Chinese begin to use negative numbers.

Zero symbol is invented.

Hypatia studies number theory, geometry, and astronomy; her death in Alexandria is followed by decline of Alexandria as center of learning.

✸

0 to 500 A.D.

✸

Jesus Christ's teachings establish new religion.

Under Emperor Claudius, Romans conquer Britain.

Roman Empire is divided into east and west.

Visigoths sack Rome.

Christianity becomes official religion of Roman Empire.

al-Khowârizmî composes key book on algebra and Hindu numerals.

Early computing algorithms are developed.

Omar Khayyam creates geometric solutions of cubic equations and calendric problems.

✸

500 A.D. to 1000 A.D.

✸

Justinian's legal code is instituted.

Hegira of Muhammad takes place.

Chinese invent compass, gunpowder, and printing.

Charlemagne is crowned Holy Roman Emperor.

Fibonacci's *Liber Abaci* advocates Hindu-Arabic numeral system, which supplants Roman system.

✸

1000 A.D. to 1500 A.D.

✸

Jerusalem is captured in First Crusade.

Genghis Khan rules.

Marco Polo travels through the East.

Universities are established at Bologna, Paris, Oxford, and Cambridge.

Bubonic plague kills one-fourth of Europe's population.

Printing with movable type is invented; rise of humanism occurs.

Columbus discovers New World.

algebra and geometry.

Pierre de Fermat develops modern number theory.

Fermat and Pascal help lay foundations for theory of probability.

Isaac Newton and Gottfried Leibniz independently discover calculus.

Bernoulli family makes numerous contributions in analysis.

Newton's *Principia Mathematica* has enormous impact throughout Europe.

✸

1500 A.D. to 1700 A.D.

✸

Johannes Kepler describes laws governing planetary movement (important to geometry and astronomy).

Works of da Vinci, Michelangelo, Raphael, Titian, and others mark the High Renaissance in Italy.

Protestant Reformation begins with Martin Luther's ninety-five theses.

Nicolas Copernicus attacks theory of geocentric universe.

Elizabeth accedes the throne. Sir Francis Drake defeats Spanish Armada.

William Shakespeare's plays are published.

Galileo invents telescope.

Mayflower lands at Plymouth Rock.

Newton formulates laws of gravity.

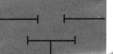

Hindu and Arabian Period 200 B.C. to 1250 A.D.
Hindu-Arabic numeral system; Arab absorption of Hindu arithmetic and Greek geometry

Modern Period (Early) 1450 A.D. to 1800 A.D.
Logarithms; modern number theory; analytic geometry; calculus; the exploitation of the calculus

Period of Transmission 1250 A.D. to 1500 A.D.
Learning preserved by Arabs slowly transmitted to Western Europe

INSTRUCTOR'S EDITION

Mathematical Ideas

TENTH EDITION AND EXPANDED TENTH EDITION

Charles D. Miller

Vern E. Heeren
American River College

John Hornsby
University of New Orleans

and

Margaret L. Morrow
Plattsburgh State University of New York
for the chapter on Graph Theory

and

Jill Van Newenhizen
Lake Forest College
for the chapter on Voting and Apportionment

PEARSON
Addison
Wesley

Boston San Francisco New York
London Toronto Sydney Tokyo Singapore Madrid
Mexico City Munich Paris Cape Town Hong Kong Montreal

To Mom and Dad, the roots of our family tree
V. E. H.

To Margaret and Shannon—like sisters to each other, like daughters to us
E. J. H.

Publisher: Greg Tobin
Sponsoring Editor: Anne Kelly
Editorial Assistant: Cecilia Stashwick
Associate Project Editor: Joanne Ha
Senior Production Supervisor: Karen Wernholm
Production Coordination: Elm Street Publishing Services, Inc.
Marketing Manager: Becky Anderson
Marketing Coordinator: Julia Coen
Associate Producer: Sara Anderson
Software Development: Sharon Humbert and Marty Wright
Senior Manufacturing Buyer: Hugh Crawford
Rights and Permissions Advisor: Dana Weightman
Text Design: Elm Street Publishing Services, Inc.
Cover Design: Barbara T. Atkinson
Composition: Beacon Publishing Services
Illustrations: Precision Graphics
Cover Photo: © Richard Cummins/CORBIS /Foreground Sculpture: Alexander Liberman. The cover photo shows an orange-red painted steel sculpture, Alexander Liberman's Olympic Iliad. Liberman's piece was installed in the Sculpture Garden at the Seattle Center in 1984. The Seattle Space Needle, which opened in 1962, rises to the height of 605 feet in the background.

For permission to use copyrighted material, grateful acknowledgment is made to the copyright holders on pages A-103–A-104, which are hereby made part of this copyright page.

Library of Congress Cataloging-in-Publication Data
Miller, Charles David
   Mathematical ideas.—10th ed./Charles D. Miller, Vern E. Heeren, John Hornsby.
      p. cm.
   Includes bibliographical references and index.
   ISBN 0-321-16808-9—ISBN 0-201-79391-1 (Expanded Edition)—ISBN 0-321-17264-7 (Instructor's Edition)
   1. Mathematics.   I. Heeren, Vern E.   II. Hornsby, E. John.   III. Title.
QA39.3.M55 2003
510—dc21

2002190901

Copyright © 2004 Pearson Education, Inc. All rights reserved. No part of this publication may be reproduced, stored in a retrieval system, or transmitted, in any form or by any means, electronic, mechanical, photocopying, recording, or otherwise, without the prior written permission of the publisher. Printed in the United States of America.

1 2 3 4 5 6 7 8 9 10—DOW—06050403

Contents

Preface ix

chapter 1 — The Art of Problem Solving 1

1.1 Solving Problems by Inductive Reasoning *2*
1.2 An Application of Inductive Reasoning: Number Patterns *10*
1.3 Strategies for Problem Solving *20*
1.4 Calculating, Estimating, and Reading Graphs *31*
Extension: Using Writing to Learn About Mathematics *40*
Collaborative Investigation: Discovering Mathematics in Pascal's Triangle *46*
Chapter 1 Test *47*

chapter 2 — The Basic Concepts of Set Theory 49

2.1 Symbols and Terminology *50*
2.2 Venn Diagrams and Subsets *56*
2.3 Set Operations and Cartesian Products *63*
2.4 Cardinal Numbers and Surveys *76*
2.5 Infinite Sets and Their Cardinalities *83*
Collaborative Investigation: A Survey of Your Class *90*
Chapter 2 Test *91*

chapter 3 — Introduction to Logic 94

3.1 Statements and Quantifiers *95*
3.2 Truth Tables and Equivalent Statements *102*
3.3 The Conditional and Circuits *113*
3.4 More on the Conditional *123*
3.5 Analyzing Arguments with Euler Diagrams *129*
Extension: Logic Puzzles *133*
3.6 Analyzing Arguments with Truth Tables *136*
Collaborative Investigation: Logic Puzzles Revisited *148*
Chapter 3 Test *149*

chapter 4: Numeration and Mathematical Systems 151

- 4.1 Historical Numeration Systems *152*
- 4.2 Arithmetic in the Hindu-Arabic System *160*
- 4.3 Conversion Between Number Bases *169*
- 4.4 Finite Mathematical Systems *179*
- 4.5 Groups *186*

Collaborative Investigation: A Perpetual Calendar Algorithm *193*
Chapter 4 Test *194*

chapter 5: Number Theory 195

- 5.1 Prime and Composite Numbers *196*
- 5.2 Selected Topics from Number Theory *206*
- 5.3 Greatest Common Factor and Least Common Multiple *212*
- 5.4 Clock Arithmetic and Modular Systems *220*
- 5.5 The Fibonacci Sequence and the Golden Ratio *232*

Extension: Magic Squares *240*
Collaborative Investigation: Investigating an Interesting Property of Number Squares *244*
Chapter 5 Test *245*

chapter 6: The Real Numbers and Their Representations 247

- 6.1 Real Numbers, Order, and Absolute Value *248*
- 6.2 Operations, Properties, and Applications of Real Numbers *257*
- 6.3 Rational Numbers and Decimal Representation *270*
- 6.4 Irrational Numbers and Decimal Representation *283*
- 6.5 Applications of Decimals and Percents *293*

Extension: Complex Numbers *306*
Collaborative Investigation: Budgeting to Buy a Car *310*
Chapter 6 Test *311*

chapter 7: The Basic Concepts of Algebra 314

- 7.1 Linear Equations *315*
- 7.2 Applications of Linear Equations *325*
- 7.3 Ratio, Proportion, and Variation *339*
- 7.4 Linear Inequalities *352*
- 7.5 Properties of Exponents and Scientific Notation *360*
- 7.6 Polynomials and Factoring *372*

7.7 Quadratic Equations and Applications *381*
Collaborative Investigation: How Does Your Walking Rate Compare to That of Olympic Race-walkers? *390*
Chapter 7 Test *391*

chapter 8 Graphs, Functions, and Systems of Equations and Inequalities 393

8.1 The Rectangular Coordinate System and Circles *394*
8.2 Lines and Their Slopes *402*
8.3 Equations of Lines and Linear Models *412*
8.4 An Introduction to Functions: Linear Functions, Applications, and Models *422*
8.5 Quadratic Functions, Applications, and Models *434*
8.6 Exponential and Logarithmic Functions, Applications, and Models *444*
8.7 Systems of Equations and Applications *456*
Extension: Using Matrix Row Operations to Solve Systems *476*
8.8 Linear Inequalities, Systems, and Linear Programming *481*
Collaborative Investigation: Living with AIDS *488*
Chapter 8 Test *489*

chapter 9 Geometry 491

9.1 Points, Lines, Planes, and Angles *492*
9.2 Curves, Polygons, and Circles *502*
Extension: Geometric Constructions *509*
9.3 Perimeter, Area, and Circumference *512*
9.4 The Geometry of Triangles: Congruence, Similarity, and the Pythagorean Theorem *523*
9.5 Space Figures, Volume, and Surface Area *536*
9.6 Transformational Geometry *545*
9.7 Non-Euclidean Geometry, Topology, and Networks *556*
9.8 Chaos and Fractal Geometry *567*
Collaborative Investigation: Generalizing the Angle Sum Concept *573*
Chapter 9 Test *574*

chapter 10 Trigonometry 576

10.1 Angles and Their Measures *577*
10.2 Trigonometric Functions of Angles *582*
10.3 Trigonometric Identities *586*
10.4 Right Triangles and Function Values *591*

10.5 Applications of Right Triangles *600*
10.6 The Laws of Sines and Cosines; Area Formulas *608*
10.7 The Unit Circle and Graphs *620*
Collaborative Investigation: Making a *Point* About Trigonometric Function Values *626*
Chapter 10 Test 627

chapter 11 Counting Methods 629

11.1 Counting by Systematic Listing *630*
11.2 Using the Fundamental Counting Principle *637*
11.3 Using Permutations and Combinations *646*
11.4 Using Pascal's Triangle and the Binomial Theorem *658*
11.5 Counting Problems Involving "Not" and "Or" *664*
Collaborative Investigation: Approximating Factorials Using Stirling's Formula *671*
Chapter 11 Test 672

chapter 12 Probability 673

12.1 Basic Concepts *674*
12.2 Events Involving "Not" and "Or" *684*
12.3 Events Involving "And" *691*
12.4 Binomial Probability *701*
12.5 Expected Value *707*
12.6 Estimating Probabilities by Simulation *716*
Collaborative Investigation: Finding Empirical Values of π *720*
Chapter 12 Test 721

chapter 13 Statistics 723

13.1 Frequency Distributions and Graphs *725*
13.2 Measures of Central Tendency *735*
13.3 Measures of Dispersion *748*
13.4 Measures of Position *756*
13.5 The Normal Distribution *763*
Extension: How to Lie with Statistics *771*
13.6 Regression and Correlation *778*
Collaborative Investigation: Combining Sets of Data *784*
Chapter 13 Test 785

CONTENTS vii

chapter 14 — Consumer Mathematics 788

14.1 The Time Value of Money *789*
Extension: Annuities *802*
14.2 Consumer Credit *806*
14.3 Truth in Lending *813*
14.4 Purchasing a House *822*
14.5 Investing *832*
Collaborative Investigation: To Buy or to Rent? *848*
Chapter 14 Test *849*

chapter 15 — Graph Theory (available only in the expanded tenth edition) 850

15.1 Basic Concepts *851*
15.2 Euler Circuits *865*
15.3 Hamilton Circuits *880*
15.4 Trees and Minimum Spanning Trees *894*
Collaborative Investigation: Finding the Number of Edges in a Complete Graph *906*
Chapter 15 Test *908*

chapter 16 — Voting and Apportionment (available only in the expanded tenth edition) 911

16.1 The Possibilities of Voting *912*
16.2 The Impossibilities of Voting *930*
16.3 The Possibilities of Apportionment *947*
16.4 The Impossibilities of Apportionment *962*
Collaborative Investigation: Class Favorites, an Election Exploration *975*
Chapter 16 Test *977*

Appendix The Metric System *A-1*
Answers to Selected Exercises *A-9*
Credits *A-103*
Index *I-1*
Index of Applications *I-19*

Preface

In 1969, Chuck Miller and Vern Heeren authored the first edition of *Mathematical Ideas,* one of the first textbooks for the liberal arts mathematics curriculum. Its goal then, as it remains today, was to serve non-physical science students through a practical coverage that vividly connects mathematics to their world. We arrive at the tenth edition, the most comprehensive ever. Flexible to evolve alongside changing trends, yet steadfast to its original objectives, *Mathematical Ideas* stays true to its pedagogy and features. We present the tenth edition with a fresh and modern feel and hope that this text continues to inspire students and instructors alike.

This edition has been designed with a variety of students in mind. It is well suited for several courses, including those geared toward the aforementioned liberal arts audience and survey courses in mathematics, finite mathematics, and mathematics for prospective and in-service elementary and middle-school teachers. Ample topics are included for a two-term course, yet the variety of topics and flexibility of sequence makes the text suitable for shorter courses as well. Our main objectives continue to be comprehensive coverage of topics, appropriate organization, clear exposition, an abundance of examples, and well-planned exercise sets with numerous applications.

We have worked diligently to incorporate the recommendations of the *Principles and Standards for School Mathematics,* prepared by the National Council of Teachers of Mathematics (NCTM), and the *Standards for Introductory College Mathematics Before Calculus,* presented by the American Mathematical Association of Two-Year Colleges (AMATYC).

Overview of Chapters

- **Chapter 1 (The Art of Problem Solving)** introduces the student to inductive reasoning, pattern recognition, and problem-solving techniques. Many of the new problems are taken from the popular monthly calendars found in the NCTM publication *Mathematics Teacher.*
- **Chapter 2 (The Basic Concepts of Set Theory) and Chapter 3 (Introduction to Logic)** give brief overviews of set theory and elementary logic. Instructors wishing to do so may cover Chapter 3 before Chapter 2.
- **Chapter 4 (Numeration and Mathematical Systems)** covers various types of numeration systems and group theory.
- **Chapter 5 (Number Theory)** presents an introduction to topics such as prime and composite numbers, the Fibonacci sequence, and magic squares and includes updated information on new developments in the field of prime numbers. We have expanded the coverage of clock arithmetic as the lead-in to modular systems.

- **Chapter 6 (The Real Numbers and Their Representations)** introduces some of the basic concepts of real numbers, their various forms of representation, and operations of arithmetic with them.

- **Chapter 7 (The Basic Concepts of Algebra)** and **Chapter 8 (Graphs, Functions, and Systems of Equations and Inequalities)** offer numerous new applications that help form the core of the text's algebra component.

- **Chapter 9 (Geometry)** has undergone moderate revision, offering a new section on transformational geometry and a new extension on constructions.

- **Chapter 10 (Trigonometry)** is entirely new. At the request of reviewers, topics include angles in standard position, right angle trigonometry, laws of sines and cosines, the unit circle, and graphs.

- **Chapter 11 (Counting Methods)** focuses on elementary counting techniques, in preparation for the chapter to follow.

- **Chapter 12 (Probability)** covers the basics of probability, odds, and expected value.

- **Chapter 13 (Statistics)** has been revised to include new data in examples and exercises.

- **Chapter 14 (Consumer Mathematics)** provides the student with the basics of the mathematics of finance as applied to inflation, consumer debt, and house buying. The chapter includes an expanded section on investing, with emphasis on stocks, bonds, and mutual funds. Examples and exercises have been updated to reflect current interest rates and investment returns.

The following chapters are available in the Expanded Edition of this text:

- **Chapter 15 (Graph Theory)** covers the basic concepts of graph theory and its applications.

- **Chapter 16 (Voting and Apportionment)** starts with an insightful new chapter opening application on the 2000 presidential election and includes additional margin notes.

Course Outline Considerations

For the most part, the chapters in the text are independent and may be covered in the order chosen by the instructor. The few exceptions are as follows: Chapter 6 contains some material dependent on the ideas found in Chapter 5; Chapter 6 should be covered before Chapter 7 if student background so dictates; Chapters 7 and 8 form an algebraic "package" and should be covered in sequential order; a thorough coverage of Chapter 12 depends on knowledge of Chapter 11 material, although probability can be covered without teaching extensive counting methods by avoiding the more difficult exercises; and the latter part of Chapter 13, on inferential statistics, depends on an understanding of probability (Chapter 12).

Features of the Tenth Edition

Enhanced: Varied Exercise Sets   We continue to present a variety of exercises that integrate drill, conceptual, and applied problems. The text contains a wealth of exercises to provide students with opportunities to practice, apply, connect, and extend the mathematical skills they are learning. We have updated the exercises that focus on real-life data and have retained their titles for easy identification. Several chapters are enriched with new applications, particularly Chapters 7, 8, 10, and 14. We continue to use graphs, tables, and charts when appropriate. Many of the graphs use a style similar to that seen by students in today's print and electronic media.

Enhanced: Margin Notes   This popular feature is a hallmark of this text and has been retained and updated where appropriate. These notes are interspersed throughout the text and deal with various subjects such as lives of mathematicians, historical vignettes, philatelic and numismatic reproductions, anecdotes on mathematics textbooks of the past, newspaper and magazine articles, and current research in mathematics. We hope that users continue to enjoy them as much as we enjoy researching and composing them.

Enhanced: Collaborative Investigations   The importance of cooperative learning is addressed in this end-of-chapter feature.

Problem Solving   Special paragraphs labeled "Problem Solving" relate the discussion of problem-solving strategies to techniques that have been learned earlier.

Optional Graphing Technology   We continue to provide sample graphing calculator screens (generated by a TI-83 Plus calculator) to show how technology can be used to support results found analytically. It is not essential, however, that a student have a graphing calculator to study from this text; *the technology component is optional.*

Flexibility   Some topics in the first six chapters require a basic knowledge of equation solving. Depending on the background of the students, the instructor may omit topics that require this skill. On the other hand, the two algebra chapters (Chapters 7 and 8) provide an excellent overview of algebra, and because of the flexibility of the text, they may be covered at almost any time.

Chapter Openers   Each chapter opens with an application of a chapter topic. Diverse and inspiring applications help students see the relevance of mathematics in our world.

Art Program   Visually magnetic, the text continues to feature a full-color design. Color is used for instructional emphasis in text discussions, examples, graphs, and figures. New and striking photos have been incorporated to enhance applications and provide visual appeal.

Examples   Numerous, carefully selected examples illustrate the concepts and skills introduced.

For Further Thought   These entries encourage students to share amongst themselves their reasoning processes in order to gain a deeper understanding of key mathematical concepts.

Emphasis on Real Data in the Form of Graphs, Charts, and Tables
We continue to use up-to-date information from magazines, newspapers, and the Internet to create real applications that are relevant and meaningful.

Chapter Tests Each chapter concludes with a chapter test so that students can check their mastery of the material.

Supplements

FOR THE STUDENT

Study Guide and Solutions Manual   ISBN 0-321-17265-5

This for-sale manual contains solutions to the odd-numbered exercises in the exercise sets and in the Extension and Appendix exercises, and solutions to all Chapter Test exercises. In addition, chapter summaries review key points in the text, provide extra examples, and enumerate major topic objectives.

Videotape Series   ISBN 0-321-17415-1

Videotapes presenting the text's topics in a chapter-by-chapter format are available. A qualified college-level mathematics instructor presents worked-out examples and uses visual aids to reinforce the mathematical concepts.

Digital Video Tutor   ISBN 0-321-17023-7

The videotape series for this text is provided on CD-ROM, making it easy and convenient for students to watch video segments from a computer at home or on campus. This complete video set, now affordable and portable for your students, is ideal for distance learning or extra instruction.

www.mathxl.com

MathXL is a Web-based program that provides online diagnostic testing, homework, and tutorial help using algorithmically-generated exercises correlated closely to the textbook. Students can take chapter tests, receive individualized study plans based on their test results, work unlimited practice problems and receive tutorial instruction for areas in which they need improvement, and take further tests to gauge their progress. With MathXL, instructors can assign and customize pre-loaded tests and homework assignments or create their own tests and homework and track all student results and practice work in the online gradebook. MathXL is free when the access code is bundled with a new Addison-Wesley text.

www.mymathlab.com

MyMathLab is a complete online course for Addison-Wesley mathematics textbooks that provides interactive, multimedia instruction correlated to the textbook content. MyMathLab is easily customizable to suit the needs of students and instructors and provides a comprehensive and efficient online course-management system that allows for diagnosis, assessment, and tracking of students' progress.

MyMathLab features

- Chapter and section folders in the online course mirror the textbooks' Tables of Contents and contain a wide range of multimedia instruction, including video lectures, tutorial software, and electronic supplements.
- The actual pages of the textbook are loaded into MyMathLab, and as students work through a section of the online text, they can link to multimedia resources—such as video and tutorial exercises—that are correlated directly to the examples and exercises in the text.
- Hyperlinks take you directly to online testing, diagnosis, tutorials, and tracking in MathXL—Addison-Wesley's tutorial and testing system for mathematics and statistics.
- Instructors can create, copy, edit, assign, and track all tests and homework for their course as well as track students' results and practice work.
- With push-button ease, instructors can remove, hide, or annotate Addison-Wesley preloaded content, add their own course documents, or change the order in which material is presented.
- Using the communication tools found in MyMathLab, instructors can hold online office hours, host a discussion board, create communication groups within their class, send e-mail, and maintain a course calendar.
- Print supplements are available online, side by side with their corresponding textbooks.
- Special InterAct Math Tutorials for use in Florida offer help in preparing for the College Level Academic Skills Test (CLAST). They include a review of arithmetic, a CLAST pretest, supplementary exercises, and a CLAST posttest.
- New to this edition, MyMathLab will also include PowerPoint presentations covering all important topics from chapters in the text.

For more information, visit our Web site at www.mymathlab.com or contact your Addison-Wesley sales representative for a live demonstration.

Addison-Wesley Math Tutor Center

The Addison-Wesley Math Tutor Center is staffed by qualified math and statistics instructors who provide students with tutoring on examples and odd-numbered exercises from the textbook. Tutoring is available via toll-free telephone, fax, e-mail, or the Internet, and White Board technology allows tutors and students to actually see the problems worked while they "talk" in real time over the Internet during tutoring sessions.

An access card is required. For more information, go to www.aw.com/tutorcenter.

FOR THE INSTRUCTOR

Instructor's Edition   ISBN 0-321-17264-7

This special edition of the text includes an answer section consisting of answers to all text non-writing exercises.

Instructor's Solutions Manual   ISBN 0-321-17027-X

This manual contains solutions to all section, Extension, Chapter Test, and appendix exercises.

Instructor's Testing Manual   ISBN 0-321-17414-3

This manual includes test forms paralleling the text's Chapter Tests.

TestGen with QuizMaster   ISBN 0-321-17024-5

TestGen enables instructors to build, edit, print, and administer tests using a computerized bank of questions developed to cover all the objectives of the text. TestGen is algorithmically based so that multiple yet equivalent versions of the same question or test can be generated at the click of a button. Instructors can also modify test bank questions or add new questions by using the built-in question editor, which allows users to create graphs, import graphics, insert math notation, and insert variable numbers or text. Tests can be printed or administered online via the Web or other network. Many questions in TestGen can be expressed in a short-answer or multiple-choice form, giving instructors greater flexibility in their test preparation. TestGen comes packaged with QuizMaster, which allows students to take tests on a local area network. The software is available on a dual-platform Windows/Macintosh CD-ROM.

www.mathxl.com

MathXL is a Web-based program that provides online diagnostic testing, homework, and tutorial help using algorithmically-generated exercises correlated closely to the textbook. Students can take chapter tests, receive individualized study plans based on their test results, work unlimited practice problems and receive tutorial instruction for areas in which they need improvement, and take further tests to gauge their progress. With MathXL, instructors can assign and customize pre-loaded tests and homework assignments or create their own tests and homework and track all student results and practice work in the online gradebook. MathXL is free when the access code is bundled with a new Addison-Wesley text.

www.mymathlab.com

MyMathLab is a complete online course for Addison-Wesley mathematics textbooks that provides interactive, multimedia instruction correlated to the textbook content. MyMathLab is easily customizable to suit the needs of students and instructors and provides a comprehensive and efficient online course-management system that allows for diagnosis, assessment, and tracking of students' progress.

MyMathLab features

- Chapter and section folders in the online course mirror the textbooks' Tables of Contents and contain a wide range of multimedia instruction, including video lectures, tutorial software, and electronic supplements.
- The actual pages of the textbook are loaded into MyMathLab, and as students work through a section of the online text, they can link to multimedia resources—such as video and tutorial exercises—that are correlated directly to the examples and exercises in the text.

- Hyperlinks take you directly to online testing, diagnosis, tutorials, and tracking in MathXL—Addison-Wesley's tutorial and testing system for mathematics and statistics.
- Instructors can create, copy, edit, assign, and track all tests and homework for their course as well as track students' results and practice work.
- With push-button ease, instructors can remove, hide, or annotate Addison-Wesley preloaded content, add their own course documents, or change the order in which material is presented.
- Using the communication tools found in MyMathLab, instructors can hold online office hours, host a discussion board, create communication groups within their class, send e-mail, and maintain a course calendar.
- Print supplements are available online, side by side with their corresponding textbooks.
- Special InterAct Math Tutorials for use in Florida offer help in preparing for the College Level Academic Skills Test (CLAST). They include a review of arithmetic, a CLAST pretest, supplementary exercises, and a CLAST posttest.
- New to this edition, MyMathLab will also include PowerPoint presentations covering all important topics from chapters in the text.

For more information, visit our Web site at www.mymathlab.com or contact your Addison-Wesley sales representative for a live demonstration.

Acknowledgments

We wish to thank the following reviewers for their helpful comments and suggestions for this and previous editions of the text. (Reviewers of the tenth edition are noted with an asterisk.)

H. Achepohl, *College of DuPage*

Shahrokh Ahmadi, *Northern Virginia Community College*

*Richard Andrews, *Florida A&M University*

Cindy Anfinson, *Palomar College*

Elaine Barber, *Germanna Community College*

Anna Baumgartner, *Carthage College*

*James E. Beamer, *Northeastern State University*

Elliot Benjamin, *Unity College*

Jaime Bestard, *Barry University*

Joyce Blair, *Belmont University*

Gus Brar, *Delaware County Community College*

Roger L. Brown, *Davenport College*

Douglas Burke, *Malcolm X College*

John Busovicki, *Indiana University of Pennsylvania*

Ann Cascarelle, *St. Petersburg Junior College*

Kenneth Chapman, *St. Petersburg Junior College*
Gordon M. Clarke, *University of the Incarnate Word*
M. Marsha Cupitt, *Durham Technical Community College*
James Curry, *American River College*
Ken Davis, *Mesa State College*
Nancy Davis, *Brunswick Community College*
George DeRise, *Thomas Nelson Community College*
Catherine Dermott, *Hudson Valley Community College*
Diana C. Dwan, *Yavapai College*
Laura Dyer, *Belleville Area College*
Jan Eardley, *Barat College*
Joe Eitel, *Folsom College*
Azin Enshai, *American River College*
Gayle Farmer, *Northeastern State University*
Michael Farndale, *Waldorf College*
Gordon Feathers, *Passaic County Community College*
Thomas Flohr, *New River Community College*
Bill Fulton, *Black Hawk College—East*
Anne Gardner, *Wenatchee Valley College*
Donald Goral, *Northern Virginia Community College*
Glen Granzow, *Idaho State University*
Larry Green, *Lake Tahoe Community College*
Arthur D. Grissinger, *Lock Haven University*
Don Hancock, *Pepperdine University*
Denis Hanson, *University of Regina*
Marilyn Hasty, *Southern Illinois University*
Shelby L. Hawthorne, *Thomas Nelson Community College*
Jeff Heiking, *St. Petersburg Junior College*
Emanuel Jinich, *Endicott College*
Karla Karstens, *University of Vermont*
Hilary Kight, *Wesleyan College*
Barbara J. Kniepkamp, *Southern Illinois University at Edwardsville*
Suda Kunyosying, *Shepherd College*
Pam Lamb, *J. Sargeant Reynolds Community College*
John W. Legge, *Pikeville College*
Leo Lusk, *Gulf Coast Community College*
Sherrie Lutsch, *Northwest Indian College*
Rhonda Macleod, *Florida State University*
Andrew Markoe, *Rider University*

Darlene Marnich, *Point Park College*

Victoria Martinez, *Okaloosa Walton Community College*

*Chris Mason, *Community College of Vermont*

Mark Maxwell, *Maryville University*

Carol McCarron, *Harrisburg Area Community College*

Delois McCormick, *Germanna Community College*

Daisy McCoy, *Lyndon State College*

*Cynthia McGinnis, *Okaloosa Walton Community College*

Vena McGrath, *Davenport College*

Robert Moyer, *Fort Valley State University*

Shai Neumann, *Brevard Community College*

Barbara Nienstedt, *Gloucester County College*

Chaitanya Nigam, *Gateway Community-Technical College*

Jean Okumura, *Windward Community College*

Bob Phillips, *Mesabi Range Community College*

Kathy Pinchback, *University of Memphis*

Priscilla Putman, *New Jersey City University*

Scott C. Radtke, *Davenport College*

John Reily, *Montclair State University*

Beth Reynolds, *Mater Dei College*

Shirley I. Robertson, *High Point University*

Andrew M. Rockett, *CW Post Campus of Long Island University*

*Kathleen Rodak, *St. Mary's College of Ave Maria University*

D. Schraeder, *McLennan Community College*

Wilfred Schulte, *Cosumnes River College*

Melinda Schulteis, *Concordia University*

Gary D. Shaffer, *Allegany College of Maryland*

Jane Sinibaldi, *York College of Pennsylvania*

Larry Smith, *Peninsula College*

Marguerite Smith, *Merced College*

Charlene D. Snow, *Lower Columbia College*

H. Jeannette Stephens, *Whatcom Community College*

Suzanne J. Stock, *Oakton Community College*

Dian Thom, *McKendree College*

Claude C. Thompson, *Hollins University*

Mark Tom, *College of the Sequoias*

Ida Umphers, *University of Arkansas at Little Rock*

Karen Villarreal, *University of New Orleans*

Wayne Wanamaker, *Central Florida Community College*

David Wasilewski, *Luzerne County Community College*

William Watkins, *California State University, Northridge*

Susan Williford, *Columbia State Community College*

Tom Witten, *Southwest Virginia Community College*

Fred Worth, *Henderson State University*

*Rob Wylie, *Carl Albert State College*

*Henry Wyzinski, *Indiana University Northwest*

A project of this magnitude cannot be accomplished without the help of many other dedicated individuals. Anne Kelly served as sponsoring editor for this edition. Amber Allen of Elm Street Publishing Services provided excellent production supervision. Jaime Bailey, Christine O'Brien, Greg Tobin, Karen Guardino, Barbara Atkinson, Becky Anderson, Karen Wernholm, and Joanne Ha of Addison-Wesley gave us their unwavering support. Terry McGinnis provided her usual excellent behind-the-scenes guidance, resulting in a book that includes one more chapter than the previous edition but is shorter in length! Thanks go to Dr. Margaret L. Morrow of Plattsburgh State University and Dr. Jill Van Newenhizen of Lake Forest College, who wrote the material on graph theory and voting/apportionment, respectively, for the Expanded Edition. Perian Herring, Deana Richmond, and Tom Wegleitner did an outstanding job of answer-checking, and Becky Troutman provided the *Index of Applications*. And finally, we thank our loyal users over the past thirty-five years for making this book one of the most successful in its market.

Vern E. Heeren
John Hornsby

chapter

1 | The Art of Problem Solving

Modern supercomputers can perform billions of calculations in a single second. With such computing power available, why do scientists need even faster computers to solve some problems?

Consider the question of how a steel beam supporting a bridge might fracture under extreme stress. To approach this problem, a new science called **molecular dynamics** is being used to model the motion of individual atoms in the metal. Modeling a rectangular slab of metal only 1000 atoms wide, 2000 atoms long, and 19 atoms thick requires over 50 hours for a supercomputer. In order to simulate the motion of 100 million atoms in a steel beam, 7.6 billion numbers must be stored in computer memory. Because an atom's movement is so rapid, scientists must calculate its velocity and location every 1 millionth of 1 billionth of a second. Each one of these steps in time takes 90 billion arithmetic calculations. Scientists have so far succeeded in modeling 600 million atoms at one time. But, a single speck of dust can contain more than a billion atoms, so only extremely small pieces of metal can currently be modeled. Modeling an entire steel beam as it fractures under stress is not within reach of today's supercomputers simply because they are too slow.

Real applications requiring high-speed computation are becoming increasingly important. From modeling the heart over long periods of time to predicting weather 24 hours in advance, real-world problems are indeed very complex. Understanding them will require not only faster computers, capable of performing trillions rather than billions of calculations per second, but also new problem-solving strategies.

Source: SIAM News, Vol. 28, No. 5, 1995.

1.1 Solving Problems by Inductive Reasoning

1.2 An Application of Inductive Reasoning: Number Patterns

1.3 Strategies for Problem Solving

1.4 Calculating, Estimating, and Reading Graphs

Extension
 Using Writing to Learn About Mathematics

Collaborative Investigation:
 Discovering Mathematics in Pascal's Triangle

Chapter 1 Test

CHAPTER 1 The Art of Problem Solving

1.1 Solving Problems by Inductive Reasoning

The Moscow papyrus, which dates back to about 1850 B.C., provides an example of inductive reasoning by the early Egyptian mathematicians. Problem 14 in the document reads:

You are given a truncated pyramid of 6 for the vertical height by 4 on the base by 2 on the top. You are to square this 4, result 16. You are to double 4, result 8. You are to square 2, result 4. You are to add the 16, the 8, and the 4, result 28. You are to take one-third of 6, result 2. You are to take 28 twice, result 56. See, it is 56. You will find it right.

What does all this mean? A *frustum* of a pyramid is that part of the pyramid remaining after its top has been cut off by a plane parallel to the base of the pyramid. The actual formula for finding the volume of the frustum of a pyramid with a square base is

$$V = \frac{1}{3} h(b^2 + bB + B^2),$$

where b is the area of the upper base, B is the area of the lower base, and h is the height (or altitude). The writer of the problem is giving a method of determining the volume of the frustum of a pyramid with square bases on the top and bottom, with bottom base side of length 4, top base side of length 2, and height equal to 6.

A truncated pyramid, or frustum of a pyramid

The development of mathematics can be traced to the Egyptian and Babylonian cultures (3000 B.C.–A.D. 260) as a necessity for problem solving. Their approach was an example of the "do thus and so" method: in order to solve a problem or perform an operation, a cookbook-like recipe was given, and it was performed over and over to solve similar problems. The classical Greek period (600 B.C.–A.D. 450) gave rise to a more formal type of mathematics, in which general concepts were applied to specific problems, resulting in a structured, logical development of mathematics.

By observing that a specific method worked for a certain type of problem, the Babylonians and the Egyptians concluded that the same method would work for any similar type of problem. Such a conclusion is called a *conjecture*. A **conjecture** is an educated guess based upon repeated observations of a particular process or pattern. The method of reasoning we have just described is called *inductive reasoning*.

Inductive Reasoning

Inductive reasoning is characterized by drawing a general conclusion (making a conjecture) from repeated observations of specific examples. The conjecture may or may not be true.

In testing a conjecture obtained by inductive reasoning, it takes only one example that does not work in order to prove the conjecture false. Such an example is called a **counterexample.** Inductive reasoning provides a powerful method of drawing conclusions, but it is also important to realize that there is no assurance that the observed conjecture will always be true. For this reason, mathematicians are reluctant to accept a conjecture as an absolute truth until it is formally proved using methods of *deductive reasoning*. Deductive reasoning characterized the development and approach of Greek mathematics, as seen in the works of Euclid, Pythagoras, Archimedes, and others.

Deductive Reasoning

Deductive reasoning is characterized by applying general principles to specific examples.

Let us now look at examples of these two types of reasoning. In this chapter we will often refer to the **natural** or **counting numbers:**

$$1, 2, 3, \ldots.$$

The three dots indicate that the numbers continue indefinitely in the pattern that has been established. The most probable rule for continuing this pattern is "add 1 to the previous number," and this is indeed the rule that we follow. Now consider the following list of natural numbers:

$$2, 9, 16, 23, 30.$$

What is the next number of this list? Most people would say that the next number is 37. Why? They probably reason something like this: What have 2 and 9 and 16 in common? What is the pattern?

After studying the numbers, we might see that $2 + 7 = 9$, and $9 + 7 = 16$. Is something similar true for the other numbers in this list? Do you add 16 and 7 to get 23? Do you add 23 and 7 to get 30? Yes; any number in the given list can be found by adding 7 to the preceding number, so the next number in the list should be $30 + 7 = 37$.

You set out to find the "next number" by reasoning from your observation of the numbers in the list. You may have jumped from these observations above ($2 + 7 = 9$, $9 + 7 = 16$, and so on) to the general statement that any number in the list is 7 more than the preceding number. This is an example of *inductive reasoning*.

By using inductive reasoning, we concluded that 37 was the next number in the list. But this is wrong. You were set up. You've been tricked into drawing an incorrect conclusion. Not that your logic was faulty; but the person making up the list has another answer in mind. The list of numbers

$$2, 9, 16, 23, 30$$

actually gives the dates of Mondays in June if June 1 falls on a Sunday. The next Monday after June 30 is July 7. With this pattern, the list continues as

$$2, 9, 16, 23, 30, 7, 14, 21, 28, \ldots.$$

See the calendar in Figure 1.

	June					
S	M	Tu	W	Th	F	S
1	2	3	4	5	6	7
8	9	10	11	12	13	14
15	16	17	18	19	20	21
22	23	24	25	26	27	28
29	30					

	July					
S	M	Tu	W	Th	F	S
		1	2	3	4	5
6	7	8	9	10	11	12
13	14	15	16	17	18	19
20	21	22	23	24	25	26
27	28	29	30	31		

FIGURE 1

The process you may have used to obtain the rule "add 7" in the list above reveals one main flaw of inductive reasoning. You can never be sure that what is true in a specific case will be true in general. Even a larger number of cases may not be enough. Inductive reasoning does not guarantee a true result, but it does provide a means of making a conjecture.

With deductive reasoning, we use general statements and apply them to specific situations. For example, one of the best-known rules in mathematics is the Pythagorean Theorem: In any right triangle, the sum of the squares of the legs (shorter sides) is equal to the square of the hypotenuse (longest side). Thus, if we know that the lengths of the shorter sides are 3 inches and 4 inches, we can find the length of the longest side. Let h represent the longest side:

$$3^2 + 4^2 = h^2 \quad \text{Pythagorean Theorem}$$
$$9 + 16 = h^2 \quad \text{Square the terms.}$$
$$25 = h^2 \quad \text{Add.}$$
$$h = 5. \quad \text{The positive square root of 25 is 5.}$$

Thus, the longest side measures 5 inches. We used the general rule (the Pythagorean Theorem) and applied it to the specific situation.

Reasoning through a problem usually requires certain *premises*. A **premise** can be an assumption, law, rule, widely held idea, or observation. Then reason inductively or deductively from the premises to obtain a **conclusion.** The premises and conclusion make up a **logical argument.**

EXAMPLE 1 Identify each premise and the conclusion in each of the following arguments. Then tell whether each argument is an example of inductive or deductive reasoning.

(a) Our house is made of redwood. Both of my next-door neighbors have redwood houses. Therefore, all houses in our neighborhood are made of redwood.

The premises are "Our house is made of redwood" and "Both of my next-door neighbors have redwood houses." The conclusion is "Therefore, all houses in our neighborhood are made of redwood." Since the reasoning goes from specific examples to a general statement, the argument is an example of inductive reasoning (although it may very well have a false conclusion).

(b) All word processors will type the symbol @. I have a word processor. I can type the symbol @.

Here the premises are "All word processors will type the symbol @" and "I have a word processor." The conclusion is "I can type the symbol @." This reasoning goes from general to specific, so deductive reasoning was used.

(c) Today is Friday. Tomorrow will be Saturday.

There is only one premise here, "Today is Friday." The conclusion is "Tomorrow will be Saturday." The fact that Saturday follows Friday is being used, even though this fact is not explicitly stated. Since the conclusion comes from general facts that apply to this special case, deductive reasoning was used. ■

The example involving dates earlier in this section illustrated how inductive reasoning may, at times, lead to false conclusions. However, in many cases inductive reasoning does provide correct results, if we look for the most *probable* answer.

EXAMPLE 2 Use inductive reasoning to determine the *probable* next number in each list below.

(a) 3, 7, 11, 15, 19, 23

Each number in the list is obtained by adding 4 to the previous number. The probable next number is $23 + 4 = 27$.

(b) 1, 1, 2, 3, 5, 8, 13, 21

Beginning with the third number in the list, each number is obtained by adding the two previous numbers in the list. That is, $1 + 1 = 2, 1 + 2 = 3, 2 + 3 = 5$, and so on. The probable next number in the list is $13 + 21 = 34$. (These are the first few terms of the famous *Fibonacci sequence,* covered in detail in a later chapter.)

(c) 1, 2, 4, 8, 16

It appears here that in order to obtain each number after the first, we must double the previous number. Therefore, the most probable next number is $16 \times 2 = 32$. ■

Inductive reasoning often can be used to predict an answer in a list of similarly constructed computation exercises, as shown in the next example.

$37 \times 3 = 111$
$37 \times 6 = 222$
$37 \times 9 = 333$
$37 \times 12 = 444$

EXAMPLE 3 Consider the list of equations in the margin. Use the list to predict the next multiplication fact in the list.

In each case, the left side of the equation has two factors, the first 37 and the second a multiple of 3, beginning with 3. The product (answer) in each case consists of three digits, all the same, beginning with 111 for 37×3. For this pattern to continue, the next multiplication fact would be $37 \times 15 = 555$, which is indeed true. (Note: You might wish to investigate what happens after 30 is reached for the right-hand factor, and make conjectures based on those products.)

FOR FURTHER THOUGHT

The following anecdote concerning inductive reasoning appears in the first volume of the *In Mathematical Circles* series by Howard Eves (PWS-KENT Publishing Company).

A scientist had two large jars before him on the laboratory table. The jar on his left contained a hundred fleas; the jar on his right was empty. The scientist carefully lifted a flea from the jar on the left, placed the flea on the table between the two jars, stepped back, and in a loud voice said, "Jump." The flea jumped and was put in the jar on the right. A second flea was carefully lifted from the jar on the left and placed on the table between the two jars. Again the scientist stepped back and in a loud voice said, "Jump." The flea jumped and was put in the jar on the right. In the same manner, the scientist treated each of the hundred fleas in the jar on the left, and each flea jumped as ordered. The two jars were then interchanged and the experiment continued with a slight difference. This time the scientist carefully lifted a flea from the jar on the left, yanked off its hind legs, placed the flea on the table between the jars, stepped back, and in a loud voice said, "Jump." The flea did not jump, and was put in the jar on the right. A second flea was carefully lifted from the jar on the left, its hind legs yanked off, and then placed on the table between the two jars. Again the scientist stepped back and in a loud voice said, "Jump." The flea did not jump, and was put in the jar on the right. In this manner, the scientist treated each of the hundred fleas in the jar on the left, and in no case did a flea jump when ordered. So the scientist recorded in his notebook the following induction: "A flea, if its hind legs are yanked off, cannot hear."

For Group Discussion

As a class, discuss examples from advertising on television, in newspapers, magazines, etc., that lead consumers to draw incorrect conclusions.

A classic example of the pitfalls in inductive reasoning involves the maximum number of regions formed when chords are constructed in a circle. When two points on a circle are joined with a line segment, a *chord* is formed. Locate a single point on a circle. Since no chords are formed, a single interior region is formed.

See Figure 2(a). Locate two points and draw a chord. Two interior regions are formed, as shown in Figure 2(b). Continue this pattern. Locate three points, and draw all possible chords. Four interior regions are formed, as shown in Figure 2(c). Four points yield 8 regions and five points yield 16 regions. See Figures 2(d) and 2(e).

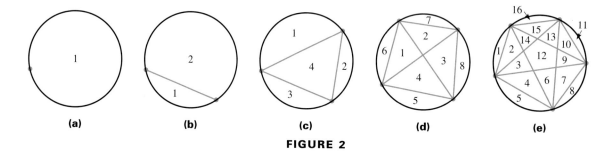

FIGURE 2

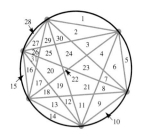

The results of the preceding observations are summarized in the chart in the margin. The pattern formed in the column headed "Number of Regions" is the same one we saw in Example 2(c), where we predicted that the next number would be 32. It seems here that for each additional point on the circle, the number of regions doubles. A reasonable inductive conjecture would be that for six points, 32 regions would be formed. But as Figure 3 indicates, there are only 31 regions!

No, a region was not "missed." It happens that the pattern of doubling ends when the sixth point is considered. Adding a seventh point would yield 57 regions. The numbers obtained here are

$$1, 2, 4, 8, 16, 31, 57.$$

For n points on the circle, the number of regions is given by the formula

$$\frac{n^4 - 6n^3 + 23n^2 - 18n + 24}{24} *.$$

We can use a graphing calculator to construct a table of values that indicates the number of regions for various numbers of points. Using X rather than n, we can define Y_1 using the expression above (see Figure 4(a)). Then, creating a table of values, as in Figure 4(b), we see how many regions (indicated by Y_1) there are for any number of points (X).

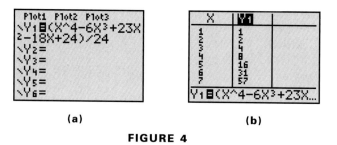

FIGURE 4

---

*For more information on this and other similar patterns, see "Counting Pizza Pieces and Other Combinatorial Problems," by Eugene Maier, in the January 1988 issue of *Mathematics Teacher*, pp. 22–26.

As indicated earlier, not until a general relationship is proved can one be sure about a conjecture since one counterexample is always sufficient to make the conjecture false.

1.1 EXERCISES

In Exercises 1–12, determine whether the reasoning is an example of deductive or inductive reasoning.

1. If the mechanic says that it will take two days to repair your car, then it will actually take four days. The mechanic says, "I figure it'll take a couple of days to fix it, ma'am." Then you can expect it to be ready four days from now.

2. If you take your medicine, you'll feel a lot better. You take your medicine. Therefore, you'll feel a lot better.

3. It has rained every day for the past five days, and it is raining today as well. So it will also rain tomorrow.

4. Natalie's first three children were boys. If she has another baby, it will be a boy.

5. Josh had 95 Pokémon trading cards. Margaret gave him 20 more for his birthday. Therefore, he now has 115 of them.

6. If the same number is subtracted from both sides of a true equation, the new equation is also true. I know that $9 + 18 = 27$. Therefore, $(9 + 18) - 12 = 27 - 12$.

7. If you build it, they will come. You build it. So, they will come.

8. All men are mortal. Socrates is a man. Therefore, Socrates is mortal.

9. It is a fact that every student who ever attended Brainchild University was accepted into medical school. Since I am attending Brainchild, I can expect to be accepted to medical school, too.

10. For the past 25 years, a rare plant has bloomed in Columbia each summer, alternating between yellow and green flowers. Last summer, it bloomed with green flowers, so this summer it will bloom with yellow flowers.

11. In the sequence $5, 10, 15, 20, \ldots$, the most probable next number is 25.

12. Britney Spears' last four single releases have reached the Top Ten list, so her current release will also reach the Top Ten.

13. Discuss the differences between inductive and deductive reasoning. Give an example of each.

14. Give an example of faulty inductive reasoning.

Determine the most probable next term in each list of numbers.

15. 6, 9, 12, 15, 18
16. 13, 18, 23, 28, 33
17. 3, 12, 48, 192, 768
18. 32, 16, 8, 4, 2
19. 3, 6, 9, 15, 24, 39
20. 1/3, 3/5, 5/7, 7/9, 9/11
21. 1/2, 3/4, 5/6, 7/8, 9/10
22. 1, 4, 9, 16, 25
23. 1, 8, 27, 64, 125
24. 2, 6, 12, 20, 30, 42
25. 4, 7, 12, 19, 28, 39
26. $-1, 2, -3, 4, -5, 6$
27. 5, 3, 5, 5, 3, 5, 5, 5, 3, 5, 5, 5, 5, 3, 5, 5, 5, 5
28. 8, 2, 8, 2, 2, 8, 2, 2, 2, 8, 2, 2, 2, 2, 8, 2, 2, 2, 2

29. Construct a list of numbers similar to those in Exercise 15 such that the most probable next number in the list is 60.

30. Construct a list of numbers similar to those in Exercise 26 such that the most probable next number in the list is 9.

In Exercises 31–42, a list of equations is given. Use the list and inductive reasoning to predict the next equation, and then verify your conjecture.

31. $(9 \times 9) + 7 = 88$
$(98 \times 9) + 6 = 888$
$(987 \times 9) + 5 = 8888$
$(9876 \times 9) + 4 = 88{,}888$

32. $(1 \times 9) + 2 = 11$
$(12 \times 9) + 3 = 111$
$(123 \times 9) + 4 = 1111$
$(1234 \times 9) + 5 = 11{,}111$

33. $3367 \times 3 = 10{,}101$
$3367 \times 6 = 20{,}202$
$3367 \times 9 = 30{,}303$
$3367 \times 12 = 40{,}404$

34. $15873 \times 7 = 111{,}111$
$15873 \times 14 = 222{,}222$
$15873 \times 21 = 333{,}333$
$15873 \times 28 = 444{,}444$

35. $34 \times 34 = 1156$
$334 \times 334 = 111{,}556$
$3334 \times 3334 = 11{,}115{,}556$

36. $11 \times 11 = 121$
$111 \times 111 = 12{,}321$
$1111 \times 1111 = 1{,}234{,}321$

37. $3 = \dfrac{3(2)}{2}$
$3 + 6 = \dfrac{6(3)}{2}$
$3 + 6 + 9 = \dfrac{9(4)}{2}$
$3 + 6 + 9 + 12 = \dfrac{12(5)}{2}$

38. $2 = 4 - 2$
$2 + 4 = 8 - 2$
$2 + 4 + 8 = 16 - 2$
$2 + 4 + 8 + 16 = 32 - 2$

39. $5(6) = 6(6 - 1)$
$5(6) + 5(36) = 6(36 - 1)$
$5(6) + 5(36) + 5(216) = 6(216 - 1)$
$5(6) + 5(36) + 5(216) + 5(1296) = 6(1296 - 1)$

40. $3 = \dfrac{3(3 - 1)}{2}$
$3 + 9 = \dfrac{3(9 - 1)}{2}$
$3 + 9 + 27 = \dfrac{3(27 - 1)}{2}$
$3 + 9 + 27 + 81 = \dfrac{3(81 - 1)}{2}$

41. $\dfrac{1}{2} = 1 - \dfrac{1}{2}$
$\dfrac{1}{2} + \dfrac{1}{4} = 1 - \dfrac{1}{4}$
$\dfrac{1}{2} + \dfrac{1}{4} + \dfrac{1}{8} = 1 - \dfrac{1}{8}$
$\dfrac{1}{2} + \dfrac{1}{4} + \dfrac{1}{8} + \dfrac{1}{16} = 1 - \dfrac{1}{16}$

42. $\dfrac{1}{1 \cdot 2} = \dfrac{1}{2}$
$\dfrac{1}{1 \cdot 2} + \dfrac{1}{2 \cdot 3} = \dfrac{2}{3}$
$\dfrac{1}{1 \cdot 2} + \dfrac{1}{2 \cdot 3} + \dfrac{1}{3 \cdot 4} = \dfrac{3}{4}$
$\dfrac{1}{1 \cdot 2} + \dfrac{1}{2 \cdot 3} + \dfrac{1}{3 \cdot 4} + \dfrac{1}{4 \cdot 5} = \dfrac{4}{5}$

A story is often told about how the great mathematician Carl Friedrich Gauss (1777–1855) at a very young age was told by his teacher to find the sum of the first 100 counting numbers. While his classmates toiled at the problem, Carl simply wrote down a single number and handed it in to his teacher. His answer was correct. When asked how he did it, the young Carl explained that he observed that there were 50 pairs of numbers that each added up to 101. (See below.) So the sum of all the numbers must be $50 \times 101 = 5050$.

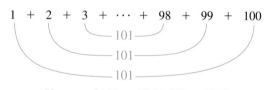

50 sums of $101 = 50 \times 101 = 5050$

Use the method of Gauss to find each of the following sums.

43. $1 + 2 + 3 + \cdots + 200$

44. $1 + 2 + 3 + \cdots + 400$

45. $1 + 2 + 3 + \cdots + 800$

46. $1 + 2 + 3 + \cdots + 2000$

47. Modify the procedure of Gauss to find the sum $1 + 2 + 3 + \cdots + 175$.

48. Explain in your own words how the procedure of Gauss can be modified to find the sum $1 + 2 + 3 + \cdots + n$, where n is an odd natural number. (An odd natural number, when divided by 2, leaves a remainder of 1.)

49. Modify the procedure of Gauss to find the sum $2 + 4 + 6 + \cdots + 100$.

50. Use the result of Exercise 49 to find the sum $4 + 8 + 12 + \cdots + 200$.

51. Find a pattern in the following figures and use inductive reasoning to predict the next figure.

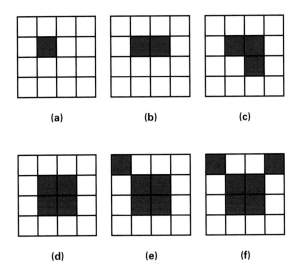

52. Consider the following table.

0	2	2	2	0	0	0	0	0
0	2	4	6	4	2	0	0	0
0	2	6	12	14	12	6	2	0
0	2	8	20	32	38	32	20	8

Find a pattern and predict the next row of the table.

53. What is the most probable next number in this list? 12, 1, 1, 1, 2, 1, 3 (*Hint:* Think about a clock.)

54. What is the next term in this list? O, T, T, F, F, S, S, E, N, T (*Hint:* Think about words and their relationship to numbers.)

55. (a) Choose any three-digit number with all different digits. Now reverse the digits, and subtract the smaller from the larger. Record your result. Choose another three-digit number and repeat this process. Do this as many times as it takes for you to see a pattern in the different results you obtain. (*Hint:* What is the middle digit? What is the sum of the first and third digits?)

(b) Write an explanation of this pattern. You may wish to use this exercise as a "number trick" to amuse your friends.

56. Choose any number, and follow these steps.
(a) Multiply by 2.
(b) Add 6.
(c) Divide by 2.
(d) Subtract the number you started with.
(e) Record your result.

Repeat the process, except in Step (b), add 8. Record your final result. Repeat the process once more, except in Step (b), add 10. Record your final result.

(f) Observe what you have done; use inductive reasoning to explain how to predict the final result. You may wish to use this exercise as a "number trick" to amuse your friends.

57. Complete the following.

$$142{,}857 \times 1 = \underline{\qquad}$$
$$142{,}857 \times 2 = \underline{\qquad}$$
$$142{,}857 \times 3 = \underline{\qquad}$$
$$142{,}857 \times 4 = \underline{\qquad}$$
$$142{,}857 \times 5 = \underline{\qquad}$$
$$142{,}857 \times 6 = \underline{\qquad}$$

What pattern exists in the successive answers? Now multiply 142,857 by 7 to obtain an interesting result.

58. Complete the following.

$$12{,}345{,}679 \times 9 = \underline{\qquad}$$
$$12{,}345{,}679 \times 18 = \underline{\qquad}$$
$$12{,}345{,}679 \times 27 = \underline{\qquad}$$

By what number would you have to multiply 12,345,679 in order to get an answer of 888,888,888?

59. Refer to Figures 2(b)–(e) and 3. Instead of counting interior regions of the circle, count the chords formed. Use inductive reasoning to predict the number of chords that would be formed if seven points were used.

60. The following number trick can be performed on one of your friends. It was provided by Dr. George DeRise of Thomas Nelson Community College.
(a) Ask your friend to write down his or her age. (Only whole numbers are allowed.)
(b) Multiply the number by 4.
(c) Add 10.
(d) Multiply by 25.
(e) Subtract the number of days in a non-leap year.
(f) Add the amount of change (less than a dollar, in cents) in his or her pocket.
(g) Ask your friend for the final answer.

If you add 115 to the answer, the first two digits are the friend's age, and the last two give the amount of change.

61. Explain how a toddler might use inductive reasoning to decide on something that will be of benefit to him or her.

62. Discuss one example of inductive reasoning that you have used recently in your life. Test your premises and your conjecture. Did your conclusion ultimately prove to be true or false?

1.2 An Application of Inductive Reasoning: Number Patterns

In the previous section we introduced inductive reasoning, and we showed how it can be applied in predicting "what comes next" in a list of numbers or equations. In this section we will continue our investigation of number patterns.

An ordered list of numbers, such as

$$3, 9, 15, 21, 27, \ldots,$$

is called a *sequence*. A **number sequence** is a list of numbers having a first number, a second number, a third number, and so on, called the **terms** of the sequence. To indicate that the terms of a sequence continue past the last term written, we use three dots (ellipsis points). The sequences in Examples 2(a) and 2(c) in the previous section are called *arithmetic* and *geometric sequences,* respectively. An arithmetic sequence has a common *difference* between successive terms, while a geometric sequence has a common *ratio* between successive terms. The Fibonacci sequence in Example 2(b) is covered in a later chapter.

Successive Differences The sequences seen in the previous section were usually simple enough for us to make an obvious conjecture about the next term. However, some sequences may provide more difficulty in making such a conjecture, and often the **method of successive differences** may be applied to determine the next term if it is not obvious at first glance. For example, consider the sequence

$$2, 6, 22, 56, 114, \ldots.$$

Since the next term is not obvious, subtract the first term from the second term, the second from the third, the third from the fourth, and so on.

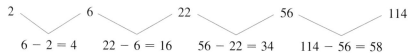

Now repeat the process with the sequence 4, 16, 34, 58 and continue repeating until the difference is a constant value, as shown in line (4):

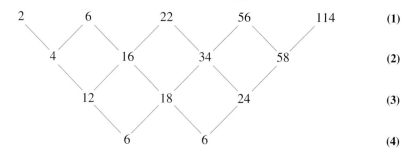

Once a line of constant values is obtained, simply work "backward" by adding until the desired term of the given sequence is obtained. Thus, for this pattern to continue, another 6 should appear in line (4), meaning that the next term in line (3) would have to be 24 + 6 = 30. The next term in line (2) would be 58 + 30 = 88. Finally, the next term in the given sequence would be 114 + 88 = 202. The final scheme of numbers is shown below.

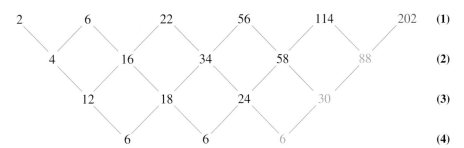

EXAMPLE 1 Use the method of successive differences to determine the next number in each sequence.

(a) 14, 22, 32, 44, …
Using the scheme described above, obtain the following:

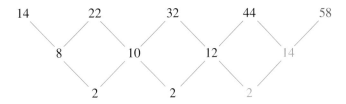

Once the row of 2s was obtained and extended, we were able to get 12 + 2 = 14, and 44 + 14 = 58, as shown above. The next number in the sequence is 58.

(b) 5, 15, 37, 77, 141, …
Proceeding as before, obtain the following diagram.

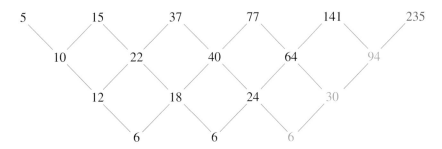

The next number in the sequence is 235.

The method of successive differences will not always work. For example, try it on the Fibonacci sequence in Example 2(b) of Section 1.1 and see what happens!

Number Patterns One of the most amazing aspects of mathematics is its seemingly endless variety of number patterns. Observe the following pattern:

$$1 = 1^2$$
$$1 + 3 = 2^2$$
$$1 + 3 + 5 = 3^2$$
$$1 + 3 + 5 + 7 = 4^2$$
$$1 + 3 + 5 + 7 + 9 = 5^2.$$

In each case, the left side of the equation is the indicated sum of the consecutive odd counting numbers beginning with 1, and the right side is the square of the number of terms on the left side. You should verify this in each case. Inductive reasoning would suggest that the next line in this pattern is

$$1 + 3 + 5 + 7 + 9 + 11 = 6^2.$$

Evaluating each side shows that each side simplifies to 36.

Can we conclude from these observations that this pattern will continue indefinitely? The answer is no, because observation of a finite number of examples does not guarantee that the pattern will continue. However, mathematicians have proved that this pattern does indeed continue indefinitely, using a method of proof called *mathematical induction*. (See any standard college algebra text.)

Any even counting number may be written in the form $2k$, where k is a counting number. It follows that the kth odd counting number is written $2k - 1$. For example, the third odd counting number, 5, can be written $2(3) - 1$. Using these ideas, we can write the result obtained above as follows.

Sum of the First n Odd Counting Numbers

If n is any counting number,

$$1 + 3 + 5 + \cdots + (2n - 1) = n^2.$$

EXAMPLE 2 In each of the following, several equations are given illustrating a suspected number pattern. Determine what the next equation would be, and verify that it is indeed a true statement.

(a)
$$1^2 = 1^3$$
$$(1 + 2)^2 = 1^3 + 2^3$$
$$(1 + 2 + 3)^2 = 1^3 + 2^3 + 3^3$$
$$(1 + 2 + 3 + 4)^2 = 1^3 + 2^3 + 3^3 + 4^3$$

The left side of each equation is the square of the sum of the first n counting numbers, while the right side is the sum of their cubes. The next equation in the pattern would be

$$(1 + 2 + 3 + 4 + 5)^2 = 1^3 + 2^3 + 3^3 + 4^3 + 5^3.$$

Each side of the above equation simplifies to 225, so the pattern is true for this equation.

(b)
$$1 = 1^3$$
$$3 + 5 = 2^3$$
$$7 + 9 + 11 = 3^3$$
$$13 + 15 + 17 + 19 = 4^3$$

The left sides of the equations contain the sum of odd counting numbers, starting with the first (1) in the first equation, the second and third (3 and 5) in the second equation, the fourth, fifth, and sixth (7, 9, and 11) in the third equation, and so on. The right side contains the cube (third power) of the number of terms on the left side in each case. Following this pattern, the next equation would be

$$21 + 23 + 25 + 27 + 29 = 5^3,$$

which can be verified by computation.

(c)
$$1 = \frac{1 \cdot 2}{2}$$
$$1 + 2 = \frac{2 \cdot 3}{2}$$
$$1 + 2 + 3 = \frac{3 \cdot 4}{2}$$
$$1 + 2 + 3 + 4 = \frac{4 \cdot 5}{2}$$

The left side of each equation gives the indicated sum of the first n counting numbers, and the right side is always of the form

$$\frac{n(n + 1)}{2}.$$

For the pattern to continue, the next equation would be

$$1 + 2 + 3 + 4 + 5 = \frac{5 \cdot 6}{2}.$$

Since each side simplifies to 15, the pattern is true for this equation. ∎

The patterns established in Examples 2(a) and 2(c) can be written as follows.

Two Special Sum Formulas

For any counting number n,

$$(1 + 2 + 3 + \cdots + n)^2 = 1^3 + 2^3 + 3^3 + \cdots + n^3$$

and

$$1 + 2 + 3 + \cdots + n = \frac{n(n + 1)}{2}.$$

Pythagoras The Greek mathematician Pythagoras lived during the sixth century B.C. He and his fellow mathematicians formed the Pythagorean brotherhood, devoted to the study of mathematics and music. The Pythagoreans investigated the figurate numbers introduced in this section. They also discovered that musical tones are related to the lengths of stretched strings by ratios of counting numbers. You can test this on a cello. Stop any string midway, so that the ratio of the whole string to the part is 2/1. If you pluck the free half of the string, you get the octave above the fundamental tone of the whole string. The ratio 3/2 gives you the fifth above the octave, and so on. The string ratio discovery was one of the first concepts of mathematical physics.

The second formula given is a generalization of the method first explained preceding Exercise 43 in the previous section, relating the story of young Carl Gauss. We can provide a general, deductive argument showing how this equation is obtained. Suppose that we let S represent the sum $1 + 2 + 3 + \cdots + n$. This sum can also be written as $S = n + (n - 1) + (n - 2) + \cdots + 1$. Now write these two equations as follows.

$$\begin{array}{l} S = 1 + 2 + 3 + \cdots + n \\ \underline{S = n + (n - 1) + (n - 2) + \cdots + 1} \\ 2S = (n + 1) + (n + 1) + (n + 1) + \cdots + (n + 1) \end{array}$$

Add the corresponding sides.

Since the right side of the equation has n terms, each of them being $(n + 1)$, we can write it as n times $(n + 1)$.

$$2S = n(n + 1)$$

$$S = \frac{n(n + 1)}{2} \quad \text{Divide both sides by 2.}$$

Now that this formula has been verified in a general manner, we can apply deductive reasoning to find the sum of the first n counting numbers for any given value of n. (See Exercises 21–24.)

Figurate Numbers The Greek mathematician Pythagoras (c. 540 B.C.) was the founder of the Pythagorean brotherhood. This group studied, among other things, numbers of geometric arrangements of points, such as *triangular numbers, square numbers,* and *pentagonal numbers.* Figure 5 illustrates the first few of each of these types of numbers.

The *figurate numbers* possess numerous interesting patterns. Every square number greater than 1 is the sum of two consecutive triangular numbers. (For example, $9 = 3 + 6$ and $25 = 10 + 15$.) Every pentagonal number can be represented as the sum of a square number and a triangular number. (For example, $5 = 4 + 1$ and $12 = 9 + 3$.)

In the expression T_n, n is called a subscript. T_n is read "T sub n," and it represents the triangular number in the nth position in the sequence. For example,

$$T_1 = 1, \quad T_2 = 3, \quad T_3 = 6, \quad \text{and} \quad T_4 = 10.$$

S_n and P_n represent the nth square and pentagonal numbers respectively.

Formulas for Triangular, Square, and Pentagonal Numbers

For any natural number n,

the nth triangular number is given by $\quad T_n = \dfrac{n(n + 1)}{2},$

the nth square number is given by $\quad S_n = n^2,$ and

the nth pentagonal number is given by $\quad P_n = \dfrac{n(3n - 1)}{2}.$

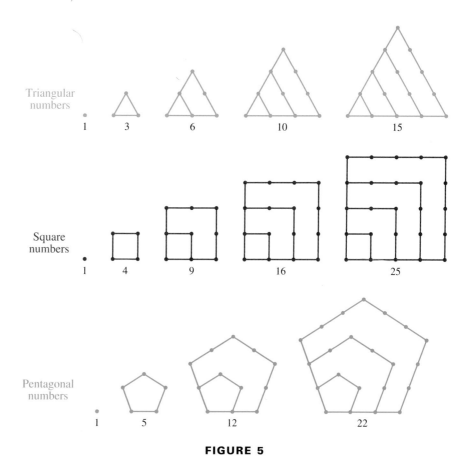

FIGURE 5

EXAMPLE 3 Use the formulas to find each of the following.

(a) the seventh triangular number

$$T_7 = \frac{7(7 + 1)}{2} = \frac{7(8)}{2} = \frac{56}{2} = 28$$

(b) the twelfth square number

$$S_{12} = 12^2 = 144$$

(c) the sixth pentagonal number

$$P_6 = \frac{6[3(6) - 1]}{2} = \frac{6(18 - 1)}{2} = \frac{6(17)}{2} = 51$$

EXAMPLE 4 Show that the sixth pentagonal number is equal to 3 times the fifth triangular number, plus 6.

From Example 3(c), $P_6 = 51$. The fifth triangular number is 15. According to the problem,

$$51 = 3(15) + 6 = 45 + 6 = 51.$$

The general relationship examined in Example 4 can be written as follows:

$$P_n = 3 \cdot T_{n-1} + n \quad (n \geq 2).$$

Other such relationships among figurate numbers are examined in the exercises of this section.

The method of successive differences, introduced at the beginning of this section, can be used to predict the next figurate number in a sequence of figurate numbers.

EXAMPLE 5 The first five pentagonal numbers are

$$1, 5, 12, 22, 35.$$

Use the method of successive differences to predict the sixth pentagonal number.

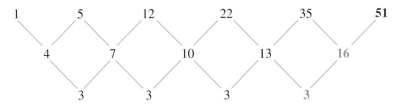

After the second line of successive differences, we can work backward to find that the sixth pentagonal number is 51, which was also found in Example 3(c).

FOR FURTHER THOUGHT

Take any three-digit number whose digits are not all the same. Arrange the digits in decreasing order, and then arrange them in increasing order. Now subtract. Repeat the process, using a 0 if necessary in the event that the difference consists of only two digits. For example, suppose that we choose the digits 1, 4, and 8.

841	963	954
−148	−369	−459
693	594	495

Notice that we have obtained the number 495, and the process will lead to 495 again. The number 495 is called a Kaprekar number, and it will eventually always be generated if this process is followed for such a three-digit number.

For Group Discussion

1. Have each student in a group of students apply the process of Kaprekar to a different two-digit number, in which the digits are not the same. (Interpret 9 as 09 if necessary.) Compare the results. What seems to be true?
2. Repeat the process for four digits, with each student in the group comparing results after several steps. What conjecture can the group make for this situation?

1.2 EXERCISES

Use the method of successive differences to determine the next number in each sequence.

1. 1, 4, 11, 22, 37, 56, …
2. 3, 14, 31, 54, 83, 118, …
3. 6, 20, 50, 102, 182, 296, …
4. 1, 11, 35, 79, 149, 251, …
5. 0, 12, 72, 240, 600, 1260, 2352, …
6. 2, 57, 220, 575, 1230, 2317, …
7. 5, 34, 243, 1022, 3121, 7770, 16799, …
8. 3, 19, 165, 771, 2503, 6483, 14409, …

9. Refer to Figures 2 and 3 in the previous section. The method of successive differences can be applied to the sequence of interior regions,

$$1, 2, 4, 8, 16, 31,$$

to find the number of regions determined by seven points on the circle. What is the next term in this sequence? How many regions would be determined by eight points? Verify this using the formula given at the end of that section.

10. Suppose that the expression $n^2 + 3n + 1$ determines the nth term in a sequence. That is, to find the first term, let $n = 1$; to find the second term, let $n = 2$, and so on.
    (a) Find the first four terms of the sequence.
    (b) Use the method of successive differences to predict the fifth term of the sequence.
    (c) Find the fifth term of the sequence by letting $n = 5$ in the expression $n^2 + 3n + 1$. Does your result agree with the one you found in part (b)?

In each of the following, several equations are given illustrating a suspected number pattern. Determine what the next equation would be, and verify that it is indeed a true statement.

11. $(1 \times 9) - 1 = 8$
    $(21 \times 9) - 1 = 188$
    $(321 \times 9) - 1 = 2888$

12. $(1 \times 8) + 1 = 9$
    $(12 \times 8) + 2 = 98$
    $(123 \times 8) + 3 = 987$

13. $999{,}999 \times 2 = 1{,}999{,}998$
    $999{,}999 \times 3 = 2{,}999{,}997$

14. $101 \times 101 = 10{,}201$
    $10{,}101 \times 10{,}101 = 102{,}030{,}201$

15. $3^2 - 1^2 = 2^3$
    $6^2 - 3^2 = 3^3$
    $10^2 - 6^2 = 4^3$
    $15^2 - 10^2 = 5^3$

16. $1 = 1^2$
    $1 + 2 + 1 = 2^2$
    $1 + 2 + 3 + 2 + 1 = 3^2$
    $1 + 2 + 3 + 4 + 3 + 2 + 1 = 4^2$

17. $2^2 - 1^2 = 2 + 1$
    $3^2 - 2^2 = 3 + 2$
    $4^2 - 3^2 = 4 + 3$

18. $1^2 + 1 = 2^2 - 2$
    $2^2 + 2 = 3^2 - 3$
    $3^2 + 3 = 4^2 - 4$

19. $1 = 1 \times 1$
    $1 + 5 = 2 \times 3$
    $1 + 5 + 9 = 3 \times 5$

20. $1 + 2 = 3$
    $4 + 5 + 6 = 7 + 8$
    $9 + 10 + 11 + 12 = 13 + 14 + 15$

Use the formula $S = \dfrac{n(n + 1)}{2}$ *derived in this section to find each of the following sums.*

21. $1 + 2 + 3 + \cdots + 300$

22. $1 + 2 + 3 + \cdots + 500$

23. $1 + 2 + 3 + \cdots + 675$

24. $1 + 2 + 3 + \cdots + 825$

Use the formula $S = n^2$ *discussed in this section to find each of the following sums. (Hint: To find n, add 1 to the last term and divide by 2.)*

25. $1 + 3 + 5 + \cdots + 101$

26. $1 + 3 + 5 + \cdots + 49$

27. $1 + 3 + 5 + \cdots + 999$

28. $1 + 3 + 5 + \cdots + 301$

29. Use the formula for finding the sum

    $$1 + 2 + 3 + \cdots + n$$

    to discover a formula for finding the sum

    $$2 + 4 + 6 + \cdots + 2n.$$

30. State in your own words the following formula discussed in this section:

    $$(1 + 2 + 3 + \cdots + n)^2 = 1^3 + 2^3 + 3^3 + \cdots + n^3.$$

31. Explain how the following diagram geometrically illustrates the formula $1 + 3 + 5 + 7 + 9 = 5^2$.

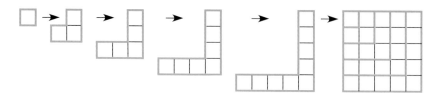

32. Explain how the following diagram geometrically illustrates the formula $1 + 2 + 3 + 4 = \dfrac{4 \times 5}{2}$.

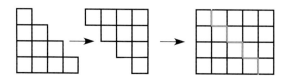

33. Use patterns to complete the table below.

Figurate Number	1st	2nd	3rd	4th	5th	6th	7th	8th
Triangular	1	3	6	10	15	21		
Square	1	4	9	16	25			
Pentagonal	1	5	12	22				
Hexagonal	1	6	15					
Heptagonal	1	7						
Octagonal	1							

34. The first five triangular, square, and pentagonal numbers may be obtained using sums of terms of sequences, as shown below.

Triangular	Square	Pentagonal
$1 = 1$	$1 = 1$	$1 = 1$
$3 = 1 + 2$	$4 = 1 + 3$	$5 = 1 + 4$
$6 = 1 + 2 + 3$	$9 = 1 + 3 + 5$	$12 = 1 + 4 + 7$
$10 = 1 + 2 + 3 + 4$	$16 = 1 + 3 + 5 + 7$	$22 = 1 + 4 + 7 + 10$
$15 = 1 + 2 + 3 + 4 + 5$	$25 = 1 + 3 + 5 + 7 + 9$	$35 = 1 + 4 + 7 + 10 + 13$

Notice the successive differences of the added terms on the right sides of the equations. The next type of figurate number is the **hexagonal** number. (A hexagon has six sides.) Use the patterns above to predict the first five hexagonal numbers.

35. Eight times any triangular number, plus 1, is a square number. Show that this is true for the first four triangular numbers.

36. Divide the first triangular number by 3 and record the remainder. Divide the second triangular number by 3 and record the remainder. Repeat this procedure several more times. Do you notice a pattern?

37. Repeat Exercise 36, but instead use square numbers and divide by 4. What pattern is determined?

38. Exercises 36 and 37 are specific cases of the following: In general, when the numbers in the sequence of n-agonal numbers are divided by n, the sequence of remainders obtained is a repeating sequence. Verify this for $n = 5$ and $n = 6$.

39. Every square number can be written as the sum of two triangular numbers. For example, 16 = 6 + 10. This can be represented geometrically by dividing a square array of dots with a line as illustrated below.

The triangular arrangement above the line represents 6, the one below the line represents 10, and the whole arrangement represents 16. Show how the square numbers 25 and 36 may likewise be geometrically represented as the sum of two triangular numbers.

40. A fraction is reduced to *lowest terms* if the greatest common factor of its numerator and its denominator is 1. For example, 3/8 is reduced to lowest terms, but 4/12 is not.
    (a) For $n = 2$ to $n = 8$, form the fractions
    $$\frac{n\text{th square number}}{(n+1)\text{th square number}}.$$
    (b) Repeat part (a), but use triangular numbers instead.
    (c) Use inductive reasoning to make a conjecture based on your results from parts (a) and (b), observing whether the fractions are reduced to lowest terms.

41. Complete the following table.

n	2	3	4	5	6	7	8
A Square of n							
B (Square of n) + n							
C One-half of Row B entry							
D (Row A entry) − n							
E One-half of Row D entry							

Use your results to answer the following, using inductive reasoning.
    (a) What kind of figurate number is obtained when you find the average of n^2 and n? (See Row C.)
    (b) If you square n and then subtract n from the result, and then divide by 2, what kind of figurate number is obtained? (See Row E.)

42. Find the smallest integer N greater than 1 such that two different figurate numbers exist for N. What are they?

In addition to the formulas for T_n, S_n, *and* P_n *shown in the text, the following formulas are true for* **hexagonal** *numbers* (H), **heptagonal** *numbers* (Hp), *and* **octagonal** *numbers* (O):

$$H_n = \frac{n(4n-2)}{2} \qquad Hp_n = \frac{n(5n-3)}{2}$$
$$O_n = \frac{n(6n-4)}{2}.$$

Use these formulas to find each of the following.

43. the sixteenth square number
44. the eleventh triangular number
45. the ninth pentagonal number
46. the seventh hexagonal number
47. the tenth heptagonal number
48. the twelfth octagonal number
49. Observe the formulas given for H_n, Hp_n, and O_n, and use patterns and inductive reasoning to predict the formula for N_n, the nth **nonagonal** number. (A nonagon has 9 sides.) Then use the fact that the sixth nonagonal number is 111 to further confirm your conjecture.
50. Use the result of Exercise 49 to find the tenth nonagonal number.

Use inductive reasoning to answer each question in Exercises 51–54.

51. If you add two consecutive triangular numbers, what kind of figurate number do you get?
52. If you add the squares of two consecutive triangular numbers, what kind of figurate number do you get?
53. Square a triangular number. Square the next triangular number. Subtract the smaller result from the larger. What kind of number do you get?
54. Choose a value of n greater than or equal to 2. Find T_{n-1}, multiply it by 3, and add n. What kind of figurate number do you get?

1.3 Strategies for Problem Solving

In the first two sections of this chapter we stressed the importance of pattern recognition and the use of inductive reasoning in solving problems. There are other useful approaches. These ideas will be used throughout the text. (Problem solving is not a topic to be covered one day and then forgotten the next!)

Probably the most famous study of problem-solving techniques was developed by George Polya (1888–1985), among whose many publications was the modern classic *How to Solve It*. In this book, Polya proposed a four-step process for problem solving.

George Polya, author of the classic *How to Solve it*, died at the age of 97 on September 7, 1985. A native of Budapest, Hungary, he was once asked why there were so many good mathematicians to come out of Hungary at the turn of the century. He theorized that it was because mathematics is the cheapest science. It does not require any expensive equipment, only pencil and paper. He authored or coauthored more than 250 papers in many languages, wrote a number of books, and was a brilliant lecturer and teacher. Yet, interestingly enough, he never learned to drive a car.

Polya's Four-Step Process for Problem Solving

1. **Understand the problem.** You cannot solve a problem if you do not understand what you are asked to find. The problem must be read and analyzed carefully. You will probably need to read it several times. After you have done so, ask yourself, "What must I find?"
2. **Devise a plan.** There are many ways to attack a problem and decide what plan is appropriate for the particular problem you are solving. See the list headed "Problem-Solving Strategies" for a number of possible approaches.
3. **Carry out the plan.** Once you know how to approach the problem, carry out your plan. You may run into "dead ends" and unforeseen roadblocks, but be persistent. If you are able to solve a problem without a struggle, it isn't much of a problem, is it?
4. **Look back and check.** Check your answer to see that it is reasonable. Does it satisfy the conditions of the problem? Have you answered all the questions the problem asks? Can you solve the problem a different way and come up with the same answer?

In Step 2 of Polya's problem-solving process, we are told to devise a plan. Here are some strategies that may prove useful.

Problem-Solving Strategies

Make a table or a chart.
Look for a pattern.
Solve a similar simpler problem.
Draw a sketch.
Use inductive reasoning.
Write an equation and solve it.

If a formula applies, use it.
Work backward.
Guess and check.
Use trial and error.
Use common sense.
Look for a "catch" if an answer seems too obvious or impossible.

A particular problem solution may involve one or more of the strategies listed here, and you should try to be creative in your problem-solving techniques. The examples that follow illustrate some of these strategies. As you read through them, keep in mind that it is one thing to read a problem solution or watch a teacher solve a problem, but it is another to be able to do it yourself. Have you ever watched your teacher solve a problem and said to yourself, "It looks easy when she does it, but I have trouble doing it myself"? Your teacher has spent many years practicing and studying problem-solving techniques. Like any skill, proficiency in problem solving requires perseverance and hard work.

FOR FURTHER THOUGHT

Various forms of the following problem have been around for many years.

In Farmer Jack's will, Jack bequeathed 1/2 of his horses to his son Johnny, 1/3 to his daughter Linda, and 1/9 to his son Jeff. Jack had 17 horses, so how were they to comply with the terms of the will? Certainly, horses cannot be divided up into fractions. Their attorney, Schwab, came to their rescue, and was able to execute the will to the satisfaction of all. How did she do it?

Here is the solution:

Schwab added one of her horses to the 17, giving a total of 18. Johnny received 1/2 of 18, or 9, Linda received 1/3 of 18, or 6, and Jeff received 1/9 of 18, or 2. That accounted for a total of 9 + 6 + 2 = 17 horses. Then Schwab took back her horse, and everyone was happy.

For Group Discussion

Discuss whether the terms of the will were actually fulfilled, according to the letter of the law.

Fibonacci (1170–1250) discovered the sequence named after him in a problem on rabbits. Fibonacci (son of Bonaccio) is one of several names for Leonardo of Pisa. His father managed a warehouse in present-day Bougie (or Bejaia), in Algeria. Thus it was that Leonardo Pisano studied with a Moorish teacher and learned the "Indian" numbers that the Moors and other Moslems brought with them in their westward drive.

Fibonacci wrote books on algebra, geometry, and trigonometry.

EXAMPLE 1 *Solving a Problem by Using a Table or a Chart* A man put a pair of rabbits in a cage. During the first month the rabbits produced no offspring but each month thereafter produced one new pair of rabbits. If each new pair thus produced reproduces in the same manner, how many pairs of rabbits will there be at the end of one year?

This problem is a famous one in the history of mathematics and first appeared in *Liber Abaci,* a book written by the Italian mathematician Leonardo Pisano (also known as Fibonacci) in the year 1202. Let us apply Polya's process to solve it.

1. **Understand the problem.** After several readings, we can reword the problem as follows: How many pairs of rabbits will the man have at the end of one year if he starts with one pair, and they reproduce this way: During the first month of life, each pair produces no new rabbits, but each month thereafter each pair produces one new pair?

2. **Devise a plan.** Since there is a definite pattern to how the rabbits will reproduce, we can construct a table as shown below. Once the table is completed, the final entry in the final column is our answer.

Month	Number of Pairs at Start	Number of New Pairs Produced	Number of Pairs at End of Month
1st			
2nd			
3rd			
4th			
5th			
6th			
7th			
8th			
9th			
10th			
11th			
12th			

> *Problem solving should be the central focus of the mathematics curriculum. As such, it is a primary goal of all mathematics instruction and an integral part of all mathematical activity. Problem solving is not a distinct topic but a process that should permeate the entire program and provide the context in which concepts and skills can be learned.*
>
> From *Curriculum and Evaluation Standards for School Mathematics*, 1989 (National Council of Teachers of Mathematics)

3. **Carry out the plan.** At the start of the first month there is only one pair of rabbits. No new pairs are produced during the first month, so there is $1 + 0 = 1$ pair present at the end of the first month. This pattern continues throughout the table. We add the number in the first column of numbers to the number in the second column to get the number in the third.

Month	Number of Pairs at Start	+	Number of New Pairs Produced	=	Number of Pairs at End of Month	
1st	1		0		1	$1 + 0 = 1$
2nd	1		1		2	$1 + 1 = 2$
3rd	2		1		3	$2 + 1 = 3$
4th	3		2		5	
5th	5		3		8	
6th	8		5		13	
7th	13		8		21	
8th	21		13		34	
9th	34		21		55	
10th	55		34		89	
11th	89		55		144	
12th	144		89		233	$144 + 89 = 233$

There will be 233 pairs of rabbits at the end of one year.

4. **Look back and check.** This problem can be checked by going back and making sure that we have interpreted it correctly, which we have. Double-check the arithmetic. We have answered the question posed by the problem, so the problem is solved. ■

The sequence shown in color in the table in Example 1 is the Fibonacci sequence, and many of its interesting properties will be investigated in a later chapter. In the remaining examples of this section, we will use Polya's process but will not list the steps specifically as we did in Example 1.

EXAMPLE 2 *Solving a Problem by Working Backward* Rob Zwettler goes to the racetrack with his buddies on a weekly basis. One week he tripled his money, but then lost $12. He took his money back the next week, doubled it, but then lost $40. The following week he tried again, taking his money back with him. He quadrupled it, and then played well enough to take that much home with him, a total of $224. How much did he start with the first week?

This problem asks us to find Rob's starting amount, given information about his winnings and losses. We also know his final amount. While we could write an algebraic equation to solve this problem, the method of working backward can be applied quite easily. Since his final amount was $224 and this represents four times the amount he started with on the third week, we *divide* $224 by 4 to find that he started the third week with $56. Before he lost $40 the second week, he had this $56 plus the $40 he lost, giving him $96. This represented double what he started with, so he started with $96 *divided by* 2, or $48, the second week. Repeating this process once more for the first week, before his $12 loss he had $48 + $12 = $60, which represents triple what he started with. Therefore, he started with $60 ÷ 3, or $20.

Augustus De Morgan (see Example 3) was an English mathematician and philosopher, who served as professor at the University of London. He wrote numerous books, one of which was *A Budget of Paradoxes*. His work in set theory and logic led to laws that bear his name and are covered in other chapters. He died in the same year as Charles Babbage.

To check our answer, $20, observe the following equations that depict the winnings and losses:

First week: $(3 \times \$20) - \$12 = \$60 - \$12 = \$48$
Second week: $(2 \times \$48) - \$40 = \$96 - \$40 = \$56$
Third week: $(4 \times \$56) = \$224$.   His final amount

EXAMPLE 3 *Solving a Problem by Trial and Error* The mathematician Augustus De Morgan lived in the nineteenth century. He once made the following statement: "I was x years old in the year x^2." In what year was De Morgan born?

We must find the year of De Morgan's birth. The problem tells us that he lived in the nineteenth century, which is another way of saying that he lived during the 1800s. One year of his life was a perfect square, so we must find a number between 1800 and 1900 that is a perfect square. Use trial and error.

$42^2 = 1764$
$43^2 = 1849$ 1849 is between 1800 and 1900.
$44^2 = 1936$

The only natural number whose square is between 1800 and 1900 is 43, since $43^2 = 1849$. Therefore, De Morgan was 43 years old in 1849. The final step in solving the problem is to subtract 43 from 1849 to find the year of his birth: $1849 - 43 = 1806$. He was born in 1806.

While the following suggestion for a check may seem unorthodox, it does work: Look up De Morgan's birth date in a book dealing with mathematics history, such as *An Introduction to the History of Mathematics,* Sixth Edition, by Howard W. Eves.

The next problem dates back to Hindu mathematics, circa 850.

EXAMPLE 4 *Solving a Problem by Guessing and Checking* One-fourth of a herd of camels was seen in the forest; twice the square root of that herd had gone to the mountain slopes; and 3 times 5 camels remained on the riverbank. What is the numerical measure of that herd of camels?

While an algebraic method of solving an equation could be used to solve this problem, we will show an alternative method. We are looking for a numerical measure of a herd of camels, so the number must be a counting number. Since the problem mentions "one-fourth of a herd" and "the square root of that herd," the number of camels must be both a multiple of 4 and a perfect square, so that no fractions will be encountered. The smallest counting number that satisfies both conditions is 4. Let us write an equation where x represents the numerical measure of the herd, and then substitute 4 for x to see if it is a solution.

"one-fourth of the herd" + "twice the square root of that herd" + "3 times 5 camels" = "the numerical measure of the herd"

$$\frac{1}{4}x + 2\sqrt{x} + 3 \cdot 5 = x$$

$$\frac{1}{4}(4) + 2\sqrt{4} + 3 \cdot 5 = 4 \quad \text{Let } x = 4.$$

$$1 + 4 + 15 = 4 \quad ?$$

$$20 \neq 4$$

Since 4 is not the solution, try 16, the next perfect square that is a multiple of 4.

$$\frac{1}{4}(16) + 2\sqrt{16} + 3 \cdot 5 = 16 \quad \text{Let } x = 16.$$
$$4 + 8 + 15 = 16 \quad ?$$
$$27 \neq 16$$

Since 16 is not a solution, try 36.

$$\frac{1}{4}(36) + 2\sqrt{36} + 3 \cdot 5 = 36 \quad \text{Let } x = 36.$$
$$9 + 12 + 15 = 36 \quad ?$$
$$36 = 36$$

We see that 36 is the numerical measure of the herd. Check in the words of the problem: "One-fourth of 36, plus twice the square root of 36, plus 3 times 5" gives 9 plus 12 plus 15, which equals 36.

EXAMPLE 5 *Solving a Problem by Considering a Similar Simpler Problem and Looking for a Pattern* The digit farthest to the right in a counting number is called the *ones* or *units* digit, since it tells how many ones are contained in the number when grouping by tens is considered. What is the ones (or units) digit in 2^{4000}?

Recall that 2^{4000} means that 2 is used as a factor 4000 times:

$$2^{4000} = \underbrace{2 \times 2 \times 2 \times \cdots \times 2}_{4000 \text{ factors}}.$$

Certainly, we are not expected to evaluate this number. In order to answer the question, we can consider examining some smaller powers of 2 and then looking for a pattern. Let us start with the exponent 1 and look at the first twelve powers of 2.

$$2^1 = 2 \qquad 2^5 = 32 \qquad 2^9 = 512$$
$$2^2 = 4 \qquad 2^6 = 64 \qquad 2^{10} = 1024$$
$$2^3 = 8 \qquad 2^7 = 128 \qquad 2^{11} = 2048$$
$$2^4 = 16 \qquad 2^8 = 256 \qquad 2^{12} = 4096$$

Notice that in each of the four rows above, the ones digit is the same. The final row, which contains the exponents 4, 8, and 12, has the ones digit 6. Each of these exponents is divisible by 4, and since 4000 is divisible by 4, we can use inductive reasoning to predict that the units digit in 2^{4000} is 6. (*Note:* The units digit for any other power can be found if we divide the exponent by 4 and consider the remainder. Then compare the result to the list of powers above. For example, to find the units digit of 2^{543}, divide 543 by 4 to get a quotient of 135 and a remainder of 3. The units digit is the same as that of 2^3, which is 8.)

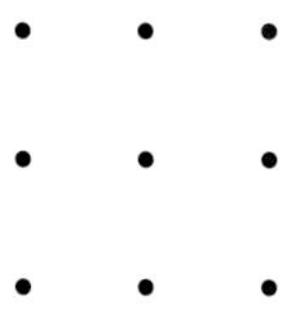

EXAMPLE 6 *Solving a Problem by Drawing a Sketch* An array of nine dots is arranged in a 3 × 3 square, as shown in Figure 6. Is it possible to join the dots with exactly four straight lines if you are not allowed to pick up your pencil from the paper and may not trace over a line that has already been drawn? If so, show how.

Figure 7 shows three attempts. In each case, something is wrong. In the first sketch, one dot is not joined. In the second, the figure cannot be drawn without picking up your pencil from the paper or tracing over a line that has already been drawn.

FIGURE 6

In the third figure, all dots have been joined, but you have used five lines as well as retraced over the figure.

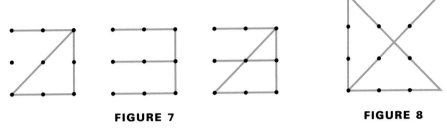

FIGURE 7   **FIGURE 8**

However, the conditions of the problem can be satisfied, as shown in Figure 8. We "went outside of the box," which was not prohibited by the conditions of the problem. This is an example of creative thinking—we used a strategy that is usually not considered at first, since our initial attempts involved "staying within the confines" of the figure.

The final example falls into a category of problems that involve a "catch" in the wording. Such problems often seem impossible to solve at first, due to preconceived notions about the matter being discussed. Some of these problems seem too easy or perhaps impossible at first, because we tend to overlook an obvious situation. We should look carefully at the use of language in such problems. And, of course, we should never forget to use common sense.

EXAMPLE 7 *Solving a Problem Using Common Sense* Two currently minted United States coins together have a total value of $1.05. One is not a dollar. What are the two coins?

Our initial reaction might be, "The only way to have two such coins with a total of $1.05 is to have a nickel and a dollar, but the problem says that one of them is not a dollar." This statement is indeed true. What we must realize here is that the one that is not a dollar is the nickel, and the *other* coin is a dollar! So the two coins are a dollar and a nickel.

1.3 EXERCISES

One of the most popular features in the journal Mathematics Teacher, *published by the National Council of Teachers of Mathematics, is the monthly calendar, which provides an interesting, unusual, or challenging problem for each day of the month. Problems are contributed by the editors of the journal, teachers, and students, and the contributors are cited in each issue. Exercises 1–25 are problems chosen from these calendars over the past sev-*

eral years, with the day, month, and year for the problem indicated. The authors wish to thank the many contributors for permission to use these problems.

Use the various problem-solving strategies to solve each of the following problems. In many cases there is more than one possible approach, so be creative.

1. *Labeling Boxes* You are working in a store that has been very careless with the stock. Three boxes of socks are each incorrectly labeled. The labels say *red socks, green socks,* and *red and green socks.* How can you relabel the boxes correctly by taking only one sock out of one box, without looking inside the boxes? (October 22, 2001)

2. *Vertical Symmetry in States' Names* (If a vertical line is drawn through the center of a figure and the left and right sides are reflections of each other across this line, the figure is said to have vertical symmetry.) When spelled with all capital letters, each letter in HAWAII has vertical symmetry. Find the name of a state whose letters all have vertical and horizontal symmetry. (September 11, 2001)

3. *Sum of Hidden Dots on Dice* Three dice with faces numbered 1 through 6 are stacked as shown. Seven of the eighteen faces are visible, leaving eleven faces hidden on the back, on the bottom, and between faces. The total number of dots not visible in this view is _____.

   A. 21
   B. 22
   C. 31
   D. 41
   E. 53
   (September 17, 2001)

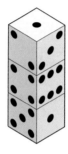

4. *Mr. Green's Age* At his birthday party, Mr. Green would not directly tell how old he was. He said, "If you add the year of my birth to this year, subtract the year of my tenth birthday and the year of my fiftieth birthday, and then add my present age, the result is eighty." How old was Mr Green? (December 14, 1997)

5. *Unfolding and Folding a Box* An unfolded box is shown below.

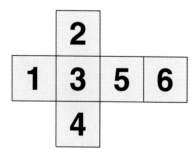

Which figure shows the box folded up? (November 7, 2001)

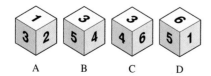

6. *Age of the Bus Driver* Today is your first day driving a city bus. When you leave downtown, you have twenty-three passengers. At the first stop, three people exit and five people get on the bus. At the second stop, eleven people exit and eight people get on the bus. At the third stop, five people exit and ten people get on. How old is the bus driver? (April 1, 2002)

7. *Matching Triangles and Squares* How can you connect each square with the triangle that has the same number? Lines cannot cross, enter a square or triangle, or go outside the diagram. (October 15, 1999)

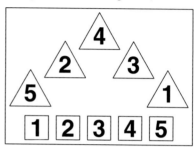

8. *Ticktacktoe Strategy* You and a friend are playing ticktacktoe, where three in a row loses. You are O. If you want to win, what must your next move be? (October 21, 2001)

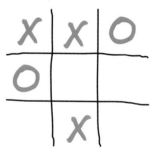

9. *Forming Perfect Square Sums* How must one place the integers from 1 to 15 in each of the spaces below in such a way that no number is repeated and the sum of the numbers in any two consecutive spaces is a perfect square? (November 11, 2001)

10. *How Old?* Pat and Chris have the same birthday. Pat is twice as old as Chris was when Pat was as old as Chris is now. If Pat is now 24 years old, how old is Chris? (December 3, 2001)

11. *Difference Triangle* Balls numbered 1 through 6 are arranged in a *difference triangle* on the next page. Note that in any row, the difference between the larger and the smaller of two successive balls is the number of the ball that appears below them.

Arrange balls numbered 1 through 10 in a *difference triangle*. (May 6, 1998)

12. *Clock Face* By drawing two straight lines, divide the face of a clock into three regions such that the numbers in the regions have the same total. (October 28, 1998)

13. *Alphametric* If a, b, and c are digits for which

$$\begin{array}{r} 7\;a\;2 \\ -4\;8\;b \\ \hline c\;7\;3, \end{array}$$

then $a + b + c =$
A. 14   **B.** 15   **C.** 16   **D.** 17   **E.** 18.
(September 22, 1999)

14. *Perfect Square* Only one of these numbers is a perfect square. Which one is it? (October 8, 1997)

$$329476 \quad 389372 \quad 964328$$
$$326047 \quad 724203$$

15. *Sleeping on the Way to Grandma's House* While traveling to his grandmother's for Christmas, George fell asleep halfway through the journey. When he awoke, he still had to travel half the distance that he had traveled while sleeping. For what part of the entire journey had he been asleep? (December 25, 1998)

16. *Counting Puzzle (Rectangles)* How many rectangles of any size are in the figure shown? (September 10, 2001)

17. *Buckets of Water* You have brought two unmarked buckets to a stream. The buckets hold 7 gallons and 3 gallons of water, respectively. How can you obtain exactly 5 gallons of water to take home? (October 19, 1997)

18. *Collecting Acorns* Chip and Dale collected thirty-two acorns on Monday and stored them with their acorn supply. After Chip fell asleep, Dale ate half the acorns. This pattern continued through Friday night, with thirty-two acorns being added and half the supply being eaten. On Saturday morning, Chip counted the acorns and found that they had only thirty-five. How many acorns had they started with on Monday morning? (March 12, 1997)

19. *Counting Puzzle (Rectangles)* How many rectangles are in the figure? (March 27, 1997)

20. *Digit Puzzle* Place each of the digits 1, 2, 3, 4, 5, 6, 7, and 8 in separate boxes so that boxes that share common corners do not contain successive digits. (November 29, 1997)

21. *Palindromic Number* (*Note:* A *palindromic* number is a number whose digits read the same left to right as right to left. For example, 383, 12321, and 9876789 are palindromic.) The odometer of the family car read 15951 when the driver noticed that the number was palindromic. "Curious," said the driver to herself. "It will be a long time before that happens again." But two hours later, the odometer showed a new palindromic number. (*Author's note:* Assume it was the next possible one.) How fast was the car driving in those two hours? (December 26, 1998)

22. *Exchange Rate* An island has no currency; it instead has the following exchange rate:

$$50 \text{ bananas} = 20 \text{ coconuts}$$
$$30 \text{ coconuts} = 12 \text{ fish}$$
$$100 \text{ fish} = 1 \text{ hammock}$$

How many bananas equal 1 hammock? (April 16, 1998)

23. *Final Digits of a Power of 7* What are the final two digits of 7^{1997}? (November 29, 1997)

24. *Brightness of a Clock Display* If a digital clock is the only light in an otherwise totally dark room, when will the room be darkest? Brightest? (May 1, 1996)

25. *Value of Coins* Which is worth more, a kilogram of $10 gold pieces or half a kilogram of $20 gold pieces? (March 20, 1995)

26. *Units Digit of a Power of 3* If you raise 3 to the 324th power, what is the units digit of the result?

27. *Units Digit of a Power of 7* What is the units digit in 7^{491}?

28. *Money Spent at a Bazaar* Alison Romike bought a book for $10 and then spent half her remaining money on a train ticket. She then bought lunch for $4 and spent half her remaining money at a bazaar. She left the bazaar with $8. How much money did she start with?

29. *Unknown Number* I am thinking of a positive number. If I square it, double the result, take half of that result, and then add 12, I get 37. What is my number?

30. *Frog Climbing up a Well* A frog is at the bottom of a 20-foot well. Each day it crawls up 4 feet but each night it slips back 3 feet. After how many days will the frog reach the top of the well?

31. *Matching Socks* A drawer contains 20 black socks and 20 white socks. If the light is off and you reach into the drawer to get your socks, what is the minimum number of socks you must pull out in order to be sure that you have a matching pair?

32. *Counting Puzzle (Squares)* How many squares are in the following figure?

33. *Counting Puzzle (Triangles)* How many triangles are in the following figure?

34. *Children in a Circle* Some children are standing in a circular arrangement. They are evenly spaced and marked in numerical order. The fourth child is standing directly opposite the twelfth child. How many children are there in the circle?

35. *Perfect Number* A *perfect number* is a counting number that is equal to the sum of all its counting number divisors except itself. For example, 28 is a perfect number, since its divisors other than itself are 1, 2, 4, 7, and 14, and $1 + 2 + 4 + 7 + 14 = 28$. What is the smallest perfect number?

36. *Naming Children* Becky's mother has three daughters. She named her first daughter Penny and her second daughter Nichole. What did she name her third daughter?

37. *Growth of a Lily Pad* A lily pad grows so that each day it doubles its size. On the twentieth day of its life, it completely covers a pond. On what day was the pond half covered?

38. *Interesting Property of a Sentence* Comment on an interesting property of this sentence: "A man, a plan, a canal, Panama." (*Hint:* See Exercise 21.)

39. *Diagram Puzzle* Draw a diagram that satisfies the following description, using the minimum number of birds: "Two birds above a bird, two birds below a bird, and a bird between two birds."

40. *Birth Year of Hornsby* Assuming that he lives that long, one of the authors of this book will be 76 years old in the year x^2, where x is a counting number. In what year was he born?

41. *High School Graduation Year of Hornsby* The same author mentioned in Exercise 40 graduated from high school in the year that satisfies these conditions: (1) The sum of the digits is 23; (2) The hundreds digit is 3 more than the tens digit; (3) No digit is an 8. In what year did he graduate?

42. *Relative Heights* Donna is taller than David but shorter than Bill. Dan is shorter than Bob. What is the first letter in the name of the tallest person?

43. *Adam and Eve's Assets* Eve said to Adam, "If you give me one dollar, then we will have the same amount of money." Adam then replied, "Eve, if you give me one dollar, I will have double the amount of money you are left with." How much does each have?

44. *Missing Digits Puzzle* In the addition problem below, some digits are missing as indicated by the blanks. If the problem is done correctly, what is the sum of the missing digits?

$$\begin{array}{r} \_\ 3\ 5 \\ 8\ \_\ 6 \\ +\ 1\ 4\ \_ \\ \hline \_\ 4\ 0\ 8 \end{array}$$

45. *Missing Digits Puzzle* Fill in the blanks so that the multiplication problem below uses all digits 0, 1, 2, 3, ..., 9 exactly once, and is correctly worked.

$$\begin{array}{r} \_\ 0\ 2 \\ \times\ \ \ \ \ 3\ \_ \\ \hline \_\ 5,\ \_\ \_\ \_ \end{array}$$

46. *Magic Square* A *magic square* is a square array of numbers that has the property that the sum of the numbers in any row, column, or diagonal is the same. Fill in the square below so that it becomes a magic square, and all digits 1, 2, 3, ..., 9 are used exactly once.

6		8
	5	
		4

47. *Magic Square* Refer to Exercise 46. Complete the magic square below so that all counting numbers 1, 2, 3, ..., 16 are used exactly once, and the sum in each row, column, or diagonal is 34.

6			9
		15	14
11		10	
16		13	

48. *Paying for a Mint* Brian Altobello has an unlimited number of cents (pennies), nickels, and dimes. In how many different ways can he pay 15¢ for a chocolate mint? (For example, one way is 1 dime and 5 pennies.)

49. *Pitches in a Baseball Game* What is the minimum number of pitches that a baseball player who pitches a complete game can make in a regulation 9-inning baseball game?

50. *Alphabetizing* What is the least natural number whose written name in the English language has its letters in alphabetical order?

51. *Crossing a River* Several soldiers must cross a deep river at a point where there is no bridge. The soldiers spot two children playing in a small rowboat. The rowboat can hold only two children or one soldier. All the soldiers get across the river. How?

52. *Weighing Coins* You have eight coins. Seven are genuine and one is a fake, which weighs a little less than the other seven. You have a balance scale, which you may use only three times. Tell how to locate the bad coin in three weighings. (Then show how to detect the bad coin in only *two* weighings.)

53. *Crossing a River* A person must take a wolf, a goat, and some cabbage across a river. The rowboat to be used has room for the person plus either the wolf, the goat, or the cabbage. If the person takes the cabbage in the boat, the wolf will eat the goat. While the wolf crosses in the boat, the cabbage will be eaten by the goat. The goat and cabbage are safe only when the person is present. Even so, the person gets everything across the river. Explain how. (This problem dates back to around the year 750.)

54. *Who's Telling the Truth?* Three women, Ms. Thompson, Ms. Johnson, and Ms. Andersen, are sitting side by side at a meeting of the neighborhood improvement group. Ms. Thompson always tells the

truth, Ms. Johnson sometimes tells the truth, and Ms. Andersen never tells the truth. The woman on the left says, "Ms. Thompson is in the middle." The woman in the middle says, "I'm Ms. Johnson," while the woman on the right says, "Ms. Andersen is in the middle." What are the correct positions of the women?

55. *Geometry Puzzle* When the diagram shown is folded to form a cube, what letter is opposite the face marked Z?

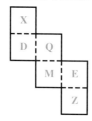

56. *Picture Puzzle* Draw a square in the following figure so that no two cats share the same region.

57. *Unknown Number* (This is an ancient Hindu problem.) Beautiful maiden with beaming eyes, tell me…which is the number that when multiplied by 3, then increased by 3/4 the product, then divided by 7, diminished by 1/3 of the quotient, multiplied by itself, diminished by 52, by the extraction of the square root, addition of 8, and division by 10 gives the number 2?

58. *Trip to the Hardware Store* At a hardware store, I can buy 1 for $.75 and I can buy 68356 for $3.75. What am I buying?

59. *Geometry Puzzle* Draw the following figure without picking up your pencil from the paper and without tracing over a line you have already drawn.

60. *Geometry Puzzle* Repeat Exercise 59 for the figure shown here.

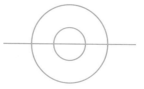

61. *Decimal Digit* What is the 100th digit in the decimal representation for 1/7?

62. *Books on a Shelf* Volumes 1 and 2 of *The Complete Works of Wally Smart* are standing in numerical order from left to right on your bookshelf. Volume 1 has 450 pages and Volume 2 has 475 pages. Excluding the covers, how many pages are between page 1 of Volume 1 and page 475 of Volume 2?

63. *Oh Brother!* The brother of the chief executive officer (CEO) of a major industrial firm died. The man who died had no brother. How is this possible?

64. *Teenager's Age* A teenager's age increased by 2 gives a perfect square. Her age decreased by 10 gives the square root of that perfect square. She is 5 years older than her brother. How old is her brother?

65. *Ages* James, Dan, Jessica, and Cathy form a pair of married couples. Their ages are 36, 31, 30, and 29. Jessica is married to the oldest person in the group. James is older than Jessica but younger than Cathy. Who is married to whom, and what are their ages?

66. *Making Change* In how many different ways can you make change for a half dollar using currently minted U.S. coins, if cents (pennies) are not allowed?

67. *Days in a Month* Some months have 30 days and some have 31 days. How many months have 28 days?

68. *Calendar Puzzle* If a year has two consecutive months with Friday the thirteenth, what months must they be?

69. *Dirt in a Hole* How much dirt is there in a cubical hole, 6 feet on each side?

70. *Fibonacci Property* Refer to Example 1, and observe the sequence of numbers in color, the Fibonacci sequence. Choose any four consecutive terms. Multiply the first one chosen by the fourth, and then multiply the two middle terms. Repeat this process a few more times. What do you notice when the two products are compared?

71. *Palindromic Greeting* The first man introduced himself to the first woman with a brief "palindromic" greeting. What was the greeting? (*Hint:* See Exercises 21, 38, and 43.)

72. *Phone Number of Heeren* Refer to Exercises 21 and 40. Another author of this book (who expects to be somewhat older than 76 in the year x^2) has a palindromic, seven-digit phone number, say *abc-defg*, satisfying these conditions:

$b = 2 \times a$

$e = f$

a, b, and c are all multiples of 4

$a + b + c + d + e + f + g = 47$.

Determine his phone number.

73. *Geometry Puzzle* What is the maximum number of small squares in which we may place crosses ($\times$) and not have any row, column, or diagonal completely filled with crosses? Illustrate your answer.

74. *Determining Operations* Place one of the arithmetic operations $+$, $-$, $\times$, or $\div$ between each pair of successive numbers on the left side of this equation to make it true. Any operation may be used more than once or not at all.

1 2 3 4 5 6 7 8 9 = 100

1.4 Calculating, Estimating, and Reading Graphs

The search for easier ways to calculate and compute has culminated in the development of hand-held calculators and computers. This text assumes that all students have access to calculators, allowing them to spend more time on the conceptual nature of mathematics and less time on computation with paper and pencil. For the general population, a calculator that performs the operations of arithmetic and a few other functions is sufficient. These are known as **four-function calculators.** Students who take higher mathematics courses (engineers, for example) usually need the added power of **scientific calculators.** **Graphing calculators,** which actually plot graphs on small screens, are also available. Since calculators differ from one manufacturer to another, remember the following.

The photograph shows the Texas Instruments TI-1795 SV, a typical four-function calculator.

Since the introduction of hand-held calculators in the early 1970s, the methods of everyday arithmetic have been drastically altered. One of the first consumer models available was the Texas Instruments SR-10, which sold for nearly $150 in 1973. It could perform the four operations of arithmetic and take square roots, but could do very little more.

Always refer to your owner's manual if you need assistance in performing an operation with your calculator. If you need further help, ask your instructor or another student who is using the same model.

Calculating Graphing calculators have become the standard in the world of advanced hand-held calculators. One of the main advantages of a graphing calculator is that both the information the user inputs into the calculator and the result

Any calculator (particularly a graphing calculator) consists of two components: the electronic "box" and the owner's manual that explains how to use it.

generated by that calculator can be viewed on the same screen. In this way, the user can verify that the information entered into the calculator is correct. While it is not necessary to have a graphing calculator to study the material presented in this text, we occasionally include graphing calculator screens to support results obtained or to provide supplemental information.*

The screens that follow illustrate some common entries and operations.

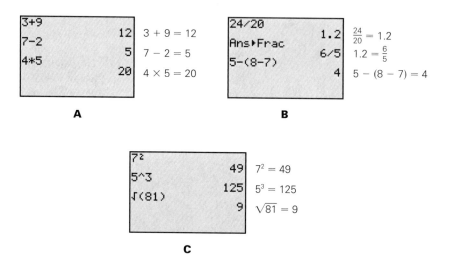

Screen A illustrates how two numbers can be added, subtracted, or multiplied. Screen B shows how two numbers can be divided, how the decimal quotient (stored in the memory cell Ans) can be converted into a fraction, and how parentheses can be used in a computation. Screen C shows how a number can be squared, how it can be cubed, and how its square root can be taken.

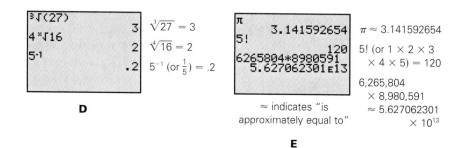

Screen D shows how higher roots (cube root and fourth root) can be found, and how the reciprocal of a number can be found using -1 as an exponent. Screen E shows how π can be accessed with its own special key, how a *factorial* (as indicated by !) can be found and how a result might be displayed in *scientific notation*. (The

---

*Because it is one of the most popular models available, we include screens generated by TI-Graph Link™ software for the Texas Instruments TI-83 Plus calculator.

"E13" following 5.627062301 means that this number is multiplied by 10^{13}. This answer is still only an approximation, because the product $6{,}265{,}804 \times 8{,}980{,}591$ contains more digits than the calculator can display.)

Estimation While calculators can make life easier when it comes to computations, many times we need only estimate an answer to a problem, and in these cases a calculator may not be necessary or appropriate.

EXAMPLE 1 A birdhouse for swallows can accommodate up to 8 nests. How many birdhouses would be necessary to accommodate 58 nests?

If we divide 58 by 8 either by hand or with a calculator, we get 7.25. Can this possibly be the desired number? Of course not, since we cannot consider fractions of birdhouses. Do we need 7 or 8 birdhouses? To provide nesting space for the nests left over after the 7 birdhouses (as indicated by the decimal fraction), we should plan to use 8 birdhouses. In this problem, we must round our answer *up* to the next counting number.

EXAMPLE 2 In 2001, David Boston of the Arizona Cardinals caught 98 passes for a total of 1598 yards. What was his approximate average number of yards per catch?

Since we are asked only to find David's approximate average, we can say that he caught about 100 passes for about 1600 yards, and his average was approximately $1600/100 = 16$ yards per catch. (A calculator shows that his average to the nearest tenth was 16.3 yards. Verify this.)

EXAMPLE 3 In a recent year there were approximately 127,000 males in the 25–29-year age bracket working on farms. This represented part of the total of 238,000 farm workers in that age bracket. Of the 331,000 farm workers in the 40–44-year age bracket, 160,000 were males. Without using a calculator, determine which age bracket had a larger proportion of males.

Here, it is best to think in terms of thousands instead of dealing with all the zeros. First, let us analyze the age bracket 25–29 years. Since there were a total of 238 thousand workers, of which 127 thousand were males, there were $238 - 127 = 111$ thousand female workers. Here, more than half of the workers were males. In the 40–44-year age bracket, of the 331 thousand workers, there were 160 thousand males, giving $331 - 160 = 171$ thousand females, meaning fewer than half were males. A comparison, then, shows that the 25–29-year age bracket had the larger proportion of males.

Reading Graphs Using graphs has become an efficient means of transmitting information in a concise way. Any issue of the newspaper *USA Today* will verify this. There are many ways to represent information using graphs and charts. Pie charts, bar graphs, and line graphs are the most common.

EXAMPLE 4 Use the graphs in Figures 9–11 to make interpretations.

The pie chart in Figure 9 shows how the price of a gallon of gasoline in California is divided among various manufacturers and agencies. The sectors (resembling slices of pie) are sized to match how the price is divided. For example, most of the price (32%) goes to the refinery, while the least portion (8%) goes for state income tax. As expected, the percents total 100%. If the price of a gallon of gasoline is $1.50, for example, the dealer's margin would be 12%, or $1.50 × .12 = $.18.

The bar graph in Figure 10 shows the relationship between the size of a room in square feet and the suggested size of air conditioner, in British thermal units (Btu). To illustrate, if we want to determine the size of air conditioner needed for an 850-square-foot room, read across at the bottom to find 850. The bar corresponding to 850 feet has a height corresponding to 16,000, found on the vertical scale at the left. We should use a 16,000 Btu air conditioner.

The line graph in Figure 11 shows the profits of CNBC from 1994 to 1999. It is read in a manner similar to the bar graph. In 1996, for example, profits were 80 million dollars, while in 1997, they were 130 million dollars. The fact that the line segments always rise from left to right indicates that profits increased each year in that time period.

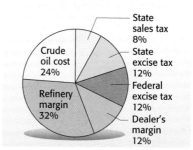

Source: California Energy Commission, 1999.

FIGURE 9

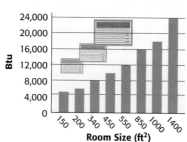

Source: Carey, Morris and James, *Home Improvement for Dummies,* IDG Books.

FIGURE 10

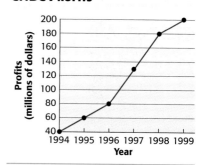

Source: Fortune, May 24, 1999, p. 142.

FIGURE 11

FOR FURTHER THOUGHT

Are You "Numerate"?

Letter is to *number* as *literacy* is to *numeracy*. In recent years much has been written about how important it is that the general population be "numerate." The essay "Quantity" by James T. Fey in *On the Shoulders of Giants: New Approaches to Numeracy* contains the following description of an approach to numeracy.

> Given the fundamental role of quantitative reasoning in applications of mathematics as well as the innate human attraction to numbers, it is not surprising that number concepts and skills form the core of school mathematics. In the earliest grades all children start on a mathematical path designed to develop computational procedures of arithmetic together with corresponding conceptual understanding that is required to solve quantitative problems and make informed decisions. Children learn many ways to describe quantitative data and relationships using numerical, graphic, and symbolic representations; to plan arithmetic and algebraic operations and to execute those plans using effective procedures; and to interpret quantitative information, to draw inferences, and to test the conclusions for reasonableness.

For Group Discussion

With calculator in hand, each student in the class should attempt to fill in the boxes with the digits 3, 4, 5, 6, 7, or 8, with each digit used at most once. Then take a poll to see who was able to come up with the closest number to the "goal number." You are allowed one minute per round. Good luck!

Round I   □ × □□□□ = 30,000

Round II   □ × □□□□ = 40,000

Round III   □ × □□□□ = 50,000

Round IV □□ × □□□□ = 30,000

Round V   □□ × □□□□ = 60,000

1.4 EXERCISES

Exercises 1–18 are designed to give you practice in learning how to do some basic operations on your calculator. Perform the indicated operations and give as many digits in your answer as shown on your calculator display. (The number of displayed digits may vary depending on the model used.)

1. $39.7 + (8.2 - 4.1)$
2. $2.8 \times (3.2 - 1.1)$
3. $\sqrt{5.56440921}$
4. $\sqrt{37.38711025}$
5. $\sqrt[3]{418.508992}$
6. $\sqrt[3]{700.227072}$
7. 2.67^2
8. 3.49^3
9. 5.76^5
10. 1.48^6
11. $\dfrac{14.32 - 8.1}{2 \times 3.11}$
12. $\dfrac{12.3 + 18.276}{3 \times 1.04}$
13. $\sqrt[5]{1.35}$
14. $\sqrt[6]{3.21}$
15. $\dfrac{\pi}{\sqrt{2}}$
16. $\dfrac{2\pi}{\sqrt{3}}$
17. $\sqrt[4]{\dfrac{2143}{22}}$
18. $\dfrac{12{,}345{,}679 \times 72}{\sqrt[3]{27}}$

19. Choose any number consisting of five digits. Multiply it by 9 on your calculator. Now add the digits in the answer. If the sum is more than 9, add the digits of this sum, and repeat until the sum is less than 10. Your answer will always be 9. Repeat the exercise with a number consisting of six digits. Does the same result hold?

20. Use your calculator to *square* the following two-digit numbers ending in 5: 15, 25, 35, 45, 55, 65, 75, 85. Write down your results, and examine the pattern that develops. Then use inductive reasoning to predict the value of 95^2. Write an explanation of how you can mentally square a two-digit number ending in 5.

By examining several similar computation problems and their answers obtained on a calculator, we can use inductive reasoning to make conjectures about certain rules, laws, properties, and definitions in mathematics. Perform each calculation and observe the answers. Then fill in the blank with the appropriate response. (Justification of these results will be discussed later in the book.)

21. $(-3) \times (-8)$; $(-5) \times (-4)$; $(-2.7) \times (-4.3)$

    Multiplying two negative numbers gives a _____ product.
    (negative/positive)

22. $5 \times (-4)$; -3×8; $2.7 \times (-4.3)$

    Multiplying a negative number and a positive number gives a _____ product.
    (negative/positive)

23. 5.6^0; π^0; 2^0; 120^0; $.5^0$

    Raising a nonzero number to the power 0 gives a result of _____.

24. 1^2; 1^3; 1^{-3}; 1^0; 1^{13}

    Raising 1 to any power gives a result of _____.

25. $1/7$; $1/(-9)$; $1/3$; $1/(-8)$

    The sign of the reciprocal of a number is _____ the sign of the
    (the same as/different from)
    number.

26. $5/0$; $9/0$; $\pi/0$; $-3/0$; $0/0$

    Dividing a number by 0 gives a(n) _____ on a calculator.

27. $0/8$; $0/2$; $0/(-3)$; $0/\pi$

    Zero divided by a nonzero number gives a quotient of _____.

28. $(-3) \times (-4) \times (-5)$; $(-3) \times (-4) \times (-5) \times (-6) \times (-7)$; $(-3) \times (-4) \times (-5) \times (-6) \times (-7) \times (-8) \times (-9)$

    Multiplying an *odd* number of negative numbers gives a _____ product.
    (positive/negative)

29. $(-3) \times (-4)$; $(-3) \times (-4) \times (-5) \times (-6)$; $(-3) \times (-4) \times (-5) \times (-6) \times (-7) \times (-8)$

    Multiplying an *even* number of negative numbers gives a _____ product.
    (positive/negative)

30. $\sqrt{-3}$; $\sqrt{-5}$; $\sqrt{-6}$; $\sqrt{-10}$

    Taking the square root of a negative number gives a(n) _____ on a calculator.

31. Find the decimal representation of 1/6 on your calculator. Following the decimal point will be a 1 and a string of 6s. The final digit will be a 7 if your calculator *rounds off* or a 6 if it *truncates*. Which kind of calculator do you have?

32. Choose any three-digit number and enter the digits into a calculator. Then enter them again to get a six-digit number. Divide this six-digit number by 7. Divide the result by 13. Divide the result by 11. What is your answer? Explain why this happens.

33. Choose any digit except 0. Multiply it by 429. Now multiply the result by 259. What is your answer? Explain why this happens.

34. Choose two natural numbers. Add 1 to the second and divide by the first to get a third. Add 1 to the third and divide by the second to get a fourth. Add 1 to the fourth and divide by the third to get a fifth. Continue this process until you discover a pattern. What is the pattern?

When a four-function or scientific calculator (not a graphing calculator, however) is turned upside down, the digits in the display correspond to letters of the English alphabet as follows:

$0 \leftrightarrow O$ $3 \leftrightarrow E$ $7 \leftrightarrow L$
$1 \leftrightarrow I$ $4 \leftrightarrow h$ $8 \leftrightarrow B$
$2 \leftrightarrow Z$ $5 \leftrightarrow S$ $9 \leftrightarrow G$.

For each of the following, perform the indicated calculation on a four-function or scientific calculator. Then turn your calculator upside down to read the word that belongs in the blank in the accompanying sentence.

35. $(100 \div 20) \times 14{,}215{,}469$

One of the biggest petroleum companies in the world is _____ .

36. $\dfrac{10 \times 10{,}609}{\sqrt{4}}$

"It's got to be the _____ ."

37. $60^2 - \dfrac{368}{4}$

The electronics manufacturer _____ produces the Wave Radio.

38. $187^2 + \sqrt{1600}$

Have you ever read *Mother* _____ nursery rhymes?

39. Make up your own exercise similar to Exercises 35–38.

40. Displayed digits on most calculators usually show some or all of the parts in the pattern shown in the figure. For the digits 0 through 9:
  (a) Which part is used most frequently?
  (b) Which part is used the least?
  (c) Which digit uses the most parts?
  (d) Which digit uses the fewest parts?

Give an appropriate counting number answer to each question in Exercises 41–44. (Find the smallest counting number that will work.)

41. *Pages to Store Trading Cards* A plastic page designed to hold trading cards will hold up to 9 cards. How many pages will be needed to store 431 cards?

42. *Drawers for Videocassettes* A sliding drawer designed to hold videocassettes has 18 compartments. If Chris wants to house his collection of 204 Disney videotapes, how many such drawers will he need?

43. *Containers for African Violets* A gardener wants to fertilize 400 African violets. Each container of fertilizer will supply up to 30 plants. How many containers will she need to do the job?

44. *Fifth-Grade Teachers Needed* Lake Harbor Middle School has 155 fifth-grade students. The principal, Cheryl Arabie, has decided that each fifth-grade teacher should have a maximum of 24 students. How many fifth-grade teachers does she need?

In Exercises 45–50, use estimation to determine the choice closest to the correct answer.

45. *Price per Acre of Land* In order to build a "millennium clock" on Mount Washington in Nevada that would tick once each year, chime once each century, and last at least 10,000 years, the nonprofit Long Now Foundation purchased 80 acres of land for $140,000. Which one of the following is the closest estimate to the price per acre?
  A. $1000   **B.** $2000   **C.** $4000   **D.** $11,200

46. *Time of a Round Trip* The distance from Seattle, Washington, to Springfield, Missouri, is 2009 miles. About how many hours would a round trip from Seattle to Springfield and back take a bus that averages 50 miles per hour for the entire trip?
  A. 60   **B.** 70   **C.** 80   **D.** 90

47. *People per Square Mile* Hale County in Texas has a population of 34,671 and covers 1005 square miles. About how many people per square mile live in Hale County?
  A. 35   **B.** 350   **C.** 3500   **D.** 35,000

48. *Revolutions of Mercury* The planet Mercury takes 88.0 Earth days to revolve around the sun once. Pluto takes 90,824.2 days to do the same. When Pluto has revolved around the sun once, about how many times will Mercury have revolved around the sun?
   A. 100,000   **B.** 10,000   **C.** 1000   **D.** 100

49. *Rushing Average* In 1998, Terrell Davis of the Denver Broncos rushed for 2008 yards in 392 attempts. His approximate number of yards gained per attempt was _____ .
   A. 1/5   **B.** 50   **C.** 4   **D.** 5

50. *Area of the Sistine Chapel* The Sistine Chapel in Vatican City measures 40.5 meters by 13.5 meters.

Which one of the following is the closest approximation to its area?
   A. 110 meters
   B. 55 meters
   C. 110 square meters
   D. 600 square meters

Satellite Television Market Share The 1999 market share for satellite television home subscribers is shown in the chart.

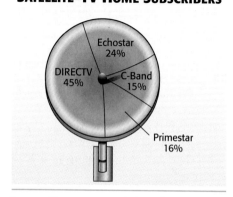

Source: Skyreport.com; *USA Today*.

The total number of users reached 12,000,000 *in August 1999. Use this information and the pie chart to work Exercises 51–54.*

51. Which provider had the largest share of the home subscriber market in August 1999? What was that share?

52. Determine the number of home subscribers to Primestar in August 1999.

53. C-Band is associated with "large dishes" while all other subscribers have "small dishes." How many subscribers had small dishes?

54. How many more subscribers did Primestar have than C-Band?

Growth of E-mail The latter half of the 1990s was characterized by incredible growth in a new method of communication: electronic mail ("e-mail"). The accompanying bar graph shows the number of e-mail boxes in North America for the years 1995–2001. Use the graph to answer the questions in Exercises 55–58.

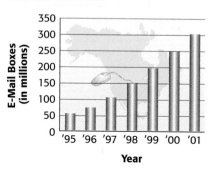

Source: IDC research.

55. How many e-mail boxes were there in 1998?

56. By how many did the number of e-mail boxes grow between 1995 and 2001?

57. In what year was the number of e-mail boxes 150 million?

58. Suppose that the number of boxes in 2002 increased the same amount from the previous year as it did in 2001. What would have been the number of boxes in 2002?

Prices for Personal Computers The accompanying line graph at the right shows average prices for personal computers (PCs) for the years 1993 through 1999. Use this information to answer the questions in Exercises 59–62.

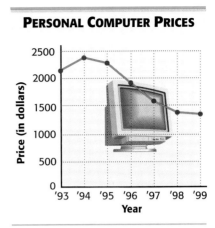

Source: CNW Marketing/Research; USA Today.

59. Between which years did the average price of a PC increase?
60. What has been the general trend in average PC prices since 1994?
61. What were the average PC prices in 1996 and in 1999?
62. About how much did PC prices decline between 1994 and 1999?

Weather Prediction The map shows the predicted weather information for a summer day in July. Use the map and the accompanying legend to answer the questions in Exercises 63–68.

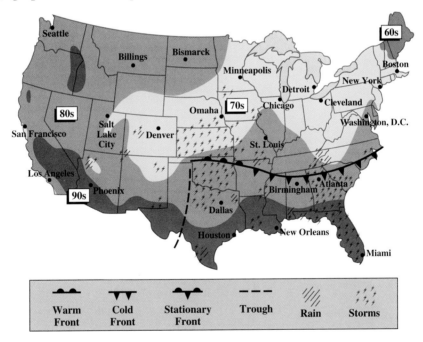

63. Which temperature range (that is, 60s, 70s, 80s, or 90s) would we expect for Detroit?
64. What type of front is moving toward Atlanta?
65. Assuming that you think anything over 80° is hot, how would you describe the weather for Miami? Use at least two descriptive words.
66. Augusta is the capital of a northeastern state. In what temperature range will Augusta be?
67. In what state is a trough located?
68. Is there a good chance that a baseball game between the Yankees and the Indians, to be played at Jacobs Field in Cleveland, will be rained out?

EXTENSION

Using Writing to Learn About Mathematics

Mathematical writing takes many forms. One of the most famous author/mathematicians was Charles Dodgson (1832–1898), who used the pen name Lewis Carroll.

Dodgson was a mathematics lecturer at Oxford University in England. Queen Victoria told Dodgson how much she enjoyed *Alice's Adventures in Wonderland* and how much she wanted to read his next book; he is said to have sent her *Symbolic Logic,* his most famous mathematical work.

The *Alice* books made Carroll famous. Late in life, however, Dodgson shunned attention and denied that he and Carroll were the same person, even though he gave away hundreds of signed copies to children and children's hospitals.

The use of writing in the mathematics curriculum has recently received increased attention, due in large part to the recommendations of the National Council of Teachers of Mathematics (N.C.T.M.). Research has indicated that the ability to express mathematical observations in writing can serve as a positive force in one's continued development as a mathematics student. The implementation of writing in the mathematics class can utilize several approaches.

Journals One way of using writing in mathematics is to keep a journal in which you spend a few minutes explaining what happened in class that day. The journal entries may be general or specific, depending on the topic covered, the degree to which you understand the topic, your interest level at the time, and so on. Journal entries are usually written in informal language, and are often an effective means of communicating to yourself, your classmates, and your instructor what feelings, perceptions, and concerns you are having at the time.

Learning Logs While journal entries are for the most part unstructured writings in which the student's thoughts are allowed to roam freely, entries in learning logs are typically more structured. An instructor may pose a specific question for a student to answer in a learning log. In this text, we intersperse writing exercises in each exercise set that are appropriate for answering in a learning log. For example, consider Exercise 13 in the exercise set for the opening section in this chapter.

> Discuss the differences between inductive and deductive reasoning. Give an example of each.

Here is a possible response to this exercise.

> Deductive reasoning occurs when you go from general ideas to specific ones. For example, I know that I can multiply both sides of $\frac{1}{2}x = 6$ by 2 to get $x = 12$, because I can multiply both sides of any equation by whatever I want (except 0). Inductive reasoning goes the other way. If I make a general conclusion from specific observations, that's inductive reasoning. Example – in the numbers 4, 8, 12, 16, and so on, I can conclude that the next number is 20, since I always add 4 to get the next number.

Reports on Articles from Mathematics Publications The motto "Publish or perish" has long been around, implying that a scholar in pursuit of an academic position must publish in a journal in his or her field. There are numerous journals that publish papers in mathematics research and/or mathematics education. In Activity 3 at the end of this section, we provide some suggestions of articles that have appeared within the last few years. A report on such an article can help you understand what mathematicians do and what ideas mathematics teachers use to convey concepts to their students.

Term Papers   A term paper in a mathematics class? In increasing numbers, professors in mathematics survey courses are requiring short term papers of their students. In this way, you can become aware of the plethora of books and articles on mathematics and mathematicians, many written specifically for the layperson. In Activities 5 and 6 at the end of this section, we provide a list of possible term paper topics.

EXTENSION ACTIVITIES

Rather than include a typical exercise set, we list some suggested activities in which writing can be used to enhance awareness and learning of mathematics.

Activity 1   Keep a journal. After each class, write for a few minutes on your perceptions about the class, the topics covered, or whatever you feel is appropriate. You may wish to use the following guidelines.

Journal Writing*

1. *WHO should write in your journal?*   You should.
2. *WHAT should you write in your journal?*   New words, ideas, formulas, or concepts; profound thoughts; wonderings, musings, problems to solve; reflections on the class; questions—both answerable and unanswerable; writing ideas
3. *WHEN should you write in your journal?*   After class each day; as you are preparing, reading, or studying for class; anytime an insight or question hits you.
4. *WHERE should you write in your journal?*   Anywhere—so keep it with you when possible.
5. *WHY should you write in your journal?*   It will help you record ideas that you might otherwise forget. It will be worthwhile for you to read later on so that you can note your growth. It will facilitate your learning, problem solving, writing, reading, and discussion in class.
6. *HOW should you write in your journal?*   In wonderful, long, flowing sentences with perfect punctuation and perfect spelling and in perfect handwriting; or in single words that express your ideas, in short phrases, in sketches, in numbers, in maps, in diagrams, in sentences (You may even prefer to organize your journal entries on your desktop, notebook, or palmtop computer.)

---

*"Journal Writing" from "No Time for Writing in Your Class?" by Margaret E. McIntosh in *Mathematics Teacher,* September 1991, p. 431. Reprinted by permission.

(continued)

Activity 2 Keep a learning log, answering at least one writing exercise from each exercise set covered in your class syllabus. Ask your teacher for suggestions of other types of specific writing assignments. For example, you might wish to choose a numbered example from a section in the text and write your own solution to the problem, or comment on the method that the authors use to solve the problem. Don't be afraid to be critical of the method used in the text.

Activity 3 The National Council of Teachers of Mathematics publishes journals in mathematics education: *Teaching Children Mathematics* (formerly called *Arithmetic Teacher*) and *Mathematics Teacher* are two of them. These journals can be found in the periodicals section of most college and university libraries. We have chosen several recent articles in each of these journals. There are thousands of other articles from which to choose. Write a short report on one of these articles according to guidelines specified by your instructor.

From *Mathematics Teacher*

1998

Andreason, Corey. "Fibonacci and Pascal Together Again: Pattern Exploration in the Fibonacci Sequence." Vol. 91, No. 3, March 1998, p. 250.

Grassl, Richard M., and Tabitha T. Y. Mingus. "Keep Counting Those Boxes—There's More." Vol. 91, No. 2, February 1998, p. 122.

Iovanelli, Robert. "Using Spreadsheets to Analyze the Historical Perspectives of Apportionment." Vol. 91, No. 2, February 1998, p. 176.

Rothbart, Andrea. "Learning to Reason from Lewis Carroll." Vol. 91, No. 1, January 1998, p. 6.

Rulf, Benjamin. "A Geometric Puzzle That Leads to Fibonacci Sequences." Vol. 91, No. 1, January 1998, p. 21.

1999

Kahan, Jeremy. "Ten Lessons from the Proof of Fermat's Last Theorem." Vol. 92, No. 6, September 1999, p. 530.

Silver, Jennifer Williams. "A Survey on the Use of Writing-to-Learn in Mathematics Classes." Vol. 92, No. 5, May 1999, p. 388.

Smith, John P. III. "Preparing Students for Modern Work: Lessons from Automobile Manufacturing." Vol. 92, No. 3, March 1999, p. 254.

2000

Kelley, Loretta. "A Mathematical History Tour." Vol. 93, No. 1, January 2000, p. 14.

Lesser, Lawrence Mark. "Sum of Songs: Making Mathematics Less Monotone!" Vol. 93, No. 5, May 2000, p. 372.

Lightner, James E. "Mathematicians Are Human Too." Vol. 93, No. 8, November 2000, p. 696.

Natsoulas, Anthula. "Group Symmetries Connect Art and History with Mathematics." Vol. 93, No. 5, May 2000, p. 364.

Simon, Marilyn K. "The Evolving Role of Women in Mathematics." Vol. 93, No. 9, December 2000, p. 782.

2001

Johnson, Craig M. "Functions of Number Theory in Music." Vol. 94, No. 8, November 2001, p. 700.

Lightner, James E. "Mathematics Didn't Just Happen." Vol. 94, No. 9, December 2001, p. 780.

McNeill, Sheila A. "The Mayan Zeros." Vol. 94, No. 7, October 2001, p. 590.

Socha, Susan. "Less Is Sometimes More." Vol. 94, No. 6, September 2001, p. 450.

2002

Houser, Don. "Roots in Music." Vol. 95, No. 1, January 2002, p. 16.

Howe, Roger. "Hermione Granger's Solution." Vol. 95, No. 2, February 2002, p. 86.

Kolpas, Sidney J. "Let Your Fingers Do the Multiplying." Vol. 95, No. 4, April 2002, p. 246.

Van Dresar, Vickie J. "Opening Young Minds to Closure Properties." Vol. 95, No. 5, May 2002, p. 326.

From *Teaching Children Mathematics*

1998

Boucher, Alfred C. "Critical Thinking through Estimation." Vol. 4, No. 7, March 1998, p. 452.

Carey, Linda M. "Parents as Math Partners: A Successful Urban Story." Vol. 4, No. 6, February 1998, p. 314.

Fennell, Francis (Skip). "Mathematics at the Mall." Vol. 4, No. 5, January 1998, p. 268.

Olson, Melfried, Lynae Sakshag, and Judith Olson. "How Many Sandwiches?" Vol. 4, No. 7, March 1998, p. 402.

Zaslavsky, Claudia. "Ethnomathematics and Multicultural Mathematics Education." Vol. 4, No. 9, May 1998, p. 502.

1999

Angerame, Shirlee S. "Math-o'-Lanterns." Vol. 5, No. 11, October 1999, p. 72.

Basile, Carol G. "Collecting Data Outdoors: Making Connections to the Real World." Vol. 5, No. 10, September 1999, p. 8.

Brahier, Daniel J., and Melfried Olsen. "The World's Largest Math Event: Promoting Mathematical Thinking." Vol. 5, No. 7, March 1999, p. 430.

Sawada, Daiyo. "Mathematics as Problem Solving: A Japanese Way." Vol. 5, No. 7, September 1999, p. 54.

2000

Glasgow, Robert, et. al. "The Decimal Dilemma." Vol. 7, No. 2, October 2000, p. 89.

Hellwig, Stacey J., Eula Ewing Monroe, and James S. Jacobs. "Making Informed Choices: Selecting Children's Trade Books for Mathematics Instruction." Vol. 7, No. 3, November 2000, p. 138.

(continued)

Lemme, Barbara. "Integrating Measurement Projects: Sand Timers." Vol. 7, No. 3, November 2000, p. 132.

Riddle, Margaret, and Bette Rodzwell. "Fractions: What Happens Between Kindergarten and the Army?" Vol. 7, No. 4, December 2000, p. 202.

2001

Karp, Karen S., and E. Todd Brown. "Geo-Dolls: Traveling in a Mathematical World." Vol. 8, No. 3, November 2001, p. 132.

Randolph, Tamela D., and Helene J. Sherman. "Alternative Algorithms: Increasing Options, Reducing Errors" Vol. 7, No. 8, April 2001, p. 480.

Sun, Wei, and Joanne Y. Zhang. "Teaching Addition and Subtraction Facts: A Chinese Perspective." Vol. 8, No. 1, September 2001, p. 28.

Whitenack, Joy W., et. al. "Second Graders Circumvent Addition and Subtraction Difficulties." Vol. 8, No. 4, December 2001, p. 228.

2002

Agosto, Melinda. "Cool Mathematics for Kids." Vol. 8, No. 7, March 2002, p. 397.

Huniker, DeAnn. "Calculators as Learning Tools for Young Children's Explorations of Number." Vol. 8, No. 6, February 2002, p. 316.

Strutchens, Marilyn E. "Multicultural Literature as a Context for Problem Solving: Children and Parents Learning Together." Vol. 8, No. 8, April 2002, p. 448.

Whitin, David J. "The Potentials and Pitfalls of Integrating Literature into the Mathematics Program." Vol. 8, No. 9, May 2002, p. 503.

Activity 4 One of the most popular mathematical films of all time is *Donald in Mathmagicland,* produced by Disney in 1959. It is available on video and may be rented at many video stores or purchased at outlets that sell video releases. Spend an entertaining half hour watching this film, and write a report on it according to the guidelines of your instructor.

Activity 5 Write a report according to the guidelines of your instructor on one of the following mathematicians, philosophers, and scientists.

Abel, N.	Cardano, G.	Gauss, C.	Noether, E.
Agnesi, M. G.	Copernicus, N.	Hilbert, D.	Pascal, B.
Agnesi, M. T.	De Morgan, A.	Kepler, J.	Plato
Al-Khowârizmi	Descartes, R.	Kronecker, L.	Polya, G.
Apollonius	Euler, L.	Lagrange, J.	Pythagoras
Archimedes	Fermat, P.	Leibniz, G.	Ramanujan, S.
Aristotle	Fibonacci	L'Hospital, G.	Riemann, G.
Babbage, C.	(Leonardo	Lobachevsky, N.	Russell, B.
Bernoulli, Jakob	of Pisa)	Mandelbrot, B.	Somerville, M.
Bernoulli,	Galileo (Galileo	Napier, J.	Tartaglia, N.
Johann	Galilei)	Nash, J.	Whitehead, A.
Cantor, G.	Galois, E.	Newton, I.	Wiles, A.

Activity 6 Write a term paper on one of the following topics in mathematics according to the guidelines of your instructor.

Babylonian mathematics	Pascal's triangle
Egyptian mathematics	The origins of probability theory
The origin of zero	Women in mathematics
Plimpton 322	Mathematical paradoxes
The Rhind papyrus	Unsolved problems in mathematics
Origins of the Pythagorean Theorem	The four-color theorem
The regular (Platonic) solids	The proof of Fermat's Last Theorem
The Pythagorean brotherhood	The search for large primes
The Golden Ratio (Golden Section)	Fractal geometry
The three famous construction problems of the Greeks	The co-inventors of calculus
The history of the approximations of π	The role of the computer in the study of mathematics
Euclid and his "Elements"	Mathematics and music
Early Chinese mathematics	Police mathematics
Early Hindu mathematics	The origins of complex numbers
Origin of the word *algebra*	Goldbach's conjecture
Magic squares	The use of the Internet in mathematics education
Figurate numbers	The development of graphing calculators
The Fibonacci sequence	Mathematics education reform movement
The Cardano/Tartaglia controversy	Multicultural mathematics
Historical methods of computation (logarithms, the abacus, Napier's rods, the slide rule, etc.)	

Activity 7 Investigate a computer program that focuses on teaching children elementary mathematics and write a critical review of it as if you were writing for a journal that contains software reviews of educational material. Be sure to address the higher-level thinking skills in addition to drill and practice.

Activity 8 The following Websites provide a fascinating list of mathematics-related topics. Go to one of them, choose a topic that interests you, and report on it, according to the guidelines of your instructor.

http://mathworld.wolfram.com/
http://world.std.com/~reinhold/mathmovies.html
http://www.mcs.surrey.ac.uk/Personal/R.Knott/
http://dir.yahoo.com/Science/Mathematics/
http://www.cut-the-knot.com/
http://www.agnesscott.edu/lriddle/women/women.htm
http://www.ics.uci.edu/~eppstein/recmath.html

COLLABORATIVE INVESTIGATION

Discovering Mathematics in Pascal's Triangle

One fascinating array of numbers, *Pascal's Triangle*, consists of rows of numbers, each of which contains one more entry than the one before. The first five rows are shown here.

To discover some of its patterns, divide the class into groups of four students each. Within each group designate one student as A, one as B, one as C, and one as D. Then perform the following activities in order.

1. Discuss among group members some of the properties of the triangle that are obvious from observing the first five rows shown.
2. It is fairly obvious that each row begins and ends with 1. Discover a method whereby the other entries in a row can be determined from the entries in the row immediately above it. (*Hint:* In the fifth row, 6 = 3 + 3.) Then, as a group, find the next four rows of the triangle, and have each member prepare his or her own copy of the entire first nine rows for later reference.
3. Now each student in the group will investigate a particular property of the triangle. In some cases, a calculator will be helpful. All students should begin working at the same time. (A discussion will follow.)
    Student A: Find the sum of the entries in each row. Notice the pattern that emerges. Now write the tenth row of the triangle.
    Student B: Investigate the successive differences in the diagonals from upper left to lower right. For example, in the diagonal that begins 1, 2, 3, 4, ..., the successive differences are all 1; in the diagonal that begins 1, 3, 6, ..., the successive differences are 2, 3, 4, and so on. Do this up through the diagonal that begins 1, 6, 21, ...
    Student C: Find the values of the first five powers of the number 11, starting with 11^0 (recall $11^0 = 1$).
    Student D: Arrange the nine rows of the triangle with all rows "flush left," and then draw lightly dashed arrows as shown:

    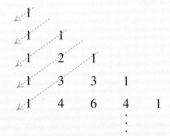

    and so on. Then add along the diagonals. Write these sums in order from left to right.
4. After all students have concluded their individual investigations in Item 3, return to a group discussion.
    (a) Have student A report the result found in Item 3, and then make a prediction concerning the sum of the entries in the tenth row.
    (b) Have student B report the successive differences discovered in the diagonals. Then have all students in the group investigate the successive differences in the diagonal that begins 1, 7, 28, ... (It may be necessary to write a few more rows of the triangle.)
    (c) Have student C report the relationship between the powers of 11 found, and then determine the value of 11^5. Why does the pattern not continue here?
    (d) Have student D report the sequence of numbers found. Then, as a group, predict what the next sum will be by observing the pattern in the sequence. Confirm your prediction by actual computation.
5. Choose a representative from each group to report to the entire class the observations made throughout this investigation.
6. Find a reference to Pascal's Triangle using a search engine of the Internet and prepare a report on the reference.

CHAPTER 1 TEST

In Exercises 1 and 2, decide whether the reasoning involved is an example of inductive or deductive reasoning.

1. Otis Taylor is a salesman for a publishing company. For the past 14 years, he has exceeded his annual sales goal, primarily by selling mathematics textbooks. Therefore, he will also exceed his annual sales goal this year.

2. For all natural numbers n, n^2 is also a natural number. 101 is a natural number. Therefore, 101^2 is a natural number.

3. What are the fourth and fifth numbers in this sequence?

    1, 4, 27, _____, _____, 46656,...

   (From *Mathematics Teacher* monthly calendar, April 25, 1994)

4. Use the list of equations and inductive reasoning to predict the next equation, and then verify your conjecture.

   $65{,}359{,}477{,}124{,}183 \times 17 = 1{,}111{,}111{,}111{,}111{,}111$

   $65{,}359{,}477{,}124{,}183 \times 34 = 2{,}222{,}222{,}222{,}222{,}222$

   $65{,}359{,}477{,}124{,}183 \times 51 = 3{,}333{,}333{,}333{,}333{,}333$

5. Use the method of successive differences to find the next term in the sequence

    3, 11, 31, 69, 131, 223, ...

6. Find the sum $1 + 2 + 3 + \cdots + 250$.

7. Consider the following equations, where the left side of each is an octagonal number.

   $1 = 1$
   $8 = 1 + 7$
   $21 = 1 + 7 + 13$
   $40 = 1 + 7 + 13 + 19$

   Use the pattern established on the right sides to predict the next octagonal number. What is the next equation in the list?

8. Use the result of Exercise 7 and the method of successive differences to find the first eight octagonal numbers. Then divide each by 4 and record the remainder. What is the pattern obtained?

9. Describe the pattern used to obtain the terms of the Fibonacci sequence 1, 1, 2, 3, 5, 8, 13, 21,....

Use problem-solving strategies to solve each of the following problems, taken from the date indicated in the monthly calendar of Mathematics Teacher.

10. *Burning Candles* One whole candle can be made from four candle "stubs" left in the candleholder after the candles burn down. If each candle burns down to a stub after one night and if you have purchased sixteen candles, then for how many nights will you have candles to burn? (October 17, 2001)

11. *Units Digit of a Power of 9* What is the units digit (ones digit) in the decimal representation of 9^{1997}? (January 27, 1997)

12. *Counting Puzzle (Triangles)* How many triangles are in this figure? (January 6, 2000)

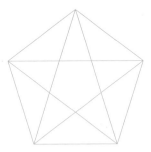

13. *Devising a Correct Addition Problem* Can you put the digits 1 through 9, each used once, in the boxes of the problem below to make an addition problem that has carrying and that is correct? If so, find a solution. If not, explain why no solution exists. (April 10, 2002)

☐☐☐ + ☐☐☐ = ☐☐☐

14. *Missing Pages in a Newspaper* A sixty-page newspaper, which consists of only one section, has the sheet with page 7 missing. What other pages are missing? (February 6, 1998)

15. *Units Digit of a Sum* Find the units digit (ones digit) of the decimal numeral representing the number $11^{11} + 14^{14} + 16^{16}$. (February 14, 1994)

16. Based on your knowledge of elementary arithmetic, describe the pattern that can be observed when the following operations are performed: 9×1, 9×2, $9 \times 3, \ldots, 9 \times 9$. (*Hint:* Add the digits in the answers. What do you notice?)

Use your calculator to evaluate each of the following. Give as many decimal places as the calculator displays.

17. $\sqrt{98.16}$ 18. 3.25^3

19. *Basketball Scoring Results* During the 1998–99 NCAA basketball season, Shawnta Rogers of George Washington University made 195 of his 503 field goal attempts. This means that for every 5 attempts, he made approximately _____ of them.
    A. 1   B. 2   C. 4   D. 3

20. *Women in Mathematics* The accompanying graph shows the number of women in mathematics or computer science professions during the past three decades.

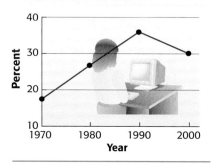

WOMEN IN MATH OR COMPUTER SCIENCE PROFESSIONS

Source: U.S. Bureau of the Census and Bureau of Labor Statistics.

(a) In what decade (10-year period) did the percent of women in math or computer science professions decrease?

(b) When did the percent of women in math or computer science professions reach a maximum?

(c) In what year was the percent of women in math or computer science professions about 27%?

chapter 2
The Basic Concepts of Set Theory

Three-dimensional computer graphics are used extensively in industrial applications involving computer-aided design of mechanical parts. Many mechanical parts are designed by a method called **constructive solid geometry**. This method uses simple elementary objects called geometric primitives to make more complex shapes. Spheres, cubes, cones, cylinders, and rectangular slabs are examples of primitives. They are conceptualized as sets of points having shape. Complex objects can be graphically designed by combining these primitive sets using operations such as union (+), difference (−), and intersection (∧). For instance, a steel washer can be created graphically by intersecting a sphere with a rectangular slab, then subtracting a cylinder. An ice cream cone is the union of a hemisphere with a cone.

2.1 Symbols and Terminology

2.2 Venn Diagrams and Subsets

2.3 Set Operations and Cartesian Products

2.4 Cardinal Numbers and Surveys

2.5 Infinite Sets and Their Cardinalities

Collaborative Investigation:
　　A Survey of Your Class

Chapter 2 Test

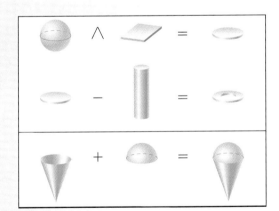

An enormous number of manufactured shapes can be created using this graphical technique. It is just one example of how the notion of a set, a fundamental concept in mathematics, is applied throughout business and science.

Source: A. Watt, *3D Computer Graphics,* Addison Wesley Publishing Company, 1993.

2.1 Symbols and Terminology

The basic ideas of set theory were developed by the German mathematician **Georg Cantor** (1845–1918) in about 1875. Some of the things he proved flew in the face of accepted mathematical beliefs of the times. Cantor created a new field of theory and at the same time continued the long debate over infinity that began in ancient times. He developed counting by one-to-one correspondence to determine how many objects are contained in a set. Infinite sets differ from finite sets by not obeying the familiar law that the whole is greater than any of its parts.

The human mind likes to create collections. Instead of seeing a group of five stars as five separate items, people tend to see them as one group of stars. The mind tries to find order and patterns. In mathematics this tendency to create collections is represented with the idea of a *set*. A **set** is a collection of objects. The objects belonging to the set are called the **elements,** or **members,** of the set.

Sets are designated in at least the following three ways: by (1) *word description,* (2) the *listing method,* and (3) *set-builder notation.* A given set may be more conveniently denoted by one method rather than another, but most sets can be given in any of the three ways. For example, the word description

the set of even counting numbers less than ten

can be expressed by listing

$$\{2, 4, 6, 8\}$$

or by set-builder notation

$$\{x \mid x \text{ is an even counting number less than } 10\}.$$

In the listing and set-builder notations, the braces at the beginning and ending indicate a set. Also, in the listing method, the commas are essential. Set-builder notation utilizes the algebraic idea of a variable. (Any symbol would do, but just as in other algebraic applications, the letter x is a common choice.) Before the vertical line we give the variable, which represents an element in general, and after the vertical line we state the criteria by which an element qualifies for membership in the set. By including *all* objects that meet the stated criteria, we generate (or build) the entire set.

Sets are commonly given names (usually capital letters) so that they can be easily referred to later in the discussion. If E is selected as a name for the set of all letters of the English alphabet, then we can write

$$E = \{a, b, c, d, e, f, g, h, i, j, k, l, m, n, o, p, q, r, s, t, u, v, w, x, y, z\}.$$

In many cases, the listing notation can be shortened by clearly establishing the pattern of elements included, and using ellipsis points (three dots) to indicate a continuation of the pattern. Thus, for example,

$$E = \{a, b, c, d, \ldots, x, y, z\} \quad \text{or} \quad E = \{a, b, c, d, e, \ldots, z\}.$$

EXAMPLE 1 Give a complete listing of all the elements of each of the following sets.

(a) the set of counting numbers between six and thirteen

This set can be denoted $\{7, 8, 9, 10, 11, 12\}$. (Notice that the word *between* excludes the endpoint values.)

(b) $\{5, 6, 7, \ldots, 13\}$

This set begins with the element 5, then 6, then 7, and so on, with each element obtained by adding 1 to the previous element in the list. This pattern stops at 13, so a complete listing is $\{5, 6, 7, 8, 9, 10, 11, 12, 13\}$.

(c) $\{x \mid x \text{ is a counting number between 6 and 7}\}$

After a little thought, we realize that there are no counting numbers between 6 and 7, so the listing for this set will contain no elements at all. We write the set as $\{\ \}$ or $\emptyset$.

A set containing no elements, such as the set in Example 1(c), is called the **empty set,** or **null set.** The symbol ∅ often is used to denote the empty set, so ∅ and { } have the same meaning. We do *not* denote the empty set with the symbol {∅}, since this notation represents a set with one element (that element being the empty set).

Example 1 referred to counting numbers (or natural numbers), which were introduced in Section 1.1. Other important categories of numbers, which will be used throughout the text, are summarized below.

Sets of Numbers

Natural or Counting numbers $\{1, 2, 3, 4, \ldots\}$

Whole numbers $\{0, 1, 2, 3, 4, \ldots\}$

Integers $\{\ldots, -3, -2, -1, 0, 1, 2, 3, \ldots\}$

Rational numbers $\{p/q \mid p \text{ and } q \text{ are integers, and } q \neq 0\}$
(Some examples of rational numbers are $3/5$, $-7/9$, 5, and 0. Any rational number may be written as a terminating decimal number, like $.25$, or a repeating decimal number, like $.666\ldots$.)

Real numbers $\{x \mid x \text{ is a number that can be written as a decimal}\}$

Irrational numbers $\{x \mid x \text{ is a real number and } x \text{ cannot be written as a quotient of integers}\}$
(Some examples of irrational numbers are $\sqrt{2}$, $\sqrt[3]{4}$, and π. Decimal representations of irrational numbers never terminate and never repeat.)

The number of elements in a set is called the **cardinal number,** or **cardinality,** of the set. The symbol $n(A)$, which is read "*n* of *A*," represents the cardinal number of set *A*.

If elements are repeated in a set listing, they should not be counted more than once when determining the cardinal number of the set. For example, the set $B = \{1, 1, 2, 2, 3\}$ has only three distinct elements, and so $n(B) = 3$.

EXAMPLE 2 Find the cardinal number of each of the following sets.

(a) $K = \{2, 4, 8, 16\}$

Set *K* contains four elements, so the cardinal number of set *K* is 4, and $n(K) = 4$.

(b) $M = \{0\}$

Set *M* contains only one element, zero, so $n(M) = 1$.

(c) $R = \{4, 5, \ldots, 12, 13\}$

There are only four elements listed, but the ellipsis indicates that there are other elements in the set. Counting them, we find that there are ten elements, so $n(R) = 10$.

(d) The empty set, ∅, contains no elements, and $n(\emptyset) = 0$.

If the cardinal number of a set is a particular whole number (0 or a counting number), as in all parts of Example 2 above, we call that set a **finite set.** Given enough time, we could finish counting all the elements of any finite set and arrive at

Infinity A close-up of a camera lens shows the infinity sign, ∞, defined as any distance greater than 1000 times the focal length of a lens.

The sign was invented by the mathematician John Wallis in 1655. Wallis used $1/\infty$ to represent an infinitely small quantity. The philosopher Voltaire described the ∞ as a loveknot, and he was skeptical about the sign making the idea of infinity any clearer.

its cardinal number. Some sets, however, are so large that we could never finish the counting process. The counting numbers themselves are such a set. Whenever a set is so large that its cardinal number is not found among the whole numbers, we call that set an **infinite set.** Cardinal numbers of infinite sets will be discussed in a later section. Infinite sets can also be designated using the three methods already mentioned.

EXAMPLE 3 Designate all odd counting numbers by the three common methods of set notation.

$$\text{The set of all odd counting numbers} \quad \text{Word description}$$
$$\{1, 3, 5, 7, 9, \ldots\} \quad \text{Listing method}$$

As indicated by the ellipsis points, there is no final element given. The listing goes on forever.

$$\{x \mid x \text{ is an odd counting number}\} \quad \text{Set-builder notation}$$

For a set to be useful, it must be *well-defined*. This means that if a particular set and some particular element are given, it must be possible to tell whether or not the element belongs to the set. For example, the preceding set E of the letters of the English alphabet is well-defined. Given the letter q, we know that q is an element of E. Given the Greek letter θ (theta), we know that it is not an element of set E.

However, given the set C of all fat chickens, and a particular chicken, Herman, it is not possible to say whether

$$\text{Herman is an element of } C \quad \text{or} \quad \text{Herman is } not \text{ an element of } C.$$

The problem is the word "fat"; how fat is fat? Since we cannot necessarily decide whether or not a given chicken belongs to set C, set C is not well-defined.

The letter q is an element of set E, where E is the set of all the letters of the English alphabet. To show this, $\in$ is used to replace the words "is an element of," or

$$q \in E,$$

which is read "q is an element of set E." The letter θ is not an element of E. To show this, $\in$ and a slash mark is used to replace the words "is not an element of," written

$$\theta \notin E.$$

This is read "θ is not an element of set E."

EXAMPLE 4 Decide whether each statement is true or false.

(a) $3 \in \{-3, -1, 5, 9, 13\}$

Since 3 is *not* an element of the set $\{-3, -1, 5, 9, 13\}$, the statement is false.

(b) $0 \in \{-3, -2, 0, 1, 2, 3\}$

Since 0 is indeed an element of the set $\{-3, -2, 0, 1, 2, 3\}$, the statement is true.

(c) $1/5 \notin \{1/3, 1/4, 1/6\}$

This statement says that 1/5 is not an element of the set $\{1/3, 1/4, 1/6\}$, which is true.

We now consider the concept of set equality.

Set Equality

Set A is **equal** to set B provided the following two conditions are met:
1. every element of A is an element of B, and
2. every element of B is an element of A.

Informally, two sets are equal if they contain exactly the same elements, regardless of order. For example,

$$\{a,b,c,d\} = \{a,c,d,b\}$$

since both sets contain exactly the same elements.

Because repetition of elements in a set listing does not add new elements, we can say that

$$\{1,0,1,2,3,3\} = \{0,1,2,3\}$$

since these sets contain exactly the same elements.

EXAMPLE 5 Are $\{-4,3,2,5\}$ and $\{-4,0,3,2,5\}$ equal sets?

Every element of the first set is an element of the second; however, 0 is an element of the second and not the first. In other words, the sets do not contain exactly the same elements, so they are not equal:

$$\{-4,3,2,5\} \neq \{-4,0,3,2,5\}.$$

Although any of the three notations can designate a given set, sometimes one will be clearer than another. For example,

$$\{x \mid x \text{ is an odd counting number between 2 and 810}\}$$

may be preferable to

$$\{3,5,7,9,\ldots,809\}$$

since it clearly states the common property that identifies all elements of the set. When working with sets, it is necessary to be able to translate from one notation to another.

EXAMPLE 6 Decide whether each statement is *true* or *false*.

(a) $\{3\} = \{x \mid x \text{ is a counting number between 1 and 5}\}$

The set on the right contains *all* counting numbers between 1 and 5, namely 2, 3, and 4, while the set on the left contains *only* the number 3. Since the sets do not contain exactly the same elements, they are not equal. The statement is false.

(b) $\{x \mid x \text{ is a negative natural number}\} = \{y \mid y \text{ is a number that is both rational and irrational}\}$

All natural numbers are positive, so the first set is $\emptyset$. By definition, if a number is rational, it cannot be irrational, so the set on the right is also $\emptyset$. Since each set is the empty set, the sets are equal. The statement is true.

2.1 EXERCISES

Match each set in Column I with the appropriate description in Column II.

I

1. $\{2, 4, 6, 8\}$
2. $\{x \mid x$ is an even integer greater than 4 and less than 6$\}$
3. $\{\ldots, -4, -3, -2, -1\}$
4. $\{\ldots, -6, -4, -2, 0, 2, 4, 6, \ldots\}$
5. $\{2, 4, 8, 16, 32\}$
6. $\{\ldots, -5, -3, -1, 1, 3, 5, \ldots\}$
7. $\{2, 4, 6, 8, 10\}$
8. $\{1, 3, 5, 7, 9\}$

II

A. the set of all even integers
B. the set of the five least positive integer powers of 2
C. the set of even positive integers less than 10
D. the set of all odd integers
E. the set of all negative integers
F. the set of odd positive integers less than 10
G. ∅
H. the set of the five least positive integer multiples of 2

List all the elements of each set. Use set notation to describe the set.

9. the set of all counting numbers less than or equal to 6
10. the set of all whole numbers greater than 8 and less than 18
11. the set of all whole numbers not greater than 4
12. the set of all counting numbers between 4 and 14
13. $\{6, 7, 8, \ldots, 14\}$
14. $\{3, 6, 9, 12, \ldots, 30\}$
15. $\{-15, -13, -11, \ldots, -1\}$
16. $\{-4, -3, -2, \ldots, 4\}$
17. $\{2, 4, 8, \ldots, 256\}$
18. $\{90, 87, 84, \ldots, 69\}$
19. $\{x \mid x$ is an even whole number less than 11$\}$
20. $\{x \mid x$ is an odd integer between -8 and 7$\}$

Denote each set by the listing method. There may be more than one correct answer.

21. the set of all counting numbers greater than 20
22. the set of all integers between -200 and 500
23. the set of Great Lakes
24. the set of United States presidents who served after Lyndon Johnson and before William Clinton (See the photograph.)

25. $\{x \mid x$ is a positive multiple of 5$\}$
26. $\{x \mid x$ is a negative multiple of 6$\}$
27. $\{x \mid x$ is the reciprocal of a natural number$\}$
28. $\{x \mid x$ is a positive integer power of 4$\}$

Denote each set by set-builder notation, using x as the variable. There may be more than one correct answer.

29. the set of all rational numbers
30. the set of all even natural numbers
31. $\{1, 3, 5, \ldots, 75\}$
32. $\{35, 40, 45, \ldots, 95\}$

Identify each set as finite *or* infinite.

33. {2, 4, 6, ..., 32}

34. {6, 12, 18}

35. {1/2, 2/3, 3/4, ...}

36. {−10, −8, −6, ...}

37. {$x \mid x$ is a natural number greater than 50}

38. {$x \mid x$ is a natural number less than 50}

39. {$x \mid x$ is a rational number}

40. {$x \mid x$ is a rational number between 0 and 1}

Find $n(A)$ for each set.

41. A = {0, 1, 2, 3, 4, 5, 6, 7}

42. A = {−3, −1, 1, 3, 5, 7, 9}

43. A = {2, 4, 6, ..., 1000}

44. A = {0, 1, 2, 3, ..., 3000}

45. A = {a, b, c, ..., z}

46. A = {$x \mid x$ is a vowel in the English alphabet}

47. A = the set of integers between −20 and 20

48. A = the set of current U.S. senators

49. A = {1/3, 2/4, 3/5, 4/6, ..., 27/29, 28/30}

50. A = {1/2, −1/2, 1/3, −1/3, ..., 1/10, −1/10}

51. Explain why it is acceptable to write the statement "x is a vowel in the English alphabet" in the set for Exercise 46, despite the fact that x is a consonant.

52. Explain how Exercise 49 can be answered without actually listing and then counting all the elements.

Identify each set as well-defined *or* not well-defined.

53. {$x \mid x$ is a real number}

54. {$x \mid x$ is a negative number}

55. {$x \mid x$ is a good athlete}

56. {$x \mid x$ is a skillful typist}

57. {$x \mid x$ is a difficult course}

58. {$x \mid x$ is a counting number less than 2}

Fill each blank with either $\in$ or $\notin$ to make the following statements true.

59. 5 _____ {2, 4, 5, 7}

60. 8 _____ {3, −2, 5, 7, 8}

61. −4 _____ {4, 7, 8, 12}

62. −12 _____ {3, 8, 12, 18}

63. 0 _____ {−2, 0, 5, 9}

64. 0 _____ {3, 4, 6, 8, 10}

65. {3} _____ {2, 3, 4, 6}

66. {6} _____ {3, 4, 5, 6, 7}

Write true *or* false *for each of the following statements.*

67. 3 ∈ {2, 5, 6, 8}

68. 6 ∈ {−2, 5, 8, 9}

69. b ∈ {h, c, d, a, b}

70. m ∈ {1, m, n, o, p}

71. 9 ∉ {6, 3, 4, 8}

72. 2 ∉ {7, 6, 5, 4}

73. {k, c, r, a} = {k, c, a, r}

74. {e, h, a, n} = {a, h, e, n}

75. {5, 8, 9} = {5, 8, 9, 0}

76. {3, 7, 12, 14} = {3, 7, 12, 14, 0}

77. {$x \mid x$ is a natural number less than 3} = {1, 2}

78. {$x \mid x$ is a natural number greater than 10} = {11, 12, 13, ...}

Write true *or* false *for each of the following statements.*

Let A = {2, 4, 6, 8, 10, 12}, B = {2, 4, 8, 10}, *and* C = {4, 10, 12}.

79. 4 ∈ A

80. 8 ∈ B

81. 4 ∉ C

82. 8 ∉ B

83. Every element of C is also an element of A.

84. Every element of C is also an element of B.

85. This section opened with the statement, "The human mind likes to create collections." Why do you suppose this is so? In explaining your thoughts, utilize one or more particular "collections," mathematical or otherwise.

86. Explain the difference between a well-defined set and a not well-defined set. Give examples and utilize terms introduced in this section.

Recall that two sets are called equal *if they contain identical elements.* On the other hand, two sets are called equivalent *if they contain the same number of elements (but not necessarily the same elements).* For each of the following conditions, give an example or explain why it is impossible.

87. two sets that are neither equal nor equivalent
88. two sets that are equal but not equivalent
89. two sets that are equivalent but not equal
90. two sets that are both equal and equivalent
91. *Volumes of Stocks* The table lists the most active stocks on the American Stock Exchange in a recent year, with share volumes (in millions).

Issue	Share Volume (millions)
1. Viacom (Class B)	223.1
2. Trans World Airlines	222.8
3. Harken Energy	214.5
4. Echo Bay Mines	188.7
5. JTS	177.0
6. Nabors Industries	172.9
7. Hasbro	159.1
8. Royal Oak Mines	140.6
9. Grey Wolf Industries	132.9
10. IVAX	132.5

(*Source:* American Stock Exchange.)

(a) List the set of issues that had a share volume of at least 188.7 million.
(b) List the set of issues that had a share volume of at most 188.7 million.

92. *Burning Calories* Jaime Bailey is health conscious, but she does like a certain chocolate bar, each of which contains 220 calories. In order to burn off unwanted calories, Jaime participates in her favorite activities, shown below, in increments of one hour and never repeats a given activity on a given day.

Activity	Symbol	Calories Burned per Hour
Volleyball	v	160
Golf	g	260
Canoeing	c	340
Swimming	s	410
Running	r	680

(a) On Monday, Jaime has time for no more than two hours of activities. List all possible sets of activities that would burn off at least the number of calories obtained from three chocolate bars.
(b) Assume that Jaime can afford up to three hours of time for activities on Saturday. List all sets of activities that would burn off at least the number of calories in five chocolate bars.

2.2 Venn Diagrams and Subsets

In every problem there is either a stated or implied **universe of discourse.** The universe of discourse includes all things under discussion at a given time. For example, in studying reactions to a proposal that a certain campus raise the minimum age of individuals to whom beer may be sold, the universe of discourse might be all the students at the school, the nearby members of the public, the board of trustees of the school, or perhaps all these groups of people.

In the mathematical theory of sets, the universe of discourse is called the **universal set.** The letter U is typically used for the universal set. The universal set might change from problem to problem.

In most areas of mathematics, our reasoning can be aided and clarified by utilizing various kinds of drawings and diagrams. In set theory, we commonly use **Venn diagrams,** developed by the logician John Venn (1834–1923). In these diagrams, the universal set is represented by a rectangle, and other sets of interest within the universal set are depicted by oval regions, or sometimes by circles or other shapes. In the Venn diagram of Figure 1, the entire region bounded by the rectangle represents the universal set U, while the portion bounded by the oval represents set A. (The size of the oval representing A is irrelevant.) The colored region inside U and outside the oval is labeled A' (read "A prime"). This set, called the *complement* of A, contains all elements that are contained in U but not contained in A.

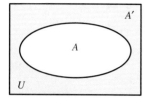

FIGURE 1

The Complement of a Set

For any set A within the universal set U, the **complement** of A, written A', is the set of elements of U that are not elements of A. That is,

$$A' = \{x | x \in U \text{ and } x \notin A\}.$$

EXAMPLE 1 Let $U = \{a, b, c, d, e, f, g, h\}$,
$M = \{a, b, e, f\}$,
$N = \{b, d, e, g, h\}$.

Find each of the following sets.

(a) M'

Set M' contains all the elements of set U that are not in set M. Since set M contains the elements a, b, e, and f, these elements will be disqualified from belonging to set M', and consequently set M' will contain c, d, g, and h, or $M' = \{c, d, g, h\}$.

(b) N'

Set N' contains all the elements of U that are not in set N, so $N' = \{a, c, f\}$.

Consider the complement of the universal set, U'. The set U' is found by selecting all the elements of U that do not belong to U. There are no such elements, so there can be no elements in set U'. This means that for any universal set U, $U' = \emptyset$.

Now consider the complement of the empty set, $\emptyset'$. Since $\emptyset' = \{x | x \in U \text{ and } x \notin \emptyset\}$ and set $\emptyset$ contains no elements, every member of the universal set U satisfies this definition. Therefore for any universal set U, $\emptyset' = U$.

Suppose that we are given the universal set $U = \{1, 2, 3, 4, 5\}$, while $A = \{1, 2, 3\}$. Every element of set A is also an element of set U. Because of this, set A is called a *subset* of set U, written

$$A \subseteq U.$$

("A is not a subset of set U" would be written $A \nsubseteq U$.) A Venn diagram showing that set M is a subset of set N is shown in Figure 2.

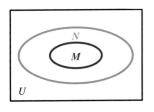

FIGURE 2

Subset of a Set

Set A is a **subset** of set B if every element of A is also an element of B. In symbols,

$$A \subseteq B.$$

EXAMPLE 2 Write $\subseteq$ or $\nsubseteq$ in each blank to make a true statement.

(a) $\{3, 4, 5, 6\}$ _____ $\{3, 4, 5, 6, 8\}$

Since every element of $\{3, 4, 5, 6\}$ is also an element of $\{3, 4, 5, 6, 8\}$, the first set is a subset of the second, so $\subseteq$ goes in the blank.

(b) $\{1, 2, 3\}$ _____ $\{2, 4, 6, 8\}$

The element 1 belongs to $\{1, 2, 3\}$ but not to $\{2, 4, 6, 8\}$. Place $\not\subseteq$ in the blank.

(c) $\{5, 6, 7, 8\}$ _____ $\{5, 6, 7, 8\}$

Every element of $\{5, 6, 7, 8\}$ is also an element of $\{5, 6, 7, 8\}$. Place $\subseteq$ in the blank.

As Example 2(c) suggests, every set is a subset of itself:

$$B \subseteq B \quad \text{for any set } B.$$

The statement of set equality in Section 2.1 can be formally presented using subset terminology.

Set Equality (Alternative definition)

If A and B are sets, then $A = B$ if $A \subseteq B$ and $B \subseteq A$.

When studying subsets of a set B, it is common to look at subsets other than set B itself. Suppose that $B = \{5, 6, 7, 8\}$ and $A = \{6, 7\}$. A is a subset of B, but A is not all of B; there is at least one element in B that is not in A. (Actually, in this case there are two such elements, 5 and 8.) In this situation, A is called a *proper subset* of B. To indicate that A is a proper subset of B, write $A \subset B$.

Proper Subset of a Set

Set A is a **proper subset** of set B if $A \subseteq B$ and $A \neq B$. In symbols,

$$A \subset B.$$

(Notice the similarity of the subset symbols, $\subset$ and $\subseteq$, to the inequality symbols from algebra, $<$ and $\leq$.)

EXAMPLE 3 Decide whether $\subset$, $\subseteq$, or both could be placed in each blank to make a true statement.

(a) $\{5, 6, 7\}$ _____ $\{5, 6, 7, 8\}$

Every element of $\{5, 6, 7\}$ is contained in $\{5, 6, 7, 8\}$, so $\subseteq$ could be placed in the blank. Also, the element 8 belongs to $\{5, 6, 7, 8\}$ but not to $\{5, 6, 7\}$, making $\{5, 6, 7\}$ a proper subset of $\{5, 6, 7, 8\}$. This means that $\subset$ could also be placed in the blank.

(b) $\{a, b, c\}$ _____ $\{a, b, c\}$

The set $\{a, b, c\}$ is a subset of $\{a, b, c\}$. Since the two sets are equal, $\{a, b, c\}$ is not a proper subset of $\{a, b, c\}$. Only $\subseteq$ may be placed in the blank.

Set A is a subset of set B if every element of set A is also an element of set B. This definition can be reworded by saying that set A is a subset of set B if there are no elements of *A* that are not also elements of *B*. This second form of the definition shows that the empty set is a subset of any set, or

$$\emptyset \subseteq B \quad \text{for any set } B.$$

This is true since it is not possible to find any elements of ∅ that are not also in *B*. (There are no elements in ∅.) The empty set ∅ is a proper subset of every set except itself:

$$\emptyset \subset B \quad \text{if } B \text{ is any set other than } \emptyset.$$

Every set (except ∅) has at least two subsets, ∅ and the set itself.

EXAMPLE 4 Find all possible subsets of each set.

(a) {7, 8}

By trial and error, the set {7, 8} has four subsets:

$$\emptyset, \{7\}, \{8\}, \{7, 8\}.$$

(b) {a, b, c}

Here trial and error leads to 8 subsets for {a, b, c}:

$$\emptyset, \{a\}, \{b\}, \{c\}, \{a, b\}, \{a, c\}, \{b, c\}, \{a, b, c\}.$$

In Example 4, the subsets of {7, 8} and the subsets of {a, b, c} were found by trial and error. An alternative method involves drawing a **tree diagram,** a systematic way of listing all the subsets of a given set. Figures 3(a) and (b) show tree diagrams for {7, 8} and {a, b, c}.

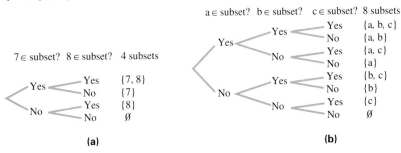

FIGURE 3

In Example 4, we determined the number of subsets of a given set by making a list of all such subsets and then counting them. The tree diagram method also produced a list of all possible subsets. In many applications, we don't need to display all the subsets but simply determine how many there are. Furthermore, the trial and error method and the tree diagram method would both involve far too much work if the original set had a very large number of elements. For these reasons, it is desirable to have a formula for the number of subsets. To obtain such a formula, we use inductive reasoning. That is, we observe particular cases to try to discover a general pattern.

Begin with the set containing the least number of elements possible—the empty set. This set, ∅, has only one subset, ∅ itself. Next, a set with one element has only two subsets, itself and ∅. These facts, together with those obtained above for sets with two and three elements, are summarized here.

Number of elements	0	1	2	3
Number of subsets	1	2	4	8

This chart suggests that as the number of elements of the set increases by one, the number of subsets doubles. This suggests that the number of subsets in each case might be a power of 2. Every number in the second row of the chart is indeed a power of 2. Add this information to the chart.

Number of elements	0	1	2	3
Number of subsets	$1 = 2^0$	$2 = 2^1$	$4 = 2^2$	$8 = 2^3$

This chart shows that the number of elements in each case is the same as the exponent on the 2. Inductive reasoning gives the following generalization.

Number of Subsets

The number of subsets of a set with n elements is 2^n.

Since the value 2^n includes the set itself, we must subtract 1 from this value to obtain the number of proper subsets of a set containing n elements.

Number of Proper Subsets

The number of proper subsets of a set with n elements is $2^n - 1$.

As shown in the chapter on problem solving, although inductive reasoning is a good way of *discovering* principles or arriving at a *conjecture,* it does not provide a proof that the conjecture is true in general. A proof must be provided by other means. The two formulas above are true, by observation, for $n = 0, 1, 2,$ or 3. (For a general proof, see Exercise 71 at the end of this section.)

EXAMPLE 5 Find the number of subsets and the number of proper subsets of each set.

(a) $\{3, 4, 5, 6, 7\}$

This set has 5 elements and $2^5 = 2 \times 2 \times 2 \times 2 \times 2 = 32$ subsets. Of these, $2^5 - 1 = 32 - 1 = 31$ are proper subsets.

(b) $\{1, 2, 3, 4, 5, 9, 12, 14\}$

This set has 8 elements. There are $2^8 = 256$ subsets and 255 proper subsets.

Powers of 2

$2^0 = 1$
$2^1 = 2$
$2^2 = 2 \times 2 = 4$
$2^3 = 2 \times 2 \times 2 = 8$
$2^4 = 2 \times 2 \times 2 \times 2 = 16$
$2^5 = 32$
$2^6 = 64$
$2^7 = 128$
$2^8 = 256$
$2^9 = 512$
$2^{10} = 1024$
$2^{11} = 2048$
$2^{12} = 4096$
$2^{15} = 32,768$
$2^{20} = 1,048,576$
$2^{25} = 33,554,432$
$2^{30} = 1,073,741,824$

The small numbers at the upper right are called *exponents.* Here the *base* is 2. Notice how quickly the values grow; this is the origin of the phrase *exponential growth.*

2.2 EXERCISES

Match each set or sets in Column I with the appropriate description in Column II.

I
1. $\{p\}, \{q\}, \{p,q\}, \emptyset$
2. $\{p\}, \{q\}, \emptyset$
3. $\{a,b\}$
4. $\emptyset$
5. U
6. $\{a\}$

II
A. the complement of $\emptyset$
B. the proper subsets of $\{p,q\}$
C. the complement of $\{c,d\}$, if $U = \{a,b,c,d\}$
D. the complement of U
E. the complement of $\{b\}$, if $U = \{a,b\}$
F. the subsets of $\{p,q\}$

Insert $\subseteq$ or $\nsubseteq$ in each blank so that the resulting statement is true.

7. $\{-2, 0, 2\}$ _____ $\{-2, -1, 1, 2\}$
8. $\{M, W, F\}$ _____ $\{S, M, T, W, Th\}$
9. $\{2, 5\}$ _____ $\{0, 1, 5, 3, 4, 2\}$
10. $\{a, n, d\}$ _____ $\{r, a, n, d, y\}$
11. $\emptyset$ _____ $\{a, b, c, d, e\}$
12. $\emptyset$ _____ $\emptyset$
13. $\{-7, 4, 9\}$ _____ $\{x \mid x \text{ is an odd integer}\}$
14. $\{2, 1/3, 5/9\}$ _____ the set of rational numbers

Decide whether $\subset$, $\subseteq$, both, or neither can be placed in each blank to make the statement true.

15. $\{B, C, D\}$ _____ $\{B, C, D, F\}$
16. $\{red, blue, yellow\}$ _____ $\{yellow, blue, red\}$
17. $\{9, 1, 7, 3, 5\}$ _____ $\{1, 3, 5, 7, 9\}$
18. $\{S, M, T, W, Th\}$ _____ $\{M, W, Th, S\}$
19. $\emptyset$ _____ $\{0\}$
20. $\emptyset$ _____ $\emptyset$
21. $\{-1, 0, 1, 2, 3\}$ _____ $\{0, 1, 2, 3, 4\}$
22. $\{5/6, 9/8\}$ _____ $\{6/5, 8/9\}$

For Exercises 23–42, tell whether each statement is true *or* false.

Let $U = \{a, b, c, d, e, f, g\}$, $A = \{a, e\}$, $B = \{a, b, e, f, g\}$, $C = \{b, f, g\}$, $D = \{d, e\}$.

23. $A \subset U$
24. $C \subset U$
25. $D \subseteq B$
26. $D \subseteq A$
27. $A \subset B$
28. $B \subseteq C$
29. $\emptyset \subset A$
30. $\emptyset \subseteq D$
31. $\emptyset \subseteq \emptyset$
32. $D \subset B$
33. $D \nsubseteq B$
34. $A \nsubseteq B$
35. There are exactly 6 subsets of C.
36. There are exactly 31 subsets of B.
37. There are exactly 3 subsets of A.
38. There are exactly 4 subsets of D.
39. There is exactly 1 subset of $\emptyset$.
40. There are exactly 127 proper subsets of U.
41. The Venn diagram below correctly represents the relationship among sets A, C, and U.
42. The Venn diagram below correctly represents the relationship among sets B, C, and U.

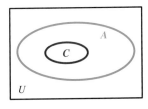
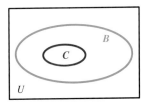

Find the number of subsets and the number of proper subsets of each of the following sets.

43. $\{1, 5, 10\}$

44. $\{8, 6, 4, 2\}$

45. $\{a, b, c, d, e, f\}$

46. the set of days of the week

47. $\{x \mid x \text{ is an odd integer between } -6 \text{ and } 4\}$

48. $\{x \mid x \text{ is an odd whole number less than } 4\}$

Let $U = \{1, 2, 3, 4, 5, 6, 7, 8, 9, 10\}$ and find the complement of each of the following sets.

49. $\{1, 4, 6, 8\}$

50. $\{2, 5, 7, 9, 10\}$

51. $\{1, 3, 4, 5, 6, 7, 8, 9, 10\}$

52. $\{1, 2, 3, 4, 6, 7, 8, 9, 10\}$

53. $\emptyset$

54. U

Vacationing in Washington, D.C. Terry McGinnis is planning to take her two sons to Washington, D.C., during their Thanksgiving vacation. In weighing her options concerning whether to fly or drive from their home in Iowa, she has listed the following characteristics.

Fly to D.C.	Drive to D.C.
Higher cost	Lower cost
Educational	Educational
More time to see the sights	Less time to see the sights
Cannot visit relatives along the way	Can visit relatives along the way

Refer to these characteristics in Exercises 55–60.

55. Find the smallest universal set U that contains all listed characteristics of both options.

Let F represent the set of characteristics of the flying option and let D represent the set of characteristics of the driving option. Use the universal set from Exercise 55.

56. Give the set F'.

57. Give the set D'.

Find the set of elements common to both sets in Exercises 58–60.

58. F and D

59. F' and D'

60. F and D'

Meeting in a Hospitality Suite Alison Romike, Bill Leonard, Charlie Dawkins, Dierdre Lynch, and Ed Moura plan to meet at the hospitality suite after the CEO makes his speech at the January sales meeting of their publishing company. Denoting these five people by A, B, C, D, and E, list all the possible sets of this group in which the given number of them can gather.

61. five people

62. four people

63. three people

64. two people

65. one person

66. no people

67. Find the total number of ways that members of this group can gather in the suite. (*Hint:* Find the total number of sets in your answers to Exercises 61–66.)

68. How does your answer in Exercise 67 compare with the number of subsets of a set of 5 elements? How can you interpret the answer to Exercise 67 in terms of subsets?

69. *Selecting Bills from a Wallet* Suppose that in your wallet you have the bills shown here.

    (a) If you must select at least one bill, and you may select up to all of the bills, how many different sums of money could you make?

    (b) In part (a), remove the condition "you must select at least one bill." Now, how many sums are possible?

70. *Selecting Coins* The photo shows a 1996 United States proof set, which contains one each of the penny, nickel, dime, quarter, and half dollar. Repeat Exercise 69, replacing "bill(s)" with "coin(s)."

71. In discovering the expression (2^n) for finding the number of subsets of a set with n elements, we observed that for the first few values of n, increasing the number of elements by one doubles the number of subsets. Here, you can prove the formula in general by showing that the same is true for any value of n. Assume set A has n elements and s subsets. Now add one additional element, say e, to the set A. (We now have a new set, say B, with $n + 1$ elements.) Divide the subsets of B into those that do not contain e and those that do.

(a) How many subsets of B do not contain e? (*Hint:* Each of these is a subset of the original set A.)

(b) How many subsets of B do contain e? (*Hint:* Each of these would be a subset of the original set A, with the element e inserted.)

(c) What is the total number of subsets of B?

(d) What do you conclude?

72. Explain why $\{\emptyset\}$ has $\emptyset$ as a subset and also has $\emptyset$ as an element.

2.3 Set Operations and Cartesian Products

Intersection of Sets Two candidates, Adelaide Boettner and David Berman, are running for a seat on the city council. A voter deciding for whom she should vote recalled the following campaign promises made by the candidates. Each promise is given a code letter.

Honest Adelaide Boettner	Determined David Berman
Spend less money, m	Spend less money, m
Emphasize traffic law enforcement, t	Crack down on crooked politicians, p
Increase service to suburban areas, s	Increase service to the city, c

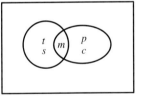

FIGURE 4

The only promise common to both candidates is promise m, to spend less money. Suppose we take each candidate's promises to be a set. The promises of Boettner give the set $\{m, t, s\}$, while the promises of Berman give $\{m, p, c\}$. The only element common to both sets is m; this element belongs to the *intersection* of the two sets $\{m, t, s\}$ and $\{m, p, c\}$, as shown in color in the Venn diagram in Figure 4. In symbols,

$$\{m, t, s\} \cap \{m, p, c\} = \{m\},$$

where the cap-shaped symbol $\cap$ represents intersection. Notice that the intersection of two sets is itself a set.

Intersection of Sets

The **intersection** of sets A and B, written $A \cap B$, is the set of elements common to both A and B, or

$$A \cap B = \{x \mid x \in A \text{ and } x \in B\}.$$

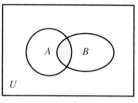

$A \cap B$

FIGURE 5

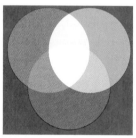

White light can be viewed as the intersection of the three primary colors.

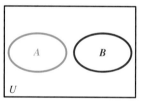

Disjoint sets

FIGURE 6

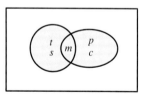

FIGURE 7

Form the intersection of sets *A* and *B* by taking all the elements included in both sets, as shown in color in Figure 5.

EXAMPLE 1 Find the intersection of the given sets.

(a) $\{3, 4, 5, 6, 7\}$ and $\{4, 6, 8, 10\}$

Since the elements common to both sets are 4 and 6,

$$\{3, 4, 5, 6, 7\} \cap \{4, 6, 8, 10\} = \{4, 6\}.$$

(b) $\{9, 14, 25, 30\}$ and $\{10, 17, 19, 38, 52\}$

These two sets have no elements in common, so

$$\{9, 14, 25, 30\} \cap \{10, 17, 19, 38, 52\} = \emptyset.$$

(c) $\{5, 9, 11\}$ and $\emptyset$

There are no elements in $\emptyset$, so there can be no elements belonging to both $\{5, 9, 11\}$ and $\emptyset$. Because of this,

$$\{5, 9, 11\} \cap \emptyset = \emptyset.$$

Examples 1(b) and 1(c) show two sets that have no elements in common. Sets with no elements in common are called **disjoint sets.** A set of dogs and a set of cats would be disjoint sets. In mathematical language, sets *A* and *B* are disjoint if $A \cap B = \emptyset$. Two disjoint sets *A* and *B* are shown in Figure 6.

Union of Sets At the beginning of this section, we showed lists of campaign promises of two candidates running for city council. Suppose a pollster wants to summarize the types of promises made by candidates for the office. The pollster would need to study *all* the promises made by *either* candidate, or the set

$$\{m, t, s, p, c\},$$

the *union* of the sets of promises made by the two candidates, as shown in color in the Venn diagram in Figure 7. In symbols,

$$\{m, t, s\} \cup \{m, p, c\} = \{m, t, s, p, c\},$$

where the cup-shaped symbol ∪ denotes set union. Be careful not to confuse this symbol with the universal set *U*. Again, the union of two sets is a set.

Union of Sets

The **union** of sets *A* and *B*, written $A \cup B$, is the set of all elements belonging to either of the sets, or

$$A \cup B = \{x | x \in A \text{ or } x \in B\}.$$

Form the union of sets *A* and *B* by taking all the elements of set *A* and then including the elements of set *B* that are not already listed, as shown in color in Figure 8.

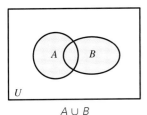

A ∪ B

FIGURE 8

EXAMPLE 2 Find the union of the given sets.

(a) {2, 4, 6} and {4, 6, 8, 10, 12}

Start by listing all the elements from the first set, 2, 4, and 6. Then list all the elements from the second set that are not in the first set, 8, 10, and 12. The union is made up of all these elements, or

$$\{2, 4, 6\} \cup \{4, 6, 8, 10, 12\} = \{2, 4, 6, 8, 10, 12\}.$$

(b) {a, b, d, f, g, h} and {c, f, g, h, k}

The union of these sets is

$$\{a, b, d, f, g, h\} \cup \{c, f, g, h, k\} = \{a, b, c, d, f, g, h, k\}.$$

(c) {3, 4, 5} and ∅

Since there are no elements in ∅, the union of {3, 4, 5} and ∅ contains only the elements 3, 4, and 5, or

$$\{3, 4, 5\} \cup \emptyset = \{3, 4, 5\}.$$

FOR FURTHER THOUGHT

The arithmetic operations of addition and multiplication, when applied to numbers, have some familiar properties. If *a*, *b*, and *c* are real numbers, then the **commutative property of addition** says that the order of the numbers being added makes no difference: $a + b = b + a$. (Is there a **commutative property of multiplication**?) The **associative property of addition** says that when three numbers are added, the grouping used makes no difference: $(a + b) + c = a + (b + c)$. (Is there an **associative property of multiplication**?) The number 0 is called the **identity element for addition** since adding it to any number does not change that number: $a + 0 = a$. (What is the **identity element for multiplication**?) Finally, the **distributive property of multiplication over addition** says that $a(b + c) = ab + ac$. (Is there a distributive property of addition over multiplication?)

For Group Discussion

Now consider the operations of union and intersection, applied to sets. By recalling definitions, or by trying examples, answer the following questions.

1. Is set union commutative? How about set intersection?
2. Is set union associative? How about set intersection?
3. Is there an identity element for set union? If so, what is it? How about set intersection?
4. Is set intersection distributive over set union? Is set union distributive over set intersection?

Recall from the previous section that *A'* represents the *complement* of set *A*. Set *A'* is formed by taking all the elements of the universal set *U* that are not in *A*.

EXAMPLE 3 Let $U = \{1, 2, 3, 4, 5, 6, 9\}$,
$$A = \{1, 2, 3, 4\},$$
$$B = \{2, 4, 6\},$$
$$C = \{1, 3, 6, 9\}.$$

Find each set.

(a) $A' \cap B$

First identify the elements of set A', the elements of U that are not in set A:
$$A' = \{5, 6, 9\}.$$

Now find $A' \cap B$, the set of elements belonging both to A' and to B:
$$A' \cap B = \{5, 6, 9\} \cap \{2, 4, 6\} = \{6\}.$$

(b) $B' \cup C' = \{1, 3, 5, 9\} \cup \{2, 4, 5\} = \{1, 2, 3, 4, 5, 9\}.$

(c) $A \cap (B \cup C')$

First find the set inside the parentheses:
$$B \cup C' = \{2, 4, 6\} \cup \{2, 4, 5\} = \{2, 4, 5, 6\}.$$

Now, find the intersection of this set with A.
$$A \cap (B \cup C') = A \cap \{2, 4, 5, 6\}$$
$$= \{1, 2, 3, 4\} \cap \{2, 4, 5, 6\}$$
$$= \{2, 4\}.$$

(d) $(A' \cup C') \cap B'$

Set $A' = \{5, 6, 9\}$ and set $C' = \{2, 4, 5\}$, with
$$A' \cup C' = \{5, 6, 9\} \cup \{2, 4, 5\} = \{2, 4, 5, 6, 9\}.$$

Set B' is $\{1, 3, 5, 9\}$, so
$$(A' \cup C') \cap B' = \{2, 4, 5, 6, 9\} \cap \{1, 3, 5, 9\} = \{5, 9\}.$$ ■

It is often said that mathematics is a "language." As such, it has the advantage of concise symbolism. For example, the set $(A \cap B)' \cup C$ is less easily expressed in words. One attempt is the following: "The set of all elements that are not in both A and B, or are in C." The key words here (*not, and, or*) will be treated more thoroughly in the chapter on logic.

EXAMPLE 4 Describe each of the following sets in words.

(a) $A \cap (B \cup C')$

This set might be described as "the set of all elements that are in A, and are in B or not in C."

(b) $(A' \cup C') \cap B'$

One possibility is "the set of all elements that are not in A or not in C, and are not in B." ■

Difference of Sets We now consider the *difference* of two sets. Suppose that $A = \{1, 2, 3, \ldots, 10\}$ and $B = \{2, 4, 6, 8, 10\}$. If the elements of B are excluded (or

2.3 Set Operations and Cartesian Products 67

taken away) from A, the set $C = \{1, 3, 5, 7, 9\}$ is obtained. C is called the difference of sets A and B.

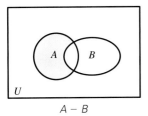

FIGURE 9

Difference of Sets

The **difference** of sets A and B, written $A - B$, is the set of all elements belonging to set A and not to set B, or

$$A - B = \{x \mid x \in A \text{ and } x \notin B\}.$$

Since $x \notin B$ has the same meaning as $x \in B'$, the set difference $A - B$ can also be described as $\{x \mid x \in A \text{ and } x \in B'\}$, or $A \cap B'$. Figure 9 illustrates the idea of set difference. The region in color represents $A - B$.

EXAMPLE 5 Let $U = \{1, 2, 3, 4, 5, 6, 7\}$,
$A = \{1, 2, 3, 4, 5, 6\}$,
$B = \{2, 3, 6\}$,
$C = \{3, 5, 7\}$.

Find each set.

(a) $A - B$

Begin with set A and exclude any elements found also in set B. So,

$$A - B = \{1, 2, 3, 4, 5, 6\} - \{2, 3, 6\} = \{1, 4, 5\}.$$

(b) $B - A$

To be in $B - A$, an element must be in set B and not in set A. But all elements of B are also in A. Thus, $B - A = \emptyset$.

(c) $(A - B) \cup C'$

From part (a), $A - B = \{1, 4, 5\}$. Also, $C' = \{1, 2, 4, 6\}$, so

$$(A - B) \cup C' = \{1, 2, 4, 5, 6\}.$$

The results in Examples 5(a) and 5(b) illustrate that, in general,

$$A - B \neq B - A.$$

Ordered Pairs When writing a set that contains several elements, the order in which the elements appear is not relevant. For example, $\{1, 5\} = \{5, 1\}$. However, there are many instances in mathematics where, when two objects are paired, the order in which the objects are written is important. This leads to the idea of the *ordered pair*. When writing ordered pairs, use parentheses (as opposed to braces, which are reserved for writing sets).

Ordered Pairs

In the **ordered pair** (a, b), a is called the **first component** and b is called the **second component**. In general, $(a, b) \neq (b, a)$.

Two ordered pairs (a, b) and (c, d) are **equal** provided that their first components are equal and their second components are equal; that is, $(a, b) = (c, d)$ if and only if $a = c$ and $b = d$.

EXAMPLE 6 Decide whether each statement is *true* or *false*.

(a) $(3, 4) = (5 - 2, 1 + 3)$

Since $3 = 5 - 2$ and $4 = 1 + 3$, the ordered pairs are equal. The statement is true.

(b) $\{3, 4\} \neq \{4, 3\}$

Since these are sets and not ordered pairs, the order in which the elements are listed is not important. Since these sets are equal, the statement is false.

(c) $(7, 4) = (4, 7)$

These ordered pairs are not equal since they do not satisfy the requirements for equality of ordered pairs. The statement is false.

Cartesian Product of Sets A set may contain ordered pairs as elements. If A and B are sets, then each element of A can be paired with each element of B, and the results can be written as ordered pairs. The set of all such ordered pairs is called the *Cartesian product* of A and B, written $A \times B$ and read "A cross B." The name comes from that of the French mathematician René Descartes.

Cartesian Product of Sets

The **Cartesian product** of sets A and B, written $A \times B$, is

$$A \times B = \{(a, b) | a \in A \text{ and } b \in B\}.$$

EXAMPLE 7 Let $A = \{1, 5, 9\}$ and $B = \{6, 7\}$. Find each set.

(a) $A \times B$

Pair each element of A with each element of B. Write the results as ordered pairs, with the element of A written first and the element of B written second. Write as a set.

$$A \times B = \{(1, 6), (1, 7), (5, 6), (5, 7), (9, 6), (9, 7)\}$$

(b) $B \times A$

Since B is listed first, this set will consist of ordered pairs that have their components interchanged when compared to those in part (a).

$$B \times A = \{(6, 1), (7, 1), (6, 5), (7, 5), (6, 9), (7, 9)\}$$

It should be noted that the order in which the ordered pairs themselves are listed is not important. For example, another way to write $B \times A$ in Example 7 would be

$$\{(6, 1), (6, 5), (6, 9), (7, 1), (7, 5), (7, 9)\}.$$

EXAMPLE 8 Let $A = \{1, 2, 3, 4, 5, 6\}$. Find $A \times A$.

In this example we take the Cartesian product of a set with *itself*. By pairing 1 with each element in the set, 2 with each element, and so on, we obtain the following set:

$$A \times A = \{(1,1), (1,2), (1,3), (1,4), (1,5), (1,6),$$
$$(2,1), (2,2), (2,3), (2,4), (2,5), (2,6),$$
$$(3,1), (3,2), (3,3), (3,4), (3,5), (3,6),$$
$$(4,1), (4,2), (4,3), (4,4), (4,5), (4,6),$$
$$(5,1), (5,2), (5,3), (5,4), (5,5), (5,6),$$
$$(6,1), (6,2), (6,3), (6,4), (6,5), (6,6)\}.$$

It is not unusual to take the Cartesian product of a set with itself, as in Example 8. In fact, the Cartesian product in Example 8 represents all possible results that are obtained when two distinguishable dice are rolled. Determining this Cartesian product is important when studying certain problems in counting techniques and probability, as we shall see in later chapters.

From Example 7 it can be seen that, in general, $A \times B \neq B \times A$, since they do not contain exactly the same ordered pairs. However, each set contains the same number of elements, six. Furthermore, $n(A) = 3$, $n(B) = 2$, and $n(A \times B) = n(B \times A) = 6$. Since $3 \times 2 = 6$, you might conclude that the cardinal number of the Cartesian product of two sets is equal to the product of the cardinal numbers of the sets. In general, this conclusion is correct.

Cardinal Number of a Cartesian Product

If $n(A) = a$ and $n(B) = b$, then **$n(A \times B) = n(B \times A) = n(A) \times n(B) = n(B) \times n(A) = ab$.**

EXAMPLE 9 Find $n(A \times B)$ and $n(B \times A)$ from the given information.

(a) $A = \{a, b, c, d, e, f, g\}$ and $B = \{2, 4, 6\}$

Since $n(A) = 7$ and $n(B) = 3$, $n(A \times B)$ and $n(B \times A)$ are both equal to 7×3, or 21.

(b) $n(A) = 24$ and $n(B) = 5$

$$n(A \times B) = n(B \times A) = 24 \times 5 = 120$$

Operations on Sets Finding intersections, unions, differences, Cartesian products, and complements of sets are examples of *set operations*. An **operation** is a rule or procedure by which one or more objects are used to obtain another object. The objects involved in an operation are usually sets or numbers. The most common operations on numbers are addition, subtraction, multiplication, and division. For example, starting with the numbers 5 and 7, the addition operation would produce the number $5 + 7 = 12$. The multiplication operation would produce $5 \times 7 = 35$.

The most common operations on sets are summarized below, along with their Venn diagrams.

Set Operations

Let A and B be any sets, with U the universal set.

The **complement** of A, written A', is
$$A' = \{x \mid x \in U \text{ and } x \notin A\}.$$

The **intersection** of A and B is
$$A \cap B = \{x \mid x \in A \text{ and } x \in B\}.$$

The **union** of A and B is
$$A \cup B = \{x \mid x \in A \text{ or } x \in B\}.$$

The **difference** of A and B is
$$A - B = \{x \mid x \in A \text{ and } x \notin B\}.$$

The **Cartesian product** of A and B is
$$A \times B = \{(x, y) \mid x \in A \text{ and } y \in B\}.$$

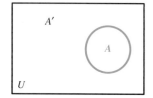

FIGURE 10

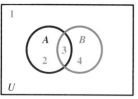

Numbering is arbitrary. The numbers indicate four regions, not cardinal numbers.

FIGURE 11

Venn Diagrams When dealing with a single set, we can use a Venn diagram as seen in Figure 10. The universal set U is divided into two regions, one representing set A and the other representing set A'.

Two sets A and B within the universal set suggest a Venn diagram as seen in Figure 11, where the four resulting regions have been numbered to provide a convenient way to refer to them. (The numbering is arbitrary.) Region 1 includes those elements outside of both set A and set B. Region 2 includes the elements belonging to A but not to B. Region 3 includes those elements belonging to both A and B. How would you describe the elements of region 4?

EXAMPLE 10 Draw a Venn diagram similar to Figure 11 and shade the region or regions representing the following sets.

(a) $A' \cap B$

Refer to Figure 11. Set A' contains all the elements outside of set A, in other words, the elements in regions 1 and 4. Set B is made up of the elements in regions 3 and 4. The intersection of sets A' and B is made up of the elements in the region common to (1 and 4) and (3 and 4), that is, region 4. Thus, $A' \cap B$ is represented by region 4, which is in color in Figure 12. This region can also be described as $B - A$.

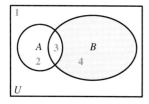

FIGURE 12

(b) $A' \cup B'$

Again, set A' is represented by regions 1 and 4, while B' is made up of regions 1 and 2. The union of A' and B', the set $A' \cup B'$, is made up of the elements belonging to the union of regions 1, 2, and 4, which are in color in Figure 13.

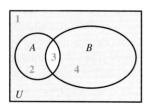

FIGURE 13

2.3 Set Operations and Cartesian Products

When the specific elements of sets A and B are known, it is sometimes useful to show where the various elements are located in the diagram.

EXAMPLE 11 Let $U = \{q, r, s, t, u, v, w, x, y, z\}$,
$A = \{r, s, t, u, v\}$,
$B = \{t, v, x\}$.

Place the elements of these sets in their proper locations on a Venn diagram.

Since $A \cap B = \{t, v\}$, elements t and v are placed in region 3 in Figure 14. The remaining elements of A, that is r, s, and u, go in region 2. The figure shows the proper placement of all other elements.

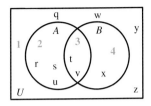

FIGURE 14

To include three sets A, B, and C in a universal set, draw a Venn diagram as in Figure 15, where again an arbitrary numbering of the regions is shown.

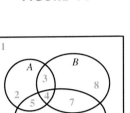

Numbering is arbitrary. The numbers indicate regions, not cardinal numbers or elements.

FIGURE 15

EXAMPLE 12 Shade the set $(A' \cap B') \cap C$ in a Venn diagram similar to the one in Figure 15.

Work first inside the parentheses. As shown in Figure 16, set A' is made up of the regions outside set A, or regions 1, 6, 7, and 8. Set B' is made up of regions 1, 2, 5, and 6. The intersection of these sets is given by the overlap of regions 1, 6, 7, 8 and 1, 2, 5, 6, or regions 1 and 6. For the final Venn diagram, find the intersection of regions 1 and 6 with set C. As seen in Figure 16, set C is made up of regions 4, 5, 6, and 7. The overlap of regions 1, 6 and 4, 5, 6, 7 is region 6, the region in color in Figure 16.

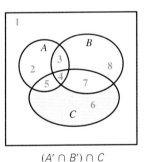

$(A' \cap B') \cap C$

FIGURE 16

EXAMPLE 13 Is the statement

$$(A \cap B)' = A' \cup B'$$

true for every choice of sets A and B?

To help decide, use the regions labeled in Figure 11. Set $A \cap B$ is made up of region 3, so that $(A \cap B)'$ is made up of regions 1, 2, and 4. These regions are in color in Figure 17(a).

To find a Venn diagram for set $A' \cup B'$, first check that A' is made up of regions 1 and 4, while set B' includes regions 1 and 2. Finally, $A' \cup B'$ is made up of regions 1 and 4, or 1 and 2, that is, regions 1, 2, and 4. These regions are in color in Figure 17(b).

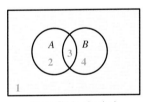

$(A \cap B)'$ is shaded.

(a)

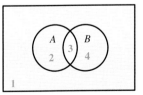

$A' \cup B'$ is shaded.

(b)

FIGURE 17

The fact that the same regions are in color in both Venn diagrams suggests that

$$(A \cap B)' = A' \cup B'.$$ ∎

An area in a Venn diagram (perhaps set off in color) may be described using set operations. When doing this, it is a good idea to translate the region into words, remembering that intersection translates as "and," union translates as "or," and complement translates as "not." There are often several ways to describe a given region.

De Morgan's Laws The result obtained in Example 13 is one of De Morgan's laws, named after the British logician Augustus De Morgan (1806–1871). De Morgan's two laws for sets follow.

De Morgan's Laws
For any sets A and B,

$$(A \cap B)' = A' \cup B' \quad \text{and} \quad (A \cup B)' = A' \cap B'.$$

The Venn diagrams in Figure 17 strongly suggest the truth of the first of De Morgan's laws. They provide a *conjecture,* as discussed in Chapter 1. Actual proofs of De Morgan's laws would require methods used in more advanced courses on set theory.

EXAMPLE 14 For each Venn diagram write a symbolic description of the area in color, using A, B, C, $\cap$, $\cup$, $-$, and $'$ as necessary.

(a)

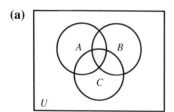

The region in color belongs to all three sets, A and B and C. Therefore, the region corresponds to $A \cap B \cap C$.

(b)

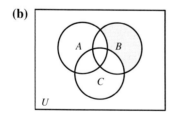

The region in color is in set B and is not in A and is not in C. Since it is not in A, it is in A', and similarly it is in C'. The region is, therefore, in B and in A' and in C', and corresponds to $B \cap A' \cap C'$.

(c) Refer to the figure in part (b) and give two additional ways of describing the region in color.

The area in color includes all of B except for the regions belonging to either A or C. This suggests the idea of set difference. The region may be described as

$$B - (A \cup C), \quad \text{or equivalently,} \quad B \cap (A \cup C)'.$$ ∎

2.3 EXERCISES

Match each term in Column I with its proper designation in Column II. Assume that A and B are sets.

I
1. the intersection of A and B
2. the union of A and B
3. the difference of A and B
4. the complement of A
5. the Cartesian product of A and B
6. the difference of B and A

II
A. the set of elements in A that are not in B
B. the set of elements common to both A and B
C. the set of elements in the universe that are not in A
D. the set of elements in B that are not in A
E. the set of ordered pairs such that each first element is from A and each second element is from B, with every element of A paired with every element of B
F. the set of elements that are in A or in B or in both A and B

Perform the indicated operations.

$$\text{Let } U = \{a, b, c, d, e, f, g\},$$
$$X = \{a, c, e, g\},$$
$$Y = \{a, b, c\},$$
$$Z = \{b, c, d, e, f\}.$$

7. $X \cap Y$
8. $X \cup Y$
9. $Y \cup Z$
10. $Y \cap Z$
11. $X \cup U$
12. $Y \cap U$
13. X'
14. Y'
15. $X' \cap Y'$
16. $X' \cap Z$
17. $X \cup (Y \cap Z)$
18. $Y \cap (X \cup Z)$
19. $(Y \cap Z') \cup X$
20. $(X' \cup Y') \cup Z$
21. $(Z \cup X')' \cap Y$
22. $(Y \cap X')' \cup Z'$
23. $X - Y$
24. $Y - X$
25. $X' - Y$
26. $Y' - X$
27. $X \cap (X - Y)$
28. $Y \cup (Y - X)$

Describe each set in words.

29. $A \cup (B' \cap C')$
30. $(A \cap B') \cup (B \cap A')$
31. $(C - B) \cup A$
32. $B \cap (A' - C)$
33. $(A - C) \cup (B - C)$
34. $(A' \cap B') \cup C'$

Adverse Effects of Alcohol and Tobacco The table lists some common adverse effects of prolonged tobacco and alcohol use.

Tobacco	Alcohol
Emphysema, e	Liver damage, l
Heart damage, h	Brain damage, b
Cancer, c	Heart damage, h

Let T be the set of listed effects of tobacco and A be the set of listed effects of alcohol. Find each set.

35. the smallest possible universal set U that includes all the effects listed
36. A'
37. T'
38. $T \cap A$
39. $T \cup A$
40. $T \cap A'$

Describe in words each set in Exercises 41–46.

Let U = the set of all tax returns,
A = the set of all tax returns with itemized deductions,
B = the set of all tax returns showing business income,
C = the set of all tax returns filed in 2003,
D = the set of all tax returns selected for audit.

41. $B \cup C$ **42.** $A \cap D$ **43.** $C - A$
44. $D \cup A'$ **45.** $(A \cup B) - D$ **46.** $(C \cap A) \cap B'$

Assuming that A and B represent any two sets, identify each of the following statements as either always true *or* not always true.

47. $A \subseteq (A \cup B)$ **48.** $A \subseteq (A \cap B)$ **49.** $(A \cap B) \subseteq A$ **50.** $(A \cup B) \subseteq A$
51. $n(A \cup B) = n(A) + n(B)$ **52.** $n(A \cap B) = n(A) - n(B)$
53. $n(A \cup B) = n(A) + n(B) - n(A \cap B)$ **54.** $n(A \cap B) = n(A) + n(B) - n(A \cup B)$

For Exercises 55–62, use your results in the earlier part(s) to answer the last part.

Let $U = \{1, 2, 3, 4, 5\}$,
$X = \{1, 3, 5\}$,
$Y = \{1, 2, 3\}$,
$Z = \{3, 4, 5\}$.

55. (a) Find $X \cup Y$. **(b)** Find $Y \cup X$.
(c) State a conjecture.

56. (a) Find $X \cap Y$. **(b)** Find $Y \cap X$.
(c) State a conjecture.

57. (a) Find $X \cup (Y \cup Z)$. **(b)** Find $(X \cup Y) \cup Z$.
(c) State a conjecture.

58. (a) Find $X \cap (Y \cap Z)$. **(b)** Find $(X \cap Y) \cap Z$.
(c) State a conjecture.

59. (a) Find $(X \cup Y)'$. **(b)** Find $X' \cap Y'$.
(c) State a conjecture.

60. (a) Find $(X \cap Y)'$. **(b)** Find $X' \cup Y'$.
(c) State a conjecture.

61. (a) Find $X \cup \emptyset$. **(b)** State a conjecture.

62. (a) Find $X \cap \emptyset$. **(b)** State a conjecture.

Decide whether each statement is true or false.

63. $(3, 2) = (5 - 2, 1 + 1)$ **64.** $(10, 4) = (7 + 3, 5 - 1)$
65. $(6, 3) = (3, 6)$ **66.** $(2, 13) = (13, 2)$
67. $\{6, 3\} = \{3, 6\}$ **68.** $\{2, 13\} = \{13, 2\}$
69. $\{(1, 2), (3, 4)\} = \{(3, 4), (1, 2)\}$ **70.** $\{(5, 9), (4, 8), (4, 2)\} = \{(4, 8), (5, 9), (4, 2)\}$

Find $A \times B$ and $B \times A$, for A and B defined as follows.

71. $A = \{2, 8, 12\}$, $B = \{4, 9\}$ **72.** $A = \{3, 6, 9, 12\}$, $B = \{6, 8\}$
73. $A = \{d, o, g\}$, $B = \{p, i, g\}$ **74.** $A = \{b, l, u, e\}$, $B = \{r, e, d\}$

For the sets specified in Exercises 75–78, use the given information to find $n(A \times B)$ and $n(B \times A)$.

75. the sets in Exercise 71   **76.** the sets in Exercise 73
77. $n(A) = 35$ and $n(B) = 6$ **78.** $n(A) = 13$ and $n(B) = 5$

Find the cardinal number specified.

79. If $n(A \times B) = 36$ and $n(A) = 12$, find $n(B)$.   **80.** If $n(A \times B) = 100$ and $n(B) = 4$, find $n(A)$.

2.3 Set Operations and Cartesian Products

Place the elements of these sets in the proper locations on the given Venn diagram.

81. Let $U = \{a, b, c, d, e, f, g\}$,
$A = \{b, d, f, g\}$,
$B = \{a, b, d, e, g\}$.

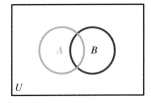

82. Let $U = \{5, 6, 7, 8, 9, 10, 11, 12, 13\}$,
$M = \{5, 8, 10, 11\}$,
$N = \{5, 6, 7, 9, 10\}$.

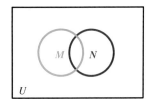

Use a Venn diagram similar to the one shown here to shade each of the following sets.

83. $B \cap A'$ **84.** $A \cup B$ **85.** $A' \cup B$
86. $A' \cap B'$ **87.** $B' \cup A$ **88.** $A' \cup A$
89. $B' \cap B$ **90.** $A \cap B'$ **91.** $B' \cup (A' \cap B')$
92. $(A \cap B) \cup B$ **93.** U' **94.** $\emptyset'$

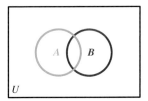

95. Let $U = \{m, n, o, p, q, r, s, t, u, v, w\}$,
$A = \{m, n, p, q, r, t\}$,
$B = \{m, o, p, q, s, u\}$,
$C = \{m, o, p, r, s, t, u, v\}$.

Place the elements of these sets in the proper location on a Venn diagram similar to the one shown here.

96. Let $U = \{1, 2, 3, 4, 5, 6, 7, 8, 9\}$,
$A = \{1, 3, 5, 7\}$,
$B = \{1, 3, 4, 6, 8\}$,
$C = \{1, 4, 5, 6, 7, 9\}$.

Place the elements of these sets in the proper location on a Venn diagram.

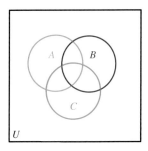

Use a Venn diagram to shade each of the following sets.

97. $(A \cap B) \cap C$ **98.** $(A \cap C') \cup B$ **99.** $(A \cap B) \cup C'$ **100.** $(A' \cap B) \cap C$
101. $(A' \cap B') \cap C$ **102.** $(A \cup B) \cup C$ **103.** $(A \cap B') \cup C$ **104.** $(A \cap C') \cap B$
105. $(A \cap B') \cap C'$ **106.** $(A' \cap B') \cup C$ **107.** $(A' \cap B') \cup C'$ **108.** $(A \cap B)' \cup C$

Write a description of each shaded area. Use the symbols A, B, C, $\cap$, $\cup$, $-$, and $'$ as necessary. More than one answer may be possible.

109.

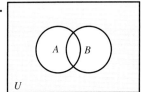

110.

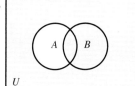

111.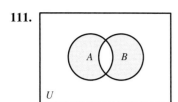
112.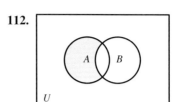
113.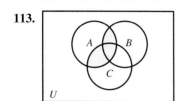
114.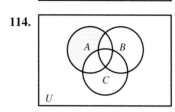
115.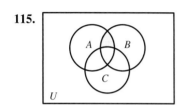
116.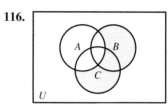

Suppose A and B are sets. Describe the conditions under which each of the following statements would be true.

117. $A = A - B$
118. $A = B - A$
119. $A = A - \emptyset$
120. $A = \emptyset - A$
121. $A \cup \emptyset = \emptyset$
122. $A \cap \emptyset = \emptyset$
123. $A \cap \emptyset = A$
124. $A \cup \emptyset = A$
125. $A \cup A = \emptyset$
126. $A \cap A = \emptyset$

For each of the following exercises, draw two appropriate Venn diagrams to decide whether the given statement is always true *or* not always true.

127. $A \cap A' = \emptyset$
128. $A \cup A' = U$
129. $(A \cap B) \subseteq A$
130. $(A \cup B) \subseteq A$
131. If $A \subseteq B$, then $A \cup B = A$.
132. If $A \subseteq B$, then $A \cap B = B$.
133. $(A \cup B)' = A' \cap B'$ (De Morgan's second law)
134. Give examples of how a language such as English, Spanish, Arabic, or Vietnamese can have an advantage over the symbolic language of mathematics.
135. If A and B are sets, is it necessarily true that $n(A - B) = n(A) - n(B)$? Explain.
136. If $Q = \{x \mid x \text{ is a rational number}\}$ and $H = \{x \mid x \text{ is an irrational number}\}$, describe each of the following sets.

   (a) $Q \cup H$ (b) $Q \cap H$

2.4 Cardinal Numbers and Surveys

Many problems involving sets of people (or other objects) require analyzing known information about certain subsets to obtain cardinal numbers of other subsets. In this section, we apply three useful problem-solving techniques to such problems: Venn diagrams, cardinal number formulas, and tables. The "known information" is quite often (though not always) obtained by administering a survey.

2.4 Cardinal Numbers and Surveys

Suppose a student group of MTV viewers on a college campus are questioned about some selected musical performers, and the following information is produced.

33 like Tim McGraw.   15 like Tim and Britney.
32 like Céline Dion.   14 like Céline and Britney.
28 like Britney Spears.   5 like all three performers.
11 like Tim and Céline.   7 like none of these performers.

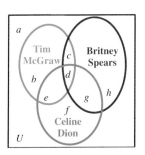

FIGURE 18

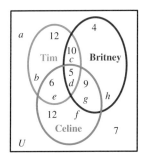

FIGURE 19

To determine the total number of students surveyed, we cannot just add the eight numbers above since there is some overlapping. For example, in Figure 18, the 33 students who like Tim McGraw should not be positioned in region b but should be distributed among regions b, c, d, and e, in a way that is consistent with all of the given data. (Region b actually contains those students who like Tim but do not like Céline and do not like Britney.)

Since, at the start, we do not know how to distribute the 33 who like Tim, we look first for some more manageable data. The smallest total listed, the 5 students who like all three singers, can be placed in region d (the intersection of the three sets). The 7 who like none of the three must go into region a. Then, the 11 who like Tim and Céline must go into regions d and e. Since region d already contains 5 students, we must place $11 - 5 = 6$ in region e. Since 15 like Tim and Britney (regions c and d), we place $15 - 5 = 10$ in region c. Now that regions c, d, and e contain 10, 5, and 6 students respectively, region b receives $33 - 10 - 5 - 6 = 12$. By similar reasoning, all regions are assigned their correct numbers, as shown in Figure 19.

EXAMPLE 1 Using the survey data on student preferences for performers, as summarized in Figure 19, answer the following questions.

(a) How many students like Britney Spears only?

A student who likes Britney only does not like Tim and does not like Céline. These students are inside the regions for Britney and outside the regions for Tim and Céline. Region h is the necessary region in Figure 19, and we see that four students like Britney only.

(b) How many students like exactly two performers?

The students in regions c, e, and g like exactly two performers. The total number of such students is $10 + 6 + 9 = 25$.

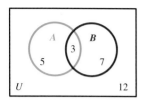

FIGURE 20

(c) How many students were surveyed?

Since each student surveyed has been placed in exactly one region of Figure 19, the total number surveyed is the sum of the numbers in all eight regions:

$$7 + 12 + 10 + 5 + 6 + 12 + 9 + 4 = 65.$$

If the numbers shown in Figure 20 are the cardinal numbers of the individual regions, then $n(A) = 5 + 3 = 8$, $n(B) = 3 + 7 = 10$, $n(A \cap B) = 3$, and $n(A \cup B) = 5 + 3 + 7 = 15$. Notice that $n(A \cup B) = n(A) + n(B) - n(A \cap B)$ since $15 = 8 + 10 - 3$. This relationship is true for any two sets A and B.

Cardinal Number Formula

For any two sets A and B,

$$n(A \cup B) = n(A) + n(B) - n(A \cap B).$$

This formula can be rearranged to find any one of its four terms when the others are known.

EXAMPLE 2 Find $n(A)$ if $n(A \cup B) = 22$, $n(A \cap B) = 8$, and $n(B) = 12$.

The formula above can be rearranged as

$$n(A) = n(A \cup B) - n(B) + n(A \cap B),$$

so
$$n(A) = 22 - 12 + 8$$
$$= 18.$$

Sometimes, even when information is presented as in Example 2, it is more convenient to fit that information into a Venn diagram as in Example 1.

EXAMPLE 3 Robert Hurst is a section chief for an electric utility company. The employees in his section cut down tall trees, climb poles, and splice wire. Hurst recently submitted the following report to the management of the utility:

My section includes 100 employees, with

T = the set of employees who can cut tall trees,
P = the set of employees who can climb poles,
W = the set of employees who can splice wire.

$n(T) = 45$ $n(P \cap W) = 20$
$n(P) = 50$ $n(T \cap W) = 25$
$n(W) = 57$ $n(T \cap P \cap W) = 11$
$n(T \cap P) = 28$ $n(T' \cap P' \cap W') = 9$

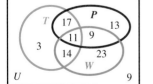

FIGURE 21

The data supplied by Hurst are reflected in Figure 21. The sum of the numbers in the diagram gives the total number of employees in the section:

$$9 + 3 + 14 + 23 + 11 + 9 + 17 + 13 = 99.$$

Hurst claimed to have 100 employees, but his data indicate only 99. The management decided that this error meant that Hurst did not qualify as section chief. He was reassigned as night-shift information operator at the North Pole. (The moral: he should have taken this course.)

Sometimes information appears in a table rather than a Venn diagram, but the basic ideas of union and intersection still apply.

EXAMPLE 4 The officer in charge of the cafeteria on a North Carolina military base wanted to know if the beverage that enlisted men and women preferred with lunch depended on their ages. On a given day, she categorized her lunch patrons according to age and preferred beverage, recording the results in a table as follows.

		Beverage			
		Cola (C)	Iced Tea (I)	Sweet Tea (S)	Totals
	18–25 (Y)	45	10	35	90
Age	26–33 (M)	20	25	30	75
	Over 33 (O)	5	30	20	55
	Totals	70	65	85	220

Using the letters in the table, find the number of people in each of the following sets.

(a) $Y \cap C$

The set Y includes all personnel represented across the top row of the table (90 in all), while C includes the 70 down the left column. The intersection of these two sets is just the upper left entry: 45 people.

(b) $O' \cup I$

The set O' excludes the bottom row, so it includes the first and second rows. The set I includes the middle column only. The union of the two sets represents

$$45 + 10 + 35 + 20 + 25 + 30 + 30 = 195 \text{ people}.$$

2.4 EXERCISES

Use the numerals representing cardinalities in the Venn diagrams to give the cardinality of each set specified.

1.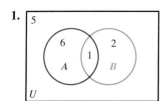

(a) $A \cap B$ (b) $A \cup B$ (c) $A \cap B'$
(d) $A' \cap B$ (e) $A' \cap B'$

2.

(a) $A \cap B$ (b) $A \cup B$ (c) $A \cap B'$
(d) $A' \cap B$ (e) $A' \cap B'$

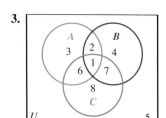

3.
(a) $A \cap B \cap C$ (b) $A \cap B \cap C'$
(c) $A \cap B' \cap C$ (d) $A' \cap B \cap C$
(e) $A' \cap B' \cap C$ (f) $A \cap B' \cap C'$
(g) $A' \cap B \cap C'$ (h) $A' \cap B' \cap C'$

4.
(a) $A \cap B \cap C$ (b) $A \cap B \cap C'$
(c) $A \cap B' \cap C$ (d) $A' \cap B \cap C$
(e) $A' \cap B' \cap C$ (f) $A \cap B' \cap C'$
(g) $A' \cap B \cap C'$ (h) $A' \cap B' \cap C'$

In Exercises 5–8, make use of an appropriate formula.

5. Evaluate $n(A \cup B)$ if $n(A) = 8$, $n(B) = 14$, and $n(A \cap B) = 5$.

6. Evaluate $n(A \cap B)$ if $n(A) = 15$, $n(B) = 12$, and $n(A \cup B) = 25$.

7. Evaluate $n(A)$ if $n(B) = 20$, $n(A \cap B) = 6$, and $n(A \cup B) = 30$.

8. Evaluate $n(B)$ if $n(A) = 35$, $n(A \cap B) = 15$, and $n(A \cup B) = 55$.

To prepare for the survey problems that follow later in this exercise set, draw an appropriate Venn diagram and use the given information to fill in the number of elements in each region.

9. $n(U) = 43$, $n(A) = 25$, $n(A \cap B) = 5$, $n(B') = 30$

10. $n(A) = 19$, $n(B) = 13$, $n(A \cup B) = 25$, $n(A') = 11$

11. $n(A \cup B) = 15$, $n(A \cap B) = 8$, $n(A) = 13$, $n(A' \cup B') = 11$

12. $n(A') = 25$, $n(B) = 28$, $n(A' \cup B') = 40$, $n(A \cap B) = 10$

13. $n(A) = 24$, $n(B) = 24$, $n(C) = 26$, $n(A \cap B) = 10$, $n(B \cap C) = 8$, $n(A \cap C) = 15$, $n(A \cap B \cap C) = 6$, $n(U) = 50$

14. $n(A) = 57$, $n(A \cap B) = 35$, $n(A \cup B) = 81$, $n(A \cap B \cap C) = 15$, $n(A \cap C) = 21$, $n(B \cap C) = 25$, $n(C) = 49$, $n(B') = 52$

15. $n(A \cap B) = 21$, $n(A \cap B \cap C) = 6$, $n(A \cap C) = 26$, $n(B \cap C) = 7$, $n(A \cap C') = 20$, $n(B \cap C') = 25$, $n(C) = 40$, $n(A' \cap B' \cap C') = 2$

16. $n(A) = 15$, $n(A \cap B \cap C) = 5$, $n(A \cap C) = 13$, $n(A \cap B') = 9$, $n(B \cap C) = 8$, $n(A' \cap B' \cap C') = 21$, $n(B \cap C') = 3$, $n(B \cup C) = 32$

Use Venn diagrams to answer the following.

17. *Compact Disc Collection* Paula Story is a fan of the music of Paul Simon and Art Garfunkel. In her collection of 22 compact discs, she has the following:

   5 on which both Simon and Garfunkel sing
   8 on which Simon sings
   7 on which Garfunkel sings
   12 on which neither Simon nor Garfunkel sings.

   (a) How many of her compact discs feature only Paul Simon?
   (b) How many of her compact discs feature only Art Garfunkel?
   (c) How many feature at least one of these two artists?

18. *Writing and Producing Music* Kent LaVoie writes and produces albums for musicians. Last year, he worked on 10 such projects.

> He wrote and produced 2 projects.
> He wrote a total of 5 projects.
> He produced a total of 7 projects.

(a) How many projects did he write but not produce?
(b) How many projects did he produce but not write?

19. *Financial Aid for Students* At the University of Louisiana, half of the 48 mathematics majors were receiving federal financial aid. Of these,

 5 had Pell Grants
14 participated in the College Work Study Program
 4 had TOPS scholarships
 2 had TOPS scholarships and participated in Work Study.

Those with Pell Grants had no other federal aid.

How many of the 48 math majors had:
(a) no federal aid?
(b) more than one of these three forms of aid?
(c) federal aid other than these three forms?
(d) a TOPS scholarship or Work Study?

20. *Viewer Response to Movies* Julianne Peterson, a sports psychologist, was planning a study of viewer response to certain aspects of the movies *The Natural, Field of Dreams,* and *The Rookie*. Upon surveying a group of 55 students, she determined the following:

17 had seen *The Natural*
17 had seen *Field of Dreams*
23 had seen *The Rookie*
 6 had seen *The Natural* and *Field of Dreams*
 8 had seen *The Natural* and *The Rookie*
10 had seen *Field of Dreams* and *The Rookie*
 2 had seen all three of these movies.

How many students had seen:
(a) exactly two of these movies?
(b) exactly one of these movies?
(c) none of these movies?
(d) *The Natural* but neither of the others?

21. *Wine Tasting* The following list shows the preferences of 102 people at a wine-tasting party:

99 like Spañada
96 like Ripple
99 like Boone's Farm Apple Wine
95 like Spañada and Ripple
94 like Ripple and Boone's
96 like Spañada and Boone's
93 like all three.

How many people like:
(a) none of the three?
(b) Spañada, but not Ripple?
(c) anything but Boone's Farm?
(d) only Ripple?
(e) exactly two kinds of wine?

22. *Cooking Habits* Robert Hurst (Example 3 in the text) was again reassigned, this time to the home economics department of the electric utility. He interviewed 140 people in a suburban shopping center to find out some of their cooking habits. He obtained the following results. There is a job opening in Siberia. Should he be reassigned yet one more time?

58 use microwave ovens
63 use electric ranges
58 use gas ranges
19 use microwave ovens and electric ranges
17 use microwave ovens and gas ranges
 4 use both gas and electric ranges
 1 uses all three
 2 cook only with solar energy.

23. *Song Themes* It was once said that Country-Western songs emphasize three basic themes: love, prison, and trucks. A survey of the local Country-Western radio station produced the following data:

12 songs about a truck driver who is in love while in prison
13 about a prisoner in love
28 about a person in love
18 about a truck driver in love
 3 about a truck driver in prison who is not in love
 2 about people in prison who are not in love and do not drive trucks
 8 about people who are out of prison, are not in love, and do not drive trucks
16 about truck drivers who are not in prison.

(a) How many songs were surveyed?

Find the number of songs about:
(b) truck drivers
(c) prisoners
(d) truck drivers in prison
(e) people not in prison
(f) people not in love.

24. *Poultry on a Farm* Old MacDonald surveyed her flock with the following results. She has:

9 fat red roosters   18 thin brown roosters
2 fat red hens   6 thin red roosters
26 fat roosters   5 thin red hens
37 fat chickens   7 thin brown hens.

Answer the following questions about the flock. [*Hint:* You need a Venn diagram with circles for fat, for male (a rooster is a male; a hen is a female), and for red (assume that brown and red are opposites in the chicken world).] How many chickens are:
(a) fat?   (b) red?
(c) male?   (d) fat, but not male?
(e) brown, but not fat?   (f) red and fat?

25. *Student Goals* Julie Ward, who sells college textbooks, interviewed freshmen on a community college campus to find out the main goals of today's students.

Let W = the set of those who want to be wealthy,
F = the set of those who want to raise a family,
E = the set of those who want to become experts in their fields.

Julie's findings are summarized here:

$n(W) = 160$ $n(E \cap F) = 90$
$n(F) = 140$ $n(W \cap F \cap E) = 80$
$n(E) = 130$ $n(E') = 95$
$n(W \cap F) = 95$ $n[(W \cup F \cup E)'] = 10$.

Find the total number of students interviewed.

26. *Hospital Patient Symptoms* Samantha Walker conducted a survey among 75 patients admitted to the cardiac unit of a Honolulu hospital during a two-week period.

Let B = the set of patients with high blood pressure,
C = the set of patients with high cholesterol levels,
S = the set of patients who smoke cigarettes.

Samantha's data are as follows:

$n(B) = 47$ $n(B \cap S) = 33$
$n(C) = 46$ $n(B \cap C) = 31$
$n(S) = 52$ $n(B \cap C \cap S) = 21$
$n[(B \cap C) \cup (B \cap S) \cup (C \cap S)] = 51$.

Find the number of these patients who:
(a) had either high blood pressure or high cholesterol levels, but not both
(b) had fewer than two of the indications listed
(c) were smokers but had neither high blood pressure nor high cholesterol levels
(d) did not have exactly two of the indications listed.

27. The figure below shows U divided into 16 regions by four sets, A, B, C, and D. Find the numbers of the regions belonging to each set.
(a) $A \cap B \cap C \cap D$
(b) $A \cup B \cup C \cup D$
(c) $(A \cap B) \cup (C \cap D)$
(d) $(A' \cap B') \cap (C \cup D)$

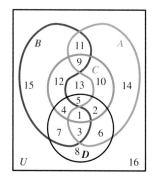

28. *Sports Viewing Habits* A survey of 130 television viewers revealed the following facts:

52 watch football
56 watch basketball
62 watch tennis
60 watch golf
21 watch football and basketball
19 watch football and tennis
22 watch basketball and tennis
27 watch football and golf
30 watch basketball and golf
21 watch tennis and golf
 3 watch football, basketball, and tennis
15 watch football, basketball, and golf
10 watch football, tennis, and golf
10 watch basketball, tennis, and golf
 3 watch all four of these sports
 5 don't watch any of these four sports.

Use a diagram like the one in Exercise 27 to answer the following.
(a) How many of these viewers watch football, basketball, and tennis, but not golf?
(b) How many watch exactly one of these four sports?
(c) How many watch exactly two of these four sports?

Solve each problem.

29. *Army Housing* A study of U.S. Army housing trends categorized personnel as commissioned officers (C), warrant officers (W), or enlisted (E), and categorized their living facilities as on-base (B), rented off-base (R), or owned off-base (O). One survey yielded the following data.

		Facilities			
		B	R	O	Totals
	C	12	29	54	95
Personnel	W	4	5	6	15
	E	374	71	285	730
	Totals	390	105	345	840

Find the number of personnel in each of the following sets.

(a) $W \cap O$ (b) $C \cup B$
(c) $R' \cup W'$ (d) $(C \cup W) \cap (B \cup R)$
(e) $(C \cap B) \cup (E \cap O)$ (f) $B \cap (W \cup R)'$

30. *Basketball Positions* Dwaine Tomlinson runs a basketball program in California. On the first day of the season, 60 young women showed up and were categorized by age level and by preferred basketball position, as shown in the following table.

		Position			
		Guard (G)	Forward (F)	Center (N)	Totals
	Junior High (J)	9	6	4	19
Age	Senior High (S)	12	5	9	26
	College (C)	5	8	2	15
	Totals	26	19	15	60

Using the set labels (letters) in the table, find the number of players in each of the following sets.

(a) $J \cap G$ (b) $S \cap N$ (c) $N \cup (S \cap F)$
(d) $S' \cap (G \cup N)$ (e) $(S \cap N') \cup (C \cap G')$
(f) $N' \cap (S' \cap C')$

31. Could the information of Example 4 have been presented in a Venn diagram similar to those in Examples 1 and 3? If so, construct such a diagram. Otherwise explain the essential difference of Example 4.

32. Explain how a cardinal number formula can be derived for the case where *three* sets occur. Specifically, give a formula relating $n(A \cup B \cup C)$ to $n(A)$, $n(B)$, $n(C)$, $n(A \cap B)$, $n(A \cap C)$, $n(B \cap C)$, and $n(A \cap B \cap C)$. Illustrate with a Venn diagram.

2.5 Infinite Sets and Their Cardinalities

As mentioned at the beginning of this chapter, most of the early work in set theory was done by Georg Cantor. He devoted much of his life to a study of the cardinal numbers of sets. Recall that the *cardinal number*, or *cardinality*, of a finite set is the number of elements that it contains. For example, the set $\{5, 9, 15\}$ contains 3 elements and has a cardinal number of 3. The cardinal number of $\emptyset$ is 0.

Cantor proved many results about the cardinal numbers of infinite sets. The proofs of Cantor are quite different from the type of proofs you may have seen in an algebra or geometry course. Because of the novelty of Cantor's methods, they were not quickly accepted by the mathematicians of his day. (In fact, some other aspects

Paradox The word in Greek originally meant "wrong opinion" as opposed to orthodox, which meant "right opinion." Over the years, the word came to mean self-contradiction. An example is the statement "This sentence is false." By assuming it is true, we get a contradiction; likewise, by assuming it is false, we get a contradiction. Thus, it's a paradox.

Before the twentieth century it was considered a paradox that any set could be placed into one-to-one correspondence with a proper subset of itself. This paradox, called Galileo's paradox after the sixteenth-century mathematician and scientist Galileo (see the picture), is now explained by saying that the ability to make such a correspondence is how we distinguish infinite sets from finite sets. What is true for finite sets is not necessarily true for infinite sets.

of Cantor's theory lead to paradoxes.) The results we will discuss here, however, are commonly accepted today.

The idea of the cardinal number of an infinite set depends on the idea of one-to-one correspondence. For example, each of the sets $\{1, 2, 3, 4\}$ and $\{9, 10, 11, 12\}$ has four elements. Corresponding elements of the two sets could be paired off in the following manner (among many other ways):

$$\{1, \quad 2, \quad 3, \quad 4\}$$
$$\updownarrow \quad \updownarrow \quad \updownarrow \quad \updownarrow$$
$$\{9, \quad 10, \quad 11, \quad 12\}.$$

Such a pairing is a **one-to-one correspondence** between the two sets. The "one-to-one" refers to the fact that each element of the first set is paired with exactly one element of the second set and each element of the second set is paired with exactly one element of the first set.

Two sets A and B which may be put in a one-to-one correspondence are said to be **equivalent.** Symbolically, this is written $A \sim B$. Do you see that the two sets shown above are equivalent but not equal?

The following correspondence between sets $\{1, 8, 12\}$ and $\{6, 11\}$,

$$\{1, \quad 8, \quad 12\}$$
$$\{6, \quad 11\},$$

is not one-to-one since the elements 8 and 12 from the first set are both paired with the element 11 from the second set. These sets are not equivalent.

It seems reasonable to say that if two non-empty sets have the same cardinal number, then a one-to-one correspondence can be established between the two sets. Also, if a one-to-one correspondence can be established between two sets, then the two sets must have the same cardinal number. These two facts are fundamental in discussing the cardinal numbers of infinite sets.

The basic set used in discussing infinite sets is the set of counting numbers, $\{1, 2, 3, 4, 5, \ldots\}$. The set of counting numbers is said to have the infinite cardinal number $\aleph_0$ (the first Hebrew letter, aleph, with a zero subscript, read "aleph-null"). Think of $\aleph_0$ as being the "smallest" infinite cardinal number. To the question "How many counting numbers are there?" answer "There are $\aleph_0$ of them."

From the discussion above, any set that can be placed in a one-to-one correspondence with the counting numbers will have the same cardinal number as the set of counting numbers, or $\aleph_0$. It turns out that many sets of numbers have cardinal number $\aleph_0$.

EXAMPLE 1 Show that the set of whole numbers $\{0, 1, 2, 3, \ldots\}$ has cardinal number $\aleph_0$.

This problem is easily stated, but not quite so easily solved. All we really know about $\aleph_0$ is that it is the cardinal number of the set of counting numbers (by definition). To show that another set, such as the whole numbers, also has $\aleph_0$ as its cardinal number, we must apparently show that set to be equivalent to the set of counting numbers. And equivalence is established by a one-to-one correspondence between the two sets. This sequence of thoughts, involving just a few basic ideas, leads us to a plan: exhibit a one-to-one correspondence between the counting numbers and the whole numbers. Our strategy will be to sketch such a correspondence, showing exactly how each counting number is paired with a unique whole number. In the correspondence

$$\{1, 2, 3, 4, 5, 6, \ldots, \quad n, \quad \ldots\} \quad \text{Counting numbers}$$
$$\updownarrow \updownarrow \updownarrow \updownarrow \updownarrow \updownarrow \quad \quad \updownarrow$$
$$\{0, 1, 2, 3, 4, 5, \ldots, n-1, \ldots\} \quad \text{Whole numbers}$$

the pairing of the counting number n with the whole number $n - 1$ continues indefinitely, with neither set containing any element not used up in the pairing process. So, even though the set of whole numbers has one more element (the number 0) than the set of counting numbers, and thus should have cardinal number $\aleph_0 + 1$, the above correspondence proves that both sets have the same cardinal number. That is,

$$\aleph_0 + 1 = \aleph_0.$$

This result shows that intuition is a poor guide for dealing with infinite sets. Intuitively, it is "obvious" that there are more whole numbers than counting numbers. However, since the sets can be placed in a one-to-one correspondence, the two sets have the same cardinal number.

The set $\{5, 6, 7\}$ is a proper subset of the set $\{5, 6, 7, 8\}$, and there is no way to place these two sets in a one-to-one correspondence. On the other hand, the set of counting numbers is a proper subset of the set of whole numbers, and Example 1 showed that these two sets *can* be placed in a one-to-one correspondence. The only way a proper subset of a set can possibly be placed in a one-to-one correspondence with the set itself is if both sets are infinite. In fact, this important property is used as the definition of an infinite set.

Infinite Set

A set is **infinite** if it can be placed in a one-to-one correspondence with a proper subset of itself.

EXAMPLE 2 Show that the set of integers $\{\ldots, -3, -2, -1, 0, 1, 2, 3, \ldots\}$ has cardinal number $\aleph_0$.

A one-to-one correspondence can be set up between the set of integers and the set of counting numbers, as follows:

$$\{1, 2, \quad 3, 4, \quad 5, 6, \quad 7, \ldots, 2n, 2n+1, \ldots\}$$
$$\updownarrow \updownarrow \quad \updownarrow \updownarrow \quad \updownarrow \updownarrow \quad \updownarrow \quad \updownarrow \quad \updownarrow$$
$$\{0, 1, -1, 2, -2, 3, -3, \ldots, \quad n, \quad -n, \ldots\}.$$

Because of this one-to-one correspondence, the cardinal number of the set of integers is the same as the cardinal number of the set of counting numbers, $\aleph_0$.

If the elements of a set can be placed in a specific order, such as the set of integers in Example 2, that set has cardinal number $\aleph_0$.

Number Lore of the Aleph-bet Aleph and other letters of the Hebrew alphabet are shown on a Kabbalistic diagram representing one of the ten emanations of God during Creation. Kabbalah, the ultramystical tradition within Judaism, arose in the fifth century and peaked in the sixteenth century in both Palestine and Poland.

Kabbalists believed that the Bible held mysteries that could be discovered in permutations, combinations, and anagrams of its very letters. They also "read" the numerical value of letters in a word by the technique called Gematria (from geometry?). This was possible since each letter in the aleph-bet has a numerical value (aleph = 1), and thus a numeration system exists. The letter Y stands for 10, so 15 should be YH (10 + 5). However, YH is a form of the Holy Name, so instead TW (9 + 6) is the symbol.

Notice that the one-to-one correspondence of Example 2 also proves that the set of integers is infinite. The set of integers was placed in a one-to-one correspondence with a proper subset of itself.

As shown by Example 2, there are just as many integers as there are counting numbers. This result is not at all intuitive. However, the next result is even less intuitive. We know that there is an infinite number of fractions between any two counting numbers. For example, there is an infinite set of fractions {1/2, 3/4, 7/8, 15/16, 31/32,...} between the counting numbers 0 and 1. This should imply that there are "more" fractions than counting numbers. It turns out, however, that there are just as many fractions as counting numbers, as shown by the next example.

EXAMPLE 3 Show that the cardinal number of the set of rational numbers is $\aleph_0$.

To show that the cardinal number of the set of rational numbers is $\aleph_0$, first show that a one-to-one correspondence may be set up between the set of nonnegative rational numbers and the counting numbers. This is done by the following ingenious scheme, devised by Georg Cantor. Look at Figure 22. The nonnegative rational numbers whose denominators are 1 are written in the first row; those whose denominators are 2 are written in the second row, and so on. Every nonnegative rational number appears in this list sooner or later. For example, 327/189 is in row 189 and column 327.

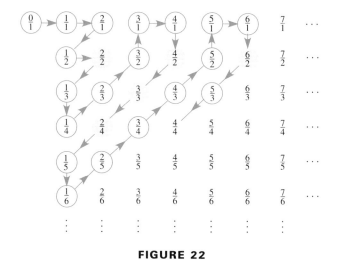

FIGURE 22

To set up a one-to-one correspondence between the set of nonnegative rationals and the set of counting numbers, follow the path drawn in Figure 22. Let 0/1 correspond to 1, let 1/1 correspond to 2, 2/1 to 3, 1/2 to 4 (skip 2/2, since 2/2 = 1/1), 1/3 to 5, 1/4 to 6, and so on. The numbers under the colored disks are omitted, since they can be reduced to lower terms, and were thus included earlier in the listing.

This procedure sets up a one-to-one correspondence between the set of nonnegative rationals and the counting numbers, showing that both of these sets have the same cardinal number, $\aleph_0$. Now by using the method of Example 2, that is, letting each negative number follow its corresponding positive number, we can extend this correspondence to include negative rational numbers as well. Thus, the set of all rational numbers has cardinal number $\aleph_0$.

2.5 Infinite Sets and Their Cardinalities

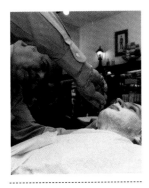

The Barber Paradox This is a version of a paradox of set theory that Bertrand Russell proposed in the early twentieth century.

1. The men in a village are of two types: men who do not shave themselves and men who do.
2. The village barber shaves all men who do not shave themselves and he shaves only those men.

But who shaves the barber? The barber cannot shave himself. If he did, he would fall into the category of men who shave themselves. However, (2) above states that the barber does not shave such men.

So the barber does not shave himself. But then he falls into the category of men who do not shave themselves. According to (2), the barber shaves all of these men; hence, the barber shaves himself, too.

We find that the barber cannot shave himself, yet the barber does shave himself—a paradox.

A set is called **countable** if it is finite, or if it has cardinal number $\aleph_0$. All the infinite sets of numbers discussed so far—the counting numbers, the whole numbers, the integers, and the rational numbers—are countable. It seems now that *every* set is countable, but this is not true. The next example shows that the set of real numbers is not countable.

EXAMPLE 4 Show that the set of all real numbers does not have cardinal number $\aleph_0$.

There are two possibilities:

1. The set of real numbers has cardinal number $\aleph_0$.
2. The set of real numbers does not have cardinal number $\aleph_0$.

Assume for the time being that the first statement is true. If the first statement is true, then a one-to-one correspondence can be set up between the set of real numbers and the set of counting numbers. We do not know what sort of correspondence this might be, but assume it can be done.

In a later chapter we show that every real number can be written as a decimal number (or simply "decimal"). Thus, in the one-to-one correspondence we are assuming, some decimal corresponds to the counting number 1, some decimal corresponds to 2, and so on. Suppose the correspondence is as follows:

$$1 \leftrightarrow .\mathbf{6}8458429006\ldots$$
$$2 \leftrightarrow .1\mathbf{3}479201038\ldots$$
$$3 \leftrightarrow .37\mathbf{2}91568341\ldots$$
$$4 \leftrightarrow .935\mathbf{2}23671611\ldots$$

and so on.

Assuming the existence of a one-to-one correspondence between the counting numbers and the real numbers means that every decimal is in the list above. Let's construct a new decimal K as follows. The first decimal in the above list has 6 as its first digit; let K start as $K = .4\ldots$. We picked 4 since $4 \neq 6$; we could have used any other digit except 6. Since the second digit of the second decimal in the list is 3, we let $K = .45\ldots$ (since $5 \neq 3$). The third digit of the third decimal is 2, so let $K = .457\ldots$ (since $7 \neq 2$). The fourth digit of the fourth decimal is 2, so let $K = .4573\ldots$ (since $3 \neq 2$). Continue K in this way.

Is K in the list that we assumed to contain all decimals? The first decimal in the list differs from K in at least the first position (K starts with 4, and the first decimal in the list starts with 6). The second decimal in the list differs from K in at least the second position, and the nth decimal in the list differs from K in at least the nth position. Every decimal in the list differs from K in at least one position, so that K cannot possibly be in the list. In summary:

We assume every decimal is in the list above.

The decimal K is not in the list.

Since these statements cannot both be true, the original assumption has led to a contradiction. This forces the acceptance of the only possible alternative to the original assumption: it is not possible to set up a one-to-one correspondence between the set of reals and the set of counting numbers; the cardinal number of the set of reals is not equal to $\aleph_0$.

The set of counting numbers is a proper subset of the set of real numbers. Because of this, it would seem reasonable to say that the cardinal number of the set of reals, commonly written c, is greater than $\aleph_0$. (The letter c here represents *continuum*.) Other, even larger, infinite cardinal numbers can be constructed. For example, the set of all subsets of the set of real numbers has a cardinal number larger than c. Continuing this process of finding cardinal numbers of sets of subsets, more and more, larger and larger infinite cardinal numbers are produced.

The six most important infinite sets of numbers were listed in an earlier section. All of them have been dealt with in this section except the irrational numbers. The irrationals have decimal representations, so they are all included among the real numbers. Since the irrationals are a subset of the reals, you might guess that the irrationals have cardinal number $\aleph_0$, just like the rationals. However, since the union of the rationals and the irrationals is all the reals, that would imply that $\aleph_0 + \aleph_0 = c$. But Example 2 showed that this is not the case. Hence, a better guess is that the cardinal number of the irrationals is c (the same as that of the reals). This is, in fact, true. The major infinite sets of numbers, with their cardinal numbers, are now summarized.

Cardinal Numbers of Infinite Number Sets

Infinite Set	Cardinal Number
Natural or Counting numbers	$\aleph_0$
Whole numbers	$\aleph_0$
Integers	$\aleph_0$
Rational numbers	$\aleph_0$
Irrational numbers	c
Real numbers	c

2.5 EXERCISES

Match each set in Column I with the set in Column II that has the same cardinality. Give the cardinal number.

I

1. $\{3\}$
2. $\{a, b, c\}$
3. $\{x \mid x \text{ is a natural number}\}$
4. $\{x \mid x \text{ is a real number}\}$
5. $\{x \mid x \text{ is an integer between 0 and 1}\}$
6. $\{x \mid x \text{ is an integer that satisfies } x^2 = 25\}$

II

A. $\{x \mid x \text{ is a rational number}\}$
B. $\{26\}$
C. $\{x \mid x \text{ is an irrational number}\}$
D. $\{x, y, z\}$
E. $\{x \mid x \text{ is a real number that satisfies } x^2 = 25\}$
F. $\{x \mid x \text{ is an integer that is both even and odd}\}$

Place each pair of sets into a one-to-one correspondence, if possible.

7. $\{I, II, III\}$ and $\{x, y, z\}$
8. $\{a, b, c, d\}$ and $\{2, 4, 6\}$
9. $\{a, d, d, i, t, i, o, n\}$ and $\{a, n, s, w, e, r\}$
10. $\{\text{Reagan, Clinton, Bush}\}$ and $\{\text{Nancy, Hillary, Laura}\}$

2.5 Infinite Sets and Their Cardinalities

Give the cardinal number of each set.

11. {a, b, c, d, ..., k}

12. {9, 12, 15, ..., 36}

13. ∅

14. {0}

15. {300, 400, 500, ...}

16. {−35, −28, −21, ..., 56}

17. {−1/4, −1/8, −1/12, ...}

18. {$x \mid x$ is an even integer}

19. {$x \mid x$ is an odd counting number}

20. {b, a, 1, 1, a, d}

21. {Jan, Feb, Mar, ..., Dec}

22. {Alabama, Alaska, Arizona, ..., Wisconsin, Wyoming}

23. Lew Lefton of Georgia Tech has revised the old song "100 Bottles of Beer on the Wall" to illustrate a property of infinite cardinal numbers. Fill in the blank in the first line of Lefton's composition:

"$\aleph_0$ bottles of beer on the wall, $\aleph_0$ bottles of beer, take one down and pass it around, _____ bottles of beer on the wall."

24. Two one-to-one correspondences are considered "different" if some elements are paired differently in one than in the other. For example:

{a, b, c}   {a, b, c}
↕ ↕ ↕ and ↕ ↕ ↕ are different,
{a, b, c}   {c, b, a}

{a, b, c}   {b, c, a}
while ↕ ↕ ↕ and ↕ ↕ ↕ are not.
{c, a, b}   {a, b, c}

(a) How many *different* correspondences can be set up between the two sets {Mike Myers, Julia Roberts, William Shatner} and {Erin Brockovich, Captain James T. Kirk, Austin Powers}?

(b) Which one of these correspondences pairs each person with the appropriate famous movie role?

Determine whether the following pairs of sets are equal, equivalent, both, *or* neither.

25. {u, v, w}, {v, u, w}   **26.** {48, 6}, {4, 86}   **27.** {X, Y, Z}, {x, y, z}   **28.** {lea}, {ale}

29. {$x \mid x$ is a positive real number}, {$x \mid x$ is a negative real number}

30. {$x \mid x$ is a positive rational number}, {$x \mid x$ is a negative real number}

Show that each set has cardinal number $\aleph_0$ by setting up a one-to-one correspondence between the given set and the set of counting numbers.

31. the set of positive even integers

32. {−10, −20, −30, −40, ...}

33. {1,000,000, 2,000,000, 3,000,000, ...}

34. the set of odd integers

35. {2, 4, 8, 16, 32, ...} (*Hint:* $4 = 2^2$, $8 = 2^3$, $16 = 2^4$, and so on)

36. {−17, −22, −27, −32, ...}

In each of Exercises 37–40, identify the given statement as always true *or* not always true. *If not always true, give a counterexample.*

37. If A and B are infinite sets, then A is equivalent to B.

38. If set A is an infinite set and set B can be put in a one-to-one correspondence with a proper subset of A, then B must be infinite.

39. If A is an infinite set and A is not equivalent to the set of counting numbers, then $n(A) = c$.

40. If A and B are both countably infinite sets, then $n(A \cup B) = \aleph_0$.

Exercises 41 and 42 are geometric applications of the concept of infinity.

41. The set of real numbers can be represented by an infinite line, extending indefinitely in both directions. Each point on the line corresponds to a unique real number, and each real number corresponds to a unique point on the line.

(a) Use the figure below, where the line segment between 0 and 1 has been bent into a semicircle and positioned above the line, to prove that

$\{x \mid x \text{ is a real number between 0 and 1}\}$ is equivalent to $\{x \mid x \text{ is a real number}\}$.

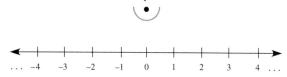

(b) What fact does part (a) establish about the set of real numbers?

42. Show that the two vertical line segments shown here both have the same number of points.

Show that each of the following sets can be placed in a one-to-one correspondence with a proper subset of itself to prove that each set is infinite.

43. $\{3, 6, 9, 12, \ldots\}$

44. $\{4, 7, 10, 13, 16, \ldots\}$

45. $\{3/4, 3/8, 3/12, 3/16, \ldots\}$

46. $\{1, 4/3, 5/3, 2, \ldots\}$

47. $\{1/9, 1/18, 1/27, 1/36, \ldots\}$

48. $\{-3, -5, -9, -17, \ldots\}$

49. Describe the difference between *equal* and *equivalent* sets.

50. Explain how the correspondence suggested in Example 4 shows that the set of real numbers between 0 and 1 is not countable.

The Paradoxes of Zeno The margin note on page 84 describes what is meant by a paradox. Other paradoxes include the famous paradoxes of Zeno, born about 496 B.C. in southern Italy. Two of them claim to show that a faster runner cannot overtake a slower one, and that, in fact, motion itself cannot even occur.

What is your explanation for the following two examples of Zeno's paradoxes?

51. Achilles, if he starts out behind a tortoise, can never overtake the tortoise even if he runs faster.

Suppose Tortoise has a head start of one meter and goes one-tenth as fast as Achilles. When Achilles reaches the point where Tortoise started, Tortoise is then one-tenth meter ahead. When Achilles reaches *that* point, Tortoise is one-hundredth meter ahead. And so on. Achilles gets closer but can never catch up.

52. Motion itself cannot occur.

You cannot travel one meter until after you have first gone a half meter. But you cannot go a half meter until after you have first gone a quarter meter. And so on. Even the tiniest motion cannot occur since a tinier motion would have to occur first.

COLLABORATIVE INVESTIGATION

A Survey of Your Class

This group activity is designed to determine the number of students present in your class without actually counting the members one-by-one. This will be accomplished by having each member of the class determine one particular set he or she belongs to, and then finding the sum of the cardinal numbers of the subsets.

For this activity, we will designate three sets: X, Y, and Z. Here are their descriptions:

X = {students in the class registered with the Republican party}

Y = {students in the class 24 years of age or younger}

Z = {students who have never been married}

Each student in the class will belong to one of the sets X, X', one of the sets Y, Y', and one of the sets Z, Z'. (Recall that the complement of a set consists of all elements in the universe (class) that are not in the set.)

As an example, suppose that a student is a 23-year-old divorced Democrat. The student belongs to the sets X', Y, and Z'. Joining these with intersection symbols, the set to which the student belongs is

$$X' \cap Y \cap Z'.$$

Now observe the Venn diagram that follows. The eight subsets are identified by lowercase letters (a) through (h). The final column will be completed when a survey is made. Each student should now determine which set he or she belongs to. (The student described earlier belongs to (g).)

Region	Description in Terms of Set Notation	Number of Class Members in the Set
(a)	$X \cap Y \cap Z$	
(b)	$X \cap Y \cap Z'$	
(c)	$X \cap Y' \cap Z$	
(d)	$X' \cap Y \cap Z$	
(e)	$X' \cap Y' \cap Z$	
(f)	$X \cap Y' \cap Z'$	
(g)	$X' \cap Y \cap Z'$	
(h)	$X' \cap Y' \cap Z'$	

The instructor will now poll the class to see how many members are in each set. *Remember that each class member will belong to one and only one set.*

After the survey is made, find the sum of the numbers in the final column. They should add up to *exactly* the number of students present. Count the class members individually to verify this.

Topics for Discussion

1. Suppose that the final column entries do not add up to the total number of class members. What might have gone wrong?
2. Why can't a class member be a member of more than one of the eight subsets listed?

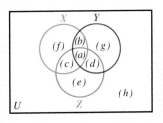

CHAPTER 2 TEST

For Exercises 1–14, let

$U = \{a, b, c, d, e, f, g, h\}$, $A = \{a, b, c, d\}$, $B = \{b, e, a, d\}$, and $C = \{a, e\}$.

Find each of the following sets.

1. $A \cup C$
2. $B \cap A$
3. B'
4. $A - (B \cap C')$

Identify each of the following statements as true or false.

5. $b \in A$
6. $C \subseteq A$
7. $B \subset (A \cup C)$
8. $c \notin C$
9. $n[(A \cup B) - C] = 4$
10. $\emptyset \subset C$
11. $A \cap B'$ is equivalent to $B \cap A'$
12. $(A \cup B)' = A' \cap B'$

Find each of the following.

13. $n(A \times C)$
14. the number of proper subsets of A

Give a word description for each of the following sets.

15. $\{-3, -1, 1, 3, 5, 7, 9\}$
16. $\{\text{January, February, March}, \ldots, \text{December}\}$

Express each of the following sets in set-builder notation.

17. $\{-1, -2, -3, -4, \ldots\}$
18. $\{24, 32, 40, 48, \ldots, 88\}$

Place $\subset$, $\subseteq$, both, or neither in each blank to make a true statement.

19. $\emptyset$ _____ $\{x \mid x \text{ is a counting number between 17 and 18}\}$
20. $\{4, 9, 16\}$ _____ $\{4, 5, 6, 7, 8, 9, 10\}$

Shade each of the following sets in an appropriate Venn diagram.

21. $X \cup Y'$
22. $X' \cap Y'$
23. $(X \cup Y) - Z$
24. $[(X \cap Y) \cup (Y \cap Z) \cup (X \cap Z)] - (X \cap Y \cap Z)$

Facts About Inventions The following table lists ten inventions, important directly or indirectly in our lives, together with other pertinent data.

Invention	Date	Inventor	Nation
Adding machine	1642	Pascal	France
Barometer	1643	Torricelli	Italy
Electric razor	1917	Schick	U.S.
Fiber optics	1955	Kapany	England
Geiger counter	1913	Geiger	Germany
Pendulum clock	1657	Huygens	Holland
Radar	1940	Watson-Watt	Scotland
Telegraph	1837	Morse	U.S.
Thermometer	1593	Galileo	Italy
Zipper	1891	Judson	U.S.

Let U = the set of all ten inventions, A = the set of items invented in the United States, and T = the set of items invented in the twentieth century. List the elements of each of the following sets.

25. $A \cap T$
26. $(A \cup T)'$
27. $A - T'$

28. State De Morgan's laws for sets in words rather than symbols.

29. The numerals in the Venn diagram indicate the number of elements in each particular subset.

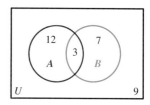

Determine the number of elements in each of the following.
(a) $A \cup B$
(b) $A \cap B'$
(c) $(A \cap B)'$

Financial Aid to College Students In one recent year, financial aid available to college students in the United States was nearly 30 billion dollars. (Much of it went unclaimed, mostly because qualified students were not aware of it, did not know how to obtain or fill out the required applications, or did not feel the results would be worth their effort.) The three major sources of aid are government grants, private scholarships, and the colleges themselves.

30. Melissa Young, Financial Aid Director of a small private Midwestern college, surveyed the records of 100 sophomores and found the following:

    49 receive government grants
    55 receive private scholarships
    43 receive aid from the college
    23 receive government grants and private scholarships
    18 receive government grants and aid from the college
    28 receive private scholarships and aid from the college
    8 receive help from all three sources.

How many of the students in the survey:
(a) have government grants only?
(b) have private scholarships but not government grants?
(c) receive financial aid from only one of these sources?
(d) receive aid from exactly two of these sources?
(e) receive no financial aid from any of these sources?
(f) receive no aid from the college or from the government?

chapter 3
Introduction to Logic

Although modern digital computers are exceedingly complex machines, their design involves only two basic components: gates and memory cells. A gate is a simple electrical switch that controls the flow of data much like a gate controls the flow of water in a canal. A memory cell can store a single bit of data. By interconnecting various gates and memory cells, a circuit can be designed to perform operations such as addition and subtraction. Since the beginning of the computer industry, there has been a constant trend toward the miniaturization of electrical circuits. In 1960 only a few gates could be reliably printed onto a silicon chip. This number doubled annually throughout the 1960s and approximately quadrupled every three years thereafter. Today over a million logic gates can be printed onto a single chip that is only a few millimeters square. Some experts believe that a silicon chip containing over a billion components may be achieved within ten years. After this, fundamental physical limitations may prevent further improvements.

How is it possible to design the interconnection of so many logic gates and have them all work together properly? You will learn the answer to this question in this chapter when you study Boolean algebra and formal logic. Without Boolean algebra and formal logic, modern computers would have been impossible. An electrical circuit is either on or off, and as a result, it can be modeled by the logical states of true or false. Instead of looking at actual connections in electrical circuits, computer engineers construct truth tables to determine the output of an electrical component for certain inputs. As a result of this abstraction, design errors can be corrected prior to making the first prototype circuit. The use of logic to create complex circuitry has propelled technology forward at an incredible pace.

Source: Stallings, W., *Computer Organization and Architecture,* Macmillan Publishing Company, New York, 1993.

3.1 Statements and Quantifiers

3.2 Truth Tables and Equivalent Statements

3.3 The Conditional and Circuits

3.4 More on the Conditional

3.5 Analyzing Arguments with Euler Diagrams

Extension: Logic Puzzles

3.6 Analyzing Arguments with Truth Tables

Collaborative Investigation: Logic Puzzles Revisited

Chapter 3 Test

3.1 Statements and Quantifiers

This section introduces the study of *symbolic logic*, which uses letters to represent statements, and symbols for words such as *and, or, not*. One of the main applications of logic is in the study of the *truth value* (that is, the truth or falsity) of statements with many parts. The truth value of these statements depends on the components that comprise them.

Many kinds of sentences occur in ordinary language, including factual statements, opinions, commands, and questions. Symbolic logic discusses only the first type of sentence, the kind that involves facts.

Statements A **statement** is defined as a declarative sentence that is either true or false, but not both simultaneously. For example, both of the following are statements:

Electronic mail provides a means of communication.

$11 + 6 = 12$.

Each one is either true or false. However, based on this definition, the following sentences are not statements:

Access the file.

Is this a great time, or what?

Nomar Garciaparra is a better baseball player than Derek Jeter.

This sentence is false.

These sentences cannot be identified as being either true or false. The first sentence is a command, and the second is a question. The third is an opinion. "This sentence is false" is a paradox; if we assume it is true, then it is false, and if we assume it is false, then it is true.

A **compound statement** may be formed by combining two or more statements. The statements making up a compound statement are called **component statements.** Various **logical connectives,** or simply **connectives,** can be used in forming compound statements. Words such as *and, or, not,* and *if … then* are examples of connectives. (While a statement such as "Today is not Tuesday" does not consist of two component statements, for convenience it is considered compound, since its truth value is determined by noting the truth value of a different statement, "Today is Tuesday.")

Gottfried Leibniz (1646–1716) was a wide-ranging philosopher and a universalist who tried to patch up Catholic-Protestant conflicts. He promoted cultural exchange between Europe and the East. Chinese ideograms led him to search for a universal symbolism.

An early inventor of symbolic logic, Leibniz hoped to build a totally objective model for human reasoning and thus reduce all reasoning errors to minor calculational mistakes. His dream has so far proved futile.

EXAMPLE 1 Decide whether each statement is compound.

(a) Shakespeare wrote sonnets and the poem exhibits iambic pentameter.

This statement is compound, since it is made up of the component statements "Shakespeare wrote sonnets" and "the poem exhibits iambic pentameter." The connective is *and*.

(b) You can pay me now or you can pay me later.

The connective here is *or*. The statement is compound.

(c) If he said it, then it must be true.

The connective here is *if … then,* discussed in more detail in a later section. The statement is compound.

The TEST menu of the TI-83 Plus calculator allows the user to test the truth or falsity of statements involving =, ≠, >, ≥, <, and ≤. If a statement is true, it returns a 1; if false, it returns a 0.

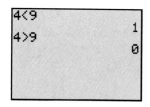

4 < 9 is true, as indicated by the 1. 4 > 9 is false, as indicated by the 0.

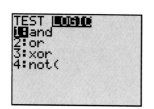

The LOGIC menu of the TI-83 Plus calculator allows the user to test truth or falsity of statements involving *and*, *or*, *exclusive or* (see Exercise 75 in the next section), and *not*.

(d) My pistol was made by Smith and Wesson.

While the word "and" is used in this statement, it is not used as a *logical* connective, since it is part of the name of the manufacturer. The statement is not compound. ■

Negations
The sentence "Tom Jones has a red car" is a statement; the **negation** of this statement is "Tom Jones does not have a red car." The negation of a true statement is false, and the negation of a false statement is true.

EXAMPLE 2 Form the negation of each statement.

(a) That state has a governor.

To negate this statement, we introduce *not* into the sentence: "That state does not have a governor."

(b) The sun is not a star.

Negation: "The sun is a star." ■

One way to detect incorrect negations is to check truth values. A negation must have the opposite truth value from the original statement.

The next example uses some of the following inequality symbols from algebra.

Symbolism	Meaning	Examples	
$a < b$	a is less than b	$4 < 9$	$\frac{1}{2} < \frac{3}{4}$
$a > b$	a is greater than b	$6 > 2$	$-5 > -11$
$a \leq b$	a is less than or equal to b	$8 \leq 10$	$3 \leq 3$
$a \geq b$	a is greater than or equal to b	$-2 \geq -3$	$-5 \geq -5$

EXAMPLE 3 Give a negation of each inequality. Do *not* use a slash symbol.

(a) $p < 9$

The negation of "p is less than 9" is "p is *not* less than 9." Since we cannot use "not," which would require writing $p \not< 9$, phrase the negation as "p is greater than or equal to 9," or $p \geq 9$.

(b) $7x + 11y \geq 77$

The negation, with no slash, is $7x + 11y < 77$. ■

Symbols
To simplify work with logic, symbols are used. Statements are represented with letters, such as p, q, or r, while several symbols for connectives are shown in the following table. The table also gives the type of compound statement having the given connective.

Connective	Symbol	Type of Statement
and	$\wedge$	Conjunction
or	$\vee$	Disjunction
not	$\sim$	Negation

The symbol ~ represents the connective *not*. If *p* represents the statement "Bill Clinton was president in 1997" then ~*p* represents "Bill Clinton was not president in 1997."

EXAMPLE 4 Let *p* represent "It is 80° today," and let *q* represent "It is Tuesday." Write each symbolic statement in words.

(a) $p \vee q$

From the table, $\vee$ symbolizes *or*; thus, $p \vee q$ represents

It is 80° today or it is Tuesday.

(b) $\sim p \wedge q$

It is not 80° today and it is Tuesday.

(c) $\sim(p \vee q)$

It is not the case that it is 80° today or it is Tuesday.

(d) $\sim(p \wedge q)$

It is not the case that it is 80° today and it is Tuesday.

Aristotle, the first to systematize the logic we use in everyday life, appears above in a detail from the painting *The School of Athens,* by Raphael. He is shown debating a point with his teacher **Plato.** Many people feel that Aristotle's logic and analysis of how things work are inseparably woven into the ways Westerners are taught to think.

The statement in part (c) of Example 4 usually is translated in English as "Neither *p* nor *q*."

Quantifiers

The words *all, each, every,* and *no(ne)* are called **universal quantifiers,** while words and phrases such as *some, there exists,* and *(for) at least one* are called **existential quantifiers.** Quantifiers are used extensively in mathematics to indicate *how many* cases of a particular situation exist. Be careful when forming the negation of a statement involving quantifiers.

The negation of a statement must be false if the given statement is true and must be true if the given statement is false, in all possible cases. Consider the statement

All girls in the group are named Mary.

Many people would write the negation of this statement as "No girls in the group are named Mary" or "All girls in the group are not named Mary." But this would not be correct. To see why, look at the three groups below.

Group I: Mary Jones, Mary Smith, Mary Jackson
Group II: Mary Johnson, Betty Parker, Margaret Westmoreland
Group III: Donna Garbarino, Paula Story, Rhonda Alessi

These groups contain all possibilities that need to be considered. In Group I, *all* girls are named Mary; in Group II, *some* girls are named Mary (and some are not); in Group III, *no* girls are named Mary. Look at the truth values in the chart on the next page and keep in mind that "some" means "at least one (and possibly all)."

Truth Value as Applied to:			
	Group I	Group II	Group III
(1) All girls in the group are named Mary. (Given)	T	F	F
(2) No girls in the group are named Mary. (Possible negation)	F	F	T
(3) All girls in the group are not named Mary. (Possible negation)	F	F	T
(4) Some girls in the group are not named Mary. (Possible negation)	F	T	T

Negation (arrows connect row 1 and row 4 in Group III column)

The negation of the given statement (1) must have opposite truth values in *all* cases. It can be seen that statements (2) and (3) do not satisfy this condition (for Group II), but statement (4) does. It may be concluded that the correct negation for "All girls in the group are named Mary" is "Some girls in the group are not named Mary." Other ways of stating the negation are:

> Not all girls in the group are named Mary.
>
> It is not the case that all girls in the group are named Mary.
>
> At least one girl in the group is not named Mary.

The following table can be used to generalize the method of finding the negation of a statement involving quantifiers.

Negations of Quantified Statements

Statement	Negation
All do.	Some do not. (Equivalently: Not all do.)
Some do.	None do. (Equivalently: All do not.)

The negation of the negation of a statement is simply the statement itself. For instance, the negations of the statements in the Negation column are simply the corresponding original statements in the Statement column. As an example, the negation of "Some do not" is "All do."

EXAMPLE 5 Write the negation of each statement.

(a) Some cats have fleas.

Since *some* means "at least one," the statement "Some cats have fleas" is really the same as "At least one cat has fleas." The negation of this is "No cat has fleas."

(b) Some cats do not have fleas.

This statement claims that at least one cat, somewhere, does not have fleas. The negation of this is "All cats have fleas."

(c) No cats have fleas.

The negation is "Some cats have fleas."

The many relationships among special sets of numbers can be expressed using universal and existential quantifiers. Earlier we introduced sets of numbers that are studied in algebra, and we repeat these in the box that follows.

Sets of Numbers

Natural or Counting numbers $\{1, 2, 3, 4, \ldots\}$

Whole numbers $\{0, 1, 2, 3, 4, \ldots\}$

Integers $\{\ldots, -3, -2, -1, 0, 1, 2, 3, \ldots\}$

Rational numbers $\{p/q \mid p \text{ and } q \text{ are integers, and } q \neq 0\}$
(Some examples of rational numbers are $3/5$, $-7/9$, 5, and 0. Any rational number may be written as a terminating decimal number, like $.25$ or a repeating decimal number, like $.666\ldots$.)

Real numbers $\{x \mid x \text{ is a number that can be written as a decimal}\}$

Irrational numbers $\{x \mid x \text{ is a real number and } x \text{ cannot be written as a quotient of integers}\}$
(Some examples of irrational numbers are $\sqrt{2}$, $\sqrt[3]{4}$, and π. Decimal representations of irrational numbers never terminate and never repeat.

EXAMPLE 6 Decide whether each of the following statements about sets of numbers involving a quantifier is *true* or *false*.

(a) There exists a whole number that is not a natural number.

Because there is such a whole number (it is 0), this statement is true.

(b) Every integer is a natural number.

This statement is false, because we can find at least one integer that is not a natural number. For example, -1 is an integer but is not a natural number. (There are infinitely many other choices we could have made.)

(c) Every natural number is a rational number.

Since every natural number can be written as a fraction with denominator 1, this statement is true.

(d) There exists an irrational number that is not real.

In order to be an irrational number, a number must first be real (see the box). Therefore, since we cannot give an irrational number that is not real, this statement is false. (Had we been able to find at least one, the statement would have then been true.)

3.1 EXERCISES

Decide whether each of the following is a statement or is not a statement.

1. December 7, 1941, was a Sunday.
2. The ZIP code for Manistee, MI, is 49660.
3. Listen, my children, and you shall hear of the midnight ride of Paul Revere.
4. Yield to oncoming traffic.
5. $5 + 8 = 13$ and $4 - 3 = 1$

6. $5 + 8 = 12$ or $4 - 3 = 2$

7. Some numbers are negative.

8. Andrew Johnson was president of the United States in 1867.

9. Accidents are the main cause of deaths of children under the age of 8.

10. *Star Wars: Episode I—The Phantom Menace* was the top-grossing movie of 1999.

11. Where are you going today?

12. Behave yourself and sit down.

13. Kevin "Catfish" McCarthy once took a prolonged continuous shower for 340 hours, 40 minutes.

14. One gallon of milk weighs more than 4 pounds.

Decide whether each of the following statements is compound.

15. I read the *Chicago Tribune* and I read the *New York Times*.

16. My brother got married in London.

17. Tomorrow is Sunday.

18. Dara Lanier is younger than 29 years of age, and so is Teri Orr.

19. Jay Beckenstein's wife loves Ben and Jerry's ice cream.

20. The sign on the back of the car read "California or bust!"

21. If Julie Ward sells her quota, then Bill Leonard will be happy.

22. If Mike is a politician, then Jerry is a crook.

Write a negation for each of the following statements.

23. Her aunt's name is Lucia.

24. The flowers are to be watered.

25. Every dog has its day.

26. No rain fell in southern California today.

27. Some books are longer than this book.

28. All students present will get another chance.

29. No computer repairman can play blackjack.

30. Some people have all the luck.

31. Everybody loves somebody sometime.

32. Everyone loves a winner.

Give a negation of each inequality. Do not use a slash symbol.

33. $y > 12$

34. $x < -6$

35. $q \geq 5$

36. $r \leq 19$

37. Try to negate the sentence "The exact number of words in this sentence is ten" and see what happens. Explain the problem that arises.

38. Explain why the negation of "$r > 4$" is not "$r < 4$."

Let p represent the statement "She has green eyes" and let q represent the statement "He is 48 years old." Translate each symbolic compound statement into words.

39. $\sim p$

40. $\sim q$

41. $p \wedge q$

42. $p \vee q$

43. $\sim p \vee q$

44. $p \wedge \sim q$

45. $\sim p \vee \sim q$

46. $\sim p \wedge \sim q$

47. $\sim(\sim p \wedge q)$

48. $\sim(p \vee \sim q)$

Let p represent the statement "Chris collects videotapes" and let q represent the statement "Jack plays the tuba." Convert each of the following compound statements into symbols.

49. Chris collects videotapes and Jack does not play the tuba.

50. Chris does not collect videotapes or Jack does not play the tuba.

51. Chris does not collect videotapes or Jack plays the tuba.

52. Jack plays the tuba and Chris does not collect videotapes.

53. Neither Chris collects videotapes nor Jack plays the tuba.

54. Either Jack plays the tuba or Chris collects videotapes, and it is not the case that both Jack plays the tuba and Chris collects videotapes.

55. Incorrect use of quantifiers often is heard in everyday language. Suppose you hear that a local electronics chain is having a 30% off sale, and the radio advertisement states "All items are not available in all stores." Do you think that, literally translated, the ad really means what it says? What do you think really is meant? Explain your answer.

56. Repeat Exercise 55 for the following: "All people don't have the time to devote to maintaining their cars properly."

Refer to the groups of art labeled A, B, *and* C, *and identify by letter the group or groups that are satisfied by the given statements involving quantifiers.*

A

B

C

57. All pictures have frames.
58. No picture has a frame.
59. At least one picture does not have a frame.
60. Not every picture has a frame.
61. At least one picture has a frame.
62. No picture does not have a frame.
63. All pictures do not have frames.
64. Not every picture does not have a frame.

Decide whether each statement in Exercises 65–74 involving a quantifier is true *or* false.

65. Every whole number is an integer.
66. Every natural number is an integer.
67. There exists a rational number that is not an integer.
68. There exists an integer that is not a natural number.
69. All rational numbers are real numbers.
70. All irrational numbers are real numbers.
71. Some rational numbers are not integers.
72. Some whole numbers are not rational numbers.
73. Each whole number is a positive number.
74. Each rational number is a positive number.

75. Explain the difference between the following statements:

All students did not pass the test.

Not all students passed the test.

76. The statement "For some real number x, $x^2 \geq 0$" is true. However, your friend does not understand why, since he claims that $x^2 \geq 0$ for *all* real numbers x (and not *some*). How would you explain his misconception to him?

77. Write the following statement using "every": There is no one here who has not done that at one time or another.

78. Only one of the following statements is true. Which one is it?
  A. For some real number x, $x \not< 0$.
  B. For all real numbers x, $x^3 > 0$.
  C. For all real numbers x less than 0, x^2 is also less than 0.
  D. For some real number x, $x^2 < 0$.

3.2 Truth Tables and Equivalent Statements

In this section, the truth values of component statements are used to find the truth values of compound statements.

Conjunctions To begin, let us decide on the truth values of the **conjunction** *p and q*, symbolized $p \wedge q$. In everyday language, the connective *and* implies the idea of "both." The statement

   Monday immediately follows Sunday and March immediately follows February

is true, since each component statement is true. On the other hand, the statement

   Monday immediately follows Sunday and March immediately follows January

is false, even though part of the statement (Monday immediately follows Sunday) is true. For the conjunction $p \wedge q$ to be true, both *p* and *q* must be true. This result is summarized by a table, called a **truth table,** which shows all four of the possible combinations of truth values for the conjunction *p and q*. The truth table for *conjunction* is shown here.

Truth Table for the Conjunction *p* and *q*

p and *q*

p	*q*	$p \wedge q$
T	T	T
T	F	F
F	T	F
F	F	F

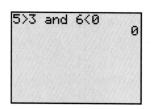

The calculator returns a "0" for 5 > 3 *and* 6 < 0, indicating that the statement is false.

EXAMPLE 1 Let *p* represent "5 > 3" and let *q* represent "6 < 0." Find the truth value of $p \wedge q$.

Here *p* is true and *q* is false. Looking in the second row of the conjunction truth table shows that $p \wedge q$ is false.

In some cases, the logical connective *but* is used in compound statements:

   He wants to go to the mountains but she wants to go to the beach.

Here, *but* is used in place of *and* to give a different sort of emphasis to the statement. In such a case, we consider the statement as we would consider the conjunction using the word *and*. The truth table for the conjunction, given above, would apply.

Disjunctions In ordinary language, the word *or* can be ambiguous. The expression "this or that" can mean either "this or that or both," or "this or that but not both." For example, the statement

   I will paint the wall or I will paint the ceiling

probably has the following meaning: "I will paint the wall or I will paint the ceiling or I will paint both." On the other hand, the statement

I will drive the Ford or the Nissan to the store

probably means "I will drive the Ford, or I will drive the Nissan, but I will not drive both."

The symbol $\vee$ normally represents the first *or* described. That is, $p \vee q$ means "*p* or *q* or both." With this meaning of *or*, $p \vee q$ is called the *inclusive disjunction*, or just the **disjunction** of *p* and *q*.

In everyday language, the disjunction implies the idea of "either." For example, the disjunction

I have a quarter or I have a dime

is true whenever I have either a quarter, a dime, or both. The only way this disjunction could be false would be if I had neither coin. A disjunction is false only if both component statements are false. The truth table for *disjunction* follows.

Truth Table for the Disjunction *p* or *q*

p or *q*

p	*q*	$p \vee q$
T	T	T
T	F	T
F	T	T
F	F	F

Statement	Reason That It Is True
$8 \geq 8$	$8 = 8$
$3 \geq 1$	$3 > 1$
$-5 \leq -3$	$-5 < -3$
$-4 \leq -4$	$-4 = -4$

The symbol $\geq$ is read "is greater than or equal to," while $\leq$ is read "is less than or equal to." If *a* and *b* are real numbers, $a \leq b$ is true if $a < b$ or $a = b$. The table in the margin shows several statements and the reasons they are true.

The **negation** of a statement *p*, symbolized $\sim p$, must have the opposite truth value from the statement *p* itself. This leads to the truth table for the negation, shown here.

Truth Table for the Negation not *p*

not *p*

p	$\sim p$
T	F
F	T

EXAMPLE 2 Suppose p is false, q is true, and r is false. What is the truth value of the compound statement $\sim p \land (q \lor \sim r)$?

Here parentheses are used to group q and $\sim r$ together. Work first inside the parentheses. Since r is false, $\sim r$ will be true. Since $\sim r$ is true and q is true, find the truth value of $q \lor \sim r$ by looking in the first row of the *or* truth table. This row gives the result T. Since p is false, $\sim p$ is true, and the final truth value of $\sim p \land (q \lor \sim r)$ is found in the top row of the *and* truth table. From the *and* truth table, when $\sim p$ is true, and $q \lor \sim r$ is true, the statement $\sim p \land (q \lor \sim r)$ is true.

The preceding paragraph may be interpreted using a short-cut symbolic method, letting T represent a true statement and F represent a false statement:

$$\sim p \land (q \lor \sim r)$$
$$\sim F \land (T \lor \sim F)$$
$$T \land (T \lor T) \quad \sim F \text{ gives T.}$$
$$T \land T \quad\quad\quad T \lor T \text{ gives T.}$$
$$T. \quad\quad\quad\quad\quad T \land T \text{ gives T.}$$

The T in the final row indicates that the compound statement is true.

FOR FURTHER THOUGHT

Raymond Smullyan is one of today's foremost writers of logic puzzles. This multitalented professor of mathematics and philosophy at City University of New York has written several books on recreational logic, including *The Lady or the Tiger?*, *What Is the Name of This Book?*, and *Alice in Puzzleland.* The title of the first of these is taken from the classic Frank Stockton short story, in which a prisoner must make a choice between two doors: behind one is a beautiful lady, and behind the other is a hungry tiger.

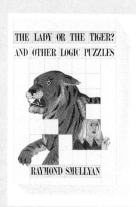

For Group Discussion

Smullyan proposes the following: What if each door has a sign, and the man knows that only one sign is true?

The sign on Door 1 reads:

  IN THIS ROOM THERE IS A LADY AND
  IN THE OTHER ROOM THERE IS A TIGER.

The sign on Door 2 reads:

  IN ONE OF THESE ROOMS THERE IS A LADY AND
  IN ONE OF THESE ROOMS THERE IS A TIGER.

With this information, the man is able to choose the correct door. Can you? (The answer is on page 106.)

Mathematical Statements We can use truth tables to determine the truth values of mathematical statements.

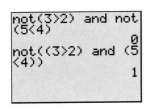

Example 3(a) explains why $\sim(3 > 2) \wedge [\sim(5 < 4)]$ is false. The calculator returns a 0. For a true statement such as $\sim[(3 > 2) \wedge (5 < 4)]$, it returns a 1.

EXAMPLE 3 Let p represent the statement $3 > 2$, q represent $5 < 4$, and r represent $3 < 8$. Decide whether the following statements are *true* or *false*.

(a) $\sim p \wedge \sim q$

Since p is true, $\sim p$ is false. By the *and* truth table, if one part of an "and" statement is false, the entire statement is false. This makes $\sim p \wedge \sim q$ false.

(b) $\sim(p \wedge q)$

First, work within the parentheses. Since p is true and q is false, $p \wedge q$ is false by the *and* truth table. Next, apply the negation. The negation of a false statement is true, making $\sim(p \wedge q)$ a true statement.

(c) $(\sim p \wedge r) \vee (\sim q \wedge \sim p)$

Here p is true, q is false, and r is true. This makes $\sim p$ false and $\sim q$ true. By the *and* truth table, the statement $\sim p \wedge r$ is false, and the statement $\sim q \wedge \sim p$ is also false. Finally,

$$(\sim p \wedge r) \vee (\sim q \wedge \sim p)$$
$$\downarrow \qquad \qquad \downarrow$$
$$F \quad \vee \quad F,$$

which is false by the *or* truth table. (For an alternative solution, see Example 7(b).)

When a quantifier is used with a conjunction or a disjunction, we must be careful in determining the truth value, as shown in the following example.

EXAMPLE 4 Identify each statement as *true* or *false*.

(a) For some real number x, $x < 5$ and $x > 2$.

Replacing x with 3 (as an example) gives $3 < 5$ and $3 > 2$. Since both $3 < 5$ and $3 > 2$ are true statements, the given statement is true by the *and* truth table. (Remember: *some* means "at least one.")

(b) For every real number b, $b > 0$ or $b < 1$.

No matter which real number might be tried as a replacement for b, at least one of the statements $b > 0$ and $b < 1$ will be true. Since an "or" statement is true if one or both component statements are true, the entire statement as given is true.

(c) For all real numbers, x, $x^2 > 0$.

Since the quantifier is a universal quantifier, we need only find one case in which the inequality is false to make the entire statement false. Can we find a real number whose square is not positive (that is, not greater than 0)? Yes, we can—0 itself is a real number (and the *only* real number) whose square is not positive. Therefore, this statement is false.

George Boole (1815–1864) grew up in poverty. His father, a London tradesman, gave him his first math lessons and taught him to make optical instruments. Boole was largely self-educated. At 16 he worked in an elementary school and by age 20 had opened his own school. He studied mathematics in his spare time. Lung disease killed him at age 49.

Boole's ideas have been used in the design of computers and telephone systems.

Truth Tables In the preceding examples, the truth value for a given statement was found by going back to the basic truth tables. In the long run, it is easier to first create a complete truth table for the given statement itself. Then final truth values can be read directly from this table. The procedure for making new truth tables is shown in the next few examples.

Answer to the Problem of *The Lady or the Tiger?*

The lady is behind Door 2. Suppose that the sign on Door 1 is true. Then the sign on Door 2 would also be true, but this is impossible. So the sign on Door 2 must be true, and the sign on Door 1 must be false. Since the sign on Door 1 says the lady is in Room 1, and this is false, the lady must be behind Door 2.

In this book we will use the following standard format for listing the possible truth values in compound statements involving two component statements.

p	q	Compound Statement
T	T	
T	F	
F	T	
F	F	

EXAMPLE 5

(a) Construct a truth table for $(\sim p \wedge q) \vee \sim q$.

Begin by listing all possible combinations of truth values for p and q, as above. Then list the truth values of $\sim p$, which are the opposite of those of p.

p	q	$\sim p$
T	T	F
T	F	F
F	T	T
F	F	T

Use only the "$\sim p$" column and the "q" column, along with the *and* truth table, to find the truth values of $\sim p \wedge q$. List them in a separate column.

p	q	$\sim p$	$\sim p \wedge q$
T	T	F	F
T	F	F	F
F	T	T	T
F	F	T	F

Next include a column for $\sim q$.

p	q	$\sim p$	$\sim p \wedge q$	$\sim q$
T	T	F	F	F
T	F	F	F	T
F	T	T	T	F
F	F	T	F	T

Finally, make a column for the entire compound statement. To find the truth values, use *or* to combine $\sim p \wedge q$ with $\sim q$.

p	q	$\sim p$	$\sim p \wedge q$	$\sim q$	$(\sim p \wedge q) \vee \sim q$
T	T	F	F	F	F
T	F	F	F	T	T
F	T	T	T	F	T
F	F	T	F	T	T

(b) Suppose both p and q are true. Find the truth value of $(\sim p \wedge q) \vee \sim q$.

Look in the first row of the final truth table above, where both p and q have truth value T. Read across the row to find that the compound statement is false. ■

3.2 Truth Tables and Equivalent Statements

EXAMPLE 6 Construct the truth table for $p \land (\sim p \lor \sim q)$.

Proceed as shown.

p	q	$\sim p$	$\sim q$	$\sim p \lor \sim q$	$p \land (\sim p \lor \sim q)$
T	T	F	F	F	F
T	F	F	T	T	T
F	T	T	F	T	F
F	F	T	T	T	F

If a compound statement involves three component statements p, q, and r, we will use the following standard format in setting up the truth table.

p	q	r	Compound Statement
T	T	T	
T	T	F	
T	F	T	
T	F	F	
F	T	T	
F	T	F	
F	F	T	
F	F	F	

Emilie, Marquise du Châtelet (1706–1749) participated in the scientific activity of the generation after Newton and Leibniz. Educated in science, music, and literature, she was studying mathematics at the time (1733) she began a long intellectual relationship with the philosopher François Voltaire (1694–1778). She and Voltaire competed independently in 1738 for a prize offered by the French Academy on the subject of fire. Although du Châtelet did not win, her dissertation was published by the academy in 1744. During the last four years of her life she translated Newton's *Principia* from Latin into French—the only French translation to date.

EXAMPLE 7

(a) Construct a truth table for $(\sim p \land r) \lor (\sim q \land \sim p)$.

This statement has three component statements, p, q, and r. The truth table thus requires eight rows to list all possible combinations of truth values of p, q, and r. The final truth table, however, can be found in much the same way as the ones above.

p	q	r	$\sim p$	$\sim p \land r$	$\sim q$	$\sim q \land \sim p$	$(\sim p \land r) \lor (\sim q \land \sim p)$
T	T	T	F	F	F	F	F
T	T	F	F	F	F	F	F
T	F	T	F	F	T	F	F
T	F	F	F	F	T	F	F
F	T	T	T	T	F	F	T
F	T	F	T	F	F	F	F
F	F	T	T	T	T	T	T
F	F	F	T	F	T	T	T

(b) Suppose p is true, q is false, and r is true. Find the truth value of $(\sim p \land r) \lor (\sim q \land \sim p)$.

By the third row of the truth table in part (a), the compound statement is false. (This is an alternative method for working part (c) of Example 3.)

Problem Solving

One strategy for problem solving is noticing a pattern and using inductive reasoning. This strategy is used in the next example.

Number of Statements	Number of Rows
1	$2 = 2^1$
2	$4 = 2^2$
3	$8 = 2^3$

EXAMPLE 8 If n is a counting number, and a logical statement is composed of n component statements, how many rows will appear in the truth table for the compound statement?

To answer this question, let us examine some of the earlier truth tables in this section. The truth table for the negation has one statement and two rows. The truth tables for the conjunction and the disjunction have two component statements, and each has four rows. The truth table in Example 7(a) has three component statements and eight rows. Summarizing these in a table, as seen in the margin, reveals a pattern seen earlier. Inductive reasoning leads us to the conjecture that, if a logical statement is composed of n component statements, it will have 2^n rows. This can be proved using more advanced concepts.

The result of Example 8 is reminiscent of the formula for the number of subsets of a set having n elements.

Number of Rows in a Truth Table

A logical statement having n component statements will have 2^n rows in its truth table.

Alternative Method for Constructing Truth Tables

After making a reasonable number of truth tables, some people prefer the shortcut method shown in Example 9, which repeats Examples 5 and 7.

EXAMPLE 9 Construct the truth table for each statement.

(a) $(\sim p \land q) \lor \sim q$

Start by inserting truth values for $\sim p$ and for q.

p	q	$(\sim p$	$\land$	$q)$	$\lor$	$\sim q$
T	T	F		T		
T	F	F		F		
F	T	T		T		
F	F	T		F		

Next, use the *and* truth table to obtain the truth values for $\sim p \land q$.

p	q	$(\sim p$	$\land$	$q)$	$\lor$	$\sim q$
T	T	F	F	T		
T	F	F	F	F		
F	T	T	T	T		
F	F	T	F	F		

Now disregard the two preliminary columns of truth values for $\sim p$ and for q, and insert truth values for $\sim q$. Finally, use the *or* truth table.

3.2 Truth Tables and Equivalent Statements

p	q	$(\sim p$	$\wedge$	$q)$	$\vee$	$\sim q$
T	T		F			F
T	F		F			T
F	T		T			F
F	F		F			T

p	q	$(\sim p$	$\wedge$	$q)$	$\vee$	$\sim q$
T	T		F		F	F
T	F		F		T	T
F	T		T		T	F
F	F		F		T	T

These steps can be summarized as follows.

p	q	$(\sim p$	$\wedge$	$q)$	$\vee$	$\sim q$
T	T	F	F	T	F	F
T	F	F	F	F	T	T
F	T	T	T	T	T	F
F	F	T	F	F	T	T
		①	②	①	④	③

The circled numbers indicate the order in which the various columns of the truth table were found.

(b) $(\sim p \wedge r) \vee (\sim q \wedge \sim p)$
Work as follows.

p	q	r	$(\sim p$	$\wedge$	$r)$	$\vee$	$(\sim q$	$\wedge$	$\sim p)$
T	T	T	F	F	T	F	F	F	F
T	T	F	F	F	F	F	F	F	F
T	F	T	F	F	T	F	T	F	F
T	F	F	F	F	F	F	T	F	F
F	T	T	T	T	T	T	F	F	T
F	T	F	T	F	F	F	F	F	T
F	F	T	T	T	T	T	T	T	T
F	F	F	T	F	F	T	T	T	T
			①	②	①	⑤	③	④	③

Equivalent Statements One application of truth tables is to show that two statements are equivalent; by definition, two statements are **equivalent** if they have the same truth value in *every* possible situation. The columns of each truth table that were the last to be completed will be exactly the same for equivalent statements.

EXAMPLE 10 Are the following statements equivalent?

$$\sim p \wedge \sim q \quad \text{and} \quad \sim(p \vee q)$$

To find out, make a truth table for each statement, with the following results.

p	q	$\sim p \wedge \sim q$
T	T	F
T	F	F
F	T	F
F	F	T

p	q	$\sim(p \vee q)$
T	T	F
T	F	F
F	T	F
F	F	T

Since the truth values are the same in all cases, as shown in the columns in color, the statements $\sim p \wedge \sim q$ and $\sim(p \vee q)$ are equivalent. Equivalence is written with a three-bar symbol, $\equiv$. Using this symbol, $\sim p \wedge \sim q \equiv \sim(p \vee q)$.

In the same way, the statements $\sim p \vee \sim q$ and $\sim(p \wedge q)$ are equivalent. We call these equivalences De Morgan's laws.

De Morgan's Laws

For any statements p and q,

$$\sim(p \vee q) \equiv \sim p \wedge \sim q$$

and

$$\sim(p \wedge q) \equiv \sim p \vee \sim q.$$

(Compare the logic statements of De Morgan's laws with the set versions.) De Morgan's laws can be used to find the negations of certain compound statements.

EXAMPLE 11 Find a negation of each statement by applying De Morgan's laws.

(a) I got an A or I got a B.

If p represents "I got an A" and q represents "I got a B," then the compound statement is symbolized $p \vee q$. The negation of $p \vee q$ is $\sim(p \vee q)$; by one of De Morgan's laws, this is equivalent to

$$\sim p \wedge \sim q,$$

or, in words,

I didn't get an A and I didn't get a B.

This negation is reasonable—the original statement says that I got either an A or a B; the negation says that I didn't get *either* grade.

(b) She won't try and he will succeed.

From one of De Morgan's laws, $\sim(p \wedge q) \equiv \sim p \vee \sim q$, so the negation becomes

She will try or he won't succeed.

(c) $\sim p \vee (q \wedge \sim p)$

Negate both component statements and change $\vee$ to $\wedge$.

$$\sim[\sim p \vee (q \wedge \sim p)] \equiv p \wedge \sim(q \wedge \sim p)$$

Now apply De Morgan's law again.

$$p \wedge \sim(q \wedge \sim p) \equiv p \wedge (\sim q \vee \sim(\sim p))$$
$$\equiv p \wedge (\sim q \vee p)$$

A truth table will show that the statements

$$\sim p \vee (q \wedge \sim p) \quad \text{and} \quad p \wedge (\sim q \vee p)$$

are negations.

3.2 EXERCISES

Use the concepts introduced in this section to answer Exercises 1–6.

1. If q is false, what must be the truth value of $(p \land \sim q) \land q$?
2. If q is true, what must be the truth value of $q \lor (q \land \sim p)$?
3. If $p \land q$ is true, and p is true, then q must be _____.
4. If $p \lor q$ is false, and p is false, then q must be _____.
5. If $\sim(p \lor q)$ is true, what must be the truth values of the component statements?
6. If $\sim(p \land q)$ is false, what must be the truth values of the component statements?

Let p represent a false statement and let q represent a true statement. Find the truth value of the given compound statement.

7. $\sim p$
8. $\sim q$
9. $p \lor q$
10. $p \land q$
11. $p \lor \sim q$
12. $\sim p \land q$
13. $\sim p \lor \sim q$
14. $p \land \sim q$
15. $\sim(p \land \sim q)$
16. $\sim(\sim p \lor \sim q)$
17. $\sim[\sim p \land (\sim q \lor p)]$
18. $\sim[(\sim p \land \sim q) \lor \sim q]$
19. Is the statement $3 \geq 1$ a conjunction or a disjunction? Why?
20. Why is the statement $6 \geq 2$ true? Why is $6 \geq 6$ true?

Let p represent a true statement, and q and r represent false statements. Find the truth value of the given compound statement.

21. $(p \land r) \lor \sim q$
22. $(q \lor \sim r) \land p$
23. $p \land (q \lor r)$
24. $(\sim p \land q) \lor \sim r$
25. $\sim(p \land q) \land (r \lor \sim q)$
26. $(\sim r \land \sim q) \lor (\sim r \land q)$
27. $\sim[(\sim p \land q) \lor r]$
28. $\sim[r \lor (\sim q \land \sim p)]$

Let p represent the statement $2 > 7$, let q represent the statement $8 \not> 6$, and let r represent the statement $19 \leq 19$. Find the truth value of the given compound statement.

29. $p \land r$
30. $p \lor \sim q$
31. $\sim q \lor \sim r$
32. $\sim p \land \sim r$
33. $(p \land q) \lor r$
34. $\sim p \lor (\sim r \lor \sim q)$
35. $(\sim r \land q) \lor \sim p$
36. $\sim(p \lor \sim q) \lor \sim r$

Give the number of rows in the truth table for each of the following compound statements.

37. $p \lor \sim r$
38. $p \land (r \land \sim s)$
39. $(\sim p \land q) \lor (\sim r \lor \sim s) \land r$
40. $[(p \lor q) \land (r \land s)] \land (t \lor \sim p)$
41. $[(\sim p \land \sim q) \land (\sim r \land s \land \sim t)] \land (\sim u \lor \sim v)$
42. $[(\sim p \land \sim q) \lor (\sim r \lor \sim s)] \lor [(\sim m \land \sim n) \land (u \land \sim v)]$

43. If the truth table for a certain compound statement has 64 rows, how many distinct component statements does it have?
44. Is it possible for the truth table of a compound statement to have exactly 48 rows? Why or why not?

Construct a truth table for each compound statement.

45. $\sim p \land q$
46. $\sim p \lor \sim q$
47. $\sim(p \land q)$
48. $p \lor \sim q$
49. $(q \lor \sim p) \lor \sim q$
50. $(p \land \sim q) \land p$

51. $\sim q \wedge (\sim p \vee q)$
52. $\sim p \vee (\sim q \wedge \sim p)$
53. $(p \vee \sim q) \wedge (p \wedge q)$
54. $(\sim p \wedge \sim q) \vee (\sim p \vee q)$
55. $(\sim p \wedge q) \wedge r$
56. $r \vee (p \wedge \sim q)$
57. $(\sim p \wedge \sim q) \vee (\sim r \vee \sim p)$
58. $(\sim r \vee \sim p) \wedge (\sim p \vee \sim q)$
59. $\sim(\sim p \wedge \sim q) \vee (\sim r \vee \sim s)$
60. $(\sim r \vee s) \wedge (\sim p \wedge q)$

Use one of De Morgan's laws to write the negation of each statement.

61. You can pay me now or you can pay me later.
62. I am not going or she is going.
63. It is summer and there is no snow.
64. $1/2$ is a positive number and -12 is less than zero.
65. I said yes but she said no.
66. Pauline Mula tried to sell the book, but she was unable to do so.
67. $5 - 1 = 4$ and $9 + 12 \neq 7$
68. $3 < 10$ or $7 \neq 2$
69. Cupid or Vixen will lead Santa's sleigh next Christmas.
70. The lawyer and the client appeared in court.

Identify each of the following statements as true *or* false.

71. For every real number y, $y < 13$ or $y > 6$.
72. For every real number t, $t > 9$ or $t < 9$.
73. For some integer p, $p \geq 4$ and $p \leq 4$.
74. There exists an integer n such that $n > 0$ and $n < 0$.
75. Complete the truth table for *exclusive disjunction*. The symbol $\underline{\vee}$ represents "one or the other is true, but not both."

p	q	$p \underline{\vee} q$
T	T	
T	F	
F	T	
F	F	

Exclusive disjunction

76. Attorneys sometimes use the phrase "and/or." This phrase corresponds to which usage of the word *or*: inclusive or exclusive?

Decide whether the following compound statements are true *or* false. *Remember that $\underline{\vee}$ is the exclusive disjunction; that is, assume "either p or q is true, but not both."*

77. $3 + 1 = 4 \underline{\vee} 2 + 5 = 7$
78. $3 + 1 = 4 \underline{\vee} 2 + 5 = 9$
79. $3 + 1 = 7 \underline{\vee} 2 + 5 = 7$
80. $3 + 1 = 7 \underline{\vee} 2 + 5 = 9$

3.3 The Conditional and Circuits

Conditionals

"If you build it, he will come."
—The Voice in the 1990 movie *Field of Dreams*

Ray Kinsella, an Iowa farmer in the movie *Field of Dreams*, heard a voice from the sky. Ray interpreted it as a promise that if he would build a baseball field in his cornfield, then the ghost of Shoeless Joe Jackson (a baseball star in the early days of the twentieth century) would come to play on it. The promise came in the form of a conditional statement. A **conditional** statement is a compound statement that uses the connective *if ... then*. For example, here are a few conditional statements.

If I read for too long, *then* I get a headache.

If looks could kill, *then* I would be dead.

If he doesn't get back soon, *then* you should go look for him.

In each of these conditional statements, the component coming after the word *if* gives a condition (but not necessarily the only condition) under which the statement coming after *then* will be true. For example, "If it is over 90°, then I'll go to the mountains" tells one possible condition under which I will go to the mountains—if the temperature is over 90°.

The conditional is written with an arrow, so "if p, then q" is symbolized

$$p \rightarrow q.$$

We read $p \rightarrow q$ as "p implies q" or "if p, then q." In the conditional $p \rightarrow q$, the statement p is the **antecedent,** while q is the **consequent.**

The conditional connective may not always be explicitly stated. That is, it may be "hidden" in an everyday expression. For example, the statement

Big girls don't cry

can be written in *if ... then* form as

If you're a big girl, then you don't cry.

As another example, the statement

It is difficult to study when you are distracted

can be written

If you are distracted, then it is difficult to study.

In the quote from the movie *Field of Dreams,* the word "then" is not stated but understood to be there from the context of the statement. In that statement, "you build it" is the antecedent and "he will come" is the consequent.

The conditional truth table is a little harder to define than the tables in the previous section. To see how to define the conditional truth table, let us analyze a statement made by a politician, Senator Bridget Terry:

If I am elected, then taxes will go down.

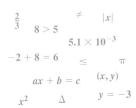

The importance of symbols was emphasized by the American philosopher-logician Charles Sanders Peirce (1839–1914), who asserted the nature of humans as symbol-using or sign-using organisms. Symbolic notation is half of mathematics, Bertrand Russell once said.

As before, there are four possible combinations of truth values for the two component statements. Let p represent "I am elected," and let q represent "Taxes will go down."

As we analyze the four possibilities, it is helpful to think in terms of the following: "Did Senator Terry lie?" If she lied, then the conditional statement is considered false; if she did not lie, then the conditional statement is considered true.

Possibility	Elected?	Taxes Go Down?	
1	Yes	Yes	p is T, q is T
2	Yes	No	p is T, q is F
3	No	Yes	p is F, q is T
4	No	No	p is F, q is F

The four possibilities are as follows:

1. In the first case assume that the senator was elected and taxes did go down (p is T, q is T). The senator told the truth, so place T in the first row of the truth table. (We do not claim that taxes went down *because* she was elected; it is possible that she had nothing to do with it at all.)
2. In the second case assume that the senator was elected and taxes did not go down (p is T, q is F). Then the senator did not tell the truth (that is, she lied). So we put F in the second row of the truth table.
3. In the third case assume that the senator was defeated, but taxes went down anyway (p is F, q is T). Senator Terry did not lie; she only promised a tax reduction if she were elected. She said nothing about what would happen if she were not elected. In fact, her campaign promise gives no information about what would happen if she lost. Since we cannot say that the senator lied, place T in the third row of the truth table.
4. In the last case assume that the senator was defeated but taxes did not go down (p is F, q is F). We cannot blame her, since she only promised to reduce taxes if elected. Thus, T goes in the last row of the truth table.

The completed truth table for the conditional is defined as follows.

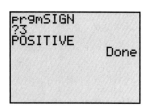

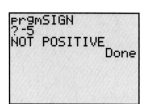

Conditional statements are useful in writing programs. The short program in the first two screens determines whether a number is positive. Notice the lines that begin with *If* and *Then*.

Truth Table for the Conditional If p, then q

If p, then q

p	q	$p \rightarrow q$
T	T	T
T	F	F
F	T	T
F	F	T

It must be emphasized that the use of the conditional connective in no way implies a cause-and-effect relationship. Any two statements may have an arrow placed between them to create a compound statement. For example,

If I pass mathematics, then the sun will rise the next day

is true, since the consequent is true. (See the box following Example 1.) There is, however, no cause-and-effect connection between my passing mathematics and the sun's rising. The sun will rise no matter what grade I get in a course.

EXAMPLE 1 Given that p, q, and r are all false, find the truth value of the statement $(p \rightarrow \sim q) \rightarrow (\sim r \rightarrow q)$.

Using the short-cut method explained in Example 2 of the previous section, we can replace p, q, and r with F (since each is false) and proceed as before, using the negation and conditional truth tables as necessary.

$$(p \rightarrow \sim q) \rightarrow (\sim r \rightarrow q)$$
$$(F \rightarrow \sim F) \rightarrow (\sim F \rightarrow F)$$
$$(F \rightarrow T) \rightarrow (T \rightarrow F) \quad \text{Use the negation truth table.}$$
$$T \rightarrow F \quad \text{Use the conditional truth table.}$$
$$F$$

The statement $(p \rightarrow \sim q) \rightarrow (\sim r \rightarrow q)$ is false when p, q, and r are all false.

The following observations come from the truth table for $p \rightarrow q$.

Special Characteristics of Conditional Statements

1. $p \rightarrow q$ is false only when the antecedent is *true* and the consequent is *false*.
2. If the antecedent is *false*, then $p \rightarrow q$ is automatically *true*.
3. If the consequent is *true*, then $p \rightarrow q$ is automatically *true*.

EXAMPLE 2 Write *true* or *false* for each statement. Here T represents a true statement, and F represents a false statement.

(a) $T \rightarrow (6 = 3)$

Since the antecedent is true, while the consequent, $6 = 3$, is false, the given statement is false by the first point mentioned above.

(b) $(5 < 2) \rightarrow F$

The antecedent is false, so the given statement is true by the second observation.

(c) $(3 \neq 2 + 1) \rightarrow T$

The consequent is true, making the statement true by the third characteristic of conditional statements.

Truth tables for compound statements involving conditionals are found using the techniques described in the previous section.

EXAMPLE 3 Construct a truth table for each statement.

(a) $(\sim p \rightarrow \sim q) \rightarrow (\sim p \wedge q)$

First insert the truth values of $\sim p$ and of $\sim q$. Then find the truth values of $\sim p \rightarrow \sim q$.

p	q	$\sim p$	$\sim q$	$\sim p \rightarrow \sim q$
T	T	F	F	T
T	F	F	T	T
F	T	T	F	F
F	F	T	T	T

Next use $\sim p$ and q to find the truth values of $\sim p \wedge q$.

p	q	$\sim p$	$\sim q$	$\sim p \rightarrow \sim q$	$\sim p \wedge q$
T	T	F	F	T	F
T	F	F	T	T	F
F	T	T	F	F	T
F	F	T	T	T	F

Now find the truth values of $(\sim p \rightarrow \sim q) \rightarrow (\sim p \wedge q)$.

p	q	$\sim p$	$\sim q$	$\sim p \rightarrow \sim q$	$\sim p \wedge q$	$(\sim p \rightarrow \sim q) \rightarrow (\sim p \wedge q)$
T	T	F	F	T	F	F
T	F	F	T	T	F	F
F	T	T	F	F	T	T
F	F	T	T	T	F	F

(b) $(p \rightarrow q) \rightarrow (\sim p \vee q)$

Go through steps similar to the ones above.

p	q	$p \rightarrow q$	$\sim p$	$\sim p \vee q$	$(p \rightarrow q) \rightarrow (\sim p \vee q)$
T	T	T	F	T	T
T	F	F	F	F	T
F	T	T	T	T	T
F	F	T	T	T	T

As the truth table in Example 3(b) shows, the statement $(p \rightarrow q) \rightarrow (\sim p \vee q)$ is always true, no matter what the truth values of the components. Such a statement is called a **tautology.** Other examples of tautologies (as can be checked by forming truth tables) include $p \vee \sim p$, $p \rightarrow p$, $(\sim p \vee \sim q) \rightarrow \sim (q \wedge p)$, and so on. By the way, the truth tables in Example 3 also could have been found by the alternative method shown in the previous section.

Negation of a Conditional Suppose that someone makes the conditional statement

"If it rains, then I take my umbrella."

When will the person have lied to you? The only case in which you would have been misled is when it rains *and* the person does *not* take the umbrella. Letting p represent "it rains" and q represent "I take my umbrella," you might suspect that the symbolic statement

$$p \wedge \sim q$$

is a candidate for the negation of $p \rightarrow q$. That is,

$$\sim(p \rightarrow q) \equiv p \wedge \sim q.$$

It happens that this is indeed the case, as the next truth table indicates.

p	q	$p \rightarrow q$	$\sim(p \rightarrow q)$	$\sim q$	$p \wedge \sim q$
T	T	T	F	F	F
T	F	F	T	T	T
F	T	T	F	F	F
F	F	T	F	T	F

Negation of $p \rightarrow q$

The negation of $p \rightarrow q$ is $p \wedge \sim q$.

Since

$$\sim(p \rightarrow q) \equiv p \wedge \sim q,$$

by negating each expression we have

$$\sim[\sim(p \rightarrow q)] \equiv \sim(p \wedge \sim q).$$

The left side of the above equivalence is $p \rightarrow q$, and one of De Morgan's laws can be applied to the right side.

$$p \rightarrow q \equiv \sim p \vee \sim(\sim q)$$
$$p \rightarrow q \equiv \sim p \vee q$$

This final row indicates that a conditional may be written as a disjunction.

Writing a Conditional as an "or" Statement

$p \rightarrow q$ is equivalent to $\sim p \vee q.$

EXAMPLE 4 Write the negation of each statement.

(a) If you build it, he will come.

If b represents "you build it" and q represents "he will come," then the given statement can be symbolized by $b \rightarrow q$. The negation of $b \rightarrow q$, as shown earlier, is $b \wedge \sim q$, so the negation of the statement is

You build it and he will not come.

(b) All dogs have fleas.

First, we must restate the given statement in *if ... then* form:

If it is a dog, then it has fleas.

Based on our earlier discussion, the negation is

It is a dog and it does not have fleas.

A common error occurs when students try to write the negation of a conditional statement as another conditional statement. As seen in Example 4, the negation of a conditional statement is written as a conjunction.

EXAMPLE 5 Write each conditional as an equivalent statement without using *if ... then*.

(a) If the Cubs win the pennant, then Gwen will be happy.

Since the conditional $p \rightarrow q$ is equivalent to $\sim p \vee q$, let p represent "The Cubs win the pennant" and q represent "Gwen will be happy." Restate the conditional as

The Cubs do not win the pennant or Gwen will be happy.

(b) If it's Borden's, it's got to be good.

If p represents "it's Borden's" and if q represents "it's got to be good," the conditional may be restated as

It's not Borden's or it's got to be good.

Circuits One of the first nonmathematical applications of symbolic logic was seen in the master's thesis of Claude Shannon in 1937. Shannon showed how logic could be used to design electrical circuits. His work was immediately used by computer designers. Then in the developmental stage, computers could be simplified and built for less money using the ideas of Shannon.

To see how Shannon's ideas work, look at the electrical switch shown in Figure 1. We assume that current will flow through this switch when it is closed and not when it is open.

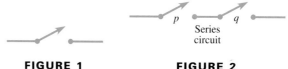

FIGURE 1   **FIGURE 2**

Figure 2 shows two switches connected in *series;* in such a circuit, current will flow only when both switches are closed. Note how closely a series circuit corresponds to the conjunction $p \wedge q$. We know that $p \wedge q$ is true only when both p and q are true.

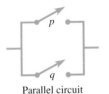

Parallel circuit

FIGURE 3

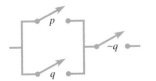

FIGURE 4

A circuit corresponding to the disjunction $p \vee q$ can be found by drawing a *parallel* circuit, as in Figure 3. Here, current flows if either p or q is closed or if both p and q are closed.

The circuit in Figure 4 corresponds to the statement $(p \vee q) \wedge \sim q$, which is a compound statement involving both a conjunction and a disjunction.

Simplifying an electrical circuit depends on the idea of equivalent statements from Section 3.2. Recall that two statements are equivalent if they have exactly the same truth table final column. The symbol $\equiv$ is used to indicate that the two statements are equivalent. Some equivalent statements are shown in the following box.

Equivalent Statements Used to Simplify Circuits

$$p \vee (q \wedge r) \equiv (p \vee q) \wedge (p \vee r) \qquad p \vee p \equiv p$$
$$p \wedge (q \vee r) \equiv (p \wedge q) \vee (p \wedge r) \qquad p \wedge p \equiv p$$
$$p \to q \equiv \sim q \to \sim p \qquad \sim(p \wedge q) \equiv \sim p \vee \sim q$$
$$p \to q \equiv \sim p \vee q \qquad \sim(p \vee q) \equiv \sim p \wedge \sim q$$

If T represents any true statement and F represents any false statement, then

$$p \vee T \equiv T \qquad p \vee \sim p \equiv T$$
$$p \wedge F \equiv F \qquad p \wedge \sim p \equiv F.$$

Circuits can be used as models of compound statements, with a closed switch corresponding to T, while an open switch corresponds to F. The method for simplifying circuits is explained in the following example.

EXAMPLE 6 Simplify the circuit of Figure 5.

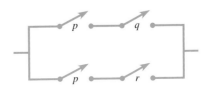

FIGURE 5

At the top of Figure 5, p and q are connected in series, and at the bottom, p and r are connected in series. These are interpreted as the compound statements $p \wedge q$ and $p \wedge r$, respectively. These two conjunctions are connected in parallel, as indicated by the figure treated as a whole. Write the disjunction of the two conjunctions:

$$(p \wedge q) \vee (p \wedge r).$$

(Think of the two switches labeled "p" as being controlled by the same handle.) By one of the pairs of equivalent statements in the preceding box,

$$(p \wedge q) \vee (p \wedge r) \equiv p \wedge (q \vee r),$$

FIGURE 6

which has the circuit of Figure 6. This circuit is logically equivalent to the one in Figure 5, and yet contains only three switches instead of four—which might well lead to a large savings in manufacturing costs.

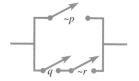

FIGURE 7

EXAMPLE 7 Draw a circuit for $p \rightarrow (q \wedge \sim r)$.

From the list of equivalent statements in the box, $p \rightarrow q$ is equivalent to $\sim p \vee q$. This equivalence gives $p \rightarrow (q \wedge \sim r) \equiv \sim p \vee (q \wedge \sim r)$, which has the circuit diagram in Figure 7.

3.3 EXERCISES

Rewrite each statement using the if ... then *connective. Rearrange the wording or add words as necessary.*

1. It must be alive if it is breathing.
2. You can believe it if you see it on the Internet.
3. Lorri Morgan visits Hawaii every summer.
4. Tom Shaffer's area code is 216.
5. Every picture tells a story.
6. All marines love boot camp.
7. No guinea pigs are scholars.
8. No koalas live in Texas.
9. Running Bear loves Little White Dove.
10. An opium-eater cannot have self-command.

In Exercises 11–18, decide whether each statement is true *or* false.

11. If the antecedent of a conditional statement is false, the conditional statement is true.
12. If the consequent of a conditional statement is true, the conditional statement is true.
13. If q is true, then $(p \wedge q) \rightarrow q$ is true.
14. If p is true, then $\sim p \rightarrow (q \vee r)$ is true.
15. The negation of "If pigs fly, I'll believe it" is "If pigs don't fly, I won't believe it."
16. The statements "If it flies, then it's a bird" and "It does not fly or it's a bird" are logically equivalent.
17. Given that $\sim p$ is true and q is false, the conditional $p \rightarrow q$ is true.
18. Given that $\sim p$ is false and q is false, the conditional $p \rightarrow q$ is true.
19. In a few sentences, explain how to determine the truth value of a conditional statement.
20. Explain why the statement "If $3 = 5$, then $4 = 6$" is true.

Tell whether each conditional is true *or* false. *Here* T *represents a true statement and* F *represents a false statement.*

21. $F \rightarrow (4 \neq 7)$
22. $T \rightarrow (6 < 3)$
23. $(6 \geq 6) \rightarrow F$
24. $F \rightarrow (3 \neq 3)$
25. $(4 = 11 - 7) \rightarrow (8 > 0)$
26. $(4^2 \neq 16) \rightarrow (4 - 4 = 8)$

3.3 The Conditional and Circuits 121

Let s represent "She has a snake for a pet," let p represent "he trains ponies," and let m represent "they raise monkeys." Express each compound statement in words.

27. $\sim m \rightarrow p$
28. $p \rightarrow \sim m$
29. $s \rightarrow (m \land p)$
30. $(s \land p) \rightarrow m$
31. $\sim p \rightarrow (\sim m \lor s)$
32. $(\sim s \lor \sim m) \rightarrow \sim p$

Let b represent "I ride my bike," let r represent "it rains," and let p represent "the play is cancelled." Write each compound statement in symbols.

33. If it rains, then I ride my bike.
34. If I ride my bike, then the play is cancelled.
35. If I do not ride my bike, then it does not rain.
36. If the play is cancelled, then it does not rain.
37. I ride my bike, or if the play is cancelled then it rains.
38. The play is cancelled, and if it rains then I do not ride my bike.
39. I'll ride my bike if it doesn't rain.
40. It rains if the play is cancelled.

Find the truth value of each statement. Assume that p and r are false, and q is true.

41. $\sim r \rightarrow q$
42. $\sim p \rightarrow \sim r$
43. $q \rightarrow p$
44. $\sim r \rightarrow p$
45. $p \rightarrow q$
46. $\sim q \rightarrow r$
47. $\sim p \rightarrow (q \land r)$
48. $(\sim r \lor p) \rightarrow p$
49. $\sim q \rightarrow (p \land r)$
50. $(\sim p \land \sim q) \rightarrow (p \land \sim r)$
51. $(p \rightarrow \sim q) \rightarrow (\sim p \land \sim r)$
52. $(p \rightarrow \sim q) \land (p \rightarrow r)$

53. Explain why, if we know that p is true, we also know that

$$[r \lor (p \lor s)] \rightarrow (p \lor q)$$

is true, even if we are not given the truth values of q, r, and s.

54. Construct a true statement involving a conditional, a conjunction, a disjunction, and a negation (not necessarily in that order), that consists of component statements p, q, and r, with all of these component statements false.

Construct a truth table for each statement. Identify any tautologies.

55. $\sim q \rightarrow p$
56. $p \rightarrow \sim q$
57. $(\sim p \rightarrow q) \rightarrow p$
58. $(\sim q \rightarrow \sim p) \rightarrow \sim q$
59. $(p \lor q) \rightarrow (q \lor p)$
60. $(p \land q) \rightarrow (p \lor q)$
61. $(\sim p \rightarrow \sim q) \rightarrow (p \land q)$
62. $r \rightarrow (p \land \sim q)$
63. $[(r \lor p) \land \sim q] \rightarrow p$
64. $(\sim r \rightarrow s) \lor (p \rightarrow \sim q)$
65. $(\sim p \land \sim q) \rightarrow (s \rightarrow r)$

66. What is the minimum number of Fs that must appear in the final column of a truth table for us to be assured that the statement is not a tautology?

Write the negation of each statement. Remember that the negation of $p \rightarrow q$ is $p \land \sim q$.

67. If that is an authentic Persian rug, I'll be surprised.
68. If Ella reaches that note, she will shatter glass.
69. If the English measures are not converted to metric measures, then the spacecraft will crash on the surface of Mars.
70. If you say "I do," then you'll be happy for the rest of your life.
71. "If you want to be happy for the rest of your life, never make a pretty woman your wife." *Jimmy Soul*
72. If loving you is wrong, I don't want to be right.

Write each statement as an equivalent statement that does not use the if ... then *connective. Remember that p → q is equivalent to ~p ∨ q.*

73. If you give your plants tender, loving care, they flourish.

74. If the check is in the mail, I'll be surprised.

75. If she doesn't, he will.

76. If I say yes, she says no.

77. All residents of Butte are residents of Montana.

78. All women were once girls.

Use truth tables to decide which of the pairs of statements are equivalent.

79. $p \to q$; $\sim p \vee q$
80. $\sim(p \to q)$; $p \wedge \sim q$
81. $p \to q$; $\sim q \to \sim p$
82. $q \to p$; $\sim p \to \sim q$
83. $p \to \sim q$; $\sim p \vee \sim q$
84. $p \to q$; $q \to p$
85. $p \wedge \sim q$; $\sim q \to \sim p$
86. $\sim p \wedge q$; $\sim p \to q$

Write a logical statement representing each of the following circuits. Simplify each circuit when possible.

87.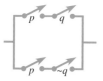

88.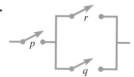

89.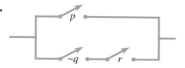

90.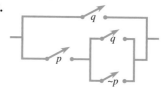

91.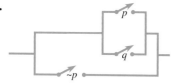

92.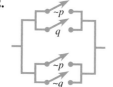

Draw circuits representing the following statements as they are given. Simplify if possible.

93. $p \wedge (q \vee \sim p)$
94. $(\sim p \wedge \sim q) \wedge \sim r$
95. $(p \vee q) \wedge (\sim p \wedge \sim q)$
96. $(\sim q \wedge \sim p) \vee (\sim p \vee q)$
97. $[(p \vee q) \wedge r] \wedge \sim p$
98. $[(\sim p \wedge \sim r) \vee \sim q] \wedge (\sim p \wedge r)$
99. $\sim q \to (\sim p \to q)$
100. $\sim p \to (\sim p \vee \sim q)$

101. Explain why the circuit

will always have exactly one open switch. What does this circuit simplify to?

102. Refer to Figures 5 and 6 in Example 6. Suppose the cost of the use of one switch for an hour is 3¢. By using the circuit in Figure 6 rather than the circuit in Figure 5, what is the savings for a year of 365 days, assuming that the circuit is in continuous use?

3.4 More on the Conditional

The conditional statement, introduced in the previous section, is one of the most important of all compound statements. Many mathematical properties and theorems are stated in *if ... then* form. Because of their usefulness, we need to study conditional statements that are related to a statement of the form $p \rightarrow q$.

Converse, Inverse, and Contrapositive Any conditional statement is made up of an antecedent and a consequent. If they are interchanged, negated, or both, a new conditional statement is formed. Suppose that we begin with the direct statement

If you stay, then I go,

and interchange the antecedent ("you stay") and the consequent ("I go"). We obtain the new conditional statement

If I go, then you stay.

This new conditional is called the **converse** of the given statement.

By negating both the antecedent and the consequent, we obtain the **inverse** of the given statement:

If you do not stay, then I do not go.

If the antecedent and the consequent are both interchanged *and* negated, the **contrapositive** of the given statement is formed:

If I do not go, then you do not stay.

These three related statements for the conditional $p \rightarrow q$ are summarized below. (Notice that the inverse is the contrapositive of the converse.)

Alfred North Whitehead (1861–1947) and Bertrand Russell worked together on *Principia Mathematica*. During that time, Whitehead was teaching mathematics at Cambridge University and had written *Universal Algebra*. In 1910 he went to the University of London, exploring not only the philosophical basis of science but also the "aims of education" (as he called one of his books). It was as philosopher that he was invited to Harvard University in 1924. Whitehead died at the age of 86 in Cambridge, Massachusetts.

Related Conditional Statements

Direct Statement	$p \rightarrow q$	(If p, then q.)
Converse	$q \rightarrow p$	(If q, then p.)
Inverse	$\sim p \rightarrow \sim q$	(If not p, then not q.)
Contrapositive	$\sim q \rightarrow \sim p$	(If not q, then not p.)

EXAMPLE 1 Given the direct statement

If I live in Miami, then I live in Florida,

write each of the following.

(a) the converse

Let p represent "I live in Miami" and q represent "I live in Florida." Then the direct statement may be written $p \rightarrow q$. The converse, $q \rightarrow p$, is

If I live in Florida, then I live in Miami.

Notice that for this statement, the converse is not necessarily true, even though the direct statement is.

(b) the inverse

The inverse of $p \rightarrow q$ is $\sim p \rightarrow \sim q$. For the given statement, the inverse is

If I don't live in Miami, then I don't live in Florida.

which is again not necessarily true.

(c) the contrapositive

The contrapositive, $\sim q \rightarrow \sim p$, is

If I don't live in Florida, then I don't live in Miami.

The contrapositive, like the direct statement, is true.

Example 1 shows that the converse and inverse of a true statement need not be true. They *can* be true, but they need not be. The relationship between the truth values of the direct statement, converse, inverse, and contrapositive is shown in the truth table that follows.

		Direct	Converse	Inverse	Contrapositive
p	q	$p \rightarrow q$	$q \rightarrow p$	$\sim p \rightarrow \sim q$	$\sim q \rightarrow \sim p$
T	T	T	T	T	T
T	F	F	T	T	F
F	T	T	F	F	T
F	F	T	T	T	T

Direct ↔ Contrapositive are Equivalent; Converse ↔ Inverse are Equivalent.

As this truth table shows, the direct statement and the contrapositive always have the same truth values, making it possible to replace any statement with its contrapositive without affecting the logical meaning. Also, the converse and inverse always have the same truth values.

This discussion is summarized as follows.

Equivalences

The direct statement and the contrapositive are equivalent, and the converse and the inverse are equivalent.

EXAMPLE 2 For the direct statement $\sim p \rightarrow q$, write each of the following.

(a) the converse

The converse of $\sim p \rightarrow q$ is $q \rightarrow \sim p$.

(b) the inverse

The inverse is $\sim(\sim p) \rightarrow \sim q$, which simplifies to $p \rightarrow \sim q$.

(c) the contrapositive

The contrapositive is $\sim q \rightarrow \sim(\sim p)$, which simplifies to $\sim q \rightarrow p$.

Bertrand Russell (1872–1970) was a student of Whitehead's before they wrote the *Principia*. Like his teacher, Russell turned toward philosophy. His works include a critique of Leibniz, analyses of mind and of matter, and a history of Western thought.

Russell became a public figure because of his involvement in social issues. Deeply aware of human loneliness, he was "passionately desirous of finding ways of diminishing this tragic isolation." During World War I he was an antiwar crusader, and he was imprisoned briefly. Again in the 1960s he championed peace. He wrote many books on social issues, winning the Nobel Prize for Literature in 1950.

Alternative Forms of "if p, then q" The conditional statement "if *p*, then *q*" can be stated in several other ways in English. For example,

> If you go to the shopping center, then you will find a place to park

can also be written

> Going to the shopping center is *sufficient* for finding a place to park.

According to this statement, going to the shopping center is enough to guarantee finding a place to park. Going to other places, such as schools or office buildings, *might* also guarantee a place to park, but at least we *know* that going to the shopping center does. Thus, $p \rightarrow q$ can be written "*p* is sufficient for *q*." Knowing that *p* has occurred is sufficient to guarantee that *q* will also occur. On the other hand,

> Turning on the set is necessary for watching television   (*)

has a different meaning. Here, we are saying that one condition that is necessary for watching television is that you turn on the set. This may not be enough; the set might be broken, for example. The statement labeled (*) could be written as

> If you watch television, then you turned on the set.

As this example suggests, $p \rightarrow q$ is the same as "*q* is necessary for *p*." In other words, if *q* doesn't happen, then neither will *p*. Notice how this idea is closely related to the idea of equivalence between the direct statement and its contrapositive.

Some common translations of $p \rightarrow q$ are summarized in the following box.

> **Common Translations of $p \rightarrow q$**
>
> The conditional $p \rightarrow q$ can be translated in any of the following ways.
>
> | If *p*, then *q*. | *p* is sufficient for *q*. |
> |---|---|
> | If *p*, *q*. | *q* is necessary for *p*. |
> | *p* implies *q*. | All *p*'s are *q*'s. |
> | *p* only if *q*. | *q* if *p*. |
>
> The translation of $p \rightarrow q$ into these various word forms does not in any way depend on the truth or falsity of $p \rightarrow q$.

For example, the statement

> If you are 18, then you can vote

can be written in any of the following ways.

> You can vote if you are 18.
>
> You are 18 only if you can vote.
>
> Being able to vote is necessary for you to be 18.
>
> Being 18 is sufficient for being able to vote.
>
> All 18-year-olds can vote.
>
> Being 18 implies that you can vote.

EXAMPLE 3 Write each statement in the form "if p, then q."

(a) You'll be sorry if I go.

If I go, then you'll be sorry.

(b) Today is Friday only if yesterday was Thursday.

If today is Friday, then yesterday was Thursday.

(c) All nurses wear white shoes.

If you are a nurse, then you wear white shoes.

FOR FURTHER THOUGHT

How many times have you heard a wise saying like "A stitch in time saves nine," "A rolling stone gathers no moss," or "Birds of a feather flock together"? In many cases, such proverbial advice can be restated as a conditional in *if ... then* form. For example, these three statements can be restated as follows.

"If you make a stitch in time, then it will save you nine (stitches)."

"If a stone rolls, then it gathers no moss."

"If they are birds of a feather, then they flock together."

For Group Discussion

1. Think of some wise sayings that have been around for a long time, and state them in *if ... then* form.

2. You have probably heard the saying "All that glitters is not gold." Do you think that what is said here is actually what is meant? If not, restate it as you think it should be stated. (*Hint:* Write the original statement in *if ... then* form.)

EXAMPLE 4 Let p represent "A triangle is equilateral," and let q represent "A triangle has three equal sides." Write each of the following in symbols.

(a) A triangle is equilateral if it has three equal sides.

$$q \rightarrow p$$

(b) A triangle is equilateral only if it has three equal sides.

$$p \rightarrow q$$

Biconditionals The compound statement p *if and only if* q (often abbreviated p *iff* q) is called a **biconditional**. It is symbolized $p \leftrightarrow q$, and is interpreted as the conjunction of the two conditionals $p \rightarrow q$ and $q \rightarrow p$. Using symbols, this conjunction is written

$$(q \rightarrow p) \wedge (p \rightarrow q)$$

so that, by definition,

$$p \leftrightarrow q \equiv (q \rightarrow p) \wedge (p \rightarrow q).$$

Using this definition, the truth table for the biconditional $p \leftrightarrow q$ can be determined.

Truth Table for the Biconditional *p* if and only if *q*

p if and only if *q*

p	q	p ↔ q
T	T	T
T	F	F
F	T	F
F	F	T

Principia Mathematica The title chosen by Whitehead and Russell was a deliberate reference to *Philosophiae naturalis principia mathematica*, or "mathematical principles of the philosophy of nature," Isaac Newton's epochal work of 1687. Newton's Principia pictured a kind of "clockwork universe" that ran via his Law of Gravitation. Newton independently invented the calculus, unaware that Leibniz had published his own formulation of it earlier. A controversy over their priority continued into the eighteenth century.

From the truth table, we see that a biconditional is true when both component statements have the same truth value. It is false when they have different truth values.

EXAMPLE 5 Tell whether each biconditional statement is *true* or *false*.

(a) $6 + 9 = 15$ if and only if $12 + 4 = 16$

Both $6 + 9 = 15$ and $12 + 4 = 16$ are true. By the truth table for the biconditional, this biconditional is true.

(b) $5 + 2 = 10$ if and only if $17 + 19 = 36$

Since the first component ($5 + 2 = 10$) is false, and the second is true, the entire biconditional statement is false.

(c) $6 = 5$ if and only if $12 \neq 12$

Both component statements are false, so by the last line of the truth table for the biconditional, the entire statement is true. (Understanding this might take some extra thought!)

In this section and in the previous two sections, truth tables have been derived for several important types of compound statements. The summary that follows describes how these truth tables may be remembered.

Summary of Basic Truth Tables

1. $\sim p$, the **negation** of p, has truth value opposite of p.
2. $p \wedge q$, the **conjunction**, is true only when both p and q are true.
3. $p \vee q$, the **disjunction**, is false only when both p and q are false.
4. $p \rightarrow q$, the **conditional**, is false only when p is true and q is false.
5. $p \leftrightarrow q$, the **biconditional**, is true only when p and q have the same truth value.

3.4 EXERCISES

For each given direct statement, write **(a)** *the converse,* **(b)** *the inverse, and* **(c)** *the contrapositive in if ... then form. In some of the exercises, it may be helpful to restate the direct statement in if ... then form.*

1. If beauty were a minute, then you would be an hour.
2. If you lead, then I will follow.
3. If it ain't broke, don't fix it.
4. If I had a nickel for each time that happened, I would be rich.
5. Walking in front of a moving car is dangerous to your health.
6. Milk contains calcium.
7. Birds of a feather flock together.
8. A rolling stone gathers no moss.
9. If you build it, he will come.
10. Where there's smoke, there's fire.
11. $p \rightarrow \sim q$
12. $\sim p \rightarrow q$
13. $\sim p \rightarrow \sim q$
14. $\sim q \rightarrow \sim p$
15. $p \rightarrow (q \vee r)$ (*Hint:* Use one of De Morgan's laws as necessary.)
16. $(r \vee \sim q) \rightarrow p$ (*Hint:* Use one of De Morgan's laws as necessary.)
17. Discuss the equivalences that exist among the direct conditional statement, the converse, the inverse, and the contrapositive.
18. State the contrapositive of "If the square of a natural number is even, then the natural number is even." The two statements must have the same truth value. Use several examples and inductive reasoning to decide whether both are true or both are false.

Write each of the following statements in the form "if p, then q."

19. If it is muddy, I'll wear my galoshes.
20. If I finish studying, I'll go to the party.
21. "17 is positive" implies that $17 + 1$ is positive.
22. "Today is Wednesday" implies that yesterday was Tuesday.
23. All integers are rational numbers.
24. All whole numbers are integers.
25. Doing crossword puzzles is sufficient for driving me crazy.
26. Being in Fort Lauderdale is sufficient for being in Florida.
27. A day's growth of beard is necessary for Greg Tobin to shave.
28. Being an environmentalist is necessary for being elected.
29. I can go from Boardwalk to Connecticut Avenue only if I pass GO.
30. The principal will hire more teachers only if the school board approves.
31. No whole numbers are not integers.
32. No integers are irrational numbers.
33. The Indians will win the pennant when their pitching improves.
34. Jesse will be a liberal when pigs fly.
35. A rectangle is a parallelogram with a right angle.
36. A parallelogram is a four-sided figure with opposite sides parallel.
37. A triangle with two sides of the same length is isosceles.
38. A square is a rectangle with two adjacent sides equal.
39. The square of a two-digit number whose units digit is 5 will end in 25.
40. An integer whose units digit is 0 or 5 is divisible by 5.
41. One of the following statements is not equivalent to all the others. Which one is it?
    A. r only if s.
    B. r implies s.
    C. If r, then s.
    D. r is necessary for s.
42. Many students have difficulty interpreting *necessary* and *sufficient*. Use the statement "Being in Canada is sufficient for being in North America" to explain why "p is sufficient for q" translates as "if p, then q."
43. Use the statement "To be an integer, it is necessary that a number be rational" to explain why "p is necessary for q" translates as "if q, then p."
44. Explain why the statement "A week has eight days if and only if December has forty days" is true.

Identify each statement as true or false.

45. $5 = 9 - 4$ if and only if $8 + 2 = 10$.
46. $3 + 1 \neq 6$ if and only if $8 \neq 8$.
47. $8 + 7 \neq 15$ if and only if $3 \times 5 \neq 9$.
48. $6 \times 2 = 14$ if and only if $9 + 7 \neq 16$.
49. Bill Clinton was president if and only if Jimmy Carter was not president.
50. Burger King sells Big Macs if and only if IBM manufactures computers.

Two statements that can both be true about the same object are **consistent.** *For example, "It is brown" and "It weighs 50 pounds" are consistent statements. Statements that cannot both be true about the same object are called* **contrary**; *"It is a Dodge" and "It is a Toyota" are contrary.*

In Exercises 51–56, label each pair of statements as either contrary *or* consistent.

51. Elvis is alive. Elvis is dead.
52. George W. Bush is a Democrat. George W. Bush is a Republican.
53. That animal has four legs. That animal is a dog.
54. That book is nonfiction. That book costs more than $70.
55. This number is an integer. This number is irrational.
56. This number is positive. This number is a natural number.
57. Make up two statements that are consistent.
58. Make up two statements that are contrary.

3.5 Analyzing Arguments with Euler Diagrams

Leonhard Euler (1707–1783) won the academy prize and edged out du Châtelet and Voltaire. That was a minor achievement, as was the invention of "Euler circles" (which antedated Venn diagrams). Euler was the most prolific mathematician of his generation despite blindness that forced him to dictate from memory.

There are two types of reasoning: inductive and deductive. With inductive reasoning we observe patterns to solve problems. Now, in this section and the next, we will study how deductive reasoning may be used to determine whether logical arguments are valid or invalid. A logical argument is made up of **premises** (assumptions, laws, rules, widely held ideas, or observations) and a **conclusion.** Together, the premises and the conclusion make up the argument. Also recall that *deductive* reasoning involves drawing specific conclusions from given general premises. When reasoning from the premises of an argument to obtain a conclusion, we want the argument to be valid.

Valid and Invalid Arguments

An argument is **valid** if the fact that all the premises are true forces the conclusion to be true. An argument that is not valid is **invalid,** or a **fallacy.**

It is very important to note that "valid" and "true" are not the same—an argument can be valid even though the conclusion is false. (See Example 4.)

Several techniques can be used to check whether an argument is valid. One of these is the visual technique based on **Euler diagrams,** illustrated by the following examples. (Another is the method of truth tables, shown in the next section.)

EXAMPLE 1 Is the following argument valid?

All dogs are animals.

Fred is a dog.

Fred is an animal.

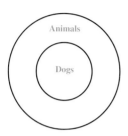

FIGURE 8

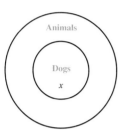

x represents Fred.

FIGURE 9

Here we use the common method of placing one premise over another, with the conclusion below a line. To begin, draw regions to represent the first premise. One is the region for "animals." Since all dogs are animals, the region for "dogs" goes inside the region for "animals," as in Figure 8.

The second premise, "Fred is a dog," suggests that "Fred" would go inside the region representing "dogs." Let *x* represent "Fred." Figure 9 shows that "Fred" is also inside the region for "animals." Therefore, if both premises are true, the conclusion that Fred is an animal must be true also. The argument is valid, as checked by Euler diagrams.

The method of Euler diagrams is especially convenient for arguments involving the quantifiers *all, some,* or *none*.

EXAMPLE 2 Is the following argument valid?

All rainy days are cloudy.
Today is not cloudy.
_____
Today is not rainy.

In Figure 10, the region for "rainy days" is drawn entirely inside the region for "cloudy days." Since "Today is *not* cloudy," place an *x* for "today" *outside* the region for "cloudy days." (See Figure 11.) Placing the *x* outside the region for "cloudy days" forces it also to be outside the region for "rainy days." Thus, if the first two premises are true, then it is also true that today is not rainy. The argument is valid.

FIGURE 10

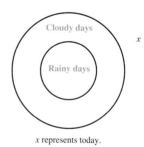

x represents today.

FIGURE 11

EXAMPLE 3 Is the following argument valid?

All banana trees have green leaves.
That plant has green leaves.
_____
That plant is a banana tree.

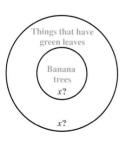

FIGURE 12

The region for "banana trees" goes entirely inside the region for "things that have green leaves." (See Figure 12.) There is a choice for locating the *x* that represents "that plant." The *x* must go inside the region for "things that have green leaves," but can go either inside or outside the region for "banana trees." Even if the premises are true, we are not forced to accept the conclusion as true. This argument is invalid; it is a fallacy.

As mentioned earlier, the validity of an argument is not the same as the truth of its conclusion. The argument in Example 3 was invalid, but the conclusion "That plant is a banana tree" may or may not be true. We cannot be sure.

EXAMPLE 4 Is the following argument valid?

All expensive things are desirable.
All desirable things make you feel good.
All things that make you feel good make you live longer.

All expensive things make you live longer.

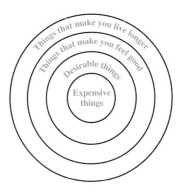

FIGURE 13

A diagram for the argument is given in Figure 13. If each premise is true, then the conclusion must be true since the region for "expensive things" lies completely within the region for "things that make you live longer." Thus, the argument is valid. (This argument is an example of the fact that a *valid* argument need *not* have a true conclusion.)

Arguments with the word "some" can be tricky.

FIGURE 14

EXAMPLE 5 Is the following argument valid?

Some students go to the beach for Spring Break.
I am a student.

I go to the beach for Spring Break.

The first premise is sketched in Figure 14. As the sketch shows, some (but not necessarily *all*) students go to the beach. There are two possibilities for *I*, as shown in Figure 15.

One possibility is that *I* go to the beach; the other is that *I* don't. Since the truth of the premises does not force the conclusion to be true, the argument is invalid.

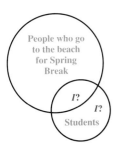

FIGURE 15

3.5 EXERCISES

Decide whether each argument is valid or invalid.

1. All boxers wear trunks.
   Chris Mader is a boxer.
   _____
   Chris Mader wears trunks.

2. All amusement parks have thrill rides.
   Great America is an amusement park.
   _____
   Great America has thrill rides.

3. All Southerners speak with an accent.
   Bill Leonard speaks with an accent.
   _____
   Bill Leonard is a Southerner.

4. All politicians lie, cheat, and steal.
   That man lies, cheats, and steals.
   _____
   That man is a politician.

5. All contractors use cell phones.
   Doug Boyle does not use a cell phone.
   _____
   Doug Boyle is not a contractor.

6. All dogs love to bury bones.
   Archie does not love to bury bones.
   _____
   Archie is not a dog.

7. All people who apply for a loan must pay for a title search.
   Cindy Herring paid for a title search.
   _____
   Cindy Herring applied for a loan.

8. All residents of Minnesota know how to live in freezing temperatures.
   Wendy Rockswold knows how to live in freezing temperatures.
   _____
   Wendy Rockswold lives in Minnesota.

9. Some philosophers are absent-minded.
   Amanda Perdaris is a philosopher.
   _____
   Amanda Perdaris is absent-minded.

10. Some dinosaurs were plant-eaters.
    Danny was a plant-eater.
    _____
    Danny was a dinosaur.

11. Some trucks have sound systems.
    Some trucks have gun racks.
    _____
    Some trucks with sound systems have gun racks.

12. Some nurses wear blue uniforms.
    Kim Falgout is a nurse.
    _____
    Kim Falgout wears a blue uniform.

13. Refer to Example 3. If the second premise and the conclusion were interchanged, would the argument then be valid?

14. Refer to Example 4. Give a different conclusion than the one given there so that the argument is still valid.

Construct a valid argument based on the Euler diagram shown.

15.

    x represents Julianne Peterson.

16.

    x represents Julie Ward.

As mentioned in the text, an argument can have a true conclusion yet be invalid. In these exercises, each argument has a true *conclusion. Identify each argument as* valid *or* invalid.

17. All cars have tires.
    All tires are rubber.
    _____
    All cars have rubber.

18. All birds fly.
    All planes fly.
    _____
    A bird is not a plane.

19. All chickens have beaks.
    All birds have beaks.
    ―――――――――――――
    All chickens are birds.

20. All chickens have beaks.
    All hens are chickens.
    ―――――――――――――
    All hens have beaks.

21. Veracruz is south of Tampico.
    Tampico is south of Monterrey.
    ―――――――――――――――――――
    Veracruz is south of Monterrey.

22. Quebec is northeast of Ottawa.
    Quebec is northeast of Toronto.
    ―――――――――――――――――――
    Ottawa is northeast of Toronto.

23. A scalene triangle has a longest side.
    A scalene triangle has a largest angle.
    ―――――――――――――――――――――――――――
    The largest angle in a scalene triangle is opposite the longest side.

24. No whole numbers are negative.
    −4 is negative.
    ―――――――――――――
    −4 is not a whole number.

In Exercises 25–30, the premises marked A, B, *and* C *are followed by several possible conclusions. Take each conclusion in turn, and check whether the resulting argument is* valid *or* invalid.

A. All people who drive contribute to air pollution.
B. All people who contribute to air pollution make life a little worse.
C. Some people who live in a suburb make life a little worse.

25. Some people who live in a suburb drive.

26. Some people who live in a suburb contribute to air pollution.

27. Some people who contribute to air pollution live in a suburb.

28. Suburban residents never drive.

29. All people who drive make life a little worse.

30. Some people who make life a little worse live in a suburb.

31. Find examples of arguments in magazine ads. Check them for validity.

32. Find examples of arguments on television commercials. Check them for validity.

EXTENSION

Logic Puzzles

Some people find that logic puzzles, which appear in periodicals such as *Official's Logic Problems*, *World-Class Logic Problems* (Penny Press), and *Logic Puzzles* (Dell), provide hours of enjoyment. They are based on deductive reasoning, and players answer questions based on clues given. The following explanation on solving such problems appeared in the Autumn 1999 issue of *World-Class Logic Problems*.

How to Solve Logic Problems Solving logic problems is entertaining and challenging. All the information you need to solve a logic problem is given in the introduction and clues, and in illustrations, when provided. If you've never solved a logic problem before, our sample should help you get started. Fill in the Sample Solving Chart as you follow our explanation. We use a "•" to signify "Yes" and an "X" to signify "No."

(continued)

134 CHAPTER 3 Introduction to Logic

Sample Logic Problem

Five couples were married last week, each on a different weekday. From the information provided, determine the woman (one is Cathy) and man (one is Paul) who make up each couple, as well as the day on which each couple was married.

1. Anne was married on Monday, but not to Wally.
2. Stan's wedding was on Wednesday. Rob was married on Friday, but not to Ida.
3. Vern (who married Fran) was married the day after Eve.

Sample Solving Chart:

	PAUL	ROB	STAN	VERN	WALLY	MONDAY	TUESDAY	WEDNESDAY	THURSDAY	FRIDAY
ANNE										
CATHY										
EVE										
FRAN										
IDA										
MONDAY										
TUESDAY										
WEDNESDAY										
THURSDAY										
FRIDAY										

Chart 1:

	PAUL	ROB	STAN	VERN	WALLY	MONDAY	TUESDAY	WEDNESDAY	THURSDAY	FRIDAY
ANNE		X	X		X	•	X	X	X	X
CATHY						X				
EVE						X				
FRAN						X				
IDA		X				X				X
MONDAY		X	X							
TUESDAY		X	X							
WEDNESDAY	X	X	•	X	X					
THURSDAY		X	X							
FRIDAY	X	•	X	X	X					

Explanation

Anne was married Mon. (1), so put a "•" at the intersection of Anne and Mon. Put "X"s in all the other days in Anne's row and all the other names in the Mon. column. (Whenever you establish a relationship, as we did here, be sure to place "X"s at the intersections of all relationships that become impossible as a result.) Anne wasn't married to Wally (1), so put an "X" at the intersection of Anne and Wally. Stan's wedding was Wed. (2), so put a "•" at the intersection of Stan and Wed. (Don't forget the "X"s.) Stan didn't marry Anne, who was married Mon., so put an "X" at the intersection of Anne and Stan. Rob was married Fri., but not to Ida (2), so put a "•" at the intersection of Rob and Fri., and "X"s at the intersections of Rob and Ida and Ida and Fri. Rob also didn't marry Anne, who was married Mon., so put an "X" at the intersection of Anne and Rob. Now your chart should look like chart 1.

Vern married Fran (3), so put a "•" at the intersection of Vern and Fran. This leaves Anne's only possible husband as Paul, so put a "•" at the intersection of Anne and Paul and Paul and Mon. Vern and Fran's wedding was the day after Eve's (3), which wasn't Mon. [Anne], so Vern's wasn't Tue. It must have been Thu. [see chart], so Eve's was Wed. (3). Put "•"s at the intersections of Vern and Thu., Fran and Thu., and Eve and Wed. Now your chart should look like chart 2.

Chart 2:

	PAUL	ROB	STAN	VERN	WALLY	MONDAY	TUESDAY	WEDNESDAY	THURSDAY	FRIDAY
ANNE	•	X	X	X	X	•	X	X	X	X
CATHY	X		X		X			X	X	
EVE	X		X			X	X	•	X	X
FRAN	X	X	X	•	X	X	X	X	•	X
IDA	X	X				X			X	X
MONDAY	•	X	X	X	X					
TUESDAY	X	X	X	X						
WEDNESDAY	X	X	•	X	X					
THURSDAY	X	X	X	•	X					
FRIDAY	X	•	X	X	X					

Chart 3:

	PAUL	ROB	STAN	VERN	WALLY	MONDAY	TUESDAY	WEDNESDAY	THURSDAY	FRIDAY
ANNE	•	X	X	X	X	•	X	X	X	X
CATHY	X	•	X	X	X	X	X	X	X	•
EVE	X	X	•	X	X	X	X	•	X	X
FRAN	X	X	X	•	X	X	X	X	•	X
IDA	X	X	X	X	•	X	•	X	X	X
MONDAY	•	X	X	X	X					
TUESDAY	X	X	X	X	•					
WEDNESDAY	X	X	•	X	X					
THURSDAY	X	X	X	•	X					
FRIDAY	X	•	X	X	X					

The chart shows that Cathy was married Fri., Ida was married Tue., and Wally was married Tue. Ida married Wally, and Cathy's wedding was Fri., so she married Rob. After this information is filled in, Eve could only have married Stan. You've completed the puzzle, and your chart should now look like chart 3.

In summary: Anne and Paul, Mon.; Cathy and Rob, Fri.; Eve and Stan, Wed.; Fran and Vern, Thu.; Ida and Wally, Tue.

In some problems, it may be necessary to make a logical guess based on facts you've established. When you do, always look for clues or other facts that disprove it. If you find that your guess is incorrect, eliminate it as a possibility.

EXTENSION EXERCISES

Follow the guidelines to solve each logic problem.

1. *Pumpkin-Patch Kids* Last Saturday, each of five children, accompanied by one of his or her parents, went to Goldman's Farm to pick out a pumpkin. Once there, however, each child (including Raven) became far more fascinated with a different aspect of the farm (in one case, feeding the animals) than he or she was in the pumpkin patch! Fortunately, each parent was able to recapture his or her child's interest just long enough for them to pick out the perfect pumpkin for their Halloween jack-o'-lantern! From the information provided, match each child with his or her parent (one is Mr. Maier) and determine the aspect of the farm each found more intriguing than pumpkin picking.

Source: World-Class Logic Problems, Autumn 1999, p. 5.

(a) Neither Lauren nor Zach was the child accompanied by Ms. Reed. Tara was more interested in shopping at the country store than in picking pumpkins.

(b) Xander and his father, Mr. Morgan, didn't go on the hay ride. Ms. Fedor's child (who isn't Zach or Lauren) was fascinated by the cider-making process.

(c) Zach is neither the child who went on the hay ride nor the one who wanted to go apple picking. Mr. Hanson's child didn't go on the hay ride.

2. *Have a Ball* With the start of the winter league just a few weeks away, the pro shop at Bowl Me Over Lanes has been packed with people picking out their new bowling balls. Just this evening, each of five bowlers purchased a ball, each a different color (one is gray) and weight (10, 12, 14, 16, or 18 pounds). After having his or her fingers sized and the holes drilled, each person's name was imprinted on his or her ball. Although their balls won't be ready for another week, these high rollers are anxious to experience life in the fast lane! From the information provided, determine the color and weight of each bowler's ball.

Source: World-Class Logic Problems, Autumn 1999, p. 5.

(a) Arlene had her name imprinted on her new orange bowling ball. Devon bought a bowling bag for his new ball, which is lighter than Silas's.

(b) Tina bowls with a 14-pound ball. The pink bowling ball is exactly 6 pounds lighter than the turquoise one.

(c) The red bowling ball, which isn't Rosetta's, weighs the most. Rosetta didn't buy the 16-pound ball.

(continued)

3. *It's a Tie* by Jenny Roberts. Four colleagues recently got into a discussion about some of the flamboyant patterns showing up on neckties these days. As a joke, each man arrived at work the next day sporting the most ridiculous tie he could find (no two men wore ties with the same pattern—one tie was decorated with smiling cupids). None of the men had to venture outside of his own closet, as each had received at least one such tie from a different relative! From the following clues, can you match each man with the pattern on his flamboyant tie, as well as determine the relative who presented each man with his tie?

(a) The tie with the grinning leprechauns wasn't a present from a daughter.
(b) Mr. Crow's tie features neither the dancing reindeer nor the yellow happy faces.
(c) Mr. Speigler's tie wasn't a present from his uncle.
(d) The tie with the yellow happy faces wasn't a gift from a sister.
(e) Mr. Evans and Mr. Speigler own the tie with the grinning leprechauns and the tie that was a present from a father-in-law, in some order.
(f) Mr. Hurley received his flamboyant tie from his sister.

Source: Dell Logic Puzzles, October 1999, p. 7.

4. *Imaginary Friends* by Keith King. Grantville's local library recently sponsored a writing contest for young children in the community. Each of four contestants (including Ralph) took on the task of bringing to life an imaginary friend in a short story. Each child selected a different type of animal (including a moose) to personify, and each described a different adventure involving this new friend (one story described how an imaginary friend had formed a rock band). From the following clues, can you match each young author with his or her imaginary friend and determine the adventure the two had together?

(a) The seal (who isn't the creation of either Joanne or Lou) neither rode to the moon in a spaceship nor took a trip around the world on a magic train.
(b) Joanne's imaginary friend (who isn't the grizzly bear) went to the circus.
(c) Winnie's imaginary friend is a zebra.
(d) The grizzly bear didn't board the spaceship to the moon.

Source: Dell Logic Puzzles, October 1999, p. 7.

3.6 Analyzing Arguments with Truth Tables

In Section 3.5 we used Euler diagrams to test the validity of arguments. While Euler diagrams often work well for simple arguments, difficulties can develop with more complex ones, since Euler diagrams require a sketch showing every possible case. In complex arguments, it is hard to be sure that all cases have been considered.

In deciding whether to use Euler diagrams to test the validity of an argument, look for quantifiers such as "all," "some," or "no." These words often indicate

3.6 Analyzing Arguments with Truth Tables

arguments best tested by Euler diagrams. If these words are absent, it may be better to use truth tables to test the validity of an argument.

As an example of this method, consider the following argument:

> If the floor is dirty, then I must mop it.
> The floor is dirty.
> _____
> I must mop it.

To test the validity of this argument, we begin by identifying the *component* statements found in the argument. They are "the floor is dirty" and "I must mop it." We assign the letters p and q to represent these statements:

> p represents "the floor is dirty";
>
> q represents "I must mop it."

Now we write the two premises and the conclusion in symbols:

> Premise 1: $p \rightarrow q$
>
> Premise 2: p
> _____
> Conclusion: q .

To decide if this argument is valid, we must determine whether the conjunction of both premises implies the conclusion for all possible cases of truth values for p and q. Therefore, write the conjunction of the premises as the antecedent of a conditional statement, and the conclusion as the consequent.

$$[\underbrace{(p \rightarrow q)}_{\text{premise}} \underbrace{\wedge}_{\text{and}} \underbrace{p}_{\text{premise}}] \underbrace{\rightarrow}_{\text{implies}} \underbrace{q}_{\text{conclusion}}$$

Finally, construct the truth table for this conditional statement, as shown below.

p	q	$p \rightarrow q$	$(p \rightarrow q) \wedge p$	$[(p \rightarrow q) \wedge p] \rightarrow q$
T	T	T	T	T
T	F	F	F	T
F	T	T	F	T
F	F	T	F	T

Since the final column, shown in color, indicates that the conditional statement that represents the argument is true for all possible truth values of p and q, the statement is a tautology. Thus, the argument is valid.

The pattern of the argument in the floor-mopping example,

> $p \rightarrow q$
>
> p
> _____
> q ,

is a common one, and is called **modus ponens,** or the *law of detachment*.

In summary, to test the validity of an argument using a truth table, follow the steps in the box.

> **Testing the Validity of an Argument with a Truth Table**
> 1. Assign a letter to represent each component statement in the argument.
> 2. Express each premise and the conclusion symbolically.
> 3. Form the symbolic statement of the entire argument by writing the *conjunction* of *all* the premises as the antecedent of a conditional statement, and the conclusion of the argument as the consequent.
> 4. Complete the truth table for the conditional statement formed in part 3 above. If it is a tautology, then the argument is valid; otherwise, it is invalid.

EXAMPLE 1 Determine whether the argument is *valid* or *invalid*.

If my check arrives in time, I'll register for the fall semester.

I've registered for the fall semester.

My check arrived in time.

Let p represent "my check arrives (arrived) in time" and let q represent "I'll register (I've registered) for the fall semester." Using these symbols, the argument can be written in the form

$$p \rightarrow q$$
$$\underline{q}$$
$$p .$$

To test for validity, construct a truth table for the statement

$$[(p \rightarrow q) \wedge q] \rightarrow p.$$

p	q	$p \rightarrow q$	$(p \rightarrow q) \wedge q$	$[(p \rightarrow q) \wedge q] \rightarrow p$
T	T	T	T	T
T	F	F	F	T
F	T	T	T	F
F	F	T	F	T

The third row of the final column of the truth table shows F, and this is enough to conclude that the argument is invalid. ∎

If a conditional and its converse were logically equivalent, then an argument of the type found in Example 1 would be valid. Since a conditional and its converse are *not* equivalent, the argument is an example of what is sometimes called the **fallacy of the converse.**

EXAMPLE 2 Determine whether the argument is *valid* or *invalid*.

If a man could be in two places at one time, I'd be with you.
I am not with you.

A man can't be in two places at one time.

If p represents "a man could be in two places at one time" and q represents "I'd be with you," the argument becomes

$$p \rightarrow q$$
$$\sim q$$
$$\overline{\sim p}$$

The symbolic statement of the entire argument is

$$[(p \rightarrow q) \wedge \sim q] \rightarrow \sim p.$$

The truth table for this argument, shown below, indicates a tautology, and the argument is valid.

p	q	$p \rightarrow q$	$\sim q$	$(p \rightarrow q) \wedge \sim q$	$\sim p$	$[(p \rightarrow q) \wedge \sim q] \rightarrow \sim p$
T	T	T	F	F	F	T
T	F	F	T	F	F	T
F	T	T	F	F	T	T
F	F	T	T	T	T	T

The pattern of reasoning of this example is called **modus tollens,** or the *law of contraposition,* or *indirect reasoning.* ■

With reasoning similar to that used to name the fallacy of the converse, the fallacy

$$p \rightarrow q$$
$$\sim p$$
$$\overline{\sim q}$$

is called the **fallacy of the inverse.** An example of such a fallacy is "If it rains, I get wet. It doesn't rain. Therefore, I don't get wet."

EXAMPLE 3 Determine whether the argument is *valid* or *invalid*.

I'll buy a car or I'll take a vacation.
I won't buy a car.

I'll take a vacation.

If p represents "I'll buy a car" and q represents "I'll take a vacation," the argument becomes

$$p \vee q$$
$$\sim p$$
$$\overline{q}$$

We must set up a truth table for

$$[(p \vee q) \wedge \sim p] \to q.$$

p	q	$p \vee q$	$\sim p$	$(p \vee q) \wedge \sim p$	$[(p \vee q) \wedge \sim p] \to q$
T	T	T	F	F	T
T	F	T	F	F	T
F	T	T	T	T	T
F	F	F	T	F	T

The statement is a tautology and the argument is valid. Any argument of this form is valid by the law of **disjunctive syllogism.**

EXAMPLE 4 Determine whether the argument is *valid* or *invalid*.

If it squeaks, then I use WD-40.
If I use WD-40, then I must go to the hardware store.
If it squeaks, then I must go to the hardware store.

Let p represent "it squeaks," let q represent "I use WD-40," and let r represent "I must go to the hardware store." The argument takes on the general form

$$p \to q$$
$$q \to r$$
$$\overline{p \to r.}$$

Make a truth table for the following statement:

$$[(p \to q) \wedge (q \to r)] \to (p \to r).$$

It will require eight rows.

p	q	r	$p \to q$	$q \to r$	$p \to r$	$(p \to q) \wedge (q \to r)$	$[(p \to q) \wedge (q \to r)] \to (p \to r)$
T	T	T	T	T	T	T	T
T	T	F	T	F	F	F	T
T	F	T	F	T	T	F	T
T	F	F	F	T	F	F	T
F	T	T	T	T	T	T	T
F	T	F	T	F	T	F	T
F	F	T	T	T	T	T	T
F	F	F	T	T	T	T	T

This argument is valid since the final statement is a tautology. The pattern of argument shown in this example is called **reasoning by transitivity,** or the *law of hypothetical syllogism.*

A summary of the valid and invalid forms of argument presented so far follows.

Valid Argument Forms

Modus Ponens	Modus Tollens	Disjunctive Syllogism	Reasoning by Transitivity
$p \rightarrow q$	$p \rightarrow q$	$p \vee q$	$p \rightarrow q$
p	$\sim q$	$\sim p$	$q \rightarrow r$
q	$\sim p$	q	$p \rightarrow r$

Invalid Argument Forms (Fallacies)

Fallacy of the Converse	Fallacy of the Inverse
$p \rightarrow q$	$p \rightarrow q$
q	$\sim p$
p	$\sim q$

When an argument contains three or more premises, it is necessary to determine the truth values of the conjunction of all of them. Remember that if *at least one* premise in a conjunction of several premises is false, then the entire conjunction is false.

EXAMPLE 5 Determine whether the argument is *valid* or *invalid*.

If Eddie goes to town, then Mabel stays at home.
If Mabel does not stay at home, then Rita will cook.
Rita will not cook.
Therefore, Eddie does not go to town.

In an argument written in this manner, the premises are given first, and the conclusion is the statement that follows the word "Therefore." Let p represent "Eddie goes to town," let q represent "Mabel stays at home," and let r represent "Rita will cook." The symbolic form of the argument is

$$p \rightarrow q$$
$$\sim q \rightarrow r$$
$$\sim r$$
$$\sim p \; .$$

To test validity, set up a truth table for the statement

$$[(p \rightarrow q) \wedge (\sim q \rightarrow r) \wedge \sim r] \rightarrow \sim p.$$

The table is shown below.

p	q	r	$p \rightarrow q$	$\sim q$	$\sim q \rightarrow r$	$\sim r$	$(p \rightarrow q) \wedge (\sim q \rightarrow r) \wedge \sim r$	$\sim p$	$[(p \rightarrow q) \wedge (\sim q \rightarrow r) \wedge \sim r] \rightarrow \sim p$
T	T	T	T	F	T	F	F	F	T
T	T	F	T	F	T	T	T	F	F
T	F	T	F	T	T	F	F	F	T
T	F	F	F	T	F	T	F	F	T
F	T	T	T	F	T	F	F	T	T
F	T	F	T	F	T	T	T	T	T
F	F	T	T	T	T	F	F	T	T
F	F	F	T	T	F	T	F	T	T

Because the final column does not contain all Ts, the statement is not a tautology. The argument is invalid. ■

Consider the following poem, which has been around for many years.

For want of a nail, the shoe was lost.
For want of a shoe, the horse was lost.
For want of a horse, the rider was lost.
For want of a rider, the battle was lost.
For want of a battle, the war was lost.
Therefore, for want of a nail, the war was lost.

Each line of the poem may be written as an *if ... then* statement. For example, the first line may be restated as "if a nail is lost, then the shoe is lost." Other statements may be worded similarly. The conclusion, "for want of a nail, the war was lost," follows from the premises since repeated use of the law of transitivity applies. Arguments used by Lewis Carroll (see the margin note in Chapter 1) often take on a similar form. The next example comes from one of his works.

Tweedlogic "I know what you're thinking about," said Tweedledum, "but it isn't so, nohow." "Contrariwise," continued Tweedledee, "if it was so, it might be; and if it were so, it would be, but as it isn't, it ain't. That's logic."

EXAMPLE 6 Supply a conclusion that yields a valid argument for the following premises.

Babies are illogical.
Nobody is despised who can manage a crocodile.
Illogical persons are despised.

First, write each premise in the form *if ... then*.

If you are a baby, then you are illogical.
If you can manage a crocodile, then you are not despised.
If you are illogical, then you are despised.

Let p be "you are a baby," let q be "you are logical," let r be "you can manage a crocodile," and let s be "you are despised." With these letters, the statements can be written symbolically as

$$p \rightarrow \sim q$$
$$r \rightarrow \sim s$$
$$\sim q \rightarrow s.$$

Now begin with any letter that appears only once. Here p appears only once. Using the contrapositive of $r \rightarrow \sim s$, which is $s \rightarrow \sim r$, rearrange the three statements as follows:

$$p \rightarrow \sim q$$
$$\sim q \rightarrow s$$
$$s \rightarrow \sim r.$$

From the three statements, repeated use of reasoning by transitivity gives the conclusion

$$p \rightarrow \sim r,$$

leading to a valid argument.

In words, the conclusion is "If you are a baby, then you cannot manage a crocodile," or, as Lewis Carroll would have written it, "Babies cannot manage crocodiles."

FOR FURTHER THOUGHT

The field of advertising is notorious for logical errors. While the conclusion reached in an advertisement may be true, the logic used to reach that conclusion may be based on an invalid argument. One fallacy that is seen time and time again is the fallacy of experts. Here, an expert (or famous figure) is quoted in a field outside of that person's field of knowledge.

In his April 8, 1997, column "When athletes endorse, why does anyone listen?" that appeared in *USA TODAY, Chicago Tribune* columnist Bob Greene made some very astute observations. He had visited Denver and noticed that Denver Bronco quarterback John Elway's face seemed to be everywhere, endorsing a variety of products.* Greene spoke to future Hall-of-Famer Elway's marketing manager, Jeff Sperbeck:

"Ever since John first came on the scene, there has been a fascination with who John Elway is," Sperbeck said. "People who have products to sell know that an association with John is a good way for people to identify with their products, because by buying the products the people feel they are making a personal connection with John."

That's the way it works in the United States, of course. When you're famous, through sports or movies or television or music, merchandisers try to sign you up for the very reason Sperbeck was talking about: so that customers, wishing to touch your life in some way, will purchase the merchandisers' products. It's a fact of American life, so ubiquitous that it's seldom even commented on: a celebrity endorsement equals big sales.

Author's note: In a halftime interview for Super Bowl XXXIV, Elway even tried to put in a free plug for a Web-based sports merchandise site.

(continued)

It might be instructive, though, for us to pause for a second or two and ask ourselves: Why?

Every consumer knows that celebrity endorsers—Elway and all the rest of them—allow their names to be used to promote products for one basic reason: The celebrities are paid for it. The expenditure of the funds is well worth it to the shoe (or shirt or hat or beverage) company. Millions of people who see a star's name attached to the marketing of the product will buy that product.

Why?

"Show me the money," that most odious of phrases from motion picture screens in recent years, was the unspoken rallying cry of professional athletes, and other celebrities, for months before a scriptwriter reduced it to painfully accurate verbal shorthand. That's what comes with sudden fame: the right to be given money for the use of one's name. This makes perfect sense for the person renting his or her name to the highest corporate bidder.

Yet why would the rest of us—all of us, any of us—purchase a product that we know a person has been paid to say he or she likes? In a court of law, an opposing attorney would leave a witness quivering, would destroy that witness in the eyes of a jury, if the attorney could establish that the witness was saying something good on behalf of someone who had paid that witness a great deal of money to say precisely such a thing. If it can be demonstrated that a person's praise is the result of a cash transaction, the person's words are made to appear worthless. In a court of law, that is.

In the court of public opinion, for some inexplicable reason, the same common-sense rules do not apply. The more a celebrity is paid to say he or she loves a product, the bigger the advertising budget seems to be—and the more consumers salivate over the product being promoted. No one ever stops to say: Wait a minute—yeah, you're saying this (shoe, drink, automobile, candy bar) is great, but why should I believe you? You've been bribed—someone gave you money to say those words.

It never comes up. In a sane world, the only endorsements that would mean anything—that would have the power to persuade anyone—would be endorsements that a famous man or woman offered for free. Those would be the sole endorsements the public would believe and accept.

Celebrity endorsements aside, a survey once found that the most effective commercials are those that contain no "live people" at all. Some of the best ones were those that featured the Eveready Bunny, the Pillsbury Doughboy, and the Keebler Elves, and although he is not an animated character, one of the most popular commercial stars in recent years is the Taco Bell Chihuahua.

Another common ploy in advertising is the fallacy of emotion, in which an appeal is made to pity, passion, brute force, snobbishness, vanity, or some other emotion. Have you seen the tire advertisement that suggests that if you don't use a certain brand of tire, your infants will be unsafe? And have you ever seen an ad for cosmetics that doesn't involve "beautiful people"?

For Group Discussion

1. Comment on the following, which appeared as part of a question addressed to columnist Sue Rusche ("Education helps control alcoholism," July 9, 1988, *Sacramento Union*):

    The simple truth is that Congress and the press seldom listen to people who actually know something about a problem. They prefer to listen instead to those whose testimony or statements, though completely uninformed, will make a publicity splash. As a British friend of mind commented, "You Americans pick an actor who has played a drug addict in some film to testify before Congress as an expert on drugs. Lord Laurence Olivier has played Hamlet at least a thousand times, but Parliament has never called on him to testify on matters Danish."

2. What are some examples of the fallacy of experts that you have seen or heard? What are some examples of the fallacy of emotion that you have seen or heard?
3. How does Michael Jordan relate to this discussion?

3.6 EXERCISES

Each of the following arguments is either valid by one of the forms of valid arguments discussed in this section, or is a fallacy by one of the forms of invalid arguments discussed. (See the summary boxes.) Decide whether the argument is valid *or a* fallacy, *and give the form that applies.*

1. If you use binoculars, then you get a glimpse of the comet.
   If you get a glimpse of the comet, then you'll be amazed.

   If you use binoculars, then you'll be amazed.

2. If Billy Joel comes to town, then I will go to the concert.
   If I go to the concert, then I'll call in sick for work.

   If Billy Joel comes to town, then I'll call in sick for work.

3. If Frank Steed sells his quota, he'll get a bonus.
   Frank Steed sells his quota.

   He gets a bonus.

4. If Amy McRee works hard enough, she will get a promotion.
   Amy McRee works hard enough.

   She gets a promotion.

5. If she buys another pair of shoes, her closet will overflow.
   Her closet will overflow.

   She buys another pair of shoes.

6. If he doesn't have to get up at 5:30 A.M., he's ecstatic.
   He's ecstatic.

   He doesn't have to get up at 5:30 A.M.

7. If Patrick Roy plays, the opponent gets shut out.
   The opponent does not get shut out.

   Patrick Roy does not play.

8. If Pedro Martinez pitches, the Red Sox win.
   The Red Sox do not win.

   Pedro Martinez does not pitch.

9. "If we evolved a race of Isaac Newtons, that would not be progress." (quote from Aldous Huxley)
   We have not evolved a race of Isaac Newtons.

   That is progress.

10. "If I have seen farther than others, it is because I stood on the shoulders of giants." (quote from Sir Isaac Newton)
    I have not seen farther than others.

    I have not stood on the shoulders of giants.

11. Alison Romike jogs or Kaare Taylor pumps iron.
    Kaare Taylor does not pump iron.

    Alison Romike jogs.

12. She uses e-commerce or she pays by credit card.
    She does not pay by credit card.

    She uses e-commerce.

Use a truth table to determine whether the argument is valid *or* invalid.

13. $p \vee q$
    p
    $\sim q$

14. $p \wedge \sim q$
    p
    $\sim q$

15. $\sim p \to \sim q$
    q
    p

16. $p \vee \sim q$
    p
    $\sim q$

17. $p \to q$
    $q \to p$
    $p \wedge q$

18. $\sim p \to q$
    p
    $\sim q$

19. $p \to \sim q$
    q
    $\sim p$

20. $p \to \sim q$
    $\sim p$
    $\sim q$

21. $(\sim p \vee q) \wedge (\sim p \to q)$
    p
    $\sim q$

22. $(p \to q) \wedge (q \to p)$
    p
    $p \vee q$

23. $(\sim p \wedge r) \to (p \vee q)$
    $\sim r \to p$
    $q \to r$

24. $(r \wedge p) \to (r \vee q)$
    $q \wedge p$
    $r \vee p$

25. Explain in a few sentences how to determine the statement for which a truth table will be constructed so that the arguments that follow in Exercises 27–36 can be analyzed for validity.

26. Earlier we showed how to analyze arguments using Euler diagrams. Refer to Example 4 in this section, restate each premise and the conclusion using a quantifier, and then draw an Euler diagram to illustrate the relationship.

Determine whether the following arguments are valid *or* invalid.

27. Jeff loves to play golf. If Joan likes to sew, then Jeff does not love to play golf. If Joan does not like to sew, then Brad sings in the choir. Therefore, Brad sings in the choir.

28. If that tree is infested with pine bark beetles, then it will die. People plant trees on Arbor Day and it will not die. Therefore, if people plant trees on Arbor Day, then that tree is not infested with pine bark beetles.

29. If the Bobble head doll craze continues, then Beanie Babies will remain popular. Barbie dolls continue to be favorites or Beanie Babies will remain popular. Barbie dolls do not continue to be favorites. Therefore, the Bobble head doll craze does not continue.

30. Christina Aguilera sings or Ricky Martin is not a teen idol. If Ricky Martin is not a teen idol, then Britney Spears does not win an American Music Award. Britney Spears wins an American Music Award. Therefore, Christina Aguilera does not sing.

31. If I've got you under my skin, then you are deep in the heart of me. If you are deep in the heart of me, then you are not really a part of me. You are deep in the heart of me or you are really a part of me. Therefore, if I've got you under my skin, then you are really a part of me.

32. The Colts will be in the playoffs if and only if Peyton leads the league in passing. Marv loves the Colts or Peyton leads the league in passing. Marv does not love the Colts. Therefore, the Colts will not be in the playoffs.

33. If Otis is a disc jockey, then he lives in Lexington. He lives in Lexington and he is a history buff. Therefore, if Otis is not a history buff, then he is not a disc jockey.

34. If I were your woman and you were my man, then I'd never stop loving you. I've stopped loving you. Therefore, I am not your woman or you are not my man.

35. All men are created equal. All people who are created equal are women. Therefore, all men are women.

36. All men are mortal. Socrates is a man. Therefore, Socrates is mortal.

37. Suppose that you ask someone for the time on your CB radio and you get the following response:

 "If I tell you the time, then we'll start chatting. If we start chatting, then you'll want to meet me at a truck stop. If we meet at a truck stop, then we'll discuss my family. If we discuss my family, then you'll find out that my daughter is available for marriage. If you find out that she is available for marriage, then you'll want to marry her. If you want to marry her, then my life will be miserable since I don't want my daughter married to some fool who can't afford a $10 watch."

Use reasoning by transitivity to draw a valid conclusion.

38. Cheryl Arabie made the following observation: "If I want to determine whether an argument leading to the statement

$$[(p \rightarrow q) \wedge \sim q] \rightarrow \sim p$$

is valid, I only need to consider the lines of the truth table which lead to T for the column headed $(p \rightarrow q) \wedge \sim q$." Cheryl was very perceptive. Can you explain why her observation was correct?

In the arguments used by Lewis Carroll, it is helpful to restate a premise in if ... then *form in order to more easily identify a valid conclusion. The following premises come from Lewis Carroll. Write each premise in* if ... then *form.*

39. All my poultry are ducks.

40. None of your sons can do logic.

41. Guinea pigs are hopelessly ignorant of music.

42. No teetotalers are pawnbrokers.

43. No teachable kitten has green eyes.

44. Opium-eaters have no self-command.

45. I have not filed any of them that I can read.

46. All of them written on blue paper are filed.

The following exercises involve premises from Lewis Carroll. Write each premise in symbols, and then in the final part, give a conclusion that yields a valid argument.

47. Let p be "it is a duck," q be "it is my poultry," r be "one is an officer," and s be "one is willing to waltz."
 (a) No ducks are willing to waltz.
 (b) No officers ever decline to waltz.
 (c) All my poultry are ducks.
 (d) Give a conclusion that yields a valid argument.

48. Let p be "one is able to do logic," q be "one is fit to serve on a jury," r be "one is sane," and s be "he is your son."
 (a) Everyone who is sane can do logic.
 (b) No lunatics are fit to serve on a jury.
 (c) None of your sons can do logic.
 (d) Give a conclusion that yields a valid argument.

49. Let p be "one is honest," q be "one is a pawnbroker," r be "one is a promise-breaker," s be "one is trustworthy," t be "one is very communicative," and u be "one is a wine-drinker."
 (a) Promise-breakers are untrustworthy.
 (b) Wine-drinkers are very communicative.
 (c) A person who keeps a promise is honest.
 (d) No teetotalers are pawnbrokers. (*Hint:* Assume "teetotaler" is the opposite of "wine-drinker.")
 (e) One can always trust a very communicative person.
 (f) Give a conclusion that yields a valid argument.

50. Let p be "it is a guinea pig," q be "it is hopelessly ignorant of music," r be "it keeps silent while the *Moonlight Sonata* is being played," and s be "it appreciates Beethoven."
 (a) Nobody who really appreciates Beethoven fails to keep silent while the *Moonlight Sonata* is being played.

- **(b)** Guinea pigs are hopelessly ignorant of music.
- **(c)** No one who is hopelessly ignorant of music ever keeps silent while the *Moonlight Sonata* is being played.
- **(d)** Give a conclusion that yields a valid argument.

51. Let p be "it begins with 'Dear Sir'," q be "it is crossed," r be "it is dated," s be "it is filed," t be "it is in black ink," u be "it is in the third person," v be "I can read it," w be "it is on blue paper," x be "it is on one sheet," and y be "it is written by Brown."
- **(a)** All the dated letters are written on blue paper.
- **(b)** None of them are in black ink, except those that are written in the third person.
- **(c)** I have not filed any of them that I can read.
- **(d)** None of them that are written on one sheet are undated.
- **(e)** All of them that are not crossed are in black ink.
- **(f)** All of them written by Brown begin with "Dear Sir."
- **(g)** All of them written on blue paper are filed.
- **(h)** None of them written on more than one sheet are crossed.
- **(i)** None of them that begin with "Dear Sir" are written in the third person.
- **(j)** Give a conclusion that yields a valid argument.

52. Let p be "he is going to a party," q be "he brushes his hair," r be "he has self-command," s be "he looks fascinating," t be "he is an opium-eater," u be "he is tidy," and v be "he wears white kid gloves."
- **(a)** No one who is going to a party ever fails to brush his hair.
- **(b)** No one looks fascinating if he is untidy.
- **(c)** Opium-eaters have no self-command.
- **(d)** Everyone who has brushed his hair looks fascinating.
- **(e)** No one wears white kid gloves unless he is going to a party. (*Hint:* "a unless b" $\equiv \sim b \rightarrow a$.)
- **(f)** A man is always untidy if he has no self-command.
- **(g)** Give a conclusion that yields a valid argument.

COLLABORATIVE INVESTIGATION

Logic Puzzles Revisited

The logic puzzles in the Extension on pages 133–136 are fairly elementary, considering the complexity of some of the other problems found in the three magazines mentioned. The problem here requires more time and reasoning skills than the ones appearing in the Extension. It is taken from *Dell Logic Puzzles*, October 1999.

The class may wish to divide up into groups and see which group can solve this problem fastest.

DUNKING BOOTH by Margaret Rounds

In order to help raise money for Dellville's annual fall festival, the coach of a local men's baseball team volunteered to sit in a dunking booth. Five of Coach Clark's players, each of whom plays a different position on the team (including first base), stepped up to the challenge of dunking him, and each player took a different number of throws (two, four, five, six, or nine) to complete his mission. From the clues below, can you determine the order in which each player (identified by full name—one surname is Winslow) dunked the coach, the number of throws it took to do so, and the position each plays?

1. The right fielder took three more throws than Josh (who wasn't the last dunker).
2. The shortstop threw immediately before Mr. Belford, who threw immediately before Geoffrey.
3. Mr. Cavallo dunked the coach with his sixth pitch.
4. The players who took an even number of throws are Scott, Mr. Hepler, and the fifth person to dunk the coach, in some order.
5. The pitcher (who wasn't the fourth dunker) took more throws than Oliver, but fewer throws than Mr. Janney.

6. The players who took an odd number of throws are Edgar and the catcher, in some order.

Order	First name	Last name	Throws	Position
___	___	___	___	___
___	___	___	___	___
___	___	___	___	___
___	___	___	___	___
___	___	___	___	___

	Edgar	Geoffrey	Josh	Oliver	Scott	Belford	Cavallo	Hepler	Janney	Winslow	Two	Four	Five	Six	Nine	Catcher	First base	Pitcher	Right field	Shortstop
First																				
Second																				
Third																				
Fourth																				
Fifth																				
Catcher																				
First base																				
Pitcher																				
Right Field																				
Shortstop																				
Two																				
Four																				
Five																				
Six																				
Nine																				
Belford																				
Cavallo																				
Hepler																				
Janney																				
Winslow																				

CHAPTER 3 TEST

Write a negation for each of the following statements.

1. $6 - 3 = 3$
2. All men are created equal.
3. Some members of the class went on the field trip.
4. If that's the way you feel, then I will accept it.
5. She passed GO and collected $200.

Let p represent "You will love me" and let q represent "I will love you." Write each of the following in symbols.

6. If you won't love me, then I will love you.
7. I will love you if you will love me.
8. I won't love you if and only if you won't love me.

Using the same statements as for Exercises 6–8, write each of the following in words.

9. $\sim p \land q$
10. $\sim(p \lor \sim q)$

In each of the following, assume that p is true and that q and r are false. Find the truth value of each statement.

11. $\sim q \land \sim r$
12. $r \lor (p \land \sim q)$
13. $r \rightarrow (s \lor r)$ (The truth value of the statement s is unknown.)
14. $p \leftrightarrow (p \rightarrow q)$
15. Explain in your own words why, if p is a statement, the biconditional $p \leftrightarrow \sim p$ must be false.
16. State the necessary conditions for
    (a) a conditional statement to be false.
    (b) a conjunction to be true.
    (c) a disjunction to be false.

Construct a truth table for each of the following.

17. $p \land (\sim p \lor q)$
18. $\sim(p \land q) \rightarrow (\sim p \lor \sim q)$

Decide whether each statement is true or false.

19. Some negative integers are whole numbers.
20. All irrational numbers are real numbers.

Write each conditional statement in if ... then *form.*

21. All integers are rational numbers.
22. Being a rhombus is sufficient for a polygon to be a quadrilateral.
23. Being divisible by 3 is necessary for a number to be divisible by 9.
24. She digs dinosaur bones only if she is a paleontologist.

For each statement, write **(a)** *the converse,* **(b)** *the inverse, and* **(c)** *the contrapositive.*

25. If a picture paints a thousand words, the graph will help me understand it.
26. $\sim p \rightarrow (q \wedge r)$ (Use one of De Morgan's laws as necessary.)
27. Use an Euler diagram to determine whether the following argument is *valid* or *invalid*.

    All members of that video club save money.
    Pat Pearson is a member of that video club.
    Pat Pearson saves money.

28. Match each argument in parts (a)–(d) with the law that justifies its validity, or the fallacy of which it is an example.
    A. Modus ponens
    B. Modus tollens
    C. Reasoning by transitivity
    D. Disjunctive syllogism
    E. Fallacy of the converse
    F. Fallacy of the inverse

    (a) If he eats liver, then he'll eat anything.
        He eats liver.
        He'll eat anything.

    (b) If you use your seat belt, you will be safer.
        You don't use your seat belt.
        You won't be safer.

    (c) If I hear *Come Saturday Morning*, I think of her.
        If I think of her, I get depressed.
        If I hear *Come Saturday Morning*, I get depressed.

    (d) She sings or she dances.
        She does not sing.
        She dances.

Use a truth table to determine whether each argument is valid *or* invalid.

29. If I write a check, it will bounce. If the bank guarantees it, then it does not bounce. The bank guarantees it. Therefore, I don't write a check.

30. $\sim p \rightarrow \sim q$
    $q \rightarrow p$
    $p \vee q$

chapter 4
Numeration and Mathematical Systems

The earliest episodes from the history of mathematics make it clear that the need for counting was the basis for the development of numeration systems. As you will see in this chapter, these systems range in concept from the simplest ones, which use tally marks, to our current place value system. One of the greatest advantages of our current system is that algorithms, or procedures, for computation are relatively easy. For example, when we add or subtract two numbers in a vertical manner, we "line them up in columns" so that digits representing the same place values are combined. It is possible that this procedure is the only one you have ever used when adding and subtracting using pencil and paper. There are, however, alternative methods for performing computations, and some of them are covered in this chapter.

One of the most interesting of these alternative algorithms is the *nines complement method* for subtracting. To use this method, we first agree that the nines complement of a digit n is $9 - n$. For example, the nines complement of 0 is 9, of 1 is 8, of 2 is 7, and so on, up to the nines complement of 9, which is 0. Now suppose we want to subtract

$$25{,}341$$
$$-8{,}496.$$

We fill in any missing place values in the subtrahend (8,496) with zeros so that there are the same number of digits in both numbers:

$$25{,}341$$
$$-08{,}496.$$

Next, replace each digit in the subtrahend with its nines complement, and then add:

$$25{,}341$$
$$+91{,}503$$
$$116{,}844.$$

Now, delete the first digit (1), and add 1 to the remaining part of the sum (16,844):

$$16{,}844 + 1 = \boxed{16{,}845}. \leftarrow \text{Answer}$$

The answer to the original subtraction problem is 16,845.

4.1 Historical Numeration Systems

4.2 Arithmetic in the Hindu-Arabic System

4.3 Conversion Between Number Bases

4.4 Finite Mathematical Systems

4.5 Groups

Collaborative Investigation:
    A Perpetual Calendar Algorithm

Chapter 4 Test

Earlier we introduced and studied the concept of a *set,* a collection of elements. A set, in itself, may have no particular structure. But when we introduce *ways of combining the elements* (called *operations*) and *ways of comparing the elements* (called *relations*), we obtain a **mathematical system.**

Mathematical System

A **mathematical system** is made up of three components:

1. a set of elements;
2. one or more operations for combining the elements;
3. one or more relations for comparing the elements.

Symbols designed to represent objects or ideas are among the oldest inventions of humans. These Indian symbols in Arizona are several hundred years old.

A familiar example of a mathematical system is the set of whole numbers $\{0, 1, 2, 3, \ldots\}$, along with the operation of addition and the relation of equality.

Historically, the earliest mathematical system to be developed involved the set of counting numbers or initially a limited subset of the "smaller" counting numbers. The various ways of symbolizing and working with the counting numbers are called **numeration systems.** The symbols of a numeration system are called **numerals.**

4.1 Historical Numeration Systems

Primitive societies have little need for large numbers. Even today, the languages of some cultures contain no words for numbers beyond "one," "two," and maybe an indefinite word suggesting "many." For example, according to UCLA physiologist Jared Diamond (*Discover,* Aug. 1987, p. 38), there are Gimi villages in New Guinea that use just two root words—*iya* for one and *rarido* for two. Slightly larger numbers are indicated using combinations of these two: for example, *rarido-rarido* is four and *rarido-rarido-iya* is five.

A practical method of keeping accounts by matching may have developed as humans established permanent settlements and began to grow crops and raise livestock. People might have kept track of the number of sheep in a flock by matching pebbles with the sheep, for example. The pebbles could then be kept as a record of the number of sheep.

A more efficient method is to keep a **tally stick.** With a tally stick, one notch or **tally** is made on a stick for each sheep. Tally sticks and tally marks have been found that appear to be many thousands of years old. Tally marks are still used today: for example, nine items are tallied by writing ⧸⧸⧸⧸ ⎮⎮⎮⎮.

Tally sticks and groups of pebbles were an important advance in counting. By these methods, the idea of *number* began to develop. Early people began to see that a group of three chickens and a group of three dogs had something in common: the idea of *three.* Gradually, people began to think of numbers separately from the things they represented. Words and symbols were developed for various numbers.

The numerical records of ancient people give us some idea of their daily lives and create a picture of them as producers and consumers. For example, Mary and Joseph went to Bethlehem to be counted in a census—a numerical record. Even earlier than that, as long as 5000 years ago, the Egyptian and Sumerian peoples were using large numbers in their government and business records. Ancient documents reveal some of their numerical methods, as well as those of the Greeks, Romans, Chinese, and Hindus. Numeration systems became more sophisticated as the need arose.

Tally sticks like this one were used by the English in about 1400 A.D. to keep track of financial transactions. Each notch stands for one pound sterling.

4.1 Historical Numeration Systems

Ancient Egyptian Numeration—Simple Grouping Early matching and tallying led to the essential ingredient of all more advanced numeration systems, that of **grouping**. Grouping allows for less repetition of symbols and also makes numerals easier to interpret. Most historical systems, including our own, have used groups of ten, indicating that people commonly learn to count by using their fingers. The size of the groupings (again, usually ten) is called the **base** of the number system. Bases of five, twenty, and sixty have also been used.

The ancient Egyptian system is an example of a simple grouping system. It utilized ten as its base, and its various symbols are shown in Table 1. The symbol for 1 (|) is repeated, in a tally scheme, for 2, 3, and so on up to 9. A new symbol is introduced for 10 (∩), and that symbol is repeated for 20, 30, and so on, up to 90. This pattern enabled the Egyptians to express numbers up to 9,999,999 with just the seven symbols shown in the table.

Much of our knowledge of **Egyptian mathematics** comes from the Rhind papyrus, from about 3800 years ago. A small portion of this papyrus, showing methods for finding the area of a triangle, is reproduced here.

TABLE 1	Early Egyptian Symbols	
Number	**Symbol**	**Description**
1			Stroke
10	∩	Heel bone
100	?	Scroll
1000	?	Lotus flower
10,000	?	Pointing finger
100,000	?	Burbot fish
1,000,000	?	Astonished person

The Egyptian symbols used denote the various **powers** of the base (ten):

$$10^0 = 1, \ 10^1 = 10, \ 10^2 = 100, \ 10^3 = 1000, \ 10^4 = 10{,}000,$$
$$10^5 = 100{,}000, \text{ and } 10^6 = 1{,}000{,}000.$$

The smaller numerals at the right of the 10s, and slightly raised, are called **exponents**.

Applied Mathematics An Egyptian tomb painting shows scribes tallying the count of a grain harvest. Egyptian mathematics was oriented more to practicality than was Greek or Babylonian mathematics, although the Egyptians did have a formula for finding the volume of a certain portion of a pyramid.

EXAMPLE 1 Write in our system the number below.

Refer to Table 1 for the values of the Egyptian symbols. Each ⌒ represents 100,000. Therefore, two ⌒s represent 2 × 100,000, or 200,000. Proceed as follows:

two	⌒	2 × 100,000 =	200,000
five	?	5 × 1000 =	5000
four	?	4 × 100 =	400
nine	∩	9 × 10 =	90
seven			7 × 1 =	7
			205,497.

The number is 205,497.

Number	Symbol
1	I
5	V
10	X
50	L
100	C
500	D
1000	M

Roman numerals still appear today, mostly for decorative purposes: on clock faces, for chapter numbers in books, and so on. The system is essentially base ten, simple grouping, but with separate symbols for the intermediate values 5, 50, and 500, as shown above. If I is positioned left of V or X, it is subtracted rather than added. Likewise for X appearing left of L or C, and for C appearing left of D or M. Thus, for example, whereas CX denotes 110, XC denotes 90.

EXAMPLE 2 Write 376,248 in Egyptian symbols.

Writing this number requires three ☒ s, seven 𓆼 s, six 𓍢 s, two 𓆐 s, four ∩s, and eight Is, or

[Egyptian numeral representation]

Notice that the position or order of the symbols makes no difference in a simple grouping system. Each of the numbers 𓆐𓆐∩∩∩IIII, IIII∩∩∩𓆐𓆐, and II∩∩𓆐𓆐∩II would be interpreted as 234. The most common order, however, is that shown in Examples 1 and 2, where like symbols are grouped together and groups of higher-valued symbols are positioned to the left.

A simple grouping system is well suited to addition and subtraction. For example, to add 𓍢𓍢𓆐𓆐∩∩II and 𓍢𓆐𓆐𓆐∩IIIIII in the early Egyptian system, work as shown. Two Is plus six Is equal to eight Is, and so on.

$$
\begin{array}{r}
𓍢𓍢 \quad 𓆐𓆐 \quad ∩∩∩ \quad II \\
+ \; 𓍢 \quad 𓆐𓆐𓆐 \quad ∩ \quad IIIIII \\
\hline
\text{Sum:} \quad 𓍢𓍢𓍢\,𓆐𓆐𓆐 \quad ∩∩ \quad IIII \\
\phantom{\text{Sum:}} \quad \,𓆐𓆐 \quad ∩∩ \quad IIII
\end{array}
$$

While we used a + sign for convenience and drew a line under the numbers, the Egyptians did not do this.

Sometimes regrouping, or "carrying," is needed as in the example below in which the answer contains more than nine heel bones. To regroup, get rid of ten heel bones from the tens group. Compensate for this by placing an extra scroll in the hundreds group.

[worked example with regrouped answer]

Subtraction is done in much the same way, as shown in the next example.

EXAMPLE 3 Subtract in each of the following.

(a) 𓆐𓆐𓆐 ∩∩ IIII
  𓆐𓆐 ∩∩ III
  −𓆐𓆐𓆐 ∩ IIII
  Difference: 𓆐𓆐 ∩∩∩ III

(b) 𓆐𓆐∩∩∩∩ II
  −𓆐 ∩∩ IIII

In part (b), to subtract four Is from two Is, "borrow" one heel bone, which is equivalent to ten Is. Finish the problem after writing ten additional Is on the right.

Greek Numerals			
1	α	60	ξ
2	β	70	o
3	γ	80	π
4	δ	90	φ
5	ε	100	ρ
6	ς	200	σ
7	ζ	300	τ
8	η	400	υ
9	θ	500	φ
10	ι	600	χ
20	κ	700	ψ
30	λ	800	ω
40	μ	900	ϡ
50	ν		

What About the Greeks?
Classical Greeks used letters of their alphabet as numerical symbols. The base of the system was the number 10, and numbers 1 through 9 were symbolized by the first nine letters of the alphabet. Rather than using repetition or multiplication, they assigned nine more letters to multiples of 10 (through 90) and more letters to multiples of 100 (through 900). This is called a ciphered system, and it sufficed for small numbers. For example, 57 would be νζ; 573 would be φογ; and 803 would be ωγ. A small stroke was used with a units symbol for multiples of 1000 (up to 9000); thus 1000 would be ‚α or 'α. Often M would indicate tens of thousands (M for myriad = 10,000) with the multiples written above M.

A procedure such as those described above is called an **algorithm**: a rule or method for working a problem. The Egyptians used an interesting algorithm for multiplication that requires only an ability to add and to double numbers, as shown in Example 4. For convenience, this example uses our symbols rather than theirs.

EXAMPLE 4 A stone used in building a pyramid has a rectangular base measuring 5 by 18 cubits. Find the area of the base.

The area of a rectangle is found by multiplying the length and the width; in this problem, we must find 5×18. To begin, build two columns of numbers, as shown below. Start the first column with 1, and the second column with 18. Each column is built downward by doubling the number above. Keep going until the first column contains numbers that can be added to equal 5. Here $1 + 4 = 5$. To find 5×18, add only those numbers from the second column that correspond to 1 and 4. Here 18 and 72 are added to get the answer 90. The area of the base of the stone is 90 square cubits.

$$1 + 4 = 5 \begin{cases} \rightarrow 1 & 18 \leftarrow \text{Corresponds to 1} \\ 2 & 36 \\ \rightarrow 4 & 72 \leftarrow \text{Corresponds to 4} \end{cases} 18 + 72 = 90$$

Finally, $5 \times 18 = 90$.

EXAMPLE 5 Use the Egyptian multiplication algorithm to find 19×70.

$$\begin{array}{rr} \rightarrow 1 & 70 \leftarrow \\ \rightarrow 2 & 140 \leftarrow \\ 4 & 280 \\ 8 & 560 \\ \rightarrow 16 & 1120 \leftarrow \end{array}$$

Form two columns, headed by 1 and by 70. Keep doubling until there are numbers in the first column that add up to 19. (Here, $1 + 2 + 16 = 19$.) Then add corresponding numbers from the second column: $70 + 140 + 1120 = 1330$, so that $19 \times 70 = 1330$.

Traditional Chinese Numeration—Multiplicative Grouping

Examples 1 through 3 above show that simple grouping, although an improvement over tallying, still requires considerable repetition of symbols. To denote 90, for example, the ancient Egyptian system must utilize nine ∩s. If an additional symbol (a "multiplier") was introduced to represent nine, say "9," then 90 could be denoted 9 ∩. All possible numbers of repetitions of powers of the base could be handled by introducing a separate multiplier symbol for each counting number less than the base. Although the ancient Egyptian system apparently did not evolve in this direction, just such a system was developed many years ago in China.

TABLE 2

Number	Symbol
1	一
2	二
3	三
4	四
5	五
6	六
7	七
8	八
9	九
10	十
100	百
1000	千
0	零

This photo is of a **quipu**. In *Ethnomathematics: A Multicultural View of Mathematical Ideas,* Marcia Ascher writes:

A quipu is an assemblage of colored knotted cotton cords. Among the Inca, cotton cloth and cordage were of great importance. Used to construct bridges, in ceremonies, for tribute, and in every phase of the life cycle from birth to death, cotton cordage and cloth were of unparalleled importance in Inca culture and, hence, not a surprising choice for its principal medium. The colors of the cords, the way the cords are connected, the relative placement of the cords, the spaces between the cords, the types of knots on the individual cords, and the relative placement of the knots are all part of the logical-numerical recording.

It was later adopted, for the most part, by the Japanese, with several versions occurring over the years. Here we show the predominant Chinese version, which used the symbols shown in Table 2. We call this type of system a **multiplicative grouping** system. In general, such a system would involve pairs of symbols, each pair containing a multiplier (with some counting number value less than the base) and then a power of the base. The Chinese numerals are read from top to bottom rather than from left to right.

Three features distinguish this system from a strictly pure multiplicative grouping system. First, the number of 1s is indicated using a single symbol rather than a pair. In effect, the multiplier $(1, 2, 3, \ldots, 9)$ is written but the power of the base (10^0) is not. Second, in the pair indicating 10s, if the multiplier is 1, then that multiplier is omitted. Just the symbol for 10 is written. Third, when a given power of the base is totally missing in a particular number, this omission is shown by the inclusion of the special zero symbol. (See Table 2.) If two or more consecutive powers are missing, just one zero symbol serves to note the total omission. The omission of 1s and 10s, and any other powers occurring at the extreme bottom of a numeral, need not be noted with a zero symbol. (Note that, for clarification in the examples that follow, we have emphasized the grouping into pairs by spacing and by using braces. These features are *not* part of the actual numeral.)

EXAMPLE 6   Interpret the Chinese numerals below.

(a)
- $3 \times 1000 = 3000$
- $1 \times 100 = 100$
- $6 \times 10 = 60$
- $4(\times 1) = \underline{4}$
- Total: 3164

(b)
- $7 \times 100 = 700$
- $0(\times 10) = 00$
- $3(\times 1) = \underline{3}$
- Total: 703

(c)
- $5 \times 1000 = 5000$
- $0(\times 100) = 000$
- $0(\times 10) = 00$
- $9(\times 1) = \underline{9}$
- Total: 5009

(d)
- $4 \times 1000 = 4000$
- $2 \times 100 = \underline{200}$
- Total: 4200

EXAMPLE 7   Write Chinese numerals for these numbers.

(a) 614

This number is made up of six 100s, one 10, and one 4, as depicted at the right.

6×100: 六百
$(1 \times)10$: 十
$4(\times 1)$: 四

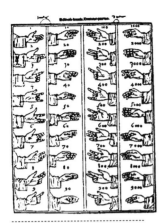

Finger Reckoning There is much evidence that early humans (in various cultures) used their fingers to represent numbers. As calculations became more complicated, *finger reckoning,* as shown in this sketch, became popular. The Romans became adept at this sort of calculating, carrying it to 10,000 or perhaps higher.

Number	Symbol
1	▼
10	◁

Babylonian numeration was positional, base sixty. But the face values within the positions were base ten simple grouping numerals, formed with the two symbols shown above. (These symbols resulted from the Babylonian method of writing on clay with a wedge-shaped stylus.) The numeral

◁◁▼▼▼◁◁◁◁▼

denotes 1421 (23 × 60 + 41 × 1).

(b) 5090

The number consists of five 1000s, no 100s, and nine 10s (no 1s).

5×1000: 五千

$0 (\times 100)$: 零

9×10: 九十

Hindu-Arabic Numeration—Positional System
A simple grouping system relies on repetition of symbols to denote the number of each power of the base. A multiplicative grouping system uses multipliers in place of repetition, which is more efficient. But the ultimate in efficiency is attained only when we proceed to the next step, a **positional** system, in which only the multipliers are used. The various powers of the base require no separate symbols, since the power associated with each multiplier can be understood by the position that the multiplier occupies in the numeral. If the Chinese system had evolved into a positional system, then the numeral for 7482 could be written

七
四   rather than   七千四百八十二
八
二

The lowest symbol is understood to represent two 1s (10^0), the next pair up denotes eight 10s (10^1), then four 100s (10^2), and finally seven 1000s (10^3). Each symbol in a numeral now has both a *face value,* associated with that particular symbol (the multiplier value), and a *place value* (a power of the base), associated with the place, or position, occupied by the symbol.

Positional Numeration

In a positional numeral, each symbol (called a **digit**) conveys two things:

1. **face value**—the inherent value of the symbol
2. **place value**—the power of the base which is associated with the position that the digit occupies in the numeral.

The place values in a Hindu-Arabic numeral, from right to left, are 1, 10, 100, 1000, and so on. The three 4s in the number 46,424 all have the same face value but different place values. The first 4, on the left, denotes four 10,000s, the next one denotes four 100s, and the one on the right denotes four 1s. Place values (in base ten) are named as shown here:

This numeral is read as eight billion, three hundred twenty-one million, four hundred fifty-six thousand, seven hundred ninety-five.

To work successfully, a positional system must have a symbol for zero to serve as a **placeholder** in case one or more powers of the base are not needed. Because of this requirement, some early numeration systems took a long time to evolve to a positional form, or never did. Although the traditional Chinese system does utilize a zero symbol, it never did incorporate all the features of a positional system, but remained essentially a multiplicative grouping system.

The one numeration system that did achieve the maximum efficiency of positional form is our own system, commonly known for historical reasons as the **Hindu-Arabic** system. It was developed over many centuries. Its symbols have been traced to the Hindus of 200 B.C. They were picked up by the Arabs and eventually transmitted to Spain, where a late tenth-century version appeared like this:

$$I\ Z\ Z\ X\ Y\ 6\ 7\ 8\ 9.$$

From Tally to Tablet The clay tablet above, despite damage, shows the durability of the mud of the Babylonians. Thousands of years after the tablets were made, Babylonian algebra problems can be worked out from the original writings.

In recent times, herdsmen make small "tokens" out of this clay as tallies of animals. Similar tokens, some 10,000 years old, have been unearthed by archaeologists in the land that was once Babylonia. Shaped like balls, disks, cones, and other regular forms, they rarely exceed 5 cm in diameter.

The earliest stages of the system evolved under the influence of navigational, trade, engineering, and military requirements. And in early modern times, the advance of astronomy and other sciences led to a structure well suited to fast and accurate computation. The purely positional form that the system finally assumed was introduced to the West by Leonardo Fibonacci of Pisa (1170–1250) early in the thirteenth century. But widespread acceptance of standardized symbols and form was not achieved until the invention of printing during the fifteenth century. Since that time, no better system of numeration has been devised, and the positional base ten Hindu-Arabic system is commonly used around the world today. (In India, where it all began, standardization still is not totally achieved, as various local systems are used today.)

In the next section we shall look in more detail at the structure of the Hindu-Arabic system and some early methods and devices for doing computation.

4.1 EXERCISES

Convert each Egyptian numeral to Hindu-Arabic form.

1.
2.
3.
4.

Convert each Hindu-Arabic numeral to Egyptian form.

5. 23,145
6. 427
7. 8,657,000
8. 306,090

Chapter 1 of the book of Numbers in the Bible describes a census of the draft-eligible men of Israel after Moses led them out of Egypt into the Desert of Sinai, about 1450 B.C. Write an Egyptian numeral for the number of available men from each tribe listed.

9. 59,300 from the tribe of Simeon

10. 46,500 from the tribe of Reuben

11. 74,600 from the tribe of Judah

12. 45,650 from the tribe of Gad

13. 62,700 from the tribe of Dan

14. 54,400 from the tribe of Issachar

Convert each Chinese numeral to Hindu-Arabic form.

15. 九百三十五

16. 二百四十六

17. 三千零七

18. 四千九百零二

Convert each Hindu-Arabic numeral to Chinese.

19. 960

20. 63

21. 7012

22. 2416

Though Chinese art forms began before written history, their highest development was achieved during four particular dynasties. Write traditional Chinese numerals for the beginning and ending dates of each dynasty listed.

23. Ming (1368 to 1644)

24. Sung (960 to 1279)

25. T'ang (618 to 907)

26. Han (202 B.C. to A.D. 220)

Work each of the following addition or subtraction problems, using regrouping as necessary. Convert each answer to Hindu-Arabic form.

27. 9 ∩∩ ∩∩ ‖
  + ∩∩∩∩ ∩∩∩ ‖‖‖‖

28. 99∩∩ ‖‖‖‖ ‖‖‖
  + 9 ∩∩∩ ∩∩ ‖‖ ‖‖‖

29. ℓℓℓ 𝄢𝄢 ∩ ‖‖‖
  + ℓℓ 𝄢 99 ∩∩∩ ‖‖‖‖
       999∩∩∩ ‖‖‖‖

30. 𝄢𝄢𝄢 99 ∩∩ ‖
       999 ∩∩
  + 𝄢𝄢 999 ∩∩ ‖‖‖
       999   ‖‖‖

31. 99∩∩∩ ‖‖‖‖
  − 9 ∩∩ ‖

32. ∩∩ ‖‖‖
   ∩∩ ‖‖‖
  − ∩∩∩ ‖‖‖‖

33. ℓ 𝄢 99 ‖‖‖‖
  − ℓℓℓ 999 ‖‖‖
     999 ‖‖‖

34. 𝄢 999 ∩∩∩ ‖‖‖
      99   ‖‖‖
  − 99 ∩∩∩   ‖‖‖
       ∩∩∩∩

Use the Egyptian algorithm to find each product.

35. 26 × 53

36. 33 × 81

37. 58 × 103

38. 67 × 115

Convert all numbers in the following problems to Egyptian numerals. Multiply using the Egyptian algorithm, and add using the Egyptian symbols. Give the final answer using a Hindu-Arabic numeral.

39. *Value of a Biblical Treasure* The book of Ezra in the Bible describes the return of the exiles to Jerusalem. When they rebuilt the temple, the King of Persia gave them the following items: thirty golden basins, a thousand silver basins, four hundred ten silver bowls, and thirty golden bowls. Find the total value of this treasure, if each gold basin is worth 3000 shekels, each silver basin is worth 500 shekels, each silver bowl is worth 50 shekels, and each golden bowl is worth 400 shekels.

40. *Total Bill for King Solomon* King Solomon told the King of Tyre (now Lebanon) that Solomon needed the best cedar for his temple, and that he would "pay you for your men whatever sum you fix." Find the total bill to Solomon if the King of Tyre used the following numbers of men: 5500 tree cutters at two shekels per week each, for a total of seven weeks; 4600 sawers of wood at three shekels per week each, for a total of 32 weeks; and 900 sailors at one shekel per week each, for a total of 16 weeks.

Explain why each of the following steps would be an improvement in the development of numeration systems.

41. progressing from carrying groups of pebbles to making tally marks on a stick

42. progressing from tallying to simple grouping

43. progressing from simple grouping to multiplicative grouping

44. progressing from multiplicative grouping to positional numeration

Recall that the ancient Egyptian system described in this section was simple grouping, used a base of ten, and contained seven distinct symbols. The largest number expressible in that system is 9,999,999. Identify the largest number expressible in each of the following simple grouping systems. (In Exercises 49–52, d can be any counting number.)

45. base ten, five distinct symbols

46. base ten, ten distinct symbols

47. base five, five distinct symbols

48. base five, ten distinct symbols

49. base ten, d distinct symbols

50. base five, d distinct symbols

51. base seven, d distinct symbols

52. base b, d distinct symbols (where b is any counting number 2 or greater)

The Hindu-Arabic system is positional and uses ten as the base. Describe any advantages or disadvantages that may have resulted in each of the following cases.

53. Suppose the base had been larger, say twelve or twenty for example.

54. Suppose the base had been smaller, maybe eight or five.

4.2 Arithmetic in the Hindu-Arabic System

The historical development of numeration culminated in positional systems, the most successful of which is the Hindu-Arabic system. As stated in the previous section, Hindu-Arabic place values are powers of the base ten. For example, 10^4 denotes the fourth power of ten. Such expressions are often called **exponential expressions**. In this case, 10 is the **base** and 4 is the **exponent**. Exponents actually indicate repeated multiplication of the base:

$$10^4 = \underbrace{10 \times 10 \times 10 \times 10}_{\text{Four factors of 10}} = 10{,}000.$$

In the same way, $10^2 = 10 \times 10 = 100$, $10^6 = 10 \times 10 \times 10 \times 10 \times 10 \times 10 = 1{,}000{,}000$, and so on. The base does not have to be 10; for example,

$$4^3 = 4 \times 4 \times 4 = 64, \quad 2^2 = 2 \times 2 = 4,$$
$$3^5 = 3 \times 3 \times 3 \times 3 \times 3 = 243,$$

and so on. Expressions of this type are defined in general as follows.

4.2 Arithmetic in the Hindu-Arabic System

Digits This Iranian stamp should remind us that counting on fingers (and toes) is an age-old practice. In fact, our word *digit,* referring to the numerals 0–9, comes from a Latin word for "finger" (or "toe"). Aristotle first noted the relationships between fingers and base ten in Greek numeration. Anthropologists go along with the notion. Some cultures, however, have used two, three, or four as number bases, for example, counting on the joints of the fingers or the spaces between them.

Exponential Expressions

For any number a and any counting number m,

$$a^m = \underbrace{a \times a \times a \times \cdots \times a}_{m \text{ factors of } a}.$$

The number a is the **base,** m is the **exponent,** and a^m is read "a to the power m."

EXAMPLE 1 Find each power.

(a) $10^3 = 10 \times 10 \times 10 = 1000$

(10^3 is read "10 cubed," or "10 to the third power.")

(b) $7^2 = 7 \times 7 = 49$

(7^2 is read "7 squared," or "7 to the second power.")

(c) $5^4 = 5 \times 5 \times 5 \times 5 = 625$

(5^4 is read "5 to the fourth power.")

To simplify work with exponents, it is agreed that $a^0 = 1$ for any nonzero number a. By this agreement, $7^0 = 1$, $52^0 = 1$, and so on. At the same time, $a^1 = a$ for any number a. For example, $8^1 = 8$, and $25^1 = 25$. The exponent 1 is usually omitted.

With the use of exponents, numbers can be written in **expanded form** in which the value of the digit in each position is made clear. For example, write 924 in expanded form by thinking of 924 as nine 100s plus two 10s plus four 1s, or

$$924 = 900 + 20 + 4$$
$$924 = (9 \times 100) + (2 \times 10) + (4 \times 1).$$

By the definition of exponents, $100 = 10^2$, $10 = 10^1$, and $1 = 10^0$. Use these exponents to write 924 in expanded form as follows:

$$924 = (9 \times 10^2) + (2 \times 10^1) + (4 \times 10^0).$$

EXAMPLE 2 Write each number in expanded form.

(a) $1906 = (1 \times 10^3) + (9 \times 10^2) + (0 \times 10^1) + (6 \times 10^0)$

Since $0 \times 10^1 = 0$, this term could be omitted, but the form is clearer with it included.

(b) $46{,}424 = (4 \times 10^4) + (6 \times 10^3) + (4 \times 10^2) + (2 \times 10^1) + (4 \times 10^0)$

EXAMPLE 3 Simplify each of the following expansions.

(a) $(3 \times 10^5) + (2 \times 10^4) + (6 \times 10^3) + (8 \times 10^2) + (7 \times 10^1) + (9 \times 10^0) = 326{,}879$

(b) $(2 \times 10^1) + (8 \times 10^0) = 28$

Expanded notation can be used to see why standard algorithms for addition and subtraction really work. The key idea behind these algorithms is based on the **distributive property,** which will be discussed more fully later in this chapter. It can be written in one form as follows.

Distributive Property

For all real numbers a, b, and c,

$$(b \times a) + (c \times a) = (b + c) \times a.$$

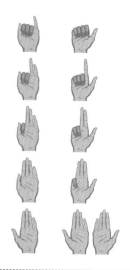

Finger Counting The first digits many people used for counting were their fingers. In Africa the Zulu used the method shown here to count to ten. They started on the left hand with palm up and fist closed. The Zulu finger positions for 1–5 are shown above on the left. The Zulu finger positions for 6–10 are shown on the right.

For example,

$$(3 \times 10^4) + (2 \times 10^4) = (3 + 2) \times 10^4$$
$$= 5 \times 10^4.$$

EXAMPLE 4 Use expanded notation to add 23 and 64.

$$\begin{aligned} 23 &= (2 \times 10^1) + (3 \times 10^0) \\ +64 &= (6 \times 10^1) + (4 \times 10^0) \\ \hline &(8 \times 10^1) + (7 \times 10^0) = 87 \end{aligned}$$

Subtraction works in much the same way.

EXAMPLE 5 Find $695 - 254$.

$$\begin{aligned} 695 &= (6 \times 10^2) + (9 \times 10^1) + (5 \times 10^0) \\ -254 &= (2 \times 10^2) + (5 \times 10^1) + (4 \times 10^0) \\ \hline &(4 \times 10^2) + (4 \times 10^1) + (1 \times 10^0) = 441 \end{aligned}$$

Expanded notation and the distributive property can also be used to show how to solve addition problems where a power of 10 ends up with a multiplier of more than one digit.

EXAMPLE 6 Use expanded notation to add 75 and 48.

$$\begin{aligned} 75 &= (7 \times 10^1) + (5 \times 10^0) \\ +48 &= (4 \times 10^1) + (8 \times 10^0) \\ \hline &(11 \times 10^1) + (13 \times 10^0) \end{aligned}$$

Since the units position (10^0) has room for only one digit, 13×10^0 must be modified:

$$\begin{aligned} 13 \times 10^0 &= (10 \times 10^0) + (3 \times 10^0) \quad \text{Distributive property} \\ &= (1 \times 10^1) + (3 \times 10^0). \end{aligned}$$

In effect, the 1 from 13 moved to the left from the units position to the tens position. This is called "carrying." Now our sum is

$$(11 \times 10^1) + (1 \times 10^1) + (3 \times 10^0)$$
$$= (12 \times 10^1) + (3 \times 10^0) \qquad \text{Distributive property}$$
$$= (10 \times 10^1) + (2 \times 10^1) + (3 \times 10^0)$$
$$= (1 \times 10^2) + (2 \times 10^1) + (3 \times 10^0)$$
$$= 123.$$

Subtraction problems often require "borrowing," which can also be clarified with expanded notation.

EXAMPLE 7 Use expanded notation to subtract 186 from 364.

$$364 = (3 \times 10^2) + (6 \times 10^1) + (4 \times 10^0)$$
$$-186 = (1 \times 10^2) + (8 \times 10^1) + (6 \times 10^0)$$

Since, in the units position, we cannot subtract 6 from 4, we modify the top expansion as follows (the units position borrows from the tens position):

$$(3 \times 10^2) + (6 \times 10^1) + (4 \times 10^0)$$
$$= (3 \times 10^2) + (5 \times 10^1) + (1 \times 10^1) + (4 \times 10^0) \qquad \text{Distributive property}$$
$$= (3 \times 10^2) + (5 \times 10^1) + (10 \times 10^0) + (4 \times 10^0)$$
$$= (3 \times 10^2) + (5 \times 10^1) + (14 \times 10^0). \qquad \text{Distributive property}$$

(We can now subtract 6 from 14 in the units position, but cannot take 8 from 5 in the tens position, so we continue the modification, borrowing from the hundreds to the tens position.)

$$(3 \times 10^2) + (5 \times 10^1) + (14 \times 10^0)$$
$$= (2 \times 10^2) + (1 \times 10^2) + (5 \times 10^1) + (14 \times 10^0) \qquad \text{Distributive property}$$
$$= (2 \times 10^2) + (10 \times 10^1) + (5 \times 10^1) + (14 \times 10^0)$$
$$= (2 \times 10^2) + (15 \times 10^1) + (14 \times 10^0) \qquad \text{Distributive property}$$

Now we can complete the subtraction.

$$(2 \times 10^2) + (15 \times 10^1) + (14 \times 10^0)$$
$$- (1 \times 10^2) + (8 \times 10^1) + (6 \times 10^0)$$
$$(1 \times 10^2) + (7 \times 10^1) + (8 \times 10^0) = 178$$

Examples 4 through 7 used expanded notation and the distributive property to clarify our usual addition and subtraction methods. In practice, our actual work for these four problems would appear as follows:

$$\begin{array}{rrrr}
 & & \overset{1}{} & \overset{2}{}\overset{15}{} \\
23 & 695 & 75 & \cancel{3}\,\cancel{6}\,4 \\
+\,64 & -254 & +\,48 & -1\,8\,6 \\
\hline
87 & 441 & 123 & 1\,7\,8.
\end{array}$$

The **Carmen de Algorismo** (opening verses shown here) by Alexander de Villa Dei, thirteenth century, popularized the new art of "algorismus":

> ... from these twice five figures 0 9 8 7 6 5 4 3 2 1 of the Indians we benefit ...

The *Carmen* related that Algor, an Indian king, invented the art. But actually, "algorism" (or "algorithm") comes in a roundabout way from the name Muhammad ibn Musa al-Khorârizmi, an Arabian mathematician of the ninth century, whose arithmetic book was translated into Latin. Furthermore, this Muhammad's book on equations, *Hisab al-jabr w'almuqâbalah*, yielded the term "algebra" in a similar way.

Palmtop computers are the latest in the development of calculating devices.

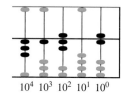

10^4 10^3 10^2 10^1 10^0

FIGURE 1

Merry Math The following two rhymes come from *Marmaduke Multiply's Merry Method of Making Minor Mathematicians,* a primer published in the late 1830s in Boston:

1. Twice 2 are 4. Pray hasten on before.
2. Five times 5 are 25. I thank my stars I'm yet alive.

The procedures seen in this section also work for positional systems with bases other than ten.

Since our numeration system is based on powers of ten, it is often called the **decimal system,** from the Latin word *decem,* meaning ten.* Over the years, many methods have been devised for speeding calculations in the decimal system. One of the oldest is the **abacus,** a device made with a series of rods with sliding beads and a dividing bar. Reading from right to left, the rods have values of 1, 10, 100, 1000, and so on. The bead above the bar has five times the value of those below. Beads moved *toward* the bar are in the "active" position, and those toward the frame are ignored. In our illustrations of abaci (plural form of abacus), such as in Figure 1, the activated beads are shown in black for emphasis.

EXAMPLE 8 The number on the abacus in Figure 1 is found as follows:

$$(3 \times 10{,}000) + (1 \times 1000) + [(1 \times 500) + (2 \times 100)] + 0$$
$$+ [(1 \times 5) + (1 \times 1)]$$
$$= 30{,}000 + 1000 + 500 + 200 + 0 + 5 + 1$$
$$= 31{,}706.$$

As paper became more readily available, people gradually switched from devices like the abacus (though these still are commonly used in some areas) to paper-and-pencil methods of calculation. One early scheme, used both in India and Persia, was the **lattice method,** which arranged products of single digits into a diagonalized lattice, as shown in the following example.

EXAMPLE 9 Find the product 38×794 by the lattice method.

Work as follows.

Step 1: Write the problem, with one number at the side and one across the top.

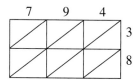

Step 2: Within the lattice, write the products of all pairs of digits from the top and side.

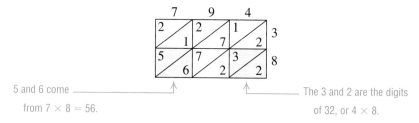

5 and 6 come from $7 \times 8 = 56$.

The 3 and 2 are the digits of 32, or 4×8.

---

*December was the tenth month in an old form of the calendar. It is interesting to note that *decem* became *dix* in the French language; a ten-dollar bill, called a "dixie," was in use in New Orleans before the Civil War. "Dixie Land" was a nickname for that city before Dixie came to refer to all the Southern states, as in Daniel D. Emmett's song, written in 1859.

Step 3: Starting at the right of the lattice add diagonally, carrying as necessary.

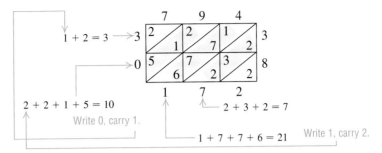

Step 4: Read the answer around the left side and bottom:

$$38 \times 794 = 30{,}172.$$

The Scottish mathematician John Napier (1550–1617) introduced a significant calculating tool called **Napier's rods,** or **Napier's bones.** Napier's invention, based on the lattice method of multiplication, is widely acknowledged as a very early forerunner of modern computers. It consisted of a set of strips, several for each digit 0 through 9, on which multiples of each digit appeared in a sort of lattice column. See Figure 2.

John Napier's most significant mathematical contribution, developed over a period of at least 20 years, was the concept of *logarithms,* which, among other things, allow multiplication and division to be accomplished with addition and subtraction. It was a great computational advantage given the state of mathematics at the time (1614).

Napier himself regarded his interest in mathematics as a recreation, his main involvements being political and religious. A supporter of John Knox and James I, he published a widely read anti-Catholic work which analyzed the Biblical book of Revelation. He concluded that the Pope was the Antichrist and that the Creator would end the world between 1688 and 1700. Napier was one of many who, over the years, have miscalculated the end of the world.

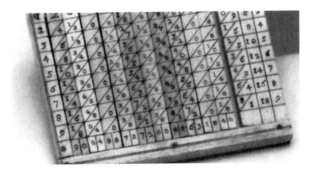

FIGURE 2

An additional strip, called the index, could be laid beside any of the others to indicate the multiplier at each level. Napier's rods were used for mechanically multiplying, dividing, and taking square roots. Figure 3 shows how to multiply 2806 by 7. Select the rods for 2, 8, 0, and 6, placing them side by side. Then using the index, locate the level for a multiplier of 7. The resulting lattice, shown at the bottom of the figure, gives the product 19,642.

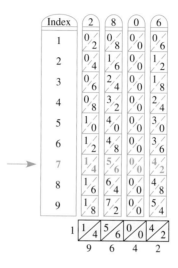

FIGURE 3

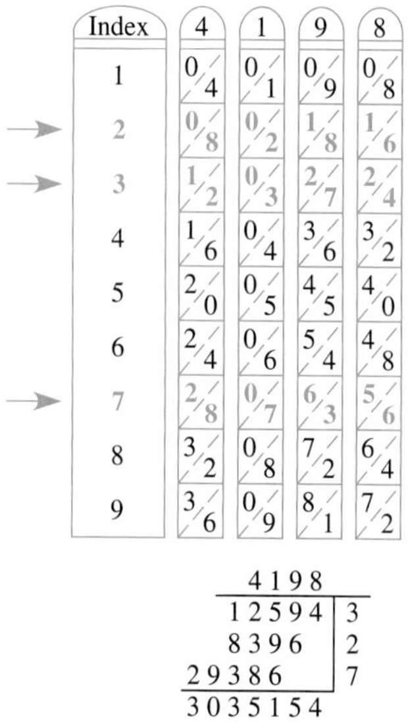

EXAMPLE 10 Use Napier's rods to find the product of 723 and 4198.

We line up the rods for 4, 1, 9, and 8 next to the index as in Figure 4. The product 3×4198 is found as described in Example 9 and written at the bottom of the figure. Then 2×4198 is found similarly and written below, shifted one place to the left. (Why?) Finally, the product 7×4198 is written shifted two places to the left. The final answer is found by addition to obtain

$$723 \times 4198 = 3,035,154.$$

One other paper-and-pencil method of multiplication is the **Russian peasant method,** which is similar to the Egyptian method of doubling explained in the previous section. (In fact both of these methods work, in effect, by expanding one of the numbers to be multiplied, but in base two rather than in base ten. Base two numerals are discussed in the next section.) To multiply 37 and 42 by the Russian peasant method, make two columns headed by 37 and 42. Form the first column by dividing 37 by 2 again and again, ignoring any remainders. Stop when 1 is obtained. Form the second column by doubling each number down the column.

FIGURE 4

	37	42
	18	84
Divide by 2, ignoring remainders.	9	168	Double each number.
	4	336
	2	672
	1	1344

Now add up only the second column numbers that correspond to odd numbers in the first column. Omit those corresponding to even numbers in the first column.

Odd numbers	→37	42 ←
	18	84
	→ 9	168 ←
	4	336
	2	672
	→ 1	1344 ←

Add these numbers.

Finally, $37 \times 42 = 42 + 168 + 1344 = 1554$.

FOR FURTHER THOUGHT

The abacus has been (and still is) used to perform very rapid calculations. A simple example is adding 526 and 362. Start with 526 on the abacus:

To add 362, start by "activating" an additional 2 on the 1s rod:

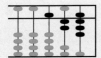

Next, activate an additional 6 on the 10s rod:

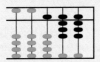

Finally, activate an additional 3 on the 100s rod:

The sum, read from the abacus, is 888.

For problems where carrying or borrowing is required, it takes a little more thought and skill. Try to obtain an actual abacus (or, otherwise, make sketches) and practice some addition and subtraction problems until you can do them quickly.

For Group Discussion

1. Use an abacus to add: $13{,}728 + 61{,}455$. Explain each step of your procedure.
2. Use an abacus to subtract: $6512 - 4816$. Again, explain each step of your procedure.

4.2 EXERCISES

Write each number in expanded form.

1. 73 **2.** 925 **3.** 3774 **4.** 12,398

5. four thousand, nine hundred twenty-four

6. fifty-two thousand, one hundred eighteen

7. fourteen million, two hundred six thousand, forty

8. two hundred twelve million, eleven thousand, nine hundred sixteen

Simplify each of the following expansions.

9. $(4 \times 10^1) + (2 \times 10^0)$

10. $(3 \times 10^2) + (5 \times 10^1) + (0 \times 10^0)$

11. $(6 \times 10^3) + (2 \times 10^2) + (0 \times 10^1) + (9 \times 10^0)$

12. $(5 \times 10^5) + (0 \times 10^4) + (3 \times 10^3) + (5 \times 10^2) + (6 \times 10^1) + (8 \times 10^0)$

13. $(7 \times 10^7) + (4 \times 10^5) + (1 \times 10^3) + (9 \times 10^0)$

14. $(3 \times 10^8) + (8 \times 10^7) + (2 \times 10^2) + (3 \times 10^0)$

In each of the following, add in expanded notation.

15. $54 + 35$

16. $782 + 413$

In each of the following, subtract in expanded notation.

17. $85 - 53$

18. $784 - 523$

Perform each addition using expanded notation.

19. $75 + 34$

20. $537 + 278$

21. $434 + 299$

22. $6755 + 4827$

Perform each subtraction using expanded notation.

23. $54 - 48$

24. $364 - 59$

25. $645 - 439$

26. $816 - 335$

Identify the number represented on each abacus.

27.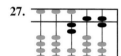

28.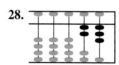

29.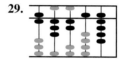

30.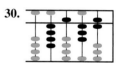

Sketch an abacus to show each number.

31. 38

32. 183

33. 2547

34. 70,163

Use the lattice method to find each product.

35. 65×29

36. 32×741

37. 525×73

38. 912×483

Refer to Example 10 where Napier's rods were used to find the product of 723 and 4198. Then complete Exercises 39 and 40.

39. Find the product of 723 and 4198 by completing the lattice process shown here.

40. Explain how Napier's rods could have been used in Example 10 to set up one complete lattice product rather than adding three individual (shifted) lattice products. Illustrate with a sketch.

Make use of Napier's rods (Figure 2) to find each product.

41. 8×62

42. 32×73

43. 26×8354

44. 526×4863

Refer to the chapter introduction, and perform each of the following subtractions using the nines complement method.

45. $283 - 41$

46. $536 - 425$

47. $50,000 - 199$

48. $40,002 - 4846$

Use the Russian peasant method to find each product.

49. 5×92

50. 41×53

51. 62×529

52. 145×63

4.3 Conversion Between Number Bases

Although the numeration systems discussed in the opening section were all base ten, other bases have occurred historically. For example, the ancient Babylonians used 60 as their base. The Mayan Indians of Central America and Mexico used 20. In this section we consider bases other than ten, but we use the familiar Hindu-Arabic symbols. We will consistently indicate bases other than ten with a spelled-out subscript, as in the numeral 43_{five}. Whenever a number appears without a subscript, it is to be assumed that the intended base is ten. It will help to be careful how you read (or verbalize) numerals here. The numeral 43_{five} is read "four three base five." (Do *not* read it as "forty-three," as that terminology implies base ten and names a totally different number.)

For reference in doing number expansions and base conversions, Table 3 gives the first several powers of some numbers used as alternative bases in this section.

TABLE 3 Selected Powers of Some Alternative Number Bases

	Fourth Power	Third Power	Second Power	First Power	Zero Power
Base two	16	8	4	2	1
Base five	625	125	25	5	1
Base seven	2401	343	49	7	1
Base eight	4096	512	64	8	1
Base sixteen	65,536	4096	256	16	1

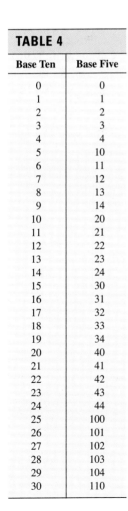

TABLE 4

Base Ten	Base Five
0	0
1	1
2	2
3	3
4	4
5	10
6	11
7	12
8	13
9	14
10	20
11	21
12	22
13	23
14	24
15	30
16	31
17	32
18	33
19	34
20	40
21	41
22	42
23	43
24	44
25	100
26	101
27	102
28	103
29	104
30	110

We begin with the base five system, which requires just five distinct symbols, 0, 1, 2, 3, and 4. Table 4 compares base five and decimal (base ten) numerals for the whole numbers 0 through 30. Notice that, while the base five system uses fewer distinct symbols, it sometimes requires more digits to denote the same number.

EXAMPLE 1 Convert 1342_{five} to decimal form.

Referring to the powers of five in Table 3, we see that this number has one 125, three 25s, four 5s, and two 1s, so

$$1342_{\text{five}} = (1 \times 125) + (3 \times 25) + (4 \times 5) + (2 \times 1)$$
$$= 125 + 75 + 20 + 2$$
$$= 222.$$

A shortcut for converting from base five to decimal form, which is *particularly useful when you use a calculator,* can be derived as follows. (We can illustrate this by repeating the conversion of Example 1.)

$$1342_{\text{five}} = (1 \times 5^3) + (3 \times 5^2) + (4 \times 5) + 2$$

Now 5 can be factored out of the three quantities in parentheses, so

$$1342_{\text{five}} = ((1 \times 5^2) + (3 \times 5) + 4) \times 5 + 2.$$

Now, factoring another five out of the two "inner" quantities, we get

$$1342_{\text{five}} = (((1 \times 5) + 3) \times 5 + 4) \times 5 + 2.$$

The inner parentheses around 1×5 are not needed since the product would be automatically done before the 3 is added. Therefore, we can write

$$1342_{\text{five}} = ((1 \times 5 + 3) \times 5 + 4) \times 5 + 2.$$

This series of products and sums is easily done as an uninterrupted sequence of operations on a calculator, with no intermediate results written down. The same method works for converting to base ten from any other base. The procedure is summarized as follows.

> **Calculator Shortcut**
>
> **To convert from another base to decimal form:** Start with the first digit on the left and multiply by the base. Then add the next digit, multiply again by the base, and so on. The last step is to add the last digit on the right. Do *not* multiply it by the base.

Exactly how you accomplish these steps depends on the type of calculator you use. With some, only the digits, the multiplications, and the additions need to be entered, in order. With others, you may need to press the $\boxed{=}$ key following each addition of a digit. If you handle grouped expressions on your calculator by actually entering parentheses, then enter the expression just as illustrated above and in the following example. (The number of left parentheses to start with will be two fewer than the number of digits in the original numeral.)

EXAMPLE 2 Use the calculator shortcut to convert 244314_{five} to decimal form.

$$244314_{\text{five}} = ((((2 \times 5 + 4) \times 5 + 4) \times 5 + 3) \times 5 + 1) \times 5 + 4$$
$$= 9334$$

EXAMPLE 3 Convert 497 from decimal form to base five.

The base five place values, starting from the right, are 1, 5, 25, 125, 625, and so on. Since 497 is between 125 and 625, it will require no 625s, but some 125s, as well as possibly some 25s, 5s, and 1s. Dividing 497 by 125 determines the proper number of 125s. The quotient is 3, with remainder 122. So we need three 125s. Next, the remainder, 122, is divided by 25 (the next place value) to find the proper number of 25s. The quotient is 4, with remainder 22, so we need four 25s. Dividing 22 by 5 yields 4, with remainder 2. So we need four 5s. Dividing 2 by 1 yields 2 (with remainder 0), so we need two 1s. Finally, we see that 497 consists of three 125s, four 25s, four 5s, and two 1s, so $497 = 3442_{\text{five}}$.

More concisely, this process can be written as follows.

$497 \div 125 = 3$	Remainder 122
$122 \div 25 = 4$	Remainder 22
$22 \div 5 = 4$	Remainder 2
$2 \div 1 = 2$	Remainder 0
$497 = 3442_{\text{five}}$	

Yin-yang The binary (base two) symbols of the *I Ching*, a 2000-year-old Chinese classic, permute into 8 elemental trigrams; 64 hexagrams are interpreted in casting oracles.

The basic symbol here is the ancient Chinese "yin-yang," in which the black and the white enfold each other, each containing a part of the other. A kind of duality is conveyed between destructive (yin) and beneficial (yang) aspects. Leibniz (1646–1716) studied Chinese ideograms in search of a universal symbolic language and promoted East-West cultural contact. He saw parallels between the trigrams and his binary arithmetic.

Niels Bohr (1885–1962), famous Danish Nobel laureate in physics (atomic theory), adopted the yin-yang symbol in his coat of arms to depict his principle of *complementarity*, which he believed was fundamental to reality at the deepest levels. Bohr also pushed for East-West cooperation.

In its 1992 edition, *The World Book Dictionary* first judged "yin-yang" to have been used enough to become a permanent part of our ever changing language, assigning it the definition, "made up of opposites."

Check: $3442_{\text{five}} = (3 \times 125) + (4 \times 25) + (4 \times 5) + (2 \times 1)$
$= 375 + 100 + 20 + 2$
$= 497.$

The calculator shortcut for converting from another base to decimal form involved repeated *multiplications* by the other base. (See Example 2.) A shortcut for converting from decimal form to another base makes use of repeated *divisions* by the other base. Just divide the original decimal numeral, and the resulting quotients in turn, by the desired base until the quotient 0 appears.

EXAMPLE 4 Repeat Example 3 using the shortcut just described.

```
                Remainder
    5 | 497
    5 |  99  ← 2
    5 |  19  ← 4
    5 |   3  ← 4
        0   ← 3
```

Read the answer from the remainder column, reading from the bottom up:

$$497 = 3442_{\text{five}}.$$

To see why this shortcut works, notice the following:

The first division shows that four hundred ninety-seven 1s are equivalent to ninety-nine 5s and two 1s. (The two 1s are set aside and account for the last digit of the answer.)

The second division shows that ninety-nine 5s are equivalent to nineteen 25s and four 5s. (The four 5s account for the next digit of the answer.)

The third division shows that nineteen 25s are equivalent to three 125s and four 25s. (The four 25s account for the next digit of the answer.)

The fourth (and final) division shows that the three 125s are equivalent to no 625s and three 125s. The remainders, as they are obtained *from top to bottom*, give the number of 1s, then 5s, then 25s, then 125s.

The methods for converting between bases ten and five, including the shortcuts, can be adapted for conversions between base ten and any other base.

EXAMPLE 5 Convert 6343_{seven} to decimal form, by expanding in powers, and by using the calculator shortcut.

$6343_{\text{seven}} = (6 \times 7^3) + (3 \times 7^2) + (4 \times 7^1) + (3 \times 7^0)$
$= (6 \times 343) + (3 \times 49) + (4 \times 7) + (3 \times 1)$
$= 2058 + 147 + 28 + 3$
$= 2236$

Calculator shortcut: $6343_{\text{seven}} = ((6 \times 7 + 3) \times 7 + 4) \times 7 + 3 = 2236.$

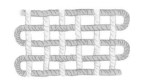

Woven fabric is a binary system of threads going lengthwise (warp threads—tan in the diagram above) and threads going crosswise (weft or woof). At any point in a fabric, either warp or weft is on top, and the variation creates the pattern.

Nineteenth-century looms for weaving operated using punched cards, "programmed" for pattern. The looms were set up with hooked needles, the hooks holding the warp. Where there were holes in cards, the needles moved, the warp lifted, and the weft passed under. Where no holes were, the warp did not lift, and the weft was on top. The system parallels the on-off system in calculators and computers. In fact, these looms were models in the development of modern calculating machinery.

Joseph Marie Jacquard (1752–1823) is credited with improving the mechanical loom so that mass production of fabric was feasible.

EXAMPLE 6 Convert 7508 to base seven.

Divide 7508 by 7, then divide the resulting quotient by 7, and so on, until a quotient of 0 results.

$$
\begin{array}{r|l}
 & \text{Remainder} \\
7\,\underline{|\,7508} & \\
7\,\underline{|\,1072} & \leftarrow 4 \\
7\,\underline{|\,153} & \leftarrow 1 \\
7\,\underline{|\,21} & \leftarrow 6 \\
7\,\underline{|\,3} & \leftarrow 0 \\
0 & \leftarrow 3
\end{array}
$$

From the remainders, reading bottom to top, $7508 = 30614_{\text{seven}}$.

Because we are accustomed to doing arithmetic in base ten, most of us would handle conversions between arbitrary bases (where neither is ten) by going from the given base to base ten and then to the desired base, as illustrated in the next example.

EXAMPLE 7 Convert 3164_{seven} to base five.

First convert to decimal form.

$$
\begin{aligned}
3164_{\text{seven}} &= (3 \times 7^3) + (1 \times 7^2) + (6 \times 7^1) + (4 \times 7^0) \\
&= (3 \times 343) + (1 \times 49) + (6 \times 7) + (4 \times 1) \\
&= 1029 + 49 + 42 + 4 \\
&= 1124
\end{aligned}
$$

Next convert this decimal result to base five.

$$
\begin{array}{r|l}
 & \text{Remainder} \\
5\,\underline{|\,1124} & \\
5\,\underline{|\,224} & \leftarrow 4 \\
5\,\underline{|\,44} & \leftarrow 4 \\
5\,\underline{|\,8} & \leftarrow 4 \\
5\,\underline{|\,1} & \leftarrow 3 \\
0 & \leftarrow 1
\end{array}
$$

From the remainders, $3164_{\text{seven}} = 13444_{\text{five}}$.

TABLE 5

Base Ten (decimal)	Base Two (binary)
0	0
1	1
2	10
3	11
4	100
5	101
6	110
7	111
8	1000
9	1001
10	1010
11	1011
12	1100
13	1101
14	1110
15	1111
16	10000
17	10001
18	10010
19	10011
20	10100

Computer Mathematics There are three alternative base systems that are most useful in computer applications. These are the **binary** (base two), **octal** (base eight), and **hexadecimal** (base sixteen) systems. Computers and handheld calculators actually use the binary system for their internal calculations since that system consists of only two symbols, 0 and 1. All numbers can then be represented by electronic "switches," of one kind or another, where "on" indicates 1 and "off" indicates 0. The octal system is used extensively by programmers who work with internal computer codes. In a computer, the CPU (central processing unit) often uses the hexadecimal system to communicate with a printer or other output device.

The binary system is extreme in that it has only two available symbols (0 and 1); because of this, representing numbers in binary form requires more digits than in any other base. Table 5 shows the whole numbers up to 20 expressed in binary form.

4.3 Conversion Between Number Bases

Conversions between any of these three special base systems (binary, octal, and hexadecimal) and the decimal system can be done by the methods already discussed, including the shortcut methods.

EXAMPLE 8 Convert 110101_{two} to decimal form, by expanding in powers, and by using the calculator shortcut.

$$110101_{two} = (1 \times 2^5) + (1 \times 2^4) + (0 \times 2^3) + (1 \times 2^2)$$
$$+ (0 \times 2^1) + (1 \times 2^0)$$
$$= (1 \times 32) + (1 \times 16) + (0 \times 8) + (1 \times 4) + (0 \times 2)$$
$$+ (1 \times 1)$$
$$= 32 + 16 + 0 + 4 + 0 + 1$$
$$= 53$$

Calculator shortcut:

$$110101_{two} = ((((1 \times 2 + 1) \times 2 + 0) \times 2 + 1) \times 2 + 0) \times 2 + 1$$
$$= 53.$$

Trick or Tree? The octal number 31 is equal to the decimal number 25. This may be written as

31 OCT = 25 DEC

Does this mean that Halloween and Christmas fall on the same day of the year?

EXAMPLE 9 Convert 9583 to octal form.
Divide repeatedly by 8, writing the remainders at the side.

```
              Remainder
     8 | 9583
     8 | 1197    ←—— 7
     8 |  149    ←—— 5
     8 |   18    ←—— 5
     8 |    2    ←—— 2
            0    ←—— 2
```

From the remainders, $9583 = 22557_{eight}$.

The hexadecimal system, having base 16, which is greater than 10, presents a new problem. Since distinct symbols are needed for all whole numbers from 0 up to one less than the base, base sixteen requires more symbols than are normally used in our decimal system. Computer programmers commonly use the letters A, B, C, D, E, and F as hexadecimal digits for the numbers ten through fifteen, respectively.

Converting Calculators A number of scientific calculators are available that will convert between decimal, binary, octal, and hexadecimal, and will also do calculations directly in all of these separate modes.

EXAMPLE 10 Convert $FA5_{sixteen}$ to decimal form.
Since the hexadecimal digits F and A represent 15 and 10, respectively,

$$FA5_{sixteen} = (15 \times 16^2) + (10 \times 16^1) + (5 \times 16^0)$$
$$= 3840 + 160 + 5$$
$$= 4005.$$

EXAMPLE 11 Convert 748 from decimal form to hexadecimal form. Use repeated division by 16.

```
                          Remainder   Hexadecimal
                                       notation
        16 | 748
        16 |  46   ←──── 12   ←──── C
        16 |   2   ←──── 14   ←──── E
               0   ←────  2   ←──── 2
```

From the remainders at the right, $748 = 2EC_{\text{sixteen}}$.

The decimal whole numbers 0 through 17 are shown in Table 6 along with their equivalents in the common computer-oriented bases (two, eight, and sixteen). Conversions among binary, octal, and hexadecimal systems can generally be accomplished by the shortcuts explained below, and are illustrated in the next several examples.

TABLE 6 Some Decimal Equivalents in the Common Computer-Oriented Bases

Decimal (Base Ten)	Hexadecimal (Base Sixteen)	Octal (Base Eight)	Binary (Base Two)
0	0	0	0
1	1	1	1
2	2	2	10
3	3	3	11
4	4	4	100
5	5	5	101
6	6	6	110
7	7	7	111
8	8	10	1000
9	9	11	1001
10	A	12	1010
11	B	13	1011
12	C	14	1100
13	D	15	1101
14	E	16	1110
15	F	17	1111
16	10	20	10000
17	11	21	10001

The binary system is the natural one for internal computer workings because of its compatibility with the two-state electronic switches. It is very cumbersome, however, for human use, because so many digits occur even in the numerals for relatively small numbers. The octal and hexadecimal systems are the choices of computer programmers mainly because of their close relationship with the binary system. *Both eight and sixteen are powers of two.* When conversions involve one

4.3 Conversion Between Number Bases

TABLE 7

Octal	Binary
0	000
1	001
2	010
3	011
4	100
5	101
6	110
7	111

base that is a power of the other, there is a quick conversion shortcut available. For example, since $8 = 2^3$, every octal digit (0 through 7) can be expressed as a 3-digit binary numeral. See Table 7.

EXAMPLE 12 Convert 473_{eight} to binary form.

Replace each octal digit with its 3-digit binary equivalent. (Leading zeros can be omitted only when they occur in the leftmost group.) Then combine all the binary equivalents into a single binary numeral.

$$\begin{array}{ccc} 4 & 7 & 3_{eight} \\ \downarrow & \downarrow & \downarrow \\ 100 & 111 & 011_{two} \end{array}$$

By this method, $473_{eight} = 100111011_{two}$.

Convert from binary form to octal form in a similar way. Start at the right and break the binary numeral into groups of three digits. (Leading zeros in the leftmost group may be omitted.)

TABLE 8

Hexadecimal	Binary
0	0000
1	0001
2	0010
3	0011
4	0100
5	0101
6	0110
7	0111
8	1000
9	1001
A	1010
B	1011
C	1100
D	1101
E	1110
F	1111

EXAMPLE 13 Convert 10011110_{two} to octal form.

Starting at the right, break the digits into groups of three. Then convert the groups to their octal equivalents.

$$\begin{array}{ccc} 10 & 011 & 110_{two} \\ \downarrow & \downarrow & \downarrow \\ 2 & 3 & 6_{eight} \end{array}$$

Finally, $10011110_{two} = 236_{eight}$.

Since $16 = 2^4$, every hexadecimal digit can be equated to a 4-digit binary numeral (see Table 8), and conversions between binary and hexadecimal forms can be done in a manner similar to that used in Examples 12 and 13.

EXAMPLE 14 Convert $8B4F_{sixteen}$ to binary form.

Each hexadecimal digit yields a 4-digit binary equivalent.

$$\begin{array}{cccc} 8 & B & 4 & F_{sixteen} \\ \downarrow & \downarrow & \downarrow & \downarrow \\ 1000 & 1011 & 0100 & 1111_{two} \end{array}$$

Combining these groups of digits, we see that

$$8B4F_{sixteen} = 1000101101001111_{two}.$$

Several games and tricks are based on the binary system. For example, Table 9 can be used to find the age of a person 31 years old or younger. The person need only tell you the columns that contain his or her age. For example, suppose Kellen Dawson says that her age appears in columns B and D only. To find her age, add the numbers from the top row of these columns:

Kellen is $2 + 8 = 10$ years old.

Do you see how this trick works? (See Exercises 68–71.)

Several years ago, the Kellogg Company featured a Magic Trick Age Detector activity on specially marked packages of *Kellogg's ® Rice Krispies ®* cereal. The trick is simply an extension of the discussion in the text.

Kellogg's ® Rice Krispies ® and characters *Snap! ® Crackle! ®* and *Pop! ®* are registered trademarks of Kellogg Company.

TABLE 9

A	B	C	D	E
1	2	4	8	16
3	3	5	9	17
5	6	6	10	18
7	7	7	11	19
9	10	12	12	20
11	11	13	13	21
13	14	14	14	22
15	15	15	15	23
17	18	20	24	24
19	19	21	25	25
21	22	22	26	26
23	23	23	27	27
25	26	28	28	28
27	27	29	29	29
29	30	30	30	30
31	31	31	31	31

4.3 EXERCISES

List the first twenty counting numbers in each of the following bases.

1. seven (Only digits 0 through 6 are used in base seven.)
2. eight (Only digits 0 through 7 are used.)
3. nine (Only digits 0 through 8 are used.)
4. sixteen (The digits 0, 1, 2, . . . , 9, A, B, C, D, E, F are used in base sixteen.)

For each of the following, write (in the same base) the counting numbers just before and just after the given number. (Do not convert to base ten.)

5. 14_{five}
6. 555_{six}
7. $\text{B6F}_{\text{sixteen}}$
8. 10111_{two}

Determine the number of distinct symbols needed in each of the following positional systems.

9. base three
10. base seven
11. base eleven
12. base sixteen

Determine, in each of the following bases, the smallest and largest four-digit numbers and their decimal equivalents.

13. three
14. sixteen

4.3 Conversion Between Number Bases

Convert each of the following to decimal form by expanding in powers and by using the calculator shortcut.

15. 24_{five}
16. 62_{seven}
17. 1011_{two}
18. 35_{eight}
19. $3BC_{\text{sixteen}}$
20. 34432_{five}
21. 2366_{seven}
22. 101101110_{two}
23. 70266_{eight}
24. $ABCD_{\text{sixteen}}$
25. 2023_{four}
26. 6185_{nine}
27. 41533_{six}
28. 88703_{nine}

Convert each of the following from decimal form to the given base.

29. 86 to base five
30. 65 to base seven
31. 19 to base two
32. 935 to base eight
33. 147 to base sixteen
34. 2730 to base sixteen
35. 36401 to base five
36. 70893 to base seven
37. 586 to base two
38. 12888 to base eight
39. 8407 to base three
40. 11028 to base four
41. 9346 to base six
42. 99999 to base nine

Make the following conversions as indicated.

43. 43_{five} to base seven
44. 27_{eight} to base five
45. 6748_{nine} to base four
46. $C02_{\text{sixteen}}$ to base seven

Convert each of the following from octal form to binary form.

47. 367_{eight}
48. 2406_{eight}

Convert each of the following from binary form to octal form.

49. 100110111_{two}
50. 11010111101_{two}

Make the following conversions as indicated.

51. DC_{sixteen} to binary
52. $F111_{\text{sixteen}}$ to binary
53. 101101_{two} to hexadecimal
54. $101111011101000_{\text{two}}$ to hexadecimal

Identify the largest number from each list in Exercises 55 and 56.

55. 42_{seven}, 37_{eight}, $1D_{\text{sixteen}}$
56. 1101110_{two}, 407_{five}, $6F_{\text{sixteen}}$

*There is a theory that twelve would be a better base than ten for general use. This is mainly because twelve has more divisors (1, 2, 3, 4, 6, 12) than ten (1, 2, 5, 10), which makes fractions easier in base twelve. The base twelve system is called the **duodecimal system**. Just as in the decimal system we speak of a one, a ten, and a hundred (and so on), in the duodecimal system we say a one, a dozen (twelve), and a gross (twelve squared, or one hundred forty-four).*

57. Otis Taylor's clients ordered 9 gross, 10 dozen, and 11 copies of *The Minnie Minoso Story* during 2002. How many copies was that in base ten?

58. Which amount is larger: 3 gross, 6 dozen or 2 gross, 19 dozen?

One very common method of converting symbols into binary digits for computer processing is called ASCII (American Standard Code of Information Interchange). The uppercase letters A through Z are assigned the numbers 65 through 90, so A has binary code 1000001 and Z has code 1011010. Lowercase letters a through z have codes 97 through 122 (that is, 1100001 through 1111010). ASCII codes, as well as other numerical computer output, normally appear without commas.

Write the binary code for each of the following letters.

59. C **60.** X **61.** k

Break each of the following into groups of seven digits and write as letters.

62. 10010001000101100110001010000 **63.** 1000011100100010101011000011001011

Translate each word into an ASCII string of binary digits. (Be sure to distinguish upper and lowercase letters.)

64. New **65.** Orleans

66. Explain why the octal and hexadecimal systems are convenient for people who code for computers.

67. There are thirty-seven counting numbers whose base eight numerals contain two digits but whose base three numerals contain four digits. Find the smallest and largest of these numbers.

Refer to Table 9 for Exercises 68–71.

68. After observing the binary forms of the numbers 1–31, identify a common property of all Table 9 numbers in each of the following columns.

   (a) Column A **(b)** Column B **(c)** Column C **(d)** Column D **(e)** Column E

69. Explain how the "trick" of Table 9 works.

70. How many columns would be needed for Table 9 to include all ages up to 63?

71. How many columns would be needed for Table 9 to include all numbers up to 127?

In our decimal system, we distinguish odd and even numbers by looking at their ones (or units) digits. If the ones digit is even (0, 2, 4, 6, or 8), the number is even. If the ones digit is odd (1, 3, 5, 7, or 9), the number is odd. For Exercises 72–79, determine whether this same criterion works for numbers expressed in the given bases.

72. two **73.** three **74.** four **75.** five
76. six **77.** seven **78.** eight **79.** nine

80. Consider all even bases. If the criterion works for all, explain why. If not, find one that does work for all even bases.

81. Consider all odd bases. If the criterion works for all, explain why. If not, find one that does work for all odd bases.

Determine whether the given base five numeral represents one that is divisible by five.

82. 3204_{five} **83.** 200_{five} **84.** 2310_{five} **85.** 342_{five}

4.4 Finite Mathematical Systems

We continue our study of mathematical systems by considering examples built upon *finite* sets. Some examples here will consist of numbers, but others will be made up of elements denoted by letters. And the operations are often represented by abstract symbols with no particular mathematical meaning. This is to make the point that a system is characterized by how its elements behave under its operations, not by the choice of symbols used.

To begin, let us introduce a finite mathematical system made up of the set of elements $\{a, b, c, d\}$, and an operation we shall write with the symbol $\star$. We give meaning to operation $\star$ by displaying an **operation table,** which shows how operation $\star$ combines any two elements from the set $\{a, b, c, d\}$. The operation table for $\star$ is shown in Table 10. To use the table to find, say, $c \star d$, first locate c on the left, and d across the top. This row and column give b, so that

$$c \star d = b.$$

TABLE 10

$\star$	a	b	c	d
a	a	b	c	d
b	b	d	a	c
c	c	a	d	b
d	d	c	b	a

The important properties to look for in a system are the following: **closure, commutative, associative, identity,** and **inverse.** Let us decide which properties are satisfied by the system made up of $\{a, b, c, d\}$ and operation $\star$.

Closure Property For this system to be closed under the operation $\star$, the answer to any possible combination of elements from the system must be in the set $\{a, b, c, d\}$. A glance at Table 10 shows that the answers in the body of the table are all elements of this set. This means that the system is closed. If an element other than a, b, c, or d had appeared in the body of the table, the system would not have been closed.

Commutative Property In order for the system to have the commutative property, it must be true that $\Gamma \star \Delta = \Delta \star \Gamma$, where Γ and Δ stand for any elements from the set $\{a, b, c, d\}$. For example,

$$c \star d = b \quad \text{and} \quad d \star c = b, \quad \text{so} \quad c \star d = d \star c.$$

TABLE 11

$\star$	a	b	c	d
a	a	b	c	d
b	b	d	a	c
c	c	a	d	b
d	d	c	b	a

To see that the same is true for *all* choices of Γ and Δ, observe that Table 11 is symmetric with respect to the diagonal line shown. This "diagonal line test" establishes that $\star$ is a commutative operation for this system.

Associative Property The system is associative if $(\Gamma \star \Delta) \star \Upsilon = \Gamma \star (\Delta \star \Upsilon)$, where Γ, Δ, and Υ represent any elements from the set $\{a, b, c, d\}$. There is no quick way to check a table for the associative property, as there is for the commutative property. All we can do is try some examples. Using the table that defines operation $\star$,

$$(a \star d) \star b = d \star b = c, \quad \text{and} \quad a \star (d \star b) = a \star c = c,$$

so that
$$(a \star d) \star b = a \star (d \star b).$$

In the same way,
$$b \star (c \star d) = (b \star c) \star d.$$

In both these examples, changing the location of parentheses did not change the answers. Since the two examples worked, we suspect that the system is associative. We cannot be sure of this, however, unless every possible choice of three letters from the set is checked. (Although we have not completely verified it here, this system does, in fact, satisfy the associative property.)

Bernard Bolzano (1781–1848) was an early exponent of rigor and precision in mathematics. Many early results in such areas as calculus were produced by the masters in the field; these masters knew what they were doing and produced accurate results. However, their sloppy arguments caused trouble in the hands of the less gifted. The work of Bolzano and others helped put mathematics on a strong footing.

Identity Property For the identity property to hold, there must be an element Δ from the set of the system such that $\Delta \star X = X$ and $X \star \Delta = X$, where X represents any element from the set $\{a, b, c, d\}$. We can see that a is such an element as follows. In Table 11, the column below a (at the top) is identical to the column at the left, and the row across from a (at the left) is identical to the row at the top. Therefore, a is in fact the identity element of the system. (It is shown in more advanced courses that if a system has an identity element, it has *only* one.)

Inverse Property We found above that a is the identity element for the system using operation $\star$. Is there any inverse in this system for, say, the element b? If Δ represents the inverse of b in this system, then

$$b \star \Delta = a \quad \text{and} \quad \Delta \star b = a \quad \text{(since } a \text{ is the identity element)}.$$

Inspecting the table for operation $\star$ shows that Δ should be replaced with c:

$$b \star c = a \quad \text{and} \quad c \star b = a.$$

We can inspect the table to see if every element of our system has an inverse in the system. We see (in Table 11) that the identity element a appears exactly once in each row, and that, in each case, the pair of elements that produces a also produces it in the opposite order. Therefore, we conclude that the system satisfies the inverse property.

In summary, the mathematical system made up of the set $\{a, b, c, d\}$ and operation $\star$ satisfies the closure, commutative, associative, identity, and inverse properties.

The basic properties that may (or may not) be satisfied by a mathematical system involving a single operation follow.

Potential Properties of a Single-Operation System

Here a, b, and c represent elements from the set of the system, and $\circ$ represents the operation of the system.

Closure The system is closed if for all elements a and b,

$$a \circ b$$

is in the set of the system.

Commutative The system has the commutative property if

$$a \circ b = b \circ a$$

for all elements a and b from the system.

Associative The system has the associative property if

$$(a \circ b) \circ c = a \circ (b \circ c)$$

for every choice of three elements a, b, and c of the system.

Identity The system has an identity element e (where e is in the set of the system) if

$$a \circ e = a \quad \text{and} \quad e \circ a = a$$

for every element a in the system.

Inverse The system satisfies the inverse property if, for every element a of the system, there is an element x in the system such that

$$a \circ x = e \quad \text{and} \quad x \circ a = e,$$

where e is the identity element of the system.

4.4 Finite Mathematical Systems

⊗	0	1	2	3	4	5
0	0	0	0	0	0	0
1	0	1	2	3	4	5
2	0	2	4	0	2	4
3	0	3	0	3	0	3
4	0	4	2	0	4	2
5	0	5	4	3	2	1

EXAMPLE 1 The table in the margin is a table for the set $\{0, 1, 2, 3, 4, 5\}$ under an operation designated $\otimes$. Which of the properties above are satisfied by this system?

All the numbers in the body of the table come from the set $\{0, 1, 2, 3, 4, 5\}$, so the system is closed. If we draw a line from upper left to lower right, we could fold the table along this line and have the corresponding elements match; the system has the commutative property.

To check for the associative property, try some examples:

$$2 \otimes (3 \otimes 5) = 2 \otimes 3 = 0 \quad \text{and} \quad (2 \otimes 3) \otimes 5 = 0 \otimes 5 = 0,$$

so that
$$2 \otimes (3 \otimes 5) = (2 \otimes 3) \otimes 5.$$

Also,
$$5 \otimes (4 \otimes 2) = (5 \otimes 4) \otimes 2.$$

Any other examples that we might try would also work. The system has the associative property.

Since the column at the left of the multiplication table is repeated under 1 in the body of the table, 1 is a candidate for the identity element in the system. To be sure that 1 is indeed the identity element here, check that the row corresponding to 1 at the left is identical with the row at the top of the table.

To find inverse elements, look for the identity element, 1, in the rows of the table. The identity element appears in the second row, $1 \otimes 1 = 1$; and in the bottom row, $5 \otimes 5 = 1$; so 1 and 5 both are their own inverses. There is no identity element in the rows opposite the numbers 0, 2, 3, and 4, so none of these elements has an inverse.

In summary, the system made up of the set $\{0, 1, 2, 3, 4, 5\}$ under this operation $\otimes$ satisfies the closure, associative, commutative, and identity properties, but not the inverse property. ∎

⊠	1	2	3	4	5	6
1	1	2	3	4	5	6
2	2	4	6	1	3	5
3	3	6	2	5	1	4
4	4	1	5	2	6	3
5	5	3	1	6	4	2
6	6	5	4	3	2	1

EXAMPLE 2 The table in the margin is a table for the set of numbers $\{1, 2, 3, 4, 5, 6\}$ under an operation designated $\boxtimes$. Which of the properties are satisfied by this system?

Notice here that 0 is not an element of this system. This is perfectly legitimate. Since we are defining the system, we can include (or exclude) whatever we wish. Check that the system satisfies the closure, commutative, associative, and identity properties, with identity element 1. Let us now check for inverses. The element 1 is its own inverse, since $1 \boxtimes 1 = 1$. In row 2, the identity element 1 appears under the number 4, so $2 \boxtimes 4 = 1$ (and $4 \boxtimes 2 = 1$), with 2 and 4 inverses of each other. Also, 3 and 5 are inverses of each other, and 6 is its own inverse. Since each number in the set of the system has an inverse, the system satisfies the inverse property. ∎

When a mathematical system has two operations, rather than just one, we can look for an additional, very important property, namely the **distributive property**. For example, when we studied Hindu-Arabic arithmetic of counting numbers, we saw that multiplication is distributive with respect to (or "over") addition.

$$3 \times (5 + 9) = 3 \times 5 + 3 \times 9$$

and
$$(5 + 9) \times 3 = 5 \times 3 + 9 \times 3.$$

In each case, the factor 3 is "distributed" over the 5 and the 9.

Although the distributive property (as well as the other properties discussed here) will be applied to the real numbers in general in a later section, we state it here just for the system of integers.

> **Distributive Property of Multiplication with Respect to Addition**
>
> For integers a, b, and c, the **distributive property** holds for multiplication with respect to addition, so
>
> $$a \times (b + c) = a \times b + a \times c$$
>
> and
>
> $$(b + c) \times a = b \times a + c \times a.$$

EXAMPLE 3 Is addition distributive over multiplication?

To find out, exchange $\times$ and $+$ in the first equation above:

$$a + (b \times c) = (a + b) \times (a + c).$$

We need to find out whether this statement is true for *every* choice of three numbers that we might make. Try an example. If $a = 3$, $b = 4$, and $c = 5$,

$$a + (b \times c) = 3 + (4 \times 5) = 3 + 20 = 23,$$

while

$$(a + b) \times (a + c) = (3 + 4) \times (3 + 5) = 7 \times 8 = 56.$$

Since $23 \neq 56$, we have $3 + (4 \times 5) \neq (3 + 4) \times (3 + 5)$. This false result is a *counterexample* (an example showing that a general statement is false). This counterexample shows that addition is *not* distributive over multiplication. ■

Because subtraction of real numbers is defined in terms of addition, the distributive property of multiplication also holds with respect to subtraction.

> **Distributive Property of Multiplication with Respect to Subtraction**
>
> For integers a, b, and c, the **distributive property** holds for multiplication with respect to subtraction, so
>
> $$a \times (b - c) = a \times b - a \times c$$
>
> and
>
> $$(b - c) \times a = b \times a - c \times a.$$

The **general form of the distributive property** appears below.

> **General Form of the Distributive Property**
>
> Let ☆ and ○ be two operations defined for elements in the same set. Then ☆ is distributive over ○ if
>
> $$a \, ☆ \, (b \circ c) = (a \, ☆ \, b) \circ (a \, ☆ \, c)$$
>
> for every choice of elements a, b, and c from the set.

The final example illustrates how the distributive property may hold for a finite system.

EXAMPLE 4 Suppose that the set $\{a,b,c,d,e\}$ has two operations ☆ and ∘ defined by the tables below.

☆	a	b	c	d	e
a	a	a	a	a	a
b	a	b	c	d	e
c	a	c	e	b	d
d	a	d	b	e	c
e	a	e	d	c	b

∘	a	b	c	d	e
a	a	b	c	d	e
b	b	c	d	e	a
c	c	d	e	a	b
d	d	e	a	b	c
e	e	a	b	c	d

The distributive property of ☆ with respect to ∘ holds in this system. Verify for the following case: $e ☆ (d ∘ b) = (e ☆ d) ∘ (e ☆ b)$.

First evaluate the left side of the equation by using the tables.

$$e ☆ (d ∘ b) = e ☆ e \quad \text{Use the ∘ table.}$$
$$= b \quad \text{Use the ☆ table.}$$

Now, evaluate the right side of the equation.

$$(e ☆ d) ∘ (e ☆ b) = c ∘ e \quad \text{Use the ☆ table twice.}$$
$$= b \quad \text{Use the ∘ table.}$$

Both times the final result is b, and the distributive property is verified for this case.

4.4 EXERCISES

For each system in Exercises 1–10, decide which of the properties of single-operation systems are satisfied. If the identity property is satisfied, give the identity element. If the inverse property is satisfied, give the inverse of each element. If the identity property is satisfied but the inverse property is not, name the elements that have no inverses.

1. $\{1, 2\}$; operation ⊗

⊗	1	2
1	1	2
2	2	1

2. $\{1, 2, 3, 4\}$; operation ⊗

⊗	1	2	3	4
1	1	2	3	4
2	2	4	1	3
3	3	1	4	2
4	4	3	2	1

3. $\{1, 2, 3, 4, 5, 6, 7\}$; operation ⊠

⊠	1	2	3	4	5	6	7
1	1	2	3	4	5	6	7
2	2	4	6	0	2	4	6
3	3	6	1	4	7	2	5
4	4	0	4	0	4	0	4
5	5	2	7	4	1	6	3
6	6	4	2	0	6	4	2
7	7	6	5	4	3	2	1

4. $\{1, 2, 3, 4, 5\}$; operation ⊠

⊠	1	2	3	4	5
1	1	2	3	4	5
2	2	4	0	2	4
3	3	0	3	0	3
4	4	2	0	4	2
5	5	4	3	2	1

184 CHAPTER 4 Numeration and Mathematical Systems

5. $\{1, 3, 5, 7, 9\}$; operation ☆

☆	1	3	5	7	9
1	1	3	5	7	9
3	3	9	5	1	7
5	5	5	5	5	5
7	7	1	5	9	3
9	9	7	5	3	1

6. $\{1, 3, 5, 7\}$; operation ☆

☆	1	3	5	7
1	1	3	5	7
3	3	1	7	5
5	5	7	1	3
7	7	5	3	1

7. $\{A, B, F\}$; operation *

*	A	B	F
A	B	F	A
B	F	A	B
F	A	B	F

8. $\{m, n, p\}$; operation J

J	m	n	p
m	n	p	n
n	p	m	n
p	n	n	m

9. $\{r, s, t, u\}$; operation Z

Z	r	s	t	u
r	u	t	r	s
s	t	u	s	r
t	r	s	t	u
u	s	r	u	t

10. $\{A, J, T, U\}$; operation #

#	A	J	T	U
A	A	J	T	U
J	J	T	U	A
T	T	U	A	J
U	U	A	J	T

The tables in the finite mathematical systems that we developed in this section can be obtained in a variety of ways. For example, let us begin with a square, as shown in the figure. Let the symbols a, b, c, and d be defined as shown in the figure.

Let *a* represent zero rotation—leave the original square as is.

Let *b* represent rotation of 90° clockwise from original position.

Let *c* represent rotation of 180° clockwise from original position.

Let *d* represent rotation of 270° clockwise from original position.

Define an operation □ for these letters as follows. To evaluate $b \square c$, for example, first perform b by rotating the square 90°. (See the figure.) Then perform operation c by rotating the square an additional 180°. The net result is the same as if we had performed d only. Thus,

$$b \square c = d.$$

Start with *a*. Perform *b*. Start with *b*, and perform *c*.

Use this method to find each of the following.

11. $b \square d$ **12.** $b \square b$ **13.** $d \square b$ **14.** $a \square b$

Solve each problem.

15. Complete the table at the right.

□	a	b	c	d
a	a	b	c	d
b	b	c		a
c	c		a	
d	d	a		

16. Which of the properties from this section are satisfied by this system?

17. Define a universal set U as the set of counting numbers. Form a new set that contains all possible subsets of U. This new set of subsets together with the operation of set intersection forms a mathematical system. Which of the properties listed in this section are satisfied by this system?

18. Replace the word "intersection" with the word "union" in Exercise 17; then answer the same question.

19. Complete the table at the right so that the result is *not* the same as operation □ of Exercise 15, but so that the five properties listed in this section still hold.

	a	b	c	d
a				
b				
c				
d				

Try examples to help you decide whether the following operations, when applied to the integers, satisfy the distributive property.

20. subtraction with respect to multiplication

21. addition with respect to subtraction

22. subtraction with respect to addition

Recall that Example 3 provided a counterexample for the general statement

$$a + (b \times c) = (a + b) \times (a + c).$$

Thus, addition is not distributive with respect to multiplication. Now work Exercises 23–26.

23. Decide if the statement above is true for each of the following sets of values.
 (a) $a = 2, b = -5, c = 4$
 (b) $a = -7, b = 5, c = 3$
 (c) $a = -8, b = 14, c = -5$
 (d) $a = 1, b = 6, c = -6$

24. Find another set of a, b, and c values that make the statement true.

25. Under what general conditions will the statement above be true?

26. Explain why, regardless of the results in Exercises 23–25, addition is still *not* distributive with respect to multiplication.

27. Give the conditions under which each of the following equations would be true.
 (a) $a + (b - c) = (a + b) - (a + c)$
 (b) $a - (b + c) = (a - b) + (a - c)$

28. (a) Find values of a, b, and c such that
$$a - (b \times c) = (a - b) \times (a - c).$$
 (b) Does this mean that subtraction is distributive with respect to multiplication? Explain.

Verify for the mathematical system of Example 4 that the distributive property holds for the following cases.

29. $c \star (d \circ e) = (c \star d) \circ (c \star e)$

30. $a \star (a \circ b) = (a \star a) \circ (a \star b)$

31. $d \star (e \circ c) = (d \star e) \circ (d \star c)$

32. $b \star (b \circ b) = (b \star b) \circ (b \star b)$

These exercises are for students who have studied sets.

33. Use Venn diagrams to show that the distributive property for union with respect to intersection holds for sets A, B, and C. That is,
$$A \cup (B \cap C) = (A \cup B) \cap (A \cup C).$$

34. Use Venn diagrams to show that *another* distributive property holds for sets A, B, and C. It is the distributive property of intersection with respect to union.
$$A \cap (B \cup C) = (A \cap B) \cup (A \cap C)$$

These exercises are for students who have studied logic.

35. Use truth tables to show that the following distributive property holds:
$$p \vee (q \wedge r) \equiv (p \vee q) \wedge (p \vee r).$$

36. Use truth tables to show that *another* distributive property holds:
$$p \wedge (q \vee r) \equiv (p \wedge q) \vee (p \wedge r).$$

4.5 Groups

We have considered some mathematical systems, most of which have satisfied some or all of the closure, associative, commutative, identity, inverse, and distributive properties. Systems are commonly classified according to which properties they satisfy. One important category is the mathematical group, which we define here.

Group

A mathematical system is called a **group** if, under its operation, it satisfies the closure, associative, identity, and inverse properties.

Niels Henrik Abel (1802–1829) of Norway was identified in childhood as a mathematical genius but never received in his lifetime the professional recognition his work deserved.

At 16, influenced by a perceptive teacher, he read the works of Newton, Euler, and Lagrange. One of Abel's achievements was the demonstration that a general formula for solving fifth-degree equations does not exist. The quadratic formula (for equations of degree 2) is well known, and formulas do exist for solving third- and fourth-degree equations. Abel's accomplishment ended a search that had lasted for years.

In the study of abstract algebra, groups which have the commutative property are referred to as **abelian** groups in honor of Abel. He died of tuberculosis at age 27.

EXAMPLE 1 Does the set $\{-1, 1\}$ under the operation of multiplication form a group?

Check the necessary four properties.

Closure The given system leads to the multiplication table below. All entries in the body of the table are either -1 or 1; the system is closed.

×	−1	1
−1	1	−1
1	−1	1

Associative Both -1 and 1 are integers, and multiplication of integers is associative.

Identity The identity for multiplication is 1, an element of the set of the system $\{-1, 1\}$.

Inverse Both -1 and 1 are their own inverses for multiplication.

All four of the properties are satisfied, so the system is a group.

EXAMPLE 2 Does the set $\{-1, 1\}$ under the operation of addition form a group?

The addition table below shows that closure is not satisfied, so there is no need to check further. The system is not a group.

+	−1	1
−1	−2	0
1	0	2

The system of Example 1 is a finite group. Let's look for an infinite group.

Amalie ("Emmy") Noether (1882–1935) was an outstanding mathematician in the field of abstract algebra. She studied and worked in Germany at a time when it was very difficult for a woman to do so. At the University of Erlangen in 1900, Noether was one of only two women. Although she could attend classes, professors could and did deny her the right to take the exams for their courses. Not until 1904 was Noether allowed to officially register. She completed her doctorate four years later.

In 1916 Emmy Noether went to Göttingen to work with David Hilbert on the general theory of relativity. But even with Hilbert's backing and prestige, it was three years before the faculty voted to make Noether a *Privatdozent,* the lowest rank in the faculty. In 1922 Noether was made an unofficial professor (or assistant). She received no pay for this post, although she was given a small stipend to lecture in algebra.

Noether's area of interest was abstract algebra, particularly structures called rings and ideals. (Groups are structures, too, with different properties.) One special type of ring bears her name; she was the first to study its properties.

EXAMPLE 3 Does the set of integers $\{\ldots, -3, -2, -1, 0, 1, 2, 3, \ldots\}$ under the operation of addition form a group?

Check the required properties.

Closure The sum of any two integers is an integer; the system is closed.

Associative Try some examples:

$$2 + (5 + 8) = 2 + 13 = 15$$

and $$(2 + 5) + 8 = 7 + 8 = 15,$$

so $$2 + (5 + 8) = (2 + 5) + 8.$$

$$-4 + (7 + 14) = -4 + 21 = 17$$

and $$(-4 + 7) + 14 = 3 + 14 = 17,$$

so $$-4 + (7 + 14) = (-4 + 7) + 14.$$

Apparently, addition of integers is associative.

Identity We know that $a + 0 = a$ and $0 + a = a$ for any integer a. The identity element for addition of integers is 0.

Inverse Given any integer a, its additive inverse, $-a$, is also an integer. For example, 5 and -5 are inverses. The system satisfies the inverse property.

Since all four properties are satisfied, this (infinite) system *is* a group. ■

Group structure applies not only to sets of numbers. One common group is the group of **symmetries of a square,** which we now develop. First, cut out a small square, and label it as shown in Figure 5.

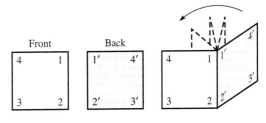

FIGURE 5

Make sure that 1 is in front of $1'$, 2 is in front of $2'$, 3 is in front of $3'$, and 4 is in front of $4'$. Let the letter M represent a clockwise rotation of 90° *about the center of the square* (marked with a dot in Figure 6). Let N represent a rotation of 180°, and so on. A list of the symmetries of a square is given in Figure 6 on the next page.

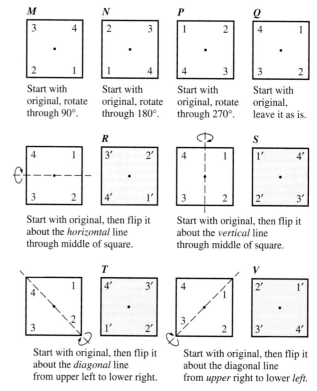

Symmetries of a square

FIGURE 6

Combine symmetries as follows: Let NP represent N followed by P. Performing N and then P is the same as performing just M, so that $NP = M$. See Figure 7.

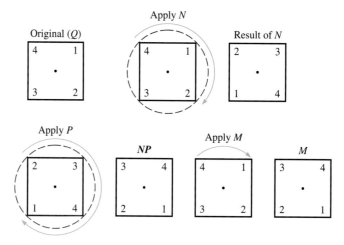

Think of N as advancing each corner two quarter turns clockwise. Thus 4 goes from upper left to lower right. To this result apply P, which advances each corner three quarter turns. Thus 2 goes from upper left to lower left. The result, NP, is the same as advancing each (original) corner one quarter turn, which M does. Thus $NP = M$.

FIGURE 7

4.5 Groups

Évariste Galois (1811–1832), as a young Frenchman, agreed to fight a duel. He had been engaged in profound mathematical research for some time. Now, anticipating the possibility of his death, he summarized the essentials of his discoveries in a letter to a friend. The next day Galois was killed. He was not yet 21 years old when he died.

It was not until 1846 that Galois's theories were published. Mathematicians began to appreciate the importance of Galois's work, which centered on solving equations by using groups. Galois found a way to derive a group which corresponds to each equation. So-called Galois groups form an important part of modern abstract algebra.

EXAMPLE 4 Find RT.

First, perform R by flipping the square about a horizontal line through the middle. Then, perform T by flipping the result of R about a diagonal from upper left to lower right. The result of RT is the same as performing only M, so that $RT = M$.

The method in Example 4 can be used to complete the following table for combining the symmetries of a square.

□	M	N	P	Q	R	S	T	V
M	N	P	Q	M	V	T	R	S
N	P	Q	M	N	S	R	V	T
P	Q	M	N	P	T	V	S	R
Q	M	N	P	Q	R	S	T	V
R	T	S	V	R	Q	N	M	P
S	V	R	T	S	N	Q	P	M
T	S	V	R	T	P	M	Q	N
V	R	T	S	V	M	P	N	Q

EXAMPLE 5 Show that the system made up of the symmetries of a square is a group.

For the system to be a group, it must satisfy the closure, associative, identity, and inverse properties.

Closure All the entries in the body of the table come from the set $\{M, N, P, Q, R, S, T, V\}$. Thus, the system is closed.

Associative Try examples:
$$P(MT) = P(R) = T.$$
Also, $\quad (PM)T = (Q)T = T,$
so that $\quad P(MT) = (PM)T.$

Other similar examples also work. (See Exercises 25–28.) Thus, the system has the associative property.

Identity The column at the left in the table is repeated under Q. Check that Q is indeed the identity element.

Inverse In the first row, Q appears under P. Check that M and P are inverses of each other. In fact, every element in the system has an inverse. (See Exercises 29–34.)

Since all four properties are satisfied, the system is a group.

□	M	N	P	Q
M	N	P	Q	M
N	P	Q	M	N
P	Q	M	N	P
Q	M	N	P	Q

EXAMPLE 6 Form a mathematical system by using only the set $\{M, N, P, Q\}$ from the group of symmetries of a square. Is this new system a group?

The table for the elements $\{M, N, P, Q\}$ (shown at the left) is just one corner of the table for the entire system. Verify that the system represented by this table satisfies all four properties and thus is a group. This new group is a *subgroup* of the original group of the symmetries of a square.

Monster Groups Much research in group theory now is devoted to "simple" groups (which are not at all simple). These groups are fundamental—other groups are built up from them. One of the largest of these simple groups has

$2^{46} \cdot 3^{20} \cdot 5^9 \cdot 7^6 \cdot 11^2 \cdot 13^3 \cdot 17 \cdot 19 \cdot 23 \cdot 29 \cdot 31 \cdot 41 \cdot 47 \cdot 59 \cdot 71$

different elements. This huge number of elements is why the group is called the "monster group." If this "monster group" contains too many elements, how about the "baby monster" simple group, with only

$2^{41} \cdot 3^{13} \cdot 5^6 \cdot 7^2 \cdot 11 \cdot 13 \cdot 17 \cdot 19 \cdot 23 \cdot 31 \cdot 47$

different elements? (For more information, see "The Classification of the Finite Simple Groups," by Michael Aschbacher in *The Mathematical Intelligencer*, Issue 2, 1981.)

Permutation Groups A very useful example of a group comes from studying the arrangements, or permutations, of a list of numbers. Start with the symbols 1-2-3, in that order.

There are several ways in which the order could be changed—for example, 2-3-1. This rearrangement is written:

$$1\text{-}2\text{-}3$$
$$2\text{-}3\text{-}1.$$

Replace 1 with 2, replace 2 with 3, and replace 3 with 1. In the same way,

$$1\text{-}2\text{-}3$$
$$3\text{-}1\text{-}2$$

means replace 1 with 3, 2 with 1, and 3 with 2, while

$$1\text{-}2\text{-}3$$
$$3\text{-}2\text{-}1$$

says to replace 1 with 3, leave the 2 unchanged, and replace 3 with 1. All possible rearrangements of the symbols 1-2-3 are listed below where, for convenience, a name has been given to each rearrangement.

A^*: 1-2-3	B^* 1-2-3	C^*: 1-2-3	D^*: 1-2-3	E^*: 1-2-3	F^*: 1-2-3
2-3-1	2-1-3	1-2-3	1-3-2	3-1-2	3-2-1

Two rearrangements can be combined as with the symmetries of a square; for example, the symbol B^*F^* means to first apply B^* to 1-2-3 and then apply F^* to the result. Rearrangement B^* changes 1-2-3 into 2-1-3. Then apply F^* to this result: 1 becomes 3, 2 is unchanged, and 3 becomes 1. In summary:

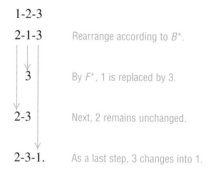

The net result of B^*F^* is to change 1-2-3 into 2-3-1, which is exactly what A^* does to 1-2-3. Therefore

$$B^*F^* = A^*.$$

4.5 Groups 191

EXAMPLE 7 Find $D*E*$.

Use the procedure described above.

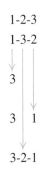

Apply $D*$. To apply $E*$, first replace 1 with 3. Next, replace 2 with 1. The last step is to replace 3 with 2. The result: $D*E*$ converts 1-2-3 into 3-2-1, as does $F*$, so

$$D*E* = F*.$$

As further examples, $A*B* = D*$ and $F*E* = B*$.

Once again, we see that we have encountered a mathematical system: the set $\{A*, B*, C*, D*, E*, F*\}$ and the operation of the combination of two rearrangements. To see whether this system is a group, check the requirements.

Closure   Combine any two rearrangements and the result is another rearrangement, so the system is closed.

Associative   Try an example:

First, $(B*D*)A* = E*A* = C*,$
while $B*(D*A*) = B*B* = C*,$
so that $(B*D*)A* = B*(D*A*).$

Since other examples will work out similarly, the system is associative.

Identity   The identity element is $C*$. If x is any rearrangement, then $xC* = C*x = x$.

Inverse   Does each rearrangement have an inverse rearrangement? Begin with the basic order 1-2-3 and then apply, say $B*$, resulting in 2-1-3. The inverse of $B*$ must convert this 2-1-3 back into 1-2-3, by changing 2 into 1 and 1 into 2. But $B*$ itself will do this. Hence $B*B* = C*$ and $B*$ is its own inverse. By the same process, $E*$ and $A*$ are inverses of each other. Also, each of $C*$, $D*$, and $F*$ is its own inverse.

Since all four requirements are satisfied, the system is a group. Rearrangements are also referred to as *permutations,* so this group is sometimes called the **permutation group on three symbols.** The total number of different permutations of a given number of symbols can be determined by methods described in the chapter on counting.

4.5 EXERCISES

What is wrong with the way in which each of the following questions is stated?

1. Do the integers form a group?
2. Does multiplication satisfy all of the group properties?

Decide whether each system is a group. If not a group, identify all properties that are not satisfied. (Recall that any system failing to satisfy the identity property automatically fails to satisfy the inverse property also.) For the finite systems, it may help to construct tables. For infinite systems, try some examples to help you decide.

3. $\{0\}$; multiplication
4. $\{0\}$; addition
5. $\{0, 1\}$; addition
6. $\{0\}$; subtraction
7. $\{-1, 1\}$; division
8. $\{0, 1\}$; multiplication
9. $\{-1, 0, 1\}$; multiplication
10. $\{-1, 0, 1\}$; addition
11. integers; subtraction
12. integers; multiplication
13. odd integers; multiplication
14. counting numbers; addition
15. rational numbers; addition
16. even integers; addition
17. prime numbers; addition
18. nonzero rational numbers; multiplication
19. Explain why a *finite* group based on the operation of ordinary addition of numbers cannot contain the element 1.
20. Explain why a group based on the operation of ordinary addition of numbers *must* contain the element 0.

Exercises 21–34 apply to the system of symmetries of a square presented in the text.
Find each of the following.

21. RN
22. PR
23. TV
24. VP

Verify each of the following statements.

25. $N(TR) = (NT)R$
26. $V(PS) = (VP)S$
27. $T(VN) = (TV)N$
28. $S(MR) = (SM)R$

Find the inverse of each element.

29. N
30. Q
31. R
32. S
33. T
34. V

A group that also satisfies the commutative property is called a commutative group *(or an* abelian group, *after Niels Henrik Abel).*

Determine whether each of the following groups is commutative.

35. the group of symmetries of a square
36. the subgroup of Example 6
37. the integers under addition
38. the permutation group on three symbols

Give illustrations to support your answers for Exercises 39–42.

39. Produce a mathematical system with two operations which is a group under one operation but not a group under the other operation.
40. Explain what property is gained when the system of counting numbers is extended to the system of whole numbers.
41. Explain what property is gained when the system of whole numbers is extended to the system of integers.
42. Explain what property is gained when the system of integers is extended to the system of rational numbers.

COLLABORATIVE INVESTIGATION

A Perpetual Calendar Algorithm

In this chapter we examined some alternative algorithms for arithmetic computations. Algorithms appear in various places in mathematics and computer science, and in each case give us a specified method of carrying out a procedure that produces a desired result. The algorithm that follows allows us to find the day of the week on which a particular date occurred or will occur.

In applying the algorithm, you will need to know whether a particular year is a leap year. In general, if a year is divisible (evenly) by 4, it is a leap year. However, there are exceptions. Century years, such as 1800 and 1900, are not leap years, despite the fact that they are divisible by 4. Furthermore, as an exception to the exception, a century year that is divisible by 400 (such as the year 2000) is a leap year.

In groups of three to five students, read the algorithm and then work the Topics for Discussion.

THE ALGORITHM

This algorithm requires several *key numbers*. Key numbers for the month, day, and century are determined by the following tables.

Month	Key
January	1 (0 if a leap year)
February	4 (3 if a leap year)
March	4
April	0
May	2
June	5
July	0
August	3
September	6
October	1
November	4
December	6

Day	Key
Saturday	0
Sunday	1
Monday	2
Tuesday	3
Wednesday	4
Thursday	5
Friday	6

Century	Key
1700s	4
1800s	2
1900s	0
2000s	6

The algorithm works as follows. (We use October 12, 1949, as an example.)

Step 1: Obtain the following five numbers.   Example
1. The number formed by the last two digits of the year   49
2. The number in Step 1, divided by 4, with the remainder ignored   12
3. The month key (1 for October in our example)   1
4. The day of the month (12 for October 12)   12
5. The century key (0 for the 1900s)   0

Step 2: Add these five numbers.   74

Step 3: Divide the sum by 7, and retain the remainder. (74/7 = 10, with remainder 4)

Step 4: Find this remainder in the day key table. (The number 4 implies that October 12, 1949 was a Wednesday.)

TOPICS FOR DISCUSSION

1. Have each person in the group determine the day of the week on which he or she was born.

(continued)

2. Among the group members, discuss whether the following poem applies. (This is all in good fun, of course.)

> Monday's child is fair of face,
> Tuesday's child is full of grace.
> Wednesday's child is full of woe,
> Thursday's child has far to go.
> Friday's child is loving and giving,
> Saturday's child works hard for a living.
> But the child that is born on the Sabbath day is bonny and good, happy and gay.

3. Determine the day of the week on which the following important historical events occurred.
   (a) December 7, 1941 (the bombing of Pearl Harbor)
   (b) November 22, 1963 (assassination of John F. Kennedy)
   (c) July 4, 1976 (bicentennial of the United States)
   (d) January 1, 2000 (the "dreaded" Y2K day)
   (e) September 11, 2001 (the terrorist attacks on the United States)

CHAPTER 4 TEST

1. For the numeral [symbol], identify the numeration system, and give the Hindu-Arabic equivalent.
2. Simplify: $(8 \times 10^3) + (3 \times 10^2) + (6 \times 10^1) + (4 \times 10^0)$.
3. Write in expanded notation: 60,923.

Perform each operation using the alternative algorithm specified.

4. 37×54 (Russian peasant or Egyptian method)
5. 236×94 (Lattice method)
6. $21,325 - 8498$ (Nines complement method)

Convert each of the following to base ten.

7. 424_{five}
8. 100110_{two}
9. $\text{A80C}_{\text{sixteen}}$

Convert as indicated.

10. 58 to base two
11. 1846 to base five
12. 10101110_{two} to base eight
13. $\text{B52}_{\text{sixteen}}$ to base two

Briefly explain each of the following.

14. the advantage of multiplicative grouping over simple grouping
15. the advantage, in a positional numeration system, of a smaller base over a larger base
16. the advantage, in a positional numeration system, of a larger base over a smaller base

Answer the questions in Exercises 17–19.

17. For addition of whole numbers, what is the identity element?
18. For multiplication of rational numbers, what is the inverse of 3?
19. For any whole numbers a, b, and c, $(a + b) + c = (b + a) + c$. What property does this illustrate?
20. Describe in general what constitutes a mathematical group.

A mathematical system with the operation V is defined by the table here.

V	a	e	i	o	u
a	o	e	u	a	i
e	u	o	a	e	i
i	e	u	o	i	a
o	a	e	i	o	u
u	i	a	e	u	o

21. (a) Is there an identity element in this system?
    (b) If so, what is it?
22. (a) Is closure satisfied by this system?
    (b) Explain.
23. (a) Is this system commutative?
    (b) Explain.
24. (a) Is the distributive property satisfied in this system?
    (b) Explain.
25. (a) Is the system with the operation V a group?
    (b) Explain.

chapter

5

Number Theory

In today's world, information is a valuable resource. Many private companies and government agencies must keep certain types of data secret. Codes are used to safely communicate classified information in a form that must be decoded before it can be understood. It is essential that a code be easy to use but difficult to break. One of the most widely used codes in industry was developed by Ronald L. Rivest, Adi Shamir, and Leonard M. Adleman. This code, called RSA for its three co-inventors, starts by changing the message to be coded into a large number M. Then M is transformed into a different number C using the formula

$$C = M^k \pmod{n}.$$

(This type of "modular" equation is discussed in Section 5.4.) The numbers n and k are made available to everyone so that anyone can code a message. Therefore RSA is called a "public key" system. However, the only known way to decode C into M is to first factor n into its two unique prime factors p and q (which are known to the one who chose n originally, but not to anyone else).

When the RSA code was first introduced in 1977, Rivest challenged researchers to decode a message using an n with 129 digits. It was estimated that this message would take approximately 20,000 years to decipher without any knowledge of p and q. However, with the aid of number theory, it took 600 mathematicians in 25 different countries only 17 years to factor n into 64- and 65-digit prime factors. The factorization is given in the margin note on page 197. The decoded message said, "The magic words are squeamish ossifrage." For more detail on RSA codes, see the *Mathematics Teacher* referred to below.

Today, ever larger numbers are being factored. But rather than involving many mathematicians, the effort now is likely to involve many computers, in a process often called "distributed computing." Many computers share the task to obtain a result in a reasonable amount of time.

Advancing techniques sometimes favor those devising codes, but can also favor those trying to decipher someone else's code. So there is a continual contest between the code makers and the code breakers.

Source: SIAM News, Vol. 27, No. 6, 1994, and *Mathematics Teacher*, November 2001, p. 689.

5.1 Prime and Composite Numbers

5.2 Selected Topics from Number Theory

5.3 Greatest Common Factor and Least Common Multiple

5.4 Clock Arithmetic and Modular Systems

5.5 The Fibonacci Sequence and the Golden Ratio

Extension: Magic Squares

Collaborative Investigation: Investigating an Interesting Property of Number Squares

Chapter 5 Test

The famous German mathematician Carl Friedrich Gauss once remarked, "Mathematics is the Queen of Science, and number theory is the Queen of Mathematics." This chapter is centered around the study of number theory. Number theory is the branch of mathematics devoted to the study of the properties of the natural numbers (that is, the infinite set of numbers 1, 2, 3, 4, ..., also called the counting numbers).

5.1 Prime and Composite Numbers

In earlier chapters we discussed the set of **natural numbers,** also called the **counting numbers** or the **positive integers:**

$$\{1, 2, 3, ...\}.$$

Number theory deals with the study of the properties of this set of numbers, and a key concept of number theory is the idea of *divisibility*. Informally, we say that one counting number is *divisible* by another if the operation of dividing the first by the second leaves a remainder 0. A formal definition follows.

Divisibility

The natural number a is **divisible** by the natural number b if there exists a natural number k such that $a = bk$. If b divides a, then we write $b\,|\,a$.

Notice that if b divides a, then the quotient a/b is a natural number. For example, 4 divides 20 since there exists a natural number k such that $20 = 4k$. The value of k here is 5, since $20 = 4 \cdot 5$. (A dot is used here to denote multiplication.) The natural number 20 is not divisible by 7, for example, since there is no natural number k satisfying $20 = 7k$. Alternatively, we think "20 divided by 7 gives quotient 2 with remainder 6" and since there is a nonzero remainder, divisibility does not hold. We write $7 \!\not|\, 20$ to indicate that 7 does not divide 20.

If the natural number a is divisible by the natural number b, then b is a **factor** (or **divisor**) of a, and a is a **multiple** of b. For example, 5 is a factor of 30, and 30 is a multiple of 5. Also, 6 is a factor of 30, and 30 is a multiple of 6. The number 30 equals $6 \cdot 5$; this product $6 \cdot 5$ is called a **factorization** of 30. Other factorizations of 30 include $3 \cdot 10$, $2 \cdot 15$, $1 \cdot 30$, and $2 \cdot 3 \cdot 5$.

The ideas of even and odd natural numbers are based on the concept of divisibility. A natural number is even if it is divisible by 2, and odd if it is not.

EXAMPLE 1 Decide whether the first number listed is divisible by the second.

(a) 45; 9

Is there a natural number k that satisfies $45 = 9k$? The answer is yes, since $45 = 9 \cdot 5$, and 5 is a natural number. Therefore, 9 divides 45, written $9\,|\,45$.

(b) 60; 7

Since the quotient $60 \div 7$ is not a natural number, 60 is not divisible by 7, written $7 \!\not|\, 60$.

Breaking Codes Finding prime factors of extremely large numbers was once considered a mere computer exercise—interesting for the improved methods of working with computers, but of no value in its own right. This has changed in recent years with new methods of *computer coding* in which very large numbers are used in an attempt to provide unbreakable codes for computer data. Just as fast as these numbers are used, other people try to find prime factors of them so that the code can be broken.

An article in *Science News,* May 7, 1994 ("Team Sieving Cracks a Huge Number," p. 292), reported that the coded message put forth by Rivest seventeen years earlier had been deciphered. (See the chapter introduction, page 195.) The key 129-digit number, known as RSA-129, had been factored by the quadratic sieve method (QS), with more than 600 computers working together. Here is the number with its two factors.

114,381,625,757,888,867,669,235,779,
976,146,612,010,218,296,721,242,362,
562,561,842,935,706,935,245,733,897,
830,597,123,563,958,705,058,989,075,
147,599,290,026,879,543,541 =
3,490,529,510,847,650,949,147,849,
619,903,898,133,417,764,638,493,387,
843,990,820,577 × 32,769,132,993,
266,709,549,961,988,190,834,461,413,
177,642,967,992,942,539,798,288,533.

(c) 19; 19

The quotient $19 \div 19$ is the natural number 1, so 19 is divisible by 19. (In fact, any natural number is divisible by itself and also by 1.)

(d) 26; 1

The quotient $26 \div 1$ is the natural number 26, so 26 is divisible by 1. (In fact, any natural number is divisible by 1. People often forget this.)

The observations made in parts (c) and (d) of Example 1 may be generalized as follows: For any natural number a, $a \mid a$ and $1 \mid a$.

EXAMPLE 2 Find all the natural number factors of each number.

(a) 36

To find the factors of 36, try to divide 36 by 1, 2, 3, 4, 5, 6, and so on. Doing this gives the following list of natural number factors of 36:

$$1, 2, 3, 4, 6, 9, 12, 18, \text{ and } 36.$$

(b) 50

The factors of 50 are 1, 2, 5, 10, 25, and 50.

(c) 11

The only natural number factors of 11 are 11 and 1.

Like the number 19 in Example 1(c), the number 11 has only two natural number factors, itself and 1. Such a natural number is called a *prime number*.

Prime and Composite Numbers

A natural number greater than 1 that has only itself and 1 as factors is called a **prime number.** A natural number greater than 1 that is not prime is called **composite.**

Mathematicians agree that the natural number 1 is neither prime nor composite. The following alternative definition of a prime number clarifies that 1 is not a prime.

Alternative Definition of a Prime Number

A **prime number** is a natural number that has *exactly* two different natural number factors.

This alternative definition excludes the possibility of 1 being prime, since 1 has only one natural number factor (namely, 1).

There is a systematic method for identifying prime numbers in a list of numbers: $2, 3, \ldots, n$. The method, known as the **Sieve of Eratosthenes,** is named after the

Greek geographer, poet, astronomer, and mathematician, who lived from about 276 to 192 B.C. To construct such a sieve, list all the natural numbers from 2 through some given natural number *n*, such as 100. The number 2 is prime, but all other multiples of 2 (4, 6, 8, 10, and so on) are composite. Circle the prime 2, and cross out all other multiples of 2. The next number not crossed out and not circled is 3, the next prime. Circle the 3, and cross out all other multiples of 3 (6, 9, 12, 15, and so on) that are not already crossed out. Circle the next prime, 5, and cross out all other multiples of 5 not already crossed out. Continue this process for all primes less than or equal to the square root of the last number in the list. For this list, we may stop with 7, since the next prime, 11, is greater than the square root of 100, which is 10. At this stage, simply circle all remaining numbers that are not crossed out. Table 1 shows the Sieve of Eratosthenes for 2, 3, 4, ... , 100, identifying the 25 primes in that range. Theoretically, such a sieve can be constructed for any value of *n*.

TABLE 1 Sieve of Eratosthenes

	②	③	4	⑤	6	⑦	8	9	10	⑪	12	⑬	14
15	16	⑰	18	⑲	20	21	22	㉓	24	25	26	27	28
㉙	30	㉛	32	33	34	35	36	㊲	38	39	40	㊁	42
㊸	44	45	46	㊼	48	49	50	51	52	㊽	54	55	56
57	58	㊾	60	㊿	62	63	64	65	66	67	68	69	70
71	72	73	74	75	76	77	78	79	80	81	82	83	84
85	86	87	88	89	90	91	92	93	94	95	96	97	98
99	100												

How to Use Up Lots of Chalk
In 1903, long before the age of the computer, the mathematician F. N. Cole presented before a meeting of the American Mathematical Society his discovery of a factorization of the number

$$2^{67} - 1.$$

He walked up to the chalkboard, raised 2 to the 67th power, and then subtracted 1. Then he moved over to another part of the board and multiplied out

193,707,721
× 761,838,257,287.

The two calculations agreed, and Cole received a standing ovation for a presentation that did not include a single word.

EXAMPLE 3 Decide whether each number is prime or composite.

(a) 97

Since the only factors of 97 are 97 and 1, the number 97 is prime.

(b) 59,872

The number 59,872 is even, so it is divisible by 2. It is composite. (There is only one even prime, the number 2 itself.)

(c) 697

For 697 to be composite, there must be a number other than 697 and 1 that divides into it with remainder 0. Start by trying 2, and then 3. Neither works. There is no need to try 4. (If 4 divides with remainder 0 into a number, then 2 will also.) Try 5. There is no need to try 6 or any succeeding even number. (Why?) Try 7. Try 11. (Why not try 9?) Try 13. Keep trying numbers until one works, or until a number is tried whose square exceeds the given number. Try 17:

$$697 \div 17 = 41.$$

The number 697 is composite. ∎

An aid in determining whether a natural number is divisible by another natural number is called a **divisibility test.** Some simple divisibility tests exist for small natural numbers, and they are given in Table 2. Divisibility tests for 7 and 11 are a bit involved, and they are discussed in the exercises for this section.

TABLE 2 Divisibility Tests

Divisible By	Test	Example
2	Number ends in 0, 2, 4, 6, or 8. (The last digit is even.)	9,489,994 ends in 4; it is divisible by 2.
3	Sum of the digits is divisible by 3.	897,432 is divisible by 3, since $8 + 9 + 7 + 4 + 3 + 2 = 33$ is divisible by 3.
4	Last two digits form a number divisible by 4.	7,693,432 is divisible by 4, since 32 is divisible by 4.
5	Number ends in 0 or 5.	890 and 7635 are divisible by 5.
6	Number is divisible by both 2 and 3.	27,342 is divisible by 6 since it is divisible by both 2 and 3.
8	Last three digits form a number divisible by 8.	1,437,816 is divisible by 8, since 816 is divisible by 8.
9	Sum of the digits is divisible by 9.	428,376,105 is divisible by 9 since sum of digits is 36, which is divisible by 9.
10	The last digit is 0.	897,463,940 is divisible by 10.
12	Number is divisible by both 4 and 3.	376,984,032 is divisible by 12.

EXAMPLE 4 Decide whether the first number listed is divisible by the second.

(a) 2,984,094; 4

The last two digits form the number 94. Since 94 is not divisible by 4, the given number is not divisible by 4.

(b) 4,119,806,514; 9

The sum of the digits is $4 + 1 + 1 + 9 + 8 + 0 + 6 + 5 + 1 + 4 = 39$, which is not divisible by 9. The given number is therefore not divisible by 9.

As mentioned earlier, a number such as 15, which can be written as a product of prime factors ($15 = 3 \cdot 5$), is called a composite number. An important theorem in mathematics states that there is only one possible way to write the prime factorization of a composite natural number. This theorem is called the *fundamental theorem of arithmetic,* a form of which was known to the ancient Greeks.*

Writing in the October 1, 1994 issue of *Science News,* Ivars Peterson gives a fascinating account of the recent discovery of a 75-year-old factoring machine ("Cranking Out Primes: Tracking Down a Long-lost Factoring Machine"). In 1989, Jeffrey Shallit of the University of Waterloo in Ontario came across an article in an obscure 1920 French journal, in which the author, Eugene Olivier Carissan, reported his invention of the factoring apparatus. Shallit and two colleagues embarked on a search for the machine. They contacted all telephone subscribers in France named Carissan and received a reply from Eugene Carissan's daughter. The machine was still in existence and in working condition, stored in a drawer at an astronomical observatory in Floirac, near Bordeaux.

Peterson explains in the article how the apparatus works. Using the machine, Carissan took just ten minutes to prove that 708,158,977 is a prime number, and he was able to factor a 13-digit number. While this cannot compare to what technology can accomplish today, it was a significant achievement for Carissan's day.

---

*A theorem is a statement that can be proved true from other statements. For a proof of this theorem, see *What Is Mathematics?* by Richard Courant and Herbert Robbins (Oxford University Press, 1941), p. 23.

The following program, written by Charles W. Gantner and provided courtesy of Texas Instruments, can be used on the TI-83 Plus calculator to list all primes less than or equal to a given natural number *N*.

PROGRAM: PRIMES
: Disp "INPUT N ≥ 2"
: Disp "TO GET"
: Disp "PRIMES ≤ N"
: Input N
: 2 → T
: Disp T
: 1 → A
: Lbl 1
: A + 2 → A
: 3 → B
: If A > N
: Stop
: Lbl 2
: If B ≤ √(A)
: Goto 3
: Disp A
: Pause
: Goto 1
: Lbl 3
: If A/B ≤ int (A/B)
: Goto 1
: B + 2 → B
: Goto 2

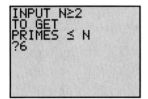

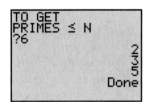

The display indicates that the primes less than or equal to 6 are 2, 3, and 5.

The Fundamental Theorem of Arithmetic

Every composite natural number can be expressed in one and only one way as a product of primes (if the order of the factors is disregarded).

This theorem is sometimes called the **unique factorization theorem** to reflect the idea that there is only one (unique) prime factorization possible for any given natural number.

A composite number can be factored into primes by using a "factor tree" as illustrated in the next example.

EXAMPLE 5 Find the prime factorization of the composite number 504.

Since 2 is a factor of 504, we can begin the first branches of the factor tree by writing 504 as 2 · 252. Then factor 252 as 2 · 126. When a composite factor appears that does not have 2 as a factor, try 3, then 5, then 7, and so on, until there is a prime number at the end of each branch.

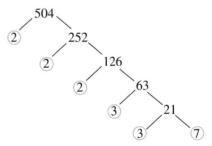

The prime factors are circled. Using exponents, the prime factorization of 504 is $2^3 \cdot 3^2 \cdot 7$. (*Note:* Because the fundamental theorem of arithmetic guarantees that this prime factorization is unique, we could have started the factor tree with 3 · 168, or 8 · 63, or several other possibilities, as long as we had continued until all factors were prime. In every case, the same final factorization would be obtained.)

As an alternative to the factor tree method, we may divide 504 by primes over and over, using the compact form shown below.

$$
\begin{array}{r|l}
2 & 504 \\
\hline
2 & 252 \\
\hline
2 & 126 \\
\hline
3 & 63 \\
\hline
3 & 21 \\
\hline
& 7
\end{array}
$$

Keep going, using as necessary the primes 2, 3, 5, 7, 11, and so on, until the last number is a prime. Read the prime factors as shown in color, to get $504 = 2^3 \cdot 3^2 \cdot 7$.

The Infinitude of Primes

Since prime numbers are, in a sense, the building blocks of the entire number system, mathematicians (amateur as well as professional) have sought for thousands of years to learn as much as possible about them. Although much still is unknown, there is one important basic result that was proved by Euclid around 300 B.C., namely that there are infinitely many primes. This means, for one thing, that no matter how large a prime we identify, there are always others even larger. Euclid's proof remains today as one of the most elegant proofs in all of mathematics. (An *elegant* mathematical proof is one that exhibits the desired result in a most direct, concise manner. Mathematicians strive for elegance in their proofs.) It is called a **proof by contradiction.**

A statement can be proved by contradiction as follows: We assume that the negation of the statement is true. The assumption that the negation is true is used to produce some sort of contradiction, or absurdity. The fact that the negation of the original statement leads to a contradiction means that the original statement must be true.

In order to better understand a particular part of the proof that there are infinitely many primes, it is helpful to examine the following argument.

Suppose that $M = 2 \cdot 3 \cdot 5 \cdot 7 + 1 = 211$. Now M is the product of the first four prime numbers, plus 1. If we divide 211 by each of the primes 2, 3, 5, and 7, the remainder is always 1.

$$
\begin{array}{cccc}
105 & 70 & 42 & 30 \\
2\overline{)211} & 3\overline{)211} & 5\overline{)211} & 7\overline{)211} \\
\underline{210} & \underline{210} & \underline{210} & \underline{210} \\
1 & 1 & 1 & 1
\end{array}
$$

So 211 is not divisible by any of the primes 2, 3, 5, and 7.

Now we are ready to prove that there are infinitely many primes. If it can be shown that *there is no largest prime number,* then there must be infinitely many primes.

Theorem There is no largest prime number.

Proof: Suppose that there is a largest prime number and call it P. Now form the number M such that

$$M = p_1 \cdot p_2 \cdot p_3 \cdots P + 1,$$

where $p_1, p_2, p_3, \ldots, P$ represent all the primes less than or equal to P. Now the number M must be either prime or composite.

1. Suppose that M is prime.
   M is obviously larger than P, so if M is prime, it is larger than the assumed largest prime P. We have reached a *contradiction.*
2. Suppose that M is composite.
   If M is composite, it must have a prime factor. But none of $p_1, p_2, p_3, \ldots, P$ are factors of M, since division by each will leave a remainder of 1. (Recall the above argument.) So if M has a prime factor, it must be greater than P. But this is a *contradiction,* since P is the assumed largest prime.

In either case 1 or 2, we reach a contradiction. The whole argument was based upon the assumption that a largest prime exists, but since this leads to contradictions, there must be no largest prime, or equivalently, there are infinitely many primes. ■

The Search for Large Primes Identifying larger and larger prime numbers and factoring large composite numbers into their prime components is of great practical importance today since it is the basis of modern **cryptography systems** (secret codes). Various codes have been used for centuries in military applications.

Prime Search At one time, $2^{11,213} - 1$ was the largest known Mersenne prime. To honor its discovery, the Urbana, Illinois, post office used the cancellation picture above. Since then, larger prime numbers have been found.

Today the security of industrial and business data also depends upon the theory of prime numbers.

As mathematicians continue to search for larger and larger primes, a formula for generating all the primes would be nice (something similar, for example, to the formula $2n$, which generates all even counting numbers for $n = 1, 2, 3, \ldots$, or the formula n^2, which generates all the perfect squares). Numbers generated by the formula $M_n = 2^n - 1$ are called **Mersenne numbers** to honor the French monk Marin Mersenne (1588–1648). It was long known that a *composite* value of n would always generate a composite Mersenne number. (See Exercises 81–86.) And some early mathematicians believed that a *prime* value of n would always generate a prime Mersenne number. That is, starting with any known prime number n, one could always produce another, larger prime number $2^n - 1$. (Although "always generate a prime" is not the same as "generate all primes," at least it would be a source of primes.)

Marin Mersenne (1588–1648), in his *Cogitata Physico-Mathematica* (1644), claimed that M_n was prime for $n = 2, 3, 5, 7, 13, 17, 19, 31, 67, 127,$ and $257,$ and composite for all other prime numbers n less than 257. Other mathematicians at the time knew that Mersenne could not have actually tested all these values, but no one else could prove or disprove them either. It was more then 300 years later before all primes up to 257 were legitimately checked out, and Mersenne was finally revealed to have made five errors:

M_{61} is prime.
M_{67} is composite.
M_{89} is prime.
M_{107} is prime.
M_{257} is composite.

EXAMPLE 6 Evaluate the Mersenne number M_n for $n = 2, 3,$ and 5.

$$M_2 = 2^2 - 1 = 3 \qquad M_3 = 2^3 - 1 = 7 \qquad M_5 = 2^5 - 1 = 31$$

Note that all three values, 3, 7, and 31, are indeed primes.

It turns out that M_7 is also a prime (see Exercise 26), but it was discovered in 1536 that $M_{11} = 2^{11} - 1 = 2047$ is not prime (since it is $23 \cdot 89$). So prime values of n do *not* always produce prime M_n. The question then became which prime values of n do produce prime Mersenne numbers (the so-called **Mersenne primes**). Since no way was ever found to identify, in general, which prime values of n result in Mersenne primes, it became a matter of checking out each prime n value individually—not an easy task given that the Mersenne numbers rapidly become very large.

The Mersenne prime search yielded results slowly. By about 1600, M_n had been verified as prime for all prime n up to 19 (except for 11, as mentioned above). The next one was M_{31}, verified by Euler in 1732. About 100 years later, M_{127} was shown to be prime. No larger Mersenne prime was discovered until 1952, when an early computer verified that M_{521}, M_{607}, M_{1279}, M_{2203}, and M_{2281} all are primes.

Over the last 50 years, most new record-breaking primes have been identified by mathematicians expert in devising algorithms and programming them to run on the most powerful computers available. In early 1996, the **Great Internet Mersenne Prime Search (GIMPS)** was launched. By early 2002, more than 130,000 individuals had signed on, receiving free software and source code to run during slack time on more than 205,000 personal computers throughout the world. The latest Mersenne prime was $M_{13,466,917} = 2^{13,466,917} - 1$, a number with 4,053,946 decimal digits. It was the 39th known Mersenne prime. (In a BBC online article it was reckoned that the accomplishment by GIMPS had utilized a combined 13,000 years of computer time.)

During the same general period that Mersenne was thinking about prime numbers, Pierre de Fermat (about 1601–1665) conjectured that the formula $2^{2^n} + 1$ would always produce a prime, for any whole number value of n. Table 3 shows how this formula generates the first four **Fermat numbers,** which are indeed all primes. The fifth Fermat number (from $n = 4$) is likewise prime, but the sixth Fermat number (from $n = 5$) turns out to be 4,294,967,297, which is *not* prime. (See Exercises 75 and 76.) To date, no more primes have been found among the Fermat numbers.

Prime Does Pay On November 14, 2001, Michael Cameron's computer, an 800 MHz AMD T-Bird, found the 39th known Mersenne prime, after just 42 days of idle time computing. The number contains 4,053,946 decimal digits. Starting with the first million-digit prime, the Electronics Frontier Foundation has paid rewards of $50,000 for record-breaking primes. The first discoverer of a ten-million-digit prime will receive $100,000.

If you would like to join the Great Internet Mersenne Prime Search, check out

http://www.mersenne.org.

TABLE 3 The Generation of Fermat Numbers

n	2^n	2^{2^n}	$2^{2^n} + 1$
0	1	2	3
1	2	4	5
2	4	16	17
3	8	256	257

Of historical note are a couple of polynomial formulas that produce primes. In 1732, Leonhard Euler offered the formula $n^2 - n + 41$, which generates primes for n up to 40 and fails at $n = 41$. In 1879, E. B. Escott produced more primes with the formula $n^2 - 79n + 1601$, which first fails at $n = 80$.

EXAMPLE 7 Find the first five numbers produced by each of the polynomial formulas of Euler and Escott.

Table 4 shows the required numbers.

TABLE 4 A Few Polynomial-Generated Prime Numbers

n	Euler formula $n^2 - n + 41$	Escott formula $n^2 - 79n + 1601$
1	41	1523
2	43	1447
3	47	1373
4	53	1301
5	61	1231

All values found here are primes. (Use Table 1 to verify the Euler values.)

Actually, it is not hard to prove that there can be no polynomial that will consistently generate primes. More complicated mathematical formulas do exist for generating primes, but none produced so far can be practically applied in a reasonable amount of time, even using the fastest computers.

5.1 EXERCISES

Decide whether the following statements are true *or* false.

1. Every natural number is divisible by 1.
2. No natural number is both prime and composite.
3. There are no even prime numbers.
4. If n is a natural number and $9 \mid n$, then $3 \mid n$.
5. If n is a natural number and $5 \mid n$, then $10 \mid n$.
6. 1 is the smallest prime number.
7. Every natural number is both a factor and a multiple of itself.
8. If 16 divides a natural number, then 2, 4, and 8 must also divide that natural number.

9. The composite number 50 has exactly two prime factorizations.

10. The prime number 53 has exactly two natural number factors.

11. The number $2^{11} - 1$ is an example of a Mersenne prime.

12. As of early 2003, only five Fermat primes had ever been found.

Find all natural number factors of each number.

13. 12 14. 18 15. 20 16. 28 17. 120 18. 172

Use divisibility tests to decide whether the given number is divisible by each of the following.

(a) 2 (b) 3 (c) 4 (d) 5 (e) 6 (f) 8 (g) 9 (h) 10 (i) 12

19. 315 20. 630 21. 25,025
22. 45,815 23. 123,456,789 24. 987,654,321

25. (a) In constructing the Sieve of Eratosthenes for 2 through 100, we said that any composite in that range had to be a multiple of some prime less than or equal to 7 (since the next prime, 11, is greater than the square root of 100). Explain.
   (b) To extend the Sieve of Eratosthenes to 200, what is the largest prime whose multiples would have to be considered?
   (c) Complete this statement: In seeking prime factors of a given number, we need only consider all primes up to and including the _____ _____ of that number, since a prime factor greater than the _____ _____ can only occur if there is at least one other prime factor less than the _____ _____.
   (d) Complete this statement: If no prime less than or equal to $\sqrt{n}$ divides n, then n is a _____ number.

26. (a) Continue the Sieve of Eratosthenes in Table 1 from 101 to 200 and list the primes between 100 and 200. How many are there?
   (b) From your list in part (a), verify that the Mersenne number M_7 is indeed prime.

27. List two primes that are consecutive natural numbers. Can there be any others?

28. Can there be three primes that are consecutive natural numbers? Explain.

29. For a natural number to be divisible by both 2 and 5, what must be true about its last digit?

30. Consider the divisibility tests for 2, 4, and 8 (all powers of 2). Use inductive reasoning to predict the divisibility test for 16. Then, use the test to show that 456,882,320 is divisible by 16.

Find the prime factorization of each composite number.

31. 240 32. 300 33. 360 34. 425 35. 663 36. 885

Here is a divisibility test for 7.

(a) *Double the last digit of the given number, and subtract this value from the given number with the last digit omitted.*
(b) *Repeat the process of part (a) as many times as necessary until the number obtained can easily be divided by 7.*
(c) *If the final number obtained is divisible by 7, then the given number also is divisible by 7. If the final number is not divisible by 7, then neither is the given number.*

Use this divisibility test to determine whether each of the following is divisible by 7.

37. 142,891 38. 409,311 39. 458,485 40. 287,824

Here is a divisibility test for 11.

(a) *Starting at the left of the given number, add together every other digit.*
(b) *Add together the remaining digits.*
(c) *Subtract the smaller of the two sums from the larger. (If they are the same, the difference is 0.)*
(d) *If the final number obtained is divisible by 11, then the given number also is divisible by 11. If the final number is not divisible by 11, then neither is the given number.*

Use this divisibility test to determine whether each of the following is divisible by 11.

41. 8,493,969 42. 847,667,942 43. 453,896,248 44. 552,749,913

45. Consider the divisibility test for the composite number 6, and make a conjecture for the divisibility test for the composite number 15.

46. Explain what is meant by a "proof by contradiction."

Determine all possible digit replacements for x so that the first number is divisible by the second. For example, 37,58x is divisible by 2 if x = 0, 2, 4, 6, or 8.

47. 398,87x; 2 **48.** 2,45x,765; 3 **49.** 64,537,84x; 4 **50.** 2,143,89x; 5

51. 985,23x; 6 **52.** 7,643,24x; 8 **53.** 4,329,7x5; 9 **54.** 23,x54,470; 10

*There is a method to determine the **number of divisors** of a composite number. To do this, write the composite number in its prime factored form, using exponents. Add 1 to each exponent and multiply these numbers. Their product gives the number of divisors of the composite number. For example, $24 = 2^3 \cdot 3 = 2^3 \cdot 3^1$. Now add 1 to each exponent: 3 + 1 = 4, 1 + 1 = 2. Multiply 4 × 2 to get 8. There are 8 divisors of 24. (Since 24 is rather small, this can be verified easily. The divisors are 1, 2, 3, 4, 6, 8, 12, and 24, a total of eight as predicted.)*

Find the number of divisors of each composite number.

55. 36 **56.** 48 **57.** 72 **58.** 144 **59.** $2^8 \cdot 3^2$ **60.** $2^4 \cdot 3^4 \cdot 5^2$

Leap years occur when the year number is divisible by 4. An exception to this occurs when the year number is divisible by 100 (that is, it ends in two zeros). In such a case, the number must be divisible by 400 in order for the year to be a leap year. Determine which of the following years are leap years.

61. 1776 **62.** 1948 **63.** 1994 **64.** 1894 **65.** 2400 **66.** 1800

67. Why is the following *not* a valid divisibility test for 8? "A number is divisible by 8 if it is divisible by both 4 and 2." Support your answer with an example.

68. Choose any three consecutive natural numbers, multiply them together, and divide the product by 6. Repeat this several times, using different choices of three consecutive numbers. Make a conjecture concerning the result.

69. Explain why the product of three consecutive natural numbers must be divisible by 6.

70. Choose any 6-digit number consisting of three digits followed by the same three digits in the same order (for example, 467,467). Divide by 13. Divide by 11. Divide by 7. What do you notice? Why do you think this happens?

71. Verify that Euler's polynomial prime-generating formula $n^2 - n + 41$ fails to produce a prime for $n = 41$.

72. Evaluate Euler's polynomial formula for **(a)** $n = 42$, and **(b)** $n = 43$.

73. Choose the correct completion: For $n > 41$, Euler's formula produces a prime
A. never. **B.** sometimes. **C.** always.
(*Hint:* If no prime less than or equal to $\sqrt{n}$ divides n, then n is prime.)

74. Recall that Escott's formula, $n^2 - 79n + 1601$, fails to produce a prime for $n = 80$. Evaluate this formula for $n = 81$ and $n = 82$ and then complete the following statement: For $n > 80$, Escott's formula produces a prime
A. never. **B.** sometimes. **C.** always.

75. (a) Evaluate the Fermat number $2^{2^n} + 1$ for $n = 4$.
(b) In seeking possible prime factors of the Fermat number of part (a), what is the largest potential prime factor that one would have to try? (As stated in the text, this "fifth" Fermat number is in fact prime.)

76. (a) Verify the value given in the text for the "sixth" Fermat number (i.e., $2^{2^5} + 1$).
(b) Divide this Fermat number by 641. (Euler discovered this factorization in 1732, proving that the sixth Fermat number is not prime.)

77. Write a short report on the Great Internet Mersenne Prime Search (GIMPS).

78. Write a short report identifying the 36th, 37th, and 38th known Mersenne primes and how, when, and by whom they were found.

79. The Mersenne margin note cites a 1644 claim that was not totally resolved for some 300 years. Find out when, and by whom, Mersenne's five errors were demonstrated.

80. In Euclid's proof that there is no largest prime, we formed a number M by taking the product of primes and adding 1. Observe the pattern below.

$M = 2 + 1 = 3$ (3 is prime)
$M = 2 \cdot 3 + 1 = 7$ (7 is prime)
$M = 2 \cdot 3 \cdot 5 + 1 = 31$ (31 is prime)
$M = 2 \cdot 3 \cdot 5 \cdot 7 + 1 = 211$ (211 is prime)
$M = 2 \cdot 3 \cdot 5 \cdot 7 \cdot 11 + 1 = 2311$ (2311 is prime)

It seems as though this pattern will always yield a prime number. Now evaluate

$$M = 2 \cdot 3 \cdot 5 \cdot 7 \cdot 11 \cdot 13 + 1.$$

Is M prime or composite? If composite, give its prime factorization.

The text stated that the Mersenne number M_n is composite whenever n is composite. The following exercises develop one way you can always find a factor of such a Mersenne number.

81. Find $M_n = 2^n - 1$ for the composite number $n = 6$.

82. Notice that $p = 3$ is a prime factor of $n = 6$. Find $2^p - 1$ for $p = 3$. Is $2^p - 1$ a factor of $2^n - 1$?

83. Complete this statement: If p is a prime factor of n, then _____ is a factor of the Mersenne number $2^n - 1$.

84. Find $M_n = 2^n - 1$ for $n = 10$.

85. Use the procedure illustrated above to find two distinct factors of M_{10}.

86. Do you think this procedure will always produce *prime* factors of M_n for composite n? (*Hint:* Consider $n = 22$ and its prime factor $p = 11$, and recall the statement following Example 6.) Explain.

5.2 Selected Topics from Number Theory

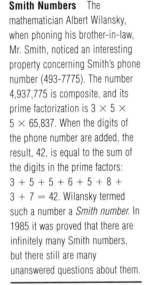

Smith Numbers The mathematician Albert Wilansky, when phoning his brother-in-law, Mr. Smith, noticed an interesting property concerning Smith's phone number (493-7775). The number 4,937,775 is composite, and its prime factorization is $3 \times 5 \times 5 \times 65{,}837$. When the digits of the phone number are added, the result, 42, is equal to the sum of the digits in the prime factors: $3 + 5 + 5 + 6 + 5 + 8 + 3 + 7 = 42$. Wilansky termed such a number a *Smith number*. In 1985 it was proved that there are infinitely many Smith numbers, but there still are many unanswered questions about them.

In an earlier chapter we introduced figurate numbers, a topic investigated by the Pythagoreans. This group of Greek mathematicians and musicians held their meetings in secret, and were led by Pythagoras. In this section we examine some of the other special numbers that fascinated the Pythagoreans and are still studied by mathematicians today.

Perfect Numbers Divisors of a natural number were covered in Section 5.1. The **proper divisors** of a natural number include all divisors of the number except the number itself. For example, the proper divisors of 8 are 1, 2, and 4. (8 is *not* a proper divisor of 8.)

Perfect Numbers

A natural number is said to be **perfect** if it is equal to the sum of its proper divisors.

Is 8 perfect? No, since $1 + 2 + 4 = 7$, and $7 \neq 8$. The smallest perfect number is 6, since the proper divisors of 6 are 1, 2, and 3, and

$$1 + 2 + 3 = 6. \quad \text{6 is perfect.}$$

EXAMPLE 1 Show that 28 is a perfect number.

The proper divisors of 28 are 1, 2, 4, 7, and 14. The sum of these is 28:

$$1 + 2 + 4 + 7 + 14 = 28.$$

By the definition, 28 is perfect.

The numbers 6 and 28 are the two smallest perfect numbers. The next two are 496 and 8128. The pattern of these first four perfect numbers led early writers to conjecture that

1. The nth perfect number contains exactly n digits.
2. The even perfect numbers end in the digits 6 and 8, alternately.

(Exercises 41–43 will help you evaluate these conjectures.)

There still are many unanswered questions about perfect numbers. Euclid showed that if $2^n - 1$ is prime, then $2^{n-1}(2^n - 1)$ is perfect, and conversely. Since the prime values of $2^n - 1$ are the Mersenne primes (discussed in the previous section), this means that for every new Mersenne prime discovered, another perfect number is automatically revealed. (Hence, as of late 2002, there were also 39 known perfect numbers.) It is also known that all even perfect numbers must take the form $2^{n-1}(2^n - 1)$ and it is strongly suspected that no odd perfect numbers exist. (Any odd one would have at least eight different prime factors and would have at least 300 decimal digits.) Therefore, Euclid and the early Greeks most likely identified the form of all perfect numbers.

Deficient and Abundant Numbers Earlier we saw that 8 is not perfect since it is greater than the sum of its proper divisors ($8 > 7$).

> **Deficient and Abundant Numbers**
>
> A natural number is **deficient** if it is greater than the sum of its proper divisors. It is **abundant** if it is less than the sum of its proper divisors.

Based on this definition, a *deficient number does not have enough* proper divisors to add up to itself, while an *abundant number has more than enough* proper divisors to add up to itself.

Weird Numbers A number is said to be *weird* if it is abundant without being equal to the sum of any set of its own proper divisors. For example, 70 is weird because it is abundant ($1 + 2 + 5 + 7 + 10 + 14 + 35 = 74 > 70$), but no set of the factors 1, 2, 5, 7, 10, 14, 35 adds up to 70.

EXAMPLE 2 Decide whether the given number is deficient or abundant.

(a) 12

The proper divisors of 12 are 1, 2, 3, 4, and 6. The sum of these is 16. Since $12 < 16$, the number 12 is abundant.

(b) 10

The proper divisors of 10 are 1, 2, and 5. Since $1 + 2 + 5 = 8$, and $10 > 8$, the number 10 is deficient.

Amicable (Friendly) Numbers Suppose that we add the proper divisors of 284:

$$1 + 2 + 4 + 71 + 142 = 220.$$

Their sum is 220. Now, add the proper divisors of 220:

$$1 + 2 + 4 + 5 + 10 + 11 + 20 + 22 + 44 + 55 + 110 = 284.$$

Notice that the sum of the proper divisors of 220 is 284, while the sum of the proper divisors of 284 is 220. Number pairs such as these are said to be *amicable* or *friendly*.

Sociable Numbers An extension of the idea of amicable numbers results in sociable numbers. In a chain of sociable numbers, the sum of the proper divisors of each number is the next number in the chain, and the sum of the proper divisors of the last number in the chain is the first number. Here is a 5-link chain of sociable numbers:

12,496
14,288
15,472
14,536
14,264.

The number 14,316 starts a 28-link chain of sociable numbers.

A Dull Number? The Indian mathematician Srinivasa Ramanujan (1887–1920) developed many ideas in number theory. His friend and collaborator on occasion was G. H. Hardy, also a number theorist and professor at Cambridge University in England.

A story has been told about Ramanujan that illustrates his genius. Hardy once mentioned to Ramanujan that he had just taken a taxicab with a rather dull number: 1729. Ramanujan countered by saying that this number isn't dull at all; it is the smallest natural number that can be expressed as the sum of two cubes in two different ways:

$1^3 + 12^3 = 1729$

and $9^3 + 10^3 = 1729$.

Show that 85 can be written as the sum of two squares in two ways.

Amicable or Friendly Numbers

The natural numbers a and b are **amicable,** or **friendly,** if the sum of the proper divisors of a is b, and the sum of the proper divisors of b is a.

The smallest pair of amicable numbers, 220 and 284, was known to the Pythagoreans, but it was not until more than 1000 years later that the next pair, 18,416 and 17,296, was discovered. Many more pairs were found over the next few decades, but it took a 16-year-old Italian boy named Nicolo Paganini to discover in the year 1866 that the pair of amicable numbers 1184 and 1210 had been overlooked for centuries!

Today, powerful computers continually extend the lists of known amicable pairs. The last time we checked, well over half a million pairs were known, some of them having over 5000 digits. It still is unknown, however, if there are infinitely many such pairs. Finally, no one has found an amicable pair without prime factors in common, but the possibility of such a pair has not been eliminated.

Goldbach's Conjecture One of the most famous unsolved problems in mathematics is Goldbach's conjecture. The mathematician Christian Goldbach (1690–1764) conjectured (guessed) that every even number greater than 2 can be written as the sum of two prime numbers. For example, $8 = 5 + 3$ and $10 = 5 + 5$ (or $10 = 7 + 3$). Mathematicians have tried, without success, to prove Goldbach's conjecture, though it has been verified for numbers up to 4×10^{14}.

EXAMPLE 3 Write each even number as the sum of two primes.

(a) 18

$18 = 5 + 13$. Another way of writing it is $7 + 11$. Notice that $1 + 17$ is *not* valid, since by definition 1 is not a prime number.

(b) 60

$60 = 7 + 53$. Can you find other ways? Why is $3 + 57$ not valid?

Twin Primes Prime numbers that differ by 2 are called **twin primes.** Some twin primes are 3 and 5, 5 and 7, 11 and 13, and so on. One unproved conjecture about prime numbers (the "twin prime conjecture") states that there are infinitely many pairs of twin primes. Like Goldbach's conjecture, this has never been proved. You may wish to verify that there are eight such pairs less than 100, using the Sieve of Eratosthenes in Table 1. As of late 2002, the largest known twin primes were $318,032,361 \cdot 2^{107,001} \pm 1$. Each contains 32,220 digits.

Recall from Section 5.1 that Euclid's proof of the infinitude of primes used numbers of the form $p_1 \cdot p_2 \cdot p_3 \cdots p_n + 1$, where all the p's are prime. It may seem that any such number must be prime, but that is not so. (See Exercise 80 of Section 5.1.) However, this form often does produce primes (as does the same form with the plus replaced by a minus). When *all* the primes up to p_n are included, the resulting primes are called **primorial primes.** They are denoted $p\# \pm 1$. For example, $5\# + 1 = 2 \cdot 3 \cdot 5 + 1 = 31$ is a primorial prime. (In late 2002, the largest known primorial prime was $392{,}113\# + 1$, a number with 169,966 digits.) The primorial primes are a popular place to look for twin primes.

Sophie Germain (1776–1831) studied at the École Polytechnique in Paris in a day when female students were not admitted. She submitted her work to the professor Joseph Louis Lagrange (profiled in a margin note in the Metric System appendix) under the pseudonym M. Leblanc. Lagrange eventually learned her true identity, and praised her as a promising young mathematician. Later in her career, she proved Fermat's last theorem for a special case involving exponents divisible by certain kinds of primes, which were therefore named in her honor. A **Sophie Germain prime** is an odd prime p for which $2p + 1$ also is prime. Lately, large Sophie Germain primes have been discovered at the rate of several per year. As of early 2003, the largest one known was $2{,}540{,}041{,}185 \cdot 2^{114{,}729} - 1$, which has 34,547 digits. *Source:* http://www.utm.edu/research/primes

EXAMPLE 4 Verify that the primorial formula $p\# \pm 1$ produces twin prime pairs for both **(a)** $p = 3$ and **(b)** $p = 5$.

(a) $3\# \pm 1 = 2 \cdot 3 \pm 1 = 6 \pm 1 = 5$ and 7 (twin primes)

(b) $5\# \pm 1 = 2 \cdot 3 \cdot 5 \pm 1 = 30 \pm 1 = 29$ and 31 (twin primes)

Fermat's Last Theorem In any right triangle with shorter sides a and b, and longest side (hypotenuse) c, the equation $a^2 + b^2 = c^2$ will hold true. This is the famous Pythagorean Theorem. For example, with $a = 3$, $b = 4$, and $c = 5$, we have

$$3^2 + 4^2 = 5^2$$
$$25 = 25.$$

There are infinitely many such triples of numbers that satisfy this equation. Are there any triples of natural numbers that satisfy $a^n + b^n = c^n$ for natural numbers $n \geq 3$? Pierre de Fermat, profiled in a margin note in the next section, thought not and wrote in the margin of a book that he had "a truly wonderful proof" for this, but that the margin was "too small to contain it." Did he indeed have a proof, or did he have an incorrect proof?

Whatever the case, Fermat's assertion was the object of more than 400 years of attempts by mathematicians to provide a suitable proof. While it was verified for many specific cases (Fermat himself proved it for $n = 3$), a proof of the general case could not be found until the Princeton mathematician Andrew Wiles announced a proof in the spring of 1993. Although some flaws were discovered in his argument, Wiles was able, by the fall of 1994, to repair and even improve the proof.

There were probably about 100 mathematicians around the world qualified to understand the Wiles proof. Many of these examined and approved it. Some portions of it were simplified and improved. Today Fermat's Last Theorem finally is regarded by the mathematics community as officially proved.

Andrew Wiles proved Fermat's Last Theorem

EXAMPLE 5 One of the theorems proved by Fermat is as follows: *Every odd prime can be expressed as the difference of two squares in one and only one way.* For each of the following odd primes, express it as the difference of two squares.

(a) 3

$$3 = 4 - 1 = 2^2 - 1^2$$

(b) 7

$$7 = 16 - 9 = 4^2 - 3^2$$

FOR FURTHER THOUGHT

One of the most remarkable books on number theory is *The Penguin Dictionary of Curious and Interesting Numbers* (1986) by David Wells. This book contains fascinating numbers and their properties, some of which are given here.

- There are only three sets of three digits which form prime numbers in all possible arrangements: {1, 1, 3}, {1, 9, 9}, and {3, 3, 7}.
- Find the sum of the cubes of the digits of 136: $1^3 + 3^3 + 6^3 = 244$. Repeat the process with the digits of 244: $2^3 + 4^3 + 4^3 = 136$. We're back to where we started.
- 635,318,657 is the smallest number that can be expressed as the sum of two fourth powers in two ways:
  $$635{,}318{,}657 = 59^4 + 158^4 = 133^4 + 134^4.$$
- The number 24,678,050 has an interesting property:
  $$24{,}678{,}050 = 2^8 + 4^8 + 6^8 + 7^8 + 8^8 + 0^8 + 5^8 + 0^8.$$
- The number 54,748 has a similar interesting property:
  $$54{,}748 = 5^5 + 4^5 + 7^5 + 4^5 + 8^5.$$
- The number 3435 has this property:
  $$3435 = 3^3 + 4^4 + 3^3 + 5^5.$$

For anyone whose curiosity is piqued by such facts, this book is for you!

For Group Discussion

Have each student in the class choose a three-digit number that is a multiple of 3. Add the cubes of the digits. Repeat the process until the same number is obtained over and over. Then, have the students compare their results. What is curious and interesting about this process?

5.2 EXERCISES

Decide whether each statement in Exercises 1–10 is true *or* false.

1. There are infinitely many prime numbers.
2. The prime numbers 2 and 3 are twin primes.
3. There is no perfect number between 496 and 8128.
4. $2^n - 1$ is prime if and only if $2^{n-1}(2^n - 1)$ is perfect.
5. Any prime number must be deficient.
6. The equation $17 + 51 = 68$ verifies Goldbach's conjecture for the number 68.
7. There are more Mersenne primes known than there are perfect numbers.
8. The number 31 cannot be represented as the difference of two squares.
9. The number $2^6(2^7 - 1)$ is perfect.
10. A natural number greater than 1 will be one and only one of the following: perfect, deficient, or abundant.
11. The proper divisors of 496 are 1, 2, 4, 8, 16, 31, 62, 124, and 248. Use this information to verify that 496 is perfect.
12. The proper divisors of 8128 are 1, 2, 4, 8, 16, 32, 64, 127, 254, 508, 1016, 2032, and 4064. Use this information to verify that 8128 is perfect.
13. As mentioned in the text, when $2^n - 1$ is prime, $2^{n-1}(2^n - 1)$ is perfect. By letting $n = 2, 3, 5$, and 7, we obtain the first four perfect numbers. Show that $2^n - 1$ is prime for $n = 13$, and then find the decimal digit representation for the fifth perfect number.
14. In 2002, the largest known prime number was $2^{13,466,917} - 1$. Use the formula in Exercise 13 to write an expression for the perfect number generated by this prime number.
15. It has been proved that the reciprocals of *all* the positive divisors of a perfect number have a sum of 2. Verify this for the perfect number 6.

16. Consider the following equations.

$$6 = 1 + 2 + 3$$
$$28 = 1 + 2 + 3 + 4 + 5 + 6 + 7$$

Show that a similar equation is valid for the third perfect number, 496.

Determine whether each number is abundant *or* deficient.

17. 36   **18.** 30   **19.** 75   **20.** 95

21. There are four abundant numbers between 1 and 25. Find them. (*Hint:* They are all even, and no prime number is abundant.)

22. Explain why a prime number must be deficient.

23. The first odd abundant number is 945. Its proper divisors are 1, 3, 5, 7, 9, 15, 21, 27, 35, 45, 63, 105, 135, 189, and 315. Use this information to verify that 945 is abundant.

24. Explain in your own words the terms *perfect number, abundant number,* and *deficient number.*

25. Nicolo Paganini's numbers 1184 and 1210 are amicable. The proper divisors of 1184 are 1, 2, 4, 8, 16, 32, 37, 74, 148, 296, and 592. The proper divisors of 1210 are 1, 2, 5, 10, 11, 22, 55, 110, 121, 242, and 605. Use the definition of amicable (friendly) numbers to show that they are indeed amicable.

26. An Arabian mathematician of the ninth century stated the following.

If the three numbers

$$x = 3 \cdot 2^{n-1} - 1,$$
$$y = 3 \cdot 2^n - 1,$$
and $$z = 9 \cdot 2^{2n-1} - 1$$

are all prime and $n \geq 2$, then $2^n xy$ and $2^n z$ are amicable numbers.

(a) Use $n = 2$, and show that the result is the pair of amicable numbers mentioned in the discussion in the text.

(b) Use $n = 4$ to obtain another pair of amicable numbers.

Write each of the following even numbers as the sum of two primes. (There may be more than one way to do this.)

27. 14   **28.** 22   **29.** 26   **30.** 32

31. Joseph Louis Lagrange (1736–1813) conjectured that every odd natural number greater than 5 can be written as a sum $a + 2b$, where a and b are both primes. Verify this for the odd natural number 11.

32. An unproved conjecture in number theory states that every natural number multiple of 6 can be written as the difference of two primes. Verify this for 6, 12, and 18.

Find one pair of twin primes between the two numbers given.

33. 65, 80   **34.** 85, 105   **35.** 125, 140

While Pierre de Fermat probably is best known for his now famous "last theorem," he did provide proofs of many other theorems in number theory. Exercises 36–40 investigate some of these theorems.

36. If p is prime and the natural numbers a and p have no common factor except 1, then $a^{p-1} - 1$ is divisible by p.
(a) Verify this for $p = 5$ and $a = 3$.
(b) Verify this for $p = 7$ and $a = 2$.

37. Every odd prime can be expressed as the difference of two squares in one and only one way.
(a) Find this one way for the prime number 5.
(b) Find this one way for the prime number 11.

38. A prime number of the form $4k + 1$ can be represented as the sum of two squares.
(a) The prime number 5 satisfies the conditions of the theorem, with $k = 1$. Verify this theorem for 5.
(b) Verify this theorem for 13 (here, $k = 3$).

39. There is only one solution in natural numbers for $a^2 + 2 = b^3$, and it is $a = 5, b = 3$. Verify this solution.

40. There are only two solutions in integers for $a^2 + 4 = b^3$, and they are $a = 2, b = 2$, and $a = 11, b = 5$. Verify these solutions.

The first four perfect numbers were identified in the text: 6, 28, 496, and 8128. The next two are 33,550,336 and 8,589,869,056. Use this information about perfect numbers to work Exercises 41–43.

41. Verify that each of these six perfect numbers ends in either 6 or 28. (In fact, this is true of all even perfect numbers.)

42. Is conjecture (1) in the text (that the nth perfect number contains exactly n digits) true or false? Explain.

43. Is conjecture (2) in the text (that the even perfect numbers end in the digits 6 and 8, alternately) true or false? Explain.

44. If the early Greeks knew the form of all even perfect numbers, namely $2^{n-1}(2^n - 1)$, then why did they not discover all the ones that are known today?

45. Explain why the primorial formula $p\# \pm 1$ does not result in a pair of twin primes for the prime value $p = 2$.

46. (a) What two numbers does the primorial formula produce for $p = 7$?
(b) Which, if either, of these numbers is prime?

47. Choose the correct completion: The primorial formula produces twin primes
A. never.   B. sometimes.   C. always.

See the margin note defining a Sophie Germain prime, and complete this table.

	p	$2p + 1$	Is p a Sophie Germain prime?
48.	2	___	___
49.	3	___	___
50.	5	___	___
51.	7	___	___
52.	11	___	___
53.	13	___	___

Yet another modern source of large prime numbers is the class of **factorial primes,** *those of the form* $n! \pm 1$ *for natural numbers n. (n! denotes "n factorial," the product of all natural numbers up to n, not just the primes as in the primorial primes.) As of late 2002, the largest verified factorial prime was* $34{,}790! - 1$, *which has 142,891 digits. Find the missing entries in this table.*

	n	$n!$	$n! - 1$	$n! + 1$	Is $n! - 1$ prime?	Is $n! + 1$ prime?
	2	2	1	3	no	yes
54.	3	___	___	___	___	___
55.	4	___	___	___	___	___
56.	5	___	___	___	___	___

57. Explain why the factorial prime formula does not give twin primes for $n = 2$.

Based on the preceding table, complete each of the following statements with one of the following: **A.** *never,* **B.** *sometimes, or* **C.** *always. When applied to particular values of n, the factorial formula* $n! \pm 1$ *produces*

58. no primes ___ .   **59.** exactly one prime ___ .   **60.** twin primes ___ .

5.3 Greatest Common Factor and Least Common Multiple

The **greatest common factor (GCF)** of a group of natural numbers is the largest natural number that is a factor of all the numbers in the group. For example, 18 is the greatest common factor of 36 and 54, since 18 is the largest natural number that divides both 36 and 54. Also, 1 is the greatest common factor of 7 and 18.

Greatest common factors can be found by using prime factorizations. To find the GCF of 36 and 54, first write the prime factorization of each number:

$$36 = 2^2 \cdot 3^2 \quad \text{and} \quad 54 = 2^1 \cdot 3^3.$$

The GCF is the product of the primes common to the factorizations, with each prime raised to the power indicated by the *smallest* exponent that it has in any factorization. Here, the prime 2 has 1 as the smallest exponent (in $54 = 2^1 \cdot 3^3$), while the prime 3 has 2 as the smallest exponent (in $36 = 2^2 \cdot 3^2$). The GCF of 36 and 54 is

$$2^1 \cdot 3^2 = 2 \cdot 9 = 18,$$

as stated earlier. We summarize as follows.

Pierre de Fermat (about 1601–1665), a government official who did not interest himself in mathematics until he was past 30, devoted leisure time to its study. He was a worthy scholar, best known for his work in number theory. His other major contributions involved certain applications in geometry and his original work in probability.

Much of Fermat's best work survived only on loose sheets or jotted, without proof, in the margins of works that he read. Mathematicians of subsequent generations have not always had an easy time verifying some of those results, though their truth has generally not been doubted.

Finding the Greatest Common Factor (Prime Factors Method)

1. Write the prime factorization of each number.
2. Choose all primes common to *all* factorizations, with each prime raised to the *smallest* exponent that it has in any factorization.
3. Form the product of all the numbers in Step 2; this product is the greatest common factor.

EXAMPLE 1 Find the greatest common factor of 360 and 2700.

Write the prime factorization of each number:

$$360 = 2^3 \cdot 3^2 \cdot 5 \quad \text{and} \quad 2700 = 2^2 \cdot 3^3 \cdot 5^2.$$

Now find the primes common to both factorizations, with each prime having as exponent the *smallest* exponent from either product: 2^2, 3^2, 5. Then form the product of these numbers.

$$\text{GCF} = 2^2 \cdot 3^2 \cdot 5 = 180$$

The greatest common factor of 360 and 2700 is 180.

The calculator shows that the greatest common divisor (factor) of 360 and 2700 is 180. Compare with Example 1.

EXAMPLE 2 Find the greatest common factor of 720, 1000, and 1800.

Write the prime factorization for each number:

$$720 = 2^4 \cdot 3^2 \cdot 5, \quad 1000 = 2^3 \cdot 5^3, \quad \text{and} \quad 1800 = 2^3 \cdot 3^2 \cdot 5^2.$$

Use the smallest exponent on each prime common to the factorizations:

$$\text{GCF} = 2^3 \cdot 5 = 40.$$

(The prime 3 is not used in the greatest common factor since it does not appear in the prime factorization of 1000.)

EXAMPLE 3 Find the greatest common factor of 80 and 63.

Start with

$$80 = 2^4 \cdot 5 \quad \text{and} \quad 63 = 3^2 \cdot 7.$$

There are no primes in common here, so the GCF is 1. The number 1 is the largest number that will divide into both 80 and 63.

Two numbers, such as 80 and 63, with a greatest common factor of 1 are called **relatively prime** numbers—that is, they are prime *relative* to one another.

Another method of finding the greatest common factor involves dividing the numbers by common prime factors.

> ### Finding the Greatest Common Factor (Dividing by Prime Factors Method)
> 1. Write the numbers in a row.
> 2. Divide each of the numbers by a common prime factor. Try 2, then try 3, and so on.
> 3. Divide the quotients by a common prime factor. Continue until no prime will divide into all the quotients.
> 4. The product of the primes in Steps 2 and 3 is the greatest common factor.

This method is illustrated in the next example.

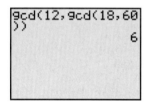

This screen uses the fact that gcd (a, b, c) = gcd (a, gcd(b, c)). Compare with Example 4.

EXAMPLE 4 Find the greatest common factor of 12, 18, and 60.

Write the numbers in a row and divide by 2.

$$\begin{array}{r|rrr} 2 & 12 & 18 & 60 \\ \hline & 6 & 9 & 30 \end{array}$$

The numbers 6, 9, and 30 are not all divisible by 2, but they are divisible by 3.

$$\begin{array}{r|rrr} 2 & 12 & 18 & 60 \\ \hline 3 & 6 & 9 & 30 \\ \hline & 2 & 3 & 10 \end{array}$$

No prime divides into 2, 3, and 10, so the greatest common factor of the numbers 12, 18, and 60 is given by the product of the primes on the left, 2 and 3.

$$\begin{array}{r|rrr} 2 & 12 & 18 & 60 \\ \hline 3 & 6 & 9 & 30 \\ \hline & 2 & 3 & 10 \end{array}$$

$$2 \cdot 3 = 6$$

The GCF of 12, 18, and 60 is 6.

There is yet another method of finding the greatest common factor of two numbers (but not more than two) that does not require factoring into primes or successively dividing by primes. It is called the **Euclidean algorithm,*** and it is illustrated in the next example.

---

*For a proof that this process does indeed give the greatest common factor, see *Elementary Introduction to Number Theory, Second Edition*, by Calvin T. Long, pp. 34–35.

EXAMPLE 5 Use the Euclidean algorithm to find the greatest common factor of 90 and 168.

Step 1: Begin by dividing the larger, 168, by the smaller, 90. Disregard the quotient, but note the remainder.

Step 2: Divide the smaller of the two numbers by the remainder obtained in Step 1. Once again, note the remainder.

Step 3: Continue dividing the successive remainders, as many times as necessary to obtain a remainder of 0.

Step 4: The *last positive remainder* in this process is the greatest common factor of 90 and 168. It can be seen that their GCF is 6.

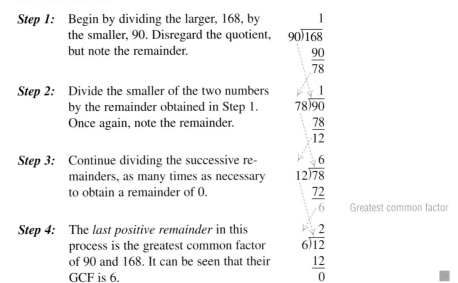

The Euclidean algorithm is particularly useful if the two numbers are difficult to factor into primes. We summarize the algorithm here.

Finding the Greatest Common Factor (Euclidean Algorithm)

To find the greatest common factor of two unequal numbers, divide the larger by the smaller. Note the remainder, and divide the previous divisor by this remainder. Continue the process until a remainder of 0 is obtained. The greatest common factor is the last positive remainder obtained in this process.

Closely related to the idea of the greatest common factor is the concept of the least common multiple. The **least common multiple (LCM)** of a group of natural numbers is the smallest natural number that is a multiple of all the numbers in the group. For example, if we wish to find the least common multiple of 15 and 10, we can specify the sets of multiples of 15 and multiples of 10.

Multiples of 15: {15, 30, 45, 60, 75, 90, 105, …}

Multiples of 10: {10, 20, 30, 40, 50, 60, 70, …}

The set of natural numbers that are multiples of *both* 15 and 10 form the set of *common multiples:*

{30, 60, 90, 120, …}.

While there are infinitely many common multiples, the *least* common multiple is observed to be 30.

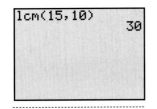

The least common multiple of 15 and 10 is 30.

A method similar to the first one given for the greatest common factor may be used to find the least common multiple of a group of numbers.

Finding the Least Common Multiple (Prime Factors Method)

1. Write the prime factorization of each number.
2. Choose all primes belonging to *any* factorization, with each prime raised to the power indicated by the *largest* exponent that it has in any factorization.
3. Form the product of all the numbers in Step 2; this product is the least common multiple.

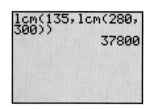

The least common multiple of 135, 280, and 300 is 37,800. Compare with Example 6.

EXAMPLE 6 Find the least common multiple of 135, 280, and 300.

Write the prime factorizations:

$$135 = 3^3 \cdot 5, \quad 280 = 2^3 \cdot 5 \cdot 7, \quad \text{and} \quad 300 = 2^2 \cdot 3 \cdot 5^2.$$

Form the product of all the primes that appear in *any* of the factorizations. Use the *largest* exponent from any factorization.

$$\text{LCM} = 2^3 \cdot 3^3 \cdot 5^2 \cdot 7 = 37{,}800$$

The smallest natural number divisible by 135, 280, and 300 is 37,800. ■

The least common multiple of a group of numbers can also be found by dividing by prime factors. The process is slightly different than that for finding the GCF.

Finding the Least Common Multiple (Dividing by Prime Factors Method)

1. Write the numbers in a row.
2. Divide each of the numbers by a common prime factor. Try 2, then try 3, and so on.
3. Divide the quotients by a common prime factor. When no prime will divide all quotients, but a prime will divide some of them, divide where possible and bring any nondivisible quotients down. Continue until no prime will divide any two quotients.
4. The product of all prime divisors from Steps 2 and 3 as well as all remaining quotients is the least common multiple.

EXAMPLE 7 Find the least common multiple of 12, 18, and 60.

Proceed just as in Example 4 to obtain the following.

$$\begin{array}{r|rrr} 2 & 12 & 18 & 60 \\ \hline 3 & 6 & 9 & 30 \\ \hline & 2 & 3 & 10 \end{array}$$

Now, even though no prime will divide 2, 3, and 10, the prime 2 will divide 2 and 10. Divide the 2 and the 10 and bring down the 3.

$$
\begin{array}{r|rrr}
2 & 12 & 18 & 60 \\
3 & 6 & 9 & 30 \\
2 & 2 & 3 & 10 \\
& 1 & 3 & 5
\end{array}
$$

$$2 \cdot 3 \cdot 2 \cdot 1 \cdot 3 \cdot 5 = 180$$

The LCM of 12, 18, and 60 is 180.

It is shown in more advanced courses that the least common multiple of two numbers m and n can be obtained by dividing their product by their greatest common factor. (This method works only for two numbers, not for more than two.)

Finding the Least Common Multiple (Formula)

The least common multiple of m and n is given by

$$\text{LCM} = \frac{m \cdot n}{\text{Greatest common factor of } m \text{ and } n}.$$

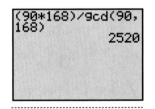

This supports the result in Example 8.

EXAMPLE 8 Use the formula to find the least common multiple of 90 and 168.

In Example 5 we used the Euclidean algorithm to find that the greatest common factor of 90 and 168 is 6. Therefore, the formula gives us

$$\text{Least common multiple of 90 and 168} = \frac{90 \cdot 168}{6} = 2520.$$

Problem Solving

Problems that deal with questions such as "How many objects will there be in each group if each group contains the same number of objects?" and "When will two events occur at the same time?" can sometimes be solved using the ideas of greatest common factor and least common multiple.

EXAMPLE 9 The King Theatre and the Star Theatre run movies continuously, and each starts its first feature at 1:00 P.M. If the movie shown at the King lasts 80 minutes and the movie shown at the Star lasts 2 hours, when will the two movies start again at the same time?

First, convert 2 hours to 120 minutes. The question can be restated as follows: "What is the smallest number of minutes it will take for the two movies to start at the same time again?" This is equivalent to saying, "What is the least common multiple of 80 and 120?" Using any of the methods described in this section, it can be shown that the least common multiple of 80 and 120 is 240. Therefore, it will take 240 minutes, or $240/60 = 4$ hours for the movies to start again at the same time. By adding 4 hours to 1:00 P.M., we find that they will start together again at 5:00 P.M.

EXAMPLE 10 Joshua Hornsby has 450 football cards and 840 baseball cards. He wants to place them in stacks on a table so that each stack has the same number of cards, and no stack has different types of cards within it. What is the largest number of cards that he can have in each stack?

Here, we are looking for the largest number that will divide evenly into 450 and 840. This is, of course, the greatest common factor of 450 and 840. Using any of the methods described in this section, we find that

$$\text{Greatest common factor of 450 and 840} = 30.$$

Therefore, the largest number of cards he can have in each stack is 30.

5.3 EXERCISES

Decide whether each of the following is true *or* false.

1. Two even natural numbers cannot be relatively prime.
2. Two different prime numbers must be relatively prime.
3. If p is a prime number, then the greatest common factor of p and p^2 is p.
4. If p is a prime number, then the least common multiple of p and p^2 is p^3.
5. There is no prime number p such that the greatest common factor of p and 2 is 2.
6. The set of all common multiples of two given natural numbers is finite.
7. Two natural numbers must have at least one common factor.
8. The least common multiple of two different primes is their product.
9. Two composite numbers may be relatively prime.
10. The set of all common factors of two given natural numbers is finite.

Use the prime factors method to find the greatest common factor of each group of numbers.

11. 70 and 120
12. 180 and 300
13. 480 and 1800
14. 168 and 504
15. 28, 35, and 56
16. 252, 308, and 504

Use the method of dividing by prime factors to find the greatest common factor of each group of numbers.

17. 60 and 84
18. 130 and 455
19. 310 and 460
20. 234 and 470
21. 12, 18, and 30
22. 450, 1500, and 432

Use the Euclidean algorithm to find the greatest common factor of each group of numbers.

23. 36 and 60
24. 25 and 70
25. 84 and 180
26. 72 and 120
27. 210 and 560
28. 150 and 480

29. Explain in your own words how to find the greatest common factor of a group of numbers.
30. Explain in your own words how to find the least common multiple of a group of numbers.

Use the prime factors method to find the least common multiple of each group of numbers.

31. 24 and 30
32. 12 and 32
33. 56 and 96
34. 28 and 70
35. 30, 40, and 70
36. 24, 36, and 48

Use the method of dividing by prime factors to find the least common multiple of each group of numbers.

37. 24 and 32
38. 35 and 56
39. 45 and 75
40. 48, 54, and 60
41. 16, 120, and 216
42. 210, 385, and 2310

Use the formula given in the text and the results of Exercises 23–28 to find the least common multiple of each group of numbers.

43. 36 and 60
44. 25 and 70
45. 84 and 180
46. 72 and 120
47. 210 and 560
48. 150 and 480

49. If p, q, and r are different primes, and a, b, and c are natural numbers such that $a > b > c$,
    (a) what is the greatest common factor of $p^a q^c r^b$ and $p^b q^a r^c$?
    (b) what is the least common multiple of $p^b q^a$, $q^b r^c$, and $p^a r^b$?

50. Find (a) the greatest common factor and (b) the least common multiple of $2^{31} \cdot 5^{17} \cdot 7^{21}$ and $2^{34} \cdot 5^{22} \cdot 7^{13}$. Leave your answers in prime factored form.

It is possible to extend the Euclidean algorithm in order to find the greatest common factor of more than two numbers. For example, if we wish to find the greatest common factor of 150, 210, and 240, we can first use the algorithm to find the greatest common factor of two of these (say, for example, 150 and 210). Then we find the greatest common factor of that result and the third number, 240. The final result is the greatest common factor of the original group of numbers.

Use the Euclidean algorithm as described above to find the greatest common factor of each group of numbers.

51. 150, 210, and 240
52. 12, 75, and 120
53. 90, 105, and 315
54. 48, 315, and 450
55. 144, 180, and 192
56. 180, 210, and 630

If we allow repetitions of prime factors, we can use Venn diagrams (Chapter 2) to find the greatest common factor and the least common multiple of two numbers. For example, consider $36 = 2^2 \cdot 3^2$ and $45 = 3^2 \cdot 5$. Their greatest common factor is $3^2 = 9$, and their least common multiple is $2^2 \cdot 3^2 \cdot 5 = 180$.

Intersection gives 3, 3.
Union gives 2, 2, 3, 3, 5.

Use this method to find (a) the greatest common factor and (b) the least common multiple of the two numbers given.

57. 12 and 18
58. 27 and 36
59. 54 and 72

60. Suppose that the least common multiple of p and q is q. What can we say about p and q?

61. Suppose that the least common multiple of p and q is pq. What can we say about p and q?

62. Suppose that the greatest common factor of p and q is p. What can we say about p and q?

63. Recall some of your early experiences in mathematics (for example, in the elementary grade classroom). What topic involving fractions required the use of the least common multiple? Give an example.

64. Recall some of your experiences in elementary algebra. What topics required the use of the greatest common factor? Give an example.

Refer to Examples 9 and 10 to solve the following problems.

65. *Inspecting Calculators* Colleen Jones and Nancy Barre work on an assembly line, inspecting electronic calculators. Colleen inspects the electronics of every sixteenth calculator, while Nancy inspects the workmanship of every thirty-sixth calculator. If they both start working at the same time, which calculator will be the first that they both inspect?

66. *Night Off for Security Guards* Paul Crockett and Cindy Herring work as security guards at a publishing company. Paul has every sixth night off, and Cindy has every tenth night off. If both are off on July 1, what is the next night that they will both be off together?

67. *Stacking Coins* Sheila Abbruzzo has 240 pennies and 288 nickels. She wants to place the pennies and nickels in stacks so that each stack has the same number of coins, and each stack contains only one denomination of coin. What is the largest number of coins that she can place in each stack?

68. *Bicycle Racing* Kathryn Campbell and Tami Dreyfus are in a bicycle race, following a circular track. If they start at the same place and travel in the same direction, and Kathryn completes a revolution every 40 seconds and Tami completes a revolution every 45 seconds, how long will it take them before they reach the starting point again simultaneously?

69. *Selling Books* John Cross sold some books at $24 each, and had enough money to buy some concert tickets at $50 each. He had no money left over after buying the tickets. What is the least amount of money he could have earned from selling the books? What is the least number of books he could have sold?

70. *Sawing Lumber* Jill Bos has some two-by-four pieces of lumber. Some are 60 inches long, and some are 72 inches long. She wishes to saw them so as to obtain equal-length pieces. What is the longest such piece she can saw so that no lumber is left over?

5.4 Clock Arithmetic and Modular Systems

Clock Arithmetic The introduction to Chapter 4 described a "mathematical system" as (1) a set of elements, along with (2) one or more operations for combining those elements, and (3) one or more relations for comparing those elements. We now consider the *12-hour clock system,* which is based on an ordinary clock face, except that 12 is replaced with 0 and only a single hand, say the hour hand, is used. See Figure 1.

FIGURE 1

The clock face yields the finite set {0, 1, 2, 3, 4, 5, 6, 7, 8, 9, 10, 11}. As an operation for this clock system, addition is defined as follows: add by moving the hour hand in a clockwise direction. For example, to add 5 and 2 on a clock, first move the hour hand to 5, as in Figure 2. Then, to add 2, move the hour hand 2 more hours in a clockwise direction. The hand stops at 7, so

$$5 + 2 = 7.$$

This result agrees with traditional addition. However, the sum of two numbers from the 12-hour clock system is not always what might be expected, as the following example shows.

Plus 2 hours
$5 + 2 = 7$

FIGURE 2

EXAMPLE 1 Find each sum in 12-hour clock arithmetic.

(a) $8 + 9$

Move the hour hand to 8, as in Figure 3. Then advance the hand clockwise through 9 more hours. It stops at 5, so

$$8 + 9 = 5.$$

(b) $11 + 3$

Proceed as shown in Figure 4. Check that

$$11 + 3 = 2.$$ ■

$8 + 9 = 5$

FIGURE 3

Since there are infinitely many whole numbers, it is not possible to write a complete table of addition facts for that set. Such a table, to show the sum of every possible pair of whole numbers, would have an infinite number of rows and columns, making it impossible to construct.

On the other hand, the 12-hour clock system uses only the whole numbers 0, 1, 2, 3, 4, 5, 6, 7, 8, 9, 10, and 11. A table of all possible sums for this system requires only 12 rows and 12 columns. The 12-hour clock **addition table** is shown in Table 5. Since the 12-hour system is built upon a finite set, it is called a **finite mathematical system.**

$11 + 3 = 2$

FIGURE 4

TABLE 5 12-Hour Clock Addition

+	0	1	2	3	4	5	6	7	8	9	10	11
0	0	1	2	3	4	5	6	7	8	9	10	11
1	1	2	3	4	5	6	7	8	9	10	11	0
2	2	3	4	5	6	7	8	9	10	11	0	1
3	3	4	5	6	7	8	9	10	11	0	1	2
4	4	5	6	7	8	9	10	11	0	1	2	3
5	5	6	7	8	9	10	11	0	1	2	3	4
6	6	7	8	9	10	11	0	1	2	3	4	5
7	7	8	9	10	11	0	1	2	3	4	5	6
8	8	9	10	11	0	1	2	3	4	5	6	7
9	9	10	11	0	1	2	3	4	5	6	7	8
10	10	11	0	1	2	3	4	5	6	7	8	9
11	11	0	1	2	3	4	5	6	7	8	9	10

Plagued by serious maritime mishaps linked to navigational difficulties, several European governments offered prizes for an effective method of determining longitude. The largest prize was 20,000 pounds (equivalent to several million dollars in today's currency) offered by the British Parliament in the Longitude Act of 1714. While famed scientists, academics, and politicians pursued an answer in the stars, John Harrison, a clock maker, set about to build a clock that could maintain accuracy at sea. This turned out to be the key, and Harrison's Chronometer eventually earned him the prize.

For a fascinating account of this drama and of Harrison's struggle to collect his prize money from the government, see the book *The Illustrated Longitude* by Dava Sobel and William J. H. Andrewes.

EXAMPLE 2 Use the 12-hour clock addition table to find each sum.

(a) $7 + 11$

Find 7 on the left of the addition table and 11 across the top. The intersection of the row headed 7 and the column headed 11 gives the number 6. Thus, $7 + 11 = 6$.

(b) Also from the table, $11 + 1 = 0$.

So far, our 12-hour clock system consists of the set $\{0, 1, 2, 3, 4, 5, 6, 7, 8, 9, 10, 11\}$, together with the operation of clock addition. Next we will check whether this sytem has the *closure, commutative, associative, identiy,* and *inverse* properties, as described in Section 4.4.

Table 5 shows that the sum of two numbers on a clock face is always a number on the clock face. That is, if a and b are any clock numbers in the set of the system, then $a + b$ is also in the set of the system. The system has the **closure property.** (The set of the system is *closed* under clock addition.)

Notice also that in this system, $5 + 9$ and $9 + 5$ both yield 2. Also $7 + 11$ and $11 + 7$ both yield 6. The order in which elements are added does not seem to matter. In fact, you can see in Table 5 that the part of the table *above* the colored diagonal line is a mirror image of the part *below* the diagonal. This shows that, for any clock numbers a and b, $a + b = b + a$. The system has the **commutative property.**

The next question is: When three elements are combined in a given order, say $a + b + c$, does it matter whether the first and second or the second and third are associated initially? In other words, is it true that, for any a, b, and c in the system, $(a + b) + c = a + (b + c)$?

EXAMPLE 3 Is 12-hour clock addition associative?

It would take lots of work to prove that the required relationship *always* holds. But a few examples should either disprove it (by revealing a *counterexample*—a case where it fails to hold), or should make it at least plausible. Using the clock numbers 4, 5, and 9, we see that

$$(4 + 5) + 9 = 9 + 9 \qquad 4 + (5 + 9) = 4 + 2$$
$$= 6 \qquad\qquad\qquad = 6.$$

Thus, $(4 + 5) + 9 = 4 + (5 + 9)$. Try another example:

$$(7 + 6) + 3 = 1 + 3 \qquad 7 + (6 + 3) = 7 + 9$$
$$= 4 \qquad\qquad\qquad = 4.$$

So $(7 + 6) + 3 = 7 + (6 + 3)$. Other examples also will work. The 12-hour clock system has the **associative property.**

Our next question is whether the clock face contains some element (number) which, when combined with any element (in either order), produces that same element. Such an element (call it e) would satisfy $a + e = a$ and $e + a = a$ for any element a of the system. Notice in Table 5 that $4 + 0 = 0 + 4 = 4$, $6 + 0 = 0 + 6 = 6$, and so on. The number 0 is the required *identity element.* The system has the **identity property.**

Generally, if a finite system has an identity element e, it can be located easily in the operation table. Check the body of Table 5 for a column that is identical to the column at the left side of the table. Since the column under 0 meets this requirement, $a + 0 = a$ holds for all elements a in the system. Thus 0 is *possibly* the identity.

Minus 5 hours

$2 - 5 = 9$

FIGURE 5

Now locate 0 at the left of the table. Since the corresponding row is identical to the row at the top of the table, $0 + a = a$ also holds for all elements a, which is the other requirement of an identity element. Hence 0 is *indeed* the identity.

Subtraction can be performed on a 12-hour clock. Subtraction may be interpreted on the clock face by a movement in the counterclockwise direction. For example, to perform the subtraction $2 - 5$, begin at 2 and move 5 hours counterclockwise, ending at 9, as shown in Figure 5. Therefore, in this system,

$$2 - 5 = 9.$$

In our usual system, subtraction may be checked by addition and this is also the case in clock arithmetic. To check that $2 - 5 = 9$, simply add $9 + 5$. The result on the clock face is 2, verifying the accuracy of this subtraction.

The *additive inverse*, $-a$, of an element a in clock arithmetic is the element that satisfies the following statement: $a + (-a) = 0$ and $(-a) + a = 0$. The next example examines this idea.

EXAMPLE 4 Determine the additive inverse of 5 in 12-hour clock arithmetic.

The additive inverse for the clock number 5 is a number x such that

$$5 + x = 0.$$

Going from 5 to 0 on the clock face requires 7 more hours, so

$$5 + 7 = 0.$$

This means that 7 is the additive inverse of 5.

The method used in Example 4 may be used to verify that *every* element of the system has an additive inverse (also in the system). So the system has the **inverse property.**

A simpler way to verify the inverse property, once you have the table, is to make sure the identity element appears exactly once in each row, and that the pair of elements that produces it also produces it in the opposite order. (This last condition is automatically true if the commutative property holds for the system.) For example, note in Table 5 that row 3 contains one 0, under the 9, so $3 + 9 = 0$, and that row 9 contains 0, under the 3, so $9 + 3 = 0$ also. Therefore, 3 and 9 are inverses.

The following chart lists the elements and their additive inverses. Notice that one element, 6, is its own inverse for addition.

Clock value a	0	1	2	3	4	5	6	7	8	9	10	11
Additive inverse $-a$	0	11	10	9	8	7	6	5	4	3	2	1

Using the additive inverse symbol, we can say that in clock arithmetic,

$$-5 = 7, \quad -11 = 1, \quad -10 = 2,$$

and so on.

We have now seen that the 12-hour clock system, with addition, has all five of the potential properties of a single-operation mathematical system, as discussed in Section 4.4. Other operations can also be introduced to the clock arithmetic system.

5.4 Clock Arithmetic and Modular Systems

Having discussed additive inverses, we can define subtraction formally. Notice that the definition is the same as for ordinary subtraction of whole numbers.

Subtraction on a Clock

If a and b are elements in clock arithmetic, then the **difference**, $a - b$, is defined as

$$a - b = a + (-b).$$

Chess Clock A double clock is used to time chess, backgammon, and Scrabble games. Push one button, and that clock stops—the other begins simultaneously. When a player's allotted time for the game has expired, that player will lose if he or she has not made the required number of moves.

Mathematics and chess both involve structured relationships and demand logical thinking. A chess player may depend more on psychological acumen and knowledge of the opponent than on the mathematical component of the game.

Emanuel Lasker achieved mastery in both fields. He was best known as a World Chess Champion for 27 years, until 1921. Lasker also was famous in mathematical circles for his work concerning the theory of primary ideals, algebraic analogies of prime numbers. An important result, the Lasker-Noether theorem, bears his name along with that of Emmy Noether. Noether extended Lasker's work. Her father had been Lasker's Ph.D. advisor.

EXAMPLE 5 Find each of the following differences.

(a) $8 - 5 = 8 + (-5)$ Use the definition of subtraction.
$\qquad = 8 + 7$ The additive inverse of 5 is 7, from the table of inverses.
$8 - 5 = 3$

This result agrees with traditional arithmetic. Check by adding 5 and 3; the sum is 8.

(b) $6 - 11 = 6 + (-11)$
$\qquad = 6 + 1$
$6 - 11 = 7$

Clock numbers can also be multiplied. For example, find the product 5×4 by adding 4 a total of 5 times:

$$5 \times 4 = 4 + 4 + 4 + 4 + 4 = 8.$$

EXAMPLE 6 Find each product, using clock arithmetic.

(a) $6 \times 9 = 9 + 9 + 9 + 9 + 9 + 9 = 6$
(b) $3 \times 4 = 4 + 4 + 4 = 0$
(c) $6 \times 0 = 0 + 0 + 0 + 0 + 0 + 0 = 0$
(d) $0 \times 8 = 0$

Some properties of the system of 12-hour clock numbers with the operation of multiplication will be investigated in Exercises 6–8.

Modular Systems We now expand the ideas of clock arithmetic to **modular systems** in general. Recall that 12-hour clock arithmetic was set up so that answers were always whole numbers less than 12. For example, $8 + 6 = 2$. The traditional sum, $8 + 6 = 14$, reflects the fact that moving the clock hand forward 8 hours from 0, and then forward another 6 hours, amounts to moving it forward 14 hours total. But since the final position of the clock is at 2, we see that 14 and 2 are, in a sense, equivalent. More formally we say that 14 and 2 are **congruent modulo** 12 (or **congruent mod** 12), which is written

$$14 \equiv 2 \pmod{12} \qquad \text{(The sign } \equiv \text{ indicates congruence.)}$$

By observing clock hand movements, you can also see that, for example,

$$26 \equiv 2 \ (\text{mod } 12), \quad 38 \equiv 2 \ (\text{mod } 12), \quad \text{and so on.}$$

In each case, the congruence is true because the difference of the two congruent numbers is a multiple of 12:

$$14 - 2 = 12 = 1 \times 12, \quad 26 - 2 = 24 = 2 \times 12, \quad 38 - 2 = 36 = 3 \times 12.$$

This suggests the following definition.

Congruence Modulo m

The integers a and b are **congruent modulo m** (where m is a natural number greater than 1 called the **modulus**) if and only if the difference $a - b$ is divisible by m. Symbolically, this congruence is written

$$a \equiv b \ (\text{mod } m).$$

Since being divisible by m is the same as being a multiple of m, we can say that

$$a \equiv b \ (\text{mod } m) \text{ if and only if } a - b = km \text{ for some integer } k.$$

EXAMPLE 7 Decide whether each statement is *true* or *false*.

(a) $16 \equiv 10 \ (\text{mod } 2)$

The difference $16 - 10 = 6$ is divisible by 2, so $16 \equiv 10 \ (\text{mod } 2)$ is true.

(b) $49 \equiv 32 \ (\text{mod } 5)$

This statement is false, since $49 - 32 = 17$, which is not divisible by 5.

(c) $30 \equiv 345 \ (\text{mod } 7)$

This statement is true, since $30 - 345 = -315$ is divisible by 7. (It doesn't matter if we find $30 - 345$ or $345 - 30$.)

There is another method of determining if two numbers, a and b, are congruent modulo m.

Criterion for Congruence

$a \equiv b \ (\text{mod } m)$ if and only if the same remainder is obtained when a and b are divided by m.

For example, we know that $27 \equiv 9 \ (\text{mod } 6)$ because $27 - 9 = 18$, which is divisible by 6. Now, if 27 is divided by 6, the quotient is 4 and the remainder is 3. Also, if 9 is divided by 6, the quotient is 1 and the remainder is 3. According to the criterion above, $27 \equiv 9 \ (\text{mod } 6)$ since both remainders are the same.

Addition, subtraction, and multiplication can be performed in any modular system just as with clock numbers. Since final answers should be whole numbers less than the modulus, we can first find an answer using ordinary arithmetic. Then, as long as the answer is nonnegative, simply divide it by the modulus and keep the remainder. This produces the smallest nonnegative integer that is congruent (modulo m) to the ordinary answer.

EXAMPLE 8 Find each of the following sums, differences, and products.

(a) $(9 + 14)(\mod 3)$

First add 9 and 14 to get 23. Then divide 23 by 3. The remainder is 2, so $23 \equiv 2 \pmod 3$ and
$$(9 + 14) \equiv 2 \pmod 3.$$

(b) $(27 - 5)(\mod 6)$

$27 - 5 = 22$. Divide 22 by 6, obtaining 4 as a remainder:
$$(27 - 5) \equiv 4 \pmod 6.$$

(c) $(50 + 34)(\mod 7)$

$50 + 34 = 84$. When 84 is divided by 7, a remainder of 0 is found:
$$(50 + 34) \equiv 0 \pmod 7.$$

(d) $(8 \times 9)(\mod 10)$

Since $8 \times 9 = 72$, and 72 leaves a remainder of 2 when divided by 10,
$$(8 \times 9) \equiv 2 \pmod{10}.$$

(e) $(12 \times 10)(\mod 5)$
$$(12 \times 10) = 120 \equiv 0 \pmod 5.$$

Problem Solving

Modular systems can often be applied to questions involving cyclical changes. For example, our method of dividing time into weeks causes the days to repeatedly cycle through the same pattern of seven. Suppose today is Sunday and we want to know what day of the week it will be 45 days from now. Since we don't care how many weeks will pass between now and then, we can discard the largest whole number of weeks in 45 days and keep the remainder. (We are finding the smallest nonnegative integer that is congruent to 45 modulo 7.) Dividing 45 by 7 leaves remainder 3, so the desired day of the week is 3 days past Sunday, or *Wednesday*.

EXAMPLE 9 If today is Thursday, November 12, and *next* year is a leap year, what day of the week will it be one year from today?

A modulo 7 system applies here, but we need to know the number of days between today and one year from today. Today's date, November 12, is unimportant except that it shows we are later in the year than the end of February and therefore the next year (starting today) will contain 366 days. (This would not be so if today were, say, January 12.) Now dividing 366 by 7 produces 52 with remainder 2. Two days past Thursday is our answer. That is, one year from today will be a Saturday.

> **Problem Solving**
>
> A modular system (mod m) allows only a fixed set of remainder values, 0, 1, 2,..., $m - 1$. One practical approach to solving modular equations, at least when m is reasonably small, is to simply try all these integers. For each solution found in this way, others can be found by adding multiples of the modulus to it.

EXAMPLE 10 Solve $(3 + x) \equiv 5 \pmod{7}$.

In a mod 7 system, any integer will be congruent to one of the integers 0, 1, 2, 3, 4, 5, or 6. So, the equation $(3 + x) \equiv 5 \pmod 7$ can be solved by trying, in turn, each of these integers as a replacement for x.

$x = 0$: Is it true that $(3 + 0) \equiv 5 \pmod 7$? No
$x = 1$: Is it true that $(3 + 1) \equiv 5 \pmod 7$? No
$x = 2$: Is it true that $(3 + 2) \equiv 5 \pmod 7$? Yes

Try $x = 3$, $x = 4$, $x = 5$, and $x = 6$ to see that none of them work. Of the integers from 0 through 6, only 2 is a solution of the equation $(3 + x) \equiv 5 \pmod 7$.

Since 2 is a solution, find other solutions to this mod 7 equation by repeatedly adding 7:

$$2 + 7 = 9, \quad 9 + 7 = 16, \quad 16 + 7 = 23, \text{ and so on.}$$

The set of all positive solutions of $(3 + x) \equiv 5 \pmod 7$ is

$$\{2, 9, 16, 23, 30, 37, ...\}.$$

EXAMPLE 11 Solve the equation $5x \equiv 4 \pmod 9$.

Because the modulus is 9, try 0, 1, 2, 3, 4, 5, 6, 7, and 8:

Is it true that $5 \times 0 \equiv 4 \pmod 9$? No
Is it true that $5 \times 1 \equiv 4 \pmod 9$? No

Continue trying numbers. You should find that none work except $x = 8$:

$$5 \times 8 = 40 \equiv 4 \pmod 9.$$

The set of all positive solutions to the equation $5x \equiv 4 \pmod 9$ is

$$\{8, 8 + 9, 8 + 9 + 9, 8 + 9 + 9 + 9, ...\} \text{ or}$$
$$\{8, 17, 26, 35, 44, 53, ...\}.$$

EXAMPLE 12 Solve the equation $6x \equiv 3 \pmod 8$.

Try the numbers 0, 1, 2, 3, 4, 5, 6, and 7. You should find that none work. Therefore, the equation $6x \equiv 3 \pmod 8$ has no solutions at all. Write the set of all solutions as the empty set, $\emptyset$.

This result is reasonable since $6x$ will always be even, no matter which whole number is used for x. Since $6x$ is even and 3 is odd, the difference $6x - 3$ will be odd, and therefore not divisible by 8.

EXAMPLE 13 Solve $8x \equiv 8 \pmod 8$.

Trying the integers 0, 1, 2, 3, 4, 5, 6, and 7 shows that *any* replacement will work. The solution set is $\{0, 1, 2, 3, ...\}$.

An equation such as $8x \equiv 8 \pmod{8}$ in Example 13 that is true for all values of the variable (x, y, and so on) is called an **identity.** Other examples of identities (in ordinary algebra) include $2x + 3x = 5x$ and $y^2 \cdot y^3 = y^5$.

Some problems can be solved by writing down two or more modular equations and finding their common solutions. The next example illustrates the process.

EXAMPLE 14 Julio wants to arrange his CD collection in equal size stacks, but after trying stacks of 4, stacks of 5, and stacks of 6, he finds that there is always 1 disc left over. Assuming Julio owns more than one CD, what is the least possible number of discs in his collection?

The given information leads to three modular equations,

$$x \equiv 1 \pmod{4}, \quad x \equiv 1 \pmod{5}, \quad x \equiv 1 \pmod{6},$$

whose sets of positive solutions are, respectively,

$\{1, 5, 9, 13, 17, 21, 25, 29, 33, 37, 41, 45, 49, 53, 57, 61, 65, 69, \ldots\}$,
$\{1, 6, 11, 16, 21, 26, 31, 36, 41, 46, 51, 56, 61, 66, 71, 76, \ldots\}$, and
$\{1, 7, 13, 19, 25, 31, 37, 43, 49, 55, 61, \ldots\}$.

The smallest common solution greater than 1 is 61, so the least possible number of discs in the collection is 61.

EXAMPLE 15 A dry-wall contractor is ordering materials to finish a 17-foot-by-45-foot room. The wallboard panels come in 4-foot widths. Show that, after uncut panels are applied, all four walls will require additional partial strips of the same width.

The width of any partial strip needed will be the remainder when the wall length is divided by 4 (the panel width). In terms of congruence, we must show that $17 \equiv 45 \pmod{4}$. By the criterion given above, we see that this is true since both 17 and 45 give the same remainder (namely 1) when divided by 4. A 1-foot partial strip will be required for each wall. (In this case four 1-foot strips can be cut from a single panel, so there will be no waste.)

FOR FURTHER THOUGHT

A Card Trick

Many card "tricks" that have been around for years are really not tricks at all, but are based on mathematical properties that allow anyone to do them with no special conjuring abilities. One of them is based on mod 14 arithmetic.

In this trick, suits play no role. Each card has a numerical value: 1 for ace, 2 for two,..., 11 for jack, 12 for queen, and 13 for king. The deck is shuffled and given to a spectator, who is instructed to place the deck of cards face up on a table, and is told to follow the procedure described: A card is laid down with its face up. (We shall call it the "starter" card.) The starter card will be at the bottom of a pile. In order to form a pile, note the value of the starter card, and then add cards on top of it while counting up to 13. For example, if the starter card is a six, pile up seven cards on top of it. If it is a jack, add two cards to it, and so on.

(continued)

When the first pile is completed, it is picked up and placed face down. The next card becomes the starter card for the next pile, and the process is repeated. This continues until all cards are used or until there are not enough cards to complete the last pile. Any cards that are left over are put aside for later use. We shall refer to these as "leftovers."

The performer then requests that a spectator choose three piles at random. The remaining piles are added to the leftovers. The spectator is then instructed to turn over any two top cards from the piles. The performer is then able to determine the value of the third top card.

The secret to the trick is that the performer adds the values of the two top cards that were turned over, and then adds 10 to this sum. The performer then counts off this number of cards from the leftovers. The number of cards remaining in the leftovers is the value of the remaining top card!

For Group Discussion

1. Obtain a deck of playing cards and perform the "trick" as described above. (As with many activities, you'll find that doing it is simpler than describing it.) Does it work?
2. Explain why this procedure works. (If you want to see how someone else explained it, using modulo 14 arithmetic, see "An Old Card Trick Revisited," by Barry C. Felps, in the December 1976 issue of the journal *The Mathematics Teacher*.)

5.4 EXERCISES

Find each of the following differences on the 12-hour clock.

1. $8 - 3$
2. $4 - 9$
3. $2 - 8$
4. $0 - 3$

5. Complete the 12-hour clock multiplication table below. You can use repeated addition and the addition table (for example, $3 \times 7 = 7 + 7 + 7 = 2 + 7 = 9$) or use mod 12 multiplication techniques, as in Example 8, parts (d) and (e).

×	0	1	2	3	4	5	6	7	8	9	10	11
0	0	0	0	0	0	0	0	0	0	0	0	0
1	0	1	2	3	4	5	6	7	8	9	10	11
2	0	2	4	6	8	10		2	4		8	
3	0	3	6	9	0	3	6		3	6		
4	0	4	8			8			4		4	8
5	0	5	10	3	8		6	11	4			
6	0	6	0		0	6	0	6		6		6
7	0	7	2	9			1			10		
8	0	8	4	0			8	4		8		4
9	0	9			0		6		0			
10	0	10	8		2							2
11	0	11										1

By referring to your table in Exercise 5, determine which of the following properties hold for the system of 12-hour clock numbers with the operation of multiplication.

6. closure
7. commutative
8. identity

A 5-hour clock system utilizes the set $\{0, 1, 2, 3, 4\}$, and relates to the clock face shown here.

9. Complete this 5-hour clock addition table.

+	0	1	2	3	4
0	0	1	2	3	4
1	1	2	3	4	
2	2	3	4		1
3	3	4			
4	4			3	

5-hour clock

Which of the following properties are satisfied by the system of 5-hour clock numbers with the operation of addition?

10. closure   **11.** commutative

12. identity (If so, what is the identity element?)

13. inverse (If so, name the inverse of each element.)

14. Complete this 5-hour clock multiplication table.

×	0	1	2	3	4
0	0	0	0	0	0
1	0	1	2	3	4
2	0	2	4		
3	0	3			
4	0	4			

Which of the following properties are satisfied by the system of 5-hour clock numbers with the operation of multiplication?

15. closure   **16.** commutative

17. identity (If so, what is the identity element?)

In clock arithmetic, as in ordinary arithmetic, $a - b = d$ is true if and only if $b + d = a$. Similarly, $a \div b = q$ if and only if $b \times q = a$.

Use the idea above and your 5-hour clock multiplication table of Exercise 14 to find the following quotients on a 5-hour clock.

18. $1 \div 3$ **19.** $3 \div 1$

20. $2 \div 3$ **21.** $3 \div 2$

22. Is division commutative on a 5-hour clock? Explain.

23. Is there an answer for $4 \div 0$ on a 5-hour clock? Find it or explain why not.

The military uses a 24-hour clock to avoid the problems of "A.M." and "P.M." For example, 1100 hours is 11 A.M., while 2100 hours is 9 P.M. (12 noon + 9 hours). In these designations, the last two digits represent minutes, and the digits before that represent hours. Find each of the following sums in the 24-hour clock system.

24. $1400 + 500$ **25.** $1300 + 1800$

26. $0750 + 1630$ **27.** $1545 + 0815$

28. Explain how the following three statements can *all* be true. (*Hint:* Think of clocks.)

$$1145 + 1135 = 2280$$
$$1145 + 1135 = 1120$$
$$1145 + 1135 = 2320$$

Answer true *or* false *for each of the following.*

29. $5 \equiv 19 \pmod{3}$ **30.** $35 \equiv 8 \pmod{9}$

31. $5445 \equiv 0 \pmod{3}$ **32.** $7021 \equiv 4202 \pmod{6}$

Work each of the following modular arithmetic problems.

33. $(12 + 7)(\bmod 4)$ **34.** $(62 + 95)(\bmod 9)$

35. $(35 - 22)(\bmod 5)$ **36.** $(82 - 45)(\bmod 3)$

37. $(5 \times 8)(\bmod 3)$ **38.** $(32 \times 21)(\bmod 8)$

39. $[4 \times (13 + 6)](\bmod 11)$

40. $[(10 + 7) \times (5 + 3)](\bmod 10)$

41. The text described how to do arithmetic mod m when the ordinary answer comes out nonnegative. Explain what to do when the ordinary answer is negative.

Work each of the following modular arithmetic problems.

42. $(3 - 27)(\bmod 5)$ **43.** $(16 - 60)(\bmod 7)$

44. $[(-8) \times 11](\bmod 3)$ **45.** $[2 \times (-23)](\bmod 5)$

In each of Exercises 46 and 47:
(a) *Complete the given addition table.*
(b) *Decide whether the closure, commutative, identity, and inverse properties are satisfied.*
(c) *If the inverse property is satisfied, give the inverse of each number.*

46. mod 4

+	0	1	2	3
0	0	1	2	3
1				
2				
3				

47. mod 7

+	0	1	2	3	4	5	6
0	0	1	2	3	4	5	6
1	1	2	3	4	5	6	
2							
3							
4							
5							
6							

In each of Exercises 48–51:
(a) *Complete the given multiplication table.*
(b) *Decide whether the closure, commutative, identity, and inverse properties are satisfied.*
(c) *Give the inverse of each nonzero number that has an inverse.*

48. mod 2

×	0	1
0	0	0
1	0	

49. mod 3

×	0	1	2
0	0	0	0
1	0	1	2
2	0	2	

50. mod 4

×	0	1	2	3
0	0	0	0	0
1	0	1	2	3
2	0	2		
3	0	3		

51. mod 9

×	0	1	2	3	4	5	6	7	8
0	0	0	0	0	0	0	0	0	0
1	0	1	2	3	4	5	6	7	8
2	0	2	4	6	8			5	
3	0	3	6	0		6		3	6
4	0	4	8		7		6		5
5	0	5	1		2		3	8	
6	0	6	3	0	6	3	0	6	3
7	0	7	5		8				2
8	0	8	7		4	3		1	

52. Complete this statement: a modular system satisfies the inverse property for multiplication only if the modulus is a(n) _____ number.

Find all positive solutions for each of the following equations. Note any identities.

53. $x \equiv 3 \pmod{7}$
54. $(2 + x) \equiv 7 \pmod{3}$
55. $6x \equiv 2 \pmod{2}$
56. $(5x - 3) \equiv 7 \pmod{4}$

Solve each problem.

57. *Odometer Readings* For many years automobile odometers showed five whole number digits and a digit for tenths of a mile. For those odometers showing just five whole number digits, totals are recorded according to what modulus?

58. *Distance Traveled by a Car* If a car's five-digit whole number odometer shows a reading of 29,306, *in theory* how many miles might the car have traveled?

59. *Piles of Ticket Stubs* Lawrence Rosenthal finds that whether he sorts his White Sox ticket stubs into piles of 10, piles of 15, or piles of 20, there are always 2 left over. What is the least number of stubs he could have (assuming he has more than 2)?

60. *Silver Spoon Collection* Roxanna Parker has a collection of silver spoons from all over the world. She finds that she can arrange her spoons in sets of 7 with 6 left over, sets of 8 with 1 left over, or sets of 15 with 3 left over. If Roxanna has fewer than 200 spoons, how many are there?

61. *Determining Day of the Week* Refer to Example 9 in the text. (Recall that *next* year is a leap year.) Assuming today was Thursday, January 12, answer the following questions.
(a) How many days would the next year (starting today) contain?
(b) What day of the week would occur one year from today?

62. *Determining a Range of Dates* Assume again, as in Example 9, that *next* year is a leap year. If the next year (starting today) does *not* contain 366 days, what is the range of possible dates for today?

63. *Flight Attendant Schedules* Robin Strang and Kristyn Wasag, flight attendants for two different airlines, are close friends and like to get together as often as possible. Robin flies a 21-day schedule (including days off), which then repeats, while Kristyn has a repeating 30-day schedule. Both of their routines include stopovers in Chicago, New Orleans, and San Francisco. The table below shows which days of each of their individual schedules they are in these cities. (Assume the first day of a cycle is day number 1.)

	Days in Chicago	Days in New Orleans	Days in San Francisco
Robin	1, 2, 8	5, 12	6, 18, 19
Kristyn	23, 29, 30	5, 6, 17	8, 10, 15, 20, 25

If today is July 1 and both are starting their schedules today (day 1), list the days during July and August that they will be able to see each other in each of the three cities.

The basis of the complex number system is the "imaginary" number i. The powers of i cycle through a repeating pattern of just 4 distinct values as shown here:

$$i^0 = 1, \quad i^1 = i, \quad i^2 = -1, \quad i^3 = -i, \quad i^4 = 1,$$
$$i^5 = i, \quad \text{and so on.}$$

Find the value of each of the following powers of i.

64. i^{16}
65. i^{47}
66. i^{98}
67. i^{137}

The following formula can be used to find the day of the week on which a given year begins.* Here y represents the year (after 1582, when our current calendar began). First calculate

$$a = y + [\![(y-1)/4]\!] - [\![(y-1)/100]\!] + [\![(y-1)/400]\!],$$

where $[\![x]\!]$ represents the greatest integer less than or equal to x. (For example, $[\![9.2]\!] = 9$, and $[\![\pi]\!] = 3$.) After finding a, find the smallest nonnegative integer b such that

$$a \equiv b \pmod{7}.$$

Then b gives the day of January 1, with $b = 0$ representing Sunday, $b = 1$ Monday, and so on.

Find the day of the week on which January 1 would occur in the following years.

68. 1812 **69.** 1865
70. 2002 **71.** 2020

Some people believe that Friday the thirteenth is unlucky. The table* below shows the months that will have a Friday the thirteenth if the first day of the year is known. A year is a leap year if it is divisible by 4. The only exception to this rule is that a century year (1900, for example) is a leap year only when it is divisible by 400.

First Day of Year	Non-leap Year	Leap Year
Sunday	Jan., Oct.	Jan., April, July
Monday	April, July	Sept., Dec.
Tuesday	Sept., Dec.	June
Wednesday	June	March, Nov.
Thursday	Feb., March, Nov.	Feb., Aug.
Friday	August	May
Saturday	May	Oct.

Use the table to determine the months that have a Friday the thirteenth for the following years.

72. 2003 **73.** 2002
74. 2004 **75.** 2200

76. Modular arithmetic can be used to create **residue designs**. For example, the designs (11, 3) and (65, 3) are shown at the top of the next column.

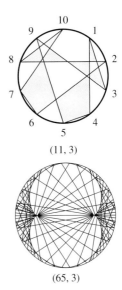

(11, 3)

(65, 3)

To see how such designs are created, construct a new design, (11, 5), by proceeding as follows.

(a) Draw a circle and divide the circumference into 10 equal parts. Label the division points as 1, 2, 3, ..., 10.

(b) Since $1 \times 5 \equiv 5 \pmod{11}$, connect 1 and 5. (We use 5 as a multiplier since we are making an (11, 5) design.)

(c) $2 \times 5 \equiv 10 \pmod{11}$
Therefore, connect 2 and _____.

(d) $3 \times 5 \equiv$ _____ (mod 11)
Connect 3 and _____.

(e) $4 \times 5 \equiv$ _____ (mod 11)
Connect 4 and _____.

(f) $5 \times 5 \equiv$ _____ (mod 11)
Connect 5 and _____.

(g) $6 \times 5 \equiv$ _____ (mod 11)
Connect 6 and _____.

(h) $7 \times 5 \equiv$ _____ (mod 11)
Connect 7 and _____.

(i) $8 \times 5 \equiv$ _____ (mod 11)
Connect 8 and _____.

(j) $9 \times 5 \equiv$ _____ (mod 11)
Connect 9 and _____.

(k) $10 \times 5 \equiv$ _____ (mod 11)
Connect 10 and _____.

(l) You might want to shade some of the regions you have found to make an interesting pattern. For more information, see "Residue Designs," by Phil Locke in *The Mathematics Teacher,* March 1972, pages 260–263.

---

*Given in "An Aid to the Superstitious," by G. L. Ritter, S. R. Lowry, H. B. Woodruff, and T. L. Isenhour. *The Mathematics Teacher,* May 1977, pp. 456–457.

Identification numbers are used in various ways for many kinds of different products.* Books, for example, are assigned International Standard Book Numbers (ISBNs). Each ISBN is a ten-digit number. It includes a check digit, which is determined on the basis of modular arithmetic. The ISBN for one version of this book is

$$0\text{-}321\text{-}16808\text{-}9.$$

The first digit, 0, identifies the book as being published in an English-speaking country. The next digits, 321, identify the publisher, while 16808 identifies this particular book. The final digit, 9, is a check digit. To find this check digit, start at the left and multiply the digits of the ISBN by 10, 9, 8, 7, 6, 5, 4, 3, and 2, respectively. Then add these products. For this book we get

$$(10 \times 0) + (9 \times 3) + (8 \times 2) + (7 \times 1) + (6 \times 1) + (5 \times 6)$$
$$+ (4 \times 8) + (3 \times 0) + (2 \times 8) = 134.$$

The check digit is the smallest number that must be added to this result to get a multiple of 11. Since $134 + 9 = 143$, *a multiple of 11, the check digit is 9. (It is possible to have a check "digit" of 10; the letter X is used instead of 10.)*

When an order for this book is received, the ISBN is entered into a computer, and the check digit evaluated. If this result does not match the check digit on the order, the order will not be processed. Which of the following ISBNs have correct check digits?

77. 0-399-13615-4   **78.** 0-691-02356-5

Find the appropriate check digit for each of the following ISBNs.

79. *A Beautiful Mind,* by Sylvia Nasar, 0-684-81906-___

80. *Brunelleschi's Dome,* by Ross King, 0-802-71366-___

81. *Empire Falls,* by Richard Russo, 0-679-43247-___

82. *Longitudes and Attitudes,* by Thomas L. Friedman, 0-374-19066-___

5.5 The Fibonacci Sequence and the Golden Ratio

The solution of Fibonacci's rabbit problem is examined in Chapter 1, pages 21–22.

The Fibonacci Sequence One of the most famous problems in elementary mathematics comes from the book *Liber Abaci,* written in 1202 by Leonardo of Pisa, a.k.a. Fibonacci. The problem is as follows:

> A man put a pair of rabbits in a cage. During the first month the rabbits produced no offspring, but each month thereafter produced one new pair of rabbits. If each new pair thus produced reproduces in the same manner, how many pairs of rabbits will there be at the end of one year?

The solution of this problem leads to a sequence of numbers known as the **Fibonacci sequence.** Here are the first fifteen terms of the Fibonacci sequence:

$$1, 1, 2, 3, 5, 8, 13, 21, 34, 55, 89, 144, 233, 377, 610.$$

Notice the pattern established in the sequence. After the first two terms (both 1), each term is obtained by adding the two previous terms. For example, the third term is obtained by adding $1 + 1$ to get 2, the fourth term is obtained by adding $1 + 2$ to get 3, and so on. This can be described by a mathematical formula known as a *recursion formula.* If F_n represents the Fibonacci number in the nth position in the sequence, then

$$F_1 = 1$$
$$F_2 = 1$$
$$F_n = F_{n-1} + F_{n-2} \quad \text{for } n \geq 3.$$

---

*For an interesting general discussion, see "The Mathematics of Identification Numbers," by Joseph A. Gallian in *The College Mathematics Journal,* May 1991, p. 194.

5.5 The Fibonacci Sequence and the Golden Ratio

The Fibonacci Association is a research organization dedicated to investigation into the Fibonacci sequence and related topics. Check your library to see if it has the journal *Fibonacci Quarterly*. The first two journals of 1963 contain a basic introduction to the Fibonacci sequence.

The Fibonacci sequence exhibits many interesting patterns, and by inductive reasoning we can make many conjectures about these patterns. However, as we have indicated many times earlier, simply observing a finite number of examples does not provide a proof of a statement. Proofs of the properties of the Fibonacci sequence often involve mathematical induction (covered in college algebra texts). Here we simply observe the patterns and do not attempt to provide such proofs.

As an example of one of the many interesting properties of the Fibonacci sequence, choose any term of the sequence after the first and square it. Then multiply the terms on either side of it, and subtract the smaller result from the larger. The difference is always 1. For example, choose the sixth term in the sequence, 8. The square of 8 is 64. Now multiply the terms on either side of 8: $5 \times 13 = 65$. Subtract 64 from 65 to get $65 - 64 = 1$. This pattern continues throughout the sequence.

EXAMPLE 1 Find the sum of the squares of the first n Fibonacci numbers for $n = 1, 2, 3, 4, 5$, and examine the pattern. Generalize this relationship.

$$1^2 = 1 = 1 \cdot 1 = F_1 \cdot F_2$$
$$1^2 + 1^2 = 2 = 1 \cdot 2 = F_2 \cdot F_3$$
$$1^2 + 1^2 + 2^2 = 6 = 2 \cdot 3 = F_3 \cdot F_4$$
$$1^2 + 1^2 + 2^2 + 3^2 = 15 = 3 \cdot 5 = F_4 \cdot F_5$$
$$1^2 + 1^2 + 2^2 + 3^2 + 5^2 = 40 = 5 \cdot 8 = F_5 \cdot F_6$$

The sum of the squares of the first n Fibonacci numbers seems to always be the product of F_n and F_{n+1}. This has been proven to be true, in general, using mathematical induction.

There are many other patterns similar to the one examined in Example 1, and some of them are discussed in the exercises of this section. An interesting property of the decimal value of the reciprocal of 89, the eleventh Fibonacci number, is examined in Example 2.

EXAMPLE 2 Observe the steps of the long-division process used to find the first few decimal places for $1/89$.

```
       .011235...
   89)1.000000
      89
      ---
      110
       89
      ---
      210
      178
      ---
      320
      267
      ---
      530
      445
      ---
      850...
```

Notice that after the 0 in the tenths place, the next five digits are the first five terms of the Fibonacci sequence. In addition, as indicated in color in the process, the digits 1, 1, 2, 3, 5, 8 appear in the division steps. Now, look at the digits next to the ones in color, beginning with the second "1"; they, too, are 1, 1, 2, 3, 5, ...

The following program for the TI-83 Plus utilizes the *Binet form* of the nth Fibonacci number (see Exercises 33–38) to determine its value.

PROGRAM: FIB
: Clr Home
: Disp "WHICH TERM"
: Disp "OF THE"
: Disp "SEQUENCE DO"
: Disp "YOU WANT?"
: Input N
: $(1 + \sqrt{(5)})/2 \to A$
: $(1 - \sqrt{(5)})/2 \to B$
: $(A^\wedge N - B^\wedge N)$
     $/\sqrt{(5)} \to F$
: Disp F

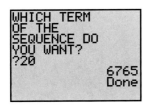

This screen indicates that the twentieth Fibonacci number is 6765.

If the division process is continued past the final step shown on the preceding page, the pattern seems to stop, since to ten decimal places, $1/89 \approx .0112359551$. (The decimal representation actually begins to repeat later in the process, since $1/89$ is a rational number.) However, the sum below indicates how the Fibonacci numbers are actually "hidden" in this decimal.

$$
\begin{aligned}
&.01\\
&.001\\
&.0002\\
&.00003\\
&.000005\\
&.0000008\\
&.00000013\\
&.000000021\\
&.0000000034\\
&.00000000055\\
&\underline{.000000000089}\\
1/89 = {}&.0112359550\ldots
\end{aligned}
$$

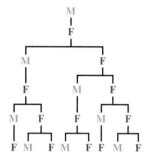

FIGURE 6

Fibonacci patterns have been found in numerous places in nature. For example, male honeybees hatch from eggs which have not been fertilized, so a male bee has only one parent, a female. On the other hand, female honeybees hatch from fertilized eggs, so a female has two parents, one male and one female. Figure 6 shows several generations of a male honeybee. (In this case, we plot generations of *ancestors* downward rather than generations of descendants.)

Notice that in the first generation there is 1 bee, in the second there is 1 bee, in the third there are 2 bees, and so on. These are the terms of the Fibonacci sequence. Furthermore, beginning with the second generation, the numbers of female bees form the sequence, and beginning with the third generation, the numbers of male bees also form the sequence.

Successive terms in the Fibonacci sequence also appear in plants. For example, the photo (below left) shows the double spiraling of a daisy head, with 21 clockwise spirals and 34 counterclockwise spirals. These numbers are successive terms in the sequence. Most pineapples (see the photo on the right) exhibit the Fibonacci sequence in the following way: Count the spirals formed by the "scales" of the cone, first counting from lower left to upper right. Then count the spirals from lower right to upper left. You should find that in one direction you get 8 spirals, and in the other you get 13 spirals, once again successive terms of the Fibonacci sequence. Many pinecones exhibit 5 and 8 spirals, and the cone of the giant sequoia has 3 and 5 spirals.

The Golden Ratio

If we consider the quotients of successive Fibonacci numbers, a pattern emerges.

$$\frac{1}{1} = 1 \qquad \frac{13}{8} = 1.625$$

$$\frac{2}{1} = 2 \qquad \frac{21}{13} \approx 1.615384615$$

$$\frac{3}{2} = 1.5 \qquad \frac{34}{21} \approx 1.619047619$$

$$\frac{5}{3} = 1.666\ldots \qquad \frac{55}{34} \approx 1.617647059$$

$$\frac{8}{5} = 1.6 \qquad \frac{89}{55} = 1.618181818\ldots$$

It seems as if these quotients are approaching some "limiting value" close to 1.618. This is indeed the case. As we go farther and farther out into the sequence, these quotients approach the number

$$\frac{1 + \sqrt{5}}{2},$$

known as the **golden ratio**, and often symbolized by ϕ, the Greek letter phi.

The golden ratio appears over and over in art, architecture, music, and nature. Its origins go back to the days of the ancient Greeks, who thought that a golden rectangle exhibited the most aesthetically pleasing proportion. A golden rectangle is defined as a rectangle whose dimensions satisfy the equation

$$\frac{\text{Length}}{\text{Width}} = \frac{\text{Length} + \text{Width}}{\text{Length}}.$$

If we let L represent length and W represent width, we have

$$\frac{L}{W} = \frac{L + W}{L}$$

$$\frac{L}{W} = \frac{L}{L} + \frac{W}{L}.$$

Since $L/W = \phi$, $L/L = 1$, and $W/L = 1/\phi$, this equation can be written

$$\phi = 1 + \frac{1}{\phi}.$$

Multiply both sides by ϕ to get

$$\phi^2 = \phi + 1.$$

Subtract ϕ and subtract 1 from both sides to get the quadratic equation

$$\phi^2 - \phi - 1 = 0.$$

Using the quadratic formula from algebra, the positive solution of this equation is found to be $(1 + \sqrt{5})/2 \approx 1.618033989$, the golden ratio.

Continuing the Discussion of the Golden Ratio A fraction such as

$$1 + \cfrac{1}{1 + \cfrac{1}{1 + \cfrac{1}{1 + \cdot}}}$$

is called a *continued fraction*. This continued fraction can be evaluated as follows.

Let $\quad x = 1 + \cfrac{1}{1 + \cfrac{1}{1 + \cdot}}$

Then $\quad x = 1 + \dfrac{1}{x}$

$\quad\quad\quad x^2 = x + 1$

$\quad\quad\quad x^2 - x - 1 = 0.$

By the quadratic formula from algebra,

$$x = \frac{1 \pm \sqrt{1 - 4(1)(-1)}}{2(1)}$$

$$x = \frac{1 \pm \sqrt{5}}{2}.$$

Notice that the positive solution

$$\frac{1 + \sqrt{5}}{2}$$

is the golden ratio.

A Golden Rectangle in Art
The rectangle outlining the figure in *St. Jerome* by Leonardo da Vinci is an example of a golden rectangle.

An example of a golden rectangle is shown in Figure 7. The Parthenon (see the photo), built on the Acropolis in ancient Athens during the fifth century B.C., is an example of architecture exhibiting many distinct golden rectangles.

FIGURE 7

To see how the terms of the Fibonacci sequence relate geometrically to the golden ratio, a rectangle that measures 89 by 55 units is constructed. (See Figure 8 on the next page.) This is a very close approximation to a golden rectangle. Within this rectangle a square is then constructed, 55 units on a side. The remaining rectangle is also approximately a golden rectangle, measuring 55 units by 34 units. Each time this process is repeated, a square and an approximate golden rectangle are formed. As indicated in the figure, vertices of the square may be joined by a smooth curve known as a *spiral*. This spiral resembles the outline of a cross section of the shell of the chambered nautilus, as shown in the photograph next to Figure 8.

FOR FURTHER THOUGHT

Or, should we say, "For Further Viewing"? The 1959 animated film *Donald in Mathmagic Land* has endured for more than 40 years as a classic. It provides a 30-minute trip with Donald Duck, led by the Spirit of Mathematics, through the world of mathematics. Several minutes of the film are devoted to the golden ratio (or, as it is termed there, the golden section).

© The Walt Disney Company

Disney provides animation to explain the golden ratio in a way that the printed word simply cannot do. The golden ratio is seen in architecture, nature, and the human body. The film is available on video and can easily be purchased or rented.

For Group Discussion

1. Verify the following Fibonacci pattern in the conifer family. Obtain a pineapple, and count spirals formed by the "scales" of the cone, first counting from lower left to upper right. Then count the spirals from lower right to upper left. What do you find?
2. Two popular sizes of index cards are 3" by 5" and 5" by 8". Why do you think that these are industry-standard sizes?
3. Divide your height by the height of your navel. Find a class average. What value does this come close to?

5.5 The Fibonacci Sequence and the Golden Ratio

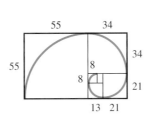

FIGURE 8

5.5 EXERCISES

Answer each of the following questions concerning the Fibonacci sequence or the golden ratio.

1. The sixteenth Fibonacci number is 987 and the seventeenth Fibonacci number is 1597. What is the eighteenth Fibonacci number?

2. Recall that F_n represents the Fibonacci number in the nth position in the sequence. What are the only two values of n such that $F_n = n$?

3. $F_{23} = 28{,}657$ and $F_{25} = 75{,}025$. What is the value of F_{24}?

4. If two successive terms of the Fibonacci sequence are both odd, is the next term even or odd?

5. What is the exact value of the golden ratio?

6. What is the approximate value of the golden ratio to the nearest thousandth?

In each of Exercises 7–14, a pattern is established involving terms of the Fibonacci sequence. Use inductive reasoning to make a conjecture concerning the next equation in the pattern, and verify it. You may wish to refer to the first few terms of the sequence given in the text.

7. $1 = 2 - 1$
   $1 + 1 = 3 - 1$
   $1 + 1 + 2 = 5 - 1$
   $1 + 1 + 2 + 3 = 8 - 1$
   $1 + 1 + 2 + 3 + 5 = 13 - 1$

8. $1 = 2 - 1$
   $1 + 3 = 5 - 1$
   $1 + 3 + 8 = 13 - 1$
   $1 + 3 + 8 + 21 = 34 - 1$
   $1 + 3 + 8 + 21 + 55 = 89 - 1$

9. $1 = 1$
   $1 + 2 = 3$
   $1 + 2 + 5 = 8$
   $1 + 2 + 5 + 13 = 21$
   $1 + 2 + 5 + 13 + 34 = 55$

10. $1^2 + 1^2 = 2$
    $1^2 + 2^2 = 5$
    $2^2 + 3^2 = 13$
    $3^2 + 5^2 = 34$
    $5^2 + 8^2 = 89$

11. $2^2 - 1^2 = 3$
    $3^2 - 1^2 = 8$
    $5^2 - 2^2 = 21$
    $8^2 - 3^2 = 55$

12. $2^3 + 1^3 - 1^3 = 8$
    $3^3 + 2^3 - 1^3 = 34$
    $5^3 + 3^3 - 2^3 = 144$
    $8^3 + 5^3 - 3^3 = 610$

13. $1 = 1^2$
    $1 - 2 = -1^2$
    $1 - 2 + 5 = 2^2$
    $1 - 2 + 5 - 13 = -3^2$
    $1 - 2 + 5 - 13 + 34 = 5^2$

14. $1 - 1 = -1 + 1$
    $1 - 1 + 2 = 1 + 1$
    $1 - 1 + 2 - 3 = -2 + 1$
    $1 - 1 + 2 - 3 + 5 = 3 + 1$
    $1 - 1 + 2 - 3 + 5 - 8 = -5 + 1$

15. Every natural number can be expressed as a sum of Fibonacci numbers, where no number is used more than once. For example, $25 = 21 + 3 + 1$. Express each of the following in this way.
    (a) 37 (b) 40 (c) 52

16. It has been shown that if m divides n, then F_m is a factor of F_n. Show that this is true for the following values of m and n.
    (a) $m = 2, n = 6$ (b) $m = 3, n = 9$
    (c) $m = 4, n = 8$

17. It has been shown that if the greatest common factor of m and n is r, then the greatest common factor of F_m and F_n is F_r. Show that this is true for the following values of m and n.
    (a) $m = 10, n = 4$
    (b) $m = 12, n = 6$
    (c) $m = 14, n = 6$

18. For any prime number p except 2 or 5, either F_{p+1} or F_{p-1} is divisible by p. Show that this is true for the following values of p.
    (a) $p = 3$
    (b) $p = 7$
    (c) $p = 11$

19. Earlier we saw that if a term of the Fibonacci sequence is squared and then the product of the terms on each side of the term is found, there will always be a difference of 1. Follow the steps below, choosing the seventh Fibonacci number, 13.
    (a) Square 13. Multiply the terms of the sequence two positions away from 13 (i.e., 5 and 34). Subtract the smaller result from the larger, and record your answer.
    (b) Square 13. Multiply the terms of the sequence three positions away from 13. Once again, subtract the smaller result from the larger, and record your answer.
    (c) Repeat the process, moving four terms away from 13.
    (d) Make a conjecture about what will happen when you repeat the process, moving five terms away. Verify your answer.

20. Here is a number trick that you can perform. Ask someone to pick any two numbers at random and to write them down. Ask the person to determine a third number by adding the first and second, a fourth number by adding the second and third, and so on, until ten numbers are determined. Then ask the person to add these ten numbers. You will be able to give the sum before the person even completes the list, because the sum will always be 11 times the seventh number in the list. Verify that this is true, by using x and y as the first two numbers arbitrarily chosen. (*Hint:* Remember the distributive property from algebra.)

Another Fibonacci-type sequence that has been studied by mathematicians is the **Lucas sequence**, *named after a French mathematician of the nineteenth century. The first ten terms of the Lucas sequence are*

$$1, 3, 4, 7, 11, 18, 29, 47, 76, 123.$$

Exercises 21–26 pertain to the Lucas sequence.

21. What is the eleventh term of the Lucas sequence?

22. Choose any term of the Lucas sequence and square it. Then multiply the terms on either side of the one you chose. Subtract the smaller result from the larger. Repeat this for a different term of the sequence. Do you get the same result? Make a conjecture about this pattern.

23. The first term of the Lucas sequence is 1. Add the first and third terms. Record your answer. Now add the first, third, and fifth terms and record your answer. Continue this pattern, each time adding another term that is in an *odd* position in the sequence. What do you notice about all of your results?

24. The second term of the Lucas sequence is 3. Add the second and fourth terms. Record your answer. Now add the second, fourth, and sixth terms and record your answer. Continue this pattern, each time adding another term that is in an *even* position of the sequence. What do you notice about all of your sums?

25. Many interesting patterns exist between the terms of the Fibonacci sequence and the Lucas sequence. Make a conjecture about the next equation that would appear in each of the lists and then verify it.
    (a) $1 \cdot 1 = 1$
    $1 \cdot 3 = 3$
    $2 \cdot 4 = 8$
    $3 \cdot 7 = 21$
    $5 \cdot 11 = 55$
    (b) $1 + 2 = 3$
    $1 + 3 = 4$
    $2 + 5 = 7$
    $3 + 8 = 11$
    $5 + 13 = 18$
    (c) $1 + 1 = 2 \cdot 1$
    $1 + 3 = 2 \cdot 2$
    $2 + 4 = 2 \cdot 3$
    $3 + 7 = 2 \cdot 5$
    $5 + 11 = 2 \cdot 8$

26. In the text we illustrate that the quotients of successive terms of the Fibonacci sequence approach the golden ratio. Make a similar observation for the terms of the Lucas sequence; that is, find the decimal approximations for the quotients

$$\frac{3}{1}, \frac{4}{3}, \frac{7}{4}, \frac{11}{7}, \frac{18}{11}, \frac{29}{18},$$

and so on, using a calculator. Then make a conjecture about what seems to be happening.

Recall the **Pythagorean theorem** *from geometry: If a right triangle has legs of lengths a and b and hypotenuse of length c, then $a^2 + b^2 = c^2$. Suppose that we choose any four successive terms of the Fibonacci sequence. Multiply the first and fourth. Double the product of the second and third. Add the squares of the second and third. The three results obtained form a* Pythagorean triple *(three numbers that satisfy the equation $a^2 + b^2 = c^2$). Find the*

Pythagorean triple obtained this way using the four given successive terms of the Fibonacci sequence.

27. 1, 1, 2, 3
28. 1, 2, 3, 5
29. 2, 3, 5, 8
30. Look at the values of the hypotenuse (*c*) in the answers to Exercises 27–29. What do you notice about each of them?
31. The following array of numbers is called **Pascal's triangle**.

$$\begin{array}{c} 1 \\ 1\ 1 \\ 1\ 2\ 1 \\ 1\ 3\ 3\ 1 \\ 1\ 4\ 6\ 4\ 1 \\ 1\ 5\ 10\ 10\ 5\ 1 \\ 1\ 6\ 15\ 20\ 15\ 6\ 1 \end{array}$$

This array is important in the study of counting techniques and probability (see later chapters) and appears in algebra in the binomial theorem. If the triangular array is written in a different form, as follows, and the sums along the diagonals as indicated by the dashed lines are found, there is an interesting occurrence. What do you find when the numbers are added?

$$\begin{array}{l} 1 \\ 1\quad 1 \\ 1\quad 2\quad 1 \\ 1\quad 3\quad 3\quad 1 \\ 1\quad 4\quad 6\quad 4\quad 1 \\ 1\quad 5\quad 10\quad 10\quad 5\quad 1 \\ 1\quad 6\quad 15\quad 20\quad 15\quad 6\quad 1 \end{array}$$

32. Write a paragraph explaining some of the occurrences of the Fibonacci sequence and the golden ratio in your everyday surroundings.

Exercises 33–38 require a scientific calculator.

33. The positive solution of the equation $x^2 - x - 1 = 0$ is $(1 + \sqrt{5})/2$, as indicated in the text. The negative solution is $(1 - \sqrt{5})/2$. Find the decimal approximations for both. What similarity do you notice between the two decimals?

34. In some cases, writers define the golden ratio to be the *reciprocal* of $(1 + \sqrt{5})/2$. Find a decimal approximation for the reciprocal of $(1 + \sqrt{5})/2$. What similarity do you notice between the decimals for $(1 + \sqrt{5})/2$ and its reciprocal?

A remarkable relationship exists between the two solutions of $x^2 - x - 1 = 0$,

$$\phi = \frac{1 + \sqrt{5}}{2} \quad \text{and} \quad \overline{\phi} = \frac{1 - \sqrt{5}}{2},$$

and the Fibonacci numbers. To find the nth Fibonacci number without using the recursion formula, evaluate

$$\frac{\phi^n - \overline{\phi}^n}{\sqrt{5}}$$

using a calculator. For example, to find the thirteenth Fibonacci number, evaluate

$$\frac{\left(\frac{1 + \sqrt{5}}{2}\right)^{13} - \left(\frac{1 - \sqrt{5}}{2}\right)^{13}}{\sqrt{5}}.$$

This form is known as the **Binet form** *of the nth Fibonacci number. Use the Binet form and a calculator to find the nth Fibonacci number for each of the following values of n.*

35. $n = 14$
36. $n = 20$
37. $n = 22$
38. $n = 25$

The equations in Exercises 7–14 are all instances of various relationships among terms of the Fibonacci sequence. For example, from the patterns in Exercises 7 and 8 we get, respectively,

$$F_1 + F_2 + \cdots + F_n = F_{n+2} - 1$$

and

$$F_2 + F_4 + \cdots + F_{2n} = F_{2n+1} - 1.$$

(You can refer to the listing of Fibonacci terms in the text to verify these patterns.) Complete the following formulas.

39. From Exercise 9:
$F_1 + F_3 + \cdots + F_{2n-1} =$ _____

40. From Exercise 10:
$F_n^2 + F_{n+1}^2 =$ _____

41. From Exercise 11:
$F_n^2 - F_{n-2}^2 =$ _____

42. From Exercise 12:
$F_n^3 + F_{n-1}^3 - F_{n-2}^3 =$ _____

EXTENSION

Magic Squares

Legend has it that in about 2200 B.C. the Chinese Emperor Yu discovered on the bank of the Yellow River a tortoise whose shell bore the diagram in Figure 9. This so-called *lo-shu* is an early example of a **magic square.** If the numbers of dots are counted and arranged in a square fashion, the array in Figure 10 is obtained. A magic square is a square array of numbers with the property that the sum along each row, column, and diagonal is the same. This common value is called the "magic sum." The **order** of a magic square is simply the number of rows (and columns) in the square. The magic square of Figure 10 is an order 3 magic square.

By using the formula for the sum of the first n terms of an arithmetic sequence, it can be shown that if a magic square of order n has entries $1, 2, 3, \ldots, n^2$, then the sum of *all entries* in the square is

$$\frac{n^2(n^2 + 1)}{2}.$$

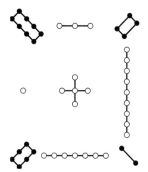

FIGURE 9

8	3	4
1	5	9
6	7	2

FIGURE 10

Since there are n rows (and columns), the magic sum of the square may be found by dividing the above expression by n. This results in the following rule for finding the magic sum.

Magic Sum Formula

If a magic square of order n has entries $1, 2, 3, \ldots, n^2$, then the magic sum MS is given by the formula

$$\mathbf{MS} = \frac{n(n^2 + 1)}{2}.$$

With $n = 3$ in this formula we find that the magic sum of the square in Figure 10, which may be verified by direct addition, is

$$\mathbf{MS} = \frac{3(3^2 + 1)}{2} = 15.$$

There is a method of constructing an odd-order magic square which is attributed to an early French envoy, *de la Loubere,* that is sometimes referred to as the "staircase method." The method is described below for an order 5 square, with entries $1, 2, 3, \ldots, 25$.

Begin by sketching a square divided into 25 cells into which the numbers 1–25 are to be entered. Proceed as described below, referring to Figures 11 and 12 for clarification.

1. Write 1 in the middle cell of the top row.
2. Always try to enter numbers in sequence in the cells by moving diagonally from lower left to upper right. There are two exceptions to this:
   (a) If you go outside of the magic square, move all the way across the row or column to enter the number. Then proceed to move diagonally.

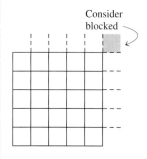

FIGURE 11

18	25	2	9	16
17	24	1	8	15
23	5	7	14	16
4	6	13	20	22
10	12	19	21	3
11	18	25	2	9

(Extended cells on right: 17, 23, 4, 10)

FIGURE 12

(b) If you run into a cell which is already occupied (that is, you are "blocked"), drop down one cell from the last entry written and enter the next number there. Then proceed to move diagonally.

3. Your last entry, 25, will be in the middle cell of the bottom row.

Figure 12 shows the completed magic square. Its magic sum is 65.

Benjamin Franklin admitted that he would amuse himself while in the Pennsylvania Assembly with magic squares or circles "or any thing to avoid Weariness." He wrote about the usefulness of mathematics in the *Gazette* in 1735, saying that no employment can be managed without arithmetic, no mechanical invention without geometry. He also thought that mathematical demonstrations are better than academic logic for training the mind to reason with exactness and distinguish truth from falsity even outside of mathematics.

The square shown here is one developed by Franklin. It has a sum of 2056 in each row and diagonal, and, in Franklin's words, has the additional property "that a four-square hole being cut in a piece of paper of such size as to take in and show through it just 16 of the little squares, when laid on the greater square, the sum of the 16 numbers so appearing through the hole, wherever it was placed on the greater square should likewise make 2056." He claimed that it was "the most magically magic square ever made by any magician."

You might wish to verify the following property of this magic square: The sum of any four numbers that are opposite each other and at equal distances from the center is 514 (which is one-fourth of the magic sum).

EXTENSION EXERCISES

Given a magic square, other magic squares may be obtained by rotating the given one. For example, starting with the magic square in Figure 10, a 90° rotation in a clockwise direction gives the magic square shown here.

6	1	8
7	5	3
2	9	4

Start with Figure 10 and give the magic square obtained by each rotation described.

1. 180° in a clockwise direction
2. 90° in a counterclockwise direction

Start with Figure 12 and give the magic square obtained by each rotation described.

3. 90° in a clockwise direction
4. 180° in a clockwise direction
5. 90° in a counterclockwise direction
6. Try to construct an order 2 magic square containing the entries 1, 2, 3, 4. What happens?

Given a magic square, other magic squares may be obtained by adding or subtracting a constant value to or from each entry, multiplying each entry by a constant value, or dividing each entry by a nonzero constant value. In Exercises 7–10, start with the magic square whose figure number is indicated, and perform the operation described to find a new magic square. Give the new magic sum.

7. Figure 10, multiply by 3
8. Figure 10, add 7
9. Figure 12, divide by 2
10. Figure 12, subtract 10

(continued)

According to a fanciful story by Charles Trigg in Mathematics Magazine *(September 1976, page 212), the Emperor Charlemagne (742–814) ordered a five-sided fort to be built at an important point in his kingdom. As good-luck charms, he had magic squares placed on all five sides of the fort. He had one restriction for these magic squares: all the numbers in them must be prime.*

The magic squares are given in Exercises 11–15, with one missing entry. Find the missing entry in each square.

11.

	71	257
47	269	491
281	467	59

12.

389		227
107	269	431
311	347	149

13.

389	227	191
71	269	
347	311	149

14.

401	227	179
47	269	491
359		137

15.

401	257	149
17		521
389	281	137

16. Compare the magic sums in Exercises 11–15. Charlemagne had stipulated that each magic sum should be the year in which the fort was built. What was that year?

Find the missing entries in each magic square.

17.

75	68	(a)
(b)	72	(c)
71	76	(d)

18.

1	8	13	(a)
(b)	14	7	2
16	9	4	(c)
(d)	(e)	(f)	15

19.

3	20	(a)	24	11
(b)	14	1	18	10
9	21	13	(c)	17
16	8	25	12	(d)
(e)	2	(f)	(g)	(h)

20.

3	36	2	35	31	4
10	12	(a)	26	7	27
21	13	17	14	(b)	22
16	(c)	23	(d)	18	15
28	30	8	(e)	25	9
(f)	1	32	5	6	34

21. Use the "staircase method" to construct a magic square of order 7, containing the entries 1, 2, 3, ..., 49.

The magic square shown in the photograph is from a woodcut by Albrecht Dürer entitled Melancholia.

The two numbers in the center of the bottom row give 1514, the year the woodcut was created. Refer to this magic square to answer Exercises 22–30.

16	3	2	13
5	10	11	8
9	6	7	12
4	15	14	1

Dürer's Magic Square

22. What is the magic sum?

23. Verify: The sum of the entries in the four corners is equal to the magic sum.

24. Verify: The sum of the entries in any 2 by 2 square at a corner of the given magic square is equal to the magic sum.

25. Verify: The sum of the entries in the diagonals is equal to the sum of the entries not in the diagonals.

26. Verify: The sum of the squares of the entries in the diagonals is equal to the sum of the squares of the entries not in the diagonals.

27. Verify: The sum of the cubes of the entries in the diagonals is equal to the sum of the cubes of the entries not in the diagonals.

28. Verify: The sum of the squares of the entries in the top two rows is equal to the sum of the squares of the entries in the bottom two rows.

29. Verify: The sum of the squares of the entries in the first and third rows is equal to the sum of the squares of the entries in the second and fourth rows.

30. Find another interesting property of Dürer's magic square and state it.

31. A magic square of order 4 may be constructed as follows. Lightly sketch in the diagonals of the blank magic square. Beginning at the upper left, move across each row from left to right, counting the cells as you go along. If the cell is on a diagonal, count it but do not enter its number. If it is not on a diagonal, enter its number. When this is completed, reverse the procedure, beginning at the bottom right and moving across from right to left. As you count the cells, enter the number if the cell is not occupied. If it is already occupied, count it but do not enter its number. You should obtain a magic square similar to the one given for Exercises 22–30. How do they differ?

With chosen values for a, b, and c, an order 3 magic square can be constructed by substituting these values in the generalized form shown here.

$a+b$	$a-b-c$	$a+c$
$a-b+c$	a	$a+b-c$
$a-c$	$a+b+c$	$a-b$

Use the given values of a, b, and c to construct an order 3 magic square, using this generalized form.

32. $a = 5, \quad b = 1, \quad c = -3$
33. $a = 16, \quad b = 2, \quad c = -6$
34. $a = 5, \quad b = 4, \quad c = -8$

35. It can be shown that if an order n magic square has least entry k, and its entries are consecutive counting numbers, then its magic sum is given by the formula

$$MS = \frac{n(2k + n^2 - 1)}{2}.$$

Construct an order 7 magic square whose least entry is 10 using the staircase method. What is its magic sum?

36. Use the formula of Exercise 35 to find the missing entries in the following order 4 magic square whose least entry is 24.

(a)	38	37	27
35	(b)	30	32
31	33	(c)	28
(d)	26	25	(e)

In a 1769 letter from Benjamin Franklin to a Mr. Peter Collinson, Franklin exhibited the following magic square of order 8.

52	61	4	13	20	29	36	45
14	3	62	51	46	35	30	19
53	60	5	12	21	28	37	44
11	6	59	54	43	38	27	22
55	58	7	10	23	26	39	42
9	8	57	56	41	40	25	24
50	63	2	15	18	31	34	47
16	1	64	49	48	33	32	17

37. What is the magic sum?

Verify the following properties of this magic square.

38. The sums in the first half of each row and the second half of each row are both equal to half the magic sum.

39. The four corner entries added to the four center entries is equal to the magic sum.

40. The "bent diagonals" consisting of eight entries, going up four entries from left to right and down four entries from left to right, give the magic sum. (For example, starting with 16, one bent diagonal sum is $16 + 63 + 57 + 10 + 23 + 40 + 34 + 17$.)

(continued)

If we use a "knight's move" (up two, right one) from chess, a variation of the staircase method gives rise to the magic square shown here. (When blocked, we move to the cell just below the previous entry.)

10	18	1	14	22
11	24	7	20	3
17	5	13	21	9
23	6	19	2	15
4	12	25	8	16

Use a similar process to construct an order 5 magic square, starting with 1 in the cell described.

41. fourth row, second column (up two, right one; when blocked, move to the cell just below the previous entry)

42. third row, third column (up one, right two; when blocked, move to the cell just to the left of the previous entry)

COLLABORATIVE INVESTIGATION

Investigating an Interesting Property of Number Squares

In the Extension at the end of this chapter, we looked at magic squares. Now in this group activity we will investigate another property of squares of numbers. Begin by dividing up the class into groups of three or four students. Each student in the group should prepare a square of numbers like the one that follows:

1	2	3	4	5
6	7	8	9	10
11	12	13	14	15
16	17	18	19	20
21	22	23	24	25

Topics for Discussion

1. Each student should do the following individually:

   Choose any number in the first row. Circle it, and cross out all entries in the column below it. (For example, if you circle 4, cross out 9, 14, 19, and 24.) Now circle any remaining number in the second row, and cross out all entries in the column below it.

   Repeat this procedure for the third and fourth rows, and then circle the final remaining number in the fifth row.

   Now each student in the group should add the circled numbers and compare his or her sum with all others in the group. What do you notice?

2. How does the sum obtained in Exercise 1 compare with the magic sum for an order 5 magic square?

3. Suppose Exercise 1 was done as shown here:

1	②	3	4	5
6	7̸	⑧	9	10
11	1̸2̸	1̸3̸	14	⑮
⑯	1̸7̸	1̸8̸	19	2̸0̸
2̸1̸	2̸2̸	2̸3̸	㉔	2̸5̸

Notice that summing the circled entries is just like summing $1 + 2 + 3 + 4 + 5$, except that

3 is replaced by $3 + 5$,

5 is replaced by $5 + 10$,

1 is replaced by $1 + 15$,

4 is replaced by $4 + 20$.

We can express this as

$$\text{sum} = (1 + 2 + 3 + 4 + 5)$$
$$+ (5 + 10 + 15 + 20)$$
$$= 15 + 50 = 65.$$

4. Explain why, whatever entries you choose to circle in the various rows, the sum is always the same.

5. Prepare a similar square of the natural numbers 1 through 36. Then repeat Exercise 1. Discuss your results. How does the sum compare with the magic sum for an order 6 magic square?

6. As a group, fill in the entries in this equation for the 6 by 6 square.

   sum = (__ + __ + __ + __ + __ + __)
   + (__ + __ + __ + __ + __)

7. As a group, predict the sum of the circled numbers in a 7 by 7 square by expressing it as follows. (Do not actually construct the square.)

   sum = (__ + __ + __ + __ + __ + __
   + __) + (__ + __ + __ + __ + __ + __)

   How does the sum compare with the magic sum for an order 7 magic square?

8. Each individual should now prepare another 5 by 5 square and repeat Exercise 1, except this time start with a number in the first *column* and cross out remaining numbers in *rows*. In your group, discuss and explain what you observe.

CHAPTER 5 TEST

In Exercises 1–5, decide whether each statement is true *or* false.

1. No two prime numbers differ by 1.
2. There are infinitely many prime numbers.
3. If a natural number is divisible by 9, then it must also be divisible by 3.
4. If p and q are different primes, 1 is their greatest common factor and pq is their least common multiple.
5. For all natural numbers n, 1 is a factor of n and n is a multiple of n.
6. Use divisibility tests to determine whether the number

   331,153,470

   is divisible by each of the following.
   - (a) 2
   - (b) 3
   - (c) 4
   - (d) 5
   - (e) 6
   - (f) 8
   - (g) 9
   - (h) 10
   - (i) 12

7. Decide whether each number is prime, composite, or neither.
   - (a) 93
   - (b) 1
   - (c) 59

8. Give the prime factorization of 1440.

9. In your own words state the Fundamental Theorem of Arithmetic.

10. Decide whether each number is perfect, deficient, or abundant.
    - (a) 17
    - (b) 6
    - (c) 24

11. Which of the following statements is false?
    - A. There are no known odd perfect numbers.
    - B. Every even perfect number must end in 6 or 28.
    - C. Goldbach's Conjecture for the number 8 is verified by the equation $8 = 7 + 1$.

12. Give a pair of twin primes between 40 and 50.

13. Find the greatest common factor of 270 and 450.

14. Find the least common multiple of 24, 36, and 60.

15. *Day Off for Fast-food Workers* Both Sherrie Firavich and Della Daniel work at a fast-food outlet. Sherrie has every sixth day off and Della has every fourth day off. If they are both off on Wednesday of this week, what will be the day of the week that they are next off together?

16. The twenty-second Fibonacci number is 17,711 and the twenty-third Fibonacci number is 28,657. What is the twenty-fourth Fibonacci number?

17. Make a conjecture about the next equation in the following list, and verify it.

$$8 - (1 + 1 + 2 + 3) = 1$$
$$13 - (1 + 2 + 3 + 5) = 2$$
$$21 - (2 + 3 + 5 + 8) = 3$$
$$34 - (3 + 5 + 8 + 13) = 5$$
$$55 - (5 + 8 + 13 + 21) = 8$$

18. Choose the correct completion of this statement: If p is a prime number, then $2^p - 1$ is prime
    A. never B. sometimes C. always.

19. Calculate: $(12 + 16)(\text{mod } 5)$.

20. Find all positive solutions for $(4x - 7) \equiv 2 \,(\text{mod } 3)$.

21. (a) Give the first eight terms of a Fibonacci-type sequence with first term 1 and second term 5.

    (b) Choose any term after the first in the sequence just formed. Square it. Multiply the two terms on either side of it. Subtract the smaller result from the larger. Now repeat the process with a different term. Make a conjecture about what this process will yield for any term of the sequence.

22. Which one of the following is the *exact* value of the golden ratio?

    A. $\dfrac{1 + \sqrt{5}}{2}$ B. $\dfrac{1 - \sqrt{5}}{2}$ C. 1.6   D. 1.618

23. Write a short paragraph explaining what Fermat's Last Theorem says and giving an account of the recent developments concerning it.

chapter

6

The Real Numbers and Their Representations

Sunbathing is a popular pastime. However, excessive exposure to ultraviolet light can be responsible for both skin and eye damage. The intensity of ultraviolet light is affected by seasonal changes in the ozone layer, cloud cover, and time of day. The table shows maximum ultraviolet intensity, measured in milliwatts per square meter for various latitudes and dates.

Latitude	Mar. 21	June 21	Sept. 21	Dec. 21
0°	325	254	325	272
10°	311	275	280	220
20°	249	292	256	143
30°	179	248	182	80
40°	99	199	127	34
50°	57	143	75	13

If a student from Seattle, located at a latitude of 47.5°, spends spring break in Hawaii, with a latitude of 20°, the sun's ultraviolet rays in Hawaii will be approximately $249/57 \approx 4.37$ times more intense than in Seattle. Hawaii's spring sun is approximately $249/143 \approx 1.74$ times stronger than Seattle's most intense sun in the summer. Because ultraviolet light is scattered through reflection, clouds and shade do not stop sunburn completely. A person sitting in the shade can receive 40 percent of the ultraviolet light occurring in sunlit areas. Thin clouds transmit 80 percent of ultraviolet light, while a person swimming $1\frac{1}{2}$ feet below the surface of the water also receives 80 percent of the sun's ultraviolet radiation. So, even sitting in the shade, our Seattle student will still feel a more intense sun in Hawaii.

Real numbers are needed to quantify the effect of ultraviolet light on humans. This chapter introduces the different types of numbers like whole numbers, decimals, fractions, and percentages which are essential in the description of real-world phenomena.

Source: Williams, J., *The USA Today Weather Almanac 1995,* Vintage Books, New York, 1994.

6.1 Real Numbers, Order, and Absolute Value

6.2 Operations, Properties, and Applications of Real Numbers

6.3 Rational Numbers and Decimal Representation

6.4 Irrational Numbers and Decimal Representation

6.5 Applications of Decimals and Percents

Extension: Complex Numbers

Collaborative Investigation: Budgeting to Buy a Car

Chapter 6 Test

As mathematics developed, it was discovered that the *counting*, or *natural*, *numbers* did not satisfy all requirements of mathematicians. Consequently, new, expanded number systems were created. The mathematician Leopold Kronecker (1823–1891) once made the statement, "God made the integers, all the rest is the work of man." In this chapter we look at those sets that, according to Kronecker, are the work of mankind.

6.1 Real Numbers, Order, and Absolute Value

The Origins of Zero The Mayan Indians of Mexico and Central America had one of the earliest numeration systems that included a symbol for zero. The very early Babylonians had a positional system, but they placed only a space between "digits" to indicate a missing power. When the Greeks absorbed Babylonian astronomy, they used the letter omicron, o, of their alphabet or $\bar{o}$ to represent "no power," or "zero." The Greek numeration system was gradually replaced by the Roman numeration system.

The Roman system was the one most commonly used in Europe from the time of Christ until perhaps 1400 A.D., when the Hindu-Arabic system began to take over. The original Hindu word for zero was *sunya*, meaning "void." The Arabs adopted this word as *sifr*, or "vacant." The word *sifr* passed into Latin as *zephirum*, which over the years became *zevero*, *zepiro*, and finally, *zero*.

Sets of Real Numbers The *natural numbers* are those numbers with which we count discrete objects. By including 0 in the set, we obtain the set of *whole numbers*.

Natural Numbers

$\{1, 2, 3, 4, \ldots\}$ is the set of **natural numbers.**

Whole Numbers

$\{0, 1, 2, 3, \ldots\}$ is the set of **whole numbers.**

These numbers, along with many others, can be represented on **number lines** like the one pictured in Figure 1. We draw a number line by locating any point on the line and calling it 0. Choose any point to the right of 0 and call it 1. The distance between 0 and 1 gives a unit of measure used to locate other points, as shown in Figure 1. The points labeled in Figure 1 and those continuing in the same way to the right correspond to the set of whole numbers.

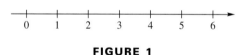

FIGURE 1

All the whole numbers starting with 1 are located to the right of 0 on the number line. But numbers may also be placed to the left of 0. These numbers, written $-1, -2, -3$, and so on, are shown in Figure 2. (The negative sign is used to show that the numbers are located to the *left* of 0.)

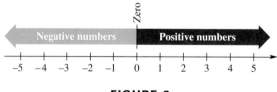

FIGURE 2

The numbers to the *left* of 0 are **negative numbers.** The numbers to the *right* of 0 are **positive numbers.** The number 0 itself is neither positive nor negative. Positive numbers and negative numbers are called **signed numbers.**

6.1 Real Numbers, Order, and Absolute Value

The Origins of Negative Numbers Negative numbers can be traced back to the Chinese between 200 B.C. and 220 A.D. Mathematicians at first found negative numbers ugly and unpleasant, even though they kept cropping up in the solutions of problems. For example, an Indian text of about 1150 A.D. gives the solution of an equation as -5 and then makes fun of anything so useless.

Leonardo of Pisa (Fibonacci), while working on a financial problem, was forced to conclude that the solution must be a negative number (that is, a financial loss). In 1545 A.D., the rules governing operations with negative numbers were published by Girolamo Cardano in his *Ars Magna* (Great Art).

There are many practical applications of negative numbers. For example, temperatures sometimes fall below zero. The lowest temperature ever recorded in meteorological records was $-128.6°F$ at Vostok, Antarctica, on July 22, 1983. Altitudes below sea level can be represented by negative numbers. The shore surrounding the Dead Sea is 1312 feet below sea level; this can be represented as -1312 feet.

The set of numbers marked on the number line in Figure 2, including positive and negative numbers and zero, is part of the set of *integers*.

Integers
$\{\ldots, -3, -2, -1, 0, 1, 2, 3, \ldots\}$ is the set of **integers**.

Not all numbers are integers. For example, $\frac{1}{2}$ is not; it is a number halfway between the integers 0 and 1. Also, $3\frac{1}{4}$ is not an integer. Several numbers that are not integers are *graphed* in Figure 3. The **graph** of a number is a point on the number line. Think of the graph of a set of numbers as a picture of the set. All the numbers in Figure 3 can be written as quotients of integers. These numbers are examples of *rational numbers*.

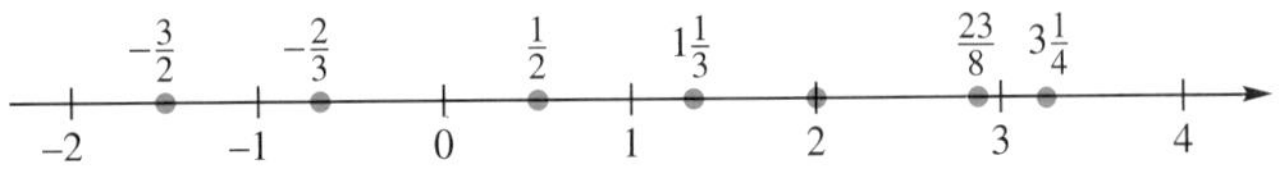

FIGURE 3

Notice that an integer, such as 2, is also a rational number; for example, $2 = 2/1$.

Rational Numbers
$\{x \mid x \text{ is a quotient of two integers, with denominator not equal to } 0\}$ is the set of **rational numbers**.
(Read the part in the braces as "the set of all numbers x such that x is a quotient of two integers, with denominator not equal to 0.")

The set symbolism used in the definition of rational numbers,

$$\{x \mid x \text{ has a certain property}\},$$

is called **set-builder notation.** This notation is convenient to use when it is not possible, or practical, to list all the elements of the set.

Although a great many numbers are rational, not all are. For example, a square that measures one unit on a side has a diagonal whose length is the square root of 2, written $\sqrt{2}$. See Figure 4. It will be shown later that $\sqrt{2}$ cannot be written as a quotient of integers. Because of this, $\sqrt{2}$ is not rational; it is *irrational*.

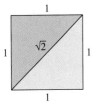

FIGURE 4

Irrational Numbers
$\{x \mid x \text{ is a number on the number line that is not rational}\}$ is the set of **irrational numbers.**

Examples of irrational numbers include $\sqrt{3}$, $\sqrt{7}$, $-\sqrt{10}$, and π, which is the ratio of the distance around a circle (its *circumference*) to the distance across it (its *diameter*).

All numbers that can be represented by points on the number line are called *real numbers*.

Real Numbers

$\{x \mid x$ is a number that can be represented by a point on the number line$\}$ is the set of **real numbers**.

Real numbers can be written as decimal numbers. Any rational number can be written as a decimal that will come to an end (terminate), or repeat in a fixed "block" of digits. For example, $2/5 = .4$ and $27/100 = .27$ are rational numbers with terminating decimals; $1/3 = .3333\ldots$ and $3/11 = .27272727\ldots$ are repeating decimals. The decimal representation of an irrational number will neither terminate nor repeat. Decimal representations of rational and irrational numbers will be discussed further later in this chapter.

Figure 5 illustrates two ways to represent the relationships among the various sets of real numbers.

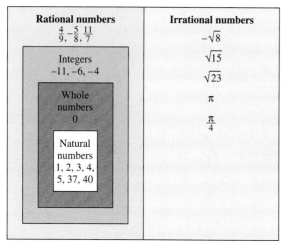

All numbers shown are real numbers.

(a)

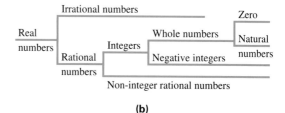

(b)

FIGURE 5

EXAMPLE 1 List the numbers in the set

$$\left\{-5, -\frac{2}{3}, 0, \sqrt{2}, \frac{13}{4}, 5, 5.8\right\}$$

that belong to each of the following sets of numbers.

(a) natural numbers

The only natural number in the set is 5.

6.1 Real Numbers, Order, and Absolute Value 251

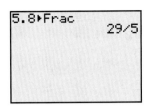

The TI-83 Plus calculator will convert a decimal to a fraction. See Example 1(d).

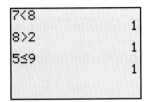

The calculator returns a 1 for these statements of inequality, signifying that each is true.

The symbol for equality, =, was first introduced by the Englishman Robert Recorde in his 1557 algebra text *The Whetstone of Witte*. He used two parallel line segments because, he claimed, no two things can be more equal.

The symbols for order relationships, < and >, were first used by Thomas Harriot (1560–1621), another Englishman. These symbols were not immediately adopted by other mathematicians.

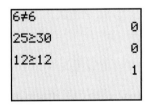

The inequalities in Example 2(a) and (d) are false, as signified by the 0. The statement in Example 2(e) is true.

(b) whole numbers

The whole numbers consist of the natural numbers and 0. So the elements of the set that are whole numbers are 0 and 5.

(c) integers

The integers in the set are -5, 0, and 5.

(d) rational numbers

The rational numbers are -5, $-2/3$, 0, $13/4$, 5, and 5.8, since each of these numbers *can* be written as the quotient of two integers. For example, $5.8 = 58/10 = 29/5$.

(e) irrational numbers

The only irrational number in the set is $\sqrt{2}$.

(f) real numbers

All the numbers in the set are real numbers.

Order Two real numbers may be compared, or ordered, using the ideas of equality and inequality. Suppose that a and b represent two real numbers. If their graphs on the number line are the same point, they are **equal.** If the graph of a lies to the left of b, a **is less than** b, and if the graph of a lies to the right of b, a **is greater than** b. We use symbols to represent these ideas.

When read from left to right, the symbol < means "is less than," so "7 is less than 8" is written

$$7 < 8.$$

The symbol > means "is greater than." We write "8 is greater than 2" as

$$8 > 2.$$

Notice that the symbol always points to the smaller number. For example, write "8 is less than 15" by pointing the symbol toward the 8:

$$8 < 15.$$

The symbol ≤ means "is less than or equal to," so

$$5 \leq 9$$

means "5 is less than or equal to 9." This statement is true, since $5 < 9$ is true. If either the < part or the = part is true, then the inequality ≤ is true.

The symbol ≥ means "is greater than or equal to." Again,

$$9 \geq 5$$

is true because $9 > 5$ is true. Also, $8 \leq 8$ is true since $8 = 8$ is true. But it is not true that $13 \leq 9$ because neither $13 < 9$ nor $13 = 9$ is true.

EXAMPLE 2 Determine whether each statement is *true* or *false*.

(a) $6 \neq 6$ The statement is false because 6 *is equal to* 6.

(b) $5 < 19$ Since 5 is indeed less than 19, this statement is true.

(c) $15 \leq 20$ The statement is true, since $15 < 20$.

(d) $25 \geq 30$ Both $25 > 30$ and $25 = 30$ are false, so $25 \geq 30$ is false.

(e) $12 \geq 12$ Since $12 = 12$, this statement is true.

Additive Inverses and Absolute Value For any real number x (except 0), there is exactly one number on the number line the same distance from 0 as x but on the opposite side of 0. For example, Figure 6 shows that the numbers 3 and -3 are both the same distance from 0 but are on opposite sides of 0. The numbers 3 and -3 are called **additive inverses,** or **opposites,** of each other.

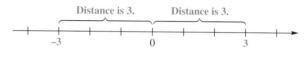

FIGURE 6

The additive inverse of the number 0 is 0 itself. This makes 0 the only real number that is its own additive inverse. Other additive inverses occur in pairs. For example, 4 and -4, and 5 and -5, are additive inverses of each other. Several pairs of additive inverses are shown in Figure 7.

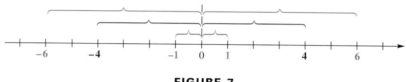

FIGURE 7

The additive inverse of a number can be indicated by writing the symbol $-$ in front of the number. With this symbol, the additive inverse of 7 is written -7. The additive inverse of -4 is written $-(-4)$, and can also be read "the opposite of -4" or "the negative of -4." Figure 7 suggests that 4 is an additive inverse of -4. Since a number can have only one additive inverse, the symbols 4 and $-(-4)$ must represent the same number, which means that

$$-(-4) = 4.$$

This idea can be generalized as follows.

Double Negative Rule

For any real number x,

$$-(-x) = x.$$

Number	Additive Inverse
-4	$-(-4)$ or 4
0	0
19	-19
$-\frac{2}{3}$	$\frac{2}{3}$

The chart shows several numbers and their additive inverses. An important property of additive inverses will be studied later in this chapter: $a + (-a) = (-a) + a = 0$ for all real numbers a.

As mentioned above, additive inverses are numbers that are the same distance from 0 on the number line. See Figure 7. This idea can also be expressed by saying that a number and its additive inverse have the same absolute value. The **absolute value** of a real number can be defined as the distance between 0 and the number on the number line. The symbol for the absolute value of the number x is $|x|$, read "the absolute value of x." For example, the distance between 2 and 0 on the number line is 2 units, so

$$|2| = 2.$$

Because the distance between -2 and 0 on the number line is also 2 units,

$$|-2| = 2.$$

Since distance is a physical measurement, which is never negative, *the absolute value of a number is never negative.* For example, $|12| = 12$ and $|-12| = 12$, since both 12 and -12 lie at a distance of 12 units from 0 on the number line. Also, since 0 is a distance of 0 units from 0,

$$|0| = 0.$$

In symbols, the absolute value of x is defined as follows.

Formal Definition of Absolute Value

$$|x| = \begin{cases} x & \text{if } x \geq 0 \\ -x & \text{if } x < 0 \end{cases}$$

By this definition, if x is a positive number or 0, then its absolute value is x itself. For example, since 8 is a positive number, $|8| = 8$. However, if x is a negative number, then its absolute value is the additive inverse of x. This means that if $x = -9$, then $|-9| = -(-9) = 9$, since the additive inverse of -9 is 9.

The formal definition of absolute value can be confusing if it is not read carefully. The "$-x$" in the second part of the definition *does not* represent a negative number. Since x is negative in the second part, $-x$ represents the opposite of a negative number, that is, a positive number. *The absolute value of a number is never negative.*

EXAMPLE 3 Simplify by finding the absolute value.

(a) $|5| = 5$

(b) $|-5| = -(-5) = 5$

(c) $-|5| = -(5) = -5$

(d) $-|-14| = -(14) = -14$

(e) $|8 - 2| = |6| = 6$

(f) $-|8 - 2| = -|6| = -6$ ■

Part (e) of Example 3 shows that absolute value bars are also grouping symbols. You must perform any operations that appear inside absolute value symbols before finding the absolute value.

Applications A table of data provides a concise way of relating information.

EXAMPLE 4 The projected annual rates of employment change (in percent) in some of the fastest growing and most rapidly declining industries from 1994 through 2005 are shown in the table.

Industry (1994–2005)	Percent Rate of Change
Health services	5.7
Computer and data processing services	4.9
Child day care services	4.3
Footware, except rubber and plastic	−6.7
Household audio and video equipment	−4.2
Luggage, handbags, and leather products	−3.3

Source: U.S. Bureau of Labor Statistics.

What industry in the list is expected to see the greatest change? the least change?

We want the greatest *change*, without regard to whether the change is an increase or a decrease. Look for the number in the list with the largest absolute value. That number is found in footware, since $|-6.7| = 6.7$. Similarly, the least change is in the luggage, handbags, and leather products industry: $|-3.3| = 3.3$.

6.1 EXERCISES

In Exercises 1–6, give a number that satisfies the given condition.

1. An integer between 3.5 and 4.5
2. A rational number between 3.8 and 3.9
3. A whole number that is not positive and is less than 1
4. A whole number greater than 4.5
5. An irrational number that is between $\sqrt{11}$ and $\sqrt{13}$
6. A real number that is neither negative nor positive

In Exercises 7–10, decide whether each statement is true or false.

7. Every natural number is positive.
8. Every whole number is positive.
9. Every integer is a rational number.
10. Every rational number is a real number.

In Exercises 11 and 12, list all numbers from each set that are **(a)** *natural numbers;* **(b)** *whole numbers;* **(c)** *integers;* **(d)** *rational numbers;* **(e)** *irrational numbers;* **(f)** *real numbers.*

11. $\left\{-9, -\sqrt{7}, -1\frac{1}{4}, -\frac{3}{5}, 0, \sqrt{5}, 3, 5.9, 7\right\}$

12. $\left\{-5.3, -5, -\sqrt{3}, -1, -\frac{1}{9}, 0, 1.2, 1.8, 3, \sqrt{11}\right\}$

13. Explain in your own words the different sets of numbers introduced in this section, and give an example of each kind.

14. What two possible situations exist for the decimal representation of a rational number?

Use an integer to express each number representing a change in the following applications.

15. *Population of Michigan* Between 1990 and 1999, the population of Michigan increased by 568,488. (*Source:* U.S. Bureau of the Census.)

16. *Height of the Sears Tower* The Sears Tower in Chicago is 1450 feet high. (*Source:* Council on Tall Buildings and Urban Habitat.)

17. *Boiling Point of Chlorine* The boiling point of chlorine is approximately 30° below 0° Fahrenheit.

18. *Height of Mt. Arenal* The height of Mt. Arenal, an active volcano in Costa Rica, is 5436 feet above sea level. (*Source: The Universal Almanac,* 1997, John W. Wright, General Editor.)

19. *Population of District of Columbia* Between 1990 and 2000, the population of the District of Columbia decreased by 34,841. (*Source:* U.S. Bureau of the Census.)

20. *Auto Production in Taiwan* In a recent year, the country of Taiwan produced 159,376 more passenger cars than commercial vehicles. (*Source:* American Automobile Manufacturers Association.)

21. *Elevation of New Orleans* The city of New Orleans lies 8 feet below sea level. (*Source:* U.S. Geological Survey, *Elevations and Distances in the United States,* 1990.)

22. *Windchill* When the wind speed is 20 miles per hour and the actual temperature is 25° Fahrenheit, the windchill factor is 3° below 0° Fahrenheit. (Give three responses.)

23. *Depths and Heights of Seas and Mountains* The chart gives selected depths and heights of bodies of water and mountains.

Bodies of Water	Average Depth in Feet (as a negative number)	Mountains	Altitude in Feet (as a positive number)
Pacific Ocean	−12,925	McKinley	20,320
South China Sea	−4802	Point Success	14,150
Gulf of California	−2375	Matlalcueyetl	14,636
Caribbean Sea	−8448	Ranier	14,410
Indian Ocean	−12,598	Steele	16,644

Source: The World Almanac and Book of Facts.

(a) List the bodies of water in order, starting with the deepest and ending with the shallowest.
(b) List the mountains in order, starting with the lowest and ending with the highest.
(c) *True or false:* The absolute value of the depth of the Pacific Ocean is greater than the absolute value of the depth of the Indian Ocean.
(d) *True or false:* The absolute value of the depth of the Gulf of California is greater than the absolute value of the depth of the Caribbean Sea.

24. *Savings for Retirement Plans* The bar graph compares average monthly savings, including retirement plans, for five countries.

(a) Which country has the largest average monthly savings? the smallest?
(b) Which countries have average monthly savings greater than $200?
(c) Estimate the average monthly savings for Japan and Italy.
(d) Find the difference between average monthly savings in Japan and Italy.

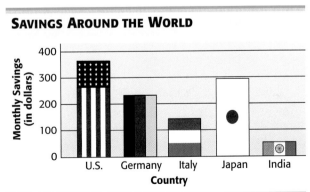

Source: Taylor Nelson-Sofres for American Express.

Graph each group of numbers on a number line.

25. $-2, -6, -4, 3, 4$

26. $-5, -3, -2, 0, 4$

27. $\frac{1}{4}, 2\frac{1}{2}, -3\frac{4}{5}, -4, -1\frac{5}{8}$

28. $5\frac{1}{4}, 4\frac{5}{9}, -2\frac{1}{3}, 0, -3\frac{2}{5}$

29. Match each expression in Column I with its value in Column II. Some choices in Column II may not be used.

    I   II

(a) $|-7|$ A. 7
(b) $-(-7)$ B. -7
(c) $-|-7|$ C. neither A nor B
(d) $-|-(-7)|$ D. both A and B

30. Fill in the blanks with the correct values: The opposite of -2 is _____, while the absolute value of -2 is _____. The additive inverse of -2 is _____, while the additive inverse of the absolute value of -2 is _____.

Find **(a)** *the additive inverse (or opposite) of each number and* **(b)** *the absolute value of each number.*

31. -2 **32.** -8 **33.** 6 **34.** 11
35. $7 - 4$ **36.** $8 - 3$ **37.** $7 - 7$ **38.** $3 - 3$

39. Look at Exercises 35 and 36 and use the results to complete the following: If $a - b > 0$, then the absolute value of $a - b$ in terms of a and b is _____.

40. Look at Exercises 37 and 38 and use the results to complete the following: If $a - b = 0$, then the absolute value of $a - b$ is _____.

Select the lesser of the two given numbers.

41. $-12, -4$ **42.** $-9, -14$ **43.** $-8, -1$
44. $-15, -16$ **45.** $3, |-4|$ **46.** $5, |-2|$
47. $|-3|, |-4|$ **48.** $|-8|, |-9|$ **49.** $-|-6|, -|-4|$
50. $-|-2|, -|-3|$ **51.** $|5 - 3|, |6 - 2|$ **52.** $|7 - 2|, |8 - 1|$

Decide whether each statement is true or false.

53. $6 > -(-2)$ **54.** $-8 > -(-2)$ **55.** $-4 \leq -(-5)$
56. $-6 \leq -(-3)$ **57.** $|-6| < |-9|$ **58.** $|-12| < |-20|$
59. $-|8| > |-9|$ **60.** $-|12| > |-15|$ **61.** $-|-5| \geq -|-9|$
62. $-|-12| \leq -|-15|$ **63.** $|6 - 5| \geq |6 - 2|$ **64.** $|13 - 8| \leq |7 - 4|$

Producer Price Index To answer the questions in Exercises 65–68, refer to the table, which gives the changes in producer price indexes for two recent years.

65. What commodity for which years represents the greatest decrease?

66. What commodity for which years represents the least change?

67. Which has the smaller absolute value, the change for video/audio equipment from 1995 to 1996 or from 1996 to 1997?

68. Which has greater absolute value, the change for apparel from 1995 to 1996 or from 1996 to 1997?

Commodity	Change from 1995 to 1996	Change from 1996 to 1997
Food	4.9	4.0
Transportation	3.9	1.3
Apparel	$-.5$	.9
Video/Audio equipment	-2.6	-2.2
Shelter	5.3	5.3

Source: U.S. Bureau of Labor Statistics.

69. *Comparing Employment Data* Refer to the table in Example 4. Of the household audio/video equipment industry and computer/data processing services, which will show the greater change (without regard to sign)?

70. Students often say "Absolute value is always positive." Is this true? If not, explain why.

Give three numbers between -6 and 6 that satisfy each given condition.

71. Positive real numbers but not integers

72. Real numbers but not positive numbers

73. Real numbers but not whole numbers

74. Rational numbers but not integers

75. Real numbers but not rational numbers

76. Rational numbers but not negative numbers

6.2 Operations, Properties, and Applications of Real Numbers

Operations The result of adding two numbers is called their **sum**.

Adding Real Numbers

Like Signs Add two numbers with the *same* sign by adding their absolute values. The sign of the sum (either $+$ or $-$) is the same as the sign of the two numbers.

Unlike Signs Add two numbers with *different* signs by subtracting the smaller absolute value from the larger. The sum is positive if the positive number has the larger absolute value. The sum is negative if the negative number has the larger absolute value.

For example, to add -12 and -8, first find their absolute values:

$$|-12| = 12 \quad \text{and} \quad |-8| = 8.$$

Since -12 and -8 have the *same* sign, add their absolute values: $12 + 8 = 20$. Give the sum the sign of the two numbers. Since both numbers are negative, the sum is negative and

$$-12 + (-8) = -20.$$

Find $-17 + 11$ by subtracting the absolute values, since these numbers have different signs.

$$|-17| = 17 \quad \text{and} \quad |11| = 11$$
$$17 - 11 = 6$$

Give the result the sign of the number with the larger absolute value.

$$-17 + 11 = -6$$

Negative since $|-17| > |11|$

Practical Arithmetic From the time of Egyptian and Babylonian merchants, practical aspects of arithmetic complemented mystical (or "Pythagorean") tendencies. This was certainly true in the time of Adam Riese (1489–1559), a "reckon master" influential when commerce was growing in Northern Europe. Riese's likeness on the stamp above comes from the title page of one of his popular books on *Rechnung* (or "reckoning"). He championed new methods of reckoning using Hindu-Arabic numerals and quill pens. (The Roman methods then in common use moved counters on a ruled board.) Riese thus fulfilled Fibonacci's efforts 300 years earlier to supplant Roman numerals and methods.

```
 -6+-3
            -9
 4+-1
            3
 -16+12
           -4
```

The calculator supports the results of Example 1(a), (c), and (e).

EXAMPLE 1 Find each of the following sums.

(a) $(-6) + (-3) = -(6 + 3) = -9$

(b) $(-12) + (-4) = -(12 + 4) = -16$

(c) $4 + (-1) = 3$ (d) $-9 + 16 = 7$ (e) $-16 + 12 = -4$

The result of subtraction is called the **difference.** Thus, the difference between 7 and 5 is 2. Compare the two statements below.

$$7 - 5 = 2$$
$$7 + (-5) = 2$$

In a similar way, $9 - 3 = 9 + (-3)$.

That is, to subtract 3 from 9, add the additive inverse of 3 to 9. These examples suggest the following rule for subtraction.

Definition of Subtraction

For all real numbers a and b,

$$a - b = a + (-b).$$

(Change the sign of the second number and add.)

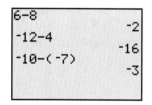

The calculator supports the results of Example 2(a), (b), and (c). Notice how the *negative* (negation) sign differs from the *minus* (subtraction) sign. There are different keys on the calculator for these purposes.

EXAMPLE 2 Perform the indicated operations.

(a) $6 - 8 = 6 + (-8) = -2$ *Change to addition. Change sign of second number.*

(b) $-12 - 4 = -12 + (-4) = -16$ *Change to addition. Sign changed.*

(c) $-10 - (-7) = -10 + [-(-7)]$ This step can be omitted.
$ = -10 + 7$
$ = -3$

(d) $15 - (-3) - 5 - 12$

Perform the additions and subtractions in order from left to right.

$$15 - (-3) - 5 - 12 = (15 + 3) - 5 - 12$$
$$= 18 - 5 - 12$$
$$= 13 - 12$$
$$= 1$$

The **product** is the result of a multiplication problem. Any rules for multiplication with negative real numbers should be consistent with the usual rules for multiplication of positive real numbers and zero. To inductively obtain a rule

Early ways of writing the four basic operation symbols were quite different from those used today. The addition symbol shown below was derived from the Italian word *più* (plus) in the sixteenth century. The + sign used today is shorthand for the Latin *et* (and).

The subtraction symbol shown below was used by Diophantus in Greece sometime during the second or third century A.D. Our subtraction bar may be derived from a bar used by medieval traders to mark differences in weights of products.

In the seventeenth century, Leibniz used the symbol below for multiplication to avoid × as too similar to the "unknown" x. The multiplication symbol × is based on St. Andrew's Cross.

The division symbol shown below was used by Gallimard in the eighteenth century. The familiar ÷ symbol may come from the fraction bar, embellished with the dots above and below.

for multiplying a positive real number and a negative real number, observe the pattern of products below.

$$4 \cdot 5 = 20$$
$$4 \cdot 4 = 16$$
$$4 \cdot 3 = 12$$
$$4 \cdot 2 = 8$$
$$4 \cdot 1 = 4$$
$$4 \cdot 0 = 0$$
$$4 \cdot (-1) = ?$$

What number must be assigned as the product $4 \cdot (-1)$ so that the pattern is maintained? The numbers just to the left of the equality signs decrease by 1 each time, and the products to the right decrease by 4 each time. To maintain the pattern, the number to the right in the bottom equation must be 4 less than 0, which is -4, so

$$4 \cdot (-1) = -4.$$

The pattern continues with

$$4 \cdot (-2) = -8$$
$$4 \cdot (-3) = -12$$
$$4 \cdot (-4) = -16,$$

and so on. In the same way,

$$-4 \cdot 2 = -8$$
$$-4 \cdot 3 = -12$$
$$-4 \cdot 4 = -16,$$

and so on. A similar observation can be made about the product of two negative real numbers. Look at the pattern that follows.

$$-5 \cdot 4 = -20$$
$$-5 \cdot 3 = -15$$
$$-5 \cdot 2 = -10$$
$$-5 \cdot 1 = -5$$
$$-5 \cdot 0 = 0$$
$$-5 \cdot (-1) = ?$$

The numbers just to the left of the equality signs decrease by 1 each time. The products on the right increase by 5 each time. To maintain the pattern, the product $-5 \cdot (-1)$ must be 5 more than 0, so

$$-5 \cdot (-1) = 5.$$

Continuing this pattern gives

$$-5 \cdot (-2) = 10$$
$$-5 \cdot (-3) = 15$$
$$-5 \cdot (-4) = 20,$$

and so on. These observations lead to the following rules for multiplication.

Multiplying Real Numbers

Like Signs Multiply two numbers with the *same* sign by multiplying their absolute values. The product is positive.

Unlike Signs Multiply two numbers with *different* signs by multiplying their absolute values. The product is negative.

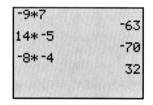

An asterisk (*) represents multiplication on this screen. The display supports the results of Example 3.

EXAMPLE 3 Find each of the following products.

(a) $-9 \cdot 7 = -63$ (b) $14 \cdot (-5) = -70$ (c) $-8 \cdot (-4) = 32$

The result obtained by dividing real numbers is called the **quotient.** For real numbers a, b, and c, where $b \neq 0$, $a/b = c$ means that $a = b \cdot c$. To illustrate this, consider the division problem

$$\frac{10}{-2}.$$

The value of this quotient is obtained by asking, "What number multiplied by -2 gives 10?" From our discussion of multiplication, the answer to this question must be "-5." Therefore,

$$\frac{10}{-2} = -5,$$

because $(-2) \cdot (-5) = 10$. Similar reasoning leads to the following results.

$$\frac{-10}{2} = -5 \quad \text{and} \quad \frac{-10}{-2} = 5$$

These facts, along with the fact that the quotient of two positive numbers is positive, lead to the following rule for division.

Dividing Real Numbers

Like Signs Divide two numbers with the *same* sign by dividing their absolute values. The quotient is positive.

Unlike Signs Divide two numbers with *different* signs by dividing their absolute values. The quotient is negative.

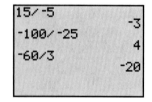

The division operation is represented by a slash (/). This screen supports the results of Example 4.

EXAMPLE 4 Find each of the following quotients.

(a) $\dfrac{15}{-5} = -3$ This is true because $-5 \cdot (-3) = 15$.

(b) $\dfrac{-100}{-25} = 4$ (c) $\dfrac{-60}{3} = -20$

If 0 is divided by a nonzero number, the quotient is 0. That is,

$$\frac{0}{a} = 0 \quad \text{for } a \neq 0.$$

This is true because $a \cdot 0 = 0$. However, we cannot divide by 0. There is a good reason for this. Whenever a division is performed, we want to obtain one and only one quotient. Now consider the division problem

$$\frac{7}{0}.$$

We must ask ourselves "What number multiplied by 0 gives 7?" There is no such number, since the product of 0 and any number is zero. On the other hand, if we consider the quotient

$$\frac{0}{0},$$

there are infinitely many answers to the question, "What number multiplied by 0 gives 0?" Since division by 0 does not yield a *unique* quotient, it is not permitted. To summarize these two situations, we make the following statement.

Dividing by zero leads to this message on the TI-83 Plus.

What result does the calculator give? The order of operations determines the answer. (See Example 5(a).)

Division by Zero
Division by 0 is undefined.

Order of Operations
Given a problem such as $5 + 2 \cdot 3$, should 5 and 2 be added first or should 2 and 3 be multiplied first? When a problem involves more than one operation, we use the following **order of operations.** (This is the order used by computers and many calculators.)

The sentence "Please excuse my dear Aunt Sally" is often used to help remember the rule for order of operations. The letters P, E, M, D, A, S are the first letters of the words of the sentence, and they stand for *parentheses, exponents, multiply, divide, add, subtract.* (Remember also that M and D have equal priority, as do A and S. Operations with equal priority are performed in order from left to right.)

Order of Operations

If parentheses or square brackets are present:

Step 1: Work separately above and below any **fraction bar.**

Step 2: Use the rules below within each set of **parentheses or square brackets.** Start with the innermost set and work outward.

If no parentheses or brackets are present:

Step 1: Apply any **exponents.**

Step 2: Do any **multiplications or divisions** in the order in which they occur, working from left to right.

Step 3: Do any **additions or subtractions** in the order in which they occur, working from left to right.

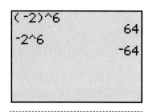

Notice the difference in the two expressions. This supports $(-2)^6 \neq -2^6$.

When evaluating an exponential expression that involves a negative sign, be aware that $(-a)^n$ and $-a^n$ do not necessarily represent the same quantity. For example, if $a = 2$ and $n = 6$,

$$(-2)^6 = (-2)(-2)(-2)(-2)(-2)(-2) = 64 \quad \text{The base is } -2.$$

while

$$-2^6 = -(2 \cdot 2 \cdot 2 \cdot 2 \cdot 2 \cdot 2) = -64. \quad \text{The base is 2.}$$

EXAMPLE 5 Use the order of operations to simplify each of the following.

(a) $5 + 2 \cdot 3$

First multiply, and then add.

$$5 + 2 \cdot 3 = 5 + 6 \quad \text{Multiply.}$$
$$= 11 \quad \text{Add.}$$

(b) $4 \cdot 3^2 + 7 - (2 + 8)$

Work inside the parentheses first.

$$4 \cdot 3^2 + 7 - (2 + 8) = 4 \cdot 3^2 + 7 - 10$$
$$= 4 \cdot 9 + 7 - 10 \quad \text{Apply the exponent.}$$
$$= 36 + 7 - 10 \quad \text{Multiply.}$$
$$= 43 - 10 \quad \text{Add.}$$
$$= 33 \quad \text{Subtract.}$$

(c) $\dfrac{2(8 - 12) - 11(4)}{5(-2) - 3} = \dfrac{2(-4) - 11(4)}{5(-2) - 3} \quad \text{Work separately above and below fraction bar.}$

$$= \dfrac{-8 - 44}{-10 - 3}$$
$$= \dfrac{-52}{-13} = 4$$

The calculator supports the results in Example 5(a), (d), and (f).

(d) $-4^4 = -(4 \cdot 4 \cdot 4 \cdot 4) = -256 \quad \text{Base is 4.}$

(e) $(-4)^4 = (-4)(-4)(-4)(-4) = 256 \quad \text{Base is } -4.$

(f) $(-8)(-3) - [4 - (3 - 6)] = (-8)(-3) - [4 - (-3)]$
$$= (-8)(-3) - [4 + 3]$$
$$= (-8)(-3) - 7$$
$$= 24 - 7$$
$$= 17$$

Properties of Addition and Multiplication of Real Numbers

Several properties of addition and multiplication of real numbers that are essential to our study in this chapter are summarized in the following box.

Properties of Addition and Multiplication

For real numbers a, b, and c, the following properties hold.

Closure Properties $\quad a + b$ and ab are real numbers.

Commutative Properties $\quad a + b = b + a \quad ab = ba$

Associative Properties $\quad (a + b) + c = a + (b + c)$
$\quad (ab)c = a(bc)$

Identity Properties $\quad$ There is a real number 0 such that
$$a + 0 = a \quad \text{and} \quad 0 + a = a.$$
There is a real number 1 such that
$$a \cdot 1 = a \quad \text{and} \quad 1 \cdot a = a.$$

Inverse Properties $\quad$ For each real number a, there is a single real number $-a$ such that
$$a + (-a) = 0 \quad \text{and} \quad (-a) + a = 0.$$
For each nonzero real number a, there is a single real number $1/a$ such that
$$a \cdot \frac{1}{a} = 1 \quad \text{and} \quad \frac{1}{a} \cdot a = 1.$$

Distributive Property of Multiplication with Respect to Addition $\quad a(b + c) = ab + ac$
$\quad (b + c)a = ba + ca$

The set of real numbers is said to be closed with respect to the operations of addition and multiplication. This means that the sum of two real numbers and the product of two real numbers are themselves real numbers. The commutative properties state that two real numbers may be added or multiplied in either order without affecting the result. The associative properties allow us to group terms or factors in any manner we wish without affecting the result. The number 0 is called the **identity element for addition,** and it may be added to any real number to obtain that real number as a sum. Similarly, 1 is called the **identity element for multiplication,** and multiplying a real number by 1 will always yield that real number. Each real number a has an **additive inverse,** $-a$, such that their sum is the additive identity element 0. Each nonzero real number a has a **multiplicative inverse,** or **reciprocal,** $1/a$, such that their product is the multiplicative identity element 1. The distributive property allows us to change certain products to sums and certains sums to products.

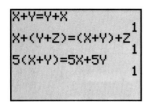

No matter what values are stored in X, Y, and Z, the commutative, associative, and distributive properties assure us that these statements are true.

EXAMPLE 6 Some specific examples of the properties of addition and multiplication of real numbers are given here.

(a) $5 + 7$ is a real number.   Closure property of addition

(b) $5 + (6 + 8) = (5 + 6) + 8$ Associative property of addition

(c) $8 + 0 = 8$ Identity property of addition

(d) $-4\left(-\dfrac{1}{4}\right) = 1$ Inverse property of multiplication

(e) $4 + (3 + 9) = 4 + (9 + 3)$ Commutative property of addition

(f) $5(x + y) = 5x + 5y$ Distributive property

Applications of Real Numbers

The usefulness of negative numbers can be seen by considering situations that arise in everyday life. For example, we need negative numbers to express the temperatures on January days in Anchorage, Alaska, where they often drop below zero. See Exercise 80, which explains the phrases "in the red" and "in the black." And, of course, haven't we all experienced a checking account balance below zero, with hopes that our deposit will make it to the bank before our outstanding checks "bounce"?

When problems deal with gains and losses, the gains may be interpreted as positive numbers and the losses as negative numbers. Temperatures below 0° are negative, and those above 0° are positive. Altitudes above sea level are considered positive and those below sea level are considered negative.

EXAMPLE 7 The Producer Price Index is the oldest continuous statistical series published by the Bureau of Labor Statistics. It measures the average changes in prices received by producers of all commodities produced in the United States. The bar graph in Figure 8 gives the Producer Price Index (PPI) for construction materials between 1993 and 2000.

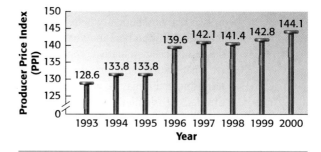

Source: U.S. Bureau of Labor Statistics, Producer Price Indexes, monthly and annual.

FIGURE 8

6.2 Operations, Properties, and Applications of Real Numbers

(a) Use a signed number to represent the change in the PPI from 1993 to 1994.

To find this change, we start with the index number from 1994 and subtract from it the index number from 1993.

$$\underbrace{133.8}_{\text{The 1994 index}} - \underbrace{128.6}_{\text{The 1993 index}} = \underbrace{5.2}_{\substack{\text{A positive number} \\ \text{indicates an increase.}}}$$

(b) Use a signed number to represent the change in the PPI from 1997 to 1998. Use the same procedure as in part (a).

$$\underbrace{141.4}_{\text{The 1998 index}} - \underbrace{142.1}_{\text{The 1997 index}} = 141.4 + (-142.1) = \underbrace{-.7}_{\substack{\text{A negative number} \\ \text{indicates a decrease.}}}$$

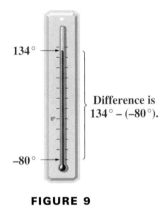

FIGURE 9

EXAMPLE 8 The record high temperature in the United States was 134° Fahrenheit, recorded at Death Valley, California, in 1913. The record low was −80°F, at Prospect Creek, Alaska, in 1971. See Figure 9. What is the difference between these highest and lowest temperatures? (*Source: The World Almanac and Book of Facts,* 2002.)

We must subtract the lower temperature from the higher temperature.

$$134 - (-80) = 134 + 80 \quad \text{Use the definition of subtraction.}$$
$$= 214 \quad \text{Add.}$$

The difference between the two temperatures is 214°F.

6.2 EXERCISES

Fill in each blank with the correct response.

1. The sum of two negative numbers will always be a _____ number.
   (positive/negative)

2. The sum of a number and its opposite will always be _____.

3. To simplify the expression $8 + [-2 + (-3 + 5)]$, I should begin by adding _____ and _____, according to the rule for order of operations.

4. If I am adding a positive number and a negative number, and the negative number has the larger absolute value, the sum will be a _____ number.
   (positive/negative)

5. Explain in words how to add signed numbers. Consider the various cases and give examples.

6. Explain in words how to multiply signed numbers.

Perform the indicated operations, using the order of operations as necessary.

7. $-12 + (-8)$
8. $-5 + (-2)$
9. $12 + (-16)$
10. $-6 + 17$
11. $-12 - (-1)$
12. $-3 - (-8)$
13. $-5 + 11 + 3$
14. $-9 + 16 + 5$
15. $12 - (-3) - (-5)$
16. $15 - (-6) - (-8)$
17. $-9 - (-11) - (4 - 6)$
18. $-4 - (-13) + (-5 + 10)$
19. $(-12)(-2)$
20. $(-3)(-5)$
21. $9(-12)(-4)(-1)3$
22. $-5(-17)(2)(-2)4$
23. $\dfrac{-18}{-3}$
24. $\dfrac{-100}{-50}$
25. $\dfrac{36}{-6}$
26. $\dfrac{52}{-13}$
27. $\dfrac{0}{12}$
28. $\dfrac{0}{-7}$
29. $-6 + [5 - (3 + 2)]$
30. $-8[4 + (7 - 8)]$
31. $-8(-2) - [(4^2) + (7 - 3)]$
32. $-7(-3) - [(2^3) - (3 - 4)]$
33. $-4 - 3(-2) + 5^2$
34. $-6 - 5(-8) + 3^2$
35. $(-8 - 5)(-2 - 1)$
36. $\dfrac{(-10 + 4) \cdot (-3)}{-7 - 2}$
37. $\dfrac{(-6 + 3) \cdot (-4)}{-5 - 1}$
38. $\dfrac{2(-5 + 3)}{-2^2} - \dfrac{(-3^2 + 2)3}{3 - (-4)}$
39. $\dfrac{2(-5) + (-3)(-2^2)}{-6 + 5 + 1}$
40. $\dfrac{3(-4) + (-5)(-2)}{2^3 - 2 + (-6)}$
41. $-\dfrac{1}{4}[3(-5) + 7(-5) + 1(-2)]$
42. $\dfrac{5 - 3\left(\dfrac{-5 - 9}{-7}\right) - 6}{-9 - 11 + 3 \cdot 7}$

43. Which of the following expressions are undefined?
    A. $\dfrac{8}{0}$ B. $\dfrac{9}{6 - 6}$ C. $\dfrac{4 - 4}{5 - 5}$ D. $\dfrac{0}{-1}$

44. If you have no money in your pocket and you divide it equally among your three siblings, how much does each get? Use this situation to explain division of zero by a nonzero number.

Identify the property illustrated by each of the following statements.

45. $6 + 9 = 9 + 6$
46. $8 \cdot 4 = 4 \cdot 8$
47. $7 + (2 + 5) = (7 + 2) + 5$
48. $(3 \cdot 5) \cdot 4 = 4 \cdot (3 \cdot 5)$
49. $9 + (-9) = 0$
50. $12 + 0 = 12$
51. $9 \cdot 1 = 9$
52. $\left(\dfrac{1}{-3}\right) \cdot (-3) = 1$
53. $0 + 283 = 283$
54. $6 \cdot (4 \cdot 2) = (6 \cdot 4) \cdot 2$
55. $2 \cdot (4 + 3) = 2 \cdot 4 + 2 \cdot 3$
56. $9 \cdot 6 + 9 \cdot 8 = 9 \cdot (6 + 8)$
57. $19 + 12$ is a real number.
58. $19 \cdot 12$ is a real number.

Exercises 59–68 are designed to explore the properties of real numbers in further detail.

59. (a) Evaluate $6 - 8$ and $8 - 6$.
    (b) By the results of part (a), we may conclude that subtraction is not a(n) _____ operation.
    (c) Are there *any* real numbers a and b for which $a - b = b - a$? If so, give an example.

60. (a) Evaluate $4 \div 8$ and $8 \div 4$.
    (b) By the results of part (a), we may conclude that division is not a(n) _____ operation.
    (c) Are there *any* real numbers a and b for which $a \div b = b \div a$? If so, give an example.

61. Many everyday occurrences can be thought of as operations that have opposites or inverses. For example, the inverse operation for "going to sleep" is "waking up." For each of the given activities, specify its inverse activity.
    (a) cleaning up your room
    (b) earning money
    (c) increasing the volume on your CD player

62. Many everyday activities are commutative; that is, the order in which they occur does not affect the outcome. For example, "putting on your shirt" and "putting on your pants" are commutative operations. Decide whether the given activities are commutative.
(a) putting on your shoes; putting on your socks
(b) getting dressed; taking a shower
(c) combing your hair; brushing your teeth

63. The following conversation actually took place between one of the authors of this text and his son, Jack, when Jack was four years old.

Daddy: "Jack, what is 3 + 0?"
Jack: "3"
Daddy: "Jack, what is 4 + 0?"
Jack: "4... and Daddy, *string* plus zero equals *string*!"
What property of addition of real numbers did Jack recognize?

64. The phrase *defective merchandise counter* is an example of a phrase that can have different meanings depending upon how the words are grouped (think of the associative properties). For example, (*defective merchandise*) *counter* is a location at which we would return an item that does not work, while *defective* (*merchandise counter*) is a broken place where items are bought and sold. For each of the following phrases, explain why the associative property does not hold.
(a) difficult test question
(b) woman fearing husband
(c) man biting dog

65. The distributive property holds for multiplication with respect to addition. Does the distributive property hold for addition with respect to multiplication? That is, is $a + (b \cdot c) = (a + b) \cdot (a + c)$ true for all values of a, b, and c? (*Hint:* Let $a = 2$, $b = 3$, and $c = 4$.)

66. Suppose someone makes the following claim: The distributive property for addition with respect to multiplication (from Exercise 65) is valid, and here's why: Let $a = 2$, $b = -4$, and $c = 3$. Then $a + (b \cdot c) = 2 + (-4 \cdot 3) = 2 + (-12) = -10$, and $(a + b) \cdot (a + c) = [2 + (-4)] \cdot [2 + 3] = -2 \cdot 5 = -10$. Since both expressions are equal, the property must be valid. How would you respond to this reasoning?

67. Suppose that a student shows you the following work.

$$-3(4 - 6) = -3(4) - 3(6) = -12 - 18 = -30$$

The student has made a very common error in applying the distributive property. Write a short paragraph explaining the student's mistake, and work the problem correctly.

68. Work the following problem in two ways, first using the order of operations, and then using the distributive property: Evaluate $9(11 + 15)$.

Each of the expressions in Exercises 69–76 is equal to either 81 *or* −81. *Decide which of these is the correct value.*

69. -3^4 **70.** $-(3^4)$ **71.** $(-3)^4$ **72.** $-(-3^4)$
73. $-(-3)^4$ **74.** $[-(-3)]^4$ **75.** $-[-(-3)]^4$ **76.** $-[-(-3^4)]$

Solve each problem by evaluating a sum or difference of real numbers.

77. *Defense Budget* The bar graph shows the federal budget outlays for national defense for the years 1990 through 2000. Use a signed number to represent the change in outlay for each of the following time periods.
(a) 1991–1992
(b) 1993–1994
(c) 1996–1997
(d) 1997–1998

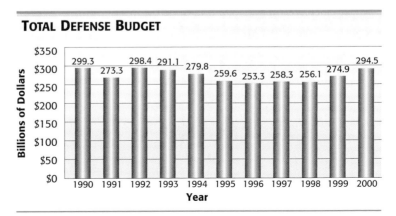

Source: U.S. Office of Management and Budget.

78. *Heights of Mountains and Depths of Trenches* The chart shows the heights in feet of some selected mountains and the depths in feet (as negative numbers) of some selected trenches.

Mountain	Height	Trench	Depth
Foraker	17,400	Philippine	−32,995
Wilson	14,246	Cayman	−24,721
Pikes Peak	14,110	Java	−23,376

Source: The World Almanac and Book of Facts, 2002.

(a) What is the difference between the height of Mt. Foraker and the depth of the Philippine Trench?
(b) What is the difference between the height of Pikes Peak and the depth of the Java Trench?
(c) How much deeper is the Cayman Trench than the Java Trench?
(d) How much deeper is the Philippine Trench than the Cayman Trench?

79. *Social Security Finances* The table shows Social Security tax revenue and cost of benefits (in billions of dollars).

Year	Tax Revenue	Cost of Benefits
2000	538	409
2010*	916	710
2020*	1479	1405
2030*	2041	2542

*Projected
Source: Social Security Board of Trustees.

(a) Find the difference between Social Security tax revenue and cost of benefits for each year shown in the table.
(b) Interpret your answer for 2030.

80. *Revenue and Expenditures of Yahoo* The table gives revenue and expenditures (both in millions of dollars) for Yahoo, Inc., for three different years. Determine the absolute value of the difference between revenue and expenditures for each year, and tell whether the company made a profit (i.e., was "in the black") or experienced a loss (i.e., was "in the red"). (These descriptions go back to the days when bookkeepers used black ink to represent gains and red ink to represent losses. This convention is still used. For example, Yahoo and online brokers display stock gains in black and losses in red.)

Year	Revenue	Expenditures
1996	19.07	21.39
1997	67.41	90.29
1998	203.20	177.61

Source: www.quote.com

(a) 1996   (b) 1997   (c) 1998

81. *Checking Account Balance* Shalita's checking account balance is $54.00. She then takes a gamble by writing a check for $89.00. What is her new balance? (Write the balance as a signed number.)

82. *Altitude of Hikers* The surface, or rim, of a canyon is at altitude 0. On a hike down into the canyon, a party of hikers stops for a rest at 130 meters below the surface. They then descend another 54 meters. What is their new altitude? (Write the altitude as a signed number.)

83. *Drastic Temperature Change* On January 23, 1943, the temperature rose 49°F in two minutes in Spearfish, South Dakota. If the starting temperature was −4°F, what was the temperature two minutes later? (*Source: The Guinness Book of World Records.*)

84. *Extreme Temperatures in Little Rock* The lowest temperature ever recorded in Little Rock, Arkansas, was −5°F. The highest temperature ever recorded there was 117°F more than the lowest. What was this highest temperature? (*Source: The World Almanac and Book of Facts,* 2002.)

85. *Low Temperatures in Chicago and Huron* The coldest temperature recorded in Chicago, Illinois, was −27°F in 1985. The record low in Huron, South Dakota, was set in 1994 and was 14°F lower than −27°F. What was the record low in Huron? (*Source: The World Almanac and Book of Facts,* 2002.)

86. *Breaching of Humpback Whales* No one knows just why humpback whales heave their 45-ton bodies out of the water, but leap they do. (This activity is called *breaching.*) Mark and Debbie, two researchers based on the island of Maui, noticed that one of their favorite whales, "Pineapple," leaped 15 feet above the surface of the ocean while her mate cruised

12 feet below the surface. What is the difference between these two levels?

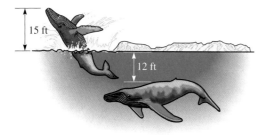

87. *Elevation Difference* The top of Mt. Whitney, visible from Death Valley, has an altitude of 14,494 feet above sea level. The bottom of Death Valley is 282 feet below sea level. Using zero as sea level, find the difference between these two elevations. (*Source: The World Almanac and Book of Facts,* 2002.)

88. *Highest Point in Louisiana* The highest point in Louisiana is Driskill Mountain, at an altitude of 535 feet. The lowest point is at Spanish Fort, 8 feet below sea level. Using zero as sea level, find the difference between these two elevations. (*Source: The World Almanac and Book of Facts,* 2002.)

89. *Weight of a Polar Bear* A female polar bear weighed 660 pounds when she entered her winter den. She lost 45 pounds during each of the first two months of hibernation, and another 205 pounds before leaving the den with her two cubs in March. How much did she weigh when she left the den?

90. *Scuba Diving Depths* Dean Baldus enjoys scuba diving. He dives to 34 feet below the surface of a lake. His partner, Jeff Balius, dives to 40 feet below the surface, but then ascends 20 feet. What is the vertical distance between Dean and Jeff?

91. *Passing Yardage* On three consecutive passing plays, Peyton Manning passed for a gain of 6 yards, was sacked for a loss of 12 yards, and passed for a gain of 43 yards. What positive or negative number represents the total net yardage for the three plays?

92. *Plane Altitude* Marty McDonald, a pilot for a major airline, announced to her passengers that their plane, currently at 34,000 feet, would descend 2500 feet to avoid turbulence, and then ascend 3000 feet once they were out of danger from the turbulence. What would their final altitude be?

93. *Triominoes Score* J. D. Patin enjoys playing Triominoes every Wednesday night. Last Wednesday, on four successive turns, his scores were -19, 28, -5, and 13. What was his final score for the four turns?

94. *Birthdate of a Greek Mathematician* A certain Greek mathematician was born in 426 B.C. His father was born 43 years earlier. In what year was his father born?

95. *MasterCard Balance* Kim Falgout owes $870.00 on her MasterCard account. She returns two items costing $35.90 and $150.00 and receives credits for these on the account. Next, she makes a purchase of $82.50, and then two more purchases of $10.00 each. She makes a payment of $500.00. She then incurs a finance charge of $37.23. How much does she still owe?

96. *Downhill Skiing Times* The 1988 Women's Olympic Downhill Skiing champion, Marina Kiehl, from West Germany, finished the course in 1 min, 25.86 sec. The winning time of the 1994 champion, Katja Seizinger of Germany, increased by 10.07 sec. Seizinger won again in 1998. Her time decreased by 7.04 sec. What was Seizinger's winning time in 1998? (*Source: World Almanac and Book of Facts,* 1999.)

6.3 Rational Numbers and Decimal Representation

Properties and Operations The set of real numbers is composed of two important mutually exclusive subsets: the rational numbers and the irrational numbers. (Two sets are *mutually exclusive* if they contain no elements in common.)

Recall from Section 6.1 that quotients of integers are called **rational numbers.** Think of the rational numbers as being made up of all the fractions (quotients of integers with denominator not equal to zero) and all the integers. Any integer can be written as the quotient of two integers. For example, the integer 9 can be written as the quotient 9/1, or 18/2, or 27/3, and so on. Also, −5 can be expressed as a quotient of integers as −5/1 or −10/2, and so on. (How can the integer 0 be written as a quotient of integers?) Since both fractions and integers can be written as quotients of integers, the set of rational numbers is defined as follows.

> ### Rational Numbers
> Rational numbers = $\{x \mid x \text{ is a quotient of two integers, with denominator not } 0\}$

A rational number is said to be in **lowest terms** if the greatest common factor of the numerator (top number) and the denominator (bottom number) is 1. Rational numbers are written in lowest terms by using the *fundamental property of rational numbers*.

> ### Fundamental Property of Rational Numbers
> If a, b, and k are integers with $b \neq 0$ and $k \neq 0$, then
> $$\frac{a \cdot k}{b \cdot k} = \frac{a}{b}.$$

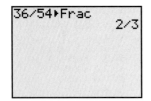

The calculator reduces 36/54 to lowest terms, as illustrated in Example 1.

EXAMPLE 1 Write 36/54 in lowest terms.

Since the greatest common factor of 36 and 54 is 18,

$$\frac{36}{54} = \frac{2 \cdot 18}{3 \cdot 18} = \frac{2}{3}.$$

In the above example, $\frac{36}{54} = \frac{2}{3}$. If we multiply the numerator of the fraction on the left by the denominator of the fraction on the right, we obtain $36 \cdot 3 = 108$. If we multiply the denominator of the fraction on the left by the numerator of the fraction on the right, we obtain $54 \cdot 2 = 108$. The result is the same in both cases. One way of determining whether two fractions are equal is to perform this test. If the product of the "extremes" (36 and 3 in this case) equals the product of the "means" (54 and 2), the fractions are equal. This test for equality of rational numbers is called the **cross-product test.**

6.3 Rational Numbers and Decimal Representation

Benjamin Banneker
(1731–1806) spent the first half of his life tending a farm in Maryland. He gained a reputation locally for his mechanical skills and abilities in mathematical problem solving. In 1772 he acquired astronomy books from a neighbor and devoted himself to learning astronomy, observing the skies, and making calculations. In 1789 Banneker joined the team that surveyed what is now the District of Columbia.

Banneker published almanacs yearly from 1792 to 1802. He sent a copy of his first almanac to Thomas Jefferson along with an impassioned letter against slavery. Jefferson subsequently championed the cause of this early African-American mathematician.

Cross-Product Test for Equality of Rational Numbers

For rational numbers a/b and c/d, $b \neq 0, d \neq 0$,

$$\frac{a}{b} = \frac{c}{d} \quad \text{if and only if} \quad a \cdot d = b \cdot c.$$

The operation of addition of rational numbers can be illustrated by the sketches in Figure 10. The rectangle at the top left is divided into three equal portions, with one of the portions in color. The rectangle at the top right is divided into five equal parts, with two of them in color.

The total of the areas in color is represented by the sum

$$\frac{1}{3} + \frac{2}{5}.$$

To evaluate this sum, the areas in color must be redrawn in terms of a common unit. Since the least common multiple of 3 and 5 is 15, redraw both rectangles with 15 parts. See Figure 11. In the figure, 11 of the small rectangles are in color, so

$$\frac{1}{3} + \frac{2}{5} = \frac{5}{15} + \frac{6}{15} = \frac{11}{15}.$$

In general, the sum

$$\frac{a}{b} + \frac{c}{d}$$

may be found by writing a/b and c/d with the common denominator bd, retaining this denominator in the sum, and adding the numerators:

$$\frac{a}{b} + \frac{c}{d} = \frac{ad}{bd} + \frac{bc}{bd} = \frac{ad + bc}{bd}.$$

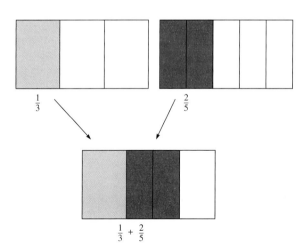

FIGURE 10

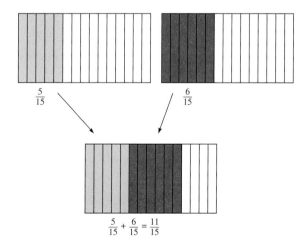

FIGURE 11

A similar case can be made for the difference between rational numbers. A formal definition of addition and subtraction of rational numbers follows.

> **Adding and Subtracting Rational Numbers**
>
> If a/b and c/d are rational numbers, then
>
> $$\frac{a}{b} + \frac{c}{d} = \frac{ad + bc}{bd} \quad \text{and} \quad \frac{a}{b} - \frac{c}{d} = \frac{ad - bc}{bd}.$$

This formal definition is seldom used in practice. In practical problems involving addition and subtraction of rational numbers, we usually rewrite the fractions with the least common multiple of their denominators, called the **least common denominator.**

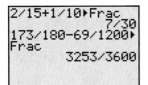

The results of Example 2 are illustrated in this screen.

EXAMPLE 2 (a) Add: $\dfrac{2}{15} + \dfrac{1}{10}$.

The least common multiple of 15 and 10 is 30. Now write 2/15 and 1/10 with denominators of 30, and then add the numerators. Proceed as follows:

Since $30 \div 15 = 2$, $\qquad \dfrac{2}{15} = \dfrac{2 \cdot 2}{15 \cdot 2} = \dfrac{4}{30},$

and since $30 \div 10 = 3$, $\qquad \dfrac{1}{10} = \dfrac{1 \cdot 3}{10 \cdot 3} = \dfrac{3}{30}.$

Thus, $\qquad \dfrac{2}{15} + \dfrac{1}{10} = \dfrac{4}{30} + \dfrac{3}{30} = \dfrac{7}{30}.$

(b) Subtract: $\dfrac{173}{180} - \dfrac{69}{1200}$.

The least common multiple of 180 and 1200 is 3600.

$$\frac{173}{180} - \frac{69}{1200} = \frac{3460}{3600} - \frac{207}{3600} = \frac{3460 - 207}{3600} = \frac{3253}{3600}$$

The product of two rational numbers is defined as follows.

> **Multiplying Rational Numbers**
>
> If a/b and c/d are rational numbers, then
>
> $$\frac{a}{b} \cdot \frac{c}{d} = \frac{ac}{bd}.$$

EXAMPLE 3   Find each of the following products.

(a) $\dfrac{3}{4} \cdot \dfrac{7}{10} = \dfrac{3 \cdot 7}{4 \cdot 10} = \dfrac{21}{40}$

(b) $\dfrac{5}{18} \cdot \dfrac{3}{10} = \dfrac{5 \cdot 3}{18 \cdot 10} = \dfrac{15}{180} = \dfrac{1 \cdot 15}{12 \cdot 15} = \dfrac{1}{12}$

To illustrate the results of Example 3, we use parentheses around the fraction factors.

In practice, a multiplication problem such as this is often solved by using slash marks to indicate that common factors have been divided out of the numerator and denominator.

$$\dfrac{\overset{1}{\cancel{5}}}{\underset{6}{\cancel{18}}} \cdot \dfrac{\overset{1}{\cancel{3}}}{\underset{2}{\cancel{10}}} = \dfrac{1}{6} \cdot \dfrac{1}{2}$$

3 is divided out of the terms 3 and 18; 5 is divided out of 5 and 10.

$$= \dfrac{1}{12}$$

FOR FURTHER THOUGHT

The Influence of Spanish Coinage on Stock Prices

Until August 28, 2000, when decimalization of the U.S. stock market began, market prices were reported with fractions having denominators with powers of 2, such as $17\dfrac{3}{4}$ and $112\dfrac{5}{8}$. Did you ever wonder why this was done?

During the early years of the United States, prior to the minting of its own coinage, the Spanish eight-reales coin, also known as the Spanish milled dollar, circulated freely in the states. Its fractional parts, the four reales, two reales, and one real, were known as **pieces of eight,** and described as such in pirate and treasure lore. When the New York Stock Exchange was founded in 1792, it chose to use the Spanish milled dollar as its price basis, rather than the decimal base as proposed by Thomas Jefferson that same year.

In the September 1997 issue of *COINage,* Tom Delorey's article "The End of 'Pieces of Eight'" gives the following account:

> As the Spanish dollar and its fractions continued to be legal tender in America alongside the decimal coins until 1857, there was no urgency to change the system—and by the time the Spanish-American money was withdrawn in 1857, pricing stocks in eighths of a dollar—and no less—was a tradition carved in stone. Being somewhat a conservative organization, the NYSE saw no need to fix what was not broken.

All prices on the U.S. stock markets are now reported in decimals. (*Source:* "Stock price tables go to decimal listings," *The Times Picayune,* June 27, 2000.)

For Group Discussion

Consider this: Have you ever heard this old cheer? "Two bits, four bits, six bits, a dollar. All for the (home team), stand up and holler." The term **two bits** refers to 25 cents. Discuss how this cheer is based on the Spanish eight-reales coin.

Early U.S. cents and half cents used fractions to denote their denominations. The half cent used $\frac{1}{200}$ and the cent used $\frac{1}{100}$. (See Exercise 18 for a photo of an interesting error coin.)

The coins shown here were part of the collection of Louis E. Eliasberg, Sr., that was auctioned by Bowers and Merena, Inc., several years ago. Louis Eliasberg was the only person ever to assemble a complete collection of United States coins. The half cent pictured sold for $506,000 and the cent sold for $27,500. The cent shown in Exercise 18 went for a mere $2970.

In a fraction, the fraction bar indicates the operation of division. Recall that, in the previous section, we defined the multiplicative inverse, or reciprocal, of the nonzero number b. The multiplicative inverse of b is $1/b$. We can now define division using multiplicative inverses.

Definition of Division

If a and b are real numbers, $b \neq 0$, then

$$\frac{a}{b} = a \cdot \frac{1}{b}.$$

You have probably heard the rule, "To divide fractions, invert the divisor and multiply." But have you ever wondered why this rule works? To illustrate it, suppose that you have $7/8$ of a gallon of milk and you wish to find how many quarts you have. Since a quart is $1/4$ of a gallon, you must ask yourself, "How many $1/4$s are there in $7/8$?" This would be interpreted as

$$\frac{7}{8} \div \frac{1}{4} \quad \text{or} \quad \frac{\frac{7}{8}}{\frac{1}{4}}.$$

The fundamental property of rational numbers discussed earlier can be extended to rational number values of a, b, and k. With $a = 7/8$, $b = 1/4$, and $k = 4$ (the reciprocal of $b = 1/4$),

$$\frac{a}{b} = \frac{a \cdot k}{b \cdot k} = \frac{\frac{7}{8} \cdot 4}{\frac{1}{4} \cdot 4} = \frac{\frac{7}{8} \cdot 4}{1} = \frac{7}{8} \cdot \frac{4}{1}.$$

Now notice that we began with the division problem $7/8 \div 1/4$ which, through a series of equivalent expressions, led to the multiplication problem $(7/8) \cdot (4/1)$. So dividing by $1/4$ is equivalent to multiplying by its reciprocal, $4/1$. By the definition of multiplication of fractions,

$$\frac{7}{8} \cdot \frac{4}{1} = \frac{28}{8} = \frac{7}{2},$$

and thus there are $7/2$ or $3\frac{1}{2}$ quarts in $\frac{7}{8}$ gallon.*

We now state the rule for dividing a/b by c/d.

Dividing Rational Numbers

If a/b and c/d are rational numbers, where $c/d \neq 0$, then

$$\frac{a}{b} \div \frac{c}{d} = \frac{a}{b} \cdot \frac{d}{c} = \frac{ad}{bc}.$$

---

*$3\frac{1}{2}$ is a **mixed number**. Mixed numbers are covered in the exercises for this section.

6.3 Rational Numbers and Decimal Representation

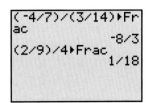

This screen supports the results in Example 4(b) and (c).

EXAMPLE 4 Find each of the following quotients.

(a) $\dfrac{3}{5} \div \dfrac{7}{15} = \dfrac{3}{5} \cdot \dfrac{15}{7} = \dfrac{45}{35} = \dfrac{9 \cdot 5}{7 \cdot 5} = \dfrac{9}{7}$

(b) $\dfrac{-4}{7} \div \dfrac{3}{14} = \dfrac{-4}{7} \cdot \dfrac{14}{3} = \dfrac{-56}{21} = \dfrac{-8 \cdot 7}{3 \cdot 7} = \dfrac{-8}{3} = -\dfrac{8}{3}$

(c) $\dfrac{2}{9} \div 4 = \dfrac{2}{9} \div \dfrac{4}{1} = \dfrac{2}{9} \cdot \dfrac{1}{4} = \dfrac{\overset{1}{\cancel{2}}}{9} \cdot \dfrac{1}{\underset{2}{\cancel{4}}} = \dfrac{1}{18}$

There is no integer between two consecutive integers, such as 3 and 4. However, a rational number can always be found between any two distinct rational numbers. For this reason, the set of rational numbers is said to be *dense*.

Density Property of the Rational Numbers

If r and t are distinct rational numbers, with $r < t$, then there exists a rational number s such that

$$r < s < t.$$

To find the **arithmetic mean,** or **average,** of n numbers, we add the numbers and then divide the sum by n. For two numbers, the number that lies halfway between them is their average.

Year	Number (in thousands)
1995	16,360
1996	16,269
1997	16,110
1998	16,211
1999	16,477
2000	16,258

Source: U.S. Bureau of Labor Statistics.

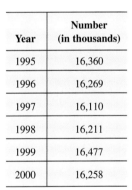

The computation in Example 5(b) is shown here.

EXAMPLE 5

(a) Find the rational number halfway between 2/3 and 5/6.

First, find their sum. $\dfrac{2}{3} + \dfrac{5}{6} = \dfrac{4}{6} + \dfrac{5}{6} = \dfrac{9}{6} = \dfrac{3}{2}$

Now divide by 2.

$$\dfrac{3}{2} \div 2 = \dfrac{3}{2} \cdot \dfrac{1}{2} = \dfrac{3}{4}$$

The number halfway between 2/3 and 5/6 is 3/4.

(b) The table in the margin shows the number of labor union or employee association members, in thousands, for the years 1995–2000. What is the average number, in thousands, for this six-year period?

To find this average, divide the sum by 6.

$$\dfrac{16{,}360 + 16{,}269 + 16{,}110 + 16{,}211 + 16{,}477 + 16{,}258}{6} = \dfrac{97{,}685}{6}$$

$$\approx 16{,}280.8$$

The average to the nearest whole number of thousands is 16,281.

Simon Stevin (1548–1620) worked as a bookkeeper in Belgium and became an engineer in the Netherlands army. He is usually given credit for the development of decimals.

Repeated application of the density property implies that between two given rational numbers are *infinitely many* rational numbers. It is also true that between any two *real* numbers there is another *real* number. Thus, we say that the set of real numbers is dense.

Decimal Form of Rational Numbers

Up to now in this section, we have discussed rational numbers in the form of quotients of integers. Rational numbers can also be expressed as decimals. Decimal numerals have place values that are powers of 10. For example, the decimal numeral 483.039475 is read "four hundred eighty-three and thirty-nine thousand, four hundred seventy-five millionths." The place values are as shown here.

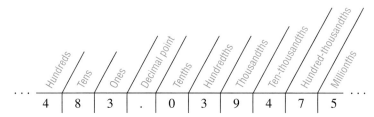

Given a rational number in the form a/b, it can be expressed as a decimal most easily by entering it into a calculator. For example, to write 3/8 as a decimal, enter 3, then enter the operation of division, then enter 8. Press the equals key to find the following equivalence.

$$\frac{3}{8} = .375$$

Of course, this same result may be obtained by long division, as shown in the margin. By this result, the rational number 3/8 is the same as the decimal .375. A decimal such as .375, which stops, is called a **terminating decimal.** Other examples of terminating decimals are

$$\frac{1}{4} = .25, \quad \frac{7}{10} = .7, \quad \text{and} \quad \frac{89}{1000} = .089.$$

Not all rational numbers can be represented by terminating decimals. For example, convert 4/11 into a decimal by dividing 11 into 4 using a calculator. The display shows

.3636363636, or perhaps .363636364.

However, we see that the long division process, shown in the margin, indicates that we will actually get .3636..., with the digits 36 repeating over and over indefinitely. To indicate this, we write a bar (called a *vinculum*) over the "block" of digits that repeats. Therefore, we can write

$$\frac{4}{11} = .\overline{36}.$$

A decimal such as $.\overline{36}$, which continues indefinitely, is called a **repeating decimal.** Other examples of repeating decimals are

$$\frac{5}{11} = .\overline{45}, \quad \frac{1}{3} = .\overline{3}, \quad \text{and} \quad \frac{5}{6} = .8\overline{3}.$$

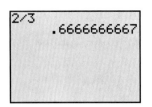

While 2/3 has a repeating decimal representation $(2/3 = .\overline{6})$, the calculator rounds off in the final decimal place displayed.

6.3 Rational Numbers and Decimal Representation 277

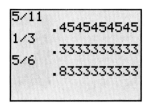

Although only ten decimal digits are shown, all three fractions have decimals that repeat endlessly.

Because of the limitations of the display of a calculator, and because some rational numbers have repeating decimals, it is important to be able to interpret calculator results accordingly when obtaining repeating decimals.

While we shall distinguish between *terminating* and *repeating* decimals in this book, some mathematicians prefer to consider all rational numbers as repeating decimals. This can be justified by thinking this way: if the division process leads to a remainder of 0, then zeros repeat without end in the decimal form. For example, we can consider the decimal form of 3/4 as follows.

$$\frac{3}{4} = .75\overline{0}$$

By considering the possible remainders that may be obtained when converting a quotient of integers to a decimal, we can draw an important conclusion about the decimal form of rational numbers. If the remainder is never zero, the division will produce a repeating decimal. This happens because each step of the division process must produce a remainder that is less than the divisor. Since the number of different possible remainders is less than the divisor, the remainders must eventually begin to repeat. This makes the digits of the quotient repeat, producing a repeating decimal.

To find a baseball player's batting average, we divide the number of hits by the number of at-bats. A surprising paradox exists concerning averages; it is possible for Player *A* to have a higher batting average than Player *B* in each of two successive years, yet for the two-year period, Player *B* can have a higher total batting average. Look at the chart.

Year	Player A	Player B
1998	$\frac{20}{40} = .500$	$\frac{90}{200} = .450$
1999	$\frac{60}{200} = .300$	$\frac{10}{40} = .250$
Two year-total	$\frac{80}{240} = .333$	$\frac{100}{240} = .417$

In both individual years, Player *A* had a higher average, but for the two-year period, Player *B* had the higher average. This is an example of *Simpson's paradox* from statistics.

Decimal Representation of Rational Numbers

Any rational number can be expressed as either a terminating decimal or a repeating decimal.

To determine whether the decimal form of a quotient of integers will terminate or repeat, we use the following rule.

Criteria for Terminating and Repeating Decimals

A rational number a/b in lowest terms results in a **terminating decimal** if the only prime factor of the denominator is 2 or 5 (or both).

A rational number a/b in lowest terms results in a **repeating decimal** if a prime other than 2 or 5 appears in the prime factorization of the denominator.

The justification of this rule is based on the fact that the prime factors of 10 are 2 and 5, and the decimal system uses ten as its base.

 Without actually dividing, determine whether the decimal form of the given rational number terminates or repeats.

(a) $\frac{7}{8}$

Since 8 factors as 2^3, the decimal form will terminate. No primes other than 2 or 5 divide the denominator.

(b) $\dfrac{13}{150}$

$150 = 2 \cdot 3 \cdot 5^2$. Since 3 appears as a prime factor of the denominator, the decimal form will repeat.

(c) $\dfrac{6}{75}$

First write the rational number in lowest terms.

$$\dfrac{6}{75} = \dfrac{2}{25}$$

Since $25 = 5^2$, the decimal form will terminate.

We have seen that a rational number will be represented by either a terminating or a repeating decimal. What about the reverse process? That is, must a terminating decimal or a repeating decimal represent a rational number? The answer is *yes*. For example, the terminating decimal .6 represents a rational number.

$$.6 = \dfrac{6}{10} = \dfrac{3}{5}$$

The results of Example 7 are supported in this screen.

EXAMPLE 7 Write each terminating decimal as a quotient of integers.

(a) $.437 = \dfrac{437}{1000}$ (b) $8.2 = 8 + \dfrac{2}{10} = \dfrac{82}{10} = \dfrac{41}{5}$

Repeating decimals cannot be converted into quotients of integers quite so quickly. The steps for making this conversion are given in the next example. (This example uses basic algebra.)

EXAMPLE 8 Find a quotient of two integers equal to $.\overline{85}$.

Step 1: Let $x = .\overline{85}$, so $x = .858585\ldots$.

Step 2: Multiply both sides of the equation $x = .858585\ldots$ by 100. (Use 100 since there are two digits in the part that repeats, and $100 = 10^2$.)

$$x = .858585\ldots$$
$$100x = 100(.858585\ldots)$$
$$100x = 85.858585\ldots$$

Step 3: Subtract the expressions in Step 1 from the final expressions in Step 2.

$$100x = 85.858585\ldots$$ (Recall that $x = 1x$ and
$$\underline{x = .858585\ldots}$$ $100x - x = 99x$.)
$$99x = 85$$

Step 4: Solve the equation $99x = 85$ by dividing both sides by 99.

$$99x = 85$$
$$\dfrac{99x}{99} = \dfrac{85}{99}$$
$$x = \dfrac{85}{99}$$
$$.\overline{85} = \dfrac{85}{99} \qquad x = .\overline{85}$$

$1 = .999999999\ldots$

Terminating or Repeating?
One of the most baffling truths of elementary mathematics is the following:

$$1 = .9999\ldots.$$

Most people believe that $.\overline{9}$ has to be less that 1, but this is not the case. The following argument shows otherwise. Let $x = .9999\ldots$. Then

$10x = 9.9999\ldots$
$\underline{x = .9999\ldots}$
$9x = 9$ Subtract.
$x = 1$.

Therefore, $1 = .9999\ldots$. Similarly, it can be shown that any terminating decimal can be represented as a repeating decimal with an endless string of 9s. For example, $.5 = .49999\ldots$ and $2.6 = 2.59999\ldots$. This is a way of justifying that any rational number may be represented as a repeating decimal.

See Exercises 95 and 96 for more on $.999\ldots = 1$.

This result may be checked with a calculator. Remember, however, that the calculator will only show a finite number of decimal places, and may round off in the final decimal place shown. ■

6.3 EXERCISES

Choose the expression(s) that is (are) equivalent to the given rational number.

1. $\dfrac{4}{8}$

   A. $\dfrac{1}{2}$ **B.** $\dfrac{8}{4}$ **C.** .5

   D. $.5\overline{0}$ **E.** $.\overline{55}$

2. $\dfrac{2}{3}$

   A. .67 **B.** $.\overline{6}$ **C.** $\dfrac{20}{30}$

   D. .666... **E.** .6

3. $\dfrac{5}{9}$

   A. .56 **B.** $.\overline{55}$ **C.** $.\overline{5}$

   D. $\dfrac{9}{5}$ **E.** $1\dfrac{4}{5}$

4. $\dfrac{1}{4}$

   A. .25 **B.** $.24\overline{9}$ **C.** $\dfrac{25}{100}$

   D. 4 **E.** $\dfrac{10}{400}$

Use the fundamental property of rational numbers to write each of the following in lowest terms.

5. $\dfrac{16}{48}$

6. $\dfrac{21}{28}$

7. $-\dfrac{15}{35}$

8. $-\dfrac{8}{48}$

Use the fundamental property to write each of the following in three other ways.

9. $\dfrac{3}{8}$

10. $\dfrac{9}{10}$

11. $-\dfrac{5}{7}$

12. $-\dfrac{7}{12}$

13. For each of the following, write a fraction in lowest terms that represents the portion of the figure that is in color.

   (a)   (b)   (c)   (d)

14. For each of the following, write a fraction in lowest terms that represents the region described.

   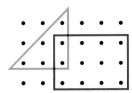

   (a) the dots in the rectangle as a part of the dots in the entire figure
   (b) the dots in the triangle as a part of the dots in the entire figure
   (c) the dots in the rectangle as a part of the dots in the union of the triangle and the rectangle
   (d) the dots in the intersection of the triangle and the rectangle as a part of the dots in the union of the triangle and the rectangle

15. Refer to the figure for Exercise 14 and write a description of the region that is represented by the fraction 1/12.

16. *Batting Averages* In a softball league, the first six games produced the following results: Greg Tobin got 8 hits in 20 at-bats, and Jason Jordan got 12 hits in 30 at-bats. Which player (if either) had the higher batting average?

17. *Batting Averages* After ten games, the following batting statistics were obtained.

Player	At-bats	Hits	Home Runs
Anne Kelly	40	9	2
Christine O'Brien	36	12	3
Brenda Bravener	11	5	1
Otis Taylor	16	8	0
Carol Britz	20	10	2

Answer each of the following, using estimation skills as necessary.
(a) Which player got a hit in exactly 1/3 of his or her at-bats?
(b) Which player got a hit in just less than 1/2 of his or her at-bats?
(c) Which player got a home run in just less than 1/10 of his or her at-bats?
(d) Which player got a hit in just less than 1/4 of his or her at-bats?
(e) Which two players got hits in exactly the same fractional parts of their at-bats? What was the fractional part, reduced to lowest terms?

18. Refer to the margin note discussing the use of common fractions on early U.S. copper coinage. The photo on the right here shows an error near the bottom that occurred on certain early cents. Discuss the error and how it represents a mathematical impossibility.

Perform the indicated operations and express answers in lowest terms. Use the order of operations as necessary.

19. $\dfrac{3}{8} + \dfrac{1}{8}$

20. $\dfrac{7}{9} + \dfrac{1}{9}$

21. $\dfrac{5}{16} + \dfrac{7}{12}$

22. $\dfrac{1}{15} + \dfrac{7}{18}$

23. $\dfrac{2}{3} - \dfrac{7}{8}$

24. $\dfrac{13}{20} - \dfrac{5}{12}$

25. $\dfrac{5}{8} - \dfrac{3}{14}$

26. $\dfrac{19}{15} - \dfrac{7}{12}$

27. $\dfrac{3}{4} \cdot \dfrac{9}{5}$

28. $\dfrac{3}{8} \cdot \dfrac{2}{7}$

29. $-\dfrac{2}{3} \cdot -\dfrac{5}{8}$

30. $-\dfrac{2}{4} \cdot \dfrac{3}{9}$

31. $\dfrac{5}{12} \div \dfrac{15}{4}$

32. $\dfrac{15}{16} \div \dfrac{30}{8}$

33. $-\dfrac{9}{16} \div -\dfrac{3}{8}$

34. $-\dfrac{3}{8} \div \dfrac{5}{4}$

35. $\left(\dfrac{1}{3} \div \dfrac{1}{2}\right) + \dfrac{5}{6}$

36. $\dfrac{2}{5} \div \left(-\dfrac{4}{5} \div \dfrac{3}{10}\right)$

37. *Recipe for Grits* The following chart appears on a package of Quaker Quick Grits.

	Microwave	Stove Top		
Servings	1	1	4	6
Water	$\frac{3}{4}$ cup	1 cup	3 cups	4 cups
Grits	3 Tbsp	3 Tbsp	$\frac{3}{4}$ cup	1 cup
Salt (optional)	dash	dash	$\frac{1}{4}$ tsp	$\frac{1}{2}$ tsp

(a) How many cups of water would be needed for 6 microwave servings?
(b) How many cups of grits would be needed for 5 stove-top servings? (*Hint:* 5 is halfway between 4 and 6.)

38. *U.S. Immigrants* More than 8 million immigrants were admitted to the United States between 1990 and 1997. The pie chart gives the fractional number from each region of birth for these immigrants.

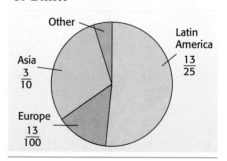

Source: U. S. Bureau of the Census.

(a) What fractional part of the immigrants were from Other regions?
(b) What fractional part of the immigrants were from Latin America or Asia?
(c) How many (in millions) were from Europe?

The **mixed number** $2\frac{5}{8}$ represents the sum $2 + \frac{5}{8}$. We can convert $2\frac{5}{8}$ to a fraction as follows:

$$2\frac{5}{8} = 2 + \frac{5}{8} = \frac{2}{1} + \frac{5}{8} = \frac{16}{8} + \frac{5}{8} = \frac{21}{8}.$$

The fraction 21/8 can be converted back to a mixed number by dividing 8 into 21. The quotient is 2, the remainder is 5, and the divisor is 8.

Convert each mixed number in the following exercises to a fraction, and convert each fraction to a mixed number.

39. $4\frac{1}{3}$ **40.** $3\frac{7}{8}$ **41.** $2\frac{9}{10}$

42. $\frac{18}{5}$ **43.** $\frac{27}{4}$ **44.** $\frac{19}{3}$

It is possible to add mixed numbers by first converting them to fractions, adding, and then converting the sum back to a mixed number. For example,

$$2\frac{1}{3} + 3\frac{1}{2} = \frac{7}{3} + \frac{7}{2} = \frac{14}{6} + \frac{21}{6} = \frac{35}{6} = 5\frac{5}{6}.$$

The other operations with mixed numbers may be performed in a similar manner.

Perform each operation and express your answer as a mixed number.

45. $3\frac{1}{4} + 2\frac{7}{8}$ **46.** $6\frac{1}{5} - 2\frac{7}{15}$

47. $-4\frac{7}{8} \cdot 3\frac{2}{3}$ **48.** $-4\frac{1}{6} \div 1\frac{2}{3}$

Solve each problem.

49. *Socket Wrench Measurements* A hardware store sells a 40-piece socket wrench set. The measure of the largest socket is $\frac{3}{4}$ in., while the measure of the smallest socket is $\frac{3}{16}$ in. What is the difference between these measures?

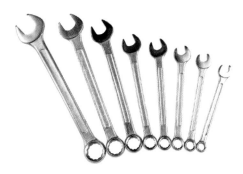

50. *Swiss Cheese Hole Sizes* Under existing standards, most of the holes in Swiss cheese must have diameters between $\frac{11}{16}$ and $\frac{13}{16}$ in. To accommodate new high-speed slicing machines, the USDA wants to reduce the minimum size to $\frac{3}{8}$ in. How much smaller is $\frac{3}{8}$ in. than $\frac{11}{16}$ in.? (*Source:* U.S. Department of Agriculture.)

A quotient of quantities containing fractions (with denominator not zero) is called a **complex fraction**. There are two methods that are used to simplify a complex fraction.

Method 1: Simplify the numerator and denominator separately. Then rewrite as a division problem, and proceed as you would when dividing fractions.

Method 2: Multiply both the numerator and denominator by the least common denominator of all the fractions found within the complex fraction. (This is, in effect, multiplying the fraction by 1, which does not change its value.) Apply the distributive property, if necessary, and simplify.

Use one of the methods above to simplify each of the following complex fractions.

51. $\dfrac{\frac{1}{2} + \frac{1}{4}}{\frac{1}{2} - \frac{1}{4}}$ **52.** $\dfrac{\frac{2}{3} + \frac{1}{6}}{\frac{2}{3} - \frac{1}{6}}$ **53.** $\dfrac{\frac{5}{8} - \frac{1}{4}}{\frac{1}{8} + \frac{3}{4}}$

54. $\dfrac{\frac{3}{16} + \frac{1}{2}}{\frac{5}{16} - \frac{1}{8}}$ **55.** $\dfrac{\frac{7}{11} + \frac{3}{10}}{\frac{1}{11} - \frac{9}{10}}$ **56.** $\dfrac{\frac{11}{15} + \frac{1}{9}}{\frac{13}{15} - \frac{2}{3}}$

*The expressions in Exercises 57 and 58 are called **continued fractions**. Write each of the following continued fractions in the form p/q, reduced to lowest terms. (Hint: Start at the bottom and work upward.)*

57. $2 + \cfrac{1}{1 + \cfrac{1}{3 + \cfrac{1}{2}}}$

58. $4 + \cfrac{1}{2 + \cfrac{1}{1 + \cfrac{1}{3}}}$

Find the rational number halfway between the two given rational numbers.

59. $\dfrac{1}{2}, \dfrac{3}{4}$ **60.** $\dfrac{1}{3}, \dfrac{5}{12}$ **61.** $\dfrac{3}{5}, \dfrac{2}{3}$ **62.** $\dfrac{7}{12}, \dfrac{5}{8}$ **63.** $-\dfrac{2}{3}, -\dfrac{5}{6}$ **64.** $-3, -\dfrac{5}{2}$

65. *Hourly Earnings in Private Industry* The table shows average hourly earnings of various fields in private industry for a recent year. Find the average of these amounts.

Private Industry Group	Hourly Earnings
Mining	$15.30
Construction	15.08
Manufacturing	12.37
Transportation, public utilities	14.23
Wholesale trade	12.43
Retail trade	7.69
Finance, insurance, real estate	12.33
Service	11.39

Source: U.S. Bureau of Labor Statistics.

66. *Average Housing Costs* The table shows the annual average housing costs in selected cities in 1998. Find the average of these amounts. (The annual rental housing costs are based on a U.S. expatriate family of two earning $75,000 and residing in a four- to six-room unit. Utilities and renter's insurance are included.)

City	1998 Housing Cost (Dollars)
Bombay, India	$63,418
Dusseldorf, Germany	26,719
Hong Kong, China	143,378
London, England	61,726
Madrid, Spain	30,638
New York, NY	58,425
Paris, France	35,551
San Francisco, CA	40,216
Singapore	73,386
Sydney, Australia	29,604
Tokyo, Japan	78,848

Source: Runzheimer International.

In the March 1973 issue of The Mathematics Teacher *there appeared an article by Laurence Sherzer, an eighth-grade mathematics teacher, that immortalized one of his students, Robert McKay. The class was studying the density property and Sherzer was explaining how to find a rational number between two given positive rational numbers by finding the average. McKay pointed out that there was no need to go to all that trouble. To find a number (not necessarily their average) between two positive rational numbers a/b and c/d, he claimed, simply "add the tops and add the bottoms." Much to Sherzer's surprise, this method really does work. For example, to find a rational number between 1/3 and 1/4, add 1 + 1 = 2 to get the numerator and 3 + 4 = 7 to get the denominator. Therefore, by* **McKay's theorem,** *2/7 is between 1/3 and 1/4. Sherzer provided a proof of this method in the article. Use* McKay's theorem *to find a rational number between the two given rational numbers.*

67. $\dfrac{5}{6}$ and $\dfrac{9}{13}$ **68.** $\dfrac{10}{11}$ and $\dfrac{13}{19}$

69. $\dfrac{4}{13}$ and $\dfrac{9}{16}$ **70.** $\dfrac{6}{11}$ and $\dfrac{13}{14}$

71. 2 and 3   **72.** 3 and 4

73. Apply McKay's theorem to any pair of consecutive integers, and make a conjecture about what happens in this case.

74. Explain in your own words how to find the rational number that is one-fourth of the way between two different rational numbers.

Convert each rational number into either a repeating or a terminating decimal. Use a calculator if your instructor so allows.

75. $\dfrac{3}{4}$ **76.** $\dfrac{7}{8}$ **77.** $\dfrac{3}{16}$ **78.** $\dfrac{9}{32}$

79. $\dfrac{3}{11}$ **80.** $\dfrac{9}{11}$ **81.** $\dfrac{2}{7}$ **82.** $\dfrac{11}{15}$

Convert each terminating decimal into a quotient of integers. Write each in lowest terms.

83. .4 **84.** .9 **85.** .85 **86.** .105 **87.** .934 **88.** .7984

Use the method of Example 6 to decide whether each of the following rational numbers would yield a repeating or a terminating decimal. (Hint: Write in lowest terms before trying to decide.)

89. $\dfrac{8}{15}$ **90.** $\dfrac{8}{35}$ **91.** $\dfrac{13}{125}$ **92.** $\dfrac{3}{24}$ **93.** $\dfrac{22}{55}$ **94.** $\dfrac{24}{75}$

95. Follow through on each part of this exercise in order.
  (a) Find the decimal for $1/3$.
  (b) Find the decimal for $2/3$.
  (c) By adding the decimal expressions obtained in parts (a) and (b), obtain a decimal expression for $1/3 + 2/3 = 3/3 = 1$.
  (d) Does your result seem bothersome? Read the margin note on terminating and repeating decimals in this section, which refers to this idea.

96. It is a fact that $1/3 = .333\ldots$. Multiply both sides of this equation by 3. Does your answer bother you? See the margin note on terminating and repeating decimals in this section.

6.4 Irrational Numbers and Decimal Representation

Tsu Ch'ung-chih (about 500 A.D.), the Chinese mathematician honored on the above stamp, investigated the digits of π. **Aryabhata,** his Indian contemporary, gave 3.1416 as the value.

The irrational number $\sqrt{2}$ was discovered by the Pythagoreans in about 500 B.C. This discovery was a great setback to their philosophy that everything is based upon the whole numbers. The Pythagoreans kept their findings secret, and legend has it that members of the group who divulged this discovery were sent out to sea, and, according to Proclus (410–485), "perished in a shipwreck, to a man."

Basic Concepts In the previous section we saw that any rational number has a decimal form that terminates or repeats. Also, every repeating or terminating decimal represents a rational number. Some decimals, however, neither repeat nor terminate. For example, the decimal

$$.1020010002000010000002\ldots$$

does not terminate and does not repeat. (It is true that there is a pattern in this decimal, but no single block of digits repeats indefinitely.)*

A number represented by a nonrepeating, nonterminating decimal is called an **irrational number.** As the name implies, it cannot be represented as a quotient of integers. The decimal number just mentioned is an irrational number. Other irrational numbers include $\sqrt{2}$, $\sqrt{7}$, and π (the ratio of the circumference of a circle to its diameter). There are infinitely many irrational numbers.

Figure 12 illustrates how a point with coordinate $\sqrt{2}$ can be located on a number line.

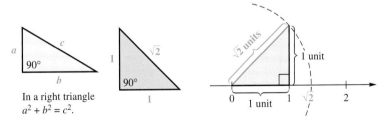

FIGURE 12

---

*In this section we will assume that the digits of a number such as this continue indefinitely in the pattern established. The next few digits would be 000000100000002, and so on.

The proof that $\sqrt{2}$ is irrational is a classic example of a **proof by contradiction**. We begin by assuming that $\sqrt{2}$ is rational, which leads to a contradiction, or absurdity. The method is also called **reductio ad absurdum** (Latin for "reduce to the absurd"). In order to understand the proof, we consider three preliminary facts:

1. When a rational number is written in lowest terms, the greatest common factor of the numerator and denominator is 1.
2. If an integer is even, then it has 2 as a factor and may be written in the form $2k$, where k is an integer.
3. If a perfect square is even, then its square root is even.

Theorem $\sqrt{2}$ is an irrational number.

Proof: Assume that $\sqrt{2}$ is a rational number. Then by definition,

$$\sqrt{2} = \frac{p}{q}, \text{ for some integers } p \text{ and } q.$$

Furthermore, assume that p/q is the form of $\sqrt{2}$ that is written in lowest terms, so the greatest common factor of p and q is 1. Squaring both sides of the equation gives

$$2 = \frac{p^2}{q^2}$$

and multiplying through by q^2 gives

$$2q^2 = p^2.$$

This last equation indicates that 2 is a factor of p^2. So p^2 is even, and thus p is even. Since p is even, it may be written in the form $2k$, where k is an integer.

Now, substitute $2k$ for p in the last equation and simplify:

$$2q^2 = (2k)^2$$
$$2q^2 = 4k^2$$
$$q^2 = 2k^2.$$

Since 2 is a factor of q^2, q^2 must be even, and thus, q must be even. This leads to a contradiction: p and q cannot both be even because they would then have a common factor of 2, although it was assumed that their greatest common factor is 1.

Therefore, since the original assumption that $\sqrt{2}$ is rational has led to a contradiction, it must follow that $\sqrt{2}$ is irrational. ∎

A calculator with a square root key can give approximations of square roots of numbers that are not perfect squares. To show that they are approximations, we use the $\approx$ symbol to indicate "is approximately equal to." Some such calculator approximations are as follows:

$$\sqrt{2} \approx 1.414213562, \quad \sqrt{6} \approx 2.449489743, \quad \text{and} \quad \sqrt{1949} \approx 44.14748011.$$

An interesting way to represent the lengths corresponding to $\sqrt{2}$, $\sqrt{3}$, $\sqrt{4}$, $\sqrt{5}$, and so on, is shown in the figure. Use the Pythagorean theorem to verify the lengths in the figure.

These are calculator approximations of irrational numbers.

> **The number *e*** is a fundamental number in our universe. For this reason, *e*, like π, is called a *universal constant*. If there are intelligent beings elsewhere, they too will have to use *e* to do higher mathematics.
>
> The letter *e* is used to honor Leonhard Euler, who published extensive results on the number in 1748. The first few digits of the decimal value of *e* are 2.7182818. Since it is an irrational number, its decimal expansion never terminates and never repeats.
>
> The properties of *e* are used in calculus and in higher mathematics extensively.

Not all square roots are irrational. For example,

$$\sqrt{4} = 2, \quad \sqrt{36} = 6, \quad \text{and} \quad \sqrt{100} = 10$$

are all rational numbers. However, if *n* is a positive integer that is not the square of an integer, then $\sqrt{n}$ is an irrational number. The chart below shows some examples of rational numbers and irrational numbers.

Rational Numbers	Irrational Numbers	
$\frac{3}{4}$	$\sqrt{2}$	
.64	.23233233323333...	
.74	$\sqrt{5}$	
$\sqrt{16}$	π	
1.618	$\frac{1 + \sqrt{5}}{2}$	The exact value of the golden ratio
2.718	*e*	An important number in higher mathematics—see the margin note.

> The symbol for π in use today is a Greek letter. It was first used in England in the 1700s. In 1859 the symbol for π shown above was proposed by Professor Benjamin Peirce of Harvard.

One of the most important irrational numbers is π, the ratio of the circumference to the diameter of a circle. (Many formulas from geometry involve π, such as the formula for area of a circle, $A = \pi r^2$.) For some 4000 years mathematicians have been finding better and better approximations for π. The ancient Egyptians used a method for finding the area of a circle that is equivalent to a value of 3.1605 for π. The Babylonians used numbers that give $3\frac{1}{8}$ for π. In the Bible (I Kings 7:23), a verse describes a circular pool at King Solomon's temple, about 1000 B.C. The pool is said to be ten cubits across, "and a line of 30 cubits did compass it round about." This implies a value of 3 for π (to the nearest whole number).

FOR FURTHER THOUGHT

In 2002, the computer scientist Yasumasa Kanada and his colleagues of the Information Techology Center at Tokyo University announced the computation of 1,241,100,000,000 decimal digits of pi, over six times their own previous record of 206,158,430,000 digits, set in 1999. According to "The π Pages" at the web site http://www.cecm.sfu.ca, "The computation of Pi is virtually the only topic from the most ancient stratum of mathematics that is still of serious interest to modern mathematical research." A visit to http://www.joyofpi.com will lead to interesting links about this fascinating number.

(continued)

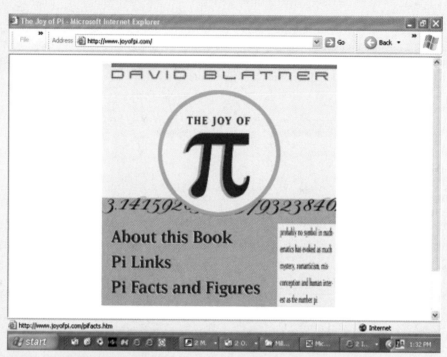

Source: David Blatner, *The Joy of Pi*, www.joyofpi.com.

One of the reasons for computing so many digits is to determine how often each digit appears and to identify any interesting patterns among the digits. Two American mathematicians, Gregory and David Chudnovsky, have spent a great deal of time and effort looking for patterns in the digits. For example, six 9s in a row appear relatively early in the decimal, within the first 800 decimal places. And past the half-billion mark appears the sequence 123456789. One conjecture about π deals with what mathematicians term "normality." The normality conjecture says that all digits appear with the same average frequency. According to Gregory, "There is absolutely no doubt that π is a 'normal' number. Yet we can't prove it. We don't even know how to *try* to prove it."

The computation of π has fascinated mathematicians and laymen for centuries. In the nineteenth century the British mathematician William Shanks spent many years of his life calculating π to 707 decimal places. It turned out that only the first 527 were correct.

In 1767 J. H. Lambert proved that π is irrational (and thus its decimal will never terminate and never repeat). Nevertheless, the 1897 Indiana state legislature considered a bill that would have *legislated* the value of π. In one part of the bill, the value was stated to be 4, and in another part, 3.2. Amazingly, the bill passed the House, but the Senate postponed action on the bill indefinitely!

The following expressions are some which may be used to compute π to more and more decimal places.

$$\frac{\pi}{2} = \frac{2 \cdot 2 \cdot 4 \cdot 4 \cdot 6 \cdot 6 \cdot 8 \ldots}{1 \cdot 3 \cdot 3 \cdot 5 \cdot 5 \cdot 7 \cdot 7 \ldots}$$

$$\frac{\pi}{4} = 1 - \frac{1}{3} + \frac{1}{5} - \frac{1}{7} + \cdots$$

$$\frac{2}{\pi} = \frac{\sqrt{2}}{2} \cdot \frac{\sqrt{2 + \sqrt{2}}}{2} \cdot \frac{\sqrt{2 + \sqrt{2 + \sqrt{2}}}}{2} \cdots$$

The fascinating history of π has been chronicled by Petr Beckman in the book *A History of Pi*.

For Group Discussion

1. Have each class member ask someone outside of class "What is π?" Then as a class, discuss the various responses obtained.
2. As with Mount Everest, some people enjoy climbing the mountain of π simply because it is there. Have you ever tackled a project for no reason other than to simply say "I did it"? Share any such experiences with the class.
3. Divide the class into three groups, and, armed with calculators, spend a few minutes calculating the expressions given previously to approximate π. Compare your results to see which one of the expressions converges toward π the fastest.
4. A **mnemonic device** is a scheme whereby one is able to recall facts by memorizing something completely unrelated to the facts. One way of learning the first few digits of the decimal for π is to memorize a sentence (or several sentences) and count the letters in each word of the sentence. For example, "See, I know a digit," will give the first 5 digits of π: "See" has 3 letters, "I" has 1 letter, "know" has 4 letters, "a" has 1 letter, and "digit" has 5 letters. So the first five digits are 3.1415.

   Verify that the following mnemonic devices work. (See the margin note on the next page for 31 decimal digits of π.)
   (a) "May I have a large container of coffee?"
   (b) "See, I have a rhyme assisting my feeble brain, its tasks ofttimes resisting."
   (c) "How I want a drink, alcoholic of course, after the heavy lectures involving quantum mechanics."

Square Roots In everyday mathematical work, nearly all of our calculations deal with rational numbers, usually in decimal form. In our study of mathematics, however, we must sometimes perform operations with irrational numbers, and in many instances the irrational numbers are square roots. Recall that $\sqrt{a}$, for $a \geq 0$, is the nonnegative number whose square is a; that is, $(\sqrt{a})^2 = a$. We will now look at some simple operations with square roots.

Notice that
$$\sqrt{4} \cdot \sqrt{9} = 2 \cdot 3 = 6$$
and
$$\sqrt{4 \cdot 9} = \sqrt{36} = 6.$$

Thus, $\sqrt{4} \cdot \sqrt{9} = \sqrt{4 \cdot 9}$. This is a particular case of the following product rule.

Product Rule for Square Roots

For nonnegative real numbers a and b,
$$\sqrt{a} \cdot \sqrt{b} = \sqrt{a \cdot b}.$$

This poem, dedicated to Archimedes ("the immortal Syracusan"), allows us to learn the first 31 digits of the decimal representation of π. By replacing each word with the number of letters it contains, with a decimal point following the initial 3, the decimal is found. The poem was written by A. C. Orr, and appeared in the *Literary Digest* in 1906.

Now I, even I, would celebrate
In rhymes unapt, the great
Immortal Syracusan, rivaled nevermore,
Who in his wondrous lore
Passed on before,
Left men his guidance
How to circles mensurate.

From this poem, we can determine these digits of π: 3.1415926535897932384626433832795.

Archimedes was able to use circles inscribed and circumscribed by polygons to find that the value of π is somewhere between 223/71 and 22/7.

```
√(27)=3√(3)
                     1
√(27)
         5.196152423
3√(3)
         5.196152423
```

The 1 after the first line indicates that the equality is true. The calculator also shows the same approximations for $\sqrt{27}$ and $3\sqrt{3}$ in the second and third answers. (See Example 1.)

Just as every rational number a/b can be written in simplest (lowest) terms (by using the fundamental property of rational numbers), every square root radical has a simplest form. A square root radical is in **simplified form** if the following three conditions are met.

Simplified Form of a Square Root Radical

1. The number under the radical (**radicand**) has no factor (except 1) that is a perfect square.
2. The radicand has no fractions.
3. No denominator contains a radical.

EXAMPLE 1 Simplify $\sqrt{27}$.

Since 9 is a factor of 27 and 9 is a perfect square, $\sqrt{27}$ is not in simplified form. The first condition in the box above is not met. We simplify as follows.

$$\sqrt{27} = \sqrt{9 \cdot 3}$$
$$= \sqrt{9} \cdot \sqrt{3} \quad \text{Use the product rule.}$$
$$= 3\sqrt{3} \quad \sqrt{9} = 3, \text{ since } 3^2 = 9.$$

Expressions such as $\sqrt{27}$ and $3\sqrt{3}$ are *exact values* of the square root of 27. If we use the square root key of a calculator, we find

$$\sqrt{27} \approx 5.196152423.$$

If we find $\sqrt{3}$ and then multiply the result by 3, we get

$$3\sqrt{3} \approx 3(1.732050808) \approx 5.196152423.$$

Notice that these approximations are the same, as we would expect. (Due to various methods of calculating, there may be a discrepancy in the final digit of the calculation.) Understand, however, that the calculator approximations do not actually *prove* that the two numbers are equal, but only strongly suggest equality. The work done in Example 1 actually provides the mathematical justification that they are indeed equal.

A rule similar to the product rule exists for quotients.

Quotient Rule for Square Roots

For nonnegative real numbers a and positive real numbers b,

$$\frac{\sqrt{a}}{\sqrt{b}} = \sqrt{\frac{a}{b}}.$$

6.4 Irrational Numbers and Decimal Representation

The radical symbol above comes from the Latin word for root, *radix*. It was first used by Leonardo da Pisa (Fibonacci) in 1220. The sixteenth-century German symbol we use today probably is also derived from the letter r.

EXAMPLE 2 Simplify each radical.

(a) $\sqrt{\dfrac{25}{9}}$

Because the radicand contains a fraction, the radical expression is not simplified. (See condition 2 in the box preceding Example 1.) Use the quotient rule as follows.

$$\sqrt{\dfrac{25}{9}} = \dfrac{\sqrt{25}}{\sqrt{9}} = \dfrac{5}{3}$$

(b) $\sqrt{\dfrac{3}{4}} = \dfrac{\sqrt{3}}{\sqrt{4}} = \dfrac{\sqrt{3}}{2}$

(c) $\sqrt{\dfrac{1}{2}} = \dfrac{\sqrt{1}}{\sqrt{2}} = \dfrac{1}{\sqrt{2}}$

This expression is not in simplified form, since condition 3 is not met. To give an equivalent expression with no radical in the denominator, we use a procedure called **rationalizing the denominator.** Multiply $1/\sqrt{2}$ by $\sqrt{2}/\sqrt{2}$, which is a form of 1, the identity element for multiplication.

$$\dfrac{1}{\sqrt{2}} = \dfrac{1}{\sqrt{2}} \cdot \dfrac{\sqrt{2}}{\sqrt{2}} = \dfrac{\sqrt{2}}{2} \qquad \sqrt{2} \cdot \sqrt{2} = 2$$

The simplified form of $\sqrt{1/2}$ is $\sqrt{2}/2$.

Is $\sqrt{4} + \sqrt{9} = \sqrt{4 + 9}$ true? Simple computation shows that the answer is "no," since $\sqrt{4} + \sqrt{9} = 2 + 3 = 5$, while $\sqrt{4 + 9} = \sqrt{13}$, and $5 \neq \sqrt{13}$. Square root radicals may be added, however, if they have the same radicand. Such radicals are **like radicals.** We add (and subtract) like radicals with the distributive property.

EXAMPLE 3 Add or subtract as indicated.

(a) $3\sqrt{6} + 4\sqrt{6}$

Since both terms contain $\sqrt{6}$, they are like radicals, and may be combined.

$$3\sqrt{6} + 4\sqrt{6} = (3 + 4)\sqrt{6} \qquad \text{Distributive property}$$
$$= 7\sqrt{6} \qquad \text{Add.}$$

(b) $\sqrt{18} - \sqrt{32}$

At first glance it seems that we cannot subtract these terms. However, if we first simplify $\sqrt{18}$ and $\sqrt{32}$, then it can be done.

$$\begin{aligned}
\sqrt{18} - \sqrt{32} &= \sqrt{9 \cdot 2} - \sqrt{16 \cdot 2} \\
&= \sqrt{9} \cdot \sqrt{2} - \sqrt{16} \cdot \sqrt{2} &&\text{Product rule} \\
&= 3\sqrt{2} - 4\sqrt{2} &&\text{Take square roots.} \\
&= (3 - 4)\sqrt{2} &&\text{Distributive property} \\
&= -1\sqrt{2} &&\text{Subtract.} \\
&= -\sqrt{2} &&-1 \cdot a = -a
\end{aligned}$$

From Example 3, we see that like radicals may be added or subtracted by adding or subtracting their coefficients (the numbers by which they are multiplied) and keeping the same radical. For example,

$$9\sqrt{7} + 8\sqrt{7} = 17\sqrt{7} \quad (\text{since } 9 + 8 = 17)$$
$$4\sqrt{3} - 12\sqrt{3} = -8\sqrt{3}, \quad (\text{since } 4 - 12 = -8)$$

and so on.

In the statements of the product and quotient rules for square roots, the radicands could not be negative. While $-\sqrt{2}$ is a real number, for example, $\sqrt{-2}$ is not: there is no real number whose square is -2. The same may be said for any negative radicand. In order to handle this situation, mathematicians have extended our number system to include *complex numbers*, discussed in the Extension at the end of this chapter.

6.4 EXERCISES

Identify each of the following as rational *or* irrational.

1. $\dfrac{4}{9}$
2. $\dfrac{7}{8}$
3. $\sqrt{10}$
4. $\sqrt{14}$
5. .37
6. .91
7. $.\overline{41}$
8. $.\overline{32}$
9. π
10. 0
11. .878778777877778...
12. .434334333433334...
13. 3.14159
14. $\dfrac{22}{7}$

15. **(a)** Find the following sum.

.272772777277772...
+.616116111611116...

(b) Based on the result of part (a), we can conclude that the sum of two _____ numbers may be a(n) _____ number.

16. **(a)** Find the following sum.

.010110111011110...
+.252552555255552...

(b) Based on the result of part (a), we can conclude that the sum of two _____ numbers may be a(n) _____ number.

Use a calculator to find a rational decimal approximation for each of the following irrational numbers. Give as many places as your calculator shows.

17. $\sqrt{39}$
18. $\sqrt{44}$
19. $\sqrt{15.1}$
20. $\sqrt{33.6}$
21. $\sqrt{884}$
22. $\sqrt{643}$
23. $\sqrt{\dfrac{9}{8}}$
24. $\sqrt{\dfrac{6}{5}}$

Use the methods of Examples 1 and 2 to simplify each of the following expressions. Then, use a calculator to approximate both the given expression and the simplified expression. (Both should be the same.)

25. $\sqrt{50}$
26. $\sqrt{32}$
27. $\sqrt{75}$
28. $\sqrt{150}$
29. $\sqrt{288}$
30. $\sqrt{200}$
31. $\dfrac{5}{\sqrt{6}}$
32. $\dfrac{3}{\sqrt{2}}$
33. $\sqrt{\dfrac{7}{4}}$
34. $\sqrt{\dfrac{8}{9}}$
35. $\sqrt{\dfrac{7}{3}}$
36. $\sqrt{\dfrac{14}{5}}$

Use the method of Example 3 to perform the indicated operations.

37. $\sqrt{17} + 2\sqrt{17}$
38. $3\sqrt{19} + \sqrt{19}$
39. $5\sqrt{7} - \sqrt{7}$
40. $3\sqrt{27} - \sqrt{27}$
41. $3\sqrt{18} + \sqrt{2}$
42. $2\sqrt{48} - \sqrt{3}$
43. $-\sqrt{12} + \sqrt{75}$
44. $2\sqrt{27} - \sqrt{300}$

Each of the following exercises deals with π. Use a calculator as necessary.

45. Move one matchstick to make the equation approximately true. (*Source:* http://www.joyofpi.com)

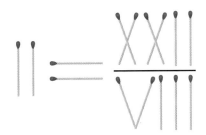

46. Find the square root of $2143/22$ using a calculator. Then find the square root of that result. Compare your result to the decimal given for π in the margin note. What do you notice?

47. Find the first eight digits in the decimal for $355/113$. Compare the result to the decimal for π given in the margin note. What do you notice?

48. You may have seen the statements "use $22/7$ for π" and "use 3.14 for π." Since $22/7$ is the quotient of two integers, and 3.14 is a terminating decimal, do these statements suggest that π is rational?

Solve each problem. Use a calculator as necessary, and give approximations to the nearest tenth unless specified otherwise.

49. *Period of a Pendulum* The period of a pendulum in seconds depends on its length, L, in feet, and is given by the formula

$$P = 2\pi \sqrt{\frac{L}{32}}.$$

If a pendulum is 5.1 feet long, what is its period? Use 3.14 for π.

50. *Radius of an Aluminum Can* The radius of the circular top or bottom of an aluminum can with surface area S and height h is given by

$$r = \frac{-h + \sqrt{h^2 + .64S}}{2}.$$

What radius should be used to make a can with height 12 inches and surface area 400 square inches?

51. *Distance to the Horizon* According to an article in *The World Scanner Report* (August 1991), the distance, D, in miles, to the horizon from an observer's point of view over water or "flat" earth is given by

$$D = \sqrt{2H},$$

where H is the height of the point of view, in feet. If a person whose eyes are 6 feet above ground level is standing at the top of a hill 28 feet above the "flat" earth, approximately how far to the horizon will she be able to see? Round your answer to the nearest tenth of a mile.

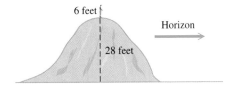

52. *Electronics Formula* The formula

$$I = \sqrt{\frac{2P}{L}}$$

relates the coefficient of self-induction L (in henrys), the energy P stored in an electronic circuit (in joules), and the current I (in amps). Find I if $P = 120$ joules and $L = 80$ henrys.

53. *Area of the Bermuda Triangle* Heron's formula gives a method of finding the area of a triangle if the lengths of its sides are known. Suppose that a, b, and c are the lengths of the sides. Let s denote one-half of the perimeter of the triangle (called the *semiperimeter*); that is,

$$s = \frac{1}{2}(a + b + c).$$

Then the area A of the triangle is given by

$$A = \sqrt{s(s-a)(s-b)(s-c)}.$$

Find the area of the Bermuda Triangle, if the "sides" of this triangle measure approximately 850 miles, 925 miles, and 1300 miles. Give your answer to the nearest thousand square miles.

54. *Area Enclosed by the Vietnam Veterans' Memorial* The Vietnam Veterans' Memorial in Washington, D.C., is in the shape of an unenclosed isosceles triangle with equal sides of length 246.75 feet. If the triangle were enclosed, the third side would have length 438.14 feet. Use Heron's formula from the previous exercise to find the area of this enclosure to the nearest hundred square feet. (*Source:* Information pamphlet obtained at the Vietnam Veterans' Memorial.)

55. *Diagonal of a Box* The length of the diagonal of a box is given by

$$D = \sqrt{L^2 + W^2 + H^2},$$

where L, W, and H are the length, the width, and the height of the box. Find the length of the diagonal, D, of a box that is 4 feet long, 3 feet high, and 2 feet wide.

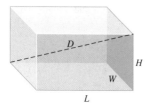

56. *Rate of Return of an Investment* If an investment of P dollars grows to A dollars in two years, the annual rate of return on the investment is given by

$$r = \frac{\sqrt{A} - \sqrt{P}}{\sqrt{P}}.$$

First rationalize the denominator and then find the annual rate of return (as a decimal) if $50,000 increases to $58,320.

57. *Accident Reconstruction* Police sometimes use the following procedure to estimate the speed at which a car was traveling at the time of an accident. A police officer drives the car involved in the accident under conditions similar to those during which the accident took place and then skids to a stop. If the car is driven at 30 miles per hour, then the speed at the time of the accident is given by

$$s = 30\sqrt{\frac{a}{p}},$$

where a is the length of the skid marks left at the time of the accident and p is the length of the skid marks in the police test. Find s for the following values of a and p.
(a) $a = 862$ feet; $p = 156$ feet
(b) $a = 382$ feet; $p = 96$ feet
(c) $a = 84$ feet; $p = 26$ feet

58. *Distance to the Horizon* A formula for calculating the distance, d, one can see from an airplane to the horizon on a clear day is

$$d = 1.22\sqrt{x},$$

where x is the altitude of the plane in feet and d is given in miles.

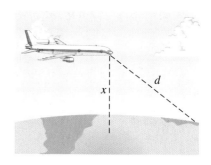

How far can one see to the horizon in a plane flying at the following altitudes? Round to the nearest tenth.
(a) 15,000 feet
(b) 18,000 feet
(c) 24,000 feet

The concept of square (second) root can be extended to **cube (third) root, fourth root,** *and so on. If $n \geq 2$ and a is a nonnegative number, $\sqrt[n]{a}$ represents the nonnegative number whose nth power is a. For example,*

$$\sqrt[3]{8} = 2 \text{ because } 2^3 = 8,$$
$$\sqrt[3]{1000} = 10 \text{ because } 10^3 = 1000,$$
$$\sqrt[4]{81} = 3 \text{ because } 3^4 = 81,$$

and so on. Find each of the following roots.

59. $\sqrt[3]{64}$ **60.** $\sqrt[3]{125}$
61. $\sqrt[3]{343}$ **62.** $\sqrt[3]{729}$
63. $\sqrt[3]{216}$ **64.** $\sqrt[3]{512}$
65. $\sqrt[4]{1}$ **66.** $\sqrt[4]{16}$
67. $\sqrt[4]{256}$ **68.** $\sqrt[4]{625}$
69. $\sqrt[4]{4096}$ **70.** $\sqrt[4]{2401}$

Use a calculator to approximate each root. Give as many places as your calculator shows.

71. $\sqrt[3]{43}$ **72.** $\sqrt[3]{87}$
73. $\sqrt[3]{198}$ **74.** $\sqrt[4]{2107}$
75. $\sqrt[3]{10,265.2}$ **76.** $\sqrt[4]{863.5}$

Solve each problem.

77. *Threshold Weight* The threshold weight, T, for a person is the weight above which the risk of death increases greatly. In one instance, a researcher found that the threshold weight in pounds for men aged 40–49 is related to height, h, in inches by the equation

$$h = 12.3\sqrt[3]{T}.$$

What height, in feet, corresponds to a threshold of 180 pounds for a 43-year-old man?

78. *Law of Tensions* In the study of sound, one version of the law of tensions is

$$f_1 = f_2 \sqrt{\frac{F_1}{F_2}}.$$

Find f_1 to the nearest unit if $F_1 = 300$, $F_2 = 60$, and $f_2 = 260$.

*Exercises 79 and 80 investigate the concept of **rational number exponents**.*

79. In algebra the expression $a^{1/2}$ is defined to be $\sqrt{a}$ for nonnegative values of a. Use a calculator with an exponential key to evaluate each of the following, and compare with the value obtained with the square root key. (Both should be the same.)
(a) $2^{1/2}$ (b) $7^{1/2}$ (c) $13.2^{1/2}$ (d) $25^{1/2}$

80. Based on Exercises 59–76, answer the following.
(a) How would you expect the expression $a^{1/3}$ to be defined as a radical?
(b) Use a calculator with a cube root key to approximate $\sqrt[3]{16}$.
(c) Use a calculator with an exponential key to approximate $16^{1/3}$.

Exercises 81 and 82 investigate the irrational number e.

81. Use a calculator with an exponential key to find values for the following: $(1.1)^{10}$, $(1.01)^{100}$, $(1.001)^{1000}$, $(1.0001)^{10,000}$, and $(1.00001)^{100,000}$. Compare your results to the approximation given for the irrational number e in the margin note in this section. What do you find?

82. Use a calculator with a key that determines powers of e to find each of the following.
(a) e^2 (b) e^3 (c) $\sqrt{e}$

6.5 Applications of Decimals and Percents

Perhaps the most frequent use of mathematics in everyday life concerns operations with decimal numbers and the concept of percent. When we use dollars and cents, we are dealing with decimal numbers. Sales tax on purchases made at the grocery store is computed using percent. The educated consumer must have a working knowledge of decimals and percent in financial matters. Look at any newspaper and you will see countless references to percent and percentages.

Operations with Decimals Because calculators have, for the most part, replaced paper-and-pencil methods for operations with decimals and percent, we will only briefly mention these latter methods. *We strongly suggest that the work in this section be done with a calculator at hand.*

Addition and Subtraction of Decimals

To add or subtract decimal numbers, line up the decimal points in a column and perform the operation.

```
.46+3.9+12.58
          16.94
12.1-8.723
          3.377
```

This screen supports the results in Example 1.

EXAMPLE 1

(a) To compute the sum $.46 + 3.9 + 12.58$, use the following method.

$$\begin{array}{r} .46 \\ 3.9 \\ +12.58 \\ \hline 16.94 \end{array}$$

Line up decimal points.

← Sum

Contrary to the fears of many, the availability of calculators and computers has expanded students' capability of performing calculations. There is no evidence to suggest that the availability of calculators makes students dependent on them for simple calculations. Students should be able to decide when they need to calculate and whether they require an exact or approximate answer. They should be able to select and use the most appropriate tool. Students should have a balanced approach to calculation, be able to choose appropriate procedures, find answers, and judge the validity of those answers.

From *Curriculum and Evaluation Standards for School Mathematics* (NCTM), p. 9.

(b) To compute the difference $12.1 - 8.723$, use this method.

$$\begin{array}{r} 12.100 \\ -8.723 \\ \hline 3.377 \end{array}$$ Attach zeros.

← Difference

Recall that when two numbers are multiplied, the numbers are called *factors* and the answer is called the *product*. When two numbers are divided, the number being divided is called the *dividend*, the number doing the dividing is called the *divisor*, and the answer is called the *quotient*.

Multiplication and Division of Decimals

1. To multiply decimals, perform the multiplication in the same manner as integers are multiplied. The number of decimal places to the right of the decimal point in the product is the *sum* of the numbers of places to the right of the decimal points in the factors.

2. To divide decimals, move the decimal point to the right the same number of places in the divisor and the dividend so as to obtain a whole number in the divisor. Perform the division in the same manner as integers are divided. The number of decimal places to the right of the decimal point in the quotient is the same as the number of places to the right in the dividend.

This screen supports the results in Example 2.

EXAMPLE 2

(a) To find the product 4.613×2.52, use the following method.

$$\begin{array}{r} 4.613 \leftarrow \text{3 decimal places} \\ \times 2.52 \leftarrow \text{2 decimal places} \\ \hline 9226 \\ 23065 \\ 9226 \\ \hline 11.62476 \leftarrow 3 + 2 = 5 \quad \text{decimal places} \end{array}$$

(b) To find the quotient $65.175 \div 8.25$, follow these steps.

$$8.25\overline{)65.175} \longrightarrow 825\overline{)6517.5}$$

$$\begin{array}{r} 7.9 \\ 825\overline{)6517.5} \\ \underline{5775} \\ 7425 \\ \underline{7425} \\ 0 \end{array}$$

Rounding Decimals Operations with decimals often result in long strings of digits in the decimal places. Since all these digits may not be needed in a practical problem, it is common to *round* a decimal to the necessary number of decimal

6.5 Applications of Decimals and Percents   295

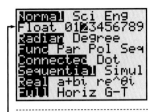

Here, the TI-83 Plus calculator is set to round the answer to two decimal places.

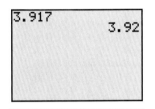

The calculator rounds 3.917 to the nearest hundredth.

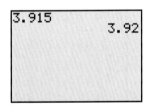

The calculator rounds 3.915 *up* to 3.92.

places. For example, in preparing federal income tax, money amounts are rounded to the nearest dollar. Round as shown in the next example.

EXAMPLE 3   Round 3.917 to the nearest hundredth.

The hundredths place in 3.917 contains the digit 1.

$$3.917$$
↑ Hundredths place

To round this decimal, locate 3.91 and 3.92 on a number line as in Figure 13.

FIGURE 13

The distance from 3.91 to 3.92 is divided into ten equal parts. The seventh of these ten parts locates the number 3.917. As the number line shows, 3.917 is closer to 3.92 than it is to 3.91, so 3.917 rounded to the nearest hundredth is 3.92. ■

If the number line method of Example 3 were used to round 3.915 to the nearest hundredth, a problem would develop—the number 3.915 is exactly halfway between 3.91 and 3.92. An arbitrary decision is then made to round *up*: 3.915 rounded to the nearest hundredth is 3.92.

Rules for Rounding Decimals

Step 1:   Locate the **place** to which the number is being rounded.

Step 2:   Look at the next **digit to the right** of the place to which the number is being rounded.

Step 3A: If this digit is **less than 5,** drop all digits to the right of the place to which the number is being rounded. Do *not change* the digit in the place to which the number is being rounded.

Step 3B: If this digit is **5 or greater,** drop all digits to the right of the place to which the number is being rounded. *Add one* to the digit in the place to which the number is being rounded.

EXAMPLE 4   Round 14.39656 to the nearest thousandth.

Step 1: Use an arrow to locate the place to which the number is being rounded.

$$14.39656$$
↑ Thousandths place

Step 2: Check to see if the first digit to the right of the arrow is 5 or greater.

Digit to the right of the arrow is 5.

Step 3: Since the digit to the right of the arrow is 5 or greater, increase by 1 the digit to which the arrow is pointing. Drop all digits to the right of the arrow.

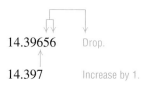

14.39656   Drop.

14.397   Increase by 1.

Finally, 14.39656 rounded to the nearest thousandth is 14.397.

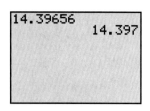

With the calculator set to round to three decimal places, the result of Example 4 is supported.

FOR FURTHER THOUGHT

Do you have the "knack"? We are all born with different abilities. That's what makes us all unique individuals. Some people have musical talents, while others can't carry a tune. Some people can repair automobiles, sew, or build houses while others can't do any of these. And some people have a knack for percents, while others are completely befuddled by them. Do you have the knack?

For example, suppose that you need to compute 20% of 50. You have the knack if you use one of these methods:

1. You think "Well, 20% means 1/5, and to find 1/5 of something I divide by 5, so 50 divided by 5 is 10. The answer is 10."
2. You think "20% is twice 10%, and to find 10% of something I move the decimal point one place to the left. So, 10% of 50 is 5, and 20% is twice 5, or 10. The answer is 10."

If you don't have the knack, you probably search for a calculator whenever you need to compute a percent, and hope like crazy that you'll work it correctly. Keep in mind one thing, however: there's nothing to be ashamed of if you don't have the knack. The methods explained in this section allow you to learn how to compute percents using tried-and-true mathematical methods. And just because you don't have the knack, it doesn't mean that you can't succeed in mathematics or can't learn other concepts. One ability does not necessarily assure success in another seemingly similar area. Case in point:

The nineteenth-century German Zacharias Dase was a lightning-swift calculator, and could do calculations like multiply 79,532,853 by 93,758,479 in his head in less than one minute, but had no concept of theoretical mathematics!

For Group Discussion

Have class members share their experiences on how they compute percents.

"I got 6 percent in math. Is that good or bad?"

6.5 Applications of Decimals and Percents

Percent One of the main applications of decimals comes from problems involving **percents.** In consumer mathematics, interest rates and discounts are often given as percents. The word *percent* means "per hundred." The symbol % represents "percent."

Percent

$$1\% = \frac{1}{100} = .01$$

EXAMPLE 5 Convert each percent to a decimal.

(a) $98\% = 98(1\%) = 98(.01) = .98$

(b) $3.4\% = .034$ (c) $.2\% = .002$

EXAMPLE 6 Convert each decimal to a percent.

(a) $.13 = 13(.01) = 13(1\%) = 13\%$

(b) $.532 = 53.2(.01) = 53.2(1\%) = 53.2\%$

(c) $2.3 = 230\%$

From Examples 5 and 6, we see that the following procedures can be used when converting between percents and decimals.

Converting Between Decimals and Percents

1. *To convert a percent to a decimal*, drop the % sign and move the decimal point two places to the left, inserting zeros as placeholders if necessary.
2. *To convert a decimal to a percent*, move the decimal point two places to the right, inserting zeros as placeholders if necessary, and attach a % sign.

EXAMPLE 7 Convert each fraction to a percent.

(a) $\frac{3}{5}$

First write 3/5 as a decimal. Dividing 5 into 3 gives $3/5 = .6 = 60\%$.

(b) $\frac{14}{25} = .56 = 56\%$

The procedure of Example 7 is summarized as follows.

Converting a Fraction to a Percent

To convert a fraction to a percent, convert the fraction to a decimal, and then convert the decimal to a percent.

The percent sign, %, probably evolved from a symbol introduced in an Italian manuscript of 1425. Instead of "per 100," "P 100," or "P cento," which were common at that time, the author used "Pr̃." By about 1650 the r̃ had become $\frac{o}{o}$, so "per $\frac{o}{o}$" was often used. Finally the "per" was dropped, leaving $\frac{o}{o}$ or %.

From *Historical Topics for the Mathematics Classroom,* the Thirty-first Yearbook of the National Council of Teachers of Mathematics, 1969.

Bicycle Mathematics
Experiments done in England, using racing cyclists on stationary bicycles, show that the most efficient saddle height is 109% of a cyclist's inside-leg measurement. You can get the most mileage out of your leg work by following these directions:

1. Stand up straight, without shoes. Have someone measure your leg on the inside (from floor to crotch bone).
2. Multiply this length by 109% (that is, by 1.09) to get a measure *R*.
3. Adjust your saddle so that the measure *R* equals the distance between the top of the saddle and the lower pedal spindle when the pedals are positioned as in the diagram above.

In the following examples involving percents, three methods are shown. The second method in each case involves using cross-products. The third method involves the percent key of a calculator.

EXAMPLE 8 Find 18% of 250.

Method 1: The key word here is "of." The word "of" translates as "times," with 18% of 250 given by

$$(18\%)(250) = (.18)(250) = 45.$$

Method 2: Think "18 is to 100 as what (x) is to 250?" This translates into the equation

$$\frac{18}{100} = \frac{x}{250}$$

$$100x = 18 \cdot 250 \qquad \text{\textit{a/b} = \textit{c/d} if and only if \textit{ad} = \textit{bc}.}$$

$$x = \frac{18 \cdot 250}{100} = 45. \qquad \text{Divide by 100 and simplify.}$$

Method 3: Use the percent key on a calculator with the following keystrokes:

With any of these methods, we find that 18% of 250 is 45.

EXAMPLE 9 What percent of 500 is 75?

Method 1: Let the phrase "what percent" be represented by $x \cdot 1\%$ or $.01x$. Again the word "of" translates as "times," while "is" translates as "equals." Thus,

$$.01x \cdot (500) = 75$$

$$5x = 75 \qquad \text{Multiply on the left side.}$$

$$x = 15. \qquad \text{Divide by 5 on both sides.}$$

Method 2: Think "What (x) is to 100 as 75 is to 500?" This translates as

$$\frac{x}{100} = \frac{75}{500}$$

$$500x = 7500$$

$$x = 15.$$

Method 3: [7] [5] [÷] [5] [0] [0] [%]   15

In each case, 15 is the percent, so we may conclude that 75 is 15% of 500.

EXAMPLE 10 38 is 5% of what number?

Method 1: $38 = .05x$

$$x = \frac{38}{.05}$$

$$x = 760$$

Method 2: Think "38 is to what number (x) as 5 is to 100?"

$$\frac{38}{x} = \frac{5}{100}$$

$$5x = 3800$$

$$x = 760$$

Method 3: [3] [8] [÷] [5] [%] 760

Each method shows us that 38 is 5% of 760.

Problem Solving

When applying percent it is often a good idea to restate the given problem as a question similar to those found in the preceding examples, and then answer that question. One strategy of problem solving deals with solving a simpler, similar problem.

EXAMPLE 11 The pie chart in Figure 14 shows the breakdown, by approximate percents, of age groups buying team sports equipment in a recent year. According to the National Sporting Goods Association, about $15,691 million was spent that year for team equipment.

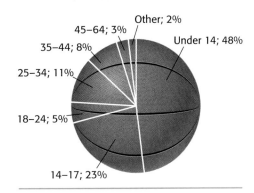

Source: National Sporting Goods Association.

FIGURE 14

How much was spent by purchasers under 14 years of age?

The portion of the circle representing the under-14 age group is 48% of the circle. Thus, we must find 48% of $15,691 million:

.48 × $15,691 million = $7351.68 million.
48%   of   Total amount   Amount spent by
                              under-14 age group

In many applications we are asked to find the percent increase or percent decrease from one quantity to another. The following guidelines summarize how to do this.

Finding Percent Increase or Decrease

1. To find the percent increase from a to b, where $b > a$, subtract a from b, and divide this result by a. Convert to a percent.

   Example: The percent increase from 4 to 7 is $\dfrac{7-4}{4} = \dfrac{3}{4} = 75\%$.

2. To find the percent decrease from a to b, where $b < a$, subtract b from a, and divide this result by a. Convert to a percent.

   Example: The percent decrease from 8 to 6 is $\dfrac{8-6}{8} = \dfrac{2}{8} = \dfrac{1}{4} = 25\%$.

EXAMPLE 12 Las Vegas, Nevada, is among the fastest-growing cities in the United States. In 1990, the population of the metropolitan area of Las Vegas was 852,646. By 1998, it had grown to 1,321,546. (*Source:* U.S. Bureau of the Census.)

(a) Estimate the percent increase over this period.

If we round the figures to 850,000 and 1,300,000, respectively, we can determine that the increase in population is approximately $1{,}300{,}000 - 850{,}000 = 450{,}000$. Then we must answer the question "What percent of 850,000 (the *original* population) is 450,000?" The fraction 450,000/850,000 simplifies to 9/17, which is a little more than 50%. Therefore, the population increased by just over 50%.

(b) Find the actual percent increase, to the nearest tenth of a percent, over this period.

We must find the difference between the two populations, and then determine what percent of 852,646 this difference comprises.

$$\underbrace{1{,}321{,}546}_{\substack{\text{Population}\\\text{in 1998}}} - \underbrace{852{,}646}_{\substack{\text{Population}\\\text{in 1990}}} = \underbrace{468{,}900}_{\substack{\text{Increase in}\\\text{population}}}$$

Now solve the problem "What percent of 852,646 is 468,900?" This is similar to the problem in Example 9. Any of the methods explained there will indicate that the answer is approximately 55.0%.

FOR FURTHER THOUGHT

Using a decimal point erroneously with a ¢ symbol is seen almost on a daily basis. Think about it. . . $.99 represents 99/100 of a dollar, or 99 cents, while 99¢ also represents 99 cents (since ¢ is the symbol for *cent*). So what does .99¢ represent? That's right, 99/100 of one cent!

Now look at the photos taken by one of the authors. At a Food Mart, a 2-liter bottle of Pepsi is advertised for .99¢. What do you think would happen if you gave the clerk a dime and asked for ten 2-liter bottles? You would most likely get a dumbfounded look. A similar response would probably be forthcoming if you asked for 24 oz of Bud, which, according to the photo, also costs .99¢. To vacuum your car, it costs .50¢. . . a mere half-cent! A color copy costs .65¢, just a bit more than one half-cent. For years, air in tires was provided as a free service; now it costs .50¢ per tire, according to the sign. And just what product is being advertised for .39¢ at the Discount Bakery Store?

It's time to end decimal point abuse.

It is fairly obvious what each of these signs actually means, but for now, assume that the products are truly being sold for the listed erroneous prices. Answer each of the following.

For Group Discussion

1. How many 2-liter bottles of Pepsi should you get for $1.00? How much change would you be owed?
2. How much should one ounce of Bud cost?
3. If you deposit two quarters to have your car vacuumed, how many times should you be able to vacuum?
4. How many color copies should you get for $6.50?
5. If you have four tires repaired and air put in all of them, how much should it total?
6. Suppose you find the item at the Discount Bakery Store that is being advertised, and you have 39 cents. How many of these items should you be able to buy?

6.5 EXERCISES

Decide whether each of the following is true or false.

1. 300% of 12 is 36.
2. 25% of a quantity is the same as 1/4 of that quantity.
3. When 759.367 is rounded to the nearest hundredth, the result is 759.40.
4. When 759.367 is rounded to the nearest hundred, the result is 759.37.
5. To find 50% of a quantity, we may simply divide the quantity by 2.
6. A soccer team that has won 12 games and lost 8 games has a winning percentage of 60%.
7. If 70% is the lowest passing grade on a quiz that has 50 items of equal value, then answering at least 35 items correctly will assure you of a passing grade.
8. 30 is more than 40% of 120.
9. .99¢ = 99 cents
10. If an item usually costs $70.00 and it is discounted 10%, then the discount price is $7.00.

Calculate each of the following using either a calculator or paper-and-pencil methods, as directed by your instructor.

11. 8.53 + 2.785
12. 9.358 + 7.2137
13. 8.74 − 12.955
14. 2.41 − 3.997
15. 25.7 × .032
16. 45.1 × 8.344
17. 1019.825 ÷ 21.47
18. −262.563 ÷ 125.03
19. $\dfrac{118.5}{1.45 + 2.3}$
20. 2.45(1.2 + 3.4 − 5.6)

Change in Population The table shows the percent change in population from 1990 through 1999 for some of the largest cities in the United States.

City	Percent Change
New York	1.4
Los Angeles	4.2
Chicago	.6
Philadelphia	−10.6
Houston	8.7
Detroit	−6.1

Source: U.S. Bureau of the Census.

21. Which city had the greatest percent change? What was this change? Was it an increase or a decline?
22. Which city had the smallest percent change? What was this change? Was it an increase or a decline?

Postage Stamp Pricing Refer to For Further Thought on decimal point abuse. At one time, the United States Postal Service sold rolls of 33-cent stamps that featured fruit berries. One such stamp is shown on the left. On the right is a photo of the pricing information found on the cellophane wrapper of such a roll.

100 STAMPS PSA
.33¢ ea. TOTAL $33.00
FRUIT BERRIES
ITEM 7757
BCA

23. Look at the second line of the pricing information. According to the price listed *per stamp*, how many stamps should you be able to purchase for one cent?
24. The total price listed is the amount the Postal Service actually charges. If you were to multiply the listed price *per stamp* by the number of stamps, what should the total price be?

(Moral of Exercises 23 and 24: Even our government is guilty of decimal point abuse.)

Solve each problem.

25. *Property Taxes* To determine their property taxes, residents of Linn County, Iowa, in the Cedar Rapids Community School District, can use the formula

$$\text{Taxes} = \dfrac{(\text{home value} \times .5485 - 4850)}{1000} \times 31.44.$$

(*Source: The Gazette,* August 19, 2000.)

Use the expression to calculate the amount of property taxes to the nearest dollar that the owner of a home with each of the following values would pay.
(a) $100,000 (b) $150,000 (c) $200,000

26. *Banking Transactions* The bank balance of Polly's Bookstore was $1856.12 on October 1. During October, Polly deposited $1742.18 received from the sale of goods, $9271.94 paid by customers on their accounts, and a $28.37 tax refund. She paid out $7195.14 for merchandise, $511.09 for salaries, and $1291.03 for other expenses.
(a) How much did Polly deposit during October?
(b) How much did she pay out?
(c) What was her bank balance at the end of October?

Exercises 27–30 are based on formulas found in Auto Math Handbook: Mathematical Calculations, Theory, and Formulas for Automotive Enthusiasts, *by John Lawlor (1991, HP Books).*

27. *Blood Alcohol Concentration* The Blood Alcohol Concentration (BAC) of a person who has been drinking is given by the formula

$$\text{BAC} = \frac{(\text{ounces} \times \text{percent alcohol} \times .075)}{\text{body weight in lb}} - (\text{hours of drinking} \times .015).$$

Suppose a policeman stops a 190-pound man who, in two hours, has ingested four 12-ounce beers, each having a 3.2 percent alcohol content. The formula would then read

$$\text{BAC} = \frac{[(4 \times 12) \times 3.2 \times .075]}{190} - (2 \times .015).$$

(a) Find this BAC.
(b) Find the BAC for a 135-pound woman who, in three hours, has drunk three 12-ounce beers, each having a 4.0 percent alcohol content.

28. *Approximate Automobile Speed* The approximate speed of an automobile in miles per hour (MPH) can be found in terms of the engine's revolutions per minute (rpm), the tire diameter in inches, and the overall gear ratio by the formula

$$\text{MPH} = \frac{\text{rpm} \times \text{tire diameter}}{\text{gear ratio} \times 336}.$$

If a certain automobile has an rpm of 5600, a tire diameter of 26 inches, and a gear ratio of 3.12, what is its approximate speed (MPH)?

29. *Engine Horsepower* Horsepower can be found from indicated mean effective pressure (mep) in pounds per square inch, engine displacement in cubic inches, and revolutions per minute (rpm) using the formula

$$\text{Horsepower} = \frac{\text{mep} \times \text{displacement} \times \text{rpm}}{792{,}000}.$$

Suppose that an engine has displacement of 302 cubic inches, and indicated mep of 195 pounds per square inch at 4000 rpm. What is its approximate horsepower?

30. *Torque Approximation* To determine the torque at a given value of rpm, the formula below applies:

$$\text{Torque} = \frac{5252 \times \text{horsepower}}{\text{rpm}}.$$

If the horsepower of a certain vehicle is 400 at 4500 rpm, what is the approximate torque?

Round each of the following numbers to the nearest (a) *tenth;* (b) *hundredth. Always round from the original number.*

31. 78.414
32. 3689.537
33. .0837
34. .0658
35. 12.68925
36. 43.99613

Convert each decimal to a percent.

37. .42
38. .87
39. .365
40. .792
41. .008
42. .0093
43. 2.1
44. 8.9

Convert each fraction to a percent.

45. $\frac{1}{5}$
46. $\frac{2}{5}$
47. $\frac{1}{100}$
48. $\frac{1}{50}$
49. $\frac{3}{8}$
50. $\frac{5}{6}$
51. $\frac{3}{2}$
52. $\frac{7}{4}$

53. Explain the difference between 1/2 of a quantity and $\frac{1}{2}$% of the quantity.

54. In Group I are some common percents, found in many everyday situations. In Group II are fractional equivalents of these percents. Match the fractions in Group I with their equivalent percents in Group II.

I		II	
(a) 25% (b) 10%		A. $\dfrac{1}{3}$ B. $\dfrac{1}{50}$	
(c) 2% (d) 20%		C. $\dfrac{3}{4}$ D. $\dfrac{1}{10}$	
(e) 75% (f) $33\dfrac{1}{3}$%		E. $\dfrac{1}{4}$ F. $\dfrac{1}{5}$	

55. Fill in each blank with the appropriate numerical response.
 (a) 5% means ____ in every 100.
 (b) 25% means 6 in every ____ .
 (c) 200% means ____ for every 4.
 (d) .5% means ____ in every 100.
 (e) ____ % means 12 for every 2.

56. The following Venn diagram shows the number of elements in the four regions formed.

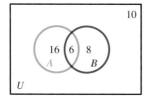

 (a) What percent of the elements in the universe are in $A \cap B$?
 (b) What percent of the elements in the universe are in A but not in B?
 (c) What percent of the elements in $A \cup B$ are in $A \cap B$?
 (d) What percent of the elements in the universe are in neither A nor B?

57. *Discount and Markup* Suppose that an item regularly costs $60.00 and it is discounted 20%. If it is then marked up 20%, is the resulting price $60.00? If not, what is it?

58. The figures in Exercise 13 of Section 6.3 are reproduced here. Express the fractional parts represented by the shaded areas as percents.

 (a) 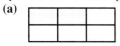 (b)

 (c) 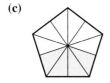 (d)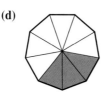

Exercises 59 and 60 deal with winning percentage in the standings of sports teams.

59. *Win-Loss Record* At the start of the play on September 25, 2002, the standings of the Central Division of the American League were as shown. Winning percentage (Pct.) is commonly expressed as a decimal rounded to the nearest thousandth. To find the winning percentage of a team, divide the number of wins (W) by the total number of games played (W + L). Find the winning percentage of each team.
 (a) Minnesota (b) Chicago (c) Cleveland

Team	W	L	Pct.
Minnesota	90	66	___
Chicago	79	78	___
Cleveland	70	87	___
Kansas City	61	96	.389
Detroit	54	102	.346

60. *Win-Loss Record* Repeat Exercise 59 for the following standings for the East Division of the National League.
 (a) Atlanta (b) Florida (c) New York

Team	W	L	Pct.
Atlanta	97	58	___
Montreal	79	78	.503
Philadelphia	79	78	.503
Florida	76	81	___
New York	74	82	___

Work each of the following problems involving percent.

61. What is 26% of 480?

62. Find 38% of 12.

63. Find 10.5% of 28.

64. What is 48.6% of 19?

65. What percent of 30 is 45?

66. What percent of 48 is 20?

67. 25% of what number is 150?

68. 12% of what number is 3600?

69. .392 is what percent of 28?

70. 78.84 is what percent of 292?

Use mental techniques to answer the questions in Exercises 71–74. Try to avoid using paper and pencil or a calculator.

71. *Allowance Increase* Dierdre Lynch's allowance was raised from $4.00 per week to $5.00 per week. What was the percent of the increase?
 A. 25% B. 20% C. 50% D. 30%

72. *Boat Purchase and Sale* Jane Gunton bought a boat five years ago for $5000 and sold it this year for $2000. What percent of her original purchase did she lose on the sale?
A. 40%   B. 50%   C. 20%   D. 60%

73. *Population of Alabama* The 2000 U.S. census showed that the population of Alabama was 4,447,000, with 26.0% represented by African-Americans. What is the best estimate of the African-American population in Alabama? (*Source:* U.S. Bureau of the Census.)
A. 500,000   B. 1,500,000
C. 1,100,000   D. 750,000

74. *Population of Hawaii* The 2000 U.S. census showed that the population of Hawaii was 1,212,000, with 21.4% of the population being of two or more races. What is the best estimate of this population of Hawaii? (*Source:* U.S. Bureau of the Census.)
A. 240,000   B. 300,000
C. 21,400   D. 24,000

Gasoline Prices The line graph shows the average price, adjusted for inflation, that Americans have paid for a gallon of gasoline for selected years since 1970.

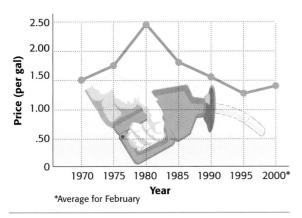

AVERAGE GASOLINE PRICES

Source: American Petroleum Institute; AP research.

75. By what percent did prices increase from 1970 to 1980?

76. By what percent did prices decrease from 1980 to 1990?

77. *Changes in Stock Prices* The following appeared in the financial section of the New Orleans *Times Picayune* on Thursday, January 20, 2000:

(By the Associated Press)

New York—The Nasdaq composite index rose to a new record Wednesday even as investors punished Microsoft for a profit report that met published estimates but failed to impress some Wall Street analysts.

The Nasdaq rose 20.48 to 4,151.29, its first new record since it closed at 4,131.15 on Jan. 3. The Dow Jones industrial average fell 71.36 to 11,489.36, a decline that came almost entirely from Microsoft.

(a) By what percent did the Nasdaq rise?
(b) By what percent did the Dow Jones industrial average fall?

78. *Union Membership* In 1999, the number of union members in the United States was 16.5 million, an increase of 265,000 from the previous year. What percent increase was this? (*Source:* U.S. Bureau of Labor Statistics.)

79. *Consumer Debt Burden* Several years ago, the nation's "consumer debt burden" was reported to be 16.4%. This means that the average American had consumer debts, such as credit card bills, auto loans, and so on, totaling 16.4% of his or her take-home pay. Suppose that Jose Valdevielso has a take-home pay of $8450 per month. What is 16.4% of his monthly take-home pay?

80. *Value of 1916-D Mercury Dime* The 1916 Mercury dime minted in Denver is quite rare. In 1979 its value in Extremely Fine condition was $625. The 2002 value had increased to $2500. What was the percent increase in the value of this coin? (*Sources: A Guide Book of United States Coins; Coin World Trends.*)

81. *Value of 1903-O Morgan Dollar* In 1963, the value of a 1903 Morgan dollar minted in New Orleans in Uncirculated condition was $1500. Due to a discovery of a large hoard of these dollars late that year, the value plummeted. Its value in 2002 was $340. What percent of its 1963 value was its 2002 value? (*Sources: A Guide Book of United States Coins; Coin World Trends.*)

82. *Price of 2002 Chevrolet Camaro* An automobile dealer had a sticker price of $26,410 on a 2002 Chevrolet Camaro. A customer was able to bargain and get 5.5% off of the sticker price. How much did the customer have to pay?

Tipping Procedure It is customary in our society to "tip" waiters and waitresses when dining in restaurants. The usual rate of tipping is 15%. A quick way of figuring a tip that will give a close approximation of 15% is as follows:

1. Round off the bill to the nearest dollar.
2. Find 10% of this amount by moving the decimal point one place to the left.
3. Take half of the amount obtained in Step 2 and add it to the result of Step 2.

This will give you approximately 15% of the bill. The amount obtained in Step 3 is 5%, and

$$10\% + 5\% = 15\%.$$

Use the method above to find an approximation of 15% for each of the following restaurant bills.

83. $29.57
84. $38.32
85. $5.15
86. $7.89

Suppose that you get extremely good service and decide to tip 20%. You can use the first two steps listed, and then in Step 3, double the amount you obtained in Step 2. Use this method to find an approximation of 20% for each of the following restaurant bills.

87. $59.96
88. $40.24
89. $180.43
90. $199.86

91. A television reporter once asked a professional wrist-wrestler what percent of his sport was physical and what percent was mental. The athlete responded "I would say it's 50% physical and 90% mental." Comment on this response.

92. We often hear the claim "(S)he gave 110%." Comment on this claim. Do you think that this is actually possible?

EXTENSION

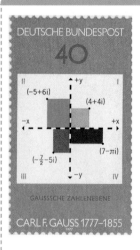

Complex Numbers

Numbers such as $\sqrt{-5}$ and $\sqrt{-16}$ were called imaginary by the early mathematicians who would not permit these numbers to be used as solutions to problems. Gradually, however, applications were found that required the use of these numbers, making it necessary to expand the set of real numbers to form the set of **complex numbers.**

Consider the equation $x^2 + 1 = 0$. It has no real number solution, since any solution must be a number whose square is -1. In the set of real numbers all squares are nonnegative numbers, because the product of either two positive numbers or two negative numbers is positive. To provide a solution for the equation $x^2 + 1 = 0$, a new number i is defined so that

$$i^2 = -1.$$

That is, i is a number whose square is -1. This definition of i makes it possible to define the square root of any negative number as follows.

For any positive real number b, $\sqrt{-b} = i\sqrt{b}$.

Extension Complex Numbers

Gauss and the Complex Numbers The stamp on the previous page honors the many contributions made by Gauss to our understanding of complex numbers. In about 1831 he was able to show that numbers of the form $a + bi$ can be represented as points on the plane (as the stamp shows) just as real numbers are. He shared this contribution with Robert Argand, a bookkeeper in Paris, who wrote an essay on the geometry of the complex numbers in 1806. This went unnoticed at the time.

EXAMPLE 1 Write each number as a product of a real number and i.

(a) $\sqrt{-100} = i\sqrt{100} = 10i$

(b) $\sqrt{-2} = \sqrt{2}i = i\sqrt{2}$

It is easy to mistake $\sqrt{2}i$ for $\sqrt{2i}$, with the i under the radical. For this reason, it is common to write $\sqrt{2}i$ as $i\sqrt{2}$.

When finding a product such as $\sqrt{-4} \cdot \sqrt{-9}$, the product rule for radicals cannot be used, since that rule applies only when both radicals represent real numbers. For this reason, always change $\sqrt{-b}$ ($b > 0$) to the form $i\sqrt{b}$ before performing any multiplications or divisions. For example,

$$\sqrt{-4} \cdot \sqrt{-9} = i\sqrt{4} \cdot i\sqrt{9} = i \cdot 2 \cdot i \cdot 3 = 6i^2.$$

Since $i^2 = -1$,

$$6i^2 = 6(-1) = -6.$$

An *incorrect* use of the product rule for radicals would give a wrong answer.

$$\sqrt{-4} \cdot \sqrt{-9} = \sqrt{(-4)(-9)} = \sqrt{36} = 6 \quad \text{Incorrect}$$

EXAMPLE 2 Multiply.

(a) $\sqrt{-3} \cdot \sqrt{-7} = i\sqrt{3} \cdot i\sqrt{7} = i^2\sqrt{3 \cdot 7} = (-1)\sqrt{21} = -\sqrt{21}$

(b) $\sqrt{-2} \cdot \sqrt{-8} = i\sqrt{2} \cdot i\sqrt{8} = i^2\sqrt{2 \cdot 8} = (-1)\sqrt{16} = (-1)4 = -4$

(c) $\sqrt{-5} \cdot \sqrt{6} = i\sqrt{5} \cdot \sqrt{6} = i\sqrt{30}$

The methods used to find products also apply to quotients, as the next example shows.

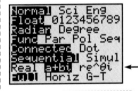

When the TI-83 Plus calculator is in complex mode, denoted by $a + bi$, it will perform complex number arithmetic.

EXAMPLE 3 Divide.

(a) $\dfrac{\sqrt{-75}}{\sqrt{-3}} = \dfrac{i\sqrt{75}}{i\sqrt{3}} = \sqrt{\dfrac{75}{3}} = \sqrt{25} = 5$

(b) $\dfrac{\sqrt{-32}}{\sqrt{8}} = \dfrac{i\sqrt{32}}{\sqrt{8}} = i\sqrt{\dfrac{32}{8}} = i\sqrt{4} = 2i$

The set of *complex numbers* are defined as follows.

```
√(-2)*√(-8)
            -4
√(-75)/√(-3)
            5
√(-32)/√(8)
            2i
```

This screen supports the results of Examples 2(b), 3(a), and 3(b).

Complex Numbers

If a and b are real numbers, then any number of the form $a + bi$ is called a **complex number.**

(continued)

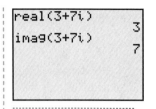

The TI-83 Plus calculator identifies the real and imaginary parts of $3 + 7i$.

In the complex number $a + bi$, the number a is called the **real part** and b is called the **imaginary part**.* When $b = 0$, $a + bi$ is a real number, so the real numbers are a subset of the complex numbers. Complex numbers with $b \neq 0$ are called **imaginary numbers.** In spite of their name, imaginary numbers are very useful in applications, particularly in work with electricity.

The relationships among the various sets of numbers discussed in this chapter are shown in Figure 15.

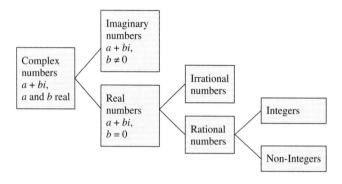

FIGURE 15

An interesting pattern emerges when we consider various powers of i. By definition, $i^0 = 1$, and $i^1 = i$. We have seen that $i^2 = -1$, and higher powers of i can be found as shown in the following list.

$$i^3 = i \cdot i^2 = i(-1) = -i \qquad i^6 = i^2 \cdot i^4 = (-1) \cdot 1 = -1$$
$$i^4 = i^2 \cdot i^2 = (-1)(-1) = 1 \qquad i^7 = i^3 \cdot i^4 = (-i) \cdot 1 = -i$$
$$i^5 = i \cdot i^4 = i \cdot 1 = i \qquad i^8 = i^4 \cdot i^4 = 1 \cdot 1 = 1$$

A few powers of i are listed here.

Powers of i

$i^1 = i$	$i^5 = i$	$i^9 = i$	$i^{13} = i$
$i^2 = -1$	$i^6 = -1$	$i^{10} = -1$	$i^{14} = -1$
$i^3 = -i$	$i^7 = -i$	$i^{11} = -i$	$i^{15} = -i$
$i^4 = 1$	$i^8 = 1$	$i^{12} = 1$	$i^{16} = 1$

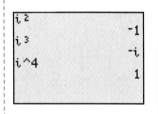

The calculator computes powers of i. Compare to the powers in the chart.

As these examples suggest, the powers of i rotate through the four numbers i, -1, $-i$, and 1. Larger powers of i can be simplified by using the fact that $i^4 = 1$. For example,

$$i^{75} = (i^4)^{18} \cdot i^3 = 1^{18} \cdot i^3 = 1 \cdot i^3 = -i.$$

*In some texts, bi is called the imaginary part.

Simplifying Large Powers of i

Step 1: Divide the exponent by 4.

Step 2: Observe the remainder obtained in Step 1. The large power of i is the same as i raised to the power determined by this remainder. Refer to the previous chart to complete the simplification. (If the remainder is 0, the power simplifies to $i^0 = 1$.)

$$\begin{array}{r} 9 \\ 4\overline{)39} \\ \underline{36} \\ 3 \end{array} \leftarrow \text{Remainder}$$

EXAMPLE 4 Find each power of i.

(a) $i^{12} = (i^4)^3 = 1^3 = 1$

(b) i^{39}

Start by dividing 39 by 4 (Step 1), as shown in the margin. The remainder is 3. So $i^{39} = i^3 = -i$ (Step 2).

Another way of simplifying i^{39} is as follows.

$$i^{39} = i^{36} \cdot i^3 = (i^4)^9 \cdot i^3 = 1^9 \cdot (-i) = -i$$

EXTENSION EXERCISES

Use the method of Examples 1–3 to write each as a real number or a product of a real number and i.

1. $\sqrt{-144}$
2. $\sqrt{-196}$
3. $-\sqrt{-225}$
4. $-\sqrt{-400}$
5. $\sqrt{-3}$
6. $\sqrt{-19}$
7. $\sqrt{-75}$
8. $\sqrt{-125}$
9. $\sqrt{-5} \cdot \sqrt{-5}$
10. $\sqrt{-3} \cdot \sqrt{-3}$
11. $\sqrt{-9} \cdot \sqrt{-36}$
12. $\sqrt{-4} \cdot \sqrt{-81}$
13. $\sqrt{-16} \cdot \sqrt{-100}$
14. $\sqrt{-81} \cdot \sqrt{-121}$
15. $\dfrac{\sqrt{-200}}{\sqrt{-100}}$
16. $\dfrac{\sqrt{-50}}{\sqrt{-2}}$
17. $\dfrac{\sqrt{-54}}{\sqrt{6}}$
18. $\dfrac{\sqrt{-90}}{\sqrt{10}}$
19. $\dfrac{\sqrt{-288}}{\sqrt{-8}}$
20. $\dfrac{\sqrt{-48} \cdot \sqrt{-3}}{\sqrt{-2}}$

21. Why is it incorrect to use the product rule for radicals to multiply $\sqrt{-3} \cdot \sqrt{-12}$?

22. In your own words describe the relationship between complex numbers and real numbers.

Use the method of Example 4 to find each power of i.

23. i^8
24. i^{16}
25. i^{42}
26. i^{86}
27. i^{47}
28. i^{63}
29. i^{101}
30. i^{141}

31. Explain the difference between $\sqrt{-1}$ and $-\sqrt{1}$. Which one of these is defined as i?

32. Is it possible to give an example of a real number that is not a complex number? Why or why not?

COLLABORATIVE INVESTIGATION

Budgeting to Buy a Car

You are shopping for a sports car and have put aside a certain amount of money each month for a car payment. Your instructor will assign this amount to you. After looking through a variety of resources, you have narrowed your choices to the cars listed in the table.

Year/Make/Model	Retail Price	Fuel Tank Size (in gallons)	Miles per Gallon (city)	Miles per Gallon (highway)
2002 Ford Mustang	$28,645	15.5	18	26
2002 Ford Thunderbird	$35,140	18.0	17	23
2002 Toyota MR2 Spyder	$24,515	12.7	25	30
2002 Mazda MX-5 Miata	$25,755	12.7	22	28
2002 Honda S2000	$32,400	13.2	20	26
2002 Chevrolet Camaro	$26,410	16.8	19	31

Source: http://autos.excite.com

As a group, work through the following steps to determine which car you can afford to buy.

A. Decide which cars you think are within your budget.

B. Select one of the cars you identified in part A. Have each member of the group calculate the monthly payment for this car using a different financing option. Use the formula given below, where P is principal, r is interest rate, and m is the number of monthly payments, along with the financing options table.

Financing Options	
Time (in years)	Interest Rate
4	7.0%
5	8.5%
6	10.0%

$$\text{Monthly Payment} = \frac{\frac{Pr}{12}}{1 - \left(\frac{12}{12+r}\right)^m}$$

C. Have each group member determine the amount of money paid in interest over the duration of the loan for his or her financing option.

D. Consider fuel expenses.

1. Assume you will travel an average of 75 miles in the city and 400 miles on the highway each week. How many gallons of gas will you need to buy each month?
2. Using typical prices for gas in your area at this time, how much money will you need to have available for buying gas?

E. Repeat parts B–D as necessary until your group can reach a consensus on the car you will buy and the financing option you will use. Write a paragraph to explain your choices.

CHAPTER 6 TEST

1. Consider the set $\{-4, -\sqrt{5}, -3/2, -.5, 0, \sqrt{3}, 4.1, 12\}$. List the elements of the set that belong to each of the following.
   (a) natural numbers   (b) whole numbers   (c) integers
   (d) rational numbers   (e) irrational numbers   (f) real numbers

2. Match the sets given in Column I with their correct set-builder notation descriptions in Column II.

   I
   (a) $\{\ldots, -4, -3, -2, -1\}$
   (b) $\{3, 4, 5, 6, \ldots\}$
   (c) $\{1, 2, 3, 4, \ldots\}$
   (d) $\{-12, \ldots, -2, -1, 0, 1, 2, \ldots, 12\}$

   II
   A. $\{x \mid x \text{ is an integer with absolute value less than or equal to } 12\}$
   B. $\{x \mid x \text{ is an integer greater than } 2.5\}$
   C. $\{x \mid x \text{ is a negative integer}\}$
   D. $\{x \mid x \text{ is a positive whole number}\}$

3. Decide whether each statement is true or false.
   (a) The absolute value of a number is always positive.
   (b) $|-7| = -(-7)$
   (c) 2/5 is an example of a real number that is not an integer.
   (d) Every real number is either positive or negative.

Perform the indicated operations. Use the rules for order of operations as necessary.

4. $6^2 - 4(9 - 1)$

5. $(-3)(-2) - [5 + (8 - 10)]$

6. $\dfrac{(-8 + 3) - (5 + 10)}{7 - 9}$

7. *California Exports* Use the graph of California exports to answer the following.

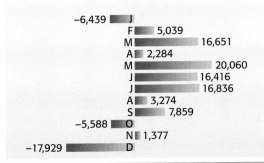

CALIFORNIA EXPORTS
Change in number of 20-foot containers exported each month in 1997 versus 1996.

Source: *USA Today*, June 23, 1998.

(a) What is the difference between the January and February changes?
(b) What is the difference between the changes in April and May?
(c) Which of the following is the best estimate of the difference between the October and November changes?
   A. -6000 B. -6500
   C. -7000 D. -7500

(d) Which of the following is the best estimate of the difference between the November and December changes?
   A. 17,000   B. 18,000
   C. 19,000   D. 20,000

8. *Altitude of a Plane* The surface of the Dead Sea has altitude 1299 feet below sea level. A stunt pilot is flying 80 feet above that surface. How much altitude must she gain to clear a 3852-foot pass by 225 feet? (Source: *The World Almanac and Book of Facts*.)

9. *Business Failures in the U.S.* The numbers of business failures in the United States for the years 1990–1996 are shown in the accompanying table. Determine, by subtraction, the change from one year to the next for each year indicated with a blank. For example, from 1991 to 1992, the change was $97{,}069 - 88{,}140 = 8929$.

Year	Number of Business Failures	Change
1990	60,747	
1991	88,140	27,393
1992	97,069	
1993	86,133	_____
1994	71,558	_____
1995	71,128	_____
1996	71,931	_____

Source: Dun & Bradstreet.

10. Match each statement in (a)–(f) with the property that justifies it in A.–F.
    (a) $7 \cdot (8 \cdot 5) = (7 \cdot 8) \cdot 5$
    (b) $3x + 3y = 3(x + y)$
    (c) $8 \cdot 1 = 1 \cdot 8 = 8$
    (d) $7 + (6 + 9) = (6 + 9) + 7$
    (e) $9 + (-9) = -9 + 9 = 0$
    (f) $5 \cdot 8$ is a real number.

    A. Distributive property
    B. Identity property
    C. Closure property
    D. Commutative property
    E. Associative property
    F. Inverse property

11. *Basketball Shot Statistics* The five starters on the local high school basketball team had the following shooting statistics after the first three games.

Player	Field Goal Attempts	Field Goals Made
Ed Moura	40	13
Jack Pritchard	10	4
Chuck Miller	20	8
Ben Whitney	6	4
Charlie Dawkins	7	2

    Answer each of the following, using estimation skills as necessary.
    (a) Which player made more than half of his attempts?
    (b) Which players made just less than 1/3 of his attempts?
    (c) Which player made exactly 2/3 of his attempts?
    (d) Which two players made the same fractional parts of their attempts? What was the fractional part, reduced to lowest terms?

Perform each operation. Reduce your answer to lowest terms.

12. $\dfrac{3}{16} + \dfrac{1}{2}$

13. $\dfrac{9}{20} - \dfrac{3}{32}$

14. $\dfrac{3}{8} \cdot \left(-\dfrac{16}{15}\right)$

15. $\dfrac{7}{9} \div \dfrac{14}{27}$

16. Convert each rational number into a repeating or terminating decimal. Use a calculator if your instructor so allows.
    (a) $\dfrac{9}{20}$ (b) $\dfrac{5}{12}$

17. Convert each decimal into a quotient of integers, reduced to lowest terms.
    (a) .72 (b) .$\overline{58}$

18. Identify each number as rational or irrational.
    (a) $\sqrt{10}$ (b) $\sqrt{16}$ (c) .01
    (d) .$\overline{01}$ (e) .0101101110…

For each of the following, (a) use a calculator to find a decimal approximation and (b) simplify the radical according to the guidelines in this chapter.

19. $\sqrt{150}$ 20. $\dfrac{13}{\sqrt{7}}$ 21. $2\sqrt{32} - 5\sqrt{128}$

22. A student using her powerful new calculator states that the *exact* value of $\sqrt{65}$ is 8.062257748. Is she correct? If not, explain.

23. Work each of the following using either a calculator or paper-and-pencil methods, as directed by your instructor.
    (a) $4.6 + 9.21$ (b) $12 - 3.725 - 8.59$
    (c) $86(.45)$ (d) $236.439 \div (-9.73)$

24. Round 9.0449 to the following place values:
    (a) hundredths (b) thousandths.

25. (a) Find 18.5% of 90.
    (b) What number is 145% of 70?

26. Consider the figure.
    (a) What percent of the total number of shapes are circles?
    (b) What percent of the total number of shapes are not stars?

27. *Sales of Books* Use estimation techniques to answer the following: In 2001, Dave Noyes sold $300,000 worth of books. In 2002, he sold $900,000. His 2002 sales were _____ of his 2001 sales.

    A. 30%   B. $33\dfrac{1}{3}\%$
    C. 200%   D. 300%

28. *Creature Comforts* From a list of "everyday items" often taken for granted, adults were recently surveyed as to those items they wouldn't want to live without. Complete the results shown in the table if 500 adults were surveyed.

Item	Percent That Wouldn't Want to Live Without	Number That Wouldn't Want to Live Without
Toilet paper	69%	___
Zipper	42%	___
Frozen Food	___	190
Self-stick note pads	___	75

(Other items included tape, hairspray, pantyhose, paper clips, and Velcro.)
Source: Market Facts for Kleenex Cottonelle.

29. *Favorite Peanuts Character* On February 13, 2000, *Peanuts* fans bid farewell to original versions of this popular comic strip. The circle graph shows the results of a survey of adults to determine their favorite Peanuts characters.
   (a) If a random sample of 2500 adults answered the survey, how many named Lucy as their favorite character?
   (b) If a certain sample resulted in 45 people naming Linus as their favorite, about how many should we expect to name Charlie Brown as their favorite? Why?

Source: USA Today/CNN/Gallup Poll conducted nationwide February 4–6, 2000.

30. *Video Rentals* The total amount spent in billions of dollars on video rentals in the United States from 1996 through 2000 is shown in the graph. By what percent did the number of rentals decrease from 1996 to 2000?

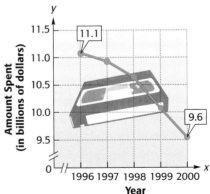

Source: Blockbuster.

chapter

7 | The Basic Concepts of Algebra

The first organized worldwide sporting event was the Summer Olympics I, held in 1896 in Athens, Greece. Three hundred eleven men from 13 nations gathered to compete. Over one hundred years later, the 2000 Summer Olympics XXVII, held in Sydney, Australia, attracted 10,651 competitors—6582 men and 4069 women, from 199 nations (*Source:* www.olympics.org). The results of an Olympic women's swimming event from 1968 to 2000 are given in the table. (Winning times are given in minutes and seconds to the nearest hundredth.)

200-Meter Freestyle	Time
1968 Deborah Meyer, U.S.	2:10.50
1972 Shane Gould, Australia	2:03.56
1976 Kornelia Ender, East Germany	1:59.26
1980 Barbara Krause, East Germany	1:58.33
1984 Mary Wayte, U.S.	1:59.23
1988 Heike Friederich, East Germany	1:57.65
1992 Nicole Haislett, U.S.	1:57.90
1996 Claudia Poll, Costa Rica	1:58.16
2000 Susie O'Neill, Australia	1:58.24

Source: World Almanac and Book of Facts, 2002.

The table gives the swimmers' times and states that the distance for the event is 200 meters. How can we find the rate of each swimmer? Using the formula relating distance, rate, and time, and using equation-solving techniques, we can answer this question. This chapter presents some of the basic ideas of algebra, including solving equations and inequalities, operations with polynomials, and factoring.

7.1 Linear Equations

7.2 Applications of Linear Equations

7.3 Ratio, Proportion, and Variation

7.4 Linear Inequalities

7.5 Properties of Exponents and Scientific Notation

7.6 Polynomials and Factoring

7.7 Quadratic Equations and Applications

Collaborative Investigation: How Does Your Walking Rate Compare to That of Olympic Racewalkers?

Chapter 7 Test

7.1 Linear Equations

Solving Linear Equations An **algebraic expression** involves the basic operations of addition, subtraction, multiplication, or division (except by 0), or raising to powers or taking roots on any collection of variables and numbers. Some examples of algebraic expressions are

$$8x + 9, \quad \sqrt{y} + 4, \quad \text{and} \quad \frac{x^3 y^8}{z}.$$

An **equation** is a statement that two algebraic expressions are equal. A linear equation in one variable involves only real numbers and one variable. Examples include

$$x + 1 = -2, \quad y - 3 = 5, \quad \text{and} \quad 2k + 5 = 10.$$

Linear Equation in One Variable

An equation in the variable x is **linear** if it can be written in the form

$$ax + b = c,$$

where a, b, and c are real numbers, with $a \neq 0$.

A linear equation in one variable is also called a **first-degree equation,** since the greatest power on the variable is one.

If the variable in an equation is replaced by a real number that makes the statement true, then that number is a **solution** of the equation. For example, 8 is a solution of the equation $x - 3 = 5$, since replacing x with 8 gives a true statement. An equation is **solved** by finding its **solution set,** the set of all solutions. The solution set of the equation $x - 3 = 5$ is $\{8\}$.

Equivalent equations are equations with the same solution set. Equations generally are solved by starting with a given equation and producing a series of simpler equivalent equations. For example,

$$8x + 1 = 17, \quad 8x = 16, \quad \text{and} \quad x = 2$$

are equivalent equations since each has the same solution set, $\{2\}$. We use the addition and multiplication properties of equality to produce equivalent equations.

Addition and Multiplication Properties of Equality

Addition Property of Equality For all real numbers a, b, and c, the equations

$$a = b \quad \text{and} \quad a + c = b + c$$

are equivalent. (The same number may be added to both sides of an equation without changing the solution set.)

Multiplication Property of Equality For all real numbers a, b, and c, where $c \neq 0$, the equations

$$a = b \quad \text{and} \quad ac = bc$$

are equivalent. (Both sides of an equation may be multiplied by the same nonzero number without changing the solution set.)

---

al-jabr, algebrista, algebra
The word *algebra* comes from the title of the work *Hisâb al-jabr w'al muquâbalah,* a ninth-century treatise by the Arab Mohammed ibn Mûsâ al-Khowârizmî. The title translates as "the science of reunion and reduction," or more generally, "the science of transposition and cancellation."

In the title of Khowârizmî's book, *jabr* ("restoration") refers to transposing negative quantities across the equals sign in solving equations. From Latin versions of Khowârizmî's text, "al-jabr" became the broad term covering the art of equation solving. (The prefix *al* means "the.")

In Spain under Moslem rule, the word *algebrista* referred to the person who restored (reset) broken bones. Signs outside barber shops read *Algebrista y Sangrador* (bonesetter and bloodletter). Such services were part of the barber's trade. The traditional red-and-white striped barber pole symbolizes blood and bandages.

Because subtraction and division are defined in terms of addition and multiplication, respectively, the addition and multiplication properties can be extended: The same number may be subtracted from both sides of an equation, and both sides may be divided by the same nonzero number without affecting the solution set.

The distributive property allows us to combine *like terms,* such as $4y$ and $2y$. For example, $4y - 2y = (4 - 2)y = 2y$.

EXAMPLE 1   Solve $4x - 2x - 5 = 4 + 6x + 3$.

First, combine like terms separately on both sides of the equation to get

$$2x - 5 = 7 + 6x.$$

Next, use the addition property to get the terms with x on the same side of the equation and the remaining terms (the numbers) on the other side. One way to do this is to first add 5 to both sides.

$$2x - 5 + 5 = 7 + 6x + 5 \quad \text{Add 5.}$$
$$2x = 12 + 6x$$

Now subtract $6x$ from both sides.

$$2x - 6x = 12 + 6x - 6x \quad \text{Subtract } 6x.$$
$$-4x = 12$$

Finally, divide both sides by -4 to obtain the term x on the left.

$$\frac{-4x}{-4} = \frac{12}{-4} \quad \text{Divide by } -4.$$
$$x = -3$$

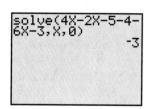

Graphing calculators can solve equations of the form

expression in X = 0.

The TI-83 Plus requires that you input the expression, the variable for which you are solving, and a "guess," separated by commas. The screen here shows how to solve the equation in Example 1.

To be sure that -3 is the solution, check by substituting back into the *original* equation (*not* an intermediate one).

$4x - 2x - 5 = 4 + 6x + 3$	Original equation
$4(-3) - 2(-3) - 5 = 4 + 6(-3) + 3$?	Let $x = -3$.
$-12 + 6 - 5 = 4 - 18 + 3$?	Multiply.
$-11 = -11$	True

Since a true statement is obtained, -3 is the solution. The solution set is $\{-3\}$. ■

The steps used to solve a linear equation in one variable are as follows. (Not all equations require all of these steps.)

Solving a Linear Equation in One Variable

Step 1: **Clear fractions.** Eliminate any fractions by multiplying both sides of the equation by a common denominator.

Step 2: **Simplify each side separately.** Simplify each side of the equation as much as possible by using the distributive property to clear parentheses and by combining like terms as needed.

(continued)

The problem-solving strategy of guessing and checking, discussed in Chapter 1, was actually used by the early Egyptians in equation solving. This method, called the **Rule of False Position,** involved making an initial guess at the solution of an equation, and then following up with an adjustment in the likely event that the guess was incorrect. For example (using our modern notation), if the equation

$$6x + 2x = 32$$

was to be solved, an initial guess might have been $x = 3$. Substituting 3 for x gives

$$6(3) + 2(3) = 32 \quad ?$$
$$18 + 6 = 32 \quad ?$$
$$24 = 32. \quad \text{False}$$

The guess, 3, gives a value (24) which is smaller than the desired value (32). Since 24 is 3/4 of 32, the guess, 3, is 3/4 of the actual solution. The actual solution, therefore, must be 4, since 3 is 3/4 of 4.

Use the methods explained in this section to verify this result.

Step 3: **Isolate the variable terms on one side.** Use the addition property of equality to get all terms with variables on one side of the equation and all numbers on the other.

Step 4: **Transform so that the coefficient of the variable is 1.** Use the multiplication property of equality to get an equation with just the variable (with coefficient 1) on one side.

Step 5: **Check.** Check by substituting back into the original equation.

In Example 1 we did not use Step 1 and the distributive property in Step 2 as given above. Many other equations, however, will require one or both of these steps.

EXAMPLE 2 Solve $2(k - 5) + 3k = k + 6$.

Since there are no fractions in this equation, Step 1 does not apply. Begin by using the distributive property to simplify and combine terms on the left side of the equation (Step 2).

$$2(k - 5) + 3k = k + 6$$
$$2k - 10 + 3k = k + 6 \qquad \text{Distributive property}$$
$$5k - 10 = k + 6 \qquad \text{Combine like terms.}$$

Next, add 10 to both sides (Step 3).

$$5k - 10 + 10 = k + 6 + 10 \qquad \text{Add 10.}$$
$$5k = k + 16$$

Now subtract k from both sides.

$$5k - k = k + 16 - k \qquad \text{Subtract } k.$$
$$4k = 16 \qquad \text{Combine like terms.}$$

Divide both sides by 4 (Step 4).

$$\frac{4k}{4} = \frac{16}{4} \qquad \text{Divide by 4.}$$
$$k = 4$$

Check that the solution set is {4} by substituting 4 for k in the original equation (Step 5). ■

In the remaining examples in this section, we will not identify the steps by number.

When fractions or decimals appear as coefficients in equations, our work can be made easier if we multiply both sides of the equation by the least common denominator of all the fractions. This is an application of the multiplication property of equality, and it produces an equivalent equation with integer coefficients.

François Viète (1540–1603) was a lawyer at the court of Henry IV of France and studied equations. Viète simplified the notation of algebra and was among the first to use letters to represent numbers. For centuries, algebra and arithmetic were expressed in a cumbersome way with words and occasional symbols. Since the time of Viète, algebra has gone beyond equation solving; the abstract nature of higher algebra depends on its symbolic language.

EXAMPLE 3 Solve $\dfrac{x+7}{6} + \dfrac{2x-8}{2} = -4$.

Start by eliminating the fractions. Multiply both sides by 6.

$$6\left(\dfrac{x+7}{6} + \dfrac{2x-8}{2}\right) = 6 \cdot (-4)$$

$$6\left(\dfrac{x+7}{6}\right) + 6\left(\dfrac{2x-8}{2}\right) = 6(-4) \quad \text{Distributive property}$$

$$x + 7 + 3(2x - 8) = -24$$

$$x + 7 + 6x - 24 = -24 \quad \text{Distributive property}$$

$$7x - 17 = -24 \quad \text{Combine terms.}$$

$$7x - 17 + 17 = -24 + 17 \quad \text{Add 17.}$$

$$7x = -7$$

$$\dfrac{7x}{7} = \dfrac{-7}{7} \quad \text{Divide by 7.}$$

$$x = -1$$

Check to see that $\{-1\}$ is the solution set.

FOR FURTHER THOUGHT

The Axioms of Equality

When we solve an equation, we must make sure that it remains "balanced"—that is, any operation that is performed on one side of an equation must also be performed on the other side in order to assure that the set of solutions remains the same.

Underlying the rules for solving equations are four axioms of equality, listed below. For all real numbers a, b, and c,

1. **Reflexive axiom** $a = a$
2. **Symmetric axiom**   If $a = b$, then $b = a$.
3. **Transitive axiom**   If $a = b$ and $b = c$, then $a = c$.
4. **Substitution axiom**   If $a = b$, then a may replace b in any statement without affecting the truth or falsity of the statement.

A relation, such as equality, which satisfies the first three of these axioms (reflexive, symmetric, and transitive), is called an equivalence relation.

For Group Discussion

1. Give an example of an everyday relation that does not satisfy the symmetric axiom.
2. Does the transitive axiom hold in sports competition, with the relation "defeats"?
3. Give an example of a relation that does not satisfy the transitive axiom.

Algebra dates back to the Babylonians of 2000 B.C. The Egyptians also worked problems in algebra, but the problems were not as complex as those of the Babylonians. In about the sixth century, the Hindus developed methods for solving problems involving interest, discounts, and partnerships.

Many Hindu and Greek works on mathematics were preserved only because Moslem scholars from about 750 to 1250 made translations of them. The Arabs took the work of the Greeks and Hindus and greatly expanded it. For example, Mohammed ibn Musa al-Khowârizmî wrote books on algebra and on the Hindu numeration system (the one we use) that had tremendous influence in Western Europe; his name is remembered today in the word *algorithm*.

Problem Solving

In the next section we solve problems involving interest rates and concentrations of solutions. These problems involve percents that are converted to decimal numbers. The equations that are used to solve such problems involve decimal coefficients. We can clear these decimals by multiplying by the largest power of 10 necessary to obtain integer coefficients. The next example shows how this is done.

EXAMPLE 4 Solve $.06x + .09(15 - x) = .07(15)$.

Since each decimal number is given in hundredths, multiply both sides of the equation by 100. (This is done by moving the decimal points two places to the right.)

$$.06x + .09(15 - x) = .07(15)$$
$$.06x + .09(15 - x) = .07(15) \quad \text{Multiply by 100.}$$
$$6x + 9(15 - x) = 7(15)$$
$$6x + 9(15) - 9x = 105 \quad \text{Distributive property}$$
$$-3x + 135 = 105 \quad \text{Combine like terms.}$$
$$-3x + 135 - 135 = 105 - 135 \quad \text{Subtract 135.}$$
$$-3x = -30$$
$$\frac{-3x}{-3} = \frac{-30}{-3} \quad \text{Divide by } -3.$$
$$x = 10$$

Check to verify that the solution set is $\{10\}$.

When multiplying the term $.09(15 - x)$ by 100 in Example 4, do not multiply both .09 and $15 - x$ by 100. This step is not an application of the distributive property, but of the associative property. The correct procedure is

$$100[.09(15 - x)] = [100(.09)](15 - x) \quad \text{Associative property}$$
$$= 9(15 - x). \quad \text{Multiply.}$$

Special Kinds of Linear Equations Each of the equations above had a solution set containing one element; for example, $2x + 1 = 13$ has solution set $\{6\}$, containing only the single number 6. An equation that has a finite (but nonzero) number of elements in its solution set is a **conditional equation.** Sometimes an equation has no solutions. Such an equation is a **contradiction** and its solution set is ∅. It is also possible for an equation to have an infinite number of solutions. An equation that is satisfied by every number for which both sides are defined is called an **identity.** The next example shows how to recognize these types of equations.

Sofia Kovalevskaya (1850–1891) was the most widely known Russian mathematician in the late nineteenth century. She did most of her work in the theory of differential equations—equations invaluable for expressing rates of change. For example, in biology, the rate of growth of a population, say of microbes, can be precisely stated by differential equations.

Kovalevskaya studied privately because public lectures were not open to women. She eventually received a degree (1874) from the University of Göttingen, Germany. In 1884 she became a lecturer at the University of Stockholm and later was appointed professor of higher mathematics.

Kovalevskaya was well known as a writer. Besides novels about Russian life, notably *The Sisters Rajevski* and *Vera Vorontzoff*, she wrote her *Recollections of Childhood*, which has been translated into English.

EXAMPLE 5 Solve each equation. Decide whether it is a conditional equation, an identity, or a contradiction.

(a) $5x - 9 = 4(x - 3)$

$$5x - 9 = 4x - 12 \quad \text{Distributive property}$$
$$5x - 9 - 4x = 4x - 12 - 4x \quad \text{Subtract } 4x.$$
$$x - 9 = -12 \quad \text{Combine like terms.}$$
$$x - 9 + 9 = -12 + 9 \quad \text{Add 9.}$$
$$x = -3$$

The solution set, $\{-3\}$, has one element, so $5x - 9 = 4(x - 3)$ is a *conditional equation*.

(b) $5x - 15 = 5(x - 3)$

$$5x - 15 = 5x - 15 \quad \text{Distributive property}$$

Both sides of the equation are *exactly the same,* so any real number would make the equation true. For this reason, the solution set is the set of all real numbers, and the equation $5x - 15 = 5(x - 3)$ is an *identity.*

(c) $5x - 15 = 5(x - 4)$

$$5x - 15 = 5x - 20 \quad \text{Distributive property}$$
$$5x - 15 - 5x = 5x - 20 - 5x \quad \text{Subtract } 5x.$$
$$-15 = -20 \quad \text{False}$$

Since the result, $-15 = -20$, is *false,* the equation has no solution. The solution set is $\emptyset$. The equation $5x - 15 = 5(x - 4)$ is a *contradiction.*

Formulas The solution of a problem in algebra often depends on the use of a mathematical statement or **formula** in which more than one letter is used to express a relationship. Examples of formulas are

$$d = rt, \quad I = prt, \quad \text{and} \quad P = 2L + 2W.$$

In some applications, the necessary formula is solved for one of its variables, which may not be the unknown number that must be found. The following examples show how to solve a formula for any one of its variables. This process is called **solving for a specified variable.** Notice how the steps used in these examples are very similar to those used in solving a linear equation. Keep in mind that *when you are solving for a specified variable, treat that variable as if it were the only one, and treat all other variables as if they were numbers.*

Solving for a Specified Variable

Step 1: Use the addition or multiplication property as necessary to get all terms containing the specified variable on one side of the equation.

Step 2: All terms not containing the specified variable should be on the other side of the equation. *(continued)*

Step 3: If necessary, use the distributive property to write the side with the specified variable as the product of that variable and a sum of terms.

In general, follow the steps given earlier for solving linear equations.

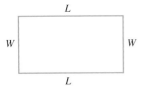

Perimeter, P, the sum of the lengths of the sides of a rectangle, is given by

$$P = 2L + 2W.$$

FIGURE 1

EXAMPLE 6 Solve the formula $P = 2L + 2W$ for W.

This formula gives the relationship between the perimeter of a rectangle, P, the length of the rectangle, L, and the width of the rectangle, W. See Figure 1.

Solve the formula for W by getting W alone on one side of the equals sign. To begin, subtract $2L$ from both sides.

$$P = 2L + 2W$$
$$P - 2L = 2L + 2W - 2L \quad \text{Subtract } 2L.$$
$$P - 2L = 2W$$
$$\frac{P - 2L}{2} = \frac{2W}{2} \quad \text{Divide by 2.}$$
$$\frac{P - 2L}{2} = W \quad \text{or} \quad W = \frac{P - 2L}{2}$$

Models Sometimes equations can be used to predict future data based on past data. Equations such as these are called **models.** Of course, models must be used with care, for often there is no guarantee that past performances will continue in the future.

EXAMPLE 7 By studying the winning times in the 500-meter speed-skating event at Olympic games back to 1900, it was found that the winning times for men were closely approximated by the equation

$$M = 46.338 - .097x,$$

where M is the time in seconds needed to win for men, with x the Olympic year and $x = 0$ corresponding to 1900. (For example, 1994 winning times would be estimated by replacing x with $1994 - 1900 = 94$.) The corresponding equation for women is

$$W = 57.484 - .196x.$$

(a) Find the year for which $M = 42.458$.

$$M = 46.338 - .097x \quad \text{Given equation}$$
$$42.458 = 46.338 - .097x \quad \text{Let } M = 42.458.$$
$$-3.88 = -.097x \quad \text{Subtract } 46.338.$$
$$x = 40 \quad \text{Divide by } -.097.$$

The year 1940 corresponds to $x = 40$.

(b) Predict the winning time for women in 2006.

The year 2006 corresponds to $x = 106$.

$$W = 57.484 - .196x \quad \text{Given equation}$$
$$W = 57.484 - .196(106) \quad \text{Let } x = 106.$$
$$W = 36.708$$

This model predicts the winning time to be just under 37 seconds.

7.1 EXERCISES

1. Which of the following equations are linear equations in x?
   A. $3x + x - 1 = 0$ **B.** $8 = x^2$
   C. $6x + 2 = 9$ **D.** $\dfrac{1}{2}x - \dfrac{1}{x} = 0$

2. Which of the equations in Exercise 1 are not linear equations in x? Explain why.

3. Decide whether 6 is a solution of $3(x + 4) = 5x$ by substituting 6 for x. If it is not a solution, explain why.

4. Use substitution to decide whether -2 is a solution of $5(x + 4) - 3(x + 6) = 9(x + 1)$. If it is not a solution, explain why.

5. If two equations are equivalent, they have the same _____ .

6. The equation $4[x + (2 - 3x)] = 2(4 - 4x)$ is an identity. Let x represent the number of letters in your last name. Is this number a solution of this equation? Check your answer.

7. The expression $.06(10 - x)(100)$ is equivalent to
   A. $.06 - .06x$ **B.** $60 - 6x$
   C. $6 - 6x$ **D.** $6 - .06x$

8. Describe in your own words the steps used to solve a linear equation.

Solve each equation.

9. $7k + 8 = 1$

10. $5m - 4 = 21$

11. $8 - 8x = -16$

12. $9 - 2r = 15$

13. $7x - 5x + 15 = x + 8$

14. $2x + 4 - x = 4x - 5$

15. $12w + 15w - 9 + 5 = -3w + 5 - 9$

16. $-4t + 5t - 8 + 4 = 6t - 4$

17. $2(x + 3) = -4(x + 1)$

18. $4(x - 9) = 8(x + 3)$

19. $3(2w + 1) - 2(w - 2) = 5$

20. $4(x - 2) + 2(x + 3) = 6$

21. $2x + 3(x - 4) = 2(x - 3)$

22. $6x - 3(5x + 2) = 4(1 - x)$

23. $6p - 4(3 - 2p) = 5(p - 4) - 10$

24. $-2k - 3(4 - 2k) = 2(k - 3) + 2$

25. $-[2z - (5z + 2)] = 2 + (2z + 7)$

26. $-[6x - (4x + 8)] = 9 + (6x + 3)$

27. $-3m + 6 - 5(m - 1) = -(2m - 4) - 5m + 5$

28. $4(k + 2) - 8k - 5 = -3k + 9 - 2(k + 6)$

29. $-[3x - (2x + 5)] = -4 - [3(2x - 4) - 3x]$

30. $2[-(x - 1) + 4] = 5 + [-(6x - 7) + 9x]$

31. $-(9 - 3a) - (4 + 2a) - 4 = -(2 - 5a) - a$

32. $(2 - 4x) - (3 - 4x) + 4 = -(-3 + 6x) + x$

33. $(2m - 6) - (3m - 4) = -(-4 + m) - 4m + 6$

34. To solve the linear equation

$$.05x + .12(x + 5000) = 940,$$

we can multiply both sides by a power of 10 so that all coefficients are integers. What is the smallest power of 10 that will accomplish this goal?

35. Suppose that in solving the equation

$$\frac{1}{3}x + \frac{1}{2}x = \frac{1}{6}x,$$

you begin by multiplying both sides by 12, rather than the *least* common denominator, 6. Should you get the correct solution anyway? Explain.

Solve each equation.

36. $\dfrac{3x}{4} + \dfrac{5x}{2} = 13$

37. $\dfrac{8x}{3} - \dfrac{2x}{4} = -13$

38. $\dfrac{x-8}{5} + \dfrac{8}{5} = -\dfrac{x}{3}$

39. $\dfrac{2r-3}{7} + \dfrac{3}{7} = -\dfrac{r}{3}$

40. $\dfrac{4t+1}{3} = \dfrac{t+5}{6} + \dfrac{t-3}{6}$

41. $\dfrac{2x+5}{5} = \dfrac{3x+1}{2} + \dfrac{-x+7}{2}$

42. $.05x + .12(x + 5000) = 940$

43. $.09k + .13(k + 300) = 61$

44. $.02(50) + .08r = .04(50 + r)$

45. $.20(14,000) + .14t = .18(14,000 + t)$

46. $.05x + .10(200 - x) = .45x$

47. $.08x + .12(260 - x) = .48x$

48. The equation $x + 2 = x + 2$ is called a(n) _____, because its solution set is {all real numbers}. The equation $x + 1 = x + 2$ is called a(n) _____, because its solution set is $\emptyset$.

49. Which one of the following is a conditional equation?
 A. $2x + 1 = 3$ **B.** $x = 3x - 2x$
 C. $3x + 1 = 3x$ **D.** $\dfrac{1}{2}x = \dfrac{1}{2}x$

50. Explain the distinctions among a conditional equation, an identity, and a contradiction.

Decide whether each equation is conditional, an identity, or a contradiction. Give the solution set.

51. $-2p + 5p - 9 = 3(p - 4) - 5$

52. $-6k + 2k - 11 = -2(2k - 3) + 4$

53. $6x + 2(x - 2) = 9x + 4$

54. $-4(x + 2) = -3(x + 5) - x$

55. $-11m + 4(m - 3) + 6m = 4m - 12$

56. $3p - 5(p + 4) + 9 = -11 + 15p$

57. $7[2 - (3 + 4r)] - 2r = -9 + 2(1 - 15r)$

58. $4[6 - (1 + 2m)] + 10m = 2(10 - 3m) + 8m$

59. When a formula is solved for a particular variable, several different equivalent forms may be possible. If we solve $A = \dfrac{1}{2}bh$ for h, one possible correct answer is

$$h = \frac{2A}{b}.$$

Which one of the following is *not* equivalent to this?

 A. $h = 2\left(\dfrac{A}{b}\right)$ **B.** $h = 2A\left(\dfrac{1}{b}\right)$

 C. $h = \dfrac{A}{\frac{1}{2}b}$ **D.** $h = \dfrac{\frac{1}{2}A}{b}$

60. One source for geometric formulas gives the formula for the perimeter of a rectangle as $P = 2L + 2W$, while another gives it as $P = 2(L + W)$. Are these equivalent? If so, what property justifies their equivalence?

Mathematical Formulas Solve each formula for the specified variable.

61. $d = rt$; for t (distance)

62. $I = prt$; for r (simple interest)

63. $A = bh$; for b (area of a parallelogram)

64. $P = 2L + 2W$; for L (perimeter of a rectangle)

65. $P = a + b + c$; for a (perimeter of a triangle)

66. $V = LWH$; for W (volume of a rectangular solid)

67. $A = \dfrac{1}{2}bh$; for h (area of a triangle)

68. $C = 2\pi r$; for r (circumference of a circle)

69. $S = 2\pi rh + 2\pi r^2$; for h (surface area of a right circular cylinder)

70. $A = \dfrac{1}{2}(B + b)h$; for B (area of a trapezoid)

71. $C = \dfrac{5}{9}(F - 32)$; for F (Fahrenheit to Celsius)

72. $F = \frac{9}{5}C + 32$; for C (Celsius to Fahrenheit)

73. $A = 2HW + 2LW + 2LH$; for H (surface area of a rectangular solid)

74. $V = \frac{1}{3}Bh$; for h (volume of a right pyramid)

Work each of the following problems. (When necessary, round up to the nearest year.)

75. *Broadway Ticket Sales* The mathematical model
$$y = .61x - 48.3$$
approximates the Broadway tickets sold from 1990 to 1999. In the model, $x = 90$ corresponds to the 1990–1991 season, $x = 91$ corresponds to the 1991–1992 season, and so on, and y is the number of tickets sold in millions. (*Source:* League of American Theaters and Producers.)

(a) Based on this model, how many tickets were sold in 1995–1996? (*Hint:* 1995–1996 corresponds to $x = 95$.)

(b) In what season did the number of tickets sold reach 7.9 million? (*Hint:* The number of tickets sold in millions corresponds to y.)

76. *Theater Ticket Sales* The chart shows the number of tickets (in millions) sold to theatergoers under age 18.

Season	Millions of Tickets
1990–1991	.5
1996–1997	1.1

Source: League of American Theaters and Producers.

If $x = 90$ corresponds to the 1990–1991 season, and so on, the model
$$y = .1x - 8.5$$
approximates the number (in millions) under age 18 attending the theater.

(a) Use the model to estimate the number of these youths attending in 1993–1994.

(b) In what season did these ticket sales amount to .75 million?

77. *Broadway Ticket Sales* This graph indicates the number of tickets sold for Broadway shows in the 1990s. For the period shown in the graph, between which two seasons did ticket sales increase the most?

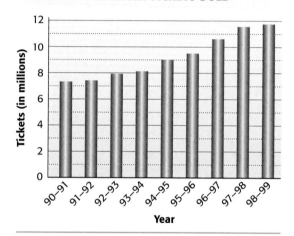

Source: League of American Theaters and Producers.

78. *Theater Ticket Sales* Use the model in Exercise 75 to find the number of tickets sold in 1994–1995. Is your answer a good approximation of the number shown in the graph in Exercise 77?

7.2 Applications of Linear Equations

In an earlier chapter we introduced Polya's problem-solving procedure. In this section we apply this procedure to solve application problems algebraically.

Translating into Algebra When algebra is used to solve practical applications, we must translate the verbal statements of the problems into mathematical statements.

Problem Solving

Usually there are key words and phrases in a verbal problem that translate into mathematical expressions involving addition, subtraction, multiplication, and division. Translations of some commonly used expressions follow.

Translation from Words to Mathematical Expressions

Verbal Expression	Mathematical Expression (where x and y are numbers)
Addition	
The **sum** of a number and 7	$x + 7$
6 **more than** a number	$x + 6$
3 **plus** 8	$3 + 8$
24 **added to** a number	$x + 24$
A number **increased by** 5	$x + 5$
The **sum** of two numbers	$x + y$
Subtraction	
2 **less than** a number	$x - 2$
12 **minus** a number	$12 - x$
A number **decreased by** 12	$x - 12$
The **difference between** two numbers	$x - y$
A number **subtracted from** 10	$10 - x$
Multiplication	
16 **times** a number	$16x$
A number **multiplied by** 6	$6x$
2/3 **of** a number (used only with fractions and percent)	$\frac{2}{3}x$
Twice (2 times) a number	$2x$
The **product** of two numbers	xy
Division	
The **quotient** of 8 and a number	$\frac{8}{x}$ ($x \neq 0$)
A number **divided by** 13	$\frac{x}{13}$
The **ratio** of two numbers or the **quotient** of two numbers	$\frac{x}{y}$ ($y \neq 0$)

The symbol of equality, $=$, is often indicated by the word *is*. In fact, since equal mathematical expressions represent different names for the same number, words that indicate the idea of "sameness" translate as $=$. For example,

If the product of a number and 12 is decreased by 7, the result is 105

translates to the mathematical equation

$$12x - 7 = 105,$$

where x represents the unknown number. (Why would $7 - 12x = 105$ be incorrect?)

Applications While there is no one method that allows us to solve all types of applied problems, the following six steps are helpful.

George Polya's problem-solving procedure can be adapted to applications of algebra as seen in the steps in the box. Steps 1 and 2 make up the first stage of Polya's procedure (*Understand the Problem*), Step 3 forms the second stage (*Devise a Plan*), Step 4 comprises the third stage (*Carry Out the Plan*), and Steps 5 and 6 form the last stage (*Look Back*).

Solving an Applied Problem

Step 1: **Read** the problem carefully until you understand what is given and what is to be found.

Step 2: **Assign a variable** to represent the unknown value, using diagrams or tables as needed. Write down what the variable represents. If necessary, express any other unknown values in terms of the variable.

Step 3: **Write an equation** using the variable expression(s).

Step 4: **Solve** the equation.

Step 5: **State the answer.** Does it seem reasonable?

Step 6: **Check** the answer in the words of the *original* problem.

The third step is often the hardest. To translate the problem into an equation, write the given phrases as mathematical expressions. Since equal mathematical expressions are names for the same number, translate any words that mean *equal* or *same* as $=$. The $=$ sign leads to an equation to be solved.

Problem Solving

A common type of problem involves finding two quantities when the sum of the quantities is known. In general, to solve such problems, choose a variable to represent one of the unknowns and then represent the other quantity in terms of the same variable, using information obtained in the problem. Then write an equation based on the words of the problem.

EXAMPLE 1 In the 2000 Olympics, U.S. contestants won 14 more gold than silver medals. They won a total of 64 gold and silver medals. How many of each type medal were won. (*Source:* United States Olympic Committee.)

Step 1: **Read** the problem. We are given information about the total number of gold and silver medals, and we are asked to find the number of each kind.

Step 2: **Assign a variable.**

Let $\quad x =$ the number of silver medals.
Then $\quad x + 14 =$ the number of gold medals.

> Here is an application of linear equations, taken from the **Greek Anthology** (about 500 A.D.), a group of 46 number problems.
>
> *Demochares has lived a fourth of his life as a boy, a fifth as a youth, a third as a man, and has spent 13 years in his dotage. How old is he?*
>
> (Answer: 60 years old)

Step 3: **Write an equation.**

$$\underset{\downarrow}{\text{The total}} \;\; \underset{\downarrow}{\text{is}} \;\; \underset{\downarrow}{\text{the number of silver}} \;\; \underset{\downarrow}{\text{plus}} \;\; \underset{\downarrow}{\text{the number of gold.}}$$
$$64 \;\; = \;\; x \;\; + \;\; (x + 14)$$

Step 4: **Solve** the equation.

$$64 = 2x + 14 \qquad \text{Combine terms.}$$
$$64 - 14 = 2x + 14 - 14 \qquad \text{Subtract 14.}$$
$$50 = 2x \qquad \text{Combine terms.}$$
$$25 = x \qquad \text{Divide by 2.}$$

Step 5: **State the answer.** Because x represents the number of silver medals, the U.S. athletes won 25 silver medals. Because $x + 14$ represents the number of gold medals, they won $25 + 14 = 39$ gold medals.

Step 6: **Check.** Since there were 39 gold and 25 silver medals, the total number of medals was $39 + 25 = 64$. Because $39 - 25 = 14$, there were 14 more gold medals than silver medals. This information agrees with what is given in the problem, so the answers check. ■

The problem in Example 1 could also have been solved by letting x represent the number of gold medals. Then $x - 14$ would represent the number of silver medals. The equation would then be

$$64 = x + (x - 14).$$

The solution of this equation is 39, which is the number of gold medals. The number of silver medals would then be $39 - 14 = 25$. The answers are the same, whichever approach is used.

EXAMPLE 2 The instructions for a woodworking project call for three pieces of wood. The longest piece must be twice the length of the middle-sized piece, and the shortest piece must be 10 inches shorter than the middle-sized piece. Maria Gonzales has a board 70 inches long that she wishes to use. How long can each piece be?

Step 1: **Read** the problem. There will be three answers.

Step 2: **Assign a variable.** Since the middle-sized piece appears in both pairs of comparisons, let x represent the length, in inches, of the middle-sized piece. We have

$$x = \text{the length of the middle-sized piece,}$$
$$2x = \text{the length of the longest piece, and}$$
$$x - 10 = \text{the length of the shortest piece.}$$

A sketch is helpful here. See Figure 2.

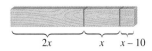

FIGURE 2

Step 3: **Write an equation.**

$$\underset{\text{Longest}}{2x} + \underset{\substack{\text{Middle-}\\\text{sized}}}{x} + \underset{\text{Shortest}}{(x-10)} \underset{\text{is}}{=} \underset{\substack{\text{Total}\\\text{length}}}{70}$$

Step 4: **Solve.**

$$4x - 10 = 70 \quad \text{Combine terms.}$$
$$4x - 10 + 10 = 70 + 10 \quad \text{Add 10 to each side.}$$
$$4x = 80 \quad \text{Combine terms.}$$
$$x = 20 \quad \text{Divide by 4 on each side.}$$

Step 5: **State the answer.** The middle-sized piece is 20 inches long, the longest piece is 2(20) = 40 inches long, and the shortest piece is 20 − 10 = 10 inches long.

Step 6: **Check.** The sum of the lengths is 70 inches. All conditions of the problem are satisfied.

Problem Solving

Percents often are used in problems involving mixing different concentrations of a substance or different interest rates. In each case, to get the amount of pure substance or the interest, we multiply.

Mixture Problems	Interest Problems (annual)
base × rate (%) = percentage	**principal × rate (%) = interest**
$b \times r = p$	$p \times r = I$

In an equation, the percent should be written as a decimal. For example, 35% is written .35, not 35, and 7% is written .07, not 7.

EXAMPLE 3

(a) If a chemist has 40 liters of a 35% acid solution, then the amount of pure acid in the solution is

$$\underset{\substack{\text{Amount of}\\\text{solution}}}{40} \times \underset{\substack{\text{Rate of}\\\text{concentration}}}{.35} = \underset{\substack{\text{Amount of}\\\text{pure acid}}}{14 \text{ liters.}}$$

(b) If $1300 is invested for one year at 7% simple interest, the amount of interest earned in the year is

$$\underset{\text{Principal}}{\$1300} \times \underset{\text{Interest rate}}{.07} = \underset{\text{Interest earned}}{\$91.}$$

7.2 Applications of Linear Equations

Problem Solving

In the examples that follow, we use tables to organize the information in the problems. A table enables us to more easily set up the equation for the problem, which is usually the most difficult step.

EXAMPLE 4 A chemist needs to mix 20 liters of 40% acid solution with some 70% acid solution to get a mixture that is 50% acid. How many liters of the 70% acid solution should be used?

Step 1: **Read** the problem. Note the percent of each solution and of the mixture.

Step 2: **Assign a variable.**

Let $x =$ the number of liters of 70% acid solution needed.

Recall from Example 3(a) that the amount of pure acid in this solution will be given by the product of the percent of strength and the number of liters of solution, or

$$\text{liters of pure acid in } x \text{ liters of 70\% solution} = .70x.$$

The amount of pure acid in the 20 liters of 40% solution is

$$\text{liters of pure acid in the 40\% solution} = .40(20) = 8.$$

The new solution will contain $(x + 20)$ liters of 50% solution. The amount of pure acid in this solution is

$$\text{liters of pure acid in the 50\% solution} = .50(x + 20).$$

Figure 3 illustrates this information, which is summarized in the table.

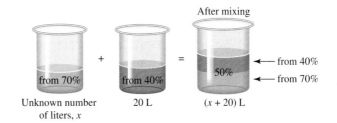

FIGURE 3

Liters of Solution	Rate (as a decimal)	Liters of Pure Acid
x	.70	$.70(x)$
20	.40	$.40(20) = 8$
$x + 20$	.50	$.50(x + 20)$

Sum must equal

Step 3: **Write an equation.** The number of liters of pure acid in the 70% solution added to the number of liters of pure acid in the 40% solution will equal the number of liters of pure acid in the final mixture, so the equation is

$$\underbrace{.70x}_{\text{Pure acid in 70\%}} + \underbrace{.40(20)}_{\text{pure acid in 40\%}} = \underbrace{.50(x+20)}_{\text{pure acid in 50\%}}.$$

Step 4: **Solve the equation.** Clear parentheses, then multiply by 100 to clear decimals.

$$.70x + .40(20) = .50x + .50(20) \quad \text{Distributive property}$$
$$70x + 40(20) = 50x + 50(20) \quad \text{Multiply by 100.}$$
$$70x + 800 = 50x + 1000$$
$$20x + 800 = 1000 \quad \text{Subtract } 50x.$$
$$20x = 200 \quad \text{Subtract 800.}$$
$$x = 10 \quad \text{Divide by 20.}$$

Step 5: **State the answer.** The chemist needs to use 10 liters of 70% solution.

Step 6: **Check.** Since

$$.70(10) + .40(20) = 7 + 8 = 15$$

and

$$.50(10 + 20) = .50(30) = 15,$$

the answer checks. ∎

The next example uses the formula for simple interest, $I = prt$. Remember that when $t = 1$, the formula becomes $I = pr$, as shown in the Problem-Solving box just before Example 3. Once again the idea of multiplying the total amount (principal) by the rate (rate of interest) gives the percentage (amount of interest).

EXAMPLE 5 Elizabeth Lemulle receives an inheritance. She plans to invest part of it at 9% and $2000 more than this amount in a less secure investment at 10%. To earn $1150 per year in interest, how much should she invest at each rate?

Step 1: **Read** the problem again.

Step 2: **Assign a variable.**

Let $\quad x =$ the amount invested at 9% (in dollars);
then $\quad x + 2000 =$ the amount invested at 10% (in dollars).

Use a table to arrange the information given in the problem.

Amount Invested in Dollars	Rate of Interest	Interest for One Year
x	.09	$.09x$
$x + 2000$	.10	$.10(x + 2000)$

As I was going to St. Ives, I met a man with seven wives. Every wife had seven sacks, Every sack had seven cats, Every cat had seven kits. Kits, cats, sacks, and wives, How many were going to St. Ives?

Do you know the "trick" answer to this nursery rhyme? The rhyme is a derivation of an old application found in the Rhind papyrus, an Egyptian manuscript that dates back to about 1650 B.C. Leonardo of Pisa (Fibonacci) also included a similar problem in *Liber Abaci* in 1202.

The answer to the question is 1. Only "I" was *going* to St. Ives.

Step 3: **Write an equation.** We multiply amount by rate to get the interest earned. Since the total interest is to be $1150, the equation is

$$\underset{\text{at 9\%}}{\text{Interest}} \quad \text{plus} \quad \underset{\text{at 10\%}}{\text{interest}} \quad \text{is} \quad \underset{\text{interest.}}{\text{total}}$$

$$.09x \quad + \quad .10(x + 2000) \quad = \quad 1150.$$

Step 4: **Solve** the equation. Clear parentheses; then clear decimals.

$$.09x + .10x + .10(2000) = 1150 \quad \text{Distributive property}$$
$$9x + 10x + 10(2000) = 115{,}000 \quad \text{Multiply by 100.}$$
$$9x + 10x + 20{,}000 = 115{,}000 \quad \text{Multiply.}$$
$$19x + 20{,}000 = 115{,}000 \quad \text{Combine terms.}$$
$$19x = 95{,}000 \quad \text{Subtract 20,000.}$$
$$x = 5000 \quad \text{Divide by 19.}$$

Step 5: **State the answer.** She should invest $5000 at 9% and $5000 + $2000 = $7000 at 10%.

Step 6: **Check.** Investing $5000 at 9% and $7000 at 10% gives total interest of .09($5000) + .10($7000) = $450 + $700 = $1150, as required in the original problem. ■

Problem Solving

Problems that involve different denominations of money or items with different monetary values are very similar to mixture and interest problems.

Money Problems

Number × Value of one = Total value

For example, if a jar contains 37 quarters, the monetary amount of the coins is

$$\underset{\text{Number of coins}}{37} \quad \times \quad \underset{\text{Denomination}}{\$.25} \quad = \quad \underset{\text{Monetary value}}{\$9.25}.$$

EXAMPLE 6 A bank teller has 25 more five-dollar bills than ten-dollar bills. The total value of the money is $200. How many of each denomination of bill does he have?

Step 1: **Read** the problem. We must find the number of each denomination of bill that the teller has.

Step 2: **Assign a variable.**

Let x = the number of ten-dollar bills;
then $x + 25$ = the number of five-dollar bills.

Organize the given information in a table.

Number of Bills	Denomination	Total Value
x	10	$10x$
$x + 25$	5	$5(x + 25)$

Step 3: **Write an equation.** Multiplying the number of bills by the denomination gives the monetary value. The value of the fives added to the value of the tens must be $200:

Value of fives plus value of tens is $200.

$$5(x + 25) + 10x = 200.$$

Step 4: **Solve.**

$5x + 125 + 10x = 200$ Distributive property
$15x + 125 = 200$ Combine terms.
$15x = 75$ Subtract 125.
$x = 5$ Divide by 15.

Step 5: **State the answer.** Since x represents the number of tens, the teller has 5 tens and $5 + 25 = 30$ fives.

Step 6: **Check.** The teller has $30 - 5 = 25$ more fives than tens, and the total value is $5(\$10) + 30(\$5) = \$200$, as required.

If an automobile travels at an average rate of 50 miles per hour for two hours, then it travels $50 \times 2 = 100$ miles. This is an example of the basic relationship between distance, rate, and time:

$$\text{distance} = \text{rate} \times \text{time},$$

given by the formula $d = rt$. By solving, in turn, for r and t in the formula, we obtain two other equivalent forms of the formula. The three forms are given below.

Distance, Rate, Time Relationship

$$d = rt \qquad r = \frac{d}{t} \qquad t = \frac{d}{r}$$

The following examples illustrate the uses of these formulas.

EXAMPLE 7

(a) The speed of sound is 1088 feet per second at sea level at 32°F. In 5 seconds under these conditions, sound travels

$$1088 \times 5 = 5440 \text{ feet.}$$
Rate × Time = Distance

Here, we found distance given rate and time, using $d = rt$.

Can we average averages?
A car travels from A to B at 40 miles per hour and returns at 60 miles per hour. What is its rate for the entire trip?

The correct answer is not 50 miles per hour, as you might expect. Remembering the distance-rate-time relationship and letting $x = $ the distance between A and B, we can simplify a complex fraction to find the correct answer.

$$\begin{aligned}
\text{Average rate for entire trip} &= \frac{\text{Total distance}}{\text{Total time}} \\
&= \frac{x + x}{\frac{x}{40} + \frac{x}{60}} \\
&= \frac{2x}{\frac{3x}{120} + \frac{2x}{120}} \\
&= \frac{2x}{\frac{5x}{120}} \\
&= 2x \cdot \frac{120}{5x} \\
&= 48
\end{aligned}$$

The average rate for the entire trip is 48 miles per hour.

(b) The winner of the first Indianapolis 500 race (in 1911) was Ray Harroun, driving a Marmon Wasp at an average speed of 74.59 miles per hour. (*Source: The Universal Almanac, 1997,* John W. Wright, General Editor.) To complete the 500 miles, it took him

$$\text{Distance} \rightarrow \frac{500}{74.59} = 6.70 \text{ hours} \quad \text{(rounded)}. \leftarrow \text{Time}$$

Here, we found time given rate and distance, using $t = d/r$. To convert .70 hour to minutes, multiply by 60 to get $.70(60) = 42$ minutes. The race took Harroun 6 hours, 42 minutes to complete.

(c) In the 2000 Olympic Games in Sydney, Australia, Dutch swimmer Inge de Bruijn won the women's 50-m freestyle swimming event in 24.32 seconds. (*Source:* www.olympics.com)

Her rate was

$$\text{Rate} = \frac{\text{Distance}}{\text{Time}} \rightarrow \frac{50}{24.32} = 2.06 \text{ meters per second} \quad \text{(rounded)}.$$

Problem Solving

The final example shows how to solve a typical application of the formula $d = rt$. We continue to use the six-step method, but now we also include a sketch along with a table in Step 2. The sketch shows what is happening in the problem and will help you set up the equation.

EXAMPLE 8 Two cars leave Baton Rouge, Louisiana, at the same time and travel east on Interstate 12. One travels at a constant speed of 55 miles per hour and the other travels at a constant speed of 63 miles per hour. In how many hours will the distance between them be 24 miles?

Step 1: **Read** the problem. We are trying to find the time when the distance between the cars will be 24 miles.

Step 2: **Assign a variable.** Since we are looking for time, let $t =$ the number of hours until the distance between them is 24 miles. The sketch in Figure 4 shows what is happening in the problem.

Now, construct a table like the one that follows. Fill in the information given in the problem, and use t for the time traveled by each car. Multiply rate by time to get the expressions for distances traveled.

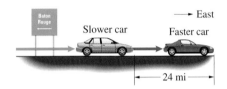

	Rate	× Time	= Distance	
Faster car	63	t	$63t$	⬅ Difference is 24 miles.
Slower car	55	t	$55t$	⬅

FIGURE 4

The quantities $63t$ and $55t$ represent the two distances. Refer to Figure 4, and notice that the difference between the larger distance and the smaller distance is 24 miles.

Step 3: **Write an equation.**
$$63t - 55t = 24$$

Step 4: **Solve.**
$$8t = 24 \quad \text{Combine like terms.}$$
$$t = 3 \quad \text{Divide by 8.}$$

Step 5: **State the answer.** It will take the cars 3 hours to be 24 miles apart.

Step 6: **Check.** After three hours the faster car will have traveled $63 \times 3 = 189$ miles, and the slower car will have traveled $55 \times 3 = 165$ miles. Since $189 - 165 = 24$, the conditions of the problem are satisfied. ■

In motion problems such as the one in Example 8, once you have filled in two pieces of information in each row of the chart, you should automatically fill in the third piece of information, using the appropriate form of the formula relating distance, rate, and time. Set up the equation based upon your sketch and the information in the table.

7.2 EXERCISES

Decide whether each of the following translates into an expression or an equation.

 1. the sum of a number and 6
 2. 75% of a number
 3. $\frac{2}{3}$ of a number is 12.
 4. 7 is 3 more than a number.
 5. the ratio of a number and 5
 6. 15 divided by a number is 3.
 7. Rework Example 6, letting the variable represent the number of five-dollar bills. Is the answer to the problem the same?
 8. Explain why $13 - x$ is *not* a correct translation of "13 less than a number."

Translate each verbal phrase into a mathematical expression. Use x to represent the unknown number.

9. a number decreased by 18
10. 12 more than a number
11. the product of 9 less than a number and 6 more than the number
12. the quotient of a number and 6
13. the ratio of 12 and a nonzero number
14. $\frac{6}{7}$ of a number
15. Write a few sentences describing the six steps for problem solving.
16. Which one of the following is *not* a valid translation of "30% of a number"?
    A. .30x **B.** .3x **C.** $\frac{3x}{10}$ **D.** .30

Let x represent the unknown, and write an equation for each sentence. Do not solve.

17. If the quotient of a number and 6 is added to twice the number, the result is 8 less than the number.
18. If the product of a number and −4 is subtracted from the number, the result is 9 more than the number.
19. When $\frac{2}{3}$ of a number is subtracted from 12, the result is 10.
20. When 75% of a number is added to 6, the result is 3 more than the number.

Use the methods of Examples 1 and 2 or your own method to solve each problem. (In Exercise 21, we outline the six-step method.)

21. *Numbers of Book Buyers* In a recent year the two most popular places where book buyers shopped were large chain bookstores and small chain/independent bookstores. In a sample of book buyers, 70 more shopped at large chain bookstores than at small chain/independent bookstores. A total of 442 book buyers shopped at these two types of stores. Complete the following problem-solving steps to find how many buyers shopped at each type of bookstore. (*Source:* Book Industry Study Group.)

    Step 1: Let x = the number of shoppers at large chain bookstores
    Step 2: Then $x - 70 = $ _____ .
    Step 3: _____ + _____ = 442
    Step 4: $x = $ _____
    Step 5: There were _____ large chain bookstore shoppers and _____ small chain/independent shoppers.
    Step 6: The number of _____ was _____ more than the number of _____ , and the total number of these two bookstore types was _____ .

22. *Concert Revenues* The rock band U2 generated top revenue on the concert circuit in 2001. U2 and second-place *NSYNC together took in $196.5 million from ticket sales. If *NSYNC took in $22.9 million less than U2, how much did each generate? (*Source:* Pollstar.)

23. *Automobile Sales* The Toyota Camry was the top-selling passenger car in the U.S. in 2000, followed by the Honda Accord. Honda Accord sales were 15 thousand less than Toyota Camry sales, and 828 thousand of these two cars were sold. How many of each make of car were sold? (*Source:* Ward's Communications.)

24. *NBA Record* In the 2001–2002 NBA regular season, the Sacramento Kings won 2 less than three times as many games as they lost. The Kings played 82 games. How many wins and losses did the team have? (*Source:* nba.com.)

25. *Hits by Ruth and Hornsby* Babe Ruth and Rogers Hornsby were two great hitters. Together they got 5803 hits in their careers. Hornsby got 57 more hits than Ruth. How many hits did each get? (*Source:* David S. Neft and Richard M. Cohen, *The Sports Encyclopedia: Baseball, 1997.* St. Martins Griffin; New York.)

26. *Votes in 2000 Presidential Campaign* In the 2000 presidential election, George W. Bush and Al Gore together received 537 electoral votes. Bush received 5 more votes than Gore. How many votes did each candidate receive? (*Source:* World Almanac and Book of Facts, 2002.)

Video Rental Revenue vs. Sales Revenue Refer to the accompanying graph to help you answer the questions in Exercises 27 and 28.

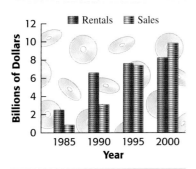

HOME VIDEO REVENUE

Source: Paul Kagan Associates, Inc.

27. In 1990, video rental revenue was $.27 billion more than twice video sales revenue. Together, these sales amounted to $9.81 billion. What was the revenue from each of these sources?

28. Video rentals accounted for $.15 billion more than video sales in a later year. At that time, combined sales were $14.83 billion. In what year was this true, and what was the revenue from each source?

29. *Drive-In Movie Screens* The number of drive-in movie screens has declined steadily in the U.S. since the 1960s. California and New York were two of the states with the most remaining drive-in screens in 2001. California had 11 more screens than New York, and there were 107 screens total in the two states. How many drive-in movie screens remained in each state? (*Source:* National Association of Theatre Owners.)

30. *Must See TV* Thursday is the most-watched night for the major broadcast TV networks (ABC, CBS, NBC, and Fox), with 20 million more viewers than Saturday, the least-watched night. The total for the two nights is 102 million viewers. How many viewers of the major networks are there on each of these nights? (*Source:* Nielsen Media Research.)

Use the method of the text discussion or your own method to solve each problem.

31. *Acid Mixture* How much pure acid is in 250 milliliters of a 14% acid solution?

32. *Alcohol Mixture* How much pure alcohol is in 150 liters of a 30% alcohol solution?

33. *Interest Earned* If $10,000 is invested for one year at 3.5% simple interest, how much interest is earned?

34. *Interest Earned* If $25,000 is invested at 3% simple interest for 2 years, how much interest is earned?

35. *Monetary Value of Coins* What is the monetary amount of 283 nickels?

36. *Monetary Value of Coins* What is the monetary amount of 35 half-dollars?

Use the method of Example 4 or your own method to solve each problem.

37. *Alcohol Mixture* In a chemistry class, 12 liters of a 12% alcohol solution must be mixed with a 20% solution to get a 14% solution. How many liters of the 20% solution are needed?

Strength	Liters of Solution	Liters of Alcohol
12%	12	
20%		
14%		

38. *Alcohol Mixture* How many liters of a 10% alcohol solution must be mixed with 40 liters of a 50% solution to get a 40% solution?

Strength	Liters of Solution	Liters of Alcohol
	x	
	40	
40%		

39. *Alcohol Mixture in First Aid Spray* A medicated first aid spray on the market is 78% alcohol by volume. If the manufacturer has 50 liters of the spray containing 70% alcohol, how much pure alcohol should be added so that the final mixture is the required 78% alcohol? (*Hint:* Pure alcohol is 100% alcohol.)

40. *Insecticide Mixture* How much water must be added to 3 gallons of a 4% insecticide solution to reduce the concentration to 3%? (*Hint:* Water is 0% insecticide.)

41. *Antifreeze Mixture* It is necessary to have a 40% antifreeze solution in the radiator of a certain car. The radiator now holds 20 liters of 20% solution. How many liters of this should be drained and replaced with 100% antifreeze to get the desired strength? (*Hint:* The number of liters drained is equal to the number of liters replaced.)

42. *Chemical Mixture* A tank holds 80 liters of a chemical solution. Currently, the solution has a strength of 30%. How much of this should be drained and replaced with a 70% solution to get a final strength of 40%?

Use the method of Example 5 or your own method to solve each problem.

43. *Investments at Different Rates* Dezneil Jones earned $12,000 last year by giving tennis lessons. He invested part at 3% simple interest and the rest at 4%. He earned a total of $440 in interest. How much did he invest at each rate?

Percent as a Decimal	Amount Invested	Interest in One Year
.03		
.04		
	12,000	440

44. *Investments at Different Rates* Cherelle Allen won $60,000 in a slot machine in Las Vegas. She invested part at 2% simple interest and the rest at 3%. She earned a total of $1600 in interest. How much was invested at each rate?

Percent as a Decimal	Amount Invested	Interest in One Year
.02	x	.02x
	60,000 − x	
		1600

45. *Investments at Different Rates* Amado Carillo invested some money at 4.5% simple interest and $1000 less than twice this amount at 3%. His total annual income from the interest was $1020. How much was invested at each rate?

46. *Investments at Different Rates* Kari Heen invested some money at 3.5% simple interest, and $5000 more than 3 times this amount at 4%. She earned $1440 in interest. How much did she invest at each rate?

47. *Investments at Different Rates* Ed Moura has $29,000 invested in stocks paying 5%. How much additional money should he invest in certificates of deposit paying 2% so that the average return on the two investments is 3%?

48. *Investments at Different Rates* Terry McGinnis placed $15,000 in an account paying 6%. How much additional money should she deposit at 4% so that the average return on the two investments is 5.5%?

Use the method of Example 6 or your own method to solve each problem.

49. *Coin Mixture* Sam Abo-zahrah has a box of coins that he uses when playing poker with his friends. The box currently contains 44 coins, consisting of pennies, dimes, and quarters. The number of pennies is equal to the number of dimes, and the total value is $4.37. How many of each denomination of coin does he have in the box?

Number of Coins	Denomination	Value
x	.01	.01x
x		
	.25	

50. *Coin Mixture* Amber Allen found some coins while looking under her sofa pillows. There were equal numbers of nickels and quarters, and twice as many half-dollars as quarters. If she found $2.60 in all, how many of each denomination of coin did she find?

Number of Coins	Denomination	Value
x	.05	.05x
x		
2x	.50	

51. *Attendance at a School Play* The school production of *Our Town* was a big success. For opening night, 410 tickets were sold. Students paid $3 each, while nonstudents paid $7 each. If a total of $1650 was collected, how many students and how many nonstudents attended?

52. *Attendance at a Concert* A total of 550 people attended a Boston Pops concert. Floor tickets cost $40 each, while balcony tickets cost $28 each. If a total of $20,800 was collected, how many of each type of ticket were sold?

53. *Attendance at a Sporting Event* At the Sacramento Monarchs home games, Row 1 seats cost $35 each and Row 2 seats cost $30 each. The 105 seats in these rows were sold out for the season. The total receipts for them were $3420. How many of each type of seat were sold? (*Source:* Sacramento Monarchs.)

54. *Coin Mixture* In the nineteenth century, the United States minted two-cent and three-cent pieces. Lee Ann Spahr has three times as many three-cent pieces as two-cent pieces, and the face value of these coins is $1.21. How many of each denomination does she have?

55. *Coin Mixture* Dave Bowers collects U.S. gold coins. He has a collection of 80 coins. Some are $10 coins and the rest are $20 coins. If the face value of the coins is $1060, how many of each denomination does he have?

56. In the nineteenth century, the United States minted half-cent coins. What decimal number (in dollars) would represent this denomination?

Automobile Racing In Exercises 57–60, find the time based on the information provided. Use a calculator and round your answers to the nearest thousandth. (*Source: The World Almanac and Book of Facts,* 2002.)

	Event and Year	Participant	Distance	Rate
57.	Indianapolis 500, 2001	Helio Castroneves (Reynard–Honda)	500 miles	131.294 mph
58.	Daytona 500, 2001	Michael Waltrip (Chevrolet)	500 miles	161.794 mph
59.	Indianapolis 500, 1980	Johnny Rutherford (Hy-Gain McLaren/Goodyear)	255 miles*	148.725 mph
60.	Indianapolis 500, 1975	Bobby Unser (Jorgensen Eagle)	435 miles*	149.213 mph

*rain-shortened

Olympics Results In Exercises 61–64, find the rate based on the information provided. Use a calculator and round your answers to the nearest hundredth. All events were at the 2000 Olympics, held in Sydney, Australia. (Source: http://espn.go.com/oly/summer00)

	Event	Participant	Distance	Time
61.	100-m hurdles, Women	Olga Shishigina, Kazakhstan	100 m	12.65 sec
62.	400-m hurdles, Women	Irina Privalova, Russia	400 m	53.02 sec
63.	400-m hurdles, Men	Angelo Taylor, USA	400 m	47.50 sec
64.	400-m dash, Men	Michael Johnson, USA	400 m	43.84 sec

65. *Distance Between Cities* A driver averaged 53 miles per hour and took 10 hours to travel from Memphis to Chicago. What is the distance between Memphis and Chicago?

66. *Distance Between Cities* A small plane traveled from Warsaw to Rome, averaging 164 miles per hour. The trip took two hours. What is the distance from Warsaw to Rome?

67. Suppose that an automobile averages 45 miles per hour, and travels for 30 minutes. Is the distance traveled $45 \times 30 = 1350$ miles? If not, explain why not, and give the correct distance.

68. Which of the following choices is the best *estimate* for the average speed of a trip of 405 miles that lasted 8.2 hours?
   A. 50 miles per hour
   B. 30 miles per hour
   C. 60 miles per hour
   D. 40 miles per hour

Use the method of Example 8 or your own method to solve each problem.

69. *Travel Times of Trains* A train leaves Little Rock, Arkansas, and travels north at 85 kilometers per hour. Another train leaves at the same time and travels south at 95 kilometers per hour. How long will it take before they are 315 kilometers apart?

	Rate	Time	Distance
First train	85	t	
Second train			

70. *Travel Times of Steamers* Two steamers leave a port on a river at the same time, traveling in opposite directions. Each is traveling 22 miles per hour. How long will it take for them to be 110 miles apart?

	Rate	Time	Distance
First steamer		t	
Second steamer	22		

71. *Travel Times of Commuters* Nancy and Mark commute to work, traveling in opposite directions. Nancy leaves the house at 8:00 A.M. and averages 35 miles per hour. Mark leaves at 8:15 A.M. and averages 40 miles per hour. At what time will they be 140 miles apart?

72. *Travel Times of Bicyclers* Jeff leaves his house on his bicycle at 8:30 A.M. and averages 5 miles per hour. His wife, Joan, leaves at 9:00 A.M., following the same path and averaging 8 miles per hour. At what time will Joan catch up with Jeff?

73. *Distance Traveled to Work* When Tri drives his car to work, the trip takes 30 minutes. When he rides the bus, it takes 45 minutes. The average speed of the bus is 12 miles per hour less than his speed when driving. Find the distance he travels to work.

74. *Distance Traveled to School* Latoya can get to school in 15 minutes if she rides her bike. It takes her 45 minutes if she walks. Her speed when walking is 10 miles per hour slower than her speed when riding. How far does she travel to school?

75. *Time Traveled by a Pleasure Boat* A pleasure boat on the Mississippi River traveled from Baton Rouge to New Orleans with a stop at White Castle. On the first part of the trip, the boat traveled at an average speed of 10 miles per hour. From White Castle to New Orleans the average speed was 15 miles per hour. The entire trip covered 100 miles. How long did the entire trip take if the two parts each took the same number of hours?

76. *Time Traveled on a Visit* Steve leaves Nashville to visit his cousin David in Napa, 80 miles away. He travels at an average speed of 50 miles per hour. One-half hour later David leaves to visit Steve, traveling at an average speed of 60 miles per hour. How long after David leaves will they meet?

7.3 Ratio, Proportion, and Variation

Ratio One of the most frequently used mathematical concepts in everyday life is ratio. A baseball player's batting average is actually a ratio. The slope, or pitch, of a roof on a building may be expressed as a ratio. Ratios provide a way of comparing two numbers or quantities.

> **Ratio**
>
> A **ratio** is a quotient of two quantities. The ratio of the number a to the number b is written
>
> $$a \text{ to } b, \quad \frac{a}{b}, \quad \text{or} \quad a:b.$$

When ratios are used in comparing units of measure, the units should be the same. This is shown in Example 1.

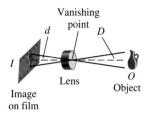

When you look a long way down a straight road or railroad track, it seems to narrow as it vanishes in the distance. The point where the sides seem to touch is called the **vanishing point.** The same thing occurs in the lens of a camera, as shown in the figure. Suppose *I* represents the length of the image, *O* the length of the object, *d* the distance from the lens to the film, and *D* the distance from the lens to the object. Then

$$\frac{\text{Image length}}{\text{Object length}} = \frac{\text{Image distance}}{\text{Object distance}}$$

or

$$\frac{I}{O} = \frac{d}{D}.$$

Given the length of the image on the film and its distance from the lens, then the length of the object determines how far away the lens must be from the object to fit on the film.

EXAMPLE 1 Write a ratio for each word phrase.

(a) the ratio of 5 hours to 3 hours

This ratio can be written as 5/3.

(b) the ratio of 5 hours to 3 days

First convert 3 days to hours: $3 \text{ days} = 3 \cdot 24 = 72 \text{ hours}$. The ratio of 5 hours to 3 days is thus 5/72.

Proportion We now define a special type of equation called a *proportion*.

Proportion

A **proportion** is a statement that says that two ratios are equal.

For example,

$$\frac{3}{4} = \frac{15}{20}$$

is a proportion that says that the ratios 3/4 and 15/20 are equal. In the proportion

$$\frac{a}{b} = \frac{c}{d},$$

a, *b*, *c*, and *d* are the **terms** of the proportion. The *a* and *d* terms are called the **extremes,** and the *b* and *c* terms are called the **means.** We can read the proportion $\frac{a}{b} = \frac{c}{d}$ as "*a* is to *b* as *c* is to *d*." Beginning with this proportion and multiplying both sides by the common denominator, *bd*, gives

$$bd \cdot \frac{a}{b} = bd \cdot \frac{c}{d}$$

$$ad = bc.$$

That is, the product of the extremes equals the product of the means. The products *ad* and *bc* can also be found by multiplying diagonally.

$$\frac{a}{b} = \frac{c}{d}$$

This is called **cross multiplication** and *ad* and *bc* are called **cross products.**

Cross Products

If $\frac{a}{b} = \frac{c}{d}$, then the cross products *ad* and *bc* are equal.

Also, if $ad = bc$, then $\frac{a}{b} = \frac{c}{d}$ (as long as $b \neq 0, d \neq 0$).

7.3 Ratio, Proportion, and Variation

From the rule given on page 340, if $\frac{a}{b} = \frac{c}{d}$ then $ad = bc$. However, if $\frac{a}{c} = \frac{b}{d}$, then $ad = cb$, or $ad = bc$. This means that the two proportions are equivalent and

the proportion $\frac{a}{b} = \frac{c}{d}$ can also be written as $\frac{a}{c} = \frac{b}{d}$.

Sometimes one form is more convenient to work with than the other.

Four numbers are used in a proportion. If any three of these numbers are known, the fourth can be found.

EXAMPLE 2 Solve each proportion.

(a) $\dfrac{63}{x} = \dfrac{9}{5}$

The cross products must be equal.

$$63 \cdot 5 = 9x \quad \text{Cross products}$$
$$315 = 9x$$
$$35 = x \quad \text{Divide by 9.}$$

The solution set is $\{35\}$.

(b) $\dfrac{8}{5} = \dfrac{12}{r}$

$$8r = 5 \cdot 12 \quad \text{Set the cross products equal.}$$
$$8r = 60$$
$$r = \frac{60}{8} = \frac{15}{2} \quad \text{Divide by 8; express in lowest terms.}$$

The solution set is $\left\{\dfrac{15}{2}\right\}$.

EXAMPLE 3 Solve the equation

$$\frac{m-2}{5} = \frac{m+1}{3}.$$

Find the cross products, and set them equal to each other.

$$3(m - 2) = 5(m + 1) \quad \text{Be sure to use parentheses.}$$
$$3m - 6 = 5m + 5 \quad \text{Distributive property}$$
$$3m = 5m + 11 \quad \text{Add 6.}$$
$$-2m = 11 \quad \text{Subtract } 5m.$$
$$m = -\frac{11}{2} \quad \text{Divide by } -2.$$

The solution set is $\left\{-\dfrac{11}{2}\right\}$.

While the cross product method is useful in solving equations of the types found in Examples 2 and 3, it cannot be used directly if there is more than one term on either side. For example, you cannot use the method directly to solve the equation

$$\frac{4}{x} + 3 = \frac{1}{9},$$

because there are two terms on the left side.

EXAMPLE 4 Biologists use algebra to estimate the number of fish in a lake. They first catch a sample of fish and mark each specimen with a harmless tag. Some weeks later, they catch a similar sample of fish from the same areas of the lake and determine the proportion of previously tagged fish in the new sample. The total fish population is estimated by assuming that the proportion of tagged fish in the new sample is the same as the proportion of tagged fish in the entire lake.

Suppose biologists tag 300 fish on May 1. When they return on June 1 and take a new sample of 400 fish, 5 of the 400 were previously tagged. Estimate the number of fish in the lake.

Let x represent the number of fish in the lake. Set up and solve a proportion.

Tagged fish on May 1 → $\dfrac{300}{x} = \dfrac{5}{400}$ ← Tagged fish in the June 1 sample
Total fish in the lake →   ← Total number in the June 1 sample

$$5x = 120{,}000$$
$$x = 24{,}000$$

There are approximately 24,000 fish in the lake.

Unit pricing—deciding which size of an item offered in different sizes produces the best price per unit—uses proportions. Suppose you buy 36 ounces of pancake syrup for $3.89. To find the price per unit, set up and solve a proportion.

$$\frac{36 \text{ ounces}}{1 \text{ ounce}} = \frac{3.89}{x}$$

$$36x = 3.89 \quad \text{Cross products}$$

$$x = \frac{3.89}{36} \quad \text{Divide by 36.}$$

$$x \approx .108 \quad \text{Use a calculator.}$$

Thus, the price for 1 ounce is $.108, or about 11 cents. Notice that the unit price is the ratio of the cost for 36 ounces, $3.89, to the number of ounces, 36, which means that the unit price for an item is found by dividing the cost by the number of units.

Size	Price
36-ounce	$3.89
24-ounce	$2.79
12-ounce	$1.89

EXAMPLE 5 Besides the 36-ounce size discussed earlier, the local supermarket carries two other sizes of a popular brand of pancake syrup, priced as shown in the table. Which size is the best buy? That is, which size has the lowest unit price?

To find the best buy, divide the price by the number of units to get the price per ounce. Each result in the table on the next page was found by using a calculator and rounding the answer to three decimal places.

Size	Unit Cost (dollars per ounce)	
36-ounce	$\dfrac{\$3.89}{36} = \$.108$	←The best buy
24-ounce	$\dfrac{\$2.79}{24} = \$.116$	
12-ounce	$\dfrac{\$1.89}{12} = \$.158$	

Since the 36-ounce size produces the lowest price per unit, it would be the best buy. (*Be careful:* Sometimes the largest container *does not* produce the lowest price per unit.) ∎

Number of Rooms	Cost of the Job
1	$ 22.50
2	$ 45.00
3	$ 67.50
4	$ 90.00
5	$112.50

Variation Suppose that a carpet cleaning service charges $22.50 per room to shampoo a carpet. The table in the margin shows the relationship between the number of rooms cleaned and the cost of the total job for 1 through 5 rooms.

If we divide the cost of the job by the number of rooms, in each case we obtain the quotient, or ratio, 22.50 (dollars per room). Suppose that we let x represent the number of rooms and y represent the cost for cleaning that number of rooms. Then the relationship between x and y is given by the equation

$$\frac{y}{x} = 22.50$$

or

$$y = 22.50x.$$

This relationship between x and y is an example of *direct variation*.

Direct Variation

y varies directly as x, or **y is directly proportional to x,** if there exists a nonzero constant k such that

$$y = kx$$

or, equivalently,

$$\frac{y}{x} = k.$$

The constant k is a numerical value called the **constant of variation.**

EXAMPLE 6 Suppose y varies directly as x, and $y = 50$ when $x = 20$. Find y when $x = 14$.

Since y varies directly as x, there exists a constant k such that $y = kx$. Find k by replacing y with 50 and x with 20.

$$y = kx$$
$$50 = k \cdot 20$$
$$\frac{5}{2} = k$$

Since $y = kx$ and $k = 5/2$,
$$y = \frac{5}{2}x.$$

Now find y when $x = 14$.
$$y = \frac{5}{2} \cdot 14 = 35$$

The value of y is 35 when $x = 14$.

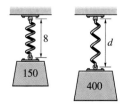

FIGURE 5

EXAMPLE 7 Hooke's law for an elastic spring states that the distance a spring stretches is directly proportional to the force applied. If a force of 150 pounds stretches a certain spring 8 centimeters, how much will a force of 400 pounds stretch the spring? See Figure 5.

If d is the distance the spring stretches and f is the force applied, then $d = kf$ for some constant k. Since a force of 150 pounds stretches the spring 8 centimeters,

$$d = kf \qquad \text{Formula}$$
$$8 = k \cdot 150 \qquad d = 8, f = 150$$
$$k = \frac{8}{150} = \frac{4}{75}, \qquad \text{Find } k.$$

and $d = \frac{4}{75}f$. For a force of 400 pounds,

$$d = \frac{4}{75}(400) = \frac{64}{3}. \qquad \text{Let } f = 400.$$

The spring will stretch 64/3 centimeters if a force of 400 pounds is applied.

In summary, follow these steps to solve a variation problem.

Solving a Variation Problem

Step 1: Write the variation equation.

Step 2: Substitute the initial values and solve for k.

Step 3: Rewrite the variation equation with the value of k from Step 2.

Step 4: Substitute the remaining values, solve for the unknown, and find the required answer.

In some cases one quantity will vary directly as a *power* of another.

Direct Variation as a Power

y **varies directly as the nth power of x** if there exists a real number k such that
$$y = kx^n.$$

An example of direct variation as a power involves the area of a circle. The formula for the area of a circle is

$$A = \pi r^2.$$

$A = \pi r^2$

Here, π is the constant of variation, and the area varies directly as the square of the radius.

EXAMPLE 8 The distance a body falls from rest varies directly as the square of the time it falls (here we disregard air resistance). If a skydiver falls 64 feet in 2 seconds, how far will she fall in 8 seconds?

Step 1: If d represents the distance the skydiver falls and t the time it takes to fall, then d is a function of t, and

$$d = kt^2$$

for some constant k.

Step 2: To find the value of k, use the fact that the object falls 64 feet in 2 seconds.

$d = kt^2$	Formula
$64 = k(2)^2$	Let $d = 64$ and $t = 2$.
$k = 16$	Find k.

Step 3: With this result, the variation equation becomes

$$d = 16t^2.$$

Step 4: Now let $t = 8$ to find the number of feet the skydiver will fall in 8 seconds.

$$d = 16(8)^2 = 1024 \quad \text{Let } t = 8.$$

The skydiver will fall 1024 feet in 8 seconds. ∎

In direct variation where $k > 0$, as x increases, y increases, and similarly as x decreases, y decreases. Another type of variation is *inverse variation*.

Inverse Variation

y varies inversely as x if there exists a real number k such that

$$y = \frac{k}{x},$$

or, equivalently,

$$xy = k.$$

Also, **y varies inversely as the nth power of x** if there exists a real number k such that

$$y = \frac{k}{x^n}.$$

Johann Kepler (1571–1630) established the importance of the ellipse in 1609, when he discovered that the orbits of the planets around the sun were elliptical, not circular. The orbits of the planets are nearly circular.

 Halley's comet, which has been studied since 467 B.C., has an elliptical orbit which is long and narrow, with one axis much longer than the other. This comet was named for the British astronomer and mathematician Edmund Halley (1656–1742), who predicted its return after observing it in 1682. The comet appears regularly every 76 years.

EXAMPLE 9 The weight of an object above the earth varies inversely as the square of its distance from the center of Earth. A space vehicle in an elliptical orbit has a maximum distance from the center of Earth (apogee) of 6700 miles. Its minimum distance from the center of Earth (perigee) is 4090 miles. See Figure 6 (not to scale). If an astronaut in the vehicle weighs 57 pounds at its apogee, what does the astronaut weigh at the perigee?

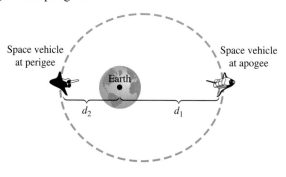

FIGURE 6

If w is the weight and d is the distance from the center of Earth, then

$$w = \frac{k}{d^2}$$

for some constant k. At the apogee the astronaut weighs 57 pounds and the distance from the center of Earth is 6700 miles. Use these values to find k.

$$57 = \frac{k}{(6700)^2} \quad \text{Let } w = 57 \text{ and } d = 6700.$$

$$k = 57(6700)^2$$

Then the weight at the perigee with $d = 4090$ miles is

$$w = \frac{57(6700)^2}{(4090)^2} \approx 153 \text{ pounds.} \quad \text{Use a calculator.}$$

It is common for one variable to depend on several others. For example, if one variable varies as the product of several other variables (perhaps raised to powers), the first variable is said to **vary jointly** as the others.

EXAMPLE 10 The strength of a rectangular beam varies jointly as its width and the square of its depth. If the strength of a beam 2 inches wide by 10 inches deep is 1000 pounds per square inch, what is the strength of a beam 4 inches wide and 8 inches deep?

If S represents the strength, w the width, and d the depth, then

$$S = kwd^2$$

for some constant k. Since $S = 1000$ if $w = 2$ and $d = 10$,

$$1000 = k(2)(10)^2. \quad \text{Let } S = 1000, w = 2, \text{ and } d = 10.$$

Solving this equation for k gives

$$1000 = k \cdot 2 \cdot 100$$
$$1000 = 200k$$
$$k = 5,$$

so
$$S = 5wd^2.$$

Find S when $w = 4$ and $d = 8$ by substitution in $S = 5wd^2$.

$$S = 5(4)(8)^2 = 1280 \qquad \text{Let } w = 4 \text{ and } d = 8.$$

The strength of the beam is 1280 pounds per square inch.

There are many combinations of direct and inverse variation. The final example shows a typical **combined variation** problem.

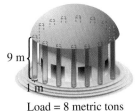

Load = 8 metric tons

FIGURE 7

EXAMPLE 11 The maximum load that a cylindrical column with a circular cross section can hold varies directly as the fourth power of the diameter of the cross section and inversely as the square of the height. A 9-meter column 1 meter in diameter will support 8 metric tons. See Figure 7. How many metric tons can be supported by a column 12 meters high and 2/3 meter in diameter?

Let L represent the load, d the diameter, and h the height. Then

$$L = \frac{kd^4}{h^2}. \qquad \begin{array}{l}\leftarrow \text{Load varies directly as the 4th power of the diameter.} \\ \leftarrow \text{Load varies inversely as the square of the height.}\end{array}$$

Now find k. Let $h = 9$, $d = 1$, and $L = 8$.

$$8 = \frac{k(1)^4}{9^2} \qquad h = 9, d = 1, L = 8$$

$$8 = \frac{k}{81}$$

$$k = 648$$

Substitute 648 for k in the first equation.

$$L = \frac{648d^4}{h^2}$$

Now find L when $h = 12$ and $d = 2/3$.

$$L = \frac{648\left(\frac{2}{3}\right)^4}{12^2} = \frac{648\left(\frac{16}{81}\right)}{144} = 648 \cdot \frac{16}{81} \cdot \frac{1}{144} = \frac{8}{9} \qquad \text{Let } h = 12, d = \frac{2}{3}.$$

The maximum load is 8/9 metric ton.

7.3 EXERCISES

Determine the ratio and write it in lowest terms.

1. 25 feet to 40 feet
2. 16 miles to 48 miles
3. 18 dollars to 72 dollars
4. 300 people to 250 people
5. 144 inches to 6 feet
6. 60 inches to 2 yards
7. 5 days to 40 hours
8. 75 minutes to 2 hours
9. Which one of the following ratios is not the same as the ratio 2 to 5?
   A. .4   B. 4 to 10   C. 20 to 50   D. 5 to 2
10. Give three ratios that are equivalent to the ratio 4 to 3.
11. Explain the distinction between *ratio* and *proportion.* Give examples.
12. Suppose that someone told you to use cross products in order to multiply fractions. How would you explain to the person what is wrong with his or her thinking?

Decide whether each proportion is true *or* false.

13. $\dfrac{5}{35} = \dfrac{8}{56}$
14. $\dfrac{4}{12} = \dfrac{7}{21}$
15. $\dfrac{120}{82} = \dfrac{7}{10}$
16. $\dfrac{27}{160} = \dfrac{18}{110}$
17. $\dfrac{\frac{1}{2}}{5} = \dfrac{1}{10}$
18. $\dfrac{\frac{1}{3}}{6} = \dfrac{1}{18}$

Solve each equation.

19. $\dfrac{k}{4} = \dfrac{175}{20}$
20. $\dfrac{49}{56} = \dfrac{z}{8}$
21. $\dfrac{x}{6} = \dfrac{18}{4}$
22. $\dfrac{z}{80} = \dfrac{20}{100}$
23. $\dfrac{3x-2}{5} = \dfrac{6x-5}{11}$
24. $\dfrac{2p+7}{3} = \dfrac{p-1}{4}$

Solve each problem by setting up and solving a proportion.

25. *Chain Saw Oil Mixture* A chain saw requires a mixture of 2-cycle engine oil and gasoline. According to the directions on a bottle of Oregon 2-cycle Engine Oil, for a 50 to 1 ratio requirement, approximately 2.5 fluid ounces of oil are required for 1 gallon of gasoline. For 2.75 gallons, how many fluid ounces of oil are required?

26. *Chain Saw Oil Mixture* The directions on the bottle mentioned in Exercise 25 indicate that if the ratio requirement is 24 to 1, approximately 5.5 ounces of oil are required for 1 gallon of gasoline. If gasoline is to be mixed with 22 ounces of oil, how much gasoline is to be used?

27. *Exchange Rate (U.S. and U.K.)* In a recent year, the average exchange rate between U.S. dollars and United Kingdom pounds was 1 pound to $1.6762. Margaret went to London and exchanged her U.S. currency for U.K. pounds, and received 400 pounds. How much in U.S. money did Margaret exchange? (*Note:* Great Britain and Switzerland are among the European nations not using euros.)

28. *Exchange Rate (U.S. and Switzerland)* If 3 U.S. dollars can be exchanged for 4.5204 Swiss francs, how many Swiss francs can be obtained for $49.20? (Round to the nearest hundredth.)

29. *Cost to Fill a Gas Tank* If 6 gallons of premium unleaded gasoline cost $3.72, how much would it cost to completely fill a 15-gallon tank?

30. *Sales Tax on a Purchase* If sales tax on a $16.00 video disc is $1.32, how much would the sales tax be on a $120.00 DVD player?

31. *Distance Between Cities* The distance between Kansas City, Missouri, and Denver is 600 miles. On a certain wall map, this is represented by a length of 2.4 feet. On the map, how many feet would there be between Memphis and Philadelphia, two cities that are actually 1000 miles apart?

32. *Distance Between Cities* The distance between Singapore and Tokyo is 3300 miles. On a certain wall map, this distance is represented by 11 inches. The actual distance between Mexico City and Cairo is 7700 miles. How far apart are they on the same map?

33. *Tagging Fish for a Population Estimate* Biologists tagged 250 fish in False River on October 5. On a later date they found 7 tagged fish in a sample of 350. Estimate the total number of fish in False River to the nearest hundred.

34. *Tagging Fish for a Population Estimate* On May 13 researchers at Argyle Lake tagged 420 fish. When they returned a few weeks later, their sample of 500 fish contained 9 that were tagged. Give an approximation of the fish population in Argyle Lake to the nearest hundred.

Olympic Funding The Olympic Committee has come to rely more and more on television rights and major corporate sponsors to finance the games. The circle graphs show the funding plans for the first Olympics in Athens and the 1996 Olympics in Atlanta, 100 years later. Use proportions and the figures to answer the questions in Exercises 35 and 36.

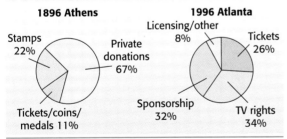

OLYMPIC GAMES FUNDING

1896 Athens: Stamps 22%, Private donations 67%, Tickets/coins/medals 11%

1996 Atlanta: Licensing/other 8%, Tickets 26%, Sponsorship 32%, TV rights 34%

Source: International Olympic Committee.

35. In the 1996 Olympics, total revenue of $350 million was raised. There were 10 major sponsors.
  (a) Write a proportion to find the amount of revenue provided by tickets. Solve it.
  (b) What amount was provided by sponsors? Assuming the sponsors contributed equally, how much was provided per sponsor?
  (c) What amount was raised by TV rights?

36. Suppose the amount of revenue raised in the 1896 Olympics was equivalent to the $350 million in 1996.
  (a) Write a proportion for the amount of revenue provided by stamps and solve it.
  (b) What amount (in dollars) would have been provided by private donations?

Merchandise Pricing A supermarket was surveyed to find the prices charged for items in various sizes. Find the best buy (based on price per unit) for each particular item.

37. Trash bags
  20 count: $3.09
  30 count: $4.59

38. Black pepper
  1-ounce size: $ .99
  2-ounce size: $1.65
  4-ounce size: $4.39

39. Breakfast cereal
  15-ounce size: $2.99
  25-ounce size: $4.49
  31-ounce size: $5.49

40. Cocoa mix
  8-ounce size: $1.39
  16-ounce size: $2.19
  32-ounce size: $2.99

41. Tomato ketchup
  14-ounce size: $ .89
  32-ounce size: $1.19
  64-ounce size: $2.95

42. Cut green beans
  8-ounce size: $ .45
  16-ounce size: $ .49
  50-ounce size: $1.59

*Two triangles are **similar** if they have the same shape (but not necessarily the same size). Similar triangles have sides that are proportional. The figure shows two similar triangles.*

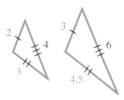

Notice that the ratios of the corresponding sides are all equal to $\frac{3}{2}$:

$$\frac{3}{2} = \frac{3}{2} \qquad \frac{4.5}{3} = \frac{3}{2} \qquad \frac{6}{4} = \frac{3}{2}.$$

If we know that two triangles are similar, we can set up a proportion to solve for the length of an unknown side.

Use a proportion to find the length x, given that the pair of triangles are similar.

43. 44.

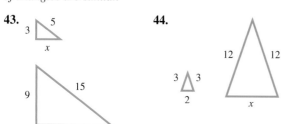

45. 46.

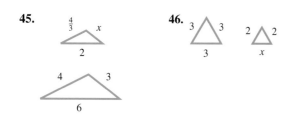

For the problems in Exercises 47 and 48, (a) draw a sketch consisting of two right triangles, depicting the situation described, and (b) solve the problem. (Source: The Guinness Book of World Records.)

47. *George Washington's Chair* An enlarged version of the chair used by George Washington at the Constitutional Convention casts a shadow 18 feet long at the same time a vertical pole 12 feet high casts a shadow 4 feet long. How tall is the chair?

48. *Candle at an Exhibition* One of the tallest candles ever constructed was exhibited at the 1897 Stockholm Exhibition. If it cast a shadow 5 feet long at the same time a vertical pole 32 feet high cast a shadow 2 feet long, how tall was the candle?

Consumer Price Index The Consumer Price Index, issued by the U.S. Bureau of Labor Statistics, provides a means of determining the purchasing power of the U.S. dollar from one year to the next. Using the period from 1982 to 1984 as a measure of 100.0, the Consumer Price Index for selected years from 1990 to 2000 is shown here.

Year	Consumer Price Index
1990	130.7
1992	140.3
1994	148.2
1996	156.9
1998	163.0
1999	166.6
2000	172.2

Source: Bureau of Labor Statistics.

To use the Consumer Price Index to predict a price in a particular year, we can set up a proportion and compare it with a known price in another year, as follows:

$$\frac{\text{Price in year } A}{\text{Index in year } A} = \frac{\text{Price in year } B}{\text{Index in year } B}.$$

Use the Consumer Price Index figures in the table to find the amount that would be charged for the use of the same amount of electricity that cost $225 in 1990. Give your answer to the nearest dollar.

49. in 1992   50. in 1994
51. in 1996   52. in 1998
53. in 1999   54. in 2000

Solve each problem involving variation.

55. If x varies directly as y, and $x = 27$ when $y = 6$, find x when $y = 2$.

56. If z varies directly as x, and $z = 30$ when $x = 8$, find z when $x = 4$.

57. If m varies directly as p^2, and $m = 20$ when $p = 2$, find m when $p = 5$.

58. If a varies directly as b^2, and $a = 48$ when $b = 4$, find a when $b = 7$.

59. If p varies inversely as q^2, and $p = 4$ when $q = \frac{1}{2}$, find p when $q = \frac{3}{2}$.

60. If z varies inversely as x^2, and $z = 9$ when $x = \frac{2}{3}$, find z when $x = \frac{5}{4}$.

61. *Interest on an Investment* The interest on an investment varies directly as the rate of interest. If the interest is $48 when the interest rate is 5%, find the interest when the rate is 4.2%.

62. *Area of a Triangle* For a given base, the area of a triangle varies directly as its height. Find the area of a triangle with a height of 6 inches, if the area is 10 square inches when the height is 4 inches.

63. *Speed of a Car* Over a specified distance, speed varies inversely with time. If a car goes a certain distance in one-half hour at 30 miles per hour, what speed is needed to go the same distance in three-fourths of an hour?

64. *Length of a Rectangle* For a constant area, the length of a rectangle varies inversely as the width. The length of a rectangle is 27 feet when the width is 10 feet. Find the length of a rectangle with the same area if the width is 18 feet.

65. *Weight of a Moose* The weight of an object on the moon varies directly as the weight of the object on Earth. According to *The Guinness Book of World Records,* "Shad," a goat owned by a couple in California, is the largest known goat, weighing 352 pounds. Shad would weigh about 59 pounds on the moon. A bull moose weighing 1800 pounds was shot in Canada and is the largest confirmed moose. How much would the moose have weighed on the moon?

66. *Voyage in a Paddle Boat* According to *The Guinness Book of World Records,* the longest recorded voyage in a paddle boat is 2226 miles in 103 days by the foot power of two boaters down the Mississippi River. Assuming a constant rate, how far would they have gone if they had traveled 120 days? (Distance varies directly as time.)

67. *Pressure Exerted by a Liquid* The pressure exerted by a certain liquid at a given point varies directly as the depth of the point beneath the surface of the liquid. The pressure at a depth of 10 feet is 50 pounds per square inch. What is the pressure at a depth of 20 feet?

68. *Pressure of a Gas in a Container* If the volume is constant, the pressure of a gas in a container varies directly as the temperature. If the pressure is 5 pounds per square inch at a temperature of 200 degrees Kelvin, what is the pressure at a temperature of 300 degrees Kelvin?

69. *Pressure of a Gas in a Container* If the temperature is constant, the pressure of a gas in a container varies inversely as the volume of the container. If the pressure is 10 pounds per square foot in a container with 3 cubic feet, what is the pressure in a container with 1.5 cubic feet?

70. *Force Required to Compress a Spring* The force required to compress a spring varies directly as the change in the length of the spring. If a force of 12 pounds is required to compress a certain spring 3 inches, how much force is required to compress the spring 5 inches?

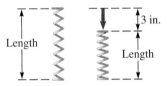

71. *Falling Body* For a body falling freely from rest (disregarding air resistance), the distance the body falls varies directly as the square of the time. If an object is dropped from the top of a tower 400 feet high and hits the ground in 5 seconds, how far did it fall in the first 3 seconds?

72. *Illumination from a Light Source* The illumination produced by a light source varies inversely as the square of the distance from the source. If the illumination produced 4 feet from a light source is 75 foot-candles, find the illumination produced 9 feet from the same source.

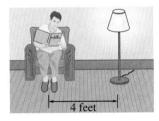

73. *Body Mass Index* Body mass index, or BMI, is used by physicians to assess a person's level of fatness. A BMI from 19 through 25 is considered desirable. BMI varies directly as an individual's weight in pounds and inversely as the square of the individual's height in inches. A person who weighs 118 pounds and is 64 inches tall has a BMI of 20. (The BMI is rounded to the nearest whole number.) Find the BMI of a person who weighs 165 pounds with a height of 70 inches.

74. *Skidding Car* The force needed to keep a car from skidding on a curve varies inversely as the radius of the curve and jointly as the weight of the car and the square of the speed. If 242 pounds of force keep a 2000-pound car from skidding on a curve of radius 500 feet at 30 miles per hour, what force would keep the same car from skidding on a curve of radius 750 feet at 50 miles per hour?

Fish Weight-Estimation Exercises 75 and 76 describe weight-estimation formulas that fishermen have used over the years. Girth *is the distance around the body of the fish.* (Source: Sacramento Bee, November 9, 2000.)

75. The weight of a bass varies jointly as its girth and the square of its length. A prize-winning bass weighed in at 22.7 pounds and measured 36 inches long with a 21-inch girth. How much would a bass 28 inches long with an 18-inch girth weigh?

76. The weight of a trout varies jointly as its length and the square of its girth. One angler caught a trout that weighed 10.5 pounds and measured 26 inches long with an 18-inch girth. Find the weight of a trout that is 22 inches long with a 15-inch girth.

7.4 Linear Inequalities

A *linear inequality in one variable* is an inequality such as

$$x + 5 < 2, \quad y - 3 \geq 5, \quad \text{or} \quad 2k + 5 \leq 10.$$

Archimedes, one of the greatest mathematicians of antiquity, is shown on this Italian stamp. He was born in the Greek city of Syracuse about 287 B.C.

A colorful story about Archimedes relates his reaction to one of his discoveries. While taking a bath, he noticed that an immersed object, if heavier than a fluid, "will, if placed in it, descend to the bottom of the fluid, and the solid will, when weighed in the fluid, be lighter than its true weight by the weight of the fluid displaced." This discovery so excited him that he ran through the streets shouting "Eureka!" ("I have found it!") without bothering to clothe himself!

Archimedes met his death at age 75 during the pillage of Syracuse. He was using a sand tray to draw geometric figures when a Roman soldier came upon him. He ordered the soldier to move clear of his "circles," and the soldier obliged by killing him.

Linear Inequality in One Variable

A **linear inequality in one variable** can be written in the form

$$ax + b < c,$$

where a, b, and c are real numbers, with $a \neq 0$.

(Throughout this section we give the definitions and rules only for $<$, but they are also valid for $>$, $\leq$, and $\geq$.)

Solving Linear Inequalities An inequality is solved by finding all numbers that make the inequality true. Usually an inequality has an infinite number of solutions. These solutions, like the solutions of equations, are found by producing a series of simpler equivalent inequalities. **Equivalent inequalities** are inequalities with the same solution set. Such inequalities are found with the addition and multiplication properties of inequality.

Addition Property of Inequality

For all real numbers a, b, and c, the inequalities

$$a < b \quad \text{and} \quad a + c < b + c$$

are equivalent. (The same number may be added to both sides of an inequality without changing the solution set.)

As with equations, the addition property can be used to *subtract* the same number from both sides of an inequality.

EXAMPLE 1 Solve $x - 7 < -12$.

$$x - 7 < -12$$
$$x - 7 + 7 < -12 + 7 \quad \text{Add 7 to both sides.}$$
$$x < -5$$

Using set-builder notation, the solution set of this inequality is written $\{x \mid x < -5\}$.

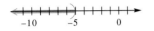

FIGURE 8

It is customary to express solution sets of inequalities with number line graphs. The solution set in Example 1 can be shown on a number line as in Figure 8. The set of numbers less than -5 is an example of an **interval** on the number line. To write intervals, we use **interval notation.** For example, using this notation, the interval of all numbers less than -5 is written $(-\infty, -5)$. The negative infinity symbol $-\infty$ does not indicate a number. It is used to show that the interval includes all real numbers less than -5. As on the number line, the parenthesis indicates that -5 is not included in the solution set.

Examples of other sets written in interval notation are shown in the following chart. In these intervals, assume that $a < b$. Note that a parenthesis is always used with the symbols $-\infty$ and ∞.

Type of Interval	Set	Interval Notation	Graph
	$\{x \mid a < x\}$	(a, ∞)	
Open interval	$\{x \mid a < x < b\}$	(a, b)	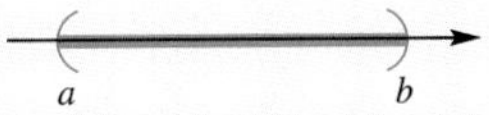
	$\{x \mid x < b\}$	$(-\infty, b)$	
	$\{x \mid a \leq x\}$	$[a, \infty)$	
Half-open interval	$\{x \mid a < x \leq b\}$	$(a, b]$	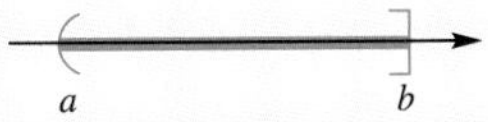
	$\{x \mid a \leq x < b\}$	$[a, b)$	
	$\{x \mid x \leq b\}$	$(-\infty, b]$	
Closed interval	$\{x \mid a \leq x \leq b\}$	$[a, b]$	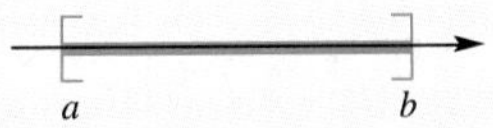

We sometimes use $(-\infty, \infty)$ to represent the set of all real numbers.

EXAMPLE 2 Solve the inequality $14 + 2m \leq 3m$. Give the solution set in interval form, and graph the solution set as well.

$$14 + 2m \leq 3m$$
$$14 + 2m - 2m \leq 3m - 2m \quad \text{Subtract } 2m.$$
$$14 \leq m \quad \text{Combine like terms.}$$

The inequality $14 \leq m$ (14 is less than or equal to m) can also be written $m \geq 14$ (m is greater than or equal to 14). Notice that in each case, the inequality symbol points to the smaller number, 14. The solution set in interval notation is $[14, \infty)$. The graph is shown in Figure 9.

FIGURE 9

Errors often occur in graphing inequalities where the variable term is on the right side. (This is probably due to the fact that we read from left to right.) To guard against such errors, it is a good idea to rewrite these inequalities so that the variable is on the left, as discussed in Example 2.

An inequality such as $3x \leq 15$ can be solved by dividing both sides by 3. This is done with the multiplication property of inequality, which is a little more involved than the corresponding property for equations. To see how this property works, start with the true statement

$$-2 < 5.$$

Multiply both sides by, say, 8.

$$-2(8) < 5(8) \quad \text{Multiply by 8.}$$
$$-16 < 40 \quad \text{True}$$

This gives a true statement. Start again with $-2 < 5$, and this time multiply both sides by -8.

$$-2(-8) < 5(-8) \quad \text{Multiply by } -8.$$
$$16 < -40 \quad \text{False}$$

The result, $16 < -40$, is false. To make it true, change the direction of the inequality symbol to get

$$16 > -40. \quad \text{True}$$

As these examples suggest, multiplying both sides of an inequality by a *negative* number forces the direction of the inequality symbol to be reversed. The same is true for dividing by a negative number, since division is defined in terms of multiplication.

Multiplication Property of Inequality

For all real numbers a, b, and c, with $c \neq 0$,

(a) the inequalities
$$a < b \quad \text{and} \quad ac < bc$$
are equivalent if $c > 0$;

(b) the inequalities
$$a < b \quad \text{and} \quad ac > bc$$
are equivalent if $c < 0$. (Both sides of an inequality may be multiplied or divided by a *positive* number without changing the direction of the inequality symbol. **Multiplying or dividing by a *negative* number requires that the inequality symbol be reversed.**)

EXAMPLE 3 Solve $-3(x + 4) + 2 \geq 8 - x$. Give the solution set in both interval and graph forms.

$$-3(x + 4) + 2 \geq 8 - x$$
$$-3x - 12 + 2 \geq 8 - x \qquad \text{Distributive property}$$
$$-3x - 10 \geq 8 - x$$
$$-3x - 10 + x \geq 8 - x + x \qquad \text{Add } x.$$
$$-2x - 10 \geq 8$$
$$-2x - 10 + 10 \geq 8 + 10 \qquad \text{Add 10.}$$
$$-2x \geq 18$$

Divide both sides by -2. Be careful. Dividing by a negative number requires changing $\geq$ to $\leq$.

FIGURE 10

$$\frac{-2x}{-2} \leq \frac{18}{-2} \qquad \text{Divide by } -2; \text{ reverse inequality symbol.}$$
$$x \leq -9$$

Figure 10 shows the graph of the solution set, $(-\infty, -9]$.

Most linear inequalities, such as the one in Example 3, require the use of both the addition and multiplication properties. The steps used in solving a linear inequality are summarized below.

Solving a Linear Inequality

Step 1: Simplify each side of the inequality as much as possible by using the distributive property to clear parentheses and by combining like terms as needed.

Step 2: Use the addition property of inequality to change the inequality so that all terms with variables are on one side and all terms without variables are on the other side.

Step 3: Use the multiplication property to change the inequality to the form $x < k$ (or $x > k$, or $x \leq k$, or $x \geq k$).

Remember: Reverse the direction of the inequality symbol only when multiplying or dividing both sides of an inequality by a negative number.

Three-part Inequalities Inequalities can be used to express the fact that a quantity lies between two other quantities. For example,

$$2 < 4 < 7$$

says that $2 < 4$ *and* $4 < 7$. This statement is a *conjunction* and is true because both parts of the statement are true. The three-part inequality

$$3 < x + 2 < 8$$

is true when the expression $x + 2$ is between 3 and 8. To solve this inequality, subtract 2 from each of the three parts of the inequality, giving

$$3 - 2 < x + 2 - 2 < 8 - 2$$
$$1 < x < 6.$$

FIGURE 11

The solution set, $(1, 6)$, is graphed in Figure 11.

When using inequalities with three parts such as the one above, it is important to have the numbers in the correct positions. It would be *wrong* to write the inequality as $8 < x + 2 < 3$, since this would imply that $8 < 3$, a false statement. In general, three-part inequalities are written so that the symbols point in the same direction, and they both point toward the smaller number.

EXAMPLE 4 Solve $-2 \leq 3k - 1 \leq 5$ and graph the solution set.

Add 1 to each of the three parts to isolate the variable term in the middle.

$$-2 + 1 \leq 3k - 1 + 1 \leq 5 + 1 \qquad \text{Add 1.}$$
$$-1 \leq 3k \leq 6$$
$$\frac{-1}{3} \leq \frac{3k}{3} \leq \frac{6}{3} \qquad \text{Divide by 3.}$$
$$-\frac{1}{3} \leq k \leq 2$$

FIGURE 12

A graph of the solution set, $[-1/3, 2]$, is shown in Figure 12.

Applications In addition to the familiar "is less than" and "is greater than," the expressions "is no more than," "is at least," and others also indicate inequalities. The table below shows how these expressions are interpreted.

Word Expression	Interpretation	Word Expression	Interpretation
a is at least b	$a \geq b$	a is at most b	$a \leq b$
a is no less than b	$a \geq b$	a is no more than b	$a \leq b$

Do not confuse a statement such as "5 is more than a number" with the phrase "5 more than a number." The first of these is a sentence, written as "$5 > x$," while the second is an expression, written with addition as "$x + 5$."

Problem Solving

The next example shows an application of algebra that is important to anyone who has ever asked himself or herself, "What score can I make on my next test and have a (particular grade) in this course?" It uses the idea of finding the average of a number of grades. In general, to find the average of n numbers, add the numbers, and divide by n.

EXAMPLE 5 Christopher Michael has test scores of 86, 88, and 78 on his first three tests in geometry. If he wants an average of at least 80 after his fourth test, what are the possible scores he can make on his fourth test?

Let $x =$ Christopher's score on his fourth test. To find his average after 4 tests, add the test scores and divide by 4.

$$\frac{86 + 88 + 78 + x}{4} \geq 80$$

Average is at least 80.

$$\frac{252 + x}{4} \geq 80 \qquad \text{Add the known scores.}$$

$$4\left(\frac{252 + x}{4}\right) \geq 4(80) \qquad \text{Multiply by 4.}$$

$$252 + x \geq 320$$

$$252 - 252 + x \geq 320 - 252 \qquad \text{Subtract 252.}$$

$$x \geq 68 \qquad \text{Combine terms.}$$

He must score 68 or more on the fourth test to have an average of *at least* 80.

EXAMPLE 6 A rental company charges $15.00 to rent a chain saw, plus $2.00 per hour. Mamie Zwettler can spend no more than $35.00 to clear some logs from her yard. What is the maximum amount of time she can use the rented saw?

Let $h =$ the number of hours she can rent the saw. She must pay $15.00, plus $2.00h$, to rent the saw for h hours, and this amount must be *no more than* $35.00.

Cost of renting is no more than 35 dollars.

$$15 + 2h \leq 35$$

$$15 + 2h - 15 \leq 35 - 15 \qquad \text{Subtract 15.}$$

$$2h \leq 20$$

$$h \leq 10 \qquad \text{Divide by 2.}$$

Mamie can use the saw for a maximum of 10 hours. (Of course, she may use it for less time, as indicated by the inequality $h \leq 10$.)

7.4 EXERCISES

Match each inequality in Exercises 1–6 with the correct graph or interval notation in A.–F. See Example 1.

1. $\{x \mid x \leq 3\}$
2. $\{x \mid x > 3\}$
3. $\{x \mid x < 3\}$
4. $\{x \mid x \geq 3\}$
5. $\{x \mid -3 \leq x \leq 3\}$
6. $\{x \mid -3 < x < 3\}$

A. ├─┼─┼─┼─┼─┤
     0   3

B. ←┼─┼─┼─┼─┼→
     0   3

C. $(3, \infty)$ D. $(-\infty, 3]$

E. $(-3, 3)$ F. $[-3, 3]$

7. Explain how to determine whether to use parentheses or brackets when graphing the solution set of an inequality.

8. Describe the steps used to solve a linear inequality. Explain when it is necessary to reverse the inequality symbol.

Solve each inequality. Give the solution set in both interval and graph forms.

9. $4x + 1 \geq 21$
10. $5t + 2 \geq 52$
11. $\dfrac{3k - 1}{4} > 5$
12. $\dfrac{5z - 6}{8} < 8$
13. $-4x < 16$
14. $-2m > 10$
15. $-\dfrac{3}{4}r \geq 30$
16. $-1.5x \leq -\dfrac{9}{2}$
17. $-1.3m \geq -5.2$
18. $-2.5x \leq -1.25$
19. $\dfrac{2k - 5}{-4} > 5$
20. $\dfrac{3z - 2}{-5} < 6$
21. $x + 4(2x - 1) \geq x$
22. $m - 2(m - 4) \leq 3m$
23. $-(4 + r) + 2 - 3r < -14$
24. $-(9 + k) - 5 + 4k \geq 4$
25. $-3(z - 6) > 2z - 2$
26. $-2(x + 4) \leq 6x + 16$
27. $\dfrac{2}{3}(3k - 1) \geq \dfrac{3}{2}(2k - 3)$
28. $\dfrac{7}{5}(10m - 1) < \dfrac{2}{3}(6m + 5)$
29. $-\dfrac{1}{4}(p + 6) + \dfrac{3}{2}(2p - 5) < 10$
30. $\dfrac{3}{5}(k - 2) - \dfrac{1}{4}(2k - 7) \leq 3$
31. $3(2x - 4) - 4x < 2x + 3$
32. $7(4 - x) + 5x < 2(16 - x)$
33. $8\left(\dfrac{1}{2}x + 3\right) < 8\left(\dfrac{1}{2}x - 1\right)$
34. $10x + 2(x - 4) < 12x - 10$

35. A student solved the inequality $5x < -20$ by dividing both sides by 5 and reversing the direction of the inequality symbol. His reasoning was that since -20 is a negative number, reversing the direction of the symbol was required. Is this correct? Explain why or why not.

36. Match each set given in interval notation with its description.
 (a) $(0, \infty)$ A. positive real numbers
 (b) $[0, \infty)$ B. negative real numbers
 (c) $(-\infty, 0]$ C. nonpositive real numbers
 (d) $(-\infty, 0)$ D. nonnegative real numbers

Solve each inequality. Give the solution set in both interval and graph forms.

37. $-4 < x - 5 < 6$
38. $-1 < x + 1 < 8$
39. $-9 \leq k + 5 \leq 15$
40. $-4 \leq m + 3 \leq 10$
41. $-6 \leq 2z + 4 \leq 16$
42. $-15 < 3p + 6 < -12$
43. $-19 \leq 3x - 5 \leq 1$
44. $-16 < 3t + 2 < -10$
45. $-1 \leq \dfrac{2x - 5}{6} \leq 5$
46. $-3 \leq \dfrac{3m + 1}{4} \leq 3$
47. $4 \leq 5 - 9x < 8$
48. $4 \leq 3 - 2x < 8$

Tornado Activity In Exercises 49–52, answer the questions based on the following graph.

Months In Which Most Tornadoes Strike

December 2.5%
November 3.6%
October 3.0%
September 4.8%
August 7.7%
July 11.1%
June 20.7%
May 22.1%
April 12.9%
March 7.0%
February 2.8%
January 1.8%

Source: The USA Today Weather Book

49. In which months did the percent of tornadoes exceed 7.7%?

50. In which months was the percent of tornadoes at least 12.9%?

51. The data used to determine the graph were based on the number of tornadoes sighted in the United States during a twenty-year period. A total of 17,252 tornadoes were reported. In which months were fewer than 1500 reported?

52. How many more tornadoes occurred during March than October? (Use the total given in Exercise 51.)

Olympic Temperature Preferences The July 14th weather forecast by time of day for the 2000 U.S. Olympic Track and Field Trials, held July 14–23, 2000, in Sacramento, California, is shown in the figure. Use this graph to work Exercises 53–56.

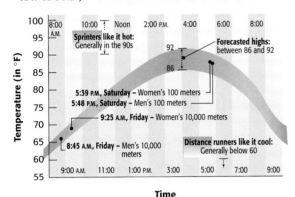

Source: Accuweather, Bcc research.

53. Sprinters prefer Fahrenheit temperatures in the 90s. Using the upper boundary of the forecast, in what time period is the temperature expected to be at least 90° F?

54. Distance runners prefer cool temperatures. During what time period are temperatures predicted to be no more than 70° F? Use the lower forecast boundary.

55. What range of temperatures is predicted for the Women's 100-meter event?

56. What range of temperatures is forecast for the Men's 10,000-meter event?

Solve each problem.

57. *Taxicab Fare* In a midwestern city, taxicabs charge $1.50 for the first $\frac{1}{5}$ mile and $.25 for each additional $\frac{1}{5}$ mile. Dantrell Davis has only $3.75 in his pocket. What is the maximum distance he can travel (not including a tip for the cabbie)?

58. *Taxicab Fare* Twelve years ago taxicab fares in the city in Exercise 57 were $.90 for the first $\frac{1}{7}$ mile and $.10 for each additional $\frac{1}{7}$ mile. Based on the information given there and the answer you found, how much farther could Dantrell Davis have traveled at that time?

59. *Grade Average* Margaret Westmoreland earned scores of 90 and 82 on her first two tests in English Literature. What score must she make on her third test to keep an average of 84 or greater?

60. *Grade Average* Susan Carsten scored 92 and 96 on her first two tests in Methods in Teaching Art. What score must she make on her third test to keep an average of 90 or greater?

61. *Car Rental* A couple wishes to rent a car for one day while on vacation. Avis wants $35.00 per day and 14¢ per mile, while Downtown Toyota wants $34.00 per day and 16¢ per mile. After how many miles would the price to rent from Downtown Toyota exceed the price to rent from Avis?

62. *Car Rental* Jane and Terry Brandsma went to Mobile, Alabama, for a week. They needed to rent a car, so they checked out two rental firms. Avis wanted $28 per day, with no mileage fee. Downtown Toyota wanted $108 per week and 14¢ per mile. How many miles would they have to drive before the Avis price is less than the Toyota price?

63. *Body Mass Index* A BMI (body mass index) between 19 and 25 is considered healthy. Use the formula

$$\text{BMI} = \frac{704 \times (\text{weight in pounds})}{(\text{height in inches})^2}$$

to find the weight range w, to the nearest pound, that gives a healthy BMI for each height. (*Source:* Washington Post.)

(a) 72 inches   (b) Your height in inches

64. *Target Heart Rate* To achieve the maximum benefit from exercising, the heart rate in beats per minute should be in the target heart rate zone (THR). For a person aged A, the formula is

$$.7(220 - A) \leq \text{THR} \leq .85(220 - A).$$

Find the THR to the nearest whole number for each age. (*Source:* Hockey, Robert V., *Physical Fitness: The Pathway to Healthful Living,* Times Mirror/Mosby College Publishing, 1989.)
(a) 35   **(b)** Your age

Profit/Cost Analysis A product will produce a profit only when the revenue R from selling the product exceeds the cost C of producing it. In Exercises 65 and 66 find the smallest whole number of units x that must be sold for the business to show a profit for the item described.

65. Peripheral Visions, Inc. finds that the cost to produce x studio quality videotapes is $C = 20x + 100$, while the revenue produced from them is $R = 24x$ (C and R in dollars).

66. Speedy Delivery finds that the cost to make x deliveries is $C = 3x + 2300$, while the revenue produced from them is $R = 5.50x$ (C and R in dollars).

7.5 Properties of Exponents and Scientific Notation

Exponents *Exponents* are used to write products of repeated factors. For example, the product $3 \cdot 3 \cdot 3 \cdot 3$ is written

$$\underbrace{3 \cdot 3 \cdot 3 \cdot 3}_{4 \text{ factors of } 3} = 3^{\overset{\leftarrow \text{Exponent}}{4}}_{\underset{\text{Base}}{\uparrow}}.$$

The number 4 shows that 3 appears as a factor four times. The number 4 is the **exponent** and 3 is the **base**. The quantity 3^4 is called an **exponential expression.** Read 3^4 as "3 to the fourth power," or "3 to the fourth." Multiplying out the four 3s gives

$$3^4 = 3 \cdot 3 \cdot 3 \cdot 3 = 81.$$

$10^{10000\cdots 0}$

The term **googol**, meaning 10^{100}, was coined by Professor Edward Kasner of Columbia University. A googol is made up of a 1 with one hundred zeros following it. This number exceeds the estimated number of electrons in the universe, which is 10^{79}.
  The Web search engine Google is named after a googol. Sergey Brin, president and cofounder of Google, Inc., was a math major. He chose the name Google to describe the vast reach of this search engine. (*Source: The Gazette,* March 2, 2001.)
  If a googol isn't big enough for you, try a **googolplex:**

$$\text{googolplex} = 10^{\text{googol}}.$$

Exponential Expression

If a is a real number and n is a natural number, then the exponential expression a^n is defined as

$$a^n = \underbrace{a \cdot a \cdot a \cdot \ldots \cdot a}_{n \text{ factors of } a}.$$

The number a is the *base* and n is the *exponent.*

EXAMPLE 1 Evaluate each exponential expression.

(a) $7^2 = 7 \cdot 7 = 49$ Read 7^2 as "7 squared."
(b) $5^3 = 5 \cdot 5 \cdot 5 = 125$ Read 5^3 as "5 cubed."
(c) $(-2)^4 = (-2)(-2)(-2)(-2) = 16$
(d) $(-2)^5 = (-2)(-2)(-2)(-2)(-2) = -32$
(e) $5^1 = 5$

In the exponential expression $3z^7$, the base of the exponent 7 is z, not $3z$. That is,

$$3z^7 = 3 \cdot z \cdot z \cdot z \cdot z \cdot z \cdot z \cdot z \quad \text{Base is } z.$$

while

$$(3z)^7 = (3z)(3z)(3z)(3z)(3z)(3z)(3z). \quad \text{Base is } 3z.$$

To evaluate $(-2)^6$, the parentheses around -2 indicate that the base is -2, so

$$(-2)^6 = (-2)(-2)(-2)(-2)(-2)(-2) = 64. \quad \text{Base is } -2.$$

In the expression -2^6, the base is 2, not -2. The $-$ sign tells us to find the negative, or additive inverse, of 2^6. It acts as a symbol for the factor -1.

$$-2^6 = -(2 \cdot 2 \cdot 2 \cdot 2 \cdot 2 \cdot 2) = -64 \quad \text{Base is } 2.$$

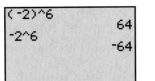

This screen supports the results preceding Example 2.

Therefore, since $64 \neq -64$, $(-2)^6 \neq -2^6$.

EXAMPLE 2 Evaluate each exponential expression.

(a) $-4^2 = -(4 \cdot 4) = -16$ (b) $-8^4 = -(8 \cdot 8 \cdot 8 \cdot 8) = -4096$

(c) $-2^4 = -(2 \cdot 2 \cdot 2 \cdot 2) = -16$

There are several useful rules that simplify work with exponents. For example, the product $2^5 \cdot 2^3$ can be simplified as follows.

$$\overset{5 + 3 = 8}{2^5 \cdot 2^3 = (2 \cdot 2 \cdot 2 \cdot 2 \cdot 2)(2 \cdot 2 \cdot 2) = 2^8}$$

This result—products of exponential expressions with the same base are found by adding exponents—is generalized as the **product rule for exponents.**

Product Rule for Exponents

If m and n are natural numbers and a is any real number, then

$$a^m \cdot a^n = a^{m+n}.$$

EXAMPLE 3 Apply the product rule for exponents in each case.

(a) $3^4 \cdot 3^7 = 3^{4+7} = 3^{11}$ (b) $5^3 \cdot 5 = 5^3 \cdot 5^1 = 5^{3+1} = 5^4$

(c) $y^3 \cdot y^8 \cdot y^2 = y^{3+8+2} = y^{13}$

(d) $(5y^2)(-3y^4) = 5(-3)y^2y^4$ Associative and commutative properties
$= -15y^{2+4}$
$= -15y^6$

(e) $(7p^3q)(2p^5q^2) = 7(2)p^3p^5qq^2 = 14p^8q^3$

So far we have discussed only positive exponents. How might we define a 0 exponent? Suppose we multiply 4^2 by 4^0. By the product rule,

$$4^2 \cdot 4^0 = 4^{2+0} = 4^2.$$

For the product rule to hold true, 4^0 must equal 1, and so we define a^0 this way for any nonzero real number a.

> **Zero Exponent**
>
> If a is any nonzero real number, then
> $$a^0 = 1.$$

The expression 0^0 **is undefined.***

EXAMPLE 4 Evaluate each expression.

(a) $12^0 = 1$
(b) $(-6)^0 = 1$ Base is -6.
(c) $-6^0 = -(6^0) = -1$ Base is 6.
(d) $5^0 + 12^0 = 1 + 1 = 2$
(e) $(8k)^0 = 1, \quad k \neq 0$

This screen supports the results in parts (b), (c), and (d) of Example 4.

How should we define a negative exponent? Using the product rule again,
$$8^2 \cdot 8^{-2} = 8^{2+(-2)} = 8^0 = 1.$$

This indicates that 8^{-2} is the reciprocal of 8^2. But $\dfrac{1}{8^2}$ is the reciprocal of 8^2, and a number can have only one reciprocal. Therefore, it is reasonable to conclude that $8^{-2} = \dfrac{1}{8^2}$. We can generalize and make the following definition.

> **Negative Exponent**
>
> For any natural number n and any nonzero real number a,
> $$a^{-n} = \frac{1}{a^n}.$$

With this definition, and the ones given earlier for positive and zero exponents, the expression a^n is meaningful for any integer exponent n and any nonzero real number a.

EXAMPLE 5 Write the following expressions with only positive exponents.

(a) $2^{-3} = \dfrac{1}{2^3} = \dfrac{1}{8}$
(b) $3^{-2} = \dfrac{1}{3^2} = \dfrac{1}{9}$
(c) $6^{-1} = \dfrac{1}{6^1} = \dfrac{1}{6}$
(d) $(5z)^{-3} = \dfrac{1}{(5z)^3}, \quad z \neq 0$

---

*In advanced treatments, 0^0 is called an *indeterminate form*.

(e) $5z^{-3} = 5\left(\dfrac{1}{z^3}\right) = \dfrac{5}{z^3}$, $z \neq 0$ (f) $(5z^2)^{-3} = \dfrac{1}{(5z^2)^3}$, $z \neq 0$

(g) $-m^{-2} = -\dfrac{1}{m^2}$, $m \neq 0$ (h) $(-m)^{-4} = \dfrac{1}{(-m)^4}$, $m \neq 0$

EXAMPLE 6   Evaluate each of the following expressions.

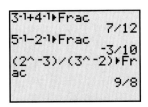

This screen supports the results in parts (a), (b), and (d) of Example 6.

(a) $3^{-1} + 4^{-1} = \dfrac{1}{3} + \dfrac{1}{4} = \dfrac{4}{12} + \dfrac{3}{12} = \dfrac{7}{12}$ $3^{-1} = \dfrac{1}{3}$ and $4^{-1} = \dfrac{1}{4}$

(b) $5^{-1} - 2^{-1} = \dfrac{1}{5} - \dfrac{1}{2} = \dfrac{2}{10} - \dfrac{5}{10} = -\dfrac{3}{10}$

(c) $\dfrac{1}{2^{-3}} = \dfrac{1}{\dfrac{1}{2^3}} = 1 \div \dfrac{1}{2^3} = 1 \cdot \dfrac{2^3}{1} = 2^3 = 8$

(d) $\dfrac{2^{-3}}{3^{-2}} = \dfrac{\dfrac{1}{2^3}}{\dfrac{1}{3^2}} = \dfrac{1}{2^3} \cdot \dfrac{3^2}{1} = \dfrac{3^2}{2^3} = \dfrac{9}{8}$

Parts (c) and (d) of Example 6 suggest the following generalizations.

Special Rules for Negative Exponents

If $a \neq 0$ and $b \neq 0$, then $\dfrac{1}{a^{-n}} = a^n$ and $\dfrac{a^{-n}}{b^{-m}} = \dfrac{b^m}{a^n}$.

A quotient, such as a^8/a^3, can be simplified in much the same way as a product. (In all quotients of this type, assume that the denominator is not 0.) Using the definition of an exponent,

$$\dfrac{a^8}{a^3} = \dfrac{a \cdot a \cdot a \cdot a \cdot a \cdot a \cdot a \cdot a}{a \cdot a \cdot a} = a \cdot a \cdot a \cdot a \cdot a = a^5.$$

Notice that $8 - 3 = 5$. In the same way,

$$\dfrac{a^3}{a^8} = \dfrac{a \cdot a \cdot a}{a \cdot a \cdot a \cdot a \cdot a \cdot a \cdot a \cdot a} = \dfrac{1}{a^5} = a^{-5}.$$

Here, $3 - 8 = -5$. These examples suggest the following **quotient rule for exponents.**

Quotient Rule for Exponents

If a is any nonzero real number and m and n are nonzero integers, then

$$\dfrac{a^m}{a^n} = a^{m-n}.$$

EXAMPLE 7 Apply the quotient rule for exponents in each case.

(a) $\dfrac{3^7}{3^2} = 3^{7-2} = 3^5$ (Numerator exponent, Denominator exponent, Minus sign)

(b) $\dfrac{p^6}{p^2} = p^{6-2} = p^4, \quad p \neq 0$

(c) $\dfrac{12^{10}}{12^9} = 12^{10-9} = 12^1 = 12$

(d) $\dfrac{7^4}{7^6} = 7^{4-6} = 7^{-2} = \dfrac{1}{7^2}$

(e) $\dfrac{k^7}{k^{12}} = k^{7-12} = k^{-5} = \dfrac{1}{k^5}, \quad k \neq 0$

EXAMPLE 8 Write each quotient using only positive exponents.

(a) $\dfrac{2^7}{2^{-3}} = 2^{7-(-3)} = 2^{10}$

(b) $\dfrac{8^{-2}}{8^5} = 8^{-2-5} = 8^{-7} = \dfrac{1}{8^7}$

(c) $\dfrac{6^{-5}}{6^{-2}} = 6^{-5-(-2)} = 6^{-3} = \dfrac{1}{6^3}$

(d) $\dfrac{4}{4^{-1}} = \dfrac{4^1}{4^{-1}} = 4^{1-(-1)} = 4^2$

(e) $\dfrac{z^{-5}}{z^{-8}} = z^{-5-(-8)} = z^3, \quad z \neq 0$

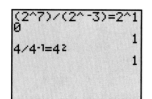

This screen supports the results in parts (a) and (d) of Example 8.

The expression $(3^4)^2$ can be simplified as $(3^4)^2 = 3^4 \cdot 3^4 = 3^{4+4} = 3^8$, where $4 \cdot 2 = 8$. This example suggests the first of the **power rules for exponents**; the other two parts can be demonstrated with similar examples.

Power Rules for Exponents

If a and b are real numbers, and m and n are integers, then

$$(a^m)^n = a^{mn}, \quad (ab)^m = a^m b^m, \quad \text{and} \quad \left(\dfrac{a}{b}\right)^m = \dfrac{a^m}{b^m} \quad (b \neq 0).$$

In the statements of rules for exponents, we always assume that zero never appears to a negative power or to the power zero.

EXAMPLE 9 Use a power rule in each case.

(a) $(p^8)^3 = p^{8 \cdot 3} = p^{24}$

(b) $\left(\dfrac{2}{3}\right)^4 = \dfrac{2^4}{3^4} = \dfrac{16}{81}$

(c) $(3y)^4 = 3^4 y^4 = 81y^4$

(d) $(6p^7)^2 = 6^2 p^{7 \cdot 2} = 6^2 p^{14} = 36p^{14}$

(e) $\left(\dfrac{-2m^5}{z}\right)^3 = \dfrac{(-2)^3 m^{5 \cdot 3}}{z^3} = \dfrac{(-2)^3 m^{15}}{z^3} = \dfrac{-8m^{15}}{z^3}, \quad z \neq 0$

Notice that

$$6^{-3} = \left(\frac{1}{6}\right)^3 = \frac{1}{216} \quad \text{and} \quad \left(\frac{2}{3}\right)^{-2} = \left(\frac{3}{2}\right)^2 = \frac{9}{4}.$$

These are examples of two special rules for negative exponents that apply when working with fractions.

Special Rules for Negative Exponents

If $a \neq 0$ and $b \neq 0$, then

$$a^{-n} = \left(\frac{1}{a}\right)^n \quad \text{and} \quad \left(\frac{a}{b}\right)^{-n} = \left(\frac{b}{a}\right)^n.$$

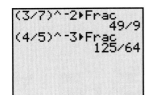

This screen supports the results of Example 10.

EXAMPLE 10 Write the following expressions with only positive exponents and then evaluate.

(a) $\left(\dfrac{3}{7}\right)^{-2} = \left(\dfrac{7}{3}\right)^2 = \dfrac{49}{9}$ (b) $\left(\dfrac{4}{5}\right)^{-3} = \left(\dfrac{5}{4}\right)^3 = \dfrac{125}{64}$

The definitions and rules of this section are summarized here.

Definitions and Rules for Exponents

For all integers m and n and all real numbers a and b, the following rules apply.

Product Rule $a^m \cdot a^n = a^{m+n}$

Quotient Rule $\dfrac{a^m}{a^n} = a^{m-n} \quad (a \neq 0)$

Zero Exponent $a^0 = 1 \quad (a \neq 0)$

Negative Exponent $a^{-n} = \dfrac{1}{a^n} \quad (a \neq 0)$

Power Rules $(a^m)^n = a^{mn} \qquad (ab)^m = a^m b^m$

$\left(\dfrac{a}{b}\right)^m = \dfrac{a^m}{b^m} \quad (b \neq 0)$

Special Rules for Negative Exponents

$\dfrac{1}{a^{-n}} = a^n \quad (a \neq 0) \qquad \dfrac{a^{-n}}{b^{-m}} = \dfrac{b^m}{a^n} \quad (a, b \neq 0)$

$a^{-n} = \left(\dfrac{1}{a}\right)^n \quad (a \neq 0)$

$\left(\dfrac{a}{b}\right)^{-n} = \left(\dfrac{b}{a}\right)^n \quad (a, b \neq 0)$

EXAMPLE 11 Simplify each expression so that no negative exponents appear in the final result. Assume all variables represent nonzero real numbers.

(a) $3^2 \cdot 3^{-5} = 3^{2+(-5)} = 3^{-3} = \dfrac{1}{3^3}$ or $\dfrac{1}{27}$

(b) $x^{-3} \cdot x^{-4} \cdot x^2 = x^{-3+(-4)+2} = x^{-5} = \dfrac{1}{x^5}$

(c) $(4^{-2})^{-5} = 4^{(-2)(-5)} = 4^{10}$

(d) $(x^{-4})^6 = x^{(-4)6} = x^{-24} = \dfrac{1}{x^{24}}$

(e) $\dfrac{x^{-4}y^2}{x^2 y^{-5}} = \dfrac{x^{-4}}{x^2} \cdot \dfrac{y^2}{y^{-5}}$

$= x^{-4-2} \cdot y^{2-(-5)}$

$= x^{-6} y^7$

$= \dfrac{y^7}{x^6}$

(f) $(2^3 x^{-2})^{-2} = (2^3)^{-2} \cdot (x^{-2})^{-2}$

$= 2^{-6} x^4$

$= \dfrac{x^4}{2^6}$ or $\dfrac{x^4}{64}$

Scientific Notation

Many of the numbers that occur in science are very large, such as the number of one-celled organisms that will sustain a whale for a few hours: 400,000,000,000,000. Other numbers are very small, such as the shortest wavelength of visible light, about .0000004 meter. Writing these numbers is simplified by using *scientific notation*.

Scientific Notation

A number is written in **scientific notation** when it is expressed in the form

$$a \times 10^n,$$

where $1 \leq |a| < 10$, and n is an integer.

As stated in the definition, scientific notation requires that the number be written as a product of a number between 1 and 10 (or -1 and -10) and some integer power of 10. (1 and -1 are allowed as values of a, but 10 and -10 are not.) For example, since

$$8000 = 8 \cdot 1000 = 8 \cdot 10^3,$$

the number 8000 is written in scientific notation as

$$8000 = 8 \times 10^3.$$

When using scientific notation, it is customary to use $\times$ instead of a dot to show multiplication.

The steps involved in writing a number in scientific notation follow. (If the number is negative, ignore the minus sign, go through these steps, and then attach a minus sign to the result.)

Converting to Scientific Notation

Step 1: **Position the decimal point.** Place a caret, ∧, to the right of the first nonzero digit, where the decimal point will be placed.

Step 2: **Determine the numeral for the exponent.** Count the number of digits from the decimal point to the caret. This number gives the absolute value of the exponent on 10.

Step 3: **Determine the sign for the exponent.** Decide whether multiplying by 10^n should make the result of Step 1 larger or smaller. The exponent should be positive to make the result larger; it should be negative to make the result smaller.

It is helpful to remember that for $n \geq 1$, $10^{-n} < 1$ and $10^n \geq 10$.

```
820000
          8.2E5
.000072
          7.2E-5
```

If a graphing calculator is set in scientific notation mode, it will give results as shown here. E5 means "times 10^5" and E-5 means "times 10^{-5}". Compare to the results of Example 12.

EXAMPLE 12 Convert each number from standard notation to scientific notation.

(a) 820,000

Place a caret to the right of the 8 (the first nonzero digit) to mark the new location of the decimal point.

$$8_\wedge 20{,}000$$

Count from the decimal point, which is understood to be after the last 0, to the caret.

$$8_\wedge 20{,}000. \leftarrow \text{Decimal point}$$
$$\text{Count 5 places.}$$

Since the number 8.2 is to be made larger, the exponent on 10 is positive.

$$820{,}000 = 8.2 \times 10^5$$

(b) .000072

Count from left to right.

$$.00007_\wedge 2$$
$$\text{5 places}$$

Since the number 7.2 is to be made smaller, the exponent on 10 is negative.

$$.000072 = 7.2 \times 10^{-5}$$

To convert a number written in scientific notation to standard notation, just work in reverse.

Converting from Scientific Notation to Standard Notation

Multiplying a number by a positive power of 10 makes the number larger, so move the decimal point to the right if n is positive in 10^n.

Multiplying by a negative power of 10 makes a number smaller, so move the decimal point to the left if n is negative.

If n is zero, leave the decimal point where it is.

EXAMPLE 13 Write each number in standard notation.

(a) $6.93 \times 10^5 = 6.93000$
    5 places

The decimal point was moved 5 places to the right. (It was necessary to attach 3 zeros.)

$$6.93 \times 10^5 = 693{,}000$$

(b) $4.7 \times 10^{-6} = 000004.7$ Attach 0s as necessary.
    6 places

The decimal point was moved 6 places to the left. Therefore,

$$4.7 \times 10^{-6} = .0000047.$$

(c) $-1.083 \times 10^0 = -1.083$

We can use scientific notation and the rules for exponents to simplify calculations.

EXAMPLE 14 Evaluate $\dfrac{1{,}920{,}000 \times .0015}{.000032 \times 45{,}000}$ by using scientific notation.

First, express all numbers in scientific notation.

$$\begin{aligned}
\frac{1{,}920{,}000 \times .0015}{.000032 \times 45{,}000} &= \frac{1.92 \times 10^6 \times 1.5 \times 10^{-3}}{3.2 \times 10^{-5} \times 4.5 \times 10^4} \\
&= \frac{1.92 \times 1.5 \times 10^6 \times 10^{-3}}{3.2 \times 4.5 \times 10^{-5} \times 10^4} && \text{Commutative and associative properties} \\
&= \frac{1.92 \times 1.5}{3.2 \times 4.5} \times 10^4 && \text{Product and quotient rules} \\
&= .2 \times 10^4 && \text{Simplify.} \\
&= (2 \times 10^{-1}) \times 10^4 \\
&= 2 \times 10^3 \\
&= 2000
\end{aligned}$$

7.5 Properties of Exponents and Scientific Notation

EXAMPLE 15 In 1990, the national health care expenditure was $695.6 billion. By 2000, this figure had risen by a factor of 1.9; that is, it almost doubled in only 10 years. (*Source:* U.S. Centers for Medicare & Medicaid Services.)

(a) Write the 1990 health care expenditure using scientific notation.

$$695.6 \text{ billion} = 695.6 \times 10^9 = (6.956 \times 10^2) \times 10^9$$
$$= 6.956 \times 10^{11} \quad \text{Product rule}$$

In 1990, the expenditure was 6.956×10^{11}.

(b) What was the expenditure in 2000?

Multiply the result in part (a) by 1.9.

$$(6.956 \times 10^{11}) \times 1.9 = (1.9 \times 6.956) \times 10^{11} \quad \text{Commutative and associative properties}$$
$$= 13.216 \times 10^{11} \quad \text{Round to three decimal places.}$$

The 2000 expenditure was $1,321,600,000,000, over $1 trillion.

7.5 EXERCISES

Match the exponential expressions in Exercises 1–6 with their equivalent expressions in Choices A.–F. Choices may be used once, more than once, or not at all.

1. $\left(\dfrac{5}{3}\right)^2$ 2. $\left(\dfrac{3}{5}\right)^2$ 3. $\left(-\dfrac{3}{5}\right)^{-2}$ 4. $\left(-\dfrac{5}{3}\right)^{-2}$ 5. $-\left(-\dfrac{3}{5}\right)^2$ 6. $-\left(-\dfrac{5}{3}\right)^2$

A. $\dfrac{25}{9}$ **B.** $-\dfrac{25}{9}$ **C.** $\dfrac{9}{25}$ **D.** $-\dfrac{9}{25}$ **E.** none of these **F.** all of these

Evaluate each exponential expression.

7. 5^4
8. 10^3
9. $(-2)^5$
10. $(-5)^4$
11. -2^3
12. -3^2
13. $-(-3)^4$
14. $-(-5)^2$
15. 7^{-2}
16. 4^{-1}
17. -7^{-2}
18. -4^{-1}
19. $\dfrac{2}{(-4)^{-3}}$
20. $\dfrac{2^{-3}}{3^{-2}}$
21. $\dfrac{5^{-1}}{4^{-2}}$
22. $\left(\dfrac{1}{2}\right)^{-3}$
23. $\left(\dfrac{1}{5}\right)^{-3}$
24. $\left(\dfrac{2}{3}\right)^{-2}$
25. $\left(\dfrac{4}{5}\right)^{-2}$
26. $3^{-1} + 2^{-1}$
27. $4^{-1} + 5^{-1}$
28. 8^0
29. 12^0
30. $(-23)^0$
31. $(-4)^0$
32. -2^0
33. $3^0 - 4^0$
34. $-8^0 - 7^0$

35. In order to raise a fraction to a negative power, we may change the fraction to its _____ and change the exponent to the _____ _____ of the original exponent.

36. Explain in your own words how to raise a power to a power.

37. Which one of the following is correct?

A. $-\dfrac{3}{4} = \left(\dfrac{3}{4}\right)^{-1}$ B. $\dfrac{3^{-1}}{4^{-1}} = \left(\dfrac{4}{3}\right)^{-1}$

C. $\dfrac{3^{-1}}{4} = \dfrac{3}{4^{-1}}$ D. $\dfrac{3^{-1}}{4^{-1}} = \left(\dfrac{3}{4}\right)^{-1}$

38. Which one of the following is incorrect?

A. $(3r)^{-2} = 3^{-2}r^{-2}$ B. $3r^{-2} = (3r)^{-2}$

C. $(3r)^{-2} = \dfrac{1}{(3r)^2}$ D. $(3r)^{-2} = \dfrac{r^{-2}}{9}$

Use the product, quotient, and power rules to simplify each expression. Write answers with only positive exponents. Assume that all variables represent nonzero real numbers.

39. $x^{12} \cdot x^4$

40. $\dfrac{x^{12}}{x^4}$

41. $\dfrac{5^{17}}{5^{16}}$

42. $\dfrac{3^{12}}{3^{13}}$

43. $\dfrac{3^{-5}}{3^{-2}}$

44. $\dfrac{2^{-4}}{2^{-3}}$

45. $\dfrac{9^{-1}}{9}$

46. $\dfrac{12}{12^{-1}}$

47. $t^5 t^{-12}$

48. $p^5 p^{-6}$

49. $(3x)^2$

50. $(-2x^{-2})^2$

51. $a^{-3}a^2a^{-4}$

52. $k^{-5}k^{-3}k^4$

53. $\dfrac{x^7}{x^{-4}}$

54. $\dfrac{p^{-3}}{p^5}$

55. $\dfrac{r^3 r^{-4}}{r^{-2} r^{-5}}$

56. $\dfrac{z^{-4} z^{-2}}{z^3 z^{-1}}$

57. $7k^2(-2k)(4k^{-5})$

58. $3a^2(-5a^{-6})(-2a)$

59. $(z^3)^{-2}z^2$

60. $(p^{-1})^3 p^{-4}$

61. $-3r^{-1}(r^{-3})^2$

62. $2(y^{-3})^4(y^6)$

63. $(3a^{-2})^3(a^3)^{-4}$

64. $(m^5)^{-2}(3m^{-2})^3$

65. $(x^{-5}y^2)^{-1}$

66. $(a^{-3}b^{-5})^2$

67. Which one of the following does *not* represent the reciprocal of x ($x \neq 0$)?

A. x^{-1} B. $\dfrac{1}{x}$ C. $\left(\dfrac{1}{x^{-1}}\right)^{-1}$ D. $-x$

68. Which one of the following is *not* in scientific notation?

A. 6.02×10^{23} B. 14×10^{-6}
C. 1.4×10^{-5} D. 3.8×10^3

Convert each number from standard notation to scientific notation.

69. 230

70. 46,500

71. .02

72. .0051

Convert each number from scientific notation to standard notation.

73. 6.5×10^3

74. 2.317×10^5

75. 1.52×10^{-2}

76. 1.63×10^{-4}

Use scientific notation to perform each of the following computations. Leave the answers in scientific notation.

77. $\dfrac{.002 \times 3900}{.000013}$

78. $\dfrac{.009 \times 600}{.02}$

79. $\dfrac{.0004 \times 56{,}000}{.000112}$

80. $\dfrac{.018 \times 20{,}000}{300 \times .0004}$

81. $\dfrac{840{,}000 \times .03}{.00021 \times 600}$

82. $\dfrac{28 \times .0045}{140 \times 1500}$

Solve each problem.

83. *NASA Budget* The projected budget of NASA (National Aeronautics and Space Administration) for 2003 is $13,757,400,000. Write this number in scientific notation. (*Source:* U.S. National Aeronautics and Space Administration.)

84. *Motor Vehicle Registrations* In 1997, there were 207,754,000 motor vehicle registrations in the United States. Write this number in scientific notation. (*Source:* U.S. Federal Highway Administration.)

85. *U.S. Budget* The U.S. budget passed $1,000,000,000 in 1917. Seventy years later in 1987 it exceeded $1,000,000,000,000 for the first time. President George W. Bush's budget request for fiscal 2003 was $2,128,000,000,000. If stacked in dollar bills, this amount would stretch 144,419 miles, almost two-thirds of the distance to the moon. Write these four numbers in scientific notation. (*Source: The Gazette*, February 5, 2002.)

86. *Wal-Mart Employees* In 1970, Wal-Mart had 1500 employees. In 1997, Wal-Mart became the largest private employer in the U.S. with 680,000 employees. In 1999, with 1,100,000 employees, Wal-Mart became the largest private employer in the world. By 2007, the company is expected to have 2,200,000 employees. Write these four numbers in scientific notation. (*Source:* Wal-Mart.)

Astronomy Data Each of the following statements comes from Astronomy! A Brief Edition *by James B. Kaler (Addison-Wesley, 1997). If the number in the statement is in scientific notation, write it in standard notation without using exponents. If the number is in standard notation, write it in scientific notation.*

87. Multiplying this view over the whole sky yields a galaxy count of more than 10 billion. (page 496)

88. The circumference of the solar orbit is … about 4.7 million km…. (in reference to the orbit of Jupiter, page 395)

89. The solar luminosity requires that 2×10^9 kg of mass be converted into energy every second. (page 327)

90. At maximum, a cosmic ray particle—a mere atomic nucleus of only 10^{-13} cm across—can carry the energy of a professionally pitched baseball. (page 445)

Solve each problem.

91. *Computer Speed* On October 28, 1998, IBM announced a computer capable of 3.9×10^8 operations per second. This was 15,000 times faster than the normal desktop computer at that time. What was the number of operations that the normal desktop could do? (*Source:* IBM.)

92. *Powerball Lottery* In the early years of the Powerball Lottery, a player had to choose five numbers from 1 through 49 and one number from 1 through 42. It can be shown that there are about 8.009×10^7 different ways to do this. Suppose that a group of 2000 people decided to purchase tickets for all these numbers and each ticket cost \$1.00. How much should each person have expected to pay? (*Source:* www.powerball.com)

93. *Distance of Uranus from the Sun* A parsec, a unit of length used in astronomy, is 1.9×10^{13} miles. The mean distance of Uranus from the sun is 1.8×10^7 miles. How many parsecs is Uranus from the sun?

94. *Number of Inches in a Mile* An inch is approximately 1.57828×10^{-5} mile. Find the reciprocal of this number to determine the number of inches in a mile.

95. *Speed of Light* The speed of light is approximately 3×10^{10} centimeters per second. How long will it take light to travel 9×10^{12} centimeters?

96. *Rocket from Earth to the Sun* The average distance from Earth to the sun is 9.3×10^7 miles. How long would it take a rocket, traveling at 2.9×10^3 miles per hour, to reach the sun?

97. *Miles in a Light-Year* A *light-year* is the distance that light travels in one year. Find the number of miles in a light-year if light travels 1.86×10^5 miles per second.

98. *Time for Light to Travel* Use the information given in the previous two exercises to find the number of minutes necessary for light from the sun to reach Earth.

99. *Rocket from Venus to Mercury* The planet Mercury has an average distance from the sun of 3.6×10^7 miles, while the mean distance of Venus from the sun is 6.7×10^7 miles. How long would it take a spacecraft traveling at 1.55×10^3 miles per hour to travel from Venus to Mercury? Assume the trip could be timed so that its start and finish would occur when the respective planets are at their average distances from the sun and the same direction from the sun. (Give your answer in hours, without scientific notation.)

100. *Distance from an Object to the Moon* When the distance between the centers of the moon and Earth is 4.60×10^8 meters, an object on the line joining the centers of the moon and Earth exerts the same gravitational force on each when it is 4.14×10^8 meters from the center of Earth. How far is the object from the center of the moon at that point?

Heart Attack Data The graph shows the estimated annual number of Americans, by age and sex, experiencing heart attacks.

Use scientific notation to represent the numbers of people for the following categories.

101. Males, 29–44

102. Females, 29–44

103. Males, 45–64

104. Females, 45–64

105. Males, 65 or older

106. Females, 65 or older

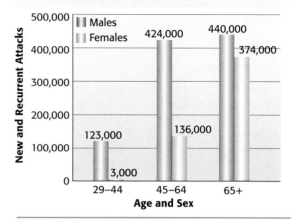

ESTIMATED ANNUAL NUMBER OF AMERICANS EXPERIENCING HEART ATTACKS

Source: American Heart Association and Framingham Heart Study, 24-year follow-up.

7.6 Polynomials and Factoring

Basic Terminology A **term**, or **monomial**, is defined to be a number, a variable, or a product of numbers and variables. A **polynomial** is a term or a finite sum or difference of terms, with only nonnegative integer exponents permitted on the variables. If the terms of a polynomial contain only the variable x, then the polynomial is called a **polynomial in x.** (Polynomials in other variables are defined similarly.) Examples of polynomials include

$$5x^3 - 8x^2 + 7x - 4, \quad 9p^5 - 3, \quad 8r^2, \quad \text{and} \quad 6.$$

The expression $9x^2 - 4x - \frac{6}{x}$ is not a polynomial because of the presence of $-\frac{6}{x}$. The terms of a polynomial cannot have variables in a denominator.

The greatest exponent in a polynomial in one variable is the **degree** of the polynomial. A nonzero constant is said to have degree 0. (The polynomial 0 has no degree.) For example, $3x^6 - 5x^2 + 2x + 3$ is a polynomial of degree 6.

A polynomial can have more than one variable. A term containing more than one variable has degree equal to the sum of all the exponents appearing on the variables in the term. For example, $-3x^4y^3z^5$ is of degree $4 + 3 + 5 = 12$. The degree of a polynomial in more than one variable is equal to the greatest degree of any term appearing in the polynomial. By this definition, the polynomial

$$2x^4y^3 - 3x^5y + x^6y^2$$

is of degree 8 because of the x^6y^2 term.

7.6 Polynomials and Factoring

A polynomial containing exactly three terms is called a **trinomial** and one containing exactly two terms is a **binomial.** For example, $7x^9 - 8x^4 + 1$ is a trinomial of degree 9. The table shows several polynomials and gives the degree and type of each.

Polynomial	Degree	Type
$9p^7 - 4p^3 + 8p^2$	7	Trinomial
$29x^{11} + 8x^{15}$	15	Binomial
$-10r^6s^8$	14	Monomial
$5a^3b^7 - 3a^5b^5 + 4a^2b^9 - a^{10}$	11	None of these

Addition and Subtraction Since the variables used in polynomials represent real numbers, a polynomial represents a real number. This means that all the properties of the real numbers mentioned in this book hold for polynomials. In particular, the distributive property holds, so

$$3m^5 - 7m^5 = (3 - 7)m^5 = -4m^5.$$

Like terms are terms that have the exact same variable factors. Thus, polynomials are added by adding coefficients of like terms; polynomials are subtracted by subtracting coefficients of like terms.

EXAMPLE 1 Add or subtract, as indicated.

(a) $(2y^4 - 3y^2 + y) + (4y^4 + 7y^2 + 6y)$
$= (2 + 4)y^4 + (-3 + 7)y^2 + (1 + 6)y$
$= 6y^4 + 4y^2 + 7y$

(b) $(-3m^3 - 8m^2 + 4) - (m^3 + 7m^2 - 3)$
$= (-3 - 1)m^3 + (-8 - 7)m^2 + [4 - (-3)]$
$= -4m^3 - 15m^2 + 7$

(c) $8m^4p^5 - 9m^3p^5 + (11m^4p^5 + 15m^3p^5) = 19m^4p^5 + 6m^3p^5$

(d) $4(x^2 - 3x + 7) - 5(2x^2 - 8x - 4)$
$= 4x^2 - 4(3x) + 4(7) - 5(2x^2) - 5(-8x) - 5(-4)$ Distributive property
$= 4x^2 - 12x + 28 - 10x^2 + 40x + 20$ Associative property
$= -6x^2 + 28x + 48$ Add like terms.

As shown in parts (a), (b), and (d) of Example 1, polynomials in one variable are often written with their terms in **descending powers;** so the term of greatest degree is first, the one with the next greatest degree is second, and so on.

Multiplication The associative and distributive properties, together with the properties of exponents, can also be used to find the product of two polynomials. For example, to find the product of $3x - 4$ and $2x^2 - 3x + 5$, treat $3x - 4$ as a single expression and use the distributive property as follows.

$$(3x - 4)(2x^2 - 3x + 5) = (3x - 4)(2x^2) - (3x - 4)(3x) + (3x - 4)(5)$$

Now use the distributive property three separate times on the right of the equality symbol to get

$$(3x - 4)(2x^2 - 3x + 5)$$
$$= (3x)(2x^2) - 4(2x^2) - (3x)(3x) - (-4)(3x) + (3x)5 - 4(5)$$
$$= 6x^3 - 8x^2 - 9x^2 + 12x + 15x - 20$$
$$= 6x^3 - 17x^2 + 27x - 20.$$

It is sometimes more convenient to write such a product vertically, as follows.

$$
\begin{array}{r}
2x^2 - 3x + 5 \\
3x - 4 \\
\hline
-8x^2 + 12x - 20 \quad \leftarrow -4(2x^2 - 3x + 5) \\
6x^3 - 9x^2 + 15x \qquad\quad \leftarrow 3x(2x^2 - 3x + 5) \\
\hline
6x^3 - 17x^2 + 27x - 20 \quad \text{Add in columns.}
\end{array}
$$

EXAMPLE 2 Multiply $(3p^2 - 4p + 1)(p^3 + 2p - 8)$.

Multiply each term of the second polynomial by each term of the first and add these products. It is most efficient to work vertically with polynomials of more than two terms, so that like terms can be placed in columns.

$$
\begin{array}{r}
3p^2 - 4p + 1 \\
p^3 + 2p - 8 \\
\hline
-24p^2 + 32p - 8 \quad \text{Multiply } 3p^2 - 4p + 1 \text{ by } -8. \\
6p^3 - 8p^2 + 2p \qquad\quad \text{Multiply } 3p^2 - 4p + 1 \text{ by } 2p. \\
3p^5 - 4p^4 + p^3 \qquad\qquad\quad \text{Multiply } 3p^2 - 4p + 1 \text{ by } p^3. \\
\hline
3p^5 - 4p^4 + 7p^3 - 32p^2 + 34p - 8 \quad \text{Add in columns.}
\end{array}
$$

The FOIL method is a convenient way to find the product of two binomials. The memory aid FOIL (for First, Outside, Inside, Last) gives the pairs of terms to be multiplied to get the product, as shown in the next examples.

EXAMPLE 3 Find each product.

$$\qquad\qquad\qquad\qquad\qquad\quad F \qquad\quad O \qquad\quad I \qquad\quad L$$
(a) $(6m + 1)(4m - 3) = (6m)(4m) + (6m)(-3) + 1(4m) + 1(-3)$
$$= 24m^2 - 14m - 3$$

(b) $(2x + 7)(2x - 7) = 4x^2 - 14x + 14x - 49$
$$= 4x^2 - 49$$

In part (a) of Example 3, the product of two binomials was a trinomial, while in part (b) the product of two binomials was a binomial. The product of two binomials of the forms $x + y$ and $x - y$ is always a binomial. Check by multiplying that the following is true.

Product of the Sum and Difference of Two Terms

$$(x + y)(x - y) = x^2 - y^2$$

The special product

$$(x + y)(x - y) = x^2 - y^2$$

can be used to solve some multiplication problems. For example,

$$51 \times 49 = (50 + 1)(50 - 1)$$
$$= 50^2 - 1^2$$
$$= 2500 - 1$$
$$= 2499$$

$$102 \times 98 = (100 + 2)(100 - 2)$$
$$= 100^2 - 2^2$$
$$= 10{,}000 - 4$$
$$= 9996.$$

Once these patterns are recognized, multiplications of this type can be done mentally.

This product is called the **difference of two squares.** Since products of this type occur frequently, it is important to be able to recognize when this pattern should be used.

EXAMPLE 4 Find each product.

(a) $(3p + 11)(3p - 11)$

Using the pattern discussed above, replace x with $3p$ and y with 11.

$$(3p + 11)(3p - 11) = (3p)^2 - 11^2 = 9p^2 - 121$$

(b) $(5m^3 - 3)(5m^3 + 3) = (5m^3)^2 - 3^2 = 25m^6 - 9$

(c) $(9k - 11r^3)(9k + 11r^3) = (9k)^2 - (11r^3)^2 = 81k^2 - 121r^6$ ■

The **squares of binomials** are also special products.

Squares of Binomials

$$(x + y)^2 = x^2 + 2xy + y^2$$
$$(x - y)^2 = x^2 - 2xy + y^2$$

EXAMPLE 5 Find each product.

(a) $(2m + 5)^2 = (2m)^2 + 2(2m)(5) + (5)^2$
$= 4m^2 + 20m + 25$

(b) $(3x - 7y^4)^2 = (3x)^2 - 2(3x)(7y^4) + (7y^4)^2$
$= 9x^2 - 42xy^4 + 49y^8$ ■

As shown in Example 5, the square of a binomial has three terms. Students often mistakenly give $x^2 + y^2$ as equivalent to the product $(x + y)^2$. Be careful to avoid that error.

The process of finding polynomials whose product equals a given polynomial is called **factoring.** For example, since $4x + 12 = 4(x + 3)$, both 4 and $x + 3$ are called **factors** of $4x + 12$. Also, $4(x + 3)$ is called the **factored form** of $4x + 12$. A polynomial that cannot be written as a product of two polynomials with integer coefficients is a **prime polynomial.** A polynomial is **factored completely** when it is written as a product of prime polynomials with integer coefficients.

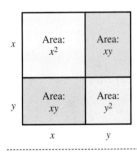

The special product

$(x + y)^2 = x^2 + 2xy + y^2$

can be illustrated geometrically using the diagram shown here. Each side of the large square has length $x + y$, so the area of the square is

$(x + y)^2$.

The large square is made up of two smaller squares and two congruent rectangles. The sum of the areas of these figures is

$x^2 + 2xy + y^2$.

Since these expressions represent the same quantity, they must be equal, thus giving us the pattern for squaring a binomial.

Factoring Out the Greatest Common Factor

Some polynomials are factored by using the distributive property. For example, to factor $6x^2y^3 + 9xy^4 + 18y^5$, we look for a monomial that is the greatest common factor of all the terms of the polynomial. For this polynomial, $3y^3$ is the greatest common factor. By the distributive property,

$$6x^2y^3 + 9xy^4 + 18y^5 = (3y^3)(2x^2) + (3y^3)(3xy) + (3y^3)(6y^2)$$
$$= 3y^3(2x^2 + 3xy + 6y^2).$$

EXAMPLE 6 Factor out the greatest common factor from each polynomial.

(a) $9y^5 + y^2 = y^2 \cdot 9y^3 + y^2 \cdot 1$ The greatest common factor is y^2.
$= y^2(9y^3 + 1)$

(b) $6x^2t + 8xt + 12t = 2t(3x^2 + 4x + 6)$

(c) $14m^4(m + 1) - 28m^3(m + 1) - 7m^2(m + 1)$

The greatest common factor is $7m^2(m + 1)$. Use the distributive property.

$14m^4(m + 1) - 28m^3(m + 1) - 7m^2(m + 1)$
$= [7m^2(m + 1)](2m^2 - 4m - 1)$
$= 7m^2(m + 1)(2m^2 - 4m - 1)$ ∎

Factoring by Grouping When a polynomial has more than three terms, it can sometimes be factored by a method called **factoring by grouping.** For example, to factor

$$ax + ay + 6x + 6y,$$

collect the terms into two groups so that each group has a common factor.

$$ax + ay + 6x + 6y = (ax + ay) + (6x + 6y)$$

Factor each group, getting

$$ax + ay + 6x + 6y = a(x + y) + 6(x + y).$$

The quantity $(x + y)$ is now a common factor, which can be factored out, producing

$$ax + ay + 6x + 6y = (x + y)(a + 6).$$

It is not always obvious which terms should be grouped. Experience and repeated trials are the most reliable tools for factoring by grouping.

EXAMPLE 7 Factor by grouping.

(a) $mp^2 + 7m + 3p^2 + 21 = (mp^2 + 7m) + (3p^2 + 21)$ Group the terms.
$= m(p^2 + 7) + 3(p^2 + 7)$ Factor each group.
$= (p^2 + 7)(m + 3)$ $p^2 + 7$ is a common factor.

(b) $2y^2 - 2z - ay^2 + az = (2y^2 - 2z) + (-ay^2 + az)$
$= 2(y^2 - z) + a(-y^2 + z)$ Factor each group.

The expression $-y^2 + z$ is the negative of $y^2 - z$, so factor out $-a$ instead of a.

$= 2(y^2 - z) - a(y^2 - z)$ Factor out $-a$.
$= (y^2 - z)(2 - a)$ Factor out $y^2 - z$. ∎

Factoring Trinomials Factoring is the opposite of multiplying. Since the product of two binomials is usually a trinomial, we can expect factorable trinomials (that have terms with no common factor) to have two binomial factors. Thus, factoring trinomials requires using FOIL backward.

EXAMPLE 8 Factor each trinomial.

(a) $4y^2 - 11y + 6$

To factor this polynomial, we must find integers a, b, c, and d such that
$$4y^2 - 11y + 6 = (ay + b)(cy + d).$$

By using FOIL, we see that $ac = 4$ and $bd = 6$. The positive factors of 4 are 4 and 1 or 2 and 2. Since the middle term is negative, we consider only negative factors of 6. The possibilities are -2 and -3 or -1 and -6. Now we try various arrangements of these factors until we find one that gives the correct coefficient of y.

$(2y - 1)(2y - 6) = 4y^2 - 14y + 6$ Incorrect
$(2y - 2)(2y - 3) = 4y^2 - 10y + 6$ Incorrect
$(y - 2)(4y - 3) = 4y^2 - 11y + 6$ Correct

The last trial gives the correct factorization.

(b) $6p^2 - 7p - 5$

Again, we try various possibilities. The positive factors of 6 could be 2 and 3 or 1 and 6. As factors of -5 we have only -1 and 5 or -5 and 1. Try different combinations of these factors until the correct one is found.

$(2p - 5)(3p + 1) = 6p^2 - 13p - 5$ Incorrect
$(3p - 5)(2p + 1) = 6p^2 - 7p - 5$ Correct

Thus, $6p^2 - 7p - 5$ factors as $(3p - 5)(2p + 1)$.

Each of the special patterns of multiplication given earlier can be used in reverse to get a pattern for factoring. Perfect square trinomials can be factored as follows.

Perfect Square Trinomials

$$x^2 + 2xy + y^2 = (x + y)^2$$
$$x^2 - 2xy + y^2 = (x - y)^2$$

EXAMPLE 9 Factor each polynomial.

(a) $16p^2 - 40pq + 25q^2$

Since $16p^2 = (4p)^2$ and $25q^2 = (5q)^2$, use the second pattern shown above with $4p$ replacing x and $5q$ replacing y to obtain
$$16p^2 - 40pq + 25q^2 = (4p)^2 - 2(4p)(5q) + (5q)^2$$
$$= (4p - 5q)^2.$$

Make sure that the middle term of the trinomial being factored, $-40pq$ here, is twice the product of the two terms in the binomial $4p - 5q$.
$$-40pq = 2(4p)(-5q)$$

(b) $169x^2 + 104xy^2 + 16y^4 = (13x + 4y^2)^2$, since $2(13x)(4y^2) = 104xy^2$.

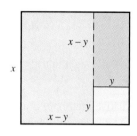

A geometric proof for the difference of squares property is shown above. (The proof is only valid for $x > y > 0$.)

$x^2 - y^2 = x(x - y) + y(x - y)$
$ = (x - y)(x + y)$

Factor out $x - y$ in the second step.

Factoring Binomials The pattern for the product of the sum and difference of two terms gives the following factorization.

Difference of Squares

$$x^2 - y^2 = (x + y)(x - y)$$

EXAMPLE 10 Factor each of the following polynomials.

(a) $4m^2 - 9$

First, recognize that $4m^2 - 9$ is the difference of squares, since $4m^2 = (2m)^2$ and $9 = 3^2$. Use the pattern for the difference of squares with $2m$ replacing x and 3 replacing y. Doing this gives

$$4m^2 - 9 = (2m)^2 - 3^2$$
$$= (2m + 3)(2m - 3).$$

(b) $256k^4 - 625m^4$

Use the difference of squares pattern twice.

$$256k^4 - 625m^4 = (16k^2)^2 - (25m^2)^2$$
$$= (16k^2 + 25m^2)(16k^2 - 25m^2)$$
$$= (16k^2 + 25m^2)(4k + 5m)(4k - 5m)$$

(c) $x^2 - 6x + 9 - y^4$

Group the first three terms to obtain a perfect square trinomial. Then use the difference of squares pattern.

$$x^2 - 6x + 9 - y^4 = (x^2 - 6x + 9) - y^4$$
$$= (x - 3)^2 - (y^2)^2$$
$$= [(x - 3) + y^2][(x - 3) - y^2]$$
$$= (x - 3 + y^2)(x - 3 - y^2)$$

Two other special results of factoring are listed below. Each can be verified by multiplying on the right side of the equation.

Sum and Difference of Cubes

Sum of Cubes $\qquad x^3 + y^3 = (x + y)(x^2 - xy + y^2)$

Difference of Cubes $\qquad x^3 - y^3 = (x - y)(x^2 + xy + y^2)$

EXAMPLE 11 Factor each polynomial.

(a) $x^3 + 27$

Notice that $27 = 3^3$, so the expression is a sum of cubes. Use the first pattern given above.

$$x^3 + 27 = x^3 + 3^3 = (x + 3)(x^2 - 3x + 9)$$

(b) $m^3 - 64n^3 = m^3 - (4n)^3$
$= (m - 4n)[m^2 + m(4n) + (4n)^2]$
$= (m - 4n)(m^2 + 4mn + 16n^2)$

(c) $8q^6 + 125p^9 = (2q^2)^3 + (5p^3)^3$
$= (2q^2 + 5p^3)[(2q^2)^2 - (2q^2)(5p^3) + (5p^3)^2]$
$= (2q^2 + 5p^3)(4q^4 - 10q^2p^3 + 25p^6)$

7.6 EXERCISES

Find each of the following sums and differences.

1. $(3x^2 - 4x + 5) + (-2x^2 + 3x - 2)$
2. $(4m^3 - 3m^2 + 5) + (-3m^3 - m^2 + 5)$
3. $(12y^2 - 8y + 6) - (3y^2 - 4y + 2)$
4. $(8p^2 - 5p) - (3p^2 - 2p + 4)$
5. $(6m^4 - 3m^2 + m) - (2m^3 + 5m^2 + 4m) + (m^2 - m)$
6. $-(8x^3 + x - 3) + (2x^3 + x^2) - (4x^2 + 3x - 1)$
7. $5(2x^2 - 3x + 7) - 2(6x^2 - x + 12)$
8. $8x^2y - 3xy^2 + 2x^2y - 9xy^2$

Find each of the following products.

9. $(x + 3)(x - 8)$
10. $(y - 3)(y - 9)$
11. $(4r - 1)(7r + 2)$
12. $(5m - 6)(3m + 4)$
13. $4x^2(3x^3 + 2x^2 - 5x + 1)$
14. $2b^3(b^2 - 4b + 3)$
15. $(2m + 3)(2m - 3)$
16. $(8s - 3t)(8s + 3t)$
17. $(4m + 2n)^2$
18. $(a - 6b)^2$
19. $(5r + 3t^2)^2$
20. $(2z^4 - 3y)^2$
21. $(2z - 1)(-z^2 + 3z - 4)$
22. $(k + 2)(12k^3 - 3k^2 + k + 1)$
23. $(m - n + k)(m + 2n - 3k)$
24. $(r - 3s + t)(2r - s + t)$
25. $(a - b + 2c)^2$
26. $(k - y + 3m)^2$

27. Which one of the following is a trinomial in descending powers, having degree 6?
   A. $5x^6 - 4x^5 + 12$ B. $6x^5 - x^6 + 4$
   C. $2x + 4x^2 - x^6$ D. $4x^6 - 6x^4 + 9x + 1$

28. Give an example of a polynomial of four terms in the variable x, having degree 5, written in descending powers, lacking a fourth degree term.

29. The exponent in the expression 6^3 is 3. Explain why the degree of 6^3 is not 3. What is its degree?

30. Explain in your own words how to square a binomial.

Factor the greatest common factor from each polynomial.

31. $8m^4 + 6m^3 - 12m^2$
32. $2p^5 - 10p^4 + 16p^3$
33. $4k^2m^3 + 8k^4m^3 - 12k^2m^4$
34. $28r^4s^2 + 7r^3s - 35r^4s^3$
35. $2(a + b) + 4m(a + b)$
36. $4(y - 2)^2 + 3(y - 2)$
37. $2(m - 1) - 3(m - 1)^2 + 2(m - 1)^3$
38. $5(a + 3)^3 - 2(a + 3) + (a + 3)^2$

Factor each of the following polynomials by grouping.

39. $6st + 9t - 10s - 15$
40. $10ab - 6b + 35a - 21$
41. $rt^3 + rs^2 - pt^3 - ps^2$
42. $2m^4 + 6 - am^4 - 3a$
43. $16a^2 + 10ab - 24ab - 15b^2$
44. $15 - 5m^2 - 3r^2 + m^2r^2$
45. $20z^2 - 8zx - 45zx + 18x^2$

46. Consider the polynomial $1 - a + ab - b$. One acceptable factored form is $(1 - a)(1 - b)$. However, there are other acceptable factored forms. Which one is *not* a factored form of this polynomial?

A. $(a - 1)(b - 1)$
B. $(-a + 1)(-b + 1)$
C. $(-1 + a)(-1 + b)$
D. $(1 - a)(b + 1)$

Factor each trinomial.

47. $x^2 - 2x - 15$ **48.** $r^2 + 8r + 12$ **49.** $y^2 + 2y - 35$ **50.** $x^2 - 7x + 6$
51. $6a^2 - 48a - 120$ **52.** $8h^2 - 24h - 320$ **53.** $3m^3 + 12m^2 + 9m$
54. $9y^4 - 54y^3 + 45y^2$ **55.** $6k^2 + 5kp - 6p^2$ **56.** $14m^2 + 11mr - 15r^2$
57. $5a^2 - 7ab - 6b^2$ **58.** $12s^2 + 11st - 5t^2$ **59.** $9x^2 - 6x^3 + x^4$
60. $30a^2 + am - m^2$ **61.** $24a^4 + 10a^3b - 4a^2b^2$ **62.** $18x^5 + 15x^4z - 75x^3z^2$

63. When a student was given the polynomial
$$4x^2 + 2x - 20$$
to factor completely on a test, she lost some credit by giving the answer $(4x + 10)(x - 2)$. She complained to her teacher that the product $(4x + 10)(x - 2)$ is indeed $4x^2 + 2x - 20$. Do you think that the teacher was justified in not giving her full credit? Explain.

64. Write an explanation as to why most people would find it more difficult to factor
$$36x^2 - 44x - 15$$
than
$$37x^2 - 183x - 10.$$

Factor each, using the method for factoring a perfect square trinomial. It may be necessary to factor out a common factor first.

65. $9m^2 - 12m + 4$ **66.** $16p^2 - 40p + 25$ **67.** $32a^2 - 48ab + 18b^2$
68. $20p^2 - 100pq + 125q^2$ **69.** $4x^2y^2 + 28xy + 49$ **70.** $9m^2n^2 - 12mn + 4$

Factor each difference of squares.

71. $x^2 - 36$ **72.** $t^2 - 64$ **73.** $y^2 - w^2$ **74.** $25 - w^2$ **75.** $9a^2 - 16$
76. $16q^2 - 25$ **77.** $25s^4 - 9t^2$ **78.** $36z^2 - 81y^4$ **79.** $p^4 - 625$ **80.** $m^4 - 81$

Factor each sum or difference of cubes.

81. $8 - a^3$ **82.** $r^3 + 27$ **83.** $125x^3 - 27$
84. $8m^3 - 27n^3$ **85.** $27y^9 + 125z^6$ **86.** $27z^3 + 729y^3$

Each of the following may be factored using one of the methods described in this section. Decide on the method, and then factor the polynomial completely.

87. $x^2 + xy - 5x - 5y$ **88.** $8r^2 - 10rs - 3s^2$ **89.** $p^4(m - 2n) + q(m - 2n)$
90. $36a^2 + 60a + 25$ **91.** $4z^2 + 28z + 49$ **92.** $6p^4 + 7p^2 - 3$
93. $1000x^3 + 343y^3$ **94.** $b^2 + 8b + 16 - a^2$ **95.** $125m^6 - 216$
96. $q^2 + 6q + 9 - p^2$ **97.** $12m^2 + 16mn - 35n^2$ **98.** $216p^3 + 125q^3$

99. The sum of squares usually cannot be factored. For example, $x^2 + y^2$ is prime. Notice that
$$x^2 + y^2 \neq (x + y)(x + y).$$
By choosing $x = 4$ and $y = 2$, show that the above inequality is true.

100. The binomial
$$9x^2 + 36$$
is a sum of squares. Can it be factored? If so, factor it.

7.7 Quadratic Equations and Applications

Quadratic Equations Recall that a linear equation is one that can be written in the form $ax + b = c$, where a, b, and c are real numbers, $a \neq 0$.

> **Quadratic Equation**
>
> An equation that can be written in the form
>
> $$ax^2 + bx + c = 0$$
>
> where a, b, and c are real numbers, with $a \neq 0$, is a **quadratic equation.**

(Why is the restriction $a \neq 0$ necessary?) A quadratic equation written in the form $ax^2 + bx + c = 0$ is in **standard form.**

The simplest method of solving a quadratic equation, but one that is not always easily applied, is by factoring. This method depends on the following property.

> **Zero-Factor Property**
>
> If $ab = 0$, then $a = 0$ or $b = 0$ or both.

EXAMPLE 1 Solve $6x^2 + 7x = 3$.

$$6x^2 + 7x = 3$$
$$6x^2 + 7x - 3 = 0 \quad \text{Standard form}$$
$$(3x - 1)(2x + 3) = 0 \quad \text{Factor.}$$

By the zero-factor property, the product $(3x - 1)(2x + 3)$ can equal 0 only if

$$3x - 1 = 0 \quad \text{or} \quad 2x + 3 = 0$$
$$3x = 1 \quad \text{or} \quad 2x = -3$$
$$x = \frac{1}{3} \quad \text{or} \quad x = -\frac{3}{2}.$$

The *solve* feature gives the two solutions of the equation in Example 1. Notice that the guess 0 yields the solution 1/3, while the guess −5 yields the solution −3/2. Compare to Example 1.

Check by first substituting $\frac{1}{3}$ and then $-\frac{3}{2}$ in the original equation. The solution set is $\left\{\frac{1}{3}, -\frac{3}{2}\right\}$.

A quadratic equation of the form $x^2 = k$, $k \geq 0$, can be solved by factoring.

$$x^2 = k$$
$$x^2 - k = 0$$
$$\left(x + \sqrt{k}\right)\left(x - \sqrt{k}\right) = 0$$
$$x + \sqrt{k} = 0 \quad \text{or} \quad x - \sqrt{k} = 0$$
$$x = -\sqrt{k} \quad \text{or} \quad x = \sqrt{k}$$

This proves the square root property for solving equations.

Square Root Property

If $k \geq 0$, then the solutions of $x^2 = k$ are $x = \pm\sqrt{k}$.

If $k > 0$, the equation $x^2 = k$ has two real solutions. If $k = 0$, there is only one solution, 0. If $k < 0$, there are no real solutions. (However, in this case, there *are* imaginary solutions. Imaginary numbers are discussed briefly in the Extension on complex numbers at the end of Chapter 6.)

Completing the square, used in deriving the quadratic formula, has important applications in algebra. To transform the expression $x^2 + kx$ into the square of a binomial, we add to it the square of half the coefficient of x; that is, $[(1/2)k]^2 = k^2/4$. We then get

$$x^2 + kx + \frac{k^2}{4} = \left(x + \frac{k}{2}\right)^2.$$

For example, to make $x^2 + 6x$ the square of a binomial, we add 9, since $9 = [(1/2)6]^2$. This results in the trinomial $x^2 + 6x + 9$, which is equal to $(x + 3)^2$.

The Greeks had a method of completing the square geometrically. For example, to complete the square for $x^2 + 6x$, begin with a square of side x. Add three rectangles of width 1 and length x to the right side and the bottom. Each rectangle has area $1x$ or x, so the total area of the figure is now $x^2 + 6x$. To fill in the corner (that is, "complete the square"), we add 9 1-by-1 squares as shown.

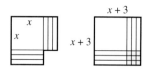

The new, larger square has sides of length $x + 3$ and area

$$(x + 3)^2 = x^2 + 6x + 9.$$

EXAMPLE 2 Use the square root property to solve each quadratic equation for real solutions.

(a) $x^2 = 25$

Since $\sqrt{25} = 5$, the solution set is $\{5, -5\}$, which may be abbreviated $\{\pm 5\}$.

(b) $r^2 = 18$

$$\begin{aligned} r &= \pm\sqrt{18} && \text{Square root property} \\ r &= \pm\sqrt{9 \cdot 2} \\ r &= \pm\sqrt{9} \cdot \sqrt{2} && \text{Product rule for square roots} \\ r &= \pm 3\sqrt{2} && \sqrt{9} = 3 \end{aligned}$$

The solution set is $\{\pm 3\sqrt{2}\}$.

(c) $z^2 = -3$

Since $-3 < 0$, there are no real roots, and the solution set is $\emptyset$.

(d) $(x - 4)^2 = 12$

Use a generalization of the square root property, working as follows.

$$\begin{aligned} (x - 4)^2 &= 12 \\ x - 4 &= \pm\sqrt{12} \\ x &= 4 \pm \sqrt{12} \\ x &= 4 \pm \sqrt{4 \cdot 3} \\ x &= 4 \pm 2\sqrt{3} \end{aligned}$$

The solution set is $\{4 \pm 2\sqrt{3}\}$.

The Quadratic Formula By using a procedure called *completing the square* (see the margin note) we can derive one of the most important formulas in algebra, the *quadratic formula*. We begin with the standard quadratic equation

$$ax^2 + bx + c = 0, \quad a \neq 0.$$

$$x^2 + \frac{b}{a}x + \frac{c}{a} = 0 \quad \text{Divide by } a.$$

$$x^2 + \frac{b}{a}x = -\frac{c}{a} \quad \text{Add } -\frac{c}{a}.$$

The polynomial on the left will be the square of a binomial if we add $b^2/(4a^2)$ to both sides of the equation.

$$x^2 + \frac{b}{a}x + \frac{b^2}{4a^2} = \frac{b^2}{4a^2} - \frac{c}{a}$$

$$\left(x + \frac{b}{2a}\right)^2 = \frac{b^2 - 4ac}{4a^2} \quad \text{Factor on the left; combine terms on the right.}$$

$$x + \frac{b}{2a} = \pm\sqrt{\frac{b^2 - 4ac}{4a^2}} \quad \text{Square root property}$$

$$x + \frac{b}{2a} = \pm\frac{\sqrt{b^2 - 4ac}}{\sqrt{4a^2}} \quad \text{Quotient rule for square roots}$$

$$x = -\frac{b}{2a} \pm \frac{\sqrt{b^2 - 4ac}}{2a} \quad \text{Subtract } \frac{b}{2a}.$$

$$x = \frac{-b \pm \sqrt{b^2 - 4ac}}{2a} \quad \text{Combine terms.}$$

For centuries mathematicians wrestled with finding a formula that could solve cubic (third-degree) equations. A story from sixteenth-century Italy concerns two main characters, **Girolamo Cardano** and **Niccolo Tartaglia**. In those days, mathematicians often participated in contests. Tartaglia had developed a method of solving a cubic equation of the form $x^3 + mx = n$ and had used it in one of these contests. Cardano begged to know Tartaglia's method and after he was told was sworn to secrecy. Nonetheless, Cardano published Tartaglia's method in his 1545 work *Ars Magna* (although he did give Tartaglia credit).

The formula for finding one real solution of the above equation is

$$x = \sqrt[3]{\frac{n}{2} + \sqrt{\left(\frac{n}{2}\right)^2 + \left(\frac{m}{3}\right)^3}}$$
$$- \sqrt[3]{-\left(\frac{n}{2}\right) + \sqrt{\left(\frac{n}{2}\right)^2 + \left(\frac{m}{3}\right)^3}}.$$

Try solving for one solution of the equation $x^3 + 9x = 26$ using this formula. (The solution given by the formula is 2.)

Quadratic Formula

The solutions of $ax^2 + bx + c = 0$, $a \neq 0$, are

$$x = \frac{-b \pm \sqrt{b^2 - 4ac}}{2a}.$$

Notice that the fraction bar in the quadratic formula extends under the $-b$ term in the numerator.

EXAMPLE 3 Solve $x^2 - 4x + 2 = 0$.

Here $a = 1$, $b = -4$, and $c = 2$. Substitute these values into the quadratic formula to obtain

$$x = \frac{-b \pm \sqrt{b^2 - 4ac}}{2a}$$

$$x = \frac{-(-4) \pm \sqrt{(-4)^2 - 4(1)2}}{2(1)} \quad a = 1, b = -4, c = 2$$

$$x = \frac{4 \pm \sqrt{16 - 8}}{2}$$

$$x = \frac{4 \pm 2\sqrt{2}}{2} \quad \sqrt{16 - 8} = \sqrt{8} = 2\sqrt{2}$$

$$x = \frac{2(2 \pm \sqrt{2})}{2} \quad \text{Factor out a 2 in the numerator.}$$

$$x = 2 \pm \sqrt{2}. \quad \text{Lowest terms}$$

The solution set is $\{2 + \sqrt{2}, 2 - \sqrt{2}\}$, abbreviated $\{2 \pm \sqrt{2}\}$.

A Radical Departure from the Other Methods of Evaluating the Golden Ratio

Recall from a previous chapter that the golden ratio is found in numerous places in mathematics, art, and nature. In a margin note there, we showed that

$$1 + \cfrac{1}{1 + \cfrac{1}{1 + \cfrac{1}{1 + \ldots}}}$$

is equal to the golden ratio, $(1 + \sqrt{5})/2$. Now consider this "nested" radical:

$$\sqrt{1 + \sqrt{1 + \sqrt{1 + \ldots}}}.$$

Let x represent this radical. Because it appears "within itself," we can write

$$x = \sqrt{1 + x}$$
$$x^2 = 1 + x \quad \text{Square both sides.}$$
$$x^2 - x - 1 = 0. \quad \text{Write in standard form.}$$

Using the quadratic formula, with $a = 1, b = -1$, and $c = -1$, it can be shown that the positive solution of this equation, and thus the value of the nested radical is ... (you guessed it!) the golden ratio.

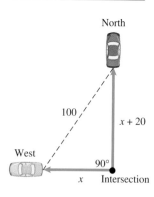

FIGURE 13

EXAMPLE 4 Solve $2x^2 = x + 4$.

To find the values of a, b, and c, first rewrite the equation in standard form as $2x^2 - x - 4 = 0$. Then $a = 2$, $b = -1$, and $c = -4$. By the quadratic formula,

$$x = \frac{-(-1) \pm \sqrt{(-1)^2 - 4(2)(-4)}}{2(2)}$$

$$x = \frac{1 \pm \sqrt{1 + 32}}{4}$$

$$x = \frac{1 \pm \sqrt{33}}{4}.$$

The solution set is $\left\{ \dfrac{1 \pm \sqrt{33}}{4} \right\}$.

Applications When solving applied problems that lead to quadratic equations, we might get a solution that does not satisfy the physical constraints of the problem. For example, if x represents a width and the two solutions of the quadratic equation are -9 and 1, the value -9 must be rejected, since a width must be a positive number.

EXAMPLE 5

Two cars left an intersection at the same time, one heading due north, and the other due west. Some time later, they were exactly 100 miles apart. The car headed north had gone 20 miles farther than the car headed west. How far had each car traveled?

Step 1: **Read** the problem carefully.

Step 2: **Assign a variable.** Let x be the distance traveled by the car headed west. Then $(x + 20)$ is the distance traveled by the car headed north. See Figure 13. The cars are 100 miles apart, so the hypotenuse of the right triangle equals 100.

Step 3: **Write an equation.** Use the Pythagorean theorem.

$$c^2 = a^2 + b^2$$
$$100^2 = x^2 + (x + 20)^2$$

Step 4: **Solve.**

$$10{,}000 = x^2 + x^2 + 40x + 400 \quad \text{Square the binomial.}$$
$$0 = 2x^2 + 40x - 9600 \quad \text{Get 0 on one side.}$$
$$0 = 2(x^2 + 20x - 4800) \quad \text{Factor out the common factor.}$$
$$0 = x^2 + 20x - 4800 \quad \text{Divide both sides by 2.}$$

Use the quadratic formula to find x.

$$x = \frac{-20 \pm \sqrt{400 - 4(1)(-4800)}}{2} \quad a = 1, b = 20, c = -4800$$

$$= \frac{-20 \pm \sqrt{19{,}600}}{2}$$

$$x = 60 \quad \text{or} \quad x = -80 \quad \text{Use a calculator.}$$

Step 5: **State the answer.** Since distance cannot be negative, discard the negative solution. The required distances are 60 miles and 60 + 20 = 80 miles.

Step 6: **Check.** Since $60^2 + 80^2 = 100^2$, the answer is correct.

EXAMPLE 6 If a rock on Earth is thrown upward from a 144-foot building with an initial velocity of 112 feet per second, its position (in feet above the ground) is given by $s = -16t^2 + 112t + 144$, where t is time in seconds after it was thrown. When does it hit the ground?

When the rock hits the ground, its distance above the ground is 0. Find t when s is 0 by solving the equation

$$0 = -16t^2 + 112t + 144. \quad \text{Let } s = 0.$$
$$0 = t^2 - 7t - 9 \quad \text{Divide both sides by } -16.$$
$$t = \frac{7 \pm \sqrt{49 + 36}}{2} \quad \text{Quadratic formula}$$
$$t = \frac{7 \pm \sqrt{85}}{2}$$
$$t \approx 8.1 \quad \text{or} \quad t \approx -1.1 \quad \text{Use a calculator.}$$

Since time cannot be negative, discard the negative solution. The rock will hit the ground about 8.1 seconds after it is thrown.

7.7 EXERCISES

Fill in each blank with the correct response.

1. For the quadratic equation $4x^2 + 5x - 9 = 0$, the values of a, b, and c are respectively ___, ___, and ___.

2. To solve the equation $3x^2 - 5x = -2$ by the quadratic formula, the first step is to add ___ to both sides of the equation.

3. When using the quadratic formula, if $b^2 - 4ac$ is positive, the equation has ___ real solution(s). (how many?)

4. If a, b, and c are integers in $ax^2 + bx + c = 0$ and $b^2 - 4ac = 17$, then the equation has ___ irrational solution(s). (how many?)

Solve each of the following equations by the zero-factor property.

5. $(x + 3)(x - 9) = 0$
6. $(m + 6)(m + 4) = 0$
7. $(2t - 7)(5t + 1) = 0$
8. $(7x - 3)(6x + 4) = 0$
9. $x^2 - x - 12 = 0$
10. $m^2 + 4m - 5 = 0$
11. $x^2 + 9x + 14 = 0$
12. $15r^2 + 7r = 2$
13. $12x^2 + 4x = 1$
14. $x(x + 3) = 4$
15. $(x + 4)(x - 6) = -16$
16. $(w - 1)(3w + 2) = 4w$

Solve each of the following by using the square root property. Give only real number solutions.

17. $x^2 = 64$
18. $w^2 = 16$
19. $x^2 = 24$
20. $x^2 = 48$
21. $r^2 = -5$
22. $x^2 = -10$
23. $(x - 4)^2 = 3$
24. $(x + 3)^2 = 11$
25. $(2x - 5)^2 = 13$
26. $(4x + 1)^2 = 19$

Solve each of the following by the quadratic formula. Give only real number solutions.

27. $4x^2 - 8x + 1 = 0$
28. $m^2 + 2m - 5 = 0$
29. $2x^2 = 2x + 1$
30. $9r^2 + 6r = 1$
31. $q^2 - 1 = q$
32. $2p^2 - 4p = 5$
33. $4k(k + 1) = 1$
34. $4r(r - 1) = 19$
35. $(g + 2)(g - 3) = 1$
36. $(x - 5)(x + 2) = 6$
37. $m^2 - 6m = -14$
38. $x^2 = 2x - 2$

39. Can the quadratic formula be used to solve the equation $2x^2 - 5 = 0$? Explain, and solve it if the answer is yes.

40. Can the quadratic formula be used to solve the equation $4x^2 + 3x = 0$? Explain, and solve it if the answer is yes.

41. Explain why the quadratic formula cannot be used to solve the equation $2x^3 + 3x - 4 = 0$.

42. A student gave the quadratic formula incorrectly as follows: $x = -b \pm \dfrac{\sqrt{b^2 - 4ac}}{2a}$. What is wrong with this?

*The expression $b^2 - 4ac$, the radicand in the quadratic formula, is called the **discriminant** of the quadratic equation $ax^2 + bx + c = 0$, $a \neq 0$. By evaluating it we can determine, without actually solving the equation, the number and nature of the solutions of the equation. Suppose that a, b, and c are integers. Then the following chart shows how the discriminant can be used to analyze the solutions.*

Discriminant	Solutions
Positive, and the square of an integer	Two different rational solutions
Positive, but not the square of an integer	Two different irrational solutions
Zero	One rational solution (a double solution)
Negative	No real solutions

*In Exercises 43–48, evaluate the discriminant, and then determine whether the equation has **(a)** two different rational solutions, **(b)** two different irrational solutions, **(c)** one rational solution (a double solution), or **(d)** no real solutions.*

43. $x^2 + 6x + 9 = 0$
44. $4x^2 + 20x + 25 = 0$
45. $6x^2 + 7x - 3 = 0$
46. $2x^2 + x - 3 = 0$
47. $9x^2 - 30x + 15 = 0$
48. $2x^2 - x + 1 = 0$

Solve each problem by using a quadratic equation. Use a calculator as necessary, and round the answer to the nearest tenth.

49. *Height of a Projectile* The Mart Hotel in Dallas, Texas, is 400 feet high. Suppose that a ball is projected upward from the top of the Mart, and its position s in feet above the ground is given by the equation $s = -16t^2 + 45t + 400$, where t is the number of seconds elapsed. How long will it take for the ball to reach a height of 200 feet above the ground? (*Source: The World Almanac and Book of Facts.*)

50. *Height of a Projectile* The Toronto Dominion Center in Winnipeg, Manitoba, is 407 feet high. Suppose that a ball is projected upward from the top of the Center, and its position s in feet above the ground is given by the equation $s = -16t^2 + 75t + 407$, where t is the number of seconds elapsed. How long will it take for the ball to reach a height of 450 feet above the ground? (*Source: The World Almanac and Book of Facts.*)

51. *Height of a Projectile* Refer to the equations in Exercises 49 and 50. Suppose that the first sentence in each problem did not give the height of the building. How could you use the equation to determine the height of the building?

52. *Position of a Searchlight Beam* A searchlight beam moves horizontally back and forth along a wall with the distance of the light from a starting point at t minutes given by $s = 100t^2 - 300t$. How long will it take before the light returns to the starting point?

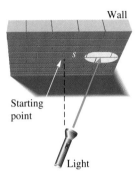

53. *Height of a Projectile* An object is projected directly upward from the ground. After t seconds its distance in feet above the ground is $s = 144t - 16t^2$.
  (a) After how many seconds will the object be 128 feet above the ground? (*Hint:* Look for a common factor before solving the equation.)
  (b) When does the object strike the ground?

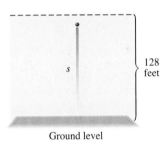

54. *Distance of a Skid* The formula $D = 100t - 13t^2$ gives the distance in feet a car going approximately 68 miles per hour will skid in t seconds. Find the time it would take for the car to skid 190 feet. (*Hint:* Your answer must be less than the time it takes the car to stop, which is 3.8 seconds.)

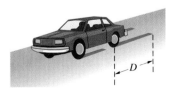

55. *Side Lengths of a Triangle* Find the lengths of the sides of the triangle.

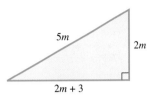

56. *Side Lengths of a Triangle* Find the lengths of the sides of the triangle.

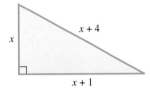

57. *Length of a Wire* Refer to Exercise 49. Suppose that a wire is attached to the top of the Mart and pulled tight. It is attached to the ground 100 feet from the base of the building, as shown in the figure. How long is the wire?

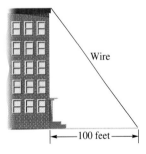

58. *Length of a Wire* Refer to Exercise 50. Suppose that a wire is attached to the top of the Center and pulled tight. The length of the wire is twice the distance between the base of the Center and the point on the ground where the wire is attached. How long is the wire?

59. *Distances Traveled by Ships* Two ships leave port at the same time, one heading due south and the other heading due east. Several hours later, they are 170 miles apart. If the ship traveling south travels 70 miles farther than the other, how many miles does each travel?

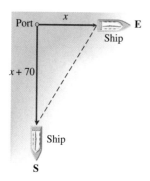

60. *Height of a Kite* Kim Hobbs is flying a kite that is 30 feet farther above her hand than its horizontal distance from her. The string from her hand to the kite is 150 feet long. How far is the kite above her hand?

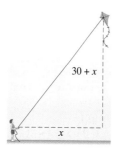

61. *Size of a Toy Piece* A toy manufacturer needs a piece of plastic in the shape of a right triangle with the longer leg 2 centimeters more than twice as long as the shorter leg, and the hypotenuse 1 centimeter more than the longer leg. How long should the three sides of the triangular piece be?

62. *Size of a Developer's Property* Michael Fuentes, a developer, owns a piece of land enclosed on three sides by streets, giving it the shape of a right triangle. The hypotenuse is 8 meters longer than the longer leg, and the shorter leg is 9 meters shorter than the hypotenuse. Find the lengths of the three sides of the property.

63. *Dimensions of Puzzle Pieces* Two pieces of a large wooden puzzle fit together to form a rectangle with a length 1 centimeter less than twice the width. The diagonal, where the two pieces meet, is 2.5 centimeters in length. Find the length and width of the rectangle.

64. *Leaning Ladder* A 13-foot ladder is leaning against a house. The distance from the bottom of the ladder to the house is 7 feet less than the distance from the top of the ladder to the ground. How far is the bottom of the ladder from the house?

65. *Dimensions of a Strip of Flooring Around a Rug* Catarina and José want to buy a rug for a room that is 15 feet by 20 feet. They want to leave an even strip of flooring uncovered around the edges of the room. How wide a strip will they have if they buy a rug with an area of 234 square feet?

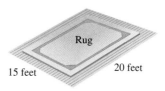

66. *Dimensions of a Border Around a Pool* A club swimming pool is 30 feet wide and 40 feet long. The club members want an exposed aggregate border in a strip of uniform width around the pool. They have enough material for 296 square feet. How wide can the strip be?

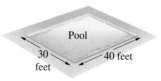

67. *Dimensions of a Garden* Arif's backyard is 20 meters by 30 meters. He wants to put a flower garden in the middle of the backyard, leaving a strip of grass of uniform width around the flower garden. Arif must have 184 square meters of grass. Under these conditions, what will the length and width of the garden be?

68. *Dimensions of a Piece of Sheet Metal* A rectangular piece of sheet metal has a length that is 4 inches less than twice the width. A square piece 2 inches on a side is cut from each corner. The sides are then turned up to form an uncovered box of volume 256 cubic inches. Find the length and width of the original piece of metal.

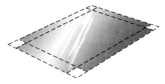

69. *Interest Rate* The formula $A = P(1 + r)^2$ gives the amount A in dollars that P dollars will grow to in 2 years at interest rate r (where r is given as a decimal), using compound interest. What interest rate will cause $2000 to grow to $2142.25 in 2 years?

70. *Cardboard Box Dimensions* If a square piece of cardboard has 3-inch squares cut from its corners and then has the flaps folded up to form an open-top box, the volume of the box is given by the formula $V = 3(x - 6)^2$, where x is the length of each side of the original piece of cardboard in inches. What original length would yield a box with a volume of 432 cubic inches?

71. *Supply and Demand for Bran Muffins* A certain bakery has found that the daily demand for bran muffins is $\frac{3200}{p}$, where p is the price of a muffin in cents. The daily supply is $3p - 200$. Find the price at which supply and demand are equal.

72. *Supply and Demand for Compact Discs* In one area the demand for compact discs is $\frac{700}{P}$ per day, where P is the price in dollars per disc. The supply is $5P - 1$ per day. At what price does supply equal demand?

Froude Number William Froude was a 19th-century naval architect who used the expression $\frac{v^2}{g\ell}$ in shipbuilding. This expression, known as the Froude number, was also used by R. McNeill Alexander in his research on dinosaurs. (See "How Dinosaurs Ran," in *Scientific American*, April 1991, pp. 130–136.)

In Exercises 73 and 74, find the value of v (in meters per second) given that $g = 9.8$ meters per second squared.

73. Rhinoceros: $\ell = 1.2$; Froude number = 2.57

74. Triceratops: $\ell = 2.8$; Froude number = .16

Recall that the corresponding sides of similar triangles are proportional. (Refer to Section 7.3 Exercises 43–46.) Use this fact to find the lengths of the indicated sides of each pair of similar triangles. Check all possible solutions in both triangles. Sides of a triangle cannot be negative.

75. Side AC

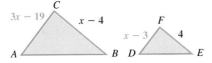

76. Side RQ

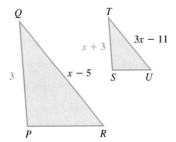

COLLABORATIVE INVESTIGATION

How Does Your Walking Rate Compare to That of Olympic Race-walkers?

Race-walking at speeds exceeding 8 miles per hour is a high fitness, long-distance competitive sport. The table below contains gold medal winners in Olympic race-walking competition.

A. Complete the table by applying the proportion given below to find the race-walker's steps per minute. **Use 10 km ≈ 6.21 miles.** (Round all answers except those for steps per minute to the nearest thousandth. Round steps per minute to the nearest whole number.)

$$\frac{70 \text{ steps per minute}}{2 \text{ miles per hour}} = \frac{x \text{ steps per minute}}{y \text{ miles per hour}}$$

Event	Gold Medal Winner	Country	Time in Hours: Minutes: Seconds	Time in Minutes	Time in Hours	y Miles per Hour	x Steps per Minute
10 km Walk, Women	Yelena Nikolayeva	Russia	0:41:49				
20 km Walk, Men	Jefferson Perez	Ecuador	1:20:07				
50 km Walk, Men	Robert Korzeniowski	Poland	3:43:30				

Source: The World Almanac and Book of Facts.

B. Using a stopwatch, take turns counting how many steps each member of the group takes in one minute while walking at a normal pace. Record the results in the table below. Then do it again at a fast pace. Record these results.

C. Use the proportion from part A to convert the numbers from part B to miles per hour and complete the chart.
  1. Find the average speed for the group at a normal pace and at a fast pace.
  2. What is the minimum number of steps per minute you would have to take to be a race-walker?
  3. At a fast pace, did anyone in the group walk fast enough to be a race-walker? Explain how you decided.

Name	Normal Pace		Fast Pace	
	x Steps per Minute	y Miles per Hour	x Steps per Minute	y Miles per Hour

CHAPTER 7 TEST

Solve each equation.

1. $5x - 3 + 2x = 3(x - 2) + 11$

2. $\dfrac{2p-1}{3} + \dfrac{p+1}{4} = \dfrac{43}{12}$

3. Decide whether the equation
   $$3x - (2 - x) + 4x = 7x - 2 - (-x)$$
   is conditional, an identity, or a contradiction. Give its solution set.

4. Solve for v: $S = vt - 16t^2$.

Solve each of the following applications.

5. *Areas of Hawaiian Islands* The three largest islands in the Hawaiian island chain are Hawaii (the Big Island), Maui, and Kauai. Together, their areas total 5300 square miles. The island of Hawaii is 3293 square miles larger than the island of Maui, and Maui is 177 square miles larger than Kauai. What is the area of each island?

6. *Chemical Mixture* How many liters of a 20% solution of a chemical should Michelle Jennings mix with 10 liters of a 50% solution to obtain a mixture that is 40% chemical?

7. *Speeds of Trains* A passenger train and a freight train leave a town at the same time and travel in opposite directions. Their speeds are 60 mph and 75 mph, respectively. How long will it take for them to be 297 miles apart?

8. *Merchandise Pricing* Which is the better buy for processed cheese slices: 8 slices for $2.19 or 12 slices for $3.30?

9. *Distance Between Cities* The distance between Milwaukee and Boston is 1050 miles. On a certain map this distance is represented by 21 inches. On the same map Seattle and Cincinnati are 46 inches apart. What is the actual distance between Seattle and Cincinnati?

10. *Current in a Circuit* The current in a simple electrical circuit is inversely proportional to the resistance. If the current is 80 amps when the resistance is 30 ohms, find the current when the resistance is 12 ohms.

Solve each inequality. Give the solution set in both interval and graph forms.

11. $-4x + 2(x - 3) \geq 4x - (3 + 5x) - 7$

12. $-10 < 3k - 4 \leq 14$

13. Which one of the following inequalities is equivalent to $x < -3$?
    A. $-3x < 9$
    B. $-3x > -9$
    C. $-3x > 9$
    D. $-3x < -9$

14. *Grade Average* Brian Kibby has scores of 83, 76, and 79 on his first three tests in Math 1031 (Survey of Mathematics). If he wants an average of at least 80 after his fourth test, what are the possible scores he can make on his fourth test?

Evaluate each exponential expression.

15. $\left(\dfrac{4}{3}\right)^2$

16. $-(-2)^6$

17. $\left(\dfrac{3}{4}\right)^{-3}$

18. $-5^0 + (-5)^0$

Use the properties of exponents to simplify each expression. Write answers with positive exponents only. Assume that all variables represent nonzero real numbers.

19. $9(4p^3)(6p^{-7})$

20. $\dfrac{m^{-2}(m^3)^{-3}}{m^{-4}m^7}$

21. Use scientific notation to evaluate
    $$\dfrac{(2{,}500{,}000)(.00003)}{(.05)(5{,}000{,}000)}.$$
    Leave the answer in scientific notation.

Solve each problem.

22. *Time Traveled for a Radio Signal* The mean distance to Earth from the planet Pluto is 4.58×10^9 kilometers. In April 1983, Pioneer 10 transmitted radio signals from Pluto to Earth at the speed of light, 3.00×10^5 kilometers per second. How long (in seconds) did it take for the signals to reach Earth?

23. *Aerospace Industry Sales* The accompanying graph depicts aerospace industry sales in current dollars. The figures at the tops of the bars represent billions of dollars. Write each figure using scientific notation.

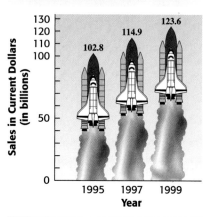

Source: U.S. Bureau of the Census, Current Industrial Reports.

Perform the indicated operations.

24. $(3k^3 - 5k^2 + 8k - 2) - (3k^3 - 9k^2 + 2k - 12)$
25. $(5x + 2)(3x - 4)$
26. $(4x^2 - 3)(4x^2 + 3)$
27. $(x + 4)(3x^2 + 8x - 9)$
28. Give an example of a polynomial in the variable t, such that it is fifth degree, in descending powers of the variable, with exactly six terms, and having a negative coefficient for its second degree term.

Factor each polynomial completely.

29. $2p^2 - 5pq + 3q^2$
30. $100x^2 - 49y^2$
31. $27y^3 - 125x^3$
32. $4x + 4y - mx - my$

Solve each quadratic equation.

33. $6x^2 + 7x - 3 = 0$
34. $x^2 - x = 7$
35. *Time an Object Has Descended* The equation

$$s = 16t^2 + 15t$$

gives the distance s in feet an object thrown off a building has descended in t seconds. Find the time t when the object has descended 25 feet. Use a calculator and round the answer to the nearest hundredth.

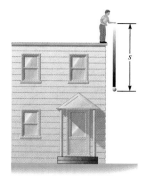

chapter 8

Graphs, Functions, and Systems of Equations and Inequalities

In 1998, Mark McGwire shattered Roger Maris's single-season home run record of 61 in 1961 by hitting 70 home runs. But McGwire's record didn't stand for long. In 2001, Barry Bonds smashed an almost unimaginable 73 homers. The all-time career home run leader is Hank Aaron with 755, followed by the immortal Babe Ruth with 714. After 17 seasons, Bonds had 613. From the graphs below, we can see how the three players' home run totals compare based on years played in the major leagues.

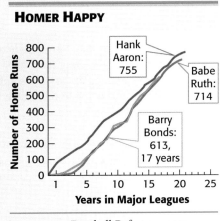

Source: www.Baseball-Reference.com.

Presenting several graphs on the same set of axes allows us to compare different trends simultaneously. This is the idea behind a *system of equations,* covered in Section 8.7. This chapter introduces the basic concepts of *graphing,* one of the fundamental procedures in the study of algebra.

8.1 The Rectangular Coordinate System and Circles

8.2 Lines and Their Slopes

8.3 Equations of Lines and Linear Models

8.4 An Introduction to Functions: Linear Functions, Applications, and Models

8.5 Quadratic Functions, Applications, and Models

8.6 Exponential and Logarithmic Functions, Applications, and Models

8.7 Systems of Equations and Applications

Extension:
  Using Matrix Row Operations to Solve Systems

8.8 Linear Inequalities, Systems, and Linear Programming

Collaborative Investigation:
  Living with AIDS

Chapter 8 Test

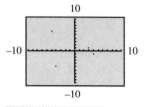

The points in Figure 2 are plotted on this calculator screen. Why is $E(-5, 0)$ not visible?

8.1 The Rectangular Coordinate System and Circles

Each of the pairs of numbers $(1, 2)$, $(-1, 5)$, and $(3, 7)$ is an example of an **ordered pair**; that is, a pair of numbers written within parentheses in which the order of the numbers is important. The two numbers are the **components** of the ordered pair. An ordered pair is graphed using two number lines that intersect at right angles at the zero points, as shown in Figure 1. The common zero point is called the **origin**. The horizontal line, the *x*-axis, represents the first number in an ordered pair, and the vertical line, the *y*-axis, represents the second. The *x*-axis and the *y*-axis make up a **rectangular** (or **Cartesian**) **coordinate system**. The axes form four **quadrants**, numbered I, II, III, and IV as shown in Figure 2. (A point on an axis is not considered to be in any of the four quadrants.)

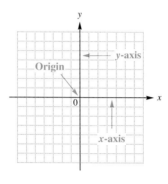
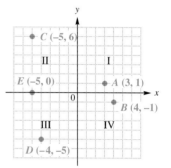

FIGURE 1 **FIGURE 2**

Double Descartes After the French postal service issued the above stamp in honor of René Descartes, sharp eyes noticed that the title of Descartes's most famous book was wrong. Thus a second stamp (see facing page) was issued with the correct title. The book in question, *Discourse on Method*, appeared in 1637. In it Descartes rejected traditional Aristotelian philosophy, outlining a universal system of knowledge that was to have the certainty of mathematics. For Descartes, method was *analysis*, going from self-evident truths step-by-step to more distant and more general truths. One of these truths is his famous statement, "I think, therefore I am." (Thomas Jefferson, also a rationalist, began the *Declaration* with the words, "We hold these truths to be self-evident.")

We locate, or **plot,** the point on the graph that corresponds to the ordered pair $(3, 1)$ by going three units from zero to the right along the *x*-axis, and then one unit up parallel to the *y*-axis. The point corresponding to the ordered pair $(3, 1)$ is labeled *A* in Figure 2. The phrase "the point corresponding to the ordered pair $(3, 1)$" often is abbreviated "the point $(3, 1)$." The numbers in an ordered pair are called the **coordinates** of the corresponding point.

The parentheses used to represent an ordered pair also are used to represent an open interval (introduced in an earlier chapter). In general, there should be no confusion between these symbols because the context of the discussion tells us whether we are discussing ordered pairs or open intervals.

Distance Formula Suppose that we wish to find the distance between two points, say $(3, -4)$ and $(-5, 3)$. The Pythagorean theorem allows us to do this. In Figure 3 on the next page, we see that the vertical line through $(-5, 3)$ and the horizontal line through $(3, -4)$ intersect at the point $(-5, -4)$. Thus, the point $(-5, -4)$ becomes the vertex of the right angle in a right triangle. By the Pythagorean Theorem, the square of the length of the hypotenuse, d, of the right triangle in Figure 3 is equal to the sum of the squares of the lengths of the two legs a and b:

$$d^2 = a^2 + b^2.$$

The length a is the distance between the endpoints of that leg. Since the *x*-coordinate of both points is -5, the side is vertical, and we can find a by finding the difference between the *y*-coordinates. Subtract -4 from 3 to get a positive value of a.

$$a = 3 - (-4) = 7$$

Similarly, find b by subtracting -5 from 3.

$$b = 3 - (-5) = 8$$

Substituting these values into the formula, we have

$$d^2 = a^2 + b^2$$
$$d^2 = 7^2 + 8^2 \quad \text{Let } a = 7 \text{ and } b = 8.$$
$$d^2 = 49 + 64$$
$$d^2 = 113$$
$$d = \sqrt{113}. \quad \text{Square root property, } d > 0.$$

Therefore, the distance between $(-5, 3)$ and $(3, -4)$ is $\sqrt{113}$.

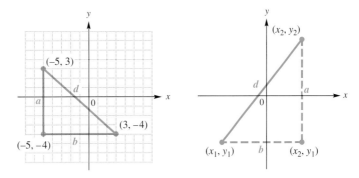

FIGURE 3   **FIGURE 4**

This result can be generalized. Figure 4 shows the two different points (x_1, y_1) and (x_2, y_2). To find a formula for the distance d between these two points, notice that the distance between (x_2, y_2) and (x_2, y_1) is given by $a = y_2 - y_1$, and the distance between (x_1, y_1) and (x_2, y_1) is given by $b = x_2 - x_1$. From the Pythagorean Theorem,

$$d^2 = (x_2 - x_1)^2 + (y_2 - y_1)^2,$$

and by using the square root property, we obtain the distance formula.

Distance Formula

The distance between the points (x_1, y_1) and (x_2, y_2) is

$$d = \sqrt{(x_2 - x_1)^2 + (y_2 - y_1)^2}.$$

This result is called the **distance formula.**

Descartes wrote his *Geometry* as an application of his method; it was published as an appendix to the *Discourse*. His attempts to unify algebra and geometry influenced the creation of what became coordinate geometry and influenced the development of calculus by Newton and Leibniz in the next generation.

In 1649 he went to Sweden to tutor Queen Christina. She preferred working in the unheated castle in the early morning; Descartes was used to staying in bed until noon. The rigors of the Swedish winter proved too much for him, and he died less than a year later.

The small numbers 1 and 2 in the ordered pairs (x_1, y_1) and (x_2, y_2) are called *subscripts*. We read x_1 as "x sub 1." Subscripts are used to distinguish between different values of a variable that have a common property. For example, in the ordered pairs $(-3, 5)$ and $(6, 4)$, -3 can be designated as x_1 and 6 as x_2. Their common property is that they are both x components of ordered pairs. This idea is used in the following example.

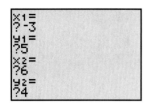

A program can be written for the distance formula. This one supports the result in Example 1, since $\sqrt{82} \approx 9.055385138$.

EXAMPLE 1 Find the distance between $(-3, 5)$ and $(6, 4)$.

When using the distance formula to find the distance between two points, designating the points as (x_1, y_1) and (x_2, y_2) is arbitrary. Let us choose $(x_1, y_1) = (-3, 5)$ and $(x_2, y_2) = (6, 4)$.

$$\begin{aligned} d &= \sqrt{(x_2 - x_1)^2 + (y_2 - y_1)^2} \\ &= \sqrt{(6 - (-3))^2 + (4 - 5)^2} \quad\quad x_2 = 6,\ y_2 = 4,\ x_1 = -3,\ y_1 = 5 \\ &= \sqrt{9^2 + (-1)^2} \\ &= \sqrt{82} \end{aligned}$$

Midpoint Formula The **midpoint** of a line segment is the point on the segment that is equidistant from both endpoints. Given the coordinates of the two endpoints of a line segment, it is not difficult to find the coordinates of the midpoint of the segment.

Midpoint Formula

The coordinates of the midpoint of the segment with endpoints (x_1, y_1) and (x_2, y_2) are

$$\left(\frac{x_1 + x_2}{2}, \frac{y_1 + y_2}{2} \right).$$

In words, the coordinates of the midpoint of a line segment are found by calculating the averages of the x- and y-coordinates of the endpoints.

EXAMPLE 2 Find the coordinates of the midpoint of the line segment with endpoints $(8, -4)$ and $(-9, 6)$.

Using the midpoint formula, we find that the coordinates of the midpoint are

$$\left(\frac{8 + (-9)}{2}, \frac{-4 + 6}{2} \right) = \left(-\frac{1}{2}, 1 \right).$$

EXAMPLE 3 Figure 5 on the next page depicts how the purchasing power of the dollar declined from 1990 to 2000, based on changes in the Consumer Price Index. In this model, the base period is 1982–1984. For example, it would have cost $1.00 to purchase in 1990 what $.766 would have purchased during the base period. Use the graph to estimate the purchasing power of the dollar in 1995, and compare it to the actual figure of $.656.

The year 1995 lies halfway between 1990 and 2000, so we must find the coordinates of the midpoint of the segment that has endpoints $(1990, .766)$ and $(2000, .581)$. This is given by

$$\left(\frac{1990 + 2000}{2}, \frac{.766 + .581}{2} \right) = (1995, .6735).$$

8.1 The Rectangular Coordinate System and Circles

Thus, based on this procedure, the purchasing power of the dollar to the nearest thousandth was $.674 in 1995. This is fairly close to the actual figure of $.656.

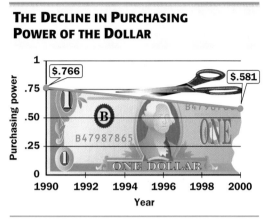

THE DECLINE IN PURCHASING POWER OF THE DOLLAR

Source: U.S. Bureau of Labor Statistics.

FIGURE 5

Circles An application of the distance formula leads to one of the most familiar shapes in geometry, the circle. A **circle** is the set of all points in a plane that lie a fixed distance from a fixed point. The fixed point is called the **center** and the fixed distance is called the **radius.**

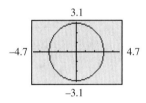

To graph $x^2 + y^2 = 9$, we solve for y to get $y_1 = \sqrt{9 - x^2}$ and $y_2 = -\sqrt{9 - x^2}$. Then we graph both in a square window.

EXAMPLE 4 Find an equation of the circle with radius 3 and center at $(0, 0)$, and graph the circle.

If the point (x, y) is on the circle, the distance from (x, y) to the center $(0, 0)$ is 3, as shown in Figure 6. By the distance formula,

$$\sqrt{(x_2 - x_1)^2 + (y_2 - y_1)^2} = d$$
$$\sqrt{(x - 0)^2 + (y - 0)^2} = 3 \qquad x_1 = 0, y_1 = 0, x_2 = x, y_2 = y$$
$$x^2 + y^2 = 9. \qquad \text{Square both sides.}$$

An equation of this circle is $x^2 + y^2 = 9$. It can be graphed by locating all points three units from the origin.

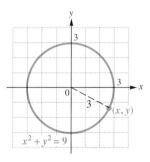

FIGURE 6

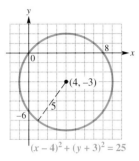

FIGURE 7

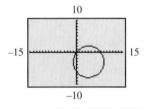

A circle can be "drawn" by using the appropriate command, entering the coordinates of the center and the radius. Compare with Example 5 and Figure 7.

EXAMPLE 5 Find an equation for the circle that has its center at $(4, -3)$ and radius 5, and graph the circle.

Again use the distance formula.

$$\sqrt{(x-4)^2 + (y+3)^2} = 5$$
$$(x-4)^2 + (y+3)^2 = 25 \quad \text{Square both sides.}$$

The graph of this circle is shown in Figure 7.

Examples 4 and 5 can be generalized to get an equation of a circle with radius r and center at (h, k). If (x, y) is a point on the circle, the distance from the center (h, k) to the point (x, y) is r. Then by the distance formula, $\sqrt{(x-h)^2 + (y-k)^2} = r$. Squaring both sides gives the following equation of a circle.

Equation of a Circle

The equation of a circle of radius r with center at (h, k) is

$$(x - h)^2 + (y - k)^2 = r^2.$$

In particular, a circle of radius r with center at the origin has equation

$$x^2 + y^2 = r^2.$$

EXAMPLE 6 Find an equation of the circle with center at $(-1, 2)$ and radius 4.

Let $h = -1$, $k = 2$, and $r = 4$ in the general equation above to get

$$(x - h)^2 + (y - k)^2 = r^2$$
$$[x - (-1)]^2 + (y - 2)^2 = 4^2$$
$$(x + 1)^2 + (y - 2)^2 = 16.$$

In the equation found in Example 5, multiplying out $(x - 4)^2$ and $(y + 3)^2$ and then combining like terms gives

$$(x - 4)^2 + (y + 3)^2 = 25$$
$$x^2 - 8x + 16 + y^2 + 6y + 9 = 25$$
$$x^2 + y^2 - 8x + 6y = 0.$$

This result suggests that an equation that has both x^2 and y^2 terms may represent a circle. The next example shows how to tell, using the method of completing the square.

EXAMPLE 7 Graph $x^2 + y^2 + 2x + 6y - 15 = 0$.

Since the equation has x^2 and y^2 terms with equal coefficients, its graph might be that of a circle. To find the center and radius, complete the squares on x and y as follows. (See the previous chapter, where completing the square is introduced.)

$$x^2 + y^2 + 2x + 6y = 15 \quad \text{Add 15 to both sides.}$$
$$(x^2 + 2x \quad) + (y^2 + 6y \quad) = 15 \quad \text{Rewrite in anticipation of completing the square.}$$
$$(x^2 + 2x + 1) + (y^2 + 6y + 9) = 15 + 1 + 9 \quad \text{Complete the squares on both } x \text{ and } y.$$
$$(x + 1)^2 + (y + 3)^2 = 25 \quad \text{Factor on the left and add on the right.}$$

The final equation shows that the graph is a circle with center at $(-1, -3)$ and radius 5. The graph is shown in Figure 8.

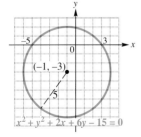

FIGURE 8

The final example in this section shows how equations of circles can be used in locating the epicenter of an earthquake.

EXAMPLE 8 Seismologists can locate the epicenter of an earthquake by determining the intersection of three circles. The radii of these circles represent the distances from the epicenter to each of three receiving stations. The centers of the circles represent the receiving stations.

Suppose receiving stations A, B, and C are located on a coordinate plane at the points $(1, 4)$, $(-3, -1)$, and $(5, 2)$. Let the distances from the earthquake epicenter to the stations be 2 units, 5 units, and 4 units, respectively. See Figure 9. Where on the coordinate plane is the epicenter located?

Graphically, it appears that the epicenter is located at $(1, 2)$. To check this algebraically, determine the equation for each circle and substitute $x = 1$ and $y = 2$.

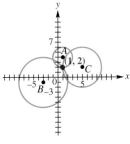

FIGURE 9

Station A:
$$(x - 1)^2 + (y - 4)^2 = 4$$
$$(1 - 1)^2 + (2 - 4)^2 = 4$$
$$0 + 4 = 4$$
$$4 = 4$$

Station B:
$$(x + 3)^2 + (y + 1)^2 = 25$$
$$(1 + 3)^2 + (2 + 1)^2 = 25$$
$$16 + 9 = 25$$
$$25 = 25$$

Station C:
$$(x - 5)^2 + (y - 2)^2 = 16$$
$$(1 - 5)^2 + (2 - 2)^2 = 16$$
$$16 + 0 = 16$$
$$16 = 16$$

Thus, we can be sure that the epicenter lies at $(1, 2)$.

8.1 EXERCISES

In Exercises 1 and 2, answer each question by locating ordered pairs on the graphs.

1. The graph shows the percent of women in math or computer science professions since 1970.
   (a) If (x, y) represents a point on the graph, what does x represent? What does y represent?
   (b) In what decade (10-year period) did the percent of women in math or computer science professions decrease?
   (c) When did the percent of women in math or computer science professions reach a maximum?
   (d) In what year was the percent of women in math or computer science professions about 27%?

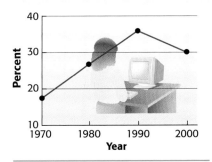

Source: U.S. Bureau of the Census and Bureau of Labor Statistics.

2. The graph indicates federal government tax revenues in billions of dollars.

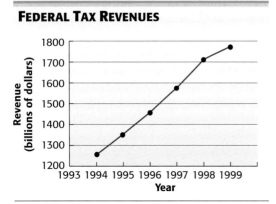

FEDERAL TAX REVENUES

Source: U.S. Office of Management and Budget.

(a) If (x, y) represents a point on the graph, what does x represent? What does y represent?
(b) What was the revenue in 1996?
(c) In what year was revenue about $1700 billion?

Fill in each blank with the correct response.

3. For any value of x, the point $(x, 0)$ lies on the _____-axis.

4. For any value of y, the point $(0, y)$ lies on the _____-axis.

5. The circle $x^2 + y^2 = 9$ has the point _____ as its center.

6. The point (_____, 0) is the center of the circle $(x - 2)^2 + y^2 = 16$.

Name the quadrant, if any, in which each point is located.

7. (a) $(1, 6)$
(b) $(-4, -2)$
(c) $(-3, 6)$
(d) $(7, -5)$
(e) $(-3, 0)$

8. (a) $(-2, -10)$
(b) $(4, 8)$
(c) $(-9, 12)$
(d) $(3, -9)$
(e) $(0, -8)$

9. Use the given information to determine the possible quadrants in which the point (x, y) must lie.
(a) $xy > 0$ (b) $xy < 0$
(c) $\dfrac{x}{y} < 0$ (d) $\dfrac{x}{y} > 0$

10. What must be true about the coordinates of any point that lies along an axis?

Locate the following points on the rectangular coordinate system, using a graph similar to Figure 2.

11. $(2, 3)$
12. $(-1, 2)$
13. $(-3, -2)$
14. $(1, -4)$
15. $(0, 5)$
16. $(-2, -4)$
17. $(-2, 4)$
18. $(3, 0)$
19. $(-2, 0)$
20. $(3, -3)$

Find (a) the distance between the pair of points, and (b) the coordinates of the midpoint of the segment having the points as endpoints.

21. $(3, 4)$ and $(-2, 1)$
22. $(-2, 1)$ and $(3, -2)$
23. $(-2, 4)$ and $(3, -2)$
24. $(1, -5)$ and $(6, 3)$
25. $(-3, 7)$ and $(2, -4)$
26. $(0, 5)$ and $(-3, 12)$

In Exercises 27–30, match each equation with the correct graph.

A.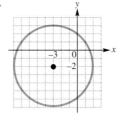

B.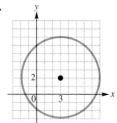

C.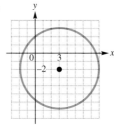

D.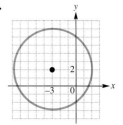

27. $(x - 3)^2 + (y - 2)^2 = 25$
28. $(x - 3)^2 + (y + 2)^2 = 25$
29. $(x + 3)^2 + (y - 2)^2 = 25$
30. $(x + 3)^2 + (y + 2)^2 = 25$

Write an equation of the circle with the given center and radius.

31. $(0, 0)$; $r = 6$
32. $(0, 0)$; $r = 5$
33. $(-1, 3)$; $r = 4$
34. $(2, -2)$; $r = 3$
35. $(0, 4)$; $r = \sqrt{3}$
36. $(-2, 0)$; $r = \sqrt{5}$

37. Suppose that a circle has an equation of the form $x^2 + y^2 = r^2, r > 0$. What is the center of the circle? What is the radius of the circle?

38. (a) How many points are there on the graph of $(x - 4)^2 + (y - 1)^2 = 0$? Explain your answer.
(b) How many points are there on the graph of $(x - 4)^2 + (y - 1)^2 = -1$? Explain your answer.

Find the center and the radius of each circle. (Hint: In Exercises 43 and 44 divide both sides by the greatest common factor.)

39. $x^2 + y^2 + 4x + 6y + 9 = 0$
40. $x^2 + y^2 - 8x - 12y + 3 = 0$
41. $x^2 + y^2 + 10x - 14y - 7 = 0$
42. $x^2 + y^2 - 2x + 4y - 4 = 0$
43. $3x^2 + 3y^2 - 12x - 24y + 12 = 0$
44. $2x^2 + 2y^2 + 20x + 16y + 10 = 0$

Graph each circle.

45. $x^2 + y^2 = 36$
46. $x^2 + y^2 = 81$
47. $(x - 2)^2 + y^2 = 36$
48. $x^2 + (y + 3)^2 = 49$
49. $(x + 2)^2 + (y - 5)^2 = 16$
50. $(x - 4)^2 + (y - 3)^2 = 25$
51. $(x + 3)^2 + (y + 2)^2 = 36$
52. $(x - 5)^2 + (y + 4)^2 = 49$

*Find **(a)** the distance between P and Q and **(b)** the coordinates of the midpoint of the segment joining P and Q.*

53.
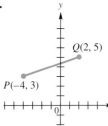

54.
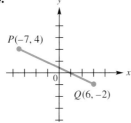

55. *(Modeling) Tuition and Fees* During the time period 1991–1999, the average annual cost (in dollars) of tuition and fees at private four-year colleges rose in an approximately linear fashion. The graph depicts this growth using a line segment. Use the midpoint formula to approximate the cost during the year 1995.

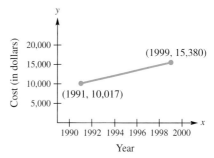

Source: The College Board.

56. *(Modeling) Two-Year College Enrollment* Enrollments in two-year colleges for 1980, 1990, and 2000 are shown in the table. Use the midpoint formula to estimate the enrollments for 1985 and 1995.

Year	Enrollment (in millions)
1980	4.5
1990	5.2
2000	5.8

Source: Statistical Abstract of the United States.

57. *(Modeling) Poverty Level Income Cutoffs* The table lists poverty level income cutoffs for a family of four since 1960. Use the midpoint formula to estimate the poverty level cutoffs in 1965 and 1985.

Year	Income (in dollars)
1960	3022
1970	3968
1980	8414
1990	13,359
1999	17,184

Source: U.S. Bureau of the Census.

58. An alternative form of the distance formula is

$$d = \sqrt{(x_1 - x_2)^2 + (y_1 - y_2)^2}.$$

Compare this to the form given in this section, and explain why the two forms are equivalent.

59. A student was asked to find the distance between the points (5, 8) and (2, 14), and wrote the following:

$$d = \sqrt{(5-8)^2 + (2-14)^2}.$$

Explain why this is incorrect.

60. A circle can be drawn on a piece of posterboard by fastening one end of a string, pulling the string taut with a pencil, and tracing a curve as shown in the figure. Explain why this method works.

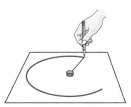

61. *Crawfish Racing* This figure shows how the crawfish race is held at the Crawfish Festival in Breaux Bridge, Louisiana. Explain why a circular "racetrack" is appropriate for such a race.

62. *Epicenter of an Earthquake* Show algebraically that if three receiving stations at $(1, 4)$, $(-6, 0)$, and $(5, -2)$ record distances to an earthquake epicenter of 4 units, 5 units, and 10 units respectively, the epicenter would lie at $(-3, 4)$.

63. *Epicenter of an Earthquake* Three receiving stations record the presence of an earthquake. The locations of the receiving stations and the distances to the epicenter are contained in the following three equations: $(x - 2)^2 + (y - 1)^2 = 25$, $(x + 2)^2 + (y - 2)^2 = 16$, and $(x - 1)^2 + (y + 2)^2 = 9$. Graph the circles and determine the location of the earthquake epicenter.

64. Without actually graphing, state whether the graphs of $x^2 + y^2 = 4$ and $x^2 + y^2 = 25$ will intersect. Explain your answer.

65. Can a circle have its center at $(2, 4)$ and be tangent to both axes? (*Tangent to* means touching in one point.) Explain.

66. Suppose that the endpoints of a line segment have coordinates (x_1, y_1) and (x_2, y_2).
    (a) Show that the distance between (x_1, y_1) and $\left(\dfrac{x_1 + x_2}{2}, \dfrac{y_1 + y_2}{2}\right)$ is the same as the distance between (x_2, y_2) and $\left(\dfrac{x_1 + x_2}{2}, \dfrac{y_1 + y_2}{2}\right)$.
    (b) Show that the sum of the distances between (x_1, y_1) and $\left(\dfrac{x_1 + x_2}{2}, \dfrac{y_1 + y_2}{2}\right)$, and (x_2, y_2) and $\left(\dfrac{x_1 + x_2}{2}, \dfrac{y_1 + y_2}{2}\right)$ is equal to the distance between (x_1, y_1) and (x_2, y_2).
    (c) From the results of parts (a) and (b), what conclusion can be made?

67. If the coordinates of one endpoint of a line segment are $(3, -8)$ and the coordinates of the midpoint of the segment are $(6, 5)$, what are the coordinates of the other endpoint?

68. Which one of the following has a circle as its graph?
    A. $x^2 - y^2 = 9$ **B.** $x^2 = 9 - y^2$
    C. $y^2 - x^2 = 9$ **D.** $-x^2 - y^2 = 9$

69. For the three choices that are not circles in Exercise 63, explain why their equations are not those of circles.

8.2 Lines and Their Slopes

Linear Equations in Two Variables In the previous chapter we studied linear equations in a single variable. The solution of such an equation is a real number. A linear equation in *two* variables will have solutions written as ordered pairs. Unlike linear equations in a single variable, equations with two variables will, in general, have an infinite number of solutions.

To find ordered pairs that satisfy the equation, select any number for one of the variables, substitute it into the equation for that variable, and then solve for the other variable. For example, suppose $x = 0$ in the equation $2x + 3y = 6$. Then

$$2x + 3y = 6$$
$$2(0) + 3y = 6 \quad \text{Let } x = 0.$$
$$0 + 3y = 6$$
$$3y = 6$$
$$y = 2,$$

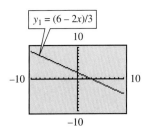

Graphing calculators can generate tables of ordered pairs. Here is an example for $2x + 3y = 6$. We must solve for y to get $Y_1 = (6 - 2X)/3$ before generating the table.

giving the ordered pair $(0, 2)$. Other ordered pairs satisfying $2x + 3y = 6$ include $(6, -2)$, $(3, 0)$, $(-3, 4)$, and $(9, -4)$.

The equation $2x + 3y = 6$ is graphed by first plotting all the ordered pairs mentioned above. These are shown in Figure 10(a). The resulting points appear to lie on a straight line. If all the ordered pairs that satisfy the equation $2x + 3y = 6$ were graphed, they would form a straight line. In fact, the graph of any first-degree equation in two variables is a straight line. The graph of $2x + 3y = 6$ is the line shown in Figure 10(b).

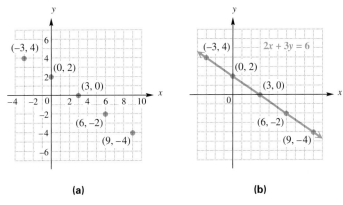

This is a calculator graph of the line shown in Figure 10(b).

FIGURE 10

Linear Equation in Two Variables

An equation that can be written in the form

$$Ax + By = C \quad (A \text{ and } B \text{ not both } 0)$$

is a **linear equation in two variables.** This form is called **standard form.**

All first-degree equations with two variables have straight-line graphs. Since a straight line is determined if any two different points on the line are known, finding two different points is enough to graph the line.

Two points that are useful for graphing lines are the x- and y-intercepts. The **x-intercept** is the point (if any) where the line crosses the x-axis, and the **y-intercept** is the point (if any) where the line crosses the y-axis. (*Note:* In many texts, the

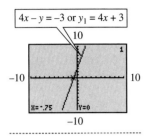

$4x - y = -3$ or $y_1 = 4x + 3$

The display at the bottom of the screen supports the fact that $(-3/4, 0)$ is the x-intercept of the line in Figure 11. We could locate the y-intercept similarly.

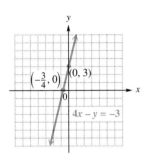

FIGURE 11

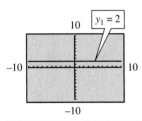

Compare this graph with the one in Figure 12(a).

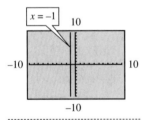

This vertical line is not an example of a *function* (see Section 8.4), so we must use a *draw* command to obtain it. Compare with Figure 12(b).

intercepts are defined as numbers, and not points. However, in this book we will refer to intercepts as points.) Intercepts can be found as follows.

Intercepts

To find the x-intercept of the graph of a linear equation, let $y = 0$.
To find the y-intercept, let $x = 0$.

EXAMPLE 1 Find the x- and y-intercepts of $4x - y = -3$, and graph the equation.

To find the x-intercept, let $y = 0$.

$4x - 0 = -3$ Let $y = 0$.
$4x = -3$
$x = -\dfrac{3}{4}$ x-intercept is $\left(-\dfrac{3}{4}, 0\right)$.

To find the y-intercept, let $x = 0$.

$4(0) - y = -3$ Let $x = 0$.
$-y = -3$
$y = 3$ y-intercept is $(0, 3)$.

The intercepts are the two points $(-3/4, 0)$ and $(0, 3)$. Use these two points to draw the graph, as shown in Figure 11.

A line may not have an x-intercept, or it may not have a y-intercept.

EXAMPLE 2 Graph each line.

(a) $y = 2$

Writing $y = 2$ as $0x + 1y = 2$ shows that any value of x, including $x = 0$, gives $y = 2$, making the y-intercept $(0, 2)$. Since y is always 2, there is no value of x corresponding to $y = 0$, and so the graph has no x-intercept. The graph, shown in Figure 12(a), is a horizontal line.

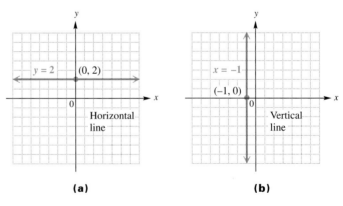

FIGURE 12

(b) $x = -1$

The form $1x + 0y = -1$ shows that every value of y leads to $x = -1$, and so no value of y makes $x = 0$. The graph, therefore, has no y-intercept. The only way a straight line can have no y-intercept is to be vertical, as shown in Figure 12(b).

8.2 Lines and Their Slopes

Slope Two different points determine a line. A line also can be determined by a point on the line and some measure of the "steepness" of the line. The measure of the steepness of a line is called the *slope* of the line. One way to get a measure of the steepness of a line is to compare the vertical change in the line (the *rise*) to the horizontal change (the *run*) while moving along the line from one fixed point to another.

Suppose that (x_1, y_1) and (x_2, y_2) are two different points on a line. Then, going along the line from (x_1, y_1) to (x_2, y_2), the y-value changes from y_1 to y_2, an amount equal to $y_2 - y_1$. As y changes from y_1 to y_2, the value of x changes from x_1 to x_2 by the amount $x_2 - x_1$. See Figure 13. The ratio of the change in y to the change in x is called the **slope** of the line. The letter m is used to denote the slope.

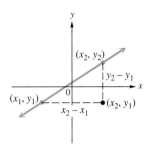

FIGURE 13

Slope

If $x_1 \neq x_2$, the slope of the line through the distinct points (x_1, y_1) and (x_2, y_2) is

$$m = \frac{\text{rise}}{\text{run}} = \frac{\text{change in } y}{\text{change in } x} = \frac{y_2 - y_1}{x_2 - x_1}.$$

EXAMPLE 3 Find the slope of the line that passes through the points $(2, -1)$ and $(-5, 3)$.

If $(2, -1) = (x_1, y_1)$ and $(-5, 3) = (x_2, y_2)$, then

$$m = \frac{y_2 - y_1}{x_2 - x_1} = \frac{3 - (-1)}{-5 - 2} = \frac{4}{-7} = -\frac{4}{7}.$$

See Figure 14. On the other hand, if $(2, -1) = (x_2, y_2)$ and $(-5, 3) = (x_1, y_1)$, the slope would be

$$m = \frac{-1 - 3}{2 - (-5)} = \frac{-4}{7} = -\frac{4}{7},$$

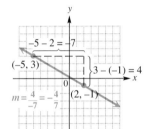

FIGURE 14

the same answer. This example suggests that the slope is the same no matter which point is considered first. Also, using similar triangles from geometry, it can be shown that the slope is the same no matter which two different points on the line are chosen. ∎

If we apply the slope formula to a vertical or a horizontal line, we find that either the numerator or denominator in the fraction is 0.

EXAMPLE 4 Find the slope, if possible, of each of the following lines.

(a) $x = -3$

By inspection, $(-3, 5)$ and $(-3, -4)$ are two points that satisfy the equation $x = -3$. Use these two points to find the slope.

$$m = \frac{-4 - 5}{-3 - (-3)} = \frac{-9}{0} \quad \text{Undefined slope}$$

Since division by zero is undefined, the slope is undefined. This is why the definition of slope includes the restriction that $x_1 \neq x_2$.

Highway slopes are measured in percent. For example, a slope of 8% means that the road gains 8 feet in altitude for each 100 feet that the road travels horizontally. Interstate highways cannot exceed a slope of 6%. While this may not seem like much of a slope, there are probably stretches of interstate highways that would be hard work for a distance runner.

(b) $y = 5$

Find the slope by selecting two different points on the line, such as $(3, 5)$ and $(-1, 5)$, and by using the definition of slope.

$$m = \frac{5 - 5}{3 - (-1)} = \frac{0}{4} = 0 \quad \text{Zero slope}$$

In Example 2, $x = -1$ has a graph that is a vertical line, and $y = 2$ has a graph that is a horizontal line. Generalizing from those results and the results of Example 4, we can make the following statements about vertical and horizontal lines.

Vertical and Horizontal Lines

A vertical line has an equation of the form $x = a$, where a is a real number, and its slope is undefined. A horizontal line has an equation of the form $y = b$, where b is a real number, and its slope is 0.

If we know the slope of a line and a point contained on the line, then we can graph the line using the method shown in the next example.

EXAMPLE 5 Graph the line that has slope $2/3$ and goes through the point $(-1, 4)$.

First locate the point $(-1, 4)$ on a graph as shown in Figure 15. Then, from the definition of slope,

$$m = \frac{\text{change in } y}{\text{change in } x} = \frac{2}{3}.$$

Move *up* 2 units in the *y*-direction and then 3 units to the *right* in the *x*-direction to locate another point on the graph (labeled *P*). The line through $(-1, 4)$ and *P* is the required graph.

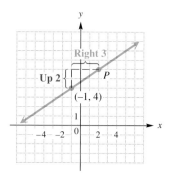

FIGURE 15

The line graphed in Figure 14 has a negative slope, $-4/7$, and the line goes down from left to right. In contrast, the line graphed in Figure 15 has a positive slope, 2/3, and it goes up from left to right. These are particular cases of a general statement that can be made about slopes. (Figure 16 shows lines of positive, zero, negative, and undefined slopes.)

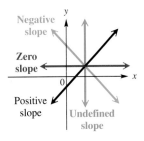

FIGURE 16

Positive and Negative Slopes

A line with a positive slope goes up (rises) from left to right, while a line with a negative slope goes down (falls) from left to right.

Parallel and Perpendicular Lines

The slopes of a pair of parallel or perpendicular lines are related in a special way. The slope of a line measures the steepness of the line. Since parallel lines have equal steepness, their slopes also must be equal. Also, lines with the same slope are parallel.

Slopes of Parallel Lines

Two nonvertical lines with the same slope are parallel; two nonvertical parallel lines have the same slope. Furthermore, any two vertical lines are parallel.

EXAMPLE 6 Are the lines L_1, through $(-2, 1)$ and $(4, 5)$, and L_2, through $(3, 0)$ and $(0, -2)$, parallel?

The slope of L_1 is

$$m_1 = \frac{5-1}{4-(-2)} = \frac{4}{6} = \frac{2}{3}.$$

The slope of L_2 is

$$m_2 = \frac{-2-0}{0-3} = \frac{-2}{-3} = \frac{2}{3}.$$

Since the slopes are equal, the lines are parallel.

Perpendicular lines are lines that meet at right angles. It can be shown that the slopes of perpendicular lines have a product of -1, provided that neither line is vertical. For example, if the slope of a line is 3/4, then any line perpendicular to it has slope $-4/3$, because $(3/4)(-4/3) = -1$.

Slopes of Perpendicular Lines

If neither is vertical, two perpendicular lines have slopes that are negative reciprocals; that is, their product is -1. Also, two lines with slopes that are negative reciprocals are perpendicular. Every vertical line is perpendicular to every horizontal line.

EXAMPLE 7 Are the lines L_1, through $(0, -3)$ and $(2, 0)$, and L_2, through $(-3, 0)$ and $(0, -2)$, perpendicular?

The slope of L_1 is

$$m_1 = \frac{0 - (-3)}{2 - 0} = \frac{3}{2}.$$

The slope of L_2 is

$$m_2 = \frac{-2 - 0}{0 - (-3)} = -\frac{2}{3}.$$

Since the product of the slopes of the two lines is $(3/2)(-2/3) = -1$, the lines are perpendicular.

Average Rate of Change

We have seen how the slope of a line is the ratio of the change in y (vertical change) to the change in x (horizontal change). This idea can be extended to real-life situations as follows: the slope gives the average rate of change of y per unit of change in x, where the value of y *is dependent upon the value of x.* The next example illustrates this idea of average rate of change. We assume a linear relationship between x and y.

EXAMPLE 8 The graph in Figure 17 approximates the percent of U.S. households owning multiple personal computers in the years 1997–2001. Find the average rate of change in percent per year for the years 1998 to 2001.

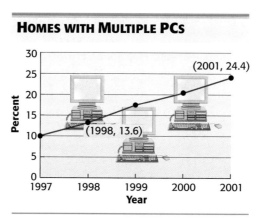

Source: The Yankee Group.

FIGURE 17

To use the slope formula, we need two pairs of data. From the graph, if we let $x = 1998$, then $y = 13.6$ and if $x = 2001$, then $y = 24.4$, so we have the ordered pairs $(1998, 13.6)$ and $(2001, 24.4)$. By the slope formula,

$$\text{average rate of change} = \frac{y_2 - y_1}{x_2 - x_1} = \frac{24.4 - 13.6}{2001 - 1998} = \frac{10.8}{3} = 3.6.$$

This means that the number of U.S. households owning multiple computers *increased* by an average of 3.6% each year in the period from 1998 to 2001.

8.2 EXERCISES

Complete the given ordered pairs for each equation. Then graph the equation.

1. $2x + y = 5$; $(0, \ ), (\ , 0), (1, \ ), (\ , 1)$
2. $3x - 4y = 24$; $(0, \ ), (\ , 0), (6, \ ), (\ , -3)$
3. $x - y = 4$; $(0, \ ), (\ , 0), (2, \ ), (\ , -1)$
4. $x + 3y = 12$; $(0, \ ), (\ , 0), (3, \ ), (\ , 6)$
5. $4x + 5y = 20$; $(0, \ ), (\ , 0), (3, \ ), (\ , 2)$
6. $2x - 5y = 12$; $(0, \ ), (\ , 0), (\ , -2), (-2, \ )$
7. $3x + 2y = 8$
8. $5x + y = 12$

x	y
0	
	0
2	
	-2

x	y
0	
	0
	-3
2	

9. Explain how to find the *x*-intercept of a linear equation in two variables.

10. Explain how to find the *y*-intercept of a linear equation in two variables.

11. Which one of the following has as its graph a horizontal line?
    A. $2y = 6$ B. $2x = 6$
    C. $x - 4 = 0$ D. $x + y = 0$

12. What is the minimum number of points that must be determined in order to graph a linear equation in two variables?

For each equation, give the x-intercept and the y-intercept. Then graph the equation.

13. $3x + 2y = 12$
14. $2x + 5y = 10$
15. $5x + 6y = 10$
16. $3y + x = 6$
17. $2x - y = 5$
18. $3x - 2y = 4$
19. $x - 3y = 2$
20. $y - 4x = 3$
21. $y + x = 0$
22. $2x - y = 0$
23. $3x = y$
24. $x = -4y$
25. $x = 2$
26. $y = -3$
27. $y = 4$
28. $x = -2$

In Exercises 29–36, match the equation with the figure that most closely resembles its graph.

A.
B.
C.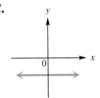
D.

29. $y + 2 = 0$
30. $y + 4 = 0$
31. $x + 3 = 0$
32. $x + 7 = 0$
33. $y - 2 = 0$
34. $y - 4 = 0$
35. $x - 3 = 0$
36. $x - 7 = 0$

37. What is the slope (or pitch) of this roof?

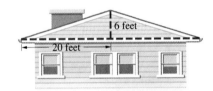

38. What is the slope (or grade) of this hill?

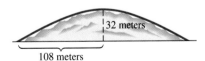

39. Use the coordinates of the indicated points to find the slope of each line.

(a) 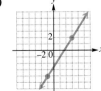   (b)

40. Tell whether the slope of the given line is positive, negative, zero, or undefined.

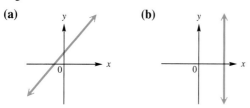

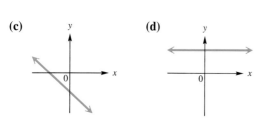

Find the slope of the line through each pair of points by using the slope formula.

41. $(-2, -3)$ and $(-1, 5)$

42. $(-4, 3)$ and $(-3, 4)$

43. $(8, 1)$ and $(2, 6)$

44. $(13, -3)$ and $(5, 6)$

45. $(2, 4)$ and $(-4, 4)$

46. $(-6, 3)$ and $(2, 3)$

47. *Public School Data* Figure A depicts public school enrollment (in thousands) in grades 9–12 in the United States. Figure B in the next column gives the (average) number of public school students per computer.

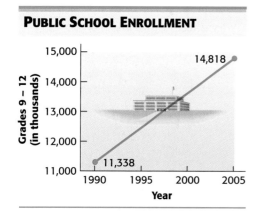

Source: Digest of Educational Statistics, annual; and *Projections of Educational Statistics*, annual.

FIGURE A

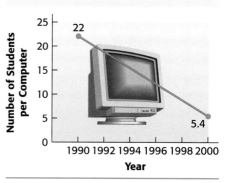

Source: Quality Education Data.

FIGURE B

(a) Use the ordered pairs (1990, 11,338) and (2005, 14,818) to find the slope of the line in Figure A.

(b) The slope of the line in Figure A is _____.
(positive/negative)
This means that during the period represented, enrollment _____.
(increased/decreased)

(c) The slope of a line represents its *rate of change*. Based on Figure A, what was the increase in students *per year* during the period shown?

(d) Use the given ordered pairs to find the slope of the line in Figure B.

(e) The slope of the line in Figure B is _____.
(positive/negative)
This means that during the period represented, the number of students per computer _____.
(increased/decreased)

(f) Based on Figure B, what was the decrease in students per computer *per year* during the period shown?

48. Use the results of Exercise 47 to make a connection between the sign of the slope of a line and the increase or decrease in the quantity represented by *y*.

Use the method of Example 5 to graph each of the following lines.

49. $m = \dfrac{1}{2}$, through $(-3, 2)$

50. $m = \dfrac{2}{3}$, through $(0, 1)$

51. $m = -\dfrac{5}{4}$, through $(-2, -1)$

52. $m = -\dfrac{3}{2}$, through $(-1, -2)$

53. $m = -2$, through $(-1, -4)$

54. $m = 3$, through $(1, 2)$

55. $m = 0$, through $(2, -5)$

56. undefined slope, through $(-3, 1)$

Determine whether the lines described are parallel, perpendicular, *or neither parallel nor perpendicular.*

57. L_1 through $(4, 6)$ and $(-8, 7)$, and L_2 through $(7, 4)$ and $(-5, 5)$

58. L_1 through $(9, 15)$ and $(-7, 12)$, and L_2 through $(-4, 8)$ and $(-20, 5)$

59. L_1 through $(2, 0)$ and $(5, 4)$, and L_2 through $(6, 1)$ and $(2, 4)$

60. L_1 through $(0, -7)$ and $(2, 3)$, and L_2 through $(0, -3)$ and $(1, -2)$

61. L_1 through $(0, 1)$ and $(2, -3)$, and L_2 through $(10, 8)$ and $(5, 3)$

62. L_1 through $(1, 2)$ and $(-7, -2)$, and L_2 through $(1, -1)$ and $(5, -9)$

Use the concept of slope to solve each problem.

63. *Steepness of an Upper Deck* The upper deck at Comiskey Park in Chicago has produced, among other complaints, displeasure with its steepness. It has been compared to a ski jump. It is 160 ft from home plate to the front of the upper deck and 250 ft from home plate to the back. The top of the upper deck is 63 ft above the bottom. What is its slope?

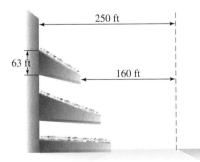

64. *Grade (Slope) of a Ramp* When designing the new FleetCenter arena in Boston to replace the old Boston Garden, architects were careful to design the ramps leading up to the entrances so that circus elephants would be able to walk up the ramps. The maximum grade (or slope) that an elephant will walk on is 13%. Suppose that such a ramp was constructed with a horizontal run of 150 ft. What would be the maximum vertical rise the architects could use?

Use the idea of average rate of change to solve each problem.

65. *Pharmaceutical Research* Merck pharmaceutical company research and development expenditures (in millions of dollars) in recent years are closely approximated by the graph.

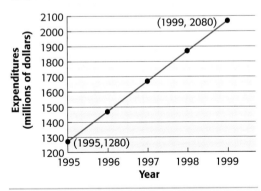

Source: Merck & Co., Inc. 1999 Annual Report.

(a) Use the given ordered pairs to determine the average rate of change in these expenditures per year.

(b) Explain how a positive rate of change is interpreted in this situation.

66. *Food Stamp Recipients* The graph provides a good approximation of the number of food stamp recipients (in millions) during the period 1994–1998.

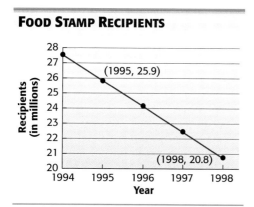

Source: U.S. Bureau of the Census.

(a) Use the given ordered pairs to find the average rate of change in food stamp recipients per year during this period.

(b) Interpret what a negative slope means in this situation.

67. *Book Sales* The table gives book publishers' net dollar sales for the period 1995–2000.

Book Publishers' Sales	
Year	Sales (in millions)
1995	19,000
1996	20,000
1997	21,000
1998	22,000
1999	23,000
2000	24,000

Source: Book Industry Study Group.

(a) Find the average rate of change for 1995–1996; 1995–1999; 1998–2000. What do you notice about your answers? What does this tell you?

(b) Calculate the rates of change in part (a) as percents. What do you notice?

68. *Cellular Telephone Subscribers* The table gives the number of cellular telephone subscribers (in thousands) from 1994 through 1999.

Cellular Telephone Subscribers	
Year	Subscribers (in thousands)
1994	24,134
1995	33,786
1996	44,043
1997	55,312
1998	69,209
1999	86,047

Source: Cellular Telecommunications Industry Association, Washington, D.C. *State of the Cellular Industry* (Annual).

(a) Find the average rate of change in subscribers for 1994–1995, 1995–1996, and so on.

(b) Is the average rate of change in successive years approximately the same? If the ordered pairs in the table were plotted, could an approximately straight line be drawn through them?

69. Explain the meaning of *slope*. Give examples showing cases of positive, negative, zero, and undefined slope.

70. Explain how the *grade* of a highway corresponds to the slope concept.

8.3 Equations of Lines and Linear Models

Equations of Lines If the slope of a line and a particular point on the line are known, it is possible to find an equation of the line. Suppose that the slope of a line is m and (x_1, y_1) is a particular point on the line. Let (x, y) be any other point on the line. Then, by the definition of slope,

$$m = \frac{y - y_1}{x - x_1}.$$

Multiplying both sides by $x - x_1$ gives the *point-slope form* of the equation of the line.

Point-Slope Form

The equation of the line through (x_1, y_1) with slope m is written in **point-slope form** as

$$y - y_1 = m(x - x_1).$$

Maria Gaetana Agnesi (1719–1799) did much of her mathematical work in coordinate geometry. She grew up in a scholarly atmosphere; her father was a mathematician on the faculty at the University of Bologna. In a larger sense she was an heir to the long tradition of Italian mathematicians.

Maria Agnesi was fluent in several languages by age 13, but she chose mathematics over literature. The curve shown below, called the *witch of Agnesi*, is studied in analytic geometry courses.

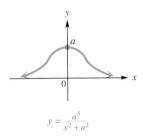

$y = \dfrac{a^3}{x^2 + a^2}$

EXAMPLE 1 Find the standard form of an equation of the line with slope $1/3$, passing through the point $(-2, 5)$.

Use the point-slope form of the equation of a line, with $(x_1, y_1) = (-2, 5)$ and $m = 1/3$.

$$y - y_1 = m(x - x_1)$$
$$y - 5 = \frac{1}{3}[x - (-2)] \quad \text{Let } y_1 = 5,\ m = \frac{1}{3},\ x_1 = -2.$$
$$y - 5 = \frac{1}{3}(x + 2)$$
$$3y - 15 = x + 2 \quad \text{Multiply by 3.}$$
$$x - 3y = -17 \quad \text{Standard form}$$

If two points on a line are known, it is possible to find an equation of the line. First, find the slope using the slope formula, and then use the slope with one of the given points in the point-slope form.

EXAMPLE 2 Find the standard form of an equation of the line passing through the points $(-4, 3)$ and $(5, -7)$.

First find the slope, using the definition.

$$m = \frac{-7 - 3}{5 - (-4)} = -\frac{10}{9}$$

Either $(-4, 3)$ or $(5, -7)$ may be used as (x_1, y_1) in the point-slope form of the equation of the line. If $(-4, 3)$ is used, then $-4 = x_1$ and $3 = y_1$.

$$y - y_1 = m(x - x_1)$$
$$y - 3 = -\frac{10}{9}[x - (-4)] \quad \text{Let } y_1 = 3,\ m = -\frac{10}{9},\ x_1 = -4.$$
$$y - 3 = -\frac{10}{9}(x + 4)$$
$$9(y - 3) = -10(x + 4) \quad \text{Multiply by 9.}$$
$$9y - 27 = -10x - 40 \quad \text{Distributive property}$$
$$10x + 9y = -13 \quad \text{Standard form}$$

Suppose that the slope m of a line is known, and the y-intercept of the line has coordinates $(0, b)$. Then substituting into the point-slope form gives

$$y - y_1 = m(x - x_1)$$
$$y - b = m(x - 0) \quad x_1 = 0, y_1 = b$$
$$y - b = mx$$
$$y = mx + b. \quad \text{Add } b \text{ to both sides.}$$

This last result is known as the *slope-intercept form* of the equation of the line.

Slope-Intercept Form

The equation of a line with slope m and y-intercept $(0, b)$ is written in **slope-intercept form** as

$$y = mx + b.$$

where m is the Slope and y-intercept is $(0, b)$.

The importance of the slope-intercept form of a linear equation cannot be overemphasized. First, every linear equation (of a nonvertical line) has a *unique* (one and only one) slope-intercept form. Second, in the next section we will study *linear functions*, where the slope-intercept form is necessary in specifying such functions.

EXAMPLE 3

(a) Write the equation of the line described in Example 1 in slope-intercept form.

We determined the standard form of the equation of the line to be

$$x - 3y = -17.$$

Solve for y to obtain the slope-intercept form.

$$-3y = -x - 17 \quad \text{Subtract } x.$$
$$y = \frac{1}{3}x + \frac{17}{3} \quad \text{Multiply by } -\frac{1}{3}.$$

The slope is $1/3$ and the y-intercept is $(0, 17/3)$.

(b) Write the equation of the line described in Example 2 in slope-intercept form.

$$10x + 9y = -13 \quad \text{Standard form}$$
$$9y = -10x - 13$$
$$y = -\frac{10}{9}x - \frac{13}{9}$$

The slope is $-10/9$ and the y-intercept is $(0, -13/9)$.

If the slope-intercept form of the equation of a line is known, the method of graphing described in Example 5 of the previous section can be used to graph the line.

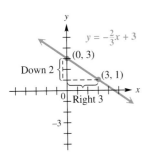

FIGURE 18

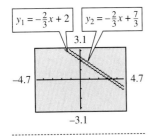

This screen gives support to the result in Example 5(a).

EXAMPLE 4 Graph the line with the equation $y = -\frac{2}{3}x + 3$.

Since the equation is given in slope-intercept form, we can easily see that the slope is $-\frac{2}{3}$ and the y-intercept is $(0, 3)$. Plot the point $(0, 3)$, and then, using the "rise over run" interpretation of slope, move *down* 2 units (because of the -2 in the numerator of the slope) and to the *right* 3 units (because of the 3 in the denominator). We arrive at the point $(3, 1)$. Plot the point $(3, 1)$, and join the two points with a line, as shown in Figure 18. (We could also have interpreted $\frac{-2}{3}$ as $\frac{2}{-3}$ and obtained a different second point; however, the line would be the same.)

As mentioned in the previous section, parallel lines have the same slope and perpendicular lines have slopes that are negative reciprocals of each other.

EXAMPLE 5 Find the slope-intercept form of the equation of each line.

(a) The line parallel to the graph of $2x + 3y = 6$, passing through the point $(-4, 5)$

The slope of the line $2x + 3y = 6$ can be found by solving for y.

$$2x + 3y = 6$$

$$3y = -2x + 6 \quad \text{Subtract } 2x.$$

$$y = -\frac{2}{3}x + 2 \quad \text{Divide by 3.}$$
$$\uparrow\!\!\text{Slope}$$

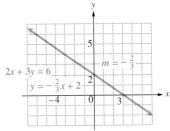

The slope is given by the coefficient of x, so $m = -\frac{2}{3}$. See the figure. The required equation of the line through $(-4, 5)$ and parallel to $2x + 3y = 6$ must also have slope $-\frac{2}{3}$. To find this equation, use the point-slope form, with $(x_1, y_1) = (-4, 5)$ and $m = -\frac{2}{3}$.

$$y - 5 = -\frac{2}{3}[x - (-4)] \quad y_1 = 5, m = -\frac{2}{3}, x_1 = -4$$

$$y - 5 = -\frac{2}{3}(x + 4)$$

$$y - 5 = -\frac{2}{3}x - \frac{8}{3} \quad \text{Distributive property}$$

$$y = -\frac{2}{3}x - \frac{8}{3} + \frac{15}{3} \quad \text{Add } 5 = \frac{15}{3}.$$

$$y = -\frac{2}{3}x + \frac{7}{3} \quad \text{Combine like terms.}$$

We did not clear fractions after the substitution step here because we want the equation in slope-intercept form—that is, solved for y. Both lines are shown in the figure.

(b) The line perpendicular to the graph of $2x + 3y = 6$, passing through the point $(-4, 5)$

To be perpendicular to the line $2x + 3y = 6$, a line must have a slope that is the negative reciprocal of $-\frac{2}{3}$, which is $\frac{3}{2}$. Use the point $(-4, 5)$ and slope $\frac{3}{2}$ in the point-slope form to obtain the equation of the perpendicular line shown in the figure.

$y - 5 = \dfrac{3}{2}[x - (-4)]$ $y_1 = 5, m = \dfrac{3}{2}, x_1 = -4$

$y - 5 = \dfrac{3}{2}(x + 4)$

$y - 5 = \dfrac{3}{2}x + 6$ Distributive property

$y = \dfrac{3}{2}x + 11$ Add 5.

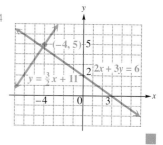

A summary of the various forms of linear equations follows.

Summary of Forms of Linear Equations

$Ax + By = C$ **Standard form**
(Neither A nor B is 0.)

$x = a$ **Vertical line**
Undefined slope; x-intercept is $(a, 0)$.

$y = b$ **Horizontal line**
Slope is 0; y-intercept is $(0, b)$.

$y = mx + b$ **Slope-intercept form**
Slope is m; y-intercept is $(0, b)$.

$y - y_1 = m(x - x_1)$ **Point-slope form**
Slope is m; Line passes through (x_1, y_1).

A Linear Model Earlier examples and exercises gave equations that described real data. Now we are able to show how such equations can be found. The process of writing an equation to fit a graph is called *curve-fitting*. The following example is an illustration of this concept for a straight line. The resulting equation is called a **linear model.**

EXAMPLE 6   Retail spending (in billions of dollars) on prescription drugs in the U.S. is shown in the graph in Figure 19 on the next page.

(a) Write an equation that models the data.

The data shown in the bar graph increase linearly; that is, a straight line could be drawn through the tops of any two bars that will pass close to the tops of all the bars. We can use the data and the point-slope form of the equation of a line to get an equation that models the relationship between year x and spending on prescription drugs y. If we let $x = 7$ represent 1997, $x = 8$ represent 1998, and so on, the given data can be written as the ordered pairs $(7, 79)$ and $(11, 155)$. The slope of the line through these two points is

$$m = \frac{155 - 79}{11 - 7} = \frac{76}{4} = 19.$$

Thus, retail spending on prescription drugs increased by about $19 billion per year. Using this slope, one of the points, say $(7, 79)$, and the point-slope form, we obtain

$y - y_1 = m(x - x_1)$ Point-slope form

$y - 79 = 19(x - 7)$ $(x_1, y_1) = (7, 79); m = 19$

$y - 79 = 19x - 133$ Distributive property

$y = 19x - 54.$ Slope-intercept form

Thus, retail spending y (in billions of dollars) on prescription drugs in the U.S. in year x can be approximated by the equation $y = 19x - 54$.

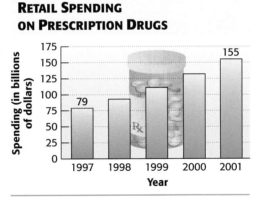
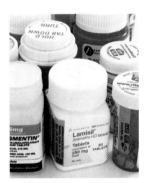

RETAIL SPENDING ON PRESCRIPTION DRUGS

Source: American Institute for Research Analysis of Scott-Levin data.

FIGURE 19

(b) Use the equation from part (a) to predict retail spending on prescription drugs in the U.S. in 2004. (Assume a constant rate of change here.)

Since $x = 7$ represents 1997 and 2004 is 7 years after 1997, $x = 14$ represents 2004. Substitute 14 for x in the equation.

$$y = 19x - 54 = 19(14) - 54 = 212$$

According to the model, $212 billion will be spent on prescription drugs in 2004.

8.3 EXERCISES

Match the correct equation in Column I with the correct description given in Column II.

I

1. $y = 4x$
2. $y = \dfrac{1}{4}x$
3. $y = -2x + 1$
4. $y - 1 = -2(x - 4)$

II

A. slope $= -2$, through the point $(4, 1)$
B. slope $= -2$, y-intercept $(0, 1)$
C. passing through the points $(0, 0)$ and $(4, 1)$
D. passing through the points $(0, 0)$ and $(1, 4)$

Use the geometric interpretation of slope (rise divided by run) to find the slope of each line. Then, by identifying the y-intercept from the graph, write the slope-intercept form of the equation of the line.

5.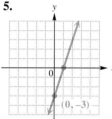

6.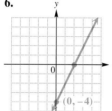

7.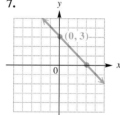

8.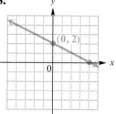

In Exercises 9–16, match each equation with the graph that it most closely resembles in Choices A.–D. (Hint: Determining the signs of m and b will help you make your decision.)

A.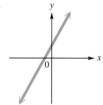

B.

C.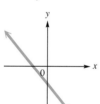

D.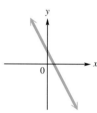

E.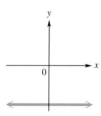

F.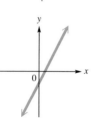

G.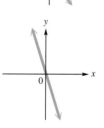

H.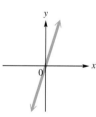

9. $y = 2x + 3$
10. $y = -2x + 3$
11. $y = -2x - 3$
12. $y = 2x - 3$
13. $y = 2x$
14. $y = -2x$
15. $y = 3$
16. $y = -3$

Write the slope-intercept form of the equation of the line satisfying the given conditions.

17. Through $(-2, 4)$; slope $-\dfrac{3}{4}$

18. Through $(-1, 6)$; slope $-\dfrac{5}{6}$

19. Through $(5, 8)$; slope -2

20. Through $(12, 10)$; slope 1

21. Through $(-5, 4)$; slope $\dfrac{1}{2}$

22. Through $(7, -2)$; slope $\dfrac{1}{4}$

23. x-intercept $(3, 0)$; slope 4

24. x-intercept $(-2, 0)$; slope -5

Write an equation that satisfies the given conditions.

25. Through $(9, 5)$; slope 0

26. Through $(-4, -2)$; slope 0

27. Through $(9, 10)$; undefined slope

28. Through $(-2, 8)$; undefined slope

29. Through $(.5, .2)$; vertical

30. Through $\left(\dfrac{5}{8}, \dfrac{2}{9}\right)$; vertical

31. Through $(-7, 8)$; horizontal

32. Through $(2, 7)$; horizontal

Write the slope-intercept form (if possible) of the equation of the line passing through the two points.

33. $(3, 4)$ and $(5, 8)$

34. $(5, -2)$ and $(-3, 14)$

35. $(6, 1)$ and $(-2, 5)$

36. $(-2, 5)$ and $(-8, 1)$

37. $\left(-\dfrac{2}{5}, \dfrac{2}{5}\right)$ and $\left(\dfrac{4}{3}, \dfrac{2}{3}\right)$

38. $\left(\dfrac{3}{4}, \dfrac{8}{3}\right)$ and $\left(\dfrac{2}{5}, \dfrac{2}{3}\right)$

39. $(2, 5)$ and $(1, 5)$

40. $(-2, 2)$ and $(4, 2)$

41. $(7, 6)$ and $(7, -8)$

42. $(13, 5)$ and $(13, -1)$

43. $(1, -3)$ and $(-1, -3)$

44. $(-4, 6)$ and $(5, 6)$

Find the equation in slope-intercept form of the line satisfying the given conditions.

45. $m = 5$; $b = 15$

46. $m = -2$; $b = 12$

47. $m = -\dfrac{2}{3}$; $b = \dfrac{4}{5}$

48. $m = -\dfrac{5}{8}$; $b = -\dfrac{1}{3}$

49. Slope $\dfrac{2}{5}$; y-intercept $(0, 5)$

50. Slope $-\dfrac{3}{4}$; y-intercept $(0, 7)$

51. Explain why the point-slope form of an equation cannot be used to find the equation of a vertical line.

52. Which one of the following equations is in standard form, according to the definition of standard form given in this text?
A. $3x + 2y - 6 = 0$ **B.** $y = 5x - 12$
C. $2y = 3x + 4$ **D.** $6x - 5y = 12$

For each equation (a) write in slope-intercept form, (b) give the slope of the line, and (c) give the y-intercept.

53. $x + y = 12$

54. $x - y = 14$

55. $5x + 2y = 20$

56. $6x + 5y = 40$

57. $2x - 3y = 10$

58. $4x - 3y = 10$

Write the equation in slope-intercept form of the line satisfying the given conditions.

59. Through $(7, 2)$; parallel to $3x - y = 8$

60. Through $(4, 1)$; parallel to $2x + 5y = 10$

61. Through $(-2, -2)$; parallel to $-x + 2y = 10$

62. Through $(-1, 3)$; parallel to $-x + 3y = 12$

63. Through $(8, 5)$; perpendicular to $2x - y = 7$

64. Through $(2, -7)$; perpendicular to $5x + 2y = 18$

65. Through $(-2, 7)$; perpendicular to $x = 9$

66. Through $(8, 4)$; perpendicular to $x = -3$

Solve each problem.

67. *Private 4-year College Costs* The table lists the average annual cost (in dollars) of tuition and fees at private 4-year colleges for selected years, where year 1 represents 1991, year 3 represents 1993, and so on.

Private College Costs	
Year	Cost (in dollars)
1	10,017
3	11,025
5	12,432
7	13,785
9	15,380

Source: The College Board.

(a) Plot the five ordered pairs (year, cost). Do the points lie in approximately a straight line?

(b) Use the ordered pairs (3, 11,025) and (9, 15,380) to find the equation of a line that approximates the data. Write the equation in slope-intercept form. (Round the slope to the nearest tenth.)

(c) Use the equation from part (b) to estimate the average annual cost at private 4-year colleges in 2003. (*Hint:* What is the value of *x* for 2003?)

68. *Nuclear Waste* The table gives the heavy metal nuclear waste (in thousands of metric tons) from spent reactor fuel now stored temporarily at reactor sites, awaiting permanent storage. (*Source:* "Burial of Radioactive Nuclear Waste under the Seabed," *Scientific American*, January 1998, p. 62.)

Heavy Metal Nuclear Waste	
Year x	Waste y
1995	32
2000*	42
2010*	61
2020*	76

*Estimates by the U.S. Department of Energy.

Let $x = 0$ represent 1995, $x = 5$ represent 2000 (since $2000 - 1995 = 5$), and so on.

(a) Plot the ordered pairs (x, y). Do the points lie approximately in a line?

(b) Use the ordered pairs (0, 32) and (25, 76) to find the equation of a line that approximates the other ordered pairs. Use the form $y = mx + b$.

(c) Use the equation from part (b) to estimate the amount of nuclear waste in 2005.

69. *U.S. Post Offices* The number of post offices in the U.S. has been declining. Use the information given on the bar graph for the years 1994 and 1998, letting $x = 0$ represent the year 1990 and y represent the number of post offices. Find a linear equation that models the data. Write it in slope-intercept form.

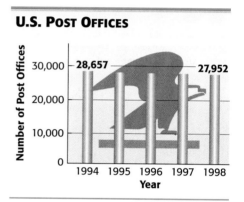

Source: U.S. Postal Service, *Annual Report of the Postmaster General.*

70. *Air Conditioner Choices* The graph shows the recommended air conditioner size (in British thermal units) for selected room sizes in square feet. Use the information given for rooms of 150 ft² and 1400 ft². Let x represent the number of square feet and y represent the corresponding Btu size air conditioner. Find a linear equation that models the data. Write it in slope-intercept form.

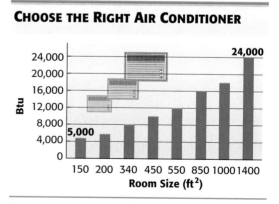

Source: Carey, Morris and James, *Home Improvement for Dummies*, IDG Books.

71. *Median Income for African-Americans* Median household income of African-Americans increased in recent years, as shown in the bar graph.

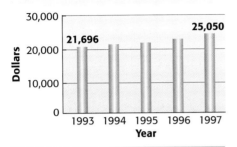

MEDIAN HOUSEHOLD INCOME FOR AFRICAN-AMERICANS

Source: U.S. Bureau of the Census.

(a) Use the information given for the years 1993 and 1997, letting $x = 3$ represent 1993, $x = 7$ represent 1997, and y represent the median income, to write an equation in slope-intercept form that models median household income.

(b) Use the equation to approximate the median income for 1995. How does your result compare to the actual value, $23,583?

72. *Median Income for Hispanics* The bar graph shows median household income for Hispanics.

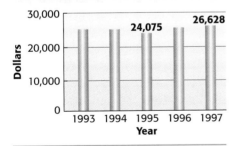

MEDIAN HOUSEHOLD INCOME FOR HISPANICS

Source: U.S. Bureau of the Census.

(a) Use the information for the years 1995 and 1997 to write an equation in slope-intercept form. Let $x = 5$ represent 1995, $x = 7$ represent 1997, and y represent the median income.

(b) Based on the graph, would you expect the equation from part (a) to give good approximations for 1993 and 1994 incomes? Would the equation give a reasonable approximation for 1998? Explain.

73. *Landfill Capacity* The Waste Management and Recycling Division of Sacramento County is responsible for managing the disposal of solid waste, including the operation of a landfill. The graph shows the remaining capacity (in millions of tons) at the county landfill at several times during the past few years. These points appear to lie close to a line.

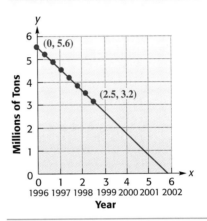

REMAINING LANDFILL CAPACITY

Source: Waste Management and Recycling Division, Sacramento County Public Works Agency, 1998 Report.

(a) Write an equation in slope-intercept form of the line shown through the points $(0, 5.6)$ and $(2.5, 3.2)$.

(b) Use the equation from part (a) to estimate the remaining capacity of the landfill in 2001.

(c) Based on the graph, when will the capacity of the landfill be used up? (*Hint:* Find the point on the graph where y equals 0; that is, the x-intercept.) Estimate the month when it will occur.

74. *Fahrenheit-Celsius Relationship* If we think of ordered pairs of the form (C, F), then the two most common methods of measuring temperature, Celsius and Fahrenheit, can be related as follows: When C = 0, F = 32, and when C = 100, F = 212. This exercise explains how this information is used to find the formula that relates the two temperature scales.

(a) There is a linear relationship between Celsius and Fahrenheit temperatures. When C = 0°, F = _____°, and when C = 100°, F = _____°.

(b) Think of ordered pairs of temperatures (C, F), where C and F represent corresponding Celsius and Fahrenheit temperatures. The equation that relates the two scales has a straight-line graph that contains the two points determined in part (a). What are these two points?

(c) Find the slope of the line described in part (b).

(d) Now think of the point-slope form of the equation in terms of C and F, where C replaces x and F replaces y. Use the slope you found in part (c) and one of the two points determined earlier, and find the equation that gives F in terms of C.

(e) To obtain another form of the formula, use the equation you found in part (d) and solve for C in terms of F.

(f) The equation found in part (d) is graphed on the graphing calculator screen shown here. Observe the display at the bottom, and interpret it in the context of this exercise.

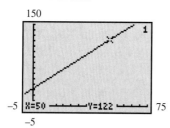

75. A table of points, generated by a graphing calculator, is shown for a line Y_1. Use any two points to find the equation of each line in slope-intercept form.

(a)

X	Y_1
-2	-.5
-1	.25
0	1
1	1.75
2	2.5
3	3.25
4	4

X = -2

(b)

X	Y_1
-4	14
-3	10
-2	6
-1	2
0	-2
1	-6
2	-10

X = -4

8.4 An Introduction to Functions: Linear Functions, Applications, and Models

We often describe one quantity in terms of another; for example, the growth of a plant is related to the amount of light it receives, the demand for a product is related to the price of the product, the cost of a trip is related to the distance traveled, and so on. To represent these corresponding quantities, we can use ordered pairs.

For example, suppose that it is time to fill up your car's tank with gasoline. At your local station, 89-octane gas is selling for $1.70 per gallon. Experience has taught you that the final price you pay is determined by the number of gallons you buy multiplied by the price per gallon (in this case, $1.70). As you pump the gas, two sets of numbers spin by: the number of gallons pumped and the price for that number of gallons. The table below uses ordered pairs to illustrate this situation.

Number of Gallons Pumped	Price for This Number of Gallons
0	$0.00 = 0($1.70)
1	$1.70 = 1($1.70)
2	$3.40 = 2($1.70)
3	$5.10 = 3($1.70)
4	$6.80 = 4($1.70)

If we let x denote the number of gallons pumped, then the price y in dollars can be found by the linear equation $y = 1.70x$. Theoretically, there are infinitely many ordered pairs (x, y) that satisfy this equation, but in this application we are limited to nonnegative values for x, since we cannot have a negative number of gallons. There also is a practical maximum value for x in this situation, which varies from one car to another. What determines this maximum value?

In this example, the total price depends on the amount of gasoline pumped. For this reason, price is called the *dependent variable,* and the number of gallons is called the *independent variable.* Generalizing, if the value of the variable y depends on the value of the variable x, then y is the **dependent variable** and x the **independent variable.**

$$\text{Independent variable} \rightarrow \quad \leftarrow \text{Dependent variable}$$
$$(x, y)$$

Relations and Functions Since related quantities can be written using ordered pairs, the concept of *relation* can be defined as follows.

Relation

A **relation** is a set of ordered pairs.

For example, the sets

$$F = \{(1, 2), (-2, 5), (3, -1)\} \quad \text{and} \quad G = \{(-4, 1), (-2, 1), (-2, 0)\}$$

both are relations. A special kind of relation, called a *function,* is very important in mathematics and its applications.

Function

A **function** is a relation in which for each value of the first component of the ordered pairs there is *exactly one value* of the second component.

Of the two examples of a relation just given, only set F is a function, because for each x-value, there is exactly one y-value. In set G, the last two ordered pairs have the same x-value paired with two different y-values, so G is a relation, but not a function.

$$F = \{(\underset{\uparrow}{1}, 2), (-2, 5), (\underset{\uparrow}{3}, -1)\} \quad \text{Function}$$
$$\text{Different } x\text{-values}$$

$$G = \{(-4, 1), (\underset{\uparrow}{-2}, 1), (\underset{\uparrow}{-2}, 0)\} \quad \text{Not a function}$$
$$\text{Same } x\text{-values}$$

In a function, there is *exactly one* value of the dependent variable, the second component, for each value of the independent variable, the first component. This is what makes functions so important in applications.

Another way to think of a functional relationship is to think of the independent variable as an input and the dependent variable as an output. A calculator is an input-output machine, for example. To find 8^2, we must input 8, press the squaring key, and see that the output is 64. Inputs and outputs also can be determined from a graph or a table.

A third way to describe a function is to give a rule that tells how to determine the dependent variable for a specific value of the independent variable. Suppose the rule is given in words as "the dependent variable is twice the independent variable." As an equation, this can be written

$$y = 2x.$$

where y is the Dependent variable and x is the Independent variable.

EXAMPLE 1 Determine the independent and dependent variables for each of the following functions. Give an example of an ordered pair belonging to the function.

(a) The 2000 Summer Olympics medal winners in men's basketball were {(gold, United States), (silver, France), (bronze, Lithuania)}. (*Source:* espn.go.com/oly/summer00/)

The independent variable (the first component in each ordered pair) is the type of medal; the dependent variable (the second component) is the recipient. Any of the three ordered pairs could be given as an example.

(b) A calculator that finds square roots

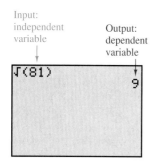

Input: independent variable
Output: dependent variable

The independent variable (the input) is a nonnegative real number, since the square root of a negative number is not a real number. The dependent variable is the nonnegative square root. For example, (81, 9) belongs to this function.

(c) The graph in Figure 20, which shows the relationship between the number of gallons of water in a small swimming pool and time in hours

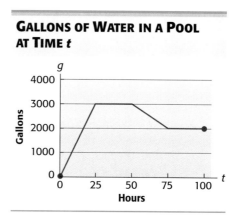

FIGURE 20

The independent variable is time, in hours, and the dependent variable is the number of the gallons of water in the pool. One ordered pair is (25, 3000).

(d) The table of petroleum imports in millions of barrels per day for selected years

U.S. Petroleum Imports	
Year	Imports
1994	9.00
1995	8.83
1996	9.40
1997	10.16
1998	10.71

Source: U.S. Energy Department.

The independent variable is the year; the dependent variable is the number of millions of barrels of petroleum. An example of an ordered pair is (1994, 9.00).

(e) $y = 3x + 4$

The independent variable is x, and the dependent variable is y. One ordered pair is $(1/3, 5)$.

Domain and Range

Domain and Range

In a relation, the set of all values of the independent variable (x) is the **domain**; the set of all values of the dependent variable (y) is the **range.**

EXAMPLE 2 Give the domain and range of each function in Example 1.

(a) The domain is the set of types of medals, {gold, silver, bronze}, and the range is the set of winning countries, {United States, France, Lithuania}.

(b) Here, the domain is restricted to nonnegative numbers: $[0, \infty)$. The range also is $[0, \infty)$.

(c) The domain is all possible values of t, the time in hours, which is the interval $[0, 100]$. The range is the number of gallons at time t, the interval $[0, 3000]$.

(d) The domain is the set of years, {1994, 1995, 1996, 1997, 1998}; the range is the set of imports (in millions of barrels per day), {9.00, 8.83, 9.40, 10.16, 10.71}.

(e) In the defining equation (or rule), $y = 3x + 4$, x can be any real number, so the domain is $\{x \mid x \text{ is a real number}\}$ or $(-\infty, \infty)$. Since every real number y can be produced by some value of x, the range also is the set $\{y \mid y \text{ is a real number}\}$, or $(-\infty, \infty)$.

Graphs of Relations The **graph of a relation** is the graph of its ordered pairs. The graph gives a picture of the relation, which can be used to determine its domain and range.

EXAMPLE 3 Give the domain and range of each relation.

(a)

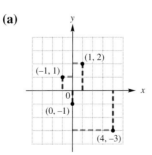

The domain is the set of x-values,

$$\{-1, 0, 1, 4\}.$$

The range is the set of y-values,

$$\{-3, -1, 1, 2\}.$$

(b)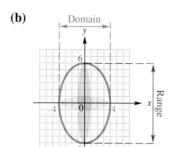

The *x*-values of the points on the graph include all numbers between −4 and 4, inclusive. The *y*-values include all numbers between −6 and 6, inclusive. Using interval notation, the domain is $[-4, 4]$; the range is $[-6, 6]$.

(c)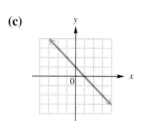

The arrowheads indicate that the line extends indefinitely left and right, as well as up and down. Therefore, both the domain and the range are the set of all real numbers, written $(-\infty, \infty)$.

(d)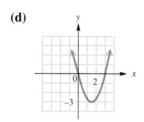

The arrowheads indicate that the graph extends indefinitely left and right, as well as upward. The domain is $(-\infty, \infty)$. Because there is a least *y*-value, −3, the range includes all real numbers greater than or equal to −3, written $[-3, \infty)$.

Relations often are defined by equations, such as $y = 2x + 3$ and $y^2 = x$. It is sometimes necessary to determine the domain of a relation from its equation. In this book, the following agreement on the domain of a relation is assumed.

Agreement on Domain

The domain of a relation is assumed to be all real numbers that produce real numbers when substituted for the independent variable.

To illustrate this agreement, since any real number can be used as a replacement for *x* in $y = 2x + 3$, the domain of this function is the set of real numbers. As another example, the function defined by $y = 1/x$ has all real numbers except 0 as domain, since *y* is undefined if $x = 0$. In general, the domain of a function defined by an algebraic expression is all real numbers, except those numbers that lead to division by 0 or an even root of a negative number.

Identifying Functions Most of the relations we have seen in the examples are functions—that is, each *x*-value corresponds to exactly one *y*-value. Now we look at ways to determine whether a given relation, defined algebraically, is a function.

8.4 An Introduction to Functions: Linear Functions, Applications, and Models

In a function each value of x leads to only one value of y, so any vertical line drawn through the graph of a function must intersect the graph in at most one point. This is the **vertical line test for a function.**

Vertical Line Test

If a vertical line intersects the graph of a relation in more than one point, then the relation is not a function.

For example, the graph shown in Figure 21(a) is not the graph of a function, since a vertical line can intersect the graph in more than one point, while the graph in Figure 21(b) does represent a function.

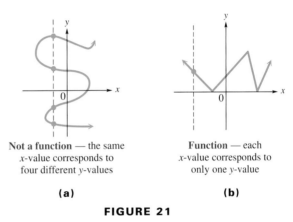

Not a function — the same x-value corresponds to four different y-values

(a)

Function — each x-value corresponds to only one y-value

(b)

FIGURE 21

The vertical line test is a simple method for identifying a function defined by a graph. It is more difficult to decide whether a relation defined by an equation is a function. The next example gives some hints that may help.

EXAMPLE 4 Decide whether each of the following defines a function and give the domain.

(a) $y = \sqrt{2x - 1}$

Here, for any choice of x in the domain, there is exactly one corresponding value for y (the radical is a nonnegative number), so this equation defines a function. Since the radicand cannot be negative,

$$2x - 1 \geq 0$$
$$2x \geq 1$$
$$x \geq \frac{1}{2}.$$

The domain is $\left[\frac{1}{2}, \infty\right)$.

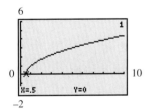

The graph of $y = \sqrt{2x - 1}$ supports the result in Example 4(a). The domain is $\left[\frac{1}{2}, \infty\right)$.

(b) $y^2 = x$

The ordered pairs $(16, 4)$ and $(16, -4)$ both satisfy this equation. Since one value of x, 16, corresponds to two values of y, 4 and -4, this equation does not define a function. Solving $y^2 = x$ for y gives $y = \sqrt{x}$ or $y = -\sqrt{x}$, which shows that two values of y correspond to each positive value of x. Because x is equal to the square of y, the values of x must always be nonnegative. The domain of the relation is $[0, \infty)$.

(c) $y \leq x - 1$

By definition, y is a function of x if every value of x leads to exactly one value of y. In this example, a particular value of x, say 1, corresponds to many values of y. The ordered pairs $(1, 0)$, $(1, -1)$, $(1, -2)$, $(1, -3)$, and so on, all satisfy the inequality. For this reason, the inequality does not define a function. Any number can be used for x, so the domain is the set of real numbers $(-\infty, \infty)$.

(d) $y = \dfrac{5}{x - 1}$

Given any value of x in the domain, we find y by subtracting 1, then dividing the result into 5. This process produces exactly one value of y for each value in the domain, so this equation defines a function. The domain includes all real numbers except those that make the denominator 0. We find these numbers by setting the denominator equal to 0 and solving for x.

$$x - 1 = 0$$
$$x = 1$$

Thus, the domain includes all real numbers except 1. In interval notation this is written as

$$(-\infty, 1) \cup (1, \infty).$$

In summary, three variations of the definition of function are given here.

> ### Variations of the Definition of Function
>
> 1. A **function** is a relation in which for each value of the first component of the ordered pairs there is exactly one value of the second component.
> 2. A **function** is a set of ordered pairs in which no first component is repeated.
> 3. A **function** is a rule or correspondence that assigns exactly one range value to each domain value.

Function Notation When a function f is defined with a rule or an equation using x and y for the independent and dependent variables, we say "y is a function of x" to emphasize that y *depends on* x. We use the notation

$$y = f(x),$$

called **function notation,** to express this and read $f(x)$ as "f of x." (In this notation the parentheses do not indicate multiplication.) The letter f stands for *function.* For example, if $y = 2x - 7$, we write

$$f(x) = 2x - 7.$$

Note that $f(x)$ is just another name for the dependent variable y. For example, if $y = f(x) = 9x - 5$, and $x = 2$, then we find y, or $f(2)$, by replacing x with 2.

$$y = f(2)$$
$$y = 9 \cdot 2 - 5$$
$$= 18 - 5$$
$$= 13.$$

The statement "if $x = 2$, then $y = 13$" is abbreviated with function notation as

$$f(2) = 13.$$

Read $f(2)$ as "f of 2" or "f at 2." Also,

$$f(0) = 9 \cdot 0 - 5 = -5, \quad \text{and} \quad f(-3) = 9(-3) - 5 = -32.$$

These ideas and the symbols used to represent them can be explained as follows.

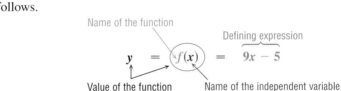

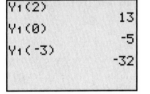

For $Y_1 = 9X - 5$, function notation capability of the TI-83 Plus supports the discussion here.

EXAMPLE 5 Let $f(x) = -x^2 + 5x - 3$. Find the following.

(a) $f(2)$

$$f(x) = -x^2 + 5x - 3$$
$$f(2) = -2^2 + 5 \cdot 2 - 3 \qquad \text{Replace } x \text{ with 2.}$$
$$= -4 + 10 - 3$$
$$= 3$$

(b) $f(-1) = -(-1)^2 + 5(-1) - 3 = -1 - 5 - 3 = -9$

(c) $f(2x) = -(2x)^2 + 5(2x) - 3 \qquad \text{Replace } x \text{ with } 2x.$
$$= -4x^2 + 10x - 3$$

Linear Functions An important type of elementary function is the *linear function.*

Linear Function

A function that can be written in the form

$$f(x) = mx + b$$

for real numbers m and b is a **linear function.**

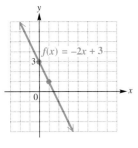

(a)

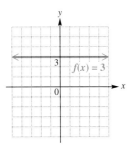

(b)

FIGURE 22

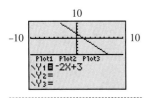

Compare with Example 6(a) and Figure 22(a).

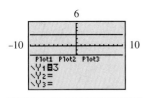

Compare with Example 6(b) and Figure 22(b).

Notice that the form $f(x) = mx + b$ defining a linear function is the same as that of the slope-intercept form of the equation of a line, first seen in the previous section. We know that the graph of $f(x) = mx + b$ will be a line with slope m and y-intercept $(0, b)$.

EXAMPLE 6 Graph each linear function.

(a) $f(x) = -2x + 3$

To graph the function, locate the y-intercept, $(0, 3)$. From this point, use the slope $-2 = \frac{-2}{1}$ to go down 2 and right 1. This second point is used to obtain the graph in Figure 22(a).

(b) $f(x) = 3$

From the previous section, we know that the graph of $y = 3$ is a horizontal line. Therefore, the graph of $f(x) = 3$ is a horizontal line with y-intercept $(0, 3)$ as shown in Figure 22(b). ■

The function defined in Example 6(b) and graphed in Figure 22(b) is an example of a constant function. A **constant function** is a linear function of the form $f(x) = b$, where b is a real number. The domain of any linear function is $(-\infty, \infty)$. The range of a nonconstant linear function (like in Example 6(a)) is also $(-\infty, \infty)$, while the range of the constant function $f(x) = b$ is $\{b\}$.

Cost and Revenue Models A company's cost of producing a product and the revenue from selling the product can be expressed as linear functions. The idea of **break-even analysis** then can be explained using the graphs of these functions. When cost is greater than revenue earned, the company loses money; when cost is less than revenue, the company makes money; and when cost equals revenue, the company breaks even.

EXAMPLE 7 Peripheral Visions, Inc., produces studio quality audiotapes of live concerts. The company places an ad in a trade newsletter. The cost of the ad is $100. Each tape costs $20 to produce, and the company charges $24 per tape.

(a) Express the cost C as a function of x, the number of tapes produced.

The *fixed cost* is $100, and for each tape produced, the *variable cost* is $20. Therefore, the cost C can be expressed as a function of x, the number of tapes produced:

$$C(x) = 20x + 100 \quad (C \text{ in dollars}).$$

(b) Express the revenue R as a function of x, the number of tapes sold.

Since each tape sells for $24, the revenue R is given by $R(x) = 24x$ (R in dollars).

(c) For what value of x does revenue equal cost?

The company will just break even (no profit and no loss) as long as revenue just equals cost, or $R(x) = C(x)$. This is true whenever

$$R(x) = C(x)$$
$$24x = 20x + 100 \quad \text{Substitute for } R(x) \text{ and } C(x).$$
$$4x = 100$$
$$x = 25.$$

If 25 tapes are produced and sold, the company will break even.

(d) Graph $C(x) = 20x + 100$ and $R(x) = 24x$ on the same coordinate system, and interpret the graph.

Figure 23 shows the graphs of the two functions. At the break-even point, we see that when 25 tapes are produced and sold, both the cost and the revenue are $600. If fewer than 25 tapes are produced and sold (that is, when $x < 25$), the company loses money. When more than 25 tapes are produced and sold (that is, when $x > 25$), there is a profit.

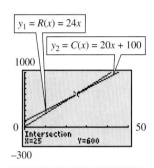

The break-even point is (25, 600), as indicated at the bottom of the screen. The calculator can find the point of intersection of the graphs.

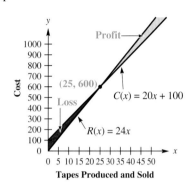

FIGURE 23

8.4 EXERCISES

1. In your own words, define *function* and give an example.
2. In your own words, define *domain of a function* and give an example.
3. In an ordered pair of a relation, is the first element a value of the independent or the dependent variable?

For each of the following relations, decide whether it is a function, and give the domain and range.

4. $\{(1, 1), (1, -1), (2, 4), (2, -4), (3, 9), (3, -9)\}$
5. $\{(2, 5), (3, 7), (4, 9), (5, 11)\}$
6. The set containing certain countries and their predicted life expectancy estimates for persons born in 2050 is $\{(U.S., 83.9), (Japan, 90.91), (Canada, 85.26), (Britain, 83.79), (France, 87.01), (Germany, 83.12), (Italy, 82.26)\}$. (*Source:* Shripad Tuljapurkar, Mountain View Research, Los Altos, California.)
7. An input-output machine accepts positive real numbers as input, and outputs both their positive and negative square roots.
8.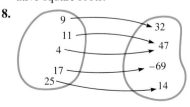

9.
U.S. Voting-age Population in 2000 (in millions)	
Hispanic	21.3
Native American	1.6
Asian American	8.2
African-American	24.6
White	152.0

Source: U.S. Bureau of the Census.

10. $\{(x, y) \mid x = |y|\}$

11.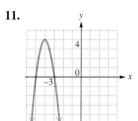

12.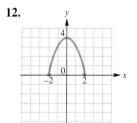

13.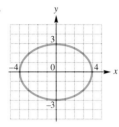
14.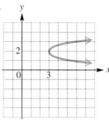

(a) Is this the graph of a function?
(b) What is the domain?
(c) Estimate the number of megawatts at 8 A.M.
(d) At what time was the most electricity used? The least electricity?

Decide whether the given relation defines y as a function of x. Give the domain.

15. $y = x^2$
16. $y = x^3$
17. $x = y^2$
18. $x = y^4$
19. $x + y < 4$
20. $x - y < 3$
21. $y = \sqrt{x}$
22. $y = -\sqrt{x}$
23. $xy = 1$
24. $xy = -3$
25. $y = \sqrt{4x + 2}$
26. $y = \sqrt{9 - 2x}$
27. $y = \dfrac{2}{x - 9}$
28. $y = \dfrac{-7}{x - 16}$

29. *Pool Water Level* Refer to Example 1, Figure 20, to answer the questions.
   (a) What numbers are possible values of the dependent variable?
   (b) For how long is the water level increasing? Decreasing?
   (c) How many gallons are in the pool after 90 hours?
   (d) Call this function f. What is $f(0)$? What does it mean in this example?

30. *Electricity Consumption* The graph shows the megawatts of electricity used on a record-breaking summer day in Sacramento, California.

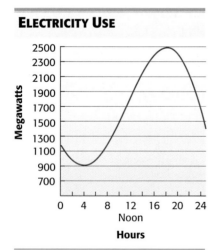

ELECTRICITY USE

Source: Sacramento Municipal Utility District.

31. Give an example of a function from everyday life. (*Hint:* Fill in the blanks: ____ depends on ____, so ____ is a function of ____.)

32. Choose the correct response. The notation $f(3)$ means
   A. the variable f times 3, or $3f$.
   B. the value of the dependent variable when the independent variable is 3.
   C. the value of the independent variable when the dependent variable is 3.
   D. f equals 3.

Let $f(x) = 3 + 2x$ and $g(x) = x^2 - 2$. Find each function value.

33. $f(1)$
34. $f(4)$
35. $g(2)$
36. $g(0)$
37. $g(-1)$
38. $g(-3)$
39. $f(-8)$
40. $f(-5)$

Sketch the graph of each linear function. Give the domain and range.

41. $f(x) = -2x + 5$
42. $g(x) = 4x - 1$
43. $h(x) = \dfrac{1}{2}x + 2$
44. $F(x) = -\dfrac{1}{4}x + 1$
45. $G(x) = 2x$
46. $H(x) = -3x$
47. $f(x) = 5$
48. $g(x) = -4$

An equation that defines y as a function of x is given.
(a) *Solve for y in terms of x, and replace y with the function notation $f(x)$.* (b) *Find $f(3)$.*

49. $y + 2x^2 = 3$
50. $y - 3x^2 = 2$
51. $4x - 3y = 8$
52. $-2x + 5y = 9$

53. Fill in the blanks with the correct responses. The equation $2x + y = 4$ has a straight ____ as its graph. One point that lies on the line is (3, ____). If we solve the equation for y and use function notation, we have a linear function $f(x) = $ ____. For this function, $f(3) = $ ____, meaning that the point (____, ____) lies on the graph of the function.

54. Which one of the following defines a linear function?
   A. $y = \dfrac{x - 5}{4}$
   B. $y = \dfrac{1}{x}$
   C. $y = x^2$
   D. $y = \sqrt{x}$

8.4 An Introduction to Functions: Linear Functions, Applications, and Models   433

55. *Taxi Fares*
(a) Suppose that a taxicab driver charges $1.50 per mile. Fill in the chart with the correct response for the price $f(x)$ she charges for a trip of x miles.

x	$f(x)$
0	
1	
2	
3	

(b) The linear function that gives a rule for the amount charged is $f(x) = $ _____ .
(c) Graph this function for the domain $\{0, 1, 2, 3\}$.

56. *Cost to Mail a Package* Suppose that a package weighing x pounds costs $f(x)$ dollars to mail to a given location, where

$$f(x) = 2.75x.$$

(a) What is the value of $f(3)$?
(b) Describe what 3 and the value $f(3)$ mean in part (a), using the terminology *independent variable* and *dependent variable*.
(c) How much would it cost to mail a 5-lb package? Write the answer using function notation.

57. *Forensic Studies* Forensic scientists use the lengths of the tibia (t), the bone from the ankle to the knee, and the femur (r), the bone from the knee to the hip socket, to calculate the height of a person. A person's height (h) is determined from the lengths of these bones using functions defined by the following formulas. All measurements are in centimeters.

For men:

$$h(r) = 69.09 + 2.24r$$

or $\quad h(t) = 81.69 + 2.39t$

For women:

$$h(r) = 61.41 + 2.32r$$

or $\quad h(t) = 72.57 + 2.53t$

(a) Find the height of a man with a femur measuring 56 centimeters.
(b) Find the height of a man with a tibia measuring 40 centimeters.
(c) Find the height of a woman with a femur measuring 50 centimeters.
(d) Find the height of a woman with a tibia measuring 36 centimeters.

58. *Pool Size for Sea Otters* Federal regulations set standards for the size of the quarters of marine mammals. A pool to house sea otters must have a volume of "the square of the sea otter's average adult length (in meters) multiplied by 3.14 and by .91 meter." If x represents the sea otter's average adult length and $f(x)$ represents the volume of the corresponding pool size, this formula can be written as

$$f(x) = (.91)(3.14)x^2.$$

Find the volume of the pool for each of the following adult lengths (in meters). Round answers to the nearest hundredth.
(a) .8   (b) 1.0   (c) 1.2   (d) 1.5

59. *Number of Post Offices* The linear function

$$f(x) = -183x + 40,034$$

is a model for the number of U.S. post offices for the period 1990–1995, where $x = 0$ corresponds to 1990, $x = 1$ corresponds to 1991, and so on. Use this model to give the approximate number of post offices during the following years. (*Source:* U.S. Postal Service, *Annual Report of the Postmaster General and Comprehensive Statement on Postal Operations.*)
(a) 1991   (b) 1993   (c) 1995
(d) The graphing calculator screen shows a portion of the graph of $y = f(x)$ with the coordinates of a point on the graph displayed at the bottom of the screen. Interpret the meaning of the display in the context of this application.

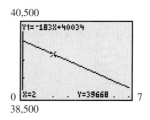

60. *U.S Defense Budget* The linear function $f(x) = -6324x + 305{,}294$ is a model for U.S. defense budgets in millions of dollars from 1992 to 1996, where $x = 0$ corresponds to 1990, $x = 2$ corresponds to 1992, and so on. Use this model to approximate the defense budget for the following years. (*Source:* U.S. Office of Management and Budget.)
 (a) 1993   (b) 1995   (c) 1996
 (d) A portion of the graph of $y = f(x)$ is shown in the graphing calculator screen, with the coordinates of a point displayed at the bottom of the screen. Interpret the meaning of the display in the context of this application.

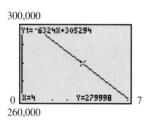

Cost and Revenue Models In each of the following, (a) express the cost C as a function of x, where x represents the quantity of items as given; (b) express the revenue R as a function of x; (c) determine the value of x for which revenue equals cost; (d) graph $y = C(x)$ and $y = R(x)$ on the same axes, and interpret the graph.

61. Perian Herring stuffs envelopes for extra income during her spare time. Her initial cost to obtain the necessary information for the job was $200.00. Each envelope costs $.02 and she gets paid $.04 per envelope stuffed. Let x represent the number of envelopes stuffed.

62. Tony Motton runs a copying service in his home. He paid $3500 for the copier and a lifetime service contract. Each sheet of paper he uses costs $.01, and he gets paid $.05 per copy he makes. Let x represent the number of copies he makes.

63. Eugene Smith operates a delivery service in a southern city. His start-up costs amounted to $2300. He estimates that it costs him (in terms of gasoline, wear and tear on his car, etc.) $3.00 per delivery. He charges $5.50 per delivery. Let x represent the number of deliveries he makes.

64. Lisa Ventura bakes cakes and sells them at county fairs. Her initial cost for the Washington Parish fair in 1996 was $40.00. She figures that each cake costs $2.50 to make, and she charges $6.50 per cake. Let x represent the number of cakes sold. (Assume that there were no cakes left over.)

8.5 Quadratic Functions, Applications, and Models

In the previous section we discussed linear functions, those that are defined by first-degree polynomials. In this section we will look at *quadratic functions,* those defined by second-degree polynomials.

Quadratic Function

A function f is a **quadratic function** if
$$f(x) = ax^2 + bx + c,$$
where a, b, and c are real numbers, with $a \neq 0$.

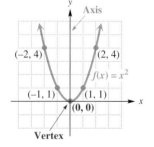

FIGURE 24

The simplest quadratic function is defined by $f(x) = x^2$. This function can be graphed by finding several ordered pairs that satisfy the equation: for example, $(0, 0)$, $(1, 1)$, $(-1, 1)$, $(2, 4)$, $(-2, 4)$, $(1/2, 1/4)$, $(-1/2, 1/4)$, $(3/2, 9/4)$, and $(-3/2, 9/4)$. Plotting these points and drawing a smooth curve through them gives the graph shown in Figure 24. This graph is called a **parabola.** Every quadratic function has a graph that is a parabola.

8.5 Quadratic Functions, Applications, and Models

Parabolas are symmetric about a line (the y-axis in Figure 24.) Intuitively, this means that if the graph were folded along the line of symmetry, the two sides would coincide. The line of symmetry for a parabola is called the **axis** of the parabola. The point where the axis intersects the parabola is the **vertex** of the parabola. The vertex is the lowest (or highest) point of a vertical parabola.

Parabolas have many practical applications. For example, the reflectors of solar ovens and flashlights are made by revolving a parabola about its axis. The **focus** of a parabola is a point on its axis that determines the curvature. See Figure 25. When the parabolic reflector of a solar oven is aimed at the sun, the light rays bounce off the reflector and collect at the focus, creating intense heat at that point. In contrast, when a lightbulb is place at the focus of a parabolic reflector, light rays reflect out parallel to the axis.

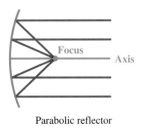

Parabolic reflector

FIGURE 25

Graphing Quadratic Functions The first example shows how the constant a affects the graph of a function of the form $g(x) = ax^2$.

EXAMPLE 1 Graph the functions defined as follows.

(a) $g(x) = -x^2$

For a given value of x, the corresponding value of $g(x)$ will be the negative of what it was for $f(x) = x^2$. (See the table of values with Figure 26(a).) Because of this, the graph of $g(x) = -x^2$ is the same shape as that of $f(x) = x^2$, but opens downward. See Figure 26(a). This is generally true; the graph of $f(x) = ax^2 + bx + c$ opens downward whenever $a < 0$.

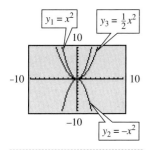

The screen illustrates the three graphs considered in Example 1 and Figures 26(a) and 26(b).

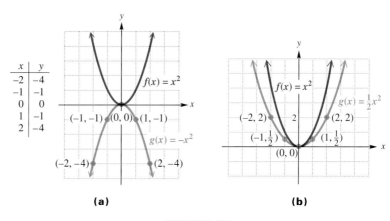

FIGURE 26

(b) $g(x) = \dfrac{1}{2}x^2$

Choose a value of x, and then find $g(x)$. The coefficient $1/2$ will cause the resulting value of $g(x)$ to be smaller than for $f(x) = x^2$, making the parabola wider than the graph of $f(x) = x^2$. See Figure 26(b). In both parabolas of this example, the axis is the vertical line $x = 0$ and the vertex is the origin $(0, 0)$.

The next few examples show the results of horizontal and vertical shifts, called **translations,** of the graph of $f(x) = x^2$.

EXAMPLE 2 Graph $g(x) = x^2 - 4$.

By comparing the tables of values for $g(x) = x^2 - 4$ and $f(x) = x^2$ shown with Figure 27, we can see that for corresponding x-values, the y-values of g are each 4 less than those for f. Thus, the graph of $g(x) = x^2 - 4$ is the same as that of $f(x) = x^2$, but translated 4 units down. See Figure 27. The vertex of this parabola (here the lowest point) is at $(0, -4)$. The axis of the parabola is the vertical line $x = 0$.

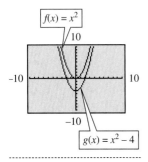

Compare with Figure 27.

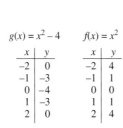

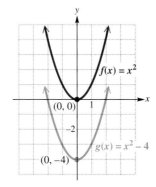

FIGURE 27

EXAMPLE 3 Graph $g(x) = (x - 4)^2$.

Comparing the tables of values shown with Figure 28 shows that the graph of $g(x) = (x - 4)^2$ is the same as that of $f(x) = x^2$, but translated 4 units to the right. The vertex is at $(4, 0)$. As shown in Figure 28, the axis of this parabola is the vertical line $x = 4$.

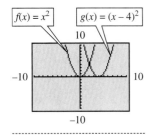

Compare with Figure 28.

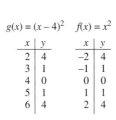

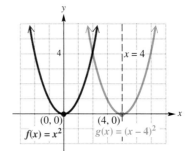

FIGURE 28

Errors frequently occur when horizontal shifts are involved. To determine the direction and magnitude of horizontal shifts, find the value that would cause the expression $x - h$ to equal 0. For example, the graph of $f(x) = (x - 5)^2$ would be shifted 5 units to the *right,* because $+5$ would cause $x - 5$ to equal 0. On the other hand, the graph of $f(x) = (x + 4)^2$ would be shifted 4 units to the *left,* because -4 would cause $x + 4$ to equal 0.

The following general principles apply for graphing functions of the form $f(x) = a(x - h)^2 + k$.

General Principles for Graphs of Quadratic Functions

1. The graph of the quadratic function defined by
   $$f(x) = a(x - h)^2 + k, \quad a \neq 0,$$
   is a parabola with vertex (h, k), and the vertical line $x = h$ as axis.
2. The graph opens upward if a is positive and downward if a is negative.
3. The graph is wider than that of $f(x) = x^2$ if $0 < |a| < 1$. The graph is narrower than that of $f(x) = x^2$ if $|a| > 1$.

EXAMPLE 4 Graph $f(x) = -2(x + 3)^2 + 4$.

The parabola opens downward (because $a < 0$), and is narrower than the graph of $f(x) = x^2$, since $a = -2$, and $|-2| > 1$. This parabola has vertex at $(-3, 4)$, as shown in Figure 29. To complete the graph, we plotted the ordered pairs $(-4, 2)$ and $(-2, 2)$.

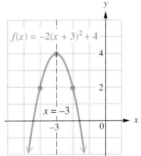

FIGURE 29

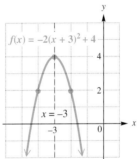

Compare with Figure 29. The vertex is $(-3, 4)$.

When the equation of a parabola is given in the form $f(x) = ax^2 + bx + c$, it is necessary to locate the vertex in order to sketch an accurate graph. This can be done in two ways. The first is by completing the square, as shown in Example 5. The second is by using a formula which can be derived by completing the square.

EXAMPLE 5 Find the vertex of the graph of $f(x) = x^2 - 4x + 5$.

To find the vertex, we need to express $x^2 - 4x + 5$ in the form $(x - h)^2 + k$. This is done by completing the square. (See the previous chapter.) To simplify the notation, replace $f(x)$ by y.

$y = x^2 - 4x + 5$

$y - 5 = x^2 - 4x$ Transform so that the constant term is on the left.

$y - 5 + 4 = x^2 - 4x + 4$ Half of -4 is -2; $(-2)^2 = 4$. Add 4 to both sides.

$y - 1 = (x - 2)^2$ Combine terms on the left and factor on the right.

$y = (x - 2)^2 + 1$ Add 1 to both sides.

Now write the original equation as $f(x) = (x - 2)^2 + 1$. As shown earlier, the vertex of this parabola is $(2, 1)$.

The trajectory of a shell fired from a cannon is a **parabola**. To reach the maximum range with a cannon, it is shown in calculus that the muzzle must be set at 45°. If the muzzle is elevated above 45°, the shell goes too high and falls too soon. If the muzzle is set below 45°, the shell is rapidly pulled to Earth by gravity.

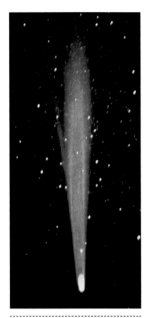

Johann Kepler (1571–1630) established the importance of a curve called an **ellipse** in 1609, when he discovered that the orbits of the planets around the sun were elliptical, not circular. The orbit of Halley's comet, shown here, also is elliptical.

See For Further Thought at the end of this section for more on ellipses.

A formula for the vertex of the graph of the quadratic function $y = ax^2 + bx + c$ can be found by completing the square for the general form of the equation. In doing so, we begin by dividing by a, since the coefficient of x^2 must be 1.

$$y = ax^2 + bx + c \quad (a \neq 0)$$

$$\frac{y}{a} = x^2 + \frac{b}{a}x + \frac{c}{a} \quad \text{Divide by } a.$$

$$\frac{y}{a} - \frac{c}{a} = x^2 + \frac{b}{a}x \quad \text{Subtract } \frac{c}{a}.$$

$$\frac{y}{a} - \frac{c}{a} + \frac{b^2}{4a^2} = x^2 + \frac{b}{a}x + \frac{b^2}{4a^2} \quad \text{Add } \frac{b^2}{4a^2}.$$

$$\frac{y}{a} + \frac{b^2 - 4ac}{4a^2} = \left(x + \frac{b}{2a}\right)^2 \quad \text{Combine terms on left and factor on the right.}$$

$$\frac{y}{a} = \left(x + \frac{b}{2a}\right)^2 - \frac{b^2 - 4ac}{4a^2} \quad \text{Transform so that the } y\text{-term is alone on the left.}$$

$$y = a\left(x + \frac{b}{2a}\right)^2 + \frac{4ac - b^2}{4a} \quad \text{Multiply by } a.$$

$$y = a\underbrace{\left[x - \left(-\frac{b}{2a}\right)\right]^2}_{h} + \underbrace{\frac{4ac - b^2}{4a}}_{k}$$

The final equation shows that the vertex (h, k) can be expressed in terms of a, b, and c. However, it is not necessary to memorize the expression for k, since it can be obtained by replacing x by $-\frac{b}{2a}$. Using function notation, if $y = f(x)$, the y-value of the vertex is $f\left(-\frac{b}{2a}\right)$.

EXAMPLE 6 Use the vertex formula to find the vertex of the graph of the function

$$f(x) = x^2 - x - 6.$$

For this function, $a = 1$, $b = -1$, and $c = -6$. The x-coordinate of the vertex of the parabola is given by

$$-\frac{b}{2a} = -\frac{(-1)}{2(1)} = \frac{1}{2}.$$

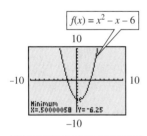

Notice the slight discrepancy when we instruct the calculator to find the vertex (a *minimum* here). This reinforces the fact that *we must understand the concepts and not totally rely on technology!*

The y-coordinate is $f\left(-\frac{b}{2a}\right) = f\left(\frac{1}{2}\right)$.

$$f\left(\frac{1}{2}\right) = \left(\frac{1}{2}\right)^2 - \frac{1}{2} - 6 = \frac{1}{4} - \frac{1}{2} - 6 = -\frac{25}{4}$$

Finally, the vertex is $\left(\frac{1}{2}, -\frac{25}{4}\right)$.

8.5 Quadratic Functions, Applications, and Models 439

A general approach to graphing quadratic functions using intercepts and the vertex is now given.

Graphing a Quadratic Function $f(x) = ax^2 + bx + c$

Step 1: **Decide whether the graph opens upward or downward.** Determine whether the graph opens upward (if $a > 0$) or opens downward (if $a < 0$) to aid in the graphing process.

Step 2: **Find the y-intercept.** Find the y-intercept by evaluating $f(0)$.

Step 3: **Find the x-intercepts.** Find the x-intercepts, if any, by solving $f(x) = 0$.

Step 4: **Find the vertex.** Find the vertex either by using the formula or by completing the square.

Step 5: **Complete the graph.** Find and plot additional points as needed, using the symmetry about the axis.

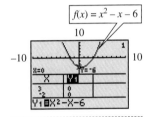

This split screen illustrates that the points $(0, -6)$, $(3, 0)$, and $(-2, 0)$ lie on the graph of $f(x) = x^2 - x - 6$. Compare with Figure 30.

EXAMPLE 7 Graph the quadratic function $f(x) = x^2 - x - 6$.

Because $a > 0$, the parabola will open upward. Now find the y-intercept.

$$f(x) = x^2 - x - 6$$
$$f(0) = 0^2 - 0 - 6 \quad \text{Find } f(0).$$
$$f(0) = -6$$

The y-intercept is $(0, -6)$. Now find any x-intercepts.

$$f(x) = x^2 - x - 6$$
$$0 = x^2 - x - 6 \quad \text{Let } f(x) = 0.$$
$$0 = (x - 3)(x + 2) \quad \text{Factor.}$$
$$x - 3 = 0 \quad \text{or} \quad x + 2 = 0 \quad \text{Set each factor equal to 0 and solve.}$$
$$x = 3 \quad \text{or} \quad x = -2$$

The x-intercepts are $(3, 0)$ and $(-2, 0)$. The vertex, found in Example 6, is $(1/2, -25/4)$. Plot the points found so far, and plot any additional points as needed. The symmetry of the graph is helpful here. The graph is shown in Figure 30.

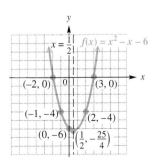

This table provides other points on the graph of

$$Y_1 = X^2 - X - 6.$$

FIGURE 30

Galileo Galilei (1564–1642) died in the year Newton was born; his work was important in Newton's development of calculus. The idea of *function* is implicit in Galileo's analysis of the parabolic path of a projectile, where height and range are functions (in our terms) of the angle of elevation and the initial velocity.

According to legend, Galileo dropped objects of different weights from the tower of Pisa to disprove the Aristotelian view that heavier objects fall faster than lighter objects. He developed a formula for freely falling objects that is described by

$$d = 16t^2,$$

where d is the distance in feet that a given object falls (discounting air resistance) in a given time t, in seconds, regardless of weight.

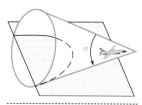

A sonic boom is a loud explosive sound caused by the shock wave that accompanies an aircraft traveling at supersonic speed. The sonic boom shock wave has the shape of a cone, and it intersects the ground in one branch of a curve known as a **hyperbola**. Everyone located along the hyperbolic curve on the ground hears the sound at the same time.

See For Further Thought at the end of this section for more on hyperbolas.

A Model for Optimization

As we have seen, the vertex of a vertical parabola is either the highest or the lowest point of the parabola. The y-value of the vertex gives the maximum or minimum value of y, while the x-value tells where that maximum or minimum occurs. Often a model can be constructed so that y can be *optimized*.

Problem Solving

In some practical problems we want to know the largest or smallest value of some quantity. When that quantity can be expressed using a quadratic function $f(x) = ax^2 + bx + c$, as in the next example, the vertex can be used to find the desired value.

EXAMPLE 8 A farmer has 120 feet of fencing. He wants to put a fence around three sides of a rectangular plot of land, with the side of a barn forming the fourth side. Find the maximum area he can enclose. What dimensions give this area?

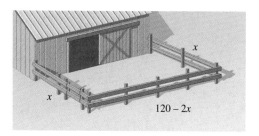

FIGURE 31

Figure 31 shows the plot. Let x represent its width. Then, since there are 120 feet of fencing,

$$x + x + \text{length} = 120 \quad \text{Sum of the three fenced sides is 120 feet.}$$
$$2x + \text{length} = 120 \quad \text{Combine terms.}$$
$$\text{length} = 120 - 2x. \quad \text{Subtract } 2x.$$

The area is modeled by the product of the length and width, or

$$A(x) = (120 - 2x)x = 120x - 2x^2.$$

To make the area (and thus $120x - 2x^2$) as large as possible, first find the vertex of the graph of the function $A(x) = 120x - 2x^2$.

$$A(x) = -2x^2 + 120x \quad \text{Standard form}$$

Here we have $a = -2$ and $b = 120$. The x-coordinate of the vertex is

$$-\frac{b}{2a} = -\frac{120}{2(-2)} = 30.$$

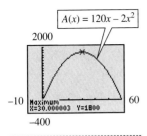

The vertex is (30, 1800), supporting the analytic result in Example 8.

The vertex is a maximum point (since $a < 0$), so the maximum area that the farmer can enclose is

$$A(30) = -2(30)^2 + 120(30) = 1800 \text{ square feet.}$$

The farmer can enclose a maximum area of 1800 square feet, when the width of the plot is 30 feet and the length is $120 - 2(30) = 60$ feet.

As seen in Example 8, be careful when interpreting the meanings of the coordinates of the vertex in problems involving maximum or minimum values. The first coordinate, x, gives the value for which the *function value* is a maximum or a minimum. Read the problem carefully to determine whether you are asked to find the value of the independent variable, the dependent variable (that is, the function value), or both.

FOR FURTHER THOUGHT

The circle, introduced in the first section of this chapter, the parabola, the ellipse, and the hyperbola are known as **conic sections**. As seen in the accompanying figure, each of these geometric shapes can be obtained by intersecting a plane and an infinite cone (made up of two *nappes*).

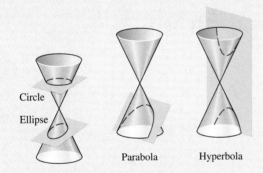

The Greek geometer Apollonius (c. 225 B.C.) was also an astronomer, and his classic work *Conic Sections* thoroughly investigated these figures. Apollonius is responsible for the names "ellipse," "parabola," and "hyperbola." The margin notes in this section show some ways that these figures appear in the world around us.

For Group Discussion

1. The terms *ellipse*, *parabola*, and *hyperbola* are similar to the terms *ellipsis*, *parable*, and *hyperbole*. What do these latter three terms mean? You might want to do some investigation as to the similarities between the mathematical terminology and these language-related terms.
2. Name some places in the world around you where conic sections are encountered.
3. The accompanying figure shows how an ellipse can be drawn using tacks and string. Have a class member volunteer to go to the board and using string and chalk, modify the method to draw a circle. Then have two class members work together to draw an ellipse. (*Hint:* Press hard!)

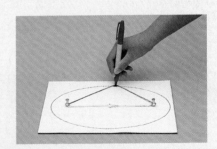

8.5 EXERCISES

In Exercises 1–6, match each equation with the figure that most closely resembles its graph.

A.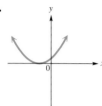

B.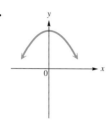

C.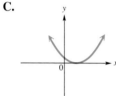

D.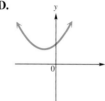

E.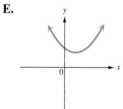

F.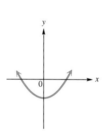

1. $g(x) = x^2 - 5$
2. $h(x) = -x^2 + 4$
3. $F(x) = (x - 1)^2$
4. $G(x) = (x + 1)^2$
5. $H(x) = (x - 1)^2 + 1$
6. $K(x) = (x + 1)^2 + 1$

7. Explain in your own words the meaning of each term.
   (a) vertex of a parabola   (b) axis of a parabola

8. Explain why the axis of the graph of a quadratic function cannot be a horizontal line.

Identify the vertex of the graph of each quadratic function.

9. $f(x) = -3x^2$
10. $f(x) = -.5x^2$
11. $f(x) = x^2 + 4$
12. $f(x) = x^2 - 4$
13. $f(x) = (x - 1)^2$
14. $f(x) = (x + 3)^2$
15. $f(x) = (x + 3)^2 - 4$
16. $f(x) = (x - 5)^2 - 8$

17. Describe how the graph of each parabola in Exercises 15 and 16 is shifted compared to the graph of $y = x^2$.

For each quadratic function, tell whether the graph opens upward or downward, and tell whether the graph is wider, narrower, or the same as the graph of $f(x) = x^2$.

18. $f(x) = -2x^2$
19. $f(x) = -3x^2 + 1$
20. $f(x) = .5x^2$
21. $f(x) = \frac{2}{3}x^2 - 4$

22. What does the value of a in $f(x) = a(x - h)^2 + k$ tell you about the graph of the function compared to the graph of $y = x^2$?

23. For $f(x) = a(x - h)^2 + k$, in what quadrant is the vertex if:
    (a) $h > 0, k > 0$;   (b) $h > 0, k < 0$;
    (c) $h < 0, k > 0$;   (d) $h < 0, k < 0$?

24. (a) What is the value of h if the graph of $f(x) = a(x - h)^2 + k$ has vertex on the y-axis?
    (b) What is the value of k if the graph of $f(x) = a(x - h)^2 + k$ has vertex on the x-axis?

Sketch the graph of each quadratic function using the methods described in this section. Indicate two points on each graph.

25. $f(x) = 3x^2$
26. $f(x) = -2x^2$
27. $f(x) = -\frac{1}{4}x^2$
28. $f(x) = \frac{1}{3}x^2$
29. $f(x) = x^2 - 1$
30. $f(x) = x^2 + 3$
31. $f(x) = -x^2 + 2$
32. $f(x) = -x^2 - 4$
33. $f(x) = 2x^2 - 2$
34. $f(x) = -3x^2 + 1$
35. $f(x) = (x - 4)^2$
36. $f(x) = (x - 3)^2$
37. $f(x) = 3(x + 1)^2$
38. $f(x) = -2(x + 1)^2$
39. $f(x) = (x + 1)^2 - 2$
40. $f(x) = (x - 2)^2 + 3$

Sketch the graph of each quadratic function. Indicate the coordinates of the vertex of the graph.

41. $f(x) = x^2 + 8x + 14$
42. $f(x) = x^2 + 10x + 23$
43. $f(x) = x^2 + 2x - 4$
44. $f(x) = 3x^2 - 9x + 8$
45. $f(x) = -2x^2 + 4x + 5$
46. $f(x) = -5x^2 - 10x + 2$

Solve each problem.

47. *Dimensions of an Exercise Run* Gina Linko has 100 meters of fencing material to enclose a rectangular exercise run for her dog. What width will give the enclosure the maximum area?

48. *Dimensions of a Parking Lot* Morgan's Department Store wants to construct a rectangular parking lot on land bordered on one side by a highway. It has 280 feet of fencing that is to be used to fence off the other three sides. What should be the dimensions of the lot if the enclosed area is to be a maximum? What is the maximum area?

49. *Height of a Propelled Object* If an object on Earth is propelled upward with an initial velocity of 32 feet per second, then its height after t seconds is given by

$$h(t) = 32t - 16t^2.$$

Find the maximum height attained by the object and the number of seconds it takes to hit the ground.

50. *Height of a Propelled Object* A projectile on Earth is fired straight upward so that its distance (in feet) above the ground t seconds after firing is given by

$$s(t) = -16t^2 + 400t.$$

Find the maximum height it reaches and the number of seconds it takes to reach that height.

51. *Height of a Propelled Object* If air resistance is neglected, a projectile on Earth shot straight upward with an initial velocity of 40 meters per second will be at a height s in meters given by the function

$$s(t) = -4.9t^2 + 40t,$$

where t is the number of seconds elapsed after projection. After how many seconds will it reach its maximum height, and what is this maximum height? Round your answers to the nearest tenth.

52. *Height of a Propelled Object* A space robot is propelled from the moon with its distance in feet given by

$$f(x) = 1.727x - .0013x^2$$

feet, where x is time in seconds. Find the maximum height the robot can reach, and the time it takes to get there. Round answers to the nearest tenth.

53. *Pharmacy Payments* The annual percent increase in the amount pharmacies paid wholesalers for drugs in the years 1990–1999 can be modeled by the quadratic function defined by

$$f(x) = .228x^2 - 2.57x + 8.97,$$

where $x = 0$ represents 1990, $x = 1$ represents 1991, and so on. (*Source: IMS Health,* Retail and Provider Perspective.)

(a) Since the coefficient of x^2 in the model is positive, the graph of this quadratic function is a parabola that opens up. Will the y-value of the vertex of this graph be a maximum or minimum?

(b) In what year was the minimum percent increase? (Round down to the nearest year.) Use the actual x-value of the vertex, to the nearest tenth, to find this increase.

54. *U.S. Oyster Catch* The U.S. domestic oyster catch (in millions) for the years 1990–1998 can be approximated by the quadratic function defined by

$$f(x) = -.566x^2 + 5.08x + 29.2,$$

where $x = 0$ represents 1990, $x = 1$ represents 1991, and so on. (*Source:* National Marine Fisheries Service.)

(a) Since the coefficient of x^2 in the model is negative, the graph of this quadratic function is a parabola that opens down. Will the y-value of the vertex of this graph be a maximum or minimum?

(b) In what year was the maximum domestic oyster catch? (Round down to the nearest year.) Use the actual x-value of the vertex, to the nearest tenth, to find this catch.

55. *Social Security Assets* The graph shows how Social Security assets are expected to change as the number of retirees receiving benefits increases.

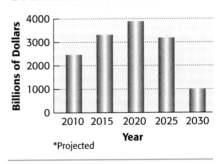

Source: Social Security Administration.

The graph suggests that a quadratic function would be a good fit to the data. The data are approximated by the function defined by

$$f(x) = -20.57x^2 + 758.9x - 3140.$$

In the model, $x = 10$ represents 2010, $x = 15$ represents 2015, and so on, and $f(x)$ is in billions of dollars.

(a) Explain why the coefficient of x^2 in the model is negative, based on the graph.

(b) Algebraically determine the vertex of the graph, with coordinates to four significant digits.

(c) Interpret the answer to part (b) as it applies to the application.

56. *Investment Portfolio Mixtures* The graph, which appears to be a portion of a parabola opening to the *right*, shows the performance of investment portfolios with different mixtures of U.S. and foreign investments for the period January 1, 1971, to December 31, 1996.

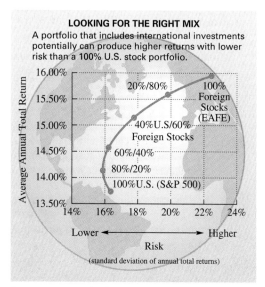

Source: Financial Ink Newsletter, Investment Management and Research, Inc., Feb. 1998. Thanks to David Van Geffen for this information.

Use the graph to answer the following questions.
(a) Is this the graph of a function? Explain.
(b) What investment mixture shown on the graph appears to represent the vertex? What relative amount of risk does this point represent? What return on investment does it provide?
(c) Which point on the graph represents the riskiest investment mixture? What return on investment does it provide?

57. *Maximum Airline Revenue* A charter flight charges a fare of $200 per person, plus $4 per person for each unsold seat on the plane. If the plane holds 100 passengers and if x represents the number of unsold seats, find the following.
(a) An expression for the total revenue $R(x)$ received for the flight (*Hint:* Multiply the number of people flying, $100 - x$, by the price per ticket.)
(b) The graph for the function of part (a)
(c) The number of unsold seats that will produce the maximum revenue
(d) The maximum revenue

58. *Maximum Bus Fare Revenue* For a trip to a resort, a charter bus company charges a fare of $48 per person, plus $2 per person for each unsold seat on the bus. If the bus has 42 seats and x represents the number of unsold seats, find the following.
(a) An expression that defines the total revenue, $R(x)$, from the trip (*Hint:* Multiply the total number riding, $42 - x$, by the price per ticket, $48 + 2x$.)
(b) The graph of the function from part (a)
(c) The number of unsold seats that produces the maximum revenue
(d) The maximum revenue

8.6 Exponential and Logarithmic Functions, Applications, and Models

Exponential Functions In this section we introduce two new types of functions. The first of these is the exponential function.

> **Exponential Function**
>
> An exponential function with base b, where $b > 0$ and $b \neq 1$, is a function of the form
>
> $$f(x) = b^x,$$
>
> where x is any real number.

8.6 Exponential and Logarithmic Functions, Applications, and Models

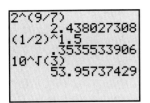

Compare with the discussion in the text.

Thus far, we have defined only integer exponents. In the definition of exponential function, we allow x to take on any real number value. By using methods not discussed in this book, expressions such as

$$2^{9/7}, \quad \left(\frac{1}{2}\right)^{1.5}, \quad \text{and} \quad 10^{\sqrt{3}}$$

can be approximated. A scientific or graphing calculator is capable of determining approximations for these numbers. See the screen in the margin.

Notice that in the definition of exponential function, the base b is restricted to positive numbers, with $b \neq 1$.

EXAMPLE 1 The graphs of $f(x) = 2^x$, $g(x) = (1/2)^x$, and $h(x) = 10^x$ are shown in Figure 32(a), (b), and (c). In each case, a table of selected points is given. The points are joined with a smooth curve, typical of the graphs of exponential functions. Notice that for each graph, the curve approaches but does not intersect the x-axis. For this reason, the x-axis is called the **horizontal asymptote** of the graph.

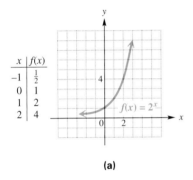
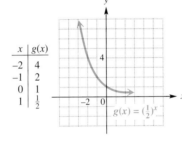
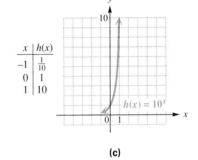

(a)   (b)   (c)

FIGURE 32

Example 1 illustrates the following facts about the graph of an exponential function.

Compare with Figure 32.

Graph of $f(x) = b^x$

1. The graph always will contain the point $(0, 1)$, since $b^0 = 1$.
2. When $b > 1$, the graph will *rise* from left to right (as in the example for $b = 2$ and $b = 10$). When $0 < b < 1$, the graph will *fall* from left to right (as in the example for $b = 1/2$).
3. The x-axis is the horizontal asymptote.
4. The domain is $(-\infty, \infty)$ and the range is $(0, \infty)$.

Probably the most important exponential function has the base e. The number e is named after Leonhard Euler (1707–1783), and is approximately 2.718281828. It is an irrational number, and its value is approached by the expression

$$\left(1 + \frac{1}{n}\right)^n$$

as n takes on larger and larger values. We write

as $n \to \infty$, $\left(1 + \dfrac{1}{n}\right)^n \to e \approx 2.718281828$.

See the table that follows.

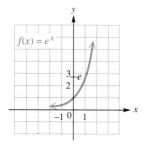

This table shows selected values for $Y_1 = \left(1 + \dfrac{1}{X}\right)^X$.

n	Approximate Value of $\left(1 + \dfrac{1}{n}\right)^n$
1	2
2	2.25
5	2.48832
10	2.59374
25	2.66584
50	2.69159
100	2.70481
500	2.71557
1000	2.71692
10,000	2.71815
1,000,000	2.71828

Powers of e can be approximated on a scientific or graphing calculator. Some powers of e obtained on a calculator are

$e^{-2} \approx .1353352832$, $e^{1.3} \approx 3.669296668$, $e^4 \approx 54.59815003$.

The graph of the function $f(x) = e^x$ is shown in Figure 33.

FIGURE 33

Applications of Exponential Functions   A real-life application of exponential functions occurs in the computation of compound interest.

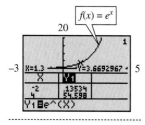

The values for e^{-2}, $e^{1.3}$, and e^4 are approximated in this split-screen graph of $f(x) = e^x$.

Compound Interest Formula

Suppose that a principal of P dollars is invested at an annual interest rate r (in percent, expressed as a decimal), compounded n times per year. Then the amount A accumulated after t years is given by the formula

$$A = P\left(1 + \dfrac{r}{n}\right)^{nt}.$$

EXAMPLE 2   Suppose that $1000 is invested at an annual rate of 8%, compounded quarterly (four times per year). Find the total amount in the account after ten years if no withdrawals are made.

Use the compound interest formula, with $P = 1000$, $r = .08$, $n = 4$, and $t = 10$.

$$A = P\left(1 + \dfrac{r}{n}\right)^{nt}$$

$$A = 1000\left(1 + \dfrac{.08}{4}\right)^{4 \cdot 10}$$

$$A = 1000(1.02)^{40}$$

8.6 Exponential and Logarithmic Functions, Applications, and Models 447

Using a calculator with an exponential key, $1.02^{40} \approx 2.20804$. Multiply by 1000 to get
$$A = 2208.04.$$
There would be $2208.04 in the account at the end of ten years. ∎

The compounding formula given earlier applies if the financial institution compounds interest for a finite number of compounding periods annually. Theoretically, the number of compounding periods per year can get larger and larger (quarterly, monthly, daily, etc.), and if n is allowed to approach infinity, we say that interest is compounded *continuously*. It can be shown that the formula for **continuous compounding** involves the number e.

Continuous Compounding Formula

Suppose that a principal of P dollars is invested at an annual interest rate r (in percent, expressed as a decimal), compounded continuously. Then the amount A accumulated after t years is given by the formula
$$A = Pe^{rt}.$$

EXAMPLE 3 Suppose that $5000 is invested at an annual rate of 6.5%, compounded continuously. Find the total amount in the account after four years if no withdrawals are made.

The continuous compounding formula applies here, with $P = 5000$, $r = .065$, and $t = 4$.
$$A = Pe^{rt}$$
$$A = 5000e^{.065(4)}$$
$$A = 5000e^{.26}$$

Use the e^x key on a calculator to find that $e^{.26} \approx 1.29693$, so $A \approx 5000(1.29693) = 6484.65$. There will be $6484.65 in the account after four years. ∎

The continuous compounding formula is an example of an **exponential growth function.** It can be shown that in situations involving growth or decay of a quantity, the amount or number present at time t can often be closely approximated by a function of the form
$$A(t) = A_0 e^{kt},$$
where A_0 represents the amount or number present at time $t = 0$, and k is a constant. If $k > 0$, there is exponential growth; if $k < 0$, there is exponential decay.

Logarithmic Functions
Consider the exponential equation
$$2^3 = 8.$$
Here we see that 3 is the exponent (or power) to which 2 must be raised in order to obtain 8. The exponent 3 is called the *logarithm* to the base 2 of 8, and this is written
$$3 = \log_2 8.$$
In general, we have the following relationship.

For $b > 0$, $b \neq 1$, if $b^y = x$, then $y = \log_b x$.

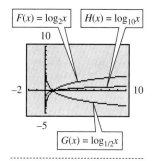

Compare with Figure 34 on the next page. Graphs of logarithmic functions with bases other than 10 and e are accomplished with the use of the change-of-base rule from algebra:

$$\log_a x = \frac{\log x}{\log a} = \frac{\ln x}{\ln a}.$$

Alternatively, they can be "drawn" by using the capability that allows the user to obtain the graph of the inverse.

The following table illustrates the relationship between exponential equations and logarithmic equations.

Exponential Equation	Logarithmic Equation
$3^4 = 81$	$4 = \log_3 81$
$(1/2)^{-4} = 16$	$-4 = \log_{1/2} 16$
$10^0 = 1$	$0 = \log_{10} 1$

The concept of inverse functions (studied in more advanced algebra courses) leads us to the definition of the logarithmic function with base b.

Logarithmic Function

A logarithmic function with base b, where $b > 0$ and $b \neq 1$, is a function of the form

$$g(x) = \log_b x,$$

where $x > 0$.

FOR FURTHER THOUGHT

In the February 5, 1989, issue of *Parade*, Carl Sagan related an oft-told legend involving **exponential growth** in the article "The Secret of the Persian Chessboard." Once upon a time, a Persian king wanted to please his executive officer, the Grand Vizier, with a gift of his choice. The Grand Vizier explained that he would like to be able to use his chessboard to accumulate wheat. A single grain of wheat would be received for the first square on the board, two grains would be received for the second square, four grains for the third, and so on, doubling the number of grains for each of the 64 squares on the board. This doubling procedure is an example of exponential growth, and is defined by the exponential function

$$f(x) = 2^{x-1},$$

where x corresponds to the number of the chessboard square, and $f(x)$ represents the number of grains of wheat corresponding to that square.

How many grains of wheat would be accumulated? As unlikely as it may seem, the number of grains would total 18.5 quintillion! Even with today's methods of production, this amount would take 150 years to produce. The Grand Vizier evidently knew his mathematics.

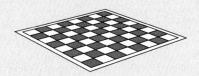

For Group Discussion

1. If a lily pad doubles in size each day, and it covers a pond after eight days of growth, when is the pond half covered? One-fourth covered?
2. Have each member of the class estimate the answer to this problem.
   If you earn 1¢ on January 1, 2¢ on January 2, 4¢ on January 3, 8¢ on January 4, and so on, doubling your salary each day, how much will you earn during the month?
As a calculator exercise, determine the answer and see who was able to estimate the answer most accurately.

8.6 Exponential and Logarithmic Functions, Applications, and Models

The graph of the function $g(x) = \log_b x$ can be found by interchanging the roles of x and y in the function $f(x) = b^x$. Geometrically, this is accomplished by reflecting the graph of $f(x) = b^x$ about the line $y = x$.

EXAMPLE 4 The graphs of $F(x) = \log_2 x$, $G(x) = \log_{1/2} x$, and $H(x) = \log_{10} x$ are shown in Figure 34(a), (b), and (c). In each case, a table of selected points is given. These points were obtained by interchanging the roles of x and y in the tables of points given in Figure 32. The points are joined with a smooth curve, typical of the graphs of logarithmic functions. Notice that for each graph, the curve approaches but does not intersect the y-axis. For this reason, the y-axis is called the **vertical asymptote** of the graph.

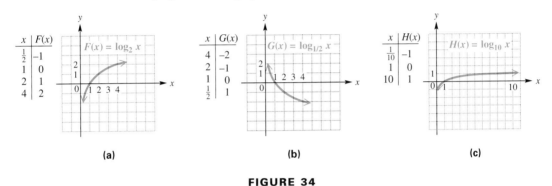

FIGURE 34

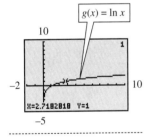

Notice that when
$x = e \approx 2.7182818$,
$y = \ln e = 1$.

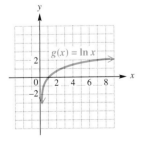

FIGURE 35

Example 4 illustrates the following facts about the graph of a logarithmic function.

Graph of $g(x) = \log_b x$

1. The graph will always contain the point $(1, 0)$, since $\log_b 1 = 0$.
2. When $b > 1$, the graph will *rise* from left to right, from the fourth quadrant to the first (as in the example for $b = 2$ and $b = 10$). When $0 < b < 1$, the graph will *fall* from left to right, from the first quadrant to the fourth (as in the example for $b = 1/2$).
3. The y-axis is the vertical asymptote.
4. The domain is $(0, \infty)$ and the range is $(-\infty, \infty)$.

One of the most important logarithmic functions is the function with base e. If we interchange the roles of x and y in the graph of $f(x) = e^x$ (Figure 33), we obtain the graph of $g(x) = \log_e x$. There is a special symbol for $\log_e x$: it is $\ln x$. That is,

$$\ln x = \log_e x.$$

Figure 35 shows the graph of $g(x) = \ln x$, called the **natural logarithmic function**.

The number e is named in honor of Leonhard Euler (1707–1783), the prolific Swiss mathematician. The value of e can be expressed as an infinite series.

$$e = 1 + \frac{1}{1} + \frac{1}{1 \cdot 2} + \frac{1}{1 \cdot 2 \cdot 3} + \frac{1}{1 \cdot 2 \cdot 3 \cdot 4} + \cdots$$

It can also be expressed in two ways using continued fractions.

$$e = 2 + \cfrac{1}{1 + \cfrac{1}{2 + \cfrac{1}{1 + \cfrac{1}{4 + \cfrac{1}{\ddots}}}}}$$

$$e = 2 + \cfrac{1}{1 + \cfrac{1}{2 + \cfrac{2}{3 + \cfrac{3}{4 + \cfrac{4}{\ddots}}}}}$$

The expression

$$\ln e^k$$

is the exponent to which the base e must be raised in order to obtain e^k. There is only one such number that will do this, and it is k itself. Therefore, we make the following important statement.

For all real numbers k,

$$\ln e^k = k.$$

This fact is used in the examples that follow.

EXAMPLE 5 Suppose that a certain amount P is invested at an annual rate of 6.5%, compounded continuously. How long will it take for the amount to triple?

We wish to find the value of t in the continuous compounding formula that will make the amount A equal to $3P$ (since we want the initial investment, P, to triple). We use the formula $A = Pe^{rt}$, with $3P$ substituted for A and .065 substituted for r.

$$A = Pe^{rt}$$
$$3P = Pe^{.065t}$$
$$3 = e^{.065t} \quad \text{Divide both sides by } P.$$

Now take the natural logarithm of both sides.

$$\ln 3 = \ln e^{.065t}$$
$$\ln 3 = .065t \quad \text{Use the fact that } \ln e^k = k.$$

Solve for t by dividing both sides by .065.

$$t = \frac{\ln 3}{.065}$$

A calculator shows that $\ln 3 \approx 1.098612289$. Dividing this by .065 gives

$$t \approx 16.9$$

to the nearest tenth. Therefore, it would take about 16.9 years for any initial investment P to triple under the given conditions.

Exponential Models

EXAMPLE 6 The *greenhouse effect* refers to the phenomenon whereby emissions of gases such as carbon dioxide, methane, and chlorofluorocarbons (CFCs) have the potential to alter the climate of the earth and destroy the ozone layer. Concentrations of CFC-12, used in refrigeration technology, in parts per billion (ppb) can be modeled by the exponential function defined by

$$f(x) = .48e^{.04x},$$

where $x = 0$ represents 1990, $x = 1$ represents 1991, and so on. Use this function to approximate the concentration in 1998.

Since $x = 0$ represents 1990, $x = 8$ represents 1998. Evaluate $f(8)$ using a calculator.

$$f(8) = .48e^{.04(8)} = .48e^{.32} \approx .66$$

In 1998, the concentration of CFC-12 was about .66 ppb. ∎

Radioactive materials disintegrate according to exponential decay functions. The **half-life** of a quantity that decays exponentially is the amount of time that it takes for any initial amount to decay to half its initial value.

EXAMPLE 7 Carbon 14 is a radioactive form of carbon that is found in all living plants and animals. After a plant or animal dies, the radiocarbon disintegrates. Scientists determine the age of the remains by comparing the amount of carbon 14 present with the amount found in living plants and animals. The amount of carbon 14 present after t years is modeled by the exponential equation

$$y = y_0 e^{-.0001216t},$$

where y_0 represents the initial amount. What is the half-life of carbon 14?

To find the half-life, let $y = \frac{1}{2}y_0$ in the equation.

$$\frac{1}{2} y_0 = y_0 e^{-.0001216t}$$

$$\frac{1}{2} = e^{-.0001216t}$$

$$\ln\left(\frac{1}{2}\right) = -.0001216t$$

$$t \approx 5700 \qquad \text{Use a calculator.}$$

The half-life of carbon 14 is about 5700 years. ∎

FOR FURTHER THOUGHT

So what does that "log" button on your scientific calculator mean? We will answer that question shortly. But let's back up to 1624 when the British mathematician Henry Briggs published what was, at that time, the greatest breakthrough ever for aiding computation: tables of common logarithms. A common, or base ten logarithm, of a positive number x is the exponent to which 10 must be raised in order to obtain x. Symbolically, we simply write log x to denote $\log_{10} x$.

Since logarithms are exponents, their properties are the same as those of exponents. These properties allowed users of tables of common logarithms to multiply by adding, divide by subtracting, raise to powers by multiplying, and take roots by dividing. For example, to multiply 534.4 by 296.7, the user would find log 534.4 in the table, then find log 296.7 in the table, and add these values. Then, by reading the table in reverse, the user would find the "antilog" of the sum in order to find the product of the original two numbers. More complicated operations were possible as well.

(continued)

The photo here shows a homework problem that someone was asked to do on May 15, 1930. It was found on a well-preserved sheet of paper stuck inside an old mathematics text purchased in a used-book store in New Orleans. Aren't you glad that we now use calculators?

So why include a base ten logarithm key on a scientific calculator? Common logarithms still are used in certain applications in science. For example, the pH of a substance is the negative of the common logarithm of the hydronium ion concentration in moles per liter. The pH value is a measure of the acidity or alkalinity of a solution. If pH > 7.0, the solution is alkaline; if pH < 7.0, it is acidic. The Richter scale, used to measure the intensity of earthquakes, is based on logarithms, and sound intensities, measured in decibels, also have a logarithmic basis.

For Group Discussion

1. Divide the class into small groups, and use the log key of a scientific calculator to multiply 458.3 by 294.6. Remember that once the logarithms have been added, it is necessary then to raise 10 to that power to obtain the product. The 10^x key will be required for this.
2. Ask someone who studied computation with logarithms in an algebra or trigonometry course before the advent of scientific calculators what it was like. Then report their recollections to the class.
3. What is a slide rule? Have someone in the class find one in an attic or closet, and bring it to class. The slide rule was a popular tool for calculation through the 1960s.

8.6 EXERCISES

Fill in each blank with the correct response.

1. For an exponential function $f(x) = a^x$, if $a > 1$, the graph _____ from left to right. If $0 < a < 1$,
   (rises/falls)
   the graph _____ from left to right.
   (rises/falls)
2. The y-intercept of the graph of $y = a^x$ is _____.
3. The graph of the exponential function $f(x) = a^x$ _____ have an x-intercept.
   (does/does not)
4. The point (2, _____) is on the graph of $f(x) = 3^{4x-3}$.
5. For a logarithmic function $g(x) = \log_a x$, if $a > 1$, the graph _____ from left to right. If
   (rises/falls)
   $0 < a < 1$, the graph _____ from left to right.
   (rises/falls)
6. The x-intercept of the graph of $y = \log_a x$ is _____.
7. The graph of the exponential function $g(x) = \log_a x$ _____ have a y-intercept.
   (does/does not)
8. The point (98, _____) lies on the graph of $g(x) = \log_{10}(x + 2)$.

Use a calculator to find an approximation for each of the following numbers. Give as many digits as the calculator displays.

9. $9^{3/7}$
10. $14^{2/7}$
11. $(.83)^{-1.2}$
12. $(.97)^{3.4}$
13. $(\sqrt{6})^{\sqrt{5}}$
14. $(\sqrt{7})^{\sqrt{3}}$
15. $\left(\dfrac{1}{3}\right)^{9.8}$
16. $\left(\dfrac{2}{5}\right)^{8.1}$

Sketch the graph of each of the following functions.

17. $f(x) = 3^x$
18. $f(x) = 5^x$
19. $f(x) = \left(\dfrac{1}{4}\right)^x$
20. $f(x) = \left(\dfrac{1}{3}\right)^x$

Use a calculator to find each of the following. Give as many digits as the calculator displays.

21. e^3
22. e^4
23. e^{-4}
24. e^{-3}

In Exercises 25–28, rewrite the exponential equation as a logarithmic equation. In Exercises 29–32, rewrite the logarithmic equation as an exponential equation.

25. $4^2 = 16$
26. $5^3 = 125$
27. $\left(\dfrac{2}{3}\right)^{-3} = \dfrac{27}{8}$
28. $\left(\dfrac{1}{10}\right)^{-4} = 10{,}000$
29. $5 = \log_2 32$
30. $3 = \log_4 64$
31. $1 = \log_3 3$
32. $0 = \log_{12} 1$

Use a calculator to find each of the following. Give as many digits as the calculator displays.

33. ln 4
34. ln 6
35. ln .35
36. ln 2.45

Global Warming The figure shown here accompanied the article "Is Our World Warming?" which appeared in the October 1990 issue of National Geographic*. It shows projected temperature increases using two graphs: one an exponential-type curve, and the other linear. From the figure, approximate the increase* **(a)** *for the exponential curve, and* **(b)** *for the linear graph for each of the following years.*

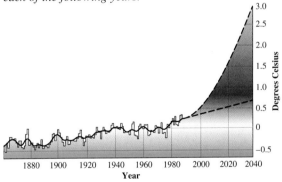

Source: "Zero Equals Average Global Temperature for the Period 1950–1979" by Dale D. Glasgow from *National Geographic*, October 1990. Reprinted by permission of the National Geographic Society.

37. 2000
38. 2010
39. 2020
40. 2040

Sketch the graph of each of the following functions. (Hint: Use the graphs of the exponential functions in Exercises 17–20 to help.)

41. $g(x) = \log_3 x$
42. $g(x) = \log_5 x$
43. $g(x) = \log_{1/4} x$
44. $g(x) = \log_{1/3} x$

Investment Determine the amount of money that will be accumulated in an account that pays compound interest, given the initial principal in each of the following.

45. $4292 at 6% compounded annually for 10 years

46. $8906.54 at 5% compounded semiannually for 9 years

47. $56,780 at 5.3% compounded quarterly for 23 quarters

48. $45,788 at 6% compounded daily (ignoring leap years) for 11 years of 365 days

49. *Investment* Karen Guardino invests a $25,000 inheritance in a fund paying 5% per year compounded continuously. What will be the amount on deposit after each of the following time periods?
(a) 1 year **(b)** 5 years **(c)** 10 years

50. *Investment* Ron Hampton, who is self-employed, wants to invest $60,000 in a pension plan. One investment offers 7% compounded quarterly. Another offers 6.75% compounded continuously. Which investment will earn more interest in 5 years? How much more will the better plan earn?

Solve each problem involving an exponential or a logarithmic function.

51. *Carbon Monoxide Emissions* Based on figures for 1970–1998, the number of worldwide carbon monoxide emissions in thousands of tons is approximated by the exponential function defined by

$$f(x) = 132{,}359(1.0124)^{-x},$$

where $x = 0$ corresponds to 1970, $x = 5$ corresponds to 1975, and so on. (*Source:* Environmental Protection Agency, *National Air Pollutant Emission Trends*.)
(a) Use this model to approximate the emissions in 1970.
(b) Repeat part (a) for 1995.
(c) In 1998, the actual amount of emissions was 89,454 million tons. How does this compare to the number that the model provides?

52. *Municipal Solid Waste* Based on figures for 1980–1999, the municipal solid waste generated in millions of tons can be approximated by the exponential function defined by

$$f(x) = 157.28(1.0204)^x,$$

where $x = 0$ corresponds to 1980, $x = 5$ corresponds to 1985, and so on. (*Source:* Environmental Protection Agency, *Characterization of Municipal Solid Waste in the United States: 1999.*)
(a) Use this model to approximate the number of tons of this waste in 1980.
(b) Use this model to approximate the number of tons of this waste in 1985.
(c) In 1999, the actual number of millions of tons of this waste was 229.9. How does this compare to the number that the model provides?

53. *Population of Brazil* The U.S. Bureau of the Census projects that the approximate population of Brazil will grow according to the function

$$f(x) = 176{,}000(2)^{.008x},$$

where $x = 0$ represents the year 2000, $x = 25$ represents 2025, and $x = 50$ represents 2050.

(a) According to this model, what was the population of Brazil in 2000?
(b) What will the population be in 2025?
(c) How will the population in 2025 compare to the population in 2000?

54. *Population of Pakistan* The U.S. Bureau of the Census projects that the approximate population of Pakistan will grow according to the function

$$f(x) = 146{,}250(2)^{.0176x},$$

where $x = 0$ represents the year 2000, $x = 25$ represents 2025, and $x = 50$ represents 2050.
(a) According to this model, what was the population of Pakistan in 2000?
(b) What will the population be in 2025?
(c) How will the population in 2025 compare to the population in 2000?

55. *Book Sales* Based on selected figures obtained during the 1980s and 1990s, consumer expenditures on all types of books in the United States can be modeled by the function defined by

$$B(x) = 8768 e^{.072x},$$

where $x = 0$ represents 1980, $x = 1$ represents 1981, and so on, and $B(x)$ is in millions of dollars. Approximate consumer expenditures for 1998. (*Source:* Book Industry Study Group.)

56. *Bachelor's Degrees Earned* Based on selected figures obtained during the 1970s, 1980s, and 1990s, the total number of bachelor's degrees earned in the United States can be modeled by the function defined by

$$D(x) = 815{,}427 e^{.0137x},$$

where $x = 1$ corresponds to 1971, $x = 10$ corresponds to 1980, and so on. Approximate the number of bachelor's degrees earned in 1994. (*Source:* U.S. National Center for Education Statistics.)

57. *Radioactive Ore Sample* The amount of radioactive material in an ore sample is given by the function defined by

$$A(t) = 100(3.2)^{-.5t},$$

where $A(t)$ is the amount present, in grams, of the sample t months after the initial measurement.
(a) How much was present at the initial measurement? (*Hint:* $t = 0$.)
(b) How much was present 2 months later?
(c) How much was present 10 months later?

58. *Value of a Copy Machine* A small business estimates that the value $V(t)$ of a copy machine is decreasing according to the function defined by

$$V(t) = 5000(2)^{-.15t},$$

where t is the number of years that have elapsed since the machine was purchased, and $V(t)$ is in dollars.
(a) What was the original value of the machine?
(b) What is the value of the machine 5 years after purchase? Give your answer to the nearest dollar.
(c) What is the value of the machine 10 years after purchase? Give your answer to the nearest dollar.

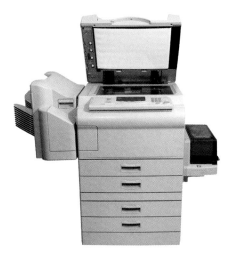

59. *Hazardous Waste Sites* According to selected figures for 1981–1995, the number of Superfund hazardous waste sites in the United States can be approximated by the function defined by

$$f(x) = 11.34 + 317.01 \log_2 x,$$

where $x = 1$ corresponds to 1981, $x = 2$ to 1982, and so on. (*Source:* Environmental Protection Agency.)
(a) Use the function to approximate the number of sites in 1984.
(b) Use the function to approximate the number of sites in 1988.

60. *Dry Natural Gas Consumption* According to selected figures for 1980–1993, the number of trillion cubic feet of dry natural gas consumed worldwide can be approximated by the function defined by

$$f(x) = 51.47 + 6.044 \log_2 x,$$

where $x = 1$ corresponds to 1980, $x = 2$ to 1981, and so on. (*Source:* Energy Information Administration.)
(a) Use the function to approximate consumption in 1980.
(b) Use the function to approximate consumption in 1987.

Earthquake Intensity In the United States, the intensity of an earthquake is rated using the Richter scale. The Richter scale rating of an earthquake of intensity x is given by

$$R = \log_{10} \frac{x}{x_0},$$

where x_0 is the intensity of an earthquake of a certain (small) size. The figure shows Richter scale ratings for major Southern California earthquakes since 1920. As the figure indicates, earthquakes "come in bunches" and the 1990s were an especially busy time.

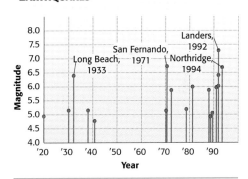

MAJOR SOUTHERN CALIFORNIA EARTHQUAKES

Source: Caltech; U.S. Geological Survey.

Writing the logarithmic equation just given in exponential form, we get

$$10^R = \frac{x}{x_0} \quad \text{or} \quad x = 10^R x_0.$$

The 1994 Northridge earthquake had a Richter scale rating of 6.7; the Landers earthquake had a rating of 7.3.

61. How many times as powerful was the Landers earthquake compared to the Northridge earthquake?
62. Compare the smallest rated earthquake in the graph (at 4.8) with the Landers quake. How many times as powerful was the Landers quake?

8.7 Systems of Equations and Applications

Linear Systems in Two Variables The worldwide personal computer market share for different manufacturers has varied, with first one, then another obtaining a larger share. As shown in Figure 36, Hewlett-Packard's share rose during 1995–1998, while Packard Bell, NEC saw its share decline. The graphs intersect at the point when the two companies had the same market share.

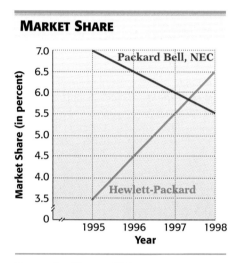

FIGURE 36

We could use a linear equation to model the graph of Hewlett-Packard's market share and another linear equation to model the graph of Packard Bell, NEC's market share. Such a set of equations is called a **system of equations.** The point where the graphs in Figure 36 intersect is a solution of each of the individual equations. It is also the solution of the system of equations.

The definition of a linear equation given earlier can be extended to more variables. Any equation of the form

$$a_1x_1 + a_2x_2 + \cdots + a_nx_n = b$$

for real numbers $a_1, a_2, \ldots, a_n$ (not all of which are 0), and b, is a **linear equation in n variables.** If all the equations in a system are linear, the system is a **system of linear equations,** or a **linear system.**

In Figure 37, the two linear equations $x + y = 5$ and $2x - y = 4$ are graphed in the same coordinate system. Notice that they intersect at the point $(3, 2)$. Because $(3, 2)$ is the only ordered pair that satisfies both equations at the same time, we say that $\{(3, 2)\}$ is the solution set of the system

$$x + y = 5$$
$$2x - y = 4.$$

Since the graph of a linear equation is a straight line, there are three possibilities for the number of solutions in the solution set of a system of two linear equations, as shown in Figure 38 on the next page.

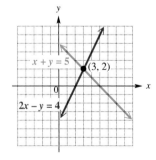

FIGURE 37

8.7 Systems of Equations and Applications   457

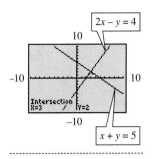

A graphing calculator supports our statement that (3, 2) is the solution of the system

$x + y = 5$
$2x - y = 4$.

Graphs of a Linear System (The Three Possibilities)

1. The two graphs intersect in a single point. The coordinates of this point give the only solution of the system. In this case, the system is **consistent,** and the equations are **independent.** This is the most common case. See Figure 38(a).

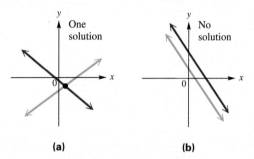

FIGURE 38

2. The graphs are parallel lines. In this case the system is **inconsistent;** that is, there is no solution common to both equations of the system, and the solution set is ∅. See Figure 38(b).

3. The graphs are the same line. In this case the equations are **dependent,** since any solution of one equation of the system is also a solution of the other. The solution set is an infinite set of ordered pairs representing the points on the line. See Figure 38(c).

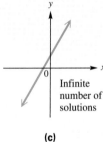

(c)

FIGURE 38

In most cases we cannot rely on graphing to solve systems. There are algebraic methods to do this, and one such method, called the **elimination method,** is explained in the following examples. The elimination method involves combining the two equations of the system so that one variable is eliminated. This is done using the following fact.

If $a = b$ and $c = d$, then $a + c = b + d$.

The general method of solving a system by elimination is summarized as follows.

The solution of systems of equations of graphs more complicated than straight lines is the principle behind the **mattang** (shown on this stamp), a stick chart used by the people of the Marshall Islands in the Pacific. A mattang is made of roots tied together with coconut fibers, and it shows the wave patterns found when approaching an island.

Solving Linear Systems by Elimination

Step 1: **Write in standard form.** Write both equations in the form
$$Ax + By = C.$$

Step 2: **Make the coefficients of one pair of variable terms opposites.** Multiply one or both equations by appropriate numbers so that the sum of the coefficients of either x or y is zero.

Step 3: **Add.** Add the new equations to eliminate a variable. The sum should be an equation with just one variable.

Step 4: **Solve.** Solve the equation from Step 3.

Step 5: **Find the other value.** Substitute the result of Step 4 into either of the given equations and solve for the other variable.

Step 6: **Find the solution set.** Check the solution in both of the given equations. Then write the solution set.

EXAMPLE 1 Solve the system.

$$5x - 2y = 4 \quad (1)$$
$$2x + 3y = 13 \quad (2)$$

Step 1: Both equations are already in standard form.

Step 2: Our goal is to add the two equations so that one of the variables is eliminated. Suppose we wish to eliminate the variable x. Since the coefficients of x are *not* opposites, we must first transform one or both equations so that the coefficients *are* opposites. Then, when we combine the equations, the term with x will have a coefficient of 0, and we will be able to solve for y. We begin by multiplying equation (1) by 2 and equation (2) by -5.

$$10x - 4y = 8 \qquad \text{2 times each side of equation (1)}$$
$$-10x - 15y = -65 \qquad \text{-5 times each side of equation (2)}$$

Step 3: Now add the two equations to eliminate x.

$$\begin{array}{r} 10x - 4y = 8 \\ -10x - 15y = -65 \\ \hline -19y = -57 \end{array} \quad \text{Add.}$$

Step 4: Solve the equation from Step 3 to get $y = 3$.

Step 5: To find x, we substitute 3 for y in either of the original equations. Substituting in equation (2) gives

$$2x + 3y = 13 \qquad (2)$$
$$2x + 3(3) = 13 \qquad \text{Let } y = 3.$$
$$2x + 9 = 13$$
$$2x = 4 \qquad \text{Subtract 9.}$$
$$x = 2. \qquad \text{Divide by 2.}$$

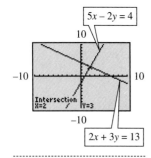

The solution in Example 1 is supported by a graphing calculator. We solve each equation for y, graph them both, and find the point of intersection of the two lines: (2, 3).

Step 6: The solution appears to be $(2, 3)$. To check, substitute 2 for x and 3 for y in both of the original equations.

$5x - 2y = 4$	(1)		$2x + 3y = 13$	(2)
$5(2) - 2(3) = 4$	?		$2(2) + 3(3) = 13$	?
$10 - 6 = 4$	?		$4 + 9 = 13$	?
$4 = 4$	True		$13 = 13$	True

The solution set is $\{(2, 3)\}$.

Linear systems can also be solved by the **substitution method.** This method is most useful for solving linear systems in which one variable has coefficient 1 or -1. As shown in more advanced algebra courses, the substitution method is the best choice for solving many *nonlinear* systems.

The method of solving a system by substitution is summarized as follows.

Solving Linear Systems by Substitution

Step 1: **Solve for one variable in terms of the other.** Solve one of the equations for either variable. (If one of the variables has coefficient 1 or -1, choose it, since the substitution method is usually easier this way.)

Step 2: **Substitute.** Substitute for that variable in the other equation. The result should be an equation with just one variable.

Step 3: **Solve.** Solve the equation from Step 2.

Step 4: **Find the other value.** Substitute the result from Step 3 into the equation from Step 1 to find the value of the other variable.

Step 5: **Find the solution set.** Check the solution in both of the given equations. Then write the solution set.

The following example illustrates this method.

EXAMPLE 2 Solve the system.

$$3x + 2y = 13 \quad (1)$$
$$4x - y = -1 \quad (2)$$

Step 1: To use the substitution method, first solve one of the equations for either x or y. Since the coefficient of y in equation (2) is -1, it is easiest to solve for y in equation (2).

$$-y = -1 - 4x \quad (2)$$
$$y = 1 + 4x$$

Step 2: Substitute $1 + 4x$ for y in equation (1) to get an equation in x.

$$3x + 2y = 13 \quad (1)$$
$$3x + 2(1 + 4x) = 13 \quad \text{Let } y = 1 + 4x.$$

Step 3: Solve for x in the equation just obtained.

$$3x + 2 + 8x = 13 \quad \text{Distributive property}$$
$$11x = 11 \quad \text{Combine terms; subtract 2.}$$
$$x = 1 \quad \text{Divide by 11.}$$

Step 4: Now solve for y. Since $y = 1 + 4x$, $y = 1 + 4(1) = 5$.

Step 5: Check to show that ordered pair $(1, 5)$ satisfies both equations. The solution set is $\{(1, 5)\}$.

The next example illustrates special cases that may result when systems are solved. (We will use the elimination method, but the same conclusions will follow when the substitution method is used.)

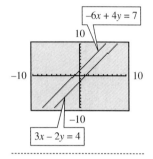

The graphs of the equations in Example 3(a) are parallel. There are no solutions.

EXAMPLE 3 **(a)** Solve the system

$$3x - 2y = 4 \quad (1)$$
$$-6x + 4y = 7. \quad (2)$$

The variable x can be eliminated by multiplying both sides of equation (1) by 2 and then adding.

$$\begin{array}{rl} 6x - 4y = & 8 \quad \text{2 times equation (1)} \\ -6x + 4y = & 7 \quad \text{(2)} \\ \hline 0 = & 15 \quad \text{False} \end{array}$$

Both variables were eliminated here, leaving the false statement $0 = 15$, indicating that these two equations have no solutions in common. The system is inconsistent, with the empty set $\emptyset$ as the solution set.

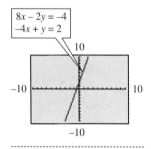

The graphs of the equations in Example 3(b) coincide. We see only one line. There are infinitely many solutions.

(b) Solve the system

$$-4x + y = 2 \quad (1)$$
$$8x - 2y = -4. \quad (2)$$

Eliminate x by multiplying both sides of equation (1) by 2 and then adding the result to equation (2).

$$\begin{array}{rl} -8x + 2y = & 4 \quad \text{2 times equation (1)} \\ 8x - 2y = & -4 \quad \text{(2)} \\ \hline 0 = & 0 \quad \text{True} \end{array}$$

This true statement, $0 = 0$, indicates that a solution of one equation is also a solution of the other, so the solution set is an infinite set of ordered pairs. The two equations are dependent.

We will write the solution set of a system of dependent equations as a set of ordered pairs by expressing x in terms of y as follows. Choose either equation and solve for x. Choosing equation (1) gives

$$-4x + y = 2$$
$$x = \frac{2 - y}{-4} = \frac{y - 2}{4}.$$

The solution set is written as

$$\left\{ \left(\frac{y-2}{4}, y \right) \right\}.$$

By selecting values for y and calculating the corresponding values for x, individual ordered pairs of the solution set can be found. For example, if $y = -2$, $x = (-2 - 2)/4 = -1$ and the ordered pair $(-1, -2)$ is a solution.

Linear Systems in Three Variables A solution of an equation in three variables, such as $2x + 3y - z = 4$, is called an **ordered triple** and is written (x, y, z). For example, the ordered triples $(1, 1, 1)$ and $(10, -3, 7)$ are both solutions of $2x + 3y - z = 4$, since the numbers in these ordered triples satisfy the equation when used as replacements for x, y, and z, respectively. The methods of solving systems of two equations in two variables can be extended to solving systems of equations in three variables such as

$$4x + 8y + z = 2$$
$$x + 7y - 3z = -14$$
$$2x - 3y + 2z = 3.$$

Theoretically, a system of this type can be solved by graphing. However, the graph of a linear equation with three variables is a *plane* and not a line. Since the graph of each equation of the system is a plane, which requires three-dimensional graphing, this method is not practical. However, it does illustrate the number of solutions possible for such systems, as Figure 39 shows.

Graphs of Linear Systems in Three Variables

1. The three planes may meet at a single, common point that forms the solution set of the system. See Figure 39(a).
2. The three planes may have the points of a line in common so that the set of points along that line is the solution set of the system. See Figure 39(b).
3. The planes may have no points common to all three so that there is no solution for the system. See Figure 39(c).
4. The three planes may coincide so that the solution set of the system is the set of all points on that plane. See Figure 39(d).

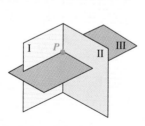

(a) A single solution

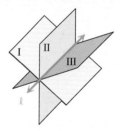

(b) Points of a line in common

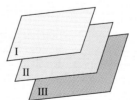

(c) One case of no points in common

(d) All points in common

FIGURE 39

There are other illustrations of the above cases. For example, two of the planes might intersect in a line, while the third plane is parallel to one of these planes, again resulting in no points common to all three planes. We give only one example of each case in Figure 39.

Since graphing to find the solution set of a system of three equations in three variables is impractical, these systems are solved with an extension of the elimination method, summarized as follows.

Solving Linear Systems in Three Variables by Elimination

Step 1: **Eliminate a variable.** Use the elimination method to eliminate any variable from any two of the given equations. The result is an equation in two variables.

Step 2: **Eliminate the same variable again.** Eliminate the *same* variable from any *other* two equations. The result is an equation in the same two variables as in Step 1.

Step 3: **Eliminate a different variable and solve.** Use the elimination method to eliminate a second variable from the two equations in two variables that result from Steps 1 and 2. The result is an equation in one variable that gives the value of that variable.

Step 4: **Find a second value.** Substitute the value of the variable found in Step 3 into either of the equations in two variables to find the value of the second variable.

Step 5: **Find a third value.** Use the values of the two variables from Steps 3 and 4 to find the value of the third variable by substituting into any of the original equations.

Step 6: **Find the solution set.** Check the solution in all of the original equations. Then write the solution set.

EXAMPLE 4 Solve the system.

$$4x + 8y + z = 2 \quad (1)$$
$$x + 7y - 3z = -14 \quad (2)$$
$$2x - 3y + 2z = 3 \quad (3)$$

Step 1: As before, the elimination method involves eliminating a variable from the sum of two equations. The choice of which variable to eliminate is arbitrary. Suppose we decide to begin by eliminating z. To do this, multiply both sides of equation (1) by 3 and then add the result to equation (2).

$$12x + 24y + 3z = 6 \quad \text{Multiply both sides of equation (1) by 3.}$$
$$\underline{x + 7y - 3z = -14} \quad (2)$$
$$13x + 31y = -8 \quad \text{Add.}$$

Step 2: The new equation has only two variables. To get another equation without z, multiply both sides of equation (1) by -2 and add the result to equation (3). It is essential at this point to *eliminate the same variable, z.*

$$-8x - 16y - 2z = -4 \quad \text{Multiply both sides of equation (1) by } -2.$$
$$\underline{2x - 3y + 2z = 3} \quad (3)$$
$$-6x - 19y = -1 \quad \text{Add.}$$

Step 3: Now solve the system of equations from Steps 1 and 2 for x and y. (This step is possible only if the *same* variable is eliminated in the first two steps.)

$$78x + 186y = -48 \quad \text{Multiply both sides of } 13x + 31y = -8 \text{ by 6.}$$
$$\underline{-78x - 247y = -13} \quad \text{Multiply both sides of } -6x - 19y = -1 \text{ by 13.}$$
$$-61y = -61 \quad \text{Add.}$$
$$y = 1$$

Step 4: Substitute 1 for y in either equation from Steps 1 and 2. Choosing $-6x - 19y = -1$ gives

$$-6x - 19y = -1$$
$$-6x - 19(1) = -1 \quad \text{Let } y = 1.$$
$$-6x - 19 = -1$$
$$-6x = 18$$
$$x = -3.$$

Step 5: Substitute -3 for x and 1 for y in any one of the three given equations to find z. Choosing equation (1) gives

$$4x + 8y + z = 2$$
$$4(-3) + 8(1) + z = 2 \quad \text{Let } x = -3 \text{ and } y = 1.$$
$$z = 6.$$

Step 6: It appears that the ordered triple $(-3, 1, 6)$ is the only solution of the system. Check that the solution satisfies all three equations of the system. We show the check here only for equation (1).

$$4x + 8y + z = 2 \quad (1)$$
$$4(-3) + 8(1) + 6 = 2 \quad ?$$
$$-12 + 8 + 6 = 2 \quad ?$$
$$2 = 2 \quad \text{True}$$

Because $(-3, 1, 6)$ also satisfies the equations (2) and (3), the solution set is $\{(-3, 1, 6)\}$.

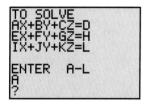
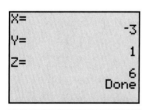

A graphing calculator can be *programmed* to solve a system such as the one in Example 4. Compare the result here to the solution in the text.

Applications of Linear Systems

Problem Solving

Many problems involve more than one unknown quantity. Although some problems with two unknowns can be solved using just one variable, many times it is easier to use two variables. To solve a problem with two unknowns, we write two equations that relate the unknown quantities; the system formed by the pair of equations then can be solved using the methods of this section.

Here is a way to find a person's age and the date of his or her birth. Ask the person to do the following.

1. Multiply the number of the month of birth by *100*.
2. Add the date of the month.
3. Multiply the sum by *2*.
4. Add *8*.
5. Multiply this result by *5*.
6. Add *4*.
7. Multiply by *10*.
8. Add *4*.
9. Add the person's age to this.

Ask the person the number obtained. You should then subtract *444* from the number you are given. The final two figures give the age, the next figure (or figures) will identify the date of the month, and the first figure (or figures) will give the number of the month.

The following steps, based on the six-step problem-solving method first introduced in Chapter 7, give a strategy for solving problems using more than one variable.

> **Solving an Applied Problem by Writing a System of Equations**
>
> *Step 1:* **Read** the problem carefully until you understand what is given and what is to be found.
>
> *Step 2:* **Assign variables** to represent the unknown values, using diagrams or tables as needed. *Write down* what each variable represents.
>
> *Step 3:* **Write a system of equations** that relates the unknowns.
>
> *Step 4:* **Solve** the system of equations.
>
> *Step 5:* **State the answer** to the problem. Does it seem reasonable?
>
> *Step 6:* **Check** the answer in the words of the original problem.

Problems about the perimeter of a geometric figure often involve two unknowns and can be solved using systems of equations.

EXAMPLE 5 Unlike football, where the dimensions of a playing field cannot vary, a rectangular soccer field may have a width between 50 and 100 yd and a length between 50 and 100 yd. Suppose that one particular field has a perimeter of 320 yd. Its length measures 40 yd more than its width. What are the dimensions of this field? (*Source: Microsoft Encarta Encyclopedia 2000.*)

Step 1: **Read the problem again.** We are asked to find the dimensions of the field.

Step 2: **Assign variables.** Let $L =$ the length and $W =$ the width. Figure 40 on the next page shows a soccer field with the length labeled L and the width labeled W.

Step 3: **Write a system of equations.** Because the perimeter is 320 yd, we find one equation by using the perimeter formula:

$$2L + 2W = 320.$$

Because the length is 40 yd more than the width, we have
$$L = W + 40.$$
The system is, therefore,
$$2L + 2W = 320 \quad (1)$$
$$L = W + 40. \quad (2)$$

Step 4: **Solve** the system of equations. Since equation (2) is solved for L, we can use the substitution method. We substitute $W + 40$ for L in equation (1), and solve for W.

$$\begin{aligned} 2L + 2W &= 320 & &(1) \\ 2(W + 40) + 2W &= 320 & &L = W + 40 \\ 2W + 80 + 2W &= 320 & &\text{Distributive property} \\ 4W + 80 &= 320 & &\text{Combine like terms.} \\ 4W &= 240 & &\text{Subtract 80.} \\ W &= 60 & &\text{Divide by 4.} \end{aligned}$$

Let $W = 60$ in the equation $L = W + 40$ to find L.
$$L = 60 + 40 = 100$$

Step 5: **State the answer.** The length is 100 yd, and the width is 60 yd. Both dimensions are within the ranges given in the problem.

Step 6: **Check.** The perimeter of this soccer field is
$$2(100) + 2(60) = 320 \text{ yd},$$
and the length, 100 yd, is indeed 40 yd more than the width, since
$$100 - 40 = 60.$$
The answer is correct.

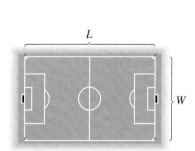

FIGURE 40

Professional sport ticket prices increase annually. Average per-ticket prices in three of the four major sports (football, basketball, and hockey) now exceed $30.00.

EXAMPLE 6 During recent National Hockey League and National Basketball Association seasons, two hockey tickets and one basketball ticket purchased at their average prices would have cost $110.40. One hockey ticket and two basketball tickets would have cost $106.32. What were the average ticket prices for the two sports? (*Source:* Team Marketing Report, Chicago.)

Step 1: **Read** the problem again. There are two unknowns.

Step 2: **Assign variables.** Let h represent the average price for a hockey ticket and b represent the average price for a basketball ticket.

Step 3: **Write a system of equations.** Because two hockey tickets and one basketball ticket cost a total of $110.40, one equation for the system is

$$2h + b = 110.40.$$

By similar reasoning, the second equation is

$$h + 2b = 106.32.$$

Therefore, the system is

$$2h + b = 110.40 \quad (1)$$
$$h + 2b = 106.32. \quad (2)$$

Step 4: **Solve** the system of equations. To eliminate h, multiply equation (2) by -2 and add.

$$\begin{aligned} 2h + b &= 110.40 & (1) \\ -2h - 4b &= -212.64 & \text{Multiply each side of (2) by } -2. \\ \hline -3b &= -102.24 & \text{Add.} \\ b &= 34.08 & \text{Divide by } -3. \end{aligned}$$

To find the value of h, let $b = 34.08$ in equation (2).

$$\begin{aligned} h + 2b &= 106.32 & (2) \\ h + 2(34.08) &= 106.32 & \text{Let } b = 34.08. \\ h + 68.16 &= 106.32 & \text{Multiply.} \\ h &= 38.16 & \text{Subtract 68.16.} \end{aligned}$$

Step 5: **State the answer.** The average price for one basketball ticket was $34.08. For one hockey ticket, the average price was $38.16.

Step 6: **Check** that these values satisfy the conditions stated in the problem. ∎

We solved mixture problems earlier using one variable. For many mixture problems, we can use more than one variable and a system of equations.

EXAMPLE 7 How many ounces each of 5% hydrochloric acid and 20% hydrochloric acid must be combined to get 10 oz of solution that is 12.5% hydrochloric acid?

Step 1: **Read** the problem. Two solutions of different strengths are being mixed together to get a specific amount of a solution with an "in-between" strength.

8.7 Systems of Equations and Applications

Problems that can be solved by writing a system of equations have been of interest historically. The following problem first appeared in a Hindu work that dates back to about A.D. 850.

The mixed price of 9 citrons [a lemonlike fruit shown in the photo] and 7 fragrant wood apples is 107; again, the mixed price of 7 citrons and 9 fragrant wood apples is 101. O you arithmetician, tell me quickly the price of a citron and the price of a wood apple here, having distinctly separated those prices well.

Use a system to solve this problem. The answer can be found at the end of the exercises for this section on page 476.

Step 2: **Assign variables.** Let x represent the number of ounces of 5% solution and y represent the number of ounces of 20% solution. Use a table to summarize the information from the problem.

Percent (as a decimal)	Ounces of Solution	Ounces of Pure Acid
5% = .05	x	.05x
20% = .20	y	.20y
12.5% = .125	10	(.125)10

Figure 41 also illustrates what is happening in the problem.

FIGURE 41

Step 3: **Write a system of equations.** When x ounces of 5% solution and y ounces of 20% solution are combined, the total number of ounces is 10, so

$$x + y = 10. \quad (1)$$

The ounces of pure acid in the 5% solution (.05x) plus the ounces of pure acid in the 20% solution (.20y) should equal the total ounces of pure acid in the mixture, which is (.125)10, or 1.25. That is,

$$.05x + .20y = 1.25. \quad (2)$$

Notice that these equations can be quickly determined by reading down in the table or using the labels in Figure 41.

Step 4: **Solve** the system of equations (1) and (2). Eliminate x by first multiplying equation (2) by 100 to clear it of decimals and then multiplying equation (1) by -5.

$$\begin{aligned} 5x + 20y &= 125 & \text{Multiply each side of (2) by 100.} \\ -5x - 5y &= -50 & \text{Multiply each side of (1) by } -5. \\ \hline 15y &= 75 & \text{Add.} \\ y &= 5 \end{aligned}$$

Because $y = 5$ and $x + y = 10$, x is also 5.

Step 5: **State the answer.** The desired mixture will require 5 ounces of the 5% solution and 5 ounces of the 20% solution.

Step 6: **Check** that these values satisfy both equations of the system.

Problems that use the distance formula $d = rt$ were first introduced in Chapter 7. In many cases, these problems can be solved with systems of two linear equations. Keep in mind that setting up a table and drawing a sketch will help you solve such problems.

François Viète, a mathematician of sixteenth-century France, did much for the symbolism of mathematics. Before his time, different symbols were often used for different powers of a quantity. Viète used the same letter with a description of the power and the coefficient. According to Howard Eves in *An Introduction to the History of Mathematics*, Viète would have written

$$5BA^2 - 2CA + A^3 = D$$

as

*B*5 in *A* quad − *C* plano 2 in *A* + *A* cub aequatur *D* solido.

EXAMPLE 8 Two executives in cities 400 mi apart drive to a business meeting at a location on the line between their cities. They meet after 4 hr. Find the speed of each car if one car travels 20 mph faster than the other.

Step 1: **Read** the problem carefully.

Step 2: **Assign variables.**

Let $x =$ the speed of the faster car,

$y =$ the speed of the slower car

We use the formula $d = rt$. Since each car travels for 4 hr, the time, t, for each car is 4. This information is shown in the table. The distance is found by using the formula $d = rt$ and the expressions already entered in the table.

	r	t	d
Faster car	x	4	$4x$
Slower car	y	4	$4y$

Find d from $d = rt$.

Draw a sketch showing what is happening in the problem. See Figure 42.

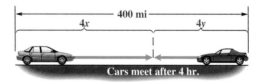

FIGURE 42

Step 3: **Write two equations.** As shown in the figure, since the total distance traveled by both cars is 400 mi, one equation is

$$4x + 4y = 400.$$

Because the faster car goes 20 mph faster than the slower car, the second equation is

$$x = 20 + y.$$

Step 4: **Solve.** This system of equations,

$$4x + 4y = 400 \quad (1)$$
$$x = 20 + y, \quad (2)$$

can be solved by substitution. Replace x with $20 + y$ in equation (1) and solve for y.

8.7 Systems of Equations and Applications 469

$$4(20 + y) + 4y = 400 \quad \text{Let } x = 20 + y.$$
$$80 + 4y + 4y = 400 \quad \text{Distributive property}$$
$$80 + 8y = 400 \quad \text{Combine like terms.}$$
$$8y = 320 \quad \text{Subtract 80.}$$
$$y = 40 \quad \text{Divide by 8.}$$

Since $x = 20 + y$, and $y = 40$,

$$x = 20 + 40 = 60.$$

Step 5: **State the answer.** The speeds of the two cars are 40 mph and 60 mph.

Step 6: **Check** the answer. Since each car travels for 4 hr, the total distance traveled is

$$4(60) + 4(40) = 240 + 160 = 400 \text{ mi},$$

as required.

The problems in Examples 5–8 also could be solved using only one variable. Many students find that the solution is simpler with two variables.

EXAMPLE 9 A company produces three color television sets, models X, Y, and Z. Each model X set requires 2 hr of electronics work, 2 hr of assembly time, and 1 hr of finishing time. Each model Y requires 1, 3, and 1 hr of electronics, assembly, and finishing time, respectively. Each model Z requires 3, 2, and 2 hr of the same work, respectively. There are 100 hr available for electronics, 100 hr available for assembly, and 65 hr available for finishing per week. How many of each model should be produced each week if all available time must be used?

Step 1: **Read** the problem again. There are three unknowns.

Step 2: **Assign variables.**

Let $x =$ the number of model X produced per week,
$y =$ the number of model Y produced per week,
and $z =$ the number of model Z produced per week.

Organize the information in a table.

	Each Model X	Each Model Y	Each Model Z	Totals
Hours of electronics work	2	1	3	100
Hours of assembly time	2	3	2	100
Hours of finishing time	1	1	2	65

Step 3: **Write a system of three equations.** The x model X sets require $2x$ hr of electronics, the y model Y sets require $1y$ (or y) hr of electronics, and the z model Z sets require $3z$ hr of electronics. Since 100 hr are available for electronics,

$$2x + y + 3z = 100. \quad (1)$$

Similarly, from the fact that 100 hr are available for assembly,
$$2x + 3y + 2z = 100, \quad (2)$$
and the fact that 65 hr are available for finishing leads to the equation
$$x + y + 2z = 65. \quad (3)$$

Again, notice the advantage of setting up a table. By reading across, we can easily determine the coefficients and constants in the equations of the system.

Step 4: **Solve** the system
$$2x + y + 3z = 100$$
$$2x + 3y + 2z = 100$$
$$x + y + 2z = 65$$
to find $x = 15$, $y = 10$, and $z = 20$.

Step 5: **State the answer.** The company should produce 15 model X, 10 model Y, and 20 model Z sets per week.

Step 6: **Check** that these values satisfy the conditions of the problem.

8.7 EXERCISES

Answer the questions in Exercises 1 and 2 by observing the graphs provided.

1. *Network News Programs* The graph shows network share (the percentage of TV sets in use) for the early evening news programs for the three major broadcast networks for 1986–2000.

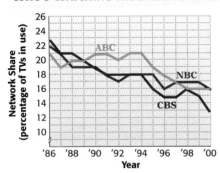

WHO'S WATCHING THE EVENING NEWS?

Source: Nielson Media Research.

(a) Between what years did the ABC early evening news dominate?

(b) During what year did ABC's dominance end? Which network equaled ABC's share that year? What was that share?

(c) During what years did ABC and CBS have equal network share? What was the share for each of these years?

(d) Which networks most recently had equal share? Write their share as an ordered pair of the form (year, share).

(e) Describe the general trend in viewership for the three major networks during these years.

2. *Music Format Sales* The graph shows how the production of vinyl LPs, audiocassettes, and compact discs (CDs) changed over the years from 1986 through 1998.

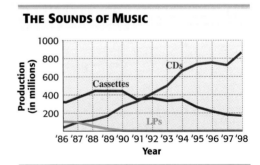

THE SOUNDS OF MUSIC

Source: Recording Industry Association of America.

(a) In what year did cassette production and CD production reach equal levels? What was that level?

(b) Express the point of intersection of the graphs of LP production and CD production as an ordered pair of the form (year, production level).

(c) Between what years did cassette production first stabilize and remain fairly constant?

(d) Describe the trend in CD production from 1986 through 1998. If a straight line were used to approximate its graph, would the line have positive, negative, or 0 slope?

(e) If a straight line were used to approximate the graph of cassette production from 1990 through 1998, would the line have positive, negative, or 0 slope? Explain.

Decide whether the ordered pair is a solution of the given system.

3. $x + y = 6$ $(5, 1)$
$x - y = 4$

4. $x - y = 17$ $(8, -9)$
$x + y = -1$

5. $2x - y = 8$ $(5, 2)$
$3x + 2y = 20$

6. $3x - 5y = -12$ $(-1, 2)$
$x - y = 1$

Solve each system by graphing.

7. $x + y = 4$
$2x - y = 2$

8. $x + y = -5$
$-2x + y = 1$

Solve each system by elimination.

9. $2x - 5y = 11$
$3x + y = 8$

10. $-2x + 3y = 1$
$-4x + y = -3$

11. $3x + 4y = -6$
$5x + 3y = 1$

12. $4x + 3y = 1$
$3x + 2y = 2$

13. $3x + 3y = 0$
$4x + 2y = 3$

14. $8x + 4y = 0$
$4x - 2y = 2$

15. $7x + 2y = 6$
$-14x - 4y = -12$

16. $x - 4y = 2$
$4x - 16y = 8$

17. $\dfrac{x}{2} + \dfrac{y}{3} = -\dfrac{1}{3}$
$\dfrac{x}{2} + 2y = -7$

18. $\dfrac{x}{5} + y = \dfrac{6}{5}$
$\dfrac{x}{10} + \dfrac{y}{3} = \dfrac{5}{6}$

19. $5x - 5y = 3$
$x - y = 12$

20. $2x - 3y = 7$
$-4x + 6y = 14$

Solve each system by substitution.

21. $4x + y = 6$
$y = 2x$

22. $2x - y = 6$
$y = 5x$

23. $3x - 4y = -22$
$-3x + y = 0$

24. $-3x + y = -5$
$x + 2y = 0$

25. $-x - 4y = -14$
$2x = y + 1$

26. $-3x - 5y = -17$
$4x = y - 8$

27. $5x - 4y = 9$
$3 - 2y = -x$

28. $6x - y = -9$
$4 + 7x = -y$

29. $x = 3y + 5$
$x = \dfrac{3}{2}y$

30. $x = 6y - 2$
$x = \dfrac{3}{4}y$

31. $\dfrac{1}{2}x + \dfrac{1}{3}y = 3$
$y = 3x$

32. $\dfrac{1}{4}x - \dfrac{1}{5}y = 9$
$y = 5x$

33. Explain what the following statement means: The solution set of the system
$2x + y + z = 3$
$3x - y + z = -2$ is $\{(-1, 2, 3)\}$.
$4x - y + 2z = 0$

34. Write a system of three linear equations in three variables that has solution set $\{(3, 1, 2)\}$. Then solve the system. (*Hint:* Start with the solution and make up three equations that are satisfied by the solution. There are many ways to do this.)

Solve each system of equations in three variables.

35. $3x + 2y + z = 8$
    $2x - 3y + 2z = -16$
    $x + 4y - z = 20$

36. $-3x + y - z = -10$
    $-4x + 2y + 3z = -1$
    $2x + 3y - 2z = -5$

37. $2x + 5y + 2z = 0$
    $4x - 7y - 3z = 1$
    $3x - 8y - 2z = -6$

38. $5x - 2y + 3z = -9$
    $4x + 3y + 5z = 4$
    $2x + 4y - 2z = 14$

39. $x + y - z = -2$
    $2x - y + z = -5$
    $-x + 2y - 3z = -4$

40. $x + 2y + 3z = 1$
    $-x - y + 3z = 2$
    $-6x + y + z = -2$

41. $2x - 3y + 2z = -1$
    $x + 2y + z = 17$
    $2y - z = 7$

42. $2x - y + 3z = 6$
    $x + 2y - z = 8$
    $2y + z = 1$

43. $4x + 2y - 3z = 6$
    $x - 4y + z = -4$
    $-x + 2z = 2$

44. $2x + 3y - 4z = 4$
    $x - 6y + z = -16$
    $-x + 3z = 8$

45. $2x + y = 6$
    $3y - 2z = -4$
    $3x - 5z = -7$

46. $4x - 8y = -7$
    $4y + z = 7$
    $-8x + z = -4$

The applications in Exercises 47–80 require solving systems with two variables, while those that follow require solving systems with three variables.

Solve each problem involving two unknowns.

47. **Win-Loss Record** During the 2000 Major League Baseball regular season, the St. Louis Cardinals played 162 games. They won 28 more games than they lost. What was their win–loss record that year?

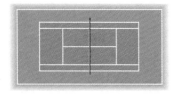

Source: www.mlb.com.

48. **Win-Loss Record** Refer to Exercise 47. During the same 162-game season, the Chicago Cubs lost 32 more games than they won. What was the team's win–loss record?

49. **Dimensions of a Tennis Court** Venus and Serena measured a tennis court and found that it was 42 feet longer than it was wide and had a perimeter of 228 feet. What were the length and the width of the tennis court?

50. **Dimensions of a Basketball Court** Shaq and Kobe found that the width of their basketball court was 44 feet less than the length. If the perimeter was 288 feet, what were the length and the width of their court?

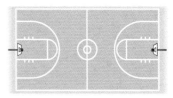

51. **Largest U.S. Companies** The two biggest U.S. companies in terms of revenue in 2000 were ExxonMobil and General Motors. ExxonMobil's revenue was $29 billion more than that of General Motors. Total revenue for the two companies was $399 billion. What was the revenue for each company? (*Source:* Bridge News, MarketGuide.com)

52. **Top U.S. Trading Partners** The top two U.S. trading partners during the first four months of 2000 were Canada and Mexico. Exports and imports with Mexico were $57 billion less than those with Canada. Total exports and imports involving these two countries were $211 billion. How much were U.S. exports and imports with each country? (*Source:* U.S. Bureau of the Census.)

53. **Dimensions of a Rectangle** The length of a rectangle is 7 feet more than the width. If the length were decreased by 3 feet and the width were increased by 2 feet, the perimeter would be 32 feet. Find the length and width of the original rectangle.

54. **Dimensions of a Square and a Triangle** The side of a square is 4 centimeters longer than the side of an equilateral triangle. The perimeter of the square is 24 centimeters more than the perimeter of the triangle. Find the lengths of a side of the square and a side of the triangle.

55. *Cost of Art Supplies* For an art project Cheryl Arabie bought 8 sheets of colored paper and 3 marker pens for $6.50. She later needed 2 sheets of colored paper and 2 marker pens. These items cost $3.00. Find the cost of 1 marker pen and 1 sheet of colored paper.

56. *Prices of Computer Monitors* Houston Community College has decided to supply its mathematics labs with color monitors. A trip to the local electronics outlet leads to the following information: 4 CGA monitors and 6 VGA monitors can be purchased for $4600, while 6 CGA monitors and 4 VGA monitors will cost $4400. What are the prices of a single CGA monitor and a single VGA monitor?

Fan Cost Index The Fan Cost Index (FCI) represents the cost of four average-price tickets, four small soft drinks, two small beers, four hot dogs, parking for one car, two game programs, and two souvenir caps to a sporting event. For example, in 1997, the FCI for Major League Baseball was $105.63. This was by far the least for the four major professional sports. (Source: Team Marketing Report, Chicago.) Use the concept of FCI in Exercises 57 and 58.

57. The FCI prices for the National Hockey League and the National Basketball Association totaled $423.12. The hockey FCI was $16.36 more than that of basketball. What were the FCIs for these sports?

58. The FCI prices for Major League Baseball and the National Football League totaled $311.03. The football FCI was $105.87 more than that of baseball. What were the FCIs for these sports?

59. *Prices at Wendy's* Andrew McGinnis works at Wendy's Old Fashioned Hamburgers. During one particular lunch hour, he sold 15 single hamburgers and 10 double hamburgers, totaling $63.25. Another lunch hour, he sold 30 singles and 5 doubles, totaling $78.65. How much did each type of burger cost? (*Source:* Wendy's Old Fashioned Hamburgers menu.)

60. *Travel Costs* Tokyo and New York are among the most expensive cities worldwide for business travelers. Using average costs per day for each city (which includes room, meals, laundry, and two taxi fares), 2 days in Tokyo and 3 days in New York cost $2015. Four days in Tokyo and 2 days in New York cost $2490. What is the average cost per day for each city? (*Source:* ECA International.)

61. *Yarn and Thread Production* A factory makes use of two basic machines, A and B, which turn out two different products, yarn and thread. Each unit of yarn requires 1 hour on machine A and 2 hours on machine B, while each unit of thread requires 1 hour on A and 1 hour on B. Machine A runs 8 hours per day, while machine B runs 14 hours per day. How many units per day of yarn and thread should the factory make to keep its machines running at capacity?

62. *Cost of Clay* For his art class, Theodis bought 2 kilograms of dark clay and 3 kilograms of light clay, paying $22 for the clay. He later needed 1 kilogram of dark clay and 2 kilograms of light clay, costing $13 altogether. What was the cost per kilogram for each type of clay?

Formulas The formulas $p = br$ (percentage = base $\times$ rate) and $I = prt$ (simple interest = principal $\times$ rate $\times$ time) are used in the applications in Exercises 67–76. To prepare to use these formulas, answer the questions in Exercises 63–66.

63. If a container of liquid contains 60 oz of solution, what is the number of ounces of pure acid if the given solution contains the following acid concentrations?
 (a) 10% (b) 25% (c) 40% (d) 50%

64. If $5000 is invested in an account paying simple annual interest, how much interest will be earned during the first year at the following rates?
 (a) 2% (b) 3% (c) 4% (d) 3.5%

65. If a pound of turkey costs $.99, how much will x pounds cost?

66. If a ticket to the movie *Chicago* costs $8 and y tickets are sold, how much is collected from the sale?

67. *Alcohol Mixture* How many gallons each of 25% alcohol and 35% alcohol should be mixed to obtain 20 gallons of 32% alcohol?

Kind of Solution	Gallons of Solution	Amount of Pure Alcohol
.25	x	.25x
.35	y	.35y
.32	20	.32(20)

68. *Acid Mixture* How many liters each of 15% acid and 33% acid should be mixed to obtain 40 liters of 21% acid?

Kind of Solution	Liters of Solution	Amount of Pure Acid
.15	x	
.33	y	
.21	40	

69. *Acid Mixture* Pure acid is to be added to a 10% acid solution to obtain 27 liters of a 20% acid solution. What amounts of each should be used?

70. *Antifreeze Mixture* A truck radiator holds 18 liters of fluid. How much pure antifreeze must be added to a mixture that is 4% antifreeze in order to fill the radiator with a mixture that is 20% antifreeze?

71. *Candy Mixture* Debbie Blanchard plans to mix pecan clusters that sell for $3.60 per pound with chocolate truffles that sell for $7.20 per pound to get a mixture that she can sell in Valentine boxes for $4.95 per pound. How much of the $3.60 clusters and the $7.20 truffles should she use to create 80 pounds of the mix?

	Number of Pounds	Price per Pound	Value of Candy
Clusters	x	3.60	3.60x
Truffles	y	7.20	7.20y
Mixture	80	4.95	4.95(80)

72. *Fruit Drink Mixture* A popular fruit drink is made by mixing fruit juices. Such a mixture with 50% juice is to be mixed with another mixture that is 30% juice to get 200 liters of a mixture that is 45% juice. How much of each should be used?

Kind of Juice	Number of Liters	Amount of Pure Juice
.50	x	.50x
.30	y	
.45		

73. *Ticket Sales* Tickets to a production of *Othello* at Nicholls State University cost $2.50 for general admission or $2.00 with student identification. If 184 people paid to see a performance and $406 was collected, how many of each type of admission were sold?

74. *Candy Mixture* A grocer plans to mix candy that sells for $1.20 per pound with candy that sells for $2.40 per pound to get a mixture that he plans to sell for $1.65 per pound. How much of the $1.20 and $2.40 candy should he use if he wants 80 pounds of the mix?

75. *Investment Mixture* A total of $3000 is invested, part at 2% simple interest and part at 4%. If the total annual return from the two investments is $100, how much is invested at each rate?

Principal	Rate	Interest
x	.02	.02x
y	.04	.04y
3000		100

76. *Investment Mixture* An investor must invest a total of $15,000 in two accounts, one paying 4% annual simple interest, and the other 3%. If he wants to earn $550 annual interest, how much should he invest at each rate?

Principal	Rate	Interest
x	.04	
y	.03	
15,000		

77. *Speeds of Trains* A freight train and an express train leave towns 390 kilometers apart, traveling toward one another. The freight train travels 30 kilometers per hour slower than the express train. They pass one another 3 hours later. What are their speeds?

78. *Speeds of Trains* A train travels 150 kilometers in the same time that a plane covers 400 kilometers. If the speed of the plane is 20 kilometers per hour less than 3 times the speed of the train, find both speeds.

79. *Speeds of Snow Speeder and Wind* Braving blizzard conditions on the planet Hoth, Luke Skywalker sets out at top speed in his snow speeder for a rebel base 3600 miles away. He travels into a steady headwind, and makes the trip in 2 hours. Returning, he finds that the trip back, still at top speed but now with a tailwind, takes only 1.5 hours. Find the top speed of Luke's snow speeder and the speed of the wind.

80. *Speeds of Boat and Current* In his motorboat, Nguyen travels upstream at top speed to his favorite fishing spot, a distance of 36 miles, in two hours. Returning, he finds that the trip downstream, still at top speed, takes only 1.5 hours. Find the speed of Nguyen's boat and the speed of the current.

Solve each problem involving three unknowns.

81. *Voter Affiliations* In a random sample of 100 Americans of voting age, 10 more Americans identify themselves as Independents than Republicans. Six fewer Americans identify themselves as Republicans than Democrats. Assuming that all of those sampled are Republican, Democrat, or Independent, how many of those in the sample identify themselves with each political affiliation? (*Source:* The Gallup Organization.)

82. *Olympic Gold Medals* In the 2000 Summer Olympics in Sydney, Australia, the United States earned 14 more gold medals than silver. The number of bronze medals earned was 17 less than twice the number of silver medals. The United States earned a total of 97 medals. How many of each kind of medal did the United States earn? (*Source: The Gazette,* October 2, 2000.)

83. *Dimensions of a Triangle* The perimeter of a triangle is 70 centimeters. The longest side is 4 centimeters less than the sum of the other two sides. Twice the shortest side is 9 centimeters less than the longest side. Find the length of each side of the triangle.

84. *Dimensions of a Triangle* The perimeter of a triangle is 56 inches. The longest side measures 4 inches less than the sum of the other two sides. Three times the shortest side is 4 inches more than the longest side. Find the lengths of the three sides.

85. *Distribution to a Wholesaler* A Mardi Gras trinket manufacturer supplies three wholesalers, A, B, and C. The output from a day's production is 320 cases of trinkets. She must send wholesaler A three times as many cases as she sends B, and she must send wholesaler C 160 cases less than she provides A and B together. How many cases should she send to each wholesaler to distribute the entire day's production to them?

86. *Hardware Production* A hardware supplier manufactures three kinds of clamps, types A, B, and C. Production restrictions require it to make 10 units more type C clamps than the total of the other types and twice as many type B clamps as type A. The shop must produce a total of 490 units of clamps per day. How many units of each type can be made per day?

87. *Globetrotter Ticket Prices* Tickets for one show on the Harlem Globetrotters' 75th Anniversary Tour cost $10, $18, or, for VIP seats, $30. So far, five times as many $18 tickets have been sold as VIP tickets. The number of $10 tickets equals the number of $18 tickets plus twice the number of VIP tickets. Sales of these tickets total $9500. How many of each kind of ticket have been sold? (*Source:* www.ticketmaster.com)

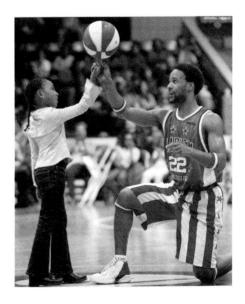

88. *Concert Ticket Prices* Three kinds of tickets are available for a Third Day concert: "up close," "in the middle," and "far out." "Up close" tickets cost $10 more than "in the middle" tickets, while "in the middle" tickets cost $10 more than "far out" tickets. Twice the cost of an "up close" ticket is $20 more than 3 times the cost of a "far out" seat. Find the price of each kind of ticket.

Answer to Margin Note problem on page 467: price of a citron: 8; price of a wood apple: 5

EXTENSION

Using Matrix Row Operations to Solve Systems

The elimination method used to solve systems introduced in the previous section can be streamlined into a systematic method by using matrices (singular: matrix). Matrices can be used to solve linear systems and matrix methods are particularly suitable for computer solutions of large systems of equations having many unknowns.

To begin, consider a system of three equations and three unknowns such as

$$a_1 x + b_1 y + c_1 z = d_1$$
$$a_2 x + b_2 y + c_2 z = d_2$$
$$a_3 x + b_3 y + c_3 z = d_3.$$

This system can be written in an abbreviated form as

$$\begin{bmatrix} a_1 & b_1 & c_1 & d_1 \\ a_2 & b_2 & c_2 & d_2 \\ a_3 & b_3 & c_3 & d_3 \end{bmatrix}.$$

Such a rectangular array of numbers enclosed by brackets is called a **matrix.** Each number in the array is an **element** or **entry.** The matrix above has three **rows** (horizontal) and four **columns** (vertical) of entries, and is called a 3 × 4 (read "3 by 4") matrix. The constants in the last column of the matrix can be set apart from the coefficients of the variables by using a vertical line, as shown in the following **augmented matrix.**

$$\text{Rows} \rightarrow \begin{bmatrix} a_1 & b_1 & c_1 & | & d_1 \\ a_2 & b_2 & c_2 & | & d_2 \\ a_3 & b_3 & c_3 & | & d_3 \end{bmatrix}$$
$$\text{Columns}$$

The rows of this augmented matrix can be treated the same as the equations of a system of equations, since the augmented matrix is actually a short form of the system. Any transformation of the matrix that will result in an equivalent system is permitted. The following **matrix row operations** produce such transformations.

Extension Using Matrix Row Operations to Solve Systems

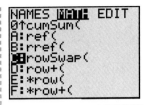

Choices C, D, E, and F provide the user of the TI-83 Plus calculator a means of performing row operations on matrices.

Matrix Row Operations

For any real number k and any augmented matrix of a system of linear equations, the following operations will produce the matrix of an equivalent system:

1. **interchanging any two rows of a matrix,**
2. **multiplying the elements of any row of a matrix by the same nonzero number k, and**
3. **adding a common multiple of the elements of one row to the corresponding elements of another row.**

If the word "row" is replaced by "equation," it can be seen that the three row operations also apply to a system of equations, so that a system of equations can be solved by transforming its corresponding matrix into the matrix of an equivalent, simpler system. The goal is a matrix in the form

$$\begin{bmatrix} 1 & 0 & | & a \\ 0 & 1 & | & b \end{bmatrix} \quad \text{or} \quad \begin{bmatrix} 1 & 0 & 0 & | & a \\ 0 & 1 & 0 & | & b \\ 0 & 0 & 1 & | & c \end{bmatrix}$$

for systems with two or three equations respectively. Notice that on the left of the vertical bar there are ones down the diagonal from upper left to lower right and zeros elsewhere in the matrices. When these matrices are rewritten as systems of equations, the values of the variables are known. The **Gauss-Jordan method** is a systematic way of using the matrix row operations to change the augmented matrix of a system into the form that shows its solution. The following examples will illustrate this method.

EXAMPLE 1 Solve the linear system.

$$3x - 4y = 1$$
$$5x + 2y = 19$$

The equations should all be in the same form, with the variable terms in the same order on the left, and the constant term on the right. Begin by writing the augmented matrix.

$$\begin{bmatrix} 3 & -4 & | & 1 \\ 5 & 2 & | & 19 \end{bmatrix}$$

The goal is to transform this augmented matrix into one in which the values of the variables will be easy to see. That is, since each column in the matrix represents the coefficients of one variable, the augmented matrix should be transformed so that it is of the form

$$\begin{bmatrix} 1 & 0 & | & k \\ 0 & 1 & | & j \end{bmatrix}$$

(continued)

for real numbers k and j. Once the augmented matrix is in this form, the matrix can be rewritten as a linear system to get

$$x = k$$
$$y = j.$$

The necessary transformations are performed as follows. It is best to work in columns beginning in each column with the element that is to become 1. In the augmented matrix,

$$\begin{bmatrix} 3 & -4 & | & 1 \\ 5 & 2 & | & 19 \end{bmatrix},$$

there is a 3 in the first row, first column position. Use row operation 2, multiplying each entry in the first row by 1/3 to get a 1 in this position. (This step is abbreviated as (1/3)R1.)

$$\begin{bmatrix} 1 & -4/3 & | & 1/3 \\ 5 & 2 & | & 19 \end{bmatrix} \quad \tfrac{1}{3}\text{R1}$$

Introduce 0 in the second row, first column by multiplying each element of the first row by -5 and adding the result to the corresponding element in the second row, using row operation 3.

$$\begin{bmatrix} 1 & -4/3 & | & 1/3 \\ 0 & 26/3 & | & 52/3 \end{bmatrix} \quad -5\text{R1} + \text{R2}$$

Obtain 1 in the second row, second column by multiplying each element of the second row by 3/26, using row operation 2.

$$\begin{bmatrix} 1 & -4/3 & | & 1/3 \\ 0 & 1 & | & 2 \end{bmatrix} \quad \tfrac{3}{26}\text{R2}$$

Finally, obtain 0 in the first row, second column by multiplying each element of the second row by 4/3 and adding the result to the corresponding element in the first row.

$$\begin{bmatrix} 1 & 0 & | & 3 \\ 0 & 1 & | & 2 \end{bmatrix} \quad \tfrac{4}{3}\text{R2} + \text{R1}$$

This last matrix corresponds to the system

$$x = 3$$
$$y = 2,$$

that has the solution set $\{(3, 2)\}$. This solution could have been read directly from the third column of the final matrix. ∎

A linear system with three equations is solved in a similar way. Row operations are used to get 1s down the diagonal from left to right and 0s above and below each 1.

EXAMPLE 2 Use the Gauss-Jordan method to solve the system.

$$x - y + 5z = -6$$
$$3x + 3y - z = 10$$
$$x + 3y + 2z = 5$$

Since the system is in proper form, begin by writing the augmented matrix of the linear system.

$$\begin{bmatrix} 1 & -1 & 5 & | & -6 \\ 3 & 3 & -1 & | & 10 \\ 1 & 3 & 2 & | & 5 \end{bmatrix}$$

The final matrix is to be of the form

$$\begin{bmatrix} 1 & 0 & 0 & | & m \\ 0 & 1 & 0 & | & n \\ 0 & 0 & 1 & | & p \end{bmatrix},$$

where m, n, and p are real numbers. This final form of the matrix gives the system $x = m$, $y = n$, and $z = p$, so the solution set is $\{(m, n, p)\}$.

There is already a 1 in the first row, first column. Get a 0 in the second row of the first column by multiplying each element in the first row by -3 and adding the result to the corresponding element in the second row, using operation 3.

$$\begin{bmatrix} 1 & -1 & 5 & | & -6 \\ 0 & 6 & -16 & | & 28 \\ 1 & 3 & 2 & | & 5 \end{bmatrix} \quad -3R1 + R2$$

Now, to change the last element in the first column to 0, use operation 3 and multiply each element of the first row by -1, then add the results to the corresponding elements of the third row.

$$\begin{bmatrix} 1 & -1 & 5 & | & -6 \\ 0 & 6 & -16 & | & 28 \\ 0 & 4 & -3 & | & 11 \end{bmatrix} \quad -1R1 + R3$$

The same procedure is used to transform the second and third columns. For both of these columns, first perform the step of getting 1 in the appropriate position of each column. Do this by multiplying the elements of the row by the reciprocal of the number in that position.

$$\begin{bmatrix} 1 & -1 & 5 & | & -6 \\ 0 & 1 & -8/3 & | & 14/3 \\ 0 & 4 & -3 & | & 11 \end{bmatrix} \quad \frac{1}{6}R2$$

$$\begin{bmatrix} 1 & 0 & 7/3 & | & -4/3 \\ 0 & 1 & -8/3 & | & 14/3 \\ 0 & 4 & -3 & | & 11 \end{bmatrix} \quad R2 + R1$$

(continued)

$$\begin{bmatrix} 1 & 0 & 7/3 & | & -4/3 \\ 0 & 1 & -8/3 & | & 14/3 \\ 0 & 0 & 23/3 & | & -23/3 \end{bmatrix} \quad -4R2 + R3$$

$$\begin{bmatrix} 1 & 0 & 7/3 & | & -4/3 \\ 0 & 1 & -8/3 & | & 14/3 \\ 0 & 0 & 1 & | & -1 \end{bmatrix} \quad \frac{3}{23}R3$$

$$\begin{bmatrix} 1 & 0 & 0 & | & 1 \\ 0 & 1 & -8/3 & | & 14/3 \\ 0 & 0 & 1 & | & -1 \end{bmatrix} \quad -\frac{7}{3}R3 + R1$$

$$\begin{bmatrix} 1 & 0 & 0 & | & 1 \\ 0 & 1 & 0 & | & 2 \\ 0 & 0 & 1 & | & -1 \end{bmatrix} \quad \frac{8}{3}R3 + R2$$

The linear system associated with this final matrix is

$$x = 1$$
$$y = 2$$
$$z = -1,$$

and the solution set is $\{(1, 2, -1)\}$.

EXTENSION EXERCISES

Use the Gauss-Jordan method to solve each system of equations.

1. $x + y = 5$
   $x - y = -1$

2. $x + 2y = 5$
   $2x + y = -2$

3. $x + y = -3$
   $2x - 5y = -6$

4. $3x - 2y = 4$
   $3x + y = -2$

5. $2x - 3y = 10$
   $2x + 2y = 5$

6. $4x + y = 5$
   $2x + y = 3$

7. $3x - 7y = 31$
   $2x - 4y = 18$

8. $5x - y = 14$
   $x + 8y = 11$

9. $x + y - z = 6$
   $2x - y + z = -9$
   $x - 2y + 3z = 1$

10. $x + 3y - 6z = 7$
    $2x - y + 2z = 0$
    $x + y + 2z = -1$

11. $2x - y + 3z = 0$
    $x + 2y - z = 5$
    $2y + z = 1$

12. $4x + 2y - 3z = 6$
    $x - 4y + z = -4$
    $-x + 2z = 2$

13. $-x + y = -1$
    $y - z = 6$
    $x + z = -1$

14. $x + y = 1$
    $2x - z = 0$
    $y + 2z = -2$

15. $2x - y + 4z = -1$
    $-3x + 5y - z = 5$
    $2x + 3y + 2z = 3$

16. $5x - 3y + 2z = -5$
    $2x + 2y - z = 4$
    $4x - y + z = -1$

8.8 Linear Inequalities, Systems, and Linear Programming

Linear Inequalities in Two Variables Linear inequalities with one variable were graphed on the number line in an earlier chapter. In this section linear inequalities in two variables are graphed in a rectangular coordinate system.

Linear Inequality in Two Variables

An inequality that can be written as

$$Ax + By < C \quad \text{or} \quad Ax + By > C,$$

where A, B, and C are real numbers and A and B are not both 0, is a **linear inequality in two variables.** The symbols $\leq$ and $\geq$ may replace $<$ and $>$ in this definition.

A line divides the plane into three regions: the line itself and the two half-planes on either side of the line. Recall that the graphs of linear inequalities in one variable are intervals on the number line that sometimes include an endpoint. The graphs of linear inequalities in two variables are *regions* in the real number plane and may include a *boundary line*. The **boundary line** for the inequality $Ax + By < C$ or $Ax + By > C$ is the graph of the *equation* $Ax + By = C$. To graph a linear inequality, we follow these steps.

Graphing a Linear Inequality

Step 1: **Draw the boundary.** Draw the graph of the straight line that is the boundary. Make the line solid if the inequality involves $\leq$ or $\geq$; make the line dashed if the inequality involves $<$ or $>$.

Step 2: **Choose a test point.** Choose any point not on the line as a test point.

Step 3: **Shade the appropriate region.** Shade the region that includes the test point if it satisfies the original inequality; otherwise, shade the region on the other side of the boundary line.

EXAMPLE 1 Graph $3x + 2y \geq 6$.

First graph the straight line $3x + 2y = 6$. The graph of this line, the boundary of the graph of the inequality, is shown in Figure 43 on the next page. The graph of the inequality $3x + 2y \geq 6$ includes the points of the line $3x + 2y = 6$, and either the points *above* the line $3x + 2y = 6$ or the points *below* that line. To decide which, select any point not on the line $3x + 2y = 6$ as a test point. The origin, $(0, 0)$, often

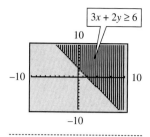

The TI-83 Plus allows us to shade the appropriate region for an inequality. Compare with Figure 43.

is a good choice. Substitute the values from the test point $(0, 0)$ for x and y in the inequality $3x + 2y \geq 6$.

$$3(0) + 2(0) \geq 6?$$
$$0 \geq 6 \quad \text{False}$$

Since the result is false, $(0, 0)$ does not satisfy the inequality, and so the solution set includes all points on the other side of the line. This region is shaded in Figure 43.

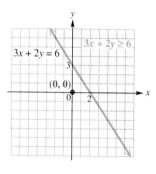

FIGURE 43

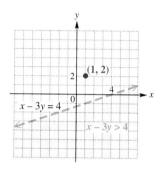

FIGURE 44

EXAMPLE 2 Graph $x - 3y > 4$.

First graph the boundary line, $x - 3y = 4$. The graph is shown in Figure 44. The points of the boundary line do not belong to the inequality $x - 3y > 4$ (since the inequality symbol is $>$ and not $\geq$). For this reason, the line is dashed. To decide which side of the line is the graph of the solution set, choose any point that is not on the line, say $(1, 2)$. Substitute 1 for x and 2 for y in the original inequality.

$$1 - 3(2) > 4?$$
$$-5 > 4 \quad \text{False}$$

Because of this false result, the solution set lies on the side of the boundary line that does *not* contain the test point $(1, 2)$. The solution set, graphed in Figure 44, includes only those points in the shaded region (not those on the line).

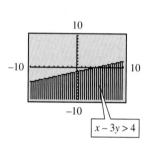

Compare with Figure 44. Can you tell from the calculator graph whether points on the boundary line are included in the solution set of the inequality?

Systems of Inequalities Methods of solving systems of *equations* were discussed in the previous section. Systems of inequalities with two variables may be solved by graphing. A system of linear inequalities consists of two or more such inequalities, and the solution set of such a system consists of all points that make all the inequalities true at the same time.

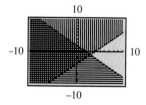

The cross-hatched region shows the solution set of the system

$$y > x - 4$$
$$y < -x + 2.$$

Graphing a System of Linear Inequalities

Step 1: **Graph each inequality in the same coordinate system.** Graph each inequality in the system, using the method described in Examples 1 and 2.

Step 2: **Find the intersection of the regions of solutions.** Indicate the intersection of the regions of solutions of the individual inequalities. This is the solution set of the system.

8.8 Linear Inequalities, Systems, and Linear Programming

EXAMPLE 3 Graph the solution set of the linear system

$$3x + 2y \leq 6$$
$$2x - 5y \geq 10.$$

Begin by graphing $3x + 2y \leq 6$. To do this, graph $3x + 2y = 6$ as a solid line. Since $(0, 0)$ makes the inequality true, shade the region containing $(0, 0)$, as shown in Figure 45.

Now graph $2x - 5y \geq 10$. The solid line boundary is the graph of $2x - 5y = 10$. Since $(0, 0)$ makes the inequality false, shade the region that does not contain $(0, 0)$, as shown in Figure 46.

The solution set of the system is given by the intersection (overlap) of the regions of the graphs in Figures 45 and 46. The solution set is the shaded region in Figure 47, and includes portions of the two boundary lines.

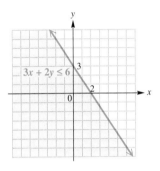

FIGURE 45

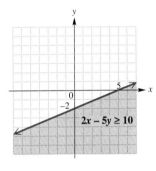

FIGURE 46

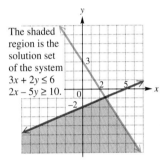

FIGURE 47

In practice, we usually do all the work in one coordinate system at the same time. In the following example, only one graph is shown.

EXAMPLE 4 Graph the solution set of the linear system

$$2x + 3y \geq 12$$
$$7x + 4y \geq 28$$
$$y \leq 6$$
$$x \leq 5.$$

The graph is obtained by graphing the four inequalities in one coordinate system and shading the region common to all four as shown in Figure 48. As the graph shows, the boundary lines are all solid.

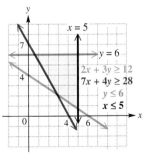

FIGURE 48

George B. Dantzig of Stanford University has been one of the key people behind **operations research** (OR). As a management science, OR is not a single discipline, but draws from mathematics, probability theory, statistics, and economics. The name given to this "multiplex" shows its historical origins in World War II, when operations of a military nature called forth the efforts of many scientists to research their fields for applications to the war effort and to solve tactical problems.

Operations research is an approach to problem solving and decision making. First of all, the problem has to be clarified. Quantities involved have to be designated as variables, and the objectives as functions. Use of **models** is an important aspect of OR.

Linear Programming A very important application of mathematics to business and social science is called **linear programming.** Linear programming is used to find an optimum value, for example, minimum cost or maximum profit. Procedures for solving linear programming problems were developed in 1947 by George Dantzig, while he was working on a problem of allocating supplies for the Air Force in a way that minimized total cost.

EXAMPLE 5   The Smartski Company makes two products, tape decks and amplifiers. Each tape deck gives a profit of $3, while each amplifier gives a profit of $7. The company must manufacture at least 1 tape deck per day to satisfy one of its customers, but no more than 5 because of production problems. Also, the number of amplifiers produced cannot exceed 6 per day. As a further requirement, the number of tape decks cannot exceed the number of amplifiers. How many of each should the company manufacture in order to obtain the maximum profit?

We translate the statements of the problem into symbols by letting

$$x = \text{number of tape decks to be produced daily}$$
$$y = \text{number of amplifiers to be produced daily}.$$

According to the statement of the problem, the company must produce at least one tape deck (one or more), so

$$x \geq 1.$$

No more than 5 tape decks may be produced:

$$x \leq 5.$$

No more than 6 amplifiers may be made in one day:

$$y \leq 6.$$

The number of tape decks may not exceed the number of amplifiers:

$$x \leq y.$$

The number of tape decks and of amplifiers cannot be negative:

$$x \geq 0 \quad \text{and} \quad y \geq 0.$$

All restrictions, or **constraints,** that are placed on production can now be summarized:

$$x \geq 1, \quad x \leq 5, \quad y \leq 6, \quad x \leq y, \quad x \geq 0, \quad y \geq 0.$$

The maximum possible profit that the company can make, subject to these constraints, is found by sketching the graph of the solution set of the system. See Figure 49. The only feasible values of x and y are those that satisfy all constraints. These values correspond to points that lie on the boundary or in the shaded region, called the **region of feasible solutions.**

Since each tape deck gives a profit of $3, the daily profit from the production of x tape decks is $3x$ dollars. Also, the profit from the production of y amplifiers will be $7y$ dollars per day. The total daily profit is thus given by the following **objective function:**

$$\text{Profit} = 3x + 7y.$$

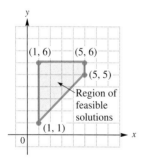

FIGURE 49

8.8 Linear Inequalities, Systems, and Linear Programming

The problem of the Smartski Company may now be stated as follows: find values of x and y in the region of feasible solutions as shown in Figure 49 that will produce the maximum possible value of $3x + 7y$. It can be shown that any optimum value (maximum or minimum) will always occur at a **vertex** (or **corner point**) of the region of feasible solutions. Locate the point (x, y) that gives the maximum profit by checking the coordinates of the vertices, shown in Figure 49 on the previous page and listed below. Find the profit that corresponds to each coordinate pair and choose the one that gives the maximum profit.

Point	Profit = $3x + 7y$
(1, 1)	$3(1) + 7(1) = 10$
(1, 6)	$3(1) + 7(6) = 45$
(5, 6)	$3(5) + 7(6) = 57$ ← Maximum
(5, 5)	$3(5) + 7(5) = 50$

The maximum profit of $57 is obtained when 5 tape decks and 6 amplifiers are produced each day.

To solve a linear programming problem in general, use the following steps.

Solving a Linear Programming Problem

Step 1: Write all necessary constraints and the objective function.
Step 2: Graph the region of feasible solutions.
Step 3: Identify all vertices.
Step 4: Find the value of the objective function at each vertex.
Step 5: The solution is given by the vertex producing the optimum value of the objective function.

It Pays to Do Your Homework

George Dantzig, profiled on the previous page, was interviewed in the September 1986 issue of the *College Mathematics Journal*. In the interview he relates the story of how he obtained his degree without actually writing a thesis.

One day Dantzig arrived late to one of his classes and on the board were two problems. Assuming that they were homework problems, he worked on them and handed them in a few days later, apologizing to his professor for taking so long to do them. Several weeks later he received an early morning visit from his professor. The problems had not been intended as homework problems; they were actually two famous *unsolved* problems in statistics! Later, when Dantzig began to think about a thesis topic, his professor told him that the two solutions would serve as his thesis.

8.8 EXERCISES

Match each system of inequalities with the correct graph from choices A.–D.

1. $x \geq 5$
    $y \leq -3$

2. $x \leq 5$
    $y \geq -3$

3. $x > 5$
    $y < -3$

4. $x < 5$
    $y > -3$

A.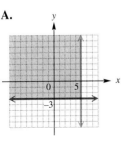

B.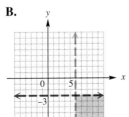

C.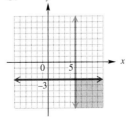

D.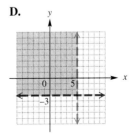

Graph each linear inequality.

5. $x + y \leq 2$
6. $x - y \geq -3$
7. $4x - y \leq 5$
8. $3x + y \geq 6$
9. $x + 3y \geq -2$
10. $4x + 6y \leq -3$
11. $x + 2y \leq -5$
12. $2x - 4y \leq 3$
13. $4x - 3y < 12$
14. $5x + 3y > 15$
15. $y > -x$
16. $y < x$

Graph each system of inequalities.

17. $x + y \leq 1$
    $x \geq 0$
18. $3x - 4y \leq 6$
    $y \geq 1$
19. $2x - y \geq 1$
    $3x + 2y \geq 6$
20. $x + 3y \geq 6$
    $3x - 4y \leq 12$
21. $-x - y < 5$
    $x - y \leq 3$
22. $6x - 4y < 8$
    $x + 2y \geq 4$

Exercises 23 and 24 show regions of feasible solutions. Find the maximum and minimum values of the given expressions.

23. $3x + 5y$

24. $40x + 75y$

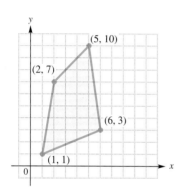

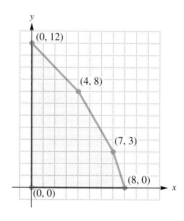

Use graphical methods to find values of x and y satisfying the given conditions. (It may be necessary to solve a system of equations in order to find vertices.) Find the value of the maximum or minimum.

25. Find $x \geq 0$ and $y \geq 0$ such that

    $2x + 3y \leq 6$
    $4x + y \leq 6$

    and $5x + 2y$ is maximized.

26. Find $x \geq 0$ and $y \geq 0$ such that

    $x + y \leq 10$
    $5x + 2y \geq 20$
    $2y \geq x$

    and $x + 3y$ is minimized.

27. Find $x \geq 2$ and $y \geq 5$ such that

    $3x - y \geq 12$
    $x + y \leq 15$

    and $2x + y$ is minimized.

28. Find $x \geq 10$ and $y \geq 20$ such that

    $2x + 3y \leq 100$
    $5x + 4y \leq 200$

    and $x + 3y$ is maximized.

Solve each linear programming problem.

29. *Refrigerator Shipping Costs* A manufacturer of refrigerators must ship at least 100 refrigerators to its two West coast warehouses. Each warehouse holds a maximum of 100 refrigerators. Warehouse A holds 25 refrigerators already, while warehouse B has 20 on hand. It costs $12 to ship a refrigerator to warehouse A and $10 to ship one to warehouse B. How many refrigerators should be shipped to each warehouse to minimize cost? What is the minimum cost?

30. *Food Supplement Costs* Bonnie, who is dieting, requires two food supplements, I and II. She can get these supplements from two different products, A and B. Product A provides 3 grams per serving of supplement I and 2 grams per serving of supplement II. Product B provides 2 grams per serving of supplement I and 4 grams per serving of supplement II. Her dietician, Dr. Dawson, has recommended that she include at least 15 grams of each supplement in her daily diet. If product A costs 25¢ per serving and product B costs 40¢ per serving, how can she satisfy her requirements most economically?

31. *Vitamin Pill Costs* Elizabeth Owens takes vitamin pills. Each day, she must have at least 16 units of Vitamin A, at least 5 units of Vitamin B_1, and at least 20 units of Vitamin C. She can choose between red pills costing 10¢ each that contain 8 units of A, 1 of B_1, and 2 of C; and blue pills that cost 20¢ each and contain 2 units of A, 1 of B_1, and 7 of C. How many of each pill should she take in order to minimize her cost and yet fulfill her daily requirements?

32. *Bolt Costs* A machine shop manufactures two types of bolts. Each can be made on any of three groups of machines, but the time required on each group differs, as shown in the table below.

		Machine Groups		
		I	II	III
Bolts	Type A	.1 min	.1 min	.1 min
	Type B	.1 min	.4 min	.5 min

Production schedules are made up one day at a time. In a day there are 240, 720, and 160 minutes available, respectively, on these machines. Type A bolts sell for 10¢ and type B bolts for 12¢. How many of each type of bolt should be manufactured per day to maximize revenue? What is the maximum revenue?

33. *Gasoline and Fuel Oil Costs* A manufacturing process requires that oil refineries manufacture at least 2 gallons of gasoline for each gallon of fuel oil. To meet the winter demand for fuel oil, at least 3 million gallons a day must be produced. The demand for gasoline is no more than 6.4 million gallons per day. If the price of gasoline is $1.90 per gallon and the price of fuel oil is $1.50 per gallon, how much of each should be produced to maximize revenue?

34. *Cake and Cookie Production* A bakery makes both cakes and cookies. Each batch of cakes requires two hours in the oven and three hours in the decorating room. Each batch of cookies needs one and a half hours in the oven and two-thirds of an hour in the decorating room. The oven is available no more than 15 hours a day, while the decorating room can be used no more than 13 hours a day. How many batches of cakes and cookies should the bakery make in order to maximize profits if cookies produce a profit of $20 per batch and cakes produce a profit of $30 per batch?

35. *Aid to Earthquake Victims* Earthquake victims in China need medical supplies and bottled water. Each medical kit measures 1 cubic foot and weighs 10 pounds. Each container of water is also 1 cubic foot and weighs 20 pounds. The plane can only carry 80,000 pounds with a total volume of 6000 cubic feet. Each medical kit will aid 6 people, while each container of water will serve 10 people. How many of each should be sent in order to maximize the number of people aided?

36. *Aid to Earthquake Victims* If each medical kit could aid 4 people instead of 6, how would the results in Exercise 35 change?

COLLABORATIVE INVESTIGATION

Living with AIDS

The graph here shows a comparison of the number of African-Americans and whites living with AIDS in the United States during 1993–2000. Form groups of 2–3 students each to work the following.

Topics for Discussion

1. The two lines were obtained by joining the data points that are of the form

   (year, number of people in thousands).

   Let $x = 0$ represent the year 1993, $x = 1$ represent 1994, and so on, and approximate the value of y for each year for African-Americans. Estimate the missing values, and fill in the table. Remember that y is in thousands.

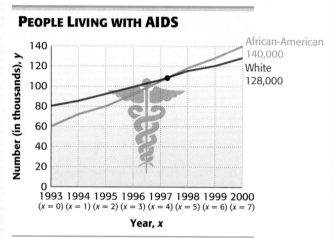

Source: U.S. Centers for Disease Control.

2. Repeat the procedure from part 1, applying the data from the line for whites living with AIDS.

Year	Number of African-Americans with AIDS (y, in thousands)
1993 ($x = 0$)	60
1994 ($x = 1$)	
1995 ($x = 2$)	
1996 ($x = 3$)	
1997 ($x = 4$)	
1998 ($x = 5$)	
1999 ($x = 6$)	
2000 ($x = 7$)	140

Year	Number of Whites with AIDS (y, in thousands)
1993 ($x = 0$)	80
1994 ($x = 1$)	
1995 ($x = 2$)	
1996 ($x = 3$)	
1997 ($x = 4$)	
1998 ($x = 5$)	
1999 ($x = 6$)	
2000 ($x = 7$)	128

Now use any two data points to find an equation of the line describing this data.

Now use any two data points to find an equation of the line describing this data.

3. The two equations from parts 1 and 2 form a system of two linear equations in two variables. Solve this system using any method you wish.

4. The x-coordinate of the solution of the system in part 3 should correspond to the year in which the two lines intersect. Look at the graph again. Does your x-value correspond the way it should?

5. Discuss why results of this activity might vary among groups performing it.

CHAPTER 8 TEST

1. Find the distance between the points $(-3, 5)$ and $(2, 1)$.

2. Find an equation of the circle whose center has coordinates $(-1, 2)$, with radius 3. Sketch its graph.

3. Find the x- and y-intercepts of the graph of $3x - 2y = 8$, and graph the equation.

4. Find the slope of the line passing through the points $(6, 4)$ and $(-1, 2)$.

5. Find the slope-intercept form of the equation of the line described.
   (a) passing through the point $(-1, 3)$, with slope $-2/5$
   (b) passing through $(-7, 2)$ and perpendicular to $y = 2x$
   (c) the line shown in the figures below (Look at the displays at the bottom.)

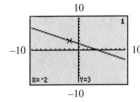
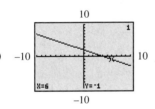

6. Which one of the following has positive slope and negative y-coordinate for its y-intercept?

   A.
   B.
   C.
   D.

7. *Income for African-Americans* Median household income of African-Americans is shown in the bar graph near the top of the next column.
   (a) Use the information given for the years 1995 and 1999, letting $x = 5$ represent 1995, $x = 9$ represent 1999, and y represent the median income, to write an equation that models median household income.
   (b) Use the equation to approximate the median income for 1997. How does your result compare to the actual value, $25,050?

MEDIAN HOUSEHOLD INCOME FOR AFRICAN-AMERICANS

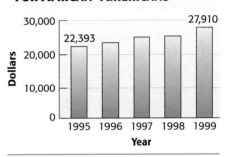

Source: U.S. Bureau of the Census.

8. *Library Fines* It costs a borrower $.05 per day for an overdue book, plus a flat $.50 charge for all books borrowed. Let x represent the number of days the book is overdue, so y represents the total fine to the tardy user. Write an equation in the form $y = mx + b$ for this situation. Then give three ordered pairs with x-values of 1, 5, and 10 that satisfy the equation.

9. Write the slope-intercept form of the equation of the line shown.

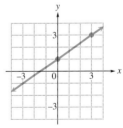

10. For the function $f(x) = x^2 - 3x + 12$,
    (a) give its domain;   (b) find $f(-2)$.

11. Give the domain of the function defined by
$$f(x) = \frac{2}{x - 3}.$$

12. *Calculator Production* If the cost to produce x units of calculators is $C(x) = 50x + 5000$ dollars, while the revenue is $R(x) = 60x$ dollars, find the number of units of calculators that must be produced in order to break even. What is the revenue at the break-even point?

13. Graph the quadratic function
$$f(x) = -(x+3)^2 + 4.$$
Give the axis, the vertex, the domain, and the range.

14. *Dimensions of a Parking Lot* Miami-Dade Community College wants to construct a rectangular parking lot on land bordered on one side by a highway. It has 320 ft of fencing with which to fence off the other three sides. What should be the dimensions of the lot if the enclosed area is to be a maximum?

15. Use a scientific calculator to find an approximation of each of the following. Give as many digits as the calculator displays.
    (a) $5.1^{4.7}$ (b) $e^{-1.85}$ (c) $\ln 23.56$

16. Which one of the following is a false statement?
    A. The domain of the function $f(x) = \log_2 x$ is $(-\infty, \infty)$.
    B. The graph of $F(x) = 3^x$ intersects the y-axis.
    C. The graph of $G(x) = \log_3 x$ intersects the x-axis.
    D. The expression $\ln x$ represents the exponent to which e must be raised in order to obtain x.

17. *Investment* Suppose that $12,000 is invested in an account that pays 4% annual interest, and is left untouched for 3 years. How much will be in the account if
    (a) interest is compounded quarterly (four times per year);
    (b) interest is compounded continuously?

18. *Decay of Plutonium-241* Suppose that the amount, in grams, of plutonium-241 present in a given sample is determined by the function defined by
$$A(t) = 2.00 e^{-.053t},$$
where t is measured in years. Find the amount present in the sample after the given number of years.
    (a) 4   (b) 10   (c) 20
    (d) What was the initial amount present?

Solve each system by using elimination, substitution, or a combination of the two methods.

19. $2x + 3y = 2$
    $3x - 4y = 20$

20. $2x + y + z = 3$
    $x + 2y - z = 3$
    $3x - y + z = 5$

21. $2x + 3y - 6z = 11$
    $x - y + 2z = -2$
    $4x + y - 2z = 7$

Solve each problem by using a system of equations.

22. *Julia Roberts' Box Office Hits* Julia Roberts is one of the biggest box-office stars in Hollywood. As of July 2001, her two top-grossing domestic films, *Pretty Woman* and *Runaway Bride,* together earned $330.7 million. If *Runaway Bride* grossed $26.1 million less than *Pretty Woman,* how much did each film gross? (*Source:* ACNielsen EDI.)

23. *Real Estate Commission* Keshon Grant sells real estate. On three recent sales, he made 10% commission, 6% commission, and 5% commission. His total commissions on these sales were $17,000, and he sold property worth $280,000. If the 5% sale amounted to the sum of the other two, what were the three sales' prices?

24. Graph the solution set of the system of inequalities.
$$x + y \leq 6$$
$$2x - y \geq 3$$

25. *Ring Sales* The Alessi company designs and sells two types of rings: the VIP and the SST. The company can produce up to 24 rings each day using up to 60 total hours of labor. It takes 3 hours to make one VIP ring, and 2 hours to make one SST ring. How many of each type of ring should be made daily in order to maximize profit, if profit on a VIP ring is $30 and profit on an SST ring is $40? What is the profit?

chapter

9

Geometry

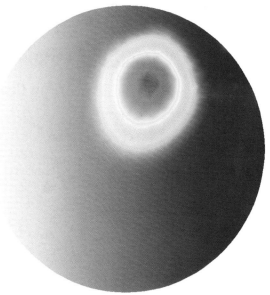

A new imaging technique called *electrical impedance tomography* (EIT) is being applied to many situations in which visualization improves diagnostic capabilities. It currently is used to measure the amount of contaminants in the soil, detect corrosive damage in aircraft, and locate valuable mineral deposits in the ground. In medicine, this technique is being researched for displaying images of internal organs.

EIT measures the distribution of impedance in a cross-section of the body. This is possible because the electrical resistivities of different body tissues vary widely, from .65 ohm m for cerebrospinal fluid to 150 ohm m for bone. Before data can be recorded, a series of electrodes are attached to a subject in a transverse plane. These are linked to a data acquisition unit which outputs data to a PC. By applying a series of small currents to the body, a set of potential difference measurements can be made from non-current carrying pairs of electrodes. Since electric currents applied to the body take the paths of least impedance, where the currents flow depends on the subject's conductivity distribution. The image reconstruction process is a non-linear problem. Therefore images can be reconstructed from the data using a variety of analytic and geometric methods.

EIT is different from X rays and CAT scans. Although both X rays and CAT scans give clearer images, EIT can display dynamic images of physiological processes, such as breathing, as they occur. In the near future researchers hope to use EIT to measure cardiac output and avoid the invasive procedure of inserting a catheter into the heart.

Sources: SIAM News, Vol. 27, No. 6, 1994.
www.geocities.com/CapeCanaveral/9710/smeit.html

9.1 Points, Lines, Planes, and Angles

9.2 Curves, Polygons, and Circles

Extension: Geometric Constructions

9.3 Perimeter, Area, and Circumference

9.4 The Geometry of Triangles: Congruence, Similarity, and the Pythagorean Theorem

9.5 Space Figures, Volume, and Surface Area

9.6 Transformational Geometry

9.7 Non-Euclidean Geometry, Topology, and Networks

9.8 Chaos and Fractal Geometry

Collaborative Investigation: Generalizing the Angle Sum Concept

Chapter 9 Test

9.1 Points, Lines, Planes, and Angles

The Geometry of Euclid

Euclid's Elements as translated by Billingsley appeared in 1570 and was the first English language translation of the text—the most influential geometry text ever written.

Unfortunately, no copy of *Elements* exists that dates back to the time of Euclid (circa 300 B.C.), and most current translations are based upon a revision of the work prepared by Theon of Alexandria.

Although *Elements* was only one of several works of Euclid, it is, by far, the most important. It ranks second only to the Bible as the most published book in history.

Let no one unversed in geometry enter here.
 —Motto over the door of Plato's Academy

To the ancient Greeks, mathematics meant geometry above all—a rigid kind of geometry from a modern-day point of view. The Greeks studied the properties of figures identical in shape and size (congruent figures) as well as figures identical in shape but not necessarily in size (similar figures). They absorbed ideas about area and volume from the Egyptians and Babylonians and established general formulas. The Greeks were the first to insist that statements in geometry be given rigorous proof.

The Greek view of geometry (and other mathematical ideas) was summarized in *Elements,* written by Euclid about 300 B.C. The influence of this book has been extraordinary; it has been studied virtually unchanged to this day as a geometry textbook and as *the* model of deductive logic.

The most basic ideas of geometry are **point, line,** and **plane.** In fact, it is not really possible to define them with other words. Euclid defined a point as "that which has no part," but this definition is so vague as to be meaningless. Do you think you could decide what a point is from this definition? But from your experience in saying "this point in time" or in sharpening a pencil, you have an idea of what he was getting at. Even though we don't try to define *point,* we do agree that, intuitively, a point has no magnitude and no size.

Euclid defined a line as "that which has breadthless length." Again, this definition is vague. Based on our experience, however, we know what Euclid meant. The drawings that we use for lines have properties of no thickness and no width, and they extend indefinitely in two directions.

What do you visualize when you read Euclid's definition of a plane: "a surface which lies evenly with the straight lines on itself"? Do you think of a flat surface, such as a tabletop or a page in a book? That is what Euclid intended.

The geometry of Euclid is a model of deductive reasoning. In this chapter, we will present geometry from an inductive viewpoint, using objects and situations found in the world around us as models for study.

Points, Lines, and Planes There are certain universally accepted conventions and symbols used to represent points, lines, planes, and angles. A capital letter usually represents a point. A line may be named by two capital letters representing points that lie on the line, or by a single (usually lowercase) letter, such as ℓ. Subscripts are sometimes used to distinguish one line from another when a lowercase letter is used. For example, ℓ_1 and ℓ_2 would represent two distinct lines. A plane may be named by three capital letters representing points that lie in the plane, or by a letter of the Greek alphabet, such as α (alpha), β (beta), or γ (gamma).

Figure 1 depicts a plane that may be represented either as α or as plane *ADE.* Contained in the plane is the line *DE* (or, equivalently, line *ED*), which is also labeled ℓ in the figure.

Selecting any point on a line divides the line into three parts: the point itself, and two **half-lines,** one on each side of the point. For example, in Figure 2, point *A* divides the line into three parts, *A* itself and two half-lines. Point *A* belongs to neither half-line. As the figure suggests, each half-line extends indefinitely in the direction opposite to the other half-line.

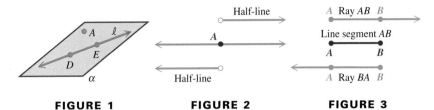

FIGURE 1 **FIGURE 2** **FIGURE 3**

Given any three points that are not in a straight line, a plane can be passed through the points. That is why camera tripods have three legs—no matter how irregular the surface, the tips of the three legs determine a plane. On the other hand, a camera support with four legs would wobble unless all four legs were carefully extended just the right amount.

Including an initial point with a half-line gives a **ray.** A ray is named with two letters, one for the initial point of the ray, and one for another point contained in the half-line. For example, in Figure 3 ray *AB* has initial point *A* and extends in the direction of *B*. On the other hand, ray *BA* has *B* as its initial point and extends in the direction of *A*.

A **line segment** includes both endpoints and is named by its endpoints. Figure 3 shows line segment *AB,* which may also be designated as line segment *BA*.

The following chart shows these figures along with the symbols used to represent them.

Name	Figure	Symbol
Line *AB* or line *BA*	←—•——•—→ *A* *B*	$\overleftrightarrow{AB}$ or $\overleftrightarrow{BA}$
Half-line *AB*	∘——•—→ *A* *B*	$\overrightarrow{AB}$ (open)
Half-line *BA*	←—•——∘ *A* *B*	$\overleftarrow{BA}$ (open)
Ray *AB*	•——•—→ *A* *B*	$\overrightarrow{AB}$
Ray *BA*	←—•——• *A* *B*	$\overleftarrow{BA}$
Segment *AB* or segment *BA*	•——• *A* *B*	$\overline{AB}$ or $\overline{BA}$

For a line, the symbol above the two letters shows two arrowheads, indicating that the line extends indefinitely in both directions. For half-lines and rays, only one arrowhead is used since these extend in only one direction. An open circle is used for a half-line to show that the endpoint is not included, while a solid circle is used for a ray to indicate the inclusion of the endpoint. Since a segment includes both endpoints and does not extend in either direction, solid circles are used to indicate endpoints of line segments.

The geometric definitions of "parallel" and "intersecting" apply to two or more lines or planes. (See Figure 4 on the next page.) **Parallel lines** lie in the same plane and never meet, no matter how far they are extended. However, **intersecting lines** do meet. If two distinct lines intersect, they intersect in one and only one point.

We use the symbol ∥ to denote parallelism. If ℓ_1 and ℓ_2 are parallel lines, as in Figure 4, then this may be indicated as $\ell_1 \parallel \ell_2$.

Parallel planes also never meet, no matter how far they are extended. Two distinct **intersecting planes** form a straight line, the one and only line they have in common. **Skew lines** do not lie in the same plane, and they never meet, no matter how far they are extended.

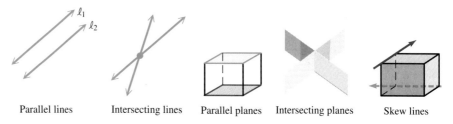

Parallel lines   Intersecting lines   Parallel planes   Intersecting planes   Skew lines

FIGURE 4

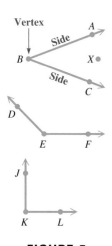

FIGURE 5

Angles An **angle** is the union of two rays that have a common endpoint, as shown in Figure 5. It is important to remember that the angle is formed by points on the rays themselves, and no other points. In Figure 5, point X is *not* a point on the angle. (It is said to be in the *interior* of the angle.) Notice that "angle" is the first basic term in this section that is actually defined, using the undefined terms *ray* and *endpoint*.

The rays forming an angle are called its **sides.** The common endpoint of the rays is the **vertex** of the angle. There are two standard ways of naming angles. If no confusion will result, an angle can be named with the letter marking its vertex. Using this method, the angles in Figure 5 can be named, respectively, angle B, angle E, and angle K. Angles also can be named with three letters: the first letter names a point on one side of the angle; the middle letter names the vertex; the third names a point on the other side of the angle. In this system, the angles in the figure can be named angle ABC, angle DEF, and angle JKL.

The symbol for representing an angle is ∡. Rather than writing "angle ABC," we may write "∡ABC."

An angle can be associated with an amount of rotation. For example, in Figure 6(a), we let $\vec{BA}$ first coincide with $\vec{BC}$—as though they were the same ray. We then rotate $\vec{BA}$ (the endpoint remains fixed) in a counterclockwise direction to form ∡ABC.

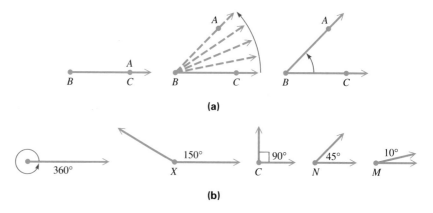

FIGURE 6

Why 360? The use of the number 360 goes back to the Babylonian culture. There are several theories regarding why 360 was chosen for the number of degrees in a complete rotation around a circle. One says that 360 was chosen because it is close to the number of days in a year, and is conveniently divisible by 2, 3, 4, 5, 6, 8, 9, 10, 12, and other numbers.

Angles are key to the study of **geodesy,** the measurement of distances on the earth's surface.

Angles are measured by the amount of rotation, using a system that dates back to the Babylonians some two centuries before Christ. Babylonian astronomers chose the number 360 to represent the amount of rotation of a ray back onto itself. Using 360 as the amount of rotation of a ray back onto itself, **one degree,** written 1°, is defined to be 1/360 of a complete rotation. Figure 6(b) shows angles of various degree measures.

Angles are classified and named with reference to their degree measures. An angle whose measure is between 0° and 90° is called an **acute angle.** Angles M and N in Figure 6(b) are acute. An angle that measures 90° is called a **right angle.** Angle C in the figure is a right angle. The squared symbol ⌐ at the vertex denotes a right angle. Angles that measure more than 90° but less than 180° are said to be **obtuse angles** (angle X, for example). An angle that measures 180° is a **straight angle.** Its sides form a straight line.

Our work in this section will be devoted primarily to angles whose measures are less than or equal to 180°. Angles whose measures are greater than 180° are studied in more detail in trigonometry courses.

A tool called a **protractor** can be used to measure angles. Figure 7 shows a protractor measuring an angle. To use a protractor, position the hole (or dot) of the protractor on the vertex of the angle. The 0-degree measure on the protractor should be placed on one side of the angle, while the other side should extend to the degree measure of the angle. The figure indicates an angle whose measure is 135°.

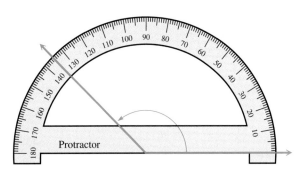

FIGURE 7

When two lines intersect to form right angles they are called **perpendicular lines.** Our sense of *vertical* and *horizontal* depends on perpendicularity.

In Figure 8, the sides of ∡NMP have been extended to form another angle, ∡RMQ. The pair ∡NMP and ∡RMQ are called **vertical angles.** Another pair of vertical angles have been formed at the same time. They are ∡NMQ and ∡PMR.

An important property of vertical angles follows.

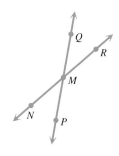

FIGURE 8

Property of Vertical Angles

Vertical angles have equal measures.

For example, ∡NMP and ∡RMQ in Figure 8 have equal measures. What other pair of angles in the figure have equal measures?

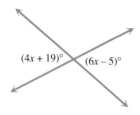

FIGURE 9

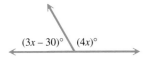

FIGURE 10

EXAMPLE 1 Refer to the appropriate figure to solve each problem.

(a) Find the measure of each marked angle in Figure 9.

Since the marked angles are vertical angles, they have the same measure. Set $4x + 19$ equal to $6x - 5$ and solve.

$$4x + 19 = 6x - 5$$
$$-4x + 4x + 19 = -4x + 6x - 5 \quad \text{Add } -4x.$$
$$19 = 2x - 5$$
$$19 + 5 = 2x - 5 + 5 \quad \text{Add 5.}$$
$$24 = 2x$$
$$12 = x \quad \text{Divide by 2.}$$

Since $x = 12$, one angle has measure $4(12) + 19 = 67$ degrees. The other has the same measure, since $6(12) - 5 = 67$ as well. Each angle measures $67°$.

(b) Find the measure of each marked angle in Figure 10.

The measures of the marked angles must add to $180°$ since together they form a straight angle. The equation to solve is

$$(3x - 30) + 4x = 180.$$
$$7x - 30 = 180 \quad \text{Combine like terms.}$$
$$7x - 30 + 30 = 180 + 30 \quad \text{Add 30.}$$
$$7x = 210$$
$$x = 30 \quad \text{Divide by 7.}$$

To find the measures of the angles, replace x with 30 in the two expressions.

$$3x - 30 = 3(30) - 30 = 90 - 30 = 60$$
$$4x = 4(30) = 120$$

The two angle measures are $60°$ and $120°$. ■

If the sum of the measures of two acute angles is $90°$, the angles are said to be **complementary,** and each is called the *complement* of the other. For example, angles measuring $40°$ and $50°$ are complementary angles, because $40° + 50° = 90°$. If two angles have a sum of $180°$, they are **supplementary.** The *supplement* of an angle whose measure is $40°$ is an angle whose measure is $140°$, because $40° + 140° = 180°$. If x represents the degree measure of an angle, $90 - x$ represents the measure of its complement, and $180 - x$ represents the measure of its supplement.

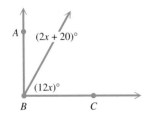

FIGURE 11

EXAMPLE 2 Find the measures of the angles in Figure 11, given that $\angle ABC$ is a right angle.

The sum of the measures of the two acute angles is $90°$ (that is, they are complementary), since they form a right angle. We add their measures to obtain a sum of 90 and solve the resulting equation.

$$(2x + 20) + 12x = 90$$
$$14x + 20 = 90$$
$$14x = 70 \quad \text{Subtract 20.}$$
$$x = 5 \quad \text{Divide by 14.}$$

The value of x is 5. Therefore, the measures of the angles are
$$2x + 20 = 2(5) + 20 = 30 \text{ degrees}$$
and
$$12x = 12(5) = 60 \text{ degrees}.$$

Moiré Patterns A set of parallel lines with equidistant spacing intersects an identical set, but at a small angle. The result is a moiré pattern, named after the fabric *moiré* ("watered") *silk*. You often see similar effects looking through window screens with bulges. Moiré patterns are related to periodic functions, which describe regular recurring phenomena (wave patterns such as heartbeats or business cycles). Moirés thus apply to the study of electromagnetic, sound, and water waves, to crystal structure, and to other wave phenomena.

EXAMPLE 3 The supplement of an angle measures 10° more than three times its complement. Find the measure of the angle.

Let $x =$ the degree measure of the angle.
Then $180 - x =$ the degree measure of its supplement,
and $90 - x =$ the degree measure of its complement.

Now use the words of the problem to write the equation.

$$\underbrace{180 - x}_{\text{Supplement}} \underbrace{=}_{\text{measures}} \underbrace{10 +}_{\text{10 more than}} \underbrace{3(90 - x)}_{\text{three times its complement.}}$$

$$180 - x = 10 + 270 - 3x \quad \text{Distributive property}$$
$$180 - x = 280 - 3x$$
$$2x = 100 \quad \text{Add } 3x\text{; subtract 180.}$$
$$x = 50 \quad \text{Divide by 2.}$$

The angle measures 50°. Since its supplement (130°) is 10° more than three times its complement (40°), that is, $130 = 10 + 3(40)$ is true, the answer checks.

Parallel lines are lines that lie in the same plane and do not intersect. Figure 12 shows parallel lines m and n. When a line q intersects two parallel lines, q is called a **transversal.** In Figure 12, the transversal intersecting the parallel lines forms eight angles, indicated by numbers. Angles 1 through 8 in the figure possess some special properties regarding their degree measures, as shown in the following chart.

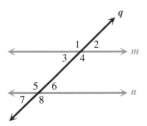

FIGURE 12

Name	Figure	Rule
Alternate interior angles	(angles 5 and 4, also 3 and 6)	Angle measures are equal.
Alternate exterior angles	(angles 1 and 8, also 2 and 7)	Angle measures are equal.
Interior angles on same side of transversal	(angles 4 and 6, also 3 and 5)	Angle measures add to 180°.
Corresponding angles	(angles 2 and 6, also 1 and 5, 3 and 7, 4 and 8)	Angle measures are equal.

The converses of the above also are true. That is, if alternate interior angles are equal, then the lines are parallel, with similar results valid for alternate exterior angles, interior angles on the same side of a transversal, and corresponding angles.

EXAMPLE 4 Find the measure of each marked angle in Figure 13, given that lines m and n are parallel.

The marked angles are alternate exterior angles, which are equal. This gives

$$3x + 2 = 5x - 40$$
$$42 = 2x \qquad \text{Subtract } 3x \text{ and add 40.}$$
$$21 = x. \qquad \text{Divide by 2.}$$

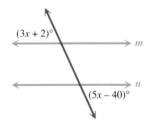

FIGURE 13

One angle has a measure of $3x + 2 = 3 \cdot 21 + 2 = 65$ degrees, and the other has a measure of $5x - 40 = 5 \cdot 21 - 40 = 65$ degrees.

9.1 EXERCISES

Fill in each blank with the correct response.

1. The sum of the measures of two complementary angles is _____ degrees.
2. The sum of the measures of two supplementary angles is _____ degrees.
3. The measures of two vertical angles are _____. (equal/not equal)
4. The measures of _____ right angles add up to the measure of a straight angle.

Decide whether each statement is true or false.

5. A line segment has two endpoints.
6. A ray has one endpoint.
7. If A and B are distinct points on a line, then ray AB and ray BA represent the same set of points.
8. If two lines intersect, they lie in the same plane.
9. If two lines are parallel, they lie in the same plane.
10. If two lines do not intersect, they must be parallel.
11. Segment AB and segment BA represent the same set of points.
12. There is no angle that is its own complement.
13. There is no angle that is its own supplement.
14. The origin of the use of the degree as a unit of measure of an angle goes back to the Egyptians.

Exercises 15–24 name portions of the line shown. For each exercise, (a) give the symbol that represents the portion of the line named, and (b) draw a figure showing just the portion named, including all labeled points.

15. line segment AB
16. ray BC
17. ray CB
18. line segment AD
19. half-line BC
20. half-line AD
21. ray BA
22. ray DA
23. line segment CA
24. line-segment DA

Match the symbol in Exercises 25–32, with the symbol in A–H that names the same set of points, based on the given figure.

25. $\overrightarrow{PQ}$
26. $\overrightarrow{QR}$
27. $\overleftrightarrow{QR}$
28. $\overrightarrow{PQ}$
29. $\overleftrightarrow{RP}$
30. $\overrightarrow{SQ}$
31. $\overleftrightarrow{PS}$
32. $\overrightarrow{PS}$

A. $\overrightarrow{QS}$
B. $\overleftrightarrow{RQ}$
C. $\overrightarrow{SR}$
D. $\overleftrightarrow{QS}$
E. $\overleftrightarrow{SP}$
F. $\overrightarrow{QP}$
G. $\overleftrightarrow{RS}$
H. none of these

*Lines, rays, half-lines, and segments may be considered sets of points. The **intersection** (symbolized ∩) of two sets is composed of all elements common to both sets, while the **union** (symbolized ∪) of two sets is composed of all elements found in at least one of the two sets. Based on the figure below, specify each of the sets given in Exercises 33–40 in a simpler way.*

33. $\overleftrightarrow{MN} \cup \overrightarrow{NO}$
34. $\overleftrightarrow{MN} \cap \overrightarrow{NO}$
35. $\overrightarrow{MO} \cap \overrightarrow{OM}$
36. $\overrightarrow{MO} \cup \overrightarrow{OM}$
37. $\overrightarrow{OP} \cap O$
38. $\overrightarrow{OP} \cup O$
39. $\overrightarrow{NP} \cap \overrightarrow{OP}$
40. $\overrightarrow{NP} \cup \overrightarrow{OP}$

Give the measure of the complement of each angle.

41. $28°$
42. $32°$
43. $89°$
44. $45°$
45. $x°$
46. $(90 - x)°$

Give the measure of the supplement of each angle.

47. 132°   **48.** 105°   **49.** 26°   **50.** 90°   **51.** $y°$ **52.** $(180-y)°$

Name all pairs of vertical angles in each figure.

53. 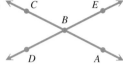   **54.**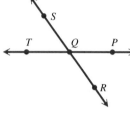

55. In Exercise 53, if ∡ABE has a measure of 52°, find the measures of the angles.
(a) ∡CBD   (b) ∡CBE

56. In Exercise 54, if ∡SQP has a measure of 126°, find the measures of the angles.
(a) ∡TQR   (b) ∡PQR

Find the measure of each marked angle.

57.    **58.**    **59.**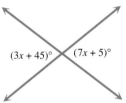

60. 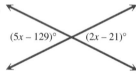   **61.** 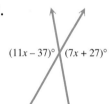   **62.**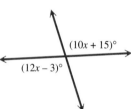

63. 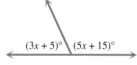   **64.** 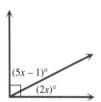   **65.**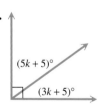

In Exercises 66–69, assume that lines m and n are parallel, and find the measure of each marked angle.

66. 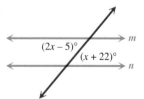   **67.**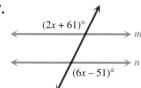

68.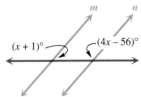

69.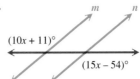

Complementary and Supplementary Angles Solve each problem in Exercises 70–73.

70. The supplement of an angle measures 25° more than twice its complement. Find the measure of the angle.

71. The complement of an angle measures 10° less than one-fifth of its supplement. Find the measure of the angle.

72. The supplement of an angle added to the complement of the angle gives 210°. What is the measure of the angle?

73. Half the supplement of an angle is 12° less than twice the complement of the angle. Find the measure of the angle.

74. The sketch shows parallel lines m and n cut by a transversal q. Using the figure, complete the steps to prove that alternate exterior angles have the same measure.

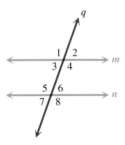

(a) Measure of $\angle 2$ = measure of $\angle$ ____, since they are vertical angles.
(b) Measure of $\angle 3$ = measure of $\angle$ ____, since they are alternate interior angles.
(c) Measure of $\angle 6$ = measure of $\angle$ ____, since they are vertical angles.
(d) By the results of parts (a), (b), and (c), the measure of $\angle 2$ must equal the measure of $\angle$ ____, showing that alternate ____ angles have equal measures.

75. Use the sketch to find the measure of each numbered angle. Assume that $m \parallel n$.

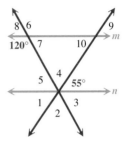

76. Complete these steps in the proof that vertical angles have equal measures. In this exercise, m($\angle$ ____) means "the measure of the angle ____." Use the figure.

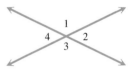

(a) m($\angle 1$) + m($\angle 2$) = ____ °
(b) m($\angle 2$) + m($\angle 3$) = ____ °
(c) Subtract the equation in part (b) from the equation in part (a) to get [m($\angle 1$) + m($\angle 2$)] − [m($\angle 2$) + m($\angle 3$)] = ____ ° − ____ °.
(d) m($\angle 1$) + m($\angle 2$) − m($\angle 2$) − m($\angle 3$) = ____ °
(e) m($\angle 1$) − m($\angle 3$) = ____ °
(f) m($\angle 1$) = m($\angle$ ____)

77. Use the approach of Exercise 74 to prove that interior angles on the same side of a transversal are supplementary.

78. Find the values of x and y in the figure, given that $x + y = 40$.

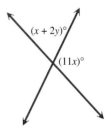

9.2 Curves, Polygons, and Circles

Curves The basic undefined term *curve* is used for describing figures in the plane. (See the examples in Figure 14.)

Simple; closed

Simple; not closed

Not simple; closed

Not simple; not closed

FIGURE 14

Simple Curve; Closed Curve

A **simple curve** can be drawn without lifting the pencil from the paper, and without passing through any point twice.

A **closed curve** has its starting and ending points the same, and is also drawn without lifting the pencil from the paper.

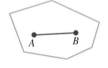
Convex
(a)

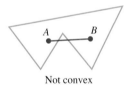
Not convex
(b)
FIGURE 15

A figure is said to be **convex** if, for any two points A and B inside the figure, the line segment AB (that is, $\overleftrightarrow{AB}$) is always completely inside the figure. Figure 15(a) shows a convex figure while (b) shows one that is not convex.

Among the most common types of curves in mathematics are those that are both simple and closed, and perhaps the most important of these are *polygons*. A **polygon** is a simple closed curve made up only of straight line segments. The line segments are called the *sides,* and the points at which the sides meet are called *vertices* (singular: *vertex*). Polygons are classified according to the number of line segments used as sides. The chart in the margin gives the special names. In general, if a polygon has n sides, and no particular value of n is specified, it is called an n-gon.

Some examples of polygons are shown in Figure 16. A polygon may or may not be convex. Polygons with all sides equal and all angles equal are **regular polygons.**

Classification of Polygons According to Number of Sides	
Number of Sides	**Name**
3	triangle
4	quadrilateral
5	pentagon
6	hexagon
7	heptagon
8	octagon
9	nonagon
10	decagon

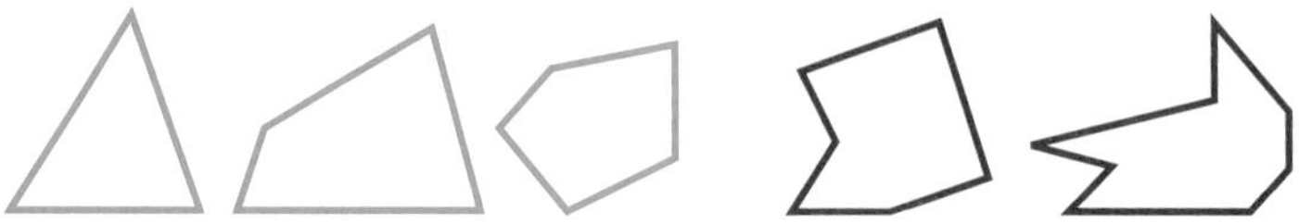

Convex   Not convex
Polygons are simple closed curves made up of straight line segments.

Regular polygons have equal sides and equal angles.

FIGURE 16

Triangles and Quadrilaterals Two of the most common types of polygons are triangles and quadrilaterals. Triangles are classified by measures of angles as well as by number of equal sides, as shown in the following box. (Notice that tick marks are used in the bottom three figures to show how side lengths are related.)

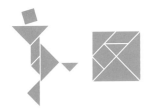

Tangrams The puzzle-game above comes from China, where it has been a popular amusement for centuries. The figure on the left is a tangram. Any tangram is composed of the same set of seven tans (the pieces making up the square are shown on the right).

Mathematicians have described various properties of tangrams. While each tan is convex, only 13 convex tangrams are possible. All others, like the figure on the left, are not convex.

Types of Triangles

	All Angles Acute	One Right Angle	One Obtuse Angle
Angles	Acute triangle	Right triangle	Obtuse triangle
	All Sides Equal	**Two Sides Equal**	**No Sides Equal**
Sides	Equilateral triangle	Isosceles triangle	Scalene triangle

Quadrilaterals are classified by sides and angles. It can be seen below that an important distinction involving quadrilaterals is whether one or more pairs of sides are parallel.

Types of Quadrilaterals

Sample Figure

Trapezoid

A **trapezoid** is a quadrilateral with one pair of parallel sides.

Parallelogram

A **parallelogram** is a quadrilateral with two pairs of parallel sides.

Rectangle

A **rectangle** is a parallelogram with a right angle (and consequently, four right angles).

Square

A **square** is a rectangle with all sides having equal length.

Rhombus

A **rhombus** is a parallelogram with all sides having equal length.

An important property of triangles that was first proved by the Greek geometers deals with the sum of the measures of the angles of any triangle.

Angle Sum of a Triangle

The sum of the measures of the angles of any triangle is 180°.

While it is not an actual proof, a rather convincing argument for the truth of this statement can be given using any size triangle cut from a piece of paper. Tear each corner from the triangle, as suggested in Figure 17(a). You should be able to rearrange the pieces so that the three angles form a straight angle, as shown in Figure 17(b).

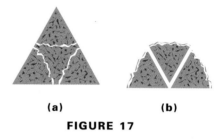

FIGURE 17

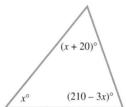

FIGURE 18

EXAMPLE 1 Find the measure of each angle in the triangle of Figure 18.

By the angle sum relationship, the three angle measures must add up to 180°. Write the equation indicating this, and then solve.

$$x + (x + 20) + (210 - 3x) = 180$$
$$-x + 230 = 180 \quad \text{Combine like terms.}$$
$$-x = -50 \quad \text{Subtract 230.}$$
$$x = 50 \quad \text{Divide by } -1.$$

One angle measures 50°, another measures $x + 20 = 50 + 20 = 70°$, and the third measures $210 - 3x = 210 - 3(50) = 60°$. Since $50° + 70° + 60° = 180°$, the answers satisfy the angle sum relationship.

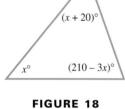

FIGURE 19

In the triangle shown in Figure 19, angles 1, 2, and 3 are called **interior angles,** while angles 4, 5, and 6 are called **exterior angles** of the triangle. Using the fact that the sum of the angle measures of any triangle is 180°, and a straight angle also measures 180°, the following property may be deduced.

Exterior Angle Measure

The measure of an exterior angle of a triangle is equal to the sum of the measures of the two opposite interior angles.

In Figure 19, the measure of angle 6 is equal to the sum of the measures of angles 1 and 2. Two other such statements can be made. What are they?

EXAMPLE 2 Find the measures of interior angles A, B, and C of the triangle in Figure 20 on the next page, and the measure of exterior angle BCD.

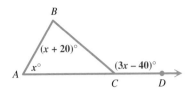

FIGURE 20

By the property concerning exterior angles, the sum of the measures of interior angles A and B must equal the measure of angle BCD. Thus,

$$x + (x + 20) = 3x - 40$$
$$2x + 20 = 3x - 40$$
$$-x = -60 \quad \text{Subtract } 3x; \text{ subtract } 20.$$
$$x = 60. \quad \text{Divide by } -1.$$

Since the value of x is 60,

Interior angle $A = 60°$
Interior angle $B = (60 + 20)° = 80°$
Interior angle $C = 180° - (60° + 80°) = 40°$
Exterior angle $BCD = [3(60) - 40]° = 140°$. ∎

Circles One of the most important plane curves is the circle. It is a simple closed curve defined as follows.

> **Circle**
>
> A **circle** is a set of points in a plane, each of which is the same distance from a fixed point.

A circle may be physically constructed with compasses, where the spike leg remains fixed and the other leg swings around to construct the circle. A string may also be used to draw a circle. For example, loop a piece of chalk on one end of a piece of string. Hold the other end in a fixed position on a chalkboard, and pull the string taut. Then swing the chalk end around to draw a circle.

A circle, along with several lines and segments, is shown in Figure 21. The points P, Q, and R lie on the circle itself. Each lies the same distance from point O, which is called the **center** of the circle. (It is the "fixed point" referred to in the definition.) $\overleftrightarrow{OP}$, $\overleftrightarrow{OQ}$, and $\overleftrightarrow{OR}$ are segments whose endpoints are the center and a point on the circle. Each is called a **radius** of the circle (plural: **radii**). $\overleftrightarrow{PQ}$ is a segment whose endpoints both lie on the circle and is an example of a **chord**. $\overleftrightarrow{PR}$ is a chord that passes through the center and is called a **diameter** of the circle. Notice that the measure of a diameter is twice that of a radius. A diameter such as $\overleftrightarrow{PR}$ in Figure 21 divides a circle into two parts of equal size, each of which is called a **semicircle.**

$\overleftrightarrow{RT}$ is a line that touches (intersects) the circle in only one point, R, and is called a **tangent** to the circle. R is the point of tangency. $\overleftrightarrow{PQ}$, which intersects the circle in two points, is called a **secant** line. (What is the distinction between a chord and a secant?)

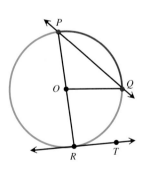

FIGURE 21

Thales made his fortune merely to prove how easy it is to become wealthy; he cornered all the oil presses during a year of an exceptionally large olive crop. Legend records that Thales studied for a time in Egypt and then introduced geometry to Greece, where he attempted to apply the principles of Greek logic to his newly learned subject.

The portion of the circle shown in red in Figure 21 is an example of an **arc** of the circle. It consists of two endpoints (P and Q) and all points on the circle "between" these endpoints. The colored portion is called arc PQ (or QP), denoted in symbols as $\overset{\frown}{PQ}$ (or $\overset{\frown}{QP}$).

The Greeks were the first to insist that all propositions, or **theorems,** about geometry be given a rigorous proof before being accepted. According to tradition, the first theorem to receive such a proof was the following.

Inscribed Angle
Any angle inscribed in a semicircle must be a right angle.

To be **inscribed** in a semicircle, the vertex of the angle must be on the circle with the sides of the angle going through the endpoints of the diameter at the base of the semicircle. (See Figure 22.) This first proof was said to have been given by the Greek philosopher Thales.

FIGURE 22

9.2 EXERCISES

Fill in each blank with the correct response.

1. A segment joining two points on a circle is called a(n) _____ .
2. A segment joining the center of a circle and a point on the circle is called a(n) _____ .
3. A regular triangle is called a(n) _____ triangle.
4. A chord that contains the center of a circle is called a(n) _____ .

Decide whether each statement is true *or* false.

5. A rhombus is an example of a regular polygon.
6. If a triangle is isosceles, then it is not scalene.
7. A triangle can have more than one obtuse angle.
8. A square is both a rectangle and a parallelogram.
9. A square must be a rhombus.
10. A rhombus must be a square.
11. In your own words, explain the distinction between a square and a rhombus.
12. What common traffic sign in the U.S. is in the shape of an octagon?

Identify each curve as simple, closed, both, *or* neither.

13.

14.

15.

16.

17.  **18.**  **19.**  **20.**

Decide whether each figure is convex *or* not convex.

21.  **22.** **23.**

24. **25.**  **26.**

Classify each triangle as either acute, right, *or* obtuse. *Also classify each as either* equilateral, isosceles, *or* scalene.

27. 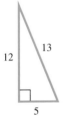 **28.** 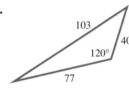 **29.** 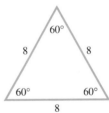 **30.**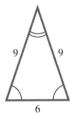

31. 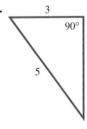 **32.**  **33.** 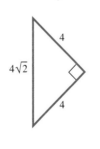 **34.**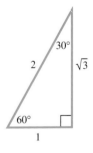

35. 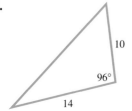 **36.** 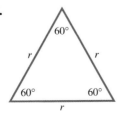 **37.** 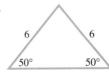 **38.**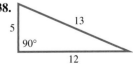

39. Write a definition of *isosceles right triangle*.

40. Explain why the sum of the lengths of any two sides of a triangle must be greater than the length of the third side.

41. Can a triangle be both right and obtuse? Explain.

42. In the classic 1939 movie *The Wizard of Oz*, the Scarecrow, upon getting a brain, says the following: "The sum of the square roots of any two sides of an isosceles triangle is equal to the square root of the remaining side." Give an example to show that his statement is incorrect.

Find the measure of each angle in triangle ABC.

43.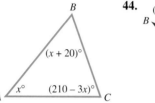

44.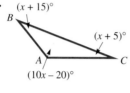

45. **46.**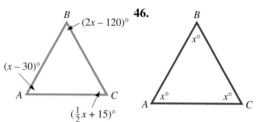

47. *Angle Measures* In triangle *ABC*, angles *A* and *B* have the same measure, while the measure of angle *C* is 24 degrees larger than the measure of each of *A* and *B*. What are the measures of the three angles?

48. *Angle Measures* In triangle *ABC*, the measure of angle *A* is 30 degrees more than the measure of angle *B*. The measure of angle *B* is the same as the measure of angle *C*. Find the measure of each angle.

In each triangle, find the measure of exterior angle BCD.

49.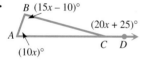

50.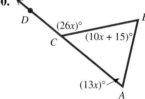

51.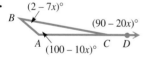

52.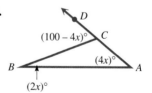

53. Using the points, segments, and lines in the figure, list all parts of the circle.
  (a) center
  (b) radii
  (c) diameters
  (d) chords
  (e) secants
  (f) tangents

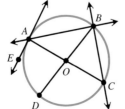

54. Refer to angles 1, 2, and 6 in the figure. Prove that the sum of the measures of angles 1 and 2 is equal to the measure of angle 6.

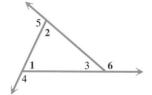

55. Go through the following argument provided by Richard Crouse in a letter to the editor of *Mathematics Teacher* in the February 1988 issue.
  (a) Place the eraser end of a pencil on vertex *A* of the triangle and let the pencil coincide with side *AC* of the triangle.
  (b) With the eraser fixed at *A*, rotate the pencil counterclockwise until it coincides with side *AB*.
  (c) With the pencil fixed at point *B*, rotate the eraser end counterclockwise until the pencil coincides with side *BC*.
  (d) With the eraser fixed at point *C*, rotate the point end of the pencil counterclockwise until the pencil coincides with side *AC*.
  (e) Notice that the pencil is now pointing in the opposite direction. What concept from this section does this exercise reinforce?

EXTENSION

Geometric Constructions

Geometria Renaissance sculptor Antonio de Pollaiolo (1433–1498) depicted Geometry in this bronze relief, one of ten reliefs on the sloping base of the bronze tomb of Sixtus IV, pope from 1471 to 1484. The important feature in this photo is the inclusion of compasses, one of two instruments allowed in classical geometry for constructions.

The Greeks did not study algebra as we do. To them geometry was the highest expression of mathematical science; their geometry was an abstract subject. Any practical application resulting from their work was nice but held no great importance. To the Greeks, a geometrical construction also needed abstract beauty. A construction could not be polluted with such practical instruments as a ruler. The Greeks permitted only two tools in geometrical construction: compasses for drawing circles and arcs of circles, and a straightedge for drawing straight line segments. The straightedge, unlike a ruler, could have no marks on it. It was not permitted to line up points by eye.

Here are four basic constructions. Their justifications are based on the *congruence properties* of Section 9.4.

Construction 1 Construct the perpendicular bisector of a given line segment.

Let the segment have endpoints *A* and *B*. Adjust the compasses for any radius greater than half the length of *AB*. Place the point of the compasses at *A* and draw an arc, then draw another arc of the same size at *B*. The line drawn through the points of intersection of these two arcs is the desired perpendicular bisector. See Figure 23.

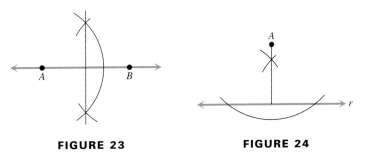

FIGURE 23   **FIGURE 24**

Construction 2 Construct a perpendicular from a point off a line to the line.

  (1) Let *A* be the point, *r* the line. Place the point of the compasses at *A* and draw an arc, cutting *r* in two points.
  (2) Swing arcs of equal radius from each of the two points on *r* which were constructed in (1). The line drawn through the intersection of the two arcs and point *A* is perpendicular to *r*. See Figure 24.

(continued)

Euclidean tools, the compasses and unmarked straightedge, proved to be sufficient for Greek geometers to accomplish a great number of geometric constructions. Basic constructions such as copying an angle, constructing the perpendicular bisector of a segment, and bisecting an angle are easily performed and verified.

There were, however, three constructions that the Greeks were not able to accomplish with these tools. Now known as the *three famous problems of antiquity,* they are:

1. To trisect an arbitrary angle;
2. To construct the length of the edge of a cube having twice the volume of a given cube;
3. To construct a square having the same area as that of a given circle.

In the nineteenth century it was learned that these constructions are, in fact, impossible to accomplish with Euclidean tools. Over the years other methods have been devised to accomplish them. For example, trisecting an arbitrary angle can be accomplished if one allows the luxury of marking on the straightedge! But this violates the rules followed by the Greeks.

Construction 3 Construct a perpendicular to a line at some given point on the line.

(1) Let r be the line and A the point. Using any convenient radius on the compasses, place the compass point at A and swing arcs which intersect r, as in Figure 25.

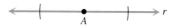

FIGURE 25

(2) Increase the radius of the compasses, place the point of the compasses on the points obtained in (1) and draw arcs. A line through A and the intersection of the two arcs is perpendicular to r. See Figure 26.

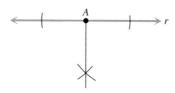

FIGURE 26

Construction 4 Copy an angle.

(1) In order to copy an angle ABC on line r, place the point of the compasses at B and draw an arc. Then place the point of the compasses on r' at some point P and draw the same arc, as in Figure 27.

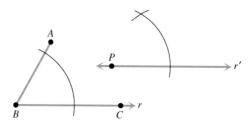

FIGURE 27

(2) Measure, with your compasses, the distance between the points where the arc intersects the angle, and transfer this distance, as shown in Figure 28. Use a straightedge to join P to the point of intersection. The angle is now copied.

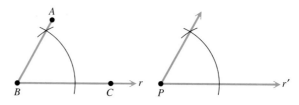

FIGURE 28

There are other basic constructions that can be found in books on plane geometry.

EXTENSION EXERCISES

Use Construction 1 to construct the perpendicular bisector of segment PQ.

1. P———Q

2. P (top), Q (bottom) on vertical line

Use Construction 2 to construct a perpendicular from P to the line r.

3. Line r (vertical), point P to the right

4. Line r (horizontal), point P below

Use Construction 3 to construct a perpendicular to the line r at P.

5. Line r (horizontal) with point P on it

6. Line r (vertical) with point P on it

Use Construction 4 to copy the given angle.

7. Angle at vertex A

8. Angle at vertex P

9. Construct a 45° angle.

10. It is impossible to trisect the general angle using only Euclidean tools. Investigate this fact, and write a short report on it. Include in your report information on the construction tool called a *tomahawk*.

9.3 Perimeter, Area, and Circumference

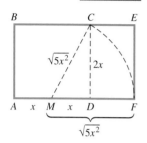

To construct a **golden rectangle**, one in which the ratio of the length to the width is equal to the ratio of the length plus the width to the length, begin with a square *ABCD*. With the point of the compasses at *M*, the midpoint of $\overrightarrow{AD}$, swing an arc of radius *MC* to intersect the extension of $\overrightarrow{AD}$ at *F*. Construct a perpendicular at *F*, and have it intersect the extension of $\overrightarrow{BC}$ at *E*. Then *ABEF* is a golden rectangle with ratio $(1 + \sqrt{5})/2$. (See Section 5.4 for more on the golden ratio.)

To verify this construction, let $AM = x$, so that $AD = 2x$. Then, by the Pythagorean theorem,

$MC = \sqrt{x^2 + (2x)^2}$
$= \sqrt{x^2 + 4x^2} = \sqrt{5x^2}$.

Since $\widehat{CF}$ is an arc of the circle with radius *MC*, $MF = MC = \sqrt{5x^2}$. Then the ratio of length *AF* to width *EF* is

$\dfrac{AF}{EF} = \dfrac{x + \sqrt{5x^2}}{2x}$
$= \dfrac{x + x\sqrt{5}}{2x}$
$= \dfrac{x(1 + \sqrt{5})}{2x}$
$= \dfrac{1 + \sqrt{5}}{2}$.

Perimeter and Area In dealing with plane figures, we are often required to find the "distance around" the figure, or the amount of surface covered by the figure. These ideas are expressed as *perimeter* and *area*.

Perimeter and Area

The **perimeter** of a plane figure composed of line segments is the sum of the measures of the line segments. The **area** of a plane figure is the measure of the surface covered by the figure. Perimeter is measured in *linear units*, while area is measured in *square units*.

The simplest polygon is a triangle. If a triangle has sides of lengths *a*, *b*, and *c*, then to find its perimeter we simply find the sum of *a*, *b*, and *c*. This can be expressed as a formula, as shown below.

Perimeter of a Triangle

The perimeter *P* of a triangle with sides of lengths *a*, *b*, and *c* is given by the formula

$$P = a + b + c.$$

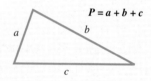

Since a rectangle is made up of two pairs of sides with each side in each pair equal in length, the formula for the perimeter of a rectangle may be stated as follows.

Perimeter of a Rectangle

The perimeter *P* of a rectangle with length ℓ and width *w* is given by the formula

$$P = 2\ell + 2w,$$

or equivalently,

$$P = 2(\ell + w).$$

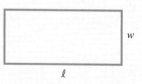

EXAMPLE 1 A plot of land is in the shape of a rectangle. If it has length 50 feet and width 26 feet, how much fencing would be needed to completely enclose the plot?

Since we must find the distance around the plot of land, the formula for the perimeter of a rectangle is needed. Substitute 50 for ℓ and 26 for *w* to find *P*.

$P = 2\ell + 2w$
$P = 2(50) + 2(26)$ $\ell = 50, w = 26$
$P = 100 + 52$
$P = 152$

The perimeter is 152 feet, so 152 feet of fencing is required.

9.3 Perimeter, Area, and Circumference

A square is a rectangle with four sides of equal length. The formula for the perimeter of a square is a special case of the formula for the perimeter of a rectangle.

Perimeter of a Square

The perimeter P of a square with all sides of length s is given by the formula

$$P = 4s.$$

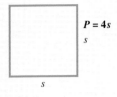

EXAMPLE 2 A square has perimeter 54 inches. What is the length of each side?

Use the formula $P = 4s$ with $P = 54$, and solve for s.

$P = 4s$
$54 = 4s$ $P = 54$
$s = 13.5$ Divide by 4.

Each side has a measure of 13.5 inches.

Problem Solving

The six-step method of solving an applied problem using algebra can be used to solve problems involving geometric figures.

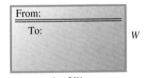

FIGURE 29

EXAMPLE 3 The length of a rectangular-shaped label is 1 centimeter more than twice the width. The perimeter is 110 centimeters. Find the length and the width.

Step 1: **Read the problem.** We must find the length and the width.

Step 2: **Assign a variable.** Let W represent the width. Then $1 + 2W$ can represent the length, since the length is 1 centimeter more than twice the width. Figure 29 shows a diagram of the label.

Step 3: **Write an equation.** In the formula $P = 2\ell + 2w$, replace w with W, ℓ with $1 + 2W$, and P with 110, since the perimeter is 110 centimeters.

$$110 = 2(1 + 2W) + 2W$$

Step 4: **Solve the equation.**

$110 = 2 + 4W + 2W$ Distributive property
$110 = 2 + 6W$ Combine terms.
$108 = 6W$ Subtract 2.
$18 = W$ Divide by 6.

Step 5: **State the answer.** Since $W = 18$, the width is 18 centimeters and the length is $1 + 2W = 1 + 2(18) = 37$ centimeters.

Step 6: **Check.** Because 37 is 1 more than twice 18, and because the perimeter $2(37) + 2(18) = 110$ centimeters, the answers are correct.

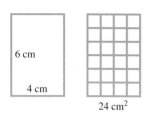

FIGURE 30

Defining the **area** of a figure requires a basic *unit of area*. One that is commonly used is the *square centimeter*, abbreviated cm². One square centimeter, or 1 cm², is the area of a square one centimeter on a side. In place of 1 cm², the basic unit of area could have been 1 in.², 1 ft², 1 m², or any appropriate unit.

As an example, we calculate the area of the rectangle shown in Figure 30. Using the basic 1 cm² unit, Figure 30 shows that 4 squares, each 1 cm on a side, can be laid off horizontally while 6 such squares can be laid off vertically. A total of $24 = 4 \cdot 6$ of the small squares are needed to cover the large rectangle. Thus, the area of the large rectangle is 24 cm².

We can generalize the above illustration to obtain a formula for the area of a rectangle.

Area of a Rectangle

The area A of a rectangle with length ℓ and width w is given by the formula

$$A = \ell w.$$

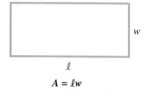

The formula for the area of a rectangle $A = \ell w$ can be used to find formulas for the areas of other figures. For example, if the letter s represents the equal lengths of the sides of the square, then $A = s \cdot s = s^2$.

Area of a Square

The area A of a square with all sides of length s is given by the formula

$$A = s^2.$$

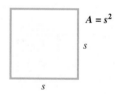

EXAMPLE 4 Figure 31 shows the floor plan of a building, made up of various rectangles. If each length given is in meters, how many square meters of carpet would be required to carpet the building?

The dashed lines in the figure break up the floor area into rectangles. The areas of the various rectangles that result are

$$10 \cdot 12 = 120 \text{ m}^2, \quad 3 \cdot 10 = 30 \text{ m}^2, \quad 3 \cdot 7 = 21 \text{ m}^2, \quad 15 \cdot 25 = 375 \text{ m}^2.$$

The total area is $120 + 30 + 21 + 375 = 546$ m², so 546 square meters of carpet would be needed.

FIGURE 31

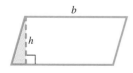

FIGURE 32

As mentioned earlier in this chapter, a **parallelogram** is a four-sided figure having both pairs of opposite sides parallel. Since a parallelogram need not be a rectangle, the formula for the area of a rectangle cannot be used directly for a parallelogram. However, this formula can be used indirectly, as shown in Figure 32. Cut off the triangle in color, and attach it at the right. The resulting figure is a rectangle having the same area as the original parallelogram.

9.3 Perimeter, Area, and Circumference 515

The *height* of the parallelogram is the perpendicular distance between two of the parallel sides and is denoted by *h* in the figure. The width of the rectangle equals the height of the parallelogram, and the length of the rectangle is the base *b* of the parallelogram, so

$$A = \text{length} \cdot \text{width} \quad \text{becomes} \quad A = \text{base} \cdot \text{height}.$$

Area of a Parallelogram

The area *A* of a parallelogram with height *h* and base *b* is given by the formula

$$A = bh.$$

(*Note:* *h* is not the length of a side.)

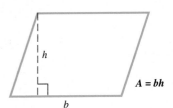

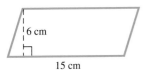

FIGURE 33

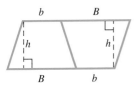

FIGURE 34

EXAMPLE 5 Find the area of the parallelogram in Figure 33.

Here the base has a length of 15 centimeters, while the height is 6 centimeters. Thus, $b = 15$ and $h = 6$.

$$A = bh = 15 \cdot 6 = 90$$

The area is 90 cm².

Figure 34 shows how we can find a formula for the area of a trapezoid. Notice that the figure as a whole is a parallelogram. It is made up of two trapezoids, each of which has height *h*, shorter base *b*, and longer base *B*. The area of the parallelogram is found by multiplying the height *h* by the base of the parallelogram, $b + B$, that is, $h(b + B)$. Since the area of the parallelogram is twice the area of each trapezoid, the area of each trapezoid is *half* the area of the parallelogram.

Area of a Trapezoid

The area *A* of a trapezoid with parallel bases *b* and *B* and height *h* is given by the formula

$$A = \frac{1}{2}h(b + B).$$

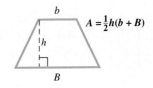

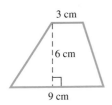

FIGURE 35

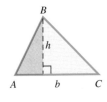

FIGURE 36

EXAMPLE 6 Find the area of the trapezoid in Figure 35.

Here $h = 6$, $b = 3$, and $B = 9$. Thus,

$$A = \frac{1}{2}h(B + b) = \frac{1}{2}(6)(9 + 3) = \frac{1}{2}(6)(12) = 36.$$

The area of the trapezoid is 36 cm².

The formula for the area of a triangle can be found from the formula for the area of a parallelogram. In Figure 36 the triangle whose vertices are at *A*, *B*, and *C* has been broken into two parts, one shown in color and one shown in gray. Repeating the part shown in color and the part in gray gives a parallelogram. The area of this parallelogram is $A = \text{base} \cdot \text{height}$, or $A = bh$. However, the parallelogram has *twice* the area of the triangle; in other words, the area of the triangle is *half* the area of the parallelogram.

In each case, $A = \frac{1}{2}bh$.

FIGURE 37

Area of a Triangle

The area A of a triangle with height h and base b is given by the formula

$$A = \frac{1}{2}bh.$$

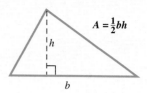

When applying the formula for the area of a triangle, remember that the height is the perpendicular distance between a vertex and the opposite side (or the extension of that side). See Figure 37.

FOR FURTHER THOUGHT

To use the formula for the area of a triangle, $A = (1/2)bh$, we must know the height of one of the sides of the triangle. Suppose that we know only the lengths of the three sides. Is there a way to determine the area from only this information?

The answer is yes, and it leads us to the formula known as **Heron's formula**. Heron of Alexandria lived during the second half of the first century A.D., and although the formula is named after him, there is evidence that it was known to Archimedes several centuries earlier.

Heron's Area Formula

Let a, b, and c be lengths of the sides of any triangle. Let $s = (1/2)(a + b + c)$ represent the semiperimeter. Then the area A of the triangle is given by the formula

$$A = \sqrt{s(s-a)(s-b)(s-c)}.$$

For example, if the sides of a triangle measure 6, 8, and 12 feet, we have $s = (1/2)(6 + 8 + 12) = 13$, and so the area of the triangle is A, where

$$A = \sqrt{13(13-6)(13-8)(13-12)}$$
$$= \sqrt{455} \approx 21.33 \text{ square feet.}$$

For Group Discussion

1. If your class is held in a rectangular-shaped room, measure the length and the width, and then multiply them to find the area. Now, measure a diagonal of the room, and use Heron's formula with the length, the width, and the diagonal to find the area of half the room. Double this result. Do your area calculations agree?

2. For a triangle to exist, the sum of the lengths of any two sides must exceed the length of the remaining side. Have half of the class try to calculate the area of a "triangle" with $a = 4$, $b = 8$, and $c = 12$, while the other half tries to calculate the area of a "triangle" with $a = 10$, $b = 20$, and $c = 34$. In both cases, use Heron's formula. Then, discuss the results, drawing diagrams on the chalkboard to support the results obtained.

3. The Vietnam Veterans' Memorial in Washington, D.C., is in the shape of an unenclosed isosceles triangle. The walls form a "V-shape," and each wall measures 246.75 feet. The distance between the ends of the walls is 438.14 feet. Use Heron's formula to find the area enclosed by the triangular shape.

Circumference of a Circle

The distance around a circle is called its **circumference** (rather than "perimeter"). To understand the formula for the circumference of a circle, use a piece of string to measure the distance around a circle. Measure its diameter and then divide the circumference by the diameter. This quotient is the same, no matter what the size of the circle. The result of this measurement is an approximation for the number π. We have

$$\pi = \frac{\text{circumference}}{\text{diameter}} = \frac{C}{d}, \quad \text{or alternatively,} \quad C = \pi d.$$

Since the diameter of a circle measures twice the radius, we have $d = 2r$. These relationships allow us to state the following formulas for the circumference of a circle.

Circumference of a Circle

The circumference C of a circle of diameter d is given by the formula

$$C = \pi d.$$

Also, the circumference C of a circle of radius r is given by the formula

$$C = 2\pi r.$$

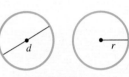

Recall that π is not a rational number. In this chapter we will use 3.14 as an approximation for π when one is required.

EXAMPLE 7

(a) A circle has diameter 12.6 cm. Find its circumference. Use $\pi \approx 3.14$.

Since $C = \pi d$,
$$C \approx 3.14(12.6) = 39.564$$

or 39.6 centimeters, rounded to the nearest tenth.

(b) The radius of a circle is 1.7 m. Find its circumference. Use $\pi \approx 3.14$.

$$C = 2\pi r \approx 2(3.14)(1.7) \approx 10.7$$

The circumference is approximately 10.7 meters.

Area of a Circle

The formula for the area of a circle can be justified as follows. Start with a circle as shown in Figure 38(a), divided into many equal pie-shaped pieces (**sectors**). Rearrange the pieces into an approximate rectangle as shown in Figure 38(b). The circle has circumference $2\pi r$, so the "length" of the approximate rectangle is one-half of the circumference, or $(1/2)(2\pi r) = \pi r$, while its "width" is r. The area of the approximate rectangle is length times width, or $(\pi r)r = \pi r^2$. By choosing smaller and smaller sectors, the figure becomes closer and closer to a rectangle, so its area becomes closer and closer to πr^2. This "limiting" procedure leads to the following formula.

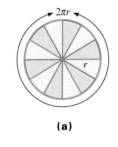

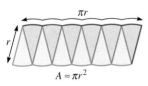

FIGURE 38

Area of a Circle

The area A of a circle with radius r is given by the formula

$$A = \pi r^2.$$

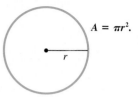

$A = \pi r^2$.

Problem Solving

The formula for the area of a circle can be used to determine the best value for the money the next time you purchase a pizza. The next example uses the idea of unit pricing.

EXAMPLE 8 Paw-Paw Johnny's delivers pizza. The price of an 8-inch diameter pepperoni pizza is $6.99, while the price of a 16-inch diameter pizza is $13.98. Which is the better buy?

To determine which pizza is the better value for the money, we must first find the area of each, and divide the price by the area to determine the price per square inch.

8-inch diameter pizza area = $\pi(4)^2 \approx 50.24$ in.^{2}
    ↑
    Radius is (1/2)(8) = 4 inches.

16-inch diameter pizza area = $\pi(8)^2 \approx 200.96$ in.^{2}
    ↑
    Radius is (1/2)(16) = 8 inches.

The price per square inch for the 8-inch pizza is $6.99/50.24 \approx 13.9$¢, while the price per square inch for the 16-inch pizza is $13.98/200.96 \approx 7.0$¢. Therefore, the 16-inch pizza is the better buy, since it costs approximately half as much per square inch.

9.3 EXERCISES

Fill in each blank with the correct response.

1. The perimeter of an equilateral triangle with side length equal to _____ inches is the same as the perimeter of a rectangle with length 10 inches and width 8 inches.

2. A square with area 16 square cm has perimeter _____ cm.

3. If the area of a certain triangle is 24 square inches, and the base measures 8 inches, then the height must measure _____ inches.

4. If the radius of a circle is doubled, then its area is multiplied by a factor of _____ .

5. The area of an equilateral triangle with side length 6 inches is _____ square inches.

6. If the length and the width of a rectangle are doubled, then the area is multiplied by a factor of _____ .

9.3 Perimeter, Area, and Circumference

Use the formulas of this section to find the area of each figure. In Exercises 19–22, use 3.14 as an approximation for π.

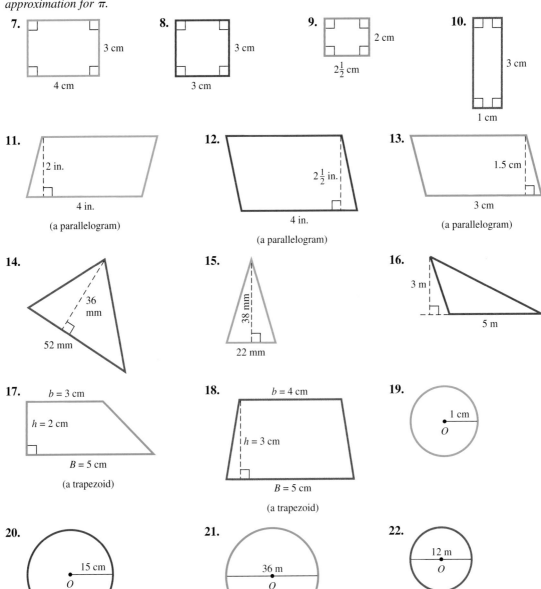

Solve each of the following problems.

23. *Window Side Length* A stained-glass window in a church is in the shape of a square. The perimeter of the square is 7 times the length of a side in meters, decreased by 12. Find the length of a side of the window.

24. *Dimensions of a Rectangle* A video rental establishment displayed a rectangular cardboard stand-up advertisement for the movie *Sweet Home Alabama*. The length was 20 in. more than the width, and the perimeter was 176 in. What were the dimensions of the rectangle?

25. *Dimensions of a Lot* A lot is in the shape of a triangle. One side is 100 ft longer than the shortest side, while the third side is 200 ft longer than the shortest side. The perimeter of the lot is 1200 ft. Find the lengths of the sides of the lot.

26. *Pennant Side Lengths* A wall pennant is in the shape of an isosceles triangle. Each of the two equal sides measures 18 in. more than the third side, and the perimeter of the triangle is 54 in. What are the lengths of the sides of the pennant?

27. *Radius of a Circular Foundation* The Peachtree Plaza Hotel in Atlanta is in the shape of a cylinder, with a circular foundation. The circumference of the foundation is 6 times the radius, increased by 12.88 ft. Find the radius of the circular foundation. (Use 3.14 as an approximation for π.)

28. *Radius of a Circle* If the radius of a certain circle is tripled, with 8.2 cm then added, the result is the circumference of the circle. Find the radius of the circle. (Use 3.14 as an approximation for π.)

29. *Area of Two Lots* The survey plat in the figure shows two lots that form a trapezoid. The measures of the parallel sides are 115.80 ft and 171.00 ft. The height of the trapezoid is 165.97 ft. Find the combined area of the two lots. Round your answer to the nearest hundredth of a square foot.

30. *Area of a Lot* Lot A in the figure is in the shape of a trapezoid. The parallel sides measure 26.84 ft and 82.05 ft. The height of the trapezoid is 165.97 ft. Find the area of Lot A. Round your answer to the nearest hundredth of a square foot.

31. *Perimeter or Area?* In order to purchase fencing to go around a rectangular yard, would you need to use perimeter or area to decide how much to buy?

32. *Perimeter or Area?* In order to purchase fertilizer for the lawn of a yard, would you need to use perimeter or area to decide how much to buy?

In the chart below, one of the values r (radius), d (diameter), C (circumference), or A (area) is given for a particular circle. Find the remaining three values. Leave π in your answers.

	r	d	C	A
33.	6 in.			
34.	9 in.			
35.		10 ft		
36.		40 ft		
37.			12π cm	
38.			18π cm	
39.				100π in.^{2}
40.				256π in.^{2}

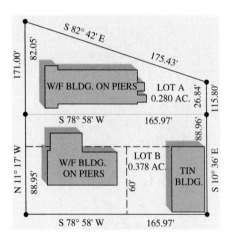

Property survey of lot in New Roads, Louisiana.

Each of the following figures has perimeter as indicated. (Figures are not necessarily to scale.) Find the value of x.

41. $P = 58$

42. $P = 42$

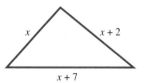

43. $P = 38$

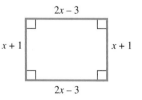

44. $P = 278$

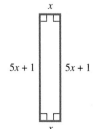

Each of the following figures has area as indicated. Find the value of x.

45. $A = 26.01$

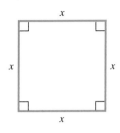

46. $A = 28$

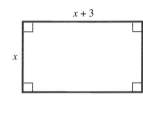

47. $A = 15$

48. $A = 30$

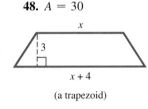

(a trapezoid)

Each of the following circles has circumference or area as indicated. Find the value of x. Use 3.14 *as an approximation for* π.

49. $C = 37.68$

50. $C = 54.95$

51. $A = 18.0864$

52. $A = 28.26$

 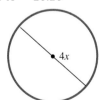

53. Work through the parts of this exercise in order, and use it to make a generalization concerning areas of rectangles.
  (a) Find the area of a rectangle 4 cm by 5 cm.
  (b) Find the area of a rectangle 8 cm by 10 cm.
  (c) Find the area of a rectangle 12 cm by 15 cm.
  (d) Find the area of a rectangle 16 cm by 20 cm.

(e) The rectangle in part (b) had sides twice as long as the sides of the rectangle in part (a). Divide the larger area by the smaller. By doubling the sides, the area increased _____ times.
(f) To get the rectangle in part (c), each side of the rectangle in part (a) was multiplied by _____. This made the larger area _____ times the size of the smaller area.
(g) To get the rectangle of part (d), each side of the rectangle of part (a) was multiplied by _____. This made the area increase to _____ times what it was originally.
(h) In general, if the length of each side of a rectangle is multiplied by n, the area is multiplied by _____.

Job Cost Use the results of Exercise 53 to solve each of the following.

54. A ceiling measuring 9 ft by 15 ft can be painted for $60. How much would it cost to paint a ceiling 18 ft by 30 ft?

55. Suppose carpet for a 10 ft by 12 ft room costs $200. Find the cost to carpet a room 20 ft by 24 ft.

56. A carpet cleaner charges $80 to do an area 31 ft by 31 ft. What would be the charge for an area 93 ft by 93 ft?

57. Use the logic of Exercise 53 to answer the following: If the radius of a circle is multiplied by n, then the area of the circle is multiplied by _____.

58. Use the logic of Exercise 53 to answer the following: If the height of a triangle is multiplied by n and the base length remains the same, then the area of the triangle is multiplied by _____.

Total Area as the Sum of Areas By considering total area as the sum of the areas of all of its parts, the area of a figure such as those in Exercises 59–62 can be determined. Find the total area of each figure. Use 3.14 *as an approximation for* π *in Exercises 61 and 62.*

59.

(a parallelogram and a triangle)

60.

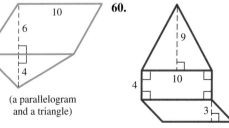

(a triangle, a rectangle, and a parallelogram)

61. **62.**

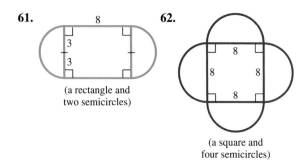

(a rectangle and two semicircles)

(a square and four semicircles)

Area of a Shaded Portion of a Plane Figure The shaded areas of the figures in Exercises 63–68 may be found by subtracting the area of the unshaded portion from the total area of the figure. Use this approach to find the area of the shaded portion. Use 3.14 as an approximation for π in Exercises 66–68, and round to the nearest hundredth.

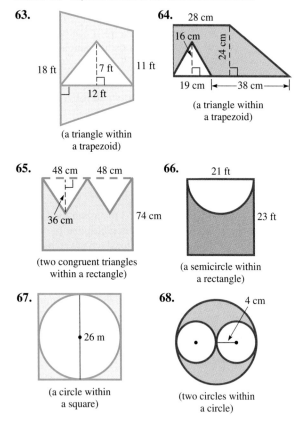

63. (a triangle within a trapezoid)

64. (a triangle within a trapezoid)

65. (two congruent triangles within a rectangle)

66. (a semicircle within a rectangle)

67. (a circle within a square)

68. (two circles within a circle)

Pizza Pricing The following exercises show prices actually charged by Maw-Maw Gigi's, a local pizzeria. In each case, the dimension is the diameter of the pizza. Find the best buy.

69. Cheese pizza: 10-in. pizza sells for $5.99, 12-in. pizza sells for $7.99, 14-in. pizza sells for $8.99.

70. Cheese pizza with two toppings: 10-in. pizza sells for $7.99, 12-in. pizza sells for $9.99, 14-in. pizza sells for $10.99.

71. All Feasts pizza: 10-in. pizza sells for $9.99, 12-in. pizza sells for $11.99, 14-in. pizza sells for $12.99.

72. Extravaganza pizza: 10-in. pizza sells for $11.99, 12-in. pizza sells for $13.99, 14-in. pizza sells for $14.99.

A polygon can be inscribed within a circle or circumscribed about a circle. In the figure, triangle ABC is inscribed within the circle, while square WXYZ is circumscribed about it. These ideas will be used in some of the remaining exercises in this section and later in this chapter.

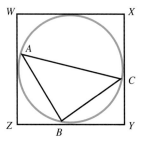

Exercises 73–80 require some ingenuity, but all may be solved using the concepts presented so far in this chapter.

73. *Diameter of a Circle* Given the circle with center O and rectangle $ABCO$, find the diameter of the circle.

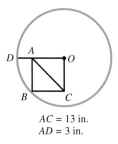

$AC = 13$ in.
$AD = 3$ in.

74. *Perimeter of a Triangle* What is the perimeter of $\triangle AEB$, if $AD = 20$ in., $DC = 30$ in., and $AC = 34$ in.?

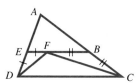

9.4 The Geometry of Triangles: Congruence, Similarity, and the Pythagorean Theorem   523

75. *Area of a Square* The area of square *PQRS* is 1250 square feet. *T*, *U*, *V*, and *W* are the midpoints of *PQ*, *QR*, *RS*, and *SP*, respectively. What is the area of square *TUVW*?

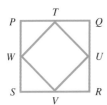

76. *Area of a Quadrilateral* The rectangle *ABCD* has length twice the width. If *P*, *Q*, *R*, and *S* are the midpoints of the sides, and the perimeter of *ABCD* is 96 in., what is the area of quadrilateral *PQRS*?

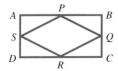

77. *Area of a Shaded Region* If *ABCD* is a square with each side measuring 36 in., what is the area of the shaded region?

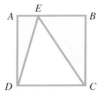

78. *Perimeter of a Polygon* Can the perimeter of the polygon shown be determined from the given information? If so, what is the perimeter?

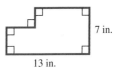

79. *Area of a Shaded Region* Express the area of the shaded region in terms of *r*, given that the circle is inscribed in the square.

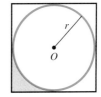

80. *Area of a Trapezoid* Find the area of trapezoid *ABCD*, given that the area of right triangle *ABE* is 30 in.^{2}.

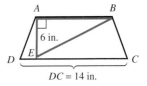

9.4 The Geometry of Triangles: Congruence, Similarity, and the Pythagorean Theorem

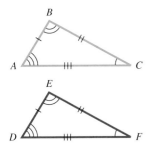

FIGURE 39

Congruent Triangles In this section we investigate special properties of triangles. Triangles that are both the same size and the same shape are called **congruent triangles.** Informally speaking, if two triangles are congruent, then it is possible to pick up one of them and place it on top of the other so that they coincide exactly. An everyday example of congruent triangles would be the triangular supports for a child's swing set, machine-produced with exactly the same dimensions each time.

Figure 39 illustrates two congruent triangles, △*ABC* and △*DEF*. The symbol ≅ denotes congruence, so △*ABC* ≅ △*DEF*. Notice how the angles and sides are marked to indicate which angles are congruent and which sides are congruent. (Using precise terminology, we refer to angles or sides as being *congruent,* while

Plimpton 322 Our knowledge of the mathematics of the Babylonians of Mesopotamia is based largely on archaeological discoveries of thousands of clay tablets. On the tablet labeled Plimpton 322, there are several columns of inscriptions that represent numbers. The far right column is simply one that serves to number the lines, but two other columns represent values of hypotenuses and legs of right triangles with integer-valued sides. Thus, it seems that while the famous theorem relating right-triangle side lengths is named for the Greek Pythagoras, the relationship was known more than 1000 years prior to the time of Pythagoras.

the *measures* of congruent angles or congruent sides are *equal*. We will often use the terms "equal angles" or "equal sides" to describe angles of equal measure or sides of equal measure.)

In geometry the following properties are used to prove that two triangles are congruent.

Congruence Properties

Side-Angle-Side (SAS) If two sides and the included angle of one triangle are equal, respectively, to two sides and the included angle of a second triangle, then the triangles are congruent.

Angle-Side-Angle (ASA) If two angles and the included side of one triangle are equal, respectively, to two angles and the included side of a second triangle, then the triangles are congruent.

Side-Side-Side (SSS) If three sides of one triangle are equal, respectively, to three sides of a second triangle, then the triangles are congruent.

Examples 1–3 show how to prove statements using these properties. We use a diagram with two columns, headed by STATEMENTS and REASONS.

STATEMENTS	REASONS

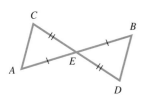

FIGURE 40

EXAMPLE 1 Refer to Figure 40.

Given: $CE = ED$
$AE = EB$

Prove: $\triangle ACE \cong \triangle BDE$

Proof:

STATEMENTS	REASONS
1. $CE = ED$	1. Given
2. $AE = EB$	2. Given
3. $\angle CEA = \angle DEB$	3. Vertical angles are equal.
4. $\triangle ACE \cong \triangle BDE$	4. SAS congruence property

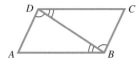

FIGURE 41

EXAMPLE 2 Refer to Figure 41.

Given: $\angle ADB = \angle CBD$
$\angle ABD = \angle CDB$

Prove: $\triangle ADB \cong \triangle CBD$

Proof:

STATEMENTS	REASONS
1. $\angle ADB = \angle CBD$	1. Given
2. $\angle ABD = \angle CDB$	2. Given
3. $DB = DB$	3. Reflexive property (a quantity is equal to itself)
4. $\triangle ADB \cong \triangle CBD$	4. ASA congruence property

9.4 The Geometry of Triangles: Congruence, Similarity, and the Pythagorean Theorem

EXAMPLE 3 Refer to Figure 42.

Given: $AD = CD$
$AB = CB$

Prove: $\triangle ABD \cong \triangle CBD$

Proof:

STATEMENTS	REASONS
1. $AD = CD$	1. Given
2. $AB = CB$	2. Given
3. $BD = BD$	3. Reflexive property
4. $\triangle ABD \cong \triangle CBD$	4. SSS congruence property

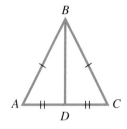

FIGURE 42

In Example 3, $\triangle ABC$ is an isosceles triangle. The results of that example allow us to make several important statements about an isosceles triangle. They are indicated symbolically in Figure 43 and stated in the following box.

FIGURE 43

Important Statements About Isosceles Triangles

If $\triangle ABC$ is an isosceles triangle with $AB = CB$, and if D is the midpoint of the base AC, then the following properties hold.

1. The base angles A and C are equal.
2. Angles ABD and CBD are equal.
3. Angles ADB and CDB are both right angles.

Similar Triangles Many of the key ideas of geometry depend on **similar triangles,** pairs of triangles that are exactly the same shape but not necessarily the same size. Figure 44 shows three pairs of similar triangles. (*Note:* The triangles do not need to be oriented in the same fashion in order to be similar.)

Suppose that a correspondence between two triangles ABC and DEF is set up as follows.

∡A corresponds to ∡D side AB corresponds to side DE
∡B corresponds to ∡E side BC corresponds to side EF
∡C corresponds to ∡F side AC corresponds to side DF

For triangle ABC to be similar to triangle DEF, the following conditions must hold.

1. Corresponding angles must have the same measure.
2. The ratios of the corresponding sides must be constant; that is, the corresponding sides are proportional.

By showing that either of these conditions holds in a pair of triangles, we may conclude that the triangles are similar.

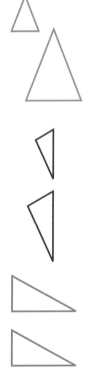

FIGURE 44

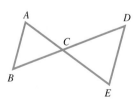

FIGURE 45

EXAMPLE 4 In Figure 45, $\overleftrightarrow{AB}$ is parallel to $\overleftrightarrow{ED}$. How can we verify that $\triangle ABC$ is similar to $\triangle EDC$?

Because $\overleftrightarrow{AB}$ is parallel to $\overleftrightarrow{ED}$, the transversal $\overleftrightarrow{BD}$ forms equal alternate interior angles ABC and EDC. Also, transversal $\overleftrightarrow{AE}$ forms equal alternate interior angles BAC and DEC. We know that ∡ACB = ∡ECD, because they are vertical angles. Because the corresponding angles have the same measures in triangles ABC and EDC, the triangles are similar.

Arizona's Navajo Bridge over the Marble Canyon of the Colorado River shows **congruent and similar triangles.** Corresponding triangles on the left and the right sides of the bridge are congruent, while the triangles are similar as they get smaller toward the center of the bridge.

Once we have shown that two angles of one triangle are equal to the two corresponding angles of a second triangle, it is not necessary to show the same for the third angle. Since, in any triangle, the sum of the angles equals 180°, we may conclude that the measures of the remaining angles *must* be equal. This leads to the following Angle-Angle similarity property.

Angle-Angle (AA) Similarity Property

If the measures of two angles of one triangle are equal to those of two corresponding angles of a second triangle, then the two triangles are similar.

EXAMPLE 5 In Figure 46, $\triangle EDF$ is similar to $\triangle CAB$. Find the unknown side lengths in $\triangle EDF$.

As mentioned above, similar triangles have corresponding sides in proportion. Use this fact to find the unknown sides in the smaller triangle. Side *DF* of the small triangle corresponds to side *AB* of the larger one, and sides *DE* and *AC* correspond. This leads to the proportion

$$\frac{8}{16} = \frac{DF}{24}.$$

Using a technique from algebra, set the two cross-products equal. (As an alternative method of solution, multiply both sides by 48, the least common multiple of 16 and 24.) Setting cross-products equal gives

$$8(24) = 16DF$$
$$192 = 16DF$$
$$12 = DF.$$

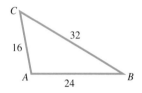

FIGURE 46

Side *DF* has length 12.

Side *EF* corresponds to side *CB*. This leads to another proportion.

$$\frac{8}{16} = \frac{EF}{32}$$
$$\frac{1}{2} = \frac{EF}{32} \qquad \frac{8}{16} = \frac{1}{2}$$
$$2EF = 32$$
$$EF = 16$$

Side *EF* has length 16.

FIGURE 47

EXAMPLE 6 Find the measures of the unknown parts of the similar triangles *STU* and *ZXY* in Figure 47.

Here angles *X* and *T* correspond, as do angles *Y* and *U*, and angles *Z* and *S*. Since angles *Z* and *S* correspond and since angle *S* is 52°, angle *Z* also must be 52°. The sum of the angles of any triangle is 180°. In the larger triangle $X = 71°$ and $Z = 52°$. To find *Y*, set up an equation and solve for *Y*.

$$X + Y + Z = 180$$
$$71 + Y + 52 = 180$$
$$123 + Y = 180$$
$$Y = 57$$

Angle Y is 57°. Since angles Y and U correspond, $U = 57°$ also.

Now find the unknown sides. Sides SU and ZY correspond, as do TS and XZ, and TU and XY, leading to the following proportions.

$$\frac{SU}{ZY} = \frac{TS}{XZ} \qquad \frac{XY}{TU} = \frac{ZY}{SU}$$
$$\frac{48}{144} = \frac{TS}{126} \qquad \frac{XY}{40} = \frac{144}{48}$$
$$\frac{1}{3} = \frac{TS}{126} \qquad \frac{XY}{40} = \frac{3}{1}$$
$$3TS = 126 \qquad XY = 120$$
$$TS = 42$$

Side TS has length 42, and side XY has length 120. ∎

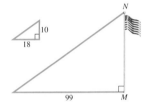

FIGURE 48

EXAMPLE 7 Lucie Wanersdorfer, the Lettsworth, LA postmaster, wants to measure the height of the office flagpole. She notices that at the instant when the shadow of the station is 18 feet long, the shadow of the flagpole is 99 feet long. The building is 10 feet high. What is the height of the flagpole?

Figure 48 shows the information given in the problem. The two triangles shown there are similar, so that corresponding sides are in proportion, with

$$\frac{MN}{10} = \frac{99}{18}$$
$$\frac{MN}{10} = \frac{11}{2}$$
$$2MN = 110$$
$$MN = 55.$$

The flagpole is 55 feet high. ∎

The Pythagorean Theorem We have used the Pythagorean theorem earlier in this book, and because of its importance in mathematics, we will investigate it further in this section on the geometry of triangles. Recall that in a right triangle, the side opposite the right angle (and consequently, the longest side) is called the **hypotenuse**. The other two sides, which are perpendicular, are called the **legs**.

The vertical distance from the base of this broken tree to the point of the break is 55 inches. The length of the broken part is 144 inches. How far along the ground is it from the base of the tree to the point where the broken part touches the ground?

Pythagorean Theorem

If the two legs of a right triangle have lengths a and b, and the hypotenuse has length c, then

$$a^2 + b^2 = c^2.$$

That is, the sum of the squares of the lengths of the legs is equal to the square of the hypotenuse.

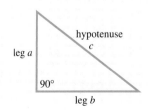

FOR FURTHER THOUGHT

Proving the Pythagorean Theorem

The Pythagorean theorem has probably been proved in more different ways than any theorem in mathematics. A book titled *The Pythagorean Proposition,* by Elisha Scott Loomis, was first published in 1927. It contained more than 250 different proofs of the theorem. It was reissued in 1968 by the National Council of Teachers of Mathematics as the first title in a series of "Classics in Mathematics Education."

One of the most popular proofs of the theorem is outlined in the group discussion exercise that follows.

For Group Discussion

Copy the accompanying figure on the board in order to prove the Pythagorean theorem. Keep in mind that the area of the large square must always be the same, no matter how it is determined. It is made up of four right triangles and a smaller square.

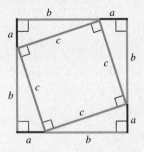

(a) The length of a side of the large square is _____ , so its area is (_____)² or _____ .

(b) The area of the large square can also be found by obtaining the sum of the areas of the four right triangles and the smaller square. The area of each right triangle is _____ , so the sum of the areas of the four right triangles is _____ . The area of the smaller square is _____ .

(c) The sum of the areas of the four right triangles and the smaller square is _____ .

(d) Since the areas in (a) and (c) represent the area of the same figure, the expressions there must be equal. Setting them equal to each other we obtain _____ = _____ .

(e) Subtract $2ab$ from each side of the equation in (d) to obtain the desired result: _____ = _____ .

FIGURE 49

Pythagoras did not actually discover the theorem that was named after him, although legend tells that he sacrificed 100 oxen to the gods in gratitude for the discovery. There is evidence that the Babylonians knew the concept quite well. The first proof, however, may have come from Pythagoras.

Figure 49 illustrates the theorem in a simple way, by using a sort of tile pattern. In the figure, the side of the square along the hypotenuse measures 5 units, while those along the legs measure 3 and 4 units. If we let $a = 3$, $b = 4$, and $c = 5$, we see that the equation of the Pythagorean theorem is satisfied.

$$a^2 + b^2 = c^2$$
$$3^2 + 4^2 = 5^2$$
$$9 + 16 = 25$$
$$25 = 25$$

The natural numbers 3, 4, 5 form a **Pythagorean triple** since they satisfy the equation of the Pythagorean theorem. There are infinitely many such triples.

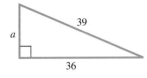

FIGURE 50

EXAMPLE 8 Find the length a in the right triangle shown in Figure 50.

Let $b = 36$ and $c = 39$. Substituting these values into the equation $a^2 + b^2 = c^2$ allows us to solve for a.

$$a^2 + b^2 = c^2$$
$$a^2 + 36^2 = 39^2$$
$$a^2 + 1296 = 1521$$
$$a^2 = 225 \quad \text{Subtract 1296 from both sides.}$$
$$a = 15 \quad \text{Choose the positive square root, since } a > 0.$$

Verify that 15, 36, 39 is a Pythagorean triple as a check.

Problem Solving

The Cairo Mathematical Papyrus is an Egyptian document that dates back to about 300 B.C. It was discovered in 1938 and examined in 1962. It contains forty problems, and nine of these deal with the Pythagorean relationship. One of these problems is solved in Example 9.

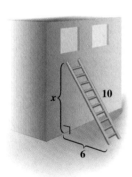

FIGURE 51

EXAMPLE 9 A ladder of length 10 cubits has its foot 6 cubits from a wall. To what height does the ladder reach?

As suggested by Figure 51, the ladder forms the hypotenuse of a right triangle, and the ground and wall form the legs. Let x represent the distance from the base of the wall to the top of the ladder. Then, by the Pythagorean theorem,

$$x^2 + 6^2 = 10^2$$
$$x^2 + 36 = 100$$
$$x^2 = 64 \quad \text{Subtract 36.}$$
$$x = 8. \quad \text{Choose the positive square root of 64, since } x \text{ represents a length.}$$

The ladder reaches a height of 8 cubits.

The statement of the Pythagorean theorem is an *if . . . then* statement. If the antecedent (the statement following the word "if") and the consequent (the statement following the word "then") are interchanged, the new statement is called the *converse* of the original one. Although the converse of a true statement may not be true, the *converse* of the Pythagorean theorem *is* also a true statement and can be used to determine if a triangle is a right triangle, given the lengths of the three sides.

Converse of the Pythagorean Theorem

If a triangle has sides of lengths a, b, and c, where c is the length of the longest side, and if $a^2 + b^2 = c^2$, then the triangle is a right triangle.

The 10 Mathematical Formulas That Changed the Face of the Earth Nicaragua issued a series of ten stamps commemorating mathematical formulas, including this one featuring the Pythagorean theorem, $a^2 + b^2 = c^2$. Notice how the compasses dominate the figure.

Problem Solving

If a carpenter is building a floor for a rectangular room, it is essential that the corners of the floor represent right angles; otherwise, problems will occur when the walls are constructed, when flooring is laid, and so on. To check that the floor is "squared off," the carpenter can use the converse of the Pythagorean theorem.

EXAMPLE 10 A carpenter has been contracted to add an 8-foot-by-12-foot laundry room onto an existing house. After the floor has been built, he finds that the length of the diagonal of the floor is 14 feet, 8 inches. Is the floor "squared off" properly?

Since 14 feet, 8 inches $= 14\frac{2}{3}$ feet, he must check to see whether the following statement is true.

$$8^2 + 12^2 = \left(14\frac{2}{3}\right)^2 \quad ?$$

$$8^2 + 12^2 = \left(\frac{44}{3}\right)^2 \quad ?$$

$$208 = \frac{1936}{9} \quad ?$$

$$208 \neq 215\frac{1}{9} \quad \text{False}$$

He has a real problem now, since his diagonal, which measures 14 feet, 8 inches, should actually measure $\sqrt{208} \approx 14.4 \approx 14$ feet, 5 inches. He must correct his error to avoid major problems later.

9.4 EXERCISES

In Exercises 1–6, provide a STATEMENTS/REASONS proof similar to the ones in Examples 1–3.

1. Given: $AC = BD$; $AD = BC$
   Prove: $\triangle ABD \cong \triangle BAC$

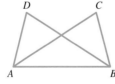

2. Given: $AC = BC$; $\angle ACD = \angle BCD$
   Prove: $\triangle ADC \cong \triangle BDC$

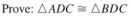

3. Given: $\overleftrightarrow{DB}$ is perpendicular to $\overleftrightarrow{AC}$; $AB = BC$
   Prove: $\triangle ABD \cong \triangle CBD$

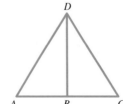

4. Given: $BC = BA$; $\angle 1 = \angle 2$
   Prove: $\triangle DBC \cong \triangle DBA$

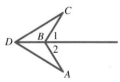

9.4 The Geometry of Triangles: Congruence, Similarity, and the Pythagorean Theorem

5. Given: ∡BAC = ∡DAC; ∡BCA = ∡DCA
   Prove: △ABC ≅ △ADC.

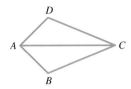

6. Given: BO = OE; $\overleftrightarrow{OB}$ is perpendicular to $\overleftrightarrow{AC}$; $\overleftrightarrow{OE}$ is perpendicular to $\overleftrightarrow{DF}$
   Prove: △AOB ≅ △FOE.

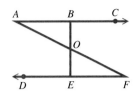

Exercises 7–10 refer to the given figure, an isosceles triangle with AB = BC.

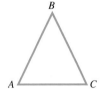

7. If ∡B measures 46°, then ∡A measures _____ and ∡C measures _____ .
8. If ∡C measures 52°, what is the measure of ∡B?
9. If BC = 12 in., and the perimeter of △ABC is 30 in., what is the length AC?
10. If the perimeter of △ABC = 40 in., and AC = 10 in., what is the length AB?
11. Explain why all equilateral triangles must be similar.
12. Explain why two congruent triangles must be similar, but two similar triangles might not be congruent.

Name the corresponding angles and the corresponding sides for each of the following pairs of similar triangles.

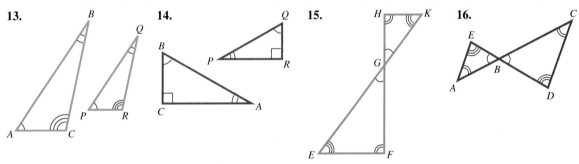

Find all unknown angle measures in each pair of similar triangles.

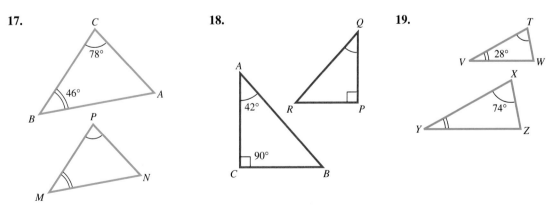

20. 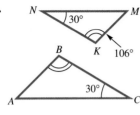 **21.** 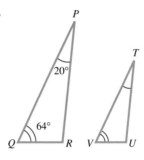 **22.**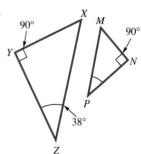

Find the unknown side lengths in each pair of similar triangles.

23. 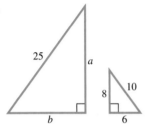 **24.** 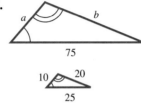 **25.**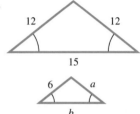

26. 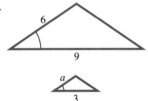 **27.** 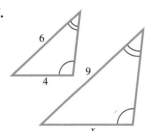 **28.**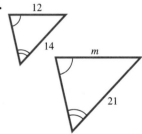

In each diagram, there are two similar triangles. Find the unknown measurement in each. (Hint: In the figure for Exercise 29, the side of length 100 in the smaller triangle corresponds to a side of length 100 + 120 = 220 in the larger triangle.)

29. 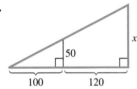 **30.**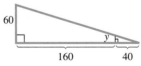

31. 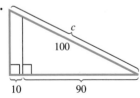 **32.**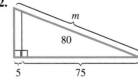

Solve the following problems.

33. *Height of a Tree* A tree casts a shadow 45 m long. At the same time, the shadow cast by a vertical 2-m stick is 3 m long. Find the height of the tree.

34. *Height of a Tower* A forest fire lookout tower casts a shadow 180 ft long at the same time that the shadow of a 9-ft truck is 15 ft long. Find the height of the tower.

35. *Lengths of Sides of a Photograph* On a photograph of a triangular piece of land, the lengths of the three sides are 4 cm, 5 cm, and 7 cm, respectively. The shortest side of the actual piece of land is 400 m long. Find the lengths of the other two sides.

36. *Height of a Lighthouse Keeper* The Santa Cruz lighthouse is 14 m tall and casts a shadow 28 m long at 7 P.M. At the same time, the shadow of the lighthouse keeper is 3.5 m long. How tall is she?

37. *Height of a Building* A house is 15 ft tall. Its shadow is 40 ft long at the same time the shadow of a nearby building is 300 ft long. Find the height of the building.

38. *Distances Between Cities* By drawing lines on a map, a triangle can be formed by the cities of Phoenix, Tucson, and Yuma. On the map, the distance between Phoenix and Tucson is 8 cm, the distance between Phoenix and Yuma is 12 cm, and the distance between Tucson and Yuma is 17 cm. The actual straight-line distance from Phoenix to Yuma is 230 km. Find the distances between the other pairs of cities.

39. *Dimensions on Mount Rushmore* Assume that Lincoln was $6\frac{1}{3}$ ft tall and his head $\frac{3}{4}$ ft long. Knowing that the carved head of Lincoln at Mount Rushmore is 60 ft tall, find out how tall his entire body would be if it were carved into the mountain.

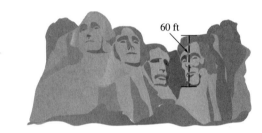

40. *Height of the World's Tallest Human* Robert Wadlow was the tallest human being ever recorded. When a 6-ft stick cast a shadow 24 in., Robert would cast a shadow 35.7 in. How tall was he?

In Exercises 41–48, a and b represent the two legs of a right triangle, while c represents the hypotenuse. Find the lengths of the unknown sides.

41.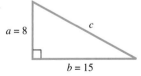

42.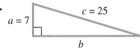

43.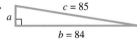

44. $a = 24$ cm; $c = 25$ cm

45. $a = 14$ m; $b = 48$ m

46. $a = 28$ km; $c = 100$ km

47. $b = 21$ in.; $c = 29$ in.

48. $b = 120$ ft; $c = 169$ ft

49. Refer to Exercise 42 in Section 9.2. Correct the Scarecrow's statement.

50. Show that if $a^2 + b^2 = c^2$, then it is not necessarily true that $a + b = c$.

There are various formulas that will generate Pythagorean triples. For example, if we choose positive integers r and s, with $r > s$, then the set of equations

$$a = r^2 - s^2, \quad b = 2rs, \quad c = r^2 + s^2$$

generates a Pythagorean triple (a, b, c). Use the values of r and s given in each of Exercises 51–56 to generate a Pythagorean triple using this method.

51. $r = 2, s = 1$ **52.** $r = 3, s = 2$
53. $r = 4, s = 3$ **54.** $r = 3, s = 1$
55. $r = 4, s = 2$ **56.** $r = 4, s = 1$

57. Show that the formula given for Exercises 51–56 actually satisfies $a^2 + b^2 = c^2$.

58. It can be shown that if $(x, x + 1, y)$ is a Pythagorean triple, then so is

$$(3x + 2y + 1, \quad 3x + 2y + 2, \quad 4x + 3y + 2).$$

Use this idea to find three more Pythagorean triples, starting with 3, 4, 5. (*Hint:* Here, $x = 3$ and $y = 5$.)

If m is an odd positive integer greater than 1, then

$$\left(m, \frac{m^2 - 1}{2}, \frac{m^2 + 1}{2} \right)$$

is a Pythagorean triple. Use this to find the Pythagorean triple generated by each value of m in Exercises 59–62.

59. $m = 3$ **60.** $m = 5$
61. $m = 7$ **62.** $m = 9$

63. Show that the expressions in the directions for Exercises 59–62 actually satisfy $a^2 + b^2 = c^2$.

64. Show why $(6, 8, 10)$ is the only Pythagorean triple consisting of consecutive even numbers.

For any integer n greater than 1,

$$(2n, n^2 - 1, n^2 + 1)$$

is a Pythagorean triple. Use this pattern to find the Pythagorean triple generated by each value of n in Exercises 65–68.

65. $n = 2$ **66.** $n = 3$
67. $n = 4$ **68.** $n = 5$

69. Show that the expressions in the directions for Exercises 65–68 actually satisfy $a^2 + b^2 = c^2$.

70. Can an isosceles right triangle have sides with integer lengths? Why or why not?

Solve each problem. (You may wish to review quadratic equations from algebra.)

71. *Side Length of a Triangle* If the hypotenuse of a right triangle is 1 m more than the longer leg, and the shorter leg is 7 m, find the length of the longer leg.

72. *Side Lengths of a Triangle* The hypotenuse of a right triangle is 1 cm more than twice the shorter leg, and the longer leg is 9 cm less than three times the shorter leg. Find the lengths of the three sides of the triangle.

73. *Height of a Tree* At a point on the ground 30 ft from the base of a tree, the distance to the top of the tree is 2 ft more than twice the height of the tree. Find the height of the tree.

74. *Dimensions of a Rectangle* The length of a rectangle is 2 in. less than twice the width. The diagonal is 5 in. Find the length and width of the rectangle.

75. *Height of a Break in Bamboo* (Problem of the broken bamboo, from the Chinese work *Arithmetic in Nine Sections* (1261)) There is a bamboo 10 ft high, the upper end of which, being broken, reaches the ground 3 ft from the stem. Find the height of the break.

76. *Depth of a Pond* (Adapted from *Arithmetic in Nine Sections*) There grows in the middle of a circular pond 10 ft in diameter a reed which projects 1 ft out of the water. When it is drawn down it just reaches the edge of the pond. How deep is the water?

Squaring Off a Floor Under Construction Imagine that you are a carpenter building the floor of a rectangular room. What must the diagonal of the room measure if your floor is to be squared off properly, given the dimensions in Exercises 77–80? Give your answer to the nearest inch.

77. 12 ft by 15 ft **78.** 14 ft by 20 ft
79. 16 ft by 24 ft **80.** 20 ft by 32 ft

81. *Garfield's Proof of the Pythagorean Theorem* James A. Garfield, the twentieth president of the United States, provided a proof of the Pythagorean theorem using the given figure. Supply the required information in each of parts (a) through (c) in order to follow his proof.
(a) Find the area of the trapezoid WXYZ using the formula for the area of a trapezoid.
(b) Find the area of each of the right triangles PWX, PZY, and PXY.
(c) Since the sum of the areas of the three right triangles must equal the area of the trapezoid, set the expression from part (a) equal to the sum of the three expressions from part (b). Simplify the equation as much as possible.

82. *Proof of the Pythagorean Theorem by Similar Triangles* In the figure, right triangles ABC, CBD, and ACD are similar. This may be used to prove the Pythagorean theorem. Fill in the blanks with the appropriate responses.

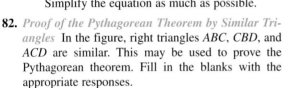

(a) By proportion, we have $c/b = $ _____ $/j$.
(b) By proportion, we also have $c/a = a/$ _____ .
(c) From part (a), $b^2 = $ _____ .
(d) From part (b), $a^2 = $ _____ .
(e) From the results of parts (c) and (d) and factoring, $a^2 + b^2 = c($ _____ $)$. But since _____ $= c$, it follows that _____ .

Exercises 83–90 require some ingenuity, but all can be solved using the concepts presented so far in this chapter.

83. *Area of a Quadrilateral* Find the area of quadrilateral ABCD, if angles A and C are right angles.

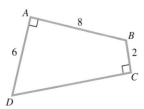

84. *Area of a Triangle* The perimeter of the isosceles triangle ABC (with $AB = BC$) is 128 in. The altitude BD is 48 in. What is the area of triangle ABC?

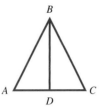

85. *Base Measure of an Isosceles Triangle* An isosceles triangle has a base of 24 and two sides of 13. What other base measure can an isosceles triangle with equal sides of 13 have and still have the same area as the given triangle?

86. *Value of a Measure in a Triangle* In right triangle ABC, if $AD = DB + 8$, what is the value of CD?

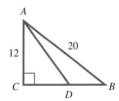

87. *Area of a Pentagon* In the figure, pentagon PQRST is formed by a square and an equilateral triangle such that $PQ = QR = RS = ST = PT$. The perimeter of the pentagon is 80. Find the area of the pentagon.

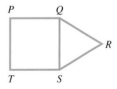

88. *Angle Measure in a Triangle* (A segment that bisects an angle divides the angle into two equal angles.) In the figure, angle A measures 50°. OB bisects angle ABC, and OC bisects angle ACB. What is the measure of angle BOC?

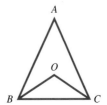

Exercises 89 and 90 refer to the given figure. The center of the circle is O.

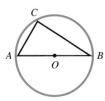

89. *Radius of a Circle* If $\overline{AC}$ measures 6 in. and $\overline{BC}$ measures 8 in., what is the radius of the circle?

90. *Lengths of Chords of a Circle* If $\overline{AB}$ measures 13 cm, and the length of $\overline{BC}$ is 7 cm more than the length of $\overline{AC}$, what are the lengths of $\overline{BC}$ and $\overline{AC}$?

Verify that the following constructions from the Extension following Section 9.2 are valid. Use a STATEMENTS/REASONS proof.

91. Construction 1

92. Construction 2

93. Construction 3

94. Construction 4

9.5 Space Figures, Volume, and Surface Area

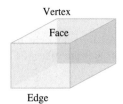

FIGURE 52

Space Figures Thus far, this chapter has discussed only **plane figures**—figures that can be drawn completely in the plane of a piece of paper. However, it takes the three dimensions of space to represent the solid world around us. For example, Figure 52 shows a "box" (a **rectangular parallelepiped** in mathematical terminology). The *faces* of a box are rectangles. The faces meet at *edges;* the "corners" are called *vertices* (plural of vertex—the same word as for the "corner" of an angle).

Boxes are one kind of space figure belonging to an important group called **polyhedrons,** the faces of which are made only of polygons. Perhaps the most interesting polyhedrons are the *regular polyhedrons*. Recall that a *regular polygon* is a polygon with all sides equal and all angles equal. A regular polyhedron is a space figure, the faces of which are only one kind of regular polygon. It turns out that there are only five different regular polyhedrons. They are shown in Figure 53. A **tetrahedron** is composed of four equilateral triangles, each three of which meet in a point. Use the figure to verify that there are four faces, four vertices, and six edges.

Tetrahedron   Hexahedron (cube)   Octahedron   Dodecahedron   Icosahedron

FIGURE 53

A regular quadrilateral is called a square. Six squares, each three of which meet at a point, form a **hexahedron,** or **cube.** Again, use Figure 53 to verify that a cube has 6 faces, 8 vertices, and 12 edges.

The three remaining regular polyhedrons are the **octahedron,** the **dodecahedron,** and the **icosahedron.** The octahedron is composed of groups of four regular triangles (i.e., equilateral) meeting at a point. The dodecahedron is formed by groups of three regular pentagons, while the icosahedron is made up of groups of five regular triangles.

FOR FURTHER THOUGHT

Many crystals and some viruses are constructed in the shapes of regular polyhedrons. Many viruses that infect animals were once thought to have spherical shapes, until X ray analysis revealed that their shells are icosahedral in nature. The sketch on the left below shows a polio virus.

Radiolaria, as shown on the right below, is a group of microorganisms. The skeletons of these single-celled animals take on beautiful geometric shapes. The one pictured is based on the tetrahedron.

The salt crystal depicted on the stamp in the next column on the left is revealed under X ray photography to have a cubic structure, with alternating ions of sodium and chlorine. The stamp was issued in honor of Sir Lawrence Bragg and his father, Sir William Henry Bragg, who jointly received a Nobel prize in 1915 for their early work in the study of X ray diffraction and solid state physics.

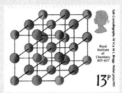

In the text we mention that a cube has 6 faces, 8 vertices, and 12 edges, while a tetrahedron has 4 faces, 4 vertices, and 6 edges. Leonhard Euler (1707–1783) investigated a remarkable relationship among these numbers. The following group discussion exercise will allow you to discover it as well.

For Group Discussion

As a class, use the figures in the text, or models brought to class by your instructor or class members, to complete the table below. Then state a conjecture concerning the value of $F + V - E$ for every polyhedron. This result, called **Euler's formula,** is given on page 545 at the end of the exercise set for this section.

Polyhedron	Faces (*F*)	Vertices (*V*)	Edges (*E*)	Value of $F + V - E$
Tetrahedron				
Hexahedron (Cube)				
Octahedron				
Dodecahedron				
Icosahedron				

Two other types of polyhedrons are familiar space figures: pyramids and prisms. **Pyramids** are made of triangular sides and a polygonal base. **Prisms** have two faces in parallel planes; these faces are congruent polygons. The remaining faces of a prism are all parallelograms. (See Figure 54(a) and (b) on the next page.) By this definition, a box is also a prism.

Tetrahedron

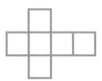
Hexahedron (cube)

Octahedron

Dodecahedron

Icosahedron

Patterns such as these may be used to actually construct three-dimensional models of the regular polyhedrons.

The circle, although a plane figure, is not a polygon. (Why?) Figure 54(c) shows space figures made up in part of circles, including *right circular cones* and *right circular cylinders*. The figure also shows how a circle can generate a *torus,* a doughnut-shaped solid that has interesting topological properties. See Section 9.7.

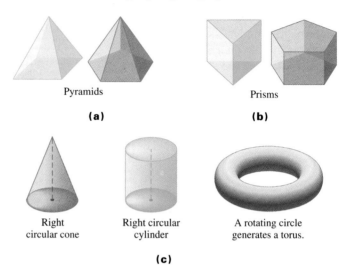

Pyramids
(a)

Prisms
(b)

Right circular cone

Right circular cylinder

A rotating circle generates a torus.

(c)

FIGURE 54

Volume and Surface Area While area is a measure of surface covered by a plane figure, **volume** is a measure of capacity of a space figure. Volume is measured in *cubic* units. For example, a cube with edge measuring 1 cm has volume 1 cubic cm, which is also written as 1 cm³, or 1 cc. The **surface area** is the total area that would be covered if the space figure were "peeled" and the peel laid flat. Surface area is measured in *square* units.

Volume and Surface Area of a Box

Suppose that a box has length ℓ, width w, and height h. Then the volume V and the surface area S are given by the formulas

$$V = \ell w h \quad \text{and} \quad S = 2\ell w + 2\ell h + 2hw.$$

In particular, if the box is a cube with edge of length s,

$$V = s^3 \quad \text{and} \quad S = 6s^2.$$

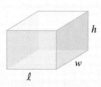

$V = \ell w h$
$S = 2\ell w + 2\ell h + 2hw$

$V = s^3$
$S = 6s^2$

9.5 Space Figures, Volume, and Surface Area

John C. Conway of Princeton University coined the term *holyhedron* to describe a polyhedron that can have a hole passing through every face and remain a polyhedron (*Science News*, Vol. 157, June 17, 2000, p. 399). He did this prior to even knowing that such a form exists. Now Jade P. Vinson, also of Princeton, has proved that such an object can exist, and has produced a monstrous holyhedron consisting of 78,585,627 faces. Conway had offered a reward for anyone producing a holyhedron: $10,000 divided by the number of faces. So for his efforts, Vinson earned $.0001. Conway suspects that someone will eventually find a holyhedron with fewer than 100 faces.

EXAMPLE 1 Find the volume V and the surface area S of the box shown in Figure 55.

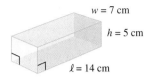

FIGURE 55

To find the volume, use the formula $V = \ell w h$ with $\ell = 14$, $w = 7$, and $h = 5$.

$$V = \ell w h = 14 \cdot 7 \cdot 5 = 490$$

Volume is measured in cubic units, so the volume of the box is 490 cubic centimeters, or 490 cm³.

To find the surface area, use the formula $S = 2\ell w + 2\ell h + 2hw$.

$$S = 2(14)(7) + 2(14)(5) + 2(5)(7) = 196 + 140 + 70 = 406$$

Like areas of plane figures, surface areas of space figures are given in square measure, so the surface area of the box is 406 square centimeters, or 406 cm².

A typical tin can is an example of a **right circular cylinder.**

Volume and Surface Area of a Right Circular Cylinder

If a right circular cylinder has height h and radius of its base equal to r, then the volume V and the surface area S are given by the formulas

$$V = \pi r^2 h$$

and $\quad S = 2\pi r h + 2\pi r^2.$

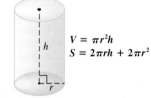

$V = \pi r^2 h$
$S = 2\pi r h + 2\pi r^2$

(In the formula for S, the areas of the top and bottom are included.)

FIGURE 56

EXAMPLE 2 In Figure 56, the right circular cylinder has surface area 288π square inches, and the radius of its base is 6 inches.

(a) Find the height of the cylinder.

Since we know that $S = 288\pi$ and $r = 6$, substitute into the formula for surface area to find h.

$$S = 2\pi r h + 2\pi r^2$$
$$288\pi = 2\pi(6)h + 2\pi(6)^2 \quad\quad S = 288\pi,\ r = 6$$
$$288\pi = 12\pi h + 72\pi$$
$$216\pi = 12\pi h \quad\quad \text{Subtract } 72\pi.$$
$$h = 18 \quad\quad \text{Divide by } 12\pi.$$

The height is 18 inches.

Ptolemy's geometric model of the solar system was fairly complex, with the earth at the center and the planets and sun moving in circular orbits and on epicycles. Ptolemy's system was accepted on authority during the Middle Ages up until the time of the Copernican Revolution, named after **Nicolas Copernicus**.

Copernicus worked out his own system with the sun at center (heliocentric) instead of Earth at center (geocentric). A small change, you think—but it marked a revolution in thinking and the beginning of the modern world.

(b) Find the volume of the cylinder.

Use the formula for volume, with $r = 6$ and $h = 18$.

$$V = \pi r^2 h = \pi(6)^2(18) = 648\pi$$

The exact volume is 648π cubic inches, or approximately 2034.72 cubic inches, using $\pi \approx 3.14$. ■

The three-dimensional analogue of a circle is a **sphere**. It is defined by replacing the word "plane" with "space" in the definition of a circle (Section 9.2).

Volume and Surface Area of a Sphere

If a sphere has radius r, then the volume V and the surface area S are given by the formulas

$$V = \frac{4}{3}\pi r^3 \quad \text{and} \quad S = 4\pi r^2.$$

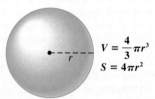

EXAMPLE 3 Suppose that a spherical tank having radius 3 meters can be filled with liquid fuel for $200. How much will it cost to fill a spherical tank of radius 6 meters with the same fuel?

We must first find the volume of the tank with radius 3 meters. Call it V_1.

$$V = \frac{4}{3}\pi r^3 \qquad \text{Formula for volume of a sphere}$$

$$V_1 = \frac{4}{3}\pi(3)^3 = \frac{4}{3}\pi(27) = 36\pi$$

Now find V_2, the volume of the tank having radius 6 meters.

$$V_2 = \frac{4}{3}\pi(6)^3 = \frac{4}{3}\pi(216) = 288\pi$$

Notice that by doubling the radius of the sphere from 3 meters to 6 meters, the volume has increased 8 times, since

$$V_2 = 288\pi = 8V_1 = 8(36\pi).$$

Therefore, the cost to fill the larger tank is eight times the cost to fill the smaller one: 8($200) = $1600. ■

FIGURE 57

The space figure shown in Figure 57 is a **right circular cone.**

Volume and Surface Area of a Right Circular Cone

If a right circular cone has height h and the radius of its circular base is r, then the volume V and the surface area S are given by the formulas

$$V = \frac{1}{3}\pi r^2 h$$

and $\quad S = \pi r \sqrt{r^2 + h^2} + \pi r^2.$

(In the formula for S, the area of the bottom is included.)

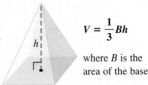

A **pyramid** is a space figure having a polygonal base and triangular sides. Figure 58 shows a pyramid with a square base.

FIGURE 58

Volume of a Pyramid

If B represents the area of the base of a pyramid, and h represents the height (that is, the perpendicular distance from the top, or apex, to the base), then the volume V is given by the formula

$$V = \frac{1}{3}Bh.$$

where B is the area of the base

The Transamerica Tower in San Francisco is a pyramid with a square base. Each side of the base has a length of 52 meters, while the height of the building is 260 meters. The formula for the volume of a pyramid indicates that the volume of the building is about 234,000 cubic meters.

EXAMPLE 4 What is the ratio of the volume of a right circular cone with radius of base r and height h to the volume of a pyramid having a square base, with each side of length r, and height h?

Using the formula for the volume of a cone, we have

$$V_1 = \text{Volume of the cone} = \frac{1}{3}\pi r^2 h.$$

Since the pyramid has a square base, the area B of its base is r^2. Using the formula for the volume of a pyramid, we get

$$V_2 = \text{Volume of the pyramid} = \frac{1}{3}Bh = \frac{1}{3}(r^2)h.$$

The ratio of the first volume to the second is

$$\frac{V_1}{V_2} = \frac{\frac{1}{3}\pi r^2 h}{\frac{1}{3}r^2 h} = \pi.$$

9.5 EXERCISES

Decide whether each of the following statements is true *or* false.

1. A cube with volume 64 cubic inches has surface area 96 square inches.
2. A tetrahedron has the same number of faces as vertices.
3. A dodecahedron can be used as a model for a calendar for a given year, where each face of the dodecahedron contains a calendar for a single month, and there are no faces left over.
4. Each face of an octahedron is an octagon.
5. If you double the length of the edge of a cube, the new cube will have a volume that is twice the volume of the original cube.
6. The numerical value of the volume of a sphere is $r/3$ times the numerical value of its surface area, where r is the measure of the radius.

Find **(a)** *the volume and* **(b)** *the surface area of each of the following space figures. When necessary, use* 3.14 *as an approximation for* π, *and round answers to the nearest hundredth.*

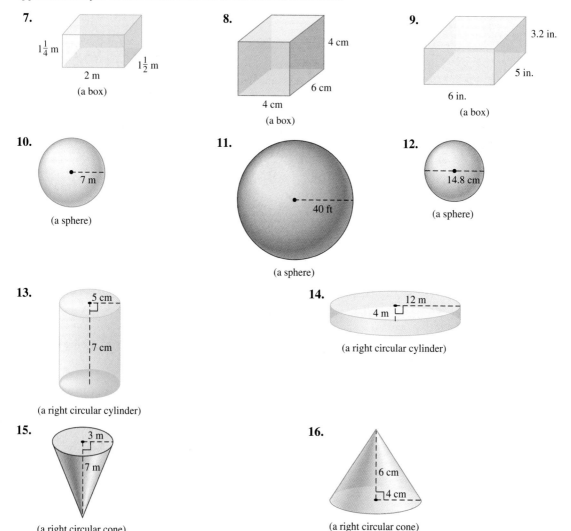

Find the volume of each pyramid. In each case, the base is a rectangle.

17.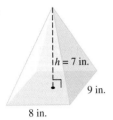
   h = 7 in., 9 in., 8 in.

18.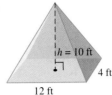
   h = 10 ft, 4 ft, 12 ft

Volumes of Common Objects Find each volume. Use 3.14 as an approximation for π when necessary.

19. a coffee can, radius 6.3 cm and height 15.8 cm
20. a soup can, radius 3.2 cm and height 9.5 cm
21. a pork-and-beans can, diameter 7.2 cm and height 10.5 cm
22. a cardboard mailing tube, diameter 2 in. and height 40 in.
23. a coffee mug, diameter 9 cm and height 8 cm
24. a bottle of typewriter correction fluid, diameter 3 cm and height 4.3 cm
25. the Great Pyramid of Cheops, near Cairo—its base is a square 230 m on a side, while the height is 137 m
26. the Peachtree Plaza Hotel in Atlanta, a cylinder with a base radius of 46 m and a height of 220 m
27. a road construction marker, a cone with height 2 m and base radius 1/2 m
28. the conical portion of a witch's hat for a Halloween costume, with height 12 in. and base radius 4 in.

In the chart below, one of the values r (radius), d (diameter), V (volume), or S (surface area) is given for a particular sphere. Find the remaining three values. Leave π in your answers.

	r	d	V	S
29.	6 in.			
30.	9 in.			
31.		10 ft		
32.		40 ft		
33.			$\frac{32}{3}\pi$ cm³	
34.			$\frac{256}{3}\pi$ cm³	
35.				4π m²
36.				144π m²

Solve each problem.

37. *Volume or Surface Area?* In order to determine the amount of liquid a spherical tank will hold, would you need to use volume or surface area?

38. *Volume or Surface Area?* In order to determine the amount of leather it would take to manufacture a basketball, would you need to use volume or surface area?

39. *Side Length of a Cube* One of the three famous construction problems of Greek mathematics required the construction of an edge of a cube with twice the volume of a given cube. If the length of each side of the given cube is x, what would be the length of each side of a cube with twice the original volume?

40. Work through the parts of this exercise in order, and use them to make a generalization concerning volumes of spheres. Leave answers in terms of π.
   (a) Find the volume of a sphere having a radius of 1 m.
   (b) Suppose the radius is doubled to 2 m. What is the volume?
   (c) When the radius was doubled, by how many times did the volume increase? (To find out, divide the answer for part (b) by the answer for part (a).)
   (d) Suppose the radius of the sphere from part (a) is tripled to 3 m. What is the volume?
   (e) When the radius was tripled, by how many times did the volume increase?
   (f) In general, if the radius of a sphere is multiplied by n, the volume is multiplied by _____ .

Cost to Fill a Spherical Tank If a spherical tank 2 m in diameter can be filled with a liquid for $300, find the cost to fill tanks of the following diameters.

41. 6 m   **42.** 8 m   **43.** 10 m

44. Use the logic of Exercise 40 to answer the following: If the radius of a sphere is multiplied by n, then the surface area of the sphere is multiplied by _____ .

Each of the following figures has volume as indicated. Find the value of x.

45. $V = 60$
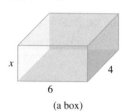
(a box)

46. $V = 450$
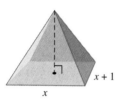
$h = 15$
Base is a rectangle.
(a pyramid)

47. $V = 36\pi$

(a sphere)

48. $V = 245\pi$

(a right circular cone)

Exercises 49–56 require some ingenuity, but all can be solved using the concepts presented so far in this chapter.

49. *Volume of a Box* The areas of the sides of a rectangular box are 30 in.², 35 in.², and 42 in.². What is the volume of the box?

50. *Ratios of Volumes* In the figure, a right circular cone is inscribed in a hemisphere. What is the ratio of the volume of the cone to the volume of the hemisphere?

51. *Volume of a Sphere* A plane intersects a sphere to form a circle as shown in the figure. If the area of the circle formed by the intersection is 576π in.², what is the volume of the sphere?

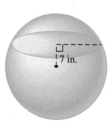

52. *Change in Volume* If the height of a right circular cylinder is halved and the diameter is tripled, how is the volume changed?

53. *Ratio of Areas* What is the ratio of the area of the circumscribed square to the area of the inscribed square?

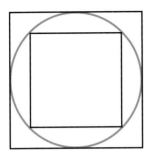

54. *Perimeter of a Square* Suppose the diameter of the circle shown is 8 in. What is the perimeter of the inscribed square $ABCD$?

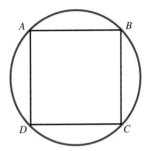

55. *Value of a Sum* In the circle shown with center O, the radius is 6. $QTSR$ is an inscribed square. Find the value of $PQ^2 + PT^2 + PR^2 + PS^2$.

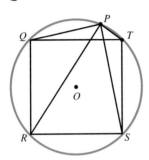

56. *Ratio of Side Lengths* The square $JOSH$ is inscribed in a semicircle. What is the ratio of x to y?

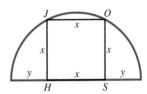

Answer to For Group Discussion on page 537: Euler's formula is $F + V - E = 2$.

9.6 Transformational Geometry

There are many branches of geometry. In this chapter we have studied concepts of Euclidean geometry. One particular branch of geometry, known as **transformational geometry**, investigates how one geometric figure can be transformed into another. In transformational geometry we are required to reflect, rotate, and change the size of figures using concepts that we now discuss.

Reflections One way to transform one geometric figure into another is by reflection. In Figure 59, line m is perpendicular to the line segment AA' and also bisects this line segment. We call point A' the **reflection image** of point A about line m. Line m is called the **line of reflection** for points A and A'. In the figure, we use a dashed line to connect points A and A' to show that these two points are images of each other under this transformation.

Point A' is the reflection image of point A only for line m; if a different line were used, A would have a different reflection image. Think of the reflection image of a point A about a line m as follows: Place a drop of ink at point A, and fold the paper along line m. The spot made by the ink on the other side of m is the reflection image of A. If A' is the image of A about line m, then A is the image of A' about the same line m.

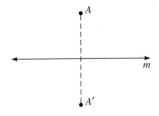

FIGURE 59

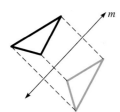

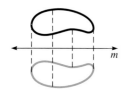

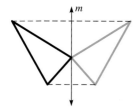

FIGURE 60

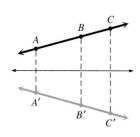

FIGURE 61

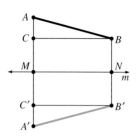

FIGURE 62

To find the reflection image of a figure, find the reflection image of each point of the figure. The set of all reflection images of the points of the original figure is called the **reflection image** of the figure. Figure 60 shows several figures (in black) and their reflection images (in color) about the lines shown.

By the definition of reflection given above, each point in a plane has exactly one reflection image point with respect to a given line of reflection. Also, each reflection image point has exactly one original point. Thus, two distinct points cannot have the same reflection image. This means there is a *1-to-1 correspondence* between the set of points of the plane and the image points with respect to a given line of reflection. Any operation, such as reflection, in which there is a 1-to-1 correspondence between the points of the plane and their image points is called a **transformation;** we can call reflection about a line the **reflection transformation.**

If a point A and its image, A', under a certain transformation are the same point, then point A is called an **invariant point** of the transformation. The only invariant points of the reflection transformation are the points of the line of reflection.

Three points that lie on the same straight line are called **collinear.** In Figure 61, points A, B, and C are collinear, and it can be shown that the reflection images A', B', and C', are also collinear. Thus, the reflection image of a line is also a line. We express this by saying that **reflection preserves collinearity.**

Distance is also preserved by the reflection transformation. Thus, in Figure 62, the distance between points A and B, written $|AB|$, is equal to the distance between the reflection images A' and B', or

$$|AB| = |A'B'|.$$

To prove this, we can use the definition of reflection image to verify that $|AM| = |MA'|$, and $|BN| = |NB'|$. Construct segments CB and $C'B'$, each perpendicular to BB'. Note that $CBB'C'$ is a rectangle. Since the opposite sides of a rectangle are equal and parallel, we have

$$|CB| = |C'B'|. \qquad \text{(Side)} \qquad \textbf{(1)}$$

Since $CBB'C'$ is a rectangle, we can also say

$$m\angle ACB = m\angle A'C'B' = 90° \qquad \text{(Angle)} \qquad \textbf{(2)}$$

where we use $m\angle ACB$ to represent the measure of angle ACB.

We know $|AM| = |MA'|$ and can show $|CM| = |MC'|$, so that $|AC|$ must equal $|A'C'|$, or

$$|AC| = |A'C'|. \qquad \text{(Side)} \qquad \textbf{(3)}$$

From statements (1), (2), and (3) above, we conclude that in triangles ABC and $A'B'C'$, two sides and the included angle of one are equal in measure to the corresponding two sides and angle of the other, and thus are congruent by SAS (Section 9.4). Since corresponding sides of congruent triangles are equal in length, we have

$$|AB| = |A'B'|,$$

which is what we wanted to show. Hence the distance between two points equals the distance between their reflection images, and thus reflection preserves distance. (The proof we have given is not really complete, since we have tacitly assumed that AB is not parallel to $A'B'$, and that A and B are on the same side of the line of reflection. Some modification would have to be made in the proof above to include these other cases.)

This playful little beagle, known as D'Artagnan, is owned by Cody and Margaret Westmoreland. Notice that the markings on his face are symmetric with respect to a vertical line through his cold, wet, little nose.

The figures shown in Figure 63 are their own reflection images about the lines of reflection shown. In this case, the line of reflection is called a **line of symmetry** for the figure. Figure 63(a) has three lines of symmetry. A circle has every line through its center as a line of symmetry.

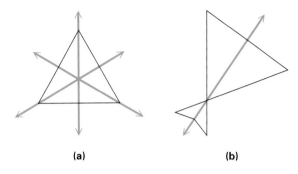

(a)   (b)

FIGURE 63

Translations and Rotations We shall use the symbol r_m to represent a reflection about line m, and let us use $r_n \cdot r_m$ to represent a reflection about line m followed by a reflection about line n. We call $r_n \cdot r_m$ the **composition,** or **product,** of the two reflections r_n and r_m. Figure 64 shows two examples of the composition of two reflections. In Figure 64(a), lines m and n are parallel, while they intersect in Figure 64(b).

An example of a reflection

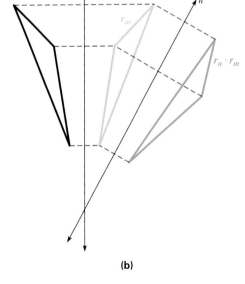

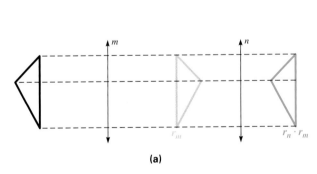

(a)   (b)

FIGURE 64

In Figure 64(a) both the original figure and its image under the composition of the two reflections appear to be oriented the same way and to have the same "tilt." In fact, it appears that the original figure could be slid along the dashed lines of Figure 64(a), with no rotation, so as to cover the image. This composite

How many kinds of symmetry do you see here?

transformation is called a **translation.** Figure 65 shows a translation, and the image can be obtained as a composition of two reflections about parallel lines. Check that the distance between a point and its image under a translation is twice the distance between the two parallel lines. The distance between a point and its image under a translation is called the **magnitude** of the translation.

A translation of magnitude 0 leaves every point of the plane unchanged, and thus is called the **identity translation.** A translation of magnitude k, followed by a similar translation of magnitude k but of opposite direction, returns a point to its original position, and thus these two translations are called **inverses** of each other. Check that there are no invariant points in a translation of magnitude $k > 0$.

A translation preserves collinearity (three points on the same line have image points that also lie on a line) and distance (the distance between two points is the same as the distance between the images of the points).

In Figure 64(b), the original figure could be rotated so as to cover the image. Hence, we call the composition of two reflections about nonparallel lines a **rotation.** The point of intersection of these two nonparallel lines is called the **center of rotation.** The black triangle of Figure 66 was reflected about line m and then reflected about line n, resulting in a rotation with center at B. The dashed lines in color represent the paths of the vertices of the triangle under the rotation. It is shown in the exercises that $m \angle ABA'$ is twice as large as $m \angle MBN$. The measure of angle ABA' is called the **magnitude** of the rotation.

Rotations also preserve collinearity and distance. The identity transformation here is a rotation of $0°$ or $360°$, and rotations of, say, $240°$ and $120°$ (or in general, $x°$ and $360° - x°$, $0 \leq x \leq 360°$) are inverses of each other. The center of rotation is the only invariant point of any rotation except the identity rotation.

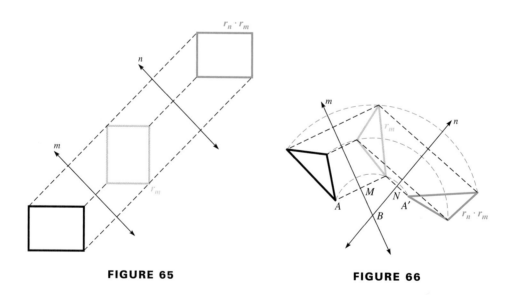

FIGURE 65   **FIGURE 66**

We have defined rotations as the composition of two reflections about nonparallel lines of reflection. We could also define a rotation by specifying its center, the angle of rotation, and a direction of rotation, as shown by the next example.

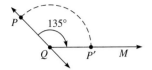

FIGURE 67

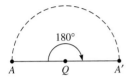

FIGURE 68

EXAMPLE 1 Find the image of a point P under a rotation transformation having center at a point Q and magnitude 135° clockwise.

To find P', the image of P, first draw angle PQM having measure 135°. Then draw an arc of a circle with center at Q and radius $|PQ|$. The point where this arc intersects side QM is P'. See Figure 67. ■

Figure 68 shows a rotation transformation having center Q and magnitude 180° clockwise. Point Q bisects the line segment from a point A to its image A', and for this reason this rotation is sometimes called a **point reflection.**

EXAMPLE 2 Find the point reflection images about point Q for each of the following figures in black.

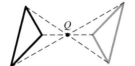 

FIGURE 69

Using the definition of point reflection, we can find the point reflection images of several sample points with the images, as shown in color in Figure 69. ■

Let r_m be a reflection about line m, and let T be a translation having nonzero magnitude and a direction parallel to m. Then the composition of T and r_m is called a **glide reflection,** as shown in Figure 70. Here a reflection followed by a translation is the same as a translation followed by a reflection, so that in this case

$$T \cdot r_m = r_m \cdot T.$$

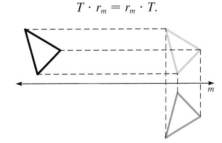

FIGURE 70

This creature exhibits *bilateral symmetry*. This kind of symmetry is often found in living organisms.

Since a translation is the composition of *two* reflections, a glide reflection is the composition of *three* reflections. Since it is required that the translation have nonzero magnitude, there is no identity glide transformation.

All the transformations of this section are **isometries,** or transformations in which the image of a figure has the same size and shape as the original figure. It can be shown that any isometry is either a reflection or the composition of two or more reflections.

Size Transformations Figure 71 shows a semicircle in black, a point M, and an image semicircle in color. Distance $A'M$ is twice the distance AM and distance $B'M$ is twice the distance BM. In fact, all the points of the image semicircle, such as C', were obtained by drawing a line through M and C, and then locating C' such that

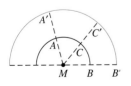

FIGURE 71

"**M.C. Escher** (1898–1972) was a Dutch graphic artist, most recognized for spatial illusions, impossible buildings, repeating geometric patterns (tessellations), and his incredible techniques in woodcutting and lithography.

Escher was a man studied and greatly appreciated by respected mathematicians, scientists, and crystallographers, yet he had no formal training in math or science. He was a humble man who considered himself neither an artist nor a mathematician.

Intricate repeating patterns, mathematically complex structures, and spatial perspectives all require a 'second look.' In Escher's work what you see the first time is most certainly not all there is to see." (*Sources:* M. C. Escher's Symmetry Drawing E66 (Winged Lions) and Waterfall © 2003 Cordon Art B.V., Baarn, Holland. All rights reserved; http://www.worldofescher.com.)

$2|MC| = |MC'|$. Such a transformation is called a **size transformation** with center M and magnitude 2. We shall assume that a size transformation can have any positive real number k as magnitude. A size transformation having magnitude $k > 1$ is called a **dilatation,** or **stretch;** while a size transformation having magnitude $k < 1$ is called a **contraction,** or **shrink.**

EXAMPLE 3 Apply a size transformation with center M and magnitude $\frac{1}{3}$ to the two triangles shown in black in Figure 72.

To find the images of these triangles, we can find the image points of some sample points. For example, if we select point A on each of the original triangles, we can find the image points by drawing a line through A and M, and locating a point A' such that $|MA'| = \frac{1}{3}|MA|$. By doing this for all points of each of the black triangles, we get the images shown in color in Figure 72.

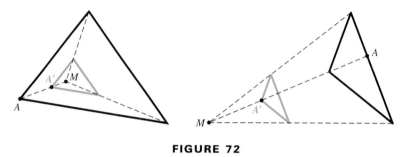

FIGURE 72

The identity transformation is a size transformation of magnitude 1, while size transformations of magnitude k and $1/k$, having the same center, are inverses of each other. The only invariant point of a size transformation of magnitude $k \neq 1$ is the center of the transformation.

EXAMPLE 4 **(a)** Does a size transformation preserve collinearity?

Figure 73 shows three collinear points, A, B, and C, and their images under two different size transformations with center at M: one of magnitude 3 and one of magnitude $\frac{1}{3}$. In each case the image points appear to be collinear, and it can be proved that they are, using similar triangles. In fact, the image of a line not through the center of the transformation is a line parallel to the original line.

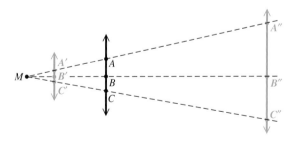

FIGURE 73

(b) Does a size transformation preserve distance?

As shown in Figure 73, $|AB| \neq |A'B'|$. Thus, a size transformation of magnitude $k \neq 1$ does not preserve distance and is not an isometry.

Summary of Transformations

	Reflection	Translation	Rotation	Glide Reflection	Size Transformation
Example					
Preserve collinearity?	Yes	Yes	Yes	Yes	Yes
Preserve distance?	Yes	Yes	Yes	Yes	No
Identity transformation?	None	Magnitude 0	Magnitude 360°	None	Magnitude 1
Inverse transformation?	None	Same magnitude; opposite direction	Same center; magnitude $360° - x°$	None	Same center; magnitude $1/k$
Composition of n reflections?	$n = 1$	$n = 2$, parallel	$n = 2$, nonparallel	$n = 3$	No
Isometry?	Yes	Yes	Yes	Yes	No
Invariant points?	Line of reflection	None	Center of rotation	None	Center of transformation

FOR FURTHER THOUGHT

The authors wish to thank Suzanne Alejandre for permission to reprint this article on tessellations, which appears at

http://mathforum.org/sum95/suzanne/whattess.html.

tessellate (verb), **tessellation** (noun): from Latin *tessera* "a square tablet" or "a die used for gambling." Latin *tessera* may have been borrowed from Greek *tessares*, meaning "four," since a square tile has four sides. The diminutive of *tessera* was *tessella*, a small, square piece of stone or a cubical tile used in mosaics. Since a mosaic extends over a given area without leaving any region uncovered, the geometric meaning of the word tessellate is "to cover the plane with a pattern in such a way as to leave no region uncovered." By extension, space or hyperspace may also be tessellated.

Definition

A dictionary will tell you that the word "tessellate" means to form or arrange small squares in a checkered or mosaic pattern. The word "tessellate" is derived from the Ionic version of the Greek word "tesseres," which in English means "four." The first tilings were made from square tiles.

A regular polygon has 3 or 4 or 5 or more sides and angles, all equal. A **regular tessellation** means a tessellation made up of congruent regular polygons. [Remember: *Regular* means that the sides of the polygon are all the same length. *Congruent* means that the polygons that you put together are all the same size and shape.]

(continued)

Only three regular polygons tessellate in the Euclidean plane: triangles, squares, or hexagons. We can't show the entire plane, but imagine that these are pieces taken from planes that have been tiled. Here are examples of

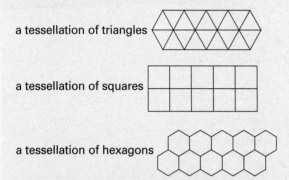

a tessellation of triangles

a tessellation of squares

a tessellation of hexagons

When you look at these three samples you can easily notice that the squares are lined up with each other while the triangles and hexagons are not. Also, if you look at 6 triangles at a time, they form a hexagon, so the tiling of triangles and the tiling of hexagons are similar and they cannot be formed by directly lining shapes up under each other—a slide (or a glide!) is involved.

You can work out the interior measure of the angles for each of these polygons:

Shape	Angle Measure in Degrees
triangle	60
square	90
pentagon	108
hexagon	120
more than six sides	more than 120 degrees

Since the regular polygons in a tessellation must fill the plane at each vertex, the interior angle must be an exact divisor of 360 degrees. This works for the triangle, square, and hexagon, and you can show working tessellations for these figures. For all the others, the interior angles are not exact divisors of 360 degrees, and therefore those figures cannot tile the plane.

Naming Conventions

A tessellation of squares is named "4.4.4.4." Here's how: choose a vertex, and then look at one of the polygons that touches that vertex. How many sides does it have?

Since it's a square, it has four sides, and that's where the first "4" comes from. Now keep going around the vertex in either direction, finding the number of sides of the polygons until you get back to the polygon you started with. How many polygons did you count?

There are four polygons, and each has four sides.

4.4.4.4

For a tessellation of regular congruent hexagons, if you choose a vertex and count the sides of the polygons that touch it, you'll see that there are three polygons and each has six sides, so this tessellation is called "6.6.6":

6.6.6

A tessellation of triangles has six polygons surrounding a vertex, and each of them has three sides: "3.3.3.3.3.3."

3.3.3.3.3.3

Semi-regular Tessellations

You can also use a variety of regular polygons to make **semi-regular tessellations**.

9.6 Transformational Geometry

A semi-regular tessellation has two properties, which are:

1. It is formed by regular polygons.
2. The arrangement of polygons at every vertex point is identical.

Here are the **eight** semi-regular tessellations:

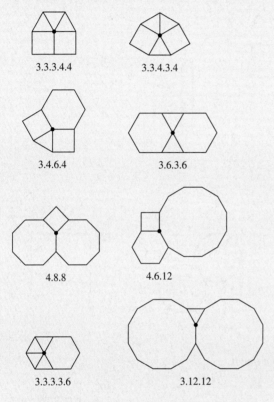

Interestingly there are other combinations that seem like they should tile the plane because the arrangements of the regular polygons fill the space around a point. For example:

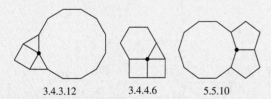

If you try tiling the plane with these units of tessellation you will find that they cannot be extended infinitely.

There is an infinite number of tessellations that can be made of patterns that do not have the same combination of angles at every vertex point. There are also tessellations made of polygons that do not share common edges and vertices.

For Group Discussion

1. Use the naming conventions to name each of these semi-regular tessellations. (These can be found at http://www.coolmath4kids.com/tesspag2.html.)

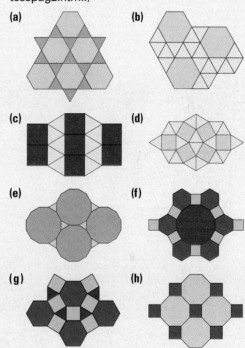

2. Why isn't this a semi-regular tessellation?

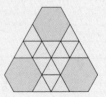

9.6 EXERCISES

Find the reflection images of the given figures about the given lines.

1.
2.
3.
4.

5.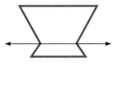
6.
7.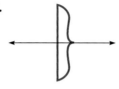
8.

Find any lines of symmetry of the given figures.

9.
10.
11.
12.

First reflect the given figure about line m. Then reflect about line n.

13.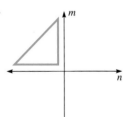
14.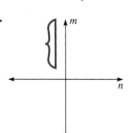
15.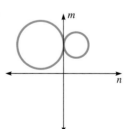
16.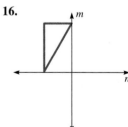

17.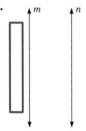
18.
19.
20.

9.6 Transformational Geometry

In Exercises 21–34, let T be a translation having magnitude $\frac{3}{4}$ inch to the right in a direction parallel to the bottom edge of the page. Let r_m be a reflection about line m, and let R_p be a rotation about point P having magnitude 60° clockwise. In each of Exercises 21–32, perform the given transformations on point A of the figure at the right.

21. r_m
22. R_P
23. T
24. $r_m \cdot r_m$
25. $T \cdot T$
26. $R_P \cdot R_P$
27. $T \cdot R_P$
28. $T \cdot r_m$
29. $r_m \cdot T$
30. $R_P \cdot r_m$
31. $r_m \cdot R_P$
32. $R_P \cdot T$
33. Is $T \cdot r_m$ a glide reflection here?
34. Does $T \cdot r_m = r_m \cdot T$ here?
35. Suppose a rotation is given by $r_m \cdot r_n$, as shown in the figure. Find the images of A, B, and C.
36. Does a glide reflection preserve
    (a) collinearity?
    (b) distance?

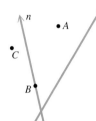

Find the point reflection images of each of the following figures with the given points as center.

37.
38.
39.
40.

Perform the indicated size transformation.

41. magnitude 2; center M

42. magnitude 1/2; center M

43. magnitude 1/2; center M

44. magnitude 2; center M

45. magnitude 1/3; center M

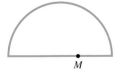

46. magnitude 1/3; center M

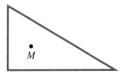

9.7 Non-Euclidean Geometry, Topology, and Networks

The *Elements* of Euclid is quite possibly the most influential mathematics book ever written. (See the margin note at the beginning of this chapter on page 492.) It begins with definitions of basic ideas such as point, line, and plane. Euclid then gives five postulates providing the foundation of all that follows.

Next, Euclid lists five axioms that he views as general truths and not just facts about geometry. (To some of the Greek writers, postulates were truths about a particlar field, while axioms were general truths. Today, "axiom" is used in either case.)

John Playfair (1748–1819) wrote his *Elements of Geometry* in 1795. Playfair's Axiom is: Given a line k and a point P not on the line, there exists one and only one line m through P that is parallel to k. Playfair was a geologist who fostered "uniformitarianism," the doctrine that geological processes long ago gave Earth its features, and processes today are the same kind as those in the past.

Euclid's Postulates	Euclid's Axioms
1. Two points determine one and only one straight line.	6. Things equal to the same thing are equal to each other.
2. A straight line extends indefinitely far in either direction.	7. If equals are added to equals, the sums are equal.
3. A circle may be drawn with any given center and any given radius.	8. If equals are subtracted from equals, the remainders are equal.
4. All right angles are equal.	9. Figures that can be made to coincide are equal.
5. Given a line k and a point P not on the line, there exists one and only one line m through P that is parallel to k.	10. The whole is greater than any of its parts.

Using only these ten statements and the basic rules of logic, Euclid was able to prove a large number of "propositions" about geometric figures.

The statement for Postulate 5 given above is actually known as Playfair's axiom on parallel lines, which is equivalent to Euclid's fifth postulate. To understand why this postulate caused trouble for so many mathematicians for so long, we must examine the original formulation.

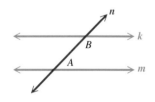

FIGURE 74

Euclid's Fifth Postulate ("Parallel Postulate") Euclid's fifth postulate states that if two lines (k and m in Figure 74) are such that a third line, n, intersects them so that the sum of the two interior angles (A and B) on one side of line n is less than two right angles, then the two lines, if extended far enough, will meet on the same side of n that has the sum of the interior angles less than two right angles.

Euclid's parallel postulate is quite different from the other nine postulates and axioms we listed. The others are simple statements that seem in complete agreement with our experience of the world around us. But the parallel postulate is long and wordy, and difficult to understand without a sketch.

The difference between the parallel postulate and the other axioms was noted by the Greeks, as well as later mathematicians. It was commonly believed that this was not a postulate at all, but a theorem to be proved. For more than 2000 years mathematicians tried repeatedly to prove it.

The most dedicated attempt came from an Italian Jesuit, Girolamo Saccheri (1667–1733). He attempted to prove the parallel postulate in an indirect way, by so-called "reduction to absurdity." He would assume the postulate to be false and then show that the assumption leads to a contradiction of something true (an absurdity). Such a contradiction would thus prove the statement true.

9.7 Non-Euclidean Geometry, Topology, and Networks

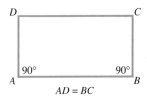

FIGURE 75

Saccheri began with a quadrilateral, as in Figure 75. He assumed angles A and B to be right angles and sides AD and BC to be equal. His plan was as follows:

1. To assume that angles C and D are obtuse angles, and to show that this leads to a contradiction.
2. To assume that angles C and D are acute angles, and to show that this also leads to a contradiction.
3. Then if C and D can be neither acute nor obtuse angles, they must be right angles.
4. If C and D are both right angles, then it can be proved that the fifth postulate is true. It thus is a theorem rather than a postulate.

Saccheri had no trouble with part 1. However, he did not actually reach a contradiction in the second part, but produced some theorems so "repugnant" that he convinced himself he had vindicated Euclid. In fact, he published a book called in English *Euclid Freed of Every Flaw.*

Today we know that the fifth postulate is indeed an axiom, and not a theorem. It is *consistent* with Euclid's other axioms.

The ten axioms of Euclid describe the world around us with remarkable accuracy. We now realize that the fifth postulate is necessary in Euclidean geometry to establish *flatness*. That is, the axioms of Euclid describe the geometry of *plane surfaces*. By changing the fifth postulate, we can describe the geometry of other surfaces. So, other geometric systems exist as much as Euclidean geometry exists, and they can even be demonstrated in our world. They are just not as familiar. A system of geometry in which the fifth postulate is changed is called a **non-Euclidean geometry.**

A song titled simply **Lobachevsky** first appeared on the LP *Songs by Tom Lehrer* (Reprise RS-6216). The liner notes indicate that the song "is a description of one way to get ahead in mathematics (which happens to be the author's own academic specialty) or any other academic field." Find a copy of the album, or the compact disc *The Remains of Tom Lehrer,* and see what Lehrer suggests!

The Origins of Non-Euclidean Geometry
One non-Euclidean system was developed by three people working separately at about the same time. Early in the nineteenth century Carl Friedrich Gauss, one of the great mathematicians, worked out a consistent geometry replacing Euclid's fifth postulate. He never published his work, however, because he feared the ridicule of people who could not free themselves from habitual ways of thinking. Gauss first used the term "non-Euclidean." Nikolai Ivanovich Lobachevski (1793–1856) published a similar system in 1830 in the Russian language. At the same time, Janos Bolyai (1802–1860), a Hungarian army officer, worked out a similar system, which he published in 1832, not knowing about Lobachevski's work. Bolyai never recovered from the disappointment of not being the first, and did no further work in mathematics.

Lobachevski replaced Euclid's fifth postulate with:

Angles C and D in the quadrilateral of Saccheri are acute angles.

This postulate of Lobachevski can be rephrased as follows:

Through a point P off a line k (Figure 76), at least two different lines can be drawn parallel to k.

Compare this form of Lobachevski's postulate to the geometry of Euclid. How many lines can be drawn through P and parallel to k in Euclidean geometry? At first glance, the postulate of Lobachevski does not agree with what we know about the world around us. But this is only because we think of our immediate surroundings as being flat.

Many of the theorems of Euclidean geometry are valid for the geometry of Lobachevski, but many are not. For example, in Euclidean geometry, the sum of the

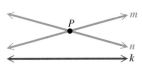

FIGURE 76

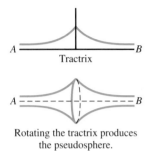

Tractrix

Rotating the tractrix produces the pseudosphere.

FIGURE 77

measures of the angles in any triangle is 180°. In Lobachevskian geometry, the sum of the measures of the angles in any triangle is *less* than 180°. Also, triangles of different sizes can never have equal angles, so similar triangles do not exist.

The geometry of Euclid can be represented on a plane. Since any portion of the earth that we are likely to see looks flat, Euclidean geometry is very useful for describing the everyday world around us. The non-Euclidean geometry of Lobachevski can be represented as a surface called a **pseudosphere.** This surface is formed by revolving a curve called a **tractrix** about the line *AB* in Figure 77.

A second non-Euclidean system was developed by Georg Riemann (1826–1866). He pointed out the difference between a line that continues indefinitely and a line having infinite length. For example, a circle on the surface of a sphere continues

FOR FURTHER THOUGHT

Artists must have a sound knowledge of geometry in order to paint realistically a world of solids in space. Our three-dimensional world must be represented in a convincing way on canvas or some other plane surface. For example, how should an artist paint a realistic view of railroad tracks going off to the horizon? In reality, the tracks are always at a constant distance apart, but they cannot be drawn that way except from overhead. The artist must make the tracks converge at a point. Only in this way will the scene look "real."

Beginning in the fifteenth century, artists led by Leone Battista Alberti, Leonardo da Vinci, and Albrecht Dürer began to study the problems of representing three dimensions in two. They found geometric methods of doing this. What artists initiated, mathematicians developed into a geometry different from that of Euclid—**projective geometry.**

Gerard Desargues (1591–1661), architect and engineer, was a technical advisor to the French government. He met Descartes at the siege of La Rochelle in 1628 and in the 1630s was a member of the Parisian group that included Descartes, Pascal, Fermat, and Mersenne. In 1636 and 1639 he published a treatise and proposals about the perspective section—Desargues had invented projective geometry. His works unfortunately were too difficult for the times; his terms were from botany and he did not use Cartesian symbolism. Desargues's geometric innovations were hidden for nearly 200 years. A manuscript by Desargues turned up in 1845, about 30 years after Jean-Victor Poncelet had rediscovered projective geometry.

For Group Discussion

Have a class member go to the board with a long straightedge and draw a figure similar to the one shown here.

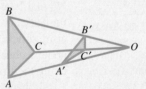

Then have the student, *very carefully,* follow these instructions.

1. Extend lines *AC* and *A'C'* of the two triangles to meet in a point *M*.
2. Extend sides *AB* and *A'B'* to meet in a point *N*.
3. Extend *BC* and *B'C'* to meet in a point *P*.

Now, as a class, make a conjecture about what has happened. This exercise is an illustration of Desargues's theorem (stated in Exercise 12 of this section).

9.7 Non-Euclidean Geometry, Topology, and Networks 559

indefinitely but does not have infinite length. Riemann developed the idea of geometry on a sphere and replaced Euclid's fifth postulate with:

Angles *C* and *D* of the Saccheri quadrilateral are obtuse angles.

In terms of parallel lines, Riemann's postulate becomes:

Through a point *P* off a line *k*, no line can be drawn that is parallel to *k*.

Riemannian geometry is important in navigation. "Lines" in this geometry are really *great circles,* or circles whose centers are at the center of the sphere. The shortest distance between two points on a sphere lies along an arc of a great circle. Great circle routes on a globe don't look at all like the shortest distance when the globe is flattened out to form a map, but this is part of the distortion that occurs when the earth is represented as a flat surface. See Figure 78. The sides of a triangle drawn on a sphere would be arcs of great circles. And, in Riemannian geometry, the sum of the measures of the angles in any triangle is *more* than 180°.

Georg Friedrich Bernhard Riemann (1826–1866) was a German mathematician. Though he lived a short time and published few papers, his work forms a basis for much modern mathematics. He made significant contributions to the theory of functions and the study of complex numbers as well as to geometry. Most calculus books today use the idea of a "Riemann sum" in defining the integral.

Riemann achieved a complete understanding of the non-Euclidean geometries of his day, expressing them on curved surfaces and showing how to extend them to higher dimensions.

FIGURE 78

Topology This chapter began by suggesting that Euclidean geometry might seem rigid from a modern-day point of view. The plane and space figures studied in the Euclidean system are carefully distinguished by differences in size, shape, angularity, and so on. For a given figure such properties are permanent, and thus we can ask sensible questions about congruence and similarity. Suppose we studied "figures" made of rubber bands, as it were, "figures" that could be stretched, bent, or otherwise distorted without tearing or scattering. **Topology,** an important twentieth-century geometry, does just that.

Topological questions concern the basic structure of objects rather than size or arrangement. For example, a typical topological question has to do with the number of holes in an object, a basic structural property that does not change during deformation. You cannot deform a rubber ball to get a rubber band without tearing it—making a hole in it. Thus the two objects are not topologically equivalent. On the other hand, a doughnut and a coffee cup are topologically equivalent, since one could be stretched so as to form the other, without changing the basic structural property.

Topology in Medicine
Trypanosome is a parasite that causes sleeping sickness in humans. The genetic material for this parasite contains a large number of linked circles. If the drug ethidium bromide is introduced into the cells, these circles become hopelessly intertwined, and the parasite cannot reproduce. Thus, topology can be used to cure sleeping sickness!

EXAMPLE 1 Decide if the figures in each pair are topologically equivalent.

(a) a football and a cereal box

If we assume that a football is made of a perfectly elastic substance such as rubber or dough, it could be twisted or kneaded into the same shape as a cereal box. Thus, the two figures are topologically equivalent.

(b) a doughnut and an unzipped coat

A doughnut has one hole, while the coat has two (the sleeve openings). Thus, a doughnut could not be stretched and twisted into the shape of the coat without tearing another hole in it. Because of this, a doughnut and the coat are not topologically equivalent.

FOR FURTHER THOUGHT

Two examples of topological surfaces are the **Möbius strip** and the **Klein bottle**. The Möbius strip is a single-sided surface named after August Ferdinand Möbius (1790–1868), a pupil of Gauss.

To construct a Möbius strip, cut out a rectangular strip of paper, perhaps 3 cm by 25 cm. Paste together the two 3-cm ends after giving the paper a half-twist. To see how the strip now has only one side, mark an *x* on the strip and then mark another *x* on what appears to be the other "side." Begin at one of the *x*'s you have drawn, and trace a path along the strip. You will eventually come to the other *x* without crossing the edge of the strip.

A branch of chemistry called chemical topology studies the structures of chemical configurations. A recent advance in this area was the synthesis of the first molecular Möbius strip, which was formed by joining the ends of a double-stranded strip of carbon and oxygen atoms.

A mathematician confided
That a Möbius strip is one-sided.
And you'll get quite a laugh
If you cut one in half,
For it stays in one piece when divided.

Möbius strip

Klein bottle

Whereas a Möbius strip results from giving a paper *strip* a half-twist and then connecting it to itself, if we could do the same thing with a paper *tube* we would obtain a Klein bottle, named after Felix Klein (1849–1925). Klein produced important results in several areas, including non-Euclidean geometry and the early beginnings of group theory. (It is not possible to construct an actual Klein bottle. However, for some interesting blown glass models, see www.kleinbottle.com.)

A mathematician named Klein
Thought the Möbius strip was divine.
Said he, "If you glue
The edges of two
You'll get a weird bottle like mine."

For Group Discussion

1. The Möbius strip has other interesting properties. With a pair of scissors, cut the strip lengthwise. Do you get two strips? Repeat the process with what you have obtained from the first cut. What happens?
2. Now construct another Möbius strip, and start cutting lengthwise about 1/3 of the way from one edge. What happens?
3. What would be the advantage of a conveyor belt with the configuration of a Möbius strip?

In topology, figures are classified according to their **genus**—that is, the number of cuts that can be made without cutting the figures into two pieces. The genus of an object is the number of holes in it. See Figure 79 on the next page.

9.7 Non-Euclidean Geometry, Topology, and Networks

Torus One of the most useful figures in topology is the torus, a doughnut-like or tire-shaped surface. Its properties are different from those of a sphere, for example. Imagine a sphere covered with hair. You cannot comb the hairs in a completely smooth way; one fixed point remains, as you can find on your own head. In the same way, on the surface of Earth the winds are not a smooth system. There is a calm point somewhere. However, the hair on a torus could be combed smooth. The winds would be blowing smoothly on a planet shaped like a torus.

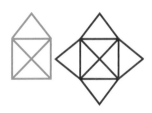

FIGURE 80

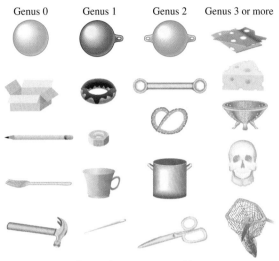

Genus of some common objects

FIGURE 79

Networks* Another branch of modern geometry is graph theory. One topic of study in graph theory is networks.

A **network** is a diagram showing the various paths (or **arcs**) between points (called **vertices,** or **nodes**). A network can be thought of as a set of arcs and vertices. Figure 80 shows two examples of networks. The study of networks began formally with the so-called Königsberg Bridge problem as solved by Leonhard Euler (1707–1783). In Königsberg, Germany, the River Pregel flowed through the middle of town. There were two islands in the river. During Euler's lifetime, there were seven bridges connecting the islands and the two banks of the river.

The people of the town loved Sunday strolls among the various bridges. Gradually, a competition developed to see if anyone could find a route that crossed each of the seven bridges exactly once. The problem concerns what topologists today call the *traversability* of a network. No one could find a solution. The problem became so famous that in 1735 it reached Euler, who was then at the court of the Russian empress Catherine the Great. In trying to solve the problem, Euler began by drawing a network representing the system of bridges, as in Figure 81.

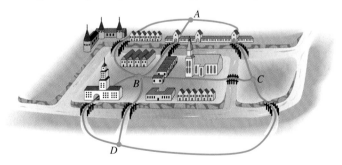

FIGURE 81

---

*An entire chapter on Graph Theory, of which networks is one topic, is available from the publisher. If this chapter is not part of this book and you are interested in more information on networks, contact Addison-Wesley.

Euler first noticed that three routes meet at vertex A. Since 3 is an odd number, he called A an **odd vertex.** Since three routes meet at A, it must be a starting or an ending point for any traverse of the network. This must be true; otherwise, when you got to A on your second trip there would be no way to get out. An **even vertex,** one where an even number of routes meet, need not be a starting or an ending point. (Why is this?) Three paths also meet at C and D, with five paths meeting at B. Thus, B, C, and D are also odd vertices. An odd vertex must be a starting or an ending point of a traverse. Thus, all four vertices A, B, C, and D must be starting or ending points. Since a network can have only two starting or ending points (one of each), this network cannot be traversed. The residents of Königsberg were trying to do the impossible.

Euler's result can be summarized as follows.

Results on Vertices and Traversability

1. The number of odd vertices of any network is *even.* (That is, a network must have $2n$ odd vertices, where $n = 0, 1, 2, 3, \ldots$.)
2. A network with no odd vertices or exactly two odd vertices can be traversed. In the case of exactly two, start at one odd vertex and end at the other.
3. A network with more than two odd vertices cannot be traversed.

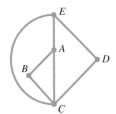

FIGURE 82

EXAMPLE 2 Decide whether the networks in Figure 82 and Figure 83 are traversable.

Since there are exactly two odd vertices (A and E) in Figure 82, it can be traversed. One way to traverse the network of the figure is to start at A, go through B to C, then back to A. (It is acceptable to go through a vertex as many times as needed.) Then go to E, to D, to C, and finally go back to E. It is traversable.

In Figure 83, because vertices A, B, C, and D are all odd (five routes meet at each of them), this network is not traversable. (One of the authors of this text, while in high school, tried for hours to traverse it, not knowing he was attempting the impossible!) ■

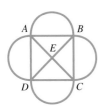

FIGURE 83

EXAMPLE 3 Figure 84 shows the floor plan of a house. Is it possible to travel through this house, going through each door exactly once?

Rooms A, C, and D have even numbers of doors, while rooms B and E have odd numbers of doors. If we think of the rooms as the vertices of a graph, then the fact that we have exactly two odd vertices means that it is possible to travel through each door of the house exactly once. (Show how—start in room B or room E.) ■

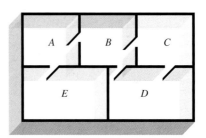

FIGURE 84

9.7 EXERCISES

The chart that follows characterizes certain properties of Euclidean and non-Euclidean geometries. Study it, and use it to respond to Exercises 1–10.

EUCLIDEAN	NON-EUCLIDEAN	
Dates back to about 300 B.C.	**Lobachevskian (about 1830)**	**Riemannian (about 1850)**
Lines have *infinite* length.		Lines have *finite* length.
Geometry on a plane	Geometry on a surface like a pseudosphere	Geometry on a sphere
Angles C and D of a Saccheri quadrilateral are *right* angles.	Angles C and D are *acute* angles.	Angles C and D are *obtuse* angles.
Given point P off line k, exactly *one* line can be drawn through P and parallel to k.	*More than one* line can be drawn through P and parallel to k.	*No* line can be drawn through P and parallel to k.
Typical triangle ABC	Typical triangle ABC	Typical triangle ABC
Two triangles can have the same size angles but different size sides (similarity as well as congruence).	Two triangles with the same size angles must have the same size sides (congruence only).	

1. In which geometry is the sum of the measures of the angles of a triangle equal to 180°?

2. In which geometry is the sum of the measures of the angles of a triangle greater than 180°?

3. In which geometry is the sum of the measures of the angles of a triangle less than 180°?

4. In a quadrilateral *ABCD* in Lobachevskian geometry, the sum of the measures of the angles must be _____ 360°.
(less than/greater than)

5. In a quadrilateral *ABCD* in Riemannian geometry, the sum of the measures of the angles must be _____ 360°.
(less than/greater than)

6. Suppose that *m* and *n* represent lines through *P* that are both parallel to *k*. In which geometry is this possible?

7. Suppose that *m* and *n* below *must* meet at a point. In which geometry is this possible?

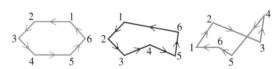

8. A globe representing the earth is a model for a surface in which geometry?

9. In which geometry is this statement possible? "Triangle *ABC* and triangle *DEF* are such that ∡*A* = ∡*D*, ∡*B* = ∡*E*, and ∡*C* = ∡*F*, and they have different areas."

10. Draw a figure (on a sheet of paper) as best you can showing the shape formed by the north pole *N* and two points *A* and *B* lying at the equator of a model of the earth.

11. Pappus, a Greek mathematician in Alexandria about A.D. 320, wrote a commentary on the geometry of the times. We will work out a theorem of his about a hexagon inscribed in two intersecting lines.

First we need to define an old word in a new way: a **hexagon** consists of any six lines in a plane, no three of which meet in the same point. As the figure shows, the vertices of several hexagons are labeled with numbers. Thus 1–2 represents a line segment joining vertices 1 and 2. Segments 1–2 and 4–5 are opposite sides of a hexagon, as are 2–3 and 5–6, and 3–4 and 1–6.
  (a) Draw an angle less than 180°.
  (b) Choose three points on one side of the angle. Label them 1, 5, 3 in that order, beginning with the point nearest the vertex.
  (c) Choose three points on the other side of the angle. Label them 6, 2, 4 in that order, beginning with the point nearest the vertex.
  (d) Draw line segments 1–6 and 3–4. Draw lines through the segments so they extend to meet in a point; call it *N*.
  (e) Let lines through 1–2 and 4–5 meet in point *M*.
  (f) Let lines through 2–3 and 5–6 meet in *P*.
  (g) Draw a straight line through points *M*, *N*, and *P*.
  (h) Write in your own words a theorem generalizing your result.

12. The following theorem comes from projective geometry:

    Theorem of Desargues in a Plane Desargues's theorem states that in a plane, if two triangles are placed so that lines joining corresponding vertices meet in a point, then corresponding sides, when extended, will meet in three collinear points. (*Collinear* points are points lying on the same line.)

    Draw a figure that illustrates this theorem.

9.7 Non-Euclidean Geometry, Topology, and Networks

In Exercises 13–20, each figure may be topologically equivalent to none or some of the objects labeled A–E. List all topological equivalences (by letter) for each figure.

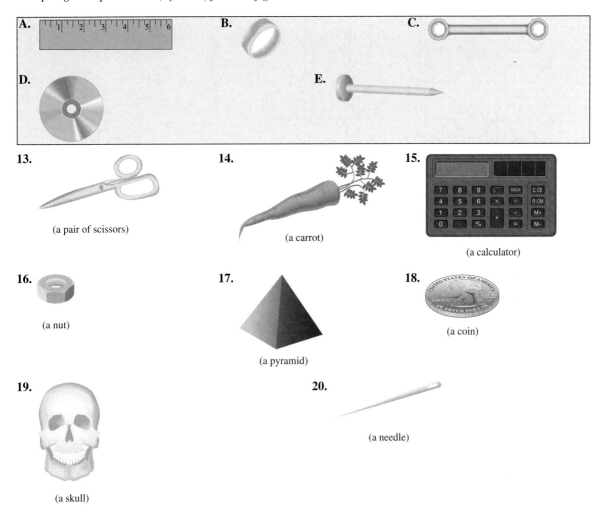

Topological Equivalence *Someone once described a topologist as "a mathematician who doesn't know the difference between a doughnut and a coffee cup." This is due to the fact that both are of genus 1—they are topologically equivalent. Based on this interpretation, would a topologist know the difference between each of the following pairs of objects?*

21. a spoon and a fork
22. a mixing bowl and a colander
23. a slice of American cheese and a slice of Swiss cheese
24. a compact disc and a phonograph record

Give the genus of each of the following objects.

25. a compact disc
26. a phonograph record
27. a sheet of loose-leaf paper made for a three-ring binder
28. a sheet of loose-leaf paper made for a two-ring binder
29. a wedding band
30. a postage stamp

For each of the following networks, decide whether each lettered vertex is even *or* odd.

31.    **32.** 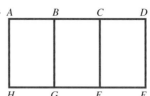   **33.**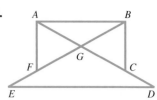

34. 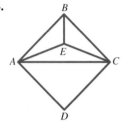   **35.** 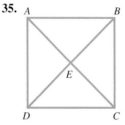   **36.**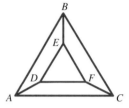

Decide whether each network is traversable. If a network is traversable, show how it can be traversed.

37.    **38.**    **39.**

40.    **41.**    **42.**

Is it possible to walk through each door of the following houses exactly once? If the answer is "yes," show how it can be done.

43. 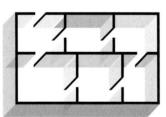   **44.**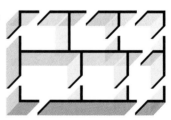

45. 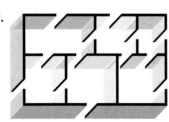   **46.**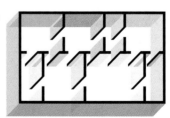

9.8 Chaos and Fractal Geometry

Chaos

Does Chaos Rule the Cosmos?
—One of the ten great unanswered questions of science, as found in the November 1992 issue of *Discover*

Consider the equation $y = kx(1 - x)$. Choosing $k = 2$ gives the equation $y = 2x(1 - x)$, which can be "iterated" by starting with an arbitrary x-value between 0 and 1, calculating the resulting y-value, substituting that y-value back in as x, calculating the resulting y-value, substituting that y-value back in as x, calculating another y-value, and so on. For example, a starting value of $x = .8$ produces the following sequence (which you can verify with a calculator):

$$.8, .32, .435, .492, .500, .500, .500, \text{ and so on.}$$

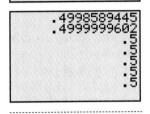

These two screens show how the TI-83 Plus calculator can produce the sequence described.

The sequence seems to begin randomly but quickly stabilizes at the value .500. A different initial x-value would produce another sequence which would also "converge" to .500. The value .500 can be called an *attractor* for the sequence generated by the equation $y = 2x(1 - x)$. The values of the sequence are "attracted" toward .500.

EXAMPLE 1 For the equation $y = kx(1 - x)$ with $k = 3$, begin with $x = .7$ and iterate with a calculator. What pattern emerges?

Using a TI-83 Plus calculator, we find the seventeenth through twentieth iterations give the sequence terms

$$.6354387337, \quad .6949690482, \quad .6359612107, \quad .6945436475.$$

This sequence apparently converges in a manner different from the initial discussion, alternating between values near .636 and .695. Therefore, for $k = 3$, the sequence tends alternately toward two distinct attractors.

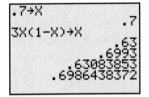
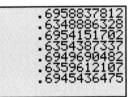

It happens that the equation in Example 1 exhibits the same behavior for any initial value of x between 0 and 1. You are asked to show this for several cases in the exercises.

EXAMPLE 2 In the equation of Example 1, change the multiplier k to 3.5, and find the forty-fourth through fifty-first terms. What pattern emerges?

Again, using a TI-83 Plus calculator, and rounding to three decimal places, we get

$$.383, .827, .501, .875, .383, .827, .501, .875.$$

This sequence seems to stabilize around four alternating attractors, .383, .827, .501, and .875.

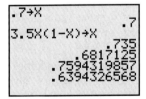
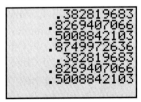

These screens support the discussion in Examples 1 and 2.

Notice that in our initial discussion, for $k = 2$, the sequence converged to *one* attractor. In Example 1, for $k = 3$, it converged to *two* attractors, and in Example 2, for $k = 3.5$, it converged to *four* attractors.

John Nash, a notable modern American mathematician (born in 1928), first came to the attention of the general public through the recent biography *A Beautiful Mind* (and the movie of the same name). In 1958 Nash narrowly lost out to René Thom (topologist and inventor of catastrophe theory, discussed on page 569) for the Fields Medal, the mathematical equivalent of the Nobel prize. Although his brilliant career was sadly interrupted by mental illness for a period of about thirty years, in 1994 Nash was awarded "the Central Bank of Sweden Prize in Economic Science in Memory of Alfred Nobel," generally regarded as equivalent to the Nobel prize. This award was for Nash's equilibrium theorem, published in his doctoral thesis in 1950. It turned out that Nash's work established a significant new way of analyzing rational conflict and cooperation in economics and other social sciences.

If k is increased further, it turns out that the number of attractors doubles over and over again, more and more often. In fact, this doubling has occurred infinitely many times before k even gets as large as 4. When we look closely at groups of these doublings we find that they are always similar to earlier groups but on a smaller scale. This is called *self-similarity,* or *scaling,* an idea that is not new, but which has taken on new significance in recent years. Somewhere before k reaches 4, the resulting sequence becomes apparently totally random, with no attractors and no stability. This type of condition is one instance of what has come to be known in the scientific community as **chaos.** This name came from an early paper by the mathematician James A. Yorke, of the University of Maryland at College Park.

The equation $y = kx(1 - x)$ does not look all that complicated, but the intricate behavior exhibited by it and similar equations has occupied some of the brightest minds (not to mention computers) in various fields—ecology, biology, physics, genetics, economics, mathematics—since about 1960. Such an equation might represent, for example, the population of some animal species where the value of k is determined by factors (such as food supply, or predators that prey on the species) that affect the increase or decrease of the population. Under certain conditions there is a long-run steady-state population (a single attractor). Under other conditions the population will eventually fluctuate between two alternating levels (two attractors), or four, or eight, and so on. But after a certain value of k, the long-term population becomes totally chaotic and unpredictable.

As long as k is small enough, there will be some number of attractors and the long-term behavior of the sequence (or population) is the same regardless of the initial x-value. But once k is large enough to cause chaos, the long-term behavior of the system will change drastically when the initial x-value is changed only slightly. For example, consider the following two sequences, both generated from $y = 4x(1 - x)$.

$$.600, .960, .154, .521, .998, .008, .032, \ldots$$
$$.610, .952, .183, .598, .962, .146, .499, \ldots$$

The fact that the two sequences wander apart from one another is partly due to roundoff errors along the way. But Yorke and others have shown that even "exact" calculations of the iterates would quickly produce divergent sequences just because of the slightly different initial values. This type of "sensitive dependence on initial conditions" was discovered (accidentally) back in the 1960s by Edward Lorenz when he was looking for an effective computerized model of weather patterns. He discerned the implication that any long-range weather predicting schemes might well be hopeless.

Patterns like those in the sequences above are more than just numerical oddities. Similar patterns apply to a great many phenomena in the physical, biological, and social sciences, many of them seemingly common natural systems that have been studied for hundreds of years. The measurement of a coastline, the description of the patterns in a branching tree, or a mountain range, or a cloud formation, or intergalactic cosmic dust, the prediction of weather patterns, the turbulent behavior of fluids of all kinds, the circulatory and neurological systems of the human body, fluctuations in populations and economic systems—these and many other phenomena remain mysteries, concealing their true nature somewhere beyond the reach of even our biggest and fastest computers.

Continuous phenomena are easily dealt with. A change in one quantity produces a predictable change in another. (For example, a little more pressure on the gas pedal produces a little more speed.) Mathematical functions that represent continuous events can be graphed by unbroken lines or curves, or perhaps smooth, gradually changing surfaces. The governing equations for such phenomena are "linear," and extensive mathematical methods of solving them have been developed. On the other hand, erratic events associated with certain other equations are harder to describe or predict. The science of chaos, made possible by modern computers, continues to open up new ways to deal with such events.

One early attempt to deal with discontinuous processes in a new way, generally acknowledged as a forerunner of chaos theory, was that of the French mathematician René Thom, who, in the 1960s, applied the methods of topology. To emphasize the feature of sudden change, Thom referred to events such as a heartbeat, a buckling beam, a stock market crash, a riot, or a tornado, as *catastrophes*. He proved that all catastrophic events (in our four-dimensional space-time) are combinations of seven elementary catastrophes. (In higher dimensions the number quickly approaches infinity.)

Each of the seven elementary catastrophes has a characteristic topological shape. Two examples are shown in Figure 85. The top figure is called a *cusp*. The bottom figure is an *elliptic umbilicus* (a belly button with an oval cross-section). Thom's work became known as **catastrophe theory.**

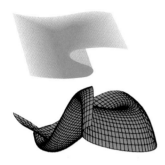

FIGURE 85

Computer graphics have been indispensable in the study of chaotic processes. The plotting of large numbers of points has revealed patterns that would otherwise have not been observed. (The underlying reasons for many of these patterns, however, have still not been explained.) The images shown in Figure 86 are created using chaotic processes.

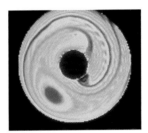 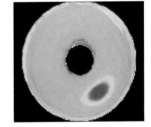

FIGURE 86

The surface of the earth, consisting of continents, mountains, oceans, valleys, and so on, has fractal dimension 2.2.

Fractals If there is one structure that has provided a key for the new study of nonlinear processes, it is **fractal geometry,** developed over a period of years mainly by the IBM mathematician Benoit Mandelbrot (1924–   ). For his work in this field, and at the recommendation of the National Science Foundation, Columbia University awarded Mandelbrot the 1985 Bernard Medal for Meritorious Service to Science.

Lines have a single dimension. Plane figures have two dimensions, and we live in a three-dimensional world. In a paper published in 1967, Mandelbrot investigated the idea of measuring the length of a coastline. He concluded that such a shape defies conventional Euclidean geometry and that rather than having a natural number dimension, it has a "fractional dimension." A coastline is an example of a *self-similar shape*—a shape that repeats itself over and over on different scales. From a distance, the bays and inlets cannot be individually observed, but as one moves closer they become more apparent. The branching of a tree, from twig to limb to trunk, also exhibits a shape that repeats itself.

FIGURE 87

FIGURE 88

In the early twentieth century, the German mathematician H. von Koch investigated the so-called Koch snowflake. It is shown in Figure 87. Starting with an equilateral triangle, each side then gives rise to another equilateral triangle. The process continues over and over, indefinitely, and a curve of infinite length is produced. The mathematics of Koch's era was not advanced enough to deal with such figures. However, using Mandelbrot's theory, it is shown that the Koch snowflake has dimension of about 1.26. This figure is obtained using a formula which involves logarithms. (Logarithms were introduced briefly in an earlier chapter.)

The theory of fractals is today being applied to many areas of science and technology. It has been used to analyze the symmetry of living forms, the turbulence of liquids, the branching of rivers, and price variation in economics. Hollywood has used fractals in the special effects found in some blockbuster movies. Figure 88 shows an example of a computer-generated fractal design.

An interesting account of the science of chaos is found in the popular 1987 book *Chaos*, by James Gleick. Mandelbrot has published two books on fractals. They are *Fractals: Form, Chance, and Dimension* (1975), and *The Fractal Geometry of Nature* (1982).

Aside from providing a geometric structure for chaotic processes in nature, fractal geometry is viewed by many as a significant new art form. (To appreciate why, see the 1986 publication *The Beauty of Fractals*, by H. O. Peitgen and P. H. Richter, which contains 184 figures, many in color.) Peitgen and others have also published *Fractals for the Classroom: Strategic Activities Volume One* (Springer-Verlag, 1991).

9.8 EXERCISES

Exercises 1–25 are taken from an issue of Student Math Notes, *published by the National Council of Teachers of Mathematics. They were written by Dr. Tami S. Martin, Mathematics Department, Illinois State University, and the authors wish to thank N.C.T.M. and Tami Martin for permission to reproduce this activity. Since the exercises should be done in numerical order, answers to all exercises (both even- and odd-numbered) appear in the answer section of this book.*

Most of the mathematical objects you have studied have dimensions that are whole numbers. For example, such solids as cubes and icosahedrons have dimension three. Squares, triangles, and many other planar figures are two-dimensional. Lines are one-dimensional, and points have dimension zero. Consider a square with side of length one. Gather several of these squares by cutting out or using patterning blocks.

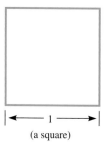

(a square)

The size of a figure is calculated by counting the number of *replicas* (*small pieces*) *that make it up.* Here, a replica is the original square with edges of length one.

1. What is the least number of these squares that can be put together edge to edge to form a larger square?

The original square is made up of one small square, so its size is one.

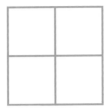

2. What is the size of the new square?
3. What is the length of each edge of the new square?

Similar figures have the same shape but are not necessarily the same size. The **scale factor** *between two similar figures can be found by calculating the ratio of corresponding edges:*

$$\frac{\text{new length}}{\text{old length}}.$$

4. What is the scale factor between the large square and the small square?
5. Find the ratio

$$\frac{\text{new size}}{\text{old size}}$$

for the two squares.

6. Form an even larger square that is three units long on each edge. Compare this square to the small square. What is the scale factor between the two squares? What is the ratio of the new size to the old size?
7. Form an even larger square that is four units long on each edge. Compare this square to the small square. What is the scale factor between the two squares? What is the ratio of the new size to the old size?
8. Complete the table for squares.

Scale Factor	2	3	4	5	6	10
Ratio of new size to old size						

9. How are the two rows in the table related?

Consider an equilateral triangle. The length of an edge of the triangle is one unit. The size of this triangle is one.

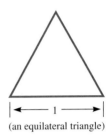

(an equilateral triangle)

10. What is the least number of equilateral triangles that can be put together edge to edge to form a similar larger triangle?
11. Complete the table for triangles.

Scale Factor	2	3	4	5	6	10
Ratio of new size to old size						

12. How does the relationship between the two rows in this table compare with the one you found in the table for squares?

One way to define the dimension, d, of a figure relates the scale factor, the new size, and the old size:

$$(\text{scale factor})^d = \frac{\text{new size}}{\text{old size}}.$$

Using a scale factor of two for squares or equilateral triangles, we can see that $2^d = 4/1$; that is, $2^d = 4$. Since $2^2 = 4$, the dimension, d, must be two. This definition of dimension confirms what we already know—that squares and equilateral triangles are two-dimensional figures.

13. Use this definition and your completed tables to confirm that the square and the equilateral triangle are two-dimensional figures for scale factors other than two.

Consider a cube, with edges of length one. Let the size of the cube be one.

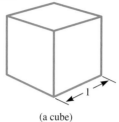

(a cube)

14. What is the least number of these cubes that can be put together face to face to form a larger cube?
15. What is the scale factor between these two cubes? What is the ratio of the new size to the old size for the two cubes?
16. Complete the table for cubes.

Scale Factor	2	3	4	5	6	10
Ratio of new size to old size						

17. How are the two rows in the table related?
18. Use the definition of dimension and a scale factor of two to verify that a cube is a three-dimensional object.

We have explored scale factors and sizes associated with two- and three-dimensional figures. Is it possible for mathematical objects to have fractional dimensions? Consider the following figure formed by replacing the middle third of a line segment of length one by one upside-down V, each of whose two sides are equal in length to the segment removed. The first four stages in the development of this figure are shown.

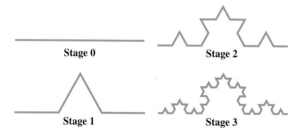

Finding the scale factor for this sequence of figures is difficult because the overall length of a representative portion of the figure remains the same while the number of pieces increases. To simplify the procedure, follow these steps.

Step 1: Start with any stage (e.g., stage 1).

Step 2: Draw the next stage (e.g., stage 2) of the sequence and "blow it up" so that it contains an exact copy of the preceding stage (in this example, stage 1).

Notice that stage 2 contains four copies, or replicas, of stage 1 and is three times as long as stage 1.

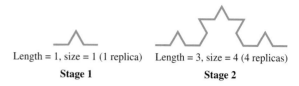

Length = 1, size = 1 (1 replica) Length = 3, size = 4 (4 replicas)
Stage 1 **Stage 2**

19. The scale factor is equal to the ratio
$$\frac{\text{new length}}{\text{old length}}$$
between any two consecutive stages. The scale factor between stage 1 and stage 2 is _____.

20. The size can be determined by counting the number of replicas of stage 1 found in stage 2. Old size = 1, new size = _____.

Use the definition of dimension to compute the dimension, d, of the figure formed by this process: $3^d = 4/1$; that is, $3^d = 4$. Since $3^1 = 3$ and $3^2 = 9$, for $3^d = 4$ the dimension of the figure must be greater than one but less than two: $1 < d < 2$.

21. Use your calculator to estimate d. Remember that d is the exponent that makes 3^d equal 4. For example, since d must be between 1 and 2, try $d = 1.5$. But $3^{1.5} = 5.196...$, which is greater than 4; thus d must be smaller than 1.5. Continue until you approximate d to three decimal places. (Use logarithms for an exact determination.)

*The original figure was a one-dimensional line segment. By iteratively adding to the line segment, an object of dimension greater than one but less than two was generated. Objects with fractional dimension are known as **fractals**. Fractals are infinitely self-similar objects formed by repeated additions to, or removals from, a figure. The object attained at the limit of the repeated procedure is the fractal.*

Next consider a two-dimensional object with sections removed iteratively. In each stage of the fractal's development, a triangle is removed from the center of each triangular region.

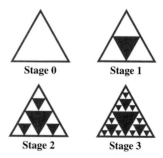

Stage 0   Stage 1
Stage 2   Stage 3

Use the process from the last example to help answer the following questions.

22. What is the scale factor of the fractal?

23. Old size = 1, new size = _____ .

24. The dimension of the fractal is between what two whole number values?

25. Use the definition of dimension and your calculator to approximate the dimension of this fractal to three decimal places.

Use a calculator to determine the pattern of attractors for the equation $y = kx(1 - x)$ for the given value of k and the given initial value of x.

26. $k = 3.25, x = .7$ **27.** $k = 3.4, x = .8$

28. $k = 3.55, x = .7$

COLLABORATIVE INVESTIGATION

Generalizing the Angle Sum Concept

In this chapter we learned that the sum of the measures of the angles of a triangle is 180°. This fact can be extended to determine a formula for the sum of the measures of the angles of any convex polygon. To begin this investigation, divide into groups of three or four students each. Prepare on a sheet of paper six figures as shown on the right.

Now we define a diagonal from vertex A to be a segment from A to a non-adjacent vertex. The triangle in Figure I has no diagonals, but the polygons in Figures II through VI do have them.

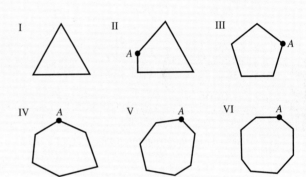

Topics for Discussion

1. Choose someone in the group to draw all possible diagonals from vertex A in each figure.
2. Now complete the following table as a group.

Polygon	Number of Sides	Number of Triangles, t	Number of Degrees in Each Triangle	Sum of the Measures of All Angles of the Polygon, $t \cdot 180°$
I	___	___	___	___
II	___	___	___	___
III	___	___	___	___
IV	___	___	___	___
V	___	___	___	___
VI	___	___	___	___

3. Based on the table you completed, answer the following in order.
   (a) As suggested by the table, the number of triangles that a convex polygon can be divided into is _____ less than the number of sides.
   (b) Thus, if a polygon has s sides, it can be divided into _____ triangles.
   (c) From the table we see that the sum of the measures of all the angles of a polygon can be found from the expression $t \cdot 180°$. Thus, if a polygon has s sides, the sum of the measures of all the angles of the polygon is given by the expression (_____ − _____) · _____°.

4. Use your discovery from Exercise 3(c) to find the sum of the measures of all the angles of
   (i) a nonagon   (ii) a decagon
   (iii) a 12-sided polygon.

CHAPTER 9 TEST

1. Consider a 38° angle. Answer each of the following.
   (a) What is the measure of its complement?
   (b) What is the measure of its supplement?
   (c) Classify it as acute, obtuse, right, or straight.

Find the measure of each marked angle.

2.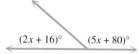

3.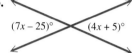

4.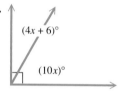

In Exercises 5 and 6, assume that lines m and n are parallel, and find the measure of each marked angle.

5.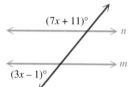

6.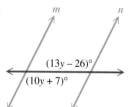

7. Explain why a rhombus must be a parallelogram, but a parallelogram might not be a rhombus.

8. Which one of the statements A.–D. is false?
   A. A square is a rhombus.
   B. The acute angles of a right triangle are complementary.
   C. A triangle may have both a right angle and an obtuse angle.
   D. A trapezoid may have nonparallel sides of the same length.

Identify each of the following curves as simple, closed, both, or neither.

9.

10.

11. Find the measure of each angle in the triangle.

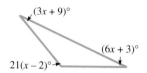

Find the area of each of the following figures.

12.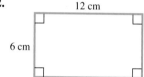

13.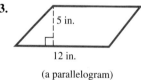
(a parallelogram)

14.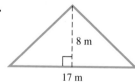

15.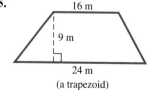
(a trapezoid)

16. If a circle has area 144π square inches, what is its circumference?

17. *Circumference of a Dome* The Skydome in Toronto, Canada, is the first stadium with a hard-shell, retractable roof. The steel dome is 630 feet in diameter. To the nearest foot, what is the circumference of this dome?

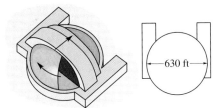

18. *Area of a Shaded Figure* What is the area of the shaded portion of the figure? Use 3.14 as an approximation for π.

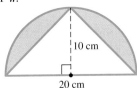

(a triangle within a semicircle)

19. Given: $\angle CAB = \angle DBA$; $DB = CA$
Prove: $\triangle ABD \cong \triangle BAC$

20. *Height of a Pole* If a 30-ft pole casts a shadow 45 ft long, how tall is a pole whose shadow is 30 ft long at the same time?

21. *Diagonal of a Rectangle* What is the measure of a diagonal of a rectangle that has width 20 m and length 21 m?

22. First, reflect the given figure about line n, and then about line m.

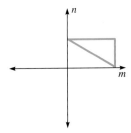

23. Find the point reflection image of the given figure with the given point as center.

Find **(a)** *the volume and* **(b)** *the surface area of each of the following space figures. When necessary, use 3.14 as an approximation for π.*

24.

(a sphere)

25.

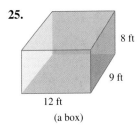

(a box)

26.

(a right circular cylinder)

27. List several main distinctions between Euclidean geometry and non-Euclidean geometry.

28. *Topological Equivalence* Are the following pairs of objects topologically equivalent?
  (a) a page of a book and the cover of the same book
  (b) a pair of glasses with the lenses removed, and the Mona Lisa

29. Decide whether it is possible to traverse the network shown. If it is possible, show how it can be done.

(a)  **(b)**

30. Use a calculator to determine the attractors for the sequence generated by the equation $y = 2.1x(1 - x)$, with initial value of $x = .6$.

chapter

10 Trigonometry

The foundations of trigonometry go back at least 3000 years. The ancient Egyptians, Babylonians, and Greeks developed trigonometry to find the lengths of the sides of triangles and the measures of their angles. In Egypt, trigonometry was used to reestablish land boundaries after the annual flood of the Nile River. In Babylonia it was used in astronomy. The very word *trigonometry* comes from the Greek words for triangle (*trigon*) and measurement (*metry*). Today trigonometry is used in electronics, surveying, and other engineering areas, and is necessary for further courses in mathematics, such as calculus.

For centuries, astronomers tried to determine the distances to various stars. It was not until 1838 that the astronomer Friedrich Bessel determined the distance to a star called 61 Cygni. He used a *parallax* method that relied on the measurement of very small angles. This measurement confirmed that the heliocentric model of Copernicus was correct and gave scientists a better understanding of the size and structure of the universe.

You observe parallax when you ride in an automobile and see a nearby object apparently moving backward with respect to more distant objects. The same is true for some stars that are relatively close to Earth. As Earth revolves around the sun, the observed angle θ (the Greek letter *theta*) of some nearby stars changes due to parallax, as shown in the figure. The table lists the size of angle θ in seconds for five stars. (One second is equal to 1/3600 of a degree.)

10.1 Angles and Their Measures

10.2 Trigonometric Functions of Angles

10.3 Trigonometric Identities

10.4 Right Triangles and Function Values

10.5 Applications of Right Triangles

10.6 The Laws of Sines and Cosines; Area Formulas

10.7 The Unit Circle and Graphs

Collaborative Investigation:
  Making a *Point* About Trigonometric Function Values

Chapter 10 Test

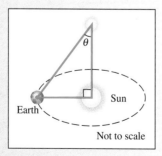

Star	θ
Alpha Centauri	.763
Barnard's Star	.546
Sirius	.377
61 Cygni	.292
Procyon	.287

Sources: Freebury, H. A., *A History of Mathematics*, MacMillan Company, 1968; Zeilik, M., S. Gregory, and E. Smith, *Introductory Astronomy and Astrophysics*, Second Edition, Saunders College Publishers, 1998.

10.1 Angles and Their Measures

Since stars are very distant objects, the parallax of a star is small and always less than one second. How can we use θ to find the distance to these stars? To solve problems like estimating distances to stars, determining the region covered by a total solar eclipse, or approximating the depth of a crater on the moon, we must understand angles, triangles, and trigonometric functions. This chapter presents a short introduction to trigonometry.

10.1 Angles and Their Measures

Basic Terminology A line may be drawn through the two distinct points A and B. This line is called **line AB**. The portion of the line between A and B, including points A and B themselves, is **segment AB**. The portion of the line AB that starts at A and continues through B, and on past B, is called **ray AB**. Point A is the endpoint of the ray. See Figure 1.

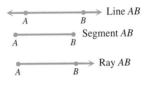

FIGURE 1

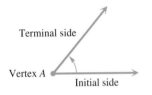

FIGURE 2

In the study of trigonometry, an **angle** is formed by rotating a ray around its endpoint. The ray in its initial position is called the **initial side** of the angle, while the ray in its location after the rotation is the **terminal side** of the angle. The endpoint of the ray is the **vertex** of the angle. Figure 2 shows the initial and terminal sides of an angle with vertex A.

If the rotation of the terminal side is counterclockwise, the angle is **positive**. If the rotation is clockwise, the angle is **negative**. Figure 3 shows two angles, one positive and one negative.

An angle can be named by using the name of its vertex. For example, the angle on the right in Figure 3 can be called angle C. Alternatively, an angle can be named using three letters, with the vertex letter in the middle. Thus, the angle on the right also could be named angle ACB or angle BCA.

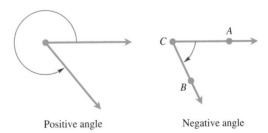

Positive angle   Negative angle

FIGURE 3

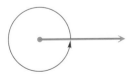

A complete rotation of a ray gives an angle whose measure is 360°.

FIGURE 4

Degree Measure There are two systems in common use for measuring the size of angles. The most common unit of measure is the **degree**. (The other common unit of measure is called the *radian*.) Degree measure was developed by the Babylonians 4000 years ago. To use degree measure we assign 360 degrees to a complete rotation of a ray. In Figure 4, notice that the terminal side of the angle corresponds to its initial side when it makes a complete rotation.

One degree, written 1°, represents 1/360 of a rotation. Therefore, 90° represents 90/360 = 1/4 of a complete rotation, and 180° represents 180/360 = 1/2 of a complete rotation. Angles of measure 1°, 90°, and 180° are shown in Figure 5.

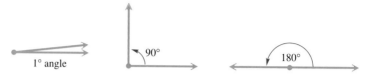

FIGURE 5

Special angles are named as shown in the following chart.

Name	Angle Measure	Example(s)
Acute angle	Between 0° and 90°	60° 82°
Right angle	Exactly 90°	90°
Obtuse angle	Between 90° and 180°	97° 138°
Straight angle	Exactly 180°	180°

> "**Trigonometry**, perhaps more than any other branch of mathematics, developed as the result of a continual and fertile interplay of supply and demand: the supply of applicable mathematical theories and techniques available at any given time and the demands of a single applied science, astronomy. So intimate was the relation that not until the thirteenth century was it useful to regard the two subjects as separate entities." (From "The History of Trigonometry" by Edward S. Kennedy, in *Historical Topics for the Mathematics Classroom,* the Thirty-first Yearbook of N.C.T.M., 1969.)

If the sum of the measures of two angles is 90°, the angles are called **complementary.** Two angles with measures whose sum is 180° are **supplementary.**

EXAMPLE 1 Give the complement and the supplement of 50°.

The complement of 50° is

$$90° - 50° = 40°,$$

while the supplement of 50° is

$$180° - 50° = 130°.$$

Do not confuse an angle with its measure. The *angle itself* consists of the vertex together with the initial and terminal sides, whereas the *measure of the angle* is the size of the rotation angle from the initial to the terminal side (commonly expressed in degrees). If angle A has a 35° rotation angle, we say that m(angle A) is 35°, where m(angle A) is read "the measure of angle A." It saves a lot of work, however, to abbreviate m(angle A) = 35° as simply angle A = 35°.

10.1 Angles and Their Measures

Traditionally, portions of a degree have been measured with minutes and seconds. One **minute,** written $1'$, is $1/60$ of a degree.

$$1' = \frac{1}{60}^\circ \quad \text{or} \quad 60' = 1^\circ$$

One **second,** $1''$, is $1/60$ of a minute.

$$1'' = \frac{1}{60}' = \frac{1}{3600}^\circ \quad \text{or} \quad 60'' = 1'$$

The measure $12°\ 42'\ 38''$ represents 12 degrees, 42 minutes, 38 seconds.

The next example shows how to perform calculations with degrees, minutes, and seconds.

EXAMPLE 2 Perform each calculation.

(a) $51°\ 29' + 32°\ 46'$

Add the degrees and the minutes separately.

$$\begin{array}{r} 51°\ 29' \\ +\ 32°\ 46' \\ \hline 83°\ 75' \end{array}$$

Since $75' = 60' + 15' = 1°\ 15'$, the sum is written

$$\begin{array}{r} 83° \\ +\ 1°\ 15' \\ \hline 84°\ 15'. \end{array}$$

(b) $90° - 73°\ 12'$

Write $90°$ as $89°\ 60'$. Then

$$\begin{array}{r} 89°\ 60' \\ -\ 73°\ 12' \\ \hline 16°\ 48'. \end{array}$$

Because calculators are an integral part of our world today, it is now common to measure angles in **decimal degrees.** For example, $12.4238°$ represents

$$12.4238° = 12\frac{4238}{10{,}000}^\circ.$$

The next example shows how to change between decimal degrees and degrees, minutes, and seconds.

EXAMPLE 3 (a) Convert $74°\ 8'\ 14''$ to decimal degrees. Round to the nearest thousandth of a degree.

Since $1' = \frac{1}{60}^\circ$ and $1'' = \frac{1}{3600}^\circ$,

$$74°\ 8'\ 14'' = 74° + \frac{8}{60}^\circ + \frac{14}{3600}^\circ$$
$$\approx 74° + .1333° + .0039°$$
$$= 74.137° \text{ (rounded).}$$

The calculations explained in Example 2 can be done with a graphing calculator capable of working with degrees, minutes, and seconds.

The conversions in Example 3 can be done on some graphing calculators. The second displayed result was obtained by setting the calculator to show only three places after the decimal point.

(b) Convert 34.817° to degrees, minutes, and seconds.

$$34.817° = 34° + .817°$$
$$= 34° + (.817)(60') \quad\quad 1° = 60'$$
$$= 34° + 49.02'$$
$$= 34° + 49' + .02'$$
$$= 34° + 49' + (.02)(60'') \quad\quad 1' = 60''$$
$$= 34° + 49' + 1.2''$$
$$= 34° \; 49' \; 1.2'' \quad\blacksquare$$

Angles in a Coordinate System An angle is in **standard position** if its vertex is at the origin of a rectangular coordinate system and its initial side lies along the positive x-axis. The two angles in Figure 6 are in standard position. An angle in standard position is said to lie in the quadrant in which its terminal side lies. For example, an acute angle is in quadrant I and an obtuse angle is in quadrant II. Angles in standard position having their terminal sides along the x-axis or y-axis, such as angles with measures 90°, 180°, 270°, and so on, are called **quadrantal angles.**

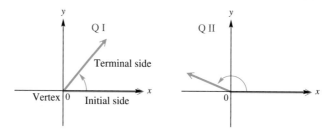

FIGURE 6

A complete rotation of a ray results in an angle of measure 360°. But there is no reason why the rotation need stop at 360°. By continuing the rotation, angles of measure larger than 360° can be produced. The angles in Figure 7(a) have measures 60° and 420°. These two angles have the same initial side and the same terminal side, but different amounts of rotation. Angles that have the same initial side and the same terminal side are called **coterminal angles.** As shown in Figure 7(b), angles with measures 110° and 830° are coterminal.

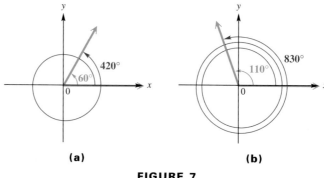

FIGURE 7

10.1 Angles and Their Measures

EXAMPLE 4 Find the angles of smallest possible positive measure coterminal with the following angles.

(a) 908°

Add or subtract 360° as many times as needed to get an angle with measure greater than 0° but less than 360°. Since 908° − 2 · 360° = 908° − 720° = 188°, an angle of 188° is coterminal with an angle of 908°. See Figure 8.

(b) −75°

Use a rotation of 360° + (−75°) = 285°. See Figure 9.

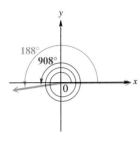

FIGURE 8

Sometimes it is necessary to find an expression that will generate all angles coterminal with a given angle. For example, suppose we wish to do this for a 60° angle. Since any angle coterminal with 60° can be obtained by adding an appropriate integer multiple of 360° to 60°, we can let n represent any integer, and the expression

$$60° + n \cdot 360°$$

will represent all such coterminal angles.

The table below shows a few possibilities.

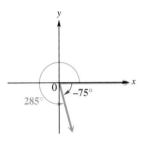

FIGURE 9

Value of n	Angle Coterminal with 60°
2	60° + 2 · 360° = 780°
1	60° + 1 · 360° = 420°
0	60° + 0 · 360° = 60° (the angle itself)
−1	60° + (−1) · 360° = −300°

10.1 EXERCISES

Give **(a)** *the complement and* **(b)** *the supplement of each angle.*

1. 30°
2. 60°
3. 45°
4. 55°
5. 89°
6. 2°

7. If an angle measures x degrees, how can we represent its complement?

8. If an angle measures x degrees, how can we represent its supplement?

Perform each calculation.

9. 62° 18′ + 21° 41′
10. 75° 15′ + 83° 32′
11. 71° 58′ + 47° 29′
12. 90° − 73° 48′
13. 90° − 51° 28′
14. 180° − 124° 51′
15. 90° − 72° 58′ 11″
16. 90° − 36° 18′ 47″

Convert each angle measure to decimal degrees. Use a calculator, and round to the nearest thousandth of a degree.

17. 20° 54′
18. 38° 42′
19. 91° 35′ 54″
20. 34° 51′ 35″
21. 274° 18′ 59″
22. 165° 51′ 9″

Convert each angle measure to degrees, minutes, and seconds. Use a calculator as necessary.

23. 31.4296°
24. 59.0854°
25. 89.9004°
26. 102.3771°
27. 178.5994°
28. 122.6853°

Find the angles of smallest positive measure coterminal with the following angles.

29. −40°   **30.** −98°   **31.** −125°   **32.** −203°
33. 539°   **34.** 699°   **35.** 850°   **36.** 1000°

Give an expression that generates all angles coterminal with the given angle. Let n represent any integer.

37. 30°   **38.** 45°   **39.** 60°   **40.** 90°

Sketch each angle in standard position. Draw an arrow representing the correct amount of rotation. Find the measure of two other angles, one positive and one negative, that are coterminal with the given angle. Give the quadrant of each angle.

41. 75°   **42.** 89°   **43.** 174°   **44.** 234°
45. 300°   **46.** 512°   **47.** −61°   **48.** −159°

10.2 Trigonometric Functions of Angles

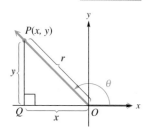

FIGURE 10

The study of trigonometry covers the six trigonometric functions defined in this section. To define these six basic functions, start with an angle θ (the Greek letter *theta*) in standard position.* Choose any point P having coordinates (x, y) on the terminal side of angle θ. (The point P must not be the vertex of the angle.) See Figure 10.

A perpendicular from P to the x-axis at point Q determines a triangle having vertices at O, P, and Q. The distance r from $P(x, y)$ to the origin, $(0, 0)$, can be found from the distance formula.

$$r = \sqrt{(x - 0)^2 + (y - 0)^2}$$
$$r = \sqrt{x^2 + y^2}$$

Notice that $r > 0$, since distance is never negative.

The six trigonometric functions of angle θ are called **sine, cosine, tangent, cotangent, secant,** and **cosecant**. In the following definitions, we use the customary abbreviations for the names of these functions.

"The founder of trigonometry is **Hipparchus,** who lived in Rhodes and Alexandria and died about 125 B.C. We know rather little about him. Most of what we do know comes from Ptolemy, who credits Hipparchus with a number of ideas in trigonometry and astronomy. We owe to him many astronomical observations and discoveries, the most influential astronomical theory of ancient times, and works on geography." (Source: *Mathematical Thought From Ancient to Modern Times,* Volume 1, by Morris Kline.)

Trigonometric Functions

Let (x, y) be a point other than the origin on the terminal side of an angle θ in standard position. The distance from the point to the origin is $r = \sqrt{x^2 + y^2}$. The six trigonometric functions of θ are:

$$\sin \theta = \frac{y}{r} \qquad \cos \theta = \frac{x}{r} \qquad \tan \theta = \frac{y}{x} \ (x \neq 0)$$

$$\csc \theta = \frac{r}{y} \ (y \neq 0) \quad \sec \theta = \frac{r}{x} \ (x \neq 0) \quad \cot \theta = \frac{x}{y} \ (y \neq 0).$$

---

*The letters of the Greek alphabet are identified in a margin note on page 586.

10.2 Trigonometric Functions of Angles   583

Although Figure 10 shows a second quadrant angle, these definitions apply to any angle θ. Because of the restrictions on the denominators in the definitions of tangent, cotangent, secant, and cosecant, some angles will have undefined function values. This will be discussed in more detail later.

EXAMPLE 1   The terminal side of an angle θ in standard position goes through the point $(8, 15)$. Find the values of the six trigonometric functions of angle θ.

Figure 11 shows angle θ and the triangle formed by dropping a perpendicular from the point $(8, 15)$ to the x-axis. The point $(8, 15)$ is 8 units to the right of the y-axis and 15 units above the x-axis, so that $x = 8$ and $y = 15$.

Since $r = \sqrt{x^2 + y^2}$,

$$r = \sqrt{8^2 + 15^2}$$
$$= \sqrt{64 + 225}$$
$$= \sqrt{289}$$
$$= 17.$$

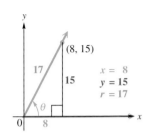

FIGURE 11

The values of the six trigonometric functions of angle θ can now be found with the definitions given above.

$$\sin \theta = \frac{y}{r} = \frac{15}{17} \qquad \cos \theta = \frac{x}{r} = \frac{8}{17} \qquad \tan \theta = \frac{y}{x} = \frac{15}{8}$$

$$\csc \theta = \frac{r}{y} = \frac{17}{15} \qquad \sec \theta = \frac{r}{x} = \frac{17}{8} \qquad \cot \theta = \frac{x}{y} = \frac{8}{15}$$

EXAMPLE 2   The terminal side of an angle θ in standard position goes through the point $(-3, -4)$. Find the values of the six trigonometric functions of θ.

As shown in Figure 12, $x = -3$ and $y = -4$. The value of r is

$$r = \sqrt{(-3)^2 + (-4)^2}$$
$$r = \sqrt{25}$$
$$r = 5.$$

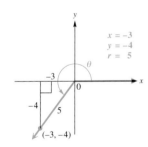

FIGURE 12

(Remember that $r > 0$.) Then by the definitions of the trigonometric functions,

$$\sin \theta = \frac{-4}{5} = -\frac{4}{5} \qquad \cos \theta = \frac{-3}{5} = -\frac{3}{5} \qquad \tan \theta = \frac{-4}{-3} = \frac{4}{3}$$

$$\csc \theta = \frac{5}{-4} = -\frac{5}{4} \qquad \sec \theta = \frac{5}{-3} = -\frac{5}{3} \qquad \cot \theta = \frac{-3}{-4} = \frac{3}{4}.$$

The six trigonometric functions can be found from *any* point on the terminal side of the angle other than the origin. To see why any point may be used, refer to Figure 13, which shows an angle θ and two distinct points on its terminal side. Point P has coordinates (x, y) and point P' (read "P-prime") has coordinates (x', y'). Let r be the length of the hypotenuse of triangle OPQ, and let r' be the length of the hypotenuse of triangle $OP'Q'$. Since corresponding sides of similar triangles are in proportion,

$$\frac{y}{r} = \frac{y'}{r'},$$

so that $\sin \theta = y/r$ is the same no matter which point is used to find it. Similar results hold for the other five functions.

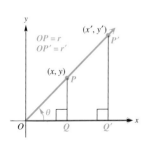

FIGURE 13

If the terminal side of an angle in standard position lies along the y-axis, any point on this terminal side has x-coordinate 0. Similarly, an angle with terminal side on the x-axis has y-coordinate 0 for any point on the terminal side. Since the values of x and y appear in the denominators of some of the trigonometric functions, and since a fraction is undefined if its denominator is 0, some of the trigonometric function values of quadrantal angles (i.e., those with terminal side on an axis) will be undefined.

EXAMPLE 3 Find the values of trigonometric functions for the following angles. Identify any that are undefined.

(a) an angle of 90°

First, select any point on the terminal side of a 90° angle. Let us select the point $(0, 1)$, as shown in Figure 14(a). Here $x = 0$ and $y = 1$. Verify that $r = 1$. Then, by the definitions of the trigonometric functions,

$$\sin 90° = \frac{1}{1} = 1 \qquad \cos 90° = \frac{0}{1} = 0 \qquad \tan 90° = \frac{1}{0} \text{ (undefined)}$$

$$\csc 90° = \frac{1}{1} = 1 \qquad \sec 90° = \frac{1}{0} \text{ (undefined)} \qquad \cot 90° = \frac{0}{1} = 0.$$

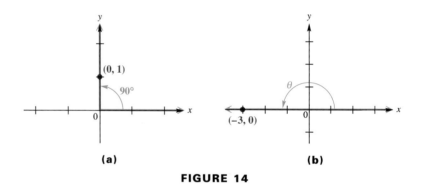

(a)   **(b)**

FIGURE 14

(b) an angle in standard position with terminal side through $(-3, 0)$

Figure 14(b) shows the angle. Here, $x = -3$, $y = 0$, and $r = 3$, so the trigonometric functions have the following values.

$$\sin \theta = \frac{0}{3} = 0 \qquad \cos \theta = \frac{-3}{3} = -1 \qquad \tan \theta = \frac{0}{-3} = 0$$

$$\csc \theta = \frac{3}{0} \text{ (undefined)} \qquad \sec \theta = \frac{3}{-3} = -1 \qquad \cot \theta = \frac{-3}{0} \text{ (undefined)}$$

The conditions under which the trigonometric function values of quadrantal angles are undefined are summarized here.

Undefined Function Values

If the terminal side of a quadrantal angle lies along the y-axis, the tangent and secant functions are undefined. If it lies along the x-axis, the cotangent and cosecant functions are undefined.

Since the most commonly used quadrantal angles are 0°, 90°, 180°, 270°, and 360°, the values of the functions of these angles are summarized in the following table. This table is for reference only; you should be able to reproduce it quickly.

Quadrantal Angles

θ	$\sin\theta$	$\cos\theta$	$\tan\theta$	$\cot\theta$	$\sec\theta$	$\csc\theta$
0°	0	1	0	Undefined	1	Undefined
90°	1	0	Undefined	0	Undefined	1
180°	0	−1	0	Undefined	−1	Undefined
270°	−1	0	Undefined	0	Undefined	−1
360°	0	1	0	Undefined	1	Undefined

10.2 EXERCISES

In Exercises 1–4, sketch an angle θ in standard position such that θ has the smallest possible positive measure, and the given point is on the terminal side of θ.

1. $(-3, 4)$ **2.** $(-4, -3)$ **3.** $(5, -12)$ **4.** $(-12, -5)$

Find the values of the trigonometric functions for the angles in standard position having the following points on their terminal sides. Identify any that are undefined. Rationalize denominators when applicable.

5. $(-3, 4)$ **6.** $(-4, -3)$ **7.** $(0, 2)$ **8.** $(-4, 0)$
9. $(1, \sqrt{3})$ **10.** $(-2\sqrt{3}, -2)$ **11.** $(3, 5)$ **12.** $(-2, 7)$
13. $(-8, 0)$ **14.** $(0, 9)$

15. For any nonquadrantal angle θ, $\sin\theta$ and $\csc\theta$ will have the same sign. Explain why this is so.

16. If $\cot\theta$ is undefined, what is the value of $\tan\theta$?

17. How is the value of r interpreted geometrically in the definitions of the sine, cosine, secant, and cosecant functions?

18. If the terminal side of an angle θ is in quadrant III, what is the sign of each of the trigonometric function values of θ?

Suppose that the point (x, y) is in the indicated quadrant. Decide whether the given ratio is positive or negative. (Hint: It may be helpful to draw a sketch.)

19. II, $\dfrac{y}{r}$ **20.** II, $\dfrac{x}{r}$ **21.** III, $\dfrac{y}{r}$ **22.** III, $\dfrac{x}{r}$

23. IV, $\dfrac{x}{r}$ **24.** IV, $\dfrac{y}{r}$ **25.** IV, $\dfrac{y}{x}$ **26.** IV, $\dfrac{x}{y}$

Use the appropriate definition to determine each function value. If it is undefined, say so.

27. $\cos 90°$ **28.** $\sin 90°$ **29.** $\tan 90°$ **30.** $\cot 90°$ **31.** $\sec 90°$
32. $\csc 90°$ **33.** $\sin 180°$ **34.** $\sin 270°$ **35.** $\tan 180°$ **36.** $\cot 270°$
37. $\sin(-270°)$ **38.** $\cos(-270°)$ **39.** $\tan 0°$ **40.** $\sec(-180°)$

10.3 Trigonometric Identities

The Reciprocal Identities The definitions of the trigonometric functions on page 582 were written so that functions above and below one another are reciprocals of each other. Since $\sin \theta = y/r$ and $\csc \theta = r/y$,

$$\sin \theta = \frac{1}{\csc \theta} \quad \text{and} \quad \csc \theta = \frac{1}{\sin \theta}.$$

Also, $\cos \theta$ and $\sec \theta$ are reciprocals, as are $\tan \theta$ and $\cot \theta$. In summary, we have the **reciprocal identities** that hold for any angle θ that does not lead to a zero denominator.

The Greek Alphabet

A	α	alpha
B	β	beta
Γ	γ	gamma
Δ	δ	delta
E	ϵ	epsilon
Z	ζ	zeta
H	η	eta
Θ	θ	theta
I	ι	iota
K	κ	kappa
Λ	λ	lambda
M	μ	mu
N	ν	nu
Ξ	ξ	xi
O	o	omicron
Π	π	pi
P	ρ	rho
Σ	σ	sigma
T	τ	tau
Y	υ	upsilon
Φ	ϕ	phi
X	χ	chi
Ψ	ψ	psi
Ω	ω	omega

Reciprocal Identities

$$\sin \theta = \frac{1}{\csc \theta} \qquad \cos \theta = \frac{1}{\sec \theta} \qquad \tan \theta = \frac{1}{\cot \theta}$$

$$\csc \theta = \frac{1}{\sin \theta} \qquad \sec \theta = \frac{1}{\cos \theta} \qquad \cot \theta = \frac{1}{\tan \theta}$$

Identities are equations that are true for all meaningful values of the variable. For example, both $(x + y)^2 = x^2 + 2xy + y^2$ and $2(x + 3) = 2x + 6$ are identities.

When studying identities, be aware that various forms exist. For example,

$$\sin \theta = \frac{1}{\csc \theta}$$

can also be written

$$\csc \theta = \frac{1}{\sin \theta} \quad \text{and} \quad (\sin \theta)(\csc \theta) = 1.$$

You should become familiar with all forms of these identities.

EXAMPLE 1 Find each function value.

(a) $\cos \theta$, if $\sec \theta = \dfrac{5}{3}$

Since $\cos \theta$ is the reciprocal of $\sec \theta$,

$$\cos \theta = \frac{1}{\sec \theta} = \frac{1}{5/3} = \frac{3}{5}.$$

(b) $\sin \theta$, if $\csc \theta = -\dfrac{\sqrt{12}}{2}$

$$\sin \theta = \frac{1}{-\sqrt{12}/2} = \frac{-2}{\sqrt{12}} = \frac{-2}{2\sqrt{3}} = \frac{-1}{\sqrt{3}} = -\frac{\sqrt{3}}{3}$$

In the definitions of the trigonometric functions, r is the distance from the origin to the point (x, y). Distance is never negative, so $r > 0$. If we choose a point (x, y)

10.3 Trigonometric Identities

in quadrant I, then both x and y will be positive. Since $r > 0$, all six of the fractions used in the definitions of the trigonometric functions will be positive, so that the values of all six functions will be positive in quadrant I.

A point (x, y) in quadrant II has $x < 0$ and $y > 0$. This makes the values of sine and cosecant positive for quadrant II angles, while the other four functions take on negative values. Similar results can be obtained for the other quadrants.

Claudius Ptolemy (c. 100–178) provided the most influential astronomical work of antiquity. Although Hipparchus is known in some circles as the founder of trigonometry, his works are lost; historians have used Ptolemy's work as their source of how the Greeks viewed trigonometry. His thirteen-book description of the Greek model of the universe, known as the *Mathematical Collection,* served as the basis for future studies both in the Islamic world and the West. Because of its importance, Islamic scientists referred to it as *al-magisti,* meaning "the greatest." Since that time, it has been called the *Almagest.* Seen here is a woodcut from one of its early printings. (*Source:* Katz, Victor J., *A History of Mathematics.*)

Signs of Function Values

θ in Quadrant	$\sin \theta$	$\cos \theta$	$\tan \theta$	$\cot \theta$	$\sec \theta$	$\csc \theta$
I	+	+	+	+	+	+
II	+	−	−	−	−	+
III	−	−	+	+	−	−
IV	−	+	−	−	+	−

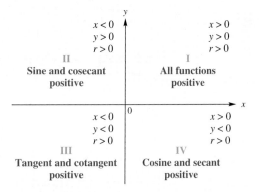

EXAMPLE 2 Identify the quadrant (or quadrants) of any angle θ that satisfies $\sin \theta > 0$, $\tan \theta < 0$.

Since $\sin \theta > 0$ in quadrants I and II, while $\tan \theta < 0$ in quadrants II and IV, both conditions are met only in quadrant II. ■

The six trigonometric functions are defined in terms of x, y, and r, where the Pythagorean theorem shows that $r^2 = x^2 + y^2$ and $r > 0$. With these relationships, knowing the value of only one function and the quadrant in which the angle lies makes it possible to find the values of all six of the trigonometric functions. This procedure is shown in the next example.

EXAMPLE 3 Suppose that angle α is in quadrant II and $\sin \alpha = 2/3$. Find the values of the other five functions.

We can choose any point on the terminal side of angle α. For simplicity, since $\sin \alpha = y/r$, we choose the point with $r = 3$. Then

$$\frac{y}{r} = \frac{2}{3}.$$

Since $r = 3$, y will be 2. To find x, use the result $x^2 + y^2 = r^2$.

$$x^2 + y^2 = r^2$$
$$x^2 + 2^2 = 3^2$$
$$x^2 + 4 = 9$$
$$x^2 = 5$$
$$x = \sqrt{5} \quad \text{or} \quad x = -\sqrt{5}$$

Since α is in quadrant II, x must be negative, as shown in Figure 15, so $x = -\sqrt{5}$. This puts the point $(-\sqrt{5}, 2)$ on the terminal side of α.

Now that the values of x, y, and r are known, the values of the remaining trigonometric functions can be found.

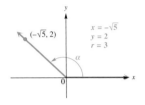

FIGURE 15

$$\cos \alpha = \frac{x}{r} = \frac{-\sqrt{5}}{5} \qquad \sec \alpha = \frac{r}{x} = \frac{3}{-\sqrt{5}} = \frac{-3\sqrt{5}}{5}$$

$$\tan \alpha = \frac{y}{x} = \frac{2}{-\sqrt{5}} = \frac{-2\sqrt{5}}{5} \qquad \cot \alpha = \frac{x}{y} = \frac{-\sqrt{5}}{2}$$

$$\csc \alpha = \frac{r}{y} = \frac{3}{2}$$

The Pythagorean Identities We can derive three very useful new identities from the relationship $x^2 + y^2 = r^2$. Dividing both sides by r^2 gives

$$\frac{x^2}{r^2} + \frac{y^2}{r^2} = \frac{r^2}{r^2},$$

or

$$\left(\frac{x}{r}\right)^2 + \left(\frac{y}{r}\right)^2 = 1.$$

Since $\sin \theta = y/r$ and $\cos \theta = x/r$, this result becomes

$$(\cos \theta)^2 + (\sin \theta)^2 = 1,$$

or, as it is often written,

$$\sin^2 \theta + \cos^2 \theta = 1.$$

Starting with $x^2 + y^2 = r^2$ and dividing through by x^2 gives

$$\frac{x^2}{x^2} + \frac{y^2}{x^2} = \frac{r^2}{x^2}$$

$$1 + \left(\frac{y}{x}\right)^2 = \left(\frac{r}{x}\right)^2$$

$$1 + (\tan \theta)^2 = (\sec \theta)^2$$

or

$$\tan^2 \theta + 1 = \sec^2 \theta.$$

On the other hand, dividing through by y^2 leads to

$$1 + \cot^2 \theta = \csc^2 \theta.$$

These three identities are called the **Pythagorean identities** since the original equation that led to them, $x^2 + y^2 = r^2$, comes from the Pythagorean theorem.

Pythagorean Identities

$$\sin^2 \theta + \cos^2 \theta = 1 \qquad \tan^2 \theta + 1 = \sec^2 \theta \qquad 1 + \cot^2 \theta = \csc^2 \theta$$

As before, we have given only one form of each identity. However, algebraic transformations can be made to get equivalent identities. For example, by subtracting $\sin^2 \theta$ from both sides of $\sin^2 \theta + \cos^2 \theta = 1$ we get the equivalent identity

$$\cos^2 \theta = 1 - \sin^2 \theta.$$

You should be able to transform these identities quickly, and also recognize their equivalent forms.

The Quotient Identities Recall that $\sin \theta = y/r$ and $\cos \theta = x/r$. Consider the quotient of $\sin \theta$ and $\cos \theta$, where $\cos \theta \neq 0$.

$$\frac{\sin \theta}{\cos \theta} = \frac{y/r}{x/r} = \frac{y}{r} \div \frac{x}{r} = \frac{y}{r} \cdot \frac{r}{x} = \frac{y}{x} = \tan \theta$$

Similarly, it can be shown that $(\cos \theta)/(\sin \theta) = \cot \theta$, for $\sin \theta \neq 0$. Thus we have two more identities, called the **quotient identities.**

Quotient Identities

$$\frac{\sin \theta}{\cos \theta} = \tan \theta \qquad \frac{\cos \theta}{\sin \theta} = \cot \theta$$

EXAMPLE 4 Find $\sin \alpha$ and $\tan \alpha$ if $\cos \alpha = -\sqrt{3}/4$ and α is in quadrant II. Start with $\sin^2 \alpha + \cos^2 \alpha = 1$, and replace $\cos \alpha$ with $-\sqrt{3}/4$.

$$\sin^2 \alpha + \left(-\frac{\sqrt{3}}{4}\right)^2 = 1 \qquad \text{Replace } \cos \alpha \text{ with } -\frac{\sqrt{3}}{4}.$$

$$\sin^2 \alpha + \frac{3}{16} = 1$$

$$\sin^2 \alpha = \frac{13}{16} \qquad \text{Subtract } \frac{3}{16}.$$

$$\sin \alpha = \pm \frac{\sqrt{13}}{4} \qquad \text{Take square roots.}$$

Since α is in quadrant II, $\sin \alpha > 0$, and

$$\sin \alpha = \frac{\sqrt{13}}{4}.$$

To find tan α, use the quotient identity $\tan \alpha = \dfrac{\sin \alpha}{\cos \alpha}$.

$$\tan \alpha = \frac{\sin \alpha}{\cos \alpha} = \frac{\dfrac{\sqrt{13}}{4}}{\dfrac{-\sqrt{3}}{4}} = \frac{\sqrt{13}}{4} \cdot \frac{4}{-\sqrt{3}} = \frac{\sqrt{13}}{-\sqrt{3}}$$

Rationalize the denominator as follows.

$$\frac{\sqrt{13}}{-\sqrt{3}} = \frac{\sqrt{13}}{-\sqrt{3}} \cdot \frac{\sqrt{3}}{\sqrt{3}}$$

$$= \frac{\sqrt{39}}{-3}$$

$$= -\frac{\sqrt{39}}{3}$$

Therefore, $\tan \alpha = -\sqrt{39}/3$.

10.3 EXERCISES

Use the appropriate reciprocal identity to find each function value. Rationalize denominators when applicable.

1. $\tan \theta$, if $\cot \theta = -3$
2. $\cot \theta$, if $\tan \theta = 5$
3. $\sin \theta$, if $\csc \theta = 3$
4. $\cos \alpha$, if $\sec \alpha = -5/2$
5. $\cot \beta$, if $\tan \beta = -1/5$
6. $\sin \alpha$, if $\csc \alpha = \sqrt{15}$
7. $\csc \alpha$, if $\sin \alpha = \sqrt{2}/4$
8. $\sec \beta$, if $\cos \beta = -1/\sqrt{7}$
9. $\tan \theta$, if $\cot \theta = -\sqrt{5}/3$
10. $\cot \theta$, if $\tan \theta = \sqrt{11}/5$
11. $\sin \theta$, if $\csc \theta = 1.5$
12. $\cos \theta$, if $\sec \theta = 7.5$

Identify the quadrant or quadrants for each angle satisfying the given conditions.

13. $\sin \alpha > 0$, $\cos \alpha < 0$
14. $\cos \beta > 0$, $\tan \beta > 0$
15. $\tan \gamma > 0$, $\sin \gamma > 0$
16. $\sin \beta < 0$, $\cos \beta > 0$
17. $\tan \omega < 0$, $\cos \omega < 0$
18. $\csc \theta < 0$, $\cos \theta < 0$
19. $\cos \beta < 0$
20. $\tan \theta > 0$

Give the signs of the six trigonometric functions for each angle.

21. 74°
22. 129°
23. 183°
24. 298°
25. 302°
26. 406°
27. 412°
28. −82°
29. −14°
30. −121°

Use identities to find the indicated value.

31. $\tan \alpha$, if $\sec \alpha = 3$, with α in quadrant IV
32. $\cos \theta$, if $\sin \theta = 2/3$, with θ in quadrant II
33. $\sin \alpha$, if $\cos \alpha = -1/4$, with α in quadrant II
34. $\csc \beta$, if $\cot \beta = -1/2$, with β in quadrant IV
35. $\tan \theta$, if $\cos \theta = 1/3$, with θ in quadrant IV
36. $\sec \theta$, if $\tan \theta = \sqrt{7}/3$, with θ in quadrant III
37. $\cos \beta$, if $\csc \beta = -4$, with β in quadrant III
38. $\sin \theta$, if $\sec \theta = 2$, with θ in quadrant IV

Find all the trigonometric function values for each angle.

39. $\tan \alpha = -15/8$, with α in quadrant II
40. $\cos \alpha = -3/5$, with α in quadrant III
41. $\cot \gamma = 3/4$, with γ in quadrant III
42. $\sin \beta = 7/25$, with β in quadrant II
43. $\tan \beta = \sqrt{3}$, with β in quadrant III
44. $\csc \theta = 2$, with θ in quadrant II
45. $\sin \beta = \sqrt{5}/7$, with $\tan \beta > 0$
46. $\cot \alpha = \sqrt{3}/8$, with $\sin \alpha > 0$
47. Derive the identity $1 + \cot^2 \theta = \csc^2 \theta$ by dividing $x^2 + y^2 = r^2$ by y^2.
48. Using a method similar to the one given in this section showing that $(\sin \theta)/(\cos \theta) = \tan \theta$, show that

$$\frac{\cos \theta}{\sin \theta} = \cot \theta.$$

10.4 Right Triangles and Function Values

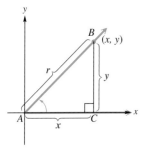

FIGURE 16

Right Triangles Figure 16 shows an acute angle A in standard position. The definitions of the trigonometric function values of angle A require x, y, and r. As drawn in Figure 16, x and y are the lengths of the two legs of the right triangle ABC, and r is the length of the hypotenuse.

The side of length y is called the **side opposite** angle A, and the side of length x is called the **side adjacent** to angle A. The lengths of these sides can be used to replace x and y in the definitions of the trigonometric functions, and r replaced by the length of the hypotenuse, to get the following right triangle-based definitions.

Right Triangle-based Definitions of Trigonometric Functions

For any acute angle A in standard position,

$$\sin A = \frac{y}{r} = \frac{\text{side opposite } A}{\text{hypotenuse}} \qquad \csc A = \frac{r}{y} = \frac{\text{hypotenuse}}{\text{side opposite } A}$$

$$\cos A = \frac{x}{r} = \frac{\text{side adjacent to } A}{\text{hypotenuse}} \qquad \sec A = \frac{r}{x} = \frac{\text{hypotenuse}}{\text{side adjacent to } A}$$

$$\tan A = \frac{y}{x} = \frac{\text{side opposite } A}{\text{side adjacent to } A} \qquad \cot A = \frac{x}{y} = \frac{\text{side adjacent to } A}{\text{side opposite } A}.$$

FIGURE 17

EXAMPLE 1 Find the values of the trigonometric functions for angles A and B in the right triangle in Figure 17.

The length of the side opposite angle A is 7. The length of the side adjacent to angle A is 24, and the length of the hypotenuse is 25. Using the relationships given above,

$$\sin A = \frac{\text{side opposite}}{\text{hypotenuse}} = \frac{7}{25} \qquad \csc A = \frac{\text{hypotenuse}}{\text{side opposite}} = \frac{25}{7}$$

$$\cos A = \frac{\text{side adjacent}}{\text{hypotenuse}} = \frac{24}{25} \qquad \sec A = \frac{\text{hypotenuse}}{\text{side adjacent}} = \frac{25}{24}$$

$$\tan A = \frac{\text{side opposite}}{\text{side adjacent}} = \frac{7}{24} \qquad \cot A = \frac{\text{side adjacent}}{\text{side opposite}} = \frac{24}{7}.$$

The length of the side opposite angle B is 24, while the length of the side adjacent to B is 7, making

$$\sin B = \frac{24}{25} \qquad \cos B = \frac{7}{25} \qquad \tan B = \frac{24}{7}$$

$$\csc B = \frac{25}{24} \qquad \sec B = \frac{25}{7} \qquad \cot B = \frac{7}{24}.$$ ∎

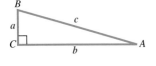

FIGURE 18

In Example 1, you may have noticed that $\sin A = \cos B$, $\cos A = \sin B$, and so on. Such relationships are always true for the two acute angles of a right triangle. Figure 18 shows a right triangle with acute angles A and B and a right angle at C. (Whenever we use A, B, and C to name the angles in a right triangle, C will be the right angle.) The length of the side opposite angle A is a, and the length of the side opposite angle B is b. The length of the hypotenuse is c.

By the definitions given above, $\sin A = a/c$. Since $\cos B$ is also equal to a/c,

$$\sin A = \frac{a}{c} = \cos B.$$

In a similar manner,

$$\tan A = \frac{a}{b} = \cot B \qquad \sec A = \frac{c}{b} = \csc B.$$

The sum of the three angles in any triangle is 180°. Since angle C equals 90°, angles A and B must have a sum of $180° - 90° = 90°$. As mentioned in Chapter 9, angles with a sum of 90° are called *complementary angles*. Since angles A and B are complementary and $\sin A = \cos B$, the functions sine and cosine are called **cofunctions**. Also, tangent and cotangent are cofunctions, as are secant and cosecant. And since the angles A and B are complementary, $A + B = 90°$, or

$$B = 90° - A,$$

giving

$$\sin A = \cos B = \cos(90° - A).$$

Similar results are true for the other trigonometric functions. We call these results the **cofunction identities.**

Cofunction Identities

For any acute angle A,

$$\sin A = \cos(90° - A) \qquad \csc A = \sec(90° - A)$$
$$\cos A = \sin(90° - A) \qquad \sec A = \csc(90° - A)$$
$$\tan A = \cot(90° - A) \qquad \cot A = \tan(90° - A).$$

These identities can be extended to *any* angle A, and not just acute angles.

EXAMPLE 2 Write each of the following in terms of cofunctions.

(a) cos 52°

Since cos A = sin(90° − A),

$$\cos 52° = \sin(90° - 52°) = \sin 38°.$$

(b) tan 71° = cot 19°

(c) sec 24° = csc 66°

Trigonometric Function Values of Special Angles

Certain special angles, such as 30°, 45°, and 60°, occur so often in trigonometry and in more advanced mathematics that they deserve special study. We can find the exact trigonometric function values of these angles by using properties of geometry and the Pythagorean theorem.

To find the trigonometric function values for 30° and 60°, we start with an equilateral triangle, a triangle with all sides of equal length. Each angle of such a triangle has a measure of 60°. While the results we will obtain are independent of the length, for convenience, we choose the length of each side to be 2 units. See Figure 19(a).

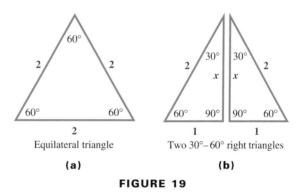

Equilateral triangle
(a)

Two 30°–60° right triangles
(b)

FIGURE 19

Bisecting one angle of this equilateral triangle leads to two right triangles, each of which has angles of 30°, 60°, and 90°, as shown in Figure 19(b). Since the hypotenuse of one of these right triangles has a length of 2, the shortest side will have a length of 1. (Why?) If x represents the length of the medium side, then, by the Pythagorean theorem,

$$2^2 = 1^2 + x^2$$
$$4 = 1 + x^2$$
$$3 = x^2$$
$$\sqrt{3} = x.$$

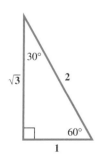

FIGURE 20

Figure 20 summarizes our results, showing a 30°–60° right triangle.

As shown in the figure, the side opposite the 30° angle has length 1; that is, for the 30° angle,

hypotenuse = 2,   side opposite = 1,   side adjacent = $\sqrt{3}$.

Using the definitions of the trigonometric functions,

$$\sin 30° = \frac{\text{side opposite}}{\text{hypotenuse}} = \frac{1}{2} \qquad \csc 30° = \frac{2}{1} = 2$$

$$\cos 30° = \frac{\text{side adjacent}}{\text{hypotenuse}} = \frac{\sqrt{3}}{2} \qquad \sec 30° = \frac{2}{\sqrt{3}} = \frac{2\sqrt{3}}{3}$$

$$\tan 30° = \frac{\text{side opposite}}{\text{side adjacent}} = \frac{1}{\sqrt{3}} = \frac{\sqrt{3}}{3} \qquad \cot 30° = \frac{\sqrt{3}}{1} = \sqrt{3}.$$

The denominator was rationalized for tan 30° and sec 30°.

In a similar manner,

$$\sin 60° = \frac{\sqrt{3}}{2} \qquad \cos 60° = \frac{1}{2} \qquad \tan 60° = \sqrt{3}$$

$$\csc 60° = \frac{2\sqrt{3}}{3} \qquad \sec 60° = 2 \qquad \cot 60° = \frac{\sqrt{3}}{3}.$$

The values of the trigonometric functions for 45° can be found by starting with a 45°–45° right triangle, as shown in Figure 21. This triangle is isosceles, and, for convenience, we choose the lengths of the equal sides to be 1 unit. (As before, the results are independent of the length of the equal sides of the right triangle.) Since the shorter sides each have length 1, if r represents the length of the hypotenuse, then

$$1^2 + 1^2 = r^2$$
$$2 = r^2$$
$$\sqrt{2} = r.$$

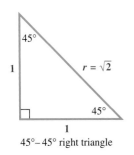

FIGURE 21

45°–45° right triangle

Using the measures indicated on the 45°–45° right triangle in Figure 21, we find

$$\sin 45° = \frac{1}{\sqrt{2}} = \frac{\sqrt{2}}{2} \qquad \cos 45° = \frac{1}{\sqrt{2}} = \frac{\sqrt{2}}{2} \qquad \tan 45° = \frac{1}{1} = 1$$

$$\csc 45° = \frac{\sqrt{2}}{1} = \sqrt{2} \qquad \sec 45° = \frac{\sqrt{2}}{1} = \sqrt{2} \qquad \cot 45° = \frac{1}{1} = 1.$$

The importance of these exact trigonometric function values of 30°, 60°, and 45° angles cannot be overemphasized, as they are used extensively. They are summarized in the chart that follows.

Function Values of Special Angles

θ	$\sin \theta$	$\cos \theta$	$\tan \theta$	$\cot \theta$	$\sec \theta$	$\csc \theta$
30°	$\frac{1}{2}$	$\frac{\sqrt{3}}{2}$	$\frac{\sqrt{3}}{3}$	$\sqrt{3}$	$\frac{2\sqrt{3}}{3}$	2
45°	$\frac{\sqrt{2}}{2}$	$\frac{\sqrt{2}}{2}$	1	1	$\sqrt{2}$	$\sqrt{2}$
60°	$\frac{\sqrt{3}}{2}$	$\frac{1}{2}$	$\sqrt{3}$	$\frac{\sqrt{3}}{3}$	2	$\frac{2\sqrt{3}}{3}$

You should be able to reproduce the function values from the preceding chart. It is not difficult to do if you remember the values of sin 30°, cos 30°, and sin 45°. Then complete the rest of the chart using the reciprocal, quotient, and cofunction identities. Another option is to visualize the appropriate triangles and use the ratios of the definitions.

Reference Angles Now that we have discussed trigonometric function values for acute angles, we can use those results to find trigonometric function values for other types of angles. Associated with every nonquadrantal angle in standard position is a positive acute angle called its *reference angle*. A **reference angle** for an angle θ, written θ', is the positive acute angle made by the terminal side of angle θ and the x-axis. Figure 22 shows several angles θ (each less than one complete counterclockwise revolution) in quadrants II, III, and IV, respectively, with the reference angle θ' also shown. In quadrant I, θ and θ' are the same. If an angle θ is negative or has measure greater than 360°, its reference angle is found by first finding its coterminal angle that is between 0° and 360°, and then using the diagram in Figure 22.

A common error is to find the reference angle by using the terminal side of θ and the y-axis. *The reference angle is always found with reference to the x-axis.*

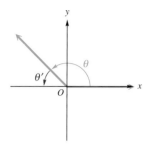

θ in quadrant II

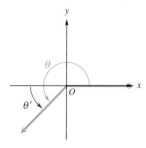

θ in quadrant III

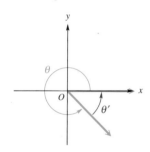

θ in quadrant IV

FIGURE 22

EXAMPLE 3 Find the reference angles for the following angles.

(a) 218°

As shown in Figure 23, the positive acute angle made by the terminal side of this angle and the x-axis is 218° − 180° = 38°. For $\theta = 218°$, the reference angle is $\theta' = 38°$.

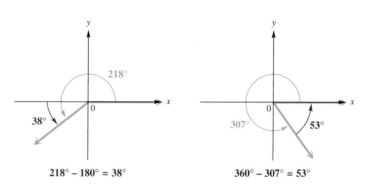

FIGURE 23   **FIGURE 24**

(b) 1387°

First find a coterminal angle between 0° and 360°. Divide 1387° by 360° to get a quotient of about 3.9. Begin by subtracting 360° three times (because of the 3 in 3.9):

$$1387° - 3 \cdot 360° = 307°.$$

The reference angle for 307° (and thus for 1387°) is 360° − 307° = 53°. See Figure 24.

The preceding example suggests the following table for finding the reference angle θ' for any angle θ between 0° and 360°.

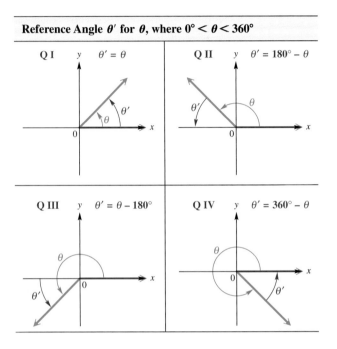

We can now find exact trigonometric function values of angles with reference angles of 30°, 60°, or 45°. In Example 4 we show how to use these function values to find the trigonometric function values for 210°.

EXAMPLE 4 Find the values of the trigonometric functions for 210°.

Even though a 210° angle is not an angle of a right triangle, the ideas mentioned earlier can still be used to find the trigonometric function values for this angle. To do so, we draw an angle of 210° in standard position, as shown in Figure 25. We choose point P on the terminal side of the angle so that the distance from the origin O to P is 2. By the results from 30°–60° right triangles, the coordinates of point P become $(-\sqrt{3}, -1)$, with $x = -\sqrt{3}$, $y = -1$, and $r = 2$. Then, by the definitions of the trigonometric functions,

$$\sin 210° = -\frac{1}{2} \quad \cos 210° = -\frac{\sqrt{3}}{2} \quad \tan 210° = \frac{\sqrt{3}}{3}$$

$$\csc 210° = -2 \quad \sec 210° = -\frac{2\sqrt{3}}{3} \quad \cot 210° = \sqrt{3}.$$

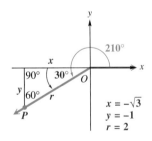

FIGURE 25

Notice in Example 4 that the trigonometric function values of 210° correspond in absolute value to those of its reference angle 30°. The signs are different for the sine, cosine, secant, and cosecant functions because 210° is a quadrant III angle. These results suggest a method for finding the trigonometric function values of a non-acute angle, using the reference angle. In Example 4, the reference angle for 210° is 30°, as shown in Figure 25. Simply by using the trigonometric function values of the reference angle, 30°, and choosing the correct signs for a quadrant III angle, we obtain the same results as found in Example 4.

The values of the trigonometric functions for any non-quadrantal angle θ can be determined by finding the function values for an angle between 0° and 90°. To do this, perform the following steps.

Historian Reviel Netz of Stanford University has been studying an ancient manuscript by the Greek mathematician Archimedes ("Unveiling the work of Archimedes," *Science News*, Vol. 157, p. 77, January 29, 2000). It is known as *Archimedes Palimpsest,* and dates from the 10th century A.D., surviving on a parchment that was later cut, scraped, and overwritten with a description of a church ritual. The use of ultraviolet photography and digital imaging makes it possible to read beneath the lines of the ritual and see Archimedes' text and diagrams. According to Netz, the diagrams suggest that Greek mathematicians emphasized qualitative relationships over quantitative accuracy.

Finding Trigonometric Function Values for Any Non-quadrantal Angle

1. If $\theta > 360°$, or if $\theta < 0°$, find a coterminal angle by adding or subtracting 360° as many times as needed to obtain an angle greater than 0° but less than 360°.
2. Find the reference angle θ'.
3. Find the necessary values of the trigonometric functions for the reference angle θ'.
4. Determine the correct signs for the values found in Step 3. (Use the table of signs in Section 10.3.) This result gives the values of the trigonometric functions for angle θ.

EXAMPLE 5 Use reference angles to find the exact value of each of the following.

(a) $\cos(-240°)$

The reference angle is 60°, as shown in Figure 26. Since the cosine is negative in quadrant II,

$$\cos(-240°) = -\cos 60° = -\frac{1}{2}.$$

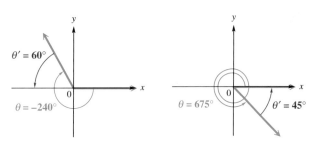

FIGURE 26   FIGURE 27

(b) $\tan 675°$

Begin by subtracting 360° to obtain a coterminal angle between 0° and 360°.

$$675° - 360° = 315°$$

As shown in Figure 27, the reference angle is $360° - 315° = 45°$. An angle of 315° is in quadrant IV, so the tangent will be negative, and

$$\tan 675° = \tan 315° = -\tan 45° = -1.$$

The ideas discussed in this section can be used inversely to find the measures of certain angles, given a trigonometric function value and an interval in which an angle θ must lie. We are most often interested in the interval $0° \leq \theta < 360°$.

EXAMPLE 6 Find all values of θ, if $0° \leq \theta < 360°$ and $\cos \theta = -\sqrt{2}/2$.

Since cosine here is negative, θ must lie in either quadrant II or III. Since the absolute value of $\cos \theta$ is $\sqrt{2}/2$, the reference angle θ' must be $45°$. The two possible angles θ are sketched in Figure 28. The quadrant II angle θ must equal $180° - 45° = 135°$, and the quadrant III angle θ must equal $180° + 45° = 225°$.

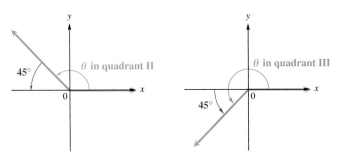

FIGURE 28

10.4 EXERCISES

In each exercise, find the values of the six trigonometric functions for angle A. Leave answers as fractions.

1.
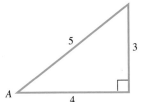

2.
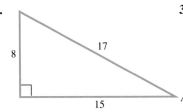

3.
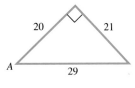

4.
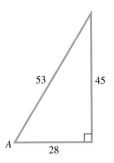

5.
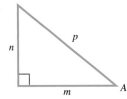

6.
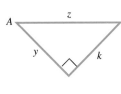

Suppose ABC is a right triangle with sides of lengths a, b, and c, with right angle C (see Figure 18). Find the unknown side length using the Pythagorean theorem, and then find the values of the six trigonometric functions for angle B. Rationalize denominators when applicable.

7. $a = 5, b = 12$ **8.** $a = 3, b = 5$ **9.** $a = 6, c = 7$ **10.** $b = 7, c = 12$

Write each of the following in terms of the cofunction.

11. tan 50° **12.** cot 73° **13.** csc 47° **14.** sec 39°
15. tan 25.4° **16.** sin 38.7° **17.** cos 13° 30′ **18.** tan 26° 10′

Refer to the discussion in this section to give the exact *trigonometric function value. Do not use a calculator.*

19. tan 30° **20.** cot 30° **21.** sin 30° **22.** cos 30°
23. csc 45° **24.** sec 45° **25.** cos 45° **26.** sin 45°
27. sin 60° **28.** cos 60° **29.** tan 60° **30.** cot 60°

Find the reference angle for each of the following.

31. 98° **32.** 212° **33.** −135° **34.** −60° **35.** 750° **36.** 480°

Use the methods of this section to find the exact values *of the six trigonometric functions for each of the following angles. Rationalize denominators when applicable.*

37. 120° **38.** 135° **39.** 150° **40.** 225° **41.** 240°
42. 300° **43.** 315° **44.** 405° **45.** 420° **46.** 480°
47. 495° **48.** 570° **49.** 750° **50.** 1305° **51.** 1500°
52. 2670° **53.** −390° **54.** −510° **55.** −1020° **56.** −1290°

Complete the following table with exact *trigonometric function values using the methods of this section.*

	θ	sin θ	cos θ	tan θ	cot θ	sec θ	csc θ
57.	30°	1/2	$\sqrt{3}/2$	___	___	$2\sqrt{3}/3$	2
58.	45°	___	___	1	1	___	___
59.	60°	___	1/2	$\sqrt{3}$	___	2	___
60.	120°	$\sqrt{3}/2$	___	$-\sqrt{3}$	___	___	$2\sqrt{3}/3$
61.	135°	$\sqrt{2}/2$	$-\sqrt{2}/2$	___	___	$-\sqrt{2}$	$\sqrt{2}$
62.	150°	___	$-\sqrt{3}/2$	$-\sqrt{3}/3$	___	___	2
63.	210°	−1/2	___	$\sqrt{3}/3$	$\sqrt{3}$	___	−2
64.	240°	$-\sqrt{3}/2$	−1/2	___	___	−2	$-2\sqrt{3}/3$

Find all values of θ, *if* $0° \leq \theta < 360°$, *and the given condition is true.*

65. sin $\theta = -1/2$ **66.** cos $\theta = -1/2$ **67.** tan $\theta = 1$
68. cot $\theta = \sqrt{3}$ **69.** sin $\theta = \sqrt{3}/2$ **70.** cos $\theta = \sqrt{3}/2$
71. sec $\theta = -2$ **72.** csc $\theta = -2$ **73.** sin $\theta = -\sqrt{2}/2$
74. cos $\theta = -\sqrt{2}/2$ **75.** tan $\theta = -\sqrt{3}$ **76.** cot $\theta = -1$

10.5 Applications of Right Triangles

Finding Calculator Approximations With the technological advances of this era in mind, the examples and exercises that follow in this chapter assume that all students have access to scientific calculators. However, since calculators differ among makes and models, students should always consult their owner's manual for specific information if questions arise concerning their use.

Thus far in this book, we have studied only one type of measure for angles—degree measure; another type of measure, radians, is studied in more theoretical work. When evaluating trigonometric functions of angles given in degrees, it is a common error to use the incorrect mode; remember that the calculator must be set in the *degree mode*. One way to avoid this problem is to get in the habit of always starting work by finding sin 90°. If the displayed answer is 1, the calculator is set for degree measure; otherwise it is not. Remember that most calculator values of trigonometric functions are approximations.

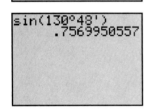

The three screens above show the results for parts (a)–(e) in Example 1. Notice that the calculator permits entering the angle measure in degrees and minutes in parts (a) and (e). In the fifth line of the first screen, Ans^{-1} tells the calculator to find the reciprocal of the answer given in the previous line.

EXAMPLE 1 Use a calculator to approximate the values of the following trigonometric functions.

(a) sin 49° 12′

Convert 49° 12′ to decimal degrees.

$$49° \ 12' = 49\frac{12°}{60} = 49.2°$$

To eight decimal places,

$$\sin 49° \ 12' = \sin 49.2° \approx .75699506.$$

(b) sec 97.977°

Calculators do not have secant keys. However,

$$\sec \theta = \frac{1}{\cos \theta}$$

for all angles θ where $\cos \theta \neq 0$. So find sec 97.977° by first finding cos 97.977° and then taking the reciprocal to get

$$\sec 97.977° \approx -7.205879213.$$

(c) cot 51.4283°

Use the identity $\cot \theta = 1/\tan \theta$.

$$\cot 51.4283° \approx .79748114$$

(d) $\sin(-246°) \approx .91354546$

(e) sin 130° 48′

130° 48′ is equal to 130.8°.

$$\sin 130° \ 48' = \sin 130.8° \approx .75699506$$

Notice that the values found in parts (a) and (e) of Example 1 are the same. The reason for this is that 49° 12′ is the reference angle for 130° 48′ and the sine function is positive for a quadrant II angle.

Finding Angles So far in this section we have used a calculator to find trigonometric function values of angles. This process can be reversed. For now we restrict our attention to angles θ in the interval $0° \leq \theta \leq 90°$. The measure of an angle can be found from one of its trigonometric function values using inverse functions as shown in the next example.

This screen supports the results of Example 2.

EXAMPLE 2 Use a calculator to find a value of θ such that $0° \leq \theta \leq 90°$, and θ satisfies each of the following. Leave answers in decimal degrees.

(a) $\sin \theta = .81815000$

We find θ using a key labeled [arc] or [INV] together with the [sin] key. Some calculators may require a key labeled [sin⁻¹] instead. Check your owner's manual to see how your calculator handles this. Again, make sure the calculator is set for degree measure. You should get

$$\theta \approx 54.900028°.$$

(b) $\sec \theta = 1.0545829$

Use the identity $\cos \theta = 1/\sec \theta$. Enter 1.0545829 and find the reciprocal. This gives $\cos \theta \approx .9482421913$. Now find θ as described in part (a) using inverse cosine. The result is

$$\theta \approx 18.514704°.$$

Compare Examples 1(b) and 2(b). Note that the reciprocal key is used *before* the inverse cosine key when finding the angle, but *after* the cosine key when finding the trigonometric function.

Significant Digits Supppose we quickly measure a room as 15 feet by 18 feet. To calculate the length of a diagonal of the room, we can use the Pythagorean theorem.

$$d^2 = 15^2 + 18^2$$
$$d^2 = 549$$
$$d = \sqrt{549} \approx 23.430749$$

Should this answer be given as the length of the diagonal of the room? Of course not. The number 23.430749 contains 6 decimal places, while the original data of 15 feet and 18 feet are only accurate to the nearest foot. Since the results of a problem can be no more accurate than the least accurate number in any calculation, we really should say that the diagonal of the 15-by-18-foot room is 23 feet.

If a wall measured to the nearest foot is 18 feet long, this actually means that the wall has length between 17.5 feet and 18.5 feet. If the wall is measured more accurately as 18.3 feet long, then its length is really between 18.25 feet and 18.35 feet. A measurement of 18.00 feet would indicate that the length of the wall is between 17.995 feet and 18.005 feet. The measurement 18 feet is said to have two *significant digits* of accuracy; 18.0 has three significant digits, and 18.00 has four.

What about the measurement 900 meters? We cannot tell whether this represents a measurement to the nearest meter, ten meters, or hundred meters. To avoid this problem, we write the number in scientific notation as 9.00×10^2 to the nearest meter, 9.0×10^2 to the nearest ten meters, or 9×10^2 to the nearest hundred meters. These three cases have three, two, and one significant digits, respectively.

A **significant digit** is a digit obtained by actual measurement. A number that represents the result of counting, or a number that results from theoretical work and is not the result of a measurement, is an **exact number.** There are 50 states in the United States, so 50 is an exact number. The number of states is not $49\frac{3}{4}$ or $50\frac{1}{4}$, nor is the number 50 used here to represent "some number between 45 and 55." In the formula for perimeter of a rectangle, $P = 2L + 2W$, the 2s are obtained from the definition of perimeter and are exact numbers.

Most values of trigonometric functions are approximations, and virtually all measurements are approximations. To perform calculations on such approximate numbers, follow the rules given below. (The rules for rounding are found in Section 6.5.)

Calculation with Significant Digits

For *adding* and *subtracting,* round the answer so that the last digit you keep is in the rightmost column in which all the numbers have significant digits.

For *multiplying* or *dividing,* round the answer to the least number of significant digits found in any of the given numbers.

For *powers* and *roots,* round the answer so that it has the same number of significant digits as the numbers whose power or root you are finding.

To **solve a triangle** means to find the measures of all the angles and sides of the triangle. When solving triangles, use the following table to determine the significant digits in angle measure.

Significant Digits for Angles

Number of Significant Digits	Angle Measure to Nearest:
2	Degree
3	Ten minutes, or nearest tenth of a degree
4	Minute, or nearest hundredth of a degree
5	Tenth of a minute, or nearest thousandth of a degree

For example, an angle measuring 52° 30′ has three significant digits (assuming that 30′ is measured to the nearest ten minutes).

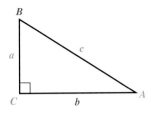

FIGURE 29

Solving Triangles In using trigonometry to solve triangles, a labeled sketch is an important aid. It is conventional to use a to represent the length of the side opposite angle A, b for the length of the side opposite angle B, and so on. As mentioned earlier, in a right triangle the letter c is reserved for the hypotenuse. Figure 29 shows the labeling of a typical right triangle.

EXAMPLE 3 Solve right triangle ABC, with $A = 34°\,30'$ and $c = 12.7$ in. See Figure 30.

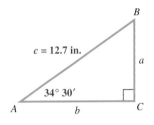

FIGURE 30

To solve the triangle, find the measures of the remaining sides and angles. The value of a can be found with a trigonometric function involving the known values of angle A and side c. Since the sine of angle A is given by the quotient of the side opposite A and the hypotenuse, use sin A.

$$\sin A = \frac{a}{c}$$

Substituting known values gives

$$\sin 34°\,30' = \frac{a}{12.7},$$

or, upon multiplying both sides by 12.7,

$$a = 12.7 \sin 34°\,30'$$
$$a = 12.7(.56640624) \quad \text{Use a calculator.}$$
$$a = 7.19 \text{ in.}$$

The value of b could be found with the Pythagorean theorem. It is better, however, to use the information given in the problem rather than a result just calculated. If a mistake were to be made in finding a, then b also would be incorrect. Also, rounding more than once may cause the result to be less accurate. Using cos A gives

$$\cos A = \frac{\text{side adjacent}}{\text{hypotenuse}} = \frac{b}{c}$$

$$\cos 34°\,30' = \frac{b}{12.7}$$

$$b = 12.7 \cos 34°\,30'$$
$$b = 10.5 \text{ in.}$$

Once b has been found, the Pythagorean theorem could be used as a check. All that remains to solve triangle ABC is to find the measure of angle B. Since $A + B = 90°$ and $A = 34° \, 30'$,

$$A + B = 90°$$
$$B = 90° - A$$
$$B = 89° \, 60' - 34° \, 30'$$
$$B = 55° \, 30'.$$

In Example 3 we could have started by finding the measure of angle B and then used the trigonometric function values of B to find the unknown sides. The process of solving a right triangle (like many problems in mathematics) can usually be done in several ways, each resulting in the correct answer. However, in order to retain as much accuracy as can be expected, always use the given information as much as possible, and avoid rounding off in intermediate steps.

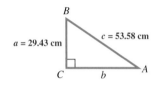

FIGURE 31

EXAMPLE 4 Solve right triangle ABC if $a = 29.43$ cm and $c = 53.58$ cm.

Draw a sketch showing the given information, as in Figure 31. One way to begin is to find the sine of angle A, and then use inverse sine.

$$\sin A = \frac{\text{side opposite}}{\text{hypotenuse}} = \frac{29.43}{53.58}$$

Using $\boxed{\text{INV}}\,\boxed{\text{sin}}$ or $\boxed{\text{sin}^{-1}}$ on a calculator, we find that $A = 33.32°$. The measure of B is $90° - 33.32° = 56.68°$.

We now find b from the Pythagorean theorem, $a^2 + b^2 = c^2$, or $b^2 = c^2 - a^2$. Since $c = 53.58$ and $a = 29.43$,

$$b^2 = 53.58^2 - 29.43^2$$

giving $\qquad b = 44.77$ cm.

Problem Solving

The process of solving right triangles is easily adapted to solving applied problems. A crucial step in such applications involves sketching the triangle and labeling the given parts correctly. Then we can use the methods described in the earlier examples to find the unknown value or values.

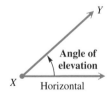

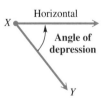

FIGURE 32

Many applications of right triangles involve the *angle of elevation* or the *angle of depression*. The **angle of elevation** from point X to point Y (above X) is the acute angle formed by ray XY and a horizontal ray with endpoint at X. The angle of elevation is always measured from the horizontal. See the angle at the top of Figure 32. The **angle of depression** from point X to point Y (below X) is the acute angle formed by ray XY and a horizontal ray with endpoint X, as shown at the bottom of Figure 32.

Errors are often made in interpreting the angle of depression. Remember that both the angle of elevation *and* the angle of depression are measured between the line of sight and the *horizontal*.

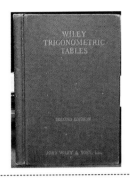

Prior to the advent of the scientific calculator in the 1970s, trigonometry students were required to use tables to find trigonometric function values. While most trigonometry texts included short versions of these tables, there were books like the one shown here that gave function values for angles, as well as information on their logarithms to make computations easier. (Authors' note: Today's students don't know how fortunate they are not to have to use such tables.)

EXAMPLE 5 Donna Garbarino knows that when she stands 123 feet from the base of a flagpole, the angle of elevation to the top is 26° 40′. If her eyes are 5.30 feet above the ground, find the height of the flagpole.

The length of the side adjacent to Donna is known and the length of the side opposite her is to be found. See Figure 33. The ratio that involves these two values is the tangent.

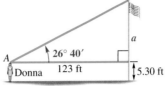

FIGURE 33

$$\tan A = \frac{\text{side opposite}}{\text{side adjacent}}$$

$$\tan 26° 40' = \frac{a}{123}$$

$$a = 123 \tan 26° 40'$$

$$a = 61.8 \text{ feet}$$

Since Donna's eyes are 5.30 feet above the ground, the height of the flagpole is

$$61.8 + 5.30 = 67.1 \text{ feet}.$$

EXAMPLE 6 The length of the shadow of a building 34.09 meters tall is 37.62 meters. Find the angle of elevation of the sun.

As shown in Figure 34, the angle of elevation of the sun is angle B. Since the side opposite B and the side adjacent to B are known, use the tangent ratio to find B.

$$\tan B = \frac{34.09}{37.62}$$

$$B = 42.18° \quad \text{Use the inverse tangent function of a calculator.}$$

The angle of elevation of the sun is 42.18°.

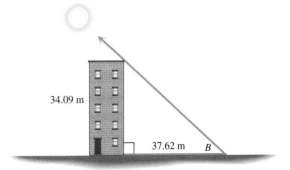

FIGURE 34

10.5 EXERCISES

Use a calculator to find a decimal approximation for each value. Give as many digits as your calculator displays.

1. tan 29° 30′
2. sin 38° 42′
3. cot 41° 24′
4. cos 27° 10′
5. sec 13° 15′
6. csc 44° 30′
7. sin 39° 40′
8. tan 17° 12′
9. csc 145° 45′
10. cot 183° 48′
11. cos 421° 30′
12. sec 312° 12′
13. tan(−80° 6′)
14. sin(−317° 36′)
15. cot(−512° 20′)
16. cos(−15′)

Find a value of θ such that 0° ≤ θ ≤ 90°, and θ satisfies the statement. Leave your answer in decimal degrees.

17. sin θ = .84802194
18. tan θ = 1.4739716
19. sec θ = 1.1606249
20. cot θ = 1.2575516
21. sin θ = .72144101
22. sec θ = 2.7496222
23. tan θ = 6.4358841
24. sin θ = .27843196

In the remaining exercises in this set, use a calculator as necessary.

Solve each right triangle.

25.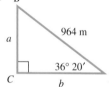

26.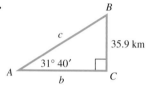

27.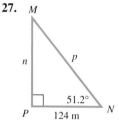

28.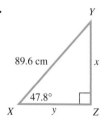

29.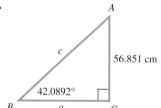

30.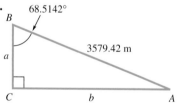

Solve each right triangle. In each case, $C = 90°$. If the angle information is given in degrees and minutes, give the answers in the same way. If given in decimal degrees, do likewise in your answers. When two sides are given, give answers in degrees and minutes.

31. $A = 28.00°$, $c = 17.4$ ft
32. $B = 46.00°$, $c = 29.7$ m
33. $B = 73.00°$, $b = 128$ in.
34. $A = 61° 00′$, $b = 39.2$ cm
35. $a = 76.4$ yd, $b = 39.3$ yd
36. $a = 958$ m, $b = 489$ m
37. $a = 18.9$ cm, $c = 46.3$ cm
38. $b = 219$ m, $c = 647$ m
39. $A = 53° 24′$, $c = 387.1$ ft
40. $A = 13° 47′$, $c = 1285$ m
41. $B = 39° 9′$, $c = .6231$ m
42. $B = 82° 51′$, $c = 4.825$ cm

Solve each problem.

43. *Ladder Leaning Against a Wall* A 13.5-meter firetruck ladder is leaning against a wall. Find the distance the ladder goes up the wall if it makes an angle of 43° 50′ with the ground.

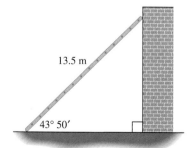

44. *Antenna Mast Guy Wire* A guy wire 77.4 meters long is attached to the top of an antenna mast that is 71.3 meters high. Find the angle that the wire makes with the ground.

45. *Guy Wire to a Tower* Find the length of a guy wire that makes an angle of 45° 30′ with the ground if the wire is attached to the top of a tower 63.0 meters high.

46. *Distance Across a Lake* To find the distance RS across a lake, a surveyor lays off $RT = 53.1$ meters, with angle $T = 32°\ 10'$ and angle $S = 57°\ 50'$. Find length RS.

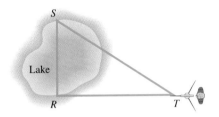

47. *Side Lengths of a Triangle* The length of the base of an isosceles triangle is 42.36 inches. Each base angle is 38.12°. Find the length of each of the two equal sides of the triangle. (*Hint:* Divide the triangle into two right triangles.)

48. *Altitude of a Triangle* Find the altitude of an isosceles triangle having a base of 184.2 cm if the angle opposite the base is 68° 44′.

49. *Cloud Ceiling* The U.S. Weather Bureau defines a *cloud ceiling* as the altitude of the lowest clouds that cover more than half the sky. To determine a cloud ceiling, a powerful searchlight projects a circle of light vertically on the bottom of the cloud. An observer sights the circle of light in the crosshairs of a tube called a *clinometer*. A pendant hanging vertically from the tube and resting on a protractor gives the angle of elevation. Find the cloud ceiling if the searchlight is located 1000 feet from the observer and the angle of elevation is 30.0° as measured with a clinometer at eye-height 6 feet. (Assume three significant digits.)

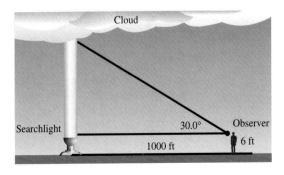

50. *Length of a Shadow* Suppose the angle of elevation of the sun is 23.4°. Find the length of the shadow cast by Cindy Newman, who is 5.75 feet tall.

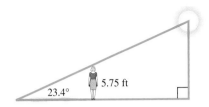

51. *Height of a Tower* The shadow of a vertical tower is 40.6 meters long when the angle of elevation of the sun is 34.6°. Find the height of the tower.

52. *Angle of Elevation of the Sun* Find the angle of elevation of the sun if a 48.6-foot flagpole casts a shadow 63.1 feet long.

53. *Distance from the Ground to the Top of a Building* The angle of depression from the top of a building to a point on the ground is 32° 30′. How far is the point on the ground from the top of the building if the building is 252 meters high?

54. *Airplane Distance* An airplane is flying 10,500 feet above the level ground. The angle of depression from the plane to the base of a tree is 13° 50′. How far horizontally must the plane fly to be directly over the tree?

55. *Height of a Building* The angle of elevation from the top of a small building to the top of a nearby taller building is 46° 40′, while the angle of depression to the bottom is 14° 10′. If the smaller building is 28.0 meters high, find the height of the taller building.

56. *Mounting a Video Camera* A video camera is to be mounted on a bank wall so as to have a good view of the head teller. Find the angle of depression that the lens should make with the horizontal.

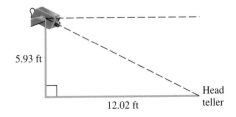

57. *Error in Measurement* A degree may seem like a very small unit, but an error of one degree in measuring an angle may be very significant. For example, suppose a laser beam directed toward the visible center of the moon misses its assigned target by 30 seconds. How far is it (in miles) from its assigned target? Take the distance from the surface of Earth to that of the moon to be 234,000 miles. (*Source: A Sourcebook of Applications of School Mathematics* by Donald Bushaw et al. Copyright © 1980 by The Mathematical Association of America.)

58. *Height of Mt. Everest* The highest mountain peak in the world is Mt. Everest, located in the Himalayas. The height of this enormous mountain was determined in 1856 by surveyors using trigonometry long before it was first climbed in 1953. This difficult measurement had to be done from a great distance. At an altitude of 14,545 feet on a different mountain, the straight line distance to the peak of Mt. Everest is 27.0134 miles and its angle of elevation is $\theta = 5.82°$. (*Source:* Dunham, W., *The Mathematical Universe*, John Wiley & Sons, 1994.)

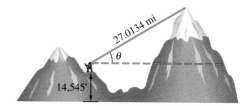

(a) Approximate the height (in feet) of Mt. Everest.
(b) In the actual measurement, Mt. Everest was over 100 miles away and the curvature of Earth had to be taken into account. Would the curvature of Earth make the peak appear taller or shorter than it actually is?

Find the exact value of each labeled part in each figure.

59.

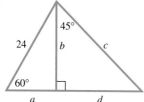

60.

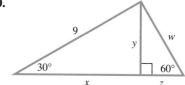

10.6 The Laws of Sines and Cosines; Area Formulas

A triangle that is not a right triangle is called an **oblique triangle**. The measures of the three sides and the three angles of a triangle can be found if at least one side and any other two measures are known. There are four possible cases.

1. One side and two angles are known.
2. Two sides and one angle not included between the two sides are known. (This case may lead to more than one triangle.)
3. Two sides and the angle included between the two sides are known.
4. Three sides are known.

If we know three angles of a triangle, we cannot find unique side lengths, since AAA assures us only of similarity, not congruence. For example, there are infinitely many triangles *ABC* with $A = 35°$, $B = 65°$, and $C = 80°$.

To solve oblique triangles, we need to use the laws of sines and cosines.

The Law of Sines To derive the law of sines, we start with an oblique triangle, such as the acute triangle in Figure 35(a) or the obtuse triangle in Figure 35(b). The following discussion applies to both triangles. First, construct the perpendicular from B to side AC or side AC extended. Let h be the length of this perpendicular. Then c is the hypotenuse of right triangle ADB, and a is the hypotenuse of right triangle BDC. Therefore,

in triangle ADB, $\quad \sin A = \dfrac{h}{c} \quad$ or $\quad h = c \sin A$,

in triangle BDC, $\quad \sin C = \dfrac{h}{a} \quad$ or $\quad h = a \sin C$.

Since $h = c \sin A$ and $h = a \sin C$,

$$a \sin C = c \sin A,$$

or, upon dividing both sides by $\sin A \sin C$,

$$\dfrac{a}{\sin A} = \dfrac{c}{\sin C}.$$

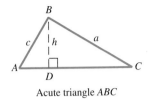

Acute triangle ABC

(a)

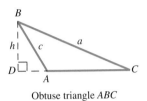

Obtuse triangle ABC

(b)

FIGURE 35

In a similar way, by constructing the perpendiculars from other vertices, it can be shown that

$$\dfrac{a}{\sin A} = \dfrac{b}{\sin B} \quad \text{and} \quad \dfrac{b}{\sin B} = \dfrac{c}{\sin C}.$$

This discussion proves the following theorem.

Law of Sines

In any triangle ABC with sides a, b, and c,

$$\dfrac{a}{\sin A} = \dfrac{b}{\sin B}, \quad \dfrac{a}{\sin A} = \dfrac{c}{\sin C}, \quad \text{and} \quad \dfrac{b}{\sin B} = \dfrac{c}{\sin C}.$$

This can be written in compact form as

$$\dfrac{a}{\sin A} = \dfrac{b}{\sin B} = \dfrac{c}{\sin C}.$$

Sometimes an alternative form of the law of sines,

$$\dfrac{\sin A}{a} = \dfrac{\sin B}{b} = \dfrac{\sin C}{c},$$

is more convenient to use.

If two angles and the side opposite one of the angles are known, the law of sines can be used directly to solve for the side opposite the other known angle. The triangle can then be solved completely, as shown in the first example.

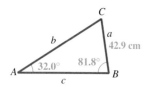

FIGURE 36

EXAMPLE 1 Solve triangle ABC if $A = 32.0°$, $B = 81.8°$, and $a = 42.9$ centimeters. See Figure 36.

Start by drawing a triangle, roughly to scale, and labeling the given parts as in Figure 36. Since the values of A, B, and a are known, use the part of the law of sines that involves these variables.

$$\frac{a}{\sin A} = \frac{b}{\sin B}$$

Substituting the known values gives

$$\frac{42.9}{\sin 32.0°} = \frac{b}{\sin 81.8°}.$$

Multiply both sides of the equation by $\sin 81.8°$.

$$b = \frac{42.9 \sin 81.8°}{\sin 32.0°}$$

When using a calculator to find b, keep intermediate answers in the calculator until the final result is found. Then round to the proper number of significant digits. In this case, find $\sin 81.8°$, and then multiply that number by 42.9. Keep the result in the calculator while you find $\sin 32.0°$, and then divide. Since the given information is accurate to three significant digits, round the value of b to get

$$b = \mathbf{80.1} \text{ centimeters.}$$

Find C from the fact that the sum of the angles of any triangle is 180°.

$$A + B + C = 180°$$
$$C = 180° - A - B$$
$$C = 180° - 32.0° - 81.8°$$
$$C = \mathbf{66.2°}$$

Now use the law of sines again to find c. (Why does the Pythagorean theorem not apply?)

$$\frac{a}{\sin A} = \frac{c}{\sin C}$$

$$\frac{42.9}{\sin 32.0°} = \frac{c}{\sin 66.2°}$$

$$c = \frac{42.9 \sin 66.2°}{\sin 32.0°}$$

$$c = \mathbf{74.1} \text{ centimeters} \qquad \blacksquare$$

Problem Solving

In applications of oblique triangles such as the one in Example 1, a correctly labeled sketch is essential in order to set up the correct equation.

EXAMPLE 2 Tri Nguyen wishes to measure the distance across the Big Muddy River. See Figure 37 on the next page. He finds that $C = 112.90°$, $A = 31.10°$, and $b = 347.6$ feet. Find the required distance.

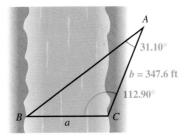

FIGURE 37

To use the law of sines, one side and the angle opposite it must be known. Since the only side whose length is given is b, angle B must be found before the law of sines can be used.

$$B = 180° - A - C$$
$$= 180° - 31.10° - 112.90° = 36.00°$$

Now use the form of the law of sines involving A, B, and b to find a, the distance across the river.

$$\frac{a}{\sin A} = \frac{b}{\sin B}$$

$$\frac{a}{\sin 31.10°} = \frac{347.6}{\sin 36.00°} \quad \text{Substitute.}$$

$$a = \frac{347.6 \sin 31.10°}{\sin 36.00°} \quad \text{Multiply by sin 31.10°.}$$

$$a = 305.5 \text{ feet} \quad \text{Use a calculator.}$$ ■

The Law of Cosines If we are given two sides and the included angle or three sides of a triangle, a unique triangle is formed. In these cases, however, we cannot begin the solution of the triangle by using the law of sines because we are not given a side and the angle opposite it. Both cases require the use of the *law of cosines*.

It will be helpful to remember the following property of triangles when applying the law of cosines.

Property of Triangle Side Lengths

In any triangle, the sum of the lengths of any two sides must be greater than the length of the remaining side.

For example, it would be impossible to construct a triangle with sides of lengths 3, 4, and 10. See Figure 38.

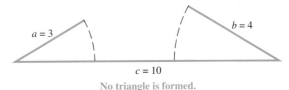

No triangle is formed.

FIGURE 38

To derive the law of cosines, let *ABC* be any oblique triangle. Choose a coordinate system so that vertex *B* is at the origin and side *BC* is along the positive *x*-axis. See Figure 39.

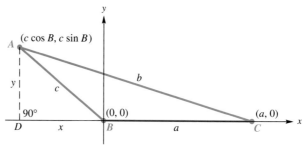

FIGURE 39

Let (x, y) be the coordinates of vertex *A* of the triangle. Verify that for angle *B*, whether obtuse or acute,

$$\sin B = \frac{y}{c} \quad \text{and} \quad \cos B = \frac{x}{c}.$$

(Here x is negative if *B* is obtuse.) From these results

$$y = c \sin B \quad \text{and} \quad x = c \cos B,$$

so that the coordinates of point *A* become

$$(c \cos B, c \sin B).$$

Point *C* has coordinates $(a, 0)$, and *AC* has length *b*. By the distance formula,

$$b = \sqrt{(c \cos B - a)^2 + (c \sin B)^2}.$$

Squaring both sides and simplifying gives

$$\begin{aligned}
b^2 &= (c \cos B - a)^2 + (c \sin B)^2 \\
&= c^2 \cos^2 B - 2ac \cos B + a^2 + c^2 \sin^2 B \\
&= a^2 + c^2 (\cos^2 B + \sin^2 B) - 2ac \cos B \\
&= a^2 + c^2(1) - 2ac \cos B \quad \text{Pythagorean identity} \\
&= a^2 + c^2 - 2ac \cos B.
\end{aligned}$$

This result is one form of the law of cosines. In the work above, we could just as easily have placed *A* or *C* at the origin. This would have given the same result, but with the variables rearranged. These various forms of the law of cosines are summarized in the following theorem.

Law of Cosines

In any triangle *ABC* with sides *a*, *b*, and *c*,

$$a^2 = b^2 + c^2 - 2bc \cos A$$
$$b^2 = a^2 + c^2 - 2ac \cos B$$
$$c^2 = a^2 + b^2 - 2ab \cos C.$$

10.6 The Laws of Sines and Cosines; Area Formulas

The law of cosines says that the square of a side of a triangle is equal to the sum of the squares of the other two sides, minus twice the product of those two sides and the cosine of the angle included between them.

If we let $C = 90°$ in the third form of the law of cosines given above, we have $\cos C = \cos 90° = 0$, and the formula becomes

$$c^2 = a^2 + b^2,$$

the familiar equation of the Pythagorean theorem. Thus, the Pythagorean theorem is a special case of the law of cosines.

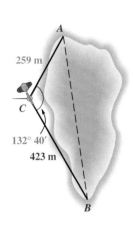

FIGURE 40

EXAMPLE 3 A surveyor wishes to find the distance between two inaccessible points A and B on opposite sides of a lake. While standing at point C, she finds that $AC = 259$ meters, $BC = 423$ meters, and angle ACB measures $132°\ 40'$. Find the distance AB. See Figure 40.

The law of cosines can be used here, since we know the lengths of two sides of the triangle and the measure of the included angle.

$AB^2 = 259^2 + 423^2 - 2(259)(423) \cos 132°\ 40'$

$AB^2 = 394{,}510.6$ Use a calculator.

$AB \approx 628$ Take the square root and round to 3 significant digits.

The distance AB is approximately 628 meters. ■

EXAMPLE 4 Solve triangle ABC if $A = 42.3°$, $b = 12.9$ meters, and $c = 15.4$ meters. See Figure 41.

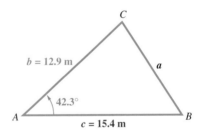

FIGURE 41

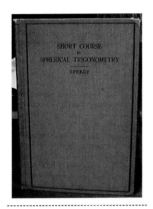

This photo shows the cover of a short text (only 60 pages) on spherical trigonometry, published in 1928. In the historical sketch provided at the end of the book, author Pauline Sperry states that Albattani, Geber of Seville, and Nassir Eddin continued the work of Ptolemy in the Arab world, while Johanne Müller (better known as **Regiomontanus**) did the same in Europe. Regiomontanus's work *De triangulis omnimodis* was "the first systematic treatment of trigonometry independent of astronomy in the western world and dominated the latter middle ages to a remarkable degree."

Start by finding a using the law of cosines.

$a^2 = b^2 + c^2 - 2bc \cos A$

$a^2 = 12.9^2 + 15.4^2 - 2(12.9)(15.4) \cos 42.3°$

$a^2 = 109.7$

$a = 10.5$ meters

We now must find the measures of angles B and C. There are several approaches that can be used at this point. We shall use the law of sines to find one of these angles. Of the two remaining angles, B must be the smaller since it is opposite the shorter of the two sides b and c. Therefore, it cannot be obtuse, and we will avoid any ambiguity when we find its sine.

$$\frac{\sin 42.3°}{10.5} = \frac{\sin B}{12.9}$$

$$\sin B = \frac{12.9 \sin 42.3°}{10.5}$$

$$B = 55.8° \qquad \text{Use the inverse sine function of a calculator.}$$

The easiest way to find C is to subtract the measures of A and B from $180°$.

$$C = 180° - A - B = 81.9°.$$

Had we chosen to use the law of sines to find C rather than B in Example 4, we would not have known whether C equals $81.9°$ or its supplement, $98.1°$.

EXAMPLE 5 Solve triangle ABC if $a = 9.47$ feet, $b = 15.9$ feet, and $c = 21.1$ feet.

We are given the lengths of three sides of the triangle, so we may use the law of cosines to solve for any angle of the triangle. Let us solve for C, the largest angle, using the law of cosines. We will be able to tell if C is obtuse if $\cos C < 0$. Use the form of the law of cosines that involves C.

$$c^2 = a^2 + b^2 - 2ab \cos C,$$

or

$$\cos C = \frac{a^2 + b^2 - c^2}{2ab}.$$

Substituting the given values leads to

$$\cos C = \frac{(9.47)^2 + (15.9)^2 - (21.1)^2}{2(9.47)(15.9)}$$

$$\cos C = -.34109402. \qquad \text{Use a calculator.}$$

Using the inverse cosine function of the calculator, we find the obtuse angle C.

$$C = 109.9°$$

We can use either the law of sines or the law of cosines to find $B = 45.1°$. (Verify this.) Since $A = 180° - B - C$,

$$A = 25.0°.$$

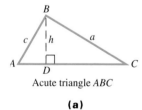

Acute triangle ABC

(a)

Obtuse triangle ABC

(b)

FIGURE 42

Area Formulas The method used to derive the law of sines can also be used to derive a useful formula to find the area of a triangle. A familiar formula for the area of a triangle is $\mathcal{A} = (1/2)bh$, where $\mathcal{A}$ represents the area, b the base, and h the height. This formula cannot always be used easily, since in practice h is often unknown. To find an alternative formula, refer to acute triangle ABC in Figure 42(a) or obtuse triangle ABC in Figure 42(b).

A perpendicular has been drawn from B to the base of the triangle (or the extension of the base). This perpendicular forms two right triangles. Using triangle ABD,

$$\sin A = \frac{h}{c},$$

or

$$h = c \sin A.$$

Substituting into the formula $\mathcal{A} = (1/2)bh$,

$$\mathcal{A} = \frac{1}{2}b(c \sin A)$$

or

$$\mathcal{A} = \frac{1}{2}bc \sin A.$$

Any other pair of sides and the angle between them could have been used, as stated in the next theorem.

Area of a Triangle

In any triangle ABC, the area $\mathcal{A}$ is given by any of the following formulas:

$$\mathcal{A} = \frac{1}{2}bc \sin A, \qquad \mathcal{A} = \frac{1}{2}ab \sin C, \qquad \mathcal{A} = \frac{1}{2}ac \sin B.$$

That is, the area is given by half the product of the lengths of two sides and the sine of the angle included between them.

EXAMPLE 6 Find the area of triangle ABC if $A = 24°\,40'$, $b = 27.3$ centimeters, and $C = 52°\,40'$.

Before we can use the formula given above, we must use the law of sines to find either a or c. Since the sum of the measures of the angles of any triangle is $180°$,

$$B = 180° - 24°\,40' - 52°\,40' = 102°\,40'.$$

Now use the form of the law of sines that relates a, b, A, and B to find a.

$$\frac{a}{\sin A} = \frac{b}{\sin B}$$

$$\frac{a}{\sin 24°\,40'} = \frac{27.3}{\sin 102°\,40'}$$

Solve for a to verify that $a = 11.7$ centimeters. Now find the area.

$$\mathcal{A} = \frac{1}{2}ab \sin C = \frac{1}{2}(11.7)(27.3) \sin 52°\,40' \approx 127$$

The area of triangle ABC is 127 square centimeters to three significant digits. ∎

The law of cosines can be used to derive a formula for the area of a triangle when only the lengths of the three sides are known. This formula is known as Heron's formula, named after the Greek mathematician Heron of Alexandria.

Heron of Alexandria lived during approximately the same period as Ptolemy and Hipparchus. He wrote extensively on both mathematical and physical subjects. The most important of his mathematical works is *Metrica*, which was discovered in Constantinople in the late 19th century. The Arabian scholar al-Biruni claims that Archimedes actually discovered the triangle area formula which bears Heron's name. (*Source:* Eves, Howard, *An Introduction to the History of Mathematics*, 6th edition.)

Heron's Area Formula

If a triangle has sides of lengths a, b, and c, and if the **semiperimeter** is

$$s = \frac{1}{2}(a + b + c),$$

then the area of the triangle is

$$\mathcal{A} = \sqrt{s(s-a)(s-b)(s-c)}.$$

EXAMPLE 7 Determine the area of the triangle having sides of lengths $a = 29.7$ feet, $b = 42.3$ feet, and $c = 38.4$ feet.

To use Heron's area formula, first find s.

$$s = \frac{1}{2}(a + b + c)$$

$$s = \frac{1}{2}(29.7 + 42.3 + 38.4)$$

$$= 55.2$$

The area is

$$\mathcal{A} = \sqrt{s(s-a)(s-b)(s-c)}$$
$$= \sqrt{55.2(55.2 - 29.7)(55.2 - 42.3)(55.2 - 38.4)}$$
$$= \sqrt{55.2(25.5)(12.9)(16.8)}$$
$$\mathcal{A} \approx 552 \text{ square feet.}$$

10.6 EXERCISES

Use the law of sines to solve each triangle.

1.

Triangle with $AB = 18$ m, angle at $A = 37°$, angle at $B = 48°$, vertex C.

2.

Triangle with $BC = 43$ cm, angle at $C = 29°$, angle at $A = 52°$.

3.
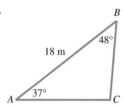

4.

5. $A = 68.41°, B = 54.23°, a = 12.75$ ft
6. $C = 74.08°, B = 69.38°, c = 45.38$ m
7. $A = 87.2°, b = 75.9$ yd, $C = 74.3°$
8. $B = 38° 40', a = 19.7$ cm, $C = 91° 40'$
9. $B = 20° 50', C = 103° 10', AC = 132$ ft
10. $A = 35.3°, B = 52.8°, AC = 675$ ft
11. $A = 39.70°, C = 30.35°, b = 39.74$ m
12. $C = 71.83°, B = 42.57°, a = 2.614$ cm
13. $B = 42.88°, C = 102.40°, b = 3974$ ft
14. $A = 18.75°, B = 51.53°, c = 2798$ yd
15. $A = 39° 54', a = 268.7$ m, $B = 42° 32'$
16. $C = 79° 18', c = 39.81$ mm, $A = 32° 57'$

Use the law of sines to solve each problem.

17. *Distance Across a River* To find the distance AB across a river, a distance $BC = 354$ meters is laid off on one side of the river. It is found that $B = 112° 10'$ and $C = 15° 20'$. Find AB.

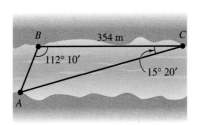

18. *Distance Across a Canyon* To determine the distance RS across a deep canyon, Joanna lays off a distance $TR = 582$ yards. She then finds that $T = 32° 50'$ and $R = 102° 20'$. Find RS.

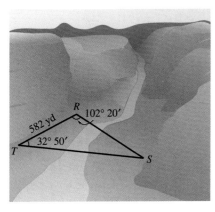

19. *Measurement of a Folding Chair* A folding chair is to have a seat 12.0 inches deep with angles as shown in the figure. How far down from the seat should the crossing legs be joined? (Find x in the figure.)

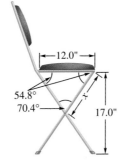

20. *Angle Formed by Radii of Gears* Three gears are arranged as shown in the figure. Find angle θ.

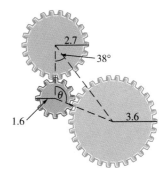

21. *Distance Between Atoms* Three atoms with atomic radii of 2.0, 3.0, and 4.5 are arranged as in the figure. Find the distance between the centers of atoms A and C.

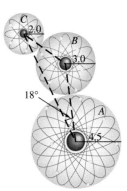

22. *Height of a Balloon* A balloonist is directly above a straight road 1.5 miles long that joins two villages. She finds that the town closer to her is at an angle of depression of 35°, and the farther town is at an angle of depression of 31°. How high above the ground is the balloon?

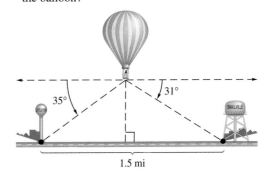

Use the law of cosines to find the length of the remaining sides of each triangle. Do not use a calculator.

23.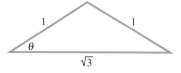

24.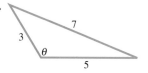

Use the law of cosines to find the value of θ in each triangle. Do not use a calculator.

25.

26.

Use the law of cosines to solve each triangle.

27. $C = 28.3°, b = 5.71$ in., $a = 4.21$ in.
28. $A = 41.4°, b = 2.78$ yd, $c = 3.92$ yd
29. $C = 45.6°, b = 8.94$ m, $a = 7.23$ m
30. $A = 67.3°, b = 37.9$ km, $c = 40.8$ km
31. $A = 80° 40', b = 143$ cm, $c = 89.6$ cm
32. $C = 72° 40', a = 327$ ft, $b = 251$ ft
33. $B = 74.80°, a = 8.919$ in., $c = 6.427$ in.
34. $C = 59.70°, a = 3.725$ mi, $b = 4.698$ mi
35. $A = 112.8°, b = 6.28$ m, $c = 12.2$ m
36. $B = 168.2°, a = 15.1$ cm, $c = 19.2$ cm
37. $a = 3.0$ ft, $b = 5.0$ ft, $c = 6.0$ ft
38. $a = 4.0$ ft, $b = 5.0$ ft, $c = 8.0$ ft
39. $a = 9.3$ cm, $b = 5.7$ cm, $c = 8.2$ cm
40. $a = 28$ ft, $b = 47$ ft, $c = 58$ ft
41. $a = 42.9$ m, $b = 37.6$ m, $c = 62.7$ m
42. $a = 187$ yd, $b = 214$ yd, $c = 325$ yd

Solve each problem, using the law of sines or the law of cosines as needed.

43. *Distance Across a Lake* Points A and B are on opposite sides of Lake Yankee. From a third point, C, the angle between the lines of sight to A and B is 46.3°. If AC is 350 meters long and BC is 286 meters long, find AB.

44. *Diagonals of a Parallelogram* The sides of a parallelogram are 4.0 cm and 6.0 cm. One angle is 58° while another is 122°. Find the lengths of the diagonals of the parallelogram.

45. *Playhouse Layout* The layout for a child's playhouse in her backyard shows the dimensions given in the figure. Find x.

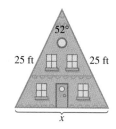

46. *Distance Between Points on a Crane* A crane with a counterweight is shown in the figure. Find the horizontal distance between points A and B.

47. *Angle Between a Beam and Cables* A weight is supported by cables attached to both ends of a horizontal beam, as shown in the figure. What angles are formed between the beam and the cables?

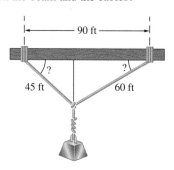

48. *Measurement Using Triangulation* Surveyors are often confronted with obstacles, such as trees, when measuring the boundary of a lot. One technique used to obtain an accurate measurement is the so-called *triangulation method.* In this technique, a triangle is constructed around the obstacle and one angle and two sides of the triangle are measured. Use this technique to find the length of the property line (the straight line between the two markers) in the figure. (*Source:* Kavanagh, B. and S. Bird, *Surveying Principles and Applications,* Fifth Edition, Prentice-Hall, 2000.)

57. $A = 56.80°, b = 32.67$ in., $c = 52.89$ in.

58. $A = 34.97°, b = 35.29$ m, $c = 28.67$ m

Find the exact area of each triangle using the formula $A = \frac{1}{2} bh,$ *and then verify that Heron's formula gives the same result.*

59.

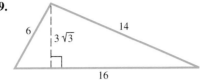

60.

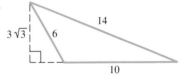

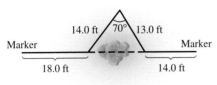

Not to scale

Find the area of each triangle using Heron's formula.

61. $a = 12$ m, $b = 16$ m, $c = 25$ m
62. $a = 22$ in., $b = 45$ in., $c = 31$ in.
63. $a = 154$ cm, $b = 179$ cm, $c = 183$ cm
64. $a = 25.4$ yd, $b = 38.2$ yd, $c = 19.8$ yd
65. $a = 76.3$ ft, $b = 109$ ft, $c = 98.8$ ft
66. $a = 15.89$ in., $b = 21.74$ in., $c = 10.92$ in.

Solve each problem.

67. *Area of a Metal Plate* A painter is going to apply a special coating to a triangular metal plate on a new building. Two sides measure 16.1 meters and 15.2 meters. She knows that the angle between these sides is 125°. What is the area of the surface she plans to cover with the coating?

68. *Area of a Triangular Lot* A real estate agent wants to find the area of a triangular lot. A surveyor takes measurements and finds the two sides are 52.1 meters and 21.3 meters, and the angle between them is 42.2°. What is the area of the triangular lot?

69. *Required Amount of Paint* A painter needs to cover a triangular region 75 meters by 68 meters by 85 meters. A can of paint covers 75 square meters of area. How many cans (to the next higher number of cans) will be needed?

70. *Area of the Bermuda Triangle* Find the area of the Bermuda Triangle if the sides of the triangle have approximate lengths 850 miles, 925 miles, and 1300 miles.

Find the exact area of each triangle using the formula $A = \frac{1}{2} bh,$ *and then verify that the formula* $A = \frac{1}{2} ab \sin C$ *gives the same result.*

49.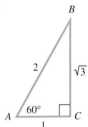

50.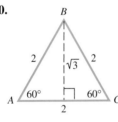

51.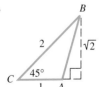

52.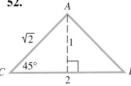

Find the area of each triangle using the formula involving the sine function (page 615).

53. $A = 42.5°, b = 13.6$ m, $c = 10.1$ m
54. $C = 72.2°, b = 43.8$ ft, $a = 35.1$ ft
55. $B = 124.5°, a = 30.4$ cm, $c = 28.4$ cm
56. $C = 142.7°, a = 21.9$ km, $b = 24.6$ km

10.7 The Unit Circle and Graphs

So far, we have defined the trigonometric functions with domains consisting of degree-measured angles. It is also possible to define them for real number domains.

Observe Figure 43. Suppose that s represents a real number. Starting at the point $(1,0)$, we lay off an arc of length $|s|$ along the circle. We go counterclockwise if s is positive and clockwise if s is negative. The endpoint of the arc is the point (x, y). The circle in Figure 43 is a **unit circle**, a circle with center at the origin and a radius of one unit (hence the name *unit circle*). Recall from algebra that the equation of this circle is

$$x^2 + y^2 = 1.$$

The value x is called the cosine of the real number s, while y is called its sine.

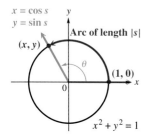

FIGURE 43

sin s and cos s

For the unit circle in Figure 43,

$$x = \cos s \quad \text{and} \quad y = \sin s.$$

Many things in daily life repeat with a predictable pattern: in warm areas electricity use goes up in the summer and down in the winter, the price of fresh fruit goes down in the summer and up in the winter, and attendance at amusement parks increases in the summer and declines in autumn. Because the sine and cosine functions repeat their values over and over in a regular pattern, they are examples of *periodic functions*.

Periodic Function

A **periodic function** is a function f such that

$$f(x) = f(x + p)$$

for every real number x in the domain of f and for some positive real number p. The smallest possible positive value of p is the **period** of the function.

The circumference of the unit circle is 2π, so the smallest value of p for which the sine and cosine functions repeat is 2π. Therefore, the sine and cosine functions are periodic functions with period 2π.

Now, use Figure 44 to verify the chart below for $y = \sin s$ and $y = \cos s$.

FIGURE 44

The unit circle $x^2 + y^2 = 1$

As s Increases from	sin s	cos s
0 to $\pi/2$	increases from 0 to 1	decreases from 1 to 0
$\pi/2$ to π	decreases from 1 to 0	decreases from 0 to -1
π to $3\pi/2$	decreases from 0 to -1	increases from -1 to 0
$3\pi/2$ to 2π	increases from -1 to 0	increases from 0 to 1

Any letter could be used instead of s for the arc length. It is customary to use x, so that we are graphing on the familiar xy-coordinate system. Selecting key values of x and finding the corresponding values of sin x give the following results. This can be done with a calculator, in **radian mode** (rather than degree mode). Decimals are rounded to the nearest tenth.

x	0	$\pi/4$	$\pi/2$	$3\pi/4$	π	$5\pi/4$	$3\pi/2$	$7\pi/4$	2π
sin x	0	.7	1	.7	0	−.7	−1	−.7	0

Plotting the points from the table of values and connecting them with a smooth curve gives the graph in Figure 45. Since $y = \sin x$ is periodic and has {all real numbers} as its domain, the graph continues in both directions indefinitely. Its range is $\{y \mid -1 \leq y \leq 1\}$. This graph is sometimes called a **sine wave** or **sinusoid**. You should learn the shape of this graph and be able to sketch it quickly. The key points of the graph are $(0, 0)$, $(\pi/2, 1)$, $(\pi, 0)$, $(3\pi/2, -1)$, and $(2\pi, 0)$. By plotting these five points and connecting them with the characteristic sine wave, you can quickly sketch the graph. The sketch can be extended if desired.

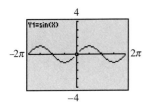

The graph of Y = sin X is shown in the standard *trig window*, having the dimensions indicated.

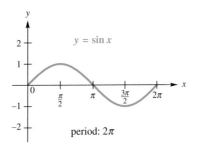

FIGURE 45

The graph of $y = \cos x$ can be found in much the same way as the graph of $y = \sin x$ was found. The domain of $y = \cos x$ is {all real numbers}, and the range is $\{y \mid -1 \leq y \leq 1\}$. A table of values is shown below for $y = \cos x$.

x	0	$\pi/4$	$\pi/2$	$3\pi/4$	π	$5\pi/4$	$3\pi/2$	$7\pi/4$	2π
cos x	1	.7	0	−.7	−1	−.7	0	.7	1

Here the key points are $(0, 1)$, $(\pi/2, 0)$, $(\pi, -1)$, $(3\pi/2, 0)$, and $(2\pi, 1)$.

The graph of $y = \cos x$, in Figure 46, has the same shape as the graph of $y = \sin x$. In fact, it is the graph of the sine function, shifted $\pi/2$ units to the left.

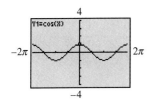

This is a calculator graph of the cosine function.

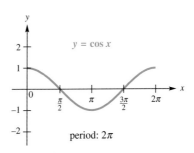

FIGURE 46

EXAMPLE 1 Graph $y = 2 \sin x$.

For a given value of x, the value of y is twice as large as it would be for $y = \sin x$, as shown in the table of values. The only change in the graph is the range, which becomes $\{y \mid -2 \leq y \leq 2\}$. See Figure 47, which also shows a graph of $y = \sin x$ for comparison, and portions of both graphs for negative x-values.

x	0	$\pi/2$	π	$3\pi/2$	2π
$\sin x$	0	1	0	-1	0
$2 \sin x$	0	2	0	-2	0

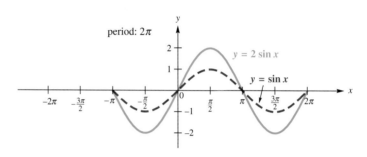

FIGURE 47

When graphing periodic functions, it is customary to first graph over one period. Then more of the graph can be sketched using the fact that the graph repeats the cycle over and over.

Generalizing from Example 1 gives the following.

Amplitude of the Sine and Cosine Functions

The graph of $y = a \sin x$ or $y = a \cos x$, with $a \neq 0$, will have the same shape as the graph of $y = \sin x$ or $y = \cos x$, respectively, except with range $\{y \mid -|a| \leq y \leq |a|\}$. The number $|a|$ is called the **amplitude.** (The amplitude of a periodic function can be interpreted as half the difference between its maximum and minimum values.)

EXAMPLE 2 Graph $y = \sin 2x$.

For this function, all function values will be found by letting $2x$ take on values between 0 and 2π. Thus, x will take on values between 0 and π. Therefore, the period is $2\pi/2 = \pi$. The graph will complete one period over the interval from 0 to π.

The endpoints are 0 and π, and the three middle points are

$$\frac{1}{4}(0 + \pi), \quad \frac{1}{2}(0 + \pi), \quad \text{and} \quad \frac{3}{4}(0 + \pi),$$

which gives the following x-values.

We now plot the points from the table of values given earlier, and join them with a smooth sinusoidal curve. More of the graph can be sketched by repeating this cycle over and over, as shown in Figure 48. Notice that the amplitude is not changed. The graph of $y = \sin x$ is included for comparison.

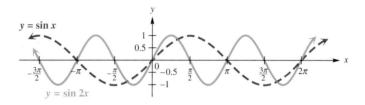

FIGURE 48

Generalizing from Example 2 leads to the following result.

Period of Sine and Cosine Functions

For $b > 0$, the graph of $y = \sin bx$ will look like that of $y = \sin x$, but with a period of $2\pi/b$. Also, the graph of $y = \cos bx$ will look like that of $y = \cos x$, but with a period of $2\pi/b$.

EXAMPLE 3 Graph $y = \cos \dfrac{2}{3} x$.

For this function the period is

$$\frac{2\pi}{2/3} = 3\pi.$$

Divide the interval from 0 to 3π into four equal parts to get the x-values that will yield minimum points, maximum points, and x-intercepts. These x-values are as follows.

$$0 \quad \frac{3\pi}{4} \quad \frac{3\pi}{2} \quad \frac{9\pi}{4} \quad 3\pi$$

These values are used to find a table of key points for one period.

x	0	$3\pi/4$	$3\pi/2$	$9\pi/4$	3π
$(2/3)x$	0	$\pi/2$	π	$3\pi/2$	2π
$\cos(2/3)x$	1	0	-1	0	1

The amplitude is 1, because the maximum value is 1, the minimum value is -1, and half of $1 - (-1) = (1/2)(2) = 1$.

Now plot these points and join them with a smooth curve. The graph is shown in Figure 49.

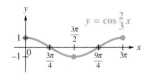

FIGURE 49

The steps used to graph $y = a \sin bx$ or $y = a \cos bx$, where $b > 0$, are given below.

Graphing the Sine and Cosine Functions

To graph $y = a \sin bx$ or $y = a \cos bx$, with $b > 0$:

1. Find the period, $2\pi/b$. Start at 0 on the x-axis and lay off a distance of $2\pi/b$.
2. Divide the interval into four equal parts.
3. Evaluate the function for each of the five x-values resulting from Step 2. The points will be maximum points, minimum points, and x-intercepts.
4. Plot the points found in Step 3, and join them with a sinusoidal curve with amplitude $|a|$.
5. Draw additional cycles of the graph, to the right and to the left, as needed.

EXAMPLE 4 Graph $y = -2 \sin 3x$.

Step 1: For this function, $b = 3$, so the period is $2\pi/3$. We will graph the function over the interval from 0 to $2\pi/3$.

Step 2: Dividing this interval into four equal parts gives the x-values 0, $\pi/6$, $\pi/3$, $\pi/2$, and $2\pi/3$.

Step 3: Make a table of points determined by the x-values resulting from Step 2.

x	0	$\pi/6$	$\pi/3$	$\pi/2$	$2\pi/3$
$3x$	0	$\pi/2$	π	$3\pi/2$	2π
$\sin 3x$	0	1	0	-1	0
$-2 \sin 3x$	0	-2	0	2	0

Step 4: Plot the points $(0,0)$, $(\pi/6, -2)$, $(\pi/3, 0)$, $(\pi/2, 2)$, and $(2\pi/3, 0)$, and join them with a sinusoidal curve with amplitude 2. See Figure 50.

Step 5: If necessary, the graph in Figure 50 can be extended by repeating the cycle over and over.

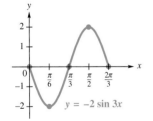

FIGURE 50

Notice the effect of the negative value of a. When a is negative, the graph of $y = a \sin bx$ will be the reflection about the x-axis of the graph of $y = |a| \sin bx$. The amplitude here is $|-2| = 2$. ■

In this section we have only discussed the graphs of the sine and cosine functions. The graphs of the other trigonometric functions are studied in standard trigonometry texts.

10.7 EXERCISES

In Exercises 1–8 match each function with its graph in choices A.–H.

1. $y = \sin x$
2. $y = \cos x$
3. $y = -\sin x$
4. $y = -\cos x$
5. $y = \sin 2x$
6. $y = \cos 2x$
7. $y = 2 \sin x$
8. $y = 2 \cos x$

A.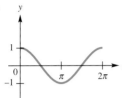
B.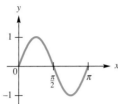
C.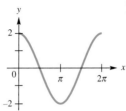
D.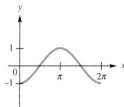

E.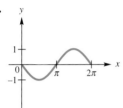
F.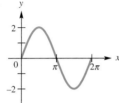
G.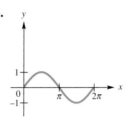
H.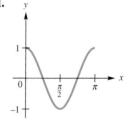

Graph each function over two periods. Give the period and the amplitude.

9. $y = 2 \cos x$
10. $y = 3 \sin x$
11. $y = \frac{2}{3} \sin x$
12. $y = \frac{3}{4} \cos x$
13. $y = -\cos x$
14. $y = -\sin x$
15. $y = -2 \sin x$
16. $y = -3 \cos x$
17. $y = \sin \frac{1}{2} x$
18. $y = \sin \frac{2}{3} x$
19. $y = \cos \frac{1}{3} x$
20. $y = \cos \frac{3}{4} x$
21. $y = \sin 3x$
22. $y = \cos 2x$
23. $y = 2 \sin \frac{1}{4} x$
24. $y = 3 \sin 2x$
25. $y = -2 \cos 3x$
26. $y = -5 \cos 2x$
27. $y = \cos \pi x$
28. $y = -\sin \pi x$

Solve each problem.

29. *Average Annual Temperature* Scientists believe that the average annual temperature in a given location is periodic. The overall temperature at a given place during a given season fluctuates as time goes on, from colder to warmer, and back to colder. The graph shows an idealized description of the temperature for the last few thousand years of a location at the same latitude as Anchorage.
    (a) Find the highest and lowest temperatures recorded.
    (b) Use these two numbers to find the amplitude. (*Hint:* An alternative definition of the amplitude is half the difference between the *y*-values of the highest and lowest points on the graph.)
    (c) Find the period of the graph.
    (d) What is the trend of the temperature now?

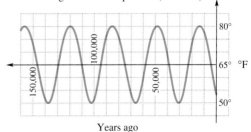

30. *Activity of a Nocturnal Animal* Many of the activities of living organisms are periodic. For example, the graph below shows the time that a certain nocturnal animal begins its evening activity.
 (a) Find the amplitude of this graph.
 (b) Find the period.

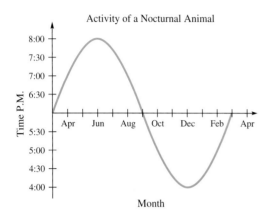

31. *Variation in Blood Pressure* The graph shown gives the variation in blood pressure for a typical person. Systolic and diastolic pressures are the upper and lower limits of the periodic changes in pressure that produce the pulse. The length of time between peaks is called the period of the pulse.
 (a) Find the amplitude of the graph.
 (b) Find the pulse rate (the number of pulse beats in one minute) for this person.

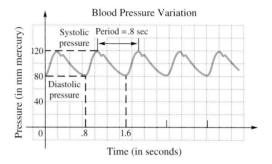

32. *Sine Waves on an Oscilloscope* Pure sounds produce single sine waves on an oscilloscope. Find the amplitude and period of each sine wave in the following graphs. On the vertical scale, each square represents .5, and on the horizontal scale each square represents $\pi/6$.

(a) 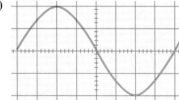   (b)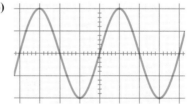

COLLABORATIVE INVESTIGATION

Making a *Point* About Trigonometric Function Values

An equation of the form $Ax + By = 0$ represents a line passing through the origin in the plane. By restricting x to take on only nonpositive or only nonnegative values, we obtain a ray with endpoint at the origin. This ray can be considered the terminal side of an angle in standard position. For example, the figure shows an angle θ in standard position whose terminal side has the equation

$$2x - y = 0, \quad x \geq 0.$$

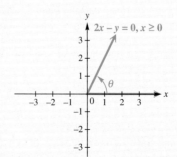

To find the trigonometric function values of θ, we can use *any* point on its terminal side except the origin. Similarly, we can find the value of θ using any point and the appropriate inverse trigonometric function. This investigation supports these statements.

Topics for Discussion

1. Divide the class into four groups: I, II, III, and IV. Each group will have a different value of x assigned to it:

   Group I: $x = 2$ Group II: $x = 4$

   Group III: $x = 6$ Group IV: $x = 8$

2. Find the coordinates of the point on the terminal side of θ corresponding to your value of x.

3. Find the corresponding value of r using the equation $r = \sqrt{x^2 + y^2}$.

4. Find the exact values of the trigonometric functions of θ. Simplify any radical expressions completely.

5. With your calculator in degree mode, find
   (a) $\sin^{-1}\left(\frac{y}{r}\right)$, (b) $\cos^{-1}\left(\frac{x}{r}\right)$, and (c) $\tan^{-1}\left(\frac{y}{x}\right)$.

6. Compare the results of Topic 4 among the different groups. Did the point chosen make any difference when finding the trigonometric function values?

7. Compare the results of Topic 5 among the different groups. Every value should be the same. What is this value? What does it represent?

8. Similar triangles have sides that are proportional, as discussed in Section 9.4. Discuss how this property justifies the results of Topic 4.

CHAPTER 10 TEST

1. Convert $74°\ 17'\ 54''$ to decimal degrees.

2. Find the angle of smallest positive measure coterminal with $-157°$.

3. If $(2, -5)$ is on the terminal side of angle θ in standard position, find $\sin\theta$, $\cos\theta$, and $\tan\theta$.

4. If $\cos\theta < 0$ and $\cot\theta > 0$, in what quadrant does θ terminate.

5. If $\cos\theta = 4/5$ and θ is in quadrant IV, find the values of the other trigonometric functions of θ.

6. Give the six trigonometric function values of angle A.

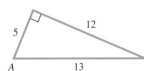

7. Give the *exact* value of each expression. If it is undefined, say so.
   (a) $\cos 60°$ (b) $\tan 45°$ (c) $\tan(-270°)$
   (d) $\sec 210°$ (e) $\csc(-180°)$ (f) $\sec 135°$

8. Use a calculator to approximate the following.
   (a) $\sin 78°\ 21'$ (b) $\tan 11.7689°$
   (c) $\sec 58.9041°$

9. Find a value of θ in the interval $0° \leq \theta \leq 90°$ in decimal degrees, if $\sin\theta = .27843196$.

10. Find two values of θ in the interval $0° \leq \theta < 360°$ that satisfy
$$\cos\theta = -\frac{\sqrt{2}}{2}.$$

11. Solve the triangle.

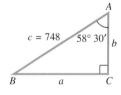

12. *Height of a Flagpole* To measure the height of a flagpole, Mario Mendoza found that the angle of elevation from a point 24.7 feet from the base to the top is $32°\ 10'$. What is the height of the flagpole?

Find the indicated part of each triangle.

13. $A = 25.2°$, $a = 6.92$ yd, $b = 4.82$ yd; find C

14. $C = 118°$, $b = 130$ km, $a = 75$ km; find c

15. $a = 17.3$ ft, $b = 22.6$ ft, $c = 29.8$ ft; find B

16. Find the area of triangle ABC in Exercise 14.

17. Find the area of a triangle having sides of lengths 22, 26, and 40.

Solve each problem.

18. *Height of a Balloon* The angles of elevation of a balloon from two points A and B on level ground are $24°\,50'$ and $47°\,20'$, respectively. As shown in the figure, points A, B, and C are in the same vertical plane and A and B are 8.4 miles apart. Approximate the height of the balloon above the ground to the nearest tenth of a mile.

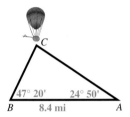

19. *Length of a Tunnel* To measure the distance through a mountain for a proposed tunnel, a point C is chosen that can be reached from each end of the tunnel. If $AC = 3800$ meters, $BC = 2900$ meters, and angle $C = 110°$, find the length of the tunnel.

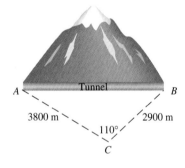

20. *Distances on a Baseball Diamond* A baseball diamond is a square 90.0 feet on a side, with home plate and the three bases as vertices. The pitcher's rubber is located 60.5 feet from home plate. Find the distance from the pitcher's rubber to each of the bases.

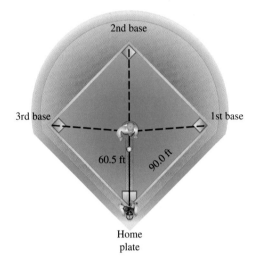

21. Graph $y = 3 \sin x$ over a one-period interval. Give the amplitude and the period.

22. Graph $y = -\cos 2x$ over a two-period interval. Give the amplitude and the period.

chapter

11

Counting Methods

One of the most profound discoveries of our time occurred in 1953 when the double helical structure of DNA (deoxyribonucleic acid) was identified. DNA is the material of heredity. Since then, scientists have been trying to understand the complicated genetic code that determines who we are. In June 2000, it was announced that highly technical research had resulted in a practically complete determination of the human genetic code. DNA testing can help determine guilt or innocence in criminal cases. Other uses of DNA profiling include archaeological studies and diagnosis (and possible prevention or treatment) of genetic diseases.

DNA is composed of four bases called adenine (A), guanine (G), thymine (T), and cytosine (C). The genetic code of an individual is made up of approximately six billion of these bases stored in a linear sequence. The nucleus of every cell contains a copy of this genetic code, tightly coiled in the shape of a double helix.

Any one of the four bases can occur anytime in the genetic code sequence. In a normal red blood cell, hemoglobin contains the genetic sequence

—ACTCCTG[A]GGAGGAGT—

whereas a person with sickle cell anemia will have exactly the same sequence except for a single base

—ACTCCTG[T]GGAGGAGT—.

Thus, one error in a sequence of six billion letters causes this disease. Since each location in the code allows four bases (A, C, G, and T), two consecutive locations allow $4^2 = 16$ choices, three allow $4^3 = 64$ choices, and so on. In 6 billion locations there are $4^{6,000,000,000}$ different sequences possible. In reality 99% to 99.9% of this genetic code is identical for all humans and only .1% to 1% is unique to the individual. It is this small percentage on which DNA profiling relies. Counting methods such as those above emphasize the uniqueness of humankind and of each individual. Small changes in the genetic code can have dramatic effects.

Sources: Easteal, S., N. McLeod, and K. Reed, *DNA PROFILING Principles, Pitfalls and Potential,* Harwood Academic Publishers, 1991; www.cnn.com (June 26, 2000).

11.1 Counting by Systematic Listing

11.2 Using the Fundamental Counting Principle

11.3 Using Permutations and Combinations

11.4 Using Pascal's Triangle and the Binomial Theorem

11.5 Counting Problems Involving "Not" and "Or"

Collaborative Investigation:
    Approximating Factorials Using Stirling's Formula

Chapter 11 Test

11.1 Counting by Systematic Listing

The methods of counting presented in this section all involve coming up with an actual list of the possible results for a given task. This approach is practical only for fairly short lists. Other methods, developed in the remainder of the chapter, will enable us to find "how many" without actually listing all the possibilities.

When listing possible results, it is extremely important to use a *systematic* approach. If we just start listing the possibilities as they happen to occur to us, we are likely to miss some of them.

Counting methods can be used to find the number of moves required to solve a Rubik's Cube. The scrambled cube must be modified so that each face is a solid color. Rubik's royalties from sales of the cube in Western countries made him Hungary's richest man.

One-Part Tasks The results for simple tasks consisting of one part can often be listed easily. For the task of tossing a single fair coin, for example, the list looks like this: *heads, tails.* There are two possible results. If the task is to roll a single fair die (a cube with faces numbered 1 through 6), the different results are 1, 2, 3, 4, 5, 6, a total of six possibilities.

EXAMPLE 1 Consider a club N with five members:

$$N = \{\text{Andy, Bill, Cathy, David, Evelyn}\},$$

or, as a shortcut, $N = \{A, B, C, D, E\}$.

In how many ways can this group select a president (assuming all members are eligible)?

The task in this case is to select one of the five members as president. There are five possible results: $A, B, C, D,$ and E. ∎

Two-Part Tasks; Using Product Tables

EXAMPLE 2 Determine the number of two-digit numbers that can be written using digits from the set $\{1, 2, 3\}$.

This task consists of two parts: choose a first digit and choose a second digit. The results for a two-part task can be pictured in a **product table** such as Table 1. From the table we obtain our list of possible results: 11, 12, 13, 21, 22, 23, 31, 32, 33. There are nine possibilities. ∎

TABLE 1

		Second Digit		
		1	2	3
First Digit	1	11	12	13
	2	21	22	23
	3	31	32	33

EXAMPLE 3 Determine the number of different possible results when two ordinary dice are rolled.

Assume the dice are easily distinguishable. Perhaps one is red and the other green. Then the task consists of two parts: roll the red die and roll the green die. Table 2 on the next page is a product table. It shows that there are thirty-six possible results. ∎

You will want to refer to Table 2 when various dice-rolling problems occur in the remainder of this chapter and the next.

TABLE 2 Rolling Two Fair Dice

		Green Die					
		1	2	3	4	5	6
Red Die	1	(1, 1)	(1, 2)	(1, 3)	(1, 4)	(1, 5)	(1, 6)
	2	(2, 1)	(2, 2)	(2, 3)	(2, 4)	(2, 5)	(2, 6)
	3	(3, 1)	(3, 2)	(3, 3)	(3, 4)	(3, 5)	(3, 6)
	4	(4, 1)	(4, 2)	(4, 3)	(4, 4)	(4, 5)	(4, 6)
	5	(5, 1)	(5, 2)	(5, 3)	(5, 4)	(5, 5)	(5, 6)
	6	(6, 1)	(6, 2)	(6, 3)	(6, 4)	(6, 5)	(6, 6)

TABLE 3 Electing Two Officers

		Secretary				
		A	B	C	D	E
President	A		AB	AC	AD	AE
	B	BA		BC	BD	BE
	C	CA	CB		CD	CE
	D	DA	DB	DC		DE
	E	EA	EB	EC	ED	

EXAMPLE 4 Find the number of ways that club N of Example 1 can elect both a president and a secretary. Assume that all members are eligible, but that no one can hold both offices.

Again, the required task has two parts: determine the president and determine the secretary. Constructing Table 3 gives us the following list (where, for example, AB denotes president A and secretary B, while BA denotes president B and secretary A):

$$AB, \quad AC, \quad AD, \quad AE, \quad BA, \quad BC, \quad BD, \quad BE, \quad CA, \quad CB,$$

$$CD, \quad CE, \quad DA, \quad DB, \quad DC, \quad DE, \quad EA, \quad EB, \quad EC, \quad ED.$$

Notice that certain entries (down the main diagonal, from upper left to lower right) are omitted from the table, since the cases AA, BB, and so on would imply one person holding both offices. Altogether, there are twenty possibilities.

Bone dice were unearthed in the remains of a Roman garrison, Vindolanda, near the border between England and Scotland. Life on the Roman frontier was occupied with gaming as well as fighting. Some of the Roman dice were loaded in favor of 6 and 1.

Life on the American frontier was reflected in cattle brands that were devised to keep alive the memories of hardships, feuds, and romances. A rancher named Ellis from Paradise Valley in Arizona designed his cattle brand in the shape of a pair of dice. You can guess that the pips were 6 and 1.

EXAMPLE 5 Find the number of ways that club N can appoint a committee of two members to represent them at an association conference.

The required task again has two parts. In fact, we can refer to Table 3 again, but this time, the order of the two letters (people) in a given pair really makes no difference. For example, BD and DB are the same committee. (In Example 4, BD and DB were different results since the two people would be holding different offices.) In the case of committees, we eliminate not only the main diagonal entries, but also all entries below the main diagonal. The resulting list contains ten possibilities:

$$AB, \quad AC, \quad AD, \quad AE, \quad BC, \quad BD, \quad BE, \quad CD, \quad CE, \quad DE.$$

Tasks with Three or More Parts; Using Tree Diagrams

Problem Solving

A task that has more than two parts is not easy to analyze with a product table. It would require a table of more than two dimensions, which is hard to construct on paper. Another helpful device is the **tree diagram,** which we use in the following examples.

EXAMPLE 6 Find the number of three-digit numbers that can be written using digits from the set {1, 2, 3}, assuming that **(a)** repeated digits are allowed, **(b)** repeated digits are not allowed.

(a) The task of constructing such a number has three parts: select the first digit, select the second digit, and select the third digit. As we move from left to right through the tree diagram in Figure 1, the tree branches at the first stage to all possibilities for the first digit. Then each first-stage branch again branches, or splits, at the second stage, to all possibilities for the second digit. Finally, the third-stage branching shows the third-digit possibilities. The list of possible results (27 of them) is shown in Figure 1.

(b) For the case of nonrepeating digits, we could construct a whole new tree diagram, as in Figure 2, or we could simply go down the list of numbers from the first tree diagram and strike out any that contain repeated digits. In either case we obtain only six possibilities.

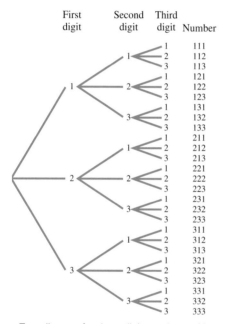

Tree diagram for three-digit numbers with digits from the set {1, 2, 3}

FIGURE 1

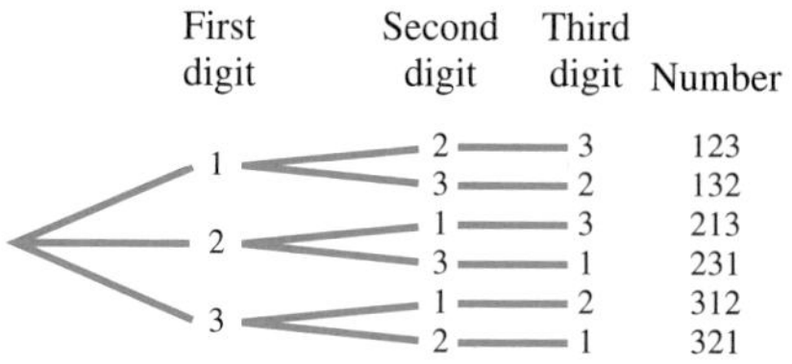

Tree diagram for nonrepeating three-digit numbers with digits from the set {1, 2, 3}

FIGURE 2

Notice the distinction between parts (a) and (b) of Example 6. There are 27 possibilities when "repetitions (of digits) are allowed," but only six possibilities when "repetitions are not allowed." Here is another way to phrase the question of Example 6: A three-digit number is to be determined by placing three slips of paper (marked 1, 2, and 3) into a hat and drawing out three slips in succession. Find the number of possible results if the drawing is done (a) with replacement, (b) without replacement. Drawing "with replacement" means drawing a slip, recording its digit, and replacing the slip into the hat so that it is again available for subsequent draws. Drawing "with replacement" has the effect of "allowing repetitions," while drawing "without replacement" has the effect of "not allowing repetitions."

The words "repetitions" and "replacement" can be important keys in the statement of a problem. But there are many other ways to express the same meaning. In Example 2, since no restrictions were stated, we can assume that *repetitions* (of digits) *are allowed,* or equivalently that digits are selected *with replacement.*

EXAMPLE 7 Michelle Clayton's computer printer allows for optional settings with a panel of four on-off switches in a row. How many different settings can she select if no two adjacent switches can both be off?

This situation is typical of user-selectable options on various devices, including computer equipment, garage door openers, and other appliances. In Figure 3 we denote "on" and "off" with 1 and 0, respectively (a common practice). The number of possible settings is seen to be eight. Notice that each time on the tree diagram that a switch is indicated as off (0), the next switch can only be on (1). This is to satisfy the restriction that no two adjacent switches can both be off.

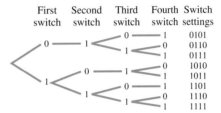

Tree diagram for printer settings

FIGURE 3

EXAMPLE 8 Aaron, Bobbette, Chuck, and Deirdre have tickets for four reserved seats in a row at a concert. In how many different ways can they seat themselves so that Aaron and Bobbette will sit next to each other?

Here we have a four-part task: assign people to the first, second, third, and fourth seats. The tree diagram in Figure 4 again avoids repetitions, since no person can occupy more than one seat. Also, once *A* or *B* appears at any stage of the tree, the other one must occur at the next stage. (Why is this?) Notice that no splitting occurs from stage three to stage four since by that time there is only one person left unassigned. The right column in the figure shows the twelve possible seating arrangements.

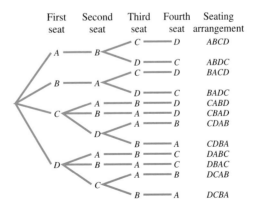

Tree diagram for concert seating

FIGURE 4

Although we have applied tree diagrams only to tasks with three or more parts, they can also be used for two-part or even simple, one-part tasks. Product tables, on the other hand, are useful only for two-part tasks.

Other Systematic Listing Methods Product tables (for two-part tasks) and tree diagrams (for tasks of any number of parts) are useful methods for picturing and listing the possible results for given tasks. There are additional systematic ways to produce complete listings of possible results.

In Example 4, we used a product table (Table 3) to list all possible president-secretary pairs for the club $N = \{A, B, C, D, E\}$. We could also systematically construct the same list using a sort of alphabetical or left-to-right approach. First, consider the results where A is president. Any of the remaining members (B, C, D, or E) could then be secretary. That gives us the pairs AB, AC, AD, and AE. Next, assume B is president. The secretary could then be A, C, D, or E. We get the pairs BA, BC, BD, and BE. Continuing in order, we get the complete list just as in Example 4:

AB, AC, AD, AE, BA, BC, BD, BE, CA, CB,
CD, CE, DA, DB, DC, DE, EA, EB, EC, ED.

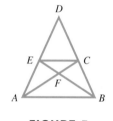

FIGURE 5

EXAMPLE 9 How many different triangles (of any size) are included in Figure 5?

One systematic approach is to label the points as shown, begin with A, and proceed in alphabetical order to write all three-letter combinations, then cross out the ones that are not triangles in the figure.

ABC, ABD, ABE, ABF, ACD, ACE, ~~ACF~~, ~~ADE~~, ~~ADF~~, AEF,
~~BCD~~, BCE, BCF, BDE, ~~BDF~~, BEF, CDE, ~~CDF~~, CEF, DEF

Finally, there are twelve different triangles in the figure. Why are ACB and CBF (and many others) not included in the list?

Another method might be first to identify the triangles consisting of a single region each: DEC, ECF, AEF, BCF, ABF. Then list those consisting of two regions each: AEC, BEC, ABE, ABC; and those with three regions each: ACD, BED. There are no triangles with four regions, but there is one with five: ABD. The total is again twelve. Can you think of other systematic ways of getting the same list? ■

Notice that in the first method shown in Example 9, the labeled points were considered in alphabetical order. In the second method, the single-region triangles were listed by using a top-to-bottom and left-to-right order. Using a definite system helps to ensure that we get a complete list.

The **"tree diagram"** on the map came from research on the feasibility of using motor-sailers (motor-driven ships with wind-sail auxiliary power) on the North Atlantic run. At the beginning of a run, weather forecasts and computer analysis are used to choose the best of the 45 million possible routes.

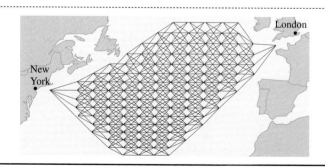

11.1 EXERCISES

Electing Officers of a Club Refer to Examples 1 and 4, involving the club

$$N = \{Andy, Bill, Cathy, David, Evelyn\}.$$

Assuming all members are eligible, but that no one can hold more than one office, list and count the different ways the club could elect the following groups of officers.

1. a president and a treasurer
2. a president and a treasurer if the president must be a female
3. a president and a treasurer if the two officers must not be the same sex
4. a president, a secretary, and a treasurer, if the president and treasurer must be women
5. a president, a secretary, and a treasurer, if the president must be a man and the other two must be women
6. a president, a secretary, and a treasurer, if the secretary must be a woman and the other two must be men

Appointing Committees List and count the ways club N could appoint a committee of three members under the following conditions.

7. There are no restrictions.
8. The committee must include more men than women.

Refer to Table 2 (the product table for rolling two dice). Of the 36 possibilities, determine the number for which the sum (for both dice) is the following.

9. 2
10. 3
11. 4
12. 5
13. 6
14. 7
15. 8
16. 9
17. 10
18. 11
19. 12
20. odd
21. even
22. from 6 through 8 inclusive
23. between 6 and 10
24. less than 5
25. Construct a product table showing all possible two-digit numbers using digits from the set

$$\{1, 2, 3, 4, 5, 6\}.$$

Of the thirty-six numbers in the product table for Exercise 25, list the ones that belong to each of the following categories.

26. odd numbers
27. numbers with repeating digits
28. multiples of 6
29. prime numbers
30. triangular numbers
31. square numbers
32. Fibonacci numbers
33. powers of 2
34. Construct a tree diagram showing all possible results when three fair coins are tossed. Then list the ways of getting the following results.
    (a) at least two heads
    (b) more than two heads
    (c) no more than two heads
    (d) fewer than two heads
35. Extend the tree diagram of Exercise 34 for four fair coins. Then list the ways of getting the following results.
    (a) more than three tails
    (b) fewer than three tails
    (c) at least three tails
    (d) no more than three tails

Determine the number of triangles (of any size) in each of the following figures.

36.
37.
38.
39.

Determine the number of squares (of any size) in each of the following figures.

40.
41.
42.
43.

Consider only the smallest individual cubes and assume solid stacks (no gaps). Determine the number of cubes in each stack that are not visible from the perspective shown.

44.   **45.**

46.   **47.**

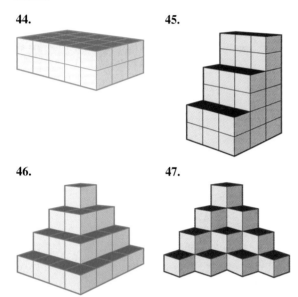

48. In the plane figure shown here, only movement that tends downward is allowed. Find the total number of paths from *A* to *B*.

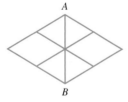

49. Find the number of paths from *A* to *B* in the figure shown here if the directions on various segments are restricted as shown.

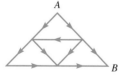

In each of Exercises 50–52, determine the number of different ways the given number can be written as the sum of two primes.

50. 30   **51.** 40   **52.** 95

53. *Shaking Hands in a Group* A group of twelve strangers sat in a circle, and each one got acquainted only with the person to the left and the person to the right. Then all twelve people stood up and each one shook hands (once) with each of the others who was still a stranger. How many handshakes occurred?

54. *Number of Games in a Chess Tournament* Fifty people enter a single-elimination chess tournament. (If you lose one game, you're out.) Assuming no ties occur, what is the number of games required to determine the tournament champion?

55. *Sums of Digits* How many of the numbers from 10 through 100 have the sum of their digits equal to a perfect square?

56. *Sums of Digits* How many three-digit numbers have the sum of their digits equal to 22?

57. *Integers Containing the Digit 2* How many integers between 100 and 400 contain the digit 2?

58. *Selecting Dinner Items* Frank Capek and friends are dining at the Bay Steamer Restaurant this evening, where a complete dinner consists of three items: (1) soup (clam chowder or minestrone) or salad (fresh spinach or shrimp), (2) sourdough rolls or bran muffin, and (3) entree (lasagna, lobster, or roast turkey). Frank selects his meal subject to the following restrictions. He cannot stomach more than one kind of seafood at a sitting. Also, whenever he tastes minestrone, he cannot resist having lasagna as well. And he cannot face the teasing he would receive from his companions if he were to order both spinach and bran. Use a tree diagram to determine the number of different choices Frank has.

Setting Options on a Computer Printer For Exercises 59–61, refer to Example 7. How many different settings could Michelle choose in each case?

59. No restrictions apply to adjacent switches.

60. No two adjacent switches can be off *and* no two adjacent switches can be on.

61. There are five switches rather than four, and no two adjacent switches can be on.

62. *Building Numbers from Sets of Digits* Determine the number of odd, nonrepeating three-digit numbers that can be written using digits from the set {0, 1, 2, 3}.

63. *Lattice Points on a Line Segment* A line segment joins the points (8, 12) and (53, 234) in the Cartesian plane. Including its endpoints, how many lattice points does this line segment contain? (A *lattice point* is a point with integer coordinates.)

64. *Lengths of Segments Joining Lattice Points* In the pattern that follows, dots are one unit apart horizontally and vertically. If a segment can join any two dots, how many segments can be drawn with each of the following lengths?

(a) 1   (b) 2   (c) 3   (d) 4   (e) 5

65. *Counting Matchsticks in a Grid* Uniform-length matchsticks are used to build a rectangular grid as shown here. If the grid is 15 matchsticks high and 28 matchsticks wide, how many matchsticks are used?

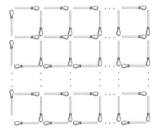

66. *Patterns in Floor Tiling* A square floor is to be tiled with square tiles as shown at the top of the next column, with blue tiles on the main diagonals and red tiles everywhere else. (In all cases, both blue and red tiles must be used and the two diagonals must have a common blue tile at the center of the floor.)

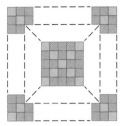

(a) If 81 blue tiles will be used, how many red tiles will be needed?
(b) For what numbers in place of 81 would this problem still be solvable?
(c) Find a formula expressing the number of red tiles required in general.

67. *Shaking Hands in a Group* Jeff Howard and his son were among four father-and-son pairs who gathered to trade baseball cards. As each person arrived, he shook hands with anyone he had not known previously. Each person ended up making a different number of new acquaintances (0–6), except Jeff and his son, who each met the same number of people. How many hands did Jeff shake?

In Exercises 68–71, restate the given counting problem in two ways, first (a) *using the word* repetition, *and then* (b) *using the word* replacement.

68. Example 2   69. Example 3
70. Example 4   71. Exercise 7

11.2 Using the Fundamental Counting Principle

In the previous section, we obtained complete lists of all possible results for various tasks. However, if the total number of possibilities is all we need to know, then an actual listing usually is unnecessary and often is difficult or tedious to obtain, especially when the list is long. In this section, we develop ways to calculate "how many" using the *fundamental counting principle*.

Figure 6 repeats Figure 2 of the previous section (for Example 6(b)) which shows all possible nonrepeating three-digit numbers with digits from the set {1, 2, 3}.

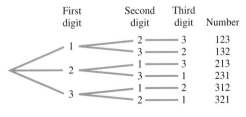

FIGURE 6

The tree diagram in Figure 6 is "uniform" in the sense that a given part of the task can be done in the same number of ways no matter which choices were selected for previous parts. For example, there are always two choices for the second digit. (If the first digit is 1, the second can be 2 or 3; if the first is 2, the second can be 1 or 3; if the first is 3, the second can be 1 or 2.) Example 6(a) of the previous section addressed the same basic question (how many three-digit numbers using the digits 1, 2, and 3), but in that case repetitions were allowed. With repetitions allowed, there were many more possibilities (27 rather than 6—see Figure 1 of Section 11.1). But the uniformity criterion mentioned above still applied. No matter what the first digit is, there are three choices for the second (1, 2, 3). And no matter what the first and second digits are, there are three choices for the third. This uniformity criterion can be stated in general as follows.

Uniformity Criterion for Multiple-Part Tasks

A multiple-part task is said to satisfy the **uniformity criterion** if the number of choices for any particular part is the same *no matter which choices were selected for previous parts.*

The uniformity criterion is not satisfied by all multiple-part counting problems. Recall Example 7 (and Figure 3) of the previous section. After the first switch (two possibilities), other switches had either one or two possible settings depending on how previous switches were set. (This "non-uniformity" arose, in that case, from the requirement that no two adjacent switches could both be off.)

In the many cases where uniformity does hold, we can avoid having to construct a tree diagram by using the **fundamental counting principle,** stated as follows.

Fundamental Counting Principle

When a task consists of k separate parts and satisfies the uniformity criterion, if the first part can be done in n_1 ways, the second part can then be done in n_2 ways, and so on through the kth part, which can be done in n_k ways, then the total number of ways to complete the task is given by the product

$$n_1 \cdot n_2 \cdot n_3 \cdot \ldots \cdot n_k.$$

Problem Solving

A problem-solving strategy suggested in Chapter 1 was: "If a formula applies, use it." The fundamental counting principle provides a formula that applies to a variety of problems. The trick is to visualize the "task" at hand as being accomplished in a sequence of two or more separate parts. A helpful technique when applying the fundamental counting principle is to write out all the separate parts of the task, with a blank for each one. Reason out how many ways each part can be done and enter these numbers in the blanks. Finally, multiply these numbers together.

11.2 Using the Fundamental Counting Principle

Richard Dedekind (1831–1916) studied at the University of Göttingen, where he was Gauss's last student. His work was not recognized during his lifetime, but his treatment of the infinite and of what constitutes a real number are influential even today.

While on vacation in Switzerland, Dedekind met Georg Cantor. Dedekind was interested in Cantor's work on infinite sets. Perhaps because both were working in new and unusual fields of mathematics, such as number theory, and because neither received the professional attention he deserved during his lifetime, the two struck up a lasting friendship.

EXAMPLE 1 How many two-digit numbers are there in our (base-ten) system of counting numbers? (While 40 is a two-digit number, 04 is not.)

Our "task" here is to select, or design, a two-digit number. We can set up the work as follows, showing the two parts to be done.

Part of task	Select first digit	Select second digit
Number of ways	_____	_____

There are nine choices for the first digit (1 through 9). Since there were no stated or implied restrictions, we assume that repetition of digits is allowed. Therefore, no matter which nonzero digit is used as the first digit, all nine choices are available for the second digit. Also, unlike the first digit, the second digit may be zero, so we have ten choices for the second digit. We can now fill in the blanks and multiply.

Part of task	Select first digit	Select second digit	
Number of ways	9 ·	10	= 90

There are 90 two-digit numbers. (As a check, notice that they are the numbers from 10 through 99, a total of $99 - 10 + 1 = 90$.) ∎

EXAMPLE 2 Find the number of two-digit numbers that do not contain repeated digits (for example, 66 is not allowed).

The basic task is again to select a two-digit number, and there are two parts: select the first digit, and select the second digit. But a new restriction applies—no repetition of digits. There are nine choices for the first digit (1 through 9). Then nine choices remain for the second digit, since one nonzero digit has been used and cannot be repeated, but zero is now available. The total number is $9 \cdot 9 = 81$. ∎

EXAMPLE 3 In how many ways can Club N of the previous section elect a president and a secretary if no one may hold more than one office and the secretary must be a man?

Recall that $N = \{A, B, C, D, E\} = \{$Andy, Bill, Cathy, David, Evelyn$\}$. Considering president first, there are five choices (no restrictions). But now we have a problem with finding the number of choices for secretary. If a woman was selected president (C or E), there are three choices for secretary (A, B, and D). If a man was selected president, only two choices (the other two men) remain for secretary. In other words, the uniformity criterion is not met and our attempt to apply the fundamental counting principle has failed.

All is not lost, however. In finding the total number of ways, there is no reason we cannot consider secretary first. There are three choices (A, B, and D). Now, no matter which man was chosen secretary, both of the other men, and both women, are available for president (four choices in every case). By considering the parts of the task in this order, we satisfy the uniformity criterion and can therefore use the fundamental counting principle. The total number of ways to elect a president and a secretary is $3 \cdot 4 = 12$. (You can check this answer by constructing a tree diagram or using some other method of listing all possibilities.) ∎

The lesson to be learned from Example 3 is this: whenever one or more parts of a task have special restrictions, try considering that part (or those parts) before other parts.

EXAMPLE 4 How many nonrepeating odd three-digit counting numbers are there?

The most restricted digit is the third, since it must be odd. There are five choices (1, 3, 5, 7, and 9). Next, consider the first digit. It can be any nonzero digit except the one already chosen as the third digit. There are eight choices. Finally, the second digit can be any digit (including 0) except for the two (nonzero) digits already used. There are eight choices. We can summarize this reasoning as follows.

Part of task	Select third digit	Select first digit	Select second digit	
Number of ways	5 ·	8 ·	8	= 320

There are 320 nonrepeating odd three-digit counting numbers.

EXAMPLE 5 In some states, auto license plates have contained three letters followed by three digits. How many such licenses are possible?

The basic task is to design a license number with three letters followed by three digits. There are six component parts to this task. Since there are no restrictions on letters or digits, the fundamental counting principle gives

$$26^3 \cdot 10^3 = 17{,}576{,}000$$

possible licenses. (In practice, a few of the possible sequences of letters are considered undesirable and are not used.)

EXAMPLE 6 A four-digit number is to be constructed using only digits from the set {1, 2, 3}.

(a) How many such numbers are possible?

To construct such a number, we must select four digits, in succession, from the given set of three digits, where the selection is done with replacement (since repetition of digits is apparently allowed). By the fundamental counting principle, the number of possibilities is $3 \cdot 3 \cdot 3 \cdot 3 = 3^4 = 81$.

(b) How many of these counting numbers are odd numbers less than 2000?

The number is less than 2000 only if the first digit is 1 (just one choice) and is odd only if the fourth digit is 1 or 3 (two choices). The second and third digits are unrestricted (three choices for each). The answer is $1 \cdot 3 \cdot 3 \cdot 2 = 18$.

Problem Solving

Two of the problem-solving strategies of Chapter 1 were to "first solve a similar simpler problem," and to "look for a pattern." In fact, a problem at hand may sometimes prove to be essentially the same, or at least fit the same pattern, as another problem already solved, perhaps in a different context.

EXAMPLE 7 Vern has four antique wood head golf clubs that he wants to give to his three sons, Mark, Chris, and Scott.

(a) How many ways can the clubs be distributed?

The task is to distribute four clubs among three sons. Consider the clubs in succession and, for each one, ask how many sons could receive it. In effect, we must select four sons, in succession, from the set {Mark, Chris, Scott}, selecting with replacement. Compare this with Example 6(a), in which we selected four digits, in succession, from the set {1, 2, 3}, selecting with replacement. In this case, we are selecting sons rather than digits, but the pattern is the same and the numbers are the same. Again our answer is $3^4 = 81$.

(b) How many choices are there if the power driver must go to Mark and the number 3 wood must go to either Chris or Scott?

Just as in Example 6(b), one part of the task is now restricted to a single choice and another part is restricted to two choices. As in that example, the number of possibilities is $1 \cdot 3 \cdot 3 \cdot 2 = 18$.

EXAMPLE 8 Rework Example 8 of the previous section, this time using the fundamental counting principle.

Recall that Aaron, Bobbette, Chuck, and Deirdre (A, B, C, and D) are to seat themselves in four adjacent seats (say 1, 2, 3, and 4) so that A and B are side-by-side. One approach to accomplish this task is to make three successive decisions as follows.

Seats available to A and B

1. Which pair of seats should A and B occupy? There are *three* choices (1 – 2, 2 – 3, 3 – 4, as illustrated in the margin).
2. Which order should A and B take? There are *two* choices (A left of B, or B left of A).
3. Which order should C and D take? There are *two* choices (C left of D, or D left of C, not necessarily right next to each other).

(Why did we not ask which two seats C and D should occupy?) The fundamental counting principle now gives the total number of choices: $3 \cdot 2 \cdot 2 = 12$, the same result found in the previous section.

Short Table of Factorials

Factorial values increase rapidly. The value of 100! is a number with 158 digits.

$$0! = 1$$
$$1! = 1$$
$$2! = 2$$
$$3! = 6$$
$$4! = 24$$
$$5! = 120$$
$$6! = 720$$
$$7! = 5040$$
$$8! = 40{,}320$$
$$9! = 362{,}880$$
$$10! = 3{,}628{,}800$$

Factorials

This section began with a discussion of nonrepeating three-digit numbers with digits from the set $\{1, 2, 3\}$. The number of possibilities was $3 \cdot 2 \cdot 1 = 6$, in keeping with the fundamental counting principle. That product can also be thought of as the total number of distinct *arrangements* of the three digits 1, 2, and 3. Similarly, the number of distinct arrangements of four objects, say A, B, C, and D, is, by the fundamental counting principle, $4 \cdot 3 \cdot 2 \cdot 1 = 24$. Since this type of product occurs so commonly in applications, we give it a special name and symbol as follows. For any counting number n, the product of *all* counting numbers from n down through 1 is called ***n* factorial**, and is denoted ***n*!**.

Factorial Formula

For any counting number n, the quantity n factorial is given by

$$n! = n(n-1)(n-2)\ldots 2 \cdot 1.$$

The first few factorial values are easily found by simple multiplication, but they rapidly become very large, as indicated in the margin. The use of a calculator is advised in most cases. (See the margin notes that follow.)

EXAMPLE 9 Evaluate the following expressions.

(a) $2! = 2 \cdot 1 = 2$
(b) $6! = 6 \cdot 5 \cdot 4 \cdot 3 \cdot 2 \cdot 1 = 720$
(c) $(6-3)! = 3! = 6$
(d) $6! - 3! = 720 - 6 = 714$
(e) $\dfrac{6!}{3!} = \dfrac{6 \cdot 5 \cdot 4 \cdot 3 \cdot 2 \cdot 1}{3 \cdot 2 \cdot 1} = 6 \cdot 5 \cdot 4 = 120$
(f) $\left(\dfrac{6}{3}\right)! = 2! = 2 \cdot 1 = 2$
(g) $15! = 1.307674368000 \times 10^{12}$
(h) $100! = 9.332621544 \times 10^{157}$

Notice the distinction between parts (c) and (d) and between parts (e) and (f) above. Parts (g) and (h) were found on an advanced scientific calculator. (Part (h) is beyond the capability of most scientific calculators.)

The results of Example 9(b), (c), and (g) are illustrated in this calculator screen.

So that factorials will be defined for all whole numbers, including zero, it is common to define 0! as follows.

Definition of Zero Factorial

$$0! = 1$$

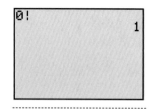

The definition $0! = 1$ is illustrated here.

(We will see later that this special definition makes other results easier to state.)

When finding the total number of ways to *arrange* a given number of distinct objects, we can use a factorial. The fundamental counting principle would do, but factorials provide a shortcut.

Arrangements of *n* Objects

The total number of different ways to arrange *n* distinct objects is ***n*!**.

EXAMPLE 10 Erika Berg has seven essays to include in her English 1A folder. In how many different orders can she arrange them?

The number of ways to arrange seven distinct objects is $7! = 5040$.

EXAMPLE 11 Lynn Damme is taking thirteen preschoolers to the park. How many ways can the children line up, in single file, to board the van?

Thirteen children can be arranged in $13! = 6,227,020,800$ different ways.

FOR FURTHER THOUGHT

A Classic Problem

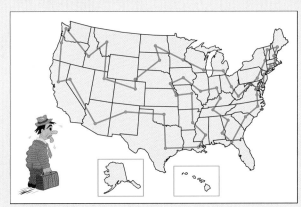

Suppose a salesperson wants to make exactly one visit to each capital city in the 48 contiguous states, starting and ending up at the same capital. What is the shortest route that would work?

The so-called **traveling salesman problem** (or **TSP**), in various versions, has baffled mathematicians for years. For the version stated above, there are 24!/3 possible routes to consider, and the huge size of this number is the reason the problem is difficult (even using the most powerful computers). In 1985, mathematician Shen Lin came up with the 10,628-mile route shown here, and even though he could not prove it was the shortest, he offered $100 to anyone who could find a shorter one.

In 1998, Rice University researchers David Applegate, Robert Bixby, and William Cook, along with Vasek Chvatal of Rutgers University, announced a breakthrough solution to the traveling salesman problem for 13,509 U.S. cities with populations of at least 500 people. Their previous record had been set in 1994, for 7397 cities.

For Group Discussion

1. Within the accuracy limits of the drawing above, try to identify other routes that may be as good or better than the one shown.
2. Realistically, what factors other than *distance* might be important to the salesperson?
3. When a chemical company schedules the steps of a complex manufacturing process, what quantities might it want to minimize?
4. Many websites provide historical information and up-to-date research data on the traveling salesman problem. For example, go to www.math.princeton.edu/tsp.

11.2 EXERCISES

1. Explain the fundamental counting principle in your own words.

2. Describe how factorials can be used in counting problems.

For Exercises 3 and 4, n and m are counting numbers. Do the following: (a) *Tell whether the given statement is true in general, and* (b) *explain your answer, using specific examples.*

3. $(n + m)! = n! + m!$

4. $(n \cdot m)! = n! \cdot m!$

Evaluate each expression without using a calculator.

5. $4!$

6. $7!$

7. $\dfrac{8!}{5!}$

8. $\dfrac{16!}{14!}$

9. $\dfrac{5!}{(5 - 2)!}$

10. $\dfrac{6!}{(6 - 4)!}$

11. $\dfrac{9!}{6!(6 - 3)!}$

12. $\dfrac{10!}{4!(10 - 4)!}$

13. $\dfrac{n!}{(n - r)!}$, where $n = 7$ and $r = 4$

14. $\dfrac{n!}{r!(n - r)!}$, where $n = 12$ and $r = 4$

Evaluate each expression using a calculator. (Some answers may not be exact.)

15. $11!$

16. $17!$

17. $\dfrac{12!}{7!}$

18. $\dfrac{15!}{9!}$

19. $\dfrac{13!}{(13 - 3)!}$

20. $\dfrac{16!}{(16 - 6)!}$

21. $\dfrac{20!}{10! \cdot 10!}$

22. $\dfrac{18!}{6! \cdot 12!}$

23. $\dfrac{n!}{(n - r)!}$, where $n = 23$ and $r = 10$

24. $\dfrac{n!}{r!(n - r)!}$, where $n = 28$ and $r = 15$

Settings on a Switch Panel A panel containing three on-off switches in a row is to be set.

25. Assuming no restrictions on individual switches, use the fundamental counting principle to find the total number of possible panel settings.

26. Assuming no restrictions, construct a tree diagram to list all the possible panel settings of Exercise 25.

27. Now assume that no two adjacent switches can both be off. Explain why the fundamental counting principle does not apply.

28. Construct a tree diagram to list all possible panel settings under the restriction of Exercise 27.

29. *Rolling Dice* Table 2 in the previous section shows that there are 36 possible outcomes when two fair dice are rolled. How many would there be if three fair dice were rolled?

30. *Counting Five-Digit Numbers* How many five-digit numbers are there in our system of counting numbers?

Matching Club Members with Tasks Recall the club
$N = \{\text{Andy, Bill, Cathy, David, Evelyn}\}$.
In how many ways could they do each of the following?

31. line up all five members for a photograph

32. schedule one member to work in the office on each of five different days, assuming members may work more than one day

33. select a male and a female to decorate for a party

34. select two members, one to open their next meeting and another to close it, given that Bill will not be present

Building Numbers from Sets of Digits In Exercises 35–38, counting numbers are to be formed using only digits from the set $\{3, 4, 5\}$. Determine the number of different possibilities for each type of number described.

35. two-digit numbers

36. odd three-digit numbers

37. four-digit numbers with one pair of adjacent 4s and no other repeated digits (*Hint:* You may want to split the task of designing such a number into three parts, such as *(1)* position the pair of 4s, *(2)* position the 3, and *(3)* position the 5.)

38. five-digit numbers beginning and ending with 3 and with unlimited repetitions allowed

Selecting Dinner Items The Casa Loma Restaurant offers four choices in the soup and salad category (two soups and two salads), two choices in the bread category, and three choices in the entree category. Find the number of dinners available in each of the following cases.

39. One item is to be included from each of the three categories.

40. Only soup and entree are to be included.

Selecting Answers on a Test Determine the number of possible ways to mark your answer sheet (with an answer for each question) for each of the following tests.

41. a six-question true-or-false test

42. a twenty-question multiple-choice test with five answer choices for each question

Selecting a College Class Schedule Tiffany Connolly is making up her class schedule for next semester, which must consist of exactly one class from each of the four categories shown here.

Category	Choices	Number of Choices
English	Medieval Literature Composition Modern Poetry	3
Mathematics	History of Mathematics College Algebra Finite Mathematics	3
Computer Information Science	Introduction to Spreadsheets Advanced Word Processing C Programming BASIC Programming	4
Sociology	Social Problems Sociology of the Middle East Aging in America Minorities in America Women in American Culture	5

For each situation in Exercises 43–48, use the table at the bottom of the left column to determine the number of different sets of classes Tiffany can take.

43. All classes shown are available.

44. She is not eligible for Modern Poetry or for C Programming.

45. All sections of Minorities in America and Women in American Culture already are filled.

46. She does not have the prerequisites for Medieval Literature, Finite Mathematics, or C Programming.

47. Funding has been withdrawn for three of the computer courses and for two of the Sociology courses.

48. She must complete English Composition and Aging in America next semester to fulfill her degree requirements.

49. *Selecting Clothing* Sean took two pairs of shoes, five pairs of pants, and six shirts on a trip. Assuming all items are compatible, how many different outfits can he wear?

50. *Selecting Music Equipment* A music equipment outlet stocks ten different guitars, four guitar cases, six amplifiers, and three effects processors, with all items mutually compatible and all suitable for beginners. How many different complete setups could Lionel choose to start his musical career?

51. *Counting ZIP Codes* Tonya's ZIP code is 85726. How many ZIP codes, altogether, could be formed using all of those same five digits?

52. *Listing Phone Numbers* John Cross keeps the phone numbers for his seven closest friends (three men and four women) in his digital phone memory. How many ways can he list them if
    (a) the men are listed before the women?
    (b) the men are all listed together?
    (c) no two men are listed next to each other?

Seating Arrangements at a Theater Aaron, Bobbette, Chuck, Deirdre, Ed, and Fran have reserved six seats in a row at the theater, starting at an aisle seat. (Refer to Example 8 in this section.)

53. In how many ways can they arrange themselves? (*Hint:* Divide the task into the series of six parts shown below, performed in order.)
    (a) If *A* is seated first, how many seats are available for him?
    (b) Now, how many are available for *B*?
    (c) Now, how many for *C*?
    (d) Now, how many for *D*?
    (e) Now, how many for *E*?
    (f) Now, how many for *F*?
    Now multiply together your six answers above.

54. In how many ways can they arrange themselves so that Aaron and Bobbette will be next to each other?

1	2	3	4	5	6
X	X	_	_	_	_
_	X	X	_	_	_
_	_	X	X	_	_
_	_	_	X	X	_
_	_	_	_	X	X

Seats available to *A* and *B*

(*Hint:* First answer the following series of questions, assuming these parts are to be accomplished in order.)
(a) How many pairs of adjacent seats can *A* and *B* occupy?
(b) Now, given the two seats for *A* and *B*, in how many orders can they be seated?
(c) Now, how many seats are available for *C*?
(d) Now, how many for *D*?
(e) Now, how many for *E*?
(f) Now, how many for *F*?
Now multiply your six answers above.

55. In how many ways can they arrange themselves if the men and women are to alternate seats and a man must sit on the aisle? (*Hint:* First answer the following series of questions.)
(a) How many choices are there for the person to occupy the first seat, next to the aisle? (It must be a man.)
(b) Now, how many choices of people may occupy the second seat from the aisle? (It must be a woman.)
(c) Now, how many for the third seat? (one of the remaining men)
(d) Now, how many for the fourth seat? (a woman)
(e) Now, how many for the fifth seat? (a man)
(f) Now, how many for the sixth seat? (a woman)
Now multiply your six answers above.

56. In how many ways can they arrange themselves if the men and women are to alternate with either a man or a woman on the aisle? (*Hint:* First answer the following series of questions.)
(a) How many choices of people are there for the aisle seat?
(b) Now, how many are there for the second seat? (This person may not be of the same sex as the person on the aisle.)
(c) Now, how many choices are there for the third seat?
(d) Now, how many for the fourth seat?
(e) Now, how many for the fifth seat?
(f) Now, how many for the sixth seat?
Now multiply your six answers above.

57. Try working Example 4 by considering digits in the order first, then second, then third. Explain what goes wrong.

58. Try working Example 4 by considering digits in the order third, then second, then first. Explain what goes wrong.

59. Repeat Example 4 but this time allow repeated digits. Does the order in which digits are considered matter in this case?

11.3 Using Permutations and Combinations

Earlier we introduced factorials as a way of counting the number of *arrangements* of a given set of objects. For example, the members of the club

$$N = \{\text{Andy, Bill, Cathy, David, Evelyn}\}$$

can arrange themselves in a row for a photograph in $5! = 120$ different ways. We have also used previous methods, like tree diagrams and the fundamental counting principle, to answer questions such as: How many ways can club *N* elect a president, a secretary, and a treasurer if no one person can hold more than one office? This again is a matter of *arrangements*. The difference is that only three, rather than all five, of the members are involved in each case. A common way to rephrase the basic question here is as follows: *How many arrangements are there of five things taken three at a time?* The answer, by the fundamental counting principle, is $5 \cdot 4 \cdot 3 = 60$. The factors begin with 5 and proceed downward, just as in a factorial product, but do not go all the way to 1. (In this example the product stops when there are three factors.) We now generalize this idea.

11.3 Using Permutations and Combinations

Permutations In the context of counting problems, arrangements are often called **permutations;** the number of permutations of n distinct things taken r at a time is written $P(n, r)$.* Since the number of objects being arranged cannot exceed the total number available, we assume for our purposes here that $r \leq n$. Applying the fundamental counting principle to arrangements of this type gives

$$P(n,r) = n(n-1)(n-2)\ldots[n-(r-1)].$$

Simplification of the last factor gives the following formula.

Permutations Formula

The number of **permutations,** or *arrangements,* of n distinct things taken r at a time, where $r \leq n$, is given by

$$P(n,r) = n(n-1)(n-2)\ldots(n-r+1).$$

The factors in this product begin at n and descend until the total number of factors is r. We now see that the number of ways in which club N can elect a president, a secretary, and a treasurer is $P(5,3) = 5 \cdot 4 \cdot 3 = 60$.

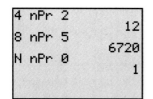

The results of Example 1(a) and (b) are supported here. For any whole number value of N, $P(N, 0) = 1$. See the final line of the display.

EXAMPLE 1 Evaluate each permutation.

(a) $P(4, 2) = 4 \cdot 3 = 12$ Begin at 4 and use two factors.

(b) $P(8, 5) = 8 \cdot 7 \cdot 6 \cdot 5 \cdot 4 = 6720$ Begin at 8 and use five factors.

(c) $P(5, 5) = 5 \cdot 4 \cdot 3 \cdot 2 \cdot 1 = 120$ Begin at 5 and use five factors.

Notice that $P(5, 5)$ is equal to 5!. It is true for all whole numbers n that $P(n, n) = n!$. (This is the number of possible arrangements of n distinct objects taken all n at a time.)

Permutations, in general, can also be related to factorials in the following way.

$P(n,r) = n(n-1)(n-2)\ldots(n-r+1)$ Permutations formula

$= \dfrac{n(n-1)(n-2)\ldots(n-r+1)(n-r)(n-r-1)\ldots 2 \cdot 1}{(n-r)(n-r-1)\ldots 2 \cdot 1}$

Multiply and divide by $(n-r)(n-r-1)\ldots 2 \cdot 1$.

$= \dfrac{n!}{(n-r)!}$ Definition of factorial

This gives us a formula for computing permutations in terms of factorials.

---

*Alternative notations are $_nP_r$ and P_r^n.

Factorial Formula for Permutations

The number of **permutations**, or *arrangements*, of n distinct things taken r at a time, where $r \leq n$, can be calculated as

$$P(n,r) = \frac{n!}{(n-r)!}.$$

If n and r are very large numbers, a calculator with a factorial key and this formula will save a lot of work when finding permutations. See the margin note.

EXAMPLE 2 Use the factorial formula for permutations to evaluate each expression.

(a) $P(9,4) = \dfrac{9!}{(9-4)!} = \dfrac{9!}{5!} = 3024$

(b) $P(35,0) = \dfrac{35!}{(35-0)!} = \dfrac{35!}{35!} = 1$

(c) $P(18,12) = \dfrac{18!}{(18-12)!} = \dfrac{18!}{6!} = 8{,}892{,}185{,}702{,}400$

```
9 nPr 4
              3024
35 nPr 0
                 1
18 nPr 12
     8.892185702E12
```

This screen supports the results of Example 2.

Concerning part (c), many calculators will not display this many digits, so you may obtain an answer such as 8.8921857×10^{12}.

Problem Solving

Permutations can be used any time we need to know the number of size-r arrangements that can be selected from a size-n set. The word *arrangement* implies an ordering, so we use permutations only in cases where the order of the items is important. Also, permutations apply only in cases where no repetition of items occurs. We summarize with the following guidelines.

Guidelines for When to Use Permutations

Permutations are applied only when

1. repetitions are not allowed, and
2. order is important.

EXAMPLE 3 How many nonrepeating three-digit numbers can be written using digits from the set $\{3, 4, 5, 6, 7, 8\}$?

Repetitions are not allowed since the numbers are to be "nonrepeating." (For example, 448 is not acceptable.) Also, order is important. (For example, 476 and 746 are *distinct* cases.) So we use permutations:

$$P(6,3) = 6 \cdot 5 \cdot 4 = 120.$$

11.3 Using Permutations and Combinations

The next example involves multiple parts, and hence calls for the fundamental counting principle, but the individual parts can be handled with permutations.

EXAMPLE 4 Suppose certain account numbers are to consist of two letters followed by four digits and then three more letters, where repetitions of letters or digits are not allowed *within* any of the three groups, but the last group of letters may contain one or both of those used in the first group. How many such accounts are possible?

The task of designing such a number consists of three parts.

1. Determine the first set of two letters.
2. Determine the set of four digits.
3. Determine the final set of three letters.

Each part requires an arrangement without repetitions, which is a permutation. Multiply together the results of the three parts.

$$P(26,2) \cdot P(10,4) \cdot P(26,3) = \underbrace{26 \cdot 25}_{\text{Part 1}} \cdot \underbrace{10 \cdot 9 \cdot 8 \cdot 7}_{\text{Part 2}} \cdot \underbrace{26 \cdot 25 \cdot 24}_{\text{Part 3}}$$
$$= 650 \cdot 5040 \cdot 15{,}600$$
$$= 51{,}105{,}600{,}000 \qquad \blacksquare$$

Change ringing, the English way of ringing church bells, combines mathematics and music. Bells are rung first in sequence, 1,2,3, ... Then the sequence is permuted ("changed"). On six bells, 720 different "changes" (different permutations of tone) can be rung: $P(6, 6) = 6!$.

Composers work out changes so that musically interesting and harmonious sequences occur regularly.

The church bells are swung by means of ropes attached to the wheels beside them. One ringer swings each bell, listening intently and watching the other ringers closely. If one ringer gets lost and stays lost, the rhythm of the ringing cannot be maintained; all the ringers have to stop.

A ringer can spend weeks just learning to keep a bell going and months learning to make the bell ring in exactly the right place. Errors of 1/4 second mean that two bells are ringing at the same time. Even errors of 1/10 second can be heard.

Combinations We introduced permutations to evaluate the number of arrangements of n things taken r at a time, where repetitions are not allowed. The order of the items was important. Recall that club $N = \{\text{Andy}, \text{Bill}, \text{Cathy}, \text{David}, \text{Evelyn}\}$ could elect three officers in $P(5,3) = 60$ different ways. With three-member committees, on the other hand, order is not important. The committees B, D, E and E, B, D are no different. The possible number of committees is not the number of arrangements of size 3. Rather, it is the number of *subsets* of size 3 (since the order of elements in a set makes no difference).

Subsets in this new context are called **combinations**. The number of combinations of n things taken r at a time (that is, the number of size r subsets, given a set of size n) is written $C(n, r)$.*

The size-3 committees (subsets) of the club (set) $N = \{A, B, C, D, E\}$ include:

$$\{A, B, C\}, \quad \{A, B, D\}, \quad \{A, B, E\}, \quad \{A, C, D\}, \quad \{A, C, E\},$$
$$\{A, D, E\}, \quad \{B, C, D\}, \quad \{B, C, E\}, \quad \{B, D, E\}, \quad \{C, D, E\}.$$

There are ten subsets of 3, so ten is the number of three-member committees possible. Just as with permutations, repetitions are not allowed. For example, $\{E, E, B\}$ is not a valid three-member subset, just as EEB is not a valid three-member arrangement.

To see how to find the number of such subsets without listing them all, notice that each size-3 subset (combination) gives rise to six size-3 arrangements (permutations). For example, the single combination ADE yields these six permutations:

$$A, D, E \quad D, A, E \quad E, A, D \quad A, E, D \quad D, E, A \quad E, D, A.$$

---

*Alternative notations are ${}_nC_r$, C_r^n, and $\binom{n}{r}$.

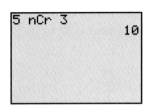

Calculators can perform operations involving combinations of *n* things taken *r* at a time. Compare this display to the result in the text.

Then there must be six times as many size-3 permutations as there are size-3 combinations, or, in other words, one-sixth as many combinations as permutations. Therefore

$$C(5,3) = \frac{P(5,3)}{6} = \frac{5 \cdot 4 \cdot 3}{6} = 10.$$

The 6 appears in the denominator because there are six different ways to arrange a set of three things (since $3! = 3 \cdot 2 \cdot 1 = 6$). Generalizing from this example, r things can be arranged in $r!$ different ways, so we obtain the following formula.

Combinations Formula

The number of **combinations,** or *subsets,* of n distinct things taken r at a time, where $r \leq n$, is given by

$$C(n,r) = \frac{P(n,r)}{r!} = \frac{n(n-1)(n-2)\ldots(n-r+1)}{r(r-1)(r-2)\ldots 2 \cdot 1}.$$

Combining this formula with the factorial formula for permutations gives a way of also evaluating combinations entirely in terms of factorials.

$$C(n,r) = \frac{P(n,r)}{r!} = \frac{\frac{n!}{(n-r)!}}{r!} = \frac{n!}{r!(n-r)!}$$

Factorial Formula for Combinations

The number of **combinations,** or *subsets,* of n distinct things taken r at a time, where $r \leq n$, can be calculated as

$$C(n,r) = \frac{P(n,r)}{r!} = \frac{n!}{r!(n-r)!}.$$

EXAMPLE 5 Use the factorial formula for combinations to evaluate each expression.

(a) $C(8,6) = \dfrac{8!}{6!\,2!} = \dfrac{40{,}320}{720 \cdot 2} = 28$ **(b)** $C(22,15) = \dfrac{22!}{15!\,7!} = 170{,}544$

"**Bilateral cipher**" (above) was invented by Francis Bacon early in the seventeenth century to code political secrets. This binary code, *a* and *b* in combinations of five, has 32 permutations. Bacon's "biformed alphabet" (bottom four rows) uses two type fonts to conceal a message in some straight text. The decoder deciphers a string of *a*s and *b*s, groups them by fives, then deciphers letters and words. This code was applied to Shakespeare's plays in efforts to prove Bacon the rightful author.

Problem Solving

The following guidelines stress that combinations have an important common property with permutations (repetitions are not allowed), but also an important distinction (order is *not* important with combinations). It is extremely important to realize this when solving problems.

Guidelines for When to Use Combinations

Combinations are applied only when

1. repetitions are not allowed, and
2. order is *not* important.

EXAMPLE 6 Find the number of different subsets of size 2 in the set $\{a, b, c, d\}$. List them to check the answer.

A subset of size 2 must have two distinct elements, so repetitions are not allowed. And since the order in which the elements of a set are listed makes no difference, we see that order is not important. Use the combinations formula with $n = 4$ and $r = 2$.

$$C(4, 2) = \frac{P(4, 2)}{2!} = \frac{4 \cdot 3}{2 \cdot 1} = 6$$

or, alternatively, $\qquad C(4, 2) = \frac{4!}{2!(4-2)!} = \frac{4!}{2!\,2!} = 6$

The six subsets of size 2 are $\{a, b\}, \{a, c\}, \{a, d\}, \{b, c\}, \{b, d\}, \{c, d\}$.

EXAMPLE 7 A common form of poker involves hands (sets) of five cards each, dealt from a standard deck consisting of 52 different cards (illustrated in the margin). How many different 5-card hands are possible?

A 5-card hand must contain five distinct cards, so repetitions are not allowed. Also, the order is not important since a given hand depends only on the cards it contains, and not on the order in which they were dealt or the order in which they are listed. Since order does not matter, use combinations (and a calculator):

$$C(52, 5) = \frac{P(52, 5)}{5!} = \frac{52 \cdot 51 \cdot 50 \cdot 49 \cdot 48}{5 \cdot 4 \cdot 3 \cdot 2 \cdot 1} = 2{,}598{,}960$$

or $\qquad C(52, 5) = \frac{52!}{5!(52-5)!} = \frac{52!}{5!\,47!} = 2{,}598{,}960.$

The set of 52 playing cards in the standard deck has four suits.

- ♠ spades   ♦ diamonds
- ♥ hearts   ♣ clubs

Ace is the unit card. Jacks, queens, and kings are "face cards." Each suit contains thirteen denominations: ace, 2, 3, . . . , 10, jack, queen, king. (In some games, ace rates above king, instead of counting as 1.)

EXAMPLE 8 Melvin wants to buy ten different books but can afford only four of them. In how many ways can he make his selection?

The four books selected must be distinct (repetitions are not allowed), and also the order of the four chosen has no bearing in this case, so we use combinations:

$$C(10, 4) = \frac{P(10, 4)}{4!} = \frac{10 \cdot 9 \cdot 8 \cdot 7}{4 \cdot 3 \cdot 2 \cdot 1} = 210 \text{ ways,}$$

or $\qquad C(10, 4) = \frac{10!}{4!(10-4)!} = \frac{10!}{4!\,6!} = 210 \text{ ways.}$

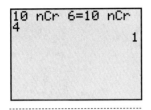

The "1" indicates that the statement

$C(10,6) = C(10,4)$

is true.

Notice that, according to our formula for combinations,

$$C(10,6) = \frac{10!}{6!(10-6)!} = \frac{10!}{6!\,4!} = 210,$$

which is the same as $C(10,4)$. In fact, Exercise 58 asks you to prove the fact that, in general, for all whole numbers n and r, with $r \leq n$,

$$C(n,r) = C(n, n-r).$$

See the margin note as well. It indicates that $C(10,6)$ is equal to $C(10,4)$.

EXAMPLE 9 How many different three-member committees could club N appoint so that exactly one woman is on the committee?

Recall that $N = \{\text{Andy, Bill, Cathy, David, Evelyn}\}$. Two members are women; three are men. Although the question mentioned only that the committee must include exactly one woman, in order to complete the committee two men must be selected as well. Therefore the task of selecting the committee members consists of two parts:

1. Choose one woman.
2. Choose two men.

One woman can be chosen in $C(2,1) = 2/1 = 2$ ways, and two men can be chosen in $C(3,2) = (3 \cdot 2)/(2 \cdot 1) = 3$ ways. Using the fundamental counting principle gives $2 \cdot 3 = 6$ different committees. This number is small enough to check by listing the possibilities:

$$\{C,A,B\},\{C,A,D\},\{C,B,D\},\{E,A,B\},\{E,A,D\},\{E,B,D\}.$$

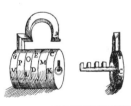

The illustration above is from the 1560s text **Logistica,** by the mathematician J. Buteo. Among other topics, the book discusses the number of possible throws of four dice and the number of arrangements of the cylinders of a combination lock. Note that "combination" is a misleading name for these locks since repetitions are allowed, and, also, order makes a difference.

EXAMPLE 10 Every member of the Alpha Beta Gamma fraternity would like to attend a special event this weekend, but only ten members will be allowed to attend. How many ways could the lucky ten be selected if there are a total of 48 members?

In this case, ten distinct men are required (repetitions are not allowed), and the order of selection makes no difference, so we use combinations. We evaluate on a calculator.

$$C(48,10) = \frac{48!}{10!\,38!} = 6{,}540{,}715{,}896$$

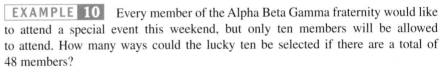

Many counting problems involve selecting some of the items from a given set of items. The particular conditions of the problem will determine which specific technique to use. The following guidelines may help.

Guidelines for Choosing a Counting Method

1. If selected items can be repeated, use the fundamental counting principle.
   Example: How many four-digit numbers are there?
   $$9 \cdot 10^3 = 9000$$

2. If selected items cannot be repeated, and order is important, use permutations.
   Example: How many ways can three of eight people line up at a ticket counter?
   $$P(8, 3) = 8 \cdot 7 \cdot 6 = 336$$

3. If selected items cannot be repeated, and order is not important, use combinations.
   Example: How many ways can a committee of three be selected from a group of twelve people?
   $$C(12, 3) = \frac{12 \cdot 11 \cdot 10}{3 \cdot 2 \cdot 1} = 220$$

Both permutations and combinations produce the number of ways of selecting r items from n items where repetitions are not allowed. Permutations apply to arrangements (order is important), while combinations apply to subsets (order is not important). These similarities and differences, as well as the appropriate formulas, are summarized in the following table.

Permutations	Combinations
Number of ways of selecting r items out of n items	
Repetitions are not allowed.	
Order is important.	Order is not important.
Arrangements of n items taken r at a time	Subsets of n items taken r at a time
$P(n, r)$ $= n(n-1)(n-2) \ldots (n-r+1)$ $= \dfrac{n!}{(n-r)!}$	$C(n, r)$ $= \dfrac{n(n-1)(n-2) \ldots (n-r+1)}{r(r-1)(r-2) \ldots 2 \cdot 1}$ $= \dfrac{n!}{r!(n-r)!}$
Clue words: arrangement, schedule, order	Clue words: set, group, sample, selection

Any problem solved with permutations (or with factorials as in the previous section) can also be solved with the fundamental counting principle. In fact, both factorials and permutations are simply ways to characterize certain product patterns that arise from applying the fundamental counting principle. And combinations are really only permutations modified for cases where order is not important.

FOR FURTHER THOUGHT

Poker Hands

In 5-card poker, played with a standard 52-card deck, 2,598,960 different hands are possible. (See Example 7.) The desirability of the various hands depends upon their relative chance of occurrence, which, in turn, depends on the number of different ways they can occur, as shown in Table 4.

TABLE 4 Categories of Hands in 5-Card Poker

Event E	Description of Event E	Number of Outcomes Favorable to E
Royal flush	Ace, king, queen, jack, and 10, all of the same suit	4
Straight flush	5 cards of consecutive denominations, all in the same suit (excluding royal flush)	36
Four of a kind	4 cards of the same denomination, plus 1 additional card	624
Full house	3 cards of one denomination, plus 2 cards of a second denomination	3744
Flush	Any 5 cards all of the same suit (excluding royal flush and straight flush)	5108
Straight	5 cards of consecutive denominations (not all the same suit)	10,200
Three of a kind	3 cards of one denomination, plus 2 cards of two additional denominations	54,912
Two pairs	2 cards of one denomination, plus 2 cards of a second denomination, plus 1 card of a third denomination	123,552
One pair	2 cards of one denomination, plus 3 additional cards of three different denominations	1,098,240
No pair	No two cards of the same denomination (and not all the same suit)	1,302,540
Total		2,598,960

For Group Discussion

As the table shows, a full house is a relatively rare occurrence. (Only four of a kind, straight flush, and royal flush are less likely.) To verify that there are 3744 different full house hands possible, carry out the following steps.

1. Explain why there are $C(4, 3)$ different ways to select three aces from the deck.
2. Explain why there are $C(4, 2)$ different ways to select two 8s from the deck.
3. If "aces and 8s" (three aces and two 8s) is one kind of full house, show that there are $P(13, 2)$ different kinds of full house altogether.
4. Multiply the expressions from Steps 1, 2, and 3 together. Explain why this product should give the total number of full house hands possible.

11.3 EXERCISES

Evaluate each of the following expressions.

1. $P(6, 4)$
2. $P(12, 2)$
3. $C(15, 3)$
4. $C(13, 6)$

Determine the number of permutations (arrangements) of each of the following.

5. 21 things taken 4 at a time
6. 14 things taken 5 at a time

Determine the number of combinations (subsets) of each of the following.

7. 10 things taken 5 at a time
8. 13 things taken 4 at a time

Use a calculator to evaluate each expression.

9. $P(25, 8)$
10. $C(36, 14)$

11. Is it possible to evaluate $P(6, 10)$? Explain.
12. Is it possible to evaluate $C(9, 12)$? Explain.
13. Explain how permutations and combinations differ.
14. Explain how factorials are related to permutations.
15. Decide whether each object is a permutation or a combination.
    (a) a telephone number
    (b) a Social Security number
    (c) a hand of cards in poker
    (d) a committee of politicians
    (e) the "combination" on a combination lock
    (f) a lottery choice of six numbers where the order does not matter
    (g) an automobile license plate number

Exercises 16–23 can be solved with permutations even though the problem statements will not always include a form of the word "permutation," or "arrangement," or "ordering."

16. *Placing in a Race* How many different ways could first-, second-, and third-place finishers occur in a race with six runners competing?

17. *Arranging New Home Models* Jeff Hubbard, a contractor, builds homes of eight different models and presently has five lots to build on. In how many different ways can he arrange homes on these lots? Assume five different models will be built.

18. *ATM PIN Numbers* An automated teller machine (ATM) requires a four-digit personal identification number (PIN), using the digits 0–9. (The first digit may be 0.) How many such PINs have no repeated digits?

19. *Electing Officers of a Club* How many ways can president and vice president be determined in a club with twelve members?

20. *Counting Prize Winners* First, second, and third prizes are to be awarded to three different people. If there are ten eligible candidates, how many outcomes are possible?

21. *Counting Prize Winners* How many ways can a teacher give five different prizes to five of her 25 students?

22. *Scheduling Security Team Visits* A security team visits 12 offices each night. How many different ways can the team order its visits?

23. *Sums of Digits* How many counting numbers have four distinct nonzero digits such that the sum of the four digits is
    (a) 10?
    (b) 11?

Exercises 24–31 can be solved with combinations even though the problem statements will not always include the word "combination" or "subset."

24. *Sampling CD Players* How many ways can a sample of five CD players be selected from a shipment of twenty-four players?

25. *Detecting Defective CD Players* If the shipment of Exercise 24 contains six defective players, how many of the size-five samples would not include any of the defective ones?

26. *Committees of U.S. Senators* How many different five-member committees could be formed from the 100 U.S. senators?

27. *Selecting Hands of Cards* Refer to the standard 52-card deck pictured in the text and notice that the

deck contains four aces, twelve face cards, thirteen hearts (all red), thirteen diamonds (all red), thirteen spades (all black), and thirteen clubs (all black). Of the 2,598,960 different five-card hands possible, decide how many would consist of the following cards.
(a) all diamonds
(b) all black cards
(c) all aces

28. *Selecting Lottery Entries* In a 7/39 lottery, you select seven distinct numbers from the set 1 through 39, where order makes no difference. How many different ways can you make your selection?

29. *Arranging New Home Models* Jeff Hubbard (the contractor) is to build six homes on a block in a new subdivision. Overhead expenses have forced him to limit his line to two different models, standard and deluxe. (All standard model homes are the same and all deluxe model homes are the same.)
    (a) How many different choices does Jeff have in positioning the six houses if he decides to build three standard and three deluxe models?
    (b) If Jeff builds only two deluxes and four standards, how many different positionings can he use?

30. *Number of Paths from Point to Point* How many paths are possible from *A* to *B* if all motion must be to the right or downward? (*Hint:* It takes ten unit steps to get from *A* to *B* and three of the ten must be downward.)

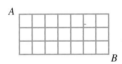

31. *Choosing a Monogram* Judy Zahrndt wants to name her new baby so that his monogram (first, middle, and last initials) will be distinct letters in alphabetical order and he will share her last name. How many different monograms could she select?

For Exercises 32–56, you may use permutations, combinations, the fundamental counting principle, or other counting methods as appropriate. (Some problems may require using more than one method.)

32. *Selecting Lottery Entries* In SuperLotto Plus, a California state lottery game, you select five distinct numbers from 1 to 47, and one MEGA number from 1 to 27, hoping that your selection will match a random list selected by lottery officials.
    (a) How many different sets of six numbers can you select?
    (b) Eileen Burke always includes her age and her husband's age as two of the first five numbers in her SuperLotto Plus selections. How many ways can she complete her list of six numbers?

33. *Drawing Cards* How many cards must be drawn (without replacement) from a standard deck of 52 to guarantee the following?
    (a) Two of the cards will be of the same suit.
    (b) Three of the cards will be of the same suit.

34. *Flush Hands in Poker* How many different 5-card poker hands would contain only cards of a single suit?

35. *Identification Numbers in Research* Subject identification numbers in a certain scientific research project consist of three letters followed by three digits and then three more letters. Assume repetitions are not allowed within any of the three groups, but letters in the first group of three may occur also in the last group of three. How many distinct identification numbers are possible?

36. *Radio Station Call Letters* Radio stations in the United States have call letters that begin with K or W (for west or east of the Mississippi River, respectively). Some have three call letters, such as WBZ in Boston, WLS in Chicago, and KGO in San Francisco. Assuming no repetition of letters, how many three-letter sets of call letters are possible?

37. *Radio Station Call Letters* Most stations that were licensed after 1927 have four call letters starting with K or W, such as WXYZ in Detroit or KRLD in Dallas. Assuming no repetitions, how many four-letter sets are possible?

38. *Scheduling Games in a Basketball League* Each team in an eight-team basketball league is scheduled to play each other team three times. How many games will be played altogether?

39. *Scheduling Batting Orders in Baseball* The Coyotes, a youth league baseball team, have seven pitchers, who only pitch, and twelve other players, all of whom can play any position other than pitcher. For Saturday's game, the coach has not yet determined which nine players to use nor what the batting order will be, except that the pitcher will bat last. How many different batting orders may occur?

40. *Ordering Performers in a Music Recital* A music class of eight girls and seven boys is having a recital. If each member is to perform once, how many ways can the program be arranged in each of the following cases?
    (a) The girls must all perform first.

(b) A girl must perform first and a boy must perform last.
(c) Elisa and Doug will perform first and last, respectively.
(d) The entire program will alternate between girls and boys.
(e) The first, eighth, and fifteenth performers must be girls.

41. *Scheduling Daily Reading* Carole begins each day by reading from one of seven inspirational books. How many ways can she choose the books for one week if the selection is done
    (a) with replacement?
    (b) without replacement?

42. *Counting Card Hands* How many of the possible 5-card hands from a standard 52-card deck would consist of the following cards?
    (a) four clubs and one non-club
    (b) two face cards and three non-face cards
    (c) two red cards, two clubs, and a spade

43. *Dividing People into Groups* In how many ways could twenty-five people be divided into five groups containing, respectively, three, four, five, six, and seven people?

44. *Points and Lines in a Plane* If any two points determine a line, how many lines are determined by seven points in a plane, no three of which are collinear?

45. *Points and Triangles in a Plane* How many triangles are determined by twenty points in a plane, no three of which are collinear?

46. *Counting Possibilities on a Combination Lock* How many different three-number "combinations" are possible on a combination lock having 40 numbers on its dial? (*Hint:* "Combination" is a misleading name for these locks since repetitions are allowed and also order makes a difference.)

47. *Selecting Drivers and Passengers for a Trip* Michael Grant, his wife and son, and four additional friends are driving, in two vehicles, to the seashore.
    (a) If all seven people are available to drive, how many ways can the two drivers be selected? (Everyone would like to drive the sports car, so it is important which driver gets which car.)
    (b) If the sports car must be driven by Michael, his wife, or their son, how many ways can the drivers now be determined?
    (c) If the sports car will accommodate only two people, and there are no other restrictions, how many ways can both drivers and passengers be assigned to both cars?

48. *Winning the Daily Double in Horse Racing* At the race track, you win the "daily double" by purchasing a ticket and selecting the winners of both of two specified races. If there are six and eight horses running in the first and second races, respectively, how many tickets must you purchase to guarantee a winning selection?

49. *Winning the Trifecta in Horse Racing* Many race tracks offer a "trifecta" race. You win by selecting the correct first-, second-, and third-place finishers. If eight horses are entered, how many tickets must you purchase to guarantee that one of them will be a trifecta winner?

50. *Selecting Committee Members* Nine people are to be distributed among three committees of two, three, and four members and a chairperson is to be selected for each committee. How many ways can this be done? (*Hint:* Break the task into the following sequence of parts.)
    (a) Select the members of the two-person committee.
    (b) Select the members of the three-person committee.
    (c) Select the chair of the two-person committee.
    (d) Select the chair of the three-person committee.
    (e) Select the chair of the four-person committee.

51. *Arranging New Home Models* Because of his good work, Jeff Hubbard gets a contract to build homes on three additional blocks in the subdivision, with six homes on each block. He decides to build nine deluxe homes on these three blocks: two on the first block, three on the second, and four on the third. The remaining nine homes will be standard.
    (a) Altogether on the three-block stretch, how many different choices does Jeff have for positioning the eighteen homes? (*Hint:* Consider the three blocks separately and use the fundamental counting principle.)
    (b) How many choices would he have if he built 2, 3, and 4 deluxe models on the three different blocks as before, but not necessarily on the first, second, and third blocks in that order?

52. *Building Numbers from Sets of Digits*
    (a) How many six-digit counting numbers can be formed using all six digits 4, 5, 6, 7, 8, and 9?
    (b) Suppose all these numbers were arranged in increasing order: 456,789; 456,798; and so on. Which number would be 364th in the list?

53. *Arranging a Wedding Reception Line* At a wedding reception, the bride and groom, the maid of honor and best man, two bridesmaids, and two ushers will form a reception line. How many ways can they be arranged in each of the following cases?
    (a) Any order will do.
    (b) The bride and groom must be the last two in line.
    (c) The groom must be last in line with the bride next to him.

54. *Assigning Student Grades* A professor teaches a class of 60 students and another class of 40 students. Five percent of the students in each class are to receive a grade of A. How many different ways can the A grades be distributed?

55. *Sums of Digits* How many counting numbers have four distinct nonzero digits such that the sum of the four digits is
    (a) 12?   (b) 13?

56. *Screening Computer Processors* A computer company will screen a shipment of 30 processors by testing a random sample of five of them. How many different samples are possible?

57. Verify that
$$C(12,9) = C(12,3).$$

58. Use the factorial formula for combinations to prove that in general,
$$C(n,r) = C(n, n-r).$$

59. (a) Use the factorial formula for permutations to evaluate, for any whole number n, $P(n,0)$.
    (b) Explain the meaning of the result in part (a).

60. (a) Use the factorial formula for combinations and the definition of 0! to evaluate, for any whole number n, $C(n,0)$.
    (b) Explain the meaning of the result in part (a).

11.4 Using Pascal's Triangle and the Binomial Theorem

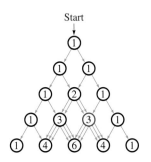

FIGURE 7

Pascal's Triangle The triangular array in Figure 7 represents what we can call "random walks" that begin at START and proceed downward according to the following rule. At each circle (branch point), a coin is tossed. If it lands heads, we go downward to the left. If it lands tails, we go downward to the right. At each point, left and right are equally likely. In each circle we have recorded the number of different routes that could bring us to that point. For example, the colored 3 can be reached as the result of three different coin-tossing sequences: htt, tht, and tth.

Another way to generate the same pattern of numbers is to begin with 1s down both diagonals and then fill in the interior entries by adding the two numbers just above a given position (to the left and right). For example, the colored 28 in Table 5 is the result of adding 7 and 21 in the row above it.

TABLE 5 Pascal's Triangle

Row Number																					Row Sum
0										1											1
1									1		1										2
2								1		2		1									4
3							1		3		3		1								8
4						1		4		6		4		1							16
5					1		5		10		10		5		1						32
6				1		6		15		20		15		6		1					64
7			1		7		21		35		35		21		7		1				128
8		1		8		28		56		70		56		28		8		1			256
9	1		9		36		84		126		126		84		36		9		1		512
10	1	10	45	120	210	252	210	120	45	10	1										1024

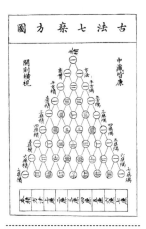

"Pascal's" triangle shown in the 1303 text *Szu-yuen Yu-chien* (*The Precious Mirror of the Four Elements*) by the Chinese mathematician Chu Shih-chieh.

By continuing to add pairs of numbers, we extend the array indefinitely downward, always beginning and ending each row with 1s. (The table shows just rows 0 through 10.) This unending "triangular" array of numbers is called **Pascal's triangle,** since Blaise Pascal wrote a treatise about it in 1653. There is evidence, though, that it was known as early as around 1100 and may have been studied in China or India still earlier. At any rate, the "triangle" possesses many interesting properties. In counting applications, the most useful property is that, in general, entry number r in row number n is equal to $C(n, r)$—the number of *combinations* of n things taken r at a time. This correspondence is shown (through row 7) in Table 6.

TABLE 6 Combination Values in Pascal's Triangle

Row Number	
0	$C(0,0)$
1	$C(1,0)\ \ C(1,1)$
2	$C(2,0)\ \ C(2,1)\ \ C(2,2)$
3	$C(3,0)\ \ C(3,1)\ \ C(3,2)\ \ C(3,3)$
4	$C(4,0)\ \ C(4,1)\ \ C(4,2)\ \ C(4,3)\ \ C(4,4)$
5	$C(5,0)\ \ C(5,1)\ \ C(5,2)\ \ C(5,3)\ \ C(5,4)\ \ C(5,5)$
6	$C(6,0)\ \ C(6,1)\ \ C(6,2)\ \ C(6,3)\ \ C(6,4)\ \ C(6,5)\ \ C(6,6)$
7	$C(7,0)\ \ C(7,1)\ \ C(7,2)\ \ C(7,3)\ \ C(7,4)\ \ C(7,5)\ \ C(7,6)\ \ C(7,7)$
	and so on

Having a copy of Pascal's triangle handy gives us another option for evaluating combinations. Any time we need to know the number of combinations of n things taken r at a time (that is, the number of subsets of size r in a set of size n), we can simply read entry number r of row number n. Keep in mind that the *first* row shown is *row number 0*. Also, the first entry of each row can be called entry number 0. This entry gives the number of subsets of size 0 (which is always 1 since there is only one empty set).

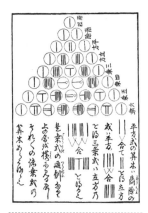

This **Japanese version** of the triangle dates from the eighteenth century. The "stick numerals" evolved from bamboo counting pieces used on a ruled board. Possibly Omar Khayyam, twelfth-century Persian mathematician and poet, may also have divined its patterns in pursuit of algebraic solutions. (The triangle lists the coefficients of the binomial expansion.)

EXAMPLE 1 A group of ten people includes six women and four men. If five of these people are randomly selected to fill out a questionnaire, how many different samples of five people are possible?

Since this is simply a matter of selecting a subset of five from a set of ten (or combinations of ten things taken five at a time), we can read $C(10, 5)$ from row 10 of Pascal's triangle in Table 5. The answer is 252.

EXAMPLE 2 Among the 252 possible samples of five people in Example 1, how many of them would consist of exactly two women and three men?

Two women can be selected from six women in $C(6, 2)$ different ways, and three men can be selected from four men in $C(4, 3)$ different ways. These combination values can be read from Pascal's triangle. Then, since the task of obtaining two women and three men requires both individual parts, the fundamental counting principle tells us to multiply the two values:

$$C(6, 2) \cdot C(4, 3) = 15 \cdot 4 = 60.$$

EXAMPLE 3 If five fair coins are tossed, in how many different ways could exactly three heads be obtained?

There are various "ways" of obtaining exactly three heads because the three heads can occur on different subsets of the coins. For example, hhtht and thhth are just two of many possibilities. When such a possibility is written down, exactly three positions are occupied by an h, the other two by a t. Each distinct way of choosing three positions from a set of five positions gives a different possibility. (Once the three positions for h are determined, each of the other two positions automatically receives a t.) So our answer is just the number of size-three subsets of a size-five set, that is, the number of combinations of five things taken three at a time. We read this answer from row 5 of Pascal's triangle: $C(5,3) = 10$. ■

Notice that row 5 of Pascal's triangle also provides answers to several other questions about tossing five fair coins. They are summarized in Table 7.

TABLE 7 Tossing Five Fair Coins

Number of Heads n	Ways of Obtaining Exactly n Heads	Listing
0	$C(5,0) = 1$	ttttt
1	$C(5,1) = 5$	htttt, thttt, tthtt, ttttt, tttth
2	$C(5,2) = 10$	hhttt, hthtt, htth, htth, thhtt, ththt, thtth, tthht, tthth, ttthh
3	$C(5,3) = 10$	hhhtt, hhtht, hhtth, hthht, hthth, htthh, thhht, thhth, ththh, tthhh
4	$C(5,4) = 5$	hhhht, hhhth, hhthh, hthhh, thhhh
5	$C(5,5) = 1$	hhhhh

The Binomial Theorem We will now look briefly at a totally different line of mathematical reasoning which also leads to Pascal's triangle, and applies it to a very important result in algebra. The "triangular" pattern of entries, it turns out, also occurs when "binomial" expressions are raised to various powers. (In algebra, any two-term expression, such as $x + y$, or $a + 2b$, or $w - 4$, is called a binomial.) The first few powers of the binomial $x + y$ are shown here, so that we can see a pattern.

$$(x + y)^0 = 1$$
$$(x + y)^1 = x + y$$
$$(x + y)^2 = x^2 + 2xy + y^2$$
$$(x + y)^3 = x^3 + 3x^2y + 3xy^2 + y^3$$
$$(x + y)^4 = x^4 + 4x^3y + 6x^2y^2 + 4xy^3 + y^4$$
$$(x + y)^5 = x^5 + 5x^4y + 10x^3y^2 + 10x^2y^3 + 5xy^4 + y^5$$

The coefficients in any one of these expansions are just the entries of one of the rows of Pascal's triangle. The expansions can be verified by direct computation, using the distributive, associative, and commutative properties of algebra. Since the coefficients of the binomial expansions give us the rows of Pascal's triangle, they are precisely the numbers we have been referring to as *combinations* in our study of counting. (In the study of algebra, they have traditionally been referred to as **binomial coefficients**, and have been denoted differently. For example, rather than

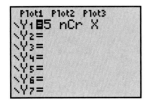

The table feature of the TI-83 Plus calculator can be used to generate a row of Pascal's triangle. Why is there a "0" for Y_1 when $X = 6$?

$C(n, r)$, you may see the notation $\binom{n}{r}$.) As "binomial coefficients," they still can be thought of as answering the question "How many?" if we reason as follows. In the expansion for $(x + y)^4$, how many are there of the expression x^2y^2? By looking in the appropriate entry of row 4 of Pascal's triangle, we see that the answer is 6.

Generalizing the pattern shown by the six expansions, we obtain the important result known as the **binomial theorem** (or sometimes known as the **general binomial expansion**).

Binomial Theorem

For any positive integer n,

$$(x + y)^n = C(n,0) \cdot x^n + C(n,1) \cdot x^{n-1}y + C(n,2) \cdot x^{n-2}y^2$$
$$+ C(n,3) \cdot x^{n-3}y^3 + \ldots$$
$$+ C(n, n-1) \cdot xy^{n-1} + C(n,n) \cdot y^n,$$

or, using the factorial formula for combinations, and the fact that $C(n, 0) = C(n, n) = 1$,

$$(x + y)^n = x^n + \frac{n!}{(n-1)!\,1!}x^{n-1}y + \frac{n!}{(n-2)!\,2!}x^{n-2}y^2$$
$$+ \frac{n!}{(n-3)!\,3!}x^{n-3}y^3 + \ldots$$
$$+ \frac{n!}{1!(n-1)!}xy^{n-1} + y^n.$$

EXAMPLE 4 Write out the binomial expansion for $(x + y)^7$.

Reading the coefficients from row 7 of Pascal's triangle (Table 5), we obtain

$$(x + y)^7 = x^7 + 7x^6y + 21x^5y^2 + 35x^4y^3 + 35x^3y^4 + 21x^2y^5 + 7xy^6 + y^7.$$

Recall that "binomial" expressions can involve terms other than x and y. The binomial theorem still applies.

EXAMPLE 5 Write out the binomial expansion for $(2a + 5)^4$.

Initially, we get coefficients from Pascal's triangle (row 4), but after all the algebra is finished the final coefficients are different numbers:

$$(2a + 5)^4 = (2a)^4 + 4(2a)^3 5 + 6(2a)^2 5^2 + 4(2a)5^3 + 5^4$$
$$= 2^4 a^4 + 4(2^3 a^3)5 + 6(2^2 a^2)5^2 + 4(2a)5^3 + 5^4$$
$$= 16a^4 + 4 \cdot 8 \cdot a^3 \cdot 5 + 6 \cdot 4 \cdot a^2 \cdot 25 + 4 \cdot 2a \cdot 125 + 625$$
$$= 16a^4 + 160a^3 + 600a^2 + 1000a + 625.$$

11.4 EXERCISES

Read the following combination values directly from Pascal's triangle.

1. $C(4,3)$
2. $C(5,2)$
3. $C(6,4)$
4. $C(7,5)$
5. $C(8,2)$
6. $C(9,4)$
7. $C(9,7)$
8. $C(10,6)$

Selecting Committees of Congressmen A committee of four Congressmen will be selected from a group of seven Democrats and three Republicans. Find the number of ways of obtaining each of the following.

9. exactly one Democrat
10. exactly two Democrats
11. exactly three Democrats
12. exactly four Democrats

Tossing Coins Suppose eight fair coins are tossed. Find the number of ways of obtaining each of the following.

13. exactly three heads
14. exactly four heads
15. exactly five heads
16. exactly six heads

Selecting Classrooms Peg Cheever, searching for an Economics class, knows that it must be in one of nine classrooms. Since the professor does not allow people to enter after the class has begun, and there is very little time left, Peg decides to try just four of the rooms at random.

17. How many different selections of four rooms are possible?
18. How many of the selections of Exercise 17 will fail to locate the class?
19. How many of the selections of Exercise 17 will succeed in locating the class?
20. What fraction of the possible selections will lead to "success"? (Give three decimal places.)

For a set of five elements, find the number of different subsets of each of the following sizes. (Use row 5 of Pascal's triangle to find the answers.)

21. 0
22. 1
23. 2
24. 3
25. 4
26. 5

27. How many subsets (of any size) are there for a set of five elements?
28. Find and explain the relationship between the row number and row sum in Pascal's triangle.

Over the years, many interesting patterns have been discovered in Pascal's triangle.* We explore a few of them in Exercises 29–35.

29. Refer to Table 5.
    (a) Choose a row whose row number is prime. Except for the 1s in this row, what is true of all the other entries?
    (b) Choose a second prime row number and see if the same pattern holds.
    (c) Continue the pattern to find row 11 in Table 5, and verify that the same pattern holds in that row. Continue the pattern to find row 11 in Table 5, and verify that the same pattern holds in that row.

30. Name the next five numbers of the diagonal sequence in the figure. What are these numbers called? (See Section 1.2.)

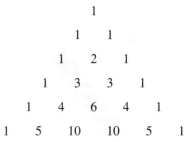

31. Complete the sequence of sums on the diagonals shown in the figure. What pattern do these sums make? What is the name of this important sequence of numbers? The presence of this sequence in the triangle apparently was not recognized by Pascal.

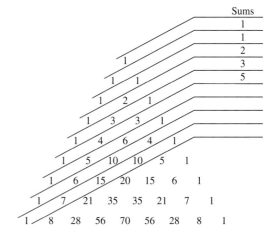

---

*For example, see the article "Serendipitous Discovery of Pascal's Triangle" by Francis W. Stanley in *The Mathematics Teacher*, February 1975.

32. Construct another "triangle" by replacing every number in Pascal's triangle (rows **0** through **5**) by its remainder when divided by 2. What special property is shared by rows **2** and **4** of this new triangle?

33. What is the next row that would have the same property as rows **2** and **4** in Exercise 32?

34. How many even numbers are there in row **256** of Pascal's triangle? (Work Exercises 32 and 33 first.)

35. The figure shows a portion of Pascal's triangle with several inverted triangular regions outlined. For any one of these regions, what can be said of the sum of the squares of the entries across its top row?

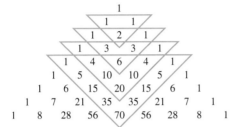

36. More than a century before Pascal's treatise on the "triangle" appeared, another work by the Italian mathematician Niccolo Tartaglia (1506–1559) came out and included the table of numbers shown here.

1	1	1	1	1	1
1	2	3	4	5	6
1	3	6	10	15	21
1	4	10	20	35	56
1	5	15	35	70	126
1	6	21	56	126	252
1	7	28	84	210	462
1	8	36	120	330	792

Explain the connection between Pascal's triangle and Tartaglia's "rectangle."

The "triangle" that Pascal studied and published in his treatise was actually more like a truncated corner of Tartaglia's rectangle, as shown here.

1	1	1	1	1	1	1	1	1	1
1	2	3	4	5	6	7	8	9	
1	3	6	10	15	21	28	36		
1	4	10	20	35	56	84			
1	5	15	35	70	126				
1	6	21	56	126					
1	7	28	84						
1	8	36							
1	9								
1									

Each number in the array can be calculated in various ways. In each of Exercises 37–40, consider the number N *to be located anywhere in the array. By checking several locations in the given array, determine how* N *is related to the sum of all entries in the shaded cells. Describe the relationship in words.*

37.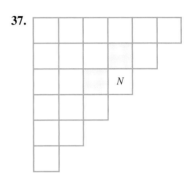

38.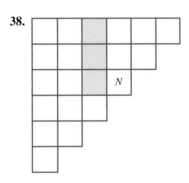

39.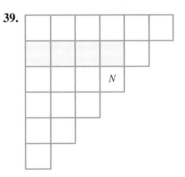

40.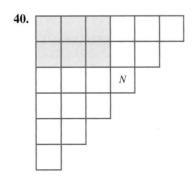

Write out the binomial expansion for each of the following powers.

41. $(x + y)^6$ **42.** $(x + y)^8$ **43.** $(z + 2)^3$

44. $(w + 3)^5$ **45.** $(2a + 5b)^4$ **46.** $(3d + 5f)^4$

47. $(b - h)^6$ (*Hint:* First change $b - h$ to $b + (-h)$.)

48. $(2n - 4m)^5$

49. How many terms appear in the binomial expansion for $(x + y)^n$?

50. Observe the pattern in this table and fill in the blanks to discover a formula for the rth term (the general term) of the binomial expansion for $(x + y)^n$.

Term Number	Coefficient	Variable Part	Term
1	1	x^n	$1\,x^n$
2	$\dfrac{n!}{(n-1)!\,1!}$	$x^{n-1}y$	$\dfrac{n!}{(n-1)!\,1!}x^{n-1}y$
3	$\dfrac{n!}{(n-2)!\,2!}$	$x^{n-2}y^2$	$\dfrac{n!}{(n-2)!\,2!}x^{n-2}y^2$
4	$\dfrac{n!}{(n-3)!\,3!}$	$x^{n-3}y^3$	$\dfrac{n!}{(n-3)!\,3!}x^{n-3}y^3$
⋮	⋮	⋮	⋮
r	$\dfrac{n!}{(n-r+1)!\,(r-1)!}$	_____	_____

Use the results of Exercise 50 to find the indicated term of each of the following expansions.

51. $(x + y)^{14}$; 5th term **52.** $(a + b)^{18}$; 16th term

53. It was stated in the text that each interior entry in Pascal's triangle can be obtained by adding the two numbers just above it (to the left and right). This fact, known as the "Pascal identity," can be written as

$$C(n, r) = C(n - 1, r - 1) + C(n - 1, r).$$

Use the factorial formula for combinations (along with some algebra) to prove the Pascal identity.

11.5 Counting Problems Involving "Not" and "Or"

When counting the number of ways that an event can occur (or that a "task" can be done), it is sometimes easier to take an indirect approach. In this section we will discuss two indirect techniques: first, the *complements principle*, where we start by counting the ways an event would *not* occur, and then the *additive principle*, where we break an event into simpler component events, and count the number of ways one component *or* another would occur.

Suppose U is the set of all possible results of some type and A is the set of all those results that satisfy a given condition. For any set S, the cardinal number of S (number of elements in S) is written $n(S)$, and the complement of S is written S'. Then Figure 8 suggests that

$$n(A) + n(A') = n(U), \quad \text{or} \quad n(A) = n(U) - n(A').$$

A summary of this principle is now stated.

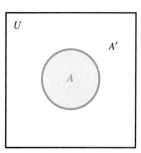

The complement of a set

FIGURE 8

Complements Principle of Counting

If A is any set within the universal set U, then

$$n(A) = n(U) - n(A').$$

By this principle, the number of ways a certain condition can be satisfied is the total number of possible results minus the number of ways the condition would **not** be satisfied. (As we have seen in Chapters 2 and 3, the arithmetic operation of *subtraction* corresponds to the set operation of *complementation* and to the logical connective of *negation,* that is "not.")

EXAMPLE 1 For the set $S = \{a, b, c, d, e, f\}$, find the number of proper subsets.

Recall that a proper subset of S is any subset with fewer than all six elements. Several subsets of different sizes would satisfy this condition. However, it is easier to consider the one subset that is not proper, namely S itself. From set theory, we know that set S has a total of $2^6 = 64$ subsets. Thus, from the complements principle, the number of proper subsets is $64 - 1 = 63$. In words, the number of subsets that *are* proper is the total number of subsets minus the number of subsets that are *not* proper.

Consider the tossing of three fair coins. Since each coin will land either heads (h) or tails (t), the possible results can be listed as follows.

hhh, hht, hth, thh, htt, tht, tth, ttt

(Even without the listing, we could have concluded that there would be eight possibilities. There are two possible outcomes for each coin, so the fundamental counting principle gives $2 \cdot 2 \cdot 2 = 2^3 = 8$.)

Now suppose we wanted the number of ways of obtaining *at least* one head. In this case, "at least one" means one or two or three. But rather than dealing with all three cases, we can note that "at least one" is the opposite (or complement) of "fewer than one" (which is zero). Since there is only one way to get zero heads (ttt), and there are a total of eight possibilities (as shown above), the complements principle gives the number of ways of getting at least one head: $8 - 1 = 7$. (The number of outcomes that include at least one head is the total number of outcomes minus the number of outcomes that do *not* include at least one head.) We find that indirect counting methods can often be applied to problems involving "at least," or "at most," or "less than," or "more than."

EXAMPLE 2 If four fair coins are tossed, in how many ways can at least one tail be obtained?

By the fundamental counting principle, $2^4 = 16$ different results are possible. Exactly one of these fails to satisfy the condition of "at least one tail" (namely, no tails, or hhhh). So our answer (from the complements principle) is $16 - 1 = 15$.

EXAMPLE 3 Carol Britz and three friends are boarding an airliner just before departure time. There are only ten seats left, three of which are aisle seats. How many ways can the four people arrange themselves in available seats so that at least one of them sits on the aisle?

The word "arrange" implies that order is important, so we shall use permutations. "At least one aisle seat" is the opposite (complement) of "no aisle seats." The total number of ways to arrange four people among ten seats is $P(10, 4) = 5040$. The number of ways to arrange four people among seven (non-aisle) seats is $P(7, 4) = 840$. Therefore, by the complements principle, the number of arrangements with at least one aisle seat is $5040 - 840 = 4200$.

The complements principle is one way of counting indirectly. Another technique is related to Figure 9. If A and B are disjoint sets (have no elements in common), then writing the union of sets A and B as $A \cup B$ gives the principle stated here.

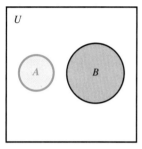

Disjoint sets

FIGURE 9

Special Additive Counting Principle

If A and B are disjoint sets, then

$$n(A \cup B) = n(A) + n(B).$$

(This principle is *special* because A and B are disjoint. It will be generalized later.) The principle states that if two conditions cannot both be satisfied together, then the number of ways that one **or** the other of them could be satisfied is found by adding the number of ways one of them could be satisfied to the number of ways the other could be satisfied. The idea is that we can sometimes analyze a set of possibilities by breaking it into a set of simpler component parts. (The arithmetic operation of *addition* corresponds to the set operation of *union* and to the logical connective of *disjunction,* that is "or.")

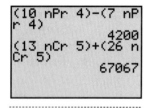

Results in Examples 3 and 4 are supported in this screen.

EXAMPLE 4 How many five-card hands consist of either all clubs or all red cards?

No hand that satisfies one of these conditions could also satisfy the other, so the two sets of possibilities (for all clubs and all red cards) are disjoint. Therefore, the special additive principle applies. Since there are 13 clubs and 26 red cards, we obtain

$$C(13, 5) + C(26, 5) = 1287 + 65{,}780$$
$$= 67{,}067.$$

The special additive principle also extends to the union of three or more disjoint sets, as illustrated by the next example.

EXAMPLE 5 Peggy Jenders needs to take twelve more specific courses for a bachelors degree, including four in math, three in physics, three in computer science, and two in business. If five courses are randomly chosen from these twelve for next semester's program, how many of the possible selections would include at least two math courses?

Of all the information given here, what is important is that there are four math courses and eight other courses to choose from, and that five of them are being

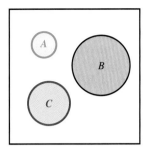

FIGURE 10 $T = A \cup B \cup C$

selected for next semester. If T denotes the set of selections that include at least two math courses, then we can write

$$T = A \cup B \cup C$$

where A = the set of selections with exactly two math courses,
B = the set of selections with exactly three math courses,
and C = the set of selections with exactly four math courses.

(In this case, *at least two* means exactly two **or** exactly three **or** exactly four.) The situation is illustrated in Figure 10.

By previous methods, we know that

$$n(A) = C(4,2) \cdot C(8,3) = 6 \cdot 56 = 336,$$
$$n(B) = C(4,3) \cdot C(8,2) = 4 \cdot 28 = 112,$$
and $$n(C) = C(4,4) \cdot C(8,1) = 1 \cdot 8 = 8,$$

so that, by the additive principle,

$$n(T) = 336 + 112 + 8 = 456.$$

Figure 11 illustrates the case in which two sets A and B are not disjoint. In this case, when the number of items in set A is added to the number in set B, any items in the intersection, $A \cap B$, get counted twice. To correct for this double inclusion, that number of items must be subtracted from the sum. The result is the following general principle. (This formula was first seen in Chapter 2, where it was called the cardinal number formula.)

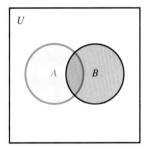

Nondisjoint sets

FIGURE 11

General Additive Counting Principle

If A and B are any two sets, disjoint or not, then

$$n(A \cup B) = n(A) + n(B) - n(A \cap B).$$

EXAMPLE 6 Table 8 categorizes a diplomatic delegation of 18 congressional members as to political party and gender. If one of the members is chosen randomly to be spokesperson for the group, in how many ways could that person be a Democrat or a woman?

TABLE 8

	Men (M)	Women (W)	Totals
Republican (R)	5	3	8
Democrat (D)	4	6	10
Totals	9	9	18

Using the general additive counting principle, we obtain

$$n(D \cup W) = n(D) + n(W) - n(D \cap W)$$
$$= 10 + 9 - 6$$
$$= 13.$$

EXAMPLE 7 How many three-digit counting numbers are multiples of 2 or multiples of 5?

A multiple of 2 must end in an even digit (0, 2, 4, 6, or 8), so there are $9 \cdot 10 \cdot 5 = 450$ three-digit multiples of 2. A multiple of 5 must end in either 0 or 5, so there are $9 \cdot 10 \cdot 2 = 180$ of those. A multiple of both 2 and 5 is a multiple of 10 and must end in 0. There are $9 \cdot 10 \cdot 1 = 90$ of those, giving

$$450 + 180 - 90 = 540$$

possible three-digit numbers that are multiples of 2 or multiples of 5. ∎

EXAMPLE 8 A single card is drawn from a standard 52-card deck.

(a) In how many ways could it be a heart or a king?

A single card can be both a heart and a king (the king of hearts), so use the general additive formula. There are thirteen hearts, four kings, and one card that is both a heart and a king:

$$13 + 4 - 1 = 16.$$

(b) In how many ways could the card be a club or a face card?

There are 13 clubs, 12 face cards, and 3 cards that are both clubs and face cards, giving

$$13 + 12 - 3 = 22.$$ ∎

EXAMPLE 9 How many subsets of a 25-element set have more than three elements?

It would be a real job to count directly all subsets of size 4, 5, 6, ..., 25. It is much easier to count those with three or fewer elements:

There is	$C(25, 0) = 1$	size-0 subset.
There are	$C(25, 1) = 25$	size-1 subsets.
There are	$C(25, 2) = 300$	size-2 subsets.
There are	$C(25, 3) = 2300$	size-3 subsets.

The total number of subsets (of all sizes, 0 through 25) is $2^{25} = 33{,}554{,}432$ (use a calculator). So the number with more than three elements must be

$$33{,}554{,}432 - (1 + 25 + 300 + 2300) = 33{,}554{,}432 - 2626$$
$$= 33{,}551{,}806.$$ ∎

In Example 9, we used both the special additive formula (to get the number of subsets with no more than three elements) and the complements principle.

Problem Solving

As you work the exercises of this section, keep in mind that indirect methods may be best, and that you may also be able to use permutations, combinations, the fundamental counting principle, or listing procedures such as product tables or tree diagrams. Also, you may want to obtain combination values, when needed, from Pascal's triangle, or find combination and permutation values on a calculator.

11.5 EXERCISES

1. Explain why the complements principle of counting is called an "indirect" method.

2. Explain the difference between the *special* and *general* additive principles of counting.

How many proper subsets are there of each of the following sets?

3. $\{w, x, y, z\}$
4. $\{A, B, C, D, E\}$

Tossing Coins If you toss seven fair coins, in how many ways can you obtain each of the following results?

5. at least one head ("At least one" is the complement of "none.")
6. at least two heads ("At least two" is the complement of "zero or one.")
7. at least two tails
8. at least one of each (a head and a tail)

Rolling Dice If you roll two fair dice (say red and green), in how many ways can you obtain each of the following? (Refer to Table 2 in Section 11.1.)

9. a 2 on the red die
10. a sum of at least 3
11. a 4 on at least one of the dice
12. a different number on each die

Identifying Properties of Counting Numbers How many two-digit counting numbers meet the following requirements?

13. not a multiple of 10
14. greater than 70 or a multiple of 10
15. *Choosing Country Music Albums* Michelle Cook's collection of eight country music albums includes *Unchained* by Johnny Cash. Michelle will choose three of her albums to play on a drive to Nashville. (Assume order is not important.)
    (a) How many different sets of three albums could she choose?
    (b) How many of these sets would not include *Unchained*?
    (c) How many of them would include *Unchained*?
16. *Choosing Broadway Hits* The ten longest Broadway runs include *Cats* and *Oh, Calcutta!*. Four of the ten are chosen randomly. (Assume order is not important.)
    (a) How many ways can the four be chosen?
    (b) How many of those groups of four would include neither of the productions mentioned?
    (c) How many of them would include at least one of the two productions mentioned?

17. *Choosing Days of the Week* How many different ways could three distinct days of the week be chosen so that at least one of them begins with the letter S? (Assume order of selection is not important.)

18. *Choosing School Assignments for Completion* Chalon Bridges has nine major assignments to complete for school this week. Two of them involve writing essays. Chalon decides to work on two of the nine assignments tonight. How many different choices of two would include at least one essay assignment? (Assume order is not important.)

Selecting Restaurants Byron Hopkins wants to dine at three different restaurants during a visit to a mountain resort. If two of eight available restaurants serve seafood, find the number of ways that at least one of the selected restaurants will serve seafood given the following conditions.

19. The order of selection is important.
20. The order of selection is not important.
21. *Seating Arrangements on an Airliner* Refer to Example 3. If one of the group decided at the last minute not to fly, then how many ways could the remaining three arrange themselves among the ten available seats so that at least one of them will sit on the aisle?
22. *Identifying Properties of Counting Numbers* Find the number of four-digit counting numbers containing at least one zero, under each of the following conditions.
    (a) Repeated digits are allowed.
    (b) Repeated digits are not allowed.
23. *Selecting Faculty Committees* A committee of four faculty members will be selected from a department of twenty-five which includes professors Fontana and Spradley. In how many ways could the committee include at least one of these two professors?
24. *Selecting Search and Rescue Teams* A Civil Air Patrol unit of twelve members includes four officers. In how many ways can four members be selected for a search and rescue mission such that at least one officer is included?

Drawing Cards If a single card is drawn from a standard 52-card deck, in how many ways could it be the following? (Use the general additive principle.)

25. a club or a jack
26. a face card or a black card

Counting Students Who Enjoy Music and Literature
Of a group of 50 students, 30 enjoy music, 15 enjoy literature, and 10 enjoy both music and literature. How many of them enjoy the following?

27. at least one of these two subjects (general additive principle)

28. neither of these two subjects (complement of "at least one")

Counting Hands of Cards Among the 2,598,960 possible 5-card poker hands from a standard 52-card deck, how many contain the following cards?

29. at least one card that is not a club (complement of "all clubs")

30. cards of more than one suit (complement of "all the same suit")

31. at least one face card (complement of "no face cards")

32. at least one diamond, but not all diamonds (complement of "no diamonds or all diamonds")

The Size of Subsets of a Set If a given set has twelve elements, how many of its subsets have the given numbers of elements?

33. at most two elements

34. at least ten elements

35. more than two elements

36. from three through nine elements

37. *Counting License Numbers* If license numbers consist of three letters followed by three digits, how many different licenses could be created having at least one letter or digit repeated? (*Hint:* Use the complements principle of counting.)

38. *Drawing Cards* If two cards are drawn from a 52-card deck without replacement (that is, the first card is not replaced in the deck before the second card is drawn), in how many different ways is it possible to obtain a king on the first draw and a heart on the second? (*Hint:* Split this event into the two disjoint components "king of hearts and then another heart" and "non-heart king and then heart." Use the fundamental counting principle on each component, then apply the special additive principle.)

Selecting National Monuments to Visit Edward Roberts is planning a long-awaited driving tour, which will take him and his family on the southern route to the West Coast. Ed is interested in seeing the twelve national monuments listed here, but will have to settle for seeing just three of them since some family members are anxious to get to Disneyland.

New Mexico	Arizona	California
Gila Cliff Dwellings	Canyon de Chelly	Devils Postpile
Petroglyph	Organ Pipe Cactus	Joshua Tree
White Sands	Saguaro	Lava Beds
Aztec Ruins		Muir Woods
		Pinnacles

In how many ways could the three monuments chosen include the following?

39. sites in only one state

40. at least one site not in California

41. sites in fewer than all three states

42. sites in exactly two of the three states

Counting Categories of Poker Hands Table 4 in this chapter (For Further Thought in Section 11.3) described the various kinds of hands in 5-card poker. Verify each statement in Exercises 43–48. (Explain all steps of your argument.)

43. There are four ways to get a royal flush.

44. There are 36 ways to get a straight flush.

45. There are 5108 ways to get a flush.

46. There are 10,200 ways to get a straight.

47. There are 624 ways to get four of a kind.

48. There are 54,912 ways to get three of a kind.

49. Extend the general additive counting principle to three overlapping sets (as in the figure) to show that
$$n(A \cup B \cup C) = n(A) + n(B) + n(C) \\ - n(A \cap B) - n(A \cap C) \\ - n(B \cap C) + n(A \cap B \cap C).$$

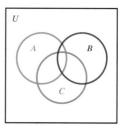

50. How many of the counting numbers 1 through 300 are *not* divisible by 2, 3, or 5? (*Hint:* Use the complements principle and the result of Exercise 49.)

COLLABORATIVE INVESTIGATION

Approximating Factorials Using Stirling's Formula

In this chapter we learned one method of computing the exact value of a factorial using a product. Because factorials increase in magnitude so quickly, theoretical mathematics provides us a formula for approximating large factorials. Interestingly enough, the formula involves the irrational numbers π and e. It is known as **Stirling's formula** and is as follows:

$$n! \approx \sqrt{2\pi n}\, n^n e^{-n}.$$

Like some other results in mathematics it is named for one person in spite of the fact that another person discovered it! According to David M. Burton, the mathematician Abraham De Moivre had just as much to do with discovering this formula as his friend, James Stirling.

As an example of an application of this formula, we observe that the exact value of 5! is 120. Using a calculator, Stirling's formula gives the approximation of 5! as 118.019168, which is off by less than 2, an error of only 1.65%.

Topics for Discussion

Divide the class into five groups, and in each group be sure that there is someone who is proficient with a calculator.

Each group should fill in one section of the table at the right. (Group 1 takes $n = 1, 2, 3$, group 2 takes $n = 4, 5, 6$, and so on.) The column values are defined as follows.

$C = n!$ (exact value, by calculator)

$S \approx n!$ (Stirling's approximation, by calculator)

D = Difference $(C - S)$

P = Percentage difference $\left(\dfrac{D}{C} \cdot 100\%\right)$

Try to obtain percentage differences accurate to two decimal places. When all groups have finished their calculations, combine the results in a large table. As a class, discuss the following question: As n grows larger and larger, does Stirling's formula become more accurate or less accurate?

n	C	S	D	P
1				
2				
3				
4				
5				
6				
7				
8				
9				
10				
11				
12				
13				
14				
15				

CHAPTER 11 TEST

If digits may be used from the set {0, 1, 2, 3, 4, 5, 6}, find the number of each of the following.

1. three-digit numbers
2. even three-digit numbers
3. three-digit numbers without repeated digits
4. three-digit multiples of five without repeated digits
5. *Counting Triangles in a Figure* Determine the number of triangles (of any size) in the figure shown here.

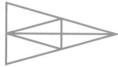

6. *Tossing Coins* Construct a tree diagram showing all possible results when a fair coin is tossed four times, if the third toss must be different than the second.
7. *Sums of Digits* How many nonrepeating four-digit numbers have the sum of their digits equal to 30?
8. *Building Numbers from Sets of Digits* Using only digits from the set {0, 1, 2}, how many three-digit numbers can be written which have no repeated odd digits?

Evaluate the following expressions.

9. 5!
10. $\dfrac{8!}{5!}$
11. $P(12,4)$
12. $C(7,3)$
13. *Building Words from Sets of Letters* How many five-letter "words" without repeated letters are possible using the English alphabet? (Assume that any five letters make a "word.")
14. *Building Words from Sets of Letters* Using the Russian alphabet (which has 32 letters), and allowing repeated letters, how many five-letter "words" are possible?

Selecting Groups of Basketball Players If there are twelve players on a basketball team, find the number of choices the coach has in selecting each of the following.

15. four players to carry the team equipment
16. two players for guard positions and two for forward positions
17. five starters and five subs
18. a set of three or more of the players

Choosing Switch Settings Determine the number of possible settings for a row of four on-off switches under each of the following conditions.

19. There are no restrictions.
20. The first and fourth switches must be on.
21. The first and fourth switches must be set the same.
22. No two adjacent switches can be off.
23. No two adjacent switches can be set the same.
24. At least two switches must be on.

Choosing Subsets of Letters Four distinct letters are to be chosen from the set

{A, B, C, D, E, F, G}.

Determine the number of ways to obtain a subset that includes each of the following.

25. the letter D
26. both A and E
27. either A or E, but not both
28. equal numbers of vowels and consonants
29. more consonants than vowels
30. Write out and simplify the binomial expansion for $(x + 2)^5$.
31. If $C(n,r) = 495$ and $C(n, r + 1) = 220$, find the value of $C(n + 1, r + 1)$.
32. If you write down the second entry of each row of Pascal's triangle (starting with row 1), what sequence of numbers do you obtain?
33. Explain why there are $r!$ times as many permutations of n things taken r at a time as there are combinations of n things taken r at a time.

chapter

12 | Probability

People are exposed to hazardous situations every day. Risk is the chance or *probability* that a harmful event will actually occur. Risk occurs in many different forms, whether it is from chemicals in the environment or behaviors that can produce detrimental consequences. The following list contains activities that carry an annual increased risk of death by 1 chance in a million.

Smoking 1.4 cigarettes
Spending 1 hour in a coal mine
Living 2 days in New York or Boston
Eating 40 teaspoons of peanut butter
Living 2 months with a cigarette smoker
Flying 1000 miles in a jet
Traveling 300 miles in a car
Riding 10 miles on a bicycle

An enormous amount of effort in our society is spent on assessing risk. Ideally, people would like to live in a world with zero risk. However, almost every action or substance exposes a person to some risk. For example, in any one year the chance of being hit by a meteorite is approximately 1 in 1,700,000,000, while the chance of dying from smoking is about 1 in 170. Meteorites will likely harm no one in the United States, whereas smoking is believed to be responsible for 420,000 deaths annually.

The game of soccer commonly involves "heading" the ball. When the issue of possible brain damage from this practice was analyzed in *Science News*, one doctor was quoted as telling soccer moms and dads that "the drive to the soccer field in their car or van is 'at least 100 times' more dangerous than the game their children will play."

Knowledge about probability allows us to put numbers to risk assessments such as those mentioned above. Armed with this knowledge, individuals are better able to keep the various risks in perspective and to make informed decisions about their lives.

Sources: Rodricks, J., *Calculated Risk*, Cambridge University Press, 1992; 1995 *Grolier Multimedia Encyclopedia; Science News,* Nov. 27, 1999.

12.1 Basic Concepts

12.2 Events Involving "Not" and "Or"

12.3 Events Involving "And"

12.4 Binomial Probability

12.5 Expected Value

12.6 Estimating Probabilities by Simulation

Collaborative Investigation:
Finding Empirical Values of π

Chapter 12 Test

"**But is it probable,**" asked Pascal, "that probability gives assurance? Nothing gives certainty but truth; nothing gives rest but the sincere search for truth." When Pascal wrote that, he had gone to live at the Jansenist convent of Port-Royal after a carriage accident in 1654.

Pascal's notes on Christianity were collected after his death in the *Pensées* (thoughts). The above quotation is included. Another develops Pascal's "rule of the wager": If you bet God exists and live accordingly, you will have gained much even if God does not exist; if you bet the opposite and God does exist, you will have lost the reason for living right—hence everything.

If you go to a supermarket and select five pounds of peaches at 89¢ per pound, you can easily predict the amount you will be charged at the checkout counter: $5 \times \$.89 = \$4.45$. The amount charged for such purchases is a **deterministic phenomenon.** It can be predicted exactly on the basis of obtainable information, namely, in this case, number of pounds and cost per pound.

On the other hand, consider the problem faced by the produce manager of the market, who must order peaches to have on hand each day without knowing exactly how many pounds customers will buy during the day. The customer demand is an example of a **random phenomenon.** It fluctuates in such a way that its value on a given day cannot be predicted exactly with obtainable information.

The study of probability is concerned with such random phenomena. Even though we cannot be certain whether a given result will occur, we often can obtain a good measure of its *likelihood,* or **probability.** This chapter discusses various ways of finding and using probabilities.

Some mathematics of probability involved in games and gambling was studied in Italy as early as the fifteenth and sixteenth centuries, but a systematic mathematical theory of chance was not begun until 1654. In that year two French mathematicians, Pierre de Fermat (about 1601–1665) and Blaise Pascal (1623–1662), corresponded with each other regarding a problem posed by the Chevalier de Méré, a gambler and member of the aristocracy. *If the two players of a game are forced to quit before the game is finished, how should the pot be divided?* Pascal and Fermat solved the problem by developing basic methods of determining each player's chance, or probability, of winning.

Nearly all published work in probability theory until about 1800 was based on dice and other games of chance. The man usually credited with being the "father" of probability theory is the French mathematician Pierre Simon de Laplace (1749–1827); he was one of the first to apply probability to matters other than gambling. Despite the work of Laplace and the urgings of a few important mathematicians in the nineteenth century, the general mathematical community failed to become interested in probability until the early twentieth century.

The development of probability as a mathematical theory was mainly due to a line of remarkable scholars in Russia, including P. L. Chebyshev (1821–1894), his student A. A. Markov (1856–1922), and finally Andrei Nikolaevich Kolmogorov (1903–1987). Kolmogorov's *Foundations of the Theory of Probability* (1933) is an axiomatic treatment.

12.1 Basic Concepts

Probability In the study of probability, we say that any observation, or measurement, of a random phenomenon is an **experiment.** The possible results of the experiment are called **outcomes,** and the set of all possible outcomes is called the **sample space.** Usually we are interested in some particular collection of the possible outcomes. Any such subset of the sample space is called an **event.** Outcomes that belong to the event are commonly referred to as "favorable outcomes," or "successes." Any time a success is observed, we say that the event has "occurred." The probability of an event, being a numerical measure of the event's likelihood, is determined in one of two ways, either *empirically* (experimentally) or *theoretically* (mathematically).

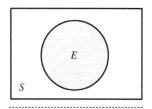

Every event is a subset of the sample space.

Christiaan Huygens
(1629–1695), a brilliant Dutch mathematician and scientist, was the first to write a formal treatise on probability. It appeared in 1657 and was based on the Pascal-Fermat correspondence.

EXAMPLE 1 If a single coin is tossed, find the probability that it will land heads up.

There is no apparent reason for one side of a coin to land up any more often than the other (in the long run), so we would normally assume that heads and tails are equally likely. This assumption can be emphasized by saying that the coin is "fair." Now the experiment here is the tossing of a single fair coin, the sample space can be denoted $S = \{h, t\}$, and the event whose probability we seek is $E = \{h\}$. Since one of two possible outcomes is a head, the probability of heads is the quotient of 1 and 2.

$$\text{Probability (heads)} = \frac{1}{2}, \quad \text{written} \quad P(h) = \frac{1}{2} \quad \text{or} \quad P(E) = \frac{1}{2}.$$

EXAMPLE 2 If a Styrofoam cup is tossed, find the probability that it will land on its top.

Intuitively, it seems that a cup will land on its side much more often than on its top or its bottom. But just how much more often is not clear. To get an idea, we performed the experiment of tossing such a cup 50 times. It landed on its side 44 times, on its top 5 times, and on its bottom just 1 time. By the frequency of "success" in this experiment, we concluded that

$$P(\text{top}) = \frac{5}{50} = \frac{1}{10}.$$

In Example 1 involving the tossing of a fair coin, the number of possible outcomes was obviously two, both were equally likely, and one of the outcomes was a head. No actual experiment was required. The desired probability was obtained *theoretically*. Theoretical probabilities apply to all kinds of games of chance (dice rolling, card games, roulette, lotteries, and so on), and also apparently to many phenomena in nature. Laplace, in his famous *Analytic Theory of Probability*, published in 1812, gave a formula that applies to any such theoretical probability, as long as the sample space S is finite and all outcomes are equally likely. (It is sometimes referred to as the *classical definition of probability*.)

Theoretical Probability Formula

If all outcomes in a sample space S are equally likely, and E is an event within that sample space, then the **theoretical probability** of event E is given by

$$P(E) = \frac{\text{number of favorable outcomes}}{\text{total number of outcomes}} = \frac{n(E)}{n(S)}.$$

On the other hand, Example 2 involved the tossing of a cup, where the likelihoods of the various outcomes were not intuitively clear. It took an actual experiment to arrive at a probability value of $1/10$. That value was found according to the *experimental*, or *empirical*, probability formula.

In 1827, **Robert Brown** (1773–1858), a Scottish physician and botanist, described the irregular motion of microscopic pollen grains suspended in water. Such "Brownian motion," as it came to be called, was not understood until 1905 when Albert Einstein explained it by treating molecular motion as a random phenomenon.

Empirical Probability Formula

If E is an event that may happen when an experiment is performed, then the **empirical probability** of event E is given by

$$P(E) = \frac{\text{number of times event } E \text{ occurred}}{\text{number of times the experiment was performed}}.$$

Usually it is clear in applications which of the two probability formulas should be used. Here are three more examples.

EXAMPLE 3 Michelle Brown wants to have exactly two daughters. Assuming that boy and girl babies are equally likely, find her probability of success in each of the following cases.

(a) She has two children altogether.

The equal likelihood assumption here allows the use of theoretical probability. But how can we determine the number of favorable outcomes and the total number of possible outcomes?

One way is to use a tree diagram (Section 11.1) to enumerate the possibilities, as shown in Figure 1. From the outcome column we obtain the sample space $S = \{gg, gb, bg, bb\}$. Only one outcome, marked with an arrow, is favorable to the event of exactly two daughters: $E = \{gg\}$. By the theoretical probability formula,

$$P(E) = \frac{n(E)}{n(S)} = \frac{1}{4}.$$

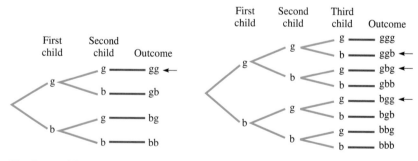

Exactly two girls among two children

FIGURE 1

Exactly two girls among three children

FIGURE 2

(b) She has three children altogether.

For three children altogether, we construct another tree diagram, as shown in Figure 2. In this case, $S = \{ggg, ggb, gbg, gbb, bgg, bgb, bbg, bbb\}$ and $E = \{ggb, gbg, bgg\}$, so $P(E) = 3/8$.

EXAMPLE 4 Find the probability of being dealt each of the following hands in five-card poker. Use a calculator to obtain answers to eight decimal places.

(a) a full house (three of one denomination and two of another)

Table 1 summarizes the various possible kinds of five-card hands. Since the 2,598,960 possible individual hands all are equally likely, we can enter the appropriate numbers from the table into the theoretical probability formula.

$$P(\text{full house}) = \frac{3744}{2{,}598{,}960} = \frac{6}{4165} \approx .00144058$$

TABLE 1 Number of Poker Hands in 5-Card Poker; Nothing Wild

Event E	Number of Outcomes Favorable to E
Royal flush	4
Straight flush	36
Four of a kind	624
Full house	3744
Flush	5108
Straight	10,200
Three of a kind	54,912
Two pairs	123,552
One pair	1,098,240
No pair	1,302,540
Total	2,598,960

(b) a royal flush (the five highest cards of a single suit)

The table shows that there are four royal flushes, one for each suit, so that

$$P(\text{royal flush}) = \frac{4}{2{,}598{,}960} = \frac{1}{649{,}740} \approx .00000154.$$

Examples 3 and 4 both utilized the theoretical probability formula because we were able to enumerate all possible outcomes and all were equally likely. In Example 3, however, the equal likelihood of girl and boy babies was *assumed*. In fact, male births typically occur a little more frequently. (At the same time, there usually are more females living at any given time, due to higher infant mortality rates among males and longer female life expectancy in general.) Example 5 shows a way of incorporating such empirical information.

EXAMPLE 5 In the year 2000, births in the United States included 2.077 million males and 1.982 million females. If a person was selected randomly from the birth records of that year, what is the probability that the person would be a male?

Since male and female births are not exactly equally likely, and we have specific experimental data to go on here, we calculate the empirical probability.

$$P(\text{male}) = \frac{\text{number of male births}}{\text{total number of births}} = \frac{2{,}077{,}000}{2{,}077{,}000 + 1{,}982{,}000} \approx .512$$

The Law of Large Numbers Now think again about the cup of Example 2. If we tossed it 50 more times, we would have 100 total tosses upon which to base an empirical probability of the cup landing on its top. The new value would likely be (at least slightly) different from what we obtained before. It would still be an empirical probability, but it would be "better" in the sense that it is based upon a larger set of outcomes.

As we increase the number of tosses, the resulting empirical probability values may approach some particular number. If so, that number can be defined as the theoretical probability of that particular cup landing on its top. This "limiting" value can occur only as the actual number of observed tosses approaches the total number of possible tosses of the cup. Since there are potentially an infinite number of possible tosses, we could never actually find the theoretical probability we want. But we can still assume such a number exists. And as the number of actual observed tosses increases, the resulting empirical probabilities should tend ever closer to the theoretical value. This very important principle is known as the **law of large numbers** (or sometimes as the "law of averages").

The **law of large numbers** also can be stated as follows. *A theoretical probability really says nothing about one, or even a few, repetitions of an experiment, but only about the proportion of successes we would expect over the long run.*

Law of Large Numbers

As an experiment is repeated more and more times, the proportion of outcomes favorable to any particular event will tend to come closer and closer to the theoretical probability of that event.

EXAMPLE 6 A fair coin was tossed 35 times, producing the following sequence of outcomes.

tthhh　ttthh　hthtt　hhthh　tthh　thttt　hhthh

Calculate the ratio of heads to total tosses after the first toss, the second toss, and so on through all 35 tosses, and plot these ratios on a graph.

The first few ratios, to two decimal places, are .00, .00, .33, .50, .60, .50, … . The graph in Figure 3 shows how the fluctuations away from .50 become smaller as the number of tosses increases, and the ratios appear to approach .50 toward the right side of the graph, in keeping with the law of large numbers.

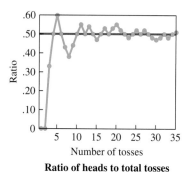

Ratio of heads to total tosses

FIGURE 3

We see then that the law of large numbers provides an important connection between empirical and theoretical probabilities. Having obtained, by experiment, an empirical probability for an event, we can then, by inductive reasoning, *estimate* that event's theoretical probability. The more repetitions the estimate is based upon, the more reliable it is.

Likewise, if we know the theoretical probability of an event, we can then, by deductive reasoning, *predict* (estimate) the fraction of times the event will occur in a series of repeated experiments. The prediction should be more accurate for larger numbers of repetitions.

Probability in Genetics Probabilities, both empirical and theoretical, have been valuable tools in many areas of science. An important early example was the work of the Austrian monk Gregor Mendel, who used the idea of randomness to help establish the study of genetics. Mendel published his results in 1865, but they were largely ignored until 1900 when others rediscovered and recognized the importance of his work.

In an effort to understand the mechanism of character transmittal from one generation to the next in plants, Mendel counted the number of occurrences of various characteristics. For example, he found that the flower color in certain pea plants obeyed this scheme:

Pure red crossed with pure white produces red.

Mendel theorized that red is "dominant" (symbolized with the capital letter R), while white is "recessive" (symbolized with the small letter r). The pure red parent carried only genes for red (R), and the pure white parent carried only genes for white (r). The offspring would receive one gene from each parent, hence one of the four combinations shown in the body of Table 2. Because every offspring receives one gene for red, that characteristic dominates and the offspring exhibits the color red.

Gregor Johann Mendel
(1822–1884) came from a peasant family who managed to send him to school. By 1847 he had been ordained and was teaching at the Abbey of St. Thomas. He finished his education at the University of Vienna and returned to the abbey to teach mathematics and natural science. Mendel began to carry out experiments on plants in the abbey garden, notably pea plants, whose distinct traits (unit characters) he had puzzled over. In 1865 he published his results. (The Czech stamp above commemorates the centennial.) His work was not appreciated at the time even though he had laid the foundation of classical genetics. Mendel had established the basic laws of heredity: laws of unit characters, of dominance, and of segregation.

TABLE 2	First to Second Generation		
		Second Parent	
		r	r
First Parent	R	Rr	Rr
	R	Rr	Rr

TABLE 3	Second to Third Generation		
		Second Parent	
		R	r
First Parent	R	RR	Rr
	r	rR	rr

Now each of these second-generation offspring, though exhibiting the color red, still carries one of each gene. So when two of them are crossed, each third-generation offspring will receive one of the gene combinations shown in Table 3. Mendel theorized that each of these four possibilities would be equally likely, and produced experimental counts that were close enough to support this hypothesis. (In more recent years, some have accused Mendel, or his assistants, of fudging the experimental data, but his conclusions have not been disputed.)

EXAMPLE 7 Referring to Table 3, determine the probability that a third-generation offspring will exhibit each of the following flower colors. Base the probabilities on the following sample space of equally likely outcomes: $S = \{RR, Rr, rR, rr\}$.

(a) red

Since red dominates white, any combination with at least one gene for red (R) will result in red flowers. Since three of the four possibilities meet this criterion, $P(\text{red}) = 3/4$.

(b) white

Only the combination rr has no gene for red, so $P(\text{white}) = 1/4$.

Due to the probabilistic laws of genetics, which began with Mendel's work, vast improvements have resulted in food supply technology, through hybrid animal and crop development. Also, the understanding and control of human diseases have been advanced by related studies.

Odds Whereas probability compares the number of favorable outcomes to the total number of outcomes, **odds** compare the number of favorable outcomes to the number of unfavorable outcomes. Odds are commonly quoted, rather than probabilities, in horse racing, lotteries, and most other gambling situations. And the odds quoted normally are odds "against" rather than odds "in favor."

> **Odds**
>
> If all outcomes in a sample space are equally likely, a of them are favorable to the event E, and the remaining b outcomes are unfavorable to E, then the **odds in favor** of E are a to b, and the **odds against** E are b to a.

An odds ratio would normally be simplified, just as a fraction. For example, it is preferable to express odds of 12 to 1 rather than 48 to 4.

EXAMPLE 8 Jennifer has been promised one of six summer jobs, three of which would be at a nearby beach resort. If she has equal chances for all six jobs, find the odds that she will land one at the resort.

Since three possibilities are favorable and three are not, the odds of working at the resort are 3 to 3, or 1 to 1. (The common factor of 3 has been divided out.) Odds of 1 to 1 are often termed "even odds," or a "50–50 chance." ∎

EXAMPLE 9 Donn Demaree has purchased six tickets for an office raffle where the winner will receive a digital phone. If 51 tickets were sold altogether and each has an equal chance of winning, what are the odds against Donn's winning the phone?

Donn has six chances to win and 45 chances to lose, so the odds against winning are 45 to 6, or 15 to 2. ∎

EXAMPLE 10 Suppose the probability of rain tomorrow is .13. Give this information in terms of odds.

We can say that

$$P(\text{rain}) = .13 = \frac{13}{100}.$$

This fraction does not mean there are necessarily 13 favorable outcomes and 100 possible outcomes, only that they would occur in this *ratio*. The corresponding number of *unfavorable* outcomes would be $100 - 13 = 87$ (total − favorable). So the odds are 13 to 87 in favor; we can say the odds are 87 to 13 against rain tomorrow. ∎

EXAMPLE 11 In a certain sweepstakes, your odds of winning are 1 to 99,999,999. What is the probability you will win?

The given odds provide a ratio of 1 favorable and 99,999,999 unfavorable outcomes, therefore $1 + 99,999,999 = 100,000,000$ total outcomes. So

$$P(\text{win}) = \frac{\text{favorable}}{\text{total}} = \frac{1}{100,000,000} = .00000001.$$ ∎

12.1 EXERCISES

In each of Exercises 1–4, give the probability that the spinner shown would land on (a) red, (b) yellow, (c) blue.

1.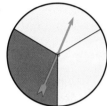

2.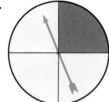

3.

4.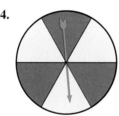

5. *Using Spinners to Generate Numbers* Suppose the spinner shown here is spun once, to determine a single-digit number, and we are interested in the event E that the resulting number is odd. Give each of the following.
   (a) the sample space
   (b) the number of favorable outcomes
   (c) the number of unfavorable outcomes
   (d) the total number of possible outcomes
   (e) the probability of an odd number
   (f) the odds in favor of an odd number

6. *Lining Up Preschool Children* Lynn Damme's group of preschool children includes eight girls and five boys. If Lynn randomly selects one child to be first in line, with E being the event that the one selected is a girl, give each of the following.
   (a) the total number of possible outcomes
   (b) the number of favorable outcomes
   (c) the number of unfavorable outcomes
   (d) the probability of event E
   (e) the odds in favor of event E

7. *Using Spinners to Generate Numbers* The spinner of Exercise 5 is spun twice in succession to determine a two-digit number. Give each of the following.
   (a) the sample space
   (b) the probability of an odd number
   (c) the probability of a number with repeated digits
   (d) the probability of a number greater than 30
   (e) the probability of a prime number

8. *Probabilities in Coin Tossing* Two fair coins are tossed (say a dime and a quarter). Give each of the following.
   (a) the sample space
   (b) the probability of heads on the dime
   (c) the probability of heads on the quarter
   (d) the probability of getting both heads
   (e) the probability of getting the same outcome on both coins

9. *Drawing Balls from an Urn* Anne Kelly randomly chooses a single ball from the urn shown here. Find the odds in favor of each of the following events.
   (a) red   (b) yellow   (c) blue

10. *Random Selection of Club Officers* Five people (Andy, Bill, Cathy, David, and Evelyn) form a club $N = \{A, B, C, D, E\}$. If they choose a president randomly, find the odds against each of the following results.
    (a) David   (b) a woman
    (c) a person whose name begins with a vowel

11. *Random Selection of Fifties Music* "Jukebox" Joe has fifty hit singles from the fifties, including exactly one by Buddy Holly, two by The Drifters, three by Bobby Darin, four by The Coasters, and five by Fats Domino. If Joe randomly selects one hit from his collection of fifty, find the probability it will be by each of the following.
    (a) Buddy Holly   (b) The Drifters
    (c) Bobby Darin   (d) The Coasters
    (e) Fats Domino

12. *Probabilities in Coin Tossing* Three fair coins are tossed. Write out the sample space, then determine the probability of each of the following events.
    (a) no heads   (b) exactly one head
    (c) exactly two heads   (d) three heads

13. *Number Sums for Rolling Two Dice* The sample space for the rolling of two fair dice appeared in Table 2 of Section 11.1. Reproduce that table, but replace each of the 36 equally likely ordered pairs with its corresponding sum (for the two dice). Then find the probability of rolling each of the following sums.
    (a) 2   (b) 3   (c) 4   (d) 5
    (e) 6   (f) 7   (g) 8   (h) 9
    (i) 10   (j) 11   (k) 12

In Exercises 14 and 15, compute answers to three decimal places.

14. *Probability of Seed Germination* In a hybrid corn research project, 200 seeds were planted, and 170 of them germinated. Find the empirical probability that any particular seed of this type will germinate.

15. *Probabilities of Boy and Girl Babies* In a certain state, 38,550 boys and 36,770 girls were born in 1996. Find the empirical probability that one of those births, chosen at random, would be
    (a) a boy   (b) a girl.

16. *Probabilities of Two Daughters Among Four Children* In Example 3, what would be Michelle's probability of having exactly two daughters if she were to have four children altogether? (You may want to use a tree diagram to construct the sample space.)

17. Explain the difference between theoretical and empirical probability.

18. Explain the difference between probability and odds.

Genetics in Snapdragons Mendel found no dominance in snapdragons (in contrast to peas) with respect to red and white flower color. When pure red and pure white parents are crossed (see Table 2), the resulting Rr combination (one of each gene) produces second-generation offspring with pink flowers. These second-generation pinks, however, still carry one red and one white gene, so when they are crossed the third generation is still governed by Table 3.

Find the following probabilities for third-generation snapdragons.

19. $P(\text{red})$ 20. $P(\text{pink})$ 21. $P(\text{white})$

Genetics in Pea Plants Mendel also investigated various characteristics besides flower color. For example, round peas are dominant over recessive wrinkled peas. First, second, and third generations can again be analyzed using Tables 2 and 3, where R represents round and r represents wrinkled.

22. Explain why crossing pure round and pure wrinkled first-generation parents will always produce round peas in the second-generation offspring.

23. When second-generation round pea plants (each of which carries both R and r genes) are crossed, find the probability that a third-generation offspring will have
    (a) round peas   (b) wrinkled peas.

Genetics of Cystic Fibrosis **Cystic fibrosis** *is one of the most common inherited diseases in North America (including the United States), occurring in about* 1 *of every* 2000 *Caucasian births and about* 1 *of every* 250,000 *non-Caucasian births. Even with modern treatment, victims usually die from lung damage by their early twenties. If we denote a cystic fibrosis gene with a* c *and a disease-free gene with a* C *(since the disease is recessive), then only a* cc *person will actually have the disease. Such persons would ordinarily die before parenting children, but a child can also inherit the disease from two* Cc *parents (who themselves are healthy, that is, have no symptoms but are "carriers" of the disease). This is like a pea plant inheriting white flowers from two red-flowered parents which both carry genes for white.*

24. Find the empirical probability (to four decimal places) that cystic fibrosis will occur in a randomly selected infant birth among U.S. Caucasians.

25. Find the empirical probability (to six decimal places) that cystic fibrosis will occur in a randomly selected infant birth among U.S. non-Caucasians.

26. Among 150,000 North American Caucasian births, about how many occurrences of cystic fibrosis would you expect?

Suppose that both partners in a marriage are cystic fibrosis carriers (a rare occurrence). Construct a chart similar to Table 3 and determine the probability of each of the following events.

27. Their first child will have the disease.

28. Their first child will be a carrier.

29. Their first child will neither have nor carry the disease.

Suppose a child is born to one cystic fibrosis carrier parent and one non-carrier parent. Find the probability of each of the following events.

30. The child will have cystic fibrosis.

31. The child will be a healthy cystic fibrosis carrier.

32. The child will neither have nor carry the disease.

Genetics of Sickle-cell Anemia **Sickle-cell anemia** occurs in about 1 of every 500 black baby births and about 1 of every 160,000 non-black baby births. It is ordinarily fatal in early childhood. There is a test to identify carriers. Unlike cystic fibrosis, which is recessive, sickle-cell anemia is **codominant.** This means that inheriting two sickle-cell genes causes the disease, while inheriting just one sickle-cell gene causes a mild (nonfatal) version (which is called **sickle-cell trait**). This is like a snapdragon plant manifesting pink flowers by inheriting one red gene and one white gene.

In Exercises 33 and 34, find the empirical probabilities of the given events.

33. A randomly selected black baby will have sickle-cell anemia. (Give your answer to three decimal places.)

34. A randomly selected non-black baby will have sickle-cell anemia. (Give your answer to six decimal places.)

35. Among 80,000 births of black babies, about how many occurrences of sickle-cell anemia would you expect?

Find the theoretical probability of each of the following conditions in a child both of whose parents have sickle-cell trait.

36. The child will have sickle-cell anemia.
37. The child will have sickle-cell trait.
38. The child will be healthy.
39. *Women's 100-meter Run* In the history of track and field, no woman has broken the 10-second barrier in the 100-meter run.
    (a) From the statement above, find the empirical probability that a woman runner will break the 10-second barrier next year.
    (b) Can you find the theoretical probability for the event of part (a)?
    (c) Is it possible that the event of part (a) will occur?
40. Is there any way a coin could fail to be "fair"? Explain.
41. On page 27 of their book *Descartes' Dream*, Philip Davis and Reuben Hersh ask the question, "Is probability real or is it just a cover-up for ignorance?" What do you think? Are some things truly random, or is everything potentially deterministic?
42. If $P(E) = .37$, find (a) the odds in favor of E, and (b) the odds against E.
43. If the odds in favor of event E are 12 to 19, find $P(E)$.
44. If the odds against event E are 10 to 3, find $P(E)$.

Probabilities of Poker Hands In 5-card poker, find the probability of being dealt each of the following. Give each answer to eight decimal places. (Refer to Table 1.)

45. a straight flush
46. two pairs
47. four queens
48. a hearts flush (*not* a royal flush or a straight flush)
49. *Probabilities in Dart Throwing* If a dart hits the square target shown here at random, what is the probability that it will hit in a colored region? (Hint: Compare the area of the colored regions to the total area of the target.)

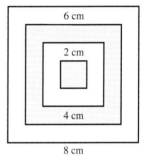

50. *Drawing Cards* When drawing cards without replacement from a standard 52-card deck, find the maximum number of cards you could possibly draw and still get
    (a) fewer than three black cards,
    (b) fewer than six spades,
    (c) fewer than four face cards,
    (d) fewer than two kings.

The remaining exercises require careful thought to determine $n(E)$ and $n(S)$. (In some cases, you may want to employ counting methods from Chapter 11, such as the fundamental counting principle, permutations, or combinations.)

Probabilities of Seating Arrangements Three married couples arrange themselves randomly in six consecutive seats in a row. Find the probability of each event in Exercises 51–54. (Hint: In each case the denominator of the probability fraction will be $6! = 720$, the total number of ways to arrange six items.)

51. Each man will sit immediately to the left of his wife.
52. Each man will sit immediately to the left of a woman.
53. The women will be in three adjacent seats.
54. The women will be in three adjacent seats, as will the men.

55. If two distinct numbers are chosen randomly from the set $\{-2, -4/3, -1/2, 0, 1/2, 3/4, 3\}$, find the probability that they will be the slopes of two perpendicular lines.

56. *Racing Bets* At most horse-racing tracks, the "trifecta" is a particular race where you win if you correctly pick the "win," "place," and "show" horses (the first-, second-, and third-place winners), in their proper order. If five horses of equal ability are entered in today's trifecta race, and you select an entry, what is the probability that you will be a winner?

57. *Probabilities of Student Course Schedules* Suppose you plan to take three courses next term. If you select them randomly from a listing of twelve courses, five of which are science courses, what is the probability that all three courses you select will be science courses?

58. *Selecting Symphony Performances* Cheryl Chechvala randomly selects three symphony performances to attend this season, choosing from a schedule of ten performances, three of which will feature works by Beethoven. Find the probability that Cheryl will select all of the Beethoven programs.

Selecting Class Reports Assuming that Ben, Jill, and Pam are three of the 36 members of the class, and that three of the class members will be chosen randomly to deliver their reports during the next class meeting, find the probability (to six decimal places) of each of the following events.

59. Ben, Jill, and Pam are selected, in that order.

60. Ben, Jill, and Pam are selected, in any order.

61. *Random Selection of Prime Numbers* If two distinct prime numbers are randomly selected from among the first eight prime numbers, what is the probability that their sum will be 24?

62. *Building Numbers from Sets of Digits* The digits 1, 2, 3, 4, and 5 are randomly arranged to form a five-digit number. Find the probability of each of the following events.
   (a) The number is even.
   (b) The first and last digits of the number both are even.

63. *Random Sums* Two integers are randomly selected from the set $\{1, 2, 3, 4, 5, 6, 7, 8, 9\}$ and are added together. Find the probability that their sum is 11 if they are selected
   (a) with replacement
   (b) without replacement.

Finding Palindromic Numbers Numbers that are **palindromes** *read the same forward and backward. For example, 30203 is a five-digit palindrome. If a single number is chosen randomly from each of the following sets, find the probability that it will be palindromic.*

64. the set of all two-digit numbers

65. the set of all three-digit numbers

12.2 Events Involving "Not" and "Or"

Remember from the previous section, that an empirical probability, based upon experimental observation, may be the best value available but still is only an approximation to the ("true") theoretical probability. For example, no human has ever been known to jump higher than eight feet vertically, so the empirical probability of such an event is zero. However, observing the rate at which high jump records have been broken, we suspect that the event is, in fact, possible and may one day occur. Hence it must have some nonzero theoretical probability, even though we have no way of assessing its exact value.

Recall also that the theoretical probability formula,

$$P(E) = \frac{n(E)}{n(S)},$$

is valid only when all outcomes in the sample space S are equally likely. For the experiment of tossing two fair coins, we can write $S = \{hh, ht, th, tt\}$ and compute correctly that

$$P(\text{both heads}) = \frac{1}{4},$$

whereas if we define the sample space with non-equally likely outcomes as $S = \{\text{both heads, both tails, one of each}\}$, we are led to

$$P(\text{both heads}) = \frac{1}{3}, \quad \text{which is } wrong.$$

(To convince yourself that $1/4$ is a better value than $1/3$, toss two fair coins 100 times or so to see what the empirical fraction seems to approach.)

Since any event E is a subset of the sample space S, we know that $0 \leq n(E) \leq n(S)$. Dividing all members of this inequality by $n(S)$ gives

$$\frac{0}{n(S)} \leq \frac{n(E)}{n(S)} \leq \frac{n(S)}{n(S)}, \quad \text{or} \quad 0 \leq P(E) \leq 1.$$

In words, the probability of any event is a number from 0 through 1, inclusive.

If event E is *impossible* (cannot happen), then $n(E)$ must be 0 (E is the empty set), so $P(E) = 0$. If event E is *certain* (cannot help but happen), then $n(E) = n(S)$, so $P(E) = n(E)/n(S) = n(S)/n(S) = 1$. These properties are summarized below.

Pierre Simon de Laplace (1749–1827) began in 1773 to solve the problem of why Jupiter's orbit seems to shrink and Saturn's orbit seems to expand. Eventually Laplace worked out the whole system. *Celestial Mechanics* resulted from almost a lifetime of work. In five volumes, it was published between 1799 and 1825 and gained for Laplace the reputation "Newton of France."

Laplace's work on probability was actually an adjunct to his celestial mechanics. He needed to demonstrate that probability is useful in interpreting scientific data. He also wrote a popular exposition of the system, which contains (in a footnote!) the "nebular hypothesis" that the sun and planets originated together in a cloud of matter, which then cooled and condensed into separate bodies.

Properties of Probability

Let E be an event from the sample space S. That is, E is a subset of S. Then the following properties hold.

1. $0 \leq P(E) \leq 1$ (The probability of an event is a number from 0 through 1, inclusive.)
2. $P(\emptyset) = 0$ (The probability of an impossible event is 0.)
3. $P(S) = 1$ (The probability of a certain event is 1.)

EXAMPLE 1 When a single fair die is rolled, find the probability of each of the following events.

(a) the number 2 is rolled

Since one of the six possibilities is a 2, $P(2) = 1/6$.

(b) a number other than 2 is rolled

There are 5 such numbers, 1, 3, 4, 5, and 6, so $P(\text{a number other than 2}) = 5/6$.

(c) the number 7 is rolled

None of the possible outcomes is 7. Thus, $P(7) = 0/6 = 0$.

(d) a number less than 7 is rolled

Since all six of the possible outcomes are less than 7, $P(\text{a number less than 7}) = 6/6 = 1$.

Mary Somerville (1780–1872) is associated with Laplace because of her brilliant exposition of his *Celestial Mechanics*. She combined a deep understanding of science with the ability to communicate its concepts to the general public.

Somerville studied Euclid thoroughly and perfected her Latin so she could read Newton's *Principia*. In about 1816 she went to London and soon became part of its literary and scientific circles. She also corresponded with Laplace and other Continental scientists.

Somerville's book on Laplace's theories came out in 1831 with great acclaim. Then followed a panoramic book, *Connection of the Physical Sciences* (1834). A statement in one of its editions suggested that irregularities in the orbit of Uranus might indicate that a more remote planet, not yet seen, existed. This caught the eye of the scientists who worked out the calculations for Neptune's orbit.

Notice that parts (c) and (d) of Example 1 illustrate probability properties 2 and 3, respectively. Also notice that the events in parts (a) and (b) are *complements* of one another, and that their probabilities add up to 1. The same is true for any two complementary events; that is, $P(E) + P(E') = 1$. By rearranging terms, we can write this equation in two other equivalent forms, the most useful of which is stated in the following rule.

Complements Rule of Probability

The probability that an event E will occur is equal to one minus the probability that it will *not* occur.

$$P(E) = 1 - P(E')$$

If you have studied counting methods, you will notice that this rule is similar to the complements principle of counting (and indeed follows from it). The complements rule is illustrated in Figure 4.

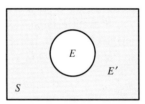

The logical connective "not" corresponds to "complement" in set theory.

$$P(\text{not } E) = P(S) - P(E)$$
$$= 1 - P(E)$$

FIGURE 4

The following example illustrates how the complements rule allows probabilities to be found indirectly when this is easier.

EXAMPLE 2 When a single card is drawn from a standard 52-card deck, what is the probability of the event of drawing a card other than a king?

It is easier to count the kings than the non-kings. Let E be the event of *not* drawing a king. Then

$$P(E) = 1 - P(E') = 1 - \frac{n(E')}{n(S)} = 1 - \frac{4}{52} = \frac{48}{52} = \frac{12}{13}.$$

We now consider some ways of combining events to obtain other events. First we look at composites involving the word "or." For event A and event B, we will want to find $P(A \text{ or } B)$. "A or B" is the event that at least one of the two individual events occurs.

EXAMPLE 3 If one number is selected randomly from the set {1, 2, 3, 4, 5, 6, 7, 8, 9, 10}, find the probability that it will be odd or a multiple of 3.

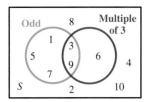

FIGURE 5

Take the sample space to be $S = \{1, 2, 3, 4, 5, 6, 7, 8, 9, 10\}$. Then, if we let A be the event that the selected number is odd and B be the event that it is a multiple of 3, we have

$$A = \{1, 3, 5, 7, 9\} \quad \text{and} \quad B = \{3, 6, 9\}.$$

Figure 5 shows the positioning of the 10 integers within the sample space, and we see that $A \cup B = \{1, 3, 5, 6, 7, 9\}$. The composite event $A \text{ or } B$ corresponds to the set $A \cup B$. So, by the theoretical probability formula,

$$P(A \text{ or } B) = \frac{6}{10} = \frac{3}{5}.$$

In general we cannot just add the individual probabilities for a composite event of the form $A \text{ or } B$. Doing so in Example 3 would have given us

$$\frac{5}{10} + \frac{3}{10} = \frac{8}{10} = \frac{4}{5}, \quad \text{which is } wrong.$$

This would have counted the outcomes in the intersection region twice. The integers 3 and 9 are both odd *and* multiples of 3. The correct general rule is stated as follows. (Since the logical connectives *or* and *and* correspond to the set operations $\cup$ (union) and $\cap$ (intersection), respectively, we can state two versions of the rule.)

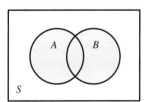

The logical connective "or" corresponds to "union" in set theory.

$P(A \text{ or } B) = P(A) + P(B) - P(A \text{ and } B)$

FIGURE 6

General Addition Rule of Probability

If A and B are any two events, then

$$P(A \text{ or } B) = P(A) + P(B) - P(A \text{ and } B),$$

or

$$P(A \cup B) = P(A) + P(B) - P(A \cap B).$$

The general addition rule is illustrated in Figure 6. This rule follows from the general additive counting principle: $n(A \cup B) = n(A) + n(B) - n(A \cap B)$. For now, we consider only events of the form "A and B" that are quite simple, as in Example 3. We deal with more involved composites using "and" in the next section.

EXAMPLE 4 Of the 20 television programs to be aired this evening, Marc plans to watch one, which he will pick at random by throwing a dart at the TV schedule. If 8 of the programs are educational, 9 are interesting, and 5 are both educational and interesting, find the probability that the show he watches will have at least one of these attributes.

If E represents "educational" and I represents "interesting," then $P(E) = 8/20$, $P(I) = 9/20$, and $P(E \text{ and } I) = 5/20$. So

$$P(E \text{ or } I) = \frac{8}{20} + \frac{9}{20} - \frac{5}{20} = \frac{12}{20} = \frac{3}{5}.$$

EXAMPLE 5 Suppose a single card is to be drawn from a standard 52-card deck. Find the probability that it will be a spade or red.

There are 13 spades in the deck and 26 red cards. But the event "spade and red" cannot possibly occur since there are no red spades. (All spades are black.) Therefore the third term in the general addition rule will be 0 (by probability property 2) and can be omitted. Hence

$$P(\text{spade } or \text{ red}) = P(\text{spade}) + P(\text{red}) = \frac{13}{52} + \frac{26}{52} = \frac{39}{52} = \frac{3}{4}.$$

In Example 5, the event "spade *and* red" had probability zero since spade and red cannot possibly both occur for a single card drawn. In general, for a particular performance of an experiment any two events that cannot both occur are called **mutually exclusive**. (As sets, such events are "disjoint.") For events A and B that are mutually exclusive, the addition rule of probability takes the simpler form stated here.

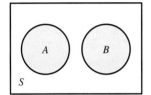

When A and B are mutually exclusive,
$P(A \text{ or } B) = P(A) + P(B)$.

FIGURE 7

Special Addition Rule of Probability

If A and B are mutually exclusive events for a given experiment, then

$$P(A \text{ or } B) = P(A) + P(B),$$

or

$$P(A \cup B) = P(A) + P(B).$$

The special addition rule is illustrated in Figure 7.

We often encounter cases in which composites of more than two events must be considered. When each event involved is mutually exclusive of all the others, an extension of the special addition rule can be applied.

x	$P(x)$
1	.05
2	.10
3	.20
4	.40
5	.10
6	.15

EXAMPLE 6 Sara figures there is only a slight chance she can finish her homework in just one hour tonight. In fact, if x represents the number of hours to be spent on homework, then she assigns probabilities to the various values of x as shown in the table at the side. (We are assuming that the actual time required will be rounded to the nearest hour and that it will not be less than 1 nor more than 6. The six time periods given are all mutually exclusive of one another. For example, if she takes 3 hours, then she will not take 2 hours or 4 hours or any of the other choices.) Find the probability that she will finish in each of the following periods of time.

(a) fewer than 3 hours

"Fewer than 3" means 1 or 2. So $P(\text{fewer than 3}) = .05 + .10 = .15$.

(b) more than 2 hours

$$P(\text{more than 2}) = P(3 \text{ or } 4 \text{ or } 5 \text{ or } 6)$$
$$= P(3) + P(4) + P(5) + P(6)$$
$$= .20 + .40 + .10 + .15 = .85$$

(c) more than 1 but no more than 5 hours

$$P(\text{more than 1 but no more than 5}) = P(2) + P(3) + P(4) + P(5)$$
$$= .10 + .20 + .40 + .10 = .80$$

(d) less than 5 hours

Although we could take a direct approach here, as in parts (a), (b), and (c), we shall combine the complements rule of probability with the special addition rule. (This would greatly improve our efficiency if the list of possibilities was very large.)

$$\begin{aligned}P(\text{less than } 5) &= 1 - P(\text{not less than } 5)\\ &= 1 - P(5 \text{ or } 6)\\ &= 1 - [P(5) + P(6)]\\ &= 1 - (.10 + .15)\\ &= 1 - .25\\ &= .75\end{aligned}$$

In Example 6, the table lists all possible time intervals so the corresponding probabilities add up to 1, a necessary condition for the way part (d) was done. The time spent on homework here is an example of a **random variable.** (It is "random" since we cannot predict which of its possible values will occur.) A table, like that of Example 6, which lists all possible values of a random variable, along with the probabilities that those values will occur, is called a **probability distribution** for that random variable. Since *all* possible values are listed, they make up the entire sample space, and so the listed probabilities must add up to 1 (by probability property 3). We discuss probability distributions further in later sections.

12.2 EXERCISES

1. *Determining Whether Two Events Are Mutually Exclusive* Julie Davis has three office assistants. If A is the event that at least two of them are men and B is the event that at least two of them are women, are A and B mutually exclusive?

2. *Attending Different Colleges* Nancy Hart earned her college degree several years ago. Consider the following four events.

   Her alma mater is in the East.
   Her alma mater is a private college.
   Her alma mater is in the Northwest.
   Her alma mater is in the South.

   Are these events all mutually exclusive of one another?

3. Explain the difference between the "general" and "special" addition rules of probability, illustrating each one with an appropriate example.

Probabilities for Rolling a Die For the experiment of rolling a single fair die, find the probability of each of the following events.

4. not less than 2
5. not prime
6. odd or less than 5
7. even or prime
8. odd or even
9. less than 3 or greater than 4

Probability and Odds for Drawing a Card For the experiment of drawing a single card from a standard 52-card deck, find **(a)** the probability, and **(b)** the odds in favor, of each of the following events.

10. not an ace
11. king or queen
12. club or heart
13. spade or face card
14. not a heart, or a 7
15. neither a heart nor a 7

Number Sums for Rolling a Pair of Dice For the experiment of rolling an ordinary pair of dice, find the probability that the sum will be each of the following. (You may want to use a table showing the sum for each of the 36 equally likely outcomes.)

16. 11 or 12
17. even or a multiple of 3
18. odd or greater than 9
19. less than 3 or greater than 9
20. Find the probability of getting a prime number in each of the following cases.
    (a) A number is chosen randomly from the set {1, 2, 3, 4, ..., 12}.
    (b) Two dice are rolled and the sum is observed.

21. Suppose, for a given experiment, A, B, C, and D are events, all mutually exclusive of one another, such that $A \cup B \cup C \cup D = S$ (the sample space). By extending the special addition rule of probability to this case, and utilizing probability property 3, what statement can you make?

Probabilities of Poker Hands If you are dealt a 5-card hand (this implies without replacement) from a standard 52-card deck, find the probability of getting each of the following. Refer to Table 1 of Section 12.1, and give answers to six decimal places.

22. a flush or three of a kind
23. a full house or a straight
24. a black flush or two pairs
25. nothing any better than two pairs

Probabilities in Golf Scoring The table gives golfer Amy Donlin's probabilities of scoring in various ranges on a par-70 course. In a given round, find the probability that her score will be each of the following.

x	$P(x)$
Below 60	.04
60–64	.06
65–69	.14
70–74	.30
75–79	.23
80–84	.09
85–89	.06
90–94	.04
95–99	.03
100 or above	.01

26. 95 or higher
27. par or above
28. in the 80s
29. less than 90
30. not in the 70s, 80s, or 90s
31. What are the odds of Amy's shooting below par?

32. *Drawing Balls from an Urn* Anne Kelly randomly chooses a single ball from the urn shown here, and x represents the color of the ball chosen. Construct a complete probability distribution for the random variable x.

33. Let x denote the sum of two distinct numbers selected randomly from the set $\{1, 2, 3, 4, 5\}$. Construct the probability distribution for the random variable x.

34. *Comparing Empirical and Theoretical Probabilities for Rolling Dice* Roll a pair of dice 50 times, keeping track of the number of times the sum is "less than 3 or greater than 9" (that is 2, 10, 11, or 12).
    (a) From your results, calculate an empirical probability for the event "less than 3 or greater than 9."
    (b) By how much does your answer differ from the *theoretical* probability of Exercise 19?

For Exercises 35–38, let A be an event within the sample space S, and let $n(A) = a$ and $n(S) = s$.

35. Use the complements principle of counting to evaluate $n(A')$.
36. Use the theoretical probability formula to express $P(A)$ and $P(A')$.
37. Evaluate, and simplify, $P(A) + P(A')$.
38. What rule have you proved?

The remaining exercises require careful thought for the determination of $n(E)$ and $n(S)$. (In some cases, you may want to employ counting methods from Chapter 11, such as the fundamental counting principle, permutations, or combinations.)

Building Numbers from Sets of Digits Suppose we want to form three-digit numbers using the set of digits $\{0, 1, 2, 3, 4, 5\}$. For example, 501 and 224 are such numbers but 035 is not.

39. How many such numbers are possible?
40. How many of these numbers are multiples of 5?
41. If one three-digit number is chosen at random from all those that can be made from the above set of digits, find the probability that the one chosen is not a multiple of 5.

42. *Multiplying Numbers Generated by Spinners* An experiment consists of spinning both spinners shown here and multiplying the resulting numbers together. Find the probability that the resulting product will be even.

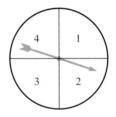 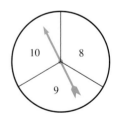

43. *Drawing Colored Marbles from Boxes* A bag contains fifty blue and fifty green marbles. Two marbles at a time are randomly selected. If both are green, they are placed in box A; if both are blue, in box B; if one is green and the other is blue, in box C. After all marbles are drawn, what is the probability that the numbers of marbles in box A and box B are the same? (This problem is borrowed with permission from the December 1992 issue of *Mathematics Teacher*, page 736.)

12.3 Events Involving "And"

In the previous section, we found the probability of an event *A or B* by adding the probabilities of event *A* and event *B,* and then subtracting the probability of the event *A and B.* The adjustment is necessary whenever *A* and *B* are not mutually exclusive.

We now consider how, in general, to find the probability of any event *A and B.*

EXAMPLE 1 If one number is selected randomly from the set {1, 2, 3, 4, 5, 6, 7, 8, 9, 10}, find the probability that it will be odd and a multiple of 3.

As in Example 3 of the previous section,

$$S = \{1, 2, 3, 4, 5, 6, 7, 8, 9, 10\}, \quad A = \{1, 3, 5, 7, 9\}, \quad \text{and} \quad B = \{3, 6, 9\}.$$

The composite event *A and B* corresponds to the set $A \cap B = \{3, 9\}$. So, by the theoretical probability formula,

$$P(A \text{ and } B) = \frac{2}{10} = \frac{1}{5}.$$

The general formula for the probability of event *A and B* will require multiplying the individual probabilities of event *A* and event *B.* But, as Example 1 shows, this must be done carefully. Although $P(A) = 5/10$ and $P(B) = 3/10$, simply multiplying these two numbers would have given

$$\frac{5}{10} \cdot \frac{3}{10} = \frac{15}{100} = \frac{3}{20}, \quad \text{which is } wrong.$$

The correct procedure is to compute the probability of the second event on the assumption that the first event has happened (or is happening, or will happen—the timing is not important). This type of probability, computed assuming some special condition, is called *conditional probability.*

Conditional Probability

The probability of event *B,* computed on the assumption that event *A* has happened, is called the **conditional probability of *B*, given *A*,** and is denoted $P(B|A)$.

The assumed condition will normally reduce the sample space to a proper subset of what it was otherwise.

In Example 1, the assumption that the selected number is odd means that we then consider only the set {1, 3, 5, 7, 9}. Of these five elements, only two, namely 3 and 9, are multiples of 3. So, using a conditional probability as the second factor, we can write

$$P(A \text{ and } B) = P(A) \cdot P(B|A) = \frac{5}{10} \cdot \frac{2}{5} = \frac{1}{5}, \quad \text{which is correct.}$$

Even **a rare occurrence** can sometimes cause widespread controversy. When Mattel Toys marketed a new talking Barbie doll in 1992, some of the Barbies were programmed to say "Math class is tough." The National Council of Teachers of Mathematics (NCTM), the American Association of University Women (AAUW), and numerous consumers voiced complaints about the damage such a message could do to the self-confidence of children and to their attitudes toward school and mathematics. Mattel subsequently agreed to erase the phrase from the microchip to be used in future doll production.

Incidentally, each Barbie was programmed to say four different statements, randomly selected from a pool of 270 prerecorded statements. Therefore, the probability of getting one that said "Math class is tough" was only

$$\frac{1 \cdot C(269, 3)}{C(270, 4)} \approx .015.$$

Other messages included in the pool were "I love school, don't you?," "I'm studying to be a doctor," and "Let's study for the quiz."

EXAMPLE 2 Given a family with two children, find the following.

(a) If it is known that at least one of the children is a girl, find the probability that both are girls. (Assume boys and girls are equally likely.)

The sample space for two children can be represented as $S = \{gg, gb, bg, bb\}$. In this case, the given condition that at least one child is a girl rules out the possibility bb, so the sample space is reduced to $\{gg, gb, bg\}$. The event "both are girls" is satisfied by one of these three equally likely outcomes, so

$$P(\text{both girls} \mid \text{at least one girl}) = \frac{1}{3}.$$

(b) If the older child is a girl, find the probability that both children are girls.

Since the older child is a girl, the sample space reduces to $\{gg, gb\}$. (We assume the children are indicated in order of birth.) The event "both are girls" is satisfied by only one of the two elements in the reduced sample space, so

$$P(\text{both girls} \mid \text{older child a girl}) = \frac{1}{2}.$$

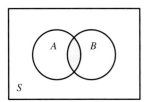

The logical connective "and" corresponds to "intersection" in set theory.

$P(A \text{ and } B) = P(A) \cdot P(B|A)$

FIGURE 8

Using a conditional probability, we can write the following *general multiplication rule of probability,* which is very similar to the fundamental counting principle of Chapter 11. It is illustrated in Figure 8.

General Multiplication Rule of Probability

If A and B are any two events, then

$$P(A \text{ and } B) = P(A) \cdot P(B|A),$$

or

$$P(A \cap B) = P(A) \cdot P(B|A).$$

EXAMPLE 3 Each year, Diane Carr adds to her book collection a number of new publications that she believes will be of lasting value and interest. She has categorized each of her twenty year 2003 acquisitions as hardcover or paperback and as fiction or nonfiction. The numbers of books in the various categories are shown in Table 4.

TABLE 4 Year 2003 Books

	Fiction (F)	Nonfiction (N)	Totals
Hardcover (H)	3	5	8
Paperback (P)	8	4	12
Totals	11	9	20

If Diane randomly chooses one of these 20 books for this evening's reading, find the probability it will be each of the following types.

(a) hardcover

Eight of the 20 books are hardcover, so $P(H) = 8/20 = 2/5$.

(b) fiction, given it is hardcover

The given condition that the book is hardcover reduces the sample space to eight books. Of those eight, just three are fiction. So $P(F|H) = 3/8$.

(c) hardcover and fiction

By the general multiplication rule of probability,

$$P(H \text{ and } F) = P(H) \cdot P(F|H) = \frac{2}{5} \cdot \frac{3}{8} = \frac{3}{20}.$$

In this case, part (c) is easier if we simply notice, directly from the table, that 3 of the 20 books are "hardcover and fiction." This verifies that the general multiplication rule of probability did give us the correct answer.

TABLE 5
Mean Distance of Planets from the Sun

Mercury	57.9
Venus	108.2
Earth	149.6
Mars	227.9
Jupiter	778.3
Saturn	1427
Uranus	2870
Neptune	4497
Pluto	5900

EXAMPLE 4 Table 5 lists the nine planets of our solar system together with their mean distance from the sun, in millions of kilometers. (Data is from the *Time Almanac 1999*.) Selina must choose two distinct planets to cover in her astronomy report. If she selects randomly, find the probability that the first one selected is closer to the sun than Mars and the second is closer than Saturn.

Define the following events.

$A:$ The first is closer than Mars.

$B:$ The second is closer than Saturn.

Then $P(A) = 3/9 = 1/3$. (Three of the original nine choices are favorable.) The planet selected first, being closer than Mars, is also closer than Saturn, and since that planet is no longer available, $P(B|A) = 4/8 = 1/2$. (Four of the remaining eight are favorable.) By the general multiplication rule, the desired probability is

$$P(A \text{ and } B) = P(A) \cdot P(B|A) = \frac{1}{3} \cdot \frac{1}{2} = \frac{1}{6} \approx .167.$$

In Example 4, the condition that A had occurred changed the probability of B, since the selection was done, in effect, without replacement. Repetitions were not allowed. We can say that event B was *dependent* upon event A. On the other hand, if the first planet selected is still available to select again (as will be the case later, in Example 5), A and B are then *independent* of one another. Independent events are defined generally as follows.

Independent Events

Two events A and B are called **independent** if knowledge about the occurrence of one of them has no effect on the probability of the other one, that is, if

$$P(B|A) = P(B).$$

Natural disasters, such as tornadoes and cyclones, earthquakes, tsunamis, volcanic eruptions, firestorms, and floods, can kill thousands of people. But a cosmic impact, the collision of a meteor, comet, or asteroid with Earth, could be at least as catastrophic as full-scale nuclear war, killing a billion or more people. So says a NASA report posted on the Internet at http://impact.arc.nasa.gov/reports/spaceguard/sg_2.html. A large enough object could even put the human species at risk of annihilation by causing drastic climate changes and destroying food crops. Many scientists now believe that impacts of the past, such as that of a 6- to 9-mile wide asteroid on the Yucatan peninsula some 65 million years ago, may have been responsible for the extinction of the dinosaurs, as well as other species.

The photo above shows damage caused by a cosmic fragment that disintegrated in the atmosphere over Tunguska, Siberia, in 1908, with an explosive energy of more than 10 megatons of TNT. Nearly 1000 square miles of uninhabited forest were flattened.

The statement $P(A|B) = P(A)$ is equivalent to the one given in the definition. So, in practice, we can establish the independence of A and B by verifying either of the two equations.

Whenever two events A and B are independent, conditional and unconditional probabilities will be no different. $P(B|A)$ will be the same as $P(B)$, so the condition can be ignored in calculating the probability. In this case, the general multiplication rule of probability reduces to the following special form.

Special Multiplication Rule of Probability

If A and B are independent events, then

$$P(A \text{ and } B) = P(A) \cdot P(B),$$

or

$$P(A \cap B) = P(A) \cdot P(B).$$

EXAMPLE 5 Selina must again select two planets, but this time one is for an oral report, the other is for a written report, and they need not be distinct. (The same planet may be selected for both.) Again find the probability that, if she selects randomly, the first is closer than Mars and the second is closer than Saturn.

Defining events A and B as in Example 4, we have $P(A) = 3/9 = 1/3$, just as before. But the selection is now done *with* replacement. Repetitions *are* allowed. In this case event B is independent of event A, so we can use the special multiplication rule, obtaining a different answer than in Example 4.

$$P(A \text{ and } B) = P(A) \cdot P(B) = \frac{1}{3} \cdot \frac{5}{9} = \frac{5}{27} \approx .185$$

EXAMPLE 6 A single card is drawn from a standard 52-card deck. Let B denote the event that the card is black, and let D denote the event that it is a diamond. Answer the following questions.

(a) Are events B and D independent?

For the unconditional probability of D, we get $P(D) = 13/52 = 1/4$. (Thirteen of the 52 cards are diamonds.) But for the conditional probability of D, given B, we have $P(D|B) = 0/26 = 0$. (None of the 26 black cards are diamonds.) Since the conditional probability $P(D|B)$ is different than the unconditional probability $P(D)$, B and D are not independent.

(b) Are events B and D mutually exclusive?

Mutually exclusive events, defined in the previous section, are events that cannot both occur for a given performance of an experiment. Since no card in the deck is both black and a diamond, B and D are mutually exclusive.

(People sometimes get the idea that "mutually exclusive" and "independent" mean the same thing. This example shows that this is not so.)

The multiplication rules of probability, both the general and special forms, can be extended to cases where more than two events are involved.

12.3 Events Involving "And"

EXAMPLE 7 Anne is still drawing balls from the same urn (shown at the side). This time she draws three balls, without replacement. Find the probability that she gets red, yellow, and blue balls, in that order.

Using appropriate letters to denote the colors, and subscripts to indicate first, second, and third draws, the event we are considering can be symbolized R_1 and Y_2 and B_3, so

$$P(R_1 \text{ and } Y_2 \text{ and } B_3) = P(R_1) \cdot P(Y_2 | R_1) \cdot P(B_3 | R_1 \text{ and } Y_2)$$
$$= \frac{4}{11} \cdot \frac{5}{10} \cdot \frac{2}{9} = \frac{4}{99} \approx .0404.$$

EXAMPLE 8 If five cards are drawn without replacement from a standard 52-card deck, find the probability that they all are hearts.

Each time a heart is drawn, the number of available cards decreases by one and the number of hearts decreases by one. The probability of drawing only hearts is

$$\frac{13}{52} \cdot \frac{12}{51} \cdot \frac{11}{50} \cdot \frac{10}{49} \cdot \frac{9}{48} = \frac{33}{66,640} \approx .000495.$$

The **search for extraterrestrial intelligence (SETI)** may have begun in earnest as early as 1961 when Dr. Frank Drake presented an equation for estimating the number of possible civilizations in the Milky Way galaxy whose communications we might detect. Over the years, the effort has been advanced by many scientists, including the late astronomer and exobiologist Carl Sagan, who popularized the issue in TV appearances and in his book *The Cosmic Connection: An Extraterrestrial Perspective* (Dell Paperback). "There must be other starfolk," said Sagan. In fact, some astronomers have estimated the odds against life on Earth being the only life in the universe at one hundred billion billion to one.

Other experts disagree. Freeman Dyson, a noted mathematical physicist and astronomer, says in his book *Disturbing the Universe* that after considering the same evidence and arguments, he believes it is just as likely as not (even odds) that there never was any other intelligent life out there.

If you studied counting methods (Chapter 11), you may prefer to solve the problem of Example 8 by using the theoretical probability formula and combinations. The total possible number of 5-card hands, drawn without replacement, is $C(52,5)$, and the number of those containing only hearts is $C(13,5)$, so the required probability is

$$\frac{C(13,5)}{C(52,5)} = \frac{\frac{13!}{5!\,8!}}{\frac{52!}{5!\,47!}} \approx .000495. \quad \text{Use a calculator.}$$

Example 9 shows how a probability problem can require the use of both addition and multiplication rules.

EXAMPLE 9 The local garage employs two mechanics, A and B, but you never know which mechanic will be working on your car. Your neighborhood consumer club has found that A does twice as many jobs as B, A does a good job three out of four times, and B does a good job only two out of five times. If you plan to take your car in for repairs, find the probability that a good job will be done.

Define the following events.

A: work done by A; B: work done by B; G: good job done

Since A does twice as many jobs as B, the (unconditional) probabilities of A and B are, respectively, $2/3$ and $1/3$. Since A does a good job three out of four times, the probability of a good job, given that A did the work, is $3/4$. And since B does well two out of five times, the probability of a good job, given that B did the work, is $2/5$. (These last two probabilities are conditional.) These four values can be summarized as $P(A) = 2/3$, $P(B) = 1/3$, $P(G|A) = 3/4$, and $P(G|B) = 2/5$. The event we are considering, G, can occur in two mutually exclusive ways: A could do the work and do a good job ($A \cap G$), or B could do the work and do a good job ($B \cap G$). Therefore

$$P(G) = P(A \cap G) + P(B \cap G) \quad \text{Special addition rule}$$
$$= P(A) \cdot P(G|A) + P(B) \cdot P(G|B) \quad \text{General multiplication rule}$$
$$= \frac{2}{3} \cdot \frac{3}{4} + \frac{1}{3} \cdot \frac{2}{5}$$
$$= \frac{1}{2} + \frac{2}{15} = \frac{19}{30} \approx .633.$$

The tree diagram in Figure 9 gives an alternative approach to finding the probability in Example 9. Use the given information to draw the tree diagram, then find the probability of a good job by adding the probabilities from the indicated branches of the tree.

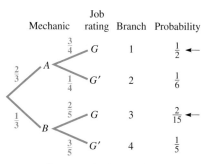

Garage mechanics experiment

FIGURE 9

Since 1999, the **search for extraterrestrial intelligence (SETI)** has been carried out mainly through **SETI@HOME**, the largest distributed computing project on Earth. Some of the world's largest radio telescopes, like the one built into a 20-acre natural bowl in Aricebo, Puerto Rico (pictured above), collect data from the most likely regions of our galaxy. The data are analyzed by personal computer users around the world. The four millionth participant registered on October 1, 2002. To learn more, or if you would like a chance at being the first to "contact" an extraterrestrial civilization, check out the website http://setiathome.ssl.berkeley.edu.

EXAMPLE 10 Michael Bolinder is among five door prize winners at a Christmas party. The five winners are asked to choose, without looking, from a bag which, they are told, contains four candy canes and one $50 gift certificate. If Michael wants the gift certificate, can he improve his chance of getting it by drawing first among the five people?

We denote candy cane by C, gift certificate by G, and first draw, second draw, and so on by subscripts $1, 2, \ldots$. Then if Michael draws first, his probability of getting the gift certificate is

$$P(G_1) = \frac{1}{5}.$$

If he draws second, his probability of getting the gift certificate is

$$P(G_2) = P(C_1 \text{ and } G_2) = P(C_1) \cdot P(G_2|C_1)$$
$$= \frac{4}{5} \cdot \frac{1}{4} = \frac{1}{5}, \quad \text{the same as before.}$$

For the third draw,

$$P(G_3) = P(C_1 \text{ and } C_2 \text{ and } G_3) = P(C_1) \cdot P(C_2|C_1) \cdot P(G_3|C_1 \text{ and } C_2)$$
$$= \frac{4}{5} \cdot \frac{3}{4} \cdot \frac{1}{3} = \frac{1}{5}, \quad \text{again the same value.}$$

Likewise, the probability of getting the gift certificate is 1/5 when drawing fourth or when drawing fifth. Therefore, the order in which Michael draws does not affect his chances.

FOR FURTHER THOUGHT

The Birthday Problem

A classic problem (with a surprising result) involves the probability that a given group of people will include at least one pair of people with the same birthday (the same day of the year, not necessarily the same year). This problem can be analyzed using the complements formula (Section 12.2) and the general multiplication rule of probability from this section. Suppose there are three people in the group. Then

P(at least one duplication of birthdays)
$= 1 - P$(no duplications)   Complements formula
$= 1 - P$(2nd is different that 1st and 3rd is different than 1st and 2nd)
$= 1 - \dfrac{364}{365} \cdot \dfrac{363}{365}$ General multiplication rule
$\approx 1 - .992$
$= .008$

(To simplify the calculations, we have assumed 365 possible birth dates, ignoring February 29.)

By doing more calculations like the one above, we find that the smaller the group, the smaller the probability of a duplication; the larger the group, the larger the probability of a duplication. The table below shows the probability of at least one duplication for numbers of people through 50.

For Group Discussion

1. Based on the data shown in the table, what are the odds in favor of a duplication in a group of 30 people?
2. Estimate from the table the least number of people for which the probability of duplication is at least 1/2.
3. How small a group is required for the probability of a duplication to be *exactly* 0?
4. How large a group is required for the probability of a duplication to be *exactly* 1?

Number of People	Probability of at Least One Duplication	Number of People	Probability of at Least One Duplication	Number of People	Probability of at Least One Duplication
2	.003	19	.379	36	.832
3	.008	20	.411	37	.849
4	.016	21	.444	38	.864
5	.027	22	.476	39	.878
6	.040	23	.507	40	.891
7	.056	24	.538	41	.903
8	.074	25	.569	42	.914
9	.095	26	.598	43	.924
10	.117	27	.627	44	.933
11	.141	28	.654	45	.941
12	.167	29	.681	46	.948
13	.194	30	.706	47	.955
14	.223	31	.730	48	.961
15	.253	32	.753	49	.966
16	.284	33	.775	50	.970
17	.315	34	.795		
18	.347	35	.814		

12.3 EXERCISES

For each of the following experiments, determine whether the two given events are independent.

1. *Tossing Coins* A fair coin is tossed twice. The events are "head on the first" and "head on the second."
2. *Rolling Dice* A pair of dice are rolled. The events are "even on the first" and "odd on the second."
3. *Comparing Planets' Mean Distances from the Sun* Two planets are selected, with replacement, from the list in Table 5. The events are "first is closer than Earth" and "second is farther than Uranus."
4. *Comparing Planets' Mean Distances from the Sun* Two planets are selected, without replacement, from the list in Table 5. The events are "first is closer than Jupiter" and "second is farther than Neptune."
5. *Guessing Answers on a Multiple-choice Test* The answers are all guessed on a twenty-question multiple-choice test. The events are "first answer is correct" and "last answer is correct."
6. *Selecting Committees of U.S. Senators* A committee of five is randomly selected from the 100 U.S. Senators. The events are "first member selected is a Republican" and "second member selected is a Republican." (Assume that there are both Republicans and non-Republicans in the Senate.)

Comparing Gender and Career Motivation of University Students One hundred college seniors attending a career fair at a major northeastern university were categorized according to gender and according to primary career motivation, as summarized here.

		Primary Career Motivation			
		Money	Allowed to be Creative	Sense of Giving to Society	Total
Gender	Male	18	21	19	58
	Female	14	13	15	42
	Total	32	34	34	100

If one of these students is to be selected at random, find the probability that the student selected will satisfy each of the following conditions.

7. female
8. motivated primarily by creativity
9. not motivated primarily by money
10. male and motivated primarily by money
11. male, given that primary motivation is a sense of giving to society
12. motivated primarily by money or creativity, given that the student is female

Selecting Pets A pet store has seven puppies, including four poodles, two terriers, and one retriever. If Rebecka and Aaron, in that order, each select one puppy at random, with replacement *(they may both select the same one)*, find the probability of each of the following events.

13. both select a poodle
14. Rebecka selects a retriever, Aaron selects a terrier
15. Rebecka selects a terrier, Aaron selects a retriever
16. both select a retriever

Selecting Pets Suppose two puppies are selected as above, but this time *without replacement (Rebecka and Aaron cannot both select the same puppy).* Find the probability of each of the following events.

17. both select a poodle
18. Aaron selects a terrier, given Rebecka selects a poodle
19. Aaron selects a retriever, given Rebecka selects a poodle
20. Rebecka selects a retriever
21. Aaron selects a retriever, given Rebecka selects a retriever
22. both select a retriever

Drawing Cards Let two cards be dealt successively, without replacement, *from a standard 52-card deck. Find the probability of each of the following events (Exercises 23–27).*

23. spade second, given spade first
24. club second, given diamond first
25. two face cards
26. no face cards
27. The first card is a jack and the second is a face card.
28. Given events A and B within the sample space S, the following sequence of steps establishes formulas that can be used to compute conditional probabilities. (The formula of Step (b) involves probabilities, while that of Step (d) involves cardinal numbers.) Justify each statement.

    (a) $P(A \text{ and } B) = P(A) \cdot P(B|A)$
    (b) Therefore, $P(B|A) = \dfrac{P(A \text{ and } B)}{P(A)}$.
    (c) Therefore, $P(B|A) = \dfrac{n(A \text{ and } B)/n(S)}{n(A)/n(S)}$.
    (d) Therefore, $P(B|A) = \dfrac{n(A \text{ and } B)}{n(A)}$.

Considering Conditions in Card Drawing Use the results of Exercise 28 to find each of the following probabilities when a single card is drawn from a standard 52-card deck.

29. $P(\text{queen}\,|\,\text{face card})$
30. $P(\text{face card}\,|\,\text{queen})$
31. $P(\text{red}\,|\,\text{diamond})$
32. $P(\text{diamond}\,|\,\text{red})$

33. If one number is chosen randomly from the integers 1 through 10, the probability of getting a number that is *odd and prime*, by the general multiplication rule, is

    $$P(\text{odd}) \cdot P(\text{prime}\,|\,\text{odd}) = \dfrac{5}{10} \cdot \dfrac{3}{5} = \dfrac{3}{10}.$$

    Compute the product $P(\text{prime}) \cdot P(\text{odd}\,|\,\text{prime})$, and compare to the product above.

34. What does Exercise 33 imply, in general, about the probability of an event of the form *A and B*?

35. *Gender in Sequences of Babies* One of the authors of this book has three sons and no daughters. Assuming boy and girl babies are equally likely, what is the probability of this event?

36. *Gender in Sequences of Babies* Under the assumptions of Exercise 35, what is the probability that three successive births will be a boy, then a girl, then a boy?

The remaining exercises, and groups of exercises, may require concepts from earlier sections, such as the complements principle and addition rules, as well as the multiplication rules of this section.

Probabilities in Warehouse Grocery Shopping Emily Falzon manages a grocery warehouse which encourages volume shopping on the part of its customers. Emily has discovered that, on any given weekday, 80 percent of the customer sales amount to more than $100. That is, any given sale on such a day has a probability of .80 of being for more than $100. (Actually, the conditional probabilities throughout the day would change slightly, depending on earlier sales, but this effect would be negligible for the first several sales of the day, so we can treat them as independent.)

Find the probability of each of the following events. (Give answers to three decimal places.)

37. The first two sales on Wednesday are both for more than $100.
38. The first three sales on Wednesday are all for more than $100.
39. None of the first three sales on Wednesday is for more than $100.
40. Exactly one of the first three sales on Wednesday is for more than $100.

Pollution from the Space Shuttle Launch Site One problem encountered by developers of the space shuttle program is air pollution in the area surrounding the launch site. A certain direction from the launch site is considered critical in terms of hydrogen chloride pollution from the exhaust cloud. It has been determined that weather conditions would cause emission cloud movement in the critical direction only 5% of the time.

Find probabilities for the following events (Exercises 41–44). Assume that probabilities for a particular launch in no way depend on the probabilities for other launches. (Give answers to two decimal places.)

41. A given launch will not result in cloud movement in the critical direction.

42. No cloud movement in the critical direction will occur during any of 5 launches.

43. Any 5 launches will result in at least one cloud movement in the critical direction.

44. Any 10 launches will result in at least one cloud movement in the critical direction.

Ordering Job Interviews Four men and three women are waiting to be interviewed for jobs. If they are all selected in random order, find the probability of each event in Exercises 45–49.

45. All the women will be interviewed first.

46. All the men will be interviewed first.

47. The first person interviewed will be a woman.

48. The second person interviewed will be a woman.

49. The last person interviewed will be a woman.

50. In Example 7, where Anne draws three balls without replacement, what would be her probability of getting one of each color, where the order does not matter?

51. *Gender in Sequences of Babies* Assuming boy and girl babies are equally likely, find the probability that it would take
    (a) at least three births to obtain two girls,
    (b) at least four births to obtain two girls,
    (c) at least five births to obtain two girls.

52. *Drawing Cards* Cards are drawn, without replacement, from an ordinary 52-card deck.
    (a) How many must be drawn before the probability of obtaining at least one face card is greater than 1/2?
    (b) How many must be drawn before the probability of obtaining at least one king is greater than 1/2?

Fair Decisions from Biased Coins Many everyday decisions, like who will drive to lunch, or who will pay for the coffee, are made by the toss of a (presumably fair) coin and using the criterion "heads, you will; tails, I will." This criterion is not quite fair, however, if the coin is biased (perhaps due to slightly irregular construction or wear). John von Neumann suggested a way to make perfectly fair decisions even with a possibly biased coin. If a coin, biased so that

$$P(h) = .5200 \quad \text{and} \quad P(t) = .4800,$$

is tossed twice, find the following probabilities. (Give answers to four decimal places.)

53. $P(hh)$ 54. $P(ht)$
55. $P(th)$ 56. $P(tt)$

57. Having completed Exercises 53–56, can you suggest what von Neumann's scheme may have been?

Programming a Garage Door Opener A certain brand of automatic garage door opener utilizes a transmitter control with six independent switches, each one set on or off. The receiver (wired to the door) must be set with the same pattern as the transmitter.* Exercises 58–61 are based on ideas similar to those of the "birthday problem" in the "For Further Thought" feature in this section.

58. How many different ways can the owner of one of these garage door openers set the switches?

59. If two residents in the same neighborhood each have one of this brand of opener, and both set the switches randomly, what is the probability to four decimal places that they are able to open each other's garage doors?

60. If five neighbors with the same type of opener set their switches independently, what is the probability of at least one pair of neighbors using the same settings? (Give your answer to four decimal places.)

61. What is the minimum number of neighbors who must use this brand of opener before the probability of at least one duplication of settings is greater than 1/2?

62. *Choosing Cards* There are three cards, one that is green on both sides, one that is red on both sides, and one that is green on one side and red on the other. One of the three cards is selected randomly and laid on the table. If it happens that the card on the table has a red side up, what is the probability that it is also red on the other side?

---

*For more information, see "Matching Garage-Door Openers," by Bonnie H. Litwiller and David R. Duncan in the March 1992 issue of *Mathematics Teacher*, p. 217.

Weather Conditions on Successive Days In November, the rain in a certain valley tends to fall in storms of several days' duration. The unconditional probability of rain on any given day of the month is .500. But the probability of rain on a day that follows a rainy day is .800, and the probability of rain on a day following a nonrainy day is .300. Find the probability of each of the following events. Give answers to three decimal places.

63. rain on two randomly selected consecutive days in November
64. rain on three randomly selected consecutive days in November
65. rain on November 1 and 2, but not on November 3
66. rain on the first four days of November, given that October 31 was clear all day

Engine Failures in a Vintage Aircraft In a certain four-engine vintage aircraft, now quite unreliable, each engine has a 10 percent chance of failure on any flight, as long as it is carrying its one-fourth share of the load. But if one engine fails, then the chance of failure increases to 20 percent for each of the other three engines. And if a second engine fails, each of the remaining two has a 30 percent chance of failure. Assuming that no two engines ever fail simultaneously, and that the aircraft can continue flying with as few as two operating engines, find each of the following for a given flight of this aircraft. (Give answers to four decimal places.)

67. the probability of no engine failures
68. the probability of exactly one engine failure (any one of four engines)
69. the probability of exactly two engine failures (any two of four engines)
70. the probability of a failed flight

One-and-one Foul Shooting in Basketball In basketball, "one-and-one" foul shooting is done as follows: if the player makes the first shot (1 point), he is given a second shot. If he misses the first shot, he is not given a second shot (see the tree diagram).

First shot	Second shot	Branch	Total points
Point	Point	1	2
Point	No point	2	1
No point		3	0

Susan Dratch, a basketball player, has a 70% foul shot record. (She makes 70% of her foul shots.) Find the probability that, on a given one-and-one foul shooting opportunity, Susan will score the following numbers of points.

71. no points   72. one point   73. two points

74. *Comparing Empirical and Theoretical Probabilities in Dice Rolling* Roll a pair of dice until a sum of seven appears, keeping track of how many rolls it took. Repeat the process a total of 50 times, each time recording the number of rolls it took to get a sum of seven.
    (a) Use your experimental data to compute an empirical probability (to two decimal places) that it would take at least three rolls to get a sum of seven.
    (b) Find the theoretical probability (to two decimal places) that it would take at least three rolls to obtain a sum of seven.

75. Go to the website mentioned in the *natural disasters* margin note in this section and write a report on the threat to humanity of cosmic impacts. Include an explanation of the abbreviation *NEO*.

12.4 Binomial Probability

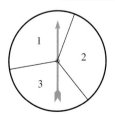

Suppose the spinner in the margin is spun twice, where we are interested in the number of times a 2 is obtained. (Assume each of the three regions contains a 120-degree arc, so that each is equally likely.) We can think of the outcome 2 as a "success," while outcomes 1 and 3 would be "failures." When the outcomes of an experiment are divided into just two categories, success and failure, the associated probabilities are called "binomial" (the prefix *bi* meaning *two*). Repeated performances of such an experiment, where the probability of success remains constant throughout all repetitions, are also known as repeated **Bernoulli trials** (after James Bernoulli).

James Bernoulli (1654–1705) is also known as Jacob or Jacques. He was charmed away from theology by the writings of Leibniz, became his pupil, and later headed the mathematics faculty at the University of Basel. His results in probability are contained in the *Art of Conjecture*, which was published in 1713, after his death, and which also included a reprint of the earlier Huygens paper. Bernoulli also made many contributions to calculus and analytic geometry.

TABLE 6
Probability Distribution for Two Spins

x	$P(x)$
0	4/9
1	4/9
2	1/9
Sum = 9/9 = 1	

If we let x denote the number of 2s obtained out of two spins, then x is a random variable, and we can construct its probability distribution. (Random variables and probability distributions were defined at the end of Section 12.2.) Apparently, x could possibly have values of 0, 1, or 2. Let's begin by listing the sample space (of equally likely outcomes) for the two spins:

$$S = \{(1,1), (1,2), (1,3), (2,1), (2,2), (2,3), (3,1), (3,2), (3,3)\}.$$

In each ordered pair, the first and second entries give the results on the first and second spins, respectively. There are nine outcomes, the number of 2s being 0 in four cases, 1 in four cases, and 2 in one case. From this information we construct the probability distribution in Table 6. In this particular case, we have an example of a **binomial probability distribution.** Notice that the sum of the probability column is 1, which agrees with property 3 of probability (Section 12.2), since all possible values of x have been listed.

In order to develop a general formula for binomial probabilities, we can consider another way to obtain the probability values in Table 6. The various spins of the spinner are independent of one another, and on each spin the probability of success (S) is 1/3 and the probability of failure (F) is 2/3. We will denote success on the first spin by S_1, failure on the second by F_2, and so on. Then, using the rules of probability, we have

$$\begin{aligned} P(x = 0) &= P(F_1 \text{ and } F_2) \\ &= P(F_1) \cdot P(F_2) \quad &\text{Special multiplication rule} \\ &= (2/3) \cdot (2/3) \\ &= 4/9, \\ P(x = 1) &= P[(S_1 \text{ and } F_2) \text{ or } (F_1 \text{ and } S_2)] \\ &= P(S_1 \text{ and } F_2) + P(F_1 \text{ and } S_2) \quad &\text{Special addition rule} \\ &= P(S_1) \cdot P(F_2) + P(F_1) \cdot P(S_2) \quad &\text{Special multiplication rule} \\ &= (1/3) \cdot (2/3) + (2/3) \cdot (1/3) \\ &= 2/9 + 2/9 \\ &= 4/9, \end{aligned}$$

and

$$\begin{aligned} P(x = 2) &= P(S_1 \text{ and } S_2) \\ &= P(S_1) \cdot P(S_2) \quad &\text{Special multiplication rule} \\ &= (1/3) \cdot (1/3) \\ &= 1/9. \end{aligned}$$

Notice the following pattern in the above calculations. There is only one way to get $x = 0$ (namely, F_1 and F_2). And there is only one way to get $x = 2$ (namely, S_1 and S_2). But there are two ways to get $x = 1$. One way is S_1 and F_2; the other is F_1 and S_2. There are two ways because the one success required can occur on the first spin or on the second spin. How many ways can exactly one success occur in two repeated trials? This question is equivalent to: How many size-one subsets are there of the set of two trials? The answer is $C(2, 1) = 2$. (The expression $C(2, 1)$ denotes "combinations of 2 things taken 1 at a time." Combinations were discussed in Chapter 11.) Each of the two ways to get exactly one success has a probability equal to $(1/3) \cdot (2/3)$, the probability of success times the probability of failure.

12.4 Binomial Probability

TABLE 7
Probability Distribution for Three Spins

x	$P(x)$
0	8/27
1	12/27
2	6/27
3	1/27

Sum = 27/27 = 1

If the same spinner is spun three times rather than two, then x, the number of successes (2s) could have values of 0, 1, 2, or 3. Then the number of ways to get exactly 1 success is $C(3, 1) = 3$. They are S_1 and F_2 and F_3, F_1 and S_2 and F_3, and F_1 and F_2 and S_3. The probability of each of these three ways is $(1/3) \cdot (2/3) \cdot (2/3) = 4/27$. So $P(x = 1) = 3 \cdot (4/27) = 12/27 = 4/9$. The tree diagram in Figure 10 shows all possibilities for three spins, and Table 7 gives the associated probability distribution. Notice in the tree diagram that the number of ways of getting two successes in three trials is also 3, in agreement with the fact that $C(3, 2) = 3$. Notice also that, again, the sum of the $P(x)$ column is 1.

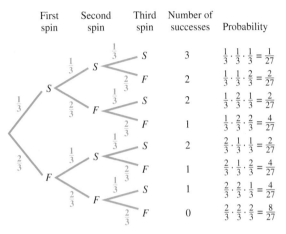

Tree diagram for three spins

FIGURE 10

Problem Solving

One of the problem-solving strategies from Chapter 1 was "Look for a pattern." Having constructed complete probability distributions for binomial experiments with 2 and 3 repeated trials (and probability of success 1/3), we can now generalize the observed pattern to any binomial experiment, as shown next.

In general, let

n = the number of repeated trials,
p = the probability of success on any given trial,
$q = 1 - p$ = the probability of failure on any given trial,

and x = the number of successes that occur.

Note that p remains fixed throughout all n trials. This means that all trials are independent of one another. The random variable x (number of successes) can have any integer value from 0 through n. In general, x successes can be assigned among n repeated trials in $C(n, x)$ different ways, since this is the number of different subsets of x positions among a set of n positions. Also, regardless of which x of the trials result in successes, there will always be x successes and $n - x$ failures, so we multiply x number of ps and $n - x$ number of qs together.

> **Binomial Probability Formula**
>
> When n independent repeated trials occur, where
>
> p = probability of success and q = probability of failure
>
> with p and q $(= 1 - p)$ remaining constant throughout all n trials, the probability of exactly x successes is given by
>
> $$P(x) = C(n,x)p^x q^{n-x}.$$

Tables of binomial probability values are commonly available in statistics texts, for various values of p, often for n ranging up to about 20. Also, computer software packages for statistics will usually do these calculations for you automatically, as will some hand-held calculators. In the following examples, we will use the formula stated above to stress the reasoning behind the calculations.

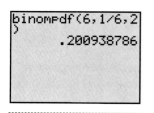

The TI-83 Plus calculator will find the probability discussed in Example 1.

EXAMPLE 1 Find the probability of obtaining exactly three heads on five tosses of a fair coin.

Let heads be "success." Then this is a binomial experiment with $n = 5$, $p = 1/2$, $q = 1/2$, and $x = 3$. By the binomial probability formula,

$$P(3) = C(5,3)\left(\frac{1}{2}\right)^3\left(\frac{1}{2}\right)^2 = 10 \cdot \frac{1}{8} \cdot \frac{1}{4} = \frac{5}{16}.$$

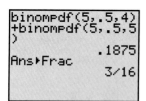

This screen supports the answer in Example 2.

EXAMPLE 2 Find the probability of obtaining exactly two 5s on six rolls of a fair die.

Let 5 be "success." Then $n = 6$, $p = 1/6$, $q = 5/6$, and $x = 2$.

$$P(2) = C(6,2)\left(\frac{1}{6}\right)^2\left(\frac{5}{6}\right)^4 = 15 \cdot \frac{1}{36} \cdot \frac{625}{1296} = \frac{3125}{15{,}552} \approx .201$$

In the case of repeated independent trials, when an event involves more than one specific number of successes, we can employ the binomial probability formula along with the complements or addition rules.

EXAMPLE 3 A couple plans to have 5 children. Find the probability they will have more than 3 girls. (Assume girl and boy babies are equally likely.)

Let a girl be "success." Then $n = 5$, $p = q = 1/2$, and $x > 3$.

$$P(x > 3) = P(x = 4 \text{ or } 5\,)$$
$$= P(4) + P(5)$$
$$= C(5,4)\left(\frac{1}{2}\right)^4\left(\frac{1}{2}\right)^1 + C(5,5)\left(\frac{1}{2}\right)^5\left(\frac{1}{2}\right)^0$$
$$= 5 \cdot \frac{1}{16} \cdot \frac{1}{2} + 1 \cdot \frac{1}{32} \cdot 1$$
$$= \frac{5}{32} + \frac{1}{32} = \frac{6}{32} = \frac{3}{16} = .1875$$

This screen supports the answer in Example 3.

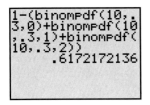

This screen supports the answer in Example 4.

EXAMPLE 4 Scott Davidson, a baseball player, has a well-established career batting average of .300. In a brief series with a rival team, Scott will bat 10 times. Find the probability that he will get more than two hits in the series.

This "experiment" involves $n = 10$ repeated Bernoulli trials, with probability of success (a hit) given by $p = .3$ (which implies $q = 1 - .3 = .7$). Since, in this case, "more than 2" means "3 or 4 or 5 or 6 or 7 or 8 or 9 or 10" (which is eight different possibilities), it will be less work to apply the complements rule.

$$\begin{aligned}
P(x > 2) &= 1 - P(x \leq 2) &&\text{Complements rule}\\
&= 1 - P(x = 0 \text{ or } 1 \text{ or } 2) &&\text{Only three different possibilities}\\
&= 1 - [P(0) + P(1) + P(2)] &&\text{Special addition rule}\\
&= 1 - [C(10,0)(.3)^0(.7)^{10} &&\text{Binomial probability formula}\\
&\quad + C(10,1)(.3)^1(.7)^9 + C(10,2)(.3)^2(.7)^8]\\
&\approx 1 - [.0282 + .1211 + .2335]\\
&= 1 - .3828\\
&= .6172
\end{aligned}$$

12.4 EXERCISES

For Exercises 1–24, give all numerical answers as common fractions reduced to lowest terms. For Exercises 25–59, give all numerical answers to three decimal places.

Coin Tossing If three fair coins are tossed, find the probability of each of the following numbers of heads.

1. 0
2. 1
3. 2
4. 3
5. 1 or 2
6. at least 1
7. no more than 1
8. fewer than 3

9. *Gender in Sequences of Babies* Assuming boy and girl babies are equally likely, find the probability that a family with three children will have exactly two boys.

10. Pascal's triangle was shown in Table 5 of Section 11.4. Explain how the probabilities in Exercises 1–4 above relate to row 3 of the "triangle." (Recall that we referred to the topmost row of the triangle as "row number 0," and to the leftmost entry of each row as "entry number 0.")

11. Generalize the pattern in Exercise 10 to complete the following statement. If n fair coins are tossed, the probability of exactly x heads is the fraction whose numerator is entry number ____ of row number ____ in Pascal's triangle, and whose denominator is the sum of the entries in row number ____ .

Binomial Probability Applied to Tossing Coins Use the pattern noted in Exercises 10 and 11 to find the probabilities of the following numbers of heads when seven fair coins are tossed.

12. 0
13. 1
14. 2
15. 3
16. 4
17. 5
18. 6
19. 7

Binomial Probability Applied to Rolling Dice A fair die is rolled three times. A 4 is considered "success," while all other outcomes are "failures." Find the probability of each of the following numbers of successes.

20. 0
21. 1
22. 2
23. 3

24. Exercises 10 and 11 established a way of using Pascal's triangle rather than the binomial probability formula to find probabilities of different numbers of successes in coin-tossing experiments. Explain why the same process would not work for Exercises 20–23.

For n repeated independent trials, with constant probability of success p for all trials, find the probability of exactly x successes in each of the following cases.

25. $n = 5$, $p = 1/3$, $x = 4$
26. $n = 10$, $p = .7$, $x = 5$
27. $n = 20$, $p = 1/8$, $x = 2$
28. $n = 30$, $p = .6$, $x = 22$

For Exercises 29–31, refer to Example 4 in the text.

29. *Batting Averages in Baseball* Does Scott's probability of a hit really remain constant at exactly .300 through all ten times at bat? Explain your reasoning.

30. *Batting Averages in Baseball* If Scott's batting average is exactly .300 going into the series, and that value is based on exactly 1200 career hits out of 4000 previous times at bat, what is the greatest his average could possibly be (to three decimal places) when he goes up to bat the tenth time of the series? What is the least his average could possibly be when he goes up to bat the tenth time of the series?

31. Do you think the use of the binomial probability formula was justified in Example 4, even though p is not strictly constant? Explain your reasoning.

Random Selection of Answers on a Multiple-choice Test Yesha Brill is taking a ten-question multiple-choice test for which each question has three answer choices, only one of which is correct. Yesha decides on answers by rolling a fair die and marking the first answer choice if the die shows 1 or 2, the second if it shows 3 or 4, and the third if it shows 5 or 6. Find the probability of each of the following events.

32. exactly four correct answers
33. exactly seven correct answers
34. fewer than three correct answers
35. at least seven correct answers

Side Effects of Prescription Drugs It is known that a certain prescription drug produces undesirable side effects in 30 percent of all patients who use it. Among a random sample of eight patients using the drug, find the probability of each of the following events.

36. None have undesirable side effects.
37. Exactly one has undesirable side effects.
38. Exactly two have undesirable side effects.
39. More than two have undesirable side effects.

Likelihood of Capable Students Attending College In a certain state, it has been shown that only 50 percent of the high school graduates who are capable of college work actually enroll in colleges. Find the probability that, among nine capable high school graduates in this state, each of the following numbers will enroll in college.

40. exactly 4
41. from 4 through 6
42. none
43. all 9

44. *Student Ownership of Personal Computers* At a large midwestern university, 80 percent of all students have their own personal computers. If five students at that university are selected at random, find the probability that exactly three of them have their own computers.

45. *Frost Survival Among Orange Trees* If it is known that 65 percent of all orange trees will survive a hard frost, then what is the probability that at least half of a group of six trees will survive such a frost?

46. *Rate of Favorable Media Coverage of an Incumbent President* An extensive survey revealed that, during a certain presidential election campaign, 64 percent of the political columns in a certain group of major newspapers were favorable to the incumbent president. If a sample of fifteen of these columns is selected at random, what is the probability that exactly ten of them will be favorable?

In the case of n independent repeated Bernoulli trials, the formula developed in this section gives the probability of exactly x successes. Sometimes, however, we are interested not in the event that exactly x successes will occur in n trials, but rather the event that the first success will occur on the xth trial. For example, consider the probability that, in a series of coin tosses, the first success (head) will occur on the fourth toss. This implies a failure first, then a second failure, then a third failure, and finally a success. Symbolically, the event is F_1 and F_2 and F_3 and S_4. The probability of this sequence of outcomes is $q \cdot q \cdot q \cdot p$, or $q^3 \cdot p$. In general, if the probability of success stays constant at p (which implies a probability of failure of $q = 1 - p$), then the probability that the first success will occur on the xth trial can be computed as follows.

$$P(F_1 \text{ and } F_2 \text{ and } \ldots \text{ and } F_{x-1} \text{ and } S_x) = q^{x-1} \cdot p$$

47. Explain why, in the formula above, there is no combination factor, such as the $C(n, x)$ in the binomial probability formula.

48. *Gender in Sequences of Babies* Assuming male and female babies are equally likely, find the probability that a family's fourth child will be their first daughter.

49. *Getting Caught when Speeding* If the probability of getting caught when you exceed the speed limit on a certain stretch of highway is .38, find the probability that the first time you will get caught is the fourth time that you exceed the speed limit.

50. *Union Members in an Industry* If 30 percent of all workers in a certain industry are union members, and workers in this industry are selected successively at random, find the probability that the first union member to occur will be on the third selection.

51. *Aborted Rocket Launches* If a certain type of rocket always has a four percent chance of an aborted launching, find the probability that the first launch to be aborted is the 20th launch.

Harvey is standing on the corner tossing a coin. He decides he will toss it 10 times, each time walking 1 block north if it lands heads up, and 1 block south if it lands tails up. In each of the following exercises, find the probability that he will end up in the indicated location. (In each case, ask how many successes, say heads, would be required and use the binomial formula.

*Some ending positions may not be possible with 10 tosses.) The random process involved here illustrates what we call a **random walk**. It is a simplified model of Brownian motion, mentioned on page 676. Further applications of the idea of a random walk are found in Section 12.6.*

52. 10 blocks north of his corner

53. 6 blocks north of his corner

54. 6 blocks south of his corner

55. 5 blocks south of his corner

56. 2 blocks north of his corner

57. at least 2 blocks north of his corner

58. at least 2 blocks from his corner

59. on his corner

12.5 Expected Value

x	$P(x)$
1	.05
2	.10
3	.20
4	.40
5	.10
6	.15

The probability distribution at the side, from Example 6 of Section 12.2, shows the probabilities assigned by Sara to the various lengths of time her homework may take on a given night. If Sara's friend Art asks her how many hours her studies will take, what would be her best guess? Six different time values are possible, with some more likely than others. One thing Sara could do is calculate a "weighted average" by multiplying each possible time value by its probability and then adding the six products.

$$1(.05) + 2(.10) + 3(.20) + 4(.40) + 5(.10) + 6(.15)$$
$$= .05 + .20 + .60 + 1.60 + .50 + .90$$
$$= 3.85$$

Thus 3.85 hours is the **expected value** (or the **mathematical expectation**) of the quantity of time to be spent. Since the original time values in the table were rounded to the nearest hour, the expected value should also be rounded, to 4 hours. A general definition of expected value follows.

> **Expected Value**
>
> If a random variable x can have any of the values $x_1, x_2, x_3, \ldots, x_n$, and the corresponding probabilities of these values occurring are $P(x_1)$, $P(x_2), P(x_3), \ldots, P(x_n)$, then the **expected value of x** is
>
> $$x_1 \cdot P(x_1) + x_2 \cdot P(x_2) + x_3 \cdot P(x_3) + \cdots + x_n \cdot P(x_n).$$

EXAMPLE 1 Find the expected number of boys for a three-child family (that is, the expected value of the number of boys).

The sample space for this experiment is

$$S = \{ggg, ggb, gbg, bgg, gbb, bgb, bbg, bbb\}.$$

The probability distribution is shown in the table below, along with the products and their sum, which gives the expected value.

Number of Boys x	Probability $P(x)$	Product $x \cdot P(x)$
0	1/8	0
1	3/8	3/8
2	3/8	6/8
3	1/8	3/8

$$\text{Expected value: } \frac{12}{8} = \frac{3}{2}$$

The expected number of boys is 3/2, or 1.5. This result seems reasonable. Since boys and girls are equally likely, "half" the children are expected to be boys.

Notice that the expected value for the number of boys in the family is itself an impossible value. So, as in this example, the expected value itself may never occur. Many times the expected value actually cannot occur; it is only a kind of long run average of the various values that could occur. (For more information on "averages," see Section 13.2.) If we recorded the number of boys in many different three-child families, then by the law of large numbers, as the number of observed families increased, the observed average number of boys should approach the expected value.

Example 1 did not involve money, but many uses of expected value do involve monetary expectations. Common applications are in gambling and in business (for example: investment decisions, contract bidding, and insurance policy pricing).

EXAMPLE 2 A player pays $3 to play the following game: He tosses three fair coins and receives back "payoffs" of $1 if he tosses no heads, $2 for one head, $3 for two heads, and $4 for three heads. Find the player's expected net winnings for this game.

Display the information in a table. (Notice that, for each possible event, "net winnings" are "gross winnings" (payoff) minus cost to play.) Probabilities are derived from the sample space

$$S = \{ttt, htt, tht, tth, hht, hth, thh, hhh\}.$$

Number of Heads	Payoff	Net Winnings x	Probability $P(x)$	Product $x \cdot P(x)$
0	$1	−$2	1/8	−$2/8
1	2	−1	3/8	−3/8
2	3	0	3/8	0
3	4	1	1/8	1/8

$$\text{Expected value: } -\$\frac{1}{2} = -\$.50$$

The expected net loss of 50 cents is a long-run average only. On any particular play of this game, the player would lose $2 or lose $1 or break even or win $1. Over a long series of plays, say 100, there would be some wins and some losses, but the total net result would likely be around a $100 \cdot (\$.50) = \$50$ *loss*.

A game in which the expected net winnings are zero is called a **fair game.** The game in Example 2 has negative expected net winnings, so it is unfair against the player. A game with positive expected net winnings is unfair in favor of the player.

EXAMPLE 3 The $3 cost to play the game of Example 2 makes the game unfair against the player (since the player's expected net winnings are negative). What cost would make this a fair game?

We already computed, in Example 2, that the $3 cost to play resulted in an expected net loss of $.50. Therefore we can conclude that the $3 cost was 50 cents too high. A fair cost to play the game would then be $3 - \$.50 = \$2.50$.

The result in Example 3 can be verified as follows. First disregard the cost to play and find the expected *gross* winnings (by summing the products of payoff times probability).

$$\$1 \cdot \frac{1}{8} + \$2 \cdot \frac{3}{8} + \$3 \cdot \frac{3}{8} + \$4 \cdot \frac{1}{8} = \frac{\$20}{8} = \$2.50$$

Since the long-run expected gross winnings (payoff) are $2.50, this amount is a fair cost to play.

EXAMPLE 4 In a certain state lottery, a player chooses three digits, in a specific order. (Leading digits may be 0, so numbers such as 028 and 003 are legitimate entries.) The lottery operators randomly select a three-digit sequence, and any player matching their selection receives a payoff of $600. What is a fair cost to play this game?

In this case, no cost has been proposed, so we have no choice but to compute expected gross winnings. The probability of selecting all three digits correctly is $(1/10) \cdot (1/10) \cdot (1/10) = 1/1000$, and the probability of not selecting all three correctly is $1 - 1/1000 = 999/1000$. The expected gross winnings are

$$\$600 \cdot \frac{1}{1000} + \$0 \cdot \frac{999}{1000} = \$.60.$$

Thus the fair cost to play this game is 60 cents. (In fact, the lottery charges $1 to play, so players should expect to lose 40 cents per play *on the average*.)

Of course, state lotteries must be unfair against players because they are designed to help fund benefits (such as the state's school system) as well as to cover administrative costs and certain other expenses. Among people's reasons for playing may be a willingness to support such causes, but most people undoubtedly play for the chance to "beat the odds" and be one of the few net winners.

Gaming casinos are major business enterprises, by no means designed to break even; the games they offer are always unfair in favor of the house. The bias does not need to be great, however, since even relatively small average losses per player multiplied by large numbers of players can result in huge profits for the house.

Roulette ("little wheel") was invented in France in the seventeenth or early eighteenth century. It has been a featured game of chance in the gambling casino of Monte Carlo.

The disk is divided into red and black alternating compartments, numbered 1 to 36 (but not in that order). There is a compartment also for 0 (and for 00 in the United States). In roulette, the wheel is set in motion, and an ivory ball is thrown into the bowl opposite to the direction of the wheel. When the wheel stops, the ball comes to rest in one of the compartments—the number and color determine who wins.

The players bet against the banker (person in charge of the pool of money) by placing money or equivalent chips in spaces on the roulette table corresponding to the wheel's colors or numbers. Bets can be made on one number or several, on odd or even, on red or black, or on combinations. The banker pays off according to the odds against the particular bet(s). For example, the classic payoff for a winning single number is $36 for each $1 bet.

EXAMPLE 5 One simple type of *roulette* is played with an ivory ball and a wheel set in motion. The wheel contains thirty-eight compartments. Eighteen of the compartments are black, eighteen are red, one is labeled "zero," and one is labeled "double zero." (These last two are neither black nor red.) In this case, assume the player places $1 on either red or black. If the player picks the correct color of the compartment in which the ball finally lands, the payoff is $2; otherwise the payoff is zero. Find the expected net winnings.

By the expected value formula, expected net winnings are

$$(\$1)\frac{18}{38} + (-\$1)\frac{20}{38} = -\$\frac{1}{19}.$$

The expected net loss here is $1/19, or about 5.3¢, per play.

In business and other areas, decisions must often be made in the face of uncertain conditions. Expected value offers one reasonable approach to these situations.

EXAMPLE 6 Michael Crenshaw, a lumber wholesaler, is considering the purchase of a (railroad) carload of varied dimensional lumber. Drawing on his considerable experience, Michael calculates that the probabilities of reselling the load for $10,000, $9000, or $8000 are .22, .33, and .45, respectively. In order to ensure an *expected* profit of at least $3000, how much can Michael afford to pay for the load?

The expected revenue (or income) from resales can be found as follows.

Income x	Probability $P(x)$	Product $x \cdot P(x)$
$10,000	.22	$2200
9000	.33	2970
8000	.45	3600

Expected revenue: $8770

In general, we have the relationship

$$\text{profit} = \text{revenue} - \text{cost}.$$

Therefore, in terms of expectations,

$$\text{expected profit} = \text{expected revenue} - \text{cost}.$$

So $3000 = $8770 − cost or cost = $8770 − $3000 = $5770. Michael can pay up to $5770 and still maintain an expected profit of at least $3000.

EXAMPLE 7 Todd Hall has $5000 to invest and will commit the whole amount, for six months, to one of three technology stocks. A number of uncertainties could affect the prices of these stocks, but Todd is confident, based on his research, that one of only several possible profit scenarios will prove true of each one at the end of the six-month period. His complete analysis is shown in the following table. (For example, stock *ABC* could lose $400, gain $800, or gain $1500.)

12.5 Expected Value

Company ABC		Company RST		Company XYZ	
Profit or Loss x	Probability $P(x)$	Profit or Loss x	Probability $P(x)$	Profit or Loss x	Probability $P(x)$
−$400	.2	$500	.8	$0	.4
800	.5	1000	.2	700	.3
1500	.3			1200	.1
				2000	.2

Find the expected profit (or loss) for each of the three stocks and select Todd's optimum choice based on these calculations.

Apply the expected value formula.

ABC: $-\$400 \cdot (.2) + \$800 \cdot (.5) + \$1500 \cdot (.3) = \$770$

RST: $\$500 \cdot (.8) + \$1000 \cdot (.2) = \$600$

XYZ: $\$0 \cdot (.4) + \$700 \cdot (.3) + \$1200 \cdot (.1) + \$2000 \cdot (.2) = \$730$

The largest expected profit is $770. By this analysis, Todd should invest the money in stock *ABC*.

Of course, by investing in stock *ABC*, Todd may in fact *lose* $400 over the six months. The "expected" return of $770 is only a long-run average over many identical situations. Since this particular investment situation may never occur again, you may argue that using expected values is not the best approach for Todd to use.

One possible alternative would be to adopt the view of an optimist, who might ignore the various probabilities and just hope for the best possibility associated with each choice and make a decision accordingly. Or, on the other hand, a pessimist may assume the worst case probably will occur, and make a decision in accordance with that view.

The first Silver Dollar Slot Machine was fashioned in 1929 by the Fey Manufacturing Company, San Francisco, inventors of the 3-reel, automatic payout machine (1895).

EXAMPLE 8 Decide which stock of Example 7 Todd would pick in each of the following cases.

(a) He is an optimist.

Disregarding the probabilities, he would focus on the best that could possibly happen with each stock. Since *ABC* could return as much as $1500, *RST* as much as $1000, and *XYZ* as much as $2000, the optimum is $2000. He would buy stock *XYZ*.

(b) He is a pessimist.

In this case, he would focus on the worst possible cases. Since *ABC* might return as little as −$400 (a $400 loss), *RST* as little as $500, and *XYZ* as little as $0, he would buy stock *RST* (the best of the three worst cases).

FOR FURTHER THOUGHT

Expected Value of Games of Chance

Slot machines are a popular game for those who want to lose their money with very little mental effort. We cannot calculate an expected value applicable to all slot machines since payoffs vary from machine to machine. But we can calculate the "typical expected value."

This Cleveland Indians fan hit four 7s in a row on a progressive nickel slot machine at the Sands Casino in Las Vegas in 1988.

A player operates a slot machine by pulling a handle after inserting a coin or coins. Reels inside the machine then rotate, and come to rest in some random order. Assume that three reels show the pictures listed in Table 8. For example, of the 20 pictures on the first reel, 2 are cherries, 5 are oranges, 5 are plums, 2 are bells, 2 are melons, 3 are bars, and 1 is the number 7.

TABLE 8

Pictures	Reels		
	1	2	3
Cherries	2	5	4
Oranges	5	4	5
Plums	5	3	3
Bells	2	4	4
Melons	2	1	2
Bars	3	2	1
7s	1	1	1
Totals	20	20	20

A picture of cherries on the first reel, but not on the second, leads to a payoff of 3 coins (*net* winnings: 2 coins); a picture of cherries on the first two reels, but not the third, leads to a payoff of 5 coins (*net* winnings: 4 coins). All other possible payoffs are as shown in Table 9.

TABLE 9 Calculating Expected Loss on a Three-reel Slot Machine

Winning Combinations	Number of Ways	Probability	Number of Coins Received	Net Winnings (in coins)	Probability Times Winnings
1 cherry (on first reel)	$2 \times 15 \times 20 = 600$	600/8000	3	2	1200/8000
2 cherries (on first two reels)	$2 \times 5 \times 16 = 160$	160/8000	5	4	640/8000
3 cherries	$2 \times 5 \times 4 = 40$	40/8000	10	9	360/8000
3 oranges	$5 \times 4 \times 5 = 100$	100/8000	10	9	900/8000
3 plums	$5 \times 3 \times 3 = 45$	45/8000	14	13	585/8000
3 bells	$2 \times 4 \times 4 = 32$	32/8000	18	17	544/8000
3 melons (jackpot)	$2 \times 1 \times 2 = 4$	4/8000	100	99	396/8000
3 bars (jackpot)	$3 \times 2 \times 1 = 6$	6/8000	200	199	1194/8000
3 7s (jackpot)	$1 \times 1 \times 1 = 1$	1/8000	500	499	499/8000
Totals	988				6318/8000

Since, according to Table 8, there are 2 ways of getting cherries on the first reel, 15 ways of *not* getting cherries on the second reel, and 20 ways of getting anything on the third reel, we have a total of $2 \times 15 \times 20 = 600$ ways of getting a net payoff of 2. Since there are 20 pictures per reel, there are a total of $20 \times 20 \times 20 = 8000$ possible outcomes. Hence, the probability of receiving a net payoff of 2 coins is 600/8000. Table 9 takes into account all *winning* outcomes, with the necessary products for computing expectation added in the last column. However, since a *nonwinning* outcome can occur in $8000 - 988 = 7012$ ways (with winnings of -1 coin), the product $(-1) \cdot 7012/8000$ must also be included. Hence, the expected value of this particular slot machine is

$$\frac{6318}{8000} + (-1) \cdot \frac{7012}{8000} \approx -.087 \text{ coin}.$$

On a machine costing one dollar per play, the expected *loss* (per play) is about

$$(.087)(1 \text{ dollar}) = 8.7 \text{ cents}.$$

Actual slot machines vary in expected loss per dollar of play. But author Hornsby was able to beat a Las Vegas slot machine in 1988. See the photo.

Table 10 comes from an article by Andrew Sterrett in *The Mathematics Teacher* (March 1967), in which he discusses rules for various games of chance and calculates their expected values. He uses expected values to find expected times it would take to lose $1000 if you played continually at the rate of $1 per play and one play per minute.

For Group Discussion

1. Explain why the entries of the "Net winnings" column of Table 9 are all one fewer than the corresponding entries of the "Number of coins received" column.
2. Verify all values in the "Number of ways" column of Table 9. (Refer to Table 8.)
3. In order to make your money last as long as possible in a casino, which game should you play?

TABLE 10 Expected Time to Lose $1000

Game	Expected Value	Days	Hours	Minutes
Roulette (with one 0)	−$.027	25	16	40
Roulette (with 0 and 00)	−$.053	13	4	40
Chuck-a-luck	−$.079	8	19	46
Keno (one number)	−$.200	3	11	20
Numbers	−$.300	2	7	33
Football pool (4 winners)	−$.375	1	20	27
Football pool (10 winners)	−$.658	1	1	19

12.5 EXERCISES

 1. Explain in words what is meant by "expected value of a random variable."

 2. Explain what a couple means by the statement, "We expect to have 1.5 sons."

3. *Tossing Coins* Five fair coins are tossed. Find the expected number of heads.

4. *Drawing Cards* Two cards are drawn, with replacement, from a standard 52-card deck. Find the expected number of diamonds.

Expected Winnings in a Die-rolling Game A certain game consists of rolling a single fair die and pays off as follows: $3 for a 6, $2 for a 5, $1 for a 4, and no payoff otherwise.

5. Find the expected winnings for this game.

6. What is a fair price to pay to play this game?

Expected Winnings in a Die-rolling Game For Exercises 7 and 8, consider a game consisting of rolling a single fair die, with payoffs as follows. If an even number of spots turns up, you receive that many dollars. But if an odd number of spots turns up, you must pay that many dollars.

7. Find the expected net winnings of this game.

8. Is this game fair, or unfair against the player, or unfair in favor of the player?

9. *Expected Winnings in a Coin-tossing Game* A certain game involves tossing 3 fair coins, and it pays 10¢ for 3 heads, 5¢ for 2 heads, and 3¢ for 1 head. Is 5¢ a fair price to pay to play this game? (That is, does the 5¢ cost to play make the game fair?)

10. *Expected Winnings in Roulette* In a form of roulette slightly different from that in Example 5, a more generous management supplies a wheel having only thirty-seven compartments, with eighteen red, eighteen black, and one zero. Find the expected net winnings if you bet on red in this game.

11. *Expected Number of Absences in a Math Class* In a certain mathematics class, the probabilities have been empirically determined for various numbers of absences on any given day. These values are shown in the table below. Find the expected number of absences on a given day. (Give the answer to two decimal places.)

Number absent	0	1	2	3	4
Probability	.12	.32	.35	.14	.07

12. *Expected Profit of an Insurance Company* An insurance company will insure a $100,000 home for its total value for an annual premium of $330. If the company spends $20 per year to service such a policy, the probability of total loss for such a home in a given year is .002, and you assume that either total loss or no loss will occur, what is the company's expected annual gain (or profit) on each such policy?

Profits from a College Foundation Raffle A college foundation raises funds by selling raffle tickets for a new car worth $36,000.

13. If 600 tickets are sold for $120 each, determine
    (a) the expected *net* winnings of a person buying one of the tickets,
    (b) the total profit for the foundation, assuming they had to purchase the car,
    (c) the total profit for the foundation, assuming the car was donated.

14. For the raffle described above, if 720 tickets are sold for $120 each, determine
    (a) the expected *net* winnings of a person buying one of the tickets,
    (b) the total profit for the foundation, assuming they had to purchase the car,
    (c) the total profit for the foundation, assuming the car was donated.

Winnings and Profits of a Raffle Five thousand raffle tickets are sold. One first prize of $1000, two second prizes of $500 each, and five third prizes of $100 each are to be awarded, with all winners selected randomly.

15. If you purchased one ticket, what are your expected winnings?

16. If you purchased two tickets, what are your expected winnings?

17. If the tickets were sold for $1 each, how much profit goes to the raffle sponsor?

18. *Expected Sales at a Theater Snack Bar* A children's theater found in a random survey that 65 customers bought one snack bar item, 40 bought two items, 26 bought three items, 14 bought four items, and 18 avoided the snack bar altogether. Use this information to find the expected number of snack bar items per customer. (Round your answer to the nearest tenth.)

19. *Expected Number of Children to Attend an Amusement Park* An amusement park, considering adding some new attractions, conducted a study over several typical days and found that, of 10,000 families entering the park, 1020 brought just one child (defined as younger than age twelve), 3370 brought two children, 3510 brought three children, 1340 brought four children, 510 brought five children, 80 brought six children, and 170 brought no

children at all. Find the expected number of children per family attending this park. (Round your answer to the nearest tenth.)

20. *Expected Sums of Randomly Selected Numbers* Five cards are numbered 1 through 5. Two of these cards are chosen randomly (without replacement), and the numbers on them are added. Find the expected value of this sum.

21. *Prospects for Electronics Jobs in a City* In a certain California city, projections for the next year are that there is a 20% chance that electronics jobs will increase by 1200, a 50% chance that they will increase by 500, and a 30% chance that they will decrease by 800. What is the expected change in the number of electronics jobs in that city in the next year?

22. *Expected Winnings in Keno* In one version of the game *keno*, the house has a pot containing 80 balls, numbered 1 through 80. A player buys a ticket for $1 and marks one number on it (from 1 to 80). The house then selects 20 of the 80 numbers at random. If the number selected by the player is among the 20 selected by the management, the player is paid $3.20. Find the expected net winnings for this game.

23. Refer to Examples 7 and 8. Considering the three different approaches (expected values, optimist, and pessimist), which one seems most reasonable to you, and why?

Contractor Decisions Based on Expected Profits Kimberli Brownlee, a commercial building contractor, will commit her company to one of three projects depending on her analysis of potential profits or losses as shown here.

Project A		Project B		Project C	
Profit or Loss x	Probability $P(x)$	Profit or Loss x	Probability $P(x)$	Profit or Loss x	Probability $P(x)$
$40,000	.10	$0	.20	$60,000	.65
180,000	.60	210,000	.35	340,000	.35
250,000	.30	290,000	.45		

Determine which project Kimberli should choose according to each of the following approaches (Exercises 24–26).

24. expected values

25. the optimist viewpoint

26. the pessimist viewpoint

Expected Winnings in a Game Show A game show contestant is offered the option of receiving a computer system worth $2300, or accepting a chance to win either a luxury vacation worth $5000 or a boat worth $8000. The contestant's probabilities of winning the vacation or the boat are .20 and .15, respectively.

27. If the contestant were to turn down the computer system and go for one of the other prizes, what would be the expected winnings?

28. Purely in terms of monetary value, what is the contestant's wiser choice?

Expected Values in Business Accounts The table below illustrates how a salesman for Levi Strauss & Co. rates his accounts by considering the existing volume of each account plus potential additional volume.*

1	2	3	4	5	6	7
Account Number	Existing Volume	Potential Additional Volume	Probability of Getting Additional Volume	Expected Value of Additional Volume	Existing Volume plus Expected Value of Additional Volume	Classification
1	$15,000	$10,000	.25	$2500	$17,500	
2	40,000	0	—	—	40,000	
3	20,000	10,000	.20	2000		
4	50,000	10,000	.10	1000		
5	5000	50,000	.50			
6	0	100,000	.60			
7	30,000	20,000	.80			

Use the table to work Exercises 29–32.

29. Compute the missing expected values in column 5.

30. Compute the missing amounts in column 6.

31. In column 7, classify each account according to this scheme: Class A if the column 6 value is $55,000 or more; Class B if the column 6 value is at least $45,000 but less than $55,000; Class C if the column 6 value is less than $45,000.

32. Considering all seven of this salesman's accounts, compute the total additional volume he can "expect" to get.

33. *Expected Winnings in Keno* Recall that in the game keno of Exercise 22, the house randomly selects 20 numbers from the counting numbers 1–80. In the variation called 6-spot keno, the player pays 60¢ for his ticket and marks 6 numbers of his choice. If the 20 numbers selected by the house contain at least 3 of those chosen by the player, he gets a payoff according to this scheme.

3 of the player's numbers among the 20	$.35
4 of the player's numbers among the 20	2.00
5 of the player's numbers among the 20	60.00
6 of the player's numbers among the 20	1250.00

Find the player's expected net winnings in this game. (*Hint:* The four probabilities required here can be found using combinations (Section 11.3), the fundamental counting principle (Section 11.2), and the theoretical probability formula (Section 12.1).)

---

*This information was provided by James McDonald of Levi Strauss & Co., San Francisco.

12.6 Estimating Probabilities by Simulation

An important area within probability theory is the process called **simulation**. It is possible to study a complicated, or unclear, phenomenon by *simulating,* or imitating, it with a simpler phenomenon involving the same basic probabilities. For example, recall from Section 12.1 Mendel's discovery that when two Rr pea plants (red-flowered but carrying both red and white genes) are crossed, the offspring will have

12.6 Estimating Probabilities by Simulation

		Second Parent	
		R	r
First Parent	R	RR	Rr
	r	rR	rr

red flowers if an R gene is received from either parent, or from both. This is because red is dominant and white is recessive. Table 3, reproduced here in the margin, shows that three of the four equally likely possibilities result in red-flowered offspring.

Now suppose we want to know (or at least approximate) the probability that three offspring in a row will have red flowers. It is much easier (and quicker) to toss coins than to cross pea plants. And the equally likely outcomes, heads and tails, can be used to simulate the transfer of the equally likely genes, R and r. If we toss two coins, say a nickel and a penny, then we can interpret the results as follows.

hh $\Rightarrow$ RR $\Rightarrow$ red gene from first parent and red gene from second parent
$\Rightarrow$ red flowers

ht $\Rightarrow$ Rr $\Rightarrow$ red gene from first parent and white gene from second parent
$\Rightarrow$ red flowers

th $\Rightarrow$ rR $\Rightarrow$ white gene from first parent and red gene from second parent
$\Rightarrow$ red flowers

tt $\Rightarrow$ rr $\Rightarrow$ white gene from first parent and white gene from second parent
$\Rightarrow$ white flowers

Simulation methods, also called "Monte Carlo" methods, have been successfully used in many areas of scientific study for nearly a century. Most practical applications require huge numbers of random digits, so computers are used to produce them. A computer, however, cannot toss coins. It must use an algorithmic process, programmed into the computer, which is called a *random number generator*. It is very difficult to avoid all nonrandom patterns in the results, so the digits produced are called "pseudorandom" numbers. They must pass a battery of tests of randomness before being "approved for use."

Computer scientists and physicists have been encountering unexpected difficulties with even the most sophisticated random number generators. As reported in the December 19, 1992, issue of *Science News* (page 422), it seems that a random number generator can pass all the tests, and work just fine for some simulation applications, but then produce faulty answers when used with a different simulation. Therefore, random number generators apparently cannot be approved for all uses in advance, but must be carefully checked along with each new simulation application proposed.

Although nothing is certain for a few tosses, the law of large numbers indicates that larger and larger numbers of tosses should become better and better indicators of general trends in the genetic process.

EXAMPLE 1 Toss two coins fifty times and use the results to approximate the probability that the crossing of Rr pea plants will produce three successive red-flowered offspring.

We actually tossed two coins 50 times and got the following sequence.

th, hh, th, tt, th, hh, ht, th, ht, th, hh, hh,
tt, th, hh, ht, ht, ht, ht, th, hh, hh, hh, tt,
ht, tt, hh, ht, ht, hh, tt, tt, tt, th, tt, tt, hh,
ht, ht, ht, hh, tt, th, hh, tt, hh, ht, tt, tt, tt

By the color interpretation described above, this gives the following sequence of flower colors in the offspring.

red–red–red–white–red–red–red–red–red–red–red–white–
red–red–red–red–red–red–red–red–red–red–white–red–white–
red–red–red–red–white–white–white–red–white–white–red–red–
red–red–red–white–red–red–white–red–red–white–white–white

We now have an experimental list of 48 sets of three successive plants, the 1st, 2nd, and 3rd entries, then the 2nd, 3rd, and 4th entries, and so on. Do you see why there are 48 in all? Now we just count up the number of these sets of three that are "red-red-red." Since there are 20 of those, our empirical probability of three successive red offspring, obtained through simulation, is $20/48 = 5/12$, or about .417. By applying the special multiplication rule (since all outcomes here are independent of one another), we find that the theoretical value is $(3/4)^3 = 27/64$, or about .422, so our approximation obtained by simulation is very close. ■

In human births boys and girls are (essentially) equally likely. Therefore, an individual birth can be simulated by tossing a fair coin, letting a head correspond to a girl and a tail to a boy.

EXAMPLE 2 A sequence of 40 actual coin tosses produced the results below.

$$\text{bbggb, gbbbg, gbgbb, bggbg, bbbbg, gbbgg, gbbgg, bgbbg}$$

(For every head we have written g, for girl; for every tail, b, for boy.) Refer to this sequence to answer the following questions.

(a) How many pairs of two successive births are represented by the above sequence?

Beginning with the 1st–2nd pair and ending with the 39th–40th pair, there are 39 pairs.

(b) How many of those pairs consist of both boys?

By observing the sequence, we count 11 pairs consisting of both boys.

(c) Find the empirical probability, based on this simulation, that two successive births both will be boys. Give your answer to three decimal places.

Utilizing parts (a) and (b), we have $11/39 \approx .282$.

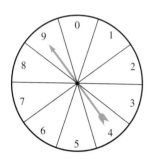

FIGURE 11

TABLE 11

51592	73219
77876	55707
36500	48007
40571	65191
04822	06772
53033	94928
92080	15709
01587	39922
36006	96365
63698	14655
17297	65587
22841	76905
91979	12369
96480	54219
74949	89329
76896	90060
47588	06975
45521	05050
02472	69774
55184	78351
40177	11464
84861	84086
86937	51497
20931	12307
22454	68009

Another way to simulate births is to generate a random sequence of digits, perhaps interpreting even digits as girls and odd digits as boys. The digits might be generated by spinning the spinner in Figure 11. It turns out that many kinds of phenomena can be simulated using random digits, so we can save lots of effort by using the spinner to obtain a table of random digits, like in Table 11, and then use that table to carry out any simulation experiment that is needed. Notice that the 250 random digits in Table 11 have been grouped conveniently so that we can easily follow down a column or across a row.

EXAMPLE 3 A couple plans to have five children. Use random number simulation to estimate the probability they will have more than three boys.

Let each sequence of five digits, as they appear in Table 11, represent a family with five children, and (arbitrarily) associate odd digits with boys, even digits with girls. (Recall that 0 is even.) Verify that, of the fifty families simulated, ten of them have more than 3 boys (4 boys or 5 boys). Therefore the estimation (empirical value) is

$$P(\text{more than 3 boys}) = 10/50 = .20.$$

The theoretical value for the probability estimated in Example 3, above, would be the same as that obtained in Example 3 of Section 12.4. It was .1875. So our estimate above was fairly close. In light of the law of large numbers, a larger sampling of random digits (more than 50 simulated families) would most likely yield a closer approximation. Very extensive tables of random digits are readily available in statistical research publications. Computers can also be programmed to generate sequences of "pseudorandom" digits, which serve the same purposes. In most simulation experiments, much larger samples than we are using here are necessary to obtain reliable results. But the examples here adequately illustrate the procedure.

EXAMPLE 4 Use random number simulation to estimate the probability that two cards drawn from a standard deck with replacement will both be of the same suit.

Use this correspondence: 0 and 1 mean clubs, 2 and 3 mean diamonds, 4 and 5 mean hearts, 6 and 7 mean spades, 8 and 9 are disregarded. Now read down the columns of Table 11. Suppose we (arbitrarily) use the first digit of each five-digit group. The first time from top to bottom gives the sequence

$$5-7-3-4-0-5-0-3-6-1-2-7-7-4-4-0-5-4-2-2.$$

(Five 8s and 9s were omitted.) Starting again at the top, we obtain

$$7-5-4-6-0-1-3-1-6-7-1-5-0-0-6-7-1-5-1-6.$$

(Again, there happened to be five 8s and 9s.) This 40-digit sequence of digits yields the sequence of suits shown next.

hearts–spades–diamonds–hearts–clubs–hearts–clubs–diamonds–spades–
clubs–diamonds–spades–spades–hearts–hearts–clubs–hearts–hearts–
diamonds–diamonds–spades–hearts–hearts–spades–clubs–clubs–
diamonds–clubs–spades–spades–clubs–hearts–clubs–clubs–spades–
spades–clubs–hearts–clubs–spades

Verify that, of the 39 successive pairs of suits (hearts–spades, spades–diamonds, diamonds–hearts, etc.), 9 of them are pairs of the same suit. This makes the estimated probability $9/39 \approx .23$. (For comparison, the theoretical value is .25.)

12.6 EXERCISES

1. *Simulating Pea Plant Reproduction with Coin Tossing* Explain why, in Example 1, fifty tosses of the coins produced only 48 sets of three successive offspring.

2. *Simulating Pea Plant Reproduction with Coin Tossing* Use the sequence of flower colors of Example 1 to approximate the probability that *four* successive offspring all will have red flowers.

3. *Comparing the Likelihoods of Girl and Boy Births* Should the probability of two successive girl births be any different from that of two successive boy births?

4. *Finding Empirical Probability* Simulate 40 births by tossing coins yourself, and obtain an empirical probability for two successive girls.

5. *Simulating Boy and Girl Children with Random Numbers* Use Table 11 to simulate fifty families with three children. Let 0–4 correspond to boys and 5–9 to girls, and use the middle grouping of three digits (159, 787, 650, and so on). Estimate the probability of exactly two boys in a family of three children. Compare the estimation with the theoretical probability of exactly two boys in a family of three children, which is $3/8 = .375$.

Simulating One-and-one Foul Shooting with Random Numbers Exercises 71–73 of Section 12.3 involved one-and-one foul shooting in basketball. Susan, who had a 70% foul-shooting record, had probabilities of scoring 0, 1, or 2 points of .30, .21, and .49, respectively. Use Table 11 (with digits 0–6 representing hit and 7–9 representing miss) to simulate 50 one-and-one shooting opportunities for Susan. Begin at the top of the first column (5, 7, 3, etc., to the bottom), then move to the second column (1, 7, 6, etc.), going until 50 one-and-one opportunities are obtained. (Notice that some "opportunities" involve one shot while others involve two shots.) Keep a tally of the numbers of times 0, 1, and 2 points are scored. From the tally, find the empirical probability that, on a given opportunity, Susan will score as follows. (Round your answers to two decimal places.)

Number of Points	Tally
0	
1	
2	

6. no points   7. 1 point   8. 2 points

Determining the Path of a Random Walk Using a Die and a Coin Exercises 52–59 of Section 12.4 illustrated a simple version of the idea of a "random walk." Atomic particles released in nuclear fission also move in a random fashion. During World War II, John von Neumann and Stanislaw Ulam used simulation with random numbers to study particle motion in nuclear reactions. Von Neumann coined the name "Monte Carlo" for the methods used, and since then the terms "Monte Carlo methods" and "simulation methods" have often been used with very little distinction.

The figure suggests a model for random motion in two dimensions. Assume that a particle moves in a series of 1-unit "jumps," each one in a random direction, any one of 12 equally likely possibilities. One way to choose directions is to roll a fair die and toss a fair coin. The die determines one of the directions 1–6, coupled with heads on the coin. Tails on the coin reverses the direction of the die, so that the die coupled with tails gives directions 7–12. Thus 3h (meaning 3 with the die and heads with the coin) gives direction 3; 3t gives direction 9 (opposite to 3); and so on.

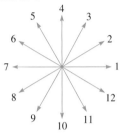

9. Simulate the motion described above with 10 rolls of a die (and tosses of a coin). Draw the 10-jump path you get. Make your drawing accurate enough so you can estimate (by measuring) how far from its starting point the particle ends up.

10. Repeat the experiment of Exercise 9 four more times. Measure distance from start to finish for each of the 5 "random trips." Add these 5 distances and divide the sum by 5, to arrive at an "expected net distance" for such a trip.

For Exercises 11 and 12, consider another two-dimensional random walk governed by the following conditions.

1. Start out from a given street corner, and travel one block north. At each intersection:
2. Turn left with probability 1/6.
3. Go straight with probability 2/6 (= 1/3).
4. Turn right with probability 3/6 (= 1/2).

(Never turn around.)

11. *A Random Walk Using a Fair Die* Explain how a fair die could be used to simulate this random walk.

12. *A Random Walk Using a Random Number Table* Use Table 11 to simulate this random walk. For every 1 encountered in the table, turn left and proceed for another block. For every 2 or 3, go straight and proceed for another block. For every 4, 5, or 6, turn right and proceed for another block. Disregard all other digits. (Do you see how this scheme satisfies the probabilities given above?) This time begin at the upper right corner of the table, running down the column 9, 7, 7, and so on, to the bottom. Then start at the top of the next column to the left, 1, 0, 0, and so on, to the bottom. When these two columns of digits are used up, stop the "walk." Describe, in terms of distance and direction, where you have ended up relative to your starting point.

COLLABORATIVE INVESTIGATION

Finding Empirical Values of π

The information in this investigation was obtained from Burton's History of Mathematics: An Introduction, Third Edition, by David M. Burton, published by Wm. C. Brown, 1995, page 440.)

The following problem was posed by Georges Louis Leclerc, Comte de Buffon (1707–1788) in his *Histoire Naturelle* in 1777.

A large plane area is ruled with equidistant parallel lines, the distance between two consecutive lines of the series being a. A thin needle of length $l < a$ is tossed randomly onto the plane. What is the probability that the needle will intersect one of these lines?

The answer to this problem is found using integral calculus, and the probability p is shown to be $p = (2l)/(\pi a)$. Solving for π gives us the formula

$$\pi = \frac{2l}{pa}, \qquad (1)$$

which can be used to approximate the value of π experimentally. This was first observed by Pierre Simon Laplace, and such an experiment was carried out by Johann Wolf, a professor of astronomy at Bern, in about 1850. In this investigation, we will perform a similar experiment.

Topics for Discussion

Divide the class into groups of 3 or 4 students each. Each group will need the following materials.

1. a sheet of paper with a series of parallel lines evenly spaced across it
2. a thin needle, or needlelike object, with a length less than the distance between adjacent parallel lines on the paper

Each group should carry out these steps:

1. Measure and record the distance between lines (a) and the length of the needle (l), using the same units for both.
2. Assign one member to drop the needle onto the paper, another to determine whether the needle "hits" a line or not, and another to keep a tally of hits and misses.
3. Discuss ways to minimize bias so that the position and orientation of the dropped needle will be as random as possible.
4. Drop the needle 100 times, and record the number of hits.
5. Calculate the probability $p =$ (number of hits)/100. Is this probability value theoretical or empirical?
6. Enter the calculated value of p and the measured values of a and l into formula (1) to obtain a value of π. Round this value to four decimal places.

Now come back together as a class and record the various values obtained for π. Discuss the following questions.

1. The correct value of π, to four decimal places, is 3.1416. Which value of π, reported by the various groups, is most accurate? How far off is it?
2. Was it necessary to drop the needle 100 times, or could more or fewer tosses have been used?
3. Wolf tossed his needle 5000 times and it hit a line 2532 times, leading to an experimental value of π equal to 3.1596. How far off was Wolf's value?
4. How could the experiment be modified to produce "better" values for π?
5. Why could different groups use different l to a ratios and still all obtain legitimate approximations for π?

CHAPTER 12 TEST

1. Explain the difference between *empirical* and *theoretical* probabilities.
2. State the *law of large numbers,* and use coin tossing to illustrate it.

Drawing Cards A single card is chosen at random from a standard 52-card deck. Find the odds against its being each of the following.

3. a heart
4. a red queen
5. a king or a black face card

Genetics of Cystic Fibrosis The chart represents genetic transmission of cystic fibrosis. C denotes a normal gene while c denotes a cystic fibrosis gene. (Normal is dominant.) Both parents in this case are Cc, which means that they inherited one of each gene, and are therefore carriers but do not have the disease.

		Second Parent	
		C	c
First Parent	C		Cc
	c		

6. Complete the chart, showing all four equally likely gene combinations.

7. Find the probability that a child of these parents will also be a carrier without the disease.

8. What are the odds against a child of these parents actually having cystic fibrosis?

Days Off for Pizza Parlor Workers The manager of a pizza parlor (which operates seven days a week) allows each of three employees to select one day off next week. Assuming the selection is done randomly, find the probability of each of the following events.

9. All three select different days.

10. All three select the same day.

11. Exactly two of them select the same day.

Building Numbers from Sets of Digits Two numbers are randomly selected without replacement from the set $\{1, 2, 3, 4, 5\}$. Find the probability of each of the following events.

12. Both numbers are even.

13. Both numbers are prime.

14. The sum of the two numbers is odd.

15. The product of the two numbers is odd.

Selecting Committees A three-member committee is selected randomly from a group consisting of three men and two women.

16. Let x denote the number of men on the committee, and complete the probability distribution table.

x	$P(x)$
0	0
1	
2	
3	

17. Find the probability that the committee members are not all men.

18. Find the expected number of men on the committee.

Rolling Dice A pair of dice are rolled. Find the following.

19. the odds against "doubles" (the same number on both dice)

20. the probability of a sum greater than 2

21. the odds against a sum of "7 or 11"

22. the probability of a sum that is even and less than 5

Making Par in Golf Greg Brueck has a .78 chance of making par on each hole of golf that he plays. Today he plans to play just three holes. Find the probability of each of the following events. Round answers to three decimal places.

23. He makes par on all three holes.

24. He makes par on exactly two of the three holes.

25. He makes par on at least one of the three holes.

26. He makes par on the first and third holes but not on the second.

Drawing Cards Two cards are drawn, without replacement, from a standard 52-card deck. Find the probability of each of the following events.

27. Both cards are red.

28. Both cards are the same color.

29. The second card is a queen, given that the first card is an ace.

30. The first card is a face card and the second is black.

Simulating Births with Random Numbers Refer to the sequence of 40 coin tosses in Example 2 of Section 12.6.

31. How many triples of 3 successive births are listed?

32. How many of those triples consist of all girls?

33. What is the empirical probability that 3 successive births will be all girls?

chapter

13

Statistics

The table below shows the aging U.S. population over the 20th century. Statistics provides a way to analyze and interpret these numerical data. Denoting the year as x and the percent 65 and over as y (and, for simplicity, letting $x = 0$ for 1990, $x = 1$ for 1910, and so on), we can plot the data in the xy-coordinate plane. The resulting points, shown below, lie roughly in a linear pattern. The statistical technique called *regression* provides the unique line that best fits the data, which also is shown in the graph. The equation of this line is

$$y = .96x + 3.27.$$

Year	1900	1910	1920	1930	1940	1950	1960	1970	1980	1990	2000
Percent 65 and over	4.1	4.3	4.7	5.4	6.8	8.1	9.2	9.8	11.3	12.5	12.4

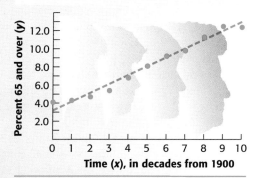

AGING U.S. POPULATION

13.1 Frequency Distributions and Graphs

13.2 Measures of Central Tendency

13.3 Measures of Dispersion

13.4 Measures of Position

13.5 The Normal Distribution

Extension: How to Lie with Statistics

13.6 Regression and Correlation

Collaborative Investigation: Combining Sets of Data

Chapter 13 Test

The slope of the regression line is .96. This shows that the proportion of the U.S. population aged 65 or over increased almost 1 percentage point per decade over the century. (Other statistical studies reveal some underlying reasons for this trend.) For more on regression, see Section 13.6.

Source: The World Almanac and Book of Facts 2002, p. 385 (data from the U.S. Census Bureau).

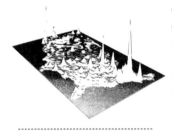

Population Densities There are many different kinds of graphs. For example, the computer-generated graph above shows U.S. population densities by peaks. The higher the peak, the denser the population in that area. The map shows the densest population to be in the New York City area. The blank areas in the West indicate fewer than two people per square mile.

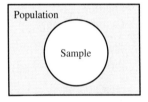

A population of 10,000

A random sample of 25

A random sample of 100

Governments collect and analyze an amazing quantity of "statistics"; the census, for example, is a vast project of gathering data. The census is not a new idea; two thousand years ago Mary and Joseph traveled to Bethlehem to be counted in a census. Long before, the Egyptians had recorded numerical information that is still being studied. For a long time, in fact, "statistics" referred to information about the government. The word itself comes from the Latin *statisticus,* meaning "of the state."

The development of current statistical theories, methods, tests, and experimental designs is due to the efforts of many people. For example, John Gaunt analyzed the weekly Bills of Mortality that recorded deaths in London in the first half of the seventeenth century. He published his *Observations* in 1662, noting that male births were more numerous than female, but eventually the numbers of both sexes came to be about equal. From these beginnings developed insurance companies, founded on the predictability of deaths as shown in mortality tables. From the field of biology, Sir Francis Galton and Karl Pearson in the nineteenth century made important contributions to statistical theory. William S. Gossett, a student of Pearson, produced some of the first results concerning small samples. He felt obliged to publish under the name "Student"; his basic findings came to be known as Student's t-test—an important statistical tool.

Statistical theory and methodology were greatly advanced during the twentieth century and become crucial in nearly all scientific, governmental, academic, social, and business fields. Important work was done by Sir Ronald Fisher, Gertrude Mary Cox, and John Tukey. Tukey's numerical and graphical methods, increasingly popular in recent years, are known as **exploratory data analysis.**

It is often important in statistics to distinguish between a **population,** which includes *all* items of interest, and a **sample,** which includes *some* (but ordinarily not all) of the items in the population. For example, to predict the outcome of an approaching presidential election, we may be interested in a population of many millions of voter preferences (those of all potential voters in the country). As a practical matter, however, even national polling organizations with considerable resources will obtain only a relatively small sample, say 2000, of those preferences. In general, a sample can be any subset of a population. See the Venn diagram at the side.

The study of statistics can be divided into two main areas. The first, **descriptive statistics,** has to do with collecting, organizing, summarizing, and presenting data (information). The main tools of descriptive statistics are tables of numbers, various kinds of graphs, and various calculated quantities, such as averages. These tools can be applied to both samples and populations.

The second main area of statistics, **inferential statistics,** has to do with drawing inferences or conclusions (making conjectures) about populations based on information from samples. It is in this area that we can best see the relationship between probability and statistics.

The Lansford Publishing Company, of San Jose, California, has produced a sampling demonstrator which consists of a large bowl containing 10,000 beads of various colors, along with several paddles for easily drawing samples of various sizes. Beneath the bowl, at the side, we show two random samples from the bowl, the first of which consists of 25 beads, 9 of which are green. From this sample we can "infer" that the bowl (the population) must contain about 9/25, or 36%, green beads, that is, about 3600. This is a matter of *inductive reasoning.* From a particular sample we have made a conjecture about the population in general. If we increase the size of our sample, then the proportion of green beads in the larger sample will most likely give a better estimate of the proportion in the population. The second sample shown contains 100 beads, 28 of which are green. So the new estimate would be about 28/100, or 28%, that is, 2800 green beads in the bowl.

In fact, the bowl is known to contain 30%, or 3000 green beads. So our larger sample estimate of 2800 was considerably more accurate than the smaller sample estimate of 3600. Knowing the true proportion of green beads in the population, we can turn the situation around, using *deductive reasoning* to predict the proportion of green beads in a particular sample. Our prediction for a sample of 100 beads would be 30% of 100, or 30. The sample, being random, may or may not end up containing exactly 30 green beads.

To summarize this discussion: If we know what a population is like, then probability theory enables us to conclude what is likely to happen in a sample (deductive reasoning); if we know what a sample is like, then inferential statistics enables us to draw inferences about the population (inductive reasoning).

13.1 Frequency Distributions and Graphs

Information that has been collected but not yet organized or processed is called **raw data.** It is often **quantitative** (or **numerical**), but can also be **qualitative** (or **nonnumerical**), as illustrated in Table 1.

TABLE 1 Examples of Raw Data
The number of siblings in ten different families (quantitative data):   3, 1, 2, 1, 5, 4, 3, 3, 8, 2
The makes of six different automobiles (qualitative data):   Toyota, Ford, Nissan, Toyota, Chevrolet, Honda

Quantitative data are generally more useful when they are **sorted,** or arranged in numerical order. In sorted form, the first list in Table 1 appears as

1, 1, 2, 2, 3, 3, 3, 4, 5, 8.

When a data set includes many repeated items, it can be organized into a **frequency distribution,** which lists the distinct data values (x) along with their frequencies (f). The frequency designates the number of times the corresponding item occurred in the data set. It is also helpful to show the **relative frequency** of each distinct item. This is the fraction, or percentage, of the data set represented by the item. If n denotes the total number of items, and a given item, x, occurred f times, then the relative frequency of x is f/n. Example 1 illustrates these ideas.

EXAMPLE 1 The 25 members of a psychology class were polled as to the number of siblings in their individual families. Construct a frequency distribution and a relative frequency distribution for their responses, which are shown here.

2, 3, 1, 3, 3, 5, 2, 3, 3, 1, 1, 4, 2, 4, 2, 5, 4, 3, 6, 5, 1, 6, 2, 2, 2

The data range from a low of 1 to a high of 6. The frequencies (obtained by inspection) and relative frequencies are shown in Table 2 on the next page.

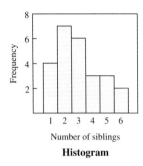

Histogram

FIGURE 1

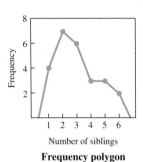

Frequency polygon

FIGURE 2

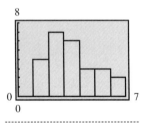

This histogram was generated using the data in Table 2. Compare with Figure 1.

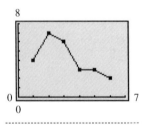

This line graph resembles the frequency polygon in Figure 2. It was generated using the data in Table 2.

TABLE 2	Frequency and Relative Frequency Distributions for Numbers of Siblings	
Number x	Frequency f	Relative Frequency f/n
1	4	4/25 = 16%
2	7	7/25 = 28%
3	6	6/25 = 24%
4	3	3/25 = 12%
5	3	3/25 = 12%
6	2	2/25 = 8%

The numerical data of Table 2 can be more easily interpreted with the aid of a **histogram.** A series of rectangles, whose lengths represent the frequencies, are placed next to one another as shown in Figure 1. On each axis, horizontal and vertical, a label and the numerical scale should be shown.

The information shown in the histogram in Figure 1 can also be conveyed by a **frequency polygon,** as in Figure 2. Simply plot a single point at the appropriate height for each frequency, connect the points with a series of connected line segments, and complete the polygon with segments that trail down to the axis beyond 1 and 6.

Grouped Frequency Distributions
Data sets containing large numbers of items are often arranged into groups, or *classes.* All data items are assigned to their appropriate classes, and then a **grouped frequency distribution** can be set up and a graph displayed. Although there are no fixed rules for establishing the classes, most statisticians agree on a few general guidelines.

Guidelines for the Classes of a Grouped Frequency Distribution

1. Make sure each data item will fit into one, and only one, class.
2. Try to make all classes the same width.
3. Make sure the classes do not overlap.
4. Use from 5 to 12 classes. (Too few or too many classes can obscure the tendencies in the data.)

EXAMPLE 2 Forty students, selected randomly in the school cafeteria on a Monday morning, were asked to estimate the number of hours they had spent studying in the past week (including both in-class and out-of-class time). Their responses are recorded here.

18	60	72	58	20	15	12	26	16	29
26	41	45	25	32	24	22	55	30	31
55	39	29	44	29	14	40	31	45	62
36	52	47	38	36	23	33	44	17	24

Tabulate a grouped frequency distribution and a grouped relative frequency distribution and construct a histogram and a frequency polygon for the given data.

Scanning the data, we see that they range from a low of 12 to a high of 72. If we let our classes run from 10 through 19, from 20 through 29, and so on up to 70 through 79, all guidelines above will be met and the common class width will be a convenient 10 units. We go through the data set, tallying each item into the appropriate class. The tally totals produce class frequencies, which in turn produce relative frequencies, as shown in Table 3. The histogram is displayed in Figure 3.

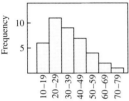

Weekly study times (in hours)
Grouped frequency histogram

FIGURE 3

TABLE 3 Grouped Frequency and Relative Frequency Distributions for Weekly Study Times

Class Limits	Tally	Frequency f	Relative Frequency f/n
10–19	|||||		6	6/40 = 15.0%
20–29	||||| |||||		11	11/40 = 27.5%
30–39	||||| ||||	9	9/40 = 22.5%
40–49	||||| ||	7	7/40 = 17.5%
50–59	||||	4	4/40 = 10.0%
60–69	||	2	2/40 = 5.0%
70–79			1	1/40 = 2.5%
	Total:	$n = 40$	

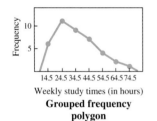

Weekly study times (in hours)
Grouped frequency polygon

FIGURE 4

In Table 3 (and Figure 3) the numbers 10, 20, 30, and so on are called the **lower class limits.** They are the smallest possible data values within the respective classes. The numbers 19, 29, 39, and so on are called the **upper class limits.** The common **class width** for the distribution is the difference of any two successive lower class limits (such as 30−20), or of any two successive upper class limits (such as 59−49). The class width for this distribution is 10.

To construct a frequency polygon, notice that, in a *grouped* frequency distribution, the data items in a given class are generally not all the same. We can obtain the "middle" value, or **class mark,** by adding the lower and upper class limits and dividing this sum by 2. We locate all the class marks along the horizontal axis and plot points above the class marks. The heights of the plotted points represent the class frequencies. The resulting points are connected just as for an ordinary (non-grouped) frequency distribution. The result, in this case, is shown in Figure 4.

TABLE 4 Grouped Frequency Distribution for Weekly Study Times

Class Limits	Frequency
10–19	6
20–29	11
30–39	9
40–49	7
50–59	4
60–69	2
70–79	1

Stem-and-Leaf Displays In Table 3 the tally marks themselves give a good impression of how the data are distributed. (It is clear, for example, that the highest frequency occurs in the class 20–29 and that the items are concentrated more heavily in the lower classes generally.) In fact, the tally marks are almost like a histogram turned on its side. Nevertheless, once the tallying is done, the tally marks are usually dropped and the grouped frequency distribution is presented as in Table 4.

The pictorial advantage of the tally marks is now lost. Furthermore, we cannot tell, from the grouped frequency distribution itself (or from the tally marks either, for that matter), what any of the original items were. We only know, for example, that there were seven items in the class 40–49. We do not know specifically what any of them were.

FOR FURTHER THOUGHT

When fair coins are tossed, the results on particular tosses cannot be reliably predicted. As more and more coins are tossed, however, the proportions of heads and tails become more predictable. This is a consequence of the "law of large numbers." For example, if five coins are tossed, then the resulting number of heads, denoted x, is a "random variable," whose possible values are 0, 1, 2, 3, 4, and 5. If the five coins are tossed repeatedly, say 64 separate times, then the binomial probability formula can be used to get **expected frequencies** (or **theoretical frequencies**), as shown in the table to the right. The first two columns of the table comprise the **expected frequency distribution** for 64 tosses of five fair coins. In an actual experiment, we could obtain **observed frequencies** (or **empirical frequencies**), which would most likely differ somewhat from the expected frequencies. But 64 repetitions of the experiment should be enough to provide fair consistency between expected and observed values.

Number of Heads x	Expected Frequency e	Observed Frequency o
0	2	
1	10	
2	20	
3	20	
4	10	
5	2	

For Group Discussion

Have members of your group toss five coins a total of 64 times, keeping a record of the results.

1. Enter your experimental results in the third column of the table above, producing an **observed frequency distribution.**
2. Construct two histograms, one from the expected frequency distribution and one from your observed frequency distribution.
3. Compare the two histograms, and explain why they are different.

To avoid the difficulties mentioned at the bottom of page 727, we can employ a tool of exploratory data analysis, the **stem-and-leaf display,** as shown in Example 3.

EXAMPLE 3 Present the study times data of Example 2 in a stem-and-leaf display.

Going back to Example 2 for the original raw data, we arrange the numbers in Table 5. The tens digits, to the left of the vertical line, are the "stems," while the corresponding ones digits are the "leaves." We have entered all items from the first row of the original data, from left to right, then the items from the second row, and so on through the fourth row.

TABLE 5 Stem-and-Leaf Display for Weekly Study Times

1	8 5 2 6 4 7
2	0 6 9 6 5 4 2 9 9 3 4
3	2 0 1 9 1 6 8 6 3
4	1 5 4 0 5 7 4
5	8 5 5 2
6	0 2
7	2

13.1 Frequency Distributions and Graphs

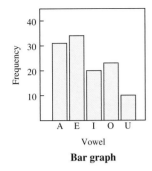

Bar graph

FIGURE 5

Notice that the stem-and-leaf display of Example 3 conveys at a glance the same pictorial impressions that a histogram would convey without the need for constructing the drawing. It also preserves the exact data values.

Bar Graphs and Pie Charts A frequency distribution of nonnumerical observations can be presented in the form of a **bar graph**, which is similar to a histogram except that the rectangles (bars) usually are not touching one another and sometimes are arranged horizontally rather than vertically. The bar graph of Figure 5 shows the frequencies of occurrence of the vowels A, E, I, O, and U in this paragraph.

A graphical alternative to the bar graph is the **circle graph**, or **pie chart**, which uses a circle to represent the total of all the categories and divides the circle into sectors, or wedges (like pieces of pie), whose sizes show the relative magnitudes of the categories. The angle around the entire circle measures 360°. For example, a category representing 20% of the whole should correspond to a sector whose central angle is 20% of 360°, that is, .20(360°) = 72°.

EXAMPLE 4 Nola Akala found that, during her first semester of college, her expenses fell into categories as shown in Table 6. Present this information in a circle graph.

The central angle of the sector for food is .30(360°) = 108°. Rent is .25(360°) = 90°. Calculate the other four angles similarly. Then draw a circle and mark off the angles with a protractor. The completed circle graph appears in Figure 6.

TABLE 6 Student Expenses

Expenses	Percent of Total
Food	30%
Rent	25%
Entertainment	15%
Clothing	10%
Books	10%
Other	10%

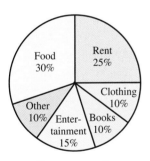

Expense categories

FIGURE 6

A circle graph shows, at a glance, the relative magnitudes of various categories. If we are interested in demonstrating how a quantity *changes*, say with respect to time, we use a **line graph**. We connect a series of line segments that rise and fall with time, according to the magnitude of the quantity being illustrated. To compare the patterns of change for two or more quantities, we can even plot multiple line graphs together. (A line graph looks somewhat like a frequency polygon, but the quantities graphed are not necessarily frequencies.)

EXAMPLE 5 Nola, from Example 4, wanted to keep track of her major expenses, food and rent, over the course of four years of college (eight semesters), in order to see how each one's budget percentage changed with time and also how the two compared. Use the data she collected (Table 7) to show this information in a line graph, and state any significant conclusions that are apparent from the graph.

TABLE 7 Food and Rent Expense Percentages

Semester	Food	Rent
First	30%	25%
Second	31	26
Third	30	28
Fourth	29	29
Fifth	28	34
Sixth	31	34
Seventh	30	37
Eighth	29	38

A comparison line graph for the given data (Figure 7) shows that the food percentage stayed fairly constant over the four years (at close to 30%), while the rent percentage, starting several points below food, rose steadily, surpassing food after the fourth semester and finishing significantly higher than food.

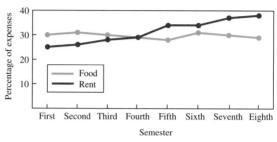

Comparison line graph

FIGURE 7

13.1 EXERCISES

In Exercises 1 and 2, use the given data to do the following:

(a) *Construct frequency and relative frequency distributions, in a table similar to Table 2.*
(b) *Construct a histogram.*
(c) *Construct a frequency polygon.*

1. *Questions Omitted on a Math Exam* The following data are the numbers of questions omitted on a math exam by the 30 members of the class.

```
1 0 3 0 0 2 1 2 2 0 0 5 1 1 3
4 2 0 2 0 1 0 1 2 3 3 4 0 1 0
```

2. *Quiz Scores in an Economics Class* The following data are quiz scores for the members of an economics class.

```
8 5 6 10 4 7 2 7 6 3  1 7 4 9
5 9 2  6 5 4 6 6 8 4 10 8 9 7
```

In each of Exercises 3–6, use the given data to do the following:

(a) *Construct grouped frequency and relative frequency distributions, in a table similar to Table 3. (In each case, follow the suggested guidelines for class limits and class width.)*
(b) *Construct a histogram.*
(c) *Construct a frequency polygon.*

3. *Heights of Baseball Players* The heights (in inches) of the 54 starting players in a baseball tournament were as follows.

```
53 51 65 62 61 55 59 52 62
64 48 54 64 57 51 67 60 49
49 59 54 52 53 60 58 60 64
52 56 56 58 66 59 62 50 58
60 63 64 52 60 58 63 53 56
58 61 55 50 65 56 61 55 54
```

Use five classes with a uniform class width of 5 inches, and use a lower limit of 45 inches for the first class.

4. *Charge Card Account Balances* The following raw data represent the monthly account balances (to the nearest dollar) for a sample of 50 brand-new charge card users.

```
138  78 175 46  79 118  90 163  88 107
126 154  85 60  42  54  62 128 114  73
129 130  81 67 119 116 145 105  96  71
100 145 117 60 125 130  94  88 136 112
118  84  74 62  81 110 108  71  85 165
```

Use seven classes with a uniform width of 20 dollars, where the lower limit of the first class is 40 dollars.

5. *Daily High Temperatures* The following data represent the daily high temperatures (in degrees Fahrenheit) for the month of June in a southwestern U.S. city.

79	84	88	96	102	104	99	97	92	94
85	92	100	99	101	104	110	108	106	106
90	82	74	72	83	107	111	102	97	94

Use nine classes with a uniform width of 5 degrees, where the lower limit of the first class is 70 degrees.

6. *IQ Scores of Tenth Graders* The following data represent IQ scores of a group of 50 tenth graders.

113	109	118	92	130	112	114	117	122	115
127	107	108	113	124	112	111	106	116	118
121	107	118	118	110	124	115	103	100	114
104	124	116	123	104	135	121	126	116	111
96	134	98	129	102	103	107	113	117	112

Use nine classes with a uniform width of 5, where the lower limit of the first class is 91.

In each of Exercises 7–10, construct a stem-and-leaf display for the given data. In each case, treat the ones digits as the leaves. For any single-digit data, use a stem of 0.

7. *Games Won in the National Basketball Association* On a certain date approaching midseason, the teams in the National Basketball Association had won the following numbers of games.

20	29	11	26	11	12	7	26	18
19	14	13	22	9	25	11	10	15
10	22	23	31	8	24	15	24	15

8. *Accumulated College Units* The students in a calculus class were asked how many college units they had accumulated to date. Their responses were as follows.

12	4	13	12	21	22	15	17	33	24
32	42	26	11	53	62	42	25	13	8
54	18	21	14	19	17	38	17	20	10

9. *Distances to School* Following are the daily round-trip distances to school (in miles) for 30 randomly chosen students at a community college in California.

12	30	10	11	18	26	34	18	8	12
26	14	5	22	4	25	9	10	6	21
18	18	9	16	44	23	4	13	36	8

10. *Yards Gained in the National Football League* The following data represent net yards gained per game by National Football League running backs who played during a given week of the season.

28	19	36	73	37	88	67	33	54	123	79
12	39	45	22	58	7	73	30	43	24	36
51	43	33	55	40	29	112	60	94	86	62
42	29	18	25	41	3	49	102	16	32	46

Network Television News Ratings Graphs of all kinds often are artistically embellished in order to catch attention or to emphasize meaning. The example here, presented in the form of a television screen, shows ratings trends for evening news broadcasts at the three major U.S. television networks. Refer to the graph for Exercises 11–15.

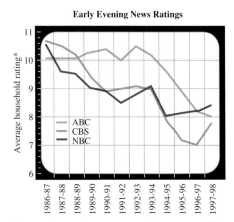

*One rating point represents 1 percent of the total U.S. households with a television.
Source: Nielsen Media Research; *1999 World Book Yearbook,* page 301. World Book, Inc., 525 W. Monroe, Chicago, IL 60661.

11. Of all U.S. households with televisions, about what total percentage (to the nearest whole percentage) watched the three networks' news broadcasts in 1986–87?

12. Of all U.S. households with televisions, about what total percentage (to the nearest whole percentage) watched the three networks' news broadcasts in 1997–98?

13. Which network's ratings dropped the most over the given period, and how much did they drop? (Estimate to the nearest tenth of a percent.)

14. In which one-year period did all three networks actually *increase* their ratings?

15. Over the entire period represented, what was the greatest spread between the highest and next highest rated network, which network attained the highest rating, and in which viewing season was it attained?

Reading Bar Graphs of Economic Indicators The bar graphs here show trends in several economic indicators over the period 1996–2001. Refer to these graphs for Exercises 16–20.

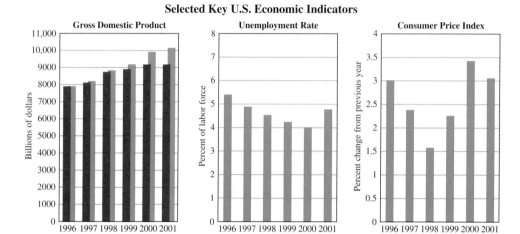

Source: U.S. Department of Commerce and U.S. Department of labor, except 2001 figures, which are estimates from The Conference Board.

Source: 2002 World Book Yearbook, page 191. World Book, Inc., 233 N. Michigan Ave., Chicago, IL 60601.

16. About what was the gross domestic product, in chained (1996) dollars, in 2001?

17. Over the six-year period, about what was the highest consumer price index, and when did it occur?

18. What was the greatest year-to-year change in the unemployment rate, and when did it occur?

19. Describe the apparent trends in both chained dollar and current dollar gross domestic product. Would it be possible for one of these to actually decrease the same year that the other increased? Why or why not?

20. Explain why the gross domestic product would generally increase when the unemployment rate decreases.

Reading Circle Graphs of Primary Energy Sources
The circle graph on the right shows the primary energy sources used to generate electricity in the U.S. in 2000. Refer to the graph for Exercises 21 and 22.

21. To the nearest degree, what is the central angle of the "Coal" sector?

22. Assuming the "Other" category includes five primary sources, what percentage (to the nearest tenth of a percent) did each of the "others" contribute on the average?

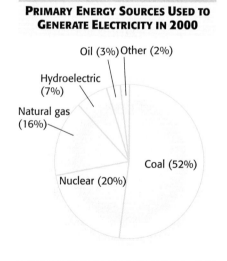

Source: 2002 World Book Yearbook, page 201.

23. *Sources of Job Training* A survey by the Bureau of Labor Statistics asked American workers how they were trained for their jobs. The percentages who responded in various categories are shown in the table below. Use the information in the table to draw a circle graph.

Principal Source of Training	Approximate Percentage of Workers
Trained in school	33%
Informal on-the-job training	25
Formal training from employers	12
Trained in military, or correspondence or other courses	10
No particular training, or could not identify any	20

24. *Correspondence Between Education and Earnings* Bureau of Labor Statistics data for a recent year showed that the average annual earnings of American workers corresponded to educational level as shown in the table below. Draw a bar graph that shows this information.

Educational Level	Average Annual Earnings
Less than 4 years of high school	$19,168
High school graduate	24,308
Four years of college	38,620

Net Worth of Retirement Savings Dave Chwalik, wishing to retire at age 60, is studying the comparison line graph here, which shows (under certain assumptions) how the net worth of his retirement savings (initially $200,000 at age 60) will change as he gets older and as he withdraws living expenses from savings. Refer to the graph for Exercises 25–28.

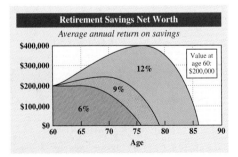

Source: *TSA Guide to Retirement Planning for California Educators,* Winter 1993.

25. Assuming Dave can maintain an average annual return of 9%, how old will he be when his money runs out?

26. If he could earn an average of 12% annually, what maximum net worth would Dave achieve? At about what age would the maximum occur?

27. Suppose Dave reaches age 70, in good health, and the average annual return has proved to be 6%.
    (a) About how much longer can he expect his money to last?
    (b) What options might he consider in order to extend that time?

28. At age 77, about how many times more will Dave's net worth be if he averages a 12% return than if he averages a 9% return?

Sample Masses in a Geology Laboratory Stem-and-leaf displays can be modified in various ways in order to obtain a reasonable number of stems. The following data, representing the measured masses (in grams) of thirty mineral samples in a geology lab, are shown in a *double-stem* display in Table 8.

60.7	41.4	50.6	39.5	46.4
58.1	49.7	38.8	61.6	55.2
47.3	52.7	62.4	59.0	44.9
35.6	36.2	40.6	56.9	42.6
34.7	48.3	55.8	54.2	33.8
51.3	50.1	57.0	42.8	43.7

TABLE 8 Stem-and-Leaf Display for Mineral Sample Masses

(30–34)	3	4.7	3.8			
(35–39)	3	9.5	8.8	5.6	6.2	
(40–44)	4	1.4	4.9	0.6	2.6	2.8	3.7
(45–49)	4	6.4	9.7	7.3	8.3	
(50–54)	5	0.6	2.7	4.2	1.3	0.1
(55–59)	5	8.1	5.2	9.0	6.9	5.8	7.0
(60–64)	6	0.7	1.6	2.4		

29. Describe how the stem-and-leaf display of Table 8 was constructed.

30. Explain why Table 8 is called a "double-stem" display.

31. In general, how many stems (total) are appropriate for a stem-and-leaf display? Explain your reasoning.

32. According to *The World Almanac and Book of Facts 2000* (page 231), the highest temperatures (in degrees Fahrenheit) ever recorded in the 50 states were as follows.

112 100 128 120 134 118 106 110 109 112
100 118 117 116 118 121 114 114 105 109
107 112 114 115 118 117 118 125 106 110
122 108 110 121 113 120 119 111 104 111
120 113 120 117 105 110 118 112 114 114

Present these data in a double-stem display.

33. *Letter Occurrence Frequencies in the English Language* The table here shows commonly accepted percentages of occurrence for the various letters in English language usage. (Code breakers have carefully analyzed these percentages as an aid in deciphering secret codes.) For example, notice that E is the most commonly occurring letter, followed by T, A, O, N, and so on. The letters Q and Z occur least often. Referring to Figure 5 in the text, would you say that the relative frequencies of occurrence of the vowels in the associated paragraph were typical or unusual? Explain your reasoning.

Letter	Percent	Letter	Percent
E	13	L	3 1/2
T	9	C, M, U	3
A, O	8	F, P, Y	2
N	7	W, G, B	1 1/2
I, R	6 1/2	V	1
S, H	6	K, X, J	1/2
D	4	Q, Z	1/5

Frequencies and Probabilities of Letter Occurrence The percentages shown in Exercise 33 are based on a very large sampling of English language text. Since they are based upon experiment, they are "empirical" rather than "theoretical." By converting each percent in that table to a decimal fraction, you can produce an **empirical probability distribution.** For example, if a single letter is randomly selected from a randomly selected passage of text, the probability that it will be an E is .13. The probability that a randomly selected letter would be a vowel (A, E, I, O, or U) is .385 (.08 + .13 + .065 + .08 + .03).

34. Rewrite the distribution shown in Exercise 33 as an empirical probability distribution. Give values to three decimal places. Note that the 26 probabilities in this distribution—one for each letter of the alphabet—should add up to 1 (except for, perhaps, a slight round-off error).

35. From your distribution of Exercise 34, construct an empirical probability distribution just for the vowels A, E, I, O, and U. (*Hint:* Divide each vowel's probability, from Exercise 34, by .385 to obtain a distribution whose five values add up to 1.) Give values to three decimal places.

36. Construct an appropriately labeled bar chart from your distribution of Exercise 35.

37. Based on the occurrences of vowels in the paragraph represented by Figure 5, construct a probability distribution for the vowels. The frequencies are: A–31, E–34, I–20, O–23, U–10. Give probabilities to three decimal places.

38. Is the probability distribution of Exercise 37 theoretical or empirical? Is it different from the distribution of Exercise 35? Which one is more accurate? Explain your reasoning.

39. *Frequencies and Probabilities of Study Times* Convert the grouped frequency distribution of Table 3 to an empirical probability distribution, using the same classes and giving probability values to three decimal places.

40. *Probabilities of Study Times* Recall that the distribution of Exercise 39 was based on weekly study times for a sample of 40 students. Suppose one of those students was chosen randomly. Using your distribution, find the probability that the study time in the past week for the student selected would have been in each of the following ranges.
(a) 30–39 hours (b) 40–59 hours
(c) fewer than 30 hours (d) at least 50 hours

Favorite Sports Among Recreation Students The 40 members of a recreation class were asked to name their favorite sports. The table below shows the numbers who responded in various ways.

Sport	Number of Class Members
Sailing	9
Hang gliding	5
Bungee jumping	7
Sky diving	3
Canoeing	12
Rafting	4

Use this information in Exercises 41–44.

41. If a member of this class is selected at random, what is the probability that the favorite sport of the person selected is bungee jumping?

42. Based on the data in the table, construct a probability distribution, giving probabilities to three decimal places.

43. (a) Is the distribution of Exercise 42 theoretical or is it empirical?
    (b) Explain your answer to part (a).

44. Explain why a frequency polygon trails down to the axis at both ends while a line graph ordinarily does not.

13.2 Measures of Central Tendency

Video Recyclers, a local business that sells "previously viewed" movies, had the following daily sales over a one-week period:

$$\$305, \quad \$285, \quad \$240, \quad \$376, \quad \$198, \quad \$264.$$

It would be desirable to have a single number to serve as a kind of representative value for this whole set of numbers—that is, some value around which all the numbers in the set tend to cluster, a kind of "middle" number or a **measure of central tendency**. Three such measures are discussed in this section.

The Mean The most common measure of central tendency is the **mean** (more properly called the **arithmetic mean**). The mean of a sample is denoted $\bar{x}$ (read "x bar"), while the mean of a complete population is denoted μ (the lower case Greek letter *mu*). Inferential statistics often involves both sample means and the population mean in the same discussion, but for our purposes here, data sets are considered to be samples, so we use $\bar{x}$.

The mean of a set of data items is found by adding up all the items and then dividing the sum by the number of items. (The mean is what most people associate with the word "average.") Since adding up, or summing, a list of items is a common procedure in statistics, we use the symbol for "summation," Σ (the capital Greek letter *sigma*). Thus the sum of n items, say $x_1, x_2, \ldots, x_n$, can be denoted $\Sigma x = x_1 + x_2 + \cdots + x_n$, and the mean is found as follows.

Many calculators find the mean automatically when a set of data items are entered. To recognize these calculators, look for a key marked $\boxed{\bar{x}}$, or perhaps $\boxed{\mu}$.

> **Mean**
>
> The **mean** of n data items $x_1, x_2, \ldots, x_n$, is given by the formula
>
> $$\bar{x} = \frac{\Sigma x}{n}.$$

```
mean({305,285,24
0,376,198,264})
                278
```

A calculator can find the mean of items in a list. This screen supports the text discussion of daily sales figures.

Now use this formula to find the central tendency of the daily sales figures for Video Recyclers:

$$\text{Mean} = \bar{x} = \frac{\Sigma x}{n} = \frac{305 + 285 + 240 + 376 + 198 + 264}{6} = \frac{1668}{6}$$
$$= 278.$$

The mean value (the "average daily sales") for the week is $278.

This screen supports the result in Example 1.

EXAMPLE 1 Last year's annual sales for eight different flower shops were

$374,910 $321,872 $242,943 $351,147
$382,740 $412,111 $334,089 $262,900.

The sum is $2,682,712. Now find the mean.

$$\bar{x} = \frac{\Sigma x}{n} = \frac{2{,}682{,}712}{8} = 335{,}339$$

The mean annual sales amount is $335,339.

Grade-point averages are very familiar to college students. For example, the table shows the units and grades earned by one student last term.

Course	Grade	Units
Mathematics	A	3
History	C	3
Chemistry	B	5
Art	B	2
PE	A	1

In one common system of finding a *grade-point average,* an A grade is assigned 4 points, with 3 points for B, 2 for C, and 1 for D. Find the grade-point average by multiplying the number of units for a course and the number assigned to each grade, and then adding these products. Finally, divide this sum by the total number of units.

Course	Grade	Grade Points	Units	(Grade Points) × (Units)
Mathematics	A	4	3	12
History	C	2	3	6
Chemistry	B	3	5	15
Art	B	3	2	6
PE	A	4	1	4
			Totals: 14	43

$$\text{Grade-point average} = \frac{43}{14} = 3.07 \quad \text{(rounded)}$$

This student earned a grade-point average of 3.07.

The calculation of a grade-point average is an example of a *weighted mean,* since the grade points for each course grade must be weighted according to the number of units of the course. (For example, five units of A is better than two units of A.) The number of units is called the *weighting factor.*

Weighted Mean

The **weighted mean** of n numbers, $x_1, x_2, \ldots, x_n$, that are weighted by the respective factors $f_1, f_2, \ldots, f_n$ is given by the formula

$$\bar{w} = \frac{\Sigma x \cdot f}{\Sigma f}.$$

13.2 Measures of Central Tendency

Salary x	Number of Employees f
$12,000	8
$16,000	11
$18,500	14
$21,000	9
$34,000	2
$50,000	1

```
mean({12000,1600
0,18500,21000,34
000,50000},{8,11
,14,9,2,1})
          18622.22222
```

In this screen supporting Example 2, the first list contains the salaries and the second list contains their frequencies.

In words, the weighted mean of a group of (weighted) items is the sum of all products of items times weighting factors, divided by the sum of all weighting factors.

The weighted mean formula is commonly used to find the mean for a frequency distribution. In this case, the weighting factors are the frequencies.

EXAMPLE 2 Find the mean salary for a small company that pays annual salaries to its employees as shown in the frequency distribution in the margin.

According to the weighted mean formula, we can set up the work as follows.

Salary x	Number of Employees f	Salary × Number $x \cdot f$
$12,000	8	$96,000
$16,000	11	$176,000
$18,500	14	$259,000
$21,000	9	$189,000
$34,000	2	$68,000
$50,000	1	$50,000
Totals:	45	$838,000

$$\text{Mean salary} = \frac{\$838,000}{45} = \$18,622 \quad \text{(rounded)}$$

FOR FURTHER THOUGHT

In baseball statistics, a player's "batting average" gives the average number of hits per time at bat. For example, a player who has gotten 84 hits in 250 times at bat has a batting average of $84/250 = .336$. This "average" can be interpreted as the empirical probability of that player's getting a hit the next time at bat.

The following are actual comparisons of hits and at-bats for two major league players in the 1989 and 1990 seasons. The numbers illustrate a puzzling statistical occurrence known as **Simpson's paradox.** (This information was reported by Richard J. Friedlander on page 845 of the November 1992 issue of the *MAA Journal*.)

For Group Discussion

1. Fill in the ten blanks in the table, giving batting averages to three decimal places.
2. Which player had a better average in 1989?
3. Which player had a better average in 1990?
4. Which player had a better average in 1989 and 1990 combined?
5. Did the results above surprise you? How can it be that one player's batting average leads another's for each of two years, and yet trails the other's for the combined years?

	Dave Justice			Andy Van Slyke		
	Hits	At-bats	Batting Average	Hits	At-bats	Batting Average
1989	12	51	_____	113	476	_____
1990	124	439	_____	140	493	_____
Combined (1989-90)	_____	_____	_____	_____	_____	_____

For some data sets the mean can be a misleading indicator of average. Consider Shady Sam who runs a small business that employs five workers at the following annual salaries.

$$\$16{,}500, \quad \$16{,}950, \quad \$17{,}800, \quad \$19{,}750, \quad \$20{,}000$$

The employees, knowing that Sam accrues vast profits to himself, decide to go on strike and demand a raise. To get public support, they go on television and tell about their miserable salaries, pointing out that the mean salary in the company is only

$$\bar{x} = \frac{\$16{,}500 + \$16{,}950 + \$17{,}800 + \$19{,}750 + \$20{,}000}{5}$$

$$= \frac{\$91{,}000}{5} = \$18{,}200.$$

The local television station schedules an interview with Sam to investigate. In preparation, Sam calculates the mean salary of *all* workers (including his own salary of $188,000):

$$\bar{x} = \frac{\$16{,}500 + \$16{,}950 + \$17{,}800 + \$19{,}750 + \$20{,}000 + \$188{,}000}{6}$$

$$= \frac{\$279{,}000}{6} = \$46{,}500.$$

When the TV crew arrives, Sam calmly assures them that there is no reason for his employees to complain since the company pays a generous mean salary of $46,500.

The employees, of course, would argue that when Sam included his own salary in the calculation, it caused the mean to be a misleading indicator of average. This was so because Sam's salary is not typical. It lies a good distance away from the general grouping of the items (salaries). An extreme value like this is referred to as an *outlier*. Since a single outlier can have a significant effect on the value of the mean, we say that the mean is "highly sensitive to extreme values."

The Median Another measure of central tendency, which is not so sensitive to extreme values, is the **median**. This is the value that divides a group of numbers into two parts, with half the numbers below the median and half above it.

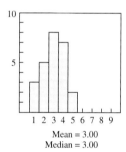

Mean = 3.00
Median = 3.00

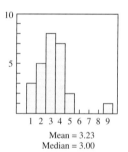

Mean = 3.23
Median = 3.00

The introduction of a single "outlier" above increased the mean by 8 percent but left the median unaffected.

Outliers should usually be considered as *possible* errors in the data.

Median

To find the **median** of a group of items:

1. Arrange the items in numerical order (from smallest to largest).
2. If the number of items is *odd,* the median is the middle item in the list.
3. If the number of items is *even,* the median is the mean of the two middle items.

For Shady Sam's business, all salaries (including Sam's), arranged in numerical order, are

$$\$16{,}500, \quad \$16{,}950, \quad \$17{,}800, \quad \$19{,}750, \quad \$20{,}000, \quad \$188{,}000,$$

so
$$\text{median} = \frac{\$17{,}800 + \$19{,}750}{2} = \frac{\$37{,}550}{2} = \$18{,}775.$$

This figure is a representative average, based on all six salaries, that the employees would probably agree is reasonable.

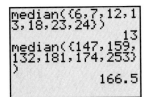

The calculator can find the median of the entries in a list. This screen supports the results in parts (a) and (c) of Example 3.

EXAMPLE 3 Find the median of each list of numbers.

(a) 6, 7, 12, 13, 18, 23, 24

This list is already in numerical order. The number of values in the list, 7, is odd, so the median is the middle value, or 13.

(b) 17, 15, 9, 13, 21, 32, 41, 7, 12

First, place the numbers in numerical order from smallest to largest.

$$7, 9, 12, 13, 15, 17, 21, 32, 41$$

The middle number can now be picked out; the median is 15.

(c) 147, 159, 132, 181, 174, 253

First write the numbers in numerical order.

$$132, 147, 159, 174, 181, 253$$

Since the list contains an even number of items, namely 6, there is no single middle item. Find the median by taking the mean of the two middle items, 159 and 174.

$$\frac{159 + 174}{2} = \frac{333}{2} = 166.5$$

In the case of a frequency distribution, locating the middle item (the median) is a bit different. First find the total number of items in the set by adding the frequencies ($n = \Sigma f$). Then the median is the item whose *position* is given by the following formula.

Position of the Median in a Frequency Distribution

$$\text{Position of median} = \frac{n+1}{2} = \frac{\Sigma f + 1}{2}$$

It is important to notice that this formula gives only the *position,* and not the actual value, of the median. The next example shows how the formula is used to find the median.

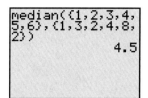

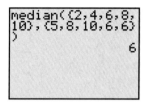

These two screens support the results in Example 4.

EXAMPLE 4 Find the medians for the following distributions.

(a)

Value	Frequency	Cumulative Frequency
1	1	1
2	3	4
3	2	6
4	4	10
5	8	18
6	2	20

Total: 20

Adding the frequencies shows that there are 20 items total, so

$$\text{position of median} = \frac{20 + 1}{2} = \frac{21}{2} = 10.5.$$

The median, then, is the average of the tenth and eleventh items. To find these items, make use of *cumulative frequencies,* which tell, for each different value, how many items have that value or a lower value. Since the value 4 has a cumulative frequency of 10 ($1 + 3 + 2 + 4 = 10$), that is, 10 items have a value of 4 or less, and 5 has a cumulative frequency of 18, the tenth item is 4 and the eleventh item is 5, making the median $(4 + 5)/2 = 9/2 = 4.5$.

(b)

Value	Frequency	Cumulative Frequency
2	5	5
4	8	13
6	10	23
8	6	29
10	6	35

Total: 35

There are 35 items total, so

$$\text{position of median} = \frac{35 + 1}{2} = \frac{36}{2} = 18.$$

From the cumulative frequency column, the fourteenth through the twenty-third items are all 6s. This means the eighteenth item is a 6, so the median is 6. ■

The Mode The third important measure of central tendency is the **mode**. If ten students earned scores on a business law examination of

74, 81, 39, 74, 82, 80, 100, 92, 74, 85,

then we notice that more students earned the score 74 than any other score. This fact makes 74 the mode of this list.

A note in *The Wall Street Journal* claimed that during a televised presentation of the Academy Awards, the average winner took 1 minute and 39 seconds to thank 7.8 friends, relatives, and loyal supporters for "making it all possible." Would these numbers refer to means, medians, or modes?

Mode

The **mode** of a data set is the value that occurs most often.

EXAMPLE 5 Find the mode for each set of data.

(a) 51, 32, 49, 49, 74, 81, 92

The number 49 occurs more often than any other. Therefore, 49 is the mode. Note that it is not necessary to place the numbers in numerical order when looking for the mode.

(b) 482, 485, 483, 485, 487, 487, 489

Both 485 and 487 occur twice. This list is said to have *two* modes, or to be *bimodal*.

(c) 10,708, 11,519, 10,972, 17,546, 13,905, 12,182

No number here occurs more than once. This list has no mode.

(d)

Value	Frequency
19	1
20	3
22	8 ← Greatest frequency
25	7
26	4
28	2

The frequency distribution shows that the most frequently occurring value (and thus the mode) is 22.

That we have included the mode (along with the mean and the median) as a measure of *central tendency* is traditional, probably because many important kinds of data sets do have their most frequently occurring values "centrally" located. However, there is no reason the mode cannot be one of the smallest values in the set or one of the largest. In such a case, the mode really is not a good measure of "central tendency."

When the data items being studied are nonnumeric, the mode may be the only usable measure of central tendency. For example, the bar graph of Figure 5 in Section 13.1 showed frequencies of occurrence of vowels in a sample paragraph. Since the vowels A, E, I, O, and U are not numbers, they cannot be added, so their mean does not exist. Furthermore, the vowels cannot be arranged in any meaningful numerical order, so their median does not exist either. The mode, however, does exist. As the bar graph shows, the mode is the letter E.

Sometimes, a distribution contains two distinct items that both occur more often than any other items such as in Example 5(b). Such a distribution is called **bimodal** (literally, "two modes"). In a large distribution, this term is commonly applied even when the two modes do not have exactly the same frequency. When a distribution has three or more different items sharing the highest frequency of occurrence, that information is not often useful. (Too many modes tend to obscure the significance of these "most frequent" items.) We say that such a distribution has *no* mode.

Central Tendency from Stem-and-Leaf Displays

As shown in Section 13.1, data are sometimes presented in a stem-and-leaf display in order to give a graphical impression of their distribution. We can also calculate measures of central tendency from a stem-and-leaf display. The median and mode are more easily identified when the "leaves" are **ranked** (arranged in numerical order) on their "stems."

In Table 9, we have rearranged the leaves of Table 5 in Section 13.1 (which showed the weekly study times from Example 2 of that section).

TABLE 9 Stem-and-Leaf Display for Weekly Study Times, with Leaves Ranked

1	2 4 5 6 7 8
2	0 2 3 4 4 5 6 6 9 9 9
3	0 1 1 2 3 6 6 8 9
4	0 1 4 4 5 5 7
5	2 5 5 8
6	0 2
7	2

EXAMPLE 6 Find the mean, median, and mode for the data in Table 9.

If you enter the data into a calculator with statistical capabilities, the mean will be automatically computed for you. Otherwise, add all items (reading from the stem-and-leaf display) and divide by $n = 40$.

$$\text{mean} = \frac{12 + 14 + 15 + \cdots + 60 + 62 + 72}{40} = \frac{1395}{40} = 34.875$$

In this case, $n = 40$ (an even number), so the median is the average of the twentieth and twenty-first items, in order. Counting leaves, we see that these will be the third and fourth items on the stem 3. So

$$\text{median} = \frac{31 + 32}{2} = 31.5.$$

By inspection, we see that 29 occurred three times and no other value occurred that often, so

$$\text{mode} = 29.$$

Symmetry in Data Sets

The most useful way to analyze a data set often depends on whether the distribution is **symmetric** or **non-symmetric**. In a "symmetric" distribution, as we move out from the central point, the pattern of frequencies is the same (or nearly so) to the left and to the right. In a "non-symmetric" distribution, the patterns to the left and right are different. On the next page, Figure 8 shows several types of symmetric distributions, while Figure 9 shows some non-symmetric distributions. A non-symmetric distribution with a tail extending out to the left is called **skewed to the left**. If the tail extends out to the right, the distribution is **skewed to the right**. Notice that a bimodal distribution may be either symmetric or non-symmetric.

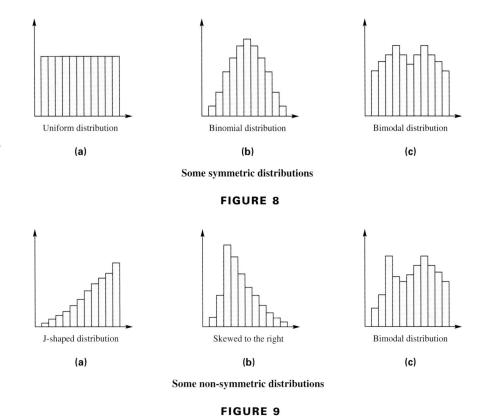

Some symmetric distributions

FIGURE 8

Some non-symmetric distributions

FIGURE 9

Summary We summarize here the measures discussed in this section.

Common Measures of Central Tendency

The **mean** of a set of numbers is found by adding all the values in the set and dividing by the number of values.

The **median** is a kind of "middle" number. To find the median, first arrange the values in numerical order. For an *odd* number of values, the median is the middle value in the list. For an *even* number of values, the median is the mean of the two middle values.

The **mode** is the value that occurs most often. Some sets of numbers have two most frequently occurring values and are **bimodal.** Other sets have no mode at all (if no value occurs more often than the others or if more than two values occur most often).

We conclude this section with a summary of the relative advantages and disadvantages of the mean, the median, and the mode as measures of central tendency. For distributions of numeric data, the mean and median will always exist, while the mode may not exist. On the other hand, for nonnumeric data, it may be that none of the three measures exists, or that only the mode exists.

Since even a single change in the data may cause the mean to change, while the median and mode may not be affected at all, the mean is the most "sensitive" measure. In a symmetric distribution, the mean, median, and mode (if a single mode exists) will all be equal. In a nonsymmetric distribution, the mean is often unduly affected by relatively few extreme values, and therefore may not be a good representative measure of central tendency. For example, distributions of salaries, or of family incomes, often include a few values that are much higher than the bulk of the items. In such cases, the median is a more useful measure. Any time there are many more items on one side of the mean than on the other, extreme non-symmetry is indicated and the median is most likely a better choice than the mean.

The mode is the only measure covered here that must always be equal to one of the data items of the distribution. In fact, more of the data items are equal to the mode than to any other number. A fashion shop planning to stock only one hat size for next season would want to know the mode (the most common) of all hat sizes among their potential customers. Likewise, a designer of family automobiles would be interested in the most common family size. In examples like these, designing for the mean or the median might not be right for anyone.

13.2 EXERCISES

For each of the following lists of data, calculate **(a)** *the mean,* **(b)** *the median, and* **(c)** *the mode or modes (if any). Round mean values to the nearest tenth.*

1. 3, 7, 12, 16, 23

2. 21, 25, 32, 48, 53, 62

3. 128, 230, 196, 224, 196, 233

4. 26, 31, 46, 31, 26, 29, 31

5. 3.1, 4.5, 6.2, 7.1, 4.5, 3.8, 6.2, 6.3

6. 14,320, 16,950, 17,330, 15,470

7. .78, .93, .66, .94, .87, .62, .74, .81

8. .53, .03, .28, .18, .39, .28, .14, .22, .04

9. 12.3, 45.6, 78.9, 1.2, 34.5, 67.8, 90.1

10. .3, .8, .4, .3, .7, .9, .2, .1, .5, .9, .6

11. 128, 131, 136, 125, 132, 128, 125, 127

12. 8.97, 5.64, 2.31, 1.02, 4.35, 7.68

Most-visited Internet Websites The table here shows the twenty most-visited Internet websites for July 2001. (*Source of data:* The World Almanac and Book of Facts 2002, *page 632.*)

Most-visited Websites, July 2001

Rank	Website	Unique visitors (in thousands)	Rank	Website	Unique visitors (in thousands)
1.	http://www.yahoo.com	62,180	11.	http://www.ebay.com	19,096
2.	http://www.msn.com	53,212	12.	http://www.netscape.com	17,538
3.	http://www.aol.com	42,779	13.	http://www.google.com	15,154
4.	http://www.x10.com	39,567	14.	http://www.fastclick.com	15,115
5.	http://www.microsoft.com	36,802	15.	http://www.excite.com	14,636
6.	http://www.passport.com	31,854	16.	http://www.about.com	14,517
7.	http://www.hotmail.com	28,656	17.	http://www.real.com	13,413
8.	http://www.lycos.com	27,359	18.	http://www.mapquest.com	13,116
9.	http://www.amazon.com	20,481	19.	http://www.tripod.com	13,057
10.	http://www.go.com	19,553	20.	http://www.cnet.com	12,057

Source: Media Metrix, Inc.

Using the table for Most-visited Websites on the previous page, find **(a)** *the mean number of visitors and* **(b)** *the median number of visitors for each of the following. (Give answers to the nearest thousand.)*

13. the sites ranked 1 through 5
14. the sites ranked 1 through 10
15. the sites ranked 11 through 20
16. the sites ranked 1 through 20

Forest Loss and Atmospheric Pollution Amid concerns about global warming and rain forest destruction, the World Resources Institute reported the five countries with the greatest average yearly forest loss for a recent five-year period, as well as the five greatest emitters of chlorofluorocarbons in a one-year period. The estimates are shown here.

Country	Average Yearly Forest Loss (in square miles)
Brazil	9700
Colombia	3400
Indonesia	2400
Mexico	2375
Ivory Coast	2000

Country	Chlorofluorocarbon Emissions (in metric tons)
United States	130,000,000
Japan	95,000,000
C.I.S.	67,000,000
Germany	34,000,000
United Kingdom	25,000,000

17. Find the mean value of yearly forest loss per country for the five countries listed. Give your answer to the nearest 100 square miles.

18. Find the mean value of chlorofluorocarbon emissions per country for the five countries listed. Give your answer to the nearest one million metric tons.

Measuring Elapsed Times While doing an experiment, a physics student recorded the following sequence of elapsed times (in seconds) in a lab notebook.

2.16, 22.2, 2.96, 2.20, 2.73, 2.28, 2.39.

19. Find the mean.   20. Find the median.

The student, when reviewing the calculations later, decided that the entry 22.2 should have been recorded as 2.22, and made that change in the listing.

21. Find the mean for the new list.
22. Find the median for the new list.
23. Which measure, the mean or the median, was affected more by correcting the error?
24. In general, which measure, mean or median, is affected less by the presence of an extreme value in the data?

Scores on Statistics Examinations Jay Beckenstein earned the following scores on his six statistics exams last semester.

79, 81, 44, 89, 79, 90

25. Find the mean, the median, and the mode for Jay's scores.
26. Which of the three averages probably is the best indicator of Jay's ability?
27. If Jay's instructor gives him a chance to replace his score of 44 by taking a "make-up" exam, what must he score on the make-up to get an overall average (mean) of 85?

Exercises 28 and 29 give frequency distributions for sets of data values. For each set find the **(a)** *mean (to the nearest hundredth),* **(b)** *median, and* **(c)** *mode or modes (if any).*

28.

Value	Frequency
2	5
4	1
6	8
8	4

29.

Value	Frequency
603	13
597	8
589	9
598	12
601	6
592	4

30. *Average Employee Salaries* A company has six employees with a salary of $19,500, eight with a salary of $23,000, four with a salary of $28,300, two with a salary of $34,500, seven with a salary of $36,900, and one with a salary of $145,500. Find the mean salary for the employees (to the nearest hundred dollars).

Grade-point Averages Find the grade-point average for each of the following students. Assume A = 4, B = 3, C = 2, D = 1, F = 0. Round to the nearest hundredth.

31.
Units	Grade
4	C
7	B
3	A
3	F

32.
Units	Grade
2	A
6	B
5	C

Area and Population in the Commonwealth of Independent States The table below gives the land area and population of the eleven founding members of the Commonwealth of Independent States. (All were former republics of the Soviet Union.) Use the information from the table for Exercises 33–36. (Source of data: The World Almanac and Book of Facts 2000, pages 768–877.)

Name	Area (square miles)	Population
Armenia	11,500	3,409,234
Azerbaijan	33,400	7,908,224
Belarus	80,200	10,401,784
Kazakhstan	1,049,200	16,824,825
Kyrgystan	76,600	4,546,055
Moldova	13,000	4,460,838
Russia	6,592,800	146,393,569
Tajikistan	55,300	6,102,854
Turkmenistan	188,500	4,366,383
Ukraine	233,100	49,811,174
Uzbekistan	172,700	24,102,473

33. Find the mean area (to the nearest 100 square miles) for these 11 states.

34. Discuss the meaningfulness of the average calculated in Exercise 33.

35. Find the mean population (to the nearest 1000) for the 11 states.

36. Discuss the meaningfulness of the average calculated in Exercise 35.

Languages Spoken by Americans The table near the top of the next column lists the ten most common languages (other than English) spoken at home by Americans. The numbers in the right column indicate the change (increase or decrease), over a recent ten-year period, in the number of Americans speaking the different languages. (Source of data: 1995 Information Please Almanac, *page 835.*) Use the table for Exercises 37–40. (The rankings are given for information only. They will not enter into your calculations.)

	Most Common Non-English Languages Among Americans	
Rank	Language	Change in Number of People
1	Spanish	5,789,839
2	French	129,901
3	German	−59,644
4	Italian	−324,631
5	Chinese	617,476
6	Tagalog	391,289
7	Polish	−102,667
8	Korean	350,766
9	Vietnamese	303,801
10	Portuguese	68,759

37. Find the mean change per language for all ten languages.

38. Find the mean change per language for the seven categories with increases.

39. Find the mean change per language for the three categories with decreases.

40. Use your answers for Exercises 38 and 39 to find the weighted mean for the change per language for all ten categories. Compare with the value calculated in Exercise 37.

41. *Triple Crown Horse Races* The table below shows the winning horse and the "value to winner" for each of the so-called Triple Crown races in 2001. (In each race, the "gross purse," or total amount distributed, was $1,000,000.) Find the average (mean) value to winner for the three races. (Source: 2002 World Book Yearbook, *page 245.*)

Race	Winning Horse	Value to Winner
Kentucky Derby	Monarchos	$812,000
Preakness Stakes	Point Given	650,000
Belmont Stakes	Point Given	600,000

Olympic Medal Standings The top ten medal-winning nations in the 2002 Winter Olympics at Salt Lake City are shown at the top of the next page. (Source: The Gazette, *February 25, 2002.*) Use the given information for Exercises 42–45.

Medal Standings for the 2002 Winter Olympics				
Nation	Gold	Silver	Bronze	Total
Germany	12	16	7	35
United States	10	13	11	34
Norway	11	7	6	24
Canada	6	3	8	17
Russia	6	6	4	16
Austria	2	4	10	16
Italy	4	4	4	12
France	4	5	2	11
Switzerland	3	2	6	11
Netherlands	3	5	0	8

Calculate each of the following for all nations shown.

42. the mean number of gold medals

43. the median number of silver medals

44. the mode, or modes, for the number of bronze medals

45. each of the following for the total number of medals
    (a) mean (b) median (c) mode or modes

In each of Exercises 46 and 47, use the given stem-and-leaf display to identify (a) the mean, (b) the median, and (c) the mode (if any) for the data represented.

46. *Auto Repair Charges* The display here represents prices (to the nearest dollar) charged by 23 different auto repair shops for a new alternator (installed). Give answers to the nearest cent.

    ```
     9 | 7
    10 | 2 4
    10 | 5 7 9
    11 | 1 3 4 4
    11 | 5 5 8 8 9
    12 | 0 4 4
    12 | 5 7 7 9
    13 | 8
    ```

47. *Scores on a Biology Exam* The display here represents scores achieved on a 100-point biology exam by the 34 members of the class.

    ```
    4 | 7
    5 | 1 3 6
    6 | 2 5 5 6 7 8 8
    7 | 0 4 5 6 7 7 8 8 8 9
    8 | 0 1 1 3 4 5 5
    9 | 0 0 0 1 6
    ```

48. *Calculating a Missing Test Score* Dinya Floyd's Business professor lost his grade book, which contained Dinya's five test scores for the course. A summary of the scores (each of which was an integer from 0 to 100) indicates that the mean was 88, the median was 87, and the mode was 92. (The data set was not bimodal.) What is the lowest possible number among the missing scores?

49. Explain what an "outlier" is and how it affects measures of central tendency.

50. *Consumer Preferences in Food Packaging* A food processing company that packages individual cups of instant soup wishes to find out the best number of cups to include in a package. In a survey of 22 consumers, they found that five prefer a package of 1, five prefer a package of 2, three prefer a package of 3, six prefer a package of 4, and three prefer a package of 6.
    (a) Calculate the mean, median, and mode values for preferred package size.
    (b) Which measure in part (a) should the food processing company use?
    (c) Explain your answer to part (b).

51. *Scores on a Math Quiz* The following are scores earned by 15 college students on a 20-point math quiz.

    0, 1, 3, 14, 14, 15, 16, 16, 17, 17, 18, 18, 18, 19, 20

    (a) Calculate the mean, median, and mode values for these scores.
    (b) Which measure in part (a) is most representative of the data?
    (c) Explain your answer to part (b).

In each of Exercises 52–55, begin a list of the given numbers, in order, starting with the smallest one. Continue the list only until the median of the listed numbers is a multiple of 4. Stop at that point and find (a) the number of numbers listed, and (b) the mean of the listed numbers (to two decimal places).

52. counting numbers   53. prime numbers

54. Fibonacci numbers   55. triangular numbers

56. Seven consecutive whole numbers add up to 147. What is the result when their mean is subtracted from their median?

57. If the mean, median, and mode are all equal for the set $\{70, 110, 80, 60, x\}$, find the value of x.

58. Michael Coons wants to include a fifth number, n, along with the numbers 2, 5, 8, and 9 so that the mean and median of the five numbers will be equal. How many choices does Michael have for the number n, and what are those choices?

59. *Average Employee Salaries* Refer to the salary data of Example 2. Explain what is wrong with simply calculating the mean salary as follows.

$$\bar{x} = \frac{\Sigma x}{n} = \frac{\$12{,}000 + \$16{,}000 + \$18{,}500 + \$21{,}000 + \$34{,}000 + \$50{,}000}{6} = \$25{,}250$$

For Exercises 60–62, refer to the grouped frequency distribution shown here.

60. Can you identify any specific data items that occurred in this sample?

61. Can you compute the actual mean for this sample?

62. Describe how you might approximate the mean for this sample. Justify your procedure.

Class Limits	Frequency f
21–25	5
26–30	3
31–35	8
36–40	12
41–45	21
46–50	38
51–55	35
56–60	20

13.3 Measures of Dispersion

The mean is a good indicator of the central tendency of a set of data values, but it does not give the whole story about the data. To see why, compare distribution A with distribution B in the table at the side.

	A	B
	5	1
	6	2
	7	7
	8	12
	9	13
Mean	7	7
Median	7	7

Both distributions of numbers have the same mean (and the same median also), but beyond that, they are quite different. In the first, 7 is a fairly typical value; but in the second, most of the values differ quite a bit from 7. What is needed here is some measure of the **dispersion,** or *spread,* of the data. Two of the most common measures of dispersion, the *range* and the *standard deviation,* are discussed in this section.

Range The **range** of a data set is a straightforward measure of dispersion defined as follows.

> **Range**
>
> For any set of data, the **range** of the set is given by
>
> Range = (greatest value in the set) − (least value in the set).

For a short list of data, calculation of the range is simple. For a more extensive list, it is more difficult to be sure you have accurately identified the greatest and least values.

In distribution A, the greatest value is 9 and the least is 5. Thus

$$\text{Range} = \text{greatest} - \text{least} = 9 - 5 = 4.$$

In distribution B,

$$\text{Range} = 13 - 1 = 12.$$

Once the data are entered, a calculator with statistical functions may actually show the range (among other things), or at least sort the data and identify the minimum and maximum items. (The associated symbols may be something like MIN Σ and MAX Σ, or minX and maxX.) Given these two values, a simple subtraction produces the range.

Dive	Mark	Myrna
1	28	27
2	22	27
3	21	28
4	26	6
5	18	27
Mean	23	23
Median	22	27
Range	10	22

The range can be misleading if it is interpreted unwisely. For example, suppose three judges for a diving contest assign points to Mark and Myrna on five different dives, as shown in the table. The ranges for the divers make it tempting to conclude that Mark is a more consistent diver than Myrna. However, Myrna is actually more consistent, with the exception of one very poor score. That score, 6, is an outlier which, if not actually recorded in error, must surely be due to some special circumstance. (Notice that the outlier does not seriously affect Myrna's median score, which is more typical of her overall performance than is her mean score.)

Standard Deviation One of the most useful measures of dispersion, the **standard deviation,** is based on *deviations from the mean* of the data values. To find how much each value deviates from the mean, first find the mean, and then subtract the mean from each data value.

EXAMPLE 1 Find the deviations from the mean for the data values

$$32, 41, 47, 53, 57.$$

Add these values and divide by the total number of values, 5. This process shows that the mean is 46. To find the deviations from the mean, subtract 46 from each data value.

Data value	32	41	47	53	57
Deviation	−14	−5	1	7	11

$$32 - 46 = -14 \qquad 57 - 46 = 11$$

To check your work, add the deviations. The sum of the deviations for a set of data is always 0. ■

It is perhaps tempting now to find a measure of dispersion by finding the mean of the deviations. However, this number always turns out to be 0 no matter how much dispersion is shown by the data; this is because the positive deviations just cancel out the negative ones. This problem of positive and negative numbers canceling each other can be avoided by *squaring* each deviation. (The square of a negative number is positive.) The following chart shows the squares of the deviations for the data above.

Data value	32	41	47	53	57
Deviation	−14	−5	1	7	11
Square of deviation	196	25	1	49	121

$$(-14) \cdot (-14) = 196 \qquad 11 \cdot 11 = 121$$

An average of the squared deviations could now be found by dividing their sum by the number of data values n (5 in this case), which we would do if our data values comprised a population. However, since we are considering the data to be a sample, we divide by $n - 1$ instead.* The average that results is itself a measure of

---

*Although the reasons cannot be explained at this level, dividing by $n - 1$ rather than n produces a sample measure that is more accurate for purposes of inference. In most cases the results using the two different divisors differ only slightly.

Most calculators find square roots, such as $\sqrt{98}$, to as many digits as you need using a key like $\boxed{\sqrt{x}}$. In this book, we normally give from two to four significant figures for such calculations.

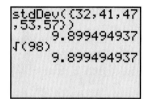

This screen supports the text discussion. Note that the standard deviation reported agrees with the approximation for $\sqrt{98}$.

dispersion, called the *variance*, but a more common measure is obtained by taking the square root of the variance. This gives, in effect, a kind of average of the deviations from the mean and is called the sample **standard deviation.** It is denoted by the letter s. (The standard deviation of a population is denoted σ, the lowercase Greek letter sigma.)

Continuing our calculations from the chart ON PAGE 749, we obtain

$$s = \sqrt{\frac{196 + 25 + 1 + 49 + 121}{4}} = \sqrt{\frac{392}{4}} = \sqrt{98} \approx 9.90.$$

The algorithm (process) described above for finding the sample standard deviation can be summarized as follows.

Calculation of Standard Deviation

Let a sample of n numbers $x_1, x_2, \ldots, x_n$ have mean $\bar{x}$. Then the **sample standard deviation, s,** of the numbers is given by

$$s = \sqrt{\frac{\Sigma (x - \bar{x})^2}{n - 1}}.$$

The individual steps involved in this calculation are as follows.

1. Calculate $\bar{x}$, the mean of the numbers.
2. Find the deviations from the mean.
3. Square each deviation.
4. Sum the squared deviations.
5. Divide the sum in Step 4 by $n - 1$.
6. Take the square root of the quotient in Step 5.

The preceding description helps show why standard deviation measures the amount of spread in a data set. For actual calculation purposes, we recommend the use of a scientific calculator, or a statistical calculator, that does all the detailed steps automatically. We illustrate both methods in the following example.

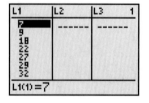

The sample in Example 2 is stored in a list. (The last entry, 40, is not shown here.)

EXAMPLE 2 Find the standard deviation of the sample

7, 9, 18, 22, 27, 29, 32, 40

by using **(a)** the step-by-step process, and **(b)** the statistical functions of a calculator.

(a) Carry out the six steps summarized above.

Step 1: Find the mean of the values.

$$\frac{7 + 9 + 18 + 22 + 27 + 29 + 32 + 40}{8} = 23$$

Step 2: Find the deviations from the mean.

Data value	7	9	18	22	27	29	32	40
Deviation	−16	−14	−5	−1	4	6	9	17

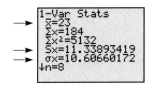

The arrows point to the mean and the sample and population standard deviations. See Example 2.

Step 3: Square each deviation.

Squares of deviations: 256 196 25 1 16 36 81 289

Step 4: Sum the squared deviations.

$$256 + 196 + 25 + 1 + 16 + 36 + 81 + 289 = 900$$

Step 5: Divide by $n - 1 = 8 - 1 = 7$.

$$900/7 \approx 128.57$$

Step 6: Take the square root.

$$\sqrt{128.57} \approx 11.3$$

(b) Enter the eight data values. (The key for entering data may look something like this: $\boxed{\Sigma +}$. Find out which key it is on your calculator.) Then press the key for standard deviation. (This one may look like

$\boxed{\text{STDEV}}$ or $\boxed{\text{SD}}$ or $\boxed{s_{n-1}}$ or $\boxed{\sigma_{n-1}}$.

If your calculator also has a key that looks like $\boxed{\sigma_n}$, it is probably for *population* standard deviation, which involves dividing by n rather than by $n - 1$, as mentioned earlier.) The result should again be 11.3. (If you *mistakenly* used the population standard deviation key, the result is 10.6 instead.)

For data given in the form of a frequency distribution, some calculators allow entry of both values and frequencies, or each value can be entered separately the number of times indicated by its frequency. Then press the standard deviation key.

The following example is included only to strengthen your understanding of frequency distributions and standard deviation, not as a practical algorithm for calculating.

EXAMPLE 3 Find the sample standard deviation for the frequency distribution shown here.

Value	Frequency
2	5
3	8
4	10
5	2

Complete the calculation by extending the table as shown below. To find the numbers in the "Deviation" column, first find the mean, and then subtract the mean from the numbers in the "Value" column.

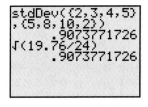

The screen supports the result in Example 3.

Value	Frequency	Value Times Frequency	Deviation	Squared Deviation	Squared Deviation Times Frequency
2	5	10	−1.36	1.8496	9.2480
3	8	24	−.36	.1296	1.0368
4	10	40	.64	.4096	4.0960
5	2	10	1.64	2.6896	5.3792
Sums	25	84			19.76

$$\bar{x} = \frac{84}{25} = 3.36 \qquad s = \sqrt{\frac{19.76}{24}} \approx \sqrt{.8233} \approx .91$$

Good average, poor consistency

Good consistency, poor average

In this case, a good average (greater central tendency) is more desirable than good consistency (lesser dispersion).

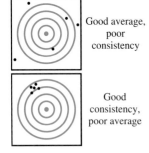

Good average, poor consistency

Good consistency, poor average

In this case, good consistency (lesser dispersion) is more desirable than a good average (central tendency).

Pafnuty Lvovich Chebyshev (1821–1894) was a Russian mathematician known mainly for his work on the theory of prime numbers. Chebyshev and French mathematician and statistician **Jules Bienaymé** (1796–1878) independently developed an important inequality of probability now known as the Bienaymé-Chebyshev inequality.

Central tendency and dispersion (or "spread tendency") are different and independent properties of a set of data. Sometimes one is more critical, sometimes the other. For example, in selecting which of two baskets of a dozen tomatoes to purchase (at equal costs), you would choose the one with the higher average weight (per tomato) rather than the one with the more consistent weight (smaller standard deviation). This maximizes the total weight of your purchase. (See the illustration at the side.)

On the other hand, a marksman is more concerned (to a certain point) with consistency than with average placement. The shooter of the bottom target at the side requires only a minor adjustment of his sights; that of the top target will require considerably more effort. (In general, consistent errors can be dealt with and corrected more easily than more dispersed errors.)

The Meaning of Dispersion Measures Standard deviation (or any other measure of dispersion) indicates the degree to which the numbers in a data set are "spread out." The larger the measure, the more spread. Beyond that, it is difficult to describe exactly what standard deviation measures. There is one useful result, however, named for the Russian mathematician Pafnuty Lvovich Chebyshev (1821–1894), which applies to all data sets.

Chebyshev's Theorem

For any set of numbers, regardless of how they are distributed, the fraction of them that lie within k standard deviations of their mean (where $k > 1$) is *at least*

$$1 - \frac{1}{k^2}.$$

Be sure to notice the words *at least* in the theorem. In certain distributions the fraction of items within k standard deviations of the mean may be more than $1 - 1/k^2$, but in no case will it ever be less. The theorem is meaningful for any value of k greater than 1 (integer or noninteger).

EXAMPLE 4 What is the minimum percentage of the items in a data set which lie within 3 standard deviations of the mean?

With $k = 3$, we calculate

$$1 - \frac{1}{3^2} = 1 - \frac{1}{9} = \frac{8}{9} \approx .889 = 88.9\%. \quad \longleftarrow \text{Minimum percentage}$$

Recall that inferential statistics involves drawing conclusions about populations based on data from samples. (See page 724.) Finding an "empirical" probability, as discussed earlier, is really a matter of drawing a statistical inference. Example 2 of Section 12.1 concluded that, since a Styrofoam cup landed on its top five times out of 50 tosses, $P(\text{top}) = 5/50 = 1/10$. That probability was an inference about the population of *all* possible tosses of that cup, based on a particular sample of 50 of those tosses.

We often use inferential techniques to compare two (or more) populations by comparing samples that come from those populations.

EXAMPLE 5 Two companies, A and B, sell 12-ounce jars of instant coffee. Five jars of each were randomly selected from markets, and the contents were carefully weighed, with the following results.

$$A: \quad 12.02, \quad 12.08, \quad 11.99, \quad 11.96, \quad 11.99$$
$$B: \quad 12.40, \quad 12.21, \quad 12.36, \quad 12.22, \quad 12.27$$

Find **(a)** which company provides more coffee in their jars and **(b)** which company fills its jars more consistently.

From the given data, we calculate the mean and standard deviation values for both samples. These values are shown in the margin.

Sample A	Sample B
$\bar{x}_A = 12.008$	$\bar{x}_B = 12.292$
$s_A = .0455$	$s_B = .0847$

(a) Since $\bar{x}_B$ is greater than $\bar{x}_A$, we *infer* that Company B most likely provides more coffee (greater mean).

(b) Since s_A is less than s_B, we *infer* that Company A seems more consistent (smaller standard deviation). ■

The conclusions drawn in Example 5 are tentative because the samples were small. We could place more confidence in our inferences if we used larger samples, for then it would be more likely that the samples were accurate representations of their respective populations. In a more detailed study of inferential statistics, you would learn techniques best suited to small samples as well as those best suited to larger samples. You would also learn to state your inferences more precisely so that the degree of uncertainty is conveyed along with the basic conclusion.

13.3 EXERCISES

1. If your calculator finds both kinds of standard deviation, the sample standard deviation and the population standard deviation, which of the two will be a larger number for a given set of data? (*Hint:* Recall the difference between how the two standard deviations are calculated.)

2. If your calculator finds only one kind of standard deviation, explain how you would determine whether it is sample or population standard deviation (assuming your calculator manual is not available).

Find **(a)** *the range, and* **(b)** *the standard deviation for each sample (Exercises 3–12). Round fractional answers to the nearest hundredth.*

3. 2, 5, 6, 8, 9, 11, 15, 19
4. 8, 5, 12, 8, 9, 15, 21, 16, 3
5. 25, 34, 22, 41, 30, 27, 31
6. 67, 83, 55, 68, 77, 63, 84, 72, 65
7. 318, 326, 331, 308, 316, 322, 310, 319, 324, 330
8. 5.7, 8.3, 7.4, 6.6, 7.4, 6.8, 7.1, 8.0, 8.5, 7.9, 7.1, 7.4, 6.9, 8.2
9. 84.53, 84.60, 84.58, 84.48, 84.72, 85.62, 85.03, 85.10, 84.96
10. 206.3, 210.4, 209.3, 211.1, 210.8, 213.5, 212.6, 210.5, 211.0, 214.2

11.

Value	Frequency
9	3
7	4
5	7
3	5
1	2

12.

Value	Frequency
14	8
16	12
18	15
20	14
22	10
24	6
26	3

Use Chebyshev's theorem for Exercises 13–28.

Find the least possible fraction of the numbers in a data set lying within the given number of standard deviations of the mean. Give answers as standard fractions reduced to lowest terms.

13. 2
14. 4
15. 7/2
16. 11/4

Find the least possible percentage (to the nearest tenth of a percent) of the items in a distribution lying within the given number of standard deviations of the mean.

17. 3
18. 5
19. 5/3
20. 5/2

In a certain distribution of numbers, the mean is 70 and the standard deviation is 8. At least what fraction of the numbers are between the following pairs of numbers? Give answers as common fractions reduced to lowest terms.

21. 54 and 86
22. 46 and 94
23. 38 and 102
24. 30 and 110

In the same distribution (mean 70 and standard deviation 8), find the largest fraction of the numbers that could meet the following requirements. Give answers as common fractions reduced to lowest terms.

25. less than 54 or more than 86
26. less than 50 or more than 90
27. less than 42 or more than 98
28. less than 52 or more than 88

Bonus Pay for a Baseball Team Leigh Jacka owns a minor league baseball team. Each time the team wins a game, Leigh pays the nine starting players, the manager, and two coaches bonuses, which are certain percentages of their regular salaries. The amounts paid are listed here.

$80, $105, $120, $175, $185, $190,
$205, $210, $215, $300, $320, $325

Use this distribution of bonuses for Exercises 29–34.

29. Find the mean of the distribution.
30. Find the standard deviation of the distribution.
31. How many of the bonus amounts are within one standard deviation of the mean?
32. How many of the bonus amounts are within two standard deviations of the mean?
33. What does Chebyshev's theorem say about the number of the amounts that are within two standard deviations of the mean?
34. Explain any discrepancy between your answers for Exercises 32 and 33.

Lifetimes of Engine Control Modules Chris Christensen manages the service department of a trucking company. Each truck in the fleet utilizes an electronic engine control module, which must be replaced when it fails. Long-lasting modules are desirable, of course, but a preventive replacement program can also avoid costly breakdowns on the highway. For this purpose it is desirable that the modules be fairly consistent in their lifetimes; that is, they should all last about the same number of miles before failure, so that the timing of preventive replacements can be done accurately. Chris tested a sample of 20 Brand A modules, and they lasted 43,560 highway miles on the average (mean), with a standard deviation of 2116 miles. The listing below shows how long each of another sample of 20 Brand B modules lasted.

50,660, 41,300, 45,680, 48,840, 47,300,
51,220, 49,100, 48,660, 47,790, 47,210,
50,050, 49,920, 47,420, 45,880, 50,110,
49,910, 47,930, 48,800, 46,690, 49,040

Use the data above for Exercises 35–37.

35. According to the sampling that was done, which brand of module has the longer average life (in highway miles)?
36. Which brand of module apparently has a more consistent (or uniform) length of life (in highway miles)?
37. If Brands A and B are the only modules available, which one should Chris purchase for the maintenance program? Explain your reasoning.

Utilize the following sample for Exercises 38–43.

13, 14, 16, 18, 20, 22, 25

38. Compute the mean and standard deviation for the sample (each to the nearest hundredth).
39. Now add 5 to each item of the given sample and compute the mean and standard deviation for the new sample.
40. Go back to the original sample. This time subtract 10 from each item, and compute the mean and standard deviation of the new sample.
41. Based on your answers for Exercises 38–40, make conjectures about what happens to the mean and standard deviation when all items of the sample have the same constant k added or subtracted.
42. Go back to the original sample again. This time multiply each item by 3, and compute the mean and standard deviation of the new sample.

13.3 Measures of Dispersion

43. Based on your answers for Exercises 38 and 42, make conjectures about what happens to the mean and standard deviation when all items of the sample are multiplied by the same constant k.

44. In Section 13.2 we showed that the mean, as a measure of central tendency, is highly sensitive to extreme values. Which measure of dispersion, covered in this section, would be more sensitive to extreme values? Illustrate your answer with one or more examples.

45. *A Cereal Marketing Survey* A food distribution company conducted a survey to determine whether a proposed premium to be included in boxes of their cereal was appealing enough to generate new sales. Four cities were used as test markets, where the cereal was distributed with the premium, and four cities as control markets, where the cereal was distributed without the premium. The eight cities were chosen on the basis of their similarity in terms of population, per capita income, and total cereal purchase volume. The results follow.

		Percent Change in Average Market Share per Month
Test cities	1	+18
	2	+15
	3	+7
	4	+10
Control cities	1	+1
	2	−8
	3	−5
	4	0

(a) Find the mean of the change in market share for the four test cities.
(b) Find the mean of the change in market share for the four control cities.
(c) Find the standard deviation of the change in market share for the test cities.
(d) Find the standard deviation of the change in market share for the control cities.
(e) Find the difference between the means of (a) and (b). This difference represents the estimate of the percent change in sales due to the premium.
(f) The two standard deviations from (c) and (d) were used to calculate an "error" of ±7.95 for the estimate in (e). With this amount of error, what are the smallest and largest estimates of the increase in sales?

On the basis of the interval estimate of part (f) the company decided to mass produce the premium and distribute it nationally.

In a skewed distribution, the mean will be farther out toward the tail than the median, as shown in the sketch.

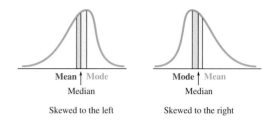

Mean | Mode
Median
Skewed to the left

Mode | Mean
Median
Skewed to the right

A common way of measuring the degree of skewness, which involves both central tendency and dispersion, is with the **skewness coefficient,** *calculated as follows.*

$$\frac{3 \times (\text{mean} - \text{median})}{\text{standard deviation}}$$

46. Under what conditions would the skewness coefficient be each of the following?
(a) positive   (b) negative

47. Explain why the mean of a skewed distribution is always farther out toward the tail than the median.

48. Suppose that the mean length of patient stay (in days) in U.S. hospitals is 2.7 days, with a standard deviation of 7.1 days. Make a sketch of how this distribution may look.

For Exercises 49–51, refer to the grouped frequency distribution shown here. (Also refer to Exercises 60–62 in Section 13.2.)

Class Limits	Frequency f
21–25	5
26–30	3
31–35	8
36–40	12
41–45	21
46–50	38
51–55	35
56–60	20

49. Can you identify any specific data items that occurred in this sample?

50. Can you compute the actual standard deviation for this sample?

51. Describe how you might approximate the standard deviation for this sample. Justify your procedure.

52. Suppose the frequency distribution of Example 3 involved 50 or 100 (or even more) distinct data values, rather than just four. Explain why the procedure of that example would then be very inefficient.

53. A "J-shaped" distribution can be skewed either to the right or to the left.
(a) In a J-shaped distribution skewed to the right, which data item would be the mode, the largest or the smallest item?
(b) In a J-shaped distribution skewed to the left, which data item would be the mode, the largest or the smallest item?
(c) Explain why the mode is a weak measure of central tendency for a J-shaped distribution.

13.4 Measures of Position

Measures of central tendency and measures of dispersion give us an effective way of characterizing an overall set of data. Central tendency has to do with where, along a number scale, the overall data set is centered. Dispersion has to do with how much the data set is spread out from the center point. And Chebyshev's theorem, stated in the previous section, tells us in a general sense what portions of the data set may be dispersed different amounts from the center point.

In many cases, we are especially interested in certain individual items in the data set, rather than in the set as a whole. We still need a way of measuring how an item fits into the collection, how it compares to other items in the collection, or even how it compares to another item in another collection. There are several common ways of creating such measures. Since they measure an item's position within the data set, they usually are called **measures of position.**

The z-score Each individual item in a sample can be assigned a *z*-score, which is defined as follows.

> **The z-score**
> If x is a data item in a sample with mean $\bar{x}$ and standard deviation s, then the **z-score** of x is given by
> $$z = \frac{x - \bar{x}}{s}.$$

Since $x - \bar{x}$ gives the amount by which x differs (or deviates) from the mean $\bar{x}$, $(x - \bar{x})/s$ gives the number of standard deviations by which x differs from $\bar{x}$. Notice that z will be positive if x is greater than $\bar{x}$ but negative if x is less than $\bar{x}$. Chebyshev's theorem assures us that, in any distribution whatsoever, at least 89% (roughly) of the items will lie within three standard deviations of the mean. That is, at least 89% of the items will have *z*-scores between -3 and 3. In fact, many common distributions, especially symmetric ones, have considerably more than 89% of their items within three standard deviations of the mean (as we will see in the next section). Hence, a *z*-score greater than 3 or less than -3 is rare.

EXAMPLE 1 Two friends, Reza and Weihua, who take different history classes, received their midterm exams back on the same day. Reza's score was 86 while Weihua's was only 78. Which student did relatively better, given the class data shown here?

	Reza	Weihua
Class mean	73	69
Class standard deviation	8	5

Calculate as follows.

$$\text{Reza: } z = \frac{86 - 73}{8} = 1.625 \qquad \text{Weihua: } z = \frac{78 - 69}{5} = 1.8$$

Since Weihua's z-score is higher, she was positioned relatively higher within her class than Reza was within his class.

Percentiles When you take the Scholastic Aptitude Test (SAT), or any other standardized test taken by large numbers of students, your raw score usually is converted to a **percentile** score, which is defined as follows.

Percentile

If approximately n percent of the items in a distribution are less than the number x, then x is the ***n*th percentile** of the distribution.

For example, if you scored at the eighty-third percentile on the SAT, it means that you outscored approximately 83% of all those who took the test. (It does *not* mean that you got 83% of the answers correct.) Since the percentile score positions an item within the data set, it is another "measure of position." The following example approximates percentiles for a fairly small collection of data.

EXAMPLE 2 The following are the numbers of dinner customers served by a restaurant on 40 consecutive days. (The numbers have been ranked lowest to highest.)

```
46  51  52  55  56  56  58  59  59  59
61  61  62  62  63  63  64  64  64  65
66  66  66  67  67  67  68  68  69  69
70  70  71  71  72  75  79  79  83  88
```

For this data set, find each of the following.

(a) The thirty-fifth percentile

The thirty-fifth percentile can be taken as the item below which 35 percent of the items are ranked. Since 35 percent of 40 is $(.35)(40) = 14$, we take the fifteenth item, or 63, as the thirty-fifth percentile.

(b) The eighty-sixth percentile

Since 86 percent of 40 is $(.86)(40) = 34.4$, we round *up* and take the eighty-sixth percentile to be the thirty-fifth item, or 72.

Technically, percentiles originally were conceived as a set of 99 values (not necessarily data items) along the scale that would divide the data set into 100 equal parts. They were computed only for very large data sets. With smaller data sets, as in Example 2 on page 757, dividing the data into 100 parts would necessarily leave many of those parts empty. However, the modern techniques of exploratory data analysis seek to apply the percentile concept to even small data sets. Thus we use approximation techniques as in Example 2. Another option is to divide the data into a lesser number of equal (or nearly equal) parts.

Deciles and Quartiles   **Deciles** are the nine values (denoted $D_1, D_2, \ldots, D_9$) along the scale that divide a data set into ten (approximately) equal parts, and **quartiles** are the three values (Q_1, Q_2, and Q_3) that divide a data set into four (approximately) equal parts. Since deciles and quartiles serve to position particular items within portions of a distribution, they also are "measures of position."

Since a decile always is equivalent to a certain percentile, it can be approximated just as in Example 2.

EXAMPLE 3   Find the fourth decile for the dinner customer data of Example 2.

Refer to the ranked data table. The fourth decile is the fortieth percentile, and 40% of 40 is $(.40)(40) = 16$. We take the fourth decile to be the seventeenth item, or 64.

Although the three quartiles also are equivalent to corresponding percentiles, notice that the second quartile, Q_2, also is equivalent to the median, a measure of central tendency introduced in Section 13.2. A common convention for computing quartiles goes back to the way we computed the median.

> **Finding Quartiles**
>
> For any set of data (ranked in order from smallest to largest):
>
> The **second quartile, Q_2,** is just the median, the middle item when the number of items is odd, or the mean of the two middle items when the number of items is even.
>
> The **first quartile, Q_1,** is the median of all items below Q_2.
>
> The **third quartile, Q_3,** is the median of all items above Q_2.

EXAMPLE 4   Find the three quartiles for the data of Example 2.

Refer to the ranked data. The two middle data items are 65 and 66, so

$$Q_2 = \frac{65 + 66}{2} = 65.5.$$

The lowest 20 items (an even number) are all below Q_2, and the two middle items in that set are 59 and 61.

$$Q_1 = \frac{59 + 61}{2} = 60$$

The highest 20 items are above Q_2.

$$Q_3 = \frac{69 + 70}{2} = 69.5$$

The Box Plot

A **box plot**, or **box-and-whisker plot**, involves the median (a measure of central tendency), the range (a measure of dispersion), and the first and third quartiles (measures of position), all incorporated into a simple visual display.

> ### Box Plot
> For a given set of data, a **box plot** (or **box-and-whisker plot**) consists of a rectangular box positioned above a numerical scale, extending from Q_1 to Q_3, with the value of Q_2 (the median) indicated within the box, and with "whiskers" (line segments) extending to the left and right from the box out to the minimum and maximum data items.

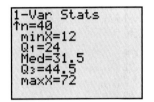

This screen supports the results of Example 5.

EXAMPLE 5 Construct a box plot for the weekly study times data of Example 2 in Section 13.1.

To determine the quartiles and the minimum and maximum values more easily, we use the stem-and-leaf display (with ranked leaves), given in Table 9 of Section 13.2:

1	2 4 5 6 7 8
2	0 2 3 4 4 5 6 6 9 9 9
3	0 1 1 2 3 6 6 8 9
4	0 1 4 4 5 5 7
5	2 5 5 8
6	0 2
7	2

The median (determined earlier in Example 6 of Section 13.2) is

$$\frac{31 + 32}{2} = 31.5.$$

From the stem-and-leaf display,

$$Q_1 = \frac{24 + 24}{2} = 24 \quad \text{and} \quad Q_3 = \frac{44 + 45}{2} = 44.5.$$

The minimum and maximum items are 12 and 72. The box plot is shown in Figure 10.

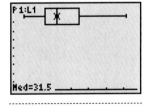

This box plot corresponds to the results of Example 5. It indicates the median in the display at the bottom. The TRACE function of the TI-83 Plus will locate the minimum, maximum, and quartile values as well.

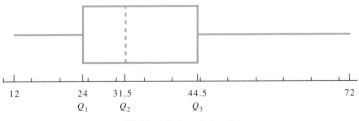

Weekly study times (in hours)

Box plot

FIGURE 10

13.4 EXERCISES

For each of Exercises 1–4, make use of z-scores.

1. *Relative Positions on Calculus Quizzes* In a calculus class, Chris Doran scored 5 on a quiz for which the class mean and standard deviation were 4.6 and 2.1, respectively. Lynn Colgin scored 6 on another quiz for which the class mean and standard deviation were 4.9 and 2.3, respectively. Relatively speaking, which student did better?

2. *Relative Performances in Track Events* In Saturday's track meet, Ryan Flahive, a high jumper, jumped 6 feet 3 inches. Conference high jump marks for the past season had a mean of 6 feet even and a standard deviation of 3.5 inches. Michael Drago, Ryan's teammate, achieved 18 feet 4 inches in the long jump. In that event the conference season average (mean) and standard deviation were 16 feet 6 inches and 1 foot 10 inches, respectively. Relative to this past season in this conference, which athlete had a better performance on Saturday?

3. *Relative Lifetimes of Tires* The lifetimes of Brand A tires are distributed with mean 45,000 miles and standard deviation 4500 miles, while Brand B tires last for only 38,000 miles on the average (mean) with standard deviation 2080 miles. Jutta's Brand A tires lasted 37,000 miles and Arvind's Brand B tires lasted 35,000 miles. Relatively speaking, within their own brands, which driver got the better wear?

4. *Relative Ratings of Fish Caught* In a certain lake, the trout average 12 inches in length with a standard deviation of 2.75 inches. The bass average 4 pounds in weight with a standard deviation of .8 pound. If Imelda caught an 18-inch trout and Timothy caught a 6-pound bass, then relatively speaking, which catch was the better trophy?

Numbers of Restaurant Customers Refer to the dinner customers data of Example 2. Approximate each of the following. (Use the methods illustrated in this section.)

5. the fifteenth percentile
6. the seventy-fifth percentile
7. the third decile
8. the eighth decile

Leading Rebounders in the National Basketball Association For Exercises 9–19, refer to the top 10 rebounders data for the 2000–2001 season of the National Basketball Association. (Source: The World Almanac and Book of Facts 2002, page 937.)

Leading Rebounders	Games Played	Offensive Rebounds	Defensive Rebounds	Total Rebounds	Average Rebounds per Game
Dikembe Mutombo, Atlanta-Philadelphia	75	307	708	1015	13.5
Ben Wallace, Detroit	80	303	749	1052	13.2
Shaquille O'Neal, L.A. Lakers	74	291	649	940	12.7
Tim Duncan, San Antonio	82	259	738	997	12.2
Antonio McDyess, Denver	70	240	605	845	12.1
Kevin Garnett, Minnesota	81	219	702	921	11.4
Chris Webber, Sacramento	70	179	598	777	11.1
Shawn Marion, Phoenix	79	220	628	848	10.7
Antonio Davis, Toronto	78	274	513	787	10.1
Elton Brand, Chicago	74	285	461	746	10.1

Compute z-scores for Exercises 9–12. (Give answers to two decimal places.)

9. Dikembe Mutombo's games played
10. Shaquille O'Neal's offensive rebounds
11. Chris Webber's average rebounds per game
12. Kevin Garnett's total rebounds

In each of Exercises 13–16, determine who occupied the given position.

13. the thirty-fifth percentile in defensive rebounds
14. the seventy-third percentile in total rebounds
15. the sixth decile in games played
16. the third decile in average rebounds per game
17. Determine who was relatively higher: Shaquille O'Neal in offensive rebounds or Tim Duncan in defensive rebounds.
18. Construct a box plot for the average rebounds per game data.
19. What does your box plot of Exercise 18 indicate about each of the following characteristics of this data?
    (a) the central tendency
    (b) the dispersion
    (c) the location of the middle half of the data items
20. The text stated that, for *any* distribution of data, at least 89% of the items will be within three standard deviations of the mean. Why couldn't we just move some items farther out from the mean to obtain a new distribution that would violate this condition?
21. Describe the basic difference between a measure of central tendency and a measure of position.

This chapter has introduced three major characteristics: central tendency, dispersion, and position, and has developed various ways of measuring them in numerical data. In each of Exercises 22–26, a new measure is described. Explain in each case which of the three characteristics you think it would measure and why.

22. Midrange = $\dfrac{\text{minimum item} + \text{maximum item}}{2}$

23. Midquartile = $\dfrac{Q_1 + Q_3}{2}$

24. Coefficient of Variation = $\dfrac{s}{\bar{x}} \cdot 100$

25. Interquartile range = $Q_3 - Q_1$

26. Semi-interquartile range = $\dfrac{Q_3 - Q_1}{2}$

27. The "skewness coefficient" was defined in the exercises of the previous section, and it is equivalent to
$$\dfrac{3 \times (\bar{x} - Q_2)}{s}.$$
Is this a measure of individual data items or of the overall distribution?

28. In a national standardized test, Jennifer scored at the ninety-second percentile. If 67,500 individuals took the test, about how many scored higher than Jennifer did?

29. Let the three quartiles (from smallest to largest) for a large population of scores be denoted Q_1, Q_2, and Q_3.
    (a) Is it necessarily true that $Q_2 - Q_1 = Q_3 - Q_2$?
    (b) Explain your answer to part (a).

In Exercises 30–33, answer yes *or* no *and explain your answer. (Consult earlier exercises for definitions of some of these measures.)*

30. Is the midquartile necessarily the same as the median?
31. Is the midquartile necessarily the same as the midrange?
32. Is the interquartile range necessarily half the range?
33. Is the semi-interquartile range necessarily half the interquartile range?

34. *Relative Positions on a Standardized Chemistry Test* Omer and Alessandro participated in the standardization process for a new statewide chemistry test. Within the large group participating, their raw scores and corresponding z-scores were as shown here.

	Raw Score	z-score
Omer	60	.69
Alessandro	72	1.67

Find the overall mean and standard deviation of the distribution of scores. (Give answers to two decimal places.)

Rating Passers in the National Football League Since the National Football League began keeping official statistics in 1932, one important aspect of the performance of quarterbacks, namely their passing effectiveness, has been rated by several different methods. The current system, adopted in 1973, is based on four performance components considered most important: completions, touchdowns, yards gained, and interceptions, as percentages of the number of passes attempted. The actual computation of the ratings seems to be poorly understood by fans and media sportscasters alike, but can be accomplished with the following formula.*

$$\text{Rating} = \frac{\left(250 \times \frac{C}{A}\right) + \left(1000 \times \frac{T}{A}\right) + \left(12.5 \times \frac{Y}{A}\right) + 6.25 - \left(1250 \times \frac{I}{A}\right)}{3},$$

where A = attempted passes, C = completed passes, Y = yards gained passing, T = touchdown passes, and I = interceptions.

In addition to the weighting factors (coefficients) appearing in the formula, the four category ratios are limited to the following maximums.

$.775$ for $\frac{C}{A}$, $.11875$ for $\frac{T}{A}$, 12.5 for $\frac{Y}{A}$, $.095$ for $\frac{I}{A}$

These limitations are intended to prevent any one component of performance from having an undue effect on the overall rating. They are not often invoked, but in special cases can have a significant effect. (See Exercises 46–49.)

The formula above rates all passers against the same performance standard and is applied, for example, after a single game, an entire season, or a career.

In 2001, the top-rated passers in the National Football Conference (N.F.C.) and the American Football Conference (A.F.C.) were Kurt Warner of the St. Louis Rams, and Rich Gannon of the Oakland Raiders, respectively. Use the data shown here for Exercises 35 and 36. Give answers to one decimal place.

35. Compute Warner's season rating.

36. Compute Gannon's season rating.

	Warner	Gannon
Attempts	546	549
Completions	375	361
Yards	4830	3828
TDs	36	27
Interceptions	22	9

Source: http://www.nfl.com.

The ratings for the ten leading passers (2001) in each conference are ranked in the table below. Find the requested measures (to one decimal place) in Exercises 37–40.

N.F.C. Passer	Rating Points	A.F.C. Passer	Rating Points
Warner, St. Louis	101.4	Gannon, Oakland	95.5
Garcia, San Francisco	94.8	McNair, Tennessee	90.2
Favre, Green Bay	94.1	Brady, New England	86.5
McNabb, Philadelphia	84.3	Manning, Indianapolis	84.1
Chandler, Atlanta	84.1	Brunell, Jacksonville	84.1
Culpepper, Minnesota	83.3	Stewart, Pittsburgh	81.7
Plummer, Arizona	79.6	Fiedler, Miami	80.3
B. Johnson, Tampa Bay	77.7	Griese, Denver	78.5
Collins, N.Y. Giants	77.1	Van Pelt, Buffalo	76.4
Batch, Detroit	76.8	Testaverde, N.Y. Jets	75.3

Source: World Almanac and Book of Facts 2003.

---

*Joseph Farabee of American River College derived the formula after sifting through a variety of descriptions of the rating system in sports columns and NFL publications.

37. the three quartiles for the N.F.C. ratings

38. the three quartiles for the A.F.C. ratings

39. the eighth decile for all ratings (the N.F.C. and A.F.C. figures combined)

40. the ninetieth percentile for all ratings (the N.F.C. and A.F.C. figures combined)

41. Construct box plots for both conferences, one above the other in the same drawing.

Refer to the box plots of Exercise 41 for Exercises 42–45. In each case, identify the conference for which the given measure is greater, and estimate about how much greater it is.

42. the median

43. the midquartile

44. the range

45. the interquartile range

Sid Luckman of the Chicago Bears, the passing champion of the 1943 season, had the following pertinent statistics in that year.

202 attempts, 110 completions, 28 touchdowns, 2194 yards, 12 interceptions

Use these numbers in the rating formula for the following exercises.

46. Compute Luckman's 1943 rating, being careful to replace each ratio in the formula with its allowed maximum if necessary.

47. Compute Luckman's rating assuming no restrictions on the ratios.

48. Which component of Luckman's passing game that year was apparently very unusual, even among high-caliber passers?

49. Compare the two ratings of Exercises 46 and 47 with Steve Young's all-time record season rating of 112.8 (achieved in 1994).

13.5 The Normal Distribution

x	$P(x)$
0	.03125
1	.15625
2	.31250
3	.31250
4	.15625
5	.03125

Probability Distribution

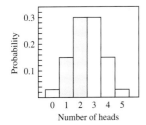

FIGURE 11

A random variable that can take on only certain fixed values is called a **discrete random variable.** For example, the number of heads in 5 tosses of a coin is discrete since its only possible values are 0, 1, 2, 3, 4, and 5. A variable whose values are not restricted in this way is a **continuous random variable.** For example, the diameter of camellia blossoms would be a continuous variable, spread over a scale perhaps from 5 to 25 centimeters. The values would not be restricted to whole numbers, or even to tenths, or hundredths, etc. Although in practice we may not measure the diameters to more accuracy than, say, tenths of centimeters, they could theoretically take on any of the infinitely many values from 5 to 25. In terms of set theory, we can say that a discrete random variable can take on only a countable number of values, whereas a continuous random variable can take on an uncountable number of values.

In contrast to most distributions discussed earlier in this chapter, which were empirical (based on observation), the distributions covered in this section and the next are theoretical (based on theoretical probabilities). A knowledge of theoretical distributions enables us to identify when actual observations are inconsistent with stated assumptions, which is the key to inferential statistics.

The theoretical probability distribution for the discrete random variable "number of heads" when 5 fair coins are tossed is shown in the margin, and Figure 11 shows the corresponding histogram. The probability values can be found using the binomial probability formula or using Pascal's triangle.

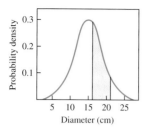

FIGURE 12

The normal curve was first developed by **Abraham De Moivre** (1667–1754), but his work went unnoticed for many years. It was independently redeveloped by Pierre Laplace (1749–1827) and Carl Friedrich Gauss (1777–1855). Gauss found so many uses for this curve that it is sometimes called the *Gaussian curve*.

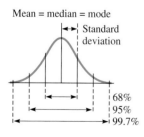

FIGURE 13

Since each rectangle in Figure 11 is 1 unit wide, the *area* of the rectangle is also equal to the probability of the corresponding number of heads. The area, and thus the probability, for the event "1 head or 2 heads" is shaded in the figure. The graph consists of 6 distinct rectangles since "number of heads" is a *discrete* variable with 6 possible values. The sum of the 6 rectangular areas is exactly 1 square unit.

In contrast to the discrete "number of heads" distribution on page 763, a probability distribution for camellia blossom diameters cannot be tabulated or graphed in quite the same way, since this variable is *continuous*. The graph would be smeared out into a "continuous" bell-shaped curve (rather than a set of rectangles) as shown in Figure 12. The vertical scale on the graph in this case shows what we call "probability density," the probability per unit along the horizontal axis.

This particular curve is highest at a diameter value of 15 cm, its center point, and drops off rapidly and equally toward a zero level in both directions. A symmetric, bell-shaped curve of this type is called a **normal curve.** Any random variable whose graph has this characteristic shape is said to have a **normal distribution.** A great many continuous random variables have this type of normal distribution, and even the distributions of discrete variables can often be approximated closely by normal curves. Thus, normal curves are well worth studying. (Of course, not all distributions are normal or approximately so—for example, income distributions usually are not.)

On a normal curve, if the quantity shown on the horizontal axis is the number of standard deviations from the mean, rather than values of the random variable itself, then we call the curve the **standard normal curve.** In that case, the horizontal dimension measures numbers of standard deviations, with the value being 0 at the mean of the distribution. The number of standard deviations, associated with a point on the horizontal axis, is the "standard score" for that point. It is the same as the *z*-score defined in the previous section.

The area under the curve along a certain interval is numerically equal to the probability that the random variable will have a value in the corresponding interval. The total area under the standard normal curve is exactly 1 square unit.

The area of the shaded region in Figure 12 is equal to the probability of a randomly chosen blossom having a diameter in the interval from exactly 16.4 cm to exactly 21.2 cm. To find that area, we first convert 16.4 and 21.2 to standard deviation units. With this approach, we can deal with any normal curve by comparing it to the standard normal curve. A single table of areas for the standard normal curve then provides probabilities for all normal curves, as will be shown later in this section.

Several properties of the normal curve are summarized below and are illustrated in Figure 13.

Properties of the Normal Curve

The graph of a normal curve is bell-shaped and symmetric about a vertical line through its center.

The mean, median, and mode of a normal curve are all equal and occur at the center of the distribution.

Empirical Rule   About 68% of all data values of a normal curve lie within 1 standard deviation of the mean (in both directions), about 95% within 2 standard deviations, and about 99.7% within 3 standard deviations.

13.5 The Normal Distribution

Close but Never Touching
When a curve approaches closer and closer to a line, without ever actually meeting it (as a normal curve approaches the horizontal axis), the line is called an **asymptote**, and the curve approaches the line **asymptotically**.

The empirical rule indicates that a very small percentage of the items in a normal distribution will lie more than 3 standard deviations from the mean (approximately .3%, divided equally between the upper and lower tails of the distribution). As we move away from the center, the curve falls rapidly, and then more and more gradually, toward the axis. But it *never* actually reaches the axis. No matter how far out we go, there is always a chance (though very small) of an item occurring even farther out. Theoretically then, the range of a true normal distribution is infinite.

It is important to realize that the percentage of items within a certain interval is equivalent to the probability that a randomly chosen item will lie in that interval. Thus one result of the empirical rule is that, if we choose a single number at random from a normal distribution, the probability that it will lie within 1 standard deviation of the mean is about .68.

EXAMPLE 1 Suppose 300 chemistry students take a midterm exam and that the distribution of their scores can be treated as normal. Find the number of scores falling into each of the following intervals.

(a) Within 1 standard deviation of the mean

By the empirical rule, 68% of all scores lie within 1 standard deviation of the mean. Since there is a total of 300 scores, the number of scores within 1 standard deviation is

$$(68\%)(300) = (.68)(300) = 204.$$

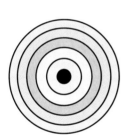

(b) Within 2 standard deviations of the mean

A total of 95% of all scores lie within 2 standard deviations of the mean. This would be

$$(.95)(300) = 285.$$

Most questions we need to answer about normal distributions involve regions other than those within 1, 2, or 3 standard deviations of the mean. For example, we might need the percentage of items within 1 1/2 or 2 1/5 standard deviations of the mean, or perhaps the area under the curve from .8 to 1.3 standard deviations above the mean. In such cases, we need more than the empirical rule. The traditional approach is to refer to a table of area values, such as Table 10, which appears on page 766. Computer software packages designed for statistical uses usually will produce the required values on command and some advanced calculators also have this capability. Those tools are recommended. As an optional approach, we illustrate the use of Table 10 here.

The table gives the fraction of all scores in a normal distribution that lie between the mean and z standard deviations from the mean. *Because of the symmetry of the normal curve, the table can be used for values above the mean or below the mean.* All of the items in the table can be thought of as corresponding to the area under the curve. The total area is arranged to be 1.000 square unit, with .500 square unit on each side of the mean. The table shows that at 3.30 standard deviations from the mean, essentially all of the area is accounted for. Whatever remains beyond is so small that it does not appear in the first three decimal places.

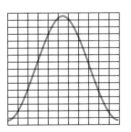

A **normal distribution** occurs in darts if the player, always aiming at the bull's-eye, tosses a fairly large number of times, and the aim on each toss is affected by independent random errors.

The column under *A* gives the proportion of the area under the entire curve that is between $z = 0$ and a positive value of *z*.

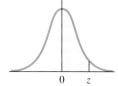

Because the curve is symmetric about the 0-value, the area between $z = 0$ and a *negative* value of *z* can be found by using the corresponding positive value of *z*.

TABLE 10 Areas Under the Standard Normal Curve

z	A	z	A	z	A	z	A	z	A	z	A
.00	.000	.56	.212	1.12	.369	1.68	.454	2.24	.487	2.80	.497
.01	.004	.57	.216	1.13	.371	1.69	.454	2.25	.488	2.81	.498
.02	.008	.58	.219	1.14	.373	1.70	.455	2.26	.488	2.82	.498
.03	.012	.59	.222	1.15	.375	1.71	.456	2.27	.488	2.83	.498
.04	.016	.60	.226	1.16	.377	1.72	.457	2.28	.489	2.84	.498
.05	.020	.61	.229	1.17	.379	1.73	.458	2.29	.489	2.85	.498
.06	.024	.62	.232	1.18	.381	1.74	.459	2.30	.489	2.86	.498
.07	.028	.63	.236	1.19	.383	1.75	.460	2.31	.490	2.87	.498
.08	.032	.64	.239	1.20	.385	1.76	.461	2.32	.490	2.88	.498
.09	.036	.65	.242	1.21	.387	1.77	.462	2.33	.490	2.89	.498
.10	.040	.66	.245	1.22	.389	1.78	.462	2.34	.490	2.90	.498
.11	.044	.67	.249	1.23	.391	1.79	.463	2.35	.491	2.91	.498
.12	.048	.68	.252	1.24	.393	1.80	.464	2.36	.491	2.92	.498
.13	.052	.69	.255	1.25	.394	1.81	.465	2.37	.491	2.93	.498
.14	.056	.70	.258	1.26	.396	1.82	.466	2.38	.491	2.94	.498
.15	.060	.71	.261	1.27	.398	1.83	.466	2.39	.492	2.95	.498
.16	.064	.72	.264	1.28	.400	1.84	.467	2.40	.492	2.96	.498
.17	.067	.73	.267	1.29	.401	1.85	.468	2.41	.492	2.97	.499
.18	.071	.74	.270	1.30	.403	1.86	.469	2.42	.492	2.98	.499
.19	.075	.75	.273	1.31	.405	1.87	.469	2.43	.492	2.99	.499
.20	.079	.76	.276	1.32	.407	1.88	.470	2.44	.493	3.00	.499
.21	.083	.77	.279	1.33	.408	1.89	.471	2.45	.493	3.01	.499
.22	.087	.78	.282	1.34	.410	1.90	.471	2.46	.493	3.02	.499
.23	.091	.79	.285	1.35	.411	1.91	.472	2.47	.493	3.03	.499
.24	.095	.80	.288	1.36	.413	1.92	.473	2.48	.493	3.04	.499
.25	.099	.81	.291	1.37	.415	1.93	.473	2.49	.494	3.05	.499
.26	.103	.82	.294	1.38	.416	1.94	.474	2.50	.494	3.06	.499
.27	.106	.83	.297	1.39	.418	1.95	.474	2.51	.494	3.07	.499
.28	.110	.84	.300	1.40	.419	1.96	.475	2.52	.494	3.08	.499
.29	.114	.85	.302	1.41	.421	1.97	.476	2.53	.494	3.09	.499
.30	.118	.86	.305	1.42	.422	1.98	.476	2.54	.494	3.10	.499
.31	.122	.87	.308	1.43	.424	1.99	.477	2.55	.495	3.11	.499
.32	.126	.88	.311	1.44	.425	2.00	.477	2.56	.495	3.12	.499
.33	.129	.89	.313	1.45	.426	2.01	.478	2.57	.495	3.13	.499
.34	.133	.90	.316	1.46	.428	2.02	.478	2.58	.495	3.14	.499
.35	.137	.91	.319	1.47	.429	2.03	.479	2.59	.495	3.15	.499
.36	.141	.92	.321	1.48	.431	2.04	.479	2.60	.495	3.16	.499
.37	.144	.93	.324	1.49	.432	2.05	.480	2.61	.495	3.17	.499
.38	.148	.94	.326	1.50	.433	2.06	.480	2.62	.496	3.18	.499
.39	.152	.95	.329	1.51	.434	2.07	.481	2.63	.496	3.19	.499
.40	.155	.96	.331	1.52	.436	2.08	.481	2.64	.496	3.20	.499
.41	.159	.97	.334	1.53	.437	2.09	.482	2.65	.496	3.21	.499
.42	.163	.98	.336	1.54	.438	2.10	.482	2.66	.496	3.22	.499
.43	.166	.99	.339	1.55	.439	2.11	.483	2.67	.496	3.23	.499
.44	.170	1.00	.341	1.56	.441	2.12	.483	2.68	.496	3.24	.499
.45	.174	1.01	.344	1.57	.442	2.13	.483	2.69	.496	3.25	.499
.46	.177	1.02	.346	1.58	.443	2.14	.484	2.70	.497	3.26	.499
.47	.181	1.03	.348	1.59	.444	2.15	.484	2.71	.497	3.27	.499
.48	.184	1.04	.351	1.60	.445	2.16	.485	2.72	.497	3.28	.499
.49	.188	1.05	.353	1.61	.446	2.17	.485	2.73	.497	3.29	.499
.50	.191	1.06	.355	1.62	.447	2.18	.485	2.74	.497	3.30	.500
.51	.195	1.07	.358	1.63	.448	2.19	.486	2.75	.497	3.31	.500
.52	.198	1.08	.360	1.64	.449	2.20	.486	2.76	.497	3.32	.500
.53	.202	1.09	.362	1.65	.451	2.21	.486	2.77	.497	3.33	.500
.54	.205	1.10	.364	1.66	.452	2.22	.487	2.78	.497	3.34	.500
.55	.209	1.11	.367	1.67	.453	2.23	.487	2.79	.497	3.35	.500

Carl Friedrich Gauss (1777–1855) was one of the greatest mathematical thinkers of history. In his *Disquisitiones arithmeticae*, published in 1798, he pulled together work by predecessors and enriched and blended it with his own into a unified whole. The book is regarded by many as the true beginning of the theory of numbers.

Of his many contributions to science, the statistical method of least squares is the most widely used today in astronomy, biology, geodesy, physics, and the social sciences. Gauss took special pride in his contributions to developing the method. Despite an aversion to teaching, he taught an annual course in the method for the last twenty years of his life.

It has been said that Gauss was the last person to have mastered all of the mathematics known in his day.

EXAMPLE 2 Use the normal curve table to find the percent of all scores that lie between the mean and the following values.

(a) One standard deviation above the mean

Here $z = 1.00$ (the number of standard deviations, written as a decimal to the nearest hundredth). Refer to Table 10. Find 1.00 in the z column. The table entry is .341, so 34.1% of all values lie between the mean and one standard deviation above the mean.

Another way of looking at this is to say that the area in color in Figure 14 represents 34.1% of the total area under the normal curve.

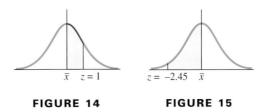

FIGURE 14 **FIGURE 15**

(b) 2.45 standard deviations below the mean

Even though we go *below* the mean here (to the left), the normal curve table still works since the normal curve is symmetrical about its mean. Find 2.45 in the z column. A total of .493, or 49.3%, of all values lie between the mean and 2.45 standard deviations below the mean. This region is shaded in Figure 15.

EXAMPLE 3 The lengths of long-distance phone calls placed through a certain company are distributed normally with mean 6 minutes and standard deviation 2 minutes. If 1 call is randomly selected from phone company records, what is the probability that it will have lasted more than 10 minutes?

Here 10 minutes is two standard deviations above the mean. The probability of such a call is equal to the shaded region in Figure 16.

From Table 10, the area between the mean and two standard deviations above is .477 ($z = 2.00$). The total area to the right of the mean is .500. Find the area from $z = 2.00$ to the right by subtracting: $.500 - .477 = .023$. So the probability of a call exceeding 10 minutes is .023.

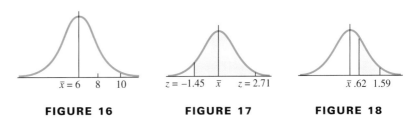

FIGURE 16 **FIGURE 17** **FIGURE 18**

EXAMPLE 4 Find the total areas indicated in the regions in color in each of Figures 17 and 18.

For Figure 17, find the area from 1.45 standard deviations below the mean to 2.71 standard deviations above the mean. From Table 10, $z = 1.45$ leads to an area of .426, while $z = 2.71$ leads to .497. The total area is the sum of these, or $.426 + .497 = .923$.

To find the indicated area in Figure 18, refer again to Table 10. From the table, $z = .62$ leads to an area of .232, while $z = 1.59$ gives .444. To get the area between these two values of z, subtract the areas: $.444 - .232 = .212$.

Examples 2–4 emphasize the *equivalence* of three quantities, as follows.

A Basic Consumer (T)issue

It all started when a reporter on consumer issues for a Midwest TV station received a complaint that rolls of Brand X toilet paper manufactured by Company Y did not have the number of sheets claimed on the wrapper. Brand X is supposed to have 375 sheets, but three rolls of it were found by reporters to have 360, 361, and 363. Shocked Company Y executives said that the **odds against** six rolls having fewer than 375 sheets each are 1 billion to 1. They counted sheets and found that several rolls of Brand X actually had 380 sheets each (machines count the sheets only in 10s). TV reporters made an independent count, and their results agreed with Company Y.

What happened the first time? Well, the reporters hadn't actually counted sheets, but had measured rolls and divided the length of a roll by the length of one sheet. Small variations in length can be expected, which add up over a roll, giving false results.

This true story perhaps points up the distinction in probability and statistics between **discrete** values and **continuous** values.

Meaning of Normal Curve Areas

In a standard normal curve, the following three quantities are equivalent.

1. **Percentage** (of total items that lie in an interval)
2. **Probability** (of a randomly chosen item lying in an interval)
3. **Area** (under the normal curve along an interval)

Which quantity we think of depends upon how a particular question is formulated. They are all evaluated by using A-values from Table 10.

In general, when we use Table 10, z is the z-score of a particular data item x. Recall from the previous section that the z-score is related to the mean $\bar{x}$ and the standard deviation s of the distribution by the formula

$$z = \frac{x - \bar{x}}{s}.$$

For example, suppose a normal curve has mean 220 and standard deviation 12. To find the number of standard deviations that a given value is from the mean, use the formula above. For the value 247,

$$z = \frac{247 - 220}{12} = \frac{27}{12} = 2.25,$$

so 247 is 2.25 standard deviations above the mean. For 204,

$$z = \frac{204 - 220}{12} = \frac{-16}{12} \approx -1.33,$$

so 204 is 1.33 standard deviations *below* the mean. (We know it is below rather than above the mean since the z-score is negative rather than positive.)

EXAMPLE 5 In one area, the average motorist drives about 1200 miles per month, with standard deviation 150 miles. Assume that the number of miles is closely approximated by a normal curve, and find the percent of all motorists driving the following distances.

(a) Between 1200 and 1600 miles per month

Start by finding how many standard deviations 1600 miles is above the mean. Use the formula for z.

$$z = \frac{1600 - 1200}{150} = \frac{400}{150} \approx 2.67$$

From Table 10, .496, or 49.6%, of all motorists drive between 1200 and 1600 miles per month.

(b) Between 1000 and 1500 miles per month

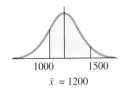

FIGURE 19

As shown in Figure 19, values of z must be found for both 1000 and 1500.

$$\text{For 1000:} \quad z = \frac{1000 - 1200}{150} = \frac{-200}{150} \approx -1.33.$$

$$\text{For 1500:} \quad z = \frac{1500 - 1200}{150} = \frac{300}{150} = 2.00.$$

From Table 10, $z = -1.33$ leads to an area of .408, while $z = 2.00$ gives .477. This means a total of $.408 + .477 = .885$, or 88.5%, of all motorists drive between 1000 and 1500 miles per month. ■

Each example thus far has given a data value and then required that z be found. The next example gives z and asks for the data value.

EXAMPLE 6 A particular normal distribution has mean $\bar{x} = 81.7$ and standard deviation $s = 5.21$. What data value from the distribution would correspond to $z = -1.35$?

Substitute the given values for z, $\bar{x}$, and s into the formula, and solve for x.

$$z = \frac{x - \bar{x}}{s}$$

$$-1.35 = \frac{x - 81.7}{5.21}$$

$$-1.35(5.21) = \frac{x - 81.7}{5.21}(5.21)$$

$$-7.0335 = x - 81.7$$

$$74.6665 = x.$$

Rounding to the nearest tenth, the required data value is 74.7. ■

13.5 EXERCISES

Note: For problems requiring the calculation of z-scores or A-values, our answers are based on Table 10. By using a calculator or computer package, you will sometimes obtain a slightly more accurate answer.

Identify each of the following variable quantities as discrete or continuous.

1. the number of heads in 50 tossed coins
2. the number of babies born in one day at a certain hospital
3. the average weight of babies born in a week
4. the heights of seedling pine trees at six months of age
5. the time as shown on a digital watch
6. the time as shown on a watch with a sweep hand

Measuring the Mass of Ore Samples Suppose 100 geology students measure the mass of an ore sample. Due to human error and limitations in the reliability of the balance, not all the readings are equal. The results are found to closely approximate a normal curve, with mean 86 g and standard deviation 1 g.

Use the symmetry of the normal curve and the empirical rule to estimate the number of students reporting readings in the following ranges.

7. more than 86 g
8. more than 85 g
9. between 85 and 87 g
10. between 84 and 87 g

Distribution of IQ Scores On standard IQ tests, the mean is 100, with a standard deviation of 15. The results come very close to fitting a normal curve. Suppose an IQ test is given to a very large group of people. Find the percent of people whose IQ scores fall into the following categories.

11. less than 100
12. greater than 115
13. between 70 and 130
14. more than 145

Find the percent of area under a normal curve between the mean and the given number of standard deviations from the mean. (Note that positive indicates above the mean, while negative indicates below the mean.)

15. 1.50
16. .92
17. −1.08
18. −2.25

Find the percent of the total area under a normal curve between the given values of z.

19. $z = 1.41$ and $z = 1.83$
20. $z = -1.74$ and $z = -1.14$
21. $z = -3.11$ and $z = 2.06$
22. $z = -1.98$ and $z = 1.02$

Find a value of z such that the following conditions are met.

23. 10% of the total area is to the right of z.
24. 4% of the total area is to the left of z.
25. 9% of the total area is to the left of z.
26. 23% of the total area is to the right of z.

Lifetimes of Lightbulbs The Better lightbulb has an average life of 600 hr, with a standard deviation of 50 hr. The length of life of the bulb can be closely approximated by a normal curve. A warehouse manager buys and installs 10,000 such bulbs. Find the total number that can be expected to last the following amounts of time.

27. at least 600 hr
28. between 600 and 675 hr
29. between 675 and 740 hr
30. between 490 and 720 hr
31. less than 740 hr
32. less than 510 hr

Weights of Chickens The chickens at Ben and Ann Rice's farm have a mean weight of 1850 g with a standard deviation of 150 g. The weights of the chickens are closely approximated by a normal curve. Find the percent of all chickens having the following weights.

33. more than 1700 g
34. less than 1800 g
35. between 1750 and 1900 g
36. between 1600 and 2000 g

Filling Cereal Boxes A certain dry cereal is packaged in 24-oz boxes. The machine that fills the boxes is set so that, on the average, a box contains 24.5 oz. The machine-filled boxes have contents weights that can be closely approximated by a normal curve. What fraction of the boxes will be underweight if the standard deviation is as follows?

37. .5 oz
38. .4 oz
39. .3 oz
40. .2 oz

41. *Recommended Daily Vitamin Allowances* In nutrition, the recommended daily allowance of vitamins is a number set by the government to guide an individual's daily vitamin intake. Actually, vitamin needs vary drastically from person to person, but the needs are closely approximated by a normal curve. To calculate the recommended daily allowance, the government first finds the average need for vitamins among people in the population and the standard deviation. The **recommended daily allowance** is then defined as the mean plus 2.5 times the standard deviation. What fraction of the population will receive adequate amounts of vitamins under this plan?

Recommended Daily Vitamin Allowances Find the recommended daily allowance for each vitamin if the mean need and standard deviation are as follows. (See Exercise 41.)

42. mean need = 1800 units; standard deviation = 140 units
43. mean need = 159 units; standard deviation = 12 units

Assume the following distributions are all normal, and use the areas under the normal curve given in Table 10 to find the appropriate areas.

44. *Filling Cartons with Milk* A machine that fills quart milk cartons is set up to average 32.2 oz per carton, with a standard deviation of 1.2 oz. What is the probability that a filled carton will contain less than 32 oz of milk?

45. *Finding Blood Clotting Times* The mean clotting time of blood is 7.47 sec, with a standard deviation of 3.6 sec. What is the probability that an individual's blood-clotting time will be less than 7 sec or greater than 8 sec?

46. *Sizes of Fish* The average length of the fish caught in Lake Amotan is 12.3 in., with a standard deviation of 4.1 in. Find the probability that a fish caught there will be longer than 18 in.

47. *Size Grading of Eggs* To be graded extra large, an egg must weigh at least 2.2 oz. If the average weight for an egg is 1.5 oz, with a standard deviation of .4 oz, how many of five dozen randomly chosen eggs would you expect to be extra large?

Distribution of Student Grades Anessa Davis teaches a course in marketing. She uses the following system for assigning grades to her students.

Grade	Score in Class
A	Greater than $\bar{x} + (3/2)s$
B	$\bar{x} + (1/2)s$ to $\bar{x} + (3/2)s$
C	$\bar{x} - (1/2)s$ to $\bar{x} + (1/2)s$
D	$\bar{x} - (3/2)s$ to $\bar{x} - (1/2)s$
F	Below $\bar{x} - (3/2)s$

From the information in the table, what percent of the students receive the following grades?

48. A **49.** B **50.** C

51. Do you think this system would be more likely to be fair in a large freshman class in psychology or in a graduate seminar of five students? Why?

Normal Distribution of Student Grades A teacher gives a test to a large group of students. The results are closely approximated by a normal curve. The mean is 75 with a standard deviation of 5. The teacher wishes to give As to the top 8% of the students and Fs to the bottom 8%. A grade of B is given to the next 15%, with Ds given similarly. All other students get Cs. Find the bottom cutoff (rounded to the nearest whole number) for the following grades. (*Hint:* Use Table 10 to find z-scores from known A-values.)

52. A **53.** B
54. C **55.** D

A normal distribution has mean 76.8 and standard deviation 9.42. Follow the method of Example 6 and find data values corresponding to the following values of z. Round to the nearest tenth.

56. $z = .72$ **57.** $z = 1.44$
58. $z = -2.39$ **59.** $z = -3.87$

60. What percentage of the items lie within 1.25 standard deviations of the mean
  (a) in any distribution (by Chebyshev's theorem)?
  (b) in a normal distribution (by Table 10)?

61. Explain the difference between the answers to parts (a) and (b) in Exercise 60.

EXTENSION

How to Lie with Statistics

The statement that there are "lies, damned lies, and statistics" is attributed to Benjamin Disraeli, Queen Victoria's prime minister. Other people have made even stronger comments about statistics. This often intense distrust of statistics has come about because of a belief that "you can prove anything with numbers." It must be admitted that there is often a conscious or unconscious distortion in many published statistics. The classic book on distortion in statistics is *How to Lie with Statistics*, by Darrell Huff (first published in 1972). This extension quotes some common methods of distortion that Huff gives in his book.

(continued)

THE SAMPLE WITH THE BUILT-IN BIAS

A house-to-house survey purporting to study magazine readership was once made in which a key question was: What magazines does your household read? When the results were tabulated and analyzed it appeared that a great many people loved *Harper's* and not very many read *True Story*. Now there were publishers' figures around at the time that showed very clearly that *True Story* had more millions of circulation than *Harper's* had hundreds of thousands. Perhaps we asked the wrong kind of people, the designers of the survey said to themselves. But no, the questions had been asked in all sorts of neighborhoods all around the country. The only reasonable conclusion then was that a good many of the respondents, as people are called when they answer such questions, had not told the truth. About all the survey had uncovered was snobbery.

In the end it was found that if you wanted to know what certain people read it was no use asking them. You could learn a good deal more by going to their houses and saying you wanted to buy old magazines and what could be had? Then all you had to do was count the *Yale Reviews* and the *Love Romances*. Even that dubious device, of course, does not tell you what people read, only what they have been exposed to.

Similarly, the next time you learn from your reading that the average American (you hear a good deal about him these days, most of it faintly improbable) brushes his teeth 1.02 times a day—a figure pulled out of the air, but it may be as good as anyone else's—ask yourself a question. How can anyone have found out such a thing? Is a woman who has read in countless advertisements that non-brushers are social offenders going to confess to a stranger that she does not brush her teeth regularly? The statistic may have meaning to one who wants to know only what people say about tooth-brushing but it does not tell a great deal about the frequency with which bristle is applied to incisor.

THE WELL-CHOSEN AVERAGE

A common trick is to use a different kind of average each time, the word "average" having a very loose meaning. It is a trick commonly used, sometimes in innocence but often in guilt, by people wishing to influence public opinion or sell advertising space. When you are told that something is an average you still don't know very much about it unless you can find out which of the common kinds of average it is—mean, median or mode.

Try your skepticism on some items from "A letter from the Publisher" in *Time* magazine. Of new subscribers it said, "Their median age is 34 years and their average family income is $7270 a year." An earlier survey of "old TIMErs" had found that their "median age was 41 years.... Average income was $9535...." The natural question is why, when median is given for ages both times, the kind of average for incomes is carefully unspecified. Could it be that the mean was used instead because it is bigger, thus seeming to dangle a richer readership before advertisers?

Be careful when reading charts or graphs—often there is no numerical scale, or no units are given. This makes the chart pretty much meaningless. Huff has a good example of this:

THE LITTLE FIGURES THAT ARE NOT THERE

Before me are wrappers from two boxes of Grape-Nuts Flakes. They are slightly different editions, as indicated by their testimonials: one cites Two-Gun Pete and the other says "If you want to be like Hoppy... you've got to eat like Hoppy!" Both offer charts to show ("Scientists *proved* it's true!") that these flakes "start giving you energy in 2 minutes!" In one case the chart hidden in these forests of exclamation points has numbers up the side; in the other case the numbers have been omitted. This is just as well, since there is no hint of what the numbers mean. Both show a steeply climbing brown line ("energy release"), but one has it starting one minute after eating Grape-Nuts Flakes, the other 2 minutes later. One line climbs about twice as fast as the other, suggesting that even the draftsman didn't think these graphs meant anything.

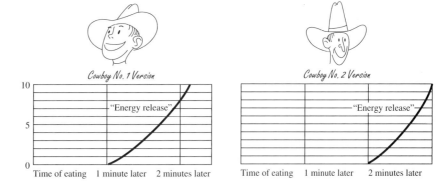

THE GEE-WHIZ GRAPH

About the simplest kind of statistical picture, or graph, is the line variety. It is very useful for showing trends, something practically everybody is interested in showing or knowing about or spotting or deploring or forecasting. We'll let our graph show how national income increased ten percent in a year.

Begin with paper ruled into squares. Name the months along the bottom. Indicate billions of dollars up the side. Plot your points and draw your line, and your graph will look like this:

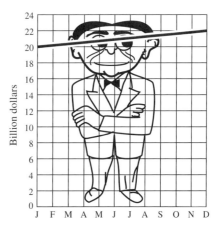

Now that's clear enough. It shows what happened during the year and it shows it month by month. He who runs may see and understand, because the whole graph is in proportion and there is a zero line at the bottom for comparison. Your ten percent looks like ten percent—an upward trend that is substantial but perhaps not overwhelming.

(continued)

Stand Up and Be Counted

According to an Associated Press release in 1988 (*The Sacramento Union,* September 28), the U.S. Census Bureau anticipated that the 1990 census would cost up to $3 billion (the cost in 1980 was $1.1 billion). Even so, significant problems were foreseen in finding the necessary workers to conduct the census and getting response from the populace. Census Director John G. Keane cited a deteriorating climate in the nation for taking censuses and surveys, due to Americans' attitude toward being counted.

That is very well if all you want to do is convey information. But suppose you wish to win an argument, shock a reader, move him into action, sell him something. For that, this chart lacks schmaltz. Chop off the bottom.

Now that's more like it. (You've saved paper too, something to point out if any carping person objects to your misleading graphics.) The figures are the same and so is the curve. It is the same graph. Nothing has been falsified—except the impression that it gives. But what the hasty reader sees now is a national income line that has climbed halfway up the paper in twelve months, all because most of the chart isn't there any more. Like the missing parts of speech in sentences that you met in grammar classes, it is "understood." Of course, the eye doesn't "understand" what isn't there, and a small rise has become, visually, a big one.

Now that you have practiced to deceive, why stop with truncating? You have a further trick available that's worth a dozen of that. It will make your modest rise of ten percent look livelier than one hundred percent is entitled to look. Simply change the proportion between the side and the bottom. There's no rule against it, and it does give your graph a prettier shape. All you have to do is let each mark up the side stand for only one-tenth as many dollars as before.

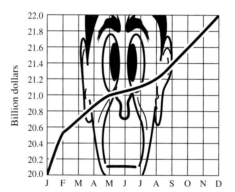

That *is* impressive, isn't it? Anyone looking at it can just feel prosperity throbbing in the arteries of the country. It is a subtler equivalent of editing "National income rose ten percent" into "… climbed a whopping ten percent." It is vastly more effective, however, because it contains no adjectives or adverbs to spoil the illusion of objectivity. There's nothing anyone can pin on you.

Suppose Diana makes twice as much money as Mike. One way to show this is with a graph using silver dollars to represent the income of each. If we used one silver dollar for Mike and two for Diana, we would be fine. But it is more common to use proportional dollars. It is common to find a dollar of one size used for Mike, with one twice as wide for Diana. This is wrong—the larger dollar actually has four times the area, giving the impression that Diana earns four times as much as Mike. Huff gives another example of this:

THE ONE-DIMENSIONAL PICTURE

Newsweek once showed how "U.S. Old Folks Grow Older" by means of a chart on which appeared two male figures, one representing the 68.2-year life expectancy of today, the other the 34-year life expectancy of 1879–1889. It was the same old story. One figure was twice as tall as the other and so would have had eight times the bulk or weight. This picture sensationalized facts in order to make a better story. It would be called a form of yellow journalism.

Cause and Effect Many people often assume that just because two things changed together one caused the other. The classic example of this is the fact that teachers' salaries and liquor consumption increased together over the last few decades. Neither of these caused the other; rather, both were caused by the same underlying growth in national prosperity. Another case of faulty cause and effect reasoning is in the claim that going to college raises your income. See Huff's comments below.

> Reams of pages of figures have been collected to show the value in dollars of a college education, and stacks of pamphlets have been published to bring these figures—and conclusions more or less based on them—to the attention of potential students. I am not quarreling with the intention. I am in favor of education myself, particularly if it includes a course in elementary statistics. Now these figures have pretty conclusively demonstrated that people who have gone to college make more money than people who have not. The exceptions are numerous, of course, but the tendency is strong and clear.
>
> The only thing wrong is that along with the figures and facts goes a totally unwarranted conclusion.... It says that these figures show that if *you* (your son, your daughter) attend college you will probably earn more money than if you decide to spend the next four years in some other manner. This unwarranted conclusion has for its basis the equally unwarranted assumption that since college-trained folks make more money, they make it because they went to college. Actually we don't know but that these are the people who would have made more money even if they had not gone to college. There are a couple of things that indicate rather strongly that this is so. Colleges get a disproportionate number of two groups of people: the bright and the rich. The bright might show good earning power without college knowledge. And as for the rich ones ... well, money breeds money in several obvious ways. Few children of rich parents are found in low-income brackets whether they go to college or not.

Extrapolation This refers to predicting the future, based only on what has happened in the past. However, it is very rare for the future to be just like the past—something will be different. One of the best examples of extrapolating incorrectly is given by Mark Twain:

> In the space of one hundred and seventy-six years the Lower Mississippi has shortened itself two hundred and forty-two miles. That is an average of a trifle over one mile and a third per year. Therefore, any calm person, who is not blind or idiotic, can see that in the Old Oölitic Silurian Period, just a million years ago next November, the Lower Mississippi River was upward of one million three hundred thousand miles long, and stuck out over the Gulf of Mexico like a fishing rod. And by the same token any person can see that seven hundred and forty-two years from now the Lower Mississippi will be only a mile and three-quarters long, and Cairo and New Orleans will have joined their streets together, and be plodding comfortably along under a single mayor and a mutual board of aldermen. There is something fascinating about science. One gets such wholesale returns of conjecture out of such a trifling investment of fact.

(continued)

EXTENSION EXERCISES

Reading Line Graphs
The Norwegian stamp at the side features two graphs.

1. What does the solid line represent?
2. Has it increased much?
3. What can you tell about the dashed line?
4. Does the graph represent a long period of time or a brief period of time?

Changing Value of the British Pound The illustrations below show the decline in the value of the British pound for a certain period.

$2.40

$1.72

5. Calculate the percent of decrease in the value by using the formula

   Percent of decrease
   $$= \frac{\text{old value} - \text{new value}}{\text{old value}}.$$

6. We estimate that the smaller banknote shown has about 50% less area than the larger one. Do you think this is close enough?

Deciphering Advertising Claims Several advertising claims are given below. Decide what further information you might need before deciding to accept the claim.

7. 98% of all Toyotas ever sold in the United States are still on the road.
8. Sir Walter Raleigh pipe tobacco is 44% fresher.
9. Eight of 10 dentists responding to a survey preferred Trident Sugarless Gum.
10. A Volvo has 2/3 the turning radius of a Continental.
11. A Ford LTD is as quiet as a glider.
12. Circulation has increased for *The Wall Street Journal*, as shown by the graph to the right.

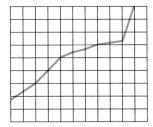

Exercises 13–16 come from Huff's book. Decide how these exercises describe possibly misleading uses of numbers.

13. *The Extent of Federal Taxes* Each of the following maps shows what portion of our national income is now being taken and spent by the federal government. The top map does this by shading the areas of most of the states *west* of the Mississippi to indicate that federal spending has become equal to the total incomes of the people of those states. The bottom map does this for states *east* of the Mississippi.

The Darkening Shadow
(Western Style)

(Eastern Style)

To show we aren't cheating, we added MD, DE, and RI for good measure

14. *Faculty-Student Marriages* Long ago, when Johns Hopkins University had just begun to admit women students, someone not particularly enamored of co-education reported a real shocker: Thirty three and one-third percent of the women at Hopkins had married faculty members!

15. *Comparative Death Rates* The death rate in the Navy during the Spanish-American War was nine per thousand. For civilians in New York City during the same period it was sixteen per thousand. Navy recruiters later used these figures to show that it was safer to be in the Navy than out of it.

16. *Reported Cases of Influenza and Pneumonia* If you should look up the latest available figures on influenza and pneumonia, you might come to the strange conclusion that these ailments are practically confined to three southern states, which account for about eighty percent of the reported cases.

Questioning the Reasonableness of Samples When a sample is selected from a population, it is important to decide whether the sample is reasonably representative of the entire population. For example, a sample of the general population that is only 20% women should make us suspicious. The same is true of a questionnaire asking, "Do you like to answer questionnaires?"

In each exercise, pick the choice that gives the most representative sample. (Sampling techniques are a major branch of inferential statistics.)

17. A factory has 10% management employees, 30% clerical employees, and 60% assembly-line workers. A sample of 50 is chosen to discuss parking.
    A. 4 management, 21 clerical, 25 assembly-line
    B. 6 management, 15 clerical, 29 assembly-line
    C. 8 management, 9 clerical, 33 assembly-line

18. A college has 35% freshmen, 28% sophomores, 21% juniors, and 16% seniors. A sample of 80 is chosen to discuss methods of electing student officers.
    A. 22 freshmen, 22 sophomores, 24 juniors, 12 seniors
    B. 24 freshmen, 20 sophomores, 22 juniors, 14 seniors
    C. 28 freshmen, 23 sophomores, 16 juniors, 13 seniors

19. A computer company has plants in Boca Raton, Jacksonville, and Tampa. The plant in Boca Raton produces 42% of all the company's output, with 27% and 31% coming from Jacksonville and Tampa, respectively. A sample of 120 parts is chosen for quality testing.
    A. 38 from Boca Raton, 39 from Jacksonville, 43 from Tampa
    B. 43 from Boca Raton, 37 from Jacksonville, 40 from Tampa
    C. 50 from Boca Raton, 31 from Jacksonville, 39 from Tampa

20. At one resort, 56% of all guests come from the Northeast, 29% from the Midwest, and 15% from Texas. A sample of 75 guests is chosen to discuss the dinner menu.
    A. 41 from the Northeast, 21 from the Midwest, 13 from Texas
    B. 45 from the Northeast, 18 from the Midwest, 12 from Texas
    C. 47 from the Northeast, 20 from the Midwest, 8 from Texas

Use the information supplied in the following problems to solve for the given variables.

21. *Employee Categories at an Insurance Agency* An insurance agency has 7 managers, 25 agents, and 18 clerical employees. A sample of 10 is chosen.
    (a) Let m be the number of managers in the sample, a the number of agents, and c the number of clerical employees. Find m, a, and c, if
    $$c = m + 2$$
    $$a = 2c.$$
    (b) To check that your answer is reasonable, calculate the numbers of the office staff that *should* be in the sample, if all groups are represented proportionately.

22. *Faculty Categories at a Small College* A small college has 12 deans, 24 full professors, 39 associate professors, and 45 assistant professors. A sample of 20 employees is chosen to discuss the graduation speaker.
    (a) Let d be the number of deans in the sample, f the number of full professors, a the number of associate professors, and s the number of assistant professors. Find d, f, a, and s if
    $$f = 2d$$
    $$a = f + d + 1$$
    $$a = s.$$
    (*Hint:* There are 2 deans.)
    (b) To check that your answer is reasonable, calculate the number of each type of employee that *should* be in the sample, if all groups are represented proportionately.

13.6 Regression and Correlation

One very important branch of inferential statistics, called **regression analysis,** is used to compare quantities or variables, to discover relationships that exist between them, and to formulate those relationships in useful ways. For example, suppose a sociologist gathers data on a few (say ten) of the residents of a small village in a remote country in order to get an idea of how annual income (in dollars) relates to age in that village. The data are shown in Table 11.

The first step in analyzing these data is to graph the results, as shown in the **scatter diagram** of Figure 20. (Graphing calculators will plot scatter diagrams.)

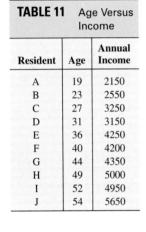

TABLE 11 Age Versus Income

Resident	Age	Annual Income
A	19	2150
B	23	2550
C	27	3250
D	31	3150
E	36	4250
F	40	4200
G	44	4350
H	49	5000
I	52	4950
J	54	5650

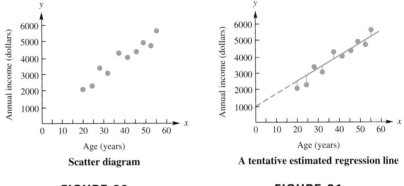

Scatter diagram

A tentative estimated regression line

FIGURE 20

FIGURE 21

Once a scatter diagram has been produced, we can draw a curve that best fits the pattern exhibited by the sample data points. This curve can have any one of many characteristic shapes, depending on how the quantities involved are related. We would like to infer a graph for the entire population of points for the related quantities, but we have available only a sample of those points, so the best-fitting curve for the sample points is called an **estimated regression curve.** If, as in the example above, the points in the scatter diagram seem to lie approximately along a straight line, the relation is assumed to be linear, and the line that best fits the data points is called the **estimated regression line.**

If we let x denote age and y denote income in the data of Table 11 and assume that the best-fitting curve is a line, then the equation of that line will take the form

$$y = ax + b,$$

where a is the slope of the line and b is the y-coordinate of the y-intercept (the y-value at which the line, if extended, would intersect the y-axis).

To completely identify the estimated regression line, we must find the values of the "regression coefficients" a and b, which requires some calculation. In Figure 21, a tentative line has been drawn through the scatter diagram.

For each x-value in the data set, the corresponding y-value usually differs from the value it would have if the data point were exactly on the line. These differences are shown in the figure by vertical segments. Choosing another line would make some of these differences greater and some lesser. The most common procedure is to choose the line where the sum of the squares of all these differences is minimized. This is called the **method of least squares,** and the resulting line is called the **least squares line.**

In the equation of the least squares line, the variable y' can be used to distinguish the *predicted* values (which would give points on the least squares line) from the *observed* values y (those occurring in the data set).

The least squares criterion mentioned above leads to specific values of a and b. We shall not give the details, which involve differential calculus, but the results are given here. (Σ—the Greek letter *sigma*—represents summation just as in earlier sections.)

Regression Coefficient Formulas

The **least squares line** $y' = ax + b$ that provides the best fit to the data points $(x_1, y_1), (x_2, y_2), \ldots, (x_n, y_n)$ has

$$a = \frac{n(\Sigma xy) - (\Sigma x)(\Sigma y)}{n(\Sigma x^2) - (\Sigma x)^2} \quad \text{and} \quad b = \frac{\Sigma y - a(\Sigma x)}{n}.$$

Francis Galton (1822–1911) learned to read at age three, was interested in mathematics and machines, but was an indifferent mathematics student at Trinity College, Cambridge. After several years as a traveling gentleman of leisure, he became interested in researching methods of predicting weather. It was during this research on weather that Galton developed early intuitive notions of correlation and regression and posed the problem of multiple regression.

Galton's key statistical work is *Natural Inheritance*. In it, he set forth his ideas on regression and correlation. He discovered the correlation coefficient while pondering Alphonse Bertillon's scheme for classifying criminals by physical characteristics. It was a major contribution to statistical method.

EXAMPLE 1 Find the equation of the least squares line for the age and income data given earlier. Graph the line.

Start with the two columns on the left in the following chart (which just repeat the original data). Then find the products $x \cdot y$, and the squares x^2.

x	y	$x \cdot y$	x^2
19	2150	40,850	361
23	2550	58,650	529
27	3250	87,750	729
31	3150	97,650	961
36	4250	153,000	1296
40	4200	168,000	1600
44	4350	191,400	1936
49	5000	245,000	2401
52	4950	257,400	2704
54	5650	305,100	2916
Sums: 375	39,500	1,604,800	15,433

From the chart, $\Sigma x = 375$, $\Sigma y = 39{,}500$, $\Sigma xy = 1{,}604{,}800$, and $\Sigma x^2 = 15{,}433$. Also, $n = 10$ since there are 10 pairs of values. Now find a with the formula given above.

$$a = \frac{10(1{,}604{,}800) - 375(39{,}500)}{10(15{,}433) - (375)^2} = \frac{1{,}235{,}500}{13{,}705} \approx 90.15$$

Finally, use this value of a to find b.

$$b = \frac{39{,}500 - 90.15(375)}{10} \approx 569.4$$

The equation of the least squares line (with coefficients rounded to whole numbers) is

$$y' = 90x + 569.$$

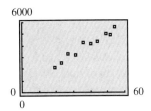

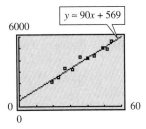

The information in Figure 22 and the accompanying discussion is supported in these screens.

Letting $x = 20$ in this equation gives $y' = 2369$, and $x = 50$ implies $y' = 5069$. The two points (20, 2369) and (50, 5069) are used to graph the regression line in Figure 22. Notice that the intercept coordinates (0, 569) also fit the extended line.

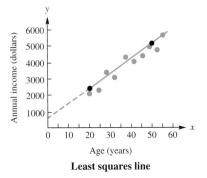

Least squares line

FIGURE 22

A computer, or a scientific, statistical, or graphing calculator is recommended for finding regression coefficients. (See the margin notes.) That way tedious calculations, such as in Example 1, can be avoided and the regression line produced automatically.

EXAMPLE 2 Use the result of Example 1 to predict the income of a village resident who is 35 years old.

Use the equation $y' = 90x + 569$ and replace x with 35.

$$y' = 90(35) + 569 = 3719$$

Based on the given data, a 35-year-old will make about $3719 per year. ■

Correlation Once an equation for the line of best fit (the least squares line) has been found, it is reasonable to ask, "Just how good is this line for predictive purposes?" If the points already observed fit the line quite closely, then future pairs of scores can be expected to do so. If the points are widely scattered about even the "best-fitting" line, then predictions are not likely to be accurate. In general, the closer the *sample* data points lie to the least squares line, the more likely it is that the entire *population* of (x, y) points really do form a line, that is, that x and y really are related linearly. Also, the better the fit, the more confidence we can have that our least squares line (based on the sample) is a good estimator of the true population line.

One common measure of the strength of the linear relationship in the sample is called the **sample correlation coefficient,** denoted r. It is calculated from the sample data according to the following formula.

Sample Correlation Coefficient Formula

In linear regression, the strength of the linear relationship is measured by the correlation coefficient

$$r = \frac{n(\Sigma xy) - (\Sigma x)(\Sigma y)}{\sqrt{n(\Sigma x^2) - (\Sigma x)^2} \cdot \sqrt{n(\Sigma y^2) - (\Sigma y)^2}}.$$

The value of r is always between -1 and 1, or perhaps equal to -1 or 1. Values of exactly 1 or -1 indicate that the least squares line goes exactly through all the data points. If r is close to 1 or -1, but not exactly equal, then the line comes "close" to fitting through all the data points, and we say that the linear correlation between x and y is "strong." If r is equal, or nearly equal, to 0, there is no linear correlation, or the correlation is weak. This would mean that the points are a totally disordered conglomeration. Or it may be that the points form an ordered pattern but one that is not linear. (More is said about this case following Example 3.) If r is neither close to 0 nor close to 1 or -1, we might describe the linear correlation as "moderate."

A positive value of r indicates a regression line with positive slope and we say the linear relationship between x and y is direct; as x increases, y also increases. A negative r-value means the line has negative slope, so there is an inverse relationship between x and y; as x increases, y decreases.

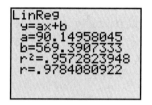

The slope *a* and *y*-intercept *b* of the regression equation, along with r^2 and r, are given. Compare with Examples 1 and 3.

EXAMPLE 3 Find r for the age and income data of Table 11.

Almost all values needed to find r were computed in Example 1.

$$n = 10 \qquad \Sigma x = 375$$
$$\Sigma y = 39{,}500 \qquad \Sigma xy = 1{,}604{,}800$$
$$\Sigma x^2 = 15{,}433$$

The only missing value is Σy^2. Squaring each y in the original data and adding the squares gives

$$\Sigma y^2 = 167{,}660{,}000.$$

Now use the formula to find r.

$$r = .98 \text{ (to two decimal places)}$$

This value of r, very close to 1, shows that age and income in this village are highly correlated. The fact that r is positive indicates that the linear relationship is direct; as age increases, income also increases. ∎

As mentioned earlier, least squares curves are not always least squares lines. If a scatter diagram reveals that the quantities represented by the data may not be linearly related, then a best-fitting curve of some other type can be sought. For example, for any set of paired data, the TI-83 Plus graphing calculator will compute best-fit curves of the types shown in Table 12 (many of which are examined in an earlier chapter of this text). Notice that an inverse square law, a common kind of relationship in nature, would be included in power functions (with $b = -2$).

TABLE 12 Curve-Fitting Options on the TI-83 Plus

Type of Curve	Equation
Linear	$y = ax + b$
Quadratic	$y = ax^2 + bx + c$
Cubic	$y = ax^3 + bx^2 + cx + d$
Quartic	$y = ax^4 + bx^3 + cx^2 + dx + e$
Logarithmic	$y = a + b \ln x$
Exponential	$y = a \cdot b^x$
Power	$y = a \cdot x^b$

The calculator provides correlation coefficients also for most types of regression curves, so it is easy to try the various types of curves to see which type will fit the sample data most closely. Computation of the various coefficients is more involved for the non-linear curves, and we do not show those formulas. For Example 4, we use the TI-83 Plus.

EXAMPLE 4 Find a reasonable curve to fit the following data, which relate the time employees have been working to the time they take to assemble a certain part.

Time on job, x (in weeks)	1	2	4	7	9	10	1	3	6	13	15	8
Time for task, y (in minutes)	8	6	5	3	2	2	10	4	2	1	1	2

The scatter diagram is shown in Figure 23. The relationship probably is not linear. We tried four possibilities, as shown in Table 13.

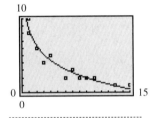

This is a calculator-generated screen corresponding to Figure 24. It uses the data of Table 13 and the corresponding logarithmic curve.

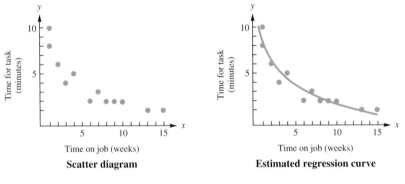

Scatter diagram

FIGURE 23

Estimated regression curve

FIGURE 24

TABLE 13 Time on Job Versus Time for a Task

Curve Type	Correlation Coefficient, r	a	b
Linear	−.857	−.536	7.360
Logarithmic	−.963	8.484	−2.974
Exponential	−.954	8.297	.855
Power	−.958	9.970	−.775

The negative r-values obtained here verify the inverse nature of the relationship (which was clear from the scatter diagram, also). All four of the curve types provide a fairly good correlation (r close to -1). But the best fit is apparently the logarithmic curve ($r = -.963$). Specifically, the best of all logarithmic curves is

$$y' = 8.484 - 2.974 \ln x.$$

The graph of this curve is superimposed upon the scatter diagram in Figure 24.

These are the values for a, b, r^2, and r, using the equation

$$y = a + b \ln x.$$

13.6 EXERCISES

Correlating Fertilizer and Corn Ear Size In a study to determine the linear relationship between the length (in decimeters) of an ear of corn (y) and the amount (in tons per acre) of fertilizer used (x), the following values were determined.

$$n = 10 \quad \Sigma xy = 75$$
$$\Sigma x = 30 \quad \Sigma x^2 = 100$$
$$\Sigma y = 24 \quad \Sigma y^2 = 80$$

1. Find an equation for the least squares line.
2. Find the coefficient of correlation.
3. If 3 tons per acre of fertilizer are used, what length (in decimeters) would the regression equation predict for an ear of corn?

Correlating Celsius and Fahrenheit Temperatures In an experiment to determine the linear relationship between temperatures on the Celsius scale (y) and on the Fahrenheit scale (x), a student got the following results.

$$n = 5 \quad \Sigma xy = 28{,}050$$
$$\Sigma x = 376 \quad \Sigma x^2 = 62{,}522$$
$$\Sigma y = 120 \quad \Sigma y^2 = 13{,}450$$

4. Find an equation for the least squares line.
5. Find the reading on the Celsius scale that corresponds to a reading of 120° Fahrenheit, using the equation of Exercise 4.
6. Find the coefficient of correlation.

Correlating Heights and Weights of Adult Men A sample of 10 adult men gave the following data on their heights and weights.

Height (inches) (x)	62	62	63	65	66
Weight (pounds) (y)	120	140	130	150	142

Height (inches) (x)	67	68	68	70	72
Weight (pounds) (y)	130	135	175	149	168

7. Find the equation of the least squares line.
8. Using the results of Exercise 7, predict the weight of a man whose height is 60 inches.
9. What would be the predicted weight of a man whose height is 70 inches?
10. Compute the coefficient of correlation.

Correlating Reading Ability and IQs The table below gives reading ability scores and IQs for a group of 10 individuals.

Reading (x)	83	76	75	85	74
IQ (y)	120	104	98	115	87

Reading (x)	90	75	78	95	80
IQ (y)	127	90	110	134	119

11. Plot a scatter diagram with reading on the horizontal axis.
12. Find the equation of a regression line.
13. Use your regression line equation to estimate the IQ of a person with a reading score of 65.

Correlating Yearly Sales of a Company Sales, in thousands of dollars, of a certain company are shown here.

Year (x)	0	1	2	3	4	5
Sales (y)	48	59	66	75	80	90

14. Find the equation of the least squares line.
15. Find the coefficient of correlation.
16. If the linear trend displayed by this data were to continue beyond year 5, what sales amount would you predict in year 7?

Comparing the Ages of Dogs and Humans It often is said that a dog's age can be multiplied by 7 to obtain the equivalent human age. A more accurate correspondence (through the first 14 years) is shown in this table from The Old Farmer's Almanac, 2000 edition, page 180.

Dog age (x)	$\frac{1}{2}$	1	2	3	4	5	6	7
Equivalent human age (y)	10	15	24	28	32	36	40	44

Dog age (x)	8	9	10	11	12	13	14
Equivalent human age (y)	48	52	56	60	64	68	70.5

17. Plot a scatter diagram for the given data.
18. Find the equation of the regression line, and graph the line on the scatter diagram of Exercise 17.
19. Describe where the data points show the most pronounced departure from the regression line, and explain why this might be so.
20. Compute the coefficient of correlation.

Statistics on the Westward Population Movement The data here (*from* The World Almanac and Book of Facts 2000, *page 532*) show the increase in the percentage of U.S. population in the West since about the time of the California Gold Rush.

Census Year	Time, in Decades from 1850 (x)	Percentage in West (y)
1850	0	.8%
1870	2	2.6
1890	4	5.0
1910	6	7.7
1930	8	10.0
1950	10	13.3
1970	12	17.1
1990	14	21.2

21. Taking x and y as indicated in the table, find the equation of the regression line.

22. Compute the coefficient of correlation.

23. Describe the degree of correlation (for example, as strong, moderate, or weak).

24. What would be the value of x in the year 2010? Use the regression line to predict the percentage in the West for that year.

Comparing State Populations with Governors' Salaries The table on the right shows the ten most populous states (as of the 2000 census) and the salaries of their governors (as of October 2001). (*Source:* The World Almanac and Book of Facts 2002, *page 380 and page 99*). Use the table for Exercises 25–28.

Rank	State	Population, in Thousands (x)	Governor's Salary, in Thousands of Dollars (y)
1	California	33,872	175
2	Texas	20,852	115
3	New York	18,976	179
4	Florida	15,982	123
5	Illinois	12,419	151
6	Pennsylvania	12,281	105
7	Ohio	11,353	126
8	Michigan	9938	177
9	New Jersey	8414	83
10	Georgia	8186	123

25. Find the equation of the regression line.

26. Compute the coefficient of correlation.

27. Describe the degree of correlation (for example, as strong, moderate, or weak).

28. What governor's salary would this linear model predict for a state with a population of 15 million citizens?

COLLABORATIVE INVESTIGATION

Combining Sets of Data

Divide your class into two separate groups, one consisting of the women and the other consisting of the men. Each group is to select a recorder to write the group's results. As a group, carry out the following tasks. (You may want to devise a way to allow the members of the groups to provide personal data anonymously.)

1. Record the number of members (n) in your group.
2. Collect shoe sizes (x) and heights in inches (y) for all members of the group.
3. Compute the mean, median, and mode(s), if any, for each of the two sets of data.
4. Compute the standard deviation of each of the two sets of data.
5. Construct a box plot for each of the two sets of data.
6. Plot a scatter diagram for the x-y data collected.
7. Find the equation of the least squares regression line ($y' = ax + b$).
8. Evaluate the correlation coefficient (r).
9. Evaluate the strength of the linear relationship between shoe size and height for your group.

Now re-combine your two groups into one. Discuss and carry out the following tasks.

1. If possible, compute the mean of the heights for the combined group, using only the means for the two individual groups and the number of members in each of the two groups. If this is not possible, explain why and describe how you *could* find the combined mean. Obtain the combined mean.
2. Do the same as in item 1 above for the median of the heights for the combined group.
3. Do the same for the mode of the heights for the combined group.
4. Fill in the table below, pertaining to heights, and discuss any apparent relationships among the computed statistics.

	Number of Members	Mean	Median	Mode
Women				
Men				
Combined				

CHAPTER 13 TEST

Statistical Trends in U.S. Farm Workers and Farms Refer to the bar graphs and line graph shown here for Exercises 1–4.

Decline in Farm Workers, 1820–1994*
Source: U.S. Department of Agriculture, Economic Research Service.

Of the approximately 2.9 mil workers in the U.S. in 1820, 71.8%, or about 2.1 mil, were employed in farm occupations. The percentage of U.S. workers in farm occupations had declined drastically by the turn of the century, and by 1994 only 2.5% of all U.S. workers were employed in farm occupations.

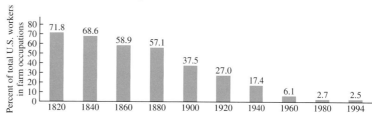

*Figures not compiled for years after 1994. Total workers for 1994 are employed workers age 15 and older; total workers for 1900 to 1960 are members of the experienced civilian labor force 14 and older; total workers for 1820 to 1880 are gainfully employed workers 10 and older.

U.S. Farms, 1940–2000
Source: National Agricultural Statistics Service, U.S. Department of Agriculture.

Since 1940, the number of farms in the U.S. has sharply declined, while the size of the average farm has grown.

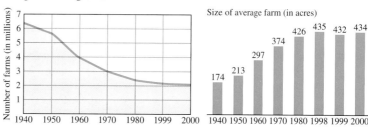

Source: The World Almanac and Book of Facts 2002, page 130.

1. From the given information, about when would you guess half of all U.S. workers were working in farm occupations?
2. Is it true that there were about 4 times as many farm workers in 1840 as in 1940? Explain your answer.
3. About how many farms were there in 1970?
4. In which year was the greater amount of land being farmed, 1940 or 2000? By about what percentage did it exceed the lesser amount?

Client Contacts of a Publisher's Representative Laura Griffin Hiller, a sales representative for a publishing company, recorded the following numbers of client contacts for the twenty-two days that she was on the road in the month of March. Use the given data for Exercises 5–7.

| 12 | 8 | 15 | 11 | 20 | 18 | 14 | 22 | 13 | 26 | 17 |
| 19 | 16 | 25 | 19 | 10 | 7 | 18 | 24 | 15 | 30 | 24 |

5. Construct grouped frequency and relative frequency distributions. Use five uniform classes of width 5 where the first class has a lower limit of 6. (Round relative frequencies to two decimal places.)

6. From your frequency distribution of Exercise 5, construct (a) a histogram and (b) a frequency polygon. Use appropriate scales and labels.

7. For the data above, how many uniform classes would be required if the first class had limits 7–9?

In Exercises 8–11, find the indicated measures for the following frequency distribution.

Value	8	10	12	14	16	18
Frequency	3	8	10	8	5	1

8. the mean
9. the median
10. the mode
11. the range

12. *Exam Scores in a Physics Class* The following data are exam scores achieved by the students in a physics class. Arrange the data into a stem-and-leaf display with leaves ranked.

79	43	65	84	77	70	52	61	80	66
68	48	55	78	71	38	45	64	67	73
77	50	67	91	84	33	49	61	79	72

Use the stem-and-leaf display shown here for Exercises 13–18.

2	3 3 4
2	6 7 8 9 9
3	0 1 1 2 3 3 3 4
3	5 6 7 8 8 9
4	1 2 2 4
4	5 7 9
5	2 4
5	8
6	0

Compute the measures required in Exercises 13–17.

13. the median
14. the mode(s), if any
15. the range
16. the third decile
17. the eighty-fifth percentile
18. Construct a box plot for the above data, showing values for the five important quantities on the numerical scale.

Test Scores in a Training Institute A certain training institute gives a standardized test to large numbers of applicants nationwide. The resulting scores form a normal distribution with mean 80 and standard deviation 5. Find the percent of all applicants with scores as follows. (Use the empirical rule.)

19. between 70 and 90
20. greater than 95 or less than 65
21. less than 75
22. between 85 and 90

Heights of Spruce Trees In a certain young forest, the heights of the spruce trees are normally distributed with mean 5.5 meters and standard deviation 2.1 meters. If a single tree is selected randomly, find the probability (to the nearest thousandth) that its height will fall in each of the following intervals.

23. less than 6.5 meters
24. between 6.2 and 9.4 meters

Season Statistics in Major League Baseball During the 2001 Major League Baseball season, most teams played a total of 162 regular season games. The tables below and on the next page show the statistics on games won for all three divisions of both leagues. In each case, n = number of teams in the division, $\bar{x}$ = average (mean) number of games won, and s = standard deviation of number of games won.

American League		
East Division	Central Division	West Division
$n = 5$	$n = 5$	$n = 4$
$\bar{x} = 76.4$	$\bar{x} = 78.0$	$\bar{x} = 91.5$
$s = 13.9$	$s = 11.8$	$s = 21.0$

Refer to the tables for Exercises 25–27.

National League		
East Division	Central Division	West Division
$n = 5$	$n = 6$	$n = 5$
$\bar{x} = 80.0$	$\bar{x} = 78.3$	$\bar{x} = 78.3$
$s = 8.1$	$s = 14.5$	$s = 14.5$

25. Overall, who had the highest winning average, the East teams, the Central teams, or the West teams?

26. Overall, where were the teams the most "consistent" in number of games won, East, Central, or West?

27. Find (to the nearest tenth) the average number of games won for all 30 teams.

Carry out the following for the paired data values shown here. (In Exercises 29–31, give all calculated values to two decimal places.)

x	1	4	6	7
y	9	7	8	1

28. Plot a scatter diagram.

29. Find the equation for the least squares regression line.

30. Use your equation from Exercise 29 to predict y when $x = 3$.

31. Find the sample correlation coefficient for the given data.

32. Evaluate the strength of the linear relationship for the above data. How confident are you in your evaluation? Explain.

33. Relate the concepts of *inferential statistics* and *inductive reasoning*.

chapter

14 | Consumer Mathematics

It is easy to put off saving for the future. Retirement may seem distant, and often there are many reasons for not starting a savings plan. Does waiting to contribute to a retirement fund have a dramatic effect on the amount of money that an individual has at age 65? Consider two retirement plans. In Plan A a person begins saving at age 20 and deposits $2000 in a retirement fund on each birthday from age 21 to age 30. After age 30 no more contributions are made. In Plan B a person waits until age 30 to start saving and then deposits $2000 in the retirement fund on every birthday from age 31 to age 65. If the retirement fund pays 8% interest, which plan will have the largest accumulation of money? The table lists the amount of money (in dollars) in the fund for each plan at various ages.

14.1 The Time Value of Money

Extension: Annuities

14.2 Consumer Credit

14.3 Truth in Lending

14.4 Purchasing a House

14.5 Investing

Collaborative Investigation:
    To Buy or to Rent?

Chapter 14 Test

Age	Plan A	Plan B
20	0	0
25	11,733	0
30	28,973	0
35	42,571	11,733
40	62,551	28,973
45	91,908	54,304
50	135,042	91,524
55	198,422	146,212
60	291,547	226,566
65	428,378	344,634

Plan A results in the accumulation of $428,378, whereas Plan B produces $344,634. Early contributions can make a substantial difference. In this chapter you will learn about compound interest. Compound interest over time is the reason for the growth of the relatively small amount, $20,000, into nearly $430,000.

As participants in the American (and international) economy, most all of us at times wish to obtain something for which we either don't presently have sufficient money or don't want to commit our own money. In such a case, a common choice is to borrow money (that is "take out a *loan*") for the purchase, thus putting ourselves into debt. Likewise, sometimes we will have more money than we need, for certain periods, in which case we may lend our money to someone else. This loan can involve depositing funds into a simple savings account or into one of many other kinds of "investments."

In the situations mentioned above, we borrow (or lend) some money, for which we expect to pay (or receive) a fee. The mathematics involved in relating that fee to the amount borrowed or lent (and evaluating certain other related costs) is included in the subject commonly called "consumer mathematics."

By studying the material in this chapter, you will gain some useful tools for managing your money, for comparing alternative sources of borrowed money, and for considering potential investments.

14.1 The Time Value of Money

To determine the value of money, we consider not only the amount (number of dollars), but also the particular point in time that the value is to be determined. If we borrow an amount of money today, we will repay a larger amount later. This increase in value is known as **interest.** The money *gains value over time.*

On the other hand, in terms of the equivalent number of goods or services that a given amount of money will buy, it is normally more today than it will be later. This decrease in value is known as **inflation.** The money *loses value over time.* These considerations are summed up in the phrase "the time value of money."

Interest The amount of a loan or a deposit is called the **principal.** The interest is usually computed as a percent of the principal. This percent is called the **rate of interest** (or the **interest rate,** or simply the **rate**). The rate of interest is always assumed to be an annual rate unless otherwise stated.

Interest calculated only on principal is called **simple interest.** Interest calculated on principal plus any previously earned interest is called **compound interest.** Generally, simple interest is used for loans of one year or less, while compound interest is used for longer loans. Banks always pay compound interest on money deposited.

Simple Interest   Simple interest is calculated according to the following formula.

> ### Simple Interest
> If P = principal, r = annual interest rate, and t = time (in years), then the **simple interest** I is given by
> $$I = Prt.$$

This is the computation required to solve Example 1.

EXAMPLE 1 Find the simple interest paid to borrow $5350 for 5 months at 6%.

Here $P = \$5350$, $r = 6\% = .06$, and $t = 5/12$ (since 5 months is 5/12 of a year). The interest is

$$I = Prt = \$5350(.06)\left(\frac{5}{12}\right) = \$133.75.$$

Future Value and Present Value In Example 1 above, at the end of 5 months the borrower would have to repay

$$\text{Principal} + \text{interest} = \$5350 + \$133.75 = \$5483.75.$$

The total amount repaid is sometimes called the **maturity value** (or simply the **value**) of the loan. We will generally refer to it as the **future value,** or **future amount,** since when a loan is being set up, repayment will be occurring in the future. We will use A to denote future amount (or value). The original principal, denoted P, can also be thought of as **present value.** The future value depends on the principal (present value) and the interest as follows.

$$A = P + I = P + Prt = P(1 + rt)$$

We summarize this relationship in the following formulas.

King Hammurabi tried to hold interest rates at 20 percent for both silver and gold, but moneylenders ignored his decrees.

Future and Present Values for Simple Interest

If a principal P is borrowed at simple interest for t years at an annual interest rate of r, then the **future value** of the loan, denoted A, is given by

$$A = P(1 + rt).$$

Likewise, the **present value,** P, is given by

$$P = \frac{A}{1 + rt}.$$

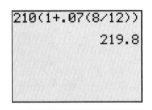

This is the computation required to solve Example 2.

EXAMPLE 2 Alex Ortega took out a simple interest loan for $210 to purchase textbooks and school supplies. If the annual interest rate is 7% and he must repay the loan after 8 months, find the future value (the maturity value) of the loan.

Applying the future value formula for simple interest,

$$A = P(1 + rt) = \$210\left[1 + .07\left(\frac{8}{12}\right)\right] = \$219.80.$$

At the end of 8 months, Alex will need to repay $219.80.

EXAMPLE 3 Suppose that Alex (Example 2) is granted an 8-month *deferral* of the $210 payment rather than a loan. That is, instead of incurring interest for 8 months, he will have to pay just the $210 at the end of that period. If he has extra money right now, and if he can earn 4% simple interest on savings, what lump sum must he deposit now in order to have the $210 available after 8 months?

14.1 The Time Value of Money

We apply the present value formula for simple interest.

$$P = \frac{A}{1+rt} = \frac{\$210}{1+(.04)(8/12)} = \$204.55$$

A deposit of $204.55 now, growing at 4% simple interest, will grow to $210 over an 8-month period.

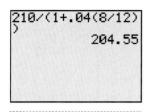

This is the computation required to solve Example 3.

Compound Interest As mentioned earlier, interest paid on principal plus interest is called compound interest. To see how this works, suppose $1000 is deposited in a savings account paying a 4% interest rate. During the first year that the money is on deposit, the account would earn interest of

$$I = Prt = \$1000\,(.04)\,(1) = \$40.$$

This amount of interest would be *credited* (added) to the account, so that at the beginning of the second year the account would contain the amount

$$\$1000 + \$40 = \$1040.$$

During the second year, the account would earn interest of

$$\$1040\,(.04)\,(1) = \$41.60.$$

At the beginning of the third year, the account would contain the amount

$$\$1040 + \$41.60 = \$1081.60.$$

The interest earned during the second year was more than the interest earned during the first year. The interest earned during the first year was added to the principal and then used in calculating the amount of interest paid during the second year. This is the way compounding of interest works.

Generalizing, we now develop a formula that will give the amount in the account at any future time, that is, the **future value** of the account. Realistically, earned interest can be credited to an account at time intervals other than one year (usually more often). For example, it can be done semiannually, quarterly, monthly, or daily. (Daily compounding is quite common.) This time interval is called the **compounding period** (or simply the **period**). Start with the following definitions.

$$P = \text{original principal deposited}$$
$$r = \text{annual interest rate}$$
$$m = \text{number of periods per year}$$
$$n = \text{total number of periods}$$

During each individual compounding period, interest is earned according to the simple interest formula, and as interest is added, the beginning principal increases from one period to the next. During the first period, the interest earned is given by

$$\text{Interest} = (\text{Principal})\,(\text{rate})\,(\text{time})$$

$$= P(r)\left(\frac{1}{m}\right) \qquad \text{One period} = \frac{1}{m} \text{ year}$$

$$= P\left(\frac{r}{m}\right). \qquad \text{Rewrite } (r)\left(\frac{1}{m}\right) \text{ as } \left(\frac{r}{m}\right).$$

> The following item shows how one community plans to use the power of compound interest to help it plan for the future. (News broadcast © CBS Inc. 1977. All rights reserved. Originally broadcast over the CBS Radio Network as part of the *Hourly News* by Charles Osgood on March 2, 1977.)
>
> "Eli Hooker, of Union City, Michigan, doesn't want his community to go broke when the tricentennial rolls around in 2076. Hooker says the local Bicentennial Committee last year didn't have enough money. He was chairman. So, to plan ahead, he has collected twenty-five dollars apiece from 42 residents of Union City, squirreling it away in a local bank. At seven-percent interest, compounded, the fund would have a million dollars by 2076.
>
> Hooker hopes that will be enough for the parades and fireworks a tricentennial celebration will require. But Eli Hooker isn't being an entirely selfless patriot about all this: he hopes one of the big events on July 4, 2076, will be the unveiling of a monument to the 42 people of Union City who contributed 99 years in advance."
>
> Unfortunately, that million dollars won't be worth as much then as we might think. If the community decides to hire people to parade around in historical costumes, it might have paid them $4 per hour in 1976. The going wage in 2076, assuming 7% annual inflation for a hundred years, would be well over $3000 per hour.

At the end of the first period, the account then contains

Ending amount = beginning amount + interest

$$= P + P\left(\frac{r}{m}\right)$$

$$= P\left(1 + \frac{r}{m}\right). \quad \text{Factor } P \text{ from both terms.}$$

Now during the second period, the interest earned is given by

Interest = [Principal](rate)(time)

$$= \left[P\left(1 + \frac{r}{m}\right)\right](r)\left(\frac{1}{m}\right)$$

$$= P\left(1 + \frac{r}{m}\right)\left(\frac{r}{m}\right),$$

so that the account ends the second period containing

Ending amount = beginning amount + interest

$$= P\left(1 + \frac{r}{m}\right) + P\left(1 + \frac{r}{m}\right)\left(\frac{r}{m}\right)$$

$$= P\left(1 + \frac{r}{m}\right)\left[1 + \frac{r}{m}\right] \quad \text{Factor } P\left(1 + \frac{r}{m}\right) \text{ from both terms.}$$

$$= P\left(1 + \frac{r}{m}\right)^2.$$

Consider one more period, namely the third. The interest earned is

Interest = [Principal](rate)(time)

$$= \left[P\left(1 + \frac{r}{m}\right)^2\right](r)\left(\frac{1}{m}\right)$$

$$= P\left(1 + \frac{r}{m}\right)^2\left(\frac{r}{m}\right),$$

so the account ends the third period containing

Ending amount = beginning amount + interest

$$= P\left(1 + \frac{r}{m}\right)^2 + P\left(1 + \frac{r}{m}\right)^2\left(\frac{r}{m}\right)$$

$$= P\left(1 + \frac{r}{m}\right)^2\left[1 + \frac{r}{m}\right] \quad \text{Factor } P\left(1 + \frac{r}{m}\right)^2 \text{ from both terms.}$$

$$= P\left(1 + \frac{r}{m}\right)^3.$$

Table 1 summarizes the preceding results. Inductive reasoning gives the bottom row of expressions for period number n.

TABLE 1 Compound Amount

Period Number	Beginning Amount	Interest Earned During Period	Ending Amount
1	P	$P\left(\dfrac{r}{m}\right)$	$P\left(1+\dfrac{r}{m}\right)$
2	$P\left(1+\dfrac{r}{m}\right)$	$P\left(1+\dfrac{r}{m}\right)\left(\dfrac{r}{m}\right)$	$P\left(1+\dfrac{r}{m}\right)^2$
3	$P\left(1+\dfrac{r}{m}\right)^2$	$P\left(1+\dfrac{r}{m}\right)^2\left(\dfrac{r}{m}\right)$	$P\left(1+\dfrac{r}{m}\right)^3$
$\vdots$	$\vdots$	$\vdots$	$\vdots$
n	$P\left(1+\dfrac{r}{m}\right)^{n-1}$	$P\left(1+\dfrac{r}{m}\right)^{n-1}\left(\dfrac{r}{m}\right)$	$P\left(1+\dfrac{r}{m}\right)^n$

The lower right entry of the table provides the following formulas.

Future and Present Values for Compound Interest

If P dollars are deposited at an annual interest rate of r, compounded m times per year, and the money is left on deposit for a total of n periods, then the **future value, A** (the final amount on deposit), is given by

$$A = P\left(1 + \frac{r}{m}\right)^n,$$

and the **present value, P,** is given by

$$P = \frac{A}{\left(1 + \dfrac{r}{m}\right)^n}.$$

In these formulas, the quantity $\left(1 + \dfrac{r}{m}\right)^n$ can be evaluated using a key such as y^x on a scientific calculator, or $\wedge$ on a graphing calculator.

EXAMPLE 4 Find the future value (final amount on deposit) and the amount of interest earned for the following deposits.

(a) $18,950 at 6% compounded quarterly for 5 years

We have $P = \$18{,}950$, $r = 6\% = .06$, $m = 4$. And over a time period of 5 years, we obtain $n = 5m = (5)(4) = 20$.

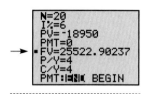

The financial functions of the TI-83 Plus calculator allow the user to solve for a missing quantity regarding the time value of money. The arrow indicates the display that supports the answer in Example 4(a).

Using the future value formula and a calculator, we get

$$A = \$18{,}950\left(1 + \frac{.06}{4}\right)^{20} = \$25{,}522.90.$$

To find the interest earned, subtract the amount originally deposited (present value) from the future value:

$$\begin{aligned}\text{Interest earned} &= \text{future value} - \text{present value} \\ &= \$25{,}522.90 - \$18{,}950 \\ &= \$6572.90.\end{aligned}$$

(b) $6429 at 2.7% compounded monthly for 30 months

Here $P = \$6429$, $r = 2.7\% = .027$, $m = 12$, and $n = 30$, so

$$A = \$6429\left(1 + \frac{.027}{12}\right)^{30} = \$6877.42.$$

$$\begin{aligned}\text{Interest earned} &= \text{future value} - \text{present value} \\ &= \$6877.42 - \$6429 \\ &= \$448.42.\end{aligned}$$

EXAMPLE 5 The Cebelinskis' daughter will need $17,000 in 5 years to help pay for her college education. What lump sum, deposited today at 7% compounded quarterly, will produce the necessary amount?

This question requires that we find the present value P based on the following.

Future value:	$A = \$17{,}000$
Annual rate:	$r = 7\% = .07$
Periods per year:	$m = 4$
Total number of periods:	$n = (5)(4) = 20$

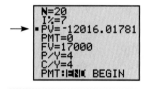

The arrow indicates the display that supports the answer in Example 5.

The present value formula gives

$$P = \frac{A}{\left(1 + \dfrac{r}{m}\right)^n} = \frac{\$17{,}000}{\left(1 + \dfrac{.07}{4}\right)^{20}} = \$12{,}016.02.$$

Assuming interest of 7% compounded quarterly can be maintained, a deposit of $12,016.02 today will grow to $17,000 five years in the future.

The next example will involve the use of logarithms. For help with those steps, see the discussion of logarithms in an earlier chapter (or see your calculator manual).

EXAMPLE 6 In a savings account paying 3% interest, compounded daily, when will the amount in the account be twice the original principal?

In the future value formula, the future value must equal two times the present value, so we substitute $2P$ for A.

$$2P = P\left(1 + \frac{r}{m}\right)^n$$

$$2 = \left(1 + \frac{r}{m}\right)^n \qquad \text{Divide both sides by } P.$$

$$2 = \left(1 + \frac{.03}{365}\right)^n \qquad \text{Substitute values of } r \text{ and } m.$$

$$\log 2 = \log\left(1 + \frac{.03}{365}\right)^n \qquad \text{Take the log of both sides.}$$

$$\log 2 = n \log\left(1 + \frac{.03}{365}\right) \qquad \text{Use the power property of logarithms.}$$

Now solving for n gives

$$n = \frac{\log 2}{\log\left(1 + \frac{.03}{365}\right)} = 8434 \quad \text{(rounding to the nearest whole number).}$$

Since n denotes the number of periods, which is days in this case, the required amount of time is 8434 days, or 23 years, 39 days (ignoring leap years).

Effective Annual Yield Banks, credit unions, and other savings institutions often give two quantities when advertising the rates they will pay. The first, the actual annualized interest rate, is the **nominal rate** (the "named" or "stated" rate). The second quantity is the equivalent rate that would produce the same final amount, or future value, at the end of one year if the interest being paid were simple rather than compound. This is called the "effective rate," or more commonly the **effective annual yield.** Since the interest is normally compounded multiple times per year, the yield will usually be somewhat higher than the nominal rate.

```
▶Eff(3.50,4)
        3.546206055
```

Financial calculators are programmed to compute effective interest rate. Compare to the result in Example 7.

EXAMPLE 7 What is the effective annual yield of an account paying a nominal rate of 3.50%, compounded quarterly?

From the given data, $r = .035$ and $m = 4$. Suppose we deposited $P = \$1$ and left it for 1 year ($n = 4$). Then the compound future value formula gives

$$A = \left(1 + \frac{.035}{4}\right)^4 \approx 1.03546.$$

The initial deposit of $1, after 1 year, has grown to $1.03546. As usual, interest earned $=$ future value $-$ present value $= 1.0355 - 1 = .0355 = 3.55\%$. A nominal rate of 3.50% results in an effective annual yield of 3.55%.

Generalizing the procedure of Example 7 gives the following formula.

Effective Annual Yield

A nominal interest rate of r, compounded m times per year, is equivalent to an **effective annual yield** of

$$Y = \left(1 + \frac{r}{m}\right)^m - 1.$$

Consumers can make practical use of the effective annual yield concept when shopping for loans or savings opportunities. A borrower should seek the lowest yield available, while a depositor should look for the highest.

TABLE 2 Consumer Price Index (CPI-U) 1970 to 2002 [Period from 1982 to 1984: 100]

Year	Average CPI-U	Percent Change in CPI-U
1970	38.8	5.7
1971	40.5	4.4
1972	41.8	3.2
1973	44.4	6.2
1974	49.3	11.0
1975	53.8	9.1
1976	56.9	5.8
1977	60.6	6.5
1978	65.2	7.6
1979	72.6	11.3
1980	82.4	13.5
1981	90.9	10.3
1982	96.5	6.2
1983	99.6	3.2
1984	103.9	4.3
1985	107.6	3.6
1986	109.6	1.9
1987	113.6	3.6
1988	118.3	4.1
1989	124.0	4.8
1990	130.7	5.4
1991	136.2	4.2
1992	140.3	3.0
1993	144.5	3.0
1994	148.2	2.6
1995	152.4	2.8
1996	156.9	3.0
1997	160.5	2.3
1998	163.0	1.6
1999	166.6	2.2
2000	172.2	3.4
2001	177.1	2.8
2002	179.9	1.6

Source: Bureau of Labor Statistics.

EXAMPLE 8 Tamara Johnson-Draper wants to deposit $2800 into a savings account and has narrowed her choices to the three institutions represented here. Which is the best choice?

Institution	Rate on Deposits of $1000 to $5000
Friendly Credit Union	2.08% annual rate, compounded monthly
Premier Savings	2.09% annual yield
Neighborhood Bank	2.05% compounded daily

Compare the effective annual yields for the three institutions:

For Friendly, $\quad Y = \left(1 + \dfrac{.0208}{12}\right)^{12} - 1 = .0210 = 2.10\%.$

For Premier, $\quad Y = 2.09\%.$

For Neighborhood, $\quad Y = \left(1 + \dfrac{.0205}{365}\right)^{365} - 1 = .0207 = 2.07\%.$

The best of the three yields, 2.10%, is offered by Friendly Credit Union.

Inflation The compounding of increases, as reflected in the future value of compound interest, also occurs in situations other than the earning or paying of interest. An important example is **inflation,** the periodic increase in the cost of living.

Inflation in an economy usually is expressed as a monthly or annual rate of price increases, estimated by government agencies in a systematic way. In the United States, the Bureau of Labor Statistics publishes **consumer price index (CPI)** figures, which reflect the prices of certain items purchased by large numbers of people. The items include such things as food, housing, automobiles, fuel, and clothing. Items such as yachts or expensive jewelry are not included.

CPI figures are reported on a monthly basis, but annual figures give adequate information for most consumers. Table 2 gives values of the primary index representing "all urban consumers" (the CPI-U). The value shown for a given year is actually the average (arithmetic mean) of the twelve monthly figures for that year. The percent change value for a given year shows the change in the CPI-U since the previous year. For example, the average CPI-U for 2002 (179.9) was 1.6% greater than the average value for 2001 (177.1).

From 1913 to the present, the record inflation rate occurred in the year 1918 and was 18.0%. The record **deflation** rate (with price levels actually decreasing from year to year) occurred in 1921 and was −10.5%. Severe deflation has not occurred since the 1930s. A period of minor deflation, normally brief and often accompanying an overall slowdown in the general economy, usually is referred to as a **recession** (rather than deflation). A business recession *can* coincide with price inflation.

Unlike account values under interest compounding, which make sudden jumps at just certain points in time (such as quarterly, monthly, or daily), price levels tend to fluctuate gradually over time. Thus it is appropriate, for inflationary estimates, to use the formula for continuous compounding (which was introduced in an earlier chapter).

Inflation depends on the prices of many goods and services—the so-called "market basket" upon which CPI calculations are based. Selecting market basket items and gathering and analyzing price and expenditure data is a huge task for the Bureau of Labor Statistics (BLS), but is crucial since many millions of workers, retirees, and program recipients are directly affected by CPI figures each year. Also pegged to the CPI are income tax exemptions and deductions and the break points between tax brackets.

The BLS constantly attempts to refine and improve the accuracy of the computed indexes. An example was the introduction, in August 2002, of a supplemental index, the Chained Consumer Price Index for all Urban Consumers, denoted C-CPI-U. This index tracks changes in relative prices within categories, and seeks to reflect the degree to which consumers shift to alternative items as prices fluctuate. It is hoped that these alternative approaches will lead to more accurate measures of the true inflation rate. Current data are published regularly online at the CPI home page, http://www.bls.gov/cpi.

Future Value for Continuous Compounding

If an initial deposit of P dollars earns continuously compounded interest at an annual rate r for a period of t years, then the **future value, A,** is given by

$$A = Pe^{rt}.$$

EXAMPLE 9 Suppose you earn a salary of $24,000 per year. About what salary would you need 20 years from now to maintain your purchasing power in case the inflation rate were to persist at each of the following levels?

(a) 2% (approximately the 1999 level)

In this case we can use the continuous compounding future value formula with $P = \$24,000$, $r = .02$, and $t = 20$. Using the $\boxed{e^x}$ key on a calculator, we find

$$A = Pe^{rt} = (\$24,000)e^{(.02)(20)} = \$35,803.79.$$

The required salary 20 years from now would be about $36,000.

(b) 13% (approximately the 1980 level)

For this level of inflation, we would have

$$A = Pe^{rt} = (\$24,000)e^{(.13)(20)} = \$323,129.71.$$

The required salary 20 years from now would be about $323,000.

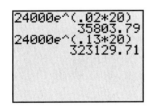

These are the computations required to solve the two parts of Example 9.

Notice that in Example 9, we stated the required salaries as rounded numbers, $36,000 and $323,000. In the first place, the given inflation rates were rounded. Also, even the more precise CPI figures published by the government are only an attempt to reflect cost of living on a very average basis. Finally, the assumption that any given rate will continue for 20 years is itself a wild speculation. (Consider the fluctuations in Table 2.) For all these reasons (and more), you should regard all inflationary projections into the future as only approximate forecasts.

Especially when working with quantities such as inflation, where continual fluctuations and inexactness prevail, we often develop rough "rules of thumb" for obtaining quick estimates. One example is the estimation of the **years to double,** which is the number of years it takes for the general level of prices to double for a given annual rate of inflation. We can derive an estimation rule as follows, beginning with the future value formula for continuous compounding.

$$A = Pe^{rt}$$

$$2P = Pe^{rt} \quad \text{Prices are to double.}$$

$$2 = e^{rt} \quad \text{Divide both sides by } P.$$

$$\ln 2 = rt \quad \text{Take the natural log of both sides.}$$

$$t = \frac{\ln 2}{r} \quad \text{Solve for } t.$$

$$t = \frac{100 \ln 2}{100r} \quad \text{Multiply numerator and denominator by 100.}$$

$$\textbf{years to double} \approx \frac{70}{\textbf{annual inflation rate}} \quad 100 \ln 2 \approx 70$$

(Since r is the inflation rate as a *decimal*, $100r$ is the inflation rate as a *percent*.)

The result above usually is called the **rule of 70**. The value it produces should be rounded *up* to the next whole number of years.

EXAMPLE 10 Estimate the years to double for an annual inflation rate of 2.3%.

By the rule of 70,

$$\text{Years to double} \approx \frac{70}{2.3} = 30.43.$$

With a sustained inflation rate of 2.3%, prices would double in about 31 years. ∎

Because 100 ln 2 ≈ 70, the two results shown here are approximately equal. see Example 10 and the discussion preceding it.

14.1 EXERCISES

In the following exercises, assume whenever appropriate that, unless otherwise known, there are 12 months per year, 30 days per month, and 365 days per year.

Find the simple interest owed for each of the following loans.

1. $1400 at 8% for 1 year
2. $8000 at 5% for 1 year
3. $650 at 6% for 9 months
4. $6000 at 7% for 4 months
5. $2675 at 8.2% for $2\frac{1}{2}$ years
6. $1460 at 7.82% for 22 months

Find the future value of each of the following deposits if the account pays (a) simple interest, and (b) interest compounded annually.

7. $700 at 3% for 6 years
8. $2000 at 4% for 5 years
9. $2500 at 2% for 3 years
10. $5000 at 8% for 5 years

Solve the following interest-related problems.

11. *Simple Interest on a Late Property Tax Payment* George Atkins was late on his property tax payment to the county. He owed $7500 and paid the tax 4 months late. The county charges a penalty of 8% simple interest. Find the amount of the penalty.

12. *Simple Interest on a Loan for Work Uniforms* Michael Bolinder bought a new supply of delivery uniforms. He paid $815 for the uniforms and agreed to pay for them in 5 months at 9% simple interest. Find the amount of interest that he will owe.

13. *Simple Interest on a Small Business Loan* Chris Campbell opened a security service on March 1. To pay for office furniture and guard dogs, Campbell borrowed $14,800 at the bank and agreed to pay the loan back in 10 months at 9% simple interest. Find the *total amount* required to repay the loan.

14. *Simple Interest on a Tax Overpayment* David Fontana is owed $260 by the Internal Revenue Service for overpayment of last year's taxes. The IRS will repay the amount at 6% simple interest. Find the *total amount* Fontana will receive if the interest is paid for 8 months.

Find the missing final amount (future value) and/or interest earned.

	Principal	Rate	Compounded	Time	Final Amount	Compound Interest
15.	$975	4%	quarterly	4 years	$1143.26	_____
16.	$1150	7%	semiannually	6 years	$1737.73	_____
17.	$480	6%	semiannually	9 years	_____	$337.17
18.	$2370	10%	quarterly	5 years	_____	_____
19.	$7500	$3\frac{1}{2}$%	annually	25 years	_____	_____
20.	$3450	2.4%	semiannually	10 years	_____	_____

For each of the following deposits, find the future value *(final amount on deposit) when compounding occurs* **(a)** *annually,* **(b)** *semiannually, and* **(c)** *quarterly.*

	Principal	Rate	Time
21.	$1000	10%	3 years
22.	$3000	5%	7 years
23.	$12,000	8%	5 years
24.	$15,000	7%	9 years

Occasionally a savings account may actually pay interest compounded continuously. For each of the following deposits, find the interest earned *if interest is compounded* **(a)** *semiannually,* **(b)** *quarterly,* **(c)** *monthly,* **(d)** *daily, and* **(e)** *continuously. (Recall that interest = future value − present value.)*

	Principal	Rate	Time
25.	$1040	7.6%	4 years
26.	$1550	2.8%	33 months (Assume 1003 days in parts (d) and (e).)

27. Describe the effect of interest being compounded more and more often. In particular, how good is continuous compounding?

Work the following interest-related problems.

28. *Finding the Amount Borrowed in a Simple Interest Loan* Chris Siragusa takes out a 7% simple interest loan today which will be repaid 15 months from now with a payoff amount of $815.63. What amount is Chris borrowing?

29. *Finding the Amount Borrowed in a Simple Interest Loan* What is the maximum amount you can borrow today if it must be repaid in 4 months with simple interest at 8% and you know that at that time you will be able to repay no more than $1500?

30. In the development of the future value formula for compound interest in the text, at least four specific problem-solving strategies were employed. Identify (name) as many of them as you can and describe their use in this case.

Find the present value for each of the following future amounts.

31. $1000 (6% compounded annually for 5 years)
32. $14,000 (4% compounded quarterly for 3 years)
33. $9860 (8% compounded semiannually for 10 years)
34. $15,080 (5% compounded monthly for 4 years)

Finding the Present Value of a Compound Interest Retirement Account Tom and Louise want to establish an account that will supplement their retirement income beginning 30 years from now. For each of the following interest rates find the lump sum they must deposit today so that $500,000 will be available at time of retirement.

35. 8% compounded quarterly
36. 6% compounded quarterly
37. 8% compounded daily
38. 6% compounded daily

Finding the Effective Annual Yield in a Savings Account Suppose a savings and loan pays a nominal rate of 5% on savings deposits. Find the effective annual yield if interest is compounded as stated in Exercises 39–45. (Give answers to the nearest thousandth of a percent.)

39. annually
40. semiannually
41. quarterly
42. monthly
43. daily
44. 1000 times per year
45. 10,000 times per year

46. Judging from Exercises 39–45, what do you suppose is the effective annual yield if a nominal rate of 5% is compounded continuously? Explain your reasoning.

Comparing Interest Rates on Certificates of Deposit For the week of December 25–31, 2002, Schools Financial Credit Union in Sacramento posted the following savings rates for "premium rate certificates" of deposit in several categories. Use this information in Exercises 47 and 48.

		Minimum Balance $1000	Minimum Balance $2500
6-month	Rate	1.50%	1.60%
	Yield	1.51%	1.61%
12-month	Rate	1.85%	1.95%
	Yield	1.86%	1.96%

47. If you put $20,000 into the 12-month option, how much would you have in 1 year?
48. How often does compounding occur in these accounts?

Solve the following interest-related problems.

49. *Finding Years to Double* How long would it take to double your money in an account paying 6% compounded quarterly? (Give your answer in years plus days, ignoring leap years.)

50. *Comparing Principal and Interest Amounts* After what time period would the interest earned equal the original principal in an account paying 6% compounded daily? (Give your answer in years plus days, ignoring leap years.)

51. *Finding the Nominal Rate of a Savings Account* At Orangevale Savings interest is compounded monthly and the effective annual yield is 2.95%. What is the nominal rate?

52. *Comparing Bank Savings Rates* In Carmichael, CA, Bank A pays a nominal rate of 3.800% compounded daily on deposits. Bank B produces the same annual yield as the first but compounds interest only quarterly and pays no interest on funds deposited for less than an entire quarter.
    (a) What nominal rate does Bank B pay (to the nearest thousandth of a percent)?
    (b) Which bank should Patty Demko choose if she has $2000 to deposit for 10 months? How much more interest will she earn than in the other bank?
    (c) Which bank should Sashaya Davis choose for a deposit of $6000 for one year? How much interest will be earned?

53. Solve the effective annual yield formula for r to obtain a general formula for nominal rate in terms of yield and the number of compounding periods per year. (Use the formula to verify your answers to Exercises 51 and 52(a).)

Estimating the Years to Double by the Rule of 70 Use the rule of 70 to estimate the years to double for the following annual inflation rates.

54. 1%
55. 2%
56. 8%
57. 9%

Estimating the Inflation Rate by the Rule of 70 Use the rule of 70 to estimate the annual inflation rate (to the nearest tenth of a percent) that would cause the general level of prices to double in the following time periods.

58. 5 years
59. 7 years
60. 16 years
61. 22 years

Work the following inflation-related problems.

62. Derive a rule for estimating the "years to triple," that is, the number of years it would take for the general levels of prices to triple for a given annual inflation rate.

63. What would you call your rule of Exercise 62? Explain your reasoning.

Estimating Future Prices for Constant Annual Inflation The year 2003 prices of several items are given below. Find the estimated future prices required to fill the blanks in the chart. (Give a number of significant figures consistent with the 2003 price figures provided.)

	Item	2003 Price	2010 Price 2% Inflation	2020 Price 2% Inflation	2010 Price 10% Inflation	2020 Price 10% Inflation
64.	House	$165,000	___	___	___	___
65.	Meal Deal	$ 4.89	___	___	___	___
66.	Gallon of gasoline	$ 1.65	___	___	___	___
67.	Small car	$ 13,500	___	___	___	___

Estimating Future Prices for Variable Annual Inflation As seen in Table 2, inflation rates do not often stay constant over a period of years. For example, from 1996 to 1997 the index increased from 156.9 to 160.5, or 2.3%, while the following year, 1997 to 1998, it increased from 160.5 to 163.0, only 1.6%. Assume that prices for the items below increased at the average annual rates shown in Table 2. Considering the increases for all appropriate years, find the missing prices in the last column of the chart. Round to the nearest dollar.

	Item	Price	Year Purchased	Price in 2002
68.	Sofa	$ 650	1999	___
69.	A dozen roses	$ 78	1996	___
70.	Skis	$ 149	1995	___
71.	Motorcycle	$6388	1994	___

Work the following interest-related problems.

72. *Finding the Present Value of a Future Real Estate Purchase* A California couple are selling their small dairy farm to a developer, but they wish to defer receipt of the money until 3 years from now, when they will be in a lower tax bracket. Find the lump sum that the developer can deposit today, at 3% compounded quarterly, so that enough will be available to pay the couple $750,000 in 3 years.

73. *Finding the Present Value of a Future Equipment Purchase* Human gene sequencing is a major research area of biotechnology. One company (PE Biosystems) leased 300 of its sequencing machines to a sibling company (Celera). If Celera was able to earn 7% compounded quarterly on invested money, what lump sum did they need to invest in order to purchase those machines 18 months later at a price of $300,000 each? (*Source: Forbes*, February 21, 2000, p. 102.)

74. Write a report on the consumer price index, relating recent and proposed changes both in the "market basket" and in the methods used to calculate the effects of various items.

EXTENSION

Annuities

Some important concepts in the mathematics of finance depend upon being able to sum the terms of a **geometric sequence.** In a geometric sequence each term, after the first, is generated by multiplying the previous term by the **common ratio,** a number remaining constant throughout the sequence. For example, a geometric sequence with first term 3 and common ratio 2 starts out as follows:

$$3, 6, 12, 24, 48, 96, \ldots.$$

In general, if the first term is denoted a, the common ratio is denoted r, and there are n terms altogether, then the complete sequence is

$$a, ar, ar^2, ar^3, \ldots, ar^{n-1}.$$

(Verify by inductive reasoning that the nth term really is ar^{n-1}.)
The sum of all n terms of the geometric sequence above can be written

$$S = a + ar + ar^2 + \cdots + ar^{n-2} + ar^{n-1}.$$

Multiply both sides of this equation by r:

$$Sr = ar + ar^2 + ar^3 + \cdots + ar^{n-1} + ar^n.$$

Now position these two equations, one below the other, and subtract one from the other, as follows.

$$
\begin{aligned}
S &= a + ar + ar^2 + \cdots + ar^{n-2} + ar^{n-1} \\
Sr &= ar + ar^2 + ar^3 + \cdots + ar^{n-1} + ar^n \\
\hline
S - Sr &= (a - ar) + (ar - ar^2) + (ar^2 - ar^3) + \cdots + (ar^{n-2} - ar^{n-1}) + (ar^{n-1} - ar^n)
\end{aligned}
$$

Now we can rearrange and regroup the terms on the right to obtain the following.

$$S - Sr = a + (ar - ar) + (ar^2 - ar^2) + \cdots + (ar^{n-1} - ar^{n-1}) - ar^n$$
$$= a + 0 + 0 + \cdots + 0 - ar^n$$

Notice that all terms on the right, except a and $-ar^n$, were arranged in pairs to cancel out. So we get the following.

$$S - Sr = a - ar^n$$
$$S(1 - r) = a(1 - r^n) \qquad \text{Factor both sides.}$$
$$S = \frac{a(1 - r^n)}{1 - r} \qquad \text{Solve for the sum } S.$$

> ### The Sum of a Geometric Sequence
> If a geometric sequence has first term a and common ratio r, and has n terms altogether, then the sum of all n terms is given by
> $$S = \frac{a(1 - r^n)}{1 - r}.$$

Now let's see how the summation formula on the previous page can be applied to systematic savings or investment plans. Consider the following situation.

> When I deposit money into a certain account, I can earn interest at an annual rate r, compounded m times per year. Assume I will make a regular deposit of R (in dollars) *at the end* of each compounding period. What will be the total value of my account after n periods?

This type of account, which receives regular periodic deposits *at the end of* each compounding period, is called an **ordinary annuity** account. The total term (time period) over which deposits are made, consisting of the n compounding periods, is called the **accumulation period** of the annuity.

As each deposit R goes into the account, it begins to earn interest and grows to some amount according to the future value formula for compound interest. In the table below we list the n deposits, starting with the last one, which is made at the very end of the nth compounding period. Since the last deposit (deposit number n) has no time to accumulate interest, its final value is just R.

Value of an Annuity Account

Deposit Number	Future Value of Deposit
n	R
$n-1$	$R\left(1 + \dfrac{r}{m}\right)$
$n-2$	$R\left(1 + \dfrac{r}{m}\right)^2$
$\vdots$	$\vdots$
3	$R\left(1 + \dfrac{r}{m}\right)^{n-3}$
2	$R\left(1 + \dfrac{r}{m}\right)^{n-2}$
1	$R\left(1 + \dfrac{r}{m}\right)^{n-1}$

(Why is the exponent in the future value of the first deposit $n - 1$ rather than n?)

The future value of the account must be the sum of the future values of all the individual deposits, which is the sum of the entries of the right-hand column of the table. But those entries comprise a geometric sequence. Its first term is R, the common ratio is $\left(1 + \dfrac{r}{m}\right)$, and the number of terms is n. So, in this case, the sum of the sequence (the **future value of the annuity**) is given by

$$\text{Sum} = \frac{R\left[1 - \left(1 + \dfrac{r}{m}\right)^n\right]}{1 - \left(1 + \dfrac{r}{m}\right)}.$$

(continued)

A little algebra produces the following formula.

Future Value of an Ordinary Annuity

If money in an annuity account earns interest at an annual rate r, compounded m times per year, and regular deposits of amount R are made into this account at the end of each compounding period, then the final value (or **future value**) of the account at the end of n periods is given by

$$V = \frac{R\left[\left(1 + \frac{r}{m}\right)^n - 1\right]}{\frac{r}{m}}.$$

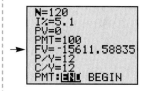

Compare the future value indicated by the arrow with the result in part (a) of the Example.

EXAMPLE Peter Harris sets up an ordinary annuity account with his credit union. Starting a month from today he will deposit $100 per month, and deposits will earn 5.1% per year, compounded monthly. Find **(a)** the value of Peter's account 10 years from today, **(b)** the total of all his deposits over the 10 years, and **(c)** the total interest earned.

(a) In this case, $R = \$100$, $r = 5.1\% = .051$, $m = 12$, and $n = 10 \times 12 = 120$. The formula for future value of an ordinary annuity gives

$$V = \frac{\$100\left[\left(1 + \frac{.051}{12}\right)^{120} - 1\right]}{\frac{.051}{12}} = \$15{,}611.59.$$

(b) Over 10 years, monthly deposits of $100 will total

$$\$100 \times 10 \times 12 = \$12{,}000.$$

(c) Interest earned = Future value − Total deposits

$$= \$15{,}611.59 - \$12{,}000$$

$$= \$3611.59.$$

The future value of Peter's account, at the end of 10 years, will be $15,611.59. This value results from a total of $12,000 in deposits which, over those 10 years, earn an additional $3611.59 in interest.

EXTENSION EXERCISES

Finding the Earnings of an Ordinary Annuity Account For each of the following ordinary annuity accounts, find **(a)** *the future value of the account (at the end of the accumulation period),* **(b)** *the total of all deposits,* and **(c)** *the total interest earned.*

	Regular Deposit	Compounded	Annual Interest Rate	Accumulation Period
1.	$ 1000	yearly	8.5%	10 years
2.	$ 2000	yearly	4.5%	20 years
3.	$50	monthly	6.0%	5 years
4.	$75	monthly	7.2%	10 years
5.	$20	weekly	5.2%	3 years
6.	$30	weekly	7.8%	5 years

An annuity account into which regular deposits are made at the beginning of each compounding period, rather than at the end, is called an **annuity due.** *(It is usually assumed that a deposit is made at the end of compounding period number n so that the future value of an annuity due is the accumulation of n + 1 deposits rather than just n.)*

7. If the interest rate, regular deposit amount, compounding frequency, and accumulation period all are the same in both cases, would the future value of an annuity due be less or more than that of an ordinary annuity? Explain.

8. Modify the derivation of the formula for future value of an ordinary annuity (presented in the text) to show that the **future value of an annuity due** is given by

$$V = \frac{R\left[\left(1 + \frac{r}{m}\right)^{n+1} - 1\right]}{\frac{r}{m}}.$$

Finding the Earnings of an Annuity Due Use the formula of Exercise 8 to find the future value of each annuity due given below.

	Regular Deposit	Compounded	Annual Interest Rate	Accumulation Period
9.	$50	monthly	6.0%	5 years
10.	$75	monthly	7.2%	10 years
11.	$20	weekly	5.2%	3 years
12.	$30	weekly	7.8%	5 years

13. Do you think it is practical for a financial institution (or insurance company) to guarantee to pay you a set interest rate for years into the future on your annuity account? Why or why not?

Interview some of your friends or acquaintances, teachers, or other employees of schools, hospitals, or other nonprofit organizations and ask them about "tax deferred annuities" (TDAs) or "tax sheltered annuities" (TSAs). Explain the following features of these accounts.

14. their basic purpose

15. their tax advantages

16. their interest rates (as compared to those of regular savings accounts)

17. penalties for excessive deposits or for early withdrawals

18. the "annuitization period" (as opposed to the accumulation period)

14.2 Consumer Credit

When you borrow your neighbor's lawnmower, you (they hope) return the whole thing all at once when you are finished with it. However, when *money* is borrowed (in any sizable amount), timely repayment is much more likely if the debt is repaid over a period of time in a series of smaller payments.

Borrowing to finance purchases, and repaying with periodic payments is called **installment buying.** In this section we discuss the two general types of consumer installment credit. The first type, **closed-end** credit, involves borrowing a set amount up front and paying a series of equal payments until the loan is paid off. Furniture, appliances, and cars commonly are financed through closed-end credit (sometimes called **fixed installment loans**).

With the second type of consumer installment credit, **open-end** credit, there is no fixed number of payments—the consumer continues paying until no balance is owed. With open-end credit, additional credit often is extended before the initial amount is paid off. Examples of open-end credit include most department store charge accounts and bank charge cards such as MasterCard and VISA.

Closed-End Credit Loans set up under closed-end credit often are based on **add-on interest.** This means that if an amount P is borrowed, the annual interest rate is to be r, and payments will extend over t years, then the required interest comes from the simple interest formula: $I = Prt$. You simply "add on" this amount of interest to the principal borrowed to arrive at the total debt (or amount to be repaid).

$$\text{Amount to be repaid} = \text{Amount borrowed} + \text{Interest due}$$
$$= P + Prt$$

This total debt is then equally divided among the payments (usually monthly) to be made over the t years.

EXAMPLE 1 The Hoangs are a newlywed couple. They buy $2700 worth of furniture and appliances to furnish their first apartment. They pay $500 down and agree to pay the balance at a 7% add-on rate for 2 years. Find the total amount they owe and their monthly payment.

They finance a total of $2700 − $500 = $2200. The interest charge is

$$I = Prt$$
$$I = \$2200 \times .07 \times 2 = \$308.$$

The total amount owed is

$$\$2200 + \$308 = \$2508,$$

with a monthly payment of

$$\frac{\$2508}{24} = \$104.50.$$

Notice that the repayment amount in Example 1 was exactly the same ($P + Prt$) as the final amount A in a savings account paying a simple interest rate r for t years. (See the previous section.) But in the case of savings, the bank keeps all of your money for the entire time period. In Example 1, the Hoangs did not keep the full principal amount for the full time period. They repaid it in 24 monthly installments. The 7% add-on rate turns out to be equivalent to a much higher "true annual interest" rate, as we shall see in the next section.

Credit card debt among students is at a record high, according to

collegecreditcounseling.com.

About 65% of the nation's nine million college students have at least one credit card. The average undergraduate student has three credit cards. And the average college student will leave school with $2500 in credit card debt.

A 2002 California law, the Student Financial Protection and Responsibility Act, prohibits the distribution of free gifts to students who apply for a credit card, requires debt education in college and university orientation programs, and requires card vendors to register with the college or university administration. (*Source:* http://horizons.eraunews.com)

Another type of fixed installment loan, the **real estate mortgage,** usually involves much larger amounts and longer repayment periods than are involved in the consumer loans covered in this section. We will discuss mortgages in Section 14.4.

Open-End Credit

With a typical department store account, or bank card, a credit limit is established initially and the consumer can make many purchases during a month (up to the credit limit). The required monthly payment can vary from a set minimum (which may depend on the account balance) up to the full balance.

For purchases made in a store or other retail location, the customer may authorize adding the purchase price to his or her charge account by signing a "charge slip." Authorization also can be given by telephone for mail orders, services, contributions, and other expenditures. Such transactions can be accomplished directly over the Internet as well, though not all consumers feel comfortable doing so.

At the end of each billing period (normally once a month), the customer receives an **itemized billing,** a statement listing purchases and cash advances, the total balance owed, the minimum payment required, and perhaps other account information. Any charges beyond cash advanced and cash prices of items purchased are called **finance charges.** Finance charges may include interest, an annual fee, credit insurance coverage, a time payment differential, or carrying charges.

A few open-end lenders still use one of the simpler methods of calculating finance charges, such as the **unpaid balance method** which is illustrated in Example 2. Typically, you would begin with the unpaid account balance at the *end of the previous month* (or *billing period*) and apply to it the current monthly interest rate (or the daily rate multiplied by the number of days in the current billing period). Any purchases or returns in the current period do not affect the finance charge calculation.

EXAMPLE 2 The table below shows Karin De Jamaer's VISA account activity for a 2-month period. If the bank charges interest of 1.1% per month on the unpaid balance, and there are no other finance charges, find the missing quantities in the table, and the total finance charges for the 2 months.

Month	Unpaid Balance at Beginning of Month	Finance Charge	Purchases During Month	Returns	Payment	Unpaid Balance at End of Month
July	$179.82	_____	$101.58	$ 0	$30	_____
August	_____	_____	$ 19.84	$42.95	$85	_____

The July finance charge is 1.1% of $179.82, which is $1.98. Then, adding any finance charge and purchases and subtracting any returns and payments, we arrive at the unpaid balance at the end of July (and the beginning of August):

$$\$179.82 + \$1.98 + \$101.58 - \$30 = \$253.38.$$

A similar calculation gives an August finance charge of

$$1.1\% \text{ of } \$253.38 = .011 \times \$253.38 = \$2.79,$$

and an unpaid balance at the end of August of

$$\$253.38 + \$2.79 + \$19.84 - \$42.95 - \$85 = \$148.06.$$

The total finance charges for the 2 months were

$$\$1.98 + \$2.79 = \$4.77.$$

Most open-end credit plans now calculate finance charges by the **average daily balance method.** From the bank's point of view, this seems fairer since it considers balances on all days of the billing period and thus comes closer to charging consumers for the credit they actually utilize. The average daily balance method is illustrated in Example 3.

EXAMPLE 3 The activity in Paige Dunbar's MasterCard account for one billing period is shown below. If the previous balance (on March 3) was $209.46, and the bank charges 1.3% per month on the average daily balance, find **(a)** the average daily balance for the next billing (April 3), **(b)** the finance charge to appear on the April 3 billing, and **(c)** the account balance on April 3.

March 3	Billing date	
March 12	Payment	$50.00
March 17	Clothes	$28.46
March 20	Mail order	$31.22
April 1	Auto parts	$59.10

Learning the following rhyme will help you in problems like the one found in Example 3.

Thirty days hath September,
April, June, and November.
All the rest have thirty-one,
Save February which has
twenty-eight days clear,
And twenty-nine in each leap year.

(a) First make a table that shows the beginning date of the billing period and the dates of all transactions in the billing period. Along with each of these, compute the running balance on that date.

Date	Running Balance
March 3	$209.46
March 12	$209.46 − $50 = $159.46
March 17	$159.46 + $28.46 = $187.92
March 20	$187.92 + $31.22 = $219.14
April 1	$219.14 + $59.10 = $278.24

Next, tabulate the running balance figures, along with the number of days until the balance changed. Multiply each balance amount by the number of days. The sum of these products gives the "sum of the daily balances."

Date	Running Balance	Number of Days Until Balance Changed	$\left(\begin{array}{c}\text{Running}\\\text{Balance}\end{array}\right) \times \left(\begin{array}{c}\text{Number}\\\text{of Days}\end{array}\right)$
March 3	$209.46	9	$ 1885.14
March 12	$159.46	5	$ 797.30
March 17	$187.92	3	$ 563.76
March 20	$219.14	12	$ 2629.68
April 1	$278.24	2	$ 556.48
	Totals:	31	$ 6432.36

$$\text{Average daily balance} = \frac{\text{Sum of daily balances}}{\text{Days in billing period}} = \frac{\$6432.36}{31} = \$207.50.$$

Dunbar will pay a finance charge based on the average daily balance of $207.50.

(b) The finance charge for the April 3 billing will be

$$1.3\% \text{ of } \$207.50 = .013 \times \$207.50 = \$2.70.$$

(c) The account balance on the April 3 billing will be the latest running balance plus the finance charge: $\$278.24 + \$2.70 = \$280.94$.

Bank card and other open-end credit accounts normally require some minimum payment each month, which depends in some way on your balance. But if you need to delay payment, you can carry most of your balance from month to month, and pay the resulting finance charges. Whatever method is used for calculating finance charges, you can avoid finance charges altogether in many cases by paying your entire new balance by the due date each month. In fact, this is a wise practice, since these types of credit accounts, although convenient, are relatively expensive. Purchases charged to your account during the month will not be billed to you until your next billing date. And even after that, most lenders allow a grace period (perhaps 20 days or more) between the billing date and the payment due date. With careful timing, you may be able to purchase an item and delay having to pay for it for up to nearly 2 months without paying interest.

Although the open-end credit examples in this section have mainly addressed the actual interest charges, there are other features of these accounts that can be at least as important as interest. For example:

1. Is an annual fee charged? If so, how much is it?
2. Is a special "introductory" rate offered? If so, how long will it last?
3. Are there other incentives, such as rebates, credits toward certain purchases, or return of interest charges for long-time use?

E-commerce, or electronic commerce, refers to the use of the Internet as a distribution channel for goods and services. It can be as simple as traditional companies selling online. But the dynamics of the Internet itself also facilitate the formation of entirely new kinds of companies and company partnerships, often marketing previously undreamed of products and services.

During the 2002 Christmas shopping season, traditional retail sales were reported to be about the same as the year before. At the same time, online sales were estimated to be around 20% higher than the year before.

14.2 EXERCISES

Round all monetary answers to the nearest cent unless directed otherwise. Assume, whenever appropriate, that unless otherwise known, there are 12 months per year, 30 days per month, and 365 days per year.

Financing an Appliance Purchase Suppose you buy appliances costing $2150 *at a store charging* 12% *add-on interest, and you make a* $500 *down payment.*

1. Find the total amount you will be financing.
2. Find the total interest if you will pay off the loan over a 2-year period.
3. Find the total amount owed.
4. The amount from Exercise 3 is to be repaid in 24 monthly installments. Find the monthly payment.

Financing a New Car Purchase Suppose you want to buy a new car that costs $16,500. *You have no cash—only your old car, which is worth* $3000 *as a trade-in.*

5. How much do you need to finance to buy the new car?
6. The dealer says the interest rate is 9% add-on for 3 years. Find the total interest.

7. Find the total amount owed.
8. Find the monthly payment.

In the following exercises, use the add-on method of calculating interest to find the total interest and the monthly payment.

	Amount of Loan	Length of Loan	Interest Rate
9.	$4500	3 years	9%
10.	$2700	2 years	8%
11.	$ 750	18 months	7.4%
12.	$2450	30 months	9.2%
13.	$1580	10 months	6%
14.	$2100	14 months	12%
15.	$ 535	16 months	11.1%
16.	$ 798	29 months	10.3%

Work the following problems. Give monetary answers to the nearest cent unless directed otherwise.

17. *Finding the Monthly Payment for an Add-on Interest Furniture Loan* The Giordanos buy $8500 worth of furniture for their new home. They pay $3000 down. The store charges 10% add-on interest. The Giordanos will pay off the furniture in 30 monthly payments ($2\frac{1}{2}$ years). Find the monthly payment.

18. *Finding the Monthly Payment for an Add-on Interest Auto Loan* Find the monthly payment required to pay off an auto loan of $9780 over 3 years if the add-on interest rate is 9.3%.

19. *Finding the Monthly Payment for an Add-on Interest Home Electronics Loan* The total purchase price of a new home entertainment system is $14,240. If the down payment is $2900 and the balance is to be financed over 48 months at 10% add-on interest, what is the monthly payment?

20. *Finding the Monthly Payment for an Add-on Interest Loan* What are the monthly payments Donna De Simone pays on a loan of $1680 for a period of 10 months if 9% add-on interest is charged?

21. *Finding the Amount Borrowed in an Add-on Interest Car Loan* Joshua Eurich has misplaced the sales contract for his car and cannot remember the amount he originally financed. He does know that the add-on interest rate was 9.8% and the loan required a total of 48 monthly payments of $314.65 each. How much did Joshua borrow (to the nearest dollar)?

22. *Finding an Add-on Interest Rate* Susan Dratch is making monthly payments of $207.31 to pay off a $3\frac{1}{2}$ year loan for $6400. What is her add-on interest rate (to the nearest tenth of a percent)?

23. *Finding the Term of an Add-on Interest Loan* How long (in years) will it take Michael Garbin to pay off an $8000 loan with monthly payments of $172.44 if the add-on interest rate is 9.2%?

24. *Finding the Number of Payments of an Add-on Interest Loan* How many monthly payments must Jawann make on a $10,000 loan if he pays $417.92 a month and the add-on interest rate is 10.15%?

Finding the Finance Charge of an Open-end Charge Account Find the finance charge on each of the following open-end charge accounts. Assume interest is calculated on the unpaid balance of the account.

	Unpaid Balance	Monthly Interest Rate
25.	$325.50	1.0%
26.	$450.25	1.2%
27.	$242.88	1.12%
28.	$655.33	1.21%

Finding the Unpaid Balance of an Open-end Charge Account Complete each of the following tables, showing the unpaid balance at the end of each month. Assume a monthly interest rate of 1.1% on the unpaid balance.

29.

Month	Unpaid Balance at Beginning of Month	Finance Charge	Purchases During Month	Returns	Payment	Unpaid Balance at End of Month
February	$319.10	_____	$86.14	0	$50	_____
March	_____	_____	109.83	$15.75	60	_____
April	_____	_____	39.74	0	72	_____
May	_____	_____	56.29	18.09	50	_____

30.

Month	Unpaid Balance at Beginning of Month	Finance Charge	Purchases During Month	Returns	Payment	Unpaid Balance at End of Month
October	$828.63	_____	$128.72	$23.15	$125	_____
November	_____	_____	291.64	0	170	_____
December	_____	_____	147.11	17.15	150	_____
January	_____	_____	27.84	139.82	200	_____

31.

Month	Unpaid Balance at Beginning of Month	Finance Charge	Purchases During Month	Returns	Payment	Unpaid Balance at End of Month
August	$684.17	_____	$155.01	$38.11	$100	_____
September	_____	_____	208.75	0	75	_____
October	_____	_____	56.30	0	90	_____
November	_____	_____	190.00	83.57	150	_____

32.

Month	Unpaid Balance at Beginning of Month	Finance Charge	Purchases During Month	Returns	Payment	Unpaid Balance at End of Month
March	$1230.30	_____	$308.13	$74.88	$250	_____
April	_____	_____	488.35	0	350	_____
May	_____	_____	134.99	18.12	175	_____
June	_____	_____	157.72	0	190	_____

Finding Finance Charges Find the finance charge for each of the following charge accounts. Assume interest is calculated on the average daily balance of the account.

	Average Daily Balance	Monthly Interest Rate
33.	$ 249.94	1.0%
34.	$ 450.21	1.06%
35.	$1073.40	1.125%
36.	$1320.42	1.375%

Finding Finance Charges and Account Balances Using the Average Daily Balance Method For the following credit card accounts, assume one month between billing dates (with the appropriate number of days) and interest of 1.2% per month on the average daily balance. Find **(a)** the average daily balance, **(b)** the monthly finance charge, and **(c)** the account balance for the next billing.

37. Previous balance: $728.36
- May 9   Billing date
- May 17   Payment $200
- May 30   Dinner $ 46.11
- June 3   Theater tickets $ 64.50

38. Previous balance: $514.79
- January 27   Billing date
- February 9   Candy $11.08
- February 13   Returns $26.54
- February 20   Payment $59
- February 25   Repairs $71.19

39. Previous balance: $462.42

June 11	Billing date	
June 15	Returns	$106.45
June 20	Jewelry	$115.73
June 24	Car rental	$ 74.19
July 3	Payment	$115
July 6	Flowers	$ 68.49

40. Previous balance: $983.25

August 17	Billing date	
August 21	Mail order	$ 14.92
August 23	Returns	$ 25.41
August 27	Beverages	$ 31.82
August 31	Payment	$108
September 9	Returns	$ 71.14
September 11	Concert tickets	$110
September 14	Cash advance	$100

Finding Finance Charges Assume no purchases or returns are made in Exercises 41 and 42.

41. At the beginning of a 31-day billing period, Margaret Dent-Dzierzanowski has an unpaid balance of $720 on her credit card. Three days before the end of the billing period, she pays $600. Find her finance charge at 1.0% per month using the following methods.
 (a) unpaid balance method
 (b) average daily balance method

42. Christine Grexa's VISA bill dated April 14 shows an unpaid balance of $1070. Five days before May 14, the end of the billing period, Grexa makes a payment of $900. Find her finance charge at 1.1% per month using the following methods.
 (a) unpaid balance method
 (b) average daily balance method

Analyzing a "90 Days Same as Cash" Offer One version of the "90 Days Same as Cash" promotion was offered by a "major purchase card," which established an account charging 1.3167% interest per month on the account balance. Interest charges are added to the balance each month, becoming part of the balance on which interest is computed the next month. If you pay off the original purchase charge within 3 months, all interest charges are cancelled. Otherwise you are liable for all the interest. Suppose you purchase $2900 worth of carpeting under this plan.

43. Find the interest charge added to the account balance at the end of
 (a) the first month,   (b) the second month,
 (c) the third month.

44. Suppose you pay off the account 1 day late (3 months plus 1 day). What total interest amount must you pay? (Do not include interest for the one extra day.)

45. Treating the 3 months as 1/4 year, find the equivalent simple interest rate for this purchase (to the nearest tenth of a percent).

Various Charges of a Bank Card Account Carole's bank card account charges 1.1% per month on the average daily balance as well as the following special fees:

Cash advance fee:	2% (not less than $2 nor more than $10)
Late payment fee:	$15
Over-the-credit-limit fee:	$5

In the month of June, Carole's average daily balance was $1846. She was on vacation during the month and did not get her account payment in on time, which resulted in a late payment and also resulted in charges accumulating to a sum above her credit limit. She also used her card for six $100 cash advances while on vacation. Find the following based on account transactions in that month.

46. interest charges to the account

47. special fees charged to the account

✏️ *Write out your responses to each of the following.*

48. Is it possible to use a bank credit card for your purchases without paying anything for credit? If so, explain how.

49. Obtain applications or descriptive brochures for several different bank card programs, compare their features (including those in fine print), and explain which deal would be best for you, and why.

50. Research and explain the difference, if any, between a "credit" card and a "debit" card.
51. Many charge card offers include the option of purchasing credit insurance coverage, which would make your monthly payments if you became disabled and could not work and/or would pay off the account balance if you died. Find out the details on at least one such offer, and discuss why you would or would not accept it.
52. Make a list of "special incentives" offered by bank cards you are familiar with, and briefly describe the pros and cons of each one.
53. One bank offered a card with a "low introductory rate" of 5.9%, good through the end of the year. And furthermore, you could receive back a percentage (up to 100%!) of all interest you pay, as shown in the table.

Use your card for:	2 years	5 years	10 years	15 years	20 years
Get back:	10%	25%	50%	75%	100%

(As soon as you take a refund, the time clock starts over.) Since you can eventually claim all your interest payments back, is this card a good deal? Explain why or why not.

54. Either recall a car-buying experience you have had, or visit a new-car dealer and interview a salesperson. Write a description of the procedure involved in purchasing a car on credit.

Comparing Bank Card Accounts Daniel and Nora Onishi are considering two bank card offers that are the same in all respects except that Bank A charges no annual fee and charges monthly interest of 1.18% on the unpaid balance while Bank B charges a $30 annual fee and monthly interest of 1.01% on the unpaid balance. From their records, the Onishis have found that the unpaid balance they tend to carry from month to month is quite consistent and averages $900.

55. Estimate their total yearly cost to use the card if they choose the card from
    (a) Bank A (b) Bank B.
56. Which card is their better choice?

14.3 Truth in Lending

In the 1960s, Congress enacted a number of measures designed to safeguard the health, safety, and well-being of American consumers. One of these was the Consumer Credit Protection Act, which was passed in 1968, took effect in 1969, and has commonly been known as the **Truth in Lending Act.**

In this section we discuss two major issues addressed in the law:

1. How can I tell the true annual interest rate a lender is charging?
2. How much of the finance charge am I entitled to save if I decide to pay off a loan sooner than originally scheduled?

Annual Percentage Rate (APR) Question 1 above arose because lenders were computing and describing the interest they charged in several different ways. For example, how does 1.5% per month at Sears compare to 9% per year add-on interest at a furniture store? Truth in Lending standardized the so-called true annual interest rate, or **annual percentage rate,** commonly denoted **APR**. All sellers (car dealers, stores, banks, insurance agents, credit card companies, and the like) must disclose the APR when you ask, and the contract must state the APR whether or not you ask. This enables a borrower to more easily compare the true costs of different loans.

Free consumer information is available in abundance from the Federal Trade Commission at http://www.ftc.gov. For example, you can find out about "Abusive Lending," "Credit," "Identity Theft," "Investments," "Privacy," and many other consumer topics. Consumers can register fraud complaints at this site using an online complaint form, or by calling the toll-free Consumer Help Line at 1-877 FTC HELP.

Theoretically, as a borrower you should not need to calculate APR yourself, but it is possible to verify the value stated by the lender if you wish. Since the formulas for finding APR are quite involved, it is easiest to use a table provided by the Federal Reserve Bank. We show an abbreviated version in Table 3. It will identify APR values to the nearest half percent from 8.0% to 14.0%, which should do in most cases. (You should be able to obtain a more complete table from your local bank.) Our table is designed to apply to loans requiring *monthly* payments and extending over the most common lengths for consumer loans from 6 to 60 months. Apart from home mortgages, which are discussed in the next section, these conditions characterize most closed-end consumer loans.

Table 3 relates the following three quantities.

APR = true annual interest rate (shown across the top)

n = total number of scheduled monthly payments (shown down the left side)

h = finance charge per $100 of amount financed (shown in the body of the table)

Later in this section, we will give the formula connected with this table. As you will see, though, it expresses h in terms of n and APR and is *not* easily solved to express APR in terms of n and h. Hence we have provided the table for determining APR values.

The use of Table 3 is illustrated in Example 1, where we find the true annual interest rate (APR) for Example 1 of the previous section.

TABLE 3 Annual Percentage Rate (APR) for Monthly Payment Loans

Number of Monthly Payments (n)	Annual Percentage Rate (APR)												
	8.0%	8.5%	9.0%	9.5%	10.0%	10.5%	11.0%	11.5%	12.0%	12.5%	13.0%	13.5%	14.0%
	(Finance charge per $100 of amount financed) (h)												
6	$2.35	$2.49	$2.64	$2.79	$2.94	$3.08	$3.23	$3.38	$3.53	$3.68	$3.83	$3.97	$4.12
12	4.39	4.66	4.94	5.22	5.50	5.78	6.06	6.34	6.62	6.90	7.18	7.46	7.74
18	6.45	6.86	7.28	7.69	8.10	8.52	8.93	9.35	9.77	10.19	10.61	11.03	11.45
24	8.55	9.09	9.64	10.19	10.75	11.30	11.86	12.42	12.98	13.54	14.10	14.66	15.23
30	10.66	11.35	12.04	12.74	13.43	14.13	14.83	15.54	16.24	16.95	17.66	18.38	19.10
36	12.81	13.64	14.48	15.32	16.16	17.01	17.86	18.71	19.57	20.43	21.30	22.17	23.04
48	17.18	18.31	19.45	20.59	21.74	22.90	24.06	25.23	26.40	27.58	28.77	29.97	31.17
60	21.66	23.10	24.55	26.01	27.48	28.96	30.45	31.96	33.47	34.99	36.52	38.06	39.61

EXAMPLE 1 Recall that the Hoangs (in Example 1 of Section 14.2) paid $500 down on a $2700 purchase and agreed to pay the balance at a 7% add-on rate for 2 years. Find the APR for their loan.

As shown previously, the total amount financed was

Purchase price − down payment = $2700 − $500 = $2200,

and the finance charge (interest) was

$I = Prt = \$2200 \times .07 \times 2 = \$308.$

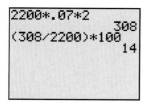

These are the computations required in the solution of Example 1.

Next find the finance charge per $100 of the amount financed. To do this, divide the finance charge by the amount financed, then multiply by $100.

$$\begin{pmatrix} \text{Finance charge per} \\ \$100 \text{ financed} \end{pmatrix} = \frac{\text{Finance charge}}{\text{Amount financed}} \times \$100$$

$$= \frac{\$308}{\$2200} \times \$100 = \$14$$

This amount, $14, represents h, the finance charge per $100 of the amount financed. Since the loan was to be paid over 24 months, look down to the "24 monthly payments" row of the table ($n = 24$). Then look across the table for the h-value closest to $14.00, which is $14.10. From that point, read up the column to find the APR, 13.0% (to the nearest half percent). In this case, a 7% add-on rate is equivalent to an APR of 13.0%.

These are the computations required in the solution of Example 2.

EXAMPLE 2 After a down payment on your new car, you still owe $7454. You agree to repay the balance in 48 monthly payments of $185 each. What is the APR on your loan?

First find the finance charge.

$$\text{Finance charge} = \text{Total payments} - \text{Amount financed}$$
$$= 48 \times \$185 - \$7454$$
$$= \$1426$$

Now find the finance charge per $100 financed as in Example 1.

$$\begin{pmatrix} \text{Finance charge per} \\ \$100 \text{ financed} \end{pmatrix} = \frac{\text{Finance charge}}{\text{Amount financed}} \times \$100$$

$$= \frac{\$1426}{\$7454} \times \$100 = \$19.13$$

Find the "48 payments" row of the table, read across to find the number closest to 19.13, which is 19.45. From there read up to find the APR, which is 9.0%.

Unearned Interest Question 2 at the beginning of the section arises when a borrower decides, for one reason or another, to pay off a closed-end loan earlier than originally scheduled. In such a case, it turns out that the lender has not loaned as much money for as long as planned and so he has not really "earned" the full finance charge originally disclosed.

If a loan is paid off early, the amount by which the original finance charge is reduced is called the **unearned interest.** We will discuss two common methods of calculating unearned interest, the **actuarial method** and the **rule of 78.** The Truth in Lending Act requires that the method for calculating this refund (or reduction) of finance charge be disclosed at the time the loan is initiated. Whichever method is used, the borrower may not, in fact, save all the unearned interest, since the lender is entitled to impose an **early payment penalty** to recover certain costs. A lender's intention to impose such a penalty in case of early payment must also be disclosed at initiation of the loan.

Consumer Information Center
Pueblo, CO 81009
www.pueblo.gsa.gov
719-948-4000

Rights and responsibilities apply to all credit accounts, and the consumer should read all disclosures provided by the lender. *The Fair Credit Billing Act* and *The Fair Credit Reporting Act* regulate, among other things, procedures for billing and for disputing bills, and for providing and disputing personal information on consumers.

If you have ever applied for a charge account, a personal loan, insurance, or a job, then information about where you work and live, how you pay your bills, and whether you've been sued, arrested, or have filed for bankruptcy appears in the files of Consumer Reporting Agencies (CRAs), which sell that information to creditors, employers, insurers, and other businesses.

For more detailed information on consumer issues, you may want to request the free *Consumer Information Catalog* from the Consumer Information Center. (The contact information is shown above.)

We will now state and illustrate both methods of computing unearned interest.

Unearned Interest—Actuarial Method

For a closed-end loan requiring *monthly* payments, which is paid off earlier than originally scheduled, let

R = regular monthly payment

k = remaining number of scheduled payments (*after* current payment)

h = finance charge per \$100, corresponding to a loan with the same APR and k monthly payments.

Then the **unearned interest**, u, is given by

$$u = kR\left(\frac{h}{\$100 + h}\right).$$

Once the unearned interest u is calculated (by any method), the amount required to actually pay off the loan early is easily found. It consists of the present regular payment due, plus k additional future payments, minus the unearned interest.

Payoff Amount

A closed-end loan requiring regular monthly payments R can be paid off early, along with the current payment. If the original loan had k additional payments scheduled (after the current payment), and the unearned interest is u, then, disregarding any possible prepayment penalty, the **payoff amount** is given by

$$\text{Payoff amount} = (k + 1)R - u.$$

EXAMPLE 3 Suppose you get an unexpected pay raise and want to pay off your car loan of Example 2 at the end of 3 years rather than paying for 4 years as originally agreed. Find **(a)** the unearned interest (the amount you will save by retiring the loan early) and **(b)** the "payoff amount" (the amount required to pay off the loan at the end of 3 years).

(a) From Example 2, recall that $R = \$185$ and APR $= 9.0\%$. The current payment is payment number 36, so $k = 48 - 36 = 12$. Use Table 3, with 12 payments and APR 9.0%, to obtain $h = \$4.94$. Finally, by the actuarial method formula,

$$u = 12 \times \$185\left(\frac{\$4.94}{\$100 + \$4.94}\right) = \$104.51.$$

By the actuarial method, you will save \$104.51 in interest by retiring the loan early.

(b) The payoff amount is found by using the appropriate formula.

$$\text{Payoff amount} = (12 + 1)\$185 - \$104.51 = \$2300.49$$

The required payoff amount at the end of 36 months is $2300.49.

The actuarial method, since it is based on the APR value, probably is the better method. However, some lenders still use the second method, the **rule of 78,** which is based on an approximating assumption. The rule of 78 works as follows.

Unearned Interest—Rule of 78

For a closed-end loan requiring *monthly* payments, which is paid off earlier than originally scheduled, let

F = original finance charge
n = number of payments originally scheduled
k = remaining number of scheduled payments (*after* current payment).

Then the **unearned interest, u,** is given by

$$u = \frac{k(k + 1)}{n(n + 1)} \times F.$$

EXAMPLE 4 Again assume that the loan in Example 2 is paid off at the time of the thirty-sixth monthly payment. This time use the rule of 78 to find **(a)** the unearned interest, and **(b)** the payoff amount.

(a) From Example 2, the original finance charge is $F = \$1426$. Also, $n = 48$ and $k = 12$. By the rule of 78,

$$u = \frac{12(12 + 1)}{48(48 + 1)} \times \$1426 = \$94.58.$$

By the rule of 78, you will save $94.58 in interest, which is $9.93 *less* than your savings by the actuarial method.

(b) In this case, the payoff amount is given by

$$\text{Payoff amount} = (12 + 1)\$185 - \$94.58 = \$2310.42.$$

The payoff amount is $2310.42, which is $9.93 *more* than the payoff amount calculated by the actuarial method in Example 3.

When the rule of 78 was first introduced into financial law (by the Indiana legislature in 1935), loans were ordinarily written for 1 year or less, interest rates were relatively low, and loan amounts were less than they tend to be today. For these reasons the rule of 78 was acceptably accurate then. Today, however, with very accurate tables and/or calculators readily available, the rule of 78 is used much less often than previously.*

---

*For an excellent discussion of the shortcomings of the rule of 78, see Alonzo F. Johnson's article, "The Rule of 78: A Rule that Outlived Its Useful Life," in *Mathematics Teacher,* September 1988.

Suppose you want to compute unearned interest accurately (so you don't trust the rule of 78), but the APR value, or the number of scheduled payments, or the number of remaining payments (or at least one of the three) is not included in Table 3. Then what? Actually, in the actuarial method, you can evaluate h (the finance charge per $100 financed) using the same formula that was used to generate Table 3.

Finance Charge per $100 Financed

If a closed-end loan requires n equal monthly payments and APR denotes the true annual interest rate for the loan (as a decimal), then h, the **finance charge per $100 financed,** is given by

$$h = \frac{n \times \frac{\text{APR}}{12} \times \$100}{1 - \left(1 + \frac{\text{APR}}{12}\right)^{-n}} - \$100.$$

```
(9*(.098/12)*100
)/(1-(1+.098/12)
^-9)-100
                4.13
9*160.39*4.13/(1
00+4.13)
               57.25
```

These are the computations required in the solution of Example 5(a).

EXAMPLE 5 Sergei borrowed $4000 to pay for music equipment for his band. His loan contract states an APR of 9.8% and stipulates 28 monthly payments of $160.39 each. Sergei decides to pay the loan in full at the time of his nineteenth payment. Find **(a)** the unearned interest, and **(b)** the payoff amount.

(a) First find h from the finance charge formula just given. (Remember to use the remaining number of payments, $28 - 19$, as the value of n.)

$$h = \frac{9 \times \frac{.098}{12} \times \$100}{1 - \left(1 + \frac{.098}{12}\right)^{-9}} - \$100 = \$4.13$$

Next use the actuarial formula for unearned interest.

Regular monthly payment: $R = \$160.39$

Remaining number of payments: $k = 28 - 19 = 9$

Finance charge per $100: $h = \$4.13$

$$u = 9 \times \$160.39 \times \frac{\$4.13}{\$100 + \$4.13} = \$57.25$$

The amount of interest Sergei will save is $57.25.

(b) Payoff amount $= (9 + 1)(\$160.39) - \$57.25 = \$1546.65$.

To pay off the loan, Sergei must pay $1546.65.

FOR FURTHER THOUGHT

To see how the rule of 78 (for unearned interest) got its name, consider a 12-month loan. First compute the sum of the natural numbers 1 through 12:

$$1 + 2 + 3 + 4 + 5 + 6 + 7 + 8 + 9 + 10 + 11 + 12 = 78.$$

Imagine the total finance charge being divided into 78 equal parts and the loan principal into 12 equal parts. *Assume* that, for the various months of the loan, the fraction of the loan principal used by the borrower and hence the fraction of the finance charge owed, would be as shown at the right.

Now suppose the loan is paid off early, say with four payments remaining. The portion (fraction) of the finance charge that should be saved can be seen by starting from the bottom of the right column of the table:

$$\frac{1}{78} + \frac{2}{78} + \frac{3}{78} + \frac{4}{78} = \frac{1+2+3+4}{78} = \frac{10}{78}.$$

Month	Fraction of Loan Principal Used by Borrower	Fraction of Finance Charge Owed
1	12/12	12/78
2	11/12	11/78
3	10/12	10/78
4	9/12	9/78
5	8/12	8/78
6	7/12	7/78
7	6/12	6/78
8	5/12	5/78
9	4/12	4/78
10	3/12	3/78
11	2/12	2/78
12	1/12	1/78
		78/78 = 1

For Group Discussion

Divide into three (or more) groups. The groups are to repeat the development above for the following loan situations. (Additional groups choose other numbers.)

(a) 10 payments total; paid in full with 3 payments remaining
(b) 18 payments total; paid in full with 5 payments remaining
(c) 24 payments total; paid in full with 8 payments remaining

1. As above, obtain for your situation the portion of the finance charge that should be saved.
2. Determine how the numerator and denominator of your fraction relate to the remaining number of payments k and the total number of payments n, respectively.
3. Have you proved the formula for the rule of 78? (*Hint:* Write the fraction of the finance charge to be saved as N/D.) Provide justification for each statement below.

$$N = 1 + 2 + \cdots + (k-1) + k$$

Also,

$$N = k + (k-1) + \cdots + 2 + 1$$
$$2N = (k+1) + (k+1) + \cdots + (k+1) + (k+1)$$
$$2N = k(k+1)$$
$$N = \frac{k(k+1)}{2}.$$

Likewise, $D = \dfrac{n(n+1)}{2}.$

Hence $\dfrac{N}{D} = \dfrac{\frac{k(k+1)}{2}}{\frac{n(n+1)}{2}} = \dfrac{k(k+1)}{n(n+1)}.$

Bring your groups back together and combine your efforts, if necessary, to justify each statement above.

4. Explain why the name "rule of 78" is not generally an appropriate name for this method of computing unearned interest.

14.3 EXERCISES

Round all monetary answers to the nearest cent unless otherwise directed.

Finding True Annual Interest Rate *Find the APR (true annual interest rate), to the nearest half percent, for each of the following.*

	Amount Financed	Finance Charge	Number of Monthly Payments
1.	$1000	$75	12
2.	$1700	$202	24
3.	$6600	$750	30
4.	$5900	$1150	48

Finding the Monthly Payment *Find the monthly payment for each of the following.*

	Purchase Price	Down Payment	Finance Charge	Number of Monthly Payments
5.	$3000	$500	$250	24
6.	$4280	$450	$700	36
7.	$3950	$300	$800	48
8.	$8400	$2500	$1300	60

Finding True Annual Interest Rate *Find the APR (true annual interest rate), to the nearest half percent, for each of the following.*

	Purchase Price	Down Payment	Add-on Interest Rate	Number of Payments
9.	$4190	$390	6%	12
10.	$3250	$750	7%	36
11.	$7480	$2200	5%	18
12.	$12,800	$4500	6%	48

Unearned Interest by the Actuarial Method *Each of the following loans was paid in full before its due date. (a) Obtain the value of h from Table 3. Then (b) use the actuarial method to find the amount of unearned interest, and (c) find the payoff amount.*

	Regular Monthly Payment	APR	Remaining Number of Scheduled Payments After Payoff
13.	$346.70	11.0%	18
14.	$783.50	8.5%	12
15.	$595.80	9.5%	6
16.	$314.50	10.0%	24

Finding Finance Charge and True Annual Interest Rate *For each of the following loans, find (a) the finance charge, and (b) the APR.*

17. Frank Barry financed a $1990 computer with 24 monthly payments of $91.50 each.

18. Pat Peterson bought a horse trailer for $5090. She paid $1240 down and paid the remainder at $152.70 per month for $2\frac{1}{2}$ years.

19. Mike Karelius still owed $2000 on his new garden tractor after the down payment. He agreed to pay monthly payments for 18 months at 6% add-on interest.

20. Phil Givant paid off a $15,000 car loan over 3 years with monthly payments of $487.54 each.

Comparing the Actuarial Method and the Rule of 78 for Unearned Interest *Each of the following loans was paid off early. Find the unearned interest by (a) the actuarial method, and (b) the rule of 78.*

	Amount Financed	Regular Monthly Payment	Total Number of Payments Scheduled	Remaining Number of Scheduled Payments After Payoff
21.	$3310	$201.85	18	6
22.	$10,230	$277.00	48	12
23.	$29,850	$641.58	60	12
24.	$16,730	$539.82	36	18

Unearned Interest by the Actuarial Method Each of the following loans was paid in full before its due date. **(a)** Obtain the value of h from the appropriate formula. Then **(b)** use the actuarial method to find the amount of unearned interest, and **(c)** find the payoff amount.

	Regular Monthly Payment	APR	Remaining Number of Scheduled Payments After Payoff
25.	$212	8.6%	4
26.	$575	9.33%	8

Comparing Loan Choices Wei-Jen Luan needs to borrow $5000 to pay for Kings season tickets for her family. She can borrow the amount from a finance company (at 6.5% add-on interest for 3 years) or from the credit union (36 monthly payments of $164.50 each). Use this information for Exercises 27–30.

27. Find the APR (to the nearest half percent) for each loan and decide which one is Wei-Jen's better choice.

28. Suppose Wei-Jen takes the credit union loan. At the time of her thirtieth payment she decides that this loan is a nuisance and decides to pay it off. If the credit union uses the rule of 78 for computing unearned interest, how much will she save by paying in full now?

29. What would Wei-Jen save in interest if she paid in full at the time of the thirtieth payment and the credit union used the actuarial method for computing unearned interest?

30. Under the conditions of Exercise 29, what amount must Wei-Jen come up with to pay off her loan?

31. Describe why, in Example 1, the APR and the add-on rate differ. Which one is more legitimate? Why?

Approximating the APR of an Add-on Rate To convert an add-on interest rate to its corresponding APR, some people recommend using the formula

$$\text{APR} = \frac{2n}{n+1} \times r,$$

where r is the add-on rate and n is the total number of payments.

32. Apply the given formula to calculate the APR (to the nearest half percent) for the loan of Example 1 ($r = .07, n = 24$).

33. Compare your APR value in Exercise 32 to the value of Example 1. What do you conclude?

The Rule of 78 with Prepayment Penalty A certain retailer's credit contract designates the rule of 78 for computing unearned interest, and also imposes a "prepayment penalty." In case of any payoff earlier than the due date, they will charge an additional 10% of the original finance charge. Find the least value of k (remaining payments after payoff) that would result in any net savings in each of the following cases.

34. 24 payments originally scheduled

35. 36 payments originally scheduled

36. 48 payments originally scheduled

The actuarial method of computing unearned interest assumes that, throughout the life of the loan, the borrower is paying interest at the rate given by APR for money actually being used by the borrower. When contemplating complete payoff along with the current payment, think of k future payments as applying to a separate loan with the same APR, and h being the finance charge per $100 of that loan. Refer to the following formula.

$$u = kR\left(\frac{h}{\$100 + h}\right)$$

37. Describe in words the quantity represented by

$$\frac{h}{\$100 + h}.$$

38. Describe in words the quantity represented by kR.

39. Explain why the product of the two quantities above represents unearned interest.

Write out your responses to the following exercises.

40. Why might a lender be justified in imposing a prepayment penalty?

41. Discuss reasons that a borrower may want to pay off a loan early.

42. Find out what federal agency you can contact if you have questions about compliance with the Truth in Lending Act. (Any bank, or retailer's credit department, should be able to help you with this, or you could try a web search.)

14.4 Purchasing a House

For many decades, home ownership has been considered a centerpiece of the "American dream." And since homes and condominiums have generally kept pace over the years with the rate of inflation, they have also been sound investments. For most people, a home represents the largest purchase of their lifetime, and it is certainly worth careful consideration.

A loan for a substantial amount, extending over a lengthy time interval (typically up to 30 years), for the purpose of buying a home or other property or real estate, and for which the property is pledged as security for the loan, is called a **mortgage**. (In some areas, a mortgage may also be called a **deed of trust** or a **security deed**.) The time until final payoff is called the **term** of the mortgage. The portion of the purchase price of the home which the buyer pays initially is called the **down payment**. The **principal amount of the mortgage** (the amount borrowed) is found by subtracting the down payment from the purchase price.

We will now discuss the common types of mortgages. We group them into two categories, **fixed-rate mortgages** and **adjustable-rate mortgages.**

Fixed-rate Mortgages

With a **fixed-rate mortgage**, the interest rate will remain constant throughout the term, and the initial principal balance, together with interest due on the loan, is repaid to the lender through regular (constant) periodic (we assume monthly) payments. This is called **amortizing** the loan. (Other items, in addition to principal and interest, which may affect the monthly payment amount, will be discussed later.)

The regular monthly payment needed to amortize a loan depends on the amount financed, the term of the loan, and the interest rate. It can be found by using the following formula.

Heating a house is another cost that may get you involved with banks and interest rates after you finally get a roof over your head. The roof you see above does more than keep off the rain. It holds glass tube solar collectors, part of the solar heating system in the building.

Regular Monthly Payment

The **regular monthly payment** required to repay a loan of P dollars, together with interest at an annual rate r, over a term of t years, is given by

$$R = \frac{P\left(\dfrac{r}{12}\right)\left(1 + \dfrac{r}{12}\right)^{12t}}{\left(1 + \dfrac{r}{12}\right)^{12t} - 1}.$$

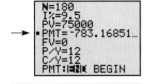

The arrow indicates a payment of $783.17, supporting the result of Example 1.

EXAMPLE 1 Find the monthly payment necessary to amortize a $75,000 mortgage at 9.5% annual interest for 15 years.

Using the formula above, we find

$$R = \frac{\$75,000\left(\dfrac{.095}{12}\right)\left(1 + \dfrac{.095}{12}\right)^{(12)(15)}}{\left(1 + \dfrac{.095}{12}\right)^{(12)(15)} - 1} = \$783.17.$$

With a programmable or financial calculator, you can store the formula above and minimize the work. Another approach is to use a tool such as Table 4, which gives payment values (per $1000 principal) for typical ranges of mortgage terms and interest rates. The entries in the table are given to five decimal places so that accuracy to the nearest cent can be obtained for normal mortgage amounts. Example 2 illustrates the use of Table 4.

EXAMPLE 2 Find the monthly payment necessary to amortize a $98,000 mortgage at 9% for 25 years.

In Table 4, read down to the 9% row and across to the column for 25 years, to find the entry 8.39196. Since this is the monthly payment amount needed to amortize a loan of $1000, and our loan is for $98,000, our required monthly payment is

$$98 \times \$8.39196 = \$822.41.$$

```
N=300.00
I%=9.00
PV=98000.00
■PMT=-822.41
FV=0.00
P/Y=12.00
C/Y=12.00
PMT:END BEGIN
```

Under the conditions of Example 2, the monthly payment is $822.41. Compare with the table method.

TABLE 4 Monthly Payments to Repay Principal and Interest on a $1000 Mortgage

Annual Rate (r)	Term of Mortgage (Years) (t)					
	5	10	15	20	25	30
5.0%	$18.87123	$10.60655	$7.90794	$6.59956	$5.84590	$5.36822
5.5%	19.10116	10.85263	8.17083	6.87887	6.14087	5.67789
6.0%	19.33280	11.10205	8.43857	7.16431	6.44301	5.99551
6.5%	19.56615	11.35480	8.71107	7.45573	6.75207	6.32068
7.0%	19.80120	11.61085	8.98828	7.75299	7.06779	6.65302
7.5%	20.03795	11.87018	9.27012	8.05593	7.38991	6.99215
8.0%	20.27639	12.13276	9.55652	8.36440	7.71816	7.33765
8.5%	20.51653	12.39857	9.84740	8.67823	8.05227	7.68913
9.0%	20.75836	12.66758	10.14267	8.99726	8.39196	8.04623
9.5%	21.00186	12.93976	10.44225	9.32131	8.73697	8.40854
10.0%	21.24704	13.21507	10.74605	9.65022	9.08701	8.77572
10.5%	21.49390	13.49350	11.05399	9.98380	9.44182	9.14739
11.0%	21.74242	13.77500	11.36597	10.32188	9.80113	9.52323
11.5%	21.99261	14.05954	11.68190	10.66430	10.16469	9.90291
12.0%	22.24445	14.34709	12.00168	11.01086	10.53224	10.28613
12.5%	22.49794	14.63762	12.32522	11.36141	10.90354	10.67258
13.0%	22.75307	14.93107	12.65242	11.71576	11.27835	11.06200
13.5%	23.00985	15.22743	12.98319	12.07375	11.65645	11.45412
14.0%	23.26825	15.52664	13.31741	12.43521	12.03761	11.84872
14.5%	23.52828	15.82868	13.65501	12.79998	12.42163	12.24556
15.0%	23.78993	16.13350	13.99587	13.16790	12.80831	12.64444

The 1986 Tax Reform Act provides another advantage to homeowners and also encourages buyers to choose minimum down payments and maximum mortgages. Interest on the amount originally borrowed for the home purchase is still tax-deductible whereas most other interest expense is not. A homeowner who has decreased his home debt over the years (through principal payments) can take out a *home equity loan* for other purposes and still deduct the interest payments (on the amount up to the original home mortgage). The rates on home equity loans (second mortgages) typically run one or two percentage points higher than on first mortgages.

So that the borrower pays interest only on the money actually owed in a month, interest on real-estate loans is computed on the decreasing balance of the loan. Each equal monthly payment is first applied toward interest for the previous month. The remainder of the payment is then applied toward reduction of the principal amount owed. Payments in the early years of a real-estate loan are mostly interest; only a small amount goes toward reducing the principal. The amount of interest decreases each month, so that larger and larger amounts of the payment will apply to the principal. During the last years of the loan, most of the monthly payment is applied toward the principal. (See Table 5 on page 825.)

An **amortization schedule** (or **repayment schedule**) shows the allotment of payments for interest and principal, and the principal balance, for each month over the entire life of the loan. Tables showing these breakdowns are available from lenders, or can be produced on a computer spreadsheet. The following steps demonstrate how the computations work. (Notice that Step 1 applies the simple interest formula $I = Prt$.)

Step 1: Interest for the month $= \begin{pmatrix} \text{Old balance} \\ \text{of principal} \end{pmatrix} \begin{pmatrix} \text{Annual} \\ \text{interest rate} \end{pmatrix} \begin{pmatrix} \dfrac{1}{12} \text{ year} \end{pmatrix}$

Step 2: Payment on principal $= \begin{pmatrix} \text{Monthly} \\ \text{payment} \end{pmatrix} - \begin{pmatrix} \text{Interest for} \\ \text{the month} \end{pmatrix}$

Step 3: New balance of principal $= \begin{pmatrix} \text{Old balance} \\ \text{of principal} \end{pmatrix} - \begin{pmatrix} \text{Payment on} \\ \text{principal} \end{pmatrix}$

This sequence of steps is done for the end of each month. The new balance obtained in Step 3 becomes the Step 1 old balance for the next month.

EXAMPLE 3 The Petersons paid for many years on a $60,000 mortgage with a term of 30 years and a 4.5% interest rate. Prepare an amortization schedule for the first two months of their mortgage.

Since Table 4 does not include interest rates as low as 4.5%, we use the regular monthly payment formula.

$$R = \dfrac{\$60{,}000 \left(\dfrac{.045}{12}\right)\left(1 + \dfrac{.045}{12}\right)^{(12)(30)}}{\left(1 + \dfrac{.045}{12}\right)^{(12)(30)} - 1} = \$304.01$$

Now apply Steps 1 through 3.

Step 1: Interest for the month $= \$60{,}000(.045)(1/12) = \$225$
Step 2: Payment on principal $= \$304.01 - \$225 = \$79.01$
Step 3: New balance of principal $= \$60{,}000 - \$79.01 = \$59{,}920.99$

Now starting with an old balance of $59,920.99, repeat the same steps for the second month.

Interest for the month $= \$59{,}920.99(.045)(1/12) = \$224.70$
Payment on principal $= \$304.01 - \$224.70 = \$79.31$
New balance of principal $= \$59{,}920.99 - \$79.31 = \$59{,}841.68$

These calculations are summarized here.

Homes and condominiums are sometimes available for a very low down payment from the inventory of HUD (the Department of Housing and Urban Development). For current information on these opportunities, consult your local real estate companies.

Amortization Schedule

Payment Number	Interest Payment	Principal Payment	Balance of Principal
			$60,000.00
1	$225.00	$79.01	$59,920.99
2	$224.70	$79.31	$59,841.68

Prevailing mortgage interest rates have varied considerably over the years. Table 5 shows portions of the amortization schedule for the Petersons' loan of Example 3, and also shows what the corresponding values would have been had their interest rate been 14.5% rather than 4.5%. Notice how much interest is involved in this home mortgage. At the (low) 4.5% rate, $49,444.03 in interest was paid along with the $60,000 principal. At a rate of 14.5%, the interest alone would total the huge sum of $204,504.88.

TABLE 5 Amortization Schedules for a $60,000, 30-year Mortgage

4.5% Interest Monthly Payment: $304.01				14.5% Interest Monthly Payment: $734.73			
Payment Number	Interest Payment	Principal Payment	Balance of Principal	Payment Number	Interest Payment	Principal Payment	Balance of Principal
Initially →			60,000.00	Initially →			60,000.00
1	225.00	79.01	59,920.99	1	725.00	9.73	59,990.27
2	224.70	79.31	59,841.68	2	724.88	9.85	59,980.42
3	224.41	79.60	59,762.08	3	724.76	9.97	59,970.45
12	221.68	82.33	59,032.06	12	723.63	11.11	59,875.11
60	205.47	98.54	54,694.75	60	714.96	19.77	59,149.53
175	152.47	151.54	40,506.65	175	656.05	78.69	54,214.82
176	151.90	152.11	40,354.54	176	655.10	79.64	54,135.18
236	113.60	190.41	30,102.65	236	571.02	163.72	47,092.76
237	112.88	191.13	29,911.52	237	569.04	165.70	46,927.07
240	110.73	193.28	29,333.83	240	562.96	171.78	46,417.87
303	59.33	244.68	15,575.66	303	368.65	366.09	30,142.47
304	58.41	245.60	15,330.05	304	364.22	370.51	29,771.96
359	2.27	301.74	302.88	359	17.44	717.29	725.96
360	1.14	302.88	0.00	360	8.77	725.96	0.00
Totals:	49,444.03	60,000.00		Totals:	204,504.88	60,000.00	

In addition to principal and interest on their mortgage, homeowners must also pay **property taxes** (usually collected by the city or county) and **homeowners insurance** (covering potential damage from fire, unusual weather, and so forth). Often these items are paid out of a **reserve account** (also called an **escrow** or an **impound account**) kept by the lender. The borrower must pay enough each month, along with amortization costs, so that the reserve account will be sufficient to pay taxes and insurance premiums when they come due. (In some cases, the homeowner pays taxes directly to the taxing authority and insurance premiums directly to an insurance company, totally separate from mortgage payments to the lender.)

EXAMPLE 4 Mr. and Mrs. Carlton have acquired a $105,000 mortgage at 8.5% for 20 years. The property taxes ($1040 per year) and homeowners insurance ($480 per year) on their new home will be paid from a lender's impound account. Find the monthly payment they must make to the lender.

The monthly amortization costs of the loan (principal and interest) can be found using the Table 4 entry 8.67823.

$$\text{Principal and interest payment} = 105 \times \$8.67823 = \$911.21$$

Monthly payments for taxes and insurance will be one-twelfth of the annual amounts.

$$\text{Tax and insurance payment} = \frac{\$1040 + \$480}{12} = \$126.67$$

Total monthly payment required:

$$\text{Principal and Interest} + \text{Taxes and Insurance} = \$911.21 + \$126.67$$
$$= \$1037.88.$$

Adjustable-rate Mortgages

The lending industry uses many variations on the basic fixed-rate mortgage to cut down on the risk to lenders and also to make home loans accessible to more people. An **adjustable-rate mortgage (ARM),** also known as a **variable-rate mortgage (VRM),** generally starts out with a lower rate than similar fixed-rate loans, but the rate changes periodically, reflecting changes in prevailing rates. To understand how ARMs work, you must know about *indexes, margins, discounts, caps, negative amortization,* and *convertibility.*

Your ARM interest rate may change every 1, 3, or 5 years (occasionally more frequently). The frequency of change in rate is called the **adjustment period.** For example, an ARM with an adjustment period of 1 year is called a 1-year ARM. When the rate changes, your payment normally changes also. These adjustments are caused by fluctuations in an **index** upon which your rate is based. A variety of indexes are used, including the 1-, 3-, and 5-year U.S. Treasury security rates. Also used are national and regional "cost of funds" indexes.

To determine your interest rate, the lender will add to the applicable index a few percentage points called the **margin.** The index and the margin are both important in determining the cost of the loan.

EXAMPLE 5 Suppose you pay $20,000 down on a $180,000 house and take out a 1-year ARM for a 30-year term. The lender uses the 1-year Treasury index (presently at 6%) and a 3% margin.

(a) Find your monthly payment for the first year.

First find the amount of your mortgage:

$$\text{Mortgage amount} = \text{Cost of house} - \text{Down payment}$$
$$= \$180,000 - \$20,000$$
$$= \$160,000.$$

Your first-year interest rate will be

$$\text{ARM interest rate} = \text{Index rate} + \text{Margin} = 6\% + 3\% = 9\%.$$

Government-backed mortgages, including FHA (Federal Housing Administration) and VA (Veterans Administration) loans, carry a government guarantee to protect the lender in case the borrower fails to repay the loan. Those who do not qualify for these loans, and obtain a conventional loan instead, usually are required to buy private mortgage insurance (PMI) as part of the loan package.

Over the second half of the twentieth century the amount of credit life insurance in force (for the purpose of covering consumer debt) increased by over 500 times, while group life insurance increased by only about 200 times, ordinary life insurance increased by about 60 times, and industrial life insurance actually decreased. (*Source:* American Council of Life Insurance.)

Now from Table 4 (using 9% over 30 years) we obtain 8.04623, so

First-year monthly payment = 160 × $8.04623 = $1287.40.

(b) Suppose that after a year the 1-year Treasury index has increased to 7.3%. Find your monthly payment for the second year.

During the first year, a small amount of the mortgage principal has been paid, so in effect we will now have a new "mortgage amount." (Also, the term will now be 1 year less than the original term.) The amortization schedule for the first year (not shown here) yields a loan balance, after the twelfth monthly payment, of $158,906.83.

Also, for the second year,

ARM interest rate = Index rate + Margin = 7.3% + 3% = 10.3%.

Since 10.3% is not included in Table 4, we resort to the regular monthly payment formula, using the new mortgage balance and also using 29 years for the remaining term.

$$\text{Second-year monthly payment} = \frac{\$158{,}906.83 \left(\frac{.103}{12}\right)\left(1 + \frac{.103}{12}\right)^{348}}{\left(1 + \frac{.103}{12}\right)^{348} - 1}$$

$$= \$1437.38$$

The first ARM interest rate adjustment has caused the second-year monthly payment to rise to $1437.38, which is an increase of $149.98 over the initial monthly payment. ■

Sometimes initial rates offered on an ARM are less than the sum of the index and the margin. Such a **discount** often is the result of the seller (a new home builder, for example) paying the lender an amount in order to reduce the buyer's initial interest rate. This arrangement is called a "seller buydown." Discounts result in lower initial monthly payments, but often are combined with higher initial fees, or even an increase in the price of the house. Also, when the discount expires, you may experience sizable increases in your monthly payments.

EXAMPLE 6 In Example 5, suppose that a seller buydown discounts your initial (first-year) rate by 2%. Find your first-year and second-year monthly payments.

The 6% index rate is discounted to 4%. Adding the 3% margin yields a net first-year rate of 7% (rather than the 9% of Example 5), so the Table 4 entry is 6.65302, and

First-year monthly payment = 160 × $6.65302 = $1064.48.

The amortization schedule shows a balance at the end of the first year of $158,374.76. The discount now expires, and the index has increased to 7.3%, so for the second year,

ARM interest rate = Index rate + Margin = 7.3% + 3% = 10.3%.

Using the monthly payment formula, with $r = 10.3\%$ and $t = 29$,

$$\text{Second-year monthly payment} = \frac{\$158{,}374.76\left(\dfrac{.103}{12}\right)\left(1 + \dfrac{.103}{12}\right)^{348}}{\left(1 + \dfrac{.103}{12}\right)^{348} - 1}$$

$$= \$1432.57.$$

The initial monthly payment of \$1064.48 looks considerably better than the \$1287.40 of Example 5, but at the start of year two, monthly payments jump by \$368.09. Such an increase may cause you to experience "payment shock." ■

To reduce the risk of payment shock, be sure the ARM you obtain includes **cap** features. An **interest rate cap** limits the amount your interest rate can increase. A **periodic cap** limits how much the rate can increase from one adjustment period to the next (typically about 1% per 6 months or 2% per year). An **overall cap** limits how much the rate can increase over the life of the loan (typically about 5% total). Most all ARMs must have overall rate caps by law and many also have periodic caps. Some ARMs also have **payment caps,** which limit how much the *payment* can increase at each adjustment. However, since such a cap limits the payment but not the interest rate, it is possible that your payment may not cover even the interest owed. In that case you could find yourself owing more principal at the end of an adjustment period than you did at the beginning. This situation is called **negative amortization.** To make up the lost interest, the lender will increase your payments more in the future, or may extend the term of the loan.

Another way to reduce risk is with a **convertibility** feature. This allows you, at certain points in time, to elect to convert the mortgage to a fixed-rate mortgage. If you elect to convert, your fixed rate will be determined by prevailing rates at the time of conversion. If you think you may want to completely, or partially, pay off the principal of the loan ahead of schedule, try to negotiate an ARM with no **prepayment penalty.** This means that you will be able to make early principal payments without paying any special fees, and your early payment(s) will go totally toward reducing the principal.

Closing Costs Besides the down payment and the periodic costs of home ownership, including principal and interest payments, property taxes, and insurance, there are some significant one-time expenses that apply to both fixed-rate and adjustable-rate mortgages and are paid when the mortgage is originally set up. Lumped together, these charges are called **closing costs** (or **settlement charges**). The **closing,** or **settlement,** occurs when all details of the transaction have been determined and the final contracts are signed. It is the formal process by which the ownership title is transferred from seller to buyer. Within 3 days after your loan application, the lender is required to deliver or mail to you certain materials that include a "good faith estimate" of your closing costs. That is the time to ask questions, clarify all proposed costs, and, if you wish, shop around for alternative providers of settlement services. At the time of the closing, it will likely be too late to avoid paying for services already provided. We illustrate typical closing costs in the following example.

Refinancing means initiating a new home mortgage and paying off the old one. In times of relatively low rates, lenders often encourage homeowners to refinance. Since setting up a new loan will involve costs, make sure you have a good reason before refinancing. Some possible reasons:

1. The new loan may be comparable to the current loan with a rate low enough to recoup the costs in a few years.
2. You may want to pay down your balance faster by switching to a shorter term loan.
3. You may want to borrow out equity to cover other major expenses, such as education for your children.
4. Your present loan may have a large balloon feature which necessitates refinancing.
5. You may be uncomfortable with your variable (hence uncertain) ARM and want to switch to a fixed-rate loan.

(*Source:* http://www.business-to-business-resources.com)

EXAMPLE 7 For a $58,000 mortgage, the borrower was charged the following closing costs.

Loan origination fee (1% of mortgage amount)	$____
Broker loan fee	1455
Lender document and underwriting fees	375
Lender tax and wire fees	205
Fee to title company	200
Title insurance fee	302
Title reconveyance fee	65
Document recording fees	35

Compute the total closing costs for this mortgage.

"Loan origination fees" are commonly referred to as **points.** Each "point" amounts to 1% of the mortgage amount. By imposing points, the lender can effectively raise the interest rate without raising monthly payments (since points are normally paid at closing rather than over the life of the loan). In this case, "one point" translates to $580. Adding this to the other amounts listed gives total closing costs of $3217.

Not all mortgage loans will involve the same closing costs. You should seek to know just what services are being provided and whether their costs are reasonable. There is considerable potential for abuse in the area of closing costs. Therefore, in 1974 Congress passed the **Real Estate Settlement Procedures Act (RESPA)** to standardize Truth in Lending practices in the area of consumer mortgages. The lender is required to provide you with clarifying information, including the APR for your loan. The APR, the cost of your loan as a yearly rate, will probably be higher than the stated interest rate since it must reflect all points and other fees paid directly for credit as well as the actual interest.

14.4 EXERCISES

Round all monetary answers to the nearest cent unless directed otherwise.

Monthly Payment on a Fixed-rate Mortgage Find the monthly payment needed to amortize principal and interest for each of the following fixed-rate mortgages. You can use either the regular monthly payment formula or Table 4, as appropriate.

	Loan Amount	Interest Rate	Term			Loan Amount	Interest Rate	Term
1.	$70,000	10.0%	20 years		5.	$227,750	12.5%	25 years
2.	$50,000	11.0%	15 years		6.	$95,450	15.5%	5 years
3.	$57,300	8.7%	25 years		7.	$132,500	7.6%	22 years
4.	$85,000	7.9%	30 years		8.	$205,000	5.5%	10 years

Amortization of a Fixed-rate Mortgage Complete the first one or two months (as required) of the following amortization schedules for fixed-rate mortgages.

9. Mortgage: $58,500
   Interest rate: 10.0%
   Term of loan: 30 years

Amortization Schedule

Payment Number	Total Payment	Interest Payment	Principal Payment	Balance of Principal
1	(a) _____	(b) _____	(c) _____	(d) _____

10. Mortgage: $87,000
    Interest rate: 8.5%
    Term of loan: 20 years

Amortization Schedule

Payment Number	Total Payment	Interest Payment	Principal Payment	Balance of Principal
1	(a) _____	(b) _____	(c) _____	(d) _____

11. Mortgage: $143,200
    Interest rate: 6.5%
    Term of loan: 15 years

Amortization Schedule

Payment Number	Total Payment	Interest Payment	Principal Payment	Balance of Principal
1	(a) _____	(b) _____	(c) _____	(d) _____
2	(e) _____	(f) _____	(g) _____	(h) _____

12. Mortgage: $124,750
    Interest rate: 9%
    Term of loan: 25 years

Amortization Schedule

Payment Number	Total Payment	Interest Payment	Principal Payment	Balance of Principal
1	(a) _____	(b) _____	(c) _____	(d) _____
2	(e) _____	(f) _____	(g) _____	(h) _____

13. Mortgage: $113,650
    Interest rate: 8.2%
    Term of loan: 10 years

Amortization Schedule

Payment Number	Total Payment	Interest Payment	Principal Payment	Balance of Principal
1	(a) _____	(b) _____	(c) _____	(d) _____
2	(e) _____	(f) _____	(g) _____	(h) _____

14. Mortgage: $150,000
    Interest rate: 6.25%
    Term of loan: 16 years

Amortization Schedule

Payment Number	Total Payment	Interest Payment	Principal Payment	Balance of Principal
1	(a) _____	(b) _____	(c) _____	(d) _____
2	(e) _____	(f) _____	(g) _____	(h) _____

Finding Monthly Mortgage Payments Find the total monthly payment, including taxes and insurance, on each of the following mortgages.

	Mortgage	Interest Rate	Term of Loan	Annual Taxes	Annual Insurance
15.	$62,300	7%	20 years	$610	$220
16.	$51,800	10%	25 years	$570	$145
17.	$89,560	6.5%	10 years	$915	$409
18.	$72,890	5.5%	15 years	$1850	$545
19.	$115,400	8.8%	20 years	$1295.16	$444.22
20.	$128,100	11.3%	30 years	$1476.53	$565.77

Comparing Total Principal and Interest on a Mortgage Suppose $140,000 is owed on a house after the down payment is made. The monthly payment for principal and interest at 8.5% for 30 years is 140 × $7.68913 = $1076.48.

21. How many monthly payments will be made?

22. What is the total amount that will be paid for principal and interest?

23. The total interest charged is the total amount paid minus the amount financed. What is the total interest?

24. Which is more—the amount financed or the total interest paid? By how much?

Long-term Effect of Interest Rates You may remember seeing home mortgage interest rates fluctuate widely in a period of not too many years. The following exercises show the effect of changing rates. Refer to Table 5, which compared the amortization of a $60,000, 30-year mortgage for rates of 4.5% and 14.5%. Give values of each of the following for (a) a 4.5% rate, and (b) a 14.5% rate.

25. monthly payments

26. percentage of first monthly payment that goes toward principal

27. balance of principal after 1 year

28. balance of principal after 20 years

29. the first monthly payment that includes more toward principal than toward interest

30. amount of interest included in final payment

The Effect of the Term on Total Amount Paid Suppose a $60,000 mortgage is to be amortized at 7.5% interest. Find the total amount of interest that would be paid for each of the following terms.

31. 10 years
32. 20 years
33. 30 years
34. 40 years

The Effect of Adjustable Rates on the Monthly Payment For each of the following adjustable-rate mortgages, find (a) the initial monthly payment, (b) the monthly payment for the second adjustment period, and (c) the change in monthly payment at the first adjustment. (The "adjusted balance" is the principal balance at the time of the first rate adjustment. Assume no caps apply.)

	Beginning Balance	Term	Initial Index Rate	Margin	Adjustment Period	Adjusted Index Rate	Adjusted Balance
35.	$75,000	20 years	6.5%	2.5%	1 year	8.0%	$73,595.52
36.	$44,500	30 years	7.2%	2.75%	3 years	6.6%	$43,669.14

The Effect of Rate Caps on Adjustable-rate Mortgages
Jeff Walenski has a 1-year ARM for $50,000 over a 20-year term. The margin is 2% and the index rate starts out at 7.5% and increases to 10.0% at the first adjustment. The balance of principal at the end of the first year is $49,119.48. The ARM includes a periodic rate cap of 2% per adjustment period. (Use this information for Exercises 37–40.)

37. Find **(a)** the interest owed and **(b)** the monthly payment due for the first month of the first year.

38. Find **(a)** the interest owed and **(b)** the monthly payment due for the first month of the second year.

39. What is the monthly payment adjustment at the end of the first year?

40. If the index rate has dropped slightly at the end of the second year, will the third-year monthly payments necessarily drop? Why or why not?

Closing Costs of a Mortgage For Exercises 41–44, refer to the following table of closing costs for the purchase of a $175,000 house requiring a 20% down payment.

Title insurance premium	$240
Document recording fee	30
Loan fee (two points)	____
Appraisal fee	225
Prorated property taxes	685
Prorated fire insurance premium	295

Find each of the following.

41. the mortgage amount
42. the loan fee
43. the total closing costs
44. the total amount of cash required of the buyer at closing (including down payment)

On the basis of material in this section, or on your own research, give brief written responses to each of the following.

45. Suppose your ARM allows conversion to a fixed-rate loan at each of the first five adjustment dates. Describe circumstances under which you would want to convert.

46. Describe each of the following types of mortgages.

    Graduated payment
    Balloon payment
    Growing equity
    Shared equity
    Partnership

47. Should a home buyer always pay the smallest down payment that will be accepted? Explain.

48. Should a borrower always choose the shortest term available in order to minimize the total interest expense? Explain.

49. Under what conditions would an ARM probably be a better choice than a fixed-rate mortgage?

50. Why are second-year monthly payments (slightly) less in Example 6 than in Example 5 even though the term, 29 years, and the interest rate, 10.3%, are the same in both cases?

51. Do you think that the discount in Example 6 actually makes the overall cost of the mortgage less? Explain.

52. Discuss the term "payment shock" mentioned at the end of Example 6.

53. Find out what is meant by the following terms and describe some of the features of each.

    FHA-backed mortgage
    VA-backed mortgage
    Conventional mortgage

14.5 Investing

In a general sense, an *investment* is a way of putting resources to work so that (hopefully) they will grow. A variety of "consumer" investments, from cash to real estate to collectibles, or even your own business, will be compared at the end of the section (in For Further Thought: Summary of Investments). But our main emphasis will be on the basic mathematical aspects of a restricted class of financial investments, namely, *stocks*, *bonds*, and *mutual funds*.

Stocks Buying stock in a corporation makes you a part owner of the corporation. You then share in any profits the company makes, and your share of profits is called a **dividend.** If the company prospers (or if increasing numbers of investors believe it will prosper in the future), your stock will be attractive to others, so that you may, if you wish, sell your shares at a profit. The profit you make by selling for more than you paid is called a **capital gain.** A negative gain, or **capital loss,** results if you sell for less than you paid. By **return on investment,** we mean the net difference between what you receive (including your sale price and any dividends received) and what you paid (your purchase price plus any other expenses of buying and selling the stock).

EXAMPLE 1 Brenda Cochrane bought 100 shares of stock in Company A on January 15, 2002, paying $30 per share. On January 15, 2003, she received a dividend of 50¢ per share, and the stock price had risen to $30.85 per share. (Ignore any costs other than the purchase price.)

(a) Find Brenda's total cost for the stock.

$$(\$30 \text{ per share}) \times (100 \text{ shares}) = \$3000$$

She paid $3000 total.

(b) Find the total dividend amount.

$$(\$.50 \text{ per share}) \times (100 \text{ shares}) = \$50$$

The dividend amount was $50.

(c) Find Brenda's capital gain if she sold the stock on January 15, 2003.

$$\text{Capital gain} = (\text{Change in price per share}) \times (\text{Number of shares})$$
$$= (\$30.85 - \$30.00) \times 100 = \$.85 \times 100 = \$85$$

The capital gain was $85.

(d) Find Brenda's total return on her one year of ownership of this stock.

$$\text{Total return} = \text{Dividends} + \text{Capital gain} = \$50 + \$85 = \$135$$

Total return on the investment was $135.

(e) Find the percentage return.

$$\text{Percentage return} = \frac{\text{Total return}}{\text{Total cost}} \times 100\% = \frac{\$135}{\$3000} \times 100\% = 4.5\%$$

The percentage return on the investment was 4.5%.

The price of a share of stock is determined by the law of supply and demand at institutions called **stock exchanges.** In the United States, the oldest and largest exchange is the New York Stock Exchange (NYSE), established in 1792 and located on Wall Street in New York City.

The public does not go directly to the exchange to buy and sell stock. Instead, members of the public buy their stock through stockbrokers, people who do have access to the exchange. Stockbrokers charge a fee for buying or selling stock.

Finding Prices Current stock prices can be found in many daily newspapers or by consulting an online computer service that provides this information. The table on the next page shows only a small portion of the activities on the New York Stock Exchange for a recent day of trading.

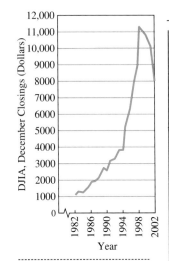

The Dow Jones Industrial Average (DJIA, or simply the "Dow") has been the most popular measure of general stock price trends since its inception in 1896. A significant time of increasing prices is a **bull market,** while an extended period of decreasing prices is a **bear market.** The Dow is a composite of 30 stocks from various sectors of the economy, which are changed occasionally to keep the list representative of the overall economy.

Stock price trends have historically been upward but with lots of variation within that general pattern. A strong bull market that began in August of 1982 saw the Dow triple over 5 years, until October 19, 1987 (known as Black Monday), when the Dow lost 22.6% of its value in a single day. The bull soon regained strength and mostly continued through the 1990s. Then another bear market set in. Starting in 2000, the Dow declined for 3 years in a row, a very rare occurrence historically. (Source: *The World Book Year Books,* 1983–2003)

The **NASDAQ (National Association of Securities Dealers Advanced Quotations),** unlike the NYSE and other exchanges which actually carry out trading at specific locations, is an electronic network of brokerages established in 1971 to facilitate the trading of over-the-counter (unlisted) stocks. Most of the high-tech company stocks, which grew very rapidly as a group in the 1980s and especially the 1990s (and many of which declined as rapidly in 2000-2002), are (or were) listed on the Nasdaq market.

The headings at the tops of the columns indicate the meanings of the numbers across the row. For example, the information for Albertsons (with stock symbol ABS) appears as follows.

YTD %CHG	52-WEEK HI	LO	STOCK (SYM)	DIV	YLD %	PE	VOL 100s	CLOSE	NET CHG
6.8	36.99	26.88	Albertsons **ABS**	.76	2.3	27	13617	33.62	0.01

The number 6.8 shows that the share value of Albertsons has increased 6.8% so far this calendar year. The next two numbers show that $36.99 is the highest price that the stock has reached in the last 52 weeks, while $26.88 is the lowest. Following the company symbol is .76, showing that the company is currently paying an annual dividend of 76¢ per share of stock. As mentioned earlier, a dividend is money paid by a company to owners of its stock. Dividends generally go up when the company is doing well and go down when business is bad. Then comes 2.3, the **current yield** on the company's stock, in percent. The dividend of 76¢ per share represents 2.3% of the current purchase price of the stock.

The 27 after the 2.3 is the **price-to-earnings ratio,** the purchase price per share divided by the earnings per share. After 27 is 13617, the number of shares sold that day, in hundreds. On the day in question, a total of $13{,}617 \times 100 = 1{,}361{,}700$ shares of Albertsons stock were sold. The next number in the listings, 33.62, shows that the stock closed for the day at $33.62 per share. That is, the final trade in Albertsons stock, prior to the closing of the market at 4 P.M. Eastern time, was executed at that price. The final number shown, 0.01, indicates that this day's closing price was 1¢ higher than that of the previous trading day.

EXAMPLE 2 Use the stock table to find the required amounts below.

(a) The highest price for the last 52 weeks for Alcoa (AA).

Find the correct line from the stock table.

1.5   45.71   27.36   Alcoa **AA**   .60   1.7   46   33587   36.07   −0.18

The highest price for this stock for the last 52 weeks (52-WEEK HI) is $45.71.

(b) The dividend for Pioneer-Standard Electronics (PIOS).

The dividend appears just after the company symbol. From the table, find the dividend for Pioneer-Standard Electronics, .12, or 12¢ per share. (Notice the special symbol ♣ preceding the company name in the listing. This means that Pioneer-Standard Electronics is one of a number of companies whose annual reports can be obtained free by readers of the *Wall Street Journal*.) ■

EXAMPLE 3 Find the cost for 100 shares of Allstate (ALL) stock, purchased at the closing price for the day.

Find the total cost of a stock purchase by multiplying the price per share times the number of shares purchased. From the stock table, the closing price for the day for Allstate was $39.08 per share.

$$\text{Total cost} = 100 \times \$39.08 = \$3908.00$$

One hundred shares of this stock would cost $3908.00, plus any broker's fees (discussed on the following pages). ■

On a given day's listings, individual stocks may carry a variety of special features. These are indicated with designated symbols which are explained where the listings are published. A few of the more prominent features are listed here.

Some Special Features in Stock Listings

1. A line listed in boldface print distinguishes a stock whose price changed by 5% or more (as long as the previous day's closing price was at least $2).
2. An underlined row in the listings indicates a stock with an especially high volume of trading for the day relative to its usual volume. In the case of the NYSE listings and the Nasdaq National Market, the 40 largest volume percentage leaders are indicated in this way.
3. A "pf" between the company name and symbol indicates a "preferred" stock, a special category of stock issued by some companies in addition to their ordinary, or "common," stock. A company must pay dividends to holders of its preferred stock before it can pay dividends to holders of its common stock. Additional advantages of preferred stocks include attractive yields, stable dividend income, and tax advantages for corporate investors.
4. The symbol ▲ at the left indicates a new 52-week high, while the symbol ▼ indicates a new 52-week low.

Finding Commissions To buy or sell a stock, it is generally necessary to use a broker (although a growing number of companies will sell their shares directly to investors, which saves brokerage expenses). A broker has representatives at the exchange who will execute a buy or sell order. The broker will charge a **commission** (the broker's fee) for executing an order. Before 1974, commission rates were set by stock exchange rules and were uniform from broker to broker. Since then, however, rates are competitive and vary considerably among brokers.

Full-service brokers, who offer research, professional opinions on buying and selling individual issues, and various other services, tend to charge the highest commissions on transactions they execute. Many **discount brokers,** on the other hand, merely buy and sell stock for their clients, offering little in the way of additional services.

During the 1990s, when it often seemed that almost any reasonable choice of stock would pay off for almost any investor, a number of online companies appeared, offering much cheaper transactions than were available through conventional brokers. Subsequently, most conventional brokerage firms, as well as the major discount brokerages, introduced their own automated (online) services. Investors today should compare a number of options before deciding which brokerage can best provide the stock investment services they need.

Commissions will normally be basically some percentage of the value of a purchase or sale (called the **principal**). Some firms will charge additional amounts in certain cases. For example, rates may depend on whether the order is for a **round lot** of shares (a multiple of 100) or an **odd lot** (any portions of an order for fewer than 100 shares). On the odd-lot portion, you may be charged an **odd-lot differential** (say, 10¢ per share). Also, it is possible to place a **limit order,** where the broker is instructed to execute a buy or sell if and when a stock reaches a predesignated price. An extra fee may be added to the commision on a limit order. One prominent discount brokerage does not charge these special fees but uses a tiered commission structure depending on the principal amount of the purchase or sale and also distinguishes between **broker-assisted** and **automated** trades, as shown here.

Typical Discount Commission Structure

Broker-assisted Trade

Principal Amount	Commission
Up to $2499.99	$35 + 1.7% of principal
$2500.00–$6249.99	$65 + .66% of principal
$6250.00–$19,999.99	$76 + .34% of principal
$20,000.00–$49,999.99	$100 + .22% of principal
$50,000.00–$499,999.99	$155 + .11% of principal
$500,000.00 or more	$255 + .09% of principal

Automated Trade

Number of Shares	Commission
Up to 1000	$29.95
More than 1000	3¢ per share

In addition to broker commissions, the Securities and Exchange Commission (SEC), a federal agency that regulates stock markets, imposes its own **SEC fee,** which goes toward funding the activities of the SEC and is assessed on stock sales only (not purchases). According to most recent law, this fee is 3.01¢ per $1000 of principal (rounded *up* to the next cent). For example, to find the fee for a sale of $1600, first divide $1600 by $1000, then multiply by 3.01¢.

$$\frac{\$1600}{\$1000} \times 3.01¢ = 4.816¢$$

Round up to obtain an SEC fee of 5¢.

EXAMPLE 4 Find the total cost (including expenses) for a broker-assisted purchase of 765 shares of Air Lease stock at $4.22 per share. Use the typical discount commission structure outlined above.

The basic cost of the stock is found by multiplying the price per share times the number of shares:

(Price per share) × (Number of shares) = $4.22 × 765 = $3228.30.

Since this principal amount falls in the second tier of the commission structure, the broker's commission is $65 plus .66% of this amount, or

Broker's commission = $65 + .0066 × $3228.30 = $65 + $21.31 = $86.31.

Find the total cost of the shares of stock as follows.

$3228.30	Basic cost
+ 86.31	Broker's commission
$3314.61	

The total cost for this purchase is $3314.61.

EXAMPLE 5 Find the amount received by a person executing an automated sale of 1500 shares of ParkerVision (PRKR) at $23.46 per share.

The basic price of the stock sold is

$$(\$23.46 \text{ per share}) \times (1500 \text{ shares}) = \$35{,}190.00.$$

Since more than 1000 shares are sold, the broker's commission is

$$(3¢ \text{ per share}) \times (1500 \text{ shares}) = \$45.00.$$

Find the SEC fee as described above.

$$\frac{\$35{,}190.00}{\$1000} \times 3.01¢ = 105.9219¢$$

Round up to obtain an SEC fee of $1.06. Proceed as follows.

$35,190.00	Basic price
− 45.00	Broker's commission
− 1.06	SEC fee
$35,143.94	

The seller receives $35,143.94.

Bonds Rather than contributing your capital (money) to a company, hoping that its employees will succeed in earning an attractive return for you and the other stockholders (owners), you may prefer merely to *lend* money to the company, receiving an agreed-upon rate of interest for the use of your money. In this case you would buy a bond from the company rather than purchase stock. The bond (loan) is issued with a stated term (life span), after which the bond "matures" and the principal (or **face value**) is paid back to you. Over the term, the company pays you a fixed rate of interest, which depends upon prevailing interest rates at the time of issue (and to some extent on the company's credit rating), rather than on the underlying value of the company.

A corporate bondholder has no stake in company profits, but is (quite) certain to receive timely interest payments. The potential return is generally less, but so is the risk of loss. If a company is unable to pay both bond interest and stock dividends, the bondholders must be paid first.

EXAMPLE 6 Chris Campbell invests $10,000 in a 5-year corporate bond paying 6% annual interest, paid semiannually. Find the total return on this investment, assuming Chris holds the bond to maturity (for the entire five years). Ignore any broker's fees.

By the simple interest formula (see Section 14.1),

$$I = Prt = \$10{,}000 \times .06 \times 5 = \$3000.$$

Notice that this amount, $3000, is the *total* return, over the 5-year period, not the annual return. Also, the fact that interest is paid twice per year (3% of the face value each time) has no bearing here since compounding does not occur.

Historically the overall rate of return over most time periods has been greater for stocks than for bonds. This fact (along with the general appeal of ownership) has led some investors to avoid bonds altogether. On the other hand, there are fewer uncertainties involved with bonds, and a bond investor's capital is subject to less risk. These facts can make bonds more attractive than stocks, especially for retired investors who depend on a steady stream of income from their investments. Many investors, however, seek to build a balanced portfolio that includes both stocks and bonds.

An abundance of information about mutual funds is available on the Internet. For example, the Investment Company Institute (ICI), the national association of the American investment company industry, provides material at the site

http://www.ici.org/.

Mutual Funds A **mutual fund** is a pool of money collected by an investment company from many individuals and/or institutions and invested in many stocks, bonds, or money market instruments (perhaps various combinations of all of these). The mutual fund industry has expanded tremendously. By the late 1990s approximately $5 trillion was invested in more than 7000 separate funds. By 2000 over 50 million U.S. households held mutual funds, and the largest category of these holdings was within retirement accounts.

Mutual Funds Versus Individual Stocks and Bonds

Advantages of Mutual Funds

1. *Simplicity*

   Let someone else do the work.

2. *Diversification*

   Being part of a large pool makes it easier to own interests in many different stocks, which decreases vulnerability to large losses in one particular stock or one particular sector of the economy.

3. *Access to New Issues*

   Initial public offerings (IPOs) can sometimes be very profitable. Whereas individual investors find it difficult to get in on these, the large overall assets of a mutual fund give its managers much more clout.

4. *Economies of Scale*

   Large stock purchases usually incur smaller expenses per dollar invested. In addition, the expenses are shared by large numbers of investors.

5. *Professional Management*

   Professional managers may have the expertise to pick stocks and time purchases to better achieve the stated objectives of the fund.

(continued)

6. *Indexing*

With minimal management, an index fund can maintain a portfolio that mimics a popular index (such as the S&P 500). This makes it easy for individual fund investors to achieve returns at least close to those of the index.

Disadvantages of Mutual Funds

1. *Impact of One-time Charges and Recurring Fees*

    Sales charges, management fees, "12b-1" fees (used to pay sales representatives), and fund expenses can mount and can be difficult to identify. And paying management fees does not guarantee getting *quality* management.

2. *Hidden Cost of Brokerage*

    Recurring fees and expenses are added to give the *expense ratio* of a fund. But commissions paid by the fund for stock purchases and sales are in addition to the expense ratio. (Since 1995, funds have been required to disclose their average commission costs in their annual reports.)

3. *Some Hidden Risks of Fund Ownership*
    (a) In the event of a market crash, getting out of the fund may mean accepting securities rather than cash, and these may be difficult to redeem for a fair price.
    (b) Managers may stray from the stated strategies of the fund.
    (c) Since tax liability is passed on to fund investors, and depends on purchases and sales made by fund managers, investors may be unable to avoid inheriting unwanted tax basis.

Source: Forbes Guide to the Markets, John Wiley & Sons, Inc., 1999.

New ways of providing financial services have emerged in the last several years. As of the turn of the century, American consumers were spending somewhere around $350 billion annually on fees and commissions for banking, insurance, and securities brokerage services. And sweeping legislation in late 1999 repealed many legal and regulatory barriers between companies operating in those three areas.

One result of the new laws was an increased number of company mergers as businesses set themselves up to offer one-stop shopping for insurance, investments, and banking. (*Source: The Sacramento Bee,* Nov 13, 1999)

Since a mutual fund owns many different stocks, each share of the fund owns a fractional interest in each of those companies. In an "open-end" fund (by far the most common type), new shares are created and issued to buyers while the fund company absorbs the shares of sellers. Each day, the value of a share in the fund (called the *net asset value,* or *NAV*) is determined as follows.

Net Asset Value of a Mutual Fund

The net asset value is given by

$$NAV = \frac{A - L}{N}$$

where A = Total fund assets,
L = Total fund liabilities, and
N = Number of shares outstanding.

EXAMPLE 7 Suppose, on a given day, a mutual fund has $500 million worth of stock, $500,000 in cash (not invested), and $300,000 in other assets. Total liabilities amount to $4 million, and there are 25 million shares outstanding. If Laura Griffin Hiller invests $50,000 in this fund, how many shares will she obtain?

We have the following.

$$A = \$500{,}000{,}000 + \$500{,}000 + \$300{,}000 = \$500{,}800{,}000,$$
$$L = \$4{,}000{,}000, \text{ and}$$
$$N = 25{,}000{,}000$$

So
$$NAV = \frac{\$500{,}800{,}000 - \$4{,}000{,}000}{25{,}000{,}000} = \$19.872.$$

Since $\frac{\$50{,}000}{\$19.872} \approx 2516$, Laura's $50,000 investment will purchase (to the nearest whole number) 2516 shares. ■

Evaluating Investment Returns Regardless of the type of investment you have, or are considering (stocks, bonds, mutual funds, or others), it is important to be able to accurately evaluate and compare returns (profits or losses). Even though past performance is never a guarantee of future returns, it is crucial information. A wealth of information on stock and bond markets and mutual funds is available in many publications (for example, the *Wall Street Journal* and *Barron's Weekly*) and on numerous Internet sources (for example, Nasdaq.com).

The most important measure of performance of an investment is the **annual rate of return.** This is not *necessarily* the same as percentage return, as calculated in Example 1, since annual rate of return depends on the time period involved. Some commonly reported periods are daily, seven-day, monthly, month-to-date, quarterly, quarter-to-date, annual, year-to-date, 2-year, 3-year, 5-year, 10-year, 20-year, and "since inception" (since the fund or other investment vehicle was begun).

Examples 8 and 9 illustrate some of the many complications that arise when one tries to evaluate and compare different investments. Yet it is important to compare different performances on an equal basis.

EXAMPLE 8 Todd Hall owns 50 shares of stock. His brokerage statement for the end of August showed that the stock closed that month at a price of $40 (per share) and the statement for the end of September showed a closing price of $40.12. For purposes of illustration, disregard any possible dividends, and assume that the money gained in a given month has no opportunity to earn additional returns. (Those earnings cannot be "reinvested.")

(a) Find the value of these shares at the end of August.

$$\text{Value} = (\text{Price per share}) \times (\text{Number of shares}) = \$40 \times 50$$
$$= \$2000$$

(b) Find Todd's monthly return on this stock.

$$\text{Return} = (\text{Change in price per share}) \times (\text{Number of shares})$$
$$= (\$40.12 - \$40) \times 50$$
$$= \$6$$

(c) Find the monthly percentage return.

$$\text{Percentage return} = \frac{\text{Total return}}{\text{Total value}} \times 100\% = \frac{\$6}{\$2000} \times 100\%$$
$$= .3\%$$

(d) Under the assumptions stated above, find the annual rate of return.

Because monthly returns do not earn more returns, we use an "arithmetic return" here. Simply add .3% twelve times (or multiply 12 times .3%), to obtain 3.6%. The annual rate of return is 3.6%.

EXAMPLE 9 Erika Gutierrez owns 80 shares of a mutual fund with a net asset value of $10 on October 1. Assume, for the sake of simplicity, that the only return that the fund earns is dividends of .4% per month, which are automatically reinvested.

(a) Find the value of Erika's holdings in this fund on October 1.

$$\text{Value} = \text{NAV} \times \text{number of shares} = \$10 \times 80 = \$800$$

(b) Find the monthly return.

$$\text{Return} = .4\% \text{ of } \$800 = \$3.20$$

(c) Find the annual rate of return.

In this case, monthly returns get reinvested. Compounding occurs. Therefore we use a "geometric" return rather than arithmetic. To make this calculation, first find the "return relative," which is 1 plus the monthly rate of return: $1 + .4\% = 1 + .004 = 1.004$. Then multiply this return relative 12 times (raise it to the 12th power) to get an annual return relative: $1.004^{12} \approx 1.049$. Finally, subtract 1 and multiply by 100 to convert back to a percentage rate. The annual percentage rate is 4.9%.

A geometric return such as the one found in Example 9 above often is called the "effective annual rate of return." (Notice that an arithmetic return calculation would have ignored the reinvestment compounding and would have understated the effective annual rate of return by .1%, since $12 \times .4\% = 4.8\%$.)

FOR FURTHER THOUGHT

Summary of Investments

The following summary of various types of investments is based on average cases and is intended only as a general comparison of investment opportunities. There are numerous exceptions to the characteristics shown. For example, though the chart shows no selling fees for mutual funds, this is really true only for the so-called no-load funds. The "loaded" funds do charge "early redemption fees" or other kinds of sales charges. That is not to say that a no-load fund is necessarily better. For example, the absence of sales fees may be offset by higher "management fees."

These various costs, as well as other important factors, will be clearly specified in a fund's prospectus and should be studied carefully before a purchase decision is made. Some expertise is not entirely unnecessary for mutual fund investing, just less so than with certain other options, like direct stock purchases, raw land, or collectibles.

To use the chart, read across the columns to find the general characteristics of the investment. "Liquidity" refers to the ability to get cash from the investment quickly.

Investment	Protection of Principal	Protection Against Inflation	Rate of Return (%)	Certainty of Return	Selling Fees	Liquidity	Long-term Growth	Requires Work from You	Expert Knowledge Required
Cash	excellent	none	0			excellent	no	no	no
Ordinary life insurance	good	poor	2–4	excellent	high	excellent	no	no	no
Savings bonds	excellent	poor	3–5	excellent	none	excellent	no	no	no
Bank savings	excellent	poor	2–4	excellent	none	excellent	no	no	no
Credit union	excellent	poor	3–4	excellent	none	excellent	no	no	no
Corporation bonds	good	poor	5–7	excellent	low	good	no	some	some
Tax-free municipal bonds	excellent	poor	2–5	excellent	medium	good	no	no	some
Corporation stock	good	good	3–12	good	medium	good	yes	some	some
Preferred stock	good	poor	4–8	excellent	medium	good	no	some	some
Mutual funds	good	good	3–12	good	none	good	yes	no	no
Your own home	good	good	4–10	good	medium	poor	yes	yes	no
Mortgages on the homes of others	fair	poor	5–8	fair	high	poor	no	yes	some
Raw land	fair	good	?	fair	medium	poor	yes	no	yes
Rental properties	good	good	4–8	fair	medium	poor	yes	yes	some
Stamps, coins, antiques	fair	good	0–10	fair	high	poor	yes	yes	yes
Your own business	fair	good	0–20	poor	high	poor	yes	yes	yes

For Group Discussion

List (a) some investment types you would recommend considering and (b) some you would recommend avoiding for a friend whose main investment objective is as follows.

1. Avoid losing money.
2. Avoid losing purchasing power.
3. Avoid having to learn about investments.
4. Avoid having to work on investments.
5. Be sure to realize a gain.
6. Be able to "cash in" at any time.
7. Get as high a return as possible.
8. Be able to "cash in" without paying fees.
9. Realize quick profits.
10. Realize growth over the long term.

844 CHAPTER 14 Consumer Mathematics

14.5 EXERCISES

Reading Stock Charts Refer to the stock table on page 834 for Exercises 1–36.

Find each of the following.

1. closing price for Agilent Technologies (A)
2. sales for the day for Airborne (ABF)
3. change from the previous day for Party City (PCTY)
4. closing price for Peet's Coffee & Tea (PEET)
5. 52-week high for Patterson Dental (PDCO)
6. 52-week low for Alaska Air Group (ALK)
7. dividend for Allegheny Energy (AYE)
8. dividend for PennFed Financial Services, preferred (PFSBP)
9. sales for the day for Perceptron (PRCP)
10. change from the previous day for Alltrista (ALC)
11. year-to-date percentage change for Alcan (AL)
12. year-to-date percentage change for PeopleSoft (PSFT)
13. price-to-earnings ratio for Patriot Transportation (PATR)
14. price-to-earnings ratio for AirNet Systems (ANS)

Finding Stock Costs Find the basic cost (ignoring any broker fees) for each of the following stock purchases, at the day's closing prices.

15. 200 shares of Albany International (AIN)
16. 400 shares of Ameren (AEE)
17. 300 shares of PC Mall (MALL)
18. 700 shares of PetsMart (PETM)

Finding Stock Costs Find the cost, at the day's closing price, for each of the following stock purchases. Include typical discount broker commissions as described in the text.

	Stock	Symbol	Number of Shares	Transaction Type
19.	Paychex	PAYX	60	broker-assisted
20.	Amerada Hess	AHC	70	broker-assisted
21.	Alltel	AT	355	automated
22.	Performance Food Group	PFGC	585	automated
23.	Agrium	AGU	2500	broker-assisted
24.	Pixelworks	PXLW	1500	automated
25.	Pathmark Stores	PTMK	2400	automated
26.	Alamosa	APS	10,000	broker-assisted

Finding Receipts for Stock Sales Find the amount received by the sellers of the following stocks (at the day's closing prices). Deduct sales expenses as described in the text.

	Stock	Symbol	Number of Shares	Transaction Type
27.	Albemarle	ALB	400	broker-assisted
28.	PartsBase	PRTS	600	broker-assisted
29.	Pharmacopeia	PCOP	500	automated
30.	Alberto Culver	ACV	700	automated
31.	PLATO Learning	TUTRs	1350	automated
32.	American Greetings	AM	2740	automated
33.	American States Water	AWR	1480	broker-assisted
34.	Photronics	PLAB	1270	broker-assisted

Finding Net Results of Combined Transactions For each of the following combined transactions (executed at the closing price of the day), find the net amount paid out or taken in. Assume typical expenses as outlined in the text.

35. Paul Borge bought 100 shares of Alpine Group (AGI) and sold 20 shares of Peak International (PEAK), both broker-assisted trades.
36. Nara Lee bought 800 shares of PETCO Animal Supplies (PETC) and sold 1200 shares of Alaska Air Group (ALK), both automated trades.

Costs and Returns of Stock Investments For each of the following stock investments, find **(a)** the total purchase price, **(b)** the total dividend amount, **(c)** the capital gain or loss, **(d)** the total return, and **(e)** the percentage return.

	Number of Shares	Purchase Price per Share	Dividend per Share	Sale Price per Share
37.	40	$20	$2	$44
38.	20	$25	$1	$22
39.	100	$12.50	$1.08	$10.15
40.	200	$ 8.80	$1.12	$11.30

14.5 Investing

Total Return on Bond Investments *Find the total return earned by each of the following bonds.*

	Face Value	Annual Interest Rate	Term to Maturity
41.	$ 1000	5.5%	5 years
42.	$ 5000	6.4%	10 years
43.	$10,000	7.11%	3 months
44.	$50,000	4.88%	6 months

Net Asset Value of a Mutual Fund *The information below applies to the day of investment in a mutual fund. In each case, find* **(a)** *the net asset value, and* **(b)** *the number of shares purchased.*

	Amount Invested	Total Fund Assets	Total Fund Liabilities	Total Shares Outstanding
45.	$ 3500	$875 million	$ 36 million	80 million
46.	$ 1800	$643 million	$102 million	50 million
47.	$25,470	$2.31 billion	$135 million	263 million
48.	$83,250	$1.48 billion	$ 84 million	112 million

Finding Monthly and Annual Investment Returns *For each of the following investments, assume that there is no opportunity for reinvestment of returns. In each case find* **(a)** *the monthly return,* **(b)** *the annual return, and* **(c)** *the annual percentage return. (Use the monthly return, rounded to the nearest cent, to find the annual return.)*

	Amount Invested	Monthly Percentage Return
49.	$ 645	1.3%
50.	$ 895	.9%
51.	$2498	2.3%
52.	$4983	1.8%

Effective Annual Rate of Return of Mutual Fund Investments *Assume that each of the following mutual fund investments earns monthly returns that are reinvested and subsequently earn at the same rate. In each case find* **(a)** *the beginning value of the investment,* **(b)** *the first monthly return, and* **(c)** *the effective annual rate of return.*

	Beginning NAV	Number of Shares Purchased	Monthly Percentage Return
53.	$ 9.63	125	1.5%
54.	$12.40	185	2.3%
55.	$11.94	350	1.83%
56.	$18.54	548	2.22%

Finding Commissions *For each of the following trades, determine the missing amounts in columns (a), (b), and (c).*

	Number of Shares	Price per Share	(a) Principal Amount	(b) Broker-assisted Commission	(c) Automated Commission
57.	10	$1.00	_____	_____	_____
58.	10	$100.00	_____	_____	_____
59.	400	$1.00	_____	_____	_____
60.	400	$100.00	_____	_____	_____
61.	4000	$1.00	_____	_____	_____
62.	4000	$100.00	_____	_____	_____

63. Referring to Exercises 57–62, fill in the blanks in the following statement.

A broker-assisted purchase is cheaper than an automated purchase only when a relatively _____ number of shares are purchased for a relatively _____ price per share.

✎ *Write a short essay in response to each of the following. Some research may be required.*

64. Describe the concept of "dollar-cost averaging," and relate it to advantage number 1 of mutual funds as listed in the text.

65. Discuss advantage number 3 of mutual funds as listed in the text.

Reading Stock Charts Refer to the stock table in the text to answer the following questions.

66. What special feature was true of Allied Capital (ALD)?
67. Give an example of a stock with all three of the following special features:
    (1) The price on this day was a new 52-week high.
    (2) The price changed by 5% or more from the previous day.
    (3) The stock was among the 40 largest volume percentage leaders on this day.

✎ 68. With respect to investing in corporate America, describe the difference between being an owner and being a lender.

Spreading Mutual Fund Investments Among Asset Classes Mutual funds (as well as other kinds of investments) normally are categorized into one of several "asset classes," according to the kinds of stocks or other securities they hold. Small capitalization funds are most aggressive, while cash is most conservative. Many investors, often with the help of an advisor, try to construct their portfolios in accordance with their stages of life. Basically, the idea is that a younger person can afford to be more aggressive (and assume more risk) while an older investor should be more conservative (assuming less risk). The investment pyramids here show percentage ranges recommended by Edward Jones, one of the largest investment firms in the country in terms of number of offices serving individual investors. In each of the following exercises, divide the given investor's money into the five categories so as to position them right at the middle of the recommended ranges.

Asset Classes	
Aggressive Growth (small cap)	■
Growth	▲
Growth & Income	◆
Income	●
Cash	+

69. Michelle Clayton, in her early investing years, with $20,000 to invest

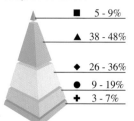

■ 5 - 9% **Early Investing Years**
▲ 38 - 48% Relatively few holdings
◆ 26 - 36% Long investing outlook
● 9 - 19%
+ 3 - 7%

70. Jeff Hubbard, in his good earning years, with $250,000 to invest

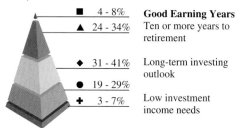

■ 4 - 8% **Good Earning Years**
▲ 24 - 34% Ten or more years to retirement
◆ 31 - 41% Long-term investing outlook
● 19 - 29%
+ 3 - 7% Low investment income needs

71. Paul Crockett, in his high income/saving years, with $400,000 to invest

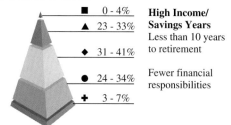

■ 0 - 4% **High Income/ Savings Years**
▲ 23 - 33% Less than 10 years to retirement
◆ 31 - 41%
● 24 - 34% Fewer financial responsibilities
+ 3 - 7%

72. Amy Donlin, in her retirement years, with $845,000 to invest

■ 0 - 4% **Retirement Years**
▲ 10 - 20% Reduced earnings power
◆ 39 - 49% High investment income needs
● 29 - 39%
+ 3 - 7%

The Effect of Taxes on Investment Returns An important consideration, not mentioned in the text, is the effect of taxation on investment returns. For example, income from municipal bonds generally is exempt from federal income tax, and bonds issued within your home state may be exempt from state tax as well. For an investor with a marginal combined state and federal tax rate of 35%, a taxable return of 6% would yield only 3.9% (65% of 6%). So at that tax rate, a tax-exempt return of 3.9% is equivalent to a taxable return of 6%. Find the tax-exempt rate of return that is equivalent to the given taxable rate of return for each of the following investors.

	Investor	Marginal Combined Tax Rate	Taxable Rate of Return
73.	Paul Altier	25%	5%
74.	Eileen Burke	30%	7%
75.	Carol Britz	35%	8%
76.	Greg Brueck	40%	10%

The Real Rate of Return When selecting long-term bonds, income investors should pay attention to the real rate of return (RRR), defined as interest rate minus inflation rate. (Since 1955, the average RRR has been 3.3%.) Find the RRR in each of the following cases. (Source: Edward Jones and Company: Income Advantage, September/October 2002, p. 11.)

	Year	Long-term Bond Rate	Inflation Rate
77.	1981	14.2%	8.9%
78.	2002	6.6%	1.3%

79. In light of Exercises 77 and 78, describe the importance to bond investors of the real rate of return.

Write a short essay in response to each of the following. Some research may be required.

80. Describe what DRIP plans are, and investigate their availability on the Internet.

81. Investigate the website www.fraud.org, and write a report on the Security and Exchange Commission's recommended questions to ask about potential investments.

82. Log onto www.napfa.org to find out about financial planners. Then report on what you would look for in a planner.

83. A great deal of research has been done in attempts to understand and predict trends in the stock market. Discuss the term *financial engineering*. Within this area of study, what is a *rocket scientist* (or *quant*)?

84. In connection with mutual funds, discuss the difference between a *front-end load* and a *contingent deferred sales charge*.

85. What is meant by the terms *growth stock* and *income stock*?

86. Describe the meaning of the graph shown here.

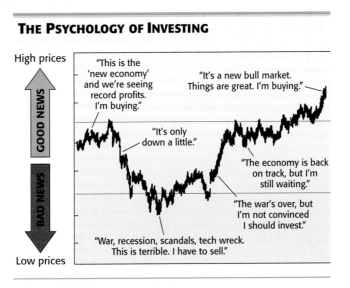

Source: Edward Jones and Company, 2002.

COLLABORATIVE INVESTIGATION

To Buy or to Rent?

Divide your class into groups of at least four students each. Every group is to first read the following.

> Scott and Aimee, a young couple with a child, Alaina, live in a rented apartment. After taxes, Scott earns $36,000 per year and Aimee's part-time job brings in $6000 per year. In the foreseeable future, their earnings probably will just keep pace with inflation. They are presently operating on the following monthly budget.
>
> | | |
> |---|---|
> | Rent | $1050 |
> | Food | 600 |
> | Daycare | 525 |
> | Clothing | 420 |
> | Utilities | 120 |
> | Entertainment | 300 |
> | Savings | 450 |
> | Other | 300 |
>
> They have accumulated $29,000 in savings.
>   Having found a house they would like to buy, and having consulted several lenders, they have discovered that, at best, purchasing the house would involve costs as shown here.
>
> | | |
> |---|---|
> | Price of house | $127,500 |
> | Down payment required (20%) | _____ |
> | Closing costs (required at closing): | |
> | Loan fee (1 point) | _____ |
> | Appraisal fee | 300 |
> | Title insurance premium | 375 |
> | Document & recording fees | 70 |
> | Prorated property taxes | 650 |
> | Prorated fire insurance premium | 270 |
> | Other | 230 |
> | Total: | _____ |
> | Fixed-rate 30-year mortgage at 7.5% | |
> | Annual taxes: | $1350 |
> | Annual fire insurance: | $540 |
>
> In order to own their own home, Scott and Aimee are willing to cut their clothing and entertainment expenditures by 30% each, and could decrease their savings allotments by 20%. (Some savings still will be necessary to provide for additional furnishings they would need and for expenses of an expanding family in the future.) They also figure that the new house and yard would require $130 monthly for maintenance and that utilities will be twice what they have been in the apartment. Other than that, their present budget allotments would remain the same.

Now divide your group into two "subgroups." The first subgroup is to answer these questions:

1. What amount would the home purchase cost Scott and Aimee in immediate expenditures?
2. Do they have enough cash on hand?

The second subgroup is to answer these questions:

3. What amount would the house cost Scott and Aimee on an ongoing monthly basis?
4. What amount will their monthly budget allow toward this cost?
5. Can they meet the monthly expenses of owning the house?

Within your group, decide whether Scott and Aimee can afford to purchase the house. Select a representative to report your findings to the class.
  Compare the evaluations of the various groups, and try to resolve any discrepancies.

CHAPTER 14 TEST

Find all monetary answers to the nearest cent. Use tables from the chapter as necessary.

Finding the Future Value of a Deposit Find the future value of each deposit.

1. $100 for 5 years at 6% simple interest
2. $50 for 2 years at 8% compounded quarterly

Solve each problem.

3. *Effective Annual Yield of an Account* Find the effective annual yield to the nearest hundredth of a percent for an account paying 3% compounded monthly.

4. *Years to Double by the Rule of 70* Use the rule of 70 to estimate the years to double at an inflation rate of 5%.

5. *Finding the Present Value of a Deposit* What amount deposited today in an account paying 4% compounded semiannually would grow to $100,000 in 10 years?

6. *Finding Bank Card Interest by the Average Daily Balance Method* Peg Cheever's MasterCard statement shows an average daily balance of $680. Find the interest due if the rate is 1.6% per month.

Analyzing a Consumer Loan Jeff Howard buys a turquoise necklace for his wife on their anniversary. He pays $4000 for the necklace with $1000 down. The dealer charges add-on interest of 8% per year. Jeff agrees to make payments for 24 months. Use this information for Exercises 7–10.

7. Find the total amount of interest he will pay.
8. Find the monthly payment.
9. Find the APR value for this loan (to the nearest half percent).
10. Find the unearned interest if he repays the loan in full with six payments remaining. Use the most accurate method available.
11. *True Annual Interest Rate in Consumer Financing* Newark Hardware wants to include financing terms in their advertising. If the price of a floor waxer is $150 and the finance charge with no down payment is $5 over a 6-month period (six equal monthly payments), find the true annual interest rate (APR).

12. Explain what a mutual fund is and discuss several of its advantages and disadvantages.

Finding the Monthly Payment on a Home Mortgage Find the monthly payment required for each of the following home loans.

13. The mortgage is for $150,000 at a fixed rate of 7% for 20 years. Amortize principal and interest only.
14. The purchase price of the home is $218,000. The down payment is 20%. Interest is at 8.5% fixed for a term of 30 years. Annual taxes are $1500 and annual insurance is $750.

Solve each problem.

15. *The Cost of Points in a Home Loan* If the lender in Exercise 14 charges two *points,* how much does that add to the cost of the loan?
16. Explain in general what *closing costs* are. Are they different from *settlement charges*?
17. *Adjusting the Rate in an Adjustable-rate Mortgage* To buy your home you obtain a 1-year ARM with 2.25% margin and a 2% periodic rate cap. The index starts at 7.85% but has increased to 10.05% by the first adjustment date. What interest rate will you pay during the second year?
18. *Reading Stock Charts* According to the stock table on page 834, how many shares of Patterson Dental (PDCO) were traded on the day represented?

Finding the Return on a Stock Investment Kathryn Campbell bought 1000 shares of stock at $12.75 per share. She received a dividend of $1.38 per share shortly after the purchase and another dividend of $1.02 per share one year later. Eighteen months after buying the stock she sold it for $10.36 per share.

19. Find Kathryn's total return on this stock.
20. Find her percentage return. Is this the *annual rate of return* on this stock transaction? Why or why not?

chapter

15

Graph Theory

"What is the diameter of the World Wide Web? The answer is not 7927 miles, even though the Web truly is worldwide. According to Albert-Laszlo Barabasi, Reka Albert, and Hawoong Joong of Notre Dame University, the diameter of the Web is 19.

The diameter in question is not a geometric distance; the concept comes from the branch of mathematics called *graph theory*. On the Web you get from place to place by clicking on hypertext links, and so it makes sense to define the distance by counting your steps through such links. The question is: If you select two Web pages at random, how many links will separate them, on average? Among the 800 million pages on the Web, there's room to wander down some very long paths, but Barabasi et al. find that if you know where you're going, you can get just about anywhere in 19 clicks of the mouse.

Barabasi's calculation reflects an interesting shift in the style and the technology of graph theory. Just a few years ago it would have been unusual to apply graph-theoretical methods to such an enormous structure as the World Wide Web. Of course, just a few years ago the Web didn't exist. Now very large netlike objects seem to be everywhere, and many of them invite graph-theoretical analysis."

In this chapter we study the kind of graphs referred to in the previous discussion.

Source: "Graph Theory in Practice: Part 1," by Brian Hayes in *American Scientist* (the magazine of Sigma Xi, the Scientific Research Society), January/February 2000, pages 9–13.

15.1 Basic Concepts

15.2 Euler Circuits

15.3 Hamilton Circuits

15.4 Trees and Minimum Spanning Trees

Collaborative Investigation: Finding the Number of Edges in a Complete Graph

Chapter 15 Test

15.1 Basic Concepts

In this chapter we study graph theory. Unlike graphs in the rectangular coordinate plane, studied earlier in the text, the "graphs" in this chapter are convenient diagrams that show relations or connections between objects in a collection. These graphs enable us to communicate and analyze complex information in a simple way. As a result, graph theory has many real-world applications.

Basic Concepts

Graphs Consider the following situation. A preschool teacher has ten children in her class: Andy, Claire, Dave, Erin, Glen, Katy, Joe, Mike, Sam, and Tim. The teacher wants to analyze the social interactions among these children. She observes with whom the children play during recess over a two-week period. Here are her observations:

Child	Played with:
Andy	no one
Claire	Dave, Erin, Glen, Katy, Sam
Dave	Claire, Erin, Glen, Katy
Erin	Claire, Dave, Katy
Glen	Claire, Dave, Sam
Katy	Claire, Dave, Erin
Joe	Mike, Tim
Mike	Joe
Sam	Claire, Glen
Tim	Joe

With the information presented like this, it is not easy to see patterns in the children's friendships. Instead, the information may be shown in a diagram, as in Figure 1. Each letter represents the first letter of a child's name. Notice that each child is represented by a dot. If two children play together, a line is drawn between the two corresponding dots to show this relationship.

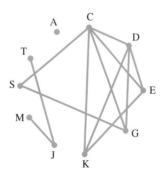

FIGURE 1

Why Are They Called Graphs?

Nineteenth-century chemists were particularly puzzled by substances that had the same chemical composition, but different physical properties. Chemist A. Crum Brown made a major breakthrough in 1864, inventing the now familiar diagrams for chemical compounds. He called these diagrams "graphic formulae." Illustrated below are two different molecules both having chemical composition C_4H_{10}. Mathematicians helped to find all possible graphic formulae corresponding to a particular chemical composition and adopted the term "graph" for their diagrams.

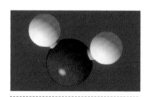

In older books on graph theory, you will find the *degree of a vertex* called the *valence* of the vertex. This terminology comes straight from chemistry; the valence of an element is the maximum number of atoms that can bond to a single atom of the element. For example, the valence of oxygen is 2, as demonstrated in the water molecule illustrated above.

A diagram such as Figure 1 is called a *graph*.* Each dot is called a **vertex** (plural **vertices**). In Figure 1, each vertex represents one of the preschool children. The lines between vertices are called the **edges** of the graph. In Figure 1, the edges show the relationship "play with each other." An edge must always begin and end at a vertex. We may have a vertex with no edges joined to it, such as A in Figure 1.

Graph

A **graph** is a collection of vertices (at least one) and edges. Each edge goes from a vertex to a vertex.

The graph in Figure 1 has 10 vertices and 12 edges. The only vertices of the graph are the dots; for example there is *not* a vertex where edges TJ and SG cross, since the crossing point does not represent one of the children. The relative positions of the vertices and the lengths of the edges have no significance; indeed, the edges do not even have to be drawn as straight lines. *Only the relation between the vertices has significance, indicated by the presence or absence of an edge between them.*

Why not draw two edges for each friendship pair? For example, why not draw one edge showing that Tim plays with Joe and another edge showing that Joe plays with Tim? Since Tim and Joe play with each other, the extra edge would show no additional information and would needlessly complicate the graph. Graphs in which there is no more than one edge between any two vertices and in which no edge goes from a vertex to the same vertex are called **simple graphs**. See Figure 2. *In this chapter, the word* graph *means simple graph, unless specifically indicated otherwise.* (The meaning of "simple" here is different from its meaning in the phrase "simple curves" discussed with geometry in Chapter 9.)

   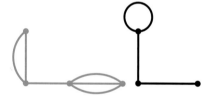

These are simple graphs.   These are not simple graphs.

FIGURE 2

From the graph in Figure 1, we can determine the size of a child's friendship circle by counting the number of edges coming from that child's vertex. The figure shows five edges coming from the vertex labeled C, indicating that Claire plays with five different children. The number of edges joined to a vertex is called the **degree of the vertex**. In the graph in Figure 1, the degree of vertex C is 5, the degree of vertex M is 1, and the degree of vertex A is 0, as no edges are joined to A. (Again, take care with terminology; this use of *degree* has nothing to do with its use in geometry in measuring the size of an angle.)

---

*In this chapter *graph* will mean finite graph. A finite graph is a graph with finitely many edges and vertices.

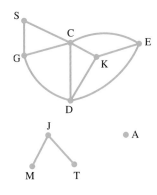

FIGURE 3

While Figure 1 clarifies the friendship patterns of the preschoolers, the graph can be drawn in a different, more informative way. In Figure 3, the vertices are in different places than in Figure 1, but the graph still has edges between the same pairs of vertices. Figure 3 shows one graph, even though it has three pieces, since it represents a single situation. By drawing the graph in this way, we make the friendship patterns more apparent. Each child still is represented by exactly one vertex, and an edge between two vertices (for example D and E) in the graph in Figure 3 is likewise an edge between the corresponding vertices in Figure 1, and vice versa. Although the graph in Figure 3 looks different from that in Figure 1, it shows the same relationships between the vertices; such graphs are said to be *isomorphic* graphs.

Isomorphic Graphs

Two graphs are **isomorphic** if there is a one-to-one correspondence between vertices of the two graphs with the property that whenever there is an edge between two vertices of either one of the graphs, there is an edge between the corresponding vertices of the other graph.

The word **isomorphism** comes from the Greek words *isos*, meaning "same," and *morphe*, meaning "form." Other words that come from the Greek root *isos* include *isobars*: lines drawn on a map connecting places with the same barometric pressure; *isosceles*: a triangle with two legs (*scelos*) of equal length; and *isomers*: chemical compounds with the same chemical composition, but different chemical structures.

A useful way to think about isomorphic graphs is to imagine a graph drawn with computer software that allows you to drag vertices without detaching any of the edges from their vertices. Any graph you can obtain by simply dragging vertices in this way will be isomorphic to the original graph.

As demonstrated with the friendship graph, replacing a graph with an isomorphic graph sometimes can convey more about the relationships being examined. For example, more features of the friendship patterns show up clearly in Figure 3, where there are no friendship links between the group consisting of Joe, Tim, and Mike, and the other children. We say that the graph is *disconnected*.

Connected and Disconnected Graphs

A graph is **connected** if one can move from each vertex of the graph to every other vertex of the graph *along edges of the graph*. If not, the graph is **disconnected.** The connected pieces of a graph are called the **components** of the graph.

Figure 3 clearly indicates that the friendship relationships among the children in the preschool class have three components. Although the graph in Figure 1 also is disconnected and includes three components, the way in which the graph is drawn does not make this as clear.

Using the graph in Figure 3 to analyze the relationships in her class, the teacher might feel concerned that Tim, Mike, and Joe do not interact with the other children. Certainly she would feel concerned about Andy's isolation. She might observe that Claire seems to get along well with other children and encourage interaction between Claire and Tim, Mike, and Joe, and she might make a special attempt to encourage the children to include Andy in their games. In summary, showing the friendship relations as a graph like that in Figure 3 made it much easier to analyze those relationships.

854　CHAPTER 15　Graph Theory

In the world of telecommunications, graphs are huge. A call graph has telephone numbers as vertices, while edges represent a call placed from one number to another. A call graph for one company for just one day contained nearly 54 million vertices and 170 million edges. It had 3.7 million components, but one of these was a giant component containing nearly 80% of the vertices.

Problem Solving

Using different colors for different components can help you determine how many components a graph has. Choose a color, begin at any vertex, and color all edges and vertices that you can get to from your starting vertex along edges of the graph. If the whole graph is not colored, then choose a different color, choose an uncolored vertex, and repeat the process. Continue until the whole graph is colored. It will then be easy to see the different components of the graph.

EXAMPLE 1　Is the graph in Figure 4 connected or disconnected? How many components does the graph have?

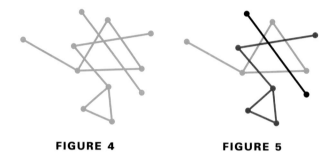

FIGURE 4　　　**FIGURE 5**

We color the graph as in Figure 5 to help solve the problem. The graph with color shows three connected components. The original graph is disconnected.

EXAMPLE 2　Are the two graphs in Figure 6 isomorphic? Justify your answer.

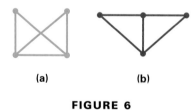

(a)　　　(b)

FIGURE 6

These graphs are isomorphic. To show this, we label corresponding vertices with the same letters and color-code the matching edges in graphs (a) and (b) of Figure 7.

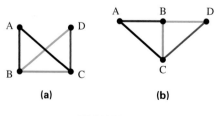

(a)　　　(b)

FIGURE 7

James Joseph Sylvester (1814–1897) was the first mathematician to use the word *graph* for these diagrams in the context of mathematics (in an article in *Nature* in 1878). Sylvester spent most of his life in Britain, but had an important influence on American mathematics during the years 1876 to 1884 as a professor at the newly founded Johns Hopkins University in Baltimore, Maryland. Sylvester was a flamboyant character. As a young man he worked as an actuary and lawyer in London, tutoring math on the side; one of his students was Florence Nightingale. His only published book, *The Laws of Verse*, was about poetry.

EXAMPLE 3   Are the two graphs in Figure 8 isomorphic?

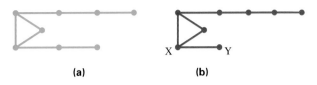

FIGURE 8

These graphs are not isomorphic. If we imagine graph (a) drawn with our special computer software, we can see that no matter how we drag the vertices we can not get this graph to look exactly like graph (b). For example, graph (b) has vertex X of degree 3 joined to vertex Y of degree one; neither of the vertices of degree 3 in graph (a) is joined to a vertex of degree one.

EXAMPLE 4   For each graph in Figure 9, determine the number of edges and the sum of the degrees of the vertices. (Note that graph (b) is not a simple graph.)

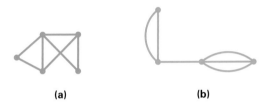

FIGURE 9

We are asked to find the *sum* of the degrees of each vertex. We start by writing the degree of each vertex next to the vertex, as in Figure 10.

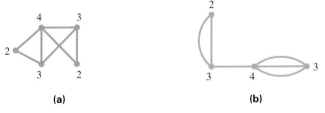

FIGURE 10

For graph (a):

Number of edges = 7

Sum of degrees of vertices
 = 2 + 3 + 2 + 3 + 4 = 14.

For graph (b):

Number of edges = 6

Sum of degrees of vertices
 = 2 + 3 + 4 + 3 = 12.

For each graph in Example 4, the sum of the degrees of the vertices is twice the number of edges. This is true for all graphs. Check that it is true for graphs in previous examples.

Sum of the Degrees Theorem

In any graph, the sum of the degrees of the vertices equals twice the number of edges.

Let us try the following thought experiment. To understand why this theorem is true, think of any graph. Imagine cutting each edge at its midpoint. (Don't add any vertices to your graph, however.) Now you have precisely twice as many half edges as you had edges in your original graph. For each vertex, count the number of half edges joined to the vertex; this is, of course, the degree of the vertex. Add these answers to determine the sum of the degrees of the vertices. You know that the number of half edges is twice the number of edges in your original graph. So, the sum of the degrees of the vertices in the graph is twice the number of edges in the graph.

EXAMPLE 5 A graph has precisely six vertices, each of degree 3. How many edges does this graph have?

The sum of the degrees of the vertices of the edges is $3 + 3 + 3 + 3 + 3 + 3 = 18$. By the theorem above, this number is twice the number of edges, so the number of edges in the graph is 18/2, or 9. ∎

Let us look at another situation in which a graph could be used to analyze and plan.

EXAMPLE 6 Suppose you decide to explore the upper Midwest next summer. You plan to travel by Greyhound bus and would like to visit Grand Forks, Fargo, Bemidji, St. Cloud, Duluth, Minneapolis, Escanaba, and Green Bay. A check of the Greyhound website shows direct bus links between destinations as indicated.

City	Direct bus links with:
Grand Forks	Bemidji, Fargo
Fargo	Grand Forks, St. Cloud, Minneapolis
Bemidji	Grand Forks, St. Cloud
St. Cloud	Bemidji, Fargo, Minneapolis
Duluth	Escanaba, Minneapolis
Minneapolis	Fargo, St. Cloud, Duluth, Green Bay
Escanaba	Duluth, Green Bay
Green Bay	Escanaba, Minneapolis

Draw a graph with vertices representing the destinations and edges representing the relation "there is a direct bus link." Is this a connected graph? Explain the significance of the graph.

Begin with eight vertices representing the eight destinations and label each vertex with the first letter of the corresponding destination, except use Y for Green Bay. Then use the bus link information to fill in the edges of the graph, as shown in Figure 11.

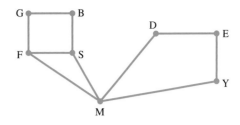

FIGURE 11

Figure 11 is a connected graph. It shows that you can travel by bus from any of your chosen destinations to any other.

Walks, Paths, and Circuits We can use Example 6 to introduce a number of important ideas. For example, what trips could you take among the eight destinations, using only direct bus links? Suppose you do not mind riding the same bus route more than once. You could then take this route: M → S → F → S → B. (Trace this with your finger on the graph in Figure 11 to check the route.) Note that you used the St. Cloud to Fargo link twice. You could not take the M → D → S route, however, since there is no direct bus link from Duluth to St. Cloud. Each possible route on the graph is called a *walk*.

Maps of subway systems in large cities usually do not faithfully represent the positions of stations, nor the distances between them. They are essentially graphs, with vertices representing stations, and edges showing subway links.

Walk

A **walk** in a graph is a sequence of vertices, each linked to the next vertex by a specified edge of the graph.

We can think of a walk as a route we can trace with a pencil without lifting the pencil from the graph. M → S → F → S → B and F → M → D → E are walks; M → D → S is not a walk, however, since there is no edge between D and S.

Suppose that the travel time for your trip is restricted, and you do not want to use any bus link more than once. A *path* is a special kind of walk.

Path

A **path** in a graph is a walk that uses no edge more than once.*

M → S → F → M, F → M → D → E, and M → S → F → M → D are paths. Trace these paths with your finger on the graph in Figure 11. Note that a path may use a *vertex* more than once. B → S → F → M → S → B is not a path, because it uses the edge BS more than once. M → D → S is not a path, as it is not even a walk.

---

*Some books call this a *trail*, and use the term *path* for a walk that visits no vertex more than once.

This diagram shows the relationships among walks, paths, and circuits.

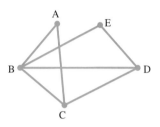

FIGURE 12

Mazes have fascinated people since earliest times. Mazes are found on coins from ancient Knossos (Greece), in the sand drawings at Nazca in Peru, on Roman pavements, on the floors of renaissance cathedrals in Europe, and, of course, as hedge mazes in gardens. The hero of a recent novel (*Larry's Party,* by Carol Shields) has a passion for garden mazes. Graphs can be used to clarify the structure of mazes. (See the exercises for this section.)

Perhaps you want to begin and end your tour in the same city (and still not use any bus route more than once). A path such as this is known as a *circuit.* Two circuits you could take are: $M \to S \to F \to M$ and $M \to S \to F \to M \to D \to E \to Y \to M$.

Circuit

A **circuit** in a graph is a path that begins and ends at the same vertex.

Notice that a circuit is a kind of path, and therefore is also a kind of walk.

EXAMPLE 7 Using the graph in Figure 12, classify each of the following sequences as a walk, a path, or a circuit.

(a) $E \to C \to D \to E$ (b) $A \to C \to D \to E \to B \to A$
(c) $B \to D \to E \to B \to C$ (d) $A \to B \to C \to D \to B \to A$

	Walk	Path	Circuit
(a)	No (no edge E to C)	No*	No*
(b)	Yes	Yes	Yes
(c)	Yes	Yes	No
(d)	Yes	No (edge AB is used twice)	No**

* If a sequence of vertices is not a walk, it cannot possibly be either a path or a circuit.
** If a sequence of vertices is not a path, it cannot possibly be a circuit, since a circuit is defined as a special kind of path.

When planning your trip around the Upper Midwest, it would be useful to know the length of time the bus ride takes. We might write the time for each bus ride on the graph, as shown in Figure 13.

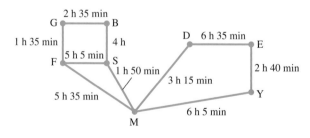

FIGURE 13

A graph with numbers on the edges as in Figure 13 is known as a **weighted graph.** The numbers on the edges are called **weights.** Weighted graphs will be important in later sections of this chapter.

15.1 Basic Concepts 859

Problem Solving

Drawing a sketch is a useful problem-solving strategy even when the original problem does not specifically refer to a geometric shape. Example 8 illustrates how drawing a sketch can simplify a problem that at first seems complicated.

FIGURE 14

EXAMPLE 8 In a round-robin tournament, every contestant plays every other contestant. The winner is the contestant who wins the most games. Suppose six tennis players compete in a round-robin tournament. How many matches will be played in the tournament?

To help solve the problem, draw a graph with six vertices to represent the six players. Then, as in Figure 14, draw an edge for each match between a pair of contestants. We can count the edges to find out how many matches must be played. Or, we can reason as follows: There are 6 vertices each of degree 5, so the sum of the degrees of the vertices in this graph is $6 \cdot 5$, which is 30. We know that the sum of the degrees is twice the number of edges. Therefore, there are 15 edges in this graph and, thus, 15 matches in the tournament.

Graphs such as that in Figure 14 are called *complete graphs*.

Complete Graph

A **complete graph** is a graph in which there is exactly one edge going from each vertex to each other vertex in the graph.

The human brain contains billions of neurons which communicate via chemical messages sent across synapses (small gaps between fibers from the neurons). Each neuron has over 1000 synapses. The image shows brain tissue with neurons. Exactly how the network of neurons produces complex thought remains a mystery. Researchers try to model thinking by using graphs. The vertices represent neurons and the edges represent synapses. Some believe that the correct graph to model a concept such as "bird" is a complete graph, with connections between all neurons for associated ideas (such as "feathers," "fly," "nest," and so on).

EXAMPLE 9 Decide whether each of the graphs in Figure 15 is complete. If a graph is not complete, explain why it is not.

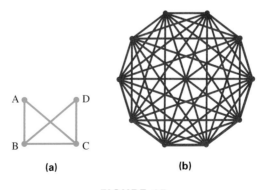

FIGURE 15

Graph (a) is not complete, since there is no edge from vertex A to vertex D. Graph (b) is complete. (Check that there is an edge from each of the ten vertices to each of the other nine vertices in the graph.)

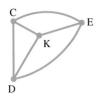

FIGURE 16

EXAMPLE 10 Find a complete graph with four vertices in the preschoolers' friendship pattern depicted in Figure 3.

Figure 16 shows a portion of the friendship graph from Figure 3. This is a complete graph with four vertices. ∎

Do you see some complete graphs with three vertices in Figure 16? In general, a graph consisting of some of the vertices of the original graph and some of the original edges between those vertices is called a **subgraph.** Vertices or edges not included in the original graph cannot be in the subgraph. (Notice the similarity to the idea of subset, discussed earlier in the text.) A subgraph may include anywhere from one to all the vertices of the original graph and anywhere from none to all the edges of the original graph.

The subgraph shown in Figure 16 is a complete graph, but a subgraph does not have to be complete. For example, graphs (a) and (b) in Figure 17 are both subgraphs of the graph in Figure 3.

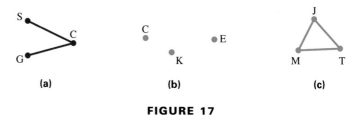

FIGURE 17

Graph (b) in Figure 17 is a subgraph even though it has no edges. (However, we could not form a subgraph by taking edges without vertices. In a graph every edge goes from a vertex to a vertex.)

Graph (c) in Figure 17 is *not* a subgraph of the graph in Figure 3, since there was no edge between M and T in the original graph.

15.1 EXERCISES

For Exercises 1–6, determine how many vertices and how many edges each graph has.

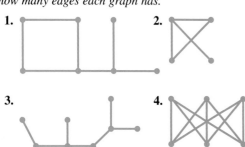

7.–10. For Exercises 7–10, refer to the graphs shown in Exercises 1–4. For each of these graphs, find the degree of each vertex in the graph. Then add the degrees to get the sum of the degrees of the vertices of the graph. What relationship do you notice between the sum of degrees and the number of edges?

For Exercises 11–16, determine whether the two graphs are isomorphic. If so, label corresponding vertices of the two graphs with the same letters and color-code corresponding edges, as in Example 2. (Note that there is more than one correct answer for many of these exercises.)

11.

(a)　　(b)

12.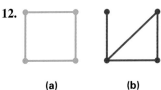

(a)　　(b)

13.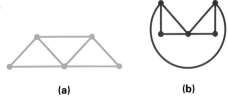

(a)　　(b)

14.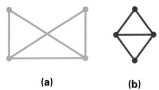

(a)　　(b)

15.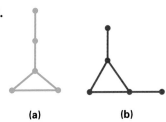

(a)　　(b)

16.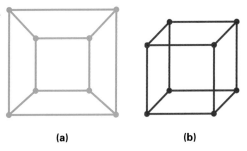

(a)　　(b)

For Exercises 17–22, determine whether the graph is connected or disconnected. Then determine how many components the graph has.

17.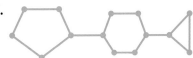

18.

19.

20.

21.

22.

For Exercises 23–26, use the theorem that relates the sum of degrees to the number of edges to determine the number of edges in the graph (without drawing the graph).

23. A graph with 4 vertices, each of degree 3

24. A graph with 8 vertices, each of degree 4

25. A graph with 5 vertices, three of degree 1, one of degree 2, and one of degree 3

26. A graph with 8 vertices, two of degree 1, three of degree 2, one of degree 3, one of degree 5, and one of degree 6

Exercises 27–29 refer to the following graph.

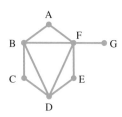

27. Which of the following are walks in the graph? If not, why not?
    (a) A → B → C
    (b) B → A → D
    (c) E → F → A → E
    (d) B → D → F → B → D
    (e) D → E
    (f) C → B → C → B

28. Which of the following are paths in the graph? If not, why not?
    (a) B → D → E → F
    (b) D → F → B → D
    (c) B → D → F → B → D
    (d) D → E → F → G → F → D
    (e) B → C → D → B → A
    (f) A → B → E → F → A

29. Which of the following are circuits in the graph? If not, why not?
    (a) A → B → C → D → E → F
    (b) A → B → D → E → F → A
    (c) C → F → E → D → C
    (d) G → F → D → E → F
    (e) F → D → F → E → D → F

Exercises 30 and 31 refer to the following graph.

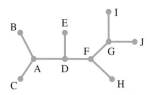

30. Which of the following are walks in the graph? If not, why not?
    (a) F → G → J → H → F
    (b) D → F
    (c) B → A → D → F → H
    (d) B → A → D → E → D → F → H
    (e) I → G → J
    (f) I → G → J → I

31. Which of the following are paths in the graph? If not, why not?
    (a) A → B → C
    (b) J → G → I → G → F
    (c) D → E → I → G → F
    (d) C → A
    (e) C → A → D → E
    (f) C → A → D → E → D → A → B

Exercises 32–37 refer to the following graph. In each case, determine whether the sequence of vertices is (i) *a walk,* (ii) *a path,* (iii) *a circuit in the graph.*

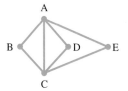

32. A → B → C → D → E
33. A → B → C
34. A → B → C → D → A
35. A → B → A → C → D → A
36. A → B → C → A → D → C → E → A
37. C → A → B → C → D → A → E

For each of Exercises 38–43, determine whether the graph is a complete graph. If not, explain why it is not complete.

38.

39.

40.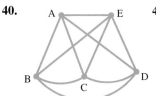

41.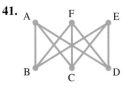

42.

43.

Solve each problem.

44. Students from two schools compete in chess. Each school has a team of four students. Each student must play one game against each student on the opposing team. Draw a graph with vertices representing the students, and edges representing the chess games. How many games must be played in the competition?

45. A chess master plays six simultaneous games with six other players. Draw a graph with vertices representing the players and edges representing the chess games. How many games are being played?

46. At a party there were four males—Tom, Joe, Chris, and Sam—and five females—Jennifer, Virginia, Mitzi, Karen, and Brenda. During the party each male danced with each female (and no female pair or male pair danced together). Draw a graph with vertices representing the people at the party and edges showing the relationship "danced with." How many edges are there in the graph?

47. There are six members on a hockey team (including the goalie). At the end of a hockey game, each member of the team shakes hands with each member of the opposing team. How many handshakes occur?

Use the theorem that the sum of the degrees of the vertices in a graph is twice the number of edges in Exercises 48 and 49.

48. There are seven people at a business meeting. One of these shakes hands with four people, four shake hands with two people, and two shake hands with three people. How many handshakes occur?

49. A lawyer is preparing his argument in a libel case. He has evidence that a libelous rumor about his client was discussed in various telephone conversations among eight people. Two of the people involved had four telephone conversations in which the rumor was discussed, one person had three, four had two, and one had one such telephone conversation. How many telephone conversations were there in which the rumor was discussed among the eight people?

50. Draw a graph with vertices representing the vertices (the corners) of a cube and edges representing the edges of the cube. In your graph, find a circuit that visits four different vertices. What figure does your circuit form on the actual cube?

51. Draw a graph with vertices representing the vertices (or corners) of a tetrahedron and edges representing the edges of the tetrahedron. In your graph, identify a circuit that visits three different vertices. What figure does your circuit form on the actual tetrahedron?

52. Mary, Erin, Sue, Jane, Katy, and Brenda are friends at college. Mary, Erin, Sue, and Jane are in the same math class. Sue, Jane, and Katy take the same English composition class.
 (a) Draw a graph with vertices representing the six students and edges representing the relation "take a common class."
 (b) Is the graph connected or disconnected? How many components does the graph have?
 (c) In your graph, identify a subgraph that is a complete graph with four vertices.
 (d) In your graph, identify three different subgraphs that are complete graphs with three vertices. (There are several correct answers.)

53. Here is another theorem about graphs: *In any graph, the number of vertices with odd degree must be even.* Explain why this theorem is true. (*Hint:* Use the theorem about the relationship between the number of edges and the sum of degrees.)

54. (a) Draw two non-isomorphic (simple) graphs with 6 vertices, with each vertex having degree 3.
 (b) Explain why your graphs are not isomorphic.

55. Draw a graph whose vertices represent the *faces* of a cube and in which an edge between two vertices shows that the corresponding faces of the actual cube share a common boundary. What is the degree of each vertex in your graph? What does the degree of any vertex in your graph tell you about the actual cube?

864 CHAPTER 15 Graph Theory

56. Graphs can be used to clarify the structure of mazes. Vertices represent entrances to the maze and points in the maze where there is either a dead end or a choice of two or more directions to proceed. Edges show how these points are connected. For example, here is the 1690 design for the hedge maze at Hampton Court in England and a graph that clarifies its structure.

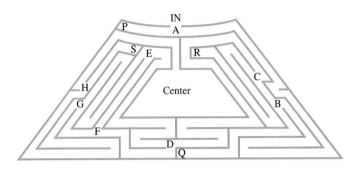

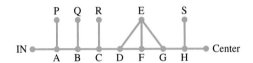

Draw a graph to clarify the underlying structure of each of the following mazes. Points that are entrances or dead ends have been labeled, but you will need to label points where there is a choice of two or more directions to proceed.

(a)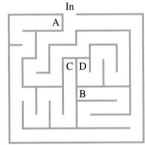

(c)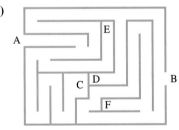

(b)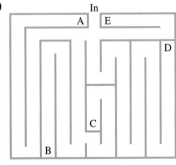

(d)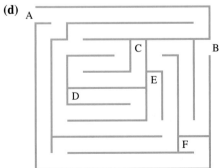

57. Write a paper on mazes. Use graphs to analyze the structures of the mazes you discuss. You may want to focus on a particular kind of maze, such as mazes in gardens, mazes in art and architecture, or mazes in ancient cultures.

58. Graphs may be used to clarify the rhyme schemes in poetry. Vertices represent words at the end of a line, edges are drawn, and a path is used to indicate the rhyme scheme. For example, here is the first stanza of *Ode to Autumn* by John Keats, with the rhyme scheme analyzed using a graph:

Season of mists and mellow fruitfulness,	A
Close bosom-friend of the maturing sun;	B
Conspiring with him how to load and bless	A
With fruit the vines that round the thatch-eves run;	B
To bend with apples the moss'd cottage-trees	C
And fill all fruit with ripeness to the core;	D
To swell the gourd, and plump the hazel shells	E
With a sweet kernel; to set budding more,	D
And still more, later flowers for the bees,	C
Until they think that warm days will never cease,	C
For Summer has o'er brimm'd their clammy cells.	E

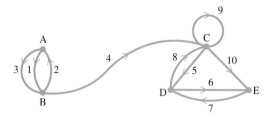

Write a paper on rhyme schemes in poetry, using graphs to illustrate. Focus either on a particular poet or on a particular type of poem.

59. Our first example in this section was a simple example of social network analysis. Research and write a paper on the use of graphs in social network analysis.

15.2 Euler Circuits

In this section we study *Euler circuits* in graphs. The ideas developed are important for planning efficient routes for snowplows and mail delivery, and for solving many other practical applications.

Königsberg Bridge Problem In the early 1700s, the city of Königsberg was the capital of East Prussia. (Königsberg is now called Kaliningrad and is in Russia.) The river Pregel ran through the city in two branches with an island between the branches. Figure 18 shows an engraving of the city as it looked in the early 1700s.

A university was founded in Königsberg in 1544. It was established as a Lutheran center of learning. Its most famous professor was the philosopher Immanual Kant, who was born in Königsberg in 1724. The university was completely destroyed during World War II.

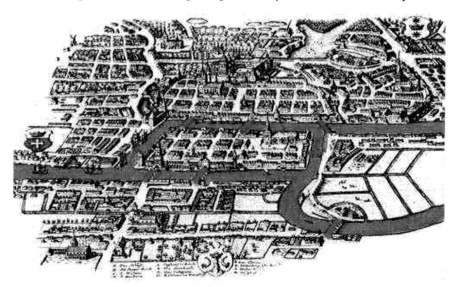

FIGURE 18

There were seven bridges joining various parts of the city. Figure 19 shows a map of the river and bridges in the city.

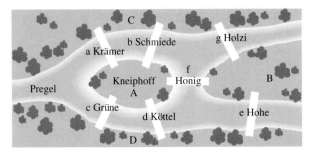

FIGURE 19

Leonhard Euler (1707–1783) was a devoted father and grandfather. He had thirteen children and frequently worked on mathematics with children playing around him. He lost the sight in his right eye at age 31, a couple of years after writing the Königsberg bridge paper. He became completely blind at age 58, but produced more than half his mathematical work after that. Euler was a prolific mathematician. His collected works consist of about a thousand books and papers.

According to Leonhard Euler (pronounced "oiler"), the following problem was well known in his time (around 1730): Is it possible for a citizen of Königsberg to take a stroll through the city, crossing each bridge exactly once, and beginning and ending at the same place?

See if you can find such a route on the map in Figure 19. (You cannot cross the river anywhere except at the bridges shown on the map.)

We can simplify the problem by drawing the map as a graph. See Figure 20. The vertices of the graph represent the land masses, and the edges of the graph represent the bridges between them. The graph focuses on the relations between the vertices; here the relation is: "there is a bridge." In this graph we have *two* edges between vertices A and C and also between vertices A and D; thus this graph is not a simple graph. We have labeled the edges with lowercase letters. Can you find a route as required in the puzzle?

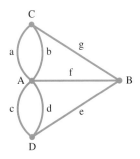

FIGURE 20

Recall that a *path* is a walk that uses no edge of the graph more than once. For example,

$$C \xrightarrow{a} A \xrightarrow{f} B \xrightarrow{g} C \xrightarrow{b} A$$

is a path in the graph in Figure 20. (We have traveled between C and A more than once, but have used different edges in doing so.) The Königsberg bridge problem demands more than simply a path: the path must use every edge *exactly* once, no more, no less. Recall that a *circuit* is a path that begins and ends at the same vertex. In the Königsberg bridge problem, we are being asked whether the graph has an *Euler circuit*.

15.2 Euler Circuits

Euler Path; Euler Circuit

An **Euler path** in a graph is a path that uses every edge of the graph exactly once. An **Euler circuit** in a graph is a circuit that uses every edge of the graph exactly once.

EXAMPLE 1

(a) Is A → B → C → D → E → F → A an Euler circuit for the graph in Figure 21? Justify your answer.

A → B → C → D → E → F → A is a circuit, but not an Euler circuit, since it does not use every edge of the graph. For example, the edge BD is not used in this circuit.

(b) Does the graph in Figure 21 have an Euler circuit?

We claim that the circuit A → B → C → D → E → F → D → B → F → A is an Euler circuit. Trace this path on the graph to check that it does use every edge of the graph exactly once. So, this graph does have an Euler circuit.

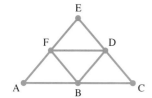

FIGURE 21

It seemed that no route of the type required in the Königsberg bridge problem could be found, but why? Leonhard Euler published a paper in 1736 that explained why the kind of walk described in the problem is impossible and provided a simple way for deciding whether any given graph has an Euler circuit.

While Euler did not use the terms *graph* and *circuit,* his paper was, in fact, the first paper on graph theory. In modern terms, Euler proved the first part of the theorem below. We call this Euler's theorem, although he proved only the first part; part 2 was not proved until 1873.

Euler's Theorem

Suppose we have a connected graph.

1. If the graph has an Euler circuit, then each vertex of the graph has even degree.
2. If each vertex of the graph has even degree, then the graph has an Euler circuit.

What does Euler's theorem predict about the graph in Figure 21? The degrees of the vertices are as follows:

A: 2   B: 4   C: 2   D: 4   E: 2   F: 4.

Each vertex has even degree, and it follows from part 2 of the theorem that the graph has an Euler circuit. (We already knew this, since we found an Euler circuit for this graph.)

What does Euler's theorem suggest about the Königsberg bridge problem? Note that the degree of vertex C in Figure 20 is 3, which is an odd number, so, by part 1

The following is part of Euler's introduction to his paper on the Königsberg bridge problem: "I was told that while some denied the possibility of doing this and others were in doubt, no one maintained that it was actually possible. On the basis of the above, I formulated the following very general problem for myself: Given any configuration of the river and the branches into which it may divide, as well as any number of bridges, to determine whether or not it is possible to cross each bridge exactly once."

(*Source:* Newman, James R. (ed.), "The Konigsberg Bridges," *Scientific American,* July 1953, in *Readings from Scientific American: Mathematics in the Modern World,* 1968.)

of the theorem, the graph cannot have an Euler circuit. (In fact, the graph has many vertices with odd degree, but the presence of *any* vertex of odd degree indicates that the graph does not have an Euler circuit.)

Why is Euler's theorem true? We will begin by considering the first part of the theorem:

If a connected graph has an Euler circuit, then each vertex has even degree.

Suppose we have a connected graph, and an Euler circuit in the graph. Suppose the circuit begins and ends at a vertex A. Imagine traveling along the Euler circuit, from A all the way back to A, putting an arrow on each edge in your direction of travel. (Of course, you travel along each edge exactly once.) Now think about any vertex B in the graph. It has edges joined to it, each with an arrow; some of the arrows point toward B, some point away from B. The number of arrows pointing toward B is the same as the number of arrows pointing away from B, for every time you visit B, you come in on one unused edge and leave via a different unused edge. In other words, you can pair off the edges joined to B, with each pair having one edge with arrow pointing toward B and one edge with arrow pointing away from B. Thus, the total number of edges joined to B must be an even number and, therefore, the degree of B is even. Check that a similar argument shows that our starting vertex, A, also has even degree.

The second part of the Euler theorem states the following:

If each vertex of a connected graph has even degree,
then the graph has an Euler circuit.

To convince you that this is true, we will provide a recipe for actually finding an Euler circuit in any such graph. (See Fleury's algorithm that follows.) First, however, we consider more examples of how Euler's theorem can be applied.

> By 1875 a new bridge had been built in Königsberg, joining the land areas we labeled C and D. This meant that it was now possible for a citizen of Königsberg to take a walk that used each bridge exactly once, provided he or she started at A and ended at B, or vice versa.

EXAMPLE 2 Decide which of the graphs in Figure 22 has an Euler circuit. Justify your answers.

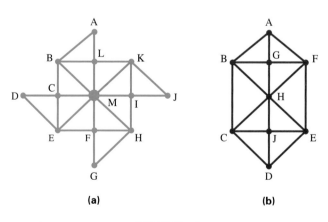

FIGURE 22

Graph (a): The graph is connected. If we write the degree next to each vertex on the graph, as in Figure 23 on the next page, we see that each vertex has even degree. Since the graph is connected, and each vertex has even degree, it follows from Euler's theorem that the graph has an Euler circuit.

15.2 Euler Circuits 869

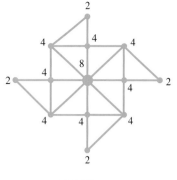

FIGURE 23

The word **algorithm** comes from the name of the Persian mathematician, **Abu Ja'far Muhammad ibn Musa Al-Khowârizmi**, who lived from about A.D. 780 to A.D. 850. One of the books he wrote was on the Hindu-Arabic system of numerals (the system we use today), and calculations using these. The Latin translation of this book was called *Algoritmi de numero Indorum*, meaning "Al-Khowârizmî on the Hindu Art of Reckoning."

Graph (b): Check the degrees of the vertices. Note that vertex A has degree three, an odd degree. We need look no further. It follows immediately from Euler's theorem that graph (b) in Figure 22 does not have an Euler circuit. (All the other vertices in this graph except for D have even degree; however, if just one vertex has odd degree, then the graph does not have an Euler circuit.) ■

Problem Solving

When solving a challenging problem, it is useful to ask yourself whether the problem is related to another problem that you do know how to solve. Example 3 illustrates this strategy.

(a)

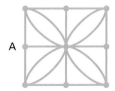

(b)

FIGURE 24

EXAMPLE 3 Beginning and ending at the same place, is it possible to trace the pattern shown in Figure 24(a) without lifting the pencil off the page and without tracing over any line or curve in the pattern twice?

The pattern in (a) is not a graph. (It has no vertices.) But what we are being asked to do is very much like being asked to find an Euler circuit. So we imagine that there are vertices at points where lines or curves of the pattern meet, obtaining the graph in Figure 24(b).

Now we can use Euler's theorem to decide whether we can trace the pattern as required. We check the degrees of the vertices and notice that vertex A has odd degree. Thus, by Euler's theorem, this graph does not have an Euler circuit. It follows that we cannot trace the original pattern in the manner required. ■

The tracing problem in Example 3 illustrates a practical problem that occurs when designing robotic arms to trace patterns. Automated machines that engrave identification tags for pets require such a mechanical arm.

Fleury's Algorithm *Fleury's algorithm* can be used to find an Euler circuit in any connected graph in which each vertex has even degree. An algorithm is like a recipe; follow the specified steps and you achieve what you need. Before introducing Fleury's algorithm, we need a definition.

Cut Edge

A **cut edge** in a graph is an edge whose removal disconnects a component of the graph.

We call such an edge a cut edge, since removing the edge *cuts* a connected piece of a graph into two pieces.*

EXAMPLE 4 Identify the cut edges in the graph in Figure 25.

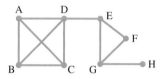

FIGURE 25

This graph has only one component, so we must look for edges whose removal would disconnect the graph.

DE is a cut edge; if we removed this edge, we would disconnect the graph into two components, obtaining graph (a) of Figure 26.

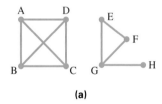 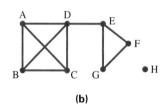 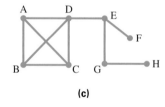

FIGURE 26

Also, HG is a cut edge; if we removed this edge, we would disconnect the graph into two components (one of which would be the single vertex H), obtaining graph (b) of Figure 26.

None of the other edges is a cut edge; we could remove any *one* of the other edges and still have a connected graph (that is, we would still have a path from each vertex of the graph to each other vertex). For example, if we remove edge GF, we end up with a graph that still is connected, as shown in Figure 26(c).

Now we are ready to describe Fleury's algorithm. Recall that this algorithm is for finding an Euler circuit in a connected graph in which each vertex has even degree.

---

*Some mathematicians use "bridge" instead of "cut edge" to describe such an edge.

Fleury's Algorithm

1. Start at any vertex. Go along any edge from this vertex to another vertex. *Remove this edge from the graph.*
2. You are now on a vertex of the revised graph. Choose any edge from this vertex, subject to only one condition: do not use a cut edge (*of the revised graph*) unless you have no other option. Go along your chosen edge. *Remove this edge from the graph.*
3. Repeat Step 2 until you have used all the edges and gotten back to the vertex at which you started.

EXAMPLE 5 Find an Euler circuit for the graph in Figure 27.

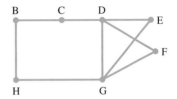

FIGURE 27

Source: http://turning.mi.
sanu.ac.yu/~jablans/tch.htm.

The Bushoong and Tshokwe peoples of Angola, Zambia, and Zaire have traditions of tracing continuous patterns in sand. The Bushoong were traditionally decorators, trading embroidered and woven objects; their children are experts at the continuous sand tracings. Amongst the Tshokwe the tracings are called *sona* and have ritual significance. Only adult men make the drawings, which are associated with stories about the gods and ancestors. The men trace the drawings as they tell the stories.

First we should check that this graph does indeed have an Euler circuit. Check that the graph is connected, and that each vertex of the graph has even degree.

We can start at any vertex; let's start at C. We could go from C to B or from C to D. Let's go to D, removing the edge CD. (We simply show it scratched out, but you must think of this edge as gone!) Our path begins: C → D. We show the revised graph in Figure 28(a).

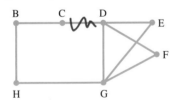 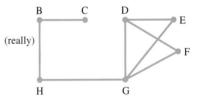

FIGURE 28(a)

Note that none of the edges DE, DF, or DG are cut edges for our current graph, so we can choose to go along any one of these. Let's go from D to G, removing edge DG. So far, our path is C → D → G. (Be sure to keep a record of your path as you go.) Figure 28(b) shows the revised graph.

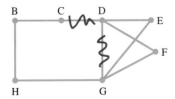 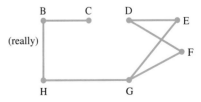

FIGURE 28(b)

The remaining edges at G are GE, GF, and GH. Note, however, that GH is a cut edge for our revised graph, and we have other options. Thus, according to Fleury's algorithm, *GH should not be used at this stage.* (Now it is apparent why this rule is in the algorithm—if you used GH now, you would never be able to get back to use the edges GF, GE, and so on.) Let's go from G to F. Our path is C → D → G → F, and we show our revised graph in Figure 28(c).

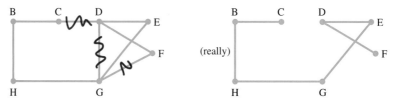

FIGURE 28(c)

Now we have only one edge to leave F, namely FD. Notice that this is a cut edge of the revised graph, but *since we have no other option* we may use this edge. Our path is C → D → G → F → D, and our revised graph is shown in Figure 28(d).

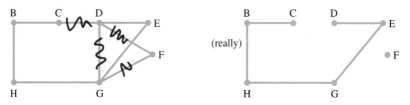

FIGURE 28(d)

We have only one edge left at D, namely DE. It is a cut edge of the remaining graph, but since we have no other option, we may use it.

It should now be clear how to complete the Euler circuit. (Be sure to return to the starting vertex.) Here is the complete Euler circuit: C → D → G → F → D → E → G → H → B → C. ∎

If we start with a connected graph in which all vertices have even degree, then Fleury's algorithm will always produce an Euler circuit for the graph. Note that a graph that has an Euler circuit always has more than one Euler circuit.

The only way to gain confidence in using an algorithm is through practice. The exercises at the end of this section include many opportunities for practicing Fleury's algorithm.

Whenever we have an algorithm for performing some task, we can program a computer to do it for us. Even for a large graph, a computer can quickly apply Fleury's algorithm. However, this is not the case for all algorithms. In the next section, for example, we shall encounter an algorithm that, when used for large graphs, requires far too much time for even the fastest computers.

As stated at the beginning of this section, Euler circuits have many practical applications. Let us consider planning a route for rural mail delivery. Figure 29 on the next page shows a map of a rural district. In rural areas, mailboxes usually are placed along the same side of a road, so the mail delivery vehicle does not have to backtrack, but can travel along the road in one direction. Suppose the roads off the main road in Figure 29 show the mail delivery region for a particular mail van.

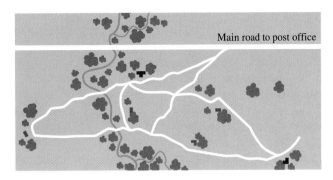

FIGURE 29

For maximum efficiency, the mail carrier should minimize the distance he or she travels. It would be best if the mail carrier could find an Euler circuit covering his or her mail delivery area. If such a circuit existed, it would certainly give the shortest traveling distance. A glance at the map in Figure 29 shows that there is no Euler circuit for this delivery area. (Imagine vertices at the road intersections; some of these vertices would have odd degree.) To minimize travel distance, the van should travel along roads already covered as little as possible. For simplicity we will assume that the time it takes to travel along any edge of the graph when not delivering mail is roughly the same as the time required to travel any other edge while not delivering mail. In Figure 30 dotted edges indicate the roads that should be traveled twice; for such a simple delivery map it is easy to see that we have added as few edges as possible.

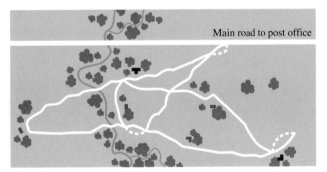

FIGURE 30

Now we can use Fleury's algorithm to find an Euler circuit on the graph in Figure 30, and this will give the most efficient route for mail delivery.

For a simple delivery area such as this, you would probably be able to choose an efficient route intuitively. For route planning in large cities, however, these techniques offer significant savings on mail delivery, mechanized street sweeping, snow-plowing, and electric meter checking. For example, in the 1980s a computerized system for planning street sweeping was implemented in an area of Washington D.C., resulting in a 20% savings. In large cities the street sweeping budget is tens of millions of dollars, so potential annual savings of 20% easily justify the initial expenditure needed to set up a computerized system for planning.

The study of Euler circuits began in the eighteenth century with a little mathematical puzzle, but the ideas developed provide methods for efficient planning in our complex modern cities.

There are additional complications when scheduling mechanical brooms for street sweeping in the real world:

1. The mechanical brooms have to travel along each curb, in the direction of traffic flow (and some streets are one-way).
2. It is time-consuming to switch the brooms from one side of the vehicle to the other.
3. Sweeping schedules have to be coordinated with city parking regulations, so that there are no parked cars where sweeping is to be done.

FOR FURTHER THOUGHT

When planning routes for mail delivery, street sweeping, or snowplowing, planners must accept that the roads in the region to be covered usually do not have an Euler circuit; it is unavoidable that certain stretches of road be traveled more than once. (This is referred to as *deadheading*.) Planners then try to find a route on which the driver will spend as little time as possible deadheading. Here we explore how this can be done.

For simplicity, we employ street grids for which the distance from corner to corner in any direction is the same. For example, suppose our street grid looks like this:

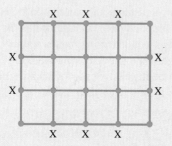

All vertices that have odd degree are marked with Xs. Now we must add edges *coinciding with existing roads* to change this into a graph that has an Euler circuit. If we add edges as shown in graph (a), we obtain a graph that has an Euler circuit. (Check that all vertices now have even degree.) However, this solution introduces 11 edges—more deadheading than is necessary.

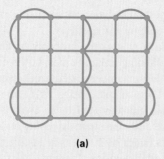

(a)

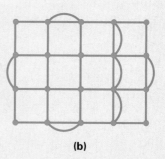

(b)

If we add edges as shown in graph (b), the addition of only 7 edges yields a graph with an Euler circuit. This is, in fact, the least number of edges that can be added (keeping to the street grid) to change the original street grid graph into a graph that has an Euler circuit. (There are different ways to add seven edges to the original graph to obtain a graph with an Euler circuit.)

A certain amount of trial and error is needed to find the least number of edges that must be added to ensure that the street grid graph has an Euler circuit.

For Group Discussion

1. In the graph below, we have added just six edges to our original street grid. Why is this not a better solution than the one with seven additional edges shown in (b) above?

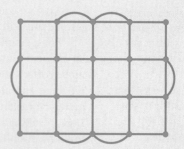

2. For each of the street grids (c), (d), and (e), add edges to obtain a street grid graph that has an Euler circuit. Try to add as few edges as possible in each case.

(c)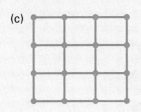

(d)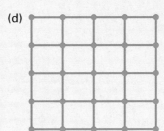

(e)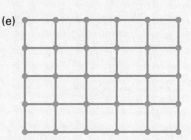

15.2 EXERCISES

For Exercises 1–3, a graph is shown and some sequences of vertices are specified. Determine which of these sequences show Euler circuits. If not, explain why not.

1. (a) A → B → C → D → A → B → C → D → A
   (b) C → B → A → D → C
   (c) A → C → D → B → A
   (d) A → B → C → D

2.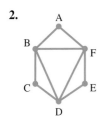

   (a) A → B → C → D → E → F → A
   (b) F → B → D → B
   (c) A → B → C → D → E → F → B → D → F → A
   (d) A → B → F → D → B → C → D → E → F → A

3.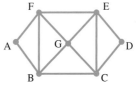

   (a) A → B → C → D → E → F → A
   (b) A → B → C → D → E → G → C → E → F → G → B → F → A
   (c) A → B → C → D → E → C → G → E → F → G → E → F → A
   (d) A → B → G → E → D → C → G → F → B → C → E → F → A

For Exercises 4–8, use Euler's theorem to decide whether the graph has an Euler circuit. (Do not actually find an Euler circuit.) Justify each answer briefly.

4.

5.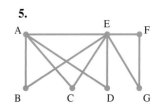

6.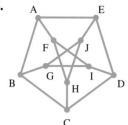

7.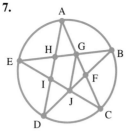

8.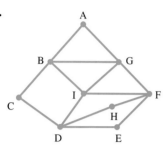

For Exercises 9 and 10, use Euler's theorem to determine whether it is possible to begin and end at the same place, trace the pattern without lifting your pencil, and trace over no line in the pattern more than once.

9.

10.

For Exercises 11–15, use Euler's theorem to determine whether the graph has an Euler circuit, justifying each answer. Then determine whether the graph has a circuit that visits each vertex exactly once, except that it returns to its starting vertex. If so, write down the circuit. (There is more than one correct answer for some of these.)

11.

12.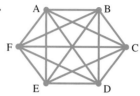

13.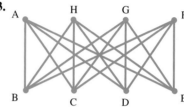

14.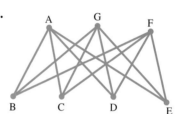

15.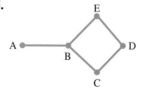

In Exercises 16–19 different floor tilings are shown. The material applied between tiles is called grout. For which of these floor tilings could the grout be applied beginning and ending at the same place, without going over any section twice, and without lifting the tool? Justify answers.

16.

17.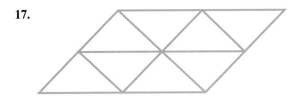

18.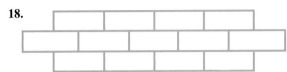

19.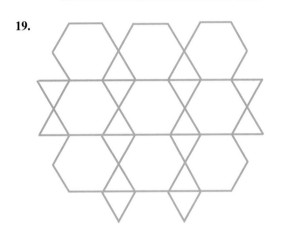

In Exercises 20–22, identify all cut edges in the graph. If there are none, say so.

20.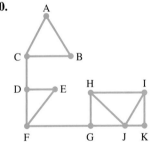

21.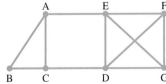

22.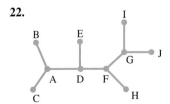

In Exercises 23–25 a graph is shown for which a student has been asked to find an Euler circuit starting at A. The student's revisions of the graph after the first few steps of Fleury's algorithm are shown, and in each case the student is now at B. For each graph, determine all edges that Fleury's algorithm permits the student to use for the next step.

23.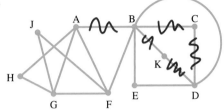

24.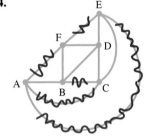

25.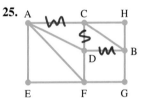

In Exercises 26–28, use Fleury's algorithm to find an Euler circuit for the graph, beginning and ending at A. There are many different correct answers.

26.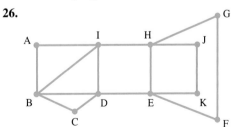

27.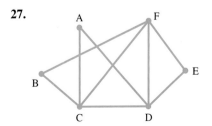

28.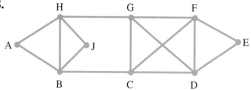

In Exercises 29–31, use Euler's theorem to determine whether the graph has an Euler circuit. If not, explain why not. If the graph does have an Euler circuit, use Fleury's algorithm to find an Euler circuit for the graph. There are many different correct answers.

29.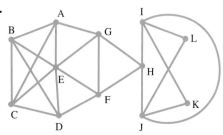

30.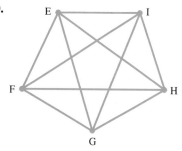

31.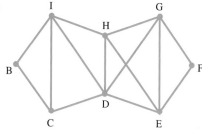

32. The graph in the next column shows the layout of the paths in a botanical garden. The edges represent the paths. Has the garden been designed in such a way that it is possible for a visitor to find a route that begins and ends at the entrance to the garden (represented by the vertex A) and that goes along each path exactly once? If so, use Fleury's algorithm to find such a route.

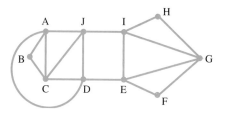

33. The map shows the roads on which parking is permitted at a national monument. This is a pay and display facility. A security guard has the task of periodically checking that all parked vehicles have a valid parking ticket displayed. He is based at the central complex, labeled A. Is there a route that he can take to walk along each of the roads exactly once, beginning and ending at A? If so, use Fleury's algorithm to find such a route.

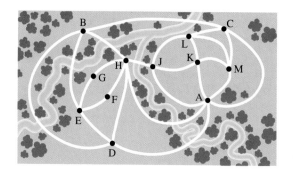

In Exercises 34–36, the floor plan of a building is shown. For which of these is it possible to start outside, walk through each door exactly once, and end up back outside? Justify each answer. (Hint: Think of the rooms and "outside" as the vertices of a graph, and the doors as the edges of the graph.)

34.

35.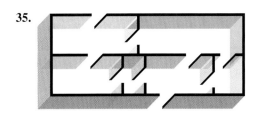

36.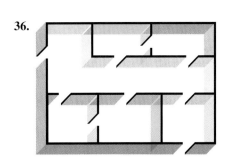

Exercises 37–44 are based on the following theorem:

1. If a graph has an Euler path that begins and ends at different vertices, then these two vertices are the only vertices with odd degree. (All the rest have even degree.)
2. If exactly two vertices in a connected graph have odd degree, then the graph has an Euler path beginning at one of these vertices and ending at the other.

For Exercises 37–40, determine whether the graph has an Euler path that begins and ends at different vertices. Justify your answer. If the graph has such a path, say at which vertices the path must begin and end.

37.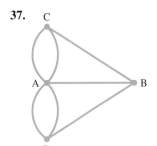

38.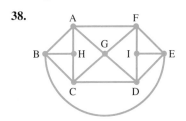

39.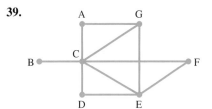

40.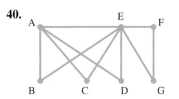

For Exercises 41–43 refer to the floor plan indicated, and determine whether it is possible to start in one of the rooms of the building, walk through each door exactly once, and end up in a different room from the one you started in. Justify your answer.

41. Refer to the floor plan shown in Exercise 34.
42. Refer to the floor plan shown in Exercise 35.
43. Refer to the floor plan shown in Exercise 36.
44. The accompanying schematic map shows a portion of the New York City area, including tunnels and bridges.

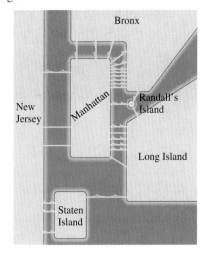

(a) Is it possible to take a drive around the New York City area using each tunnel and bridge exactly once, beginning and ending in the same place? Justify your answer.
(b) Is it possible to take a drive around the New York City area using each tunnel and bridge exactly once, beginning and ending in different places? If so, where must your drive begin and end? Justify your answer.

In Exercises 45–47 the graph does not have an Euler circuit. For each graph find a circuit that uses as many edges as possible. (There is more than one correct answer in each case.) How many edges did you use in the circuit?

45.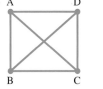

46.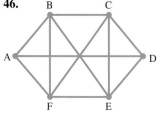

47.

48. There are just five regular polyhedra: the cube, the tetrahedron, the octahedron, the dodecahedron, and the icosahedron. (See Chapter 9 on Geometry.) For which of these can you trace your finger along each edge exactly once, beginning and ending at the same vertex? Justify your answer carefully, using the ideas developed in this section.

49. Which *complete graphs* have Euler circuits? Justify your answer carefully.

50. Write a paper on the sand tracings of the Bushoong or Tshokwe people. Include discussion of the role these drawings play in the culture.

51. Write a paper on continuous tracings in Danish folk culture.

52. Write a paper on the life of Leonhard Euler. (In addition to other sources, check the website http://www-groups.dcs.st-and.ac.uk/~history/.)

15.3 Hamilton Circuits

In this section we examine Hamilton circuits in graphs. The story of Hamilton circuits is a story of very large numbers and of unsolved problems in both mathematics and computer science.

Consider a game, called the Icosian game, invented by Irish mathematician William Hamilton in the mid-nineteenth century. It uses a wooden board marked with the graph shown in Figure 31.

Dodecahedron

Hamilton sold the **Icosian game** idea to a games dealer for 25 British pounds. It was put on the market in 1859, but it was not a huge success. A later version of the game, *A Voyage Around the World,* consisted of a regular dodecahedron with pegs at the vertices. The vertices had names such as Brussels, Delhi, and Zanzibar. The aim was to travel to all vertices along edges of the dodecahedron, visiting each vertex exactly once. We can show the vertices and edges of a dodecahedron precisely as in Figure 31.

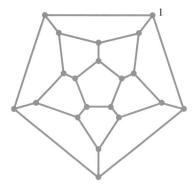

The Icosian Game

FIGURE 31

The game includes 20 pegs numbered 1 through 20. (The word *Icosian* comes from the Greek word for 20.) At each vertex of the graph there is a hole for a peg.

15.3 Hamilton Circuits   881

William Rowan Hamilton (1805–1865) spent most of his years in Dublin, Ireland. Much of his life was overshadowed by the fact that the great love of his life, Catherine Disney, was persuaded by her parents to marry a much wealthier man. He met the poet William Wordsworth while touring England and Scotland, and they became firm friends. Hamilton tried his hand at writing poems, but Wordsworth told him tactfully that his talents were in mathematics rather than poetry.

The simplest version of the game is as follows: put the pegs into the holes in order, following along the edges of the graph, in such a way that peg 20 ends up in a hole that is joined by an edge to the hole of peg 1. Try this now, numbering the vertices in the graph in Figure 31, starting at vertex 1.

The Icosian game asks you to find a circuit in the graph, but the circuit need not be an Euler circuit. The circuit must visit each *vertex* exactly once, except for returning to the starting vertex to complete the circuit. (The circuit may or may not travel all edges of the graph.) Circuits such as this are called *Hamilton circuits*.

Hamilton Circuit

A **Hamilton circuit** in a graph is a circuit that visits each vertex exactly once (returning to the starting vertex to complete the circuit).

We leave the Icosian game as a challenge for you, and continue our study of Hamilton circuits.

EXAMPLE 1 Which of the following are Hamilton circuits for the graph in Figure 32? Justify your answers briefly.

$$A \to B \to E \to D \to C \to F \to A$$
$$A \to B \to C \to D \to E \to F \to C \to E \to B \to F \to A$$
$$B \to C \to D \to E \to F \to B$$

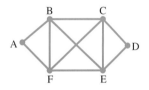

FIGURE 32

$A \to B \to E \to D \to C \to F \to A$ is a Hamilton circuit for the graph. It visits each vertex of the graph exactly once, except for returning to the starting vertex. (Trace this circuit on the graph to check it.)

$A \to B \to C \to D \to E \to F \to C \to E \to B \to F \to A$ is not a Hamilton circuit, since it visits vertex B (and vertices C, E, and F) more than once. (This is, however, an Euler circuit for the graph.)

$B \to C \to D \to E \to F \to B$ is not a Hamilton circuit since it does not visit all vertices in the graph.

Some graphs, such as the graph in Figure 33, do not have a Hamilton circuit. There is no way to visit all the vertices and return to the starting vertex without visiting vertex I more times than allowed.

It can be difficult to determine whether a particular graph has a Hamilton circuit, since there is no convenient theorem that gives necessary and sufficient conditions for a Hamilton circuit to exist. This remains an unsolved problem in Hamilton circuits. (Contrast the simple method provided by Euler's theorem for checking whether a graph has an Euler circuit.)

FIGURE 33

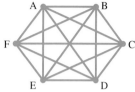

FIGURE 34

Fortunately, in many real-world applications of Hamilton circuits we are dealing with complete graphs, and any complete graph with three or more vertices does have a Hamilton circuit. (Recall our discussion of complete graphs in Section 15.1.) For example, A → B → C → D → E → F → A is a Hamilton circuit for the complete graph shown in Figure 34. We could form a Hamilton circuit for any complete graph with three or more vertices in a similar way.

Hamilton Circuits for Complete Graphs
Any complete graph with three or more vertices has a Hamilton circuit.

The graph in Figure 34 in fact has many Hamilton circuits. Try finding some that are different from the one given above.

How many Hamilton circuits does a complete graph have? Before we count, we need an agreement about when two lists of vertices will be considered to be *different* Hamilton circuits. For example, the Hamilton circuits B → C → D → E → F → A → B and A → B → C → D → E → F → A visit the vertices in essentially the same order in the graph in Figure 34, although the lists start with different vertices. If you mark the circuit on the graph, you can describe it starting at any vertex. For our purposes, it is convenient to consider the two lists above as representing the same Hamilton circuit.

When Hamilton Circuits Are the Same
Hamilton circuits that differ *only* in their starting points will be considered to be the same circuit.

The Tutte Graph The four-color theorem is the most famous theorem of graph theory. It says, roughly, that with just four different colors, we can color any map in such a way that no countries with a common border have the same color. (Cartographers seem to have known this for a long time.) Peter Tate gave a proof in 1880 based on the assumption that every connected graph that can be drawn with edges crossing only at vertices, and in which each vertex has degree 3, has a Hamilton circuit. But in 1946 Tutte found the graph shown above, and proved that it has *no Hamilton circuit*. Tate's assumption was wrong. The four-color theorem was not proved until 1976, and then only by a computer proof so long that no one can check it completely.

Problem Solving

Tree diagrams are a convenient strategy for counting when the task involved has three or more parts.

We begin by counting the number of Hamilton circuits in a complete graph with four vertices, as shown in Figure 35. We can use the same starting point for all the Hamilton circuits in the graph. We choose A as the starting point. From A we can go to any of the three remaining vertices (B, C, or D). No matter which vertex we choose, we then have two unvisited vertices from which to choose next. Then there is only one way to complete the Hamilton circuit. We illustrate this counting procedure in the tree diagram shown in Figure 36.

FIGURE 35

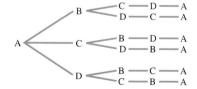

FIGURE 36

15.3 Hamilton Circuits

Large Numbers 70! is the first factorial larger than 10^{100}. Most calculators will calculate numbers up to only about 10^{99}, so if you try calculating 70! you probably will get an error message. (Try it.) 10^{100} is called one googol, a name invented by a nine-year-old child when asked to name a very, very large number.

These numbers are unimaginably big! Try these for size:

$$25! \approx 2 \times 10^{25}$$
$$50! \approx 3 \times 10^{64}$$
$$100! \approx 9 \times 10^{157}$$

Estimate of the number of *seconds* since the creation of the universe: a mere 10^{17}.

Estimate of the number of *inches* to the farthest object in the universe: a mere 10^{27}.

Approximate number of molecules we would have if we filled a sphere the size of the earth with water: just 10^{49}.

This tree diagram shows that we have $3 \cdot 2 \cdot 1$ Hamilton circuits in the graph. We write this as 3! (read "3 factorial"). Refer to Chapter 11 on counting methods for more on factorial notation.

Returning to Figure 34, we can count the number of Hamilton circuits in the complete graph with 6 vertices. Start at vertex A. From A we can proceed to any of the five remaining vertices. No matter which we choose, we then have four choices for the next vertex along the circuit, three choices for the next, and two for the next. (We could draw a counting tree for this, but it would be rather large.) This means that we have $5 \cdot 4 \cdot 3 \cdot 2 \cdot 1$ different Hamilton circuits in all. We write 5! for this number.

Likewise, we would find that a complete graph with 10 vertices has 9! Hamilton circuits. (A nice pattern is emerging here.) In general, for a complete graph the number of Hamilton circuits in the graph can be obtained by calculating the factorial of the number that is one less than the number of vertices.

Number of Hamilton Circuits in a Complete Graph

A complete graph with n vertices has $(n - 1)!$ Hamilton circuits.

As the number of vertices in a complete graph increases, the number of Hamilton circuits for that graph increases very quickly. Previously you considered how quickly exponential functions increase; factorials, however, increase much faster. For example, $25! = 15,511,210,043,330,985,984,000,000$, or approximately 1.6×10^{25}—more than a trillion trillion.

So where would we come across real-life problems involving Hamilton circuits for such large graphs? Consider, for example, that on a typical day, a UPS van might have to make deliveries to 100 different locations. For simplicity, assume that none are priority deliveries, so they can be made in any order. UPS would like to minimize the time that the driver takes to make these deliveries. Consider the 100 locations and the UPS distribution center as the vertices of a complete graph with 101 vertices. In principle, the van can go from any of these locations directly to any other, shown by having an edge joining each pair of vertices. Suppose we estimate the travel time between each pair of locations (of course, this would depend on distance and traffic conditions along the route). These estimates provide weights on the edges of the graph. Notice that the objective is to visit each location exactly once, to begin and end at the same place, and to take as little time as possible. In graph theory terms, we need a Hamilton circuit for the graph that has *smallest possible total weight*. The **total weight** of a circuit is the sum of the weights on the edges in the circuit. We call such a circuit a *minimum Hamilton circuit* for the graph.

Minimum Hamilton Circuit

In a weighted graph, a **minimum Hamilton circuit** is a Hamilton circuit with smallest possible total weight.

Problems whose solution requires us to find a minimum Hamilton circuit for a complete, weighted graph often are called *traveling salesman problems* (or TSPs).

Traveling Salesman Problems and Archaeology

Archaeologists often excavate sites in which there are no inscriptions or chemical evidence to show the sequence of the deposits. Some archaeologists have used minimum Hamilton circuits to solve the problem. For example, if a site consists of a number of burials, they consider a complete graph with vertices representing the various burials, and weights on the edges corresponding roughly to how dissimilar the burials are. A minimum Hamilton circuit in this graph gives the best guess for the order in which the deposits were made.

FIGURE 37

To understand why, think of the vertices of the complete graph as the cities that a salesperson must visit, and the weights on the edges as the cost of traveling directly between the cities. The salesperson can visit the cities in any order, but he or she would like to keep travel costs as low as possible. To achieve this the salesperson needs a minimum Hamilton circuit for the graph.

Traveling salesman problems occur throughout business and industry. For example, to manufacture integrated circuit silicon chips, many lines have to be etched on a silicon wafer. Minimizing production time involves deciding the order in which to etch the lines, a traveling salesman problem. Likewise, to manufacture circuit boards for integrated circuits, laser-drilled holes must be made for connections. Again, the order for drilling the holes is a critical factor in production time. Traveling salesman problems also are relevant to efficient routing of telephone calls and Internet connections.

Suppose we have been given a complete, weighted graph. How can we find a minimum Hamilton circuit for the graph?

One way is to systematically list all the Hamilton circuits in the graph, find the total weight of each, and choose the circuit with smallest total weight. In fact, it is sufficient to add up the weights on the edges for just half of the Hamilton circuits, since the total weight of a circuit is the same as the total weight of the circuit that uses the same edges in reverse order. This method is sometimes called the *brute force* algorithm since we grind out the solution by checking *all* the Hamilton circuits.

Brute Force Algorithm

1. Choose a starting point.
2. List all the Hamilton circuits with that starting point.
3. Find the total weight of each circuit.
4. Choose a Hamilton circuit with smallest total weight.

EXAMPLE 2 Find a minimum Hamilton circuit for the complete, weighted graph shown in Figure 37.

Choose a starting point, A. Then list all the Hamilton circuits for this graph. Since this is a complete graph with 4 vertices, we know there are 3! circuits: $3! = 3 \cdot 2 \cdot 1 = 6$. Thus, we must find 6 Hamilton circuits.

We need a systematic way of writing down all the Hamilton circuits. The counting tree of Figure 36 provides a guide. We start by finding all the Hamilton circuits that begin A → B, then include those that begin A → C, and finally add those that begin A → D, as shown below. Once all the Hamilton circuits are listed, pair those that visit the vertices in precisely opposite orders, since circuits in these pairs have the same total weight. (This will save some adding.) Finally, determine the sum of the weights of the edges in each circuit.

Circuit	Total weight of the circuit
1. A → B → C → D → A	$3 + 6 + 3 + 1 = 13$
2. A → B → D → C → A	$3 + 4 + 3 + 1 = 11$
3. A → C → B → D → A	$1 + 6 + 4 + 1 = 12$
4. A → C → D → B → A (opposite of 2)	11
5. A → D → B → C → A (opposite of 3)	12
6. A → D → C → B → A (opposite of 1)	13

We can now see that A → B → D → C → A is a minimum Hamilton circuit for the graph. The weight of this circuit is 11.

In principle, the brute force algorithm provides a way to find a minimum Hamilton circuit in any complete, weighted graph. In practice, however, there's a problem with this method. How many total weights would we have to calculate for a complete, weighted graph with 7 vertices? There are 6! or 720 Hamilton circuits in the graph, and we would have to calculate the total weight for half of these. That means we would have to do 360 separate calculations, in addition to listing the Hamilton circuits. This would take a long time. We could program a computer to do this for a complete graph with seven vertices, which would be much faster.

Computer Speeds The speed of supercomputers is measured in gigaflops. One gigaflop is approximately one billion calculations per second. The fastest computers in 1999 had a speed of about 2000 gigaflops. IBM plans to develop a new supercomputer with a speed of one million gigaflops by the year 2005.

Most real-world problems involve a lot more than 7 vertices. Manufacturing a large integrated circuit can involve a graph with almost one million vertices; likewise, telecommunications companies routinely deal with graphs with millions of vertices. As the number of vertices in the graph increases, the task of finding a minimum Hamilton circuit using the brute force algorithm soon becomes too time-consuming for even our fastest computers. For example, if one of today's supercomputers had started using the brute force algorithm on a 100-vertex traveling salesman problem when the universe was created, it would still be far from done. In contrast, Fleury's algorithm for finding an Euler circuit in a graph does not take too long for our computers, even for rather large graphs. Algorithms that do not take too much computer time are called **efficient algorithms.** Computer scientists have so far been unable to find an efficient algorithm for the traveling salesman problem, and they suspect that it is simply impossible to create such an algorithm.

The traveling salesman problem is just one of a collection of problems for which there is no known efficient algorithm. Many of these unsolved problems are related in such a way that an efficient algorithm for solving any one of them could be adapted to solve all of them. Probably the most important unsolved problem in computer science is this: either to find an efficient algorithm for the traveling salesman problem, or to explain why no one could create such an algorithm. (You might find this referred to as the P = NP problem.)

DNA Computers Researchers are developing computers based on DNA instead of silicon chips. Chemical reactions replace electric signals, and since these happen simultaneously and very quickly, computation is much faster. Programming a DNA computer is more difficult than programming a conventional computer, however, since it involves synthesizing DNA molecules to represent the specific problem. Although DNA computers have so far solved only small problems that are easily solved by a conventional computer, some believe that DNA computers will soon be sufficiently developed to solve in four months a problem that would take 10,000 years on a conventional computer.

For the traveling salesman problem there are some algorithms that do not take too much computer time, and that give *reasonably good* solutions *most of the time*. Such algorithms are called **approximate algorithms,** since they give an approximate solution to the problem. We shall consider one such algorithm for the traveling salesman problem, the **nearest neighbor algorithm.** The underlying idea of this algorithm is that from each vertex we proceed to a "nearest" available vertex.

Nearest Neighbor Algorithm

1. Choose a starting point for the circuit; let's call this vertex A.
2. Check all the edges joined to A, and choose one that has smallest weight; proceed along this edge to the next vertex.
3. At each vertex you reach, check the edges from there *to vertices not yet visited.* Choose one with smallest weight. Proceed along this edge to the next vertex.
4. Repeat Step 3 until you have visited all the vertices.
5. Return to the starting vertex.

EXAMPLE 3 A courier is based at the head office (A), and must deliver documents to four other offices (B, C, D, and E). The estimated time of travel (in minutes) between each of these offices is shown on the graph in Figure 38. The courier wants to visit the locations in an order that takes the least time. Use the nearest neighbor algorithm to find an approximate solution to this problem. Calculate the total time required to cover the chosen route.

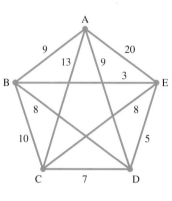

FIGURE 38   **FIGURE 39**

1. Choose a starting point: Let's start at A.
2. Choose an edge with smallest weight joined to A. Both AB and AD have weight 9, which is less than the weights of the other two edges joined to A (13 and 20). Let's choose AD.* We keep a record of our circuit as we form it:

$$A \xrightarrow{9} D$$

To help ensure that we do not visit a vertex twice, we number the vertices as we visit them and color the edges used. Our diagram now appears as in Figure 39.

Continue to number the vertices as we work through the rest of the solution.

3. Now check edges joined to D, excluding DA (since we have already been to A). DC has weight 7, DB has weight 8, and DE has weight 5. DE has smallest weight, so proceed along DE to E. Our circuit begins:

$$A \xrightarrow{9} D \xrightarrow{5} E$$

4. Now repeat the process at E. Check all edges from E that go to a vertex not yet visited. EC has weight 8 and EB has weight 3. Proceed along EB to B. Our circuit so far:

$$A \xrightarrow{9} D \xrightarrow{5} E \xrightarrow{3} B.$$

---

*If using a computer to do this, we would instruct the computer to make a random choice between AB and AD.

5. We have one vertex not yet visited, C. We go next to C and now have visited all the vertices. We return to our starting point, A. Our Hamilton circuit is

$$A \xrightarrow{9} D \xrightarrow{5} E \xrightarrow{3} B \xrightarrow{10} C \xrightarrow{13} A.$$

Its total weight is $9 + 5 + 3 + 10 + 13 = 40$.

Our advice to the courier would thus be to visit the places in the order shown in this circuit. This route will take about 40 minutes. ■

You might think that you could get a quicker route for the courier just by looking at the graph, rather than following the rules of the algorithm. You would be right. The circuit we found in Example 3 is not the minimum Hamilton circuit for the graph in Figure 39. (See below.) However, the point of an approximate algorithm is that a computer can implement it without taking too much time, and most of the time it will give a reasonably good solution to the problem. Computer scientists work hard at developing approximate algorithms to solve real-world problems.

If we had a computer performing the nearest neighbor algorithm for us, we could make a small adjustment that would give better results without taking too much longer: We could repeat the nearest neighbor algorithm for all possible starting points, and then choose from these the Hamilton circuit with smallest weight. If we do this for the example above, we find the following:

Records for Traveling Salesman Problems There is a race on to set records for exact solutions to traveling salesman problems. Computer scientists do this by using parallel processing and more sophisticated algorithms (all still exponential time algorithms). The illustration shows the exact solution to a traveling salesman problem with 3038 vertices on a printed circuit board, obtained in 1993.

Here are some of the other records, showing the year each was set and the number of vertices in the graph.

  1980: 318
  1987: 666
  1994: 7397
  1998: 13,509
  2001: 15,112

(*Source:* http://www.math.princeton.edu/tsp)

Starting Vertex	Circuit Using Nearest Neighbor	Total Weight
A	A → D → E → B → C → A	$9 + 5 + 3 + 10 + 13 = 40$
B	B → E → D → C → A → B	$3 + 5 + 7 + 13 + 9 = 37$
C	C → D → E → B → A → C	$7 + 5 + 3 + 9 + 13 = 37$
D	D → E → B → A → C → D	$5 + 3 + 9 + 13 + 7 = 37$
E	E → B → D → C → A → E	$3 + 8 + 7 + 13 + 20 = 51$

Using this variant of the nearest neighbor algorithm, we would recommend to the courier that he use a circuit with total weight 37, for example the circuit $D \to E \to B \to A \to C \to D$. This does not force him to start his journey at D. If he followed the route $A \to C \to D \to E \to B \to A$, he would be using the same edges as in $D \to E \to B \to A \to C \to D$, and his traveling time still would be 37 minutes.

Can we now be sure that we have found the minimum Hamilton circuit for the graph in Figure 39? No. This is still an *approximate* solution. If we check the total weights of all possible Hamilton circuits in the graph, we find the Hamilton circuit

$$A \xrightarrow{9} B \xrightarrow{3} E \xrightarrow{8} C \xrightarrow{7} D \xrightarrow{9} A$$

with total weight just 36 minutes. The nearest neighbor algorithm simply will not find this minimum Hamilton circuit for the graph. However, all we ask of an approximate algorithm is that it give a reasonably good solution for the problem.

FOR FURTHER THOUGHT

How do computer scientists classify the speed of algorithms? First, they write the number of steps a computer using the algorithm takes to solve a problem as a function of the "size" of the problem. We can think of the size of the problem as the number of vertices in the graph. Let's suppose that our graph has n vertices. For some algorithms the number of steps is a *polynomial function of n*, for example $n^4 + 2n$. For other algorithms, the number of steps is an *exponential function of n*, for example 2^n. (See Chapters 7 and 8 for more on polynomial and exponential functions.) As n increases, the number of steps increases much faster for exponential functions than for polynomial functions. Functions that involve factorials increase even faster. In the table below we show approximate values for functions of the three types for different values of n. (We use scientific notation to make it easier to compare the sizes of the numbers.)

n	n^4	2^n	$n!$
5	6.3×10^2	3.2×10	1.2×10^2
10	1.0×10^4	1.0×10^3	3.6×10^6
15	5.1×10^4	3.3×10^4	1.3×10^{12}
20	1.6×10^5	1.0×10^6	2.4×10^{18}
25	3.9×10^5	3.4×10^7	1.6×10^{25}
30	8.1×10^5	1.1×10^9	2.7×10^{32}
40	2.6×10^6	1.1×10^{12}	8.2×10^{47}
50	6.3×10^6	1.1×10^{15}	3.0×10^{64}

In the first line of the table (where $n = 5$), n^4 gives the largest value. But as n gets larger, 2^n grows much faster than n^4, while $n!$ grows ridiculously fast. For example, if we double the size of n from 15 to 30, n^4 becomes a little more than 10 times as large, 2^n becomes about 3×10^4 or 30,000 times as large, while $n!$ becomes approximately 2×10^{20} times as large. 10^{20} is more than a billion billion.

Algorithms whose time functions grow no faster than a polynomial function are called *polynomial time algorithms*. These are *efficient* algorithms; they do not take too much computer time.

Algorithms whose time functions grow faster than any polynomial function are called *exponential time algorithms*. These are *not efficient* algorithms. They are by nature too time-consuming for our computers. Our silicon chip computers are getting faster each year, but even this increase in speed hardly puts a dent in the time required for a computer to implement an exponential time algorithm for a very large graph.

Since the brute force algorithm for the traveling salesman problem has a time function involving $n!$, which grows faster than an exponential function of n, this algorithm is an exponential time algorithm.

For Group Discussion

Use the table to help answer these questions:

1. By approximately what factor does n^4 grow if we double n from 10 to 20?
2. By approximately what factor does 2^n grow if we double n from 10 to 20?
3. By approximately what factor does $n!$ grow if we double n from 10 to 20?
4. Repeat Exercises 1–3 if we double n from 25 to 50.
5. Suppose it would take 2^{30} years using computers at their present speeds to solve a certain problem. (2^{30} is a little over one billion.) If we assume that computers double in speed each year, how long would we have to wait before we could solve the problem in 1 year?

15.3 EXERCISES

For Exercises 1 and 2, a graph is shown, and some paths in the graph are specified. Determine which paths are Hamilton circuits for the graph. If not, say why not.

1. (a) A → E → C → D → E → B → A
   (b) A → E → C → D → B → A
   (c) D → B → E → A → B
   (d) E → D → C → B → E

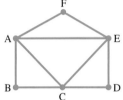

2.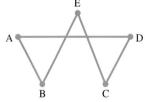

   (a) A → B → C → D → E → C → A → E → F → A
   (b) A → C → D → E → F → A
   (c) F → A → C → E → F
   (d) C → D → E → F → A → B

For Exercises 3 and 4, a graph is shown, and a string of vertices is specified. Determine whether the string is a circuit, whether it is an Euler circuit, and whether it is a Hamilton circuit. Justify your answers briefly.

3.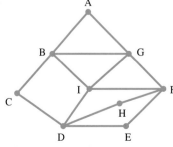

   (a) A → B → C → D → E → A
   (b) B → E → C → D → A → B
   (c) E → B → A → D → A → D → C → E

4.

   (a) A → B → C → D → E → F → G → A
   (b) B → I → G → F → E → D → H → F → I → D → C → B → G → A → B
   (c) A → B → C → D → E → F → G → H → I → A

For Exercises 5–10, determine whether the graph has a Hamilton circuit. If so, find one. (There are many different correct answers.)

5.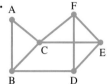

6.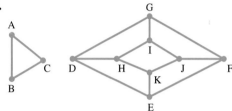

7.

8.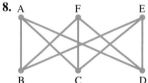

9.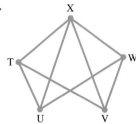

10.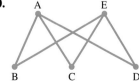

11. Draw a graph that has a Hamilton circuit, but no Euler circuit. Specify the Hamilton circuit, and explain why the graph has no Euler circuit. (There are many different correct answers.)

12. Draw a graph that has an Euler circuit, but no Hamilton circuit. Specify an Euler circuit in your graph. (There are many different correct answers.)

13. Draw a graph that has both an Euler circuit and a Hamilton circuit. Specify these circuits. (There are many different correct answers.)

14. Decide whether each of the following statements is true or false. If the statement is false, give an example to show that it is false.
    (a) A Hamilton circuit for a graph must visit each vertex in the graph.
    (b) An Euler circuit for a graph must visit each vertex in the graph.
    (c) A Hamilton circuit for a graph must use each edge in the graph.
    (d) An Euler circuit for a graph must use each edge in the graph.
    (e) A circuit cannot be both a Hamilton circuit and an Euler circuit.
    (f) An Euler circuit must visit no vertex more than once, except the vertex where the circuit begins and ends.

For Exercises 15–20, determine whether an Euler circuit, a Hamilton circuit, or neither would solve the problem.

15. The vertices of a graph represent bandstands at a festival and the edges represent paths between the bandstands. A visitor wants to visit each bandstand exactly once, returning to her starting point when she is finished.

16. The vertices of a complete graph represent the members of a five-person relay team. The team manager wants a circuit that will show the order in which the team members will run. (He will decide later who will start.)

17. The vertices of a graph represent places where paths in a botanical garden cross and the edges represent the paths. A visitor wants to walk along each path in the garden exactly once, returning to his starting point when finished.

18. The vertices of a graph represent the countries on the continent (Western Europe), with edges representing border crossings between the countries. A traveler wants to travel over each border crossing exactly once, returning to the first country visited for his flight home to the United States.

19. Vertices represent countries in sub-Saharan Africa, with an edge between two vertices if those countries have a common border. A traveler wants to visit each country exactly once, returning to the first country visited for her flight home to the United States.

20. In using X rays to analyze the structure of crystals, an X ray diffractometer measures the intensity of reflected radiation from the crystal in thousands of different positions. Consider the complete graph with vertices representing the positions where measurements must be taken. The researcher must decide the order in which to take these readings, with the diffractometer returning to its starting point when finished.

For Exercises 21–24, use your calculator, if necessary, to find the value.

21. 4!   22. 6!   23. 9!   24. 14!

For Exercises 25–28, determine how many Hamilton circuits there are in a complete graph with this number of vertices. (Leave answers in factorial notation.)

25. 10 vertices   26. 15 vertices
27. 18 vertices   28. 60 vertices

29. List all Hamilton circuits in the graph which start at P.

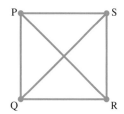

Exercises 30–36 refer to the following graph. List all Hamilton circuits in the graph that start with the indicated vertices.

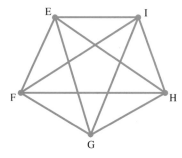

30. Starting E → F → G   31. Starting E → H → I
32. Starting E → I → H   33. Starting E → F
34. Starting E → I   35. Starting E → G
36. Starting at E

37. List all Hamilton circuits in the graph which start at A.

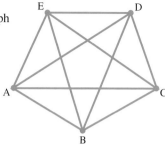

In Exercises 38–41, use the brute force algorithm to find a minimum Hamilton circuit for the graph. In each case determine the total weight of the minimum Hamilton circuit.

38.
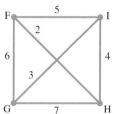

39.
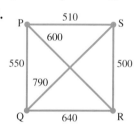

40.
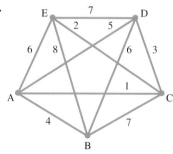

41.
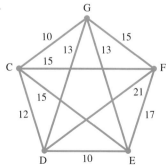

In Exercises 42–44, use the nearest neighbor algorithm starting at each of the indicated vertices to determine an approximate solution to the problem of finding a minimum Hamilton circuit for the graph. In each case, find the total weight of the circuit you have found.

42.
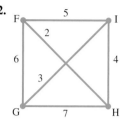

(a) Starting at F
(b) Starting at G
(c) Starting at H
(d) Starting at I

43.
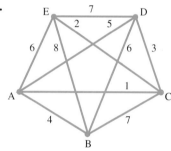

(a) Starting at A
(b) Starting at C
(c) Starting at D
(d) Starting at E

44.
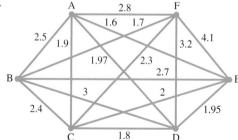

(a) Starting at A
(b) Starting at B
(c) Starting at C
(d) Starting at D
(e) Starting at E
(f) Starting at F

45. Refer to the accompanying graph. Complete parts (a)–(c) in order.

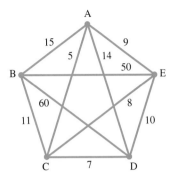

(a) Use the nearest neighbor algorithm starting at each of the vertices in turn to determine an approximate solution to the problem of finding a minimum Hamilton circuit for the graph. In each case, find the total weight of the circuit you have found.

(b) Which of the circuits that you found in part (a) gives the best solution to the problem of finding a minimum Hamilton circuit for the graph?

(c) Just by looking carefully at the graph, find a Hamilton circuit in the graph that has lower total weight than any of the circuits you found in part (a).

46. A graph is called a *complete bipartite graph* if the vertices can be separated into two groups in such a way that there are no edges between vertices in the same group, and there is an edge between each vertex in the first group and each vertex in the second group. An example is shown below.

The notation we use for these graphs is $K_{m,n}$ where m and n are the numbers of vertices in each of the groups. For example, the graph shown above is $K_{4,3}$ (or $K_{3,4}$). Answer the following in order:

(a) Draw $K_{2,2}$, $K_{2,3}$, $K_{2,4}$, $K_{3,3}$, and $K_{4,4}$.

(b) For each of the graphs you have drawn in part (a), find a Hamilton circuit, or say that the graph has no Hamilton circuit.

(c) Make a conjecture: What must be true about m and n for $K_{m,n}$ to have a Hamilton circuit?

(d) What must be true about m and n for $K_{m,n}$ to have an Euler circuit? Justify your answer.

For Exercises 47–50, find all Hamilton circuits in the graph, starting at A. (Suggestion: Use counting trees.)

47.

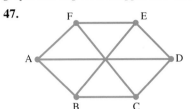

48.

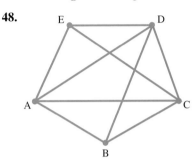

49.

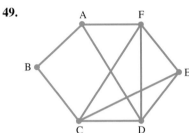

50.

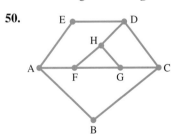

51. Paul A. M. Dirac proved the following theorem in 1952: Suppose G is a (simple) graph with n vertices, $n \geq 3$. If the degree of each vertex is greater than or equal to $\dfrac{n}{2}$, then the graph has a Hamilton circuit. Refer to the graphs following part (e) on the next page as you answer the different parts in order:

(a) Which of the graphs satisfy the condition that the degree of each vertex is greater than or equal to $\dfrac{n}{2}$?

(b) For which of the graphs can we conclude *from Dirac's theorem* that the graph has a Hamilton circuit?

(c) If a graph does *not* satisfy the condition that the degree of each vertex is greater than or equal to $\frac{n}{2}$, can we be sure that the graph does *not* have a Hamilton circuit? Justify your answer. (*Hint:* Study the accompanying graphs.)

(d) Is Dirac's theorem still true if $n < 3$? Justify your answer.

(e) Use Dirac's theorem to write a convincing argument that any complete graph with 3 or more vertices has a Hamilton circuit.

The graph below shows the Icosian game (described in the text) with the vertices labeled. For Exercises 52–54 find a Hamilton circuit for the graph in the specified version of the game. (We suggest you write numbers on the graph in the order in which you visit the vertices. There are different correct answers for these exercises.)

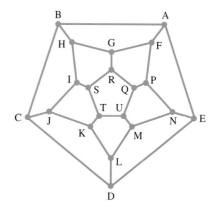

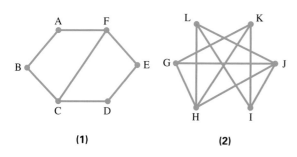

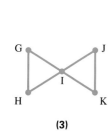

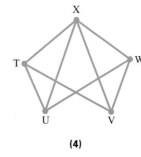

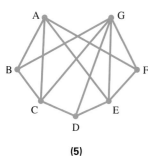

52. The circuit begins at A.

53. The circuit begins A → F.

54. The circuit begins A → B → H.

55. Refer to Exercise 58 in Section 15.1 for how one can analyze the rhyme scheme in poetry using graphs. Use this idea to write a paper on sestinas. Be sure to discuss what Hamilton circuits have to do with sestinas.

56. Find out what NP complete problems are, and write a paper on them. Include a careful description of at least one NP complete problem that is different from the traveling salesman problem.

57. Write a paper on very large numbers. Be sure to include some good examples of actual situations that involve very, very large numbers.

15.4 Trees and Minimum Spanning Trees

Trees In the previous two sections, we explored graphs with special kinds of circuits. In this section we examine graphs (called *trees*) that have no circuits.

Consider a problem that Wanda Johnson has in her garden. She wants to install an underground irrigation system. The system must connect to the faucet (F). Wanda wants outlets at various points in the garden, namely at her rose bed (R), perennial bed (P), daffodil bed (D), annuals bed (A), berry patch (B), vegetable patch (V), cut flower bed (C), and shrub bed (S). These points are shown as the vertices of the graph in Figure 40. The irrigation system will have pipes connecting these points. The outlets will be fitted with sprinklers that can be selectively switched off without disrupting the flow of water past an outlet to other outlets.

Forests, Leaves, and Pruning in Graph Theory In graph theory, the botanical analogy goes beyond trees. A *forest* is a graph in which all components are trees, as shown here. Later in this section you will encounter the words *root* and *leaf*. The terms *pruning* and *separating* occur when minimum spanning trees are applied to computer image analysis.

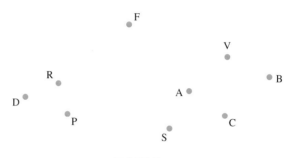

FIGURE 40

For the irrigation system to work well, there should not be more pipes than necessary to connect all the outlets. Think about this situation in terms of a graph with edges representing pipes. Wanda wants water to flow from the faucet to each of the outlets; this means that we need to create a connected graph that includes all the vertices in Figure 40.

We discussed connected graphs in Section 15.1. Now that you are familiar with paths, we can give the definition for a *connected* graph in a slightly different form that is more useful for our purposes here:

Connected Graph

A **connected graph** is one in which there is *at least one path* between each pair of vertices.

To see that this definition agrees with our earlier definition (on page 853), observe that if we select any pair of vertices in the connected graph in Figure 41 (for example, A and C) then there is at least one path between them. (In fact, there are many paths between A and C; A → B → E → C and A → F → C are two such paths.)

Wanda needs a *connected* graph that includes all the vertices of the original graph. Yet she must not use more pipes than necessary to connect all the outlets to the system. This means that the connected graph *must not contain any circuits*. If the graph had a circuit, we could remove one of the pipes in the circuit, and still have a connected system of pipes.

Figure 42 on the next page shows one possible solution to Wanda's problem.

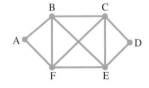

FIGURE 41

It seems mathematician **Arthur Cayley** (see page 897) invented the term "tree" in graph theory in about 1857 in connection with problems related to counting all trees of certain types. Later he realized the relevancy of his work to nineteenth-century organic chemistry, and in the 1870s published a note on this. Of course, not all molecules have treelike structures. Friedrich Kekulé's realization that the structure of benzene is *not* a tree is considered one of the most brilliant breakthroughs in organic chemistry.

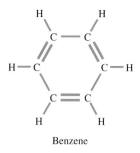
Benzene

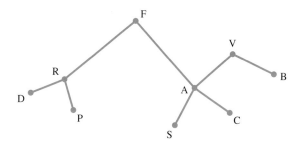

FIGURE 42

There is a name for graphs that are connected and contain no circuits.

> **Tree**
>
> We call a graph a **tree** if the graph is *connected* and contains *no circuits*.

The graphs shown in Figure 43 all are trees.

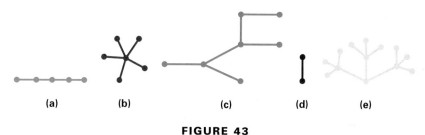

FIGURE 43

In contrast, none of the graphs shown in Figure 44 is a tree. Graph (a) is not a tree, since it is not connected. Graphs (b) and (c) are not trees, since each contains at least one circuit.

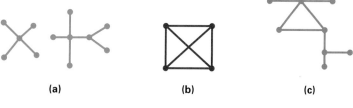

FIGURE 44

To get a better feeling for the concept of a tree, we consider an important property of trees that follows directly from the definition of a tree.

> **Property of Trees**
>
> In a tree there is always **exactly one path** from each vertex in the graph to any other vertex in the graph.

For example, starting with vertices S and B in the tree in Figure 42, there is exactly one path from vertex S to vertex B, namely S → A → V → B.

To see how this property follows from the definition of a tree, notice that since a tree is connected, there is always at least one path between each pair of vertices. Also, if the graph is a tree, there cannot be two different paths between a pair of vertices; if there were, we could join together appropriate pieces of these to create a circuit.

FOR FURTHER THOUGHT

The unique path property of a tree can be used to set up a code. Here we set up a **binary code,** that is, a code with strings of 0s and 1s representing letters. (Recall from Chapter 4 that when we represent numbers in binary form we use only the symbols 0 and 1.)

To set up the binary code we use a special kind of tree (called a binary tree) like that shown in the figure. This tree is a **directed graph** (there are arrows on the edges). The vertex at the top (with no arrows pointing toward it) is called the **root** of the directed tree; the vertices with no arrows pointing away from them are called **leaves** of the directed tree. We label each leaf with a letter we want to encode. The diagram shown provides an encoding for only 8 letters, but we could easily draw a bigger binary tree with more leaves to represent more letters.

We now write a 0 on each branch extending to the left and a 1 on each branch extending to the right.

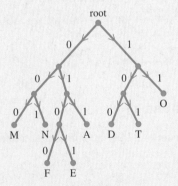

To show how the encoding works, let us write the word MAD using the code. Follow the unique path from the root of the tree down to the appropriate leaf, noting in order the labels on the edges.

M is written 000.
A is written 011.
D is written 100.

So MAD is written 000011100.

We can easily translate this code using our tree. Let us see how we could decode 000011100. Referring to the tree, we can see that there is only one path from the root to a leaf that can give rise to those first three 0s, and that is the path leading to M. So we can begin to separate the code word into letters: 000–011100. Again following down from the root, the path 011 leads us unambiguously to A. So we have 000-011-100. The path 100 leads unambiguously to D.

The reason we can translate the string of 0s and 1s back to letters without ambiguity is that no letter has a code that is the start of the code for a different letter.

For Group Discussion

1. Use the binary encoding tree to write the binary code for each of the following words: FEET, ANT
2. Use the encoding tree to find the word represented by each of the following codes: 1000101001101   0001111001
3. Decode the following message (commas are inserted to show separation of words). 00001010101101, 0000101, 011101, 1010101011

15.4 Trees and Minimum Spanning Trees   897

It is this unique path property of trees that makes them so important in real-world applications. This property also is the reason trees are not a very useful model for telephone networks; each edge of a tree is a cut edge, so failure along one link would disrupt the service. Thus telephone companies rely on networks with many circuits to provide alternate routes for calls.

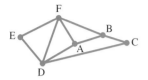

FIGURE 45

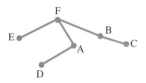

FIGURE 46

EXAMPLE 1 Brett has an old underground irrigation system in his garden, with connections at the faucet (F), outlets A through E in the garden, and existing underground pipes as shown in the graph in Figure 45. He wants to renovate this system by retaining only some of the pipes. Water must still be able to flow to each of the original outlets and, like Wanda Johnson did, Brett wants the graph of his irrigation system to be a tree. Design an irrigation system that meets Brett's objectives.

In graph theory terms, since the renovated irrigation system will consist of some of the existing pipes, we need a *subgraph* of the graph in Figure 45. Since water still must be able to flow to each of the original outlets, we need a connected subgraph that *includes all the vertices of the original graph.* Finally, this subgraph must be a *tree.* Thus, we must remove edges (but no vertices) from the graph, without disconnecting the graph, to obtain a subgraph that is a tree. One way to achieve this is to remove edges ED, FD, AB, and DC from the original graph, obtaining the subgraph shown in Figure 46.

A subgraph of the type shown in Figure 46 in Example 1 is called a *spanning tree* for the graph. This kind of subgraph "spans" the graph in the sense that it connects all the vertices of the original graph into a subgraph.

Spanning Tree

A **spanning tree** for a graph is a subgraph that includes every vertex of the original, and is a tree.

Arthur Cayley (1821–1895) was a close friend of James Sylvester. (See the margin note on page 855.) They worked together as lawyers at the courts of Lincolns Inn in London and discussed mathematics with each other during work hours. Cayley considered his work as a lawyer simply a means of support so that he could do mathematics; in the fourteen years he worked as a lawyer, he published 250 mathematical papers. At age 42, he took a drastic reduction in income to become a professor at Cambridge University.

We can find a spanning tree for any connected graph (think about that) and, if the original graph has at least one circuit, it will have a number of different spanning trees. The following example illustrates this.

EXAMPLE 2 Consider the graph shown in Figure 47. Find two different ways of removing edges from the graph (without removing any vertices) to obtain spanning trees for the graph.

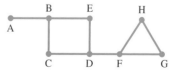

FIGURE 47

The graph has two circuits, B → C → D → E → B and F → G → H → F. To obtain a subgraph that is a tree, we must break these two circuits. That is, we must remove a single edge from each of these circuits. (Removing more than one edge from either of these circuits would disconnect the graph, and the remaining graph would no longer be a tree.) There are a number of choices for which edge to remove from each circuit. We could remove the edges CD and FH, leaving the subgraph shown in Figure 48(a). Alternatively, we could remove the edges ED and FG, leaving the subgraph shown in Figure 48(b). We now have two different subgraphs that are trees, as required. (Note that there are other subgraphs that are correct solutions for this example.)

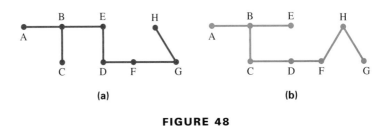

FIGURE 48

Note that if two spanning trees use different *labeled* edges, we consider them to be different spanning trees. This means that different spanning trees for a graph may in fact be isomorphic graphs. For example, the subgraphs shown in Figure 48(a) and (b) are *different* spanning trees, even though they happen to be isomorphic graphs. (Did you notice they were isomorphic? To see this, change the labels in graph (a) as follows: change E to C and C to E, then change G to H, and H to G. Now imagine graph (a) drawn with our special computer software. By dragging vertices without detaching edges, the graph could be made to look exactly like (b).)

Finding techniques for successful **computer analysis of images** is a very active field of research, particularly in medicine. (Computer analysis of images refers to having a computer actually "look" at an image to detect relevant features.) Some of this research uses minimum spanning trees. For example, these have been used to identify fibers of a particular protein in photographs of cells. The computer detects some points lying on the fibers and then determines a minimum spanning tree to deduce the overall structure of the fiber. Some studies have used minimum spanning trees to analyze cancerous tissue. Minimum spanning trees also have been used to identify road networks in satellite photographs.

Minimum Spanning Trees Consider this situation. There are 6 villages in a rural district. At present all the roads in the district are gravel roads. There is not enough funding to pave all the roads, but there is a pressing need to have good paved roads so that emergency vehicles can travel easily between the villages. The vertices in the graph in Figure 49 represent the villages in the district, and the edges represent the existing gravel roads. Estimated costs for paving the different roads are shown, in millions of dollars. (These estimates take into account the distances, as well as features of the terrain, that would make it more or less expensive to pave a given road.)

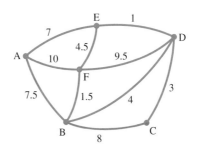

FIGURE 49

The district council wants to pave just enough roads so that emergency vehicles will be able to travel from any village to any other along paved roads, though possibly by a roundabout route. Of course, the district planners would like to achieve this at *lowest possible cost*.

The problem calls for exactly one path from each vertex in the graph to any other, so it calls for a spanning tree. In addition, this spanning tree should have minimum total weight. (The total weight of a spanning tree is the sum of the weights of the edges in the tree.)

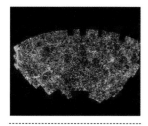

Minimum Spanning Trees in Astronomy The structure of the universe may hold clues about the formation of the universe. The image shows the large-scale structure of the universe in a region comprising approximately 10% of the sky. It shows the distribution of about 2 million galaxies. Galaxies are not uniformly distributed throughout the universe; instead, they seem to be arranged in clusters, which are grouped together into superclusters, which themselves seem to be arranged in vast chains called filaments. As the image shows, there are voids between the filaments. Some astronomers are using spanning trees to help map the filamentary structure of the universe.

Minimum Spanning Tree

A spanning tree that has minimum total weight is called a **minimum spanning tree** for the graph.

Fortunately for the district planners, there is a good algorithm for finding a minimum spanning tree for any connected, weighted graph. It is called *Kruskal's algorithm*. (The algorithm is an efficient algorithm; that is, it does not take too much computer time for even very large graphs.)

Kruskal's Algorithm

Choose edges for the spanning tree as follows:

1. Choose any edge with minimum weight.
2. Find an edge with minimum weight *among those not yet selected* and choose this as the next edge. (It does not matter at this stage if the subgraph looks disconnected.)
3. Continue to choose edges of minimum weight from those not yet selected, but *do not select any edge that creates a circuit* in the subgraph.
4. Repeat Step 3 until the tree connects all vertices of the original graph.

Algorithms such as Kruskal's often are called *greedy* algorithms; can you explain why?

Kruskal's algorithm specifically avoids forming circuits in the subgraph. Also, we stop only when the subgraph has reached all the vertices of the original graph. Thus we certainly end up with a spanning tree for the graph. Further, the algorithm always will give us a *minimum* spanning tree for the graph, but since this argument is beyond the realm of this course, we ask you to take this fact on trust.

EXAMPLE 3 Use Kruskal's algorithm to find a minimum spanning tree for the graph in Figure 49 on the previous page.

Note that the graph is connected, so Kruskal's algorithm applies. Choose an edge with minimum weight; ED has smallest weight of all the edges, so it is selected first. We color the edge a different color to show that it has been selected. See Figure 50(a).

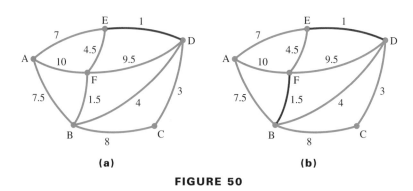

FIGURE 50

> Weather data are complex, involving numerous readings of temperature, pressure, and wind velocity over huge areas, making accurate weather prediction difficult at best. Some current research uses minimum spanning trees to help computers interpret these data; for example, minimum spanning trees are used to identify developing cold fronts.

Of the remaining edges, BF has smallest weight, so it is the second edge chosen, as shown in Figure 50(b). Note that it does not matter that at this stage the subgraph is not connected.

Continuing, the next edge chosen is CD, and then BD, as shown in Figure 50(c). Of the edges remaining, EF has the smallest weight; however, EF cannot be selected because *this would create a circuit* (E → D → B → F → E) in the subgraph. The subgraph would not be a tree. Therefore ignore EF and note that, of the edges remaining, AE has the smallest weight. Thus, AE is included in the spanning tree, as shown in Figure 50(d).

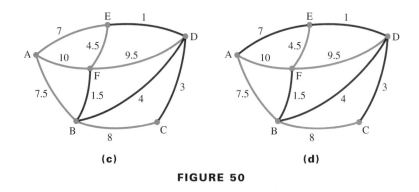

FIGURE 50

We now have a subgraph of the original that connects all vertices of the original graph into a tree. This subgraph is the required minimum spanning tree for the original graph. Its total weight is 1 + 1.5 + 3 + 4 + 7, which is 16.5. Thus, the district council can achieve its objective at a cost of 16.5 million dollars, by paving only the roads represented by edges in the minimum spanning tree. ∎

There is an interesting relation between the number of vertices and the number of edges in a tree. To see this, consider the graphs shown in Figure 51 on the next page. Information about these graphs is summarized in the table below the figure.

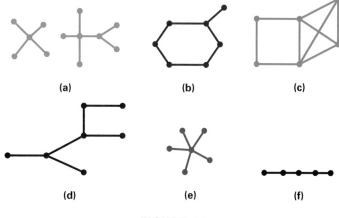

FIGURE 51

Graph	Tree?	Number of Vertices	Number of Edges
(a)	No	12	10
(b)	No	7	7
(c)	No	6	9
(d)	Yes	7	6
(e)	Yes	6	5
(f)	Yes	5	4

The first three graphs listed are not trees; there is no uniform relation between the number of edges and the number of vertices for these graphs. A graph that is not a tree may have more edges than vertices, may have fewer edges than vertices, or may have a like number of edges and vertices. In contrast, the last three graphs listed are trees. What do you notice about the number of edges compared to the number of vertices for these trees? The same relation between the number of edges and the number of vertices holds for any tree.

Number of Vertices and Edges in a Tree

If a graph is a tree, then the number of edges in the graph is one less than the number of vertices. So a tree with n vertices has $n-1$ edges.

James Sylvester was known for his flowery language. This is how he described the connection between graphs and chemistry: "Chemistry has the same quickening and suggestive influence upon the algebraist as a visit to the Royal Academy, or the old masters may be supposed to have had on a Browning or a Tennyson."

It is not hard to understand why this relation should hold for a tree. (See Exercise 56 of this section.) In addition, provided we work only with connected graphs, we have a converse for this theorem: For a connected graph, if the number of edges is one less than the number of vertices, then the graph is a tree.

EXAMPLE 4 A chemist has synthesized a new chemical compound. She knows from her analyses that a molecule of the compound contains 54 atoms and that the molecule has a treelike structure. How many chemical bonds are there in the molecule?

We can think of the atoms in the molecule as the vertices of a tree, and the bonds as edges of the tree. This tree has 54 vertices. Since the number of edges in a tree is one less than the number of vertices, the molecule must have 53 bonds.

Problem Solving

Sometimes trial and error is a good problem-solving strategy. Simply choose an arbitrary solution to the problem and check whether it works. If the arbitrary solution does not work, we may gain some insight into what to try next.

EXAMPLE 5 Suppose a graph is a tree and has 15 vertices. What is the largest number of vertices of degree 5 that this graph could have? Draw such a tree.

Since the graph is a tree with 15 vertices, it must have 14 edges. Do we know anything about graphs that will help us relate the number of edges to the degrees of vertices? Recall that the sum of the degrees of the vertices of a graph is twice the number of edges. Thus, the sum of the degrees of the vertices in the tree must be $2 \cdot 14$, which is 28.

Now we use trial and error. We want the largest possible number of vertices with degree 5. Four vertices with degree 5 would contribute 20 to the total degree sum. This would leave $28 - 20 = 8$ as the degree sum of the remaining 11 vertices, but this would mean that some of those vertices would have no edges joined to them. The graph would not be connected and would not be a tree.

Let's try 3 vertices with degree 5. That would contribute 15 to the total degree sum; this would leave $28 - 15 = 13$ as the degree sum for the remaining 12 vertices. After a little thought we see that we can draw a graph with these specifications. See Figure 52. Thus, the largest possible number of vertices of degree 5 in a tree with 15 vertices is 3.

FIGURE 52

15.4 EXERCISES

For Exercises 1–7, determine whether the graph is a tree. If not, explain why it is not.

1.  2. 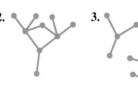 3.

4. 5.

6. a complete graph with 6 vertices
7. a connected graph with all vertices having even degree

For Exercises 8–10, add edges (no vertices) to the graph to change the graph into a tree, or explain why this is not possible. (Note: There may be more than one correct answer.)

8.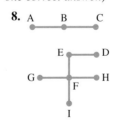

9.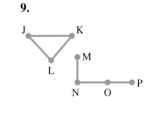

10.

For Exercises 11–13, determine whether the graph described must be a tree.

11. A sociologist is investigating the spread of a rumor. He finds that 18 people know of the rumor. One of these people must have started the rumor, and each must have heard the rumor for the first time from one of the others. He draws a graph with vertices representing the 18 people and edges showing from whom each person first heard the rumor.

12. You are using the "Excite" search engine to search the Web for information on a topic for a paper. You start at the Excite page and follow links you find, sometimes going back to a site you've already visited. To keep a record of the sites you visit, you show each site as a vertex of a graph; you draw an edge each time you connect for the *first* time to a *new* site.

13. A patient has a highly infectious disease. An employee of the Centers for Disease Control is trying to quarantine all people who have had contact over the past week with either the patient or with someone already included in the contact network. The employee draws a graph with vertices representing the patient and all people who have had contact as described. The edges of the graph represent the relation "the two people have had contact."

For Exercises 14–17, determine whether the statement is true or false. If the statement is false, draw an example to show that it is false.

14. Every graph with no circuits is a tree.

15. Every connected graph in which each edge is a cut edge is a tree.

16. Every graph in which there is a path between each pair of vertices is a tree.

17. Every graph in which each edge is a cut edge is a tree.

For Exercises 18–20, find three different spanning trees for each graph. (There are many different correct answers.)

18.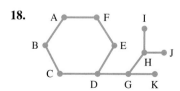

19.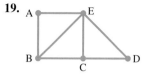

20.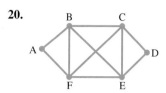

For Exercises 21–23, find all spanning trees for the graph.

21.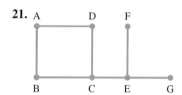

22.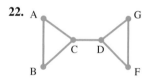

23.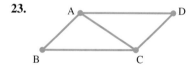

For Exercises 24–26, determine how many spanning trees the graph has.

24.

25.

26.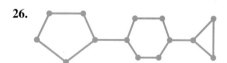

27. What is a general principle about the number of spanning trees in graphs such as those in Exercises 24–26?

28. Complete the parts of this exercise in order.
 (a) Find all the spanning trees of each of the following graphs.

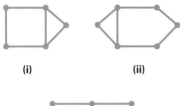

(i)   (ii)

(iii)

(b) For each of the following graphs, determine how many spanning trees the graph has.

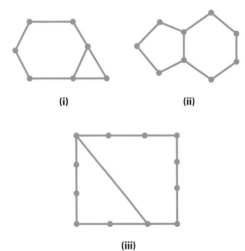

(i)   (ii)

(iii)

(c) What is a general principle about the number of spanning trees in graphs of the kind shown in (a) and (b)?

For Exercises 29–32, use Kruskal's algorithm to find a minimum spanning tree for the graph. Find the total weight of this minimum spanning tree.

29.

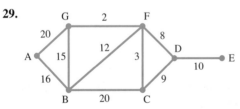

30.

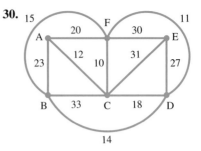

31.

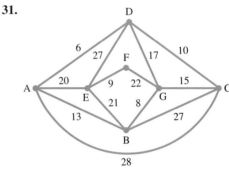

32.

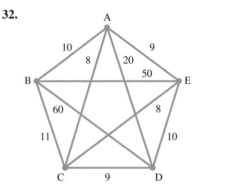

33. A school consists of 6 separate buildings, represented by the vertices in the following graph.

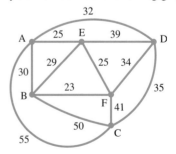

There are paths between some of the buildings as shown. The graph also shows the length in feet of each path. School administrators want to cover some of these paths with roofs so that students will be able to walk between buildings without getting wet when it rains. To minimize cost, they must select paths to be covered such that the total length to be covered is as small as possible. Use Kruskal's algorithm to determine which paths to cover. Also determine the total length of pathways that must be covered under your plan.

34. A town council is planning to provide town water to an area that previously relied on private wells. Water will be fed into the area at the point represented by the vertex labeled A in the accompanying graph.

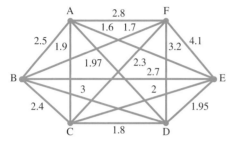

Water must be piped to each of five main distribution points, represented by the vertices labeled B through F in the accompanying graph. Town engineers have estimated the cost of laying the pipes to carry the water between each pair of points in millions of dollars, as indicated in the graph. They must now select which pipes should be laid, so that there is exactly one route for the water to be pumped from A to any one of the five distribution points (possibly via another distribution point), and they want to achieve this at minimum cost. Use Kruskal's algorithm to decide which pipes should be laid. Find the total cost of laying the pipes you select.

35. How many edges are there in a tree with 34 vertices?
36. How many vertices are there in a tree with 40 edges?
37. How many edges are there in a spanning tree for a complete graph with 63 vertices?
38. A connected graph has 27 vertices and 43 edges. How many edges must be removed to form a spanning tree for the graph?
39. We have said that we can always find at least one spanning tree for a connected graph, and usually more than one. Could it happen that different spanning trees for the same graph have different numbers of edges? Justify your answer.
40. Suppose we have a tree with 9 vertices.
    (a) Determine the number of edges in the graph.
    (b) Determine the sum of the degrees of the vertices in the graph.
    (c) Determine the smallest possible number of vertices of degree 1 in this graph.
    (d) Determine the largest possible number of vertices of degree 1 in this graph.
    (e) Answer parts (a) through (d) for a tree with n vertices.
41. Suppose we have a tree with 10 vertices.
    (a) Determine the number of edges in the graph.
    (b) Determine the sum of the degrees of the vertices in the graph.
    (c) Determine the smallest possible number of vertices of degree 4 in this graph.
    (d) Determine the largest possible number of vertices of degree 4 in this graph. Draw a graph to illustrate your answer.
42. Suppose we have a tree with 17 vertices.
    (a) Determine the number of edges in the graph.
    (b) Determine the sum of the degrees of the vertices in the graph.
    (c) Determine the largest possible number of vertices of degree 5 in this graph. Draw a graph to illustrate your answer.
43. A business has 23 employees, each with his or her own desk computer, all working in the same office. The managers want to network the computers. To do this they need to install cables between individual computers so that every computer is linked into the network. Because of the way the office is laid out, it is not convenient to simply connect all the computers in one long line. Determine the smallest number of cables the managers need to install to achieve their objective.
44. Maria Jimenez has 12 vegetable and flower beds in her garden and wants to build flagstone paths between the beds so that she can get from each bed along flagstone paths to every other bed. She also wants a path linking her front door to one of the beds. Determine the minimum number of paths she must build to achieve this.

Exercises 45–47 refer to the following situation: We start with a tree and then draw in extra edges and vertices. For each, say whether it is possible to draw in the number of edges and vertices specified to end up with a connected graph. If it is possible, determine whether the resulting graph must be a tree, may be a tree, or cannot possibly be a tree. Justify each answer briefly.

45. Draw in the same number of vertices as edges.
46. Draw in more vertices than edges.
47. Draw in more edges than vertices.
48. Starting with a tree, is it possible to draw in the same number of vertices as edges and end up with a disconnected graph? If so, give an example. If not, justify briefly.

Exercises 49–51 require the following theorem, proved by Cayley in 1889: A complete graph with n vertices has n^{n-2} spanning trees.

49. How many spanning trees are there for a complete graph with 3 vertices? Draw a complete graph with 3 vertices and find all the spanning trees.
50. How many spanning trees are there for a complete graph with 4 vertices? Draw a complete graph with 4 vertices and find all the spanning trees.
51. How many spanning trees are there for a complete graph with 5 vertices?
52. Find all non-isomorphic trees with 4 vertices. How many are there?
53. Find all non-isomorphic trees with 5 vertices. How many are there?
54. Find all non-isomorphic trees with 6 vertices. How many are there?
55. Find all non-isomorphic trees with 7 vertices. How many are there?
56. In this exercise, we explore why the number of edges in a tree is one less than the number of vertices. Since the statement is clearly true for a tree with only one vertex, we will consider a tree *with more than one vertex.*

    Answer parts (a)–(e) in order:
    (a) How many components does the tree have?
    (b) Why must the tree have at least one edge?
    (c) Remove one edge from the tree. How many components does the resulting graph have?
    (d) You have not created any new circuits by removing the edge, so each of the components of the resulting graph is a tree. If the remaining graph still has edges, choose any edge and remove it. (You have now removed 2 edges from the original tree.) Altogether, how many components remain?
    (e) Repeat the procedure described in (d). If you remove 3 edges from the original tree, how many components remain? If you remove 4 edges from your original tree, how many components remain?
    (f) Repeat the procedure in (d) until you have removed all the edges from the tree. If you have to remove *n* edges to achieve this, determine an expression involving *n* for the number of components remaining.
    (g) What *are* the components that remain when you have removed all the edges from the tree?
    (h) What can you conclude about the number of vertices in a tree with *n* edges?

57. Write a paper on the arrangement of galaxies and the filamentary structure of the universe. Include discussion of how minimum spanning trees are used to analyze this.

58. Write a paper on James Sylvester and Arthur Cayley. Include discussion of their work in graph theory. (In addition to other sources, check the website http://www-groups.dcs.st-and.ac.uk/~history/.)

COLLABORATIVE INVESTIGATION

Finding the Number of Edges in a Complete Graph

In this investigation we explore two short methods for finding the number of edges in a complete graph with a specified number of vertices. By comparing the methods, we obtain a special sum formula that you encountered in Chapter 1.

Divide into groups of 4 or 5 students. Assign to each group member a different complete graph, with 4, 5, 6, or 7 vertices. Each member should sketch his or her complete graph and count the number of edges. Next, make a table showing the number of vertices and the number of edges in complete graphs with 3 through 7 vertices.

For Group Discussion

Do you notice a pattern in the number of edges for these complete graphs? Guess how many edges there would be in a complete graph with 8 vertices. How many are in a complete graph with 10 vertices?

If you spotted a pattern in the table, you could find the number of vertices in a complete graph with, for example, 100 vertices. Of course, completing the table all the way up to 100 would take a long time! A short way of determining the number of edges in a complete graph with any specified number of vertices can be used instead. Divide your group into two groups, A and B, and complete the following tasks. When both groups have finished, come back together to share what you have learned.

Group A

1. Start by studying the number of edges in a complete graph with eight vertices. Proceed as follows.

   Have someone in the group draw the eight vertices (no edges yet).

   (a) Choose any one of the vertices and draw in all necessary edges *from this vertex*. (Remember, there must be an edge to every other vertex in the graph.) Record the number of edges you drew from this first vertex:

   (a) _____

   (b) Now choose a different vertex—it is already joined to your first vertex by an edge. Draw in all necessary additional edges *from this vertex*. Record the number of *additional* edges you drew in from this vertex: (b) _____

   Continue this process, determining how many *additional* edges you must draw from a third vertex, from a fourth vertex, and so on until the graph is complete. Record the number of additional edges in succession:

   (c) _____ (d) _____ (e) _____
   (f) _____ (g) _____

   Now add your answers from (a) through (g) to find the total number of edges:

   Total number of edges = _____ + _____ + _____
   + _____ + _____ + _____ + _____ = _____

2. Repeat the procedure described above to find the number of edges in a complete graph with 10 vertices.

3. As in Step 1 above, write down the numbers you must add to find the number of edges in a complete graph with 100 vertices. (Since there are so many numbers, write only the first three numbers, ellipsis points to indicate the omission, and then the last three numbers.)

   Total number of edges = _____ + _____ + _____
   + ... + _____ + _____ + _____

   To add these numbers by hand (or even with a calculator) would take a long time. Perhaps you remember a formula from a previous chapter? You will discuss this further when you rejoin Group B.

Group B

1. Start by studying the number of edges in a complete graph with eight vertices. Proceed as follows.

   Have someone draw a graph with 8 vertices. Then answer the following:

   (a) Each vertex in your graph has the same degree. What is it? (a) _____

   (b) Use your answer from (a) to find the sum of the degrees of the vertices in your graph.

   (b) _____

   (c) Find the number of edges in the complete graph, by using the theorem from Section 15.1 that relates the sum of degrees to the number of edges in the graph. (c) _____

2. Repeat the procedure described above to find the number of edges in a complete graph with 10 vertices.

3. Repeat the procedure described above to find the number of edges in a complete graph with 100 vertices.

Group B should now rejoin Group A. The members of Group A must explain their method to the whole group. Then the members of Group B must explain their method to the whole group. Use your combined knowledge to answer the following questions.

1. By comparing the work on Step 3 of Group A and Group B, find the sum: $99 + 98 + 97 + \ldots + 3 + 2 + 1 =$ _____

2. Determine the number of edges in a complete graph with 210 vertices.

3. Write a formula for the number of edges in a complete graph with m vertices.

4. Determine the sum of the following numbers (*Hint:* Regard the sum as the number of edges in a complete graph.):

   $155 + 154 + 153 + \ldots + 3 + 2 + 1 =$ _____

CHAPTER 15 TEST

Exercises 1–4 refer to the following graph.

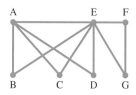

1. Determine how many vertices the graph has.
2. Determine the sum of the degrees of the vertices.
3. Determine how many edges the graph has.
4. Which of the following are paths in the graph? If not, why not?
   (a) B → A → C → E → B → A
   (b) A → B → E → A
   (c) A → C → D → E
5. Which of the following are circuits in the graph? If not, why not?
   (a) A → B → E → D → A
   (b) A → B → C → D → E → F → G → A
   (c) A → B → E → F → G → E → D → A → E → C → A
6. Draw a graph that has 2 components.
7. A graph has 10 vertices, 3 of degree 4, and the rest of degree 2. Use the theorem that relates the sum of degrees to the number of edges to determine the number of edges in the graph (without drawing the graph).
8. Determine whether graphs (a) and (b) are isomorphic. If they are, justify this by labeling corresponding vertices of the two graphs with the same letters, and color-coding the corresponding edges.

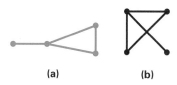

(a) (b)

9. Julia is planning to invite some friends for dinner. She plans to invite John, Adam, Bill, Tina, Nicole, and Rita. John, Nicole, and Tina all know each other. Adam knows John and Tina. Bill also knows Tina, and he knows Rita. Draw a graph with vertices representing the six friends whom Julia plans to invite to dinner and edges representing the relationship "know each other." Is your graph connected or disconnected? Which of the guests knows the largest number of other guests?

10. There are 8 contestants in a chess competition. Each contestant plays one chess game against every other contestant. The winner is the contestant who wins the most games. How many chess games will be played in the competition?

11. Is the accompanying graph a complete graph? Justify your answer.

12. Refer to the graph for Exercises 1–4. Which of the following are Euler circuits for the graph? If not, why not?
    (a) A → B → E → D → A
    (b) A → B → C → D → E → F → G → A
    (c) A → B → E → F → G → E → D → A → E → C → A

For Exercises 13 and 14, use Euler's theorem to decide whether the graph has an Euler circuit. (Do not actually find an Euler circuit.) Justify your answer briefly.

13.

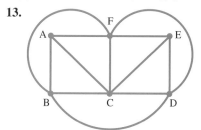

14.

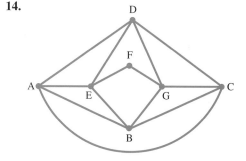

15. The floor plan of a building is shown. Is it possible to start outside, walk through each door exactly once, and end up back outside? Justify your answer.

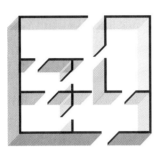

16. Use Fleury's algorithm to find an Euler circuit for the accompanying graph, beginning F → B.

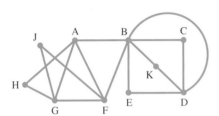

17. Refer to the graph for Exercises 1–4. Which of the following are Hamilton circuits for the graph? If not, why not?
 (a) A → B → E → D → A
 (b) A → B → C → D → E → F → G → A
 (c) A → B → E → F → G → E → D → A → E → C → A

18. List all Hamilton circuits in the graph that start F → G. How many are there?

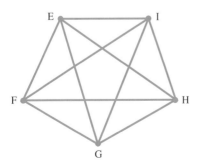

19. Use the brute force algorithm to find a minimum Hamilton circuit for the accompanying graph. Determine the total weight of the minimum Hamilton circuit.

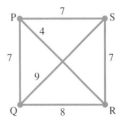

20. Use the nearest neighbor algorithm starting at A to find an approximate solution to the problem of finding a minimum Hamilton circuit for the graph. Find the total weight of this circuit.

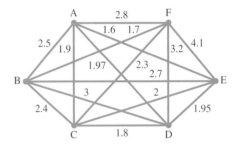

21. How many Hamilton circuits are there in a complete graph with 25 vertices? (Leave your answer as a factorial.)

22. The agent for a rock band based in Milwaukee is planning a tour for the band. He plans for the band to visit Minneapolis, Santa Barbara, Orlando, Phoenix, and St. Louis. Since the band members want to minimize the time they are away from home, they do not want to go to any of these cities more than once on the tour. Consider the complete graph with vertices representing the 6 cities mentioned. Does the problem of planning the tour require an Euler circuit, a Hamilton circuit, or neither for its solution?

23. Draw three non-isomorphic trees with 7 vertices.

For Exercises 24–26, decide whether the statement is true or false.

24. Every tree has a Hamilton circuit.

25. In a tree each edge is a cut edge.

26. Every tree is connected.

27. Find all the different spanning trees for the following graph. How many different spanning trees are there?

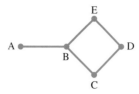

28. Use Kruskal's algorithm to find a minimum spanning tree for the following graph.

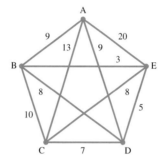

29. Determine the number of edges in a tree with 50 vertices.

chapter

16 Voting and Apportionment

March 13, 2000
  To all households:
  This is your official form for the United States Census 2000. It is used to count every person living in this house or apartment—people of all ages, citizens and non-citizens. Your answers are important. First, the number of representatives each state has in Congress depends on the number of people living in each state. . . .

In the 1888 U.S. presidential election, Democrat Grover Cleveland outpolled Republican Benjamin Harrison by a slim margin, 5,540,309 to 5,439,853. But Harrison won the election that counts, winning the Electoral College 233 to 168 to become the nation's first "minority president." In 2000 it happened again: Out of 105 million votes cast, George W. Bush recorded about 550,000 fewer popular votes than Al Gore but Bush won with 271 electoral votes to Gore's 266.

These elections generated heated argument about the fairness of our voting system. The founding fathers hoped that the Electoral College would discourage party politics, yet protect the individual states from excessive federal power. The constitution provides that each state have as many electors as it has congressional seats. Thus, a state's electoral vote depends on the size of its House delegation, which is roughly proportional to population, modified by adding 2 votes corresponding to its two Senate seats. This actually gives small states electoral power disproportionate to their populations. For example, Kentucky, with a 1990 census population of 3.7 million, has 8 electoral votes, while Idaho has 4 electoral votes but only 27% of Kentucky's population.

In this chapter we study different systems of voting and different ways of apportioning representatives in a legislative body like the U.S. House of Representatives. As you will learn, the design of a completely fair apportionment method or voting system is an impossible dream: every solution can give undesirable results in certain circumstances. While we can argue about the "best" system, we will have to work with an imperfect one.

Source: U.S. National Archives & Records Administration; William C. Kimberling, "The Electoral College."

- **16.1** The Possibilities of Voting
- **16.2** The Impossibilities of Voting
- **16.3** The Possibilities of Apportionment
- **16.4** The Impossibilities of Apportionment

Collaborative Investigation:
  Class Favorites, An Election Exploration

Chapter 16 Test

CHAPTER 16 Voting and Apportionment

16.1 The Possibilities of Voting

Harold Washington won the 1983 Chicago Democratic mayoral primary with only 36% of the vote. He eventually became Chicago's first black mayor. He died of a heart attack while in office on Nov. 15, 1987.

If a group of people want to make one choice from two alternatives, then each member of the group indicates which of the two alternatives he or she prefers. The alternative awarded the majority of the votes is the choice of the group. Clearly, selecting one alternative from two is a simple, straightforward process. The outcome, ignoring ties, is uncontestable. The essentially two-party political system of the United States means nearly all elections for public office at the state and federal level are of the straightforward variety. Most other democratic nations have three or more major political parties. Elections in these multiple-party democracies are far from straightforward.

Suppose a group of one hundred people want to select one alternative from among three. The group decides each member should indicate his or her first preference, and the alternative with the most votes will be selected as the choice of the group. It is possible that none of the alternatives gets a **majority** (more than half) of the votes. For example, the outcome of the vote may be

33 votes for Alternative a,
31 votes for Alternative b,

and
36 votes for Alternative c.

While Alternative c does *not* have a majority of the votes, it does have a **plurality** (the greatest number) of the votes. Alternative c is declared the people's choice, in spite of being disliked by 64 of 100 voters.

Situations such as this one are not rare. Whenever a group of people are required to select one alternative from three or more, an alternative favored by far fewer than a majority of the people may be selected.

Over the years, a variety of alternative voting methods have been developed. Most of these methods require each voter to give a complete ranking of the candidates/alternatives. In other words, it is necessary to know which candidate each voter ranks first, second, third, and so on. The number of possible rankings of n candidates is calculated by the factorial formula $n!$.

Problem Solving

Many techniques covered in Chapter 11 on Counting Methods have direct application in the discussion of voting.

Suppose, for example, that there are twelve voters and three candidates/alternatives. Let the three candidates be a, b, and c. There are

$$3! = 3 \cdot 2 \cdot 1 \quad \text{or} \quad 6 \text{ possible ways}$$

to rank the 3 candidates. The notation $a > b > c$ indicates the ranking in which Candidate a is the voter's first choice, Candidate b is the voter's second choice, and Candidate c is his or her third choice. Suppose 3 of the 12 voters chose the ranking $a > b > c$, 3 voters chose the ranking $c > a > b$, 4 voters chose the ranking $b > a > c$, and 2 voters chose the ranking $c > b > a$. This **profile of voters** is displayed in the table at the left. Although there are six possible rankings, only four are reflected in this voter profile.

Number of Voters	Ranking
3	$a > b > c$
3	$c > a > b$
4	$b > a > c$
2	$c > b > a$

The following discussion introduces six alternative voting methods. We do not consider final election outcomes that result in ties.

16.1 The Possibilities of Voting

The Plurality Method The *plurality method* is the most common voting method. It is the process used by the one hundred voters in the opening discussion, and in nearly all state and federal elections in the United States.

The Plurality Method

In the **plurality method** of voting, each voter gives one vote to his or her top-ranked candidate. The candidate with the most votes, a plurality of the votes, wins the election.

The plurality method does not require a voter to rank the candidates completely. Only the voter's first choice is important in the tally of votes. When an election involves only two alternatives, the plurality of the votes is the majority of votes.

Number of Voters	Ranking
6	b > p > c
5	c > p > b
4	p > c > b

EXAMPLE 1 The Culinary Arts Club is planning its annual banquet. Entree choices are beef, chicken, or pork. The 15 members of the club rank the choices according to the profile in the table. Let b represent beef, c chicken, and p pork.

Each voter's complete ranking is shown, but *only the top-ranked candidate matters in a plurality method election.* Beef receives 6 votes, chicken receives 5, and pork receives 4. The winner is beef because it receives a plurality of the votes. ∎

Number of Voters	Ranking
3	a > b > c
3	c > a > b
4	b > a > c
2	c > b > a

EXAMPLE 2 Twelve voters rank the candidates for chairperson in their school board election according to the profile in the table, where a represents Alice Adams, b represents Bobby Brown, and c represents Cathy Coutts.

The top-ranked candidate will be elected. Candidate Alice Adams gets 3 votes and Bobby Brown gets 4 votes. Cathy Coutts gets 3 votes plus 2 more votes, for a total of 5 votes. Thus, Cathy Coutts is elected chairperson of the school board because she has a plurality of the votes. ∎

The Pairwise Comparison Method The *pairwise comparison method* is more involved than the plurality method. It requires voters to make a choice between every pair of candidates, and all possible pairs of candidates must be compared to find the outcome of the election. Pairwise comparison method elections are similar to round-robin tournaments in sports.

The Pairwise Comparison Method

In the **pairwise comparison method,** each voter gives a complete ranking of the candidates. For *each pair* of candidates a and b, the number of voters preferring Candidate a is compared with the number of voters preferring Candidate b. The candidate receiving more votes is awarded one point. If the two candidates receive an equal number of votes, each is awarded a half point. The candidate with the most points is the winner of the election.

If there are n candidates, then the number of pairwise comparisons needed for a pairwise comparison method election is calculated by the combinations formula $C(n, 2)$.

A voter's complete ranking of the candidates allows the direct comparison of any pair of candidates. All candidate positions in the rankings are ignored, except the positions of the two candidates being considered. The voter's preferred candidate of the pair is the candidate ranking higher in the order. For example, a voter ranking the six candidates s > u > n > d > a > y prefers Candidate s to Candidate a, in a direct pairwise comparison. The voter prefers Candidate u in direct comparisons with any of Candidates n, d, a, or y.

Number of Voters	Ranking
6	b > p > c
5	c > p > b
4	p > c > b

EXAMPLE 3 Some members of the Culinary Arts Club are disappointed with the beef outcome of the plurality method election from Example 1. They request that the election be redone, using the pairwise comparison method. The profile of how the 15 club members rank beef, pork, and chicken is repeated in the table.

There are $n = 3$ candidates, so

$$C(n, 2) = C(3, 2) = 3.$$

Thus, three pairs must be examined:

p and c,   p and b,   c and b.

Compare pork and chicken. Pork is ranked higher than chicken in the first and third rankings; thus, $6 + 4 = 10$ voters prefer pork to chicken. Chicken is ranked higher than pork by the 5 voters with the second ranking. Pork is preferred to chicken by a margin of 10 to 5 and receives one point.

Compare pork and beef. Pork is ranked higher than beef in the second and third rankings; this means $5 + 4 = 9$ voters prefer pork to beef. The 6 voters with the first ranking rank beef higher than pork. Pork is preferred to beef by a margin of 9 to 6 and receives another point.

Compare chicken and beef. Chicken outranks beef in the second and third rankings; these $5 + 4 = 9$ voters prefer chicken to beef. Beef is ranked higher than chicken by the 6 voters with the first ranking. Chicken is preferred to beef by a margin of 9 to 6 and receives one point.

All three pairs have been considered and the results are pork with 2 points, chicken with 1 point, and beef with 0 points. The winner of the entree election, using the pairwise comparison method, is pork. ■

Number of Voters	Ranking
3	a > b > c > d
2	c > a > b > d
5	d > a > b > c
6	c > b > d > a
4	b > a > d > c

EXAMPLE 4 The twenty-member executive board of the College Art Association is meeting to decide the location of the next yearly meeting. The board members are considering four cities:

Atlanta a,   Boston b,   Chicago c,   and   Dallas d.

Each executive board member gives a complete ranking of the four cities. The profile is organized in the table. Only 5 of the $4! = 24$ possible rankings are used. The five rankings keep the profile size manageable, but interesting.

There are 4 candidates, so $C(n, 2) = C(4, 2) = 6$. Thus, six pairs of cities must be compared:

a and b,   a and c,   a and d,   b and c,   b and d,   c and d.

Compare a and b. Atlanta is preferred to Boston by the $3 + 2 + 5 = 10$ voters with the first three rankings. The remaining $6 + 4 = 10$ voters with the last two rankings prefer Boston to Atlanta. The tie for this pair means **Atlanta** and **Boston** each receive $\frac{1}{2}$ **point**.

Atlanta wins the comparison with Chicago by a margin of
$3 + 5 + 4 = 12$ to $2 + 6 = 8$. *Atlanta, 1 point*
Dallas wins the comparison with Atlanta by a margin of
$5 + 6 = 11$ to $3 + 2 + 4 = 9$. *Dallas, 1 point*
Boston wins when compared with Chicago,
$3 + 5 + 4 = 12$ to $2 + 6 = 8$. *Boston, 1 point*
Boston also wins when compared with Dallas,
$3 + 2 + 6 + 4 = 15$ to 5. *Boston, 1 point*
Chicago wins the comparison with Dallas,
$3 + 2 + 6 = 11$ to $5 + 4 = 9$. *Chicago, 1 point*

All six pairwise comparisons are complete. Boston wins two comparisons and ties in a third, receiving 2.5 points. Atlanta wins one comparison and ties in another, receiving 1.5 points. Chicago and Dallas each win one comparison and each receive 1 point. The host city selected by the pairwise comparison method is Boston.

Notice that the choice of voting method can lead to different outcomes. Beef is the plurality method selection of the Culinary Arts Club in Example 1, but Example 3 shows that pork wins a pairwise comparison method election. If the plurality method is used to decide the outcome of the election in Example 4, then Chicago, with a total of 8 first-place votes, is selected instead of Boston. (Atlanta has 3 first-place votes, Boston has 4, and Dallas has 5.)

The Borda Method What is it about pork in the Culinary Arts Club profile that allows it to fare so well in the pairwise comparison method, while losing to beef in a plurality method election? Every Culinary Arts Club member ranks pork first or second, and never last. The profile for the executive board of the College Art Association shows Boston also is highly ranked by many voters. This allows Boston to win the pairwise comparison method election, although Boston does not win a plurality method election.

The *Borda method* was first proposed in 1781 by French military man Jean-Charles de Borda (1733–1799). Winners of many popular music awards and the winner of the annual Heisman Trophy are decided with the Borda method, or a version of that method. Unlike the plurality method, which considers only the candidate a voter ranks first, the Borda method uses a voter's complete ranking of the candidates. The pairwise comparison method also uses a voter's complete ranking, but the outcome depends on many individual comparisons. Indeed, the number of comparisons necessary makes the pairwise comparison method difficult to use for contests or elections in which there are more than 4 or 5 candidates. In contrast, the Borda method computes a weighted sum for each candidate to decide the outcome of the election.

Jean-Charles de Borda (1733–1799), a Frenchman of diverse talents, was a military man with skills on land and sea, serving as both a cavalry officer and a naval captain. In addition to voting theory, he studied and wrote about other topics in mathematics and physics. Borda was fully aware that his voting method is easy for an insincere voter to manipulate. Such a voter could misrepresent his real ranking in order to help a favorite candidate. In reply to his detractors, Borda said he assumed his method would be used by honest men.

The Borda Method

In the **Borda method,** each voter must give a complete ranking of the candidates. Let n be the number of candidates. Each first-place vote a candidate receives is worth $n - 1$ points. Each second-place vote a candidate receives is worth $n - 2$ points. A third-place vote earns $n - 3$ points, fourth place earns $n - 4$ points, ..., and last place earns $n - n = 0$ points. Points are tallied separately for each candidate. The candidate with the highest tally of points is the winner.

The Jesus Seminar, a collection of biblical scholars, uses a Borda-like voting method to decide on the credibility of the Gospels. Did Jesus really deliver the Sermon on the Mount? Following a proper scholarly debate, members vote by dropping a colored bead in the voting box. Depositing a red bead means the scholar believes the Sermon on the Mount is definitely a true historical event, a pink bead means it is likely true, a gray bead means maybe it is true, a black bead means the scholar rejects any possibility of the event being factual. A weighted mathematical formula is used to compile the results of the vote and render a decision on the validity of the event under discussion.

EXAMPLE 5 The Culinary Arts Club decides to use the Borda method to select the entree for the annual banquet. A first-place vote earns 2 points, because there are $n = 3$ choices: beef, pork, and chicken. A second-place vote earns 1 point and a third-place vote earns 0 points. The profile is given in the table, with additional columns to aid the calculations.

Number of Voters	Ranking	First Place Earns 2 Points	Second Place Earns 1 Point	Third Place Earns 0 Points
6	$b > p > c$	b	p	c
5	$c > p > b$	c	p	b
4	$p > c > b$	p	c	b

Beef earns $6 \cdot 2 = 12$ points from the six voters with the first ranking, $5 \cdot 0 = 0$ points from the five voters with the second ranking, and $4 \cdot 0 = 0$ points from the four voters with the third ranking.

The tally for beef is $12 + 0 + 0 = 12$ points.

Chicken earns $6 \cdot 0 = 0$ points from the six voters with the first ranking, $5 \cdot 2 = 10$ points from the five voters with the second ranking, and $4 \cdot 1 = 4$ points from the four voters with the last ranking.

The tally for chicken is $0 + 10 + 4 = 14$ points.

Pork earns $6 \cdot 1 = 6$ points from the six voters with the first ranking, $5 \cdot 1 = 5$ points from the five voters with the second ranking, and $4 \cdot 2 = 8$ points from the four voters with the third ranking.

The tally for pork is $6 + 5 + 8 = $ **19 points.** ← Winner

The winner of the Borda method entree election is pork. Pork is the winner of the Borda method and the pairwise comparison method entree elections. Both methods take into account the overall popularity of pork. ∎

EXAMPLE 6 Midwestern members of the College Art Association executive board (see Example 4) are disappointed that Chicago was not selected by the pairwise comparison method as the site of the next yearly meeting, so they suggest a Borda method election. The modified voter profile is given in the table. The $n = 4$ choices, Atlanta, Boston, Chicago, and Dallas, mean each first-place vote earns 3 points, second place earns 2 points, third place earns 1 point, and fourth place earns 0 points.

Number of Voters	Ranking	First Place (3 points)	Second Place (2 points)	Third Place (1 point)	Fourth Place (0 points)
3	$a > b > c > d$	a	b	c	d
2	$c > a > b > d$	c	a	b	d
5	$d > a > b > c$	d	a	b	c
6	$c > b > d > a$	c	b	d	a
4	$b > a > d > c$	b	a	d	c

Atlanta earns $(3 \cdot 3) + (2 \cdot 2) + (5 \cdot 2) + (6 \cdot 0) + (4 \cdot 2) = 31$ points.
Boston earns $(3 \cdot 2) + (2 \cdot 1) + (5 \cdot 1) + (6 \cdot 2) + (4 \cdot 3) = 37$ points. ← Winner
Chicago earns $(3 \cdot 1) + (2 \cdot 3) + (5 \cdot 0) + (6 \cdot 3) + (4 \cdot 0) = 27$ points.
Dallas earns $(3 \cdot 0) + (2 \cdot 0) + (5 \cdot 3) + (6 \cdot 1) + (4 \cdot 1) = 25$ points.

The winner of the Borda method election is Boston.

Pork, the Borda method selection in Example 5, also is the selection using the pairwise comparison method for the same profile in Example 3. Boston, the Borda method selection in Example 6, also is the pairwise comparison method selection for the same profile in Example 4. The next example shows that the Borda and pairwise comparison methods do not *always* select the same winner.

EXAMPLE 7 Fifteen members of Miss Green's kindergarten class are voting on a class pet. The choices are a hamster h, a fish f, a turtle t, or a canary c. The profile in the table shows remarkable agreement among the students. It uses only two of the $4! = 24$ possible rankings.

The number of class pet candidates is $n = 4$. If the Borda method is used for the class selection, the same points are assigned to places in the rankings as in Example 6: first place earns 3 points, second place 2 points, third place 1 point, and fourth place 0 points.

A hamster earns $(10 \cdot 3) + (5 \cdot 0) = 30$ points.
A fish earns $(10 \cdot 2) + (5 \cdot 3) = 35$ points. ← Winner
A turtle earns $(10 \cdot 1) + (5 \cdot 2) = 20$ points.
A canary earns $(10 \cdot 0) + (5 \cdot 1) = 5$ points.

The winner of the Borda method election is a fish, with 35 Borda points.

If the pairwise comparison method is used, a hamster wins each pairwise comparison with a fish, a turtle, and a canary, by a margin of 10 to 5. A fish wins each comparison with a turtle and a canary, by a margin of 15 to 0. A turtle wins in the comparison with a canary, by a margin of 15 to 0.

A hamster earns 3 points. ← Winner
A fish earns 2 points.
A turtle earns 1 point.
A canary earns 0 points.

All pairwise comparisons are complete. The winner using the pairwise comparison method is a hamster, with 3 pairwise comparison points.

The Hare Method
The *Hare method* was first proposed in 1861 by British political theorist Thomas Hare (1806–1891). It is a variation of the plurality method in which candidates are eliminated in sequential rounds of voting. Votes are transferred from eliminated candidates to remaining candidates. The president of France and public officials in Australia are selected by the Hare method. It also is known as the *plurality with elimination method,* or the *single transferable vote system.*

Hare elections do not require the voters initially to rank the complete list of candidates. Instead, voters need to decide only which candidate they prefer most in each round. This is considered an advantage of the method.

In a traditional *n*-candidate Borda method election, a first-place vote earns $n - 1$ points, second place earns $n - 2$ points, third place earns $n - 3$ points, and so on, with last place earning 0 points. The important defining feature is that the points for the various places in a voter's ranking must be equally spaced. In other words, the difference between the points given for any two consecutive rankings must be equal.

Instead of using the traditional 2-, 1-, 0-point scheme, try finding the outcome of the Culinary Arts Club Borda method vote using 1 point for first place, 0 points for second, and −1 points for third place. Try again using a 17-, 10-, 3-point scheme.

Number of Voters	Ranking
10	h > f > t > c
5	f > t > c > h

Thomas Hare (1806–1891) was a British lawyer. His reason for devising his voting method was to secure proportional representation of all classes in the United Kingdom, including minorities, in the House of Commons and other elected groups. The Hare method was considered to be "among the greatest improvements yet made in the theory and practice of government" by Hare's contemporary, John Stuart Mill.

The Hare Method

In a **Hare method** election, each voter gives one vote to his or her favorite candidate in the first round. If a candidate receives a *majority* of the votes, he or she is declared the winner. If no candidate receives a majority of the votes, the candidate (or candidates) with the fewest votes is (are) eliminated, and a second election is conducted on the remaining candidates.

Each voter gives one vote to his or her favorite remaining candidate in the second round. If a candidate receives a *majority* of the votes in the second election, he or she is declared the winner. If no candidate receives a majority, the candidate (or candidates) with the fewest votes is (are) eliminated, and a third election is conducted on the remaining candidates.

The rounds of voting continue to eliminate candidates until one candidate receives a *majority* of the votes.

It may take many rounds of voting to learn the winner of a Hare method election. To simulate Hare elections in the following examples, it is assumed each voter has given a complete, initial ranking of the candidates. It also is assumed that when a candidate is eliminated from a voter's ranking, the voter votes for the next candidate in the ranking. For example, if the voter initially ranks the candidates $s > u > n > d > a > y$ and Candidate s is eliminated in the first round of voting, the voter will vote for u in the second round of voting. If s, u, and n are eliminated, the voter will vote for d in the next round.

Number of Voters	Ranking
6	$b > p > c$
5	$c > p > b$
4	$p > c > b$

EXAMPLE 8 Chicken entree supporters of the Culinary Arts Club suggest the vote be redone using the Hare method. The 15-member club profile is repeated in the table. Club members vote for their favorite of the three entrees in the first round. The results of the vote between beef, chicken, and pork are familiar.

Beef gets 6 votes.

Chicken gets 5 votes.

Pork gets 4 votes.

No entree receives a majority (8 or more) of the votes in the first round. Pork receives the fewest votes and is eliminated.

Club members vote for their favorite of the remaining entrees, beef or chicken, in the second round of Hare method voting. The 4 club members with the third ranking vote for chicken, because pork is eliminated in the first round and they prefer chicken to beef. The other 11 voters vote as they did in the first round.

Beef gets 6 votes.

Chicken gets $5 + 4 = 9$ votes. ← Winner

Chicken receives a majority of the votes in the second round and is declared the entree selection by the Hare method.

16.1 The Possibilities of Voting

The following example shows that the mechanics of a Hare method vote are identical to the mechanics of a plurality method vote when a candidate enjoys a majority of first-place rankings.

Number of Voters	Ranking
3	a > b > o
3	b > a > o
4	a > o > b
2	o > b > a

EXAMPLE 9 The D'Amico Family Fruit Company is deciding which fruit the company will donate to the city's new organic foods school lunch program. The 12 members of the D'Amico family rank the choices apples a, bananas b, and oranges o, as shown in the profile. Each of the 12 members votes for the fruit he or she ranks highest of the three in the first round.

Apples get **7 votes**. ← Winner

Bananas get 3 votes.

Oranges get 2 votes.

Apples have 7 of 12 votes, a majority, at the end of the first round. Apples are selected by the Hare method election. ∎

EXAMPLE 10 Chicago supporters still are grumbling about being able to win a plurality method election, but lose both the pairwise comparison and the Borda method elections to Boston. They convince the College Art Association to vote again using the Hare method. The twenty-member executive board profile is repeated in the table.

Number of Voters	Ranking
3	a > b > c > d
2	c > a > b > d
5	d > a > b > c
6	c > b > d > a
4	b > a > d > c

Atlanta gets 3 votes.
Boston gets 4 votes.
Chicago gets 2 + 6 = 8 votes.
Dallas gets 5 votes.

No city receives a majority (11 or more) of the votes in the first round of voting between Atlanta, Boston, Chicago, and Dallas. Atlanta has the fewest votes in the first round. It is eliminated.

Boston, Chicago, and Dallas are voted on in the second round of the Hare election. The 3 voters with the first ranking vote for Boston, because Atlanta was eliminated. Other voters vote as they did in the first round.

Boston gets 4 + 3 = 7 votes.

Chicago gets 8 votes.

Dallas gets 5 votes.

No city receives a majority of the votes in the second round of voting. Dallas has the fewest votes and is eliminated.

The third-round vote is between Boston and Chicago. The 5 voters with the middle profile prefer Boston to Chicago; with both southern cities eliminated, they transfer their votes to Boston. The other 15 voters vote as they did in the second round.

Boston gets $7 + 5 = 12$ votes. ← Winner

Chicago gets 8 votes.

Boston receives a majority of the votes and is the winner of the Hare method election. ∎

EXAMPLE 11 Nineteen members of the Outer Banks Film Confederation meet to decide which of four *Star Wars* movies will be the theme of a charity costume ball. They must choose from *The Empire Strikes Back* e, *Return of the Jedi* r, *The Phantom Menace* p, and *Attack of the Clones* a. The profile of complete rankings for the 19 members of the confederation allows a simulated Hare method election. Each member votes for his or her favorite of the four movies in the first round of the Hare election. No movie gets a majority (10 or more) of the votes. *The Empire Strikes Back* has the fewest votes and is eliminated.

Number of Voters	Ranking
1	$e > a > r > p$
6	$a > e > p > r$
7	$r > p > e > a$
3	$p > e > r > a$
2	$p > e > a > r$

The Empire Strikes Back gets 1 vote.
Attack of the Clones gets 6 votes.
Return of the Jedi gets 7 votes.
The Phantom Menace gets $3 + 2 = 5$ votes.

Each member votes for his or her favorite among *Return of the Jedi, The Phantom Menace,* or *Attack of the Clones* in the second round. The voter with the first ranking now votes for *Attack of the Clones,* because e is eliminated. The other 18 voters vote as they did in the first round.

Attack of the Clones gets $6 + 1 = 7$ votes.

Return of the Jedi gets 7 votes.

The Phantom Menace gets 5 votes.

No movie gets a majority of the votes in the second round. *The Phantom Menace* has the fewest votes and is eliminated.

The final round of voting is between *Attack of the Clones* and *Return of the Jedi*. The 5 voters with *The Phantom Menace* as their first choice do not transfer their 5 votes in a single block. The 3 members with the fourth ranking prefer *Return of the Jedi* to *Attack of the Clones*. The 2 members with the bottom profile prefer *Attack of the Clones* to *Return of the Jedi*.

Attack of the Clones gets $7 + 2 = 9$

Return of the Jedi gets $7 + 3 = 10$ votes. ← Winner

Return of the Jedi wins a majority of the votes in the final round and is selected by the Hare method as the theme of the Outer Banks Film Confederation costume ball.

The Sequential Pairwise Comparison Method The *sequential pairwise comparison method* is another variation of the plurality method. Decisions of the United States Congress often are made by the sequential pairwise comparison method, because amendments to bills are introduced and then voted on in sequential order. Voters are not required initially to rank a complete list of the candidates, but from a profile of complete rankings the necessary information can be gathered. Voting goes in rounds similar to the Hare method.

> **The Sequential Pairwise Comparison Method**
>
> In the **sequential pairwise comparison method,** the entire field of candidates are compared two at a time, in a predetermined order. The first and second candidates in the predetermined order are compared. The candidate with fewer votes is eliminated, and the candidate with more votes advances to challenge the third candidate in the predetermined order. The candidate with fewer votes in the next comparison is eliminated and the candidate with more votes advances to challenge the fourth candidate in the predetermined order.
>
> The process continues until two candidates are left. In the final pairwise comparison, the candidate with the majority of the votes is the winner.

Predetermined orders will be written as a sequence of candidates. For example, the sequence of candidates s, u, n, d, a, y denotes the predetermined order in which s is first, u is second, n is third, d is fourth, a is fifth, and y is sixth. The sequential pairwise comparison method would begin with a comparison of s and u. The winner of this pair would then face a comparison with n, and so on.

The next example makes a dramatic point. If the sequential pairwise comparison method is used, the predetermined order of the candidates can radically change the outcome of an election.

EXAMPLE 12 All members of the Intergalactic Society give a complete ranking of the candidates for the site of the next intergalactic meeting. The choices are the planets Saturn s, Jupiter j, Mars m, and Venus v. The executive board uses a computer to figure out the winner of each possible pairwise comparison. No voter profile is given for this example—only the results of the individual pairwise comparisons are shown.

> Saturn beats Jupiter.
> Saturn beats Mars.
> Venus beats Saturn.
> Jupiter beats Mars.
> Jupiter beats Venus.
> Mars beats Venus.

The executive board decides to use the sequential pairwise comparison method and the predetermined order

$$s, \quad j, \quad m, \quad v,$$

to find the winner of the election. Saturn and Jupiter are compared. Saturn wins the first vote, and is compared with Mars, the next candidate in the order. Saturn wins the second vote over Mars, and is compared with Venus, the last candidate in the order. Venus wins the third vote over Saturn, and the election.

The executive board members (who secretly hope Mars will be the site of the next intergalactic meeting) decide to try another predetermined order:

$$j, \quad m, \quad v, \quad s.$$

Jupiter and Mars are compared. Jupiter wins. Mars is out after the first vote.

Not a group to give up easily, the board decides to try the following predetermined order:

$$v, \quad s, \quad m, \quad j.$$

Venus and Saturn are compared, and Venus wins. Venus then faces Mars, the next candidate in the order. Mars wins over Venus. Mars and the last candidate in the order, Jupiter, are compared. Jupiter wins the vote and the election.

The board decides to try yet another predetermined order:

$$s, \quad j, \quad v, \quad m.$$

Saturn and Jupiter are compared, and Saturn wins. Saturn faces Venus, and Venus wins. Venus faces Mars. Mars wins the vote and, finally, an election. ∎

The Approval Voting Method The *approval voting method* was introduced in the 1970s. It is another variation of the plurality method and is used to elect the secretary-general of the United Nations. Approval voting is useful for elections involving numerous candidates. The strength of the method is that voters do not need to rank the list of candidates completely when using approval voting. Voters need to decide only which candidates they like and which they dislike.

> **The Approval Voting Method**
>
> In the **approval voting method,** each voter is allowed to give one vote to as many candidates as he or she wishes. The candidate with the most approval votes is the winner of the election.

Many emerging democracies around the globe need a voting system that is easy for their citizens to use. Approval voting is a good alternative, but it is very temperamental, as the next example shows. Voters are not required to rank the candidates completely for an approval election, but for the following voting simulation a complete ranking profile is used.

EXAMPLE 13 If three candidates are in an approval method election, then each voter can decide to vote for his or her top-ranked candidate only, or for both of his or her top-ranked candidates. (If the voter voted for all three candidates, the vote would be meaningless because it would fail to make a distinction between the candidates.) The familiar profile for the Culinary Arts Club is repeated in the table, with an additional column of information necessary for an approval voting method election simulation.

Number of Voters	Ranking	Number of Voters Voting for Two of the Three Candidates
6	b > p > c	x, $0 \leq x \leq 6$
5	c > p > b	y, $0 \leq y \leq 5$
4	p > c > b	z, $0 \leq z \leq 4$

The variable x in $0 \leq x \leq 6$ in the first row represents how many of the 6 voters with this ranking vote for both b, their top-ranked candidate, and p, their second-ranked candidate. The variable y in $0 \leq y \leq 5$ in the second row represents how many of the 5 voters with the second ranking vote for both top-ranked c and second-ranked p. The variable z in $0 \leq z \leq 4$ represents how many of the 4 voters with the third ranking vote for both top-ranked p and second-ranked c.

Approval voting method results of this profile can be generalized. Beef always receives 6 approval votes from the voters with the top ranking. Beef cannot get any additional votes, however, because it is not ranked second by anyone. Chicken always receives 5 votes from the middle ranking. Chicken also can get up to $z = 4$ additional votes from the voters with the third ranking who rank it second. Pork always receives 4 votes from the third ranking. Pork also can get up to $x = 6$ and $y = 5$ additional votes from the 11 voters with the top two rankings who rank pork second. The range of values the variables can assume makes it possible for any of the three candidates to win the election.

Beef gets 6 approval votes.

Chicken gets $5 + z$ approval votes, $0 \leq z \leq 4$.

Pork gets $4 + x + y$ approval votes, $0 \leq x \leq 6, 0 \leq y \leq 5$.

The outcome of the election depends on the values of the variables x, y, and z. The values of the variables depend on the whims of the individual voters.

- If each voter distrusts the new system and votes only for his or her top-ranked candidate, the approval method election is a plurality method election. Beef, with 6 votes, wins.
- If 3 of the 6 voters with the top ranking decide to approve of pork and beef, then $x = 3$ additional votes go to pork. Suppose the other voters approve only their top-ranked candidate.

Beef gets 6 approval votes.

Chicken gets 5 approval votes.

Pork gets $4 + 3 = 7$ approval votes.

Pork receives the most approval votes and is the selected entree.

- If 3 of the 4 voters with the bottom ranking decide to approve of both pork and chicken, then $z = 3$ additional votes go to chicken. If 1 of the 5 voters with the middle ranking decides to approve of both chicken and pork, then $y = 1$

additional vote goes to pork. If 2 of the 6 voters with the top ranking decide to approve of both beef and pork, then $x = 2$ additional votes go to pork.

Beef gets 6 approval votes.

Chicken gets $5 + 3 = 8$ approval votes.

Pork gets $4 + 2 + 1 = 7$ approval votes.

Chicken receives the most approval votes.

The distinct election outcomes in the previous example are not the result of using different voting methods. Three different outcomes, for the same profile of voters, suggest the capricious nature of approval voting. The voters *did not* change the way they ranked the candidates. They simply changed how many candidates in the ranking received an approval vote.

16.1 EXERCISES

1. A local animal shelter is choosing a poster dog. The choices are an Australian shepherd a, a boxer b, a cocker spaniel c, or a dalmatian d. Thirteen staff members completely rank the four dogs on their ballots. The thirteen individual ballots are the columns in the table.

Voter	#1	#2	#3	#4	#5	#6	#7	#8	#9	#10	#11	#12	#13
1st place	b	b	a	c	d	d	c	c	b	a	a	a	a
2nd place	c	c	d	d	c	a	a	a	c	c	d	b	b
3rd place	a	a	c	b	b	b	d	d	a	d	c	c	c
4th place	d	d	b	a	a	c	b	b	d	b	b	d	d

(a) How many different ways can a staff member complete his or her ballot?
(b) Write a voter profile for this election.
(c) Use the plurality method to determine the shelter's poster dog.

2. Repeat Exercise 1 using the following thirteen ballots.

Voter	#1	#2	#3	#4	#5	#6	#7	#8	#9	#10	#11	#12	#13
1st place	a	b	d	c	a	a	b	c	d	a	a	a	a
2nd place	b	c	b	b	d	d	c	b	b	b	b	b	b
3rd place	c	a	c	a	c	c	d	d	c	c	d	d	c
4th place	d	d	a	d	b	b	a	a	a	d	c	c	d

3. If there are $n = 5$ candidates, then how many different rankings of the candidates are possible? What if there are $n = 7$ candidates?

4. If there are $n = 6$ candidates, how many different rankings of the candidates are possible? What if there are $n = 8$ candidates?

5. Comment on the implications of the number of different rankings for elections involving more than $n = 4$ candidates.

6. How many pairwise comparisons are needed to learn the outcome of an election involving $n = 5$ candidates? What if there are 7 candidates?

7. How many pairwise comparisons are needed to learn the outcome of an election involving $n = 6$ candidates? What if there are 8 candidates?

8. Comment on the implications of the number of pairwise comparisons needed to learn the outcome of an election involving more than $n = 4$ candidates.

9. Comment on the logistics of using the pairwise comparison method to determine the outcome of an election involving $n = 8$ candidates.

10. A 13-member committee is selecting a chairperson. The 3 candidates are Albert Werner a, Barbara Hightower b, and Charles Smith c. Each committee member completely ranked the candidates on a separate ballot. The voter profile of the committee is summarized in the table.

Number of Voters	Ranking
4	a > b > c
2	b > c > a
4	b > a > c
3	c > a > b

Determine the chairperson using the
(a) plurality method
(b) pairwise comparison method
(c) Borda method
(d) Hare method
(e) sequential pairwise comparison method and the predetermined order c, b, a of the candidates.

11. Repeat Exercise 10 using the voter profile given here.

Number of Voters	Ranking
3	a > c > b
4	c > b > a
2	b > a > c
4	b > c > a

12. A 13-member committee is selecting a new company logo from 3 alternatives a, b, and c. Each committee member completely ranked the possible logos on a separate ballot. The voter profile of the committee is summarized in the table at the top of the next column.

Determine the new company logo using the
(a) plurality method
(b) pairwise comparison method
(c) Borda method
(d) Hare method
(e) sequential pairwise comparison method and the predetermined order a, c, b.

Number of Voters	Ranking
2	a > c > b
3	c > b > a
4	b > a > c
4	b > c > a

13. Repeat Exercise 12 using the following voter profile. For part (e) use the predetermined order b, c, a.

Number of Voters	Ranking
6	a > b > c
1	b > c > a
3	b > a > c
3	c > a > b

14. A senator invited one person from each of the 21 counties in her state to a weekend workshop. The senator asked the attendees to rank the issues of job creation j, education e, health care h, and gun control g in order of importance to themselves and their counties. The 21 members each gave a complete ranking of the issues on a separate ballot. The voter profile of the workshop is summarized in the table.

Number of Voters	Ranking
3	e > h > g > j
6	h > e > g > j
5	j > g > e > h
4	g > e > h > j
3	j > e > h > g

Determine which issue workshop members felt was the highest priority using the
(a) plurality method
(b) pairwise comparison method
(c) Borda method
(d) Hare method
(e) sequential pairwise comparison method and the predetermined order h, j, e, g.

15. Repeat Exercise 14 using the following voter profile.

Number of Voters	Ranking
6	h > j > g > e
5	e > g > j > h
4	g > j > h > e
3	j > h > g > e
3	e > j > h > g

16. For the profile of voters found in Exercise 2(b), determine the animal shelter's poster dog using the following voting methods.
    (a) pairwise comparison method
    (b) Borda method
    (c) Hare method

17. Repeat Exercise 16 using the profile of voters in Exercise 1(b).

18. Members of the Adventure Club are selecting an activity for their yearly vacation. Choices are hiking in the desert h, rock climbing c, white-water kayaking k, bungee jumping b, or just sitting by a mountain lake and tanning t. The voter profile of the 55 members is shown.

Number of Voters	Ranking
18	t > k > h > b > c
12	c > h > k > b > t
10	b > c > h > k > t
9	k > b > h > c > t
4	h > c > k > b > t
2	h > b > k > c > t

Determine the selected activity using the
(a) plurality method
(b) pairwise comparison method
(c) Borda method
(d) Hare method
(e) sequential pairwise comparison method and the predetermined order b, t, c, k, h.

19. Repeat Exercise 18 using the following voter profile.

Number of Voters	Ranking
18	t > b > h > k > c
12	c > h > b > k > t
10	k > c > h > b > t
9	b > k > h > c > t
4	h > c > b > k > t
2	h > k > b > c > t

20. One common solution to an election in which no candidate receives a majority of first-place votes is to have a runoff between the two candidates with the most first-place votes. For the Adventure Club voter profile in Exercise 18, what activity is selected if a runoff is held between the two candidates with the most first-place votes?

21. Repeat Exercise 20 using the voter profile in Exercise 19.

22. For the workshop voter profile in Exercise 14, what issue is the highest priority if a runoff is held between the two candidates with the most first-place votes?

23. For the workshop voter profile in Exercise 15, what issue is the highest priority if a runoff is held between the two candidates with the most first-place votes?

24. Another solution to an election in which no candidate receives a majority of first-place votes is to have a runoff between the candidates that rank second and third in first-place votes. The winner of that election faces the candidate with the most first-place votes to decide the final outcome of the election. Use this process to select an activity for the Adventure Club voter profile in Exercise 18.

25. Repeat Exercise 24 using the voter profile in Exercise 19.

26. Use the runoff method described in Exercise 24 to determine which logo the voters profiled in Exercise 12 select.

27. Use the runoff method described in Exercise 24 to determine which logo the voters profiled in Exercise 13 select.

16.1 The Possibilities of Voting

28. The members of a soccer team are electing their captain using the approval method. Candidates include team members Joan, Lori, Mary, and Alison. The 13 ballots have been marked as shown; an x indicates an approval vote for the candidate named in the first column.

Candidate	#1	#2	#3	#4	#5	#6	#7	#8	#9	#10	#11	#12	#13
Joan	x	x	x		x	x	x	x	x				
Lori	x	x		x				x	x		x		x
Mary			x		x		x		x	x	x		
Alison	x		x	x		x						x	

(a) Which candidate is selected as team captain by the approval method?

(b) If the soccer team decides to name the two candidates with the most approval votes as their co-captains, which two candidates will be selected?

29. Repeat Exercise 28 using the following thirteen ballots.

Candidate	#1	#2	#3	#4	#5	#6	#7	#8	#9	#10	#11	#12	#13
Joan	x		x		x	x			x	x	x		x
Lori		x	x	x	x		x	x				x	
Mary	x	x	x		x	x	x		x	x	x	x	
Alison	x	x		x		x	x	x	x		x		x

30. Suppose 7 candidates (a, b, c, d, e, f, g) are involved in a pairwise comparison method election. The number of pairwise comparisons each candidate wins is shown in the table for 6 of the 7 candidates.

Candidate	a	b	c	d	e	f	g
Number of pairwise comparisons won	4	6	3	1	5	1	?

(a) How many comparisons does candidate g win?

(b) Which candidate wins the election, and how many pairwise points does the winner have?

31. Suppose 7 candidates (a, b, c, d, e, f, g) are involved in a pairwise comparison method election. The number of pairwise comparisons each candidate wins is shown in the table for 6 of the 7 candidates.

Candidate	a	b	c	d	e	f	g
Number of pairwise comparisons won	3	5	7	1	2	?	1

(a) How many comparisons does candidate f win?

(b) Which candidate wins the election, and how many pairwise points does the winner have?

32. Suppose 8 candidates (a, b, c, d, e, f, g, h) are involved in a pairwise comparison method election. The number of pairwise comparisons each candidate wins is shown in the table for 7 of the 8 candidates.

Candidate	a	b	c	d	e	f	g	h
Number of pairwise comparisons won	4	5	2	?	3	1	2	3

(a) How many comparisons does candidate d win?
(b) Which candidate wins the election, and how many pairwise points does the winner have?

33. Suppose 8 candidates (a, b, c, d, e, f, g, h) are involved in a pairwise comparison method election. The number of pairwise comparisons each candidate wins is shown in the table for 7 of the 8 candidates.

Candidate	a	b	c	d	e	f	g	h
Number of pairwise comparisons won	2	6	3	4	?	2	2	2

(a) How many comparisons does candidate e win?
(b) Which candidate wins the election, and how many pairwise points does the winner have?

34. Fifteen voters are using the Borda method to select one of three candidates. Let the candidates be a, b, and c. If Candidate a receives 17 Borda points and Candidate b receives 14 Borda points, how many Borda points does Candidate c receive? Who wins the Borda method election?

35. Fifteen voters are using the Borda method to select one of three candidates. Let the candidates be a, b, and c. If Candidate a receives 15 Borda points and Candidate b receives 14 Borda points, how many Borda points does Candidate c receive? Who wins the Borda method election?

36. Twenty voters are using the Borda method to select one of five candidates a, b, c, d, or e. The number of Borda points each candidate receives is in the table, except for Candidate a. How many Borda points does Candidate a receive? Who wins the Borda method election?

Candidate	a	b	c	d	e
Number of Borda points	?	50	25	40	55

37. Twenty voters are using the Borda method to select one of five candidates a, b, c, d, or e. The number of Borda points each candidate receives is in the table, except for Candidate c. How many Borda points does Candidate c receive? Who wins the Borda method election?

Candidate	a	b	c	d	e
Number of Borda points	35	40	?	40	30

The Coombs Method The Coombs method of voting is a variation of the Hare method. Voting takes place in rounds, but instead of eliminating the candidate with the fewest first-place votes, the candidate with the most last-place votes is eliminated.

38. Use the Coombs method to determine which issue the senator's workshop members profiled in Exercise 14 felt was the highest priority.

39. Use the Coombs method to determine which issue the senator's workshop members profiled in Exercise 15 felt was the highest priority.

How a New Bill Is Processed The sequential pairwise method is used to decide the outcome of a new bill in Congress. If a representative votes for a new bill, then he or she wants it to replace the existing law. If a representative votes against a new bill, then he or she wants to retain the existing law. Before a vote is taken on the new bill, representatives can try to amend it. Amendments change the new bill. They are individually proposed and then a vote is taken. The new bill can face many sequential, one-on-one comparisons with the amendments. It may survive intact and ultimately be compared one last time with the existing law. Or, it may be altered along the way by amendments, possibly so much that the content of the new bill barely remains. If one or more amendments are approved, the amended bill is compared one-on-one with the existing law.

Suppose the existing law (e) is that you must be 21 years old to purchase alcohol. A new bill (n) proposed changes the existing law and raises the legal age to 22 years old. One amendment (a_x) alters the new bill by allowing 21-year-olds to purchase beer and wine only.

Another amendment (a_y) alters the new bill by allowing active members of the Armed Services who are 21 years old to purchase any alcohol product. The results of all possible pairwise comparisons are in the table.

Pair	How Representatives Rank the Pair
n and a_x	$n > a_x$
n and a_y	$a_y > n$
n and e	$n > e$
a_x and a_y	$a_x > a_y$
a_x and e	$e > a_x$
a_y and e	$a_y > e$

40. If the new bill faces the existing law one-on-one, which is the choice of the representatives?

41. If only Amendment a_x is proposed, the order of the sequential pairwise comparisons needed to determine the legislative outcome is n, a_x, e. What is the choice of the representatives in this case?

42. If only Amendment a_y is proposed, the order of the sequential pairwise comparisons needed to determine the legislative outcome is n, a_y, e. What is the choice of the representatives in this case?

43. If Amendment a_x is proposed and then Amendment a_y is proposed, the order of the sequential pairwise comparisons needed to determine the legislative outcome is n, a_x, a_y, e. What is the choice of the representatives in this case?

44. If Amendment a_y is proposed and then Amendment a_x is proposed, the order of the sequential pairwise comparisons needed to determine the legislative outcome is n, a_y, a_x, e. What is the choice of the representatives in this case?

45. Given the following voter profile for a 4-candidate election, arrange a total of 21 voters in such a way that candidate d is selected if the plurality method, the pairwise comparison method, the Borda method, or the Hare method is used to determine the outcome of the election.

Number of Voters	Ranking
?	$a > b > d > c$
?	$b > c > d > a$
?	$c > d > a > b$
?	$d > a > b > c$
?	$a > d > c > b$

46. Given the following voter profile for a 4-candidate election, arrange a total of 21 voters in such a way that candidate a is selected if the plurality method, the pairwise comparison method, the Borda method, or the Hare method is used to determine the outcome of the election.

Number of Voters	Ranking
?	$d > b > a > c$
?	$b > c > a > d$
?	$c > a > d > b$
?	$a > d > b > c$
?	$d > a > c > b$

47. Which voting method—plurality, pairwise comparison, Borda, or Hare—do you personally think is the best way to select one alternative from among three or more alternatives? Why do you think it is the best method?

48. Suppose only two candidates are running for office. Show that the pairwise comparison method, the Borda method, and the Hare method all select the same winner. Explain why all three methods reduce to a simple plurality method vote of giving one vote to your top-ranked candidate.

49. Comment on the power of the Intergalactic Society executive board in Example 12. What is it about the results of the individual pairwise comparisons that allows the executive board to get away with the mischief they do?

50. Discuss the strengths and weaknesses of the approval voting method. Argue that the ambiguous nature of the outcome of an approval vote is a strength of the method. Argue that it is a weakness.

16.2 The Impossibilities of Voting

Different voting methods applied to the same voter profile can show remarkable agreement. If the pairwise comparison, Borda, or Hare method is used, then Boston wins the College Art Association (CAA) city selection in the previous section. Boston does lose a plurality method CAA election, because Chicago has a plurality of first-place rankings.

The Culinary Arts Club entree selection shows how different voting methods applied to the same voter profile can result in totally different outcomes. Beef is selected by the plurality method. Pork is selected by the pairwise comparison method and the Borda method. The Hare method selects chicken.

The voting methods presented in the previous section are considered fair and reasonable. All are used in real-world decision making, but none is perfect. Approval voting and the sequential pairwise comparison method have the same type of flaw: different outcomes are possible for the same profile of voters. The voters in an approval method election have many ways to vote for their initial ranking of the candidates. The Culinary Arts Club approval method election demonstrates that this freedom can make an election outcome ambiguous. Members of the Intergalactic Society did not rearrange their initial rankings of the planets. Executive board members had the freedom to pick any predetermined order for candidate comparisons. Naturally they picked one that gave the desired result. Thus, given enough information about a voter profile, the outcome in a sequential pairwise comparison method election can be influenced.

Defects in the plurality method, the pairwise comparison method, the Borda method, and the Hare method are revealed by considering the following four desirable attributes for any voting method: the *majority criterion,* the *Condorcet criterion,* the *monotonicity criterion,* and the *independence of irrelevant alternatives* (IIA) *criterion.* The first two criteria concern desirable qualities for a voting method when it is used a single time to determine a winner. The third and fourth criteria concern desirable qualities of a voting method when the method is to be used twice in the election procedure.

The Majority Criterion
If there are only two candidates, the candidate with a majority (more than half) of first-place rankings in the voter profile wins the election. Example 1 shows that a candidate with a majority of first-place rankings in an election involving three or more candidates may not win the election.

EXAMPLE 1 The College Art Association has a new executive board. Its members are meeting to select a city to host the yearly meeting. The sites under consideration are Portland p, Boston b, Chicago c, and Oakland o. The voter profile for the new executive board shows how the 20 members rank the cities. Board members agree to use the Borda method to decide the outcome of the vote.

Number of Votes	Ranking	First Place (3 points)	Second Place (2 points)	Third Place (1 point)	Fourth Place (0 points)
8	c > p > b > o	c	p	b	o
4	c > b > o > p	c	b	o	p
4	b > p > o > c	b	p	o	c
4	b > o > p > c	b	o	p	c

One argument offered against the electoral college system is that it is unstable under very small, even accidental changes in the vote. The 2000 presidential election illustrated this risk. In Palm Beach County, Florida, as many as 2000 voters confused by the controversial "butterfly" ballot may have voted for Reform Party candidate Patrick Buchanan instead of their actual choice, Al Gore. Since George W. Bush won the Florida popular vote (and thus all 25 of Florida's electoral votes) by a margin of only 537, the effect of this voting accident was magnified to a national scale. Of course, votes were misrecorded or miscounted in other counties; indeed any voting process is subject to this kind of error. And nationwide only about 51% of the voting-age population actually cast votes in the 2000 presidential election. Determining the will of the people is an elusive goal.

Source: Wand, Jonathan N., Kenneth W. Shotts, Jasjeet S. Sekhon, Walter R. Mebane, Jr., Michael C. Herron, Henry E. Brady, 2001. "The Butterfly Did It: The Aberrant Vote for Buchanan in Palm Beach County Florida." (http://elections.fas.harvard.edu/wssmhb/)

Chicago supporters are delighted to discover they have a *majority* of first-place rankings. They are confident that, with 12 of 20 first-place votes, Chicago will be selected as the meeting site. The Borda tallies are calculated.

Portland earns $(8 \cdot 2) + (4 \cdot 0) + (4 \cdot 2) + (4 \cdot 1) = 28$ points.

Oakland earns $(8 \cdot 0) + (4 \cdot 1) + (4 \cdot 1) + (4 \cdot 2) = 16$ points.

Chicago earns $(8 \cdot 3) + (4 \cdot 3) + (4 \cdot 0) + (4 \cdot 0) = 36$ points.

Boston earns $(8 \cdot 1) + (4 \cdot 2) + (4 \cdot 3) + (4 \cdot 3) = \mathbf{40}$ **points.** ← Winner

The Borda method selects Boston as the site for next year's meeting.

When a voter profile includes a candidate with a majority of first-place rankings, it seems reasonable to expect that candidate to be elected. It is an attribute many people consider a standard or criterion of election fairness.

The Majority Criterion

If a candidate has a majority of first-place rankings in a voter profile, then that candidate should be the winner of the election.

The profile of CAA voters shows that the Borda method does not satisfy the majority criterion in Example 1. The Borda method does not select Chicago, the city with the majority of first-place rankings; instead, it selects Boston. This failure to select the candidate with the majority is a defect in the Borda method.

Do not misunderstand what it means when a voting method fails to satisfy a criterion. For some voter profiles, if a candidate has a majority of first-place rankings, the Borda method *does* select that candidate; however, it takes only one contradictory profile to show that a voting system violates a criterion. Since the Borda method *does not* select the majority candidate for the CAA voter profile in Example 1, the Borda method does not satisfy the majority criterion.

The plurality method, the Hare method, and the pairwise comparison method do satisfy the majority criterion. Despite the voter profile, if a majority candidate exists, these methods always select it. For example, a candidate with a majority of first-place rankings also has a plurality of the votes, and is selected by the plurality method. The Hare method is defined to select a candidate with a majority of first-place rankings in the first round of the voting process.

If a candidate has a majority of first-place rankings in a voter profile, then the pairwise comparison method automatically selects that candidate. For example, when there are n candidates in a pairwise comparison election, each candidate is involved in $n - 1$ pairwise comparisons. A Candidate x with a majority of first-place rankings is placed above every other candidate in more than half the voters' rankings. Consequently, that candidate wins all pairwise comparisons involving it and accumulates $n - 1$ pairwise points. The pairwise point total for any other candidate cannot exceed $n - 2$, because every candidate loses to x. Thus, Candidate x earns the most pairwise points and is selected by the pairwise comparison method.

The Condorcet Criterion If a candidate can win a one-on-one comparison with each and every other candidate, then it is natural to expect that candidate will be selected when the group votes. Expecting such an outcome is natural, but unless there are only two candidates, there is no guarantee it will happen.

Number of Voters	Ranking
6	$b > p > c$
5	$c > p > b$
4	$p > c > b$

EXAMPLE 2 The voter profile of the Culinary Arts Club, repeated here, shows that pork wins pairwise comparisons with both beef and chicken.

The 6 voters with the top ranking prefer beef to pork, but the other 9 voters rank pork higher than beef. Pork wins the comparison with beef by a margin of 9 to 6. The 5 voters with the middle ranking prefer chicken to pork, but the other 10 voters rank pork higher than chicken. Pork also wins the comparison with chicken by a margin of 10 to 5.

If the plurality method is used by the club to select the entree, then beef wins because it is ranked first by 6 (a plurality) of the members. Beef is selected in spite of the popularity of pork displayed by the pairwise comparisons. ■

A candidate who can win a pairwise comparison with every other candidate is called a **Condorcet candidate,** in honor of French aristocrat Marquis de Condorcet (1743–1794). Pork is a Condorcet candidate in Example 2. Not every voter profile yields a candidate who can win each pairwise comparison, however. It is not unusual for a voter profile to include a Condorcet candidate, but neither is it commonplace. Another standard of election fairness calls for a Condorcet candidate, if one exists, to be selected as the choice of the group.

The Marquis de Condorcet (1743–1794) was a French aristocrat. His interests included mathematics, social science, economics, and politics. Sadly, Condorcet's political passions at the end of the French Revolution sent him to prison, where he died of questionable causes. It is hard to understand how France's counterpart to America's Thomas Jefferson could have come to such a tragic end. Imagine how different the government of the United States would be today if Jefferson's great mind had been eliminated from American history at the end of the American Revolution.

The Condorcet Criterion

If a Condorcet candidate exists for a voter profile, then the Condorcet candidate should be the winner of the election.

The pairwise comparison method was created to satisfy the Condorcet criterion. The method always selects an existing Condorcet candidate, because that candidate naturally wins the most pairwise comparisons. Suppose that there are n candidates in an election. The Condorcet candidate earns $n - 1$ pairwise points; it can beat every other candidate. No other candidate's point total can exceed $n - 2$; all candidates lose to the Condorcet candidate.

The Culinary Arts Club voter profile shows that the plurality method does not satisfy the Condorcet criterion. In Example 2, the plurality method selects beef, not pork, the Condorcet candidate. The Hare method does not respect the Condorcet criterion either. For the Culinary Arts Club voter profile, the Hare method selects chicken, not pork.

The Borda method does select the Condorcet candidate, pork, as the entree for the Culinary Arts Club. For other profiles, however, the Borda method can fail to satisfy the Condorcet criterion, because it can fail to respect the majority criterion. Since a candidate with a majority of first-place rankings must win all the pairwise comparisons, a majority candidate is also a Condorcet candidate. In Example 1, Chicago is a majority candidate and, therefore, a Condorcet candidate, but the Borda method selects Boston. Thus, the Borda method can fail to respect the Condorcet criterion.

The pairwise comparison method always satisfies the Condorcet criterion. The Borda method, the Hare method, and the plurality method might select a Condorcet candidate, but they can also fail to honor the criterion.

The majority criterion and the Condorcet criterion are desirable attributes of a voting method when it is used in an election. The next two criteria involve a voting method's desirable attributes when a second election is held using the same method.

The Monotonicity Criterion Suppose the outcome of a first election is not binding, as in a straw poll. Voters may change their rankings before the next election. Suppose those who rearrange their rankings move the winner of the first election to the top of their rankings. It is reasonable to expect the winner of the first election, who now enjoys even more support, to win the second election also. This desirable attribute is known as the *monotonicity criterion*.

> **The Monotonicity Criterion**
>
> If Candidate x wins an election and, before a second election, the voters who rearrange their rankings move Candidate x to the top of their rankings, then Candidate x should win the second election.

The plurality method respects the monotonicity criterion. If Candidate x has a plurality of votes in the first election and only x can get more first-place votes, then x must have a plurality in the second vote. The Borda method and the pairwise comparison method can fail the monotonicity criterion as it is stated. They can fail because, while a rearranging voter is required to move winning Candidate x to the top of his or her ranking, how that voter shifts the *other* candidates is not restricted. The unrestricted movement of the other candidates is not obvious in the statement of the criterion, but it is implied. It allows the Borda point tallies and the pairwise comparison points of all candidates to increase, which may prevent the original winner from winning again in the second election. The Hare method, too, can violate the monotonicity criterion.

EXAMPLE 3 The College Art Association increased its executive board to 29 members. Voting for the site of the yearly meeting is among Miami m, Boston b, Chicago c, and Seattle s. The board members agree to use the Hare method to make their selection. They decide to hold a preliminary, nonbinding vote Friday afternoon and to meet again Saturday morning for the official vote. The Friday afternoon voter profile of the 29 members is given here.

Number of Voters	Ranking
7	m > b > s > c
8	b > c > m > s
10	c > s > m > b
2	m > c > s > b
2	m > s > b > c

Round One results are as follows:

Miami gets $7 + 2 + 2 = 11$ votes.
Boston gets 8 votes.
Chicago gets 10 votes.
Seattle gets 0 votes.

No city has a majority (15 or more) of the 29 votes. Seattle has the fewest votes and is eliminated.

Round Two results are as follows:

> Miami gets 11 votes.
> Boston gets 8 votes.
> Chicago gets 10 votes.

No city has a majority of the votes. Boston has the fewest votes and is eliminated. The 8 voters with the second ranking vote for Chicago in the next round, because Boston is eliminated.

Round Three results are as follows:

> Miami gets 11 votes.
> Chicago gets $10 + 8 = 18$ votes.

Friday afternoon, Chicago is selected by the preliminary Hare method election.

Friday evening, Chicago supporters wine and dine the 4 voters with the two bottom rankings. They convince the 4 voters to rearrange their rankings, placing Chicago first. The voter profile for the official vote on Saturday morning shows the 4 voters with the bottom rankings move Chicago into first place. The 2 voters with the bottom ranking were seriously swayed. The other 25 voters do not rearrange their rankings. Chicago supporters believe the 4 additional first-place rankings will work in their favor.

Number of Voters	Ranking
7	m > b > s > c
8	b > c > m > s
10	c > s > m > b
2	c > m > s > b
2	c > s > b > m

Round One results are as follows:

Miami gets 7 votes.
Boston gets 8 votes.
Chicago gets $10 + 2 + 2 = 14$ votes.
Seattle gets 0 votes.

No city has a majority of the 29 votes. Seattle has the fewest votes, and is eliminated.

Round Two results are as follows:

> Miami gets 7 votes.
> Boston gets 8 votes.
> Chicago gets 14 votes.

No city has a majority of the votes. Miami has the fewest votes and is eliminated. The 7 Miami supporters with the top ranking will vote for Boston in the next round, because Miami is eliminated.

Round Three results are as follows:

> **Boston** gets $8 + 7 = 15$ votes. ← Winner
> Chicago gets 14 votes.

Boston wins the official Saturday morning Hare method election by a single vote in the final round. Chicago loses the official vote, although Chicago has more first-place votes than it did in the preliminary round.

The conditions of the monotonicity criterion require the four voters who rearrange their rankings in Example 3 to move Chicago into first place. The voters also are permitted to reshuffle the other candidates in their rankings. To make the example obvious and dramatic, Chicago was simply exchanged with the top-ranked candidate in the two bottom rankings and the other candidates were not moved.

An alternative version of the monotonicity criterion exists. *If a voter rearranges his or her ranking before the second vote, the alternative version allows only the exchange of the winning Candidate x and the candidate immediately above it in a voter's ranking.*

The pairwise comparison method respects the alternative version of the criterion. A voter's ranking can be rearranged only by exchanging winning Candidate x and the candidate immediately above x in his or her ranking. Even if the two are exchanged, the initial order of the set of candidates ranked below original winner Candidate x and the candidate immediately above x must remain the same. The initial order of the set of candidates ranked above the pair also must remain the same. This means the outcomes of all the pairwise comparisons that do not involve x do not change. The outcomes of the pairwise comparisons involving x might change, but the outcomes must change in favor of x because only the original winner x is allowed to move up in the rankings. The number of pairwise comparisons x wins in the second pairwise comparison election cannot decrease, and the points for the other candidates cannot increase, so the original winner x also wins a second pairwise comparison method election.

The Borda method, the Hare method, and the pairwise comparison method can fail to satisfy the monotonicity criterion (as originally stated). The criterion is upheld by the plurality method.

The Independence of Irrelevant Alternatives Criterion

Suppose, that for some reason, the outcome of an original election is not considered the final outcome. If a candidate who lost the first election drops out before a second vote, expecting the winner of the first election also to win the second is reasonable. This voting method attribute is known as the *independence of irrelevant alternatives criterion*.

> ### The Independence of Irrelevant Alternatives (IIA) Criterion
> If Candidate x wins a first election and one (or more) of the losing alternatives drops out before a second vote, the winner x of the first election should win the second election.

Number of Voters	Ranking
6	b > p > c
5	c > p > b
4	p > c > b

The Culinary Arts Club profile, repeated here, shows that the plurality method and the Hare method can fail to satisfy the independence of irrelevant alternatives (IIA) criterion. Beef, with 6 first-place votes, is selected by the plurality method as the entree for the banquet. If pork is removed from the ballot, then, in a second election between beef and chicken, the former pork supporters vote for chicken. Chicken receives a majority of the votes and wins the second plurality election. (If chicken is removed from the ballot after one plurality vote, then pork wins the second decision between it and beef.)

The Hare method selects chicken as the Culinary Arts Club entree. When beef is removed from the ballot, the second Hare election is reduced to a simple majority election between pork and chicken. The former beef supporters vote for pork, causing it to win a second Hare election.

The Borda method can fail to respect the IIA criterion. When an alternative is removed from a voter's ranking, the candidates formerly ranked below it move up in the ranking. Moving up in the rankings can allow an original Borda loser to gain points and defeat the original winner in a second election. Suppose a voter has the ranking $p > l > a > c > e$, and p is the original Borda winner. If l drops out before a second vote, p still is ranked first, a and c move up, and e remains ranked last: $p > a > c > e$. Relative Borda tallies of a and c increase for this voter's ranking; the relative tally for the winner p does not change. Candidate p might lose a second Borda election, if enough rankings are shifted.

The pairwise comparison method also can fail to satisfy the IIA criterion. The outcome of a pairwise comparison election depends on how many individual comparisons each candidate wins. When a candidate is dropped from an election, the pairwise comparisons involving the candidate are no longer considered in the voting process. The next example shows how the decreased number of comparisons in a second vote can change a pairwise comparison method election outcome.

EXAMPLE 4 The conductor, along with the music director and a small group of trustees, are selecting a percussionist to replace their beloved timpani player who is retiring after 40 years with the world-famous American Orchestra. Five hopeful percussionists, v, w, x, y, and z, perform audition pieces. The committee agrees to use the pairwise comparison method for their selection, because it allows direct, one-on-one comparisons of the five musicians. Only the results of the ten possible pairwise comparisons are shown; the actual voter profile for the conductor, director, and trustees is not given.

Percussionist v ties with percussionist w; v gets $\frac{1}{2}$ point and w gets $\frac{1}{2}$ point.
Percussionist v beats percussionist x; v gets 1 point.
Percussionist y beats percussionist v; y gets 1 point.
Percussionist v beats percussionist z; v gets 1 point.
Percussionist x beats percussionist w; x gets 1 point.
Percussionist y beats percussionist w; y gets 1 point.
Percussionist z ties with percussionist w; z gets $\frac{1}{2}$ point and w gets $\frac{1}{2}$ point.
Percussionist x beats percussionist y; x gets 1 point.
Percussionist x beats percussionist z; x gets 1 point.
Percussionist z beats percussionist y; z gets 1 point.

The pairwise comparison point total for each candidate is gathered from the ten individual results.

Percussionist x gets **3 points**. ← Winner
Percussionist v gets $2\frac{1}{2}$ points.
Percussionist y gets 2 points.
Percussionist z gets $1\frac{1}{2}$ points.
Percussionist w gets 1 point.

Percussionist x wins the most comparisons and is selected by the pairwise comparison method as the next timpani player of the American Orchestra.

The conductor is about to call and congratulate x when he gets an e-mail from y. Sensing she had not played well enough to win the audition, percussionist y accepted an offer to join a small, yet popular, European orchestra. The conductor and the selection committee feel they should revisit their decision, now that y no longer is a candidate. They agree to vote again on the four remaining candidates v, w, x, and z, using the pairwise comparison method.

The only change in the voter profile (not shown) is that the committee members remove y from their rankings for the second vote. Consequently, the results of the second round of comparisons agree with the previous results, but there are no longer any comparisons with percussionist y. Only six pairwise comparisons are needed for four candidates.

Percussionists v and w tie; v gets $\frac{1}{2}$ point and w gets $\frac{1}{2}$ point.

Percussionist v beats percussionist x; v gets 1 point.

Percussionist v beats percussionist z; v gets 1 point.

Percussionist x beats percussionist w; x gets 1 point.

Percussionists z and w tie; z gets $\frac{1}{2}$ point and w gets $\frac{1}{2}$ point.

Percussionist x beats percussionist z; x gets 1 point.

The results gathered from the six individual comparisons surprise the voters.

Percussionist v gets $2\frac{1}{2}$ **points.** ← Winner

Percussionist x gets 2 points.

Percussionist w gets 1 point.

Percussionist z gets $\frac{1}{2}$ point.

The second pairwise comparison method vote selects percussionist v as the next timpani player of the American Orchestra.

The independence of irrelevant alternatives criterion seems like a reasonable expectation of a voting method. The wording of the monotonicity criterion disguises the ability to reshuffle candidates. The IIA criterion does not have any hidden features. It simply says removing a loser from the ballot should not make the original winner unable to win a second election. Surprisingly, the plurality method, the Borda method, the Hare method, and the pairwise comparison method can fail to satisfy the IIA criterion.

Arrow's Impossibility Theorem

The previous section introduced six ways to select one alternative from a set of alternatives; however, each of the six methods is inherently flawed. Approval voting and the sequential pairwise comparison method permit multiple outcomes for the same profile of voters. Each of the remaining four voting methods fails to satisfy at least one of the four criteria considered desirable attributes of a selection process. The table on the next page shows which criteria are satisfied and which criteria are not satisfied by the plurality, pairwise comparison, Hare, and Borda voting methods.

Voting theory history is being made right now by living Americans. **Steven J. Brams** of New York University and **Peter C. Fishburn** of AT&T are working hard to educate voters about the advantages of approval voting. In spite of its somewhat sensitive nature, the method has many good qualities such as encouraging voter turnout by allowing voters to more easily express their feelings.

Donald G. Saari of Northwestern University is using highly sophisticated mathematical methods to shed light on the paradoxical nature of voting methods. Saari's studies offer political and social scientists a solid mathematical foundation for their work in the new millennium.

	Voting Method			
Criterion	Plurality Method	Pairwise Comparison Method	Borda Method	Hare Method
Majority criterion	satisfied	satisfied	not satisfied (Example 1)	satisfied
Condorcet criterion	not satisfied (Example 2)	satisfied	not satisfied	not satisfied
Monotonicity criterion	satisfied	not satisfied	not satisfied	not satisfied (Example 3)
Independence of irrelevant alternatives criterion	not satisfied	not satisfied (Example 4)	not satisfied	not satisfied

Plurality voting considers only top-ranked candidates, so it manages to satisfy the majority criterion and monotonicity criterion. The Hare method uses a series of plurality elections, but only satisfies the majority criterion. The Borda method considers how a voter ranks a complete list of candidates, and although using it is easy, the Borda method fails to satisfy any of the criteria.

The pairwise comparison method also considers a voter's complete ranking of the candidates and satisfies the majority criterion and Condorcet criterion. While this method does not satisfy either of the desired criteria for a second election, it is the only method to meet both desired attributes of a single election. This should make the pairwise comparison method the ideal choice if only one vote is taken.

Unfortunately, the pairwise comparison method is difficult to use for elections with 6 or more candidates because of the large number of comparisons required to determine the outcome. The next example reveals another difficulty with the pairwise comparison method—it often fails to select a winner.

EXAMPLE 5 A family of opera enthusiasts wants to name their new parrot after a female opera character. They narrow the choices down to Tosca, Aida, Carmen, and Lulu. The family agrees to use the pairwise comparison method to select the parrot's name. Results of the six pairwise comparisons are given. (The family voting profile is not shown.)

Tosca beats Aida; Tosca gets 1 point.
Tosca beats Carmen; Tosca gets 1 point.
Lulu beats Tosca; Lulu gets 1 point.
Aida beats Carmen; Aida gets 1 point.
Aida beats Lulu; Aida gets 1 point.
Lulu beats Carmen; Lulu gets 1 point.

The pairwise comparison points for each name are calculated.

Tosca gets 2 points.
Aida gets 2 points.
Lulu gets 2 points.
Carmen gets 0 points.

16.2 The Impossibilities of Voting   939

Kenneth J. Arrow (1921–) is an American economist. In 1972, he was awarded the Nobel Memorial Prize in Economics for his work in the theory of general economic equilibrium. His famous impossibility theorem about voting (1951) resulted from discussions he had while working at the RAND Corporation. Arrow and his colleagues were applying the then-new mathematical science of game theory to the age-old science of military and diplomatic affairs. Arrow's discovery of unavoidable imperfection ended the first era of voting theory. The focus has shifted from the question of what voting method is perfect to much more mathematically challenging questions such as what can be expected of a voting method, and what can be done to help minimize the possibility of a particular paradox occurring.

The outcome is inconclusive, since no alternative has more points than all the others. Tosca, Aida, and Lulu each get 2 points. The pairwise comparison method fails to select an opera name for the family's new parrot.

All of the voting methods presented in the previous section have shortcomings. Why wasn't a better voting method presented? Keep in mind that the four fairness criteria are reasonable expectations. Wanting to elect a majority or Condorcet candidate is reasonable. Wanting the original winner to win a second election is reasonable. It is especially reasonable if the winning candidate has moved up in the voters' rankings (monotonicity) or if an original loser drops out (IIA) before the second election. These criteria may be reasonable individually, but together they are impossible to satisfy.

Economist Kenneth Arrow (see the margin note) set out to invent a better voting system. He discovered his goal was mathematically impossible. A perfect voting system cannot exist. This famous result is known as **Arrow's Impossibility Theorem.** It is said that the theorem took Arrow less than a week to prove and ultimately ended the search for a perfect voting system that had started nearly 200 years earlier.

Arrow's Impossibility Theorem

There does not exist and never will exist any voting method that simultaneously satisfies the majority criterion, the Condorcet criterion, the monotonicity criterion, and the independence of irrelevant alternatives criterion.

Arrow proved that for any voting method there will be some voter profile for which one or more of the criteria are not satisfied. The CAA voter profile in Example 1 shows that the Borda method does not always satisfy the majority criterion. In Example 2, the Culinary Arts Club profile shows that the plurality method does not always satisfy the Condorcet criterion. The monotonicity criterion is violated by the Hare method for a different CAA voter profile in Example 3. The pairwise comparison method does not satisfy the independence of irrelevant alternatives criterion for the orchestra selection committee profile in Example 4. Arrow's theorem also says that a contradictory profile of voters will exist for any new voting system yet to be invented.

16.2 EXERCISES

1. Eleven voters rank three alternatives a, b, and c according to the voter profile at the right.

   (a) Which alternative has a majority of first-place votes?
   (b) Use the Borda method to determine which alternative the voters prefer.
   (c) Does the Borda method violate the majority criterion for the voter profile?

Number of Voters	Ranking
6	a > b > c
3	b > c > a
2	c > b > a

2. Repeat Exercise 1 using the following voter profile.

Number of Voters	Ranking
6	b > a > c
3	a > c > b
2	c > a > b

3. Thirty-six voters rank four alternatives a, b, c, and d according to the following voter profile.

Number of Voters	Ranking
20	a > b > c > d
6	b > c > d > a
5	c > b > d > a
5	d > b > a > c

(a) Which alternative has a majority of first-place votes?
(b) Use the Borda method to determine which alternative the voters prefer.
(c) Does the Borda method violate the majority criterion for the voter profile?

4. Repeat Exercise 3 using the following voter profile.

Number of Voters	Ranking
4	b > c > a > d
9	c > a > d > b
4	a > c > b > d
19	d > c > b > a

5. Thirty voters rank five alternatives a, b, c, d, and e according to the voter profile at the top of the next column.
(a) Which alternative has a majority of first-place votes?
(b) Use the Borda method to determine which alternative the voters prefer.
(c) Does the Borda method violate the majority criterion for the voter profile?

Number of Voters	Ranking
16	a > b > c > d > e
3	b > c > d > e > a
5	c > d > b > e > a
3	d > b > c > a > e
3	e > c > d > a > b

6. Repeat Exercise 5 using the following voter profile.

Number of Voters	Ranking
16	b > e > c > d > a
3	a > c > b > d > e
6	c > a > e > b > d
2	d > a > c > e > b
3	e > c > a > d > b

7. In Exercise 10 of the previous section, a 13-member committee is selecting a chairperson. The three candidates are Albert Werner a, Barbara Hightower b, and Charles Smith c. The committee members ranked the candidates according to the following voter profile.

Number of Voters	Ranking
4	a > b > c
2	b > c > a
4	b > a > c
3	c > a > b

(a) Which candidate is a Condorcet candidate?
(b) Which candidate is selected by the plurality method?
(c) Which candidate is selected by the Hare method?
(d) Which candidate is selected by the Borda method?
(e) Which voting method(s)—plurality, Hare, or Borda—violate(s) the Condorcet criterion for this profile of voters?

8. Repeat Exercise 7 using the following voter profile from Exercise 11 of the previous section.

Number of Voters	Ranking
3	$a > c > b$
4	$c > b > a$
2	$b > a > c$
4	$b > c > a$

9. In Exercise 14 of the previous section, a senator is holding a workshop. The senator asked the 21 workshop members to rank the issues of job creation j, education e, health care h, and gun control g in the order of importance to themselves and the counties they represent. The workshop members rank the issues according to the following voter profile.

Number of Voters	Ranking
3	$e > h > g > j$
6	$h > e > g > j$
5	$j > g > e > h$
4	$g > e > h > j$
3	$j > e > h > g$

(a) Which issue is a Condorcet candidate?

Determine which issue the workshop members felt was the highest priority using the
(b) plurality method
(c) Hare method
(d) Borda method.
(e) Which voting method(s)—plurality, Hare, or Borda—violate(s) the Condorcet criterion for this profile of voters?

10. Repeat Exercise 9 using the following voter profile from Exercise 15 of the previous section.

Number of Voters	Ranking
6	$h > j > g > e$
5	$e > g > j > h$
4	$g > j > h > e$
3	$j > h > g > e$
3	$e > j > h > g$

11. In Exercise 18 of the previous section, members of the Adventure Club are selecting an activity for their yearly vacation. Choices are hiking in the desert h, rock climbing c, white-water kayaking k, bungee jumping b, or just sitting by a mountain lake and tanning t. The 55 Adventure Club members ranked the choices according to the following voter profile.

Number of Voters	Ranking
18	$t > k > h > b > c$
12	$c > h > k > b > t$
10	$b > c > h > k > t$
9	$k > b > h > c > t$
4	$h > c > k > b > t$
2	$h > b > k > c > t$

(a) Which activity is a Condorcet candidate?

Select one activity for the yearly vacation of the Adventure Club using the
(b) plurality method
(c) Hare method
(d) Borda method.
(e) Which voting method—plurality, Hare, or Borda—violates the Condorcet criterion for this profile of voters?

12. Repeat Exercise 11 using the following voter profile from Exercise 19 of the previous section.

Number of Voters	Ranking
18	$t > b > h > k > c$
12	$c > h > b > k > t$
10	$k > c > h > b > t$
9	$b > k > h > c > t$
4	$h > c > b > k > t$
2	$h > k > b > c > t$

13. A 14-member committee is selecting a site for its next meeting. The choices are Montreal m, Chicago c, San Francisco s, and Boston b.
(a) The committee members decide to use the pairwise comparison method to select a site in a nonbinding decision. Prior to any discussion, the

14 members rank the choices according to the following voter profile.

Number of Voters	Ranking
5	m > c > s > b
4	b > s > c > m
3	b > s > m > c
2	c > m > s > b

Show that Montreal is selected by the pairwise comparison method in the preliminary nonbinding decision.

(b) The 2 committee members with the bottom ranking in the table rearrange their ranking after listening to the discussions. The other 12 committee members stick with their original rankings of the cities. For the official vote the 14 members rank the choices according to the following voter profile.

Number of Voters	Ranking
5	m > c > s > b
4	b > s > c > m
3	b > s > m > c
2	m > b > c > s

Use the pairwise comparison method to determine the site selection of the committee.

(c) Does the pairwise comparison method violate the monotonicity criterion in this selection process?

14. Repeat Exercise 13 using the following pair of voter profiles. In part (a), instead of Montreal, show that Chicago is selected by the pairwise comparison method in the preliminary nonbinding decision. Prior to any discussion, the 14 members rank the choices according to the following voter profile.

Number of Voters	Ranking
5	c > m > b > s
4	s > b > m > c
3	s > b > c > m
2	m > c > b > s

For the official vote, the 14 members rank the choices according to the following voter profile.

Number of Voters	Ranking
5	c > m > b > s
4	s > b > m > c
3	s > b > c > m
2	c > s > m > b

15. A 19-member committee is selecting a site for the next meeting. The choices are Dallas d, Chicago c, San Francisco s, and Boston b.

(a) The committee decides to use the Borda method to select a site in a nonbinding decision. Prior to any discussion, the 19 members rank the choices according to the following voter profile.

Number of Voters	Ranking
7	s > b > c > d
5	b > d > c > s
3	d > s > c > b
4	c > s > d > b

Show that San Francisco is selected by the Borda method in the preliminary nonbinding decision.

(b) The 7 committee members with the bottom two rankings rearrange their rankings after listening to the discussions. The other 12 committee members stick with their original rankings of the cities. For the official vote, the 19 members rank the choices according to the following voter profile.

Number of Voters	Ranking
7	s > b > c > d
5	b > d > c > s
3	s > b > d > c
4	s > b > c > d

Use the Borda method to determine the site selection of the committee.

(c) Does the Borda method violate the monotonicity criterion in this selection process?

16. Repeat Exercise 15 using the following pair of voter profiles. In part (a), instead of San Francisco, show that Chicago is selected by the Borda method in the preliminary nonbinding decision. Prior to any discussion, the 19 members rank the choices according to the following voter profile.

Number of Voters	Ranking
7	c > d > s > b
5	d > b > s > c
3	b > c > s > d
4	s > c > b > d

For the official vote, the 19 members rank the choices according to the following voter profile.

Number of Voters	Ranking
7	c > d > s > b
5	d > b > s > c
3	c > d > b > s
4	c > d > s > b

17. A 17-member committee is selecting a site for the next meeting. The choices are Dallas d, Chicago c, Atlanta a, and Boston b.

(a) The committee decides to use the Hare method to select a site in a nonbinding decision. Prior to any discussion, the 17 members rank the choices according to the following voter profile.

Number of Voters	Ranking
6	a > c > b > d
5	b > d > a > c
4	c > d > b > a
2	c > a > b > d

Show that Atlanta is selected by the Hare method in the preliminary nonbinding decision.

(b) The 2 committee members with the bottom ranking rearrange their rankings after listening to the discussions. The other 15 committee members stick with their original rankings of the cities. For the official vote, the 17 members rank the choices according to the following voter profile.

Number of Voters	Ranking
6	a > c > b > d
5	b > d > a > c
4	c > d > b > a
2	a > c > b > d

Use the Hare method to determine the site selection of the committee.

(c) Does the Hare method violate the monotonicity criterion in this selection process?

18. Repeat Exercise 17 using the following pair of voter profiles. In part (a), instead of Atlanta, show that Boston is selected by the Hare method in the preliminary nonbinding decision. Prior to any discussion, the 17 members rank the choices according to the following voter profile.

Number of Voters	Ranking
6	b > c > a > d
5	a > d > b > c
4	c > d > a > b
2	c > b > a > d

For the official vote, the 17 members rank the choices according to the following voter profile.

Number of Voters	Ranking
6	b > c > a > d
5	a > d > b > c
4	c > d > a > b
2	b > c > a > d

19. Thirteen voters ranked three candidates a, b, and c according to the following voter profile.

Number of Voters	Ranking
6	a > b > c
5	b > c > a
2	c > b > a

(a) Show that Candidate a is selected if the plurality method is used to determine the outcome of the election.
(b) If losing Candidate c drops out, which candidate is selected by a second plurality method election?
(c) Does the plurality method violate the independence of irrelevant alternatives criterion in this election process?

20. Seventeen voters ranked three candidates a, b, and c according to the following voter profile.

Number of Voters	Ranking
6	a > c > b
8	b > a > c
3	c > a > b

(a) Show that Candidate b is selected if the plurality method is used to determine the outcome of the election.
(b) If losing Candidate a drops out, which candidate is selected by a second plurality method election?
(c) Does the plurality method violate the independence of irrelevant alternatives criterion in this election process?

21. Four candidates, a, b, c, and d, are ranked by 175 voters according to the following voter profile.

Number of Voters	Ranking
75	a > c > d > b
50	c > a > b > d
30	b > c > d > a
20	d > b > c > a

(a) Show that Candidate a is selected if the plurality method is used to determine the outcome of the election.
(b) If losing Candidate b drops out, which candidate is selected by a second plurality vote?
(c) Does the plurality method violate the independence of irrelevant alternatives criterion in this election process?

22. Four candidates, a, b, c, and d, are ranked by 95 voters according to the following voter profile. Repeat parts a, b, and c of Exercise 21.

Number of Voters	Ranking
40	a > c > d > b
30	c > a > b > d
15	b > c > d > a
10	d > b > c > a

23. A subcommittee of the senator's workshop is asked to include the issue of military spending m in its breakout session discussion of the issues job creation j, education e, health care h, and gun control g. The 11 subcommittee members rank the five issues in the order of importance to themselves and their counties according to the following voter profile.

Number of Voters	Ranking
5	e > j > h > g > m
2	m > j > e > h > g
2	h > m > e > g > j
2	g > m > e > j > h

(a) If the pairwise comparison method is used, show that the subcommittee felt the issue of education was the highest priority.
(b) The subcommittee decided to delete the losing issues of health care and gun control from further discussion. Use the pairwise comparison method to determine which of the remaining issues of education, job creation, or military spending the subcommittee felt was the highest priority.
(c) Does the pairwise comparison method violate the independence of irrelevant alternatives criterion in this selection process?

24. A subcommittee of the senator's workshop is asked to include the issue of military spending m in its breakout session discussion of the issues job creation j, education e, health care h, and gun control g. The 11 subcommittee members rank the five issues in the order of importance to themselves and their counties according to the voter profile in Exercise 23.
    (a) If the pairwise comparison method is used, show that the subcommittee felt the issue of education was the highest priority.
    (b) The subcommittee decided to delete the losing issues of health care, job creation, and gun control from further discussion. Use the pairwise comparison method to determine which of the remaining issues of education or military spending the subcommittee felt was the highest priority.
    (c) Does the pairwise comparison method violate the independence of irrelevant alternatives criterion in this selection process?

25. The conductor, music director, and a small group of trustees select a new percussionist using the pairwise comparison method in Example 4. When Percussionist y drops out of the audition process, the pairwise comparison method violates the independence of irrelevant alternatives criterion by selecting Percussionist v instead of original winner Percussionist x in a second pairwise comparison. If it had been Percussionist w, not Percussionist y, to drop out of the audition process, would the IIA criterion have been violated in a second pairwise comparison?

26. Consider the situation described in Exercise 25. If it had been Percussionist z, not Percussionist y, to drop out of the audition process, would the IIA criterion have been violated in a second pairwise comparison?

27. Twenty-five voters rank three candidates a, b, and c according to the following voter profile.

Number of Voters	Ranking
13	c > b > a
8	b > a > c
4	b > c > a

    (a) Show that Candidate b wins a Borda method election.
    (b) Which candidate wins a second Borda method election, if losing Candidate a drops out?
    (c) Does the Borda method violate the independence of irrelevant alternatives criterion in this election process?

28. Seventeen voters rank three candidates a, b, and c according to the following voter profile.

Number of Voters	Ranking
9	b > a > c
6	a > c > b
2	a > b > c

    (a) Show that Candidate a wins a Borda method election.
    (b) Which candidate wins a second Borda method election, if losing Candidate c drops out?
    (c) Does the Borda method violate the independence of irrelevant alternatives criterion in this election process?

29. Thirty-four voters rank three candidates a, b, and c according to the following voter profile.

Number of Voters	Ranking
12	a > c > b
10	b > a > c
8	c > b > a
4	c > a > b

    (a) Show that Candidate a wins a Hare method election.
    (b) If losing Candidate c drops out of the election, which of the remaining candidates wins a Hare method election?
    (c) Does the Hare method violate the independence of irrelevant alternatives criterion in this election process?

30. Fifty-four voters rank three candidates a, b, and c according to the following voter profile. Repeat parts (a), (b), and (c) of Exercise 29.

Number of Voters	Ranking
17	a > c > b
15	b > a > c
13	c > b > a
9	c > a > b

31. For the profile of voters in Exercise 17(a), Altanta is selected by the Hare method in the preliminary nonbinding decision.
    (a) If the losing city, Chicago, withdraws after the nonbinding vote, which city wins the official vote? Use the voter profile in Exercise 17(a) with Chicago deleted for the official Hare method election.
    (b) Does the Hare method violate the independence of irrelevant alternatives criterion in this selection process?

32. For the profile of voters in Exercise 18(a), Boston is selected by the Hare method in the preliminary nonbinding decision. Repeat Exercise 31 using the profile of voters in Exercise 18(a).

33. In a short paragraph, explain in your own words why a violation of the majority criterion is an automatic violation of the Condorcet criterion.

34. What is it about the departure for Europe of Percussionist y in Example 4 that hurts Percussionist x, the original winner, and benefits Percussionist v in the second pairwise comparison method selection?

35. Construct a voter profile for 19 voters and 6 candidates that has a majority candidate. Show that the majority candidate must win 5 pairwise comparison points, but that another candidate can win 4 pairwise points. (*Hint:* Only two of the 6! possible rankings are needed for the profile.)

36. Construct a voter profile for 40 voters and 4 candidates that shows the Borda method violates the majority criterion. Do this by assigning the remaining 19 voters to the rankings in the given incomplete voter profile in such a way that the majority candidate does not win a Borda method election.

Number of Voters	Ranking
21	d > b > c > a
?	b > c > a > d
?	c > b > d > a
?	a > c > b > d

37. Repeat Exercise 36 using the following incomplete voter profile.

Number of Voters	Ranking
21	a > b > c > d
?	b > c > a > d
?	c > b > d > a
?	d > c > b > a

38. Construct a voter profile for 13 voters and 4 candidates that has a Condorcet candidate that fails to be elected by the Borda method, but is selected by both the Hare method and the plurality method.

39. Construct a voter profile for 13 voters and 4 candidates that has a Condorcet candidate that fails to be elected by the Borda method and the plurality method, but is selected by the Hare method.

40. Construct a voter profile for 18 voters and 4 candidates that shows the pairwise comparison method violates the monotonicity criterion. The original voter profile for the 18 voters is given.

Number of Voters	Ranking
6	a > x > y > z
5	z > y > x > a
4	z > y > a > x
3	x > a > y > z

   (a) Show that Candidate a wins the pairwise comparison method election for the original given profile.
   (b) Rearrange the ranking of the 3 voters in the bottom row in such a way that for the altered voter profile Candidate z wins a second pairwise comparison method election. Candidate a must be moved to first place in the new ranking.

41. Construct a voter profile for 14 voters and 4 candidates that shows the Borda method violates the monotonicity criterion. The original voter profile for the 14 voters is given on the next page.

Number of Voters	Ranking
5	m > b > s > c
4	b > s > c > m
3	s > m > c > b
2	c > m > s > b

(a) Show that Candidate m wins the Borda method election for the original given profile.
(b) Rearrange the rankings of the 5 voters in the bottom two rows in such a way that for the altered voter profile, Candidate b wins a second Borda method election. Candidate m must be moved to first place in the new ranking.

42. Construct a voter profile for 34 voters and 3 candidates that shows the Hare method violates the monotonicity criterion. (*Hint:* Tweak the profile given in Exercise 29.)

43. Delete one of the losing candidates from the given voter profile in such a way that the plurality method violates the independence of irrelevant alternatives criterion in a second election.

Number of Voters	Ranking
15	a > b > c > d
8	b > a > c > d
9	c > b > a > d
6	d > b > a > c

44. Delete 2 or 3 of the losing candidates from the given voter profile in such a way that the pairwise comparison method violates the independence of irrelevant alternatives criterion in a second election.

Number of Voters	Ranking
15	a > c > b > d > e
6	e > c > a > b > d
6	b > e > a > d > c
6	d > e > a > c > b

45. Construct a voter profile for 41 voters and 3 candidates g, j, and e in such a way that the Borda method violates the independence of irrelevant alternatives criterion in a second election. Arrange the voter profile so that j wins the first Borda election with the most Borda points, and e has the fewest Borda points. Delete e from the second Borda method vote.

46. The discussion on the monotonicity criterion explains why the pairwise comparison method satisfies the alternative monotonicity criterion.

    (a) Explain why the plurality method satisfies the alternative monotonicity criterion.
    (b) Explain why the Borda method satisfies the alternative monotonicity criterion.

47. Now that you know the pros and cons of the plurality, pairwise comparison, Borda, and Hare methods, have you changed your mind about which one you personally think is the best method? Explain why you changed your mind or why you are staying with your original choice.

48. If you can have only two of the four criteria discussed in this section satisfied by a voting method, which would you choose? Why do you think the two criteria you selected are important?

16.3 The Possibilities of Apportionment

The Highwood School District received a generous gift of 109 computers from a local manufacturer. The benefactor stipulated that the division of the identical machines must be based on the individual enrollments at the various schools in the district. In other words, the larger the school's enrollment, the more machines it is entitled to receive. Naturally, each school must receive an integer number of computers because fractional parts of computers are worthless. **Apportionment** is a division or partition of identical, indivisible things according to some plan or proportion. The Highwood School District's task is to use the enrollments at its various schools to determine the apportionment of the 109 computers.

The story of apportionment began as the dust of the American Revolution settled at the Constitutional Convention in 1787 in Philadelphia. America's founding fathers invented a system of government with three branches—presidential, judicial, and legislative. The *Great Compromise* found in Article I, Sections 2 and 3, created a legislative branch consisting of the Senate and the House of Representatives. Each state in the union would be represented by two senators in the Senate. The number of representatives a state has in the House of Representatives would be determined by the size of the state's population.

The founding fathers did not specify exactly how to decide the number of representatives for each state—only that the number must be based on state population. America's rapidly growing population made it impossible to fix the number of seats in the House of Representatives when the Constitution was written. However, the founding fathers did specify that each congressional district represented by a seat in the House should have a population of at least 30,000. The increasing and shifting population of the United States, and the addition of each new state to the union, meant that the apportionment issue would be revisited often.

The Constitution requires a census of the population of the United States every ten years and that the House of Representatives be reapportioned according to the current census data. The first census was conducted in 1790, but a debate about what apportionment method to use postponed the first House apportionment until 1794. The House has been reapportioned two years after each subsequent census, except in the 1920s. Although six methods have been proposed, surprisingly, only four methods have ever been used to apportion seats in the House.

The first method used to determine the number of representatives apportioned to each state was proposed and championed by the great statesman Thomas Jefferson. The Jefferson method was used for the first apportionment of the House in 1794 and until the 1842 apportionment. Problems with the 1822 and 1832 Jefferson method house apportionments caused the method to fall from favor. In 1842, the Jefferson method was replaced by a method proposed by Daniel Webster. The Webster method, by its nature, has many of the same flaws as the Jefferson method, however. The Webster method was immediately replaced in 1852 by a method originally proposed years before by Alexander Hamilton. It was the Hamilton method that Jefferson's method had vanquished for the original 1794 House apportionment. Flaws discovered in the Hamilton method late in the 1800s caused the method to be replaced by the Webster method after the 1900 census.

The method currently used to apportion the House of Representatives is known as the Hill-Huntington method. It replaced the Webster method in 1941. The Hill-Huntington method is covered in the exercises for this section because it depends on some nontrivial mathematics. The Hamilton, Jefferson, and Webster methods are presented in this section. Problems and paradoxes associated with these methods will be discussed in the following section.

The U.S. House of Representatives is not the only place where apportionment is an issue. Many countries have Parliamentary governments in which legislative seats are apportioned based on the percent of the total vote received by a political party in a general election. Transportation boards decide, on the basis of ridership, how many trains are put on each rail line and how many buses are assigned to each route. Based on student enrollments, faculty positions at a college or university are divided among various departments and teaching assistants are meted out to professors.

The Hill-Huntington method has been challenged on Constitutional grounds as recently as 1991. The state of Montana challenged the reapportionment of the House after the 1990 census. Montana lost 50% of its seats when its apportionment decreased from 2 seats to 1 seat. Montana asked the Supreme Court to require the Census Bureau to recompute the apportionments using an apportionment method that minimizes the differences in district populations. If successful, Montana would have regained its lost seat at the expense of Washington state. The Court unanimously rejected Montana's challenge and ruled that, since congressional districts cannot cross state lines, the districts are bound to be lopsided occasionally.

16.3 The Possibilities of Apportionment

Alexander Hamilton
(1755–1804) was born in the British West Indies and came to New York in 1773. He was General Washington's aide-de-camp in the American Revolutionary War, an author of the influential *Federalist Papers*, and the first secretary of the treasury under President Washington. He is among the least understood of the founding fathers because of his aristocratic views and often is regarded as the patron saint of capitalism. Hamilton distrusted Aaron Burr, and the two argued publicly for years before their fateful duel in 1804 that took Hamilton's life.

The Hamilton Method The Hamilton method was among the earliest methods considered for apportioning the original House of Representatives after the first census in 1790. This method is sensible and easy to calculate, and no "magic divisors" are required. Congress sent a bill to President Washington adopting the Hamilton method, but he rejected it with the very first presidential veto. Congress could not override the veto and eventually adopted the Jefferson method.

The Hamilton method has four steps. The first step determines how many people in the overall population are to be represented by a single seat in the House. This is called the *standard divisor*. The next step produces what is called a *standard quota* for each state, which is the number of seats (not necessarily an integer number) to which a state is entitled, based on its population. Fractional parts of seats do not exist, so the third step is a practical way to ensure that the total number of seats promised is equal to or smaller than the total number of seats available. The fourth step offers a pragmatic scheme for distributing any seats that still are available to the states Hamilton considered most deserving. It is this fourth step in the Hamilton method that caused President Washington to reject it.

The Hamilton Method

1. Compute the **standard divisor,**

   $d =$ (total population)/(total number of seats).

2. Compute the **standard quota** for each state,

   $Q =$ (state's population)/d.

3. Round each state's standard quota Q *down* to the nearest integer. Each state will get at least this many seats, but must get at least one.
4. Give any additional seats one at a time (until no seats are left) to the states with the largest fractional parts of their standard quotas.

EXAMPLE 1 The Highwood School District decides to apportion the gift of 109 computers by using the Hamilton method. In legislative lingo, the computers are congressional seats, the schools are states, and the school enrollments are state populations. The district has five schools: Applegate, enrollment 335; Bayshore, enrollment 456; Claypool, enrollment 298; Delmar, enrollment 567; and Edgewater, enrollment 607. The total enrollment for the Highwood School District is

$$335 + 456 + 298 + 567 + 607 = 2263.$$

1. The standard divisor d equals (total enrollment)/(number of computers), so $2263/109 = 20.76147$. The standard divisor represents the number of students there are for each of the 109 computers. In other words, there is one computer for every 20.76147 students in the Highwood School District.
2. The standard quota for Applegate School is $Q = 335/20.76147$.
   The standard quota for Bayshore School is $Q = 456/20.76147$.
   The standard quota for Claypool School is $Q = 298/20.76147$.
   The standard quota for Delmar School is $Q = 567/20.76147$.
   The standard quota for Edgewater School is $Q = 607/20.76147$.

Each standard quota Q is given to two decimal places in the third column of the table. Usually, two decimal places are sufficient.

School	Enrollment	Standard Quota Q	Rounded-down Q
Applegate	335	16.14	16
Bayshore	456	21.96	21
Claypool	298	14.35	14
Delmar	567	27.31	27
Edgewater	607	29.24	29
Totals	2263	109	107

3. The standard quotas are rounded down to the nearest integer in the fourth column of the table. Each school receives at least this number of computers.

4. The total number of promised computers is only 107; thus, two additional computers need to be distributed. The Hamilton method distributes these two computers one at a time to the schools with the largest fractional parts of their standard quotas. The fractional part of the Bayshore School standard quota is .96 which is greater than all the other fractional parts of the other schools' standard quotas, so the first of the two extra computers is given to Bayshore School. The remaining computer is given to Claypool School because the fractional part of its standard quota, .35, is greater than the fractional parts of the standard quotas for Applegate (.14), Delmar (.31), and Edgewater (.24).

The apportionment of the 109 computers the Highwood School District received as a gift is shown below as an additional column. The bold entries of **22** and **15** in the final column for the Bayshore and Claypool Schools respectively indicate that these schools received more computers than their rounded-down standard quota Q.

School	Enrollment	Standard Quota Q	Rounded-down Q	Computers Apportioned
Applegate	335	16.14	16	16
Bayshore	456	21.96	21	**22**
Claypool	298	14.35	14	**15**
Delmar	567	27.31	27	27
Edgewater	607	29.24	29	29
Totals	2263	109	107	109

Consider Delmar School in Example 1. The fractional part of its standard quota is .31, which is not drastically smaller than the fraction (.35) that gained Claypool School an additional computer. In practice, the fractional parts of two states' standard quotas might need to be calculated to four or five decimal places to decide which is larger. Indeed, a state with a quota of $Q = 14.8678$ could get the last additional seat available (for a total of 15 seats), and a state with a quota of $Q = 14.8677$, if there was one, would get only its rounded-down quota of 14 seats. The Hamilton method has much more serious and subtle problems than the obvious one discussed here.

16.3 The Possibilities of Apportionment

It may seem that a more mathematical and direct approach to apportionment would be to replace the practical rounding-down scheme in Step 3 of the Hamilton method with a traditional rounding scheme. If the fractional part of Q is greater than or equal to .5, then Q is rounded up to the nearest integer. If the fractional part of Q is less than .5, then Q is rounded down to the nearest integer. If a traditional rounding scheme is used on the values of Q from the Highwood School District example, only 108 of the 109 computers are apportioned as shown in the table below. In some cases, traditional rounding will result in too many objects being apportioned rather than too few.

School	Standard Quota Q	Q Rounded to Nearest Integer
Applegate	16.14	16
Bayshore	21.96	22
Claypool	14.35	14
Delmar	27.31	27
Edgewater	29.24	29
Totals	109	108

The next example shows how the seats of the first House of Representatives would have been apportioned in 1794, if the Hamilton method rather than the Jefferson method had been used.

EXAMPLE 2 The 1790 census determined that the population of the United States was 3,615,920. The Union had 15 states, and it was decided that the original House would have 105 seats. The Hamilton method is used to apportion the 105 seats.

1. Compute the standard divisor $d = 3{,}615{,}920/105 = 34{,}437.33$.
2. Virginia had the largest population with 630,560 residents. The standard quota for Virginia is $Q = 630{,}560/34{,}437.33 = 18.310$. Using Q to three places is necessary for this example. The populations of the other states and their standard quotas are shown in the table.

State	Population	Standard Quota Q	Rounded-down Q	Number of Seats
Virginia	630,560	18.310	18	18
Massachusetts	475,327	13.803	13	14
Pennsylvania	432,879	12.570	12	13
North Carolina	353,523	10.266	10	10
New York	331,589	9.629	9	10
Maryland	278,514	8.088	8	8
Connecticut	236,841	6.877	6	7
South Carolina	206,236	5.989	5	6
New Jersey	179,570	5.214	5	5
New Hampshire	141,822	4.118	4	4
Vermont	85,533	2.484	2	2
Georgia	70,835	2.057	2	2
Kentucky	68,705	1.995	1	2
Rhode Island	68,446	1.988	1	2
Delaware	55,540	1.613	1	2
Totals	3,615,920	105	97	105

3. The values of Q are rounded down in the fourth column of the table. The total of this column, the number of seats promised, is 97. Eight more seats must be assigned.

4. The apportionment of the seats by the Hamilton method is in the final column. The bold entries indicate the states that receive more seats than their rounded-down values of Q. They are the eight states with the largest fractional parts of their standard quotas.

The Hamilton apportionment of the original House gives Delaware 2 seats. This actually is not a Constitutionally correct apportionment. The Constitution specifically requires that each seat represent at least 30,000 people. Delaware's population is 4460 people short of deserving 2 House seats.

The state of Virginia, where Thomas Jefferson made his home, would have received 18 seats in the first House if the Hamilton method of apportionment had been used. The actual 1794 House apportionment used the Jefferson method shown later in Example 4.

The only specific mandates given by the Constitution for apportioning seats in the House of Representatives are that apportionment be based on state population and that each seat represent at least 30,000 people. The theoretical Hamilton apportionment of the House in Example 2 can be used to illustrate one of the more serious and subtle problems with the method. This is a problem of which President Washington was certainly aware, when he vetoed the bill that would have made the Hamilton method the law of the land.

In Example 2, the additional 8 seats are assigned based on fractional parts of the standard quotas, in the order shown in the table at the side. It may be argued that these 8 additional seats are not awarded based on population when the fourth step of the Hamilton method is executed. The state with the next largest fractional part of its standard quota is Vermont, with $Q = 2.484$. The fraction .570 for Pennsylvania, the last state to receive an additional seat, is unarguably larger than Vermont's fraction of .484. Simple division shows that the fractional part .484 of Vermont's value of Q is approximately 20% (.484/2.484) of its total value of $Q = 2.484$, while the fractional part .570 of Pennsylvania's value of Q is only about 5% (.570/12.570) of its total value of $Q = 12.570$. Yet, it is Pennsylvania, not Vermont, that gets an extra seat when Step 4 of Hamilton's method is used to assign the remaining 8 seats. The original detractors of the Hamilton method argued that this kind of situation showed that the method did not follow the Constitutional mandate of apportionment by population. President Washington evidently felt their argument was valid.

State	Standard Quota Q
Kentucky	1.**995**
South Carolina	5.**989**
Rhode Island	1.**988**
Connecticut	6.**877**
Massachusetts	13.**803**
New York	9.**629**
Delaware	1.**613**
Pennsylvania	12.**570**

The Jefferson Method Unable to override Washington's veto of the Hamilton method, Congress adopted the method Thomas Jefferson was promoting. It was used to apportion the original House in 1794, based on the 1790 census. The Jefferson method eliminates the pragmatic, yet problematic and perhaps unconstitutional, Step 4 of the Hamilton method. It replaces the Hamilton method's straightforward standard divisor d with a modified, slightly magical divisor, md. Each state's population is divided by the modified divisor md to produce its modified quota mQ. Each state receives as many seats as the integer part of its modified quota mQ. The beauty of Jefferson's method is that the sum of the integer parts of all the values of mQ is equal to the total number of seats to be apportioned. No additional seats are left to be assigned. How is that possible? The modified, slightly magical divisor md is carefully and painstakingly chosen to force it to happen.

16.3 The Possibilities of Apportionment

Thomas Jefferson (1743–1826) is perhaps the most important man in American history. He shaped the very foundation of the country, helping to write both the Declaration of Independence and the Constitution of the United States. Many of the ideas in these important documents are the fruits of Jefferson's brilliant philosophical mind. Jefferson was secretary of state under President Washington and later was elected president himself. Some of his other accomplishments include devising the decimal monetary system and nearly doubling the land area of the United States with the Louisiana Purchase in 1803.

The Jefferson Method

1. Compute *md*, the **modified divisor**.
2. Compute *mQ*, the **modified quota** for each state.

$$mQ = (\text{state's population})/md$$

3. Round each state's modified quota *mQ* *down* to the nearest integer.
4. Give each state this integer number of seats.

A standard quota Q is computed by dividing a state's population by the standard divisor d. The values of Q must, by definition, sum to the total number of seats to be apportioned. Rounded down, the values of Q may sum to less than the total number of seats to be apportioned. If the rounded-down values of Q actually sum to the number of things to be apportioned, the Jefferson apportionment simply uses $md = d$. Otherwise, for the Jefferson method to work, the values of the modified quotas must be larger than the values of Q so that when the values of mQ are rounded down they sum to exactly the number of seats to be apportioned, not less. If the values of mQ must be slightly larger than the values of Q, then md, the divisor used to compute the values of mQ, must be slightly smaller than the standard divisor d used to compute the values of Q. In other words, since smaller divisors produce larger quotients, the Jefferson method produces bigger modified quotas for each state by dividing the state populations by a number smaller than the standard divisor d.

Algorithms exist to find the values of md needed for Jefferson method apportionments. The algorithms are technical and beyond the range of this discussion. The values of md in the following examples were found by trial and error, slowly decreasing the value of the standard divisor d, until the rounded values of mQ summed to the exact number of rooms or seats to be apportioned.

EXAMPLE 3 Sea Isle Vacations operates six secluded resorts in the Caribbean. They recently purchased 30 new sailboats. Executives at the company headquarters in Virginia decide to use the Jefferson method to apportion the new boats, basing the apportionment on the number of rooms at each resort. The six individual resorts are named after hurricanes that spared them. The number of rooms at each location is in the second column of the table in Step 2 (on the next page). The total number of rooms at the six resorts is 2013.

1. The standard divisor $d = (\text{total number of rooms})/(\text{number of new boats}) = 2013/30 = 67.1$. The value of the standard divisor means that, theoretically, for every 67.1 rooms at a resort, that resort is entitled to one of the new boats. The modified divisor md was found by slowly decreasing the value of d. For this example, when $md = 60$ the Jefferson method works. In fact, values of md greater than 58.5 but less than or equal to 60 also work.

2. Resort Anna has a modified quota of $mQ = 345/60$.
Resort Bob has a modified quota of $mQ = 234/60$.
Resort Cathy has a modified quota of $mQ = 420/60$.
Resort David has a modified quota of $mQ = 330/60$.
Resort Ellen has a modified quota of $mQ = 289/60$.
Resort Floyd has a modified quota of $mQ = 395/60$.

Each resort's modified quota mQ is given to two places in the fourth column of the table. The third column contains the standard quotas for comparison.

Resort	Number of Rooms	Standard Quota Q	Modified Quota mQ
Anna	345	5.14	5.75
Bob	234	3.49	3.90
Cathy	420	6.26	7.00
David	330	4.92	5.50
Ellen	289	4.31	4.82
Floyd	395	5.89	6.58
Totals	2013	30.00	33.55

3. The modified quotas are rounded down in the final column. The rounded-down values of mQ for the six resorts sum to exactly 30. This is guaranteed by the choice of $md = 60$. The rounded-down values of Q are given for comparison.

Resort	Number of Rooms	Standard Quota Q	Rounded-down Q	Modified Quota mQ	Rounded-down mQ
Anna	345	5.14	5	5.75	5
Bob	234	3.49	3	3.90	3
Cathy	420	6.26	6	7.00	7
David	330	4.92	4	5.50	5
Ellen	289	4.31	4	4.82	4
Floyd	395	5.89	5	6.58	6
Totals	2013	30.00	27	33.55	30

4. Each resort gets the number of new sailboats in the final column. The Jefferson method has automatically apportioned the exact number of new sailboats available. No boats remain unallocated.

If the Hamilton method is used to apportion the sailboats, the results are different from the results in Example 3. Rounding down the values of Q leaves three sailboats unassigned, as the total in the fourth column shows. The three boats go to resorts David, Floyd, and Bob, the resorts with the largest fractional portions of their values of Q. The table compares the apportionments of the sailboats by the Hamilton and Jefferson methods.

Resort	Number of Rooms	Rounded-down Standard Quota Q	Number of Boats Using Hamilton	Number of Boats Using Jefferson
Anna	345	5	5	5
Bob	234	3	4	3
Cathy	420	6	6	7
David	330	4	5	5
Ellen	289	4	4	4
Floyd	395	5	6	6
Totals	2013	27	30	30

Bold entries in each of the two final columns indicate the resorts that received more boats than their rounded-down values of Q. Both methods increase the number of boats for resorts David and Floyd. The Hamilton method increases the apportionment for the smallest resort, Bob, while the Jefferson method increases the apportionment for the largest resort, Cathy.

It is not a coincidence that the Jefferson method favored the largest resort. The Jefferson method always shows a preference for larger states. This bias helped Jefferson's large home state of Virginia in the actual apportionment of the House in 1794, as the next example shows.

EXAMPLE 4 The Jefferson method apportionment of the 1794 House of Representatives is based on the 1790 census. Census workers calculated the total population of the existing fifteen states as 3,615,920. The number of seats had been set at 105. The standard divisor $d = 34{,}437.33$ was computed in Example 2.

1. The modified divisor used by Congress for the actual Jefferson method apportionment was $md = 33{,}000$. Considering the hours of manual calculations it probably required to test each guess of the value of md, it must have been a welcome surprise that when $md = 33{,}000$, the Jefferson method works.

2. The modified quota for Virginia is $630{,}560/33{,}000 = 19.108$. The modified quotas for the other states are in the table below.

3. The rounded-down modified quotas mQ are in the final column of the table. The rounded-down values sum to the required number of 105 seats, because of the choice of the value of md.

4. Each state gets the number of seats shown in the final column of the table, that is, the integer part of its modified quota mQ.

The U.S. Department of the Treasury began issuing commemorative quarters for each state starting in 1999. The quarters are being phased into circulation in the same order that the states they represent joined the Union. The next time you find a new state quarter in your pocket, spend a minute or two reflecting on how the addition of that state changed the way the seats in the House of Representatives were apportioned and the mathematics involved in the process.

State	Population	Modified Quota mQ	Rounded-down mQ
Virginia	630,560	19.108	19
Massachusetts	475,327	14.404	14
Pennsylvania	432,879	13.118	13
North Carolina	353,523	10.713	10
New York	331,589	10.048	10
Maryland	278,514	8.440	8
Connecticut	236,841	7.177	7
South Carolina	206,236	6.250	6
New Jersey	179,570	5.442	5
New Hampshire	141,822	4.298	4
Vermont	85,533	2.592	2
Georgia	70,835	2.147	2
Kentucky	68,705	2.082	2
Rhode Island	68,446	2.074	2
Delaware	55,540	1.683	1
Totals	3,615,920	109.576	105

The numbers in the final column represent the way the House of Representatives was actually apportioned in 1794. The largest state, Jefferson's home state of Virginia, received 19 seats in the 1794 apportionment of the House of Representatives. Delaware, the smallest state in the Union, received only one seat when the Jefferson method was used.

The Jefferson method is considered a **divisor method,** because it uses a modified divisor *md* to produce modified quotas. Values of *mQ* are rounded down in the Jefferson method. Other divisor methods, such as the Webster method discussed next, use different rounding schemes. When President Washington rejected the Hamilton method on Constitutional grounds, supporters of other apportionment methods questioned whether the modified divisor in the Jefferson method was Constitutionally sound. Jefferson argued that the Constitution required only that the House apportionment be based on population, and not on a standard divisor. Jefferson method supporters convinced skeptics that the modified divisor *md* still produced quotients that reflect state populations and, since the same divisor is used for all states, the method was on firm Constitutional ground.

The Webster Method

The Jefferson method was used for decades without incident, until the 1820s. The apportionment of the House based on the 1820 census revealed a major flaw in this method that is discussed later. The flaw might have been dismissed as a freak incident, if the same anomaly had not occurred again following the 1830 census. The Jefferson method was replaced by a method proposed and championed by Daniel Webster. No one ever suspected it was flawed in the same way as the Jefferson method. The Webster method was used to apportion the House following the 1840 census, then again from 1900 until it was replaced by the Hill-Huntington method in 1941.

The appeal of the Webster method is simple. Standard quotas are not rounded down to the nearest integer. Instead, they are rounded using a traditional rounding scheme. A quota with a fractional part greater than or equal to .5 is rounded up to the next highest integer, and a quota with a fractional part less than .5 is rounded down to the next lowest integer. Discussion following Example 1 concerning the Highwood School District computer apportionment shows apportionment does not necessarily work when a traditional rounding scheme is used on the standard quota values of *Q*. However, the Webster method, based on a traditional rounding scheme, does work, because, like the Jefferson method, it is a divisor method. The Webster method uses a modified divisor *md*. The value of *md* is selected so that when the modified quotas (mQ = state population/*md*) are traditionally rounded, they sum to exactly the number of seats to be apportioned.

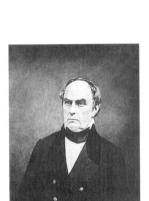

Daniel Webster (1782–1852) was an important constitutional lawyer, a brilliant speaker, and a folk hero. He served in Congress and the Senate and was secretary of state under Presidents Harrison, Tyler, and Fillmore. Webster spent his life keeping a watchful eye on the power of individual states and was an ardent supporter of the national government. He disliked the institution of slavery, but he supported the Compromise of 1850 on constitutional grounds and his support helped it pass. The Compromise of 1850 delayed the American Civil War for ten years. In that time, the free states gained enough economic power to abolish slavery in America.

The Webster Method

1. Compute *md,* the **modified divisor.**
2. Compute *mQ,* the **modified quota** for each state.

$$mQ = (\text{state's population})/md$$

3. Round each state's modified quota *mQ* *up* to the nearest integer if its fractional part is greater than or equal to .5 and *down* to the nearest integer if its fractional part is less than .5.
4. Give each state this integer number of seats.

The Webster-modified divisor *md* is a bit harder to find than the modified divisor for the Jefferson method because rounding the values of *mQ* can go either way in the traditional rounding scheme of the Webster method. The Jefferson method always rounds the values of mQ down to the nearest integer. Rounding the values of mQ down allows the modified divisor for the Jefferson method to be found by slowly decreasing the value of the standard divisor d, to produce slightly larger modified quotas with a larger sum. Traditional rounding of the standard quota values of Q can cause their sum to be less than, equal to, or even greater than the actual number of seats to be apportioned.

If the rare case of equality occurs, then simply let $md = d$ for the Webster apportionment. If the sum of the traditionally rounded values of Q is less than the number of seats to be apportioned, then slowly decrease the value of d to find the modified Webster divisor md, as in the Jefferson method. If the values of Q sum to more than the number of objects to be apportioned when traditionally rounded, then the value of md for the Webster apportionment is found by slowly increasing the value of d, the standard divisor. This is because bigger divisors make smaller quotients, and therefore smaller sums.

EXAMPLE 5 If the Webster method is used to apportion the 109 computers given to the Highwood School District, the apportionment is different from that given in Example 1, in which the Hamilton method is used to apportion the computers. The total enrollment in the district is 2263.

1. The value of the standard divisor $d = 20.76147$ and the standard quotas for the five schools are computed in Example 1. Discussion following Example 1 shows that the values of Q when traditionally rounded sum to 108, and 109 computers are to be apportioned. Here the value of md is found by slowly *decreasing* the value of d by trial and error to produce a slightly larger sum of mQ values. When $md = 20.6$, the Webster method works.

2. The modified quota for Applegate School is $mQ = 335/20.6 = 16.26$. The modified quotas for the other schools are in the table below.

3. The five values of mQ in the fourth column are traditionally rounded. The rounded values sum to exactly 109. This is guaranteed by the choice of $md = 20.6$.

School	Enrollment	Modified Quota mQ	Traditionally Rounded mQ
Applegate	335	16.26	16
Bayshore	456	22.14	22
Claypool	298	14.47	14
Delmar	567	27.52	28
Edgewater	607	29.47	29
Totals	2263		109

4. Each school gets the number of computers in the final column. The Webster method automatically apportioned exactly 109 computers.

The Webster apportionment agrees with the Hamilton apportionment (Example 1) for the Applegate, Bayshore, and Edgewater schools, but not for the other two schools. The Webster method gives Delmar School 28 computers and Claypool School 14. The Hamilton method apportions 27 computers to Delmar and 15 to Claypool.

EXAMPLE 6 A newly formed nation on a small island in the Pacific adopted a constitution modeled on the Constitution of the United States. Unlike the United States, the new nation, Timmu, decided to have just one legislative assembly with 131 seats. Like the U.S. House of Representatives, the seats of Timmu's legislature are to be apportioned based on the populations of the nation's individual states. The Timmu Constitution does not specify an apportionment method, but the leadership of the island nation voted to use the Webster method.

A census found that the population of Timmu is currently 47,841. A house size of 131 seats means that the standard divisor $d = 47{,}841/131 = 365.1985$. The small nation is divided into six states. The table shows the state populations, the values of the standard quotas to three places, and the traditionally rounded values of the standard quotas.

State	Population	Standard Quota Q	Traditionally Rounded Q
Abo	5672	15.531	16
Boa	8008	21.928	22
Cio	2400	6.572	7
Dao	6789	18.590	19
Effo	4972	13.615	14
Foti	20,000	54.765	55
Totals	47,841		133

1. The sum of the traditionally rounded values of Q is 133, which is greater than the number of seats in the Timmu house. The value of the modified divisor md needed for the Webster method is found by slowly *increasing* the value of the standard divisor $d = 365.1985$, which produces smaller quotas with a smaller sum. Using a computer spreadsheet program, the leadership of Timmu discovers that an increased value of $md = 366.9724$ produces a set of modified quota values that sum to 132 seats, after traditional rounding. An increased value of $md = 366.9730$ produces values of mQ that sum to 130 seats. The values of $md = 366.9725, 366.9726, 366.9727, 366.9728,$ and 366.9729 all produce values of mQ that sum to exactly 131 seats. The leaders use the value $md = 366.9729$.

2. The modified quota for Abo is found by dividing its population by the value of the modified divisor, $mQ = 5672/366.9729 = 15.4562$, to four places. Modified quotas to four places for the other five states are given in the table in the next step.

3. The six values of mQ are traditionally rounded in the fourth column of the table. The value of $md = 366.9729$ guarantees the traditionally rounded values of mQ sum to 131 exactly.

State	Population	Modified Quota mQ	Traditionally Rounded mQ
Abo	5672	15.4562	15
Boa	8008	21.8218	22
Cio	2400	6.5400	7
Dao	6789	18.5000	19
Effo	4972	13.5487	14
Foti	20,000	54.4999	54
Totals	47,841		131

4. Each state receives the number of seats in the Timmu House of Representatives shown in the final column of the table.

16.3 EXERCISES

1. The schoolchildren of Wisconsin began a statewide effort last year to raise money to purchase new trees for five state parks. Altogether, the students raised enough money to purchase 239 trees. A committee decided that the trees should be apportioned based on the number of acres of land in each of the five parks. The area of each of the five parks is shown in the table.

State park	a	b	c	d	e
Acres	1429	8639	7608	6660	5157

   (a) Find the total number of acres in the five state parks and compute the standard divisor for the apportionment of the 239 trees.
   (b) Use the Hamilton method to apportion the trees.
   (c) Use the Jefferson method to apportion the trees. As always, the modified divisor needed for the Jefferson apportionment is found by slowly decreasing the value of the standard divisor computed in part (a).
   (d) Round the values of all of the parks' standard quotas, using a traditional rounding scheme. Is the sum of the traditionally rounded values less than, equal to, or greater than the number of trees to be apportioned?
   (e) Should the value of the modified divisor for the Webster method apportionment be less than, equal to, or greater than the standard divisor computed in part (a)? Why?
   (f) Use the Webster method to apportion the trees.
   (g) Which of the three tree apportionments are the same? Which is different?

2. School enrollments for the Highwood School District, the standard divisor, and the Hamilton apportionment of its 109 new computers are in Example 1. The Webster apportionment is in Example 5.
   (a) Use the Jefferson method to apportion the 109 computers the Highwood School District received.
   (b) Which of the apportionments of the 109 computers are the same? Which is different?
   (c) The students at Claypool School studied the various apportionment methods and launched a massive letter-writing campaign to convince the Highwood School District to use the Hamilton method, rather than the Jefferson or Webster method, to apportion the new computers. Why?

3. The English department at Oaks College has the faculty to offer 11 sections in any combination of the four courses: Fiction Writing, Poetry, Short Story, or Multicultural Literature. The number of sections apportioned to each course is based on the enrollments in the courses. The enrollment figures are in the table.

Course	Fiction	Poetry	Short Story	Multicultural
Enrollment	56	35	78	100

   (a) Find the total enrollment for the four courses and the standard divisor for the apportionment of the 11 sections.
   (b) Use the Hamilton method to apportion the sections.
   (c) Use the Jefferson method to apportion the sections.
   (d) Round the values of all of the courses' standard quotas, using a traditional rounding scheme. Is the sum of the traditionally rounded values less than, equal to, or greater than the number of sections to be apportioned?
   (e) Should the value of the modified divisor for the Webster method apportionment be less than, equal to, or greater than the standard divisor computed in part (a)? Why?
   (f) Use the Webster method to apportion the sections.
   (g) Which of the three section apportionments are the same? Which is different?
   (h) If you were a student enrolled in the poetry course at Oaks College, which apportionment method would you hope was used to apportion the sections? Why?
   (i) If you were one of the 100 students enrolled in the multicultural literature course, which apportionment method would you hope was used to apportion the sections? Why?

4. The number of rooms at each Sea Isle resort and the standard divisor are in Example 3.
   (a) Use the Webster method to apportion the 30 new sailboats. Follow the process in Exercise 1 parts (d) and (e) to find md.
   (b) The Jefferson apportionment of the sailboats is in Example 3. The Hamilton apportionment follows Example 3. Which sailboat apportionments are the same? Which is different?

5. The information for apportioning the Timmu House of Representatives and the Webster apportionment of the 131 seats are in Example 6.
   (a) Use the Hamilton method to apportion the seats.
   (b) Use the Jefferson method to apportion the seats.
   (c) Which Timmu House of Representatives apportionments are the same? Which are different?

6. If the leadership of Timmu uses $md = 366.9724$ for the Webster method apportionment of its legislative seats, show that 132 seats, not 131, are apportioned.

7. If the leadership of Timmu uses $md = 366.9730$ for the Webster method apportionment of its legislative seats, show that 130 seats, not 131, are apportioned. Use six decimal places for mQ.

8. Show that the Webster method apportionment of the original 1794 United States House of Representatives agrees with the Hamilton method apportionment in Example 2.

9. Forty nurses from North Carolina have volunteered to help at 5 hospitals treating the victims of an earthquake in a remote mountainous region of Mexico. The Mexican government decides to apportion the nurses based on the number of beds at each of the hospitals.

Hospital	A	B	C	D	E
Number of beds	137	237	337	455	555

   (a) Find the total number of beds at the 5 hospitals and compute the standard divisor for the apportionment of the 40 nurses.
   (b) Use the Hamilton method to apportion the nurses.
   (c) Use the Jefferson method to apportion the nurses.
   (d) Round the values of the hospitals' standard quotas, using a traditional rounding scheme. Is the sum less than, equal to, or greater than the number of nurses to be apportioned?
   (e) Should the value of the modified divisor for the Webster method apportionment of the nurses be less than, equal to, or greater than the standard divisor computed in part (a)? Why?
   (f) Use the Webster method to apportion the nurses.
   (g) What do you notice about the three apportionments of the 40 nurses?

10. The governor of Wisconsin challenged the schoolchildren of his state to raise enough money for an additional 200 trees. He also decided to postpone the purchase and apportionment of the trees until the children met his challenge, which took them only six months. The trees were then apportioned based on the number of acres at each of the five parks, as given in Exercise 1.
    (a) Find the total number of acres in the parks and compute the standard divisor for the apportionment of 439 trees.
    (b) Use the Hamilton method to apportion the trees.
    (c) Use the Jefferson method to apportion the trees.
    (d) Round the values of the parks' standard quotas using a traditional rounding scheme. Is the sum of the traditionally rounded values less than, equal to, or greater than the number of trees to be apportioned?
    (e) Should the value of the modified divisor for the Webster method apportionment be less than, equal to, or greater than the value of the standard divisor computed in part (a)? Why?
    (f) Use the Webster method to apportion the trees.
    (g) What do you notice about the three apportionments of 439 trees? Comment on the political wisdom of Wisconsin's governor.

11. Create a population profile for 5 states with a total population of 1000 for which both of the following conditions are true: the Hamilton method, the Jefferson method, and the Webster method apportionments of 100 legislative seats all agree; the modified divisors for the Jefferson and Webster apportionments equal the standard divisor.

12. Create an enrollment profile for 4 science courses with a total enrollment of 198 students for which both of the following conditions are true: the Hamilton method and Webster method apportionments of 9 teaching assistants to the 4 courses are the same, and the Jefferson method apportionment is different; the modified divisor for the Webster apportionment equals the standard divisor.

13. Create a ridership profile for 5 school bus routes that must service 949 students for which the following condition is true: the Hamilton method, the Jefferson method, and the Webster method apportionments of 16 school buses to the 5 routes all are different.

The Adams Method An apportionment method considered by Congress but never adopted for use was proposed by John Quincy Adams, the sixth president of the United States. The Adams Method is a divisor method. The method uses a modified divisor selected to produce modified quotas that when rounded up to the nearest integer produce a sum that is equal to the number of seats being apportioned.

> **The Adams Method**
> 1. Compute *md*, the **modified divisor.**
> 2. Compute *mQ*, the **modified quota** for each state.
> 3. Round each state's modified quota *mQ* up to the nearest integer.
> 4. Give each state this integer number of seats.

The standard quotas rounded up to the nearest integer can automatically sum to the number of seats to be apportioned. In these cases, the Adams modified divisor md equals d, the standard divisor, and the apportionment is done. Otherwise, the modified divisor needed for an Adams method apportionment is always found by slowly increasing the value of the standard divisor.

14. Determine why the modified divisor needed for an Adams method apportionment is always found by slowly increasing the value of the standard divisor.

15. (a) Use the Adams method to apportion the 11 sections the English department at Oaks College is offering. See Exercise 3 for the enrollment numbers. The standard divisor is computed in Exercise 3(a).
    (b) Compare the Adams method apportionment of the 11 sections with the Hamilton, Jefferson, and Webster apportionments.

16. (a) Use the Adams method to apportion the 40 volunteer nurses to the five hospitals in Mexico. See Exercise 9. The standard divisor is computed in Exercise 9(a).
    (b) What do you notice about the Adams method apportionment of the nurses?

17. (a) Use the Adams method to apportion the 131 seats in Timmu's House of Representatives. See Example 6.
    (b) What do you notice about the four different apportionments of Timmu's 131 legislative seats? See Exercise 5 and Example 6.

The Hill-Huntington Method Congress currently uses the Hill-Huntington method to apportion the U.S. House of Representatives. The Hill-Huntington method is also a divisor method. The method uses a modified divisor to produce modified quotas that, when rounded by the unique Hill-Huntington rounding scheme, sum to exactly the number of seats being apportioned.

Executing the Hill-Huntington method of apportionment is very similar to executing the Webster method. First, the standard quotas Q are rounded by the Hill-Huntington rounding scheme and summed. This sum can be less than, equal to, or greater than the number of seats being apportioned. Based on the sum of the rounded values of Q, a modified divisor for the Hill-Huntington method is selected. The modified divisor is used to produce modified quotas that are rounded according to the Hill-Huntington rounding scheme and summed. If the sum of modified quotas is equal to the number of seats being apportioned, then the apportionment is complete and each state receives its rounded integer number of seats. If the sum is not equal to the number of seats being apportioned, the modified divisor is tweaked until the sum is equal to the number of seats being apportioned.

> **The Hill-Huntington Method**
> 1. Compute *md*, the **modified divisor.**
> 2. Compute *mQ*, the **modified quota** for each state.
> 3. Round each state's modified quota *mQ* to the nearest integer using the Hill-Huntington rounding scheme.
> 4. Give each state this integer number of seats.

The Hill-Huntington unique rounding scheme uses the geometric mean *of two numbers. The geometric mean of two non-negative numbers a and b is the square root of the product of the two numbers, $\sqrt{a \cdot b}$. The Hill-Huntington method uses the geometric mean of $a =$ the integer part of a modified quota and $b =$ the integer part of the modified quota $+1$ as the cutoff point for rounding the modified quota up to the nearest integer, instead of down to the nearest integer.*

For example, suppose a state has a modified quota of $mQ = 6.49$. The Hill-Huntington method cutoff point for rounding this modified quota up to 7, the nearest integer, is the geometric mean of 6 and $6 + 1 = 7$. The geometric mean of 6 and 7 is equal to $\sqrt{6 \cdot 7} = \sqrt{42} = 6.481$. The value $mQ = 6.49$ is *greater than 6.481, the geometric mean of 6 and 7, so the modified quota is rounded up to 7. Notice the modified quota $mQ = 6.49$ would not have been rounded up to 7 using a traditional rounding scheme, but is rounded up to 7 by the Hill-Huntington rounding scheme.*

18. Find the Hill-Huntington method cutoff point for rounding a modified quota of $mQ = 5.470$ up to 6, the nearest integer.

19. Find the Hill-Huntington cutoff point for rounding up $mQ = 56.498$.

20. Find the Hill-Huntington cutoff point for rounding up $mQ = 11.71$.

21. Find the Hill-Huntington cutoff point for rounding up $mQ = 32.497$.

22. If initially the Hill-Huntington rounded values of Q sum to less than the number of objects being apportioned, then how is the standard divisor d modified to produce the modified divisor md necessary for the Hill-Huntington method? Why?

23. If initially the Hill-Huntington rounded values of Q sum to more than the number of objects being apportioned, then how is the standard divisor d modified to produce the modified divisor md necessary for the Hill-Huntington method? Why?

24. If initially the Hill-Huntington rounded values of Q sum to exactly the number of objects being apportioned, then how is the standard divisor d modified to produce the modified divisor md necessary for the Hill-Huntington method? Why?

25. (a) Use the Hill-Huntington method to apportion the 11 sections the English department at Oaks College is offering. See Exercise 3. The standard divisor is computed in Exercise 3(a).
    (b) Compare the Hill-Huntington apportionment of the 11 sections with the Hamilton, Jefferson, Webster, and Adams apportionments.

26. (a) Use the Hill-Huntington Method to apportion the 40 volunteer nurses to the five hospitals in Mexico. See Exercise 9. The standard divisor is computed in Exercise 9(a).
    (b) Compare the Hill-Huntington apportionment of the nurses with the Hamilton, Jefferson, Webster, and Adams apportionments.

27. (a) Use the Hill-Huntington method to apportion the 131 seats in Timmu's House of Representatives. See Example 6.
    (b) Compare the Hill-Huntington apportionment of the 131 seats with the Hamilton, Jefferson, Webster, and Adams apportionments.

28. Which apportionment method do you think is best? Give three reasons to support your choice.

29. Suppose the total population is 1000 and there are 100 seats to be apportioned to five states. What simple condition guarantees that the Hamilton, Jefferson, and Webster apportionments are all the same and the modified divisors for the Jefferson and Webster methods equal the standard divisor? Explain why the condition guarantees what it does.

30. What do you think is the problem with the final step of the Hamilton method of apportionment? Explain.

31. Why does the Jefferson method of apportionment favor large populations?

16.4 The Impossibilities of Apportionment

The Hamilton, Jefferson, and Webster methods of apportionment have all been used to apportion the seats in the U.S. House of Representatives. Over time, all three of these methods fell from favor. What happened in the 1822 and 1832 apportionments that made Congress abandon the Jefferson method and replace it with the Webster method? Why was the Webster method immediately replaced by the Hamilton method? How did the Hamilton method fall from favor at the end of the 1800s? Why did the Webster method make a comeback, only to be replaced by the Hill-Huntington method that is currently in favor?

These questions can be answered by considering potential pitfalls of each apportionment method. For example, the modified divisors in both the Jefferson and Webster methods can produce apportionments that violate a basic measure of

In 1842, the House actually decreased in size. The apportionment for 1832 had divided 240 seats; the apportionment of 1842 divided only 223 seats. The House has had 435 seats since 1912.

Notwithstanding a direct Constitutional mandate, the House was not reapportioned following the 1920 census. Technically, this means that none of the business conducted by the House in the 1920s was completely honorable.

fairness called the *quota rule*. This is what happened in the 1822 and 1832 Jefferson apportionments. Also, three quantities vary in the apportionment process: the number of objects being apportioned; the population on which the apportionment is based; and the number of parties getting a part of the apportionment. A change in any of the three quantities can unpredictably affect the Hamilton method apportionment. In legislative lingo, if the number of seats, the state populations, or the number of states changes, then the Hamilton method can produce apportionments that are paradoxical. All three of these quantities were dramatically changing during the last half of the 1800s. The changes naturally revealed the serious problems of the Hamilton method.

The Quota Rule The Jefferson method apportionment of the House of Representatives following the 1820 census caused concern. New York had a standard quota of $Q = 32.503$. The value of New York's standard quota rounded up to the nearest integer is 33. However, when the seats were apportioned, New York received 34 seats because the modified divisor used for the Jefferson method gave New York a modified quota mQ that rounded (down) to 34. This caused some alarm (since New York received more seats than its value of Q rounded up to the nearest integer), but not enough to cause rejection of the Jefferson method. According to the 1830 census, New York had a standard quota of $Q = 38.593$, which, when rounded up to the nearest integer, is 39. The Jefferson apportionment calculated a modified quota mQ for New York that rounded (down) to 40. Thus, two consecutive House apportionments using the Jefferson method had awarded the state of New York more House seats than many people believed was fair.

The Jefferson method was known to naturally favor large states such as New York. The method's partiality for large states alone had not been enough to condemn it. However, the second time the Jefferson method apportioned more seats to a state than the state's value of Q rounded up to the nearest integer, it was rejected and replaced by the Webster method. The Jefferson method failed to pass the basic test of fairness known as the *quota rule*.

> **The Quota Rule**
>
> The integer number of objects apportioned to each state must be either the standard quota Q rounded down to the nearest integer, or the standard quota Q rounded up to the nearest integer.

A **quota method** of apportionment uses the standard divisor d and apportions seats to each state equal to the standard quota Q either rounded up or rounded down to the nearest integer. The quota rule is satisfied by any quota method of apportionment by definition.

The Hamilton method is a quota method and satisfies the quota rule. The method computes the standard divisor and uses it to produce standard quotas. The values of Q are then rounded down and each state is guaranteed at least this number of seats. This means that no state can receive fewer seats than its value of Q rounded down to the nearest integer. Any additional seats are awarded one at a time to the states with the largest fractional parts of their values of Q. A state cannot receive more than one additional seat in the fourth step of the Hamilton method because there are never more additional seats than states. If a state does receive an additional seat, then its apportionment of seats equals its rounded-down value of Q plus one. The rounded-down value of Q plus one is equal to the value of Q rounded up to the nearest integer. This means no state receives more seats than its value of Q rounded up to the nearest integer.

Historical House of Representatives apportionments of the 1820s and 1830s reveal that the Jefferson method does not satisfy the quota rule since a state can receive more seats than its value of Q rounded up to the nearest integer. A state can also receive fewer seats than its value of Q rounded down to the nearest integer. The Webster method also fails to satisfy the quota rule. Both the Jefferson and Webster methods are **divisor methods** of apportionment; they use modified divisors, not the standard divisor. Although the modified divisors of the Jefferson and Webster methods have withstood Constitutional debate, the divisors are the reason the two methods and all divisor methods of apportionment fail to satisfy the quota rule.

The Hill-Huntington method, adopted in 1941 and currently used to apportion the House, is named for Joseph Hill and Edward V. Huntington. Hill was chief statistician of the Bureau of the Census. He proposed the method in 1911. Huntington, a professor of mathematics at Harvard, gave it his endorsement.

The method was studied by the National Academy of Sciences. The Academy concluded in 1929 that the Hill-Huntington method was more neutral than the other method under consideration in its treatment of both large and small states. In 1980, Balinski and Young (see the margin note on p. 971) showed that the Academy's conclusion was false. The method does have a slight bias for small states, because of its liberal rounding scheme. Nonetheless, Hill-Huntington remains the law.

EXAMPLE 1 The Nexus Motor Company currently produces 20,638 automobiles per week at six different factories. Business has been very good for Nexus and 500 additional workers have just been hired. The CEO and the board decide to apportion the 500 new employees, using the Jefferson method, according to the weekly automobile production at each facility. Legislatively speaking, the newly hired employees are seats, the six factories are states, and the production at each factory is a state population.

1. The standard divisor $d =$ (total weekly production)/(number of new employees) $= 20{,}638/500 = 41.276$. The value of the modified divisor md is found by trial and error, slowly decreasing the value of the standard divisor d. Here, $md = 41.055$ works in the Jefferson method.

2. The modified quotas for the six factories are in the table below. The mQ values are calculated by dividing the weekly production of each factory by the modified divisor md. The standard quota values are given for comparison.

3. The rounded-down modified quotas are in the final column of the table. They sum to exactly 500. The rounded-down values of the standard quotas are given for comparison.

Factory	Production of Cars per Week	Standard Quota Q	Rounded-down Q	Modified Quota mQ	Rounded-down mQ
A	1429	34.621	34	34.807	34
B	3000	72.681	72	73.073	73
C	1642	39.781	39	39.995	39
D	9074	**219.837**	**219**	221.021	**221**
E	4382	106.163	106	106.735	106
F	1111	26.916	26	27.061	27
Totals	20,628	500	496		500

4. Each factory gets the number of new employees shown in the final column of the table.

The Jefferson method violates the quota rule in this apportionment. The standard quota of the largest factory, Factory D, is $Q = 219.837$. The quota rule states that the number of employees apportioned to Factory D must either be 219 or 220. So the apportionment of 221 new employees to Factory D violates the quota rule.

The large weekly production of Factory D in Example 1 is naturally favored by Jefferson's method. Decreasing the value of the standard divisor to find the value of *md* allows the modified quota of the large production factory to grow more rapidly than the modified quotas of the other factories.

The same kind of paradoxical apportionment is also possible when the Webster method is used. The standard divisor is always decreased to find the value of *md* required for the Jefferson method. This means the Jefferson method violates the upper bound of the quota rule if it fails to satisfy the rule. The Webster method can require the value of the standard divisor to be increased to find the value of the modified divisor. Increasing the value of the standard divisor can cause the modified quota of a small state to decrease more rapidly than the modified quotas of other states. The rapid decrease can mean the small state receives an apportionment that is smaller than its value of Q rounded down. It is this possibility of a quota rule violation that taints the Jefferson and Webster methods.

The Alabama Paradox The required tweaking of the standard divisor in any divisor method is an open invitation for quota violations, especially if one state's population is large (or small) in relation to other state populations. The Hamilton method and all quota methods automatically satisfy the quota rule by their definition. The Hamilton method would be the method of choice if the quota rule was the sole test of fairness for an apportionment method, but it is not the only such benchmark.

As the population of the United States grew and additional states joined the union, the size of the U.S. House of Representatives slowly increased from the original 105 seats to the current 435 seats. Preparations for increasing the size of the House following the 1880 census revealed a surprising flaw in the Hamilton method. Government workers calculated the different Hamilton method apportionments for all the different House sizes from 275 to 350 seats. This was quite an amazing feat, considering spreadsheet programs were not available and all calculations were done manually. The sample apportionments showed that if the House had 299 seats, then Alabama would receive 8 seats. However, if the House size was 300, then Alabama would receive 7 seats. This odd result is known as the *Alabama paradox.*

The Alabama Paradox

The situation in which an increase in the number of objects being apportioned actually forces a state to lose one of those objects is known as the **Alabama paradox.**

EXAMPLE 2 A longtime resident of Highwood has given one additional computer to the school district. The gift brings the total number of computers to be apportioned by the Hamilton method to 110.

1. Total enrollment in the school district is 2263, and now 110 computers are to be apportioned. The standard divisor $d = $ (total enrollment)/(number of computers) $= 2263/110 = 20.57273$.

2. Standard quotas for each school are computed by dividing the enrollment for the school by d. Enrollments and the standard quotas for the five schools are shown in the first table below.

3. The standard quotas are rounded down to the nearest integer in the fourth column. Each school receives at least this integer number of computers. The values of Q sum to 108, so two additional computers are to be apportioned in the fourth step of Hamilton's method.

4. The two additional computers are apportioned one at a time to the two schools with the largest fractional parts of their standard quotas. Delmar School gets the first additional computer, and Edgewater School gets the other. The fractional parts of the standard quotas for these two schools, .56 and .51 respectively, are greater than the fractional parts of the standard quotas of the other schools. The final apportionment of the 110 computers is in the last column of the table. Notice that Claypool School receives 14 computers when 110 are apportioned.

School	Enrollment	Standard Quota Q	Rounded-down Q	Computers Apportioned
Applegate	335	16.28	16	16
Bayshore	456	22.17	22	22
Claypool	298	14.49	14	14
Delmar	567	27.56	27	28
Edgewater	607	29.51	29	30
Totals	2263	110	108	110

Example 1 of the previous section uses the Hamilton method to apportion the original gift of 109 computers to the five schools in the Highwood School District. The standard divisor for apportioning 109 computers is $d = 2263/109 = 20.76147$. The original apportionment of 109 computers is repeated here.

School	Enrollment	Standard Quota Q	Rounded-down Q	Computers Apportioned
Applegate	335	16.14	16	16
Bayshore	456	21.96	21	22
Claypool	298	14.35	14	15
Delmar	567	27.31	27	27
Edgewater	607	29.24	29	29
Totals	2263	109	107	109

16.4 The Impossibilities of Apportionment

The gift of an additional computer causes the Alabama paradox. Using the Hamilton method, Claypool School receives 15 computers when 109 computers are apportioned and only 14 when 110 are apportioned. The increase in the number of computers being apportioned actually forces Claypool School to forfeit one of its machines. ∎

The paradoxical apportionment in Example 2 is caused by the pragmatic fourth step in the Hamilton method. Additional seats are distributed to the states with the largest parts of their standard quotas. If the number of seats to be apportioned increases, the value of d = (total population)/(total number of seats) decreases. A decreased value of d means the values of Q = (state population)/d increase. Example 2 shows the increased values of Q when 110 instead of 109 computers are apportioned. The problem is that the values of Q do not increase by the same amount for all states. If the number of seats is increased, then a state that originally has a large enough fractional part of its value of Q to earn an additional seat may not continue to have a large enough fractional part of Q to do so. Increases in the fractional parts of the values of Q are erratic. The Hamilton method uses the fractional values for apportioning the additional seats, and their erratic nature can cause strange phenomena such as the Alabama paradox.

The Population Paradox   The quota rule is an extremely important measure of fairness for an apportionment method. The Hamilton method always satisfies the quota rule, even if it does allow the possibility of the Alabama paradox. The Alabama paradox can occur only when the number of objects being apportioned *increases*. If the number of objects being apportioned is permanently *fixed*, then the Hamilton method appears to shine again. The size of the U.S. House of Representatives has been fixed at 435 seats for many years. Why are the seats still apportioned with a divisor method that invites quota rule violations? What keeps the Hamilton method from really shining as the apportionment method of choice? The Hamilton method allows for paradoxical outcomes other than the Alabama paradox. Even with a fixed number of objects being apportioned, another type of paradox can occur. Discovered around 1900, this paradox is known as the *population paradox*.

> **The Population Paradox**
>
> The **population paradox** occurs when, based on updated population figures, a reapportionment of a fixed number of seats causes a state to lose a seat to another state, although the percent increase in the population of the state that loses the seat is bigger than the percent increase in the population of the state that gains the seat.

EXAMPLE 3   Eleven students in a graduate school education class plan to interview 370 educators at 3 different grammar schools. They are interested in how art and music education impact scores on standardized mathematics exams. The students decide to apportion themselves into three interview teams using the Hamilton method, based on the number of educators to be interviewed at each school. In legislative lingo, the interviewers are comparable to House seats; they are what is being apportioned. The three schools are states and the number of interviews to be conducted at each school is a state population.

1. Eleven graduate students are to interview 370 educators. The standard divisor $d =$ (total number of interviews)/(number of interviewers) $= 370/11 = 33.6364$.

2. Standard quotas are computed by dividing the number of interviews at each school by $d = 33.6364$. The values of Q are shown in the table below.

3. The values of Q are rounded down in the fourth column. At least this many graduate students will conduct the interviews at each school. The sum of the rounded Q values is 9, so two additional interviewers are to be assigned in the next step of the Hamilton apportionment.

4. The two schools with the largest fractional parts of their values of Q are St. Francis (.7162) and Marshall (.6486). The final apportionment of the interviewers is in the last column of the table. Bold numbers in the last column indicate that St. Francis and Marshall get more interviewers than their value of Q rounded down.

School	Number of Interviews	Standard Quota Q	Rounded-down Q	Interviewers Apportioned
Lincoln	55	1.6351	1	1
St. Francis	125	3.7162	3	**4**
Marshall	190	5.6486	5	**6**
Totals	370	11	9	11

A month before the interviews are scheduled, the graduate students decide to include additional interviews with retired educators from the three schools. Five of the retired educators are from Lincoln, 20 are from St. Francis, and 20 are from Marshall. The interviewers must now talk with $370 + 45 = 415$ people. They reapportion themselves using the Hamilton method.

1. The new standard divisor $d = 415/11 = 37.727$.

2. The values of Q are recalculated by dividing the new number of interviews at each school by the new standard divisor $d = 37.727$. The revised values of Q are shown in the table below.

3. The values of Q are rounded down in the fourth column. They sum to 9, leaving two additional interviewers to be apportioned in the final step.

4. The two schools with the largest fractional parts of their Q values are St. Francis (.8434) and Lincoln (.5904). Bold entries in the last column indicate these two schools get more interviewers than their values of Q rounded down. The revised apportionment of the interviewers is in the last column.

School	Number of Interviews	Standard Quota Q	Rounded-down Q	Interviewers Apportioned
Lincoln	60	1.5904	1	**2**
St. Francis	145	3.8434	3	**4**
Marshall	210	5.5663	5	5
Totals	415	11	9	11

The original Hamilton apportionment gives one additional interviewer to each of the St. Francis and Marshall interview teams. An additional interviewer still is assigned to the St. Francis team by the revised apportionment, but the second additional interviewer is assigned to the Lincoln interview team, rather than Marshall.

Comparing the growth rates of the number of interviews at Lincoln and Marshall reveals that the team of graduate student interviewers has suffered the population paradox. The addition of the retired educators to the interview list causes the number of interviews at Marshall to increase by $(210 - 190)/190 = 10.53\%$, and the number of interviews at Lincoln to increase by $(60 - 55)/55 = 9.09\%$. The percent increase in the number of interviews to be conducted at Marshall is bigger than the percent increase in the number of interviews at Lincoln, yet in the final Hamilton method apportionment the Marshall interview team loses a member to the Lincoln interview team. ■

The New States Paradox The Hamilton method of apportionment allows the population paradox for the same reason it allows the Alabama paradox. Both paradoxes are triggered by the pragmatic fourth step of the method, which awards the unapportioned seats based on the fractional parts of the values of Q, the standard quotas. Paradoxes occur because the changes in the fractional parts of Q, due to an increased number of seats or revised populations, are erratic at best. If the surplus seats are distributed based on erratic values, then unsettling results are bound to occur. Another paradox triggered by the pragmatic fourth step of the Hamilton method is called the *new states paradox*. The new states paradox was discovered by Balinski and Young. (See p. 971.) As its name suggests, this paradox occurs when seats are reapportioned following the addition of a new state.

If a new state is added to the existing states, then the number of seats or objects being apportioned is usually increased to prevent a decrease in the existing apportionments. The purpose of the increase is to keep all existing apportionments the same, while adding the new state and its fair share of seats to the mix. The new states paradox may result from such an addition.

The new states paradox occurred in 1907, when Oklahoma joined the Union as the 46th state. Remember this fact when you find a new commemorative quarter for Oklahoma among the coins in your pocket. It will be issued in 2008.

The New States Paradox

The **new states paradox** occurs when a reapportionment of an increased number of seats, necessary due to the addition of a new state, causes a shift in the apportionment of the original states.

EXAMPLE 4 Several years ago the town of Lake Bluff annexed a small unincorporated subdivision. The population of the original town was 8500 and the annexed subdivision had a population of 1671. The new town was governed by 100 council members. The Hamilton method was used to apportion council seats to the original town and the annexed subdivision based on their populations.

1. The total population of the new community was $8500 + 1671 = 10{,}171$, and 100 council seats were available. This means the standard divisor $d =$ (total population)/(number of seats) $= 10{,}171/100 = 101.71$.

2. The standard quotas for the original and annexed populations are in the first table on the next page.

> Congress decided to divide 283 seats following the 1870 census. They selected this number of seats because with it the official apportionment done by the Hamilton method agreed with the theoretical outcome of the Webster method. A power grab on the floor of the House resulted in the apportionment of 292 seats, without using the lawful Hamilton method. It was this unconstitutional House that gave Rutherford B. Hayes enough electoral votes to become president. If the Hamilton method had been used, Samuel J. Tilden, who had won a majority of the national votes, would have had enough electoral votes to become president.

3. The rounded-down values of Q sum to 99, leaving one seat to be apportioned in the fourth step.

4. The additional council seat is awarded to the original town. Its fractional part of Q is .57, which is greater than .43, the annexed population's fractional part of Q. The apportionment of the 100 seats is in the final column of the table.

Community	Population	Standard Quota Q	Rounded-down Q	Council Seats Apportioned
Original	8500	83.57	83	84
Annexed	1671	16.43	16	16
Totals	10,171	100	99	100

During the following years, council members from the annexed subdivision lobbied intensely and tirelessly. They finally convinced the town to annex a second small unincorporated subdivision with a population of 545. Populations of the original town and the first annexed subdivision did not change. The town council voted to increase the number of council seats from 100 to 105. They reasoned that if the established standard divisor of $d = 101.71$ was used, then a population of 545 would have a standard quota of $Q = 545/101.71 = 5.36$, which rounds down to 5.

An updated Hamilton method apportionment of the 105 council seats shows why the first annexed subdivision lobbied so intensely for the annexation of another subdivision. The updated apportionment was done with a revised value of the standard divisor d, reflecting the increased total population of Lake Bluff and the increased number of council seats.

1. The updated total population was $8500 + 1671 + 545 = 10,716$. The revised standard divisor $d = 10,716/105 = 102.057$.

2. The updated values of Q, the standard quotas, are in the table below.

3. The rounded-down values of Q sum to 104, leaving one additional council seat to be apportioned.

4. The fractional part of Q for the first annexed subdivision is .37, which is greater than the fractional parts of the values of Q for the other two populations. It receives the additional available seat. The reapportionment of the 105 council seats is in the final column of the table.

Community	Population	Standard Quota Q	Rounded-down Q	Council Seats Apportioned
Original	8500	83.29	83	83
1st Annexed	1671	16.37	16	17
2nd Annexed	545	5.34	5	5
Totals	10,716	105	104	105

Comparing the two apportionments shows that annexing a second subdivision with a population of 545 and adding new seats to the council causes the new states paradox to occur. The shift in the apportionment for the original two communities favors the first annexed subdivision. The first apportionment splits 100 council seats, giving 84 seats to the original town and 16 seats to the first annexed subdivision. The second apportionment again splits 100 seats between the original town and the first annexed subdivision, but the apportionment shifts, giving only 83 seats to the original town and 17 seats to the first annexed subdivision.

The Balinski and Young Impossibility Theorem

As mentioned previously, the Jefferson and Webster methods are examples of divisor methods. All divisor methods are afflicted by the possibility of a quota rule violation, since they depend on a modified standard divisor—not the standard divisor. Modifying the standard divisor is necessary to force the modified values of the standard quotas, rounded according to the prescribed scheme, to sum to exactly the number of seats being apportioned.

The Hamilton method is an example of a quota method. The Hamilton method uses the standard divisor, so it avoids the possibility of a quota rule violation. However, it allows puzzling apportionments. The pragmatic final step of the Hamilton method permits the possibility of the Alabama paradox, the population paradox, or the new states paradox. In other words, increases in the number of seats to be apportioned, changes in population figures, or increases in the number of states can cause the Hamilton method to go awry. An unexpected 1980 discovery by Michel Balinski and Peyton Young is hinted at in the summary table.

Michel L. Balinski is a mathematician at CNRS and Ecole Polytechnique (Paris). **H. Peyton Young** is a mathematical economist at Johns Hopkins University. In addition to their important impossibility theorem and results about the Hill-Huntington method, they have shown that the Webster method is the only divisor method of apportionment that is not biased to either large or small states.

Method	Quota Rule	Fairness Standard		
		Alabama Paradox	Population Paradox	New States Paradox
Hamilton method	Satisfied	Can occur	Can occur	Can occur
Jefferson method	Can violate	Cannot occur	Cannot occur	Cannot occur
Webster method	Can violate	Cannot occur	Cannot occur	Cannot occur

Scholars and politicians have always expected that mathematicians would eventually find the perfect apportionment method. Instead, mathematicians Balinski and Young showed that an ultimate method was impossible.

The Balinski and Young Impossibility Theorem

Any apportionment method that always satisfies the quota rule will, by its nature, permit the possibility of a paradoxical apportionment. Likewise, any apportionment method that does not permit the possibility of a paradoxical apportionment will, by its nature, fail to always satisfy the quota rule.

Balinski and Young, like Kenneth Arrow (see the margin note on p. 939), answered an age-old question of great social importance. Unfortunately, even very talented mathematicians cannot find solutions that do not exist. Fortunately, they can tell the world when perfection does not exist and can begin developing ways of coping with unavoidable imperfection.

16.4 EXERCISES

1. Show that if the Jefferson method is used to apportion 132 legislative seats to four states with the given populations, then a violation of the quota rule occurs.

State	a	b	c	d
Population	17,179	7500	49,400	5824

2. Show that if the Jefferson method is used to apportion 200 legislative seats to four states with the given populations, then a violation of the quota rule occurs.

State	a	b	c	d
Population	67,000	35,000	15,000	9900

3. Show that if the Jefferson method is used to apportion the 131 seats in the Timmu House of Representatives, then a violation of the quota rule occurs.

State	Abo	Boa	Cio	Dao	Effo	Foti
Population	5672	8008	2400	6789	4972	20,000

4. Show that if the Jefferson method is used to apportion 200 legislative seats to five states with the given populations, then a violation of the quota rule occurs.

State	a	b	c	d	e
Population	4589	1515	2013	1118	1111

5. Show that if the Jefferson method is used to apportion 290 legislative seats to five states with the given populations, then a violation of the quota rule occurs.

State	a	b	c	d	e
Population	2567	1500	8045	950	1099

6. Show that if the Jefferson method is used to apportion 150 legislative seats to five states with the given populations, then a violation of the quota rule occurs.

State	a	b	c	d	e
Population	1720	3363	6960	24,223	8800

7. Show that if the Hamilton method is used to apportion legislative seats to four states with the given populations, then the Alabama paradox occurs when the number of seats being apportioned increases from $n = 204$ to $n = 205$.

State	a	b	c	d
Population	3462	7470	4265	5300

8. Repeat Exercise 7 using the following populations. Increase n from 71 to 72.

State	a	b	c	d
Population	1050	2040	3060	4050

9. Show that if the Hamilton method is used to apportion legislative seats to five states with the given populations, then the Alabama paradox occurs when the number of seats being apportioned increases from $n = 126$ to $n = 127$.

State	a	b	c	d	e
Population	263	808	931	781	676

10. Repeat Exercise 9 using the following populations. Increase n from 45 to 46.

State	a	b	c	d	e
Population	309	289	333	615	465

11. Repeat Exercise 9 using the following populations. Increase n from 149 to 150.

State	a	b	c	d	e
Population	5552	8260	5968	6256	5150

12. Repeat Exercise 7 using the following populations. Increase n from 160 to 161.

State	a	b	c	d
Population	5613	6958	7434	5085

13. Use the Hamilton method to apportion 11 legislative seats to three states with the given initial populations. Show that the population paradox occurs when the Hamilton method is used to apportion the 11 legislative seats again, after the population figures are revised by a census. The populations are given in thousands.

State	a	b	c
Initial population	55	125	190
Revised population	61	148	215

14. Repeat Exercise 13 using the following populations.

State	a	b	c
Initial population	55	125	190
Revised population	62	150	218

15. Use the Hamilton method to apportion 13 legislative seats to three states with the given initial populations. Show that the population paradox occurs when the Hamilton method is used to apportion the 13 legislative seats again, after the population figures are revised by a census. The populations are given in thousands.

State	a	b	c
Initial population	930	738	415
Revised population	975	750	421

16. Repeat Exercise 15 using the following populations.

State	a	b	c
Initial population	930	738	415
Revised population	975	752	422

17. Repeat Exercise 15 using the following populations.

State	a	b	c
Initial population	89	125	225
Revised population	97	145	247

18. Repeat Exercise 15 using the following populations.

State	a	b	c
Initial population	89	125	225
Revised population	98	145	249

19. Use the Hamilton method to apportion 16 legislative seats to two states with the given populations. Show that the new states paradox occurs if a new state with the indicated population and the appropriate number of additional seats is included in a second Hamilton method apportionment. The appropriate number of seats to add for the second apportionment is the state's standard quota of the new state, computed using the original two-state standard divisor, rounded down to the nearest integer. The populations are given in thousands.

State	Original State a	Original State b	New State c
Population	134	52	38

20. Repeat Exercise 19 using the following populations.

State	Original State a	Original State b	New State c
Population	134	52	39

21. Repeat Exercise 19, but apportion 75 legislative seats using the following populations. The populations are given in hundreds.

State	Original State a	Original State b	New State c
Population	3184	8475	330

22. Repeat Exercise 21 using the following populations. Apportion 75 legislative seats.

State	Original State a	Original State b	New State c
Population	3184	8475	350

23. Repeat Exercise 19, but apportion 83 legislative seats using the following populations. The populations are given in hundreds.

State	Original State a	Original State b	New State c
Population	7500	9560	1500

24. Repeat Exercise 23 using the following populations. Apportion 83 legislative seats.

State	Original State a	Original State b	New State c
Population	7500	9560	1510

25. Find an enrollment profile for four classes with a total enrollment of 335 for which the following condition is true: The Jefferson method apportionment of 25 teaching assistants produces a violation of the quota rule.

26. Find a population profile for four counties with a total population of 87,580 for which the following condition is true: The Jefferson method apportionment of 50 council seats produces a violation of the quota rule.

27. The five-state population profile in Exercise 5 produces a violation of the quota rule when the Jefferson method is used to apportion 290 legislative seats. Using the same population profile, find a different number of legislative seats for which the Jefferson method of apportionment produces a violation of the quota rule. (*Hint:* Another number of seats that produces a quota violation is between 194 and 204 seats.)

28. The five-state population profile in Exercise 6 produces a violation of the quota rule when the Jefferson method is used to apportion 150 legislative seats. Using the same population profile, find a different number of legislative seats for which the Jefferson method of apportionment produces a violation of the quota rule. (*Hint:* Another number of seats that produces a quota violation is between 95 and 105 seats.)

29. Find a number of seats n that, when increased to $n + 1$ seats, causes the Alabama paradox to occur if the Hamilton method is used to apportion the seats based on the given population profile for four states at the top of the next column. Populations are given in thousands. (*Hint:* The number n is between 45 and 55.)

State	a	b	c	d
Population	465	552	385	251

30. Repeat Exercise 29 using the following populations. (*Hint:* The number n is between 14 and 24.)

State	a	b	c	d
Population	777	418	664	280

31. Use the Hamilton method to apportion 11 teaching assistants to 3 courses with the given preliminary enrollment figures. The actual enrollment for one course is given. Find actual enrollments for the other two courses such that when the Hamilton method is used to apportion the 11 teaching assistants, based on actual enrollments, the population paradox occurs. That is, find enrollment figures for the other two courses that cause a course to lose a teaching assistant to a course with a smaller percent increase in enrollment.

Course	a	b	c
Preliminary enrollment	55	125	190
Actual enrollment	?	145	?

32. Repeat Exercise 31. Instead of 11 teaching assistants, apportion 13 teaching assistants using the following enrollments.

Course	a	b	c
Preliminary enrollment	89	125	225
Actual enrollment	?	145	?

33. In Example 4, council members from the first subdivision annexed to the town of Lake Bluff lobbied for the annexation of a second unincorporated subdivision with a population of 545 because they had realized it would cause the new states paradox to occur. The first annexed subdivision took one of the original 100 council seats away from the original town when 5 new seats were included in an updated Hamilton method apportionment. If, by strange coincidence, the population of the second subdivision had been only 531, why would it have been prudent for the first subdivision to wait until a baby was born and the population increased to 532?

34. Consider the situation described in Exercise 33. If, by strange coincidence, the population of the second subdivision had been 548, why would it have been urgent for the first subdivision to act before a baby was born and the population increased to 549?

35. Show that if the Adams method is used to apportion 220 legislative seats to the five states with the population profile given in Exercise 6, then a violation of the quota rule occurs.

36. Show that if the Adams method is used to apportion 219 legislative seats to the five states with the population profile given in Exercise 6, then a violation of the quota rule occurs.

37. Now that you know the pros and cons of the Hamilton, Jefferson, and Webster methods, have you changed your mind about which apportionment method you think is the best? Explain why you changed your mind, or why you are backing your original choice.

38. Which impossibility of apportionment—a quota rule violation, the Alabama paradox, the population paradox, or the new states paradox—do you find most disturbing? Explain.

39. Which impossibility of apportionment—a quota rule violation, the Alabama paradox, the population paradox, or the new states paradox—do you find least disturbing? Explain.

40. In your own words, describe the problem with the final step of the Hamilton method of apportionment.

41. Why is the Adams method of apportionment tainted by the possibility of a quota rule violation? What kind of a quota rule violation can occur if the Adams method is used? Why?

42. Is the Hill-Huntington method of apportionment tainted by the possibility of a quota rule violation? Explain. If quota rule violations can occur, then describe them.

COLLABORATIVE INVESTIGATION

Class Favorites, an Election Exploration

Part One: To be completed following Section 16.1: The Possibilities of Voting

Divide the class into groups of six students each. Within each group designate three teams A, B, and C of two students each. Do the following activities in order.

1. Each group should select a different collection of 4 or 5 candidates/alternatives, from which they want to learn the class favorite. Suggestions include candidates for a student government post, movies nominated for "Best Picture," songs nominated in any of the Grammy categories, the best *Star Wars* movie, or candidates in a presidential primary.

2. Groups should prepare ballots for a class vote on their selected collection of candidates. The ballots must provide a list of all of the $n!$ possible rankings of the $n = 4$ or 5 candidates. Groups should distribute a ballot to each member of the class.

3. Each member of the class should mark the complete ranking representing his or her arrangement of the particular collection of candidates. Ballots should then be returned to the appropriate groups.

4. Groups should create a voter profile of the class based on the returned ballots. To do this, simply tally how many members of the class selected each of the $n!$ rankings. Organize the profile into a table showing how many voters preferred each ranking. The voter profile will be needed in Part Two of this Collaborative Investigation.

5. Teams should determine the outcome of the election using different voting methods.
   Team A: Find the outcome of the election using the plurality method. Also, find the outcome of the election using the Borda method.

(continued)

Team B: Find the outcome of the election using the pairwise comparison method. Keep a careful record of the pairwise results; it will be useful for Part Two of the Collaborative Investigation.
Team C: Find the outcome of the election using the Hare method.

6. When the three teams have completed their individual tasks in Item 5, reassemble as a 6-student group.
   - Team A should report the winner of the plurality election and the winner of the Borda election. Team B should report the winner of the pairwise comparison method election. Team C should report the outcome of the Hare election.
   - Prepare a chart showing which candidate was selected by each method: plurality, pairwise comparison, Borda, and Hare.
   - If the outcomes of the various elections show agreement, study the voter profile and try to explain why the methods agree. If the outcomes of the various elections do not agree, try to decide what it is about the profile of voters that causes the disagreement.

7. Choose a representative from each group to report to the entire class on the election the group conducted.

Part Two: To be completed following Section 16.2: The Impossibilities of Voting

Reassemble the original groups of 6 students (divided into three teams). Do the following activities in order.

1. Study the voter profile to find out if there is a majority candidate. Use the notes from Team B to decide if a Condorcet candidate exists for the voter profile.

2. Determine which of the plurality, pairwise comparison, Borda, and Hare methods satisfy the majority criterion and Condorcet criterion for the group's voter profile. In other words, which methods elect the majority candidate, if one exists, and which do not? Do the same for the Condorcet candidate, if one exists.

3. Now investigate the outcomes of a second election, after a losing candidate has been removed from the group's collection. Team members should consider the independence of irrelevant alternatives criterion.
Team A: Remove a candidate who did not win the plurality election from the collection. Determine the winner of the remaining candidates, using the plurality method. Remove a candidate who did not win the Borda election from the collection. Determine the winner of the remaining candidates using the Borda method. If there is extra time, try removing a different loser from an election or remove more than one loser.
Team B: Remove a candidate who did not win the pairwise comparison election from the collection. Determine the winner of the remaining candidates using the pairwise comparison method. Notes from Part One are helpful. If time permits, remove a different loser and recalculate the pairwise winner.
Team C: Remove a candidate who did not win the Hare election from the collection.
Determine the winner of the remaining candidates using the Hare method. If time permits, remove a different loser and conduct another Hare election.

4. When all the team members have completed their individual tasks in Item 3, return to a group discussion.
   - Team A should report whether it can demonstrate that the plurality method does not satisfy the independence of irrelevant alternatives (IIA) criterion. It also should report whether it can show that the Borda method does not satisfy the IIA criterion.
   - Team B should report whether it can show that the pairwise comparison method does not satisfy the IIA criterion.
   - Team C should report whether it can show that the IIA criterion is not satisfied by the Hare method.

5. Choose a representative from each group to report to the class what the group showed concerning the majority criterion, the Condorcet criterion, and the IIA criterion.

CHAPTER 16 TEST

Twenty-seven members of a sorority are debating where to spend spring break. They have narrowed the list of possibilities to Aruba a, the Bahamas b, Cancún c, or the Dominican Republic d. Each of the 27 sisters gave a complete ranking of the four choices on a separate ballot. The voter profile of the sorority members is summarized in the table.

Number of Voters	Ranking
5	a > b > d > c
6	b > c > a > d
5	c > d > b > a
7	d > a > b > c
4	c > a > d > b

Determine the spring break destination of the sorority using the

1. plurality method

2. Borda method

3. Hare method

4. sequential pairwise comparison method and the predetermined order a, c, b, d of the alternatives.

5. Suppose you are the president of the sorority, and this gives you the power to dictate which voting method will be used to determine the spring break destination. Suppose further that only you and the sorority secretary have seen the actual voter profile. Which voting method will you select to determine the spring break destination of the sorority? Why?

6. Use the approval voting method to determine the spring break destination of the sorority. Suppose each sister votes for both her first- and second-choice destinations, but no other.

7. Use the approval voting method to determine the spring break destination of the sorority. Suppose each of the 5 sisters with the first ranking gives an approval vote to all the destinations except Cancún; they went to Cancún last year. Suppose all the other sisters vote for both their first- and second-choice destinations, but no other.

8. Why would the pairwise comparison method be a poor choice of voting method for the sorority destination decision?

9. For the sorority voter profile, find a predetermined order for a sequential pairwise comparison method vote that selects Cancún as the spring break destination. Use the information about each of the pairwise comparisons discovered in Exercise 8.

10. What is the majority criterion?

11. What is the Condorcet criterion?

12. Explain the difference between the monotonicity criterion and the alternative version of the monotonicity criterion. In your opinion, which is a stricter criterion?

13. Why is the independence of irrelevant alternatives criterion an important measure of fairness in a two-stage election process?

14. Show that the Borda method violates the majority criterion for the given voter profile.

Number of Voters	Ranking
16	a > b > c
8	b > c > a
7	c > b > a

15. Use the pairwise comparison method to determine which alternative the committee with the following voter profile prefers.

Number of Voters	Ranking
5	a > c > b
6	c > b > a
4	b > a > c
6	b > c > a

16. Which alternative is a Condorcet candidate for the profile of voters in Exercise 15? Find the preferred alternative of the committee in Exercise 15 by using each of the plurality, Borda, and Hare methods. For the voter profile in Exercise 15, which voting methods violate the Condorcet criterion and which uphold it?

17. Use the pairwise comparison method to determine the preferred alternative for the following profile of voters.

Number of Voters	Ranking
8	$c > m > b > s$
7	$s > b > m > c$
6	$s > b > c > m$
5	$m > c > b > s$

Use the pairwise comparison method to redetermine the preferred alternative for the voter profile, if the five voters with the last ranking rearrange their ranking to be $c > s > m > b$. What do the outcomes of the two pairwise comparison method selections show about the voting method?

18. Use the Hare method to determine the preferred alternative for the following profile of voters.

Number of Voters	Ranking
9	$a > c > b > d$
8	$b > d > a > c$
7	$c > d > b > a$
5	$c > a > b > d$

Use the Hare method to redetermine the preferred alternative for the voter profile, if the five voters with the last ranking rearrange their ranking to be $a > c > b > d$. What do the outcomes of the two Hare method selections show about the voting method?

19. Use the plurality method to determine the preferred alternative for the following profile of voters.

Number of Voters	Ranking
10	$a > b > c$
7	$b > c > a$
5	$c > b > a$

Use the plurality method to redetermine the preferred alternative for the voter profile, if alternative c is dropped from the selection process. What do the outcomes of the two plurality method selections show about the voting method?

20. Use the Borda method to determine the preferred alternative for the following profile of voters.

Number of Voters	Ranking
7	$a > b > c > d$
5	$b > c > d > a$
4	$d > c > a > b$

Use the Borda method to redetermine the preferred alternative for the voter profile, if alternative d is dropped from the selection process. What do the outcomes of the two Borda method selections show about the voting method?

21. Use the Hare method to determine the preferred alternative for the profile of voters in Exercise 20. Use the Hare method to redetermine the preferred alternative for the voter profile, if alternative b is dropped from the selection process. What do the outcomes of the two Hare method selections show about the voting method?

22. Who is Kenneth J. Arrow? Explain in your own words what Arrow's impossibility theorem says about voting methods.

The city of Smithapolis, population 2,600,700, is divided into five wards and governed by a city council of 195 representatives. The 195 council seats are apportioned to the five wards based on their populations. A recent city census reported the following populations (in hundreds) for the five wards.

Ward	1st	2nd	3rd	4th	5th
Population	1429	8639	7608	6660	1671

Apportion the 195 council seats of Smithapolis using the

23. Hamilton method
24. Jefferson method
25. Webster method.
26. What is the quota rule?
27. Explain the Alabama paradox.
28. Explain the population paradox.
29. Explain the new states paradox.

30. Show that if the Jefferson method is used to apportion 100 legislative seats to four states with the given populations, then a violation of the quota rule occurs. The populations are given in hundreds.

State	a	b	c	d
Population	2354	4500	5598	23,000

31. Show that if the Hamilton method is used to apportion legislative seats to five states with the given populations, then the Alabama paradox occurs when the number of seats being apportioned increases from $n = 126$ to $n = 127$. The populations are given in thousands.

State	a	b	c	d	e
Population	263	809	931	781	676

32. Use the Hamilton method to apportion 11 legislative seats to three states with the given initial populations. Show that the population paradox occurs when the Hamilton method is used to apportion the 11 legislative seats again, after the population figures are revised by a census. The populations are given in thousands.

State	a	b	c
Initial population	55	125	190
Revised population	63	150	220

33. Use the Hamilton method to apportion 100 legislative seats to two states with the given populations. Show that the new states paradox occurs if a new state with the indicated population and the appropriate number of additional seats is included in the second Hamilton method apportionment. The appropriate number of seats to add for the second apportionment is the state's standard quota of the new state, computed using the original two-state standard divisor, rounded down to the nearest integer. The populations are given in thousands.

State	Original State a	Original State b	New State c
Population	49	160	32

34. Who are Balinski and Young? In your own words, explain what the Balinski and Young impossibility theorem says about apportionment methods.

APPENDIX
The Metric System

Joseph Louis Lagrange (1736–1813) was born in Turin, Italy, and became a professor at age 19. In 1776 he came to Berlin at the request of Frederick the Great to take the position Euler left. A decade later Lagrange settled permanently in Paris. Napoleon was among many who admired and honored him.

Lagrange's greatest work was in the theory and application of calculus. He carried forward Euler's work of putting calculus on firm algebraic ground in his theory of functions. His *Analytic Mechanics* (1788) applied calculus to the motion of objects. Lagrange's contributions to algebra had great influence on Galois and hence the theory of groups. He also wrote on number theory; he proved, for example, that every integer is the sum of at most four squares. His study of the moon led to methods for finding longitude.

The metric system of weights and measures is based on decimals and powers of ten. For this reason, calculations and changes of units in the metric system are much easier to do than in the English system that we now use. The metric system was developed by a committee of the French Academy just after the French Revolution of 1789. The president of the committee was the mathematician Joseph Louis Lagrange.

The advantages of the metric system can be seen when compared to our English system. In the English system, one inch is one-twelfth of a foot, while one foot is one-third of a yard. One mile is equivalent to 5280 feet, or 1760 yards. Obviously, there is no consistency in subdivisions. In the metric system, prefixes are used to indicate multiplications or divisions by powers of ten. For example, the basic unit of length in the metric system is the *meter* (which is a little longer than one yard). To indicate one thousand meters, attach the prefix "kilo-" to get kilometer. To indicate one one-hundredth of a meter, use the prefix "centi-" to obtain centimeter. These prefixes are examples of those used for lengths, weights, and volumes in the metric system. A complete list of the prefixes of the metric system is shown in the table below, with the most commonly used prefixes appearing in heavy type.

Metric Prefixes

Prefix	Multiple	Prefix	Multiple
exa	1,000,000,000,000,000,000	deci	.1
peta	1,000,000,000,000,000	**centi**	.01
tera	1,000,000,000,000	**milli**	.001
giga	1,000,000,000	micro	.000001
mega	1,000,000	nano	.000000001
kilo	1000	pico	.000000000001
hecto	100	femto	.000000000000001
deka	10	atto	.000000000000000001

Length, Area, and Volume Lagrange urged the committee devising the metric system to find some natural measure for length from which weight and volume measures could be derived. It was decided that one **meter (m)** would be the basic unit of length, with a meter being defined as one ten-millionth of the distance from the equator to the North Pole. According to Paul G. Hewitt, in *Conceptual Physics*, 7th edition (HarperCollins, 1993):

This distance was thought at the time to be close to 10,000 kilometers. One ten-millionth of this, the meter, was carefully determined and marked off by means of scratches on a bar of platinum-iridium alloy. This bar is kept at the International Bureau of Weights and Measures in France. The standard meter in France has since been calibrated in terms of the wavelength of light—it is 1,650,763.73 times the wavelength of orange light emitted by the atoms of the gas krypton-86. The meter is now defined as being the length of the path traveled by light in a vacuum during a time interval of 1/299,792,458 of a second.

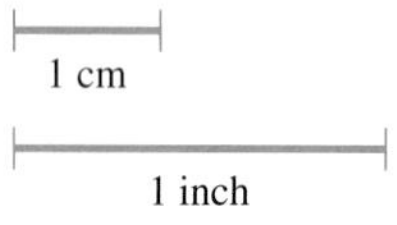

To obtain measures longer than one meter, Greek prefixes were added. For measures smaller than a meter, Latin prefixes were used. A meter is a little longer than a yard (about 39.37 inches). Shorter lengths commonly are measured in micrometers, centimeters, and millimeters. A **centimeter** is one one-hundredth of a meter and is about 2/5 of an inch. (See Figure 1.) Centimeter is symbolized cm.

Because the metric system is based on decimals and powers of ten, conversions within the system are very easy. They simply involve multiplying and dividing by powers of ten. For example, to convert 2.5 m to centimeters, multiply 2.5 by 100 (since 100 cm = 1 m) to obtain 250 cm. On the other hand, to convert 18.6 cm to meters, divide by 100 to obtain .186 m. Other conversions are made in the same manner, using the meanings of the prefixes. Why is 42 m equal to 42,000 millimeters (mm)?

Long distances usually are measured in kilometers. A **kilometer** is 1000 meters. (According to a popular dictionary, the word *kilometer* may be pronounced with the accent on either the first or the second syllable. Scientists usually stress the second syllable.) A kilometer (symbolized km) is equal to about .6 mile. Figure 2 shows the ratio of 1 kilometer to 1 mile.

Conversions from meters to kilometers, and vice versa, are made by multiplying or dividing by 1000 as necessary. For example, 37 kilometers equals 37,000 meters, while 583 meters equals .583 km.

The area of a figure can be measured in square metric units. Figure 3 shows a square that is 1 cm on each side; thus, it is a **square centimeter (cm²)**. One square meter (m²) is the area of a square with sides one meter long. How many cm² are in one m²?

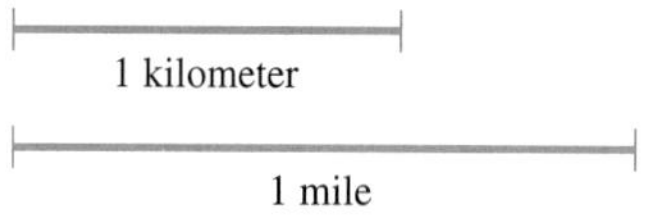

1 cm

1 inch

FIGURE 1

1 kilometer

1 mile

FIGURE 2

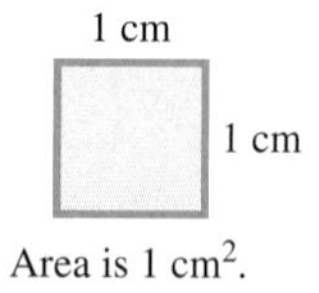

Area is 1 cm².

FIGURE 3

Volume is 1 cm³.

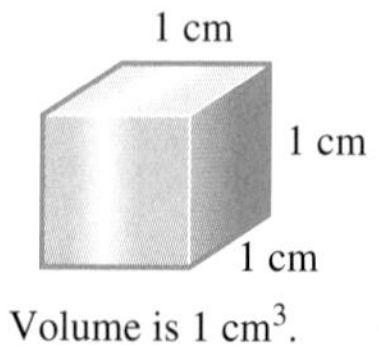

FIGURE 4

In September 1999, the Mars *Climate Orbiter* was lost due to "a failure to use metric units in the coding of a ground software file...." According to Edward Weiler, associate administrator for NASA's Office of Space Science, the metric conversion error that led to the loss "should have been caught five ways to Sunday," but it wasn't. (*Source:* "NASA's Mars Losses Spark Anger and Opportunity" by Leonard David, Washington Contributing Editor to www.space.com.)

The volume of a three-dimensional figure is measured in cubic units. If, for example, the dimensions are given in centimeters, the volume may be determined by the appropriate formula from geometry, and it will be in **cubic centimeters (cm³)**. See Figure 4 for a sketch of a box whose volume is one cm³.

Volume and Weight In the metric system, one **liter (l)** is the quantity assigned to the volume of a box that is 10 cm on a side. (See Figure 5 on the next page.) A liter is a little more than a quart. (See Figure 6 on the next page.) Notice the advantage of this definition over the equivalent one in the English system—using a ruler marked in centimeters, a volume of 1 liter (symbolized 1 L) can be constructed. On the other hand, given a ruler marked in inches, it would be difficult to construct a volume of 1 quart.

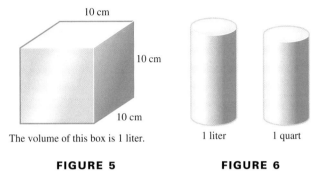

The volume of this box is 1 liter.   1 liter   1 quart

FIGURE 5   **FIGURE 6**

A Comparison of Distances	
Length in Meters	Approximate Related Distances
10^{19}	Distance to the North Star
10^{12}	Distance of Saturn from the sun
10^{11}	Distance of Venus from the sun
10^9	Diameter of the sun
10^8	Diameter of Jupiter
10^7	Diameter of Earth; distance from Washington, D.C. to Tokyo
10^6	Distance from Chicago to Wichita, Kansas
10^5	Average distance across Lake Michigan
10^4	Average width of the Grand Canyon
10^3	Length of the Golden Gate Bridge
10^2	Length of a football field
10^1	Average height of a two-story house
10^0	Width of a door
10^{-1}	Width of your hand
10^{-2}	Diameter of a piece of chalk
10^{-3}	Thickness of a dime
10^{-4}	Thickness of a piece of paper
10^{-5}	Diameter of a red blood cell
10^{-7}	Thickness of a soap bubble
10^{-8}	Average distance between molecules of air in a room
10^{-9}	Diameter of a molecule of oil
10^{-14}	Diameter of an atomic nucleus
10^{-15}	Diameter of a proton

The prefixes mentioned earlier are used throughout the metric system, so one **milliliter (ml)** is one one-thousandth of a liter, one **centiliter (cl)** is one one-hundredth of a liter, one **kiloliter (kl)** is 1000 liters, and so on. Milliliters are used extensively in science and medicine. Many beverages now are sold by milliliters and by liters. For example, 750 ml is a common size for wine bottles, and many soft drinks now are sold in 1- and 2-liter bottles.

Because of the way a liter is defined as the volume of a box 10 cm on a side,

$$1 \text{ L} = 10 \text{ cm} \times 10 \text{ cm} \times 10 \text{ cm} = 1000 \text{ cm}^3$$

or

$$\frac{1}{1000} \text{L} = 1 \text{ cm}^3.$$

Since $\frac{1}{1000}$ L = 1 ml, we have the following relationship.

$$1 \text{ ml} = 1 \text{ cm}^3$$

For example, the volume of a box which is 8 cm by 6 cm by 5 cm may be given as 240 cm³ or as 240 ml.

The box in Figure 4 is 1 cm by 1 cm by 1 cm. The volume of this box is 1 cm³, or 1 ml. By definition, the mass of the water that fills such a box is **1 gram (g)**. A nickel five-cent piece has a mass close to 5 grams, or 5 g. The volume of water used to define a gram is very small, so a gram is a very small mass. For everyday use, a **kilogram (kg)**, or one thousand grams, is more practical. A kilogram weighs about 2.2 pounds. A common abbreviation for kilogram is the word **kilo.**

Extremely small masses can be measured with **milligrams (mg)** and **centigrams (cg).** These measures are so small that they, like centiliters and milliliters, are used mainly in science and medicine.

Temperature In the metric system temperature is measured in **degrees Celsius.** On the Celsius temperature scale, water freezes at 0° and boils at 100°. These two

Anders Celsius and his scale for measuring temperature are honored on this Swedish stamp. The original scale had the freezing point of water at 100° and the boiling point at 0°, but biologist Carl von Linne inverted the scale, giving us the familiar Celsius scale of today.

Switching a country to the metric system requires a tremendous educational campaign. The stamps shown here were issued as part of one such campaign. They show the various metric units and items measured in those units.

numbers are easier to remember than the corresponding numbers on the Fahrenheit scale, 32° and 212°.

The thermometer in Figure 7 shows some typical temperatures in both Fahrenheit and Celsius. For example, room temperature is 22°C, or 70°F, and body temperature is about 37°C, or 99°F.

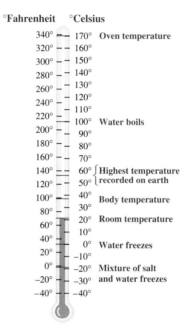

FIGURE 7

The formulas given below can be used to convert between Celsius and Fahrenheit temperatures.

Celsius-Fahrenheit Conversion Formulas

To convert a reading from Fahrenheit to Celsius, use the formula

$$C = \frac{5}{9}(F - 32).$$

To convert from Celsius to Fahrenheit, use the formula

$$F = \frac{9}{5}C + 32.$$

Metric Conversions Due to legislation enacted by Congress, the metric system is used in the United States, and an ultimate goal is for the two systems to be in use, side-by-side, with public acceptance of both systems. Many industries now use the metric system. In particular, industries that export a great many goods are using the metric system, since this is compatible with most of the countries with which they trade.

Further information on the metric system can be obtained from the U.S. Metric Association, 10245 Andasol Avenue, Northridge, CA 91325-1504.

Some scientific calculators are programmed to do conversions between the English and metric systems. Approximate conversions can be made with the aid of the tables below.

Metric to English		
To Convert from	To	Multiply by
meters	yards	1.0936
meters	feet	3.2808
meters	inches	39.37
kilometers	miles	.6214
grams	pounds	.0022
kilograms	pounds	2.20
liters	quarts	1.0567
liters	gallons	.2642

English to Metric		
To Convert from	To	Multiply by
yards	meters	.9144
feet	meters	.3048
inches	meters	.0254
miles	kilometers	1.609
pounds	grams	454
pounds	kilograms	.454
quarts	liters	.9464
gallons	liters	3.785

Appendix Exercises

Perform the following conversions by multiplying or dividing by the appropriate powers of 10.

1. 8 m to millimeters
2. 14.76 m to centimeters
3. 8500 cm to meters
4. 250 mm to meters
5. 68.9 cm to millimeters
6. 3.25 cm to millimeters
7. 59.8 mm to centimeters
8. 3.542 mm to centimeters
9. 5.3 km to meters
10. 9.24 km to meters
11. 27,500 m to kilometers
12. 14,592 m to kilometers

Use a metric ruler to perform the following measurements, first in centimeters, then in millimeters.

13. ⊢―――⊣
14. ⊢――――――⊣
15. ⊢―――――――――⊣

16. Based on your measurement of the line segment in Exercise 13, one inch is about how many centimeters? how many millimeters?

Perform the following conversions by multiplying or dividing by the appropriate powers of 10.

17. 6 L to centiliters
18. 4.1 L to milliliters
19. 8.7 L to milliliters
20. 12.5 L to centiliters
21. 925 cl to liters
22. 412 ml to liters
23. 8974 ml to liters
24. 5639 cl to liters
25. 8000 g to kilograms
26. 25,000 g to kilograms
27. 5.2 kg to grams
28. 12.42 kg to grams
29. 4.2 g to milligrams
30. 3.89 g to centigrams
31. 598 mg to grams
32. 7634 cg to grams

Use the formulas given in the text to perform the following conversions. If necessary, round to the nearest degree.

33. 86°F to Celsius
34. 536°F to Celsius
35. −114°F to Celsius
36. −40°F to Celsius
37. 10°C to Fahrenheit
38. 25°C to Fahrenheit
39. −40°C to Fahrenheit
40. −15°C to Fahrenheit

Solve each of the following problems. Refer to geometry formulas as necessary.

41. One nickel weighs 5 g. How many nickels are in 1 kg of nickels?

42. Sea water contains about 3.5 g salt per 1000 ml of water. How many grams of salt would be in one liter of sea water?

43. Helium weighs about .0002 g per milliliter. How much would one liter of helium weigh?

44. About 1500 g sugar can be dissolved in a liter of warm water. How much sugar could be dissolved in one milliliter of warm water?

45. Northside Foundry needed seven metal strips, each 67 cm long. Find the total cost of the strips, if they sell for $8.74 per meter.

46. Uptown Dressmakers bought fifteen pieces of lace, each 384 mm long. The lace sold for $54.20 per meter. Find the cost of the fifteen pieces.

47. Imported marble for desktops costs $174.20 per square meter. Find the cost of a piece of marble 128 cm by 174 cm.

48. A special photographic paper sells for $63.79 per square meter. Find the cost to buy 80 pieces of the paper, each 9 cm by 14 cm.

49. An importer received some special coffee beans in a box 82 cm by 1.1 m by 1.2 m. Give the volume of the box, both in cubic centimeters and cubic meters.

50. A fabric center receives bolts of woolen cloth in crates 1.5 m by 74 cm by 97 cm. Find the volume of a crate, both in cubic centimeters and cubic meters.

51. A medicine is sold in small bottles holding 800 ml each. How many of these bottles can be filled from a vat holding 160 L of the medicine?

52. How many 2-liter bottles of soda pop would be needed for a wedding reception if 80 people are expected, and each drinks 400 ml of soda?

Perform the following conversions. Use a calculator and/or the table in the text as necessary.

53. 982 yd to meters
54. 12.2 km to miles
55. 125 mi to kilometers
56. 1000 mi to kilometers
57. 1816 g to pounds
58. 1.42 lb to grams
59. 47.2 lb to grams
60. 7.68 kg to pounds
61. 28.6 L to quarts
62. 59.4 L to quarts
63. 28.2 gal to liters
64. 16 qt to liters

Metric measures are very common in medicine. Since we convert among metric measures by moving the decimal point, errors in locating the decimal point in medical doses are not unknown. Decide whether the following doses of medicine seem reasonable *or* unreasonable.

65. Take 2 kg of aspirin three times a day.
66. Take 4 L of liquid Mylanta every evening at bedtime.
67. Take 25 ml of cough syrup daily.
68. Soak your feet in 6 L of hot water.
69. Inject 1/2 L of insulin every morning.
70. Apply 40 g of salve to a cut on your finger.

Select the most reasonable choice for each of the following.

71. length of an adult cow
    A. 1 m B. 3 m C. 5 m
72. length of a Cadillac
    A. 1 m B. 3 m C. 5 m
73. distance from Seattle to Miami
    A. 500 km B. 5000 km C. 50,000 km
74. length across an average nose
    A. 3 cm B. 30 cm C. 300 cm
75. distance across a page of a book
    A. 1.93 mm B. 19.3 mm C. 193 mm
76. weight of a book
    A. 1 kg B. 10 kg C. 1000 kg
77. weight of a large automobile
    A. 1300 kg B. 130 kg C. 13 kg
78. volume of a 12-ounce bottle of beverage
    A. 35 ml B. 355 ml C. 3550 ml

79. height of a person
    A. 180 cm   B. 1800 cm   C. 18 cm
80. diameter of the earth
    A. 130 km   B. 1300 km   C. 13,000 km
81. length of a long freight train
    A. 8 m   B. 80 m   C. 800 m
82. volume of a grapefruit
    A. 1 L   B. 4 L   C. 8 L
83. the length of a pair of Levis
    A. 70 cm   B. 700 cm   C. 7 cm
84. a person's weight
    A. 700 kg   B. 7 kg   C. 70 kg
85. diagonal measure of the picture tube of a table model TV set
    A. 5 cm   B. 50 cm   C. 500 cm
86. width of a standard bedroom door
    A. 1 m   B. 3 m   C. 5 m
87. thickness of a standard audiotape cassette
    A. .9 mm   B. 9 mm   C. 90 mm
88. length of an edge of a record album
    A. 300 mm   B. 30 mm   C. 3000 mm
89. the temperature at the surface of a frozen lake
    A. 0°C   B. 10°C   C. 32°C
90. the temperature in the middle of Death Valley on a July afternoon
    A. 25°C   B. 40°C   C. 65°C
91. surface temperature of desert sand on a hot summer day
    A. 30°C   B. 60°C   C. 90°C
92. boiling water
    A. 100°C   B. 120°C   C. 150°C
93. air temperature on a day when you need a sweater
    A. 30°C   B. 20°C   C. 10°C
94. air temperature when you go swimming
    A. 30°C   B. 15°C   C. 10°C
95. temperature when baking a cake
    A. 120°C   B. 170°C   C. 300°C
96. temperature of bath water
    A. 35°C   B. 50°C   C. 65°C

Answers to Selected Exercises

CHAPTER 1 The Art of Problem Solving

1.1 Exercises (Page 7)

1. deductive **2.** deductive **3.** inductive **4.** inductive **5.** deductive **6.** deductive **7.** deductive
8. deductive **9.** inductive **10.** inductive **11.** inductive **12.** inductive **13.** Answers will vary.
14. Answers will vary. **15.** 21 **16.** 38 **17.** 3072 **18.** 1 **19.** 63 **20.** $\frac{11}{13}$ **21.** $\frac{11}{12}$ **22.** 36
23. 216 **24.** 56 **25.** 52 **26.** -7 **27.** 5 **28.** 2 **29.** Answers will vary. One such list is
10, 20, 30, 40, 50, ... **30.** Answers will vary. One such list is 1, -2, 3, -4, 5, -6, 7, -8, ...
31. $(98,765 \times 9) + 3 = 888,888$ **32.** $(12,345 \times 9) + 6 = 111,111$ **33.** $3367 \times 15 = 50,505$
34. $15,873 \times 35 = 555,555$ **35.** $33,334 \times 33,334 = 1,111,155,556$ **36.** $11,111 \times 11,111 = 123,454,321$
37. $3 + 6 + 9 + 12 + 15 = \frac{15(6)}{2}$ **38.** $2 + 4 + 8 + 16 + 32 = 64 - 2$
39. $5(6) + 5(36) + 5(216) + 5(1296) + 5(7776) = 6(7776 - 1)$ **40.** $3 + 9 + 27 + 81 + 243 = \frac{3(243 - 1)}{2}$
41. $\frac{1}{2} + \frac{1}{4} + \frac{1}{8} + \frac{1}{16} + \frac{1}{32} = 1 - \frac{1}{32}$ **42.** $\frac{1}{1 \cdot 2} + \frac{1}{2 \cdot 3} + \frac{1}{3 \cdot 4} + \frac{1}{4 \cdot 5} + \frac{1}{5 \cdot 6} = \frac{5}{6}$ **43.** 20,100
44. 80,200 **45.** 320,400 **46.** 2,001,000 **47.** 15,400 **49.** 2550 **50.** 5100

51.   **52.** 0 2 10 30 60 90 102 90 60 (To find any term, choose the term directly above it and add to it the two preceding terms. If there are fewer than two terms, add as many as there are.)

53. 1 (These are the numbers of chimes a clock rings, starting with 12 o'clock, if it rings the number of hours on the hour, and 1 chime on the half-hour.) **54.** E (One, Two, Three, and so on) **55.** (a) The middle digit is always 9, and the sum of the first and third digits is always 9 (considering 0 as the first digit if the difference has only two digits).
56. (f) The final result will always be half of the number added in Step (b). **57.** 142,857; 285,714; 428,571; 571,428; 714,285; 857,142. Each result consists of the same six digits, but in a different order. $142,857 \times 7 = 999,999$
58. 111,111,111; 222,222,222; 333,333,333; 72 **59.** 21 **60.** For example, suppose the age is 50 and the amount is 35 cents. Then the numbers obtained are 50, 200, 210, 5250, 4885, 4920, 5035. The final number here, 5035, provides the information. **61.** Answers will vary. **62.** Answers will vary.

1.2 Exercises (Page 16)

1. 79 **2.** 159 **3.** 450 **4.** 391 **5.** 4032 **6.** 3992 **7.** 32,758 **8.** 28,675 **9.** 57; 99
10. (a) 5, 11, 19, 29 (b) Add 12 to 29 to obtain 41. (c) $5^2 + 3(5) + 1 = 41$; The results agree.
11. $(4321 \times 9) - 1 = 38,888$ **12.** $(1234 \times 8) + 4 = 9876$ **13.** $999,999 \times 4 = 3,999,996$
14. $1,010,101 \times 1,010,101 = 1,020,304,030,201$ **15.** $21^2 - 15^2 = 6^3$
16. $1 + 2 + 3 + 4 + 5 + 4 + 3 + 2 + 1 = 5^2$ **17.** $5^2 - 4^2 = 5 + 4$ **18.** $4^2 + 4 = 5^2 - 5$
19. $1 + 5 + 9 + 13 = 4 \times 7$ **20.** $16 + 17 + 18 + 19 + 20 = 21 + 22 + 23 + 24$ **21.** 45,150 **22.** 125,250

23. 228,150   **24.** 340,725   **25.** 2601   **26.** 625   **27.** 250,000   **28.** 22,801   **29.** $S = n(n+1)$
33. row 1: 28, 36;   row 2: 36, 49, 64;   row 3: 35, 51, 70, 92;   row 4: 28, 45, 66, 91, 120;   row 5: 18, 34, 55, 81, 112, 148;   row 6: 8, 21, 40, 65, 96, 133, 176   **34.** 1, 6, 15, 28, 45   **35.** $8(1) + 1 = 9 = 3^2$; $8(3) + 1 = 25 = 5^2$; $8(6) + 1 = 49 = 7^2$; $8(10) + 1 = 81 = 9^2$ **36.** The pattern is 1, 0, 0, 1, 0, 0, ...   **37.** The pattern is 1, 0, 1, 0, 1, 0, ...   **38.** For $n = 5$, the pattern is 1, 0, 2, 2, 0, 1, 0, 2, 2, 0, ... For $n = 6$, the pattern is 1, 0, 3, 4, 3, 0, 1, 0, 3, 4, 3, 0, ...   **39.** [figure]   **40.** (a) $\dfrac{4}{9}, \dfrac{9}{16}, \dfrac{16}{25}, \dfrac{25}{36}, \dfrac{36}{49}, \dfrac{49}{64}, \dfrac{64}{81}$ (b) $\dfrac{3}{6}, \dfrac{6}{10}, \dfrac{10}{15}, \dfrac{15}{21}, \dfrac{21}{28}, \dfrac{28}{36}, \dfrac{36}{45}$

(c) The fraction formed by two consecutive square numbers is always reduced to lowest terms, while the fraction formed by two consecutive triangular numbers is never reduced to lowest terms.   **41.** (a) a triangular number   (b) a triangular number   **42.** $N = 6$; They are the third triangular number and the second hexagonal number.
43. 256   **44.** 66   **45.** 117   **46.** 91   **47.** 235   **48.** 408   **49.** $N_n = \dfrac{n(7n - 5)}{2}$ **50.** 325
51. a square number   **52.** a triangular number   **53.** a perfect cube   **54.** a pentagonal number

1.3 Exercises (page 25)

1. You should choose a sock from the box labeled *red and green socks*. Since it is mislabeled, it contains only red socks or only green socks, determined by the sock you choose. If the sock is green, relabel this box *green socks*. Since the other two boxes were mislabeled, switch the remaining label to the other box and place the label that says *red and green socks* on the unlabeled box. No other choice guarantees a correct relabeling, since you can remove only one sock.
2. OHIO   **3.** D   **4.** 70   **5.** A   **6.** Since you are the bus driver, the answer is *your* age.
7. One example of a solution follows: **8.** You must place the O in the bottom-left square. No other choice guarantees you a win.

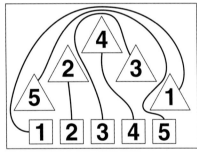

9. | 9 | 7 | 2 | 14 | 11 | 5 | 4 | 12 | 13 | 3 | 6 | 10 | 15 | 1 | 8 | **10.** 18   **11.** Here is one solution.   **12.** Each region has a sum of 26.
(or the same arrangement reading right to left)

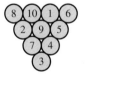

13. D   **14.** 329476   **15.** $\dfrac{1}{3}$ **16.** 31   **17.** One possible sequence is shown here. The numbers represent the number of gallons in each bucket in each successive step.

Big	7	4	4	1	1	0	7	5	5
Small	0	3	0	3	0	1	1	3	0

18. 128   **19.** 90   **20.** Here is one possible solution.   **21.** 55 miles per hour   **22.** 625 bananas   **23.** 07

	3	5	
7	1	8	2
	4	6	

24. It will be darkest at 1:11 and brightest at 10:08. See the diagram.

`1:11`   `10:08`

25. A kilogram of $10 gold pieces is worth twice as much as half a kilogram of $20 gold pieces. (The denomination has nothing to do with the answer!)

26. 1   **27.** 3   **28.** $50   **29.** 5   **30.** 17 days   **31.** 3 socks   **32.** 14   **33.** 35   **34.** 16   **35.** 6
36. Becky   **37.** the nineteenth day   **38.** It is a *palindrome,* since it reads the same backward and forward.
39.   **40.** 1949   **41.** 1967   **42.** B, since either Bill or Bob must be the tallest person.
43. Eve has $5 and Adam has $7.   **44.** 14   **45.** The correct problem is

The correct problem is
435
826
+147
1408.

402
× 39
15,678.

46.

6	1	8
7	5	3
2	9	4

47.

6	12	7	9
1	15	4	14
11	5	10	8
16	2	13	3

48. 6 ways   **49.** 25 pitches (The visiting team's pitcher retires 24 consecutive batters through the first eight innings, using only one pitch per batter. His team does not score either. Going into the bottom of the ninth tied 0–0, the first batter for the home team hits his first pitch for a home run. The pitcher threw 25 pitches and loses the game by a score of 1–0.)

50. forty   **51.** The two children row across. One stays on the opposite bank and the other returns. One soldier rows across, and the child on the opposite bank then rows back. The two children row across. One stays and the other returns. Now another soldier rows across. This process continues until all the soldiers are across.   **52.** For three weighings, first balance four against four. Of the lighter four, balance two against the other two. Finally, of the lighter two, balance them one against the other. To find the bad coin in two weighings, divide the eight coins into groups of 3, 3, 2. Weigh the groups of three against each other on the scale. If the groups weigh the same, the fake is in the two left out and can be found in one additional weighing. If the two groups of three do not weigh the same, pick the lighter group. Choose any two of the coins and weigh them. If one of these is lighter, it is the fake; if they weigh the same, then the third coin is the fake.
53. The person takes the goat across and returns alone. On the second trip, the person takes the wolf across and returns with the goat. On the third trip, the goat is left on the first side while the person takes the cabbage across. Then the person returns alone and brings the goat back across.   **54.** From left to right, they are Johnson, Andersen, and Thompson.
55. Q   **56.**    **57.** 28   **58.** house address numbers   **59.**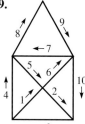

60. 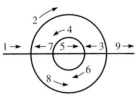   **61.** 8   **62.** none   **63.** The CEO is a woman.   **64.** 9 years old

65. Dan (36) is married to Jessica (29); James (30) is married to Cathy (31).   **66.** 10   **67.** 12; All months have 28 days.   **68.** They must be February and March in a non-leap year.   **69.** None, since there is no dirt in a hole.
70. The products will always differ by 1.   **71.** "Madam, I'm Adam."   **72.** 488-7884   **73.** 6
74. One solution is $1 + 2 + 3 + 4 + 5 + 6 + 7 + (8 \times 9) = 100$

One of several possibilities

1.4 Exercises (page 35)
1. 43.8   **2.** 5.88   **3.** 2.3589   **4.** 6.1145   **5.** 7.48   **6.** 8.88   **7.** 7.1289   **8.** 42.508549
9. 6340.338097   **10.** 10.50921537   **11.** 1   **12.** 9.8   **13.** 1.061858759   **14.** 1.214555893
15. 2.221441469   **16.** 3.627598728   **17.** 3.141592653   **18.** 296296296   **19.** yes   **20.** $95^2 = 9025$
21. positive   **22.** negative   **23.** 1   **24.** 1   **25.** the same as   **26.** error message   **27.** 0
28. negative   **29.** positive   **30.** error message   **31.** Answers will vary.   **34.** A cycle will form that has its sixth number equal to the first number chosen. For example, choosing 5 and 6 gives 5, 6, 1.4, .4, 1, 5, 6 as the first seven terms.   **35.** ShELLOIL   **36.** ShOES   **37.** BOSE   **38.** GOOSE   **39.** Answers will vary.
40. (a) (b) 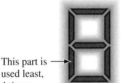   (c) 8 (all seven parts)   (d) 1 (two parts)

← This part is used most frequently, 9 times.

This part is → used least, 4 times.

41. 48   **42.** 12   **43.** 14   **44.** 7   **45.** B   **46.** C   **47.** A   **48.** C   **49.** D   **50.** D
51. DIRECTV; 45%   **52.** 1,920,000   **53.** 10,200,000   **54.** 120,000   **55.** 150,000,000
56. just less than 250,000,000   **57.** 1998   **58.** approximately 350,000,000   **59.** between 1993 and 1994
60. The line graph falls from left to right between 1994 and 1999, so PC prices have been decreasing over these years.
61. 1996: about $2000; 1999: about $1300   **62.** about $1100   **63.** 70s   **64.** a cold front
65. hot and stormy   **66.** 60s   **67.** Texas   **68.** probably not, since rain is not predicted for Cleveland

Chapter 1 Test (page 47)
1. inductive   **2.** deductive   **3.** 256, 3125 (The nth term of the sequence is n^n.)
4. $65,359,477,124,183 \times 68 = 4,444,444,444,444,444$ **5.** 351   **6.** 31,375
7. 65; $65 = 1 + 7 + 13 + 19 + 25$ **8.** 1, 8, 21, 40, 65, 96, 133, 176; The pattern is 1, 0, 1, 0, 1, 0, 1, 0, ...
9. The first two terms are both 1. Each term after the second is found by adding the two previous terms.   **10.** 21
11. 9   **12.** 35   **13.** $629 + 154 = 783$ is one of several solutions   **14.** 8, 53, and 54   **15.** 3
16. The sum of the digits is always 9.   **17.** 9.907572861 (Answers may vary due to the model of calculator used.)
18. 34.328125   **19.** B   **20.** (a) 1990–2000   (b) 1990   (c) 1980

CHAPTER 2 The Basic Concepts of Set Theory

2.1 Exercises (Page 54)

1. C   2. G   3. E   4. A   5. B   6. D   7. H   8. F   9. $\{1,2,3,4,5,6\}$
10. $\{9,10,11,12,13,14,15,16,17\}$ 11. $\{0,1,2,3,4\}$ 12. $\{5,6,7,8,9,10,11,12,13\}$
13. $\{6,7,8,9,10,11,12,13,14\}$ 14. $\{3,6,9,12,15,18,21,24,27,30\}$ 15. $\{-15,-13,-11,-9,-7,-5,-3,-1\}$
16. $\{-4,-3,-2,-1,0,1,2,3,4\}$ 17. $\{2,4,8,16,32,64,128,256\}$ 18. $\{90,87,84,81,78,75,72,69\}$
19. $\{0,2,4,6,8,10\}$ 20. $\{-7,-5,-3,-1,1,3,5\}$ 21. $\{21,22,23,\ldots\}$ 22. $\{-199,-198,-197,\ldots,499\}$
23. {Lake Erie, Lake Huron, Lake Michigan, Lake Ontario, Lake Superior}
24. {George Bush, Ronald Reagan, Jimmy Carter, Gerald Ford, Richard Nixon}   25. $\{5,10,15,20,25,\ldots\}$
26. $\{-6,-12,-18,-24,-30,\ldots\}$ 27. $\{1,\frac{1}{2},\frac{1}{3},\frac{1}{4},\frac{1}{5},\ldots\}$ 28. $\{4,16,64,256,1024,\ldots\}$
In Exercises 29–32, there are other ways to describe the sets.
29. $\{x \mid x \text{ is a rational number}\}$ 30. $\{x \mid x \text{ is an even natural number}\}$ 31. $\{x \mid x \text{ is an odd natural number less than 76}\}$
32. $\{x \mid x \text{ is a multiple of 5 between 30 and 100}\}$ 33. finite   34. finite   35. infinite   36. infinite
37. infinite   38. finite   39. infinite   40. infinite   41. 8   42. 7   43. 500   44. 3001
45. 26   46. 5   47. 39   48. 100   49. 28   50. 18   53. well-defined   54. well-defined
55. not well-defined   56. not well-defined   57. not well-defined   58. well-defined   59. $\in$ 60. $\in$
61. $\notin$ 62. $\notin$ 63. $\in$ 64. $\notin$ 65. $\notin$ 66. $\notin$ 67. false   68. false   69. true   70. true
71. true   72. true   73. true   74. true   75. false   76. false   77. true   78. true   79. true
80. true   81. false   82. false   83. true   84. false   87. $\{2\}$ and $\{3,4\}$ (Other examples are possible.)
88. This is impossible. If they are equal, they have the same number of elements and must be equivalent.
89. $\{a,b\}$ and $\{a,c\}$ (Other examples are possible.)   90. $\{5\}$ and $\{4+1\}$ (Other examples are possible.)
91. **(a)** {Viacom (Class B), Trans World Airlines, Harken Energy, Echo Bay Mines}
(b) {Echo Bay Mines, JTS, Nabors Industries, Hasbro, Royal Oak Mines, Grey Wolf Industries, IVAX}
92. **(a)** $\{r\}, \{g,s\}, \{c,s\}, \{v,r\}, \{g,r\}, \{c,r\}, \{s,r\}$ **(b)** $\{v,g,r\}, \{v,c,r\}, \{v,s,r\}, \{g,c,r\}, \{g,s,r\}, \{c,s,r\}$

2.2 Exercises (Page 61)

1. F   2. B   3. C   4. D   5. A   6. E   7. $\not\subseteq$ 8. $\not\subseteq$ 9. $\subseteq$ 10. $\subseteq$ 11. $\subseteq$ 12. $\subseteq$
13. $\not\subseteq$ 14. $\subseteq$ 15. both   16. $\subseteq$ 17. $\subseteq$ 18. neither   19. both   20. $\subseteq$ 21. neither
22. neither   23. true   24. true   25. false   26. false   27. true   28. false   29. true   30. true
31. true   32. false   33. true   34. false   35. false   36. false   37. false   38. true   39. true
40. true   41. false   42. true   43. 8; 7   44. 16; 15   45. 64; 63   46. 128; 127   47. 32; 31
48. 4; 3   49. $\{2,3,5,7,9,10\}$ 50. $\{1,3,4,6,8\}$ 51. $\{2\}$ 52. $\{5\}$ 53. $\{1,2,3,4,5,6,7,8,9,10\}$ or U
54. $\emptyset$ 55. {Higher cost, Lower cost, Educational, More time to see the sights, Less time to see the sights, Cannot visit relatives along the way, Can visit relatives along the way}   56. {Lower cost, Less time to see the sights, Can visit relatives along the way}   57. {Higher cost, More time to see the sights, Cannot visit relatives along the way}
58. {Educational}   59. $\emptyset$ 60. {Higher cost, More time to see the sights, Cannot visit relatives along the way}
61. $\{A,B,C,D,E\}$ (All are present.)   62. $\{A,B,C,D\}, \{A,B,C,E\}, \{A,B,D,E\}, \{A,C,D,E\}, \{B,C,D,E\}$
63. $\{A,B,C\}, \{A,B,D\}, \{A,B,E\}, \{A,C,D\}, \{A,C,E\}, \{A,D,E\}, \{B,C,D\}, \{B,C,E\}, \{B,D,E\}, \{C,D,E\}$
64. $\{A,B\}, \{A,C\}, \{A,D\}, \{A,E\}, \{B,C\}, \{B,D\}, \{B,E\}, \{C,D\}, \{C,E\}, \{D,E\}$ 65. $\{A\}, \{B\}, \{C\}, \{D\}, \{E\}$
66. $\emptyset$ (No one shows up.)   67. 32   68. They are the same: $32 = 2^5$. The number of ways that people from a group of five can gather is the same as the number of subsets there are of a set of five elements.
69. **(a)** 15   **(b)** 16; It is now possible to select *no* bills.   70. **(a)** 31   **(b)** 32   71. **(a)** s **(b)** s **(c)** $2s$
(d) Adding one more element will always double the number of subsets, so the expression 2^n is true in general.

2.3 Exercises (Page 73)

1. B   2. F   3. A   4. C   5. E   6. D   7. $\{a,c\}$ 8. $\{a,b,c,e,g\}$ 9. $\{a,b,c,d,e,f\}$
10. $\{b,c\}$ 11. $\{a,b,c,d,e,f,g\}$ 12. $\{a,b,c\}$ 13. $\{b,d,f\}$ 14. $\{d,e,f,g\}$ 15. $\{d,f\}$
16. $\{b,d,f\}$ 17. $\{a,b,c,e,g\}$ 18. $\{a,b,c\}$ 19. $\{a,c,e,g\}$ 20. $\{b,c,d,e,f,g\}$ 21. $\{a\}$
22. $\{a,c,d,e,f,g\}$ 23. $\{e,g\}$ 24. $\{b\}$ 25. $\{d,f\}$ 26. $\{d,f\}$ 27. $\{e,g\}$ 28. $\{a,b,c\}$

A-14 Answers to Selected Exercises

In Exercises 29–34, there may be other acceptable descriptions.
29. the set of all elements that either are in A, or are not in B and not in C **30.** the set of all elements that are in A and not in B, or are in B and not in A **31.** the set of all elements that are in C but not in B, or are in A **32.** the set of all elements that are in B and are not in A and not in C **33.** the set of all elements that are in A but not in C, or in B but not in C **34.** the set of all elements that are not in A and not in B, or are not in C **35.** $\{e,h,c,l,b\}$ **36.** $\{e,c\}$ **37.** $\{l,b\}$ **38.** $\{h\}$ **39.** $\{e,h,c,l,b\}$ **40.** $\{e,c\}$ **41.** the set of all tax returns showing business income or filed in 2003   **42.** the set of all tax returns with itemized deductions and selected for audit   **43.** the set of all tax returns filed in 2003 without itemized deductions   **44.** the set of all tax returns selected for audit or without itemized deductions   **45.** the set of all tax returns with itemized deductions or showing business income, but not selected for audit   **46.** the set of all tax returns filed in 2003 with itemized deductions but not showing business income   **47.** always true   **48.** not always true   **49.** always true   **50.** not always true   **51.** not always true   **52.** not always true   **53.** always true   **54.** always true   **55. (a)** $\{1,3,5,2\}$ **(b)** $\{1,2,3,5\}$ **(c)** For any sets X and Y, $X \cup Y = Y \cup X$.   **56. (a)** $\{1,3\}$ **(b)** $\{1,3\}$ **(c)** For any sets X and Y, $X \cap Y = Y \cap X$.   **57. (a)** $\{1,3,5,2,4\}$ **(b)** $\{1,3,5,2,4\}$ **(c)** For any sets X, Y, and Z, $X \cup (Y \cup Z) = (X \cup Y) \cup Z$.   **58. (a)** $\{3\}$ **(b)** $\{3\}$ **(c)** For any sets X, Y, and Z, $X \cap (Y \cap Z) = (X \cap Y) \cap Z$.   **59. (a)** $\{4\}$ **(b)** $\{4\}$ **(c)** For any sets X and Y, $(X \cup Y)' = X' \cap Y'$.   **60. (a)** $\{2,4,5\}$ **(b)** $\{2,4,5\}$ **(c)** For any sets X and Y, $(X \cap Y)' = X' \cup Y'$.   **61. (a)** $\{1,3,5\}$ **(b)** For any set X, $X \cup \emptyset = X$.   **62. (a)** $\emptyset$ **(b)** For any set X, $X \cap \emptyset = \emptyset$.   **63.** true   **64.** true   **65.** false   **66.** false   **67.** true   **68.** true   **69.** true   **70.** true   **71.** $A \times B = \{(2,4),(2,9),(8,4),(8,9),(12,4),(12,9)\}$; $B \times A = \{(4,2),(4,8),(4,12),(9,2),(9,8),(9,12)\}$
72. $A \times B = \{(3,6),(3,8),(6,6),(6,8),(9,6),(9,8),(12,6),(12,8)\}$; $B \times A = \{(6,3),(6,6),(6,9),(6,12),(8,3),(8,6),(8,9),(8,12)\}$
73. $A \times B = \{(d,p),(d,i),(d,g),(o,p),(o,i),(o,g),(g,p),(g,i),(g,g)\}$; $B \times A = \{(p,d),(p,o),(p,g),(i,d),(i,o),(i,g),(g,d),(g,o),(g,g)\}$
74. $A \times B = \{(b,r),(b,e),(b,d),(l,r),(l,e),(l,d),(u,r),(u,e),(u,d),(e,r),(e,e),(e,d)\}$; $B \times A = \{(r,b),(r,l),(r,u),(r,e),(e,b),(e,l),(e,u),(e,e),(d,b),(d,l),(d,u),(d,e)\}$ **75.** $n(A \times B) = 6; n(B \times A) = 6$
76. $n(A \times B) = 9; n(B \times A) = 9$ **77.** $n(A \times B) = 210; n(B \times A) = 210$ **78.** $n(A \times B) = 65; n(B \times A) = 65$
79. 3   **80.** 25   **81.** 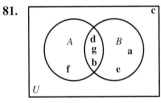   **82.**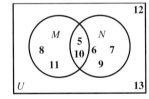

83. 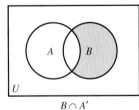   **84.** 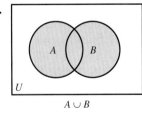   **85.**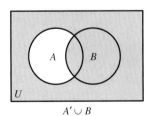

$B \cap A'$ $A \cup B$ $A' \cup B$

86. 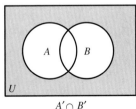   **87.** 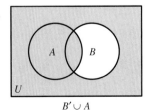   **88.**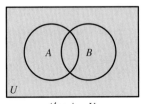

$A' \cap B'$ $B' \cup A$ $A' \cup A = U$

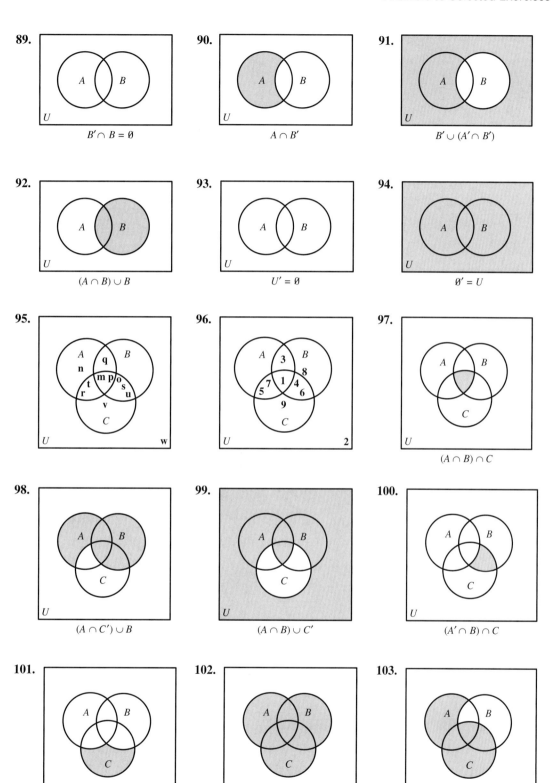

104.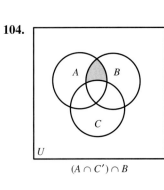
$(A \cap C') \cap B$

105.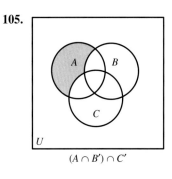
$(A \cap B') \cap C'$

106.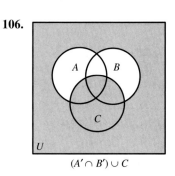
$(A' \cap B') \cup C$

107.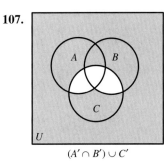
$(A' \cap B') \cup C'$

108.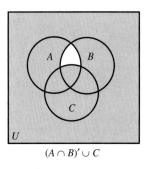
$(A \cap B)' \cup C$

109. $A' \cap B'$ or $(A \cup B)'$

110. $A' \cup B$ or $(A - B)'$ **111.** $(A \cup B) \cap (A \cap B)'$ or $(A \cup B) - (A \cap B)$
112. $A \cap B'$ or $A - B$ **113.** $(A \cap B) \cup (A \cap C)$ or $A \cap (B \cup C)$ **114.** $A \cap (B \cup C)'$ or $A \cap (B' \cap C')$ **115.** $(A \cap B) \cap C'$ or $(A \cap B) - C$ **116.** $(B \cup C) \cap A'$ or $(B \cup C) - A$
117. $A \cap B = \emptyset$ **118.** $A = B = \emptyset$ **119.** This statement is true for any set A. **120.** $A = \emptyset$
121. $A = \emptyset$ **122.** This statement is true for any set A. **123.** $A = \emptyset$ **124.** This statement is true for any set A. **125.** $A = \emptyset$ **126.** $A = \emptyset$ **127.** always true **128.** always true **129.** always true
130. not always true **131.** not always true **132.** not always true **133.** always true
136. (a) $\{x \mid x \text{ is a real number}\}$ **(b)** $\emptyset$

2.4 Exercises (Page 79)

1. (a) 1 **(b)** 9 **(c)** 6 **(d)** 2 **(e)** 5 **2. (a)** 6 **(b)** 8 **(c)** 0 **(d)** 2 **(e)** 9 **3. (a)** 1 **(b)** 2 **(c)** 6 **(d)** 7 **(e)** 8 **(f)** 3 **(g)** 4 **(h)** 5 **4. (a)** 1 **(b)** 3 **(c)** 4 **(d)** 0 **(e)** 2 **(f)** 10 **(g)** 2 **(h)** 5 **5.** 17 **6.** 2
7. 16 **8.** 35

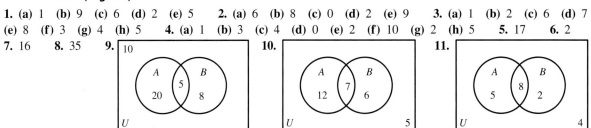

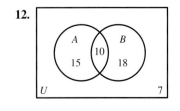

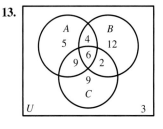

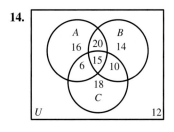

15. 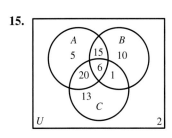 **16.** 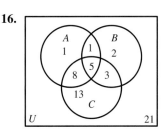 **17. (a)** 3 **(b)** 2 **(c)** 10

18. (a) 3 **(b)** 5 **19. (a)** 24 **(b)** 2 **(c)** 3 **(d)** 16 **20. (a)** 18 **(b)** 15 **(c)** 20 **(d)** 5 **21. (a)** 0 **(b)** 4 **(c)** 3 **(d)** 0 **(e)** 6 **22.** Yes, he should be reassigned to the job in Siberia. His figures add up to 142 people, not 140 as they should. **23. (a)** 51 **(b)** 31 **(c)** 18 **(d)** 15 **(e)** 33 **(f)** 23 **24. (a)** 37 **(b)** 22 **(c)** 50 **(d)** 11 **(e)** 25 **(f)** 11 **25.** 225 **26. (a)** 31 **(b)** 24 **(c)** 11 **(d)** 45 **27. (a)** 1 **(b)** 1, 2, 3, 4, 5, 6, 7, 8, 9, 10, 11, 12, 13, 14, 15 **(c)** 1, 2, 3, 4, 5, 9, 11 **(d)** 5, 8, 13 **28. (a)** none **(b)** 52 **(c)** 44 **29. (a)** 6 **(b)** 473 **(c)** 835 **(d)** 50 **(e)** 297 **(f)** 386 **30. (a)** 9 **(b)** 9 **(c)** 20 **(d)** 20 **(e)** 27 **(f)** 15

2.5 Exercises (Page 88)

1. B; 1 **2.** D; 3 **3.** A; $\aleph_0$ **4.** C; c **5.** F; 0 **6.** E; 2 **7.** (Other correspondences are possible.)

$$\{I, \ II, \ III\}$$
$$\updownarrow \ \updownarrow \ \updownarrow$$
$$\{x, \ y, \ z\}$$

8. not possible **9.** (Other correspondences are possible.)

$$\{a, \ d, \ i, \ t, \ o, \ n\}$$
$$\updownarrow \ \updownarrow \ \updownarrow \ \updownarrow \ \updownarrow \ \updownarrow$$
$$\{a, \ n, \ s, \ w, \ e, \ r\}$$

10. (Other correspondences are possible. This particular one pairs each president with his wife.)

$$\{\text{Reagan}, \ \text{Clinton}, \ \text{Bush}\}$$
$$\updownarrow \ \updownarrow \ \updownarrow$$
$$\{\text{Nancy}, \ \text{Hillary}, \ \text{Laura}\}$$

11. 11 **12.** 10 **13.** 0 **14.** 1 **15.** $\aleph_0$ **16.** 14 **17.** $\aleph_0$ **18.** $\aleph_0$ **19.** $\aleph_0$ **20.** 4 **21.** 12 **22.** 50 **23.** $\aleph_0$ **24. (a)** 6 **(b)**

$$\{\text{Mike Myers}, \ \text{Julia Roberts}, \ \text{William Shatner}\}$$
$$\updownarrow \ \updownarrow \ \updownarrow$$
$$\{\text{Austin Powers}, \ \text{Erin Brockovich}, \ \text{Captain James T. Kirk}\}$$

25. both **26.** equivalent **27.** equivalent **28.** equivalent **29.** equivalent **30.** neither

31. $\{2, \ 4, \ 6, \ 8, \ \ldots, \ 2n, \ \ldots\}$
   $\ \ \ \updownarrow \ \updownarrow \ \updownarrow \ \updownarrow \ \ \ \ \ \updownarrow$
   $\{1, \ 2, \ 3, \ 4, \ \ldots, \ n, \ \ldots\}$

32. $\{-10, \ -20, \ -30, \ -40, \ \ldots, \ -10n, \ \ldots\}$
   $\ \ \ \ \updownarrow \ \ \ \updownarrow \ \ \ \updownarrow \ \ \ \updownarrow \ \ \ \ \ \ \ \updownarrow$
   $\{1, \ 2, \ 3, \ 4, \ \ldots, \ n, \ \ldots\}$

33. $\{1{,}000{,}000, \ 2{,}000{,}000, \ 3{,}000{,}000, \ \ldots, \ 1{,}000{,}000n, \ \ldots\}$
   $\ \ \ \ \ \ \updownarrow \ \ \ \ \ \ \ \ \updownarrow \ \ \ \ \ \ \ \ \updownarrow \ \ \ \ \ \ \ \ \ \ \ \updownarrow$
   $\{ \ \ 1, \ \ \ \ \ \ \ \ 2, \ \ \ \ \ \ \ \ 3, \ \ \ldots, \ \ \ \ \ n, \ \ \ \ldots\}$

34. $\{1, \ -1, \ 3, \ -3, \ 5, \ -5, \ \ldots, \ a, \ \ldots\}$
   $\ \ \updownarrow \ \ \updownarrow \ \updownarrow \ \ \updownarrow \ \updownarrow \ \ \updownarrow \ \ \ \ \ \updownarrow$ where $a = n$ if n is odd and $a = 1 - n$ if n is even
   $\{1, \ 2, \ 3, \ 4, \ 5, \ 6, \ \ldots, \ n, \ \ldots\}$

35. $\{2, \ 4, \ 8, \ 16, \ 32, \ \ldots, \ 2^n, \ \ldots\}$
   $\ \ \updownarrow \ \updownarrow \ \updownarrow \ \updownarrow \ \ \updownarrow \ \ \ \ \ \updownarrow$
   $\{1, \ 2, \ 3, \ 4, \ 5, \ \ldots, \ n, \ \ldots\}$

36. $\{-17, \ -22, \ -27, \ -32, \ \ldots, \ -5n - 12, \ \ldots\}$
   $\ \ \ \updownarrow \ \ \ \updownarrow \ \ \ \updownarrow \ \ \ \updownarrow \ \ \ \ \ \ \ \ \updownarrow$
   $\{ \ 1, \ \ \ 2, \ \ \ 3, \ \ \ 4, \ \ldots, \ \ \ \ \ n, \ \ \ \ldots\}$

37. This statement is not always true. For example, let A = the set of counting numbers, B = the set of real numbers. **38.** This statement is not always true. For example, let A = the set of counting numbers, $B = \{a, b, c\}$. **39.** This statement is not always true. For example, A could be the set of all subsets of the set of reals. Then $n(A)$ would be an infinite number *greater* than c. **40.** This statement is always true.

41. (a) Rays emanating from point P will establish a geometric pairing of the points on the semicircle with the points on the line.

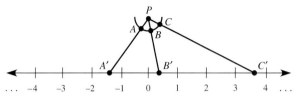

(b) The set of real numbers is infinite, having been placed in a one-to-one correspondence with a proper subset of itself.

42.

43. $\{3, 6, 9, 12, \ldots, 3n, \ldots\}$
$\{6, 9, 12, 15, \ldots, 3n+3, \ldots\}$

44. $\{4, 7, 10, 13, \ldots, 3n+1, \ldots\}$
$\{7, 10, 13, 16, \ldots, 3n+4, \ldots\}$

45. $\{3/4, 3/8, 3/12, 3/16, \ldots, 3/(4n), \ldots\}$
$\{3/8, 3/12, 3/16, 3/20, \ldots, 3/(4n+4), \ldots\}$

46. $\{1, 4/3, 5/3, 2, \ldots, (n+2)/3, \ldots\}$
$\{4/3, 5/3, 2, 7/3, \ldots, (n+3)/3, \ldots\}$

47. $\{1/9, 1/18, 1/27, \ldots, 1/(9n), \ldots\}$
$\{1/18, 1/27, 1/36, \ldots, 1/(9n+9), \ldots\}$

48. $\{-3, -5, -9, -17, \ldots, -(2^n+1), \ldots\}$
$\{-5, -9, -17, -33, \ldots, -(2^{n+1}+1), \ldots\}$

Chapter 2 Test (Page 91)

1. $\{a, b, c, d, e\}$ **2.** $\{a, b, d\}$ **3.** $\{c, f, g, h\}$ **4.** $\{a, c\}$ **5.** true **6.** false **7.** true **8.** true
9. false **10.** true **11.** true **12.** true **13.** 8 **14.** 15

Answers may vary for Exercises 15–18.
15. the set of odd integers between -4 and 10 **16.** the set of months of the year **17.** $\{x \mid x \text{ is a negative integer}\}$
18. $\{x \mid x \text{ is a multiple of 8 between 20 and 90}\}$ **19.** $\subseteq$ **20.** neither **21.**

22.

23. $(X \cup Y) - Z$

24. $[(X \cap Y) \cup (Y \cap Z) \cup (X \cap Z)] - (X \cap Y \cap Z)$

25. {Electric razor} **26.** {Adding machine, Barometer, Pendulum clock, Thermometer} **27.** {Electric razor}
29. (a) 22 **(b)** 12 **(c)** 28 **30. (a)** 16 **(b)** 32 **(c)** 33 **(d)** 45 **(e)** 14 **(f)** 26

CHAPTER 3 Introduction to Logic

3.1 Exercises (Page 99)

1. statement **2.** statement **3.** not a statement **4.** not a statement **5.** statement **6.** statement **7.** statement **8.** statement **9.** statement **10.** statement **11.** not a statement **12.** not a statement **13.** statement **14.** statement **15.** compound **16.** not compound **17.** not compound **18.** compound **19.** not compound **20.** not compound **21.** compound **22.** compound **23.** Her aunt's name is not Lucia. **24.** The flowers are not to be watered. **25.** At least one dog does not have its day. **26.** Some rain fell in southern California today. **27.** No book is longer than this book. **28.** At least one student present will not get another chance. **29.** At least one computer repairman can play blackjack. **30.** No people have all the luck. **31.** Someone does not love somebody sometime. **32.** Someone does not love a winner. **33.** $y \leq 12$ **34.** $x \geq -6$ **35.** $q < 5$ **36.** $r > 19$ **39.** She does not have green eyes. **40.** He is not 48 years old. **41.** She has green eyes and he is 48 years old. **42.** She has green eyes or he is 48 years old. **43.** She does not have green eyes or he is 48 years old. **44.** She has green eyes and he is not 48 years old. **45.** She does not have green eyes or he is not 48 years old. **46.** She does not have green eyes and he is not 48 years old. **47.** It is not the case that she does not have green eyes and he is 48 years old. **48.** It is not the case that she has green eyes or he is not 48 years old. **49.** $p \wedge \sim q$ **50.** $\sim p \vee \sim q$ **51.** $\sim p \vee q$ **52.** $q \wedge \sim p$ **53.** $\sim(p \vee q)$ or, equivalently, $\sim p \wedge \sim q$ **54.** $(q \vee p) \wedge [\sim(q \wedge p)]$ **57.** C **58.** B **59.** A, B **60.** A, B **61.** A, C **62.** C **63.** B **64.** A, C **65.** true **66.** true **67.** true **68.** true **69.** true **70.** true **71.** true **72.** false **73.** false **74.** false **77.** Everyone here has done that at one time or another. **78.** A

3.2 Exercises (Page 111)

1. false **2.** true **3.** true **4.** false **5.** They must both be false. **6.** They must both be true. **7.** T **8.** F **9.** T **10.** F **11.** F **12.** T **13.** T **14.** F **15.** T **16.** F **17.** T **18.** T **19.** It is a disjunction, because it means "3 > 1 or 3 = 1." **20.** $6 \geq 2$ is true because 6 > 2. $6 \geq 6$ is true because 6 = 6. **21.** T **22.** T **23.** F **24.** T **25.** T **26.** T **27.** T **28.** T **29.** F **30.** T **31.** T **32.** F **33.** T **34.** T **35.** T **36.** F **37.** 4 **38.** 8 **39.** 16 **40.** 32 **41.** 128 **42.** 256 **43.** six **44.** No, because 48 is not a whole number power of 2.

In Exercises 45–60, we give the truth values found in the order in which they appear in the final column of the truth table, when the truth table is constructed in the format described in the section.

45. FFTF **46.** FTTT **47.** FTTT **48.** TFFT **49.** TTTT **50.** FTFF **51.** FFFT **52.** FFTT **53.** TFFF **54.** TFTT **55.** FFFFTFFF **56.** TFTTTFTF **57.** FTFTTTTT **58.** FFFTTTTT **59.** TTTTTTTTTTTTFTTT **60.** FFFFFFFFTFTTFFFF **61.** You can't pay me now and you can't pay me later. **62.** I am going and she is not going. **63.** It is not summer or there is snow. **64.** 1/2 is not a positive number or -12 is greater than or equal to zero. **65.** I did not say yes or she did not say no. **66.** Pauline Mula did not try to sell the book, or she was able to do so. **67.** $5 - 1 \neq 4$ or $9 + 12 = 7$ **68.** $3 \geq 10$ and $7 = 2$ **69.** Neither Cupid nor Vixen will lead Santa's sleigh next Christmas. **70.** The lawyer did not appear in court or the client did not appear in court. **71.** T **72.** F **73.** T **74.** F **75.** FTTF **76.** inclusive disjunction **77.** F **78.** T **79.** T **80.** F

3.3 Exercises (Page 120)

1. If it is breathing, then it must be alive. **2.** If you see it on the Internet, then you can believe it. **3.** If it is summer, then Lorri Morgan visits Hawaii. **4.** If the person is Tom Shaffer, then his area code is 216. **5.** If it is a picture, then it tells a story. **6.** If the soldier is a marine, then the soldier loves boot camp. **7.** If it is a guinea pig, then it is not a scholar. **8.** If it is a koala, then it does not live in Texas. **9.** If he is Running Bear, then he loves Little White Dove. **10.** If it is an opium-eater, then it has no self-command. **11.** true **12.** true **13.** true **14.** true **15.** false **16.** true **17.** true **18.** false **21.** T **22.** F **23.** F **24.** T **25.** T **26.** T **27.** If they do not raise monkeys, then he trains ponies. **28.** If he trains ponies, then they do not raise monkeys. **29.** If she has a snake for a pet, then they raise monkeys and he trains ponies.

30. If she has a snake for a pet and he trains ponies, then they raise monkeys. **31.** If he does not train ponies, then they do not raise monkeys or she has a snake for a pet. **32.** If she does not have a snake for a pet or they do not raise monkeys, then he does not train ponies. **33.** $r \to b$ **34.** $b \to p$ **35.** $\sim b \to \sim r$ **36.** $p \to \sim r$ **37.** $b \lor (p \to r)$ **38.** $p \land (r \to \sim b)$ **39.** $\sim r \to b$ **40.** $p \to r$ **41.** T **42.** T **43.** F **44.** F **45.** T **46.** T **47.** F **48.** F **49.** T **50.** T **51.** T **52.** T
54. Answers will vary. One example is $p \to [(\sim q \land r) \lor p]$.
In Exercises 55–65, we give the truth values found in the order in which they appear in the final column of the truth table, when the truth table is constructed in the format described in Section 3.2.
55. TTTF **56.** FTTT **57.** TTFT **58.** FTFT **59.** TTTT; tautology **60.** TTTT; tautology
61. TFTF **62.** FTTTFTFT **63.** TTTTTTFT **64.** TTTFTTTTTTTTTTTT **65.** TTTTTTTTTTTTTFT
66. one **67.** That is an authentic Persian rug and I am not surprised. **68.** Ella reaches that note and she will not shatter glass. **69.** The English measures are not converted to metric measures and the spacecraft does not crash on the surface of Mars. **70.** Say "I do" and you are not happy for the rest of your life. **71.** You want to be happy for the rest of your life and you make a pretty woman your wife. **72.** Loving you is wrong and I want to be right. **73.** You do not give your plants tender, loving care or they flourish. **74.** The check is not in the mail or I'll be surprised. **75.** She does or he will. **76.** I do not say yes or she says no. **77.** The person is not a resident of Butte or is a resident of Montana. **78.** The person is not a woman or was once a girl.
79. equivalent **80.** equivalent **81.** equivalent **82.** equivalent **83.** equivalent **84.** not equivalent
85. not equivalent **86.** not equivalent **87.** $(p \land q) \lor (p \land \sim q)$; The statement simplifies to p.
88. $p \land (r \lor q)$ **89.** $p \lor (\sim q \land r)$ **90.** $q \lor [p \land (q \lor \sim p)]$; The statement simplifies to q.
91. $\sim p \lor (p \lor q)$; The statement simplifies to T. **92.** $(\sim p \lor q) \lor (\sim p \lor \sim q)$; The statement simplifies to T.
93. The statement simplifies to $p \land q$. **94.** 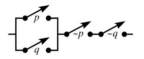 **95.** The statement simplifies to F.

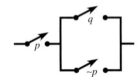

96. The statement simplifies to $\sim p \lor q$. **97.** The statement simplifies to $(r \land \sim p) \land q$.

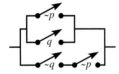 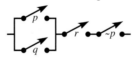

98. The statement simplifies to $(\sim p \land r) \land \sim q$. **99.** The statement simplifies to $p \lor q$.

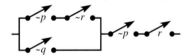

100. The statement simplifies to T. **102.** $262.80

3.4 Exercises (Page 128)

1. (a) If you were an hour, then beauty would be a minute. **(b)** If beauty were not a minute, then you would not be an hour. **(c)** If you were not an hour, then beauty would not be a minute. **2. (a)** If I follow, then you lead. **(b)** If you do not lead, then I will not follow. **(c)** If I do not follow, then you do not lead.
3. (a) If you don't fix it, then it ain't broke. **(b)** If it's broke, then fix it. **(c)** If you fix it, then it's broke.

4. (a) If I were rich, then I would have a nickel for each time that happened. **(b)** If I did not have a nickel for each time that happened, then I would not be rich. **(c)** If I were not rich, then I would not have a nickel for each time that happened. **5. (a)** If it is dangerous to your health, then you walk in front of a moving car. **(b)** If you do not walk in front of a moving car, then it is not dangerous to your health. **(c)** If it is not dangerous to your health, then you do not walk in front of a moving car. **6. (a)** If it contains calcium, then it's milk. **(b)** If it's not milk, then it does not contain calcium. **(c)** If it does not contain calcium, then it's not milk. **7. (a)** If they flock together, then they are birds of a feather. **(b)** If they are not birds of a feather, then they do not flock together. **(c)** If they do not flock together, then they are not birds of a feather. **8. (a)** If it gathers no moss, then it is a rolling stone. **(b)** If it is not a rolling stone, then it gathers moss. **(c)** If it gathers moss, then it is not a rolling stone. **9. (a)** If he comes, then you built it. **(b)** If you don't build it, then he won't come. **(c)** If he doesn't come, then you didn't build it. **10. (a)** If there's fire, then there's smoke. **(b)** If there's no smoke, then there's no fire. **(c)** If there's no fire, then there's no smoke. **11. (a)** $\sim q \to p$ **(b)** $\sim p \to q$ **(c)** $q \to \sim p$ **12. (a)** $q \to \sim p$ **(b)** $p \to \sim q$ **(c)** $\sim q \to p$ **13. (a)** $\sim q \to \sim p$ **(b)** $p \to q$ **(c)** $q \to p$ **14. (a)** $\sim p \to \sim q$ **(b)** $q \to p$ **(c)** $p \to q$ **15. (a)** $(q \lor r) \to p$ **(b)** $\sim p \to (\sim q \land \sim r)$ **(c)** $(\sim q \land \sim r) \to \sim p$ **16. (a)** $p \to (r \lor \sim q)$ **(b)** $(\sim r \land q) \to \sim p$ **(c)** $\sim p \to (\sim r \land q)$ **19.** If it is muddy, then I'll wear my galoshes. **20.** If I finish studying, then I'll go to the party. **21.** If 17 is positive, then $17 + 1$ is positive. **22.** If today is Wednesday, then yesterday was Tuesday. **23.** If a number is an integer, then it is a rational number. **24.** If a number is a whole number, then it is an integer. **25.** If I do crossword puzzles, then I am driven crazy. **26.** If you are in Fort Lauderdale, then you are in Florida. **27.** If Greg Tobin is to shave, then he must have a day's growth of beard. **28.** If one is elected, then one is an environmentalist. **29.** If I go from Boardwalk to Connecticut Avenue, then I pass GO. **30.** If the principal hires more teachers, then the school board approves. **31.** If a number is a whole number, then it is an integer. **32.** If a number is an integer, then it is rational. **33.** If their pitching improves, then the Indians will win the pennant. **34.** If pigs fly, then Jesse will be a liberal. **35.** If the figure is a rectangle, then it is a parallelogram with a right angle. **36.** If the figure is a parallelogram, then it is a four-sided figure with opposite sides parallel. **37.** If a triangle has two sides of the same length, then it is isosceles. **38.** If the figure is a square, then it is a rectangle with two adjacent sides equal. **39.** If a two-digit number whose units digit is 5 is squared, then it will end in 25. **40.** If an integer has a units digit of 0 or 5, then it is divisible by 5. **41.** D **45.** true **46.** false **47.** false **48.** true **49.** false **50.** false **51.** contrary **52.** contrary **53.** consistent **54.** consistent **55.** contrary **56.** consistent **57.** Answers will vary. One example is: That man is Kent Merrill; That man sells books. **58.** Answers will vary. One example is: Jennifer Kerber is 35 years old; Jennifer Kerber is less than 35 years old.

3.5 Exercises (Page 132)

1. valid **2.** valid **3.** invalid **4.** invalid **5.** valid **6.** valid **7.** invalid **8.** invalid **9.** invalid **10.** invalid **11.** invalid **12.** invalid **13.** yes **14.** One possible conclusion is "All expensive things make you feel good." **15.** People who have major surgery must go to the hospital.
<u>Julianne Peterson is having major surgery.</u>
Julianne Peterson must go to the hospital.
16. All people with blue eyes have blond hair. **17.** valid **18.** invalid **19.** invalid **20.** valid **21.** valid
<u>Julie Ward does not have blond hair.</u>
Julie Ward does not have blue eyes.
22. invalid **23.** invalid **24.** valid **25.** invalid **26.** invalid **27.** invalid **28.** invalid **29.** valid **30.** valid

Extension Exercises (Page 135)

1. Lauren, Mr. Maier, hay ride; Raven, Ms. Fedor, cider making; Tara, Ms. Reed, country store; Xander, Mr. Morgan, apple picking; Zach, Mr. Hanson, feeding animals **2.** Arlene, orange, 12; Devon, turquoise, 16; Rosetta, pink, 10; Silas, red, 18; Tina, gray, 14 **3.** Mr. Crow, cupids, daughter; Mr. Evans, leprechauns, uncle; Mr. Hurley, reindeer, sister; Mr. Speigler, happy faces, father-in-law **4.** Joanne, moose, circus; Lou, grizzly bear, train; Ralph, seal, rock band; Winnie, zebra, spaceship

3.6 Exercises (Page 145)

1. valid by reasoning by transitivity   **2.** valid by reasoning by transitivity   **3.** valid by modus ponens   **4.** valid by modus ponens   **5.** fallacy by fallacy of the converse   **6.** fallacy by fallacy of the converse   **7.** valid by modus tollens   **8.** valid by modus tollens   **9.** fallacy by fallacy of the inverse   **10.** fallacy by fallacy of the inverse   **11.** valid by disjunctive syllogism   **12.** valid by disjunctive syllogism   **13.** invalid   **14.** valid   **15.** valid   **16.** invalid   **17.** invalid   **18.** invalid   **19.** valid   **20.** invalid   **21.** invalid   **22.** valid   **23.** invalid   **24.** valid   **26.** Every time something squeaks, I use WD-40.
Every time I use WD-40, I go to the hardware store.
Every time something squeaks, I go to the hardware store.

27. valid   **28.** valid   **29.** invalid   **30.** invalid   **31.** invalid   **32.** invalid   **33.** valid   **34.** valid   **35.** valid   **36.** valid   **37.** If I tell you the time, then my life will be miserable.
Answers in Exercises 39–46 may be replaced by their contrapositives.
39. If it is my poultry, then it is a duck.   **40.** If he is your son, then he cannot do logic.   **41.** If it is a guinea pig, then it is hopelessly ignorant of music.   **42.** If the person is a teetotaler, then the person is not a pawnbroker.   **43.** If it is a teachable kitten, then it does not have green eyes.   **44.** If it is an opium-eater, then it has no self-command.   **45.** If I can read it, then I have not filed it.   **46.** If it is written on blue paper, then it is filed.
47. (a) $p \rightarrow \sim s$ (b) $r \rightarrow s$ (c) $q \rightarrow p$ (d) None of my poultry are officers.   **48.** (a) $r \rightarrow p$ (b) $\sim r \rightarrow \sim q$ (c) $s \rightarrow \sim p$ (d) Your sons are not fit to serve on a jury.   **49.** (a) $r \rightarrow \sim s$ (b) $u \rightarrow t$ (c) $\sim r \rightarrow p$ (d) $\sim u \rightarrow \sim q$ (e) $t \rightarrow s$ (f) All pawnbrokers are honest.   **50.** (a) $s \rightarrow r$ (b) $p \rightarrow q$ (c) $q \rightarrow \sim r$ (d) Guinea pigs don't appreciate Beethoven.   **51.** (a) $r \rightarrow w$ (b) $\sim u \rightarrow \sim t$ (c) $v \rightarrow \sim s$ (d) $x \rightarrow r$ (e) $\sim q \rightarrow t$ (f) $y \rightarrow p$ (g) $w \rightarrow s$ (h) $\sim x \rightarrow \sim q$ (i) $p \rightarrow \sim u$ (j) I can't read any of Brown's letters.   **52.** (a) $p \rightarrow q$ (b) $\sim u \rightarrow \sim s$ (c) $t \rightarrow \sim r$ (d) $q \rightarrow s$ (e) $v \rightarrow p$ (f) $\sim r \rightarrow \sim u$ (g) Opium-eaters do not wear white kid gloves.

Chapter 3 Test (Page 149)

1. $6 - 3 \neq 3$ **2.** Some men are not created equal.   **3.** No members of the class went on the field trip.   **4.** That's the way you feel and I won't accept it.   **5.** She did not pass GO or did not collect $200.   **6.** $\sim p \rightarrow q$ **7.** $p \rightarrow q$ **8.** $\sim q \leftrightarrow \sim p$ **9.** You won't love me and I will love you.   **10.** It is not the case that you will love me or I will not love you. (Equivalently: You won't love me and I will love you.)   **11.** T   **12.** T   **13.** T   **14.** F   **16.** (a) The antecedent must be true and the consequent must be false.   (b) Both component statements must be true.   (c) Both component statements must be false.
In each of Exercises 17 and 18, we give the truth values found in the order in which they appear in the final column of the truth table, when the truth table is constructed in the format described in the chapter.
17. TFFF   **18.** TTTT (tautology)   **19.** false   **20.** true
Wording may vary in the answers in Exercises 21–26.
21. If the number is an integer, then it is a rational number.   **22.** If a polygon is a rhombus, then it is a quadrilateral.   **23.** If a number is divisible by 9, then it is divisible by 3.   **24.** If she digs dinosaur bones, then she is a paleontologist.   **25.** (a) If the graph helps me understand it, then a picture paints a thousand words.   (b) If a picture doesn't paint a thousand words, then the graph won't help me understand it.   (c) If the graph doesn't help me understand it, then a picture doesn't paint a thousand words.   **26.** (a) $(q \wedge r) \rightarrow \sim p$ (b) $p \rightarrow (\sim q \vee \sim r)$ (c) $(\sim q \vee \sim r) \rightarrow p$ **27.** valid   **28.** (a) A (b) F (c) C (d) D   **29.** valid   **30.** invalid

CHAPTER 4 Numeration and Mathematical Systems

4.1 Exercises (Page 158)

1. 13,036   2. 2412   3. 7,630,729   4. 3,005,231   5. (Egyptian numerals)   6. (Egyptian numerals)
7. (Egyptian numerals)   8. (Egyptian numerals)   9. (Egyptian numerals)
10. (Babylonian numerals)   11. (Babylonian numerals)   12. (Babylonian numerals)   13. (Babylonian numerals)
14. (Babylonian numerals)   15. 935   16. 246   17. 3007   18. 4902
19. (Chinese numerals)   20. (Chinese numerals)   21. (Chinese numerals)   22. (Chinese numerals)   23. (Chinese numerals)   24. (Chinese numerals)   25. (Chinese numerals)
26. (Chinese numerals)   27. 216   28. 392   29. 53,601   30. 6168   31. 113   32. 22   33. 7598
34. 1263   35. 1378   36. 2673   37. 5974   38. 7705   39. 622,500 shekels   40. 533,000 shekels
45. 99,999   46. 9,999,999,999   47. 3124   48. 9,765,624   49. $10^d - 1$ 50. $5^d - 1$
51. $7^d - 1$ 52. $b^d - 1$

4.2 Exercises (Page 167)

1. $(7 \times 10^1) + (3 \times 10^0)$ 2. $(9 \times 10^2) + (2 \times 10^1) + (5 \times 10^0)$
3. $(3 \times 10^3) + (7 \times 10^2) + (7 \times 10^1) + (4 \times 10^0)$
4. $(1 \times 10^4) + (2 \times 10^3) + (3 \times 10^2) + (9 \times 10^1) + (8 \times 10^0)$
5. $(4 \times 10^3) + (9 \times 10^2) + (2 \times 10^1) + (4 \times 10^0)$
6. $(5 \times 10^4) + (2 \times 10^3) + (1 \times 10^2) + (1 \times 10^1) + (8 \times 10^0)$
7. $(1 \times 10^7) + (4 \times 10^6) + (2 \times 10^5) + (0 \times 10^4) + (6 \times 10^3) + (0 \times 10^2) + (4 \times 10^1) + (0 \times 10^0)$
8. $(2 \times 10^8) + (1 \times 10^7) + (2 \times 10^6) + (0 \times 10^5) + (1 \times 10^4) + (1 \times 10^3) + (9 \times 10^2) + (1 \times 10^1) + (6 \times 10^0)$ 9. 42   10. 350   11. 6209   12. 503,568   13. 70,401,009
14. 380,000,203   15. 89   16. 1195   17. 32   18. 261   19. 109   20. 815   21. 733
22. 11,582   23. 6   24. 305   25. 206   26. 481   27. 256   28. 23   29. 63,259
30. 4536   31. 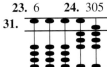   32. 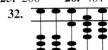   33. 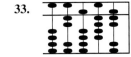   34.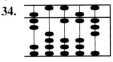
35. 1885   36. 23,712   37. 38,325   38. 440,496   39. 3,035,154   41. 496   42. 2336
43. 217,204   44. 2,557,938   45. 242   46. 111   47. 49,801   48. 35,156   49. 460   50. 2173
51. 32,798   52. 9135

4.3 Exercises (Page 176)

1. 1, 2, 3, 4, 5, 6, 10, 11, 12, 13, 14, 15, 16, 20, 21, 22, 23, 24, 25, 26   **2.** 1, 2, 3, 4, 5, 6, 7, 10, 11, 12, 13, 14, 15, 16, 17, 20, 21, 22, 23, 24   **3.** 1, 2, 3, 4, 5, 6, 7, 8, 10, 11, 12, 13, 14, 15, 16, 17, 18, 20, 21, 22   **4.** 1, 2, 3, 4, 5, 6, 7, 8, 9, A, B, C, D, E, F, 10, 11, 12, 13, 14   **5.** 13_{five}; 20_{five} **6.** 554_{six}; 1000_{six} **7.** $B6E_{sixteen}$; $B70_{sixteen}$ **8.** 10110_{two}; 11000_{two} **9.** 3   **10.** 7   **11.** 11   **12.** 16   **13.** smallest: $1000_{three} = 27$; largest: $2222_{three} = 80$ **14.** smallest: $1000_{sixteen} = 4096$; largest: $FFFF_{sixteen} = 65535$ **15.** 14   **16.** 44   **17.** 11   **18.** 29   **19.** 956   **20.** 2492   **21.** 881   **22.** 366   **23.** 28,854   **24.** 43,981   **25.** 139   **26.** 4532   **27.** 5601   **28.** 58,890   **29.** 321_{five} **30.** 122_{seven} **31.** 10011_{two} **32.** 1647_{eight} **33.** $93_{sixteen}$ **34.** $AAA_{sixteen}$ **35.** 2131101_{five} **36.** 413454_{seven} **37.** 1001001010_{two} **38.** 31130_{eight} **39.** 102112101_{three} **40.** 2230110_{four} **41.** 111134_{six} **42.** 162150_{nine} **43.** 32_{seven} **44.** 43_{five} **45.** 1031321_{four} **46.** 11651_{seven} **47.** 11110111_{two} **48.** 10100000110_{two} **49.** 467_{eight} **50.** 3275_{eight} **51.** 11011100_{two} **52.** 1111000100010001_{two} **53.** $2D_{sixteen}$ **54.** $5EE8_{sixteen}$ **55.** 37_{eight} **56.** $6F_{sixteen}$ **57.** 1427   **58.** 2 gross, 19 dozen   **59.** 1000011_{two} **60.** 1011000_{two} **61.** 1101011_{two} **62.** HELP   **63.** CHUCK   **64.** $1001110110010111101 11_{two}$ **65.** $10011111100101101100110010111000011101101110011_{two}$ **67.** 27 and 63   **68. (a)** The binary ones digit is 1.   **(b)** The binary twos digit is 1.   **(c)** The binary fours digit is 1.   **(d)** The binary eights digit is 1.   **(e)** The binary sixteens digit is 1.   **70.** 6   **71.** 7   **72.** yes   **73.** no   **74.** yes   **75.** no   **76.** yes   **77.** no   **78.** yes   **79.** no   **82.** no   **83.** yes   **84.** yes   **85.** no

4.4 Exercises (Page 183)

1. all properties; 1 is the identity element; 1 is its own inverse, as is 2.   **2.** all properties; 1 is the identity element; 1 is its own inverse, as is 4; 2 and 3 are inverses.   **3.** closure commutative, associative, and identity properties; 1 is the identity element; 2, 4, and 6 have no inverses.   **4.** closure commutative, associative, and identity properties; 1 is the identity element; 2, 3, and 4 have no inverses.   **5.** all properties except the inverse; 1 is the identity element; 5 has no inverse.   **6.** all properties; 1 is the identity element; 1, 3, 5, and 7 are their own inverses.   **7.** all properties; F is the identity element; A and B are inverses; F is its own inverse.   **8.** closure and commutative properties   **9.** all properties; t is the identity element; s and r are inverses; t and u are their own inverses.   **10.** all properties; A is the identity element; J and U are inverses; A and T are their own inverses.   **11.** a **12.** c **13.** a **14.** b **15.** row b: d; row c: d, b; row d: b, c **16.** all   **17.** associative, commutative, identity (U), closure   **18.** associative, commutative, identity ($\emptyset$), closure   **19.** Answers may vary. One possibility is as follows.

	a	b	c	d
a	a	b	c	d
b	b	a	d	c
c	c	d	a	b
d	d	c	b	a

20. no   **21.** no   **22.** no   **23. (a)** true   **(b)** true   **(c)** true   **(d)** true   **24.** Answers will vary.   **25.** $a + b + c = 1$ or $a = 0$ **27. (a)** $a = 0$ **(b)** $a = 0$ **28. (a)** Answers will vary.   **29.** Each side simplifies to e.   **30.** Each side simplifies to a.   **31.** Each side simplifies to d.   **32.** Each side simplifies to c.

33.
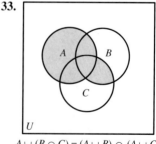
$A \cup (B \cap C) = (A \cup B) \cap (A \cup C)$

34.
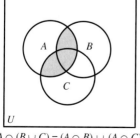
$A \cap (B \cup C) = (A \cap B) \cup (A \cap C)$

35. Both final columns read TTTTTFFF, when set up in the manner described in Chapter 3.

36. Both final columns read TTTFFFFF, when set up in the manner described in Chapter 3.

4.5 Exercises (Page 192)

1. No operation is specified.   **2.** No set of elements is specified.   **3.** yes   **4.** yes   **5.** no; closure, inverse   **6.** yes   **7.** yes   **8.** no; inverse   **9.** no; inverse   **10.** no; closure   **11.** no; associative, identity, inverse   **12.** no; inverse   **13.** no; inverse   **14.** no; identity, inverse   **15.** yes   **16.** yes   **17.** no; closure, identity, inverse   **18.** yes   **21.** S **22.** T **23.** N **24.** S **25.** Each side is equal to M.   **26.** Each side is equal to Q.   **27.** Each side is equal to Q.   **28.** Each side is equal to M.   **29.** N **30.** Q **31.** R **32.** S **33.** T **34.** V **35.** no   **36.** yes   **37.** yes   **38.** no

Chapter 4 Test (Page 194)

1. ancient Egyptian; 2536   **2.** 8364   **3.** $(6 \times 10^4) + (0 \times 10^3) + (9 \times 10^2) + (2 \times 10^1) + (3 \times 10^0)$ **4.** 1998   **5.** 22,184   **6.** 12,827   **7.** 114   **8.** 38   **9.** 43,020   **10.** 111010_{two} **11.** 24341_{five} **12.** 256_{eight} **13.** 101101010010_{two} **14.** There is less repetition of symbols.   **15.** There are fewer symbols to learn.   **16.** There are fewer digits in the numerals.   **17.** 0   **18.** $\frac{1}{3}$ **19.** commutative   **21. (a)** yes **(b)** o **22. (a)** yes   **23. (a)** no   **24. (a)** no   **25. (a)** no

CHAPTER 5 Number Theory

5.1 Exercises (Page 203)

1. true   **2.** true   **3.** false   **4.** true   **5.** false   **6.** false   **7.** true   **8.** true   **9.** false   **10.** true   **11.** false   **12.** true   **13.** 1, 2, 3, 4, 6, 12   **14.** 1, 2, 3, 6, 9, 18   **15.** 1, 2, 4, 5, 10, 20   **16.** 1, 2, 4, 7, 14, 28   **17.** 1, 2, 3, 4, 5, 6, 8, 10, 12, 15, 20, 24, 30, 40, 60, 120   **18.** 1, 2, 4, 43, 86, 172   **19. (a)** no **(b)** yes **(c)** no **(d)** yes **(e)** no **(f)** no **(g)** yes **(h)** no **(i)** no   **20. (a)** yes **(b)** yes **(c)** no **(d)** yes **(e)** yes **(f)** no **(g)** yes **(h)** yes **(i)** no   **21. (a)** no **(b)** no **(c)** no **(d)** yes **(e)** no **(f)** no **(g)** no **(h)** no **(i)** no   **22. (a)** no **(b)** no **(c)** no **(d)** yes **(e)** no **(f)** no **(g)** no **(h)** no **(i)** no   **23. (a)** no **(b)** yes **(c)** no **(d)** no **(e)** no **(f)** no **(g)** yes **(h)** no **(i)** no   **24. (a)** no **(b)** yes **(c)** no **(d)** no **(e)** no **(f)** no **(g)** yes **(h)** no **(i)** no   **25. (b)** 13 **(c)** square root; square root; square root **(d)** prime   **26. (a)** 101, 103, 107, 109, 113, 127, 131, 137, 139, 149, 151, 157, 163, 167, 173, 179, 181, 191, 193, 197, 199; There are 21. **(b)** $M_7 = 2^7 - 1 = 127$, which is in the list.   **27.** 2, 3; no   **29.** It must be 0.   **30.** The last four digits must form a number divisible by 16. The number 456,882,320 is divisible by 16, since 2320 is divisible by 16; $2320 = 16 \cdot 145$.   **31.** $2^4 \cdot 3 \cdot 5$ **32.** $2^2 \cdot 3 \cdot 5^2$ **33.** $2^3 \cdot 3^2 \cdot 5$ **34.** $5^2 \cdot 17$ **35.** $3 \cdot 13 \cdot 17$ **36.** $3 \cdot 5 \cdot 59$ **37.** yes   **38.** yes   **39.** no   **40.** no   **41.** yes   **42.** yes   **43.** no   **44.** no   **45.** The number must be divisible by both 3 and 5. That is, the sum of the digits must be divisible by 3, and the last digit must be 5 or 0.   **47.** 0, 2, 4, 6, 8   **48.** 1, 4, 7   **49.** 0, 4, 8   **50.** 0, 5   **51.** 0, 6   **52.** 0, 8   **53.** 6   **54.** any digit   **55.** 9   **56.** 10   **57.** 12   **58.** 15   **59.** 27   **60.** 75   **61.** leap year   **62.** leap year   **63.** not a leap year   **64.** not a leap year   **65.** leap year   **66.** not a leap year   **68.** The product of any three consecutive natural numbers is divisible by 6.   **70.** In each case, the number is divisible by 13, 11, and 7. This happens because such a number is divisible by 1001, and $1001 = 13 \cdot 11 \cdot 7$.   **71.** $41^2 - 41 + 41 = 41^2$, and 41^2 is not a prime.   **72. (a)** 1763 **(b)** 1847   **73.** B   **74.** B   **75. (a)** 65,537 **(b)** 251   **76. (a)** 4,294,967,297 **(b)** $4{,}294{,}967{,}297 \div 641 = 6{,}700{,}417$ **80.** composite; $M = 30{,}031 = 59 \cdot 509$ **81.** 63   **82.** 7; yes   **83.** $2^p - 1$ **84.** 1023   **85.** 3 and 31

5.2 Exercises (Page 210)

1. true   **2.** false   **3.** true   **4.** true   **5.** true   **6.** false   **7.** false   **8.** false   **9.** true   **10.** true   **11.** The sum of the proper divisors is 496: $1 + 2 + 4 + 8 + 16 + 31 + 62 + 124 + 248 = 496$.   **12.** The sum of the proper divisors is 8128: $1 + 2 + 4 + 8 + 16 + 32 + 64 + 127 + 254 + 508 + 1016 + 2032 + 4064 = 8128$.   **13.** 8191 is prime; 33,550,336   **14.** $2^{13,466,916}(2^{13,466,917} - 1)$ **15.** $1 + 1/2 + 1/3 + 1/6 = 2$ **16.** $496 = 1 + 2 + 3 + \cdots + 31$ **17.** abundant   **18.** abundant   **19.** deficient   **20.** deficient   **21.** 12, 18, 20, 24   **23.** $1 + 3 + 5 + 7 + 9 + 15 + 21 + 27 + 35 + 45 + 63 + 105 + 135 + 189 + 315 = 975$, and $975 > 945$, so 945 is abundant.

25. $1 + 2 + 4 + 8 + 16 + 32 + 37 + 74 + 148 + 296 + 592 = 1210$ and
$1 + 2 + 5 + 10 + 11 + 22 + 55 + 110 + 121 + 242 + 605 = 1184$ **26.** (a) $2^2 \cdot 5 \cdot 11 = 220$ and
$2^2 \cdot 71 = 284.$ (b) 17,296 and 18,416 **27.** $3 + 11$ **28.** $11 + 11$ **29.** $3 + 23$ **30.** $13 + 19$
31. Let $a = 5$ and $b = 3$; $11 = 5 + 2 \cdot 3$ **32.** $6 = 11 - 5$; $12 = 17 - 5$; $18 = 23 - 5$ **33.** 71 and 73
34. 101 and 103 **35.** 137 and 139 **36.** (a) $3^4 - 1 = 80$ is divisible by 5. (b) $2^6 - 1 = 63$ is divisible by 7.
37. (a) $5 = 9 - 4 = 3^2 - 2^2$ (b) $11 = 36 - 25 = 6^2 - 5^2$ **38.** (a) $5 = 1 + 4 = 1^2 + 2^2$
(b) $13 = 4 + 9 = 2^2 + 3^2$ **39.** $5^2 + 2 = 27 = 3^3$ **40.** $2^2 + 4 = 8 = 2^3$ and $11^2 + 4 = 125 = 5^3$
42. No; the fifth contains 8 digits and the sixth contains 10 digits. **43.** No; for the first six, the sequence is
6, 8, 6, 8, 6, 6. **46.** (a) 209 and 211 (b) 211 only **47.** B **48.** 5; yes **49.** 7; yes **50.** 11; yes
51. 15; no **52.** 23; yes **53.** 27; no **54.** 6; 5; 7; yes; yes **55.** 24; 23; 25; yes; no
56. 120; 119; 121; no; no **58.** B **59.** B **60.** B

5.3 Exercises (Page 218)

1. true **2.** true **3.** true **4.** false **5.** false **6.** false **7.** true **8.** true **9.** true
10. true **11.** 10 **12.** 60 **13.** 120 **14.** 168 **15.** 7 **16.** 28 **17.** 12 **18.** 65 **19.** 10
20. 2 **21.** 6 **22.** 6 **23.** 12 **24.** 5 **25.** 12 **26.** 24 **27.** 70 **28.** 30 **31.** 120
32. 96 **33.** 672 **34.** 140 **35.** 840 **36.** 144 **37.** 96 **38.** 280 **39.** 225 **40.** 2160
41. 2160 **42.** 2310 **43.** 180 **44.** 350 **45.** 1260 **46.** 360 **47.** 1680 **48.** 2400
49. (a) $p^b q^c r^c$ (b) $p^a q^a r^b$ **50.** (a) $2^{31} \cdot 5^{17} \cdot 7^{13}$ (b) $2^{34} \cdot 5^{22} \cdot 7^{21}$ **51.** 30 **52.** 3 **53.** 15
54. 3 **55.** 12 **56.** 30 **57.** (a) 6 (b) 36 **58.** (a) 9 (b) 108 **59.** (a) 18 (b) 216
60. p is a factor of q (or, q is a multiple of p). **61.** p and q are relatively prime. **62.** p is a factor of q (or, q
is a multiple of p). **65.** 144th **66.** July 31 **67.** 48 **68.** 360 seconds **69.** $600; 25 books
70. 12 inches

5.4 Exercises (Page 228)

1. 5 **2.** 7 **3.** 6 **4.** 9 **5.** row 2: 0, 6, 10; row 3: 9, 0, 9; row 4: 0, 4, 0, 8, 0; row 5: 1, 9, 2, 7;
row 6: 6, 0, 0; row 7: 4, 11, 6, 8, 3, 5; row 8: 8, 4, 0, 0; row 9: 6, 3, 9, 3, 9, 6, 3; row 10: 6, 4, 0, 10, 8, 6, 4;
row 11: 10, 9, 8, 7, 6, 5, 4, 3, 2 **6.** yes **7.** yes **8.** yes (1 is the identity element.) **9.** row 1: 0;
row 2: 0; row 3: 0, 1, 2; row 4: 0, 1, 2 **10.** yes **11.** yes **12.** yes (0 is the identity element.)
13. yes (0 is its own inverse, 1 and 4 are inverses of each other, and 2 and 3 are inverses of each other.)
14. row 2: 1, 3; row 3: 1, 4, 2; row 4: 3, 2, 1 **15.** yes **16.** yes **17.** yes (1 is the identity element.)
18. 2 **19.** 3 **20.** 4 **21.** 4 **24.** 1900 **25.** 0700 **26.** 0020 **27.** 0000
28. first answer: ordinary whole number arithmetic; second answer: 12-hour clock arithmetic; third answer: 24-hour
clock arithmetic **29.** false **30.** true **31.** true **32.** false **33.** 3 **34.** 4 **35.** 3 **36.** 1
37. 1 **38.** 0 **39.** 10 **40.** 6 **42.** 1 **43.** 5 **44.** 2 **45.** 4 **46.** (a) row 1: 1, 2, 3, 0;
row 2: 2, 3, 0, 1; row 3: 3, 0, 1, 2 (b) All four properties are satisfied. (c) 0 is its own inverse, 2 is its own inverse,
and 1 and 3 are inverses of each other. **47.** (a) row 1: 0; row 2: 2, 3, 4, 5, 6, 0, 1; row 3: 3, 4, 5, 6, 0, 1, 2;
row 4: 4, 5, 6, 0, 1, 2, 3; row 5: 5, 6, 0, 1, 2, 3, 4; row 6: 6, 0, 1, 2, 3, 4, 5 (b) All four properties are satisfied.
(c) 0 is its own inverse, 1 and 6 are inverses of each other, 2 and 5 are inverses of each other, and 3 and 4 are
inverses of each other. **48.** (a) 1 (b) All properties are satisfied. (c) 1 is its own inverse. **49.** (a) 1
(b) All properties are satisfied. (c) 1 is its own inverse, and 2 is its own inverse. **50.** (a) row 2: 0, 2;
row 3: 2, 1 (b) no inverse property (c) 1 is its own inverse as is 3; 2 has no inverse. **51.** (a) row 2: 1, 3, 7;
row 3: 3, 0; row 4: 3, 2, 1; row 5: 6, 7, 4; row 7: 3, 1, 6, 4; row 8: 6, 5, 2 (b) no inverse property (c) 1 is its
own inverse, as is 8; 2 and 5 are inverses of each other, as are 4 and 7; 3 has no inverse, and 6 has no inverse.
52. prime **53.** {3, 10, 17, 24, 31, 38, . . .} **54.** {2, 5, 8, 11, 14, 17, . . .} **55.** identity
56. {2, 6, 10, 14, 18, 22, . . .} **57.** 100,000 **58.** $29,306 + 100,000k$, where k is any nonnegative integer
59. 62 **60.** 153 **61.** (a) 365 (b) Friday **62.** January 1 through February 28 **63.** Chicago: July 23
and 29; New Orleans: July 5 and August 16; San Francisco: August 9 **64.** 1 **65.** $-i$ **66.** -1 **67.** i
68. Wednesday **69.** Sunday **70.** Tuesday **71.** Wednesday **72.** June **73.** September, December
74. February, August **75.** June **76.** (c) 10 (d) 4, 4 (e) 9, 9 (f) 3, 3 (g) 8, 8 (h) 2, 2 (i) 7, 7

(j) 1, 1 **(k)** 6, 6 **(l)**

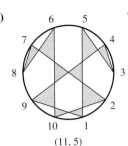

(11, 5)

77. incorrect **78.** correct **79.** 6 **80.** 1 **81.** 7 **82.** 6

5.5 Exercises (Page 237)

1. 2584 **2.** 1 and 5 **3.** 46,368 **4.** even **5.** $\dfrac{1+\sqrt{5}}{2}$ **6.** 1.618
7. $1+1+2+3+5+8 = 21 - 1$; Each expression is equal to 20.
8. $1+3+8+21+55+144 = 233 - 1$; Each expression is equal to 232.
9. $1+2+5+13+34+89 = 144$; Each expression is equal to 144.
10. $8^2 + 13^2 = 233$; Each expression is equal to 233. **11.** $13^2 - 5^2 = 144$; Each expression is equal to 144.
12. $13^3 + 8^3 - 5^3 = 2584$; Each expression is equal to 2584. **13.** $1 - 2 + 5 - 13 + 34 - 89 = -8^2$; Each expression is equal to -64. **14.** $1 - 1 + 2 - 3 + 5 - 8 + 13 = 8 + 1$; Each expression is equal to 9.
15. (There are other ways to do this.) **(a)** $37 = 34 + 3$ **(b)** $40 = 34 + 5 + 1$ **(c)** $52 = 34 + 13 + 5$
16. (a) 2 divides 6, and $F_2 = 1$ is a factor of $F_6 = 8$. **(b)** 3 divides 9, and $F_3 = 2$ is a factor of $F_9 = 34$.
(c) 4 divides 8, and $F_4 = 3$ is a factor of $F_8 = 21$. **17. (a)** The greatest common factor of 10 and 4 is 2, and the greatest common factor of $F_{10} = 55$ and $F_4 = 3$ is $F_2 = 1$. **(b)** The greatest common factor of 12 and 6 is 6, and the greatest common factor of $F_{12} = 144$ and $F_6 = 8$ is $F_6 = 8$. **(c)** The greatest common factor of 14 and 6 is 2, and the greatest common factor of $F_{14} = 377$ and $F_6 = 8$ is $F_2 = 1$. **18. (a)** $F_{3+1} = F_4 = 3$ is divisible by 3.
(b) $F_{7+1} = F_8 = 21$ is divisible by 7. **(c)** $F_{11-1} = F_{10} = 55$ is divisible by 11. **19. (a)** $5 \cdot 34 - 13^2 = 1$
(b) $13^2 - 3 \cdot 55 = 4$ **(c)** $2 \cdot 89 - 13^2 = 9$ **(d)** The difference will be 25, since we are obtaining the squares of the terms of the Fibonacci sequence. $13^2 - 1 \cdot 144 = 25 = 5^2$. **20.** The sum is $55x + 88y = 11(5x + 8y)$. Since $5x + 8y$ is the seventh number in the list, the process is verified. **21.** 199 **22.** In each case, the difference is 5.
23. Each sum is 2 less than a Lucas number. **24.** Each sum is 1 less than a Lucas number.
25. (a) $8 \cdot 18 = 144$; Each expression is equal to 144. **(b)** $8 + 21 = 29$; Each expression is equal to 29.
(c) $8 + 18 = 2 \cdot 13$; Each expression is equal to 26. **26.** The quotients seem to approach the golden ratio. (This is indeed the case.) **27.** 3, 4, 5 **28.** 5, 12, 13 **29.** 16, 30, 34 **30.** Each is a term of the Fibonacci sequence. **31.** The sums are 1, 1, 2, 3, 5, 8, 13. They are terms of the Fibonacci sequence.
33. $(1 + \sqrt{5})/2 \approx 1.618033989$ and $(1 - \sqrt{5})/2 \approx -.618033989$. After the decimal point, the digits are the same.
34. $2/(1 + \sqrt{5}) \approx .618033989$. After the decimal point, the expressions for $(1 + \sqrt{5})/2$ and its reciprocal have the same digits. **35.** 377 **36.** 6765 **37.** 17,711 **38.** 75,025 **39.** F_{2n} **40.** F_{2n+1} **41.** F_{2n-2}
42. F_{3n-3}

Extension Exercises (Page 241)

1.

2	7	6
9	5	1
4	3	8

2.

4	9	2
3	5	7
8	1	6

3.

11	10	4	23	17
18	12	6	5	24
25	19	13	7	1
2	21	20	14	8
9	3	22	16	15

4.

9	2	25	18	11
3	21	19	12	10
22	20	13	6	4
16	14	7	5	23
15	8	1	24	17

5.

15	16	22	3	9
8	14	20	21	2
1	7	13	19	25
24	5	6	12	18
17	23	4	10	11

6. Such a magic square is impossible to construct.

7.

24	9	12
3	15	27
18	21	6

Magic sum is 45.

8.

15	10	11
8	12	16
13	14	9

Magic sum is 36.

9.

$\frac{17}{2}$	12	$\frac{1}{2}$	4	$\frac{15}{2}$
$\frac{23}{2}$	$\frac{5}{2}$	$\frac{7}{2}$	7	8
2	3	$\frac{13}{2}$	10	11
5	6	$\frac{19}{2}$	$\frac{21}{2}$	$\frac{3}{2}$
$\frac{11}{2}$	9	$\frac{25}{2}$	1	$\frac{9}{2}$

Magic sum is $32\frac{1}{2}$.

10.

7	14	−9	−2	5
13	−5	−3	4	6
−6	−4	3	10	12
0	2	9	11	−7
1	8	15	−8	−1

Magic sum is 15.

11. 479   **12.** 191   **13.** 467
14. 311   **15.** 269   **16.** 807
17. (a) 73 (b) 70 (c) 74 (d) 69
18. (a) 12 (b) 11 (c) 5 (d) 6 (e) 3 (f) 10
19. (a) 7 (b) 22 (c) 5 (d) 4 (e) 15 (f) 19
(g) 6 (h) 23   **20.** (a) 29 (b) 24 (c) 19
(d) 20 (e) 11 (f) 33

21.

30	39	48	1	10	19	28
38	47	7	9	18	27	29
46	6	8	17	26	35	37
5	14	16	25	34	36	45
13	15	24	33	42	44	4
21	23	32	41	43	3	12
22	31	40	49	2	11	20

22. 34   **23.** Each sum is equal to 34.   **24.** Each sum is equal to 34.
25. Each sum is equal to 68.   **26.** Each sum is equal to 748.
27. Each sum is equal to 9248.   **28.** Each sum is equal to 748.
29. Each sum is equal to 748.

31.

16	2	3	13
5	11	10	8
9	7	6	12
4	14	15	1

The second and third columns are interchanged.

32.

6	7	2
1	5	9
8	3	4

33.

18	20	10
8	16	24
22	12	14

34.

9	9	−3
−7	5	17
13	1	1

35.

39	48	57	10	19	28	37
47	56	16	18	27	36	38
55	15	17	26	35	44	46
14	23	25	34	43	45	54
22	24	33	42	51	53	13
30	32	41	50	52	12	21
31	40	49	58	11	20	29

Magic sum is 238.

36. (a) 24 (b) 29 (c) 34 (d) 36 (e) 39   **37.** 260
38. Each sum is equal to 130.
39. 52 + 45 + 16 + 17 + 54 + 43 + 10 + 23 = 260
40. 50 + 8 + 7 + 54 + 43 + 26 + 25 + 47 = 260, and so on.

41.

5	13	21	9	17
6	19	2	15	23
12	25	8	16	4
18	1	14	22	10
24	7	20	3	11

42.

9	3	22	16	15
21	20	14	8	2
13	7	1	25	19
5	24	18	12	6
17	11	10	4	23

Chapter 5 Test (Page 245)

1. false **2.** true **3.** true **4.** true **5.** true **6. (a)** yes **(b)** yes **(c)** no **(d)** yes **(e)** yes **(f)** no **(g)** yes **(h)** yes **(i)** no **7. (a)** composite **(b)** neither **(c)** prime **8.** $2^5 \cdot 3^2 \cdot 5$ **10. (a)** deficient **(b)** perfect **(c)** abundant **11.** C **12.** 41 and 43 **13.** 90 **14.** 360 **15.** Monday **16.** 46,368 **17.** $89 - (8 + 13 + 21 + 34) = 13$; Each expression is equal to 13. **18.** B **19.** 3 **20.** $\{3, 6, 9, 12, \ldots\}$ **21. (a)** 1, 5, 6, 11, 17, 28, 45, 73 **(b)** The process will yield 19 for any term chosen. **22.** A

CHAPTER 6 The Real Numbers and Their Representations

6.1 Exercises (Page 254)

1. 4 **2.** 3.85 (There are others.) **3.** 0 **4.** 5 (There are others.) **5.** $\sqrt{12}$ (There are others.) **6.** 0 **7.** true **8.** false **9.** true **10.** true **11. (a)** 3, 7 **(b)** 0, 3, 7 **(c)** $-9, 0, 3, 7$ **(d)** $-9, -1\frac{1}{4}, -\frac{3}{5}, 0, 3, 5.9, 7$ **(e)** $-\sqrt{7}, \sqrt{5}$ **(f)** All are real numbers. **12. (a)** 3 **(b)** 0, 3 **(c)** $-5, -1, 0, 3$ **(d)** $-5.3, -5, -1, -\frac{1}{9}, 0, 1.2, 1.8, 3$ **(e)** $-\sqrt{3}, \sqrt{11}$ **(f)** All are real numbers. **14.** The decimal representation of a rational number will either terminate or repeat. **15.** 568,488 **16.** 1450 **17.** $-30°$ **18.** 5436 **19.** $-34,841$ **20.** 159,376 **21.** -8 **22.** 20; 25°; $-3°$ **23. (a)** Pacific Ocean, Indian Ocean, Caribbean Sea, South China Sea, Gulf of California **(b)** Point Success, Ranier, Matlalcueyetl, Steele, McKinley **(c)** true **(d)** false **24. (a)** largest: U.S.; smallest: India **(b)** U.S., Germany, Japan **(c)** Japan: about \$300; Italy: about \$150 **(d)** about \$150 **25.** [number line from −6 to 4] **26.** [number line from −6 to 4] **27.** $-3\frac{4}{5}, -1\frac{5}{8}, \frac{1}{4}, 2\frac{1}{3}$ **28.** $-3\frac{2}{5}, -2\frac{1}{3}, \frac{4}{9}, 5\frac{1}{4}$ **29. (a)** A **(b)** A **(c)** B **(d)** B **30.** 2, 2, 2, -2 **31. (a)** 2 **(b)** 2 **32. (a)** 8 **(b)** 8 **33. (a)** -6 **(b)** 6 **34. (a)** -11 **(b)** 11 **35. (a)** -3 **(b)** 3 **36. (a)** -5 **(b)** 5 **37. (a)** 0 **(b)** 0 **38. (a)** 0 **(b)** 0 **39.** $a - b$ **40.** 0 **41.** -12 **42.** -14 **43.** -8 **44.** -16 **45.** 3 **46.** $|-2|$ or 2 **47.** $|-3|$ or 3 **48.** $|-8|$ or 8 **49.** $-|-6|$ or -6 **50.** $-|-3|$ or -3 **51.** $|5 - 3|$ or 2 **52.** $|7 - 2|$ or 5 **53.** true **54.** false **55.** true **56.** true **57.** true **58.** true **59.** false **60.** false **61.** true **62.** false **63.** false **64.** false **65.** video/audio equipment from 1995 to 1996 **66.** apparel from 1995 to 1996 **67.** 1996 to 1997 **68.** 1996 to 1997 **69.** computer/data processing services

Answers will vary in Exercises 71–76.

71. $\frac{1}{2}, \frac{5}{8}, 1\frac{3}{4}$ **72.** $-1, -\frac{3}{4}, -5$ **73.** $-3\frac{1}{2}, -\frac{2}{3}, \frac{3}{7}$ **74.** $\frac{1}{2}, -\frac{2}{3}, \frac{2}{7}$ **75.** $\sqrt{5}, \pi, -\sqrt{3}$ **76.** $\frac{2}{3}, \frac{5}{6}, \frac{5}{2}$

6.2 Exercises (Page 265)

1. negative **2.** zero **3.** $-3; 5$ **4.** negative **7.** -20 **8.** -7 **9.** -4 **10.** 11 **11.** -11 **12.** 5 **13.** 9 **14.** 12 **15.** 20 **16.** 29 **17.** 4 **18.** 14 **19.** 24 **20.** 15 **21.** -1296 **22.** -1360 **23.** 6 **24.** 2 **25.** -6 **26.** -4 **27.** 0 **28.** 0 **29.** -6 **30.** -24 **31.** -4 **32.** 12 **33.** 27 **34.** 43 **35.** 39 **36.** -2 **37.** -2 **38.** 4 **39.** not a real number

40. not a real number **41.** 13 **42.** −7 **43.** A, B, C **45.** commutative property of addition **46.** commutative property of multiplication **47.** associative property of addition **48.** commutative property of multiplication **49.** inverse property of addition **50.** identity property of addition **51.** identity property of multiplication **52.** inverse property of multiplication **53.** identity property of addition **54.** associative property of multiplication **55.** distributive property **56.** distributive property **57.** closure property of addition **58.** closure property of multiplication **59. (a)** −2, 2 **(b)** commutative **(c)** yes; choose $a = b$. For example, $a = b = 2: 2 − 2 = 2 − 2$. **60. (a)** $\frac{1}{2}, 2$ **(b)** commutative **(c)** yes; choose $|a| = |b| \neq 0$. For example, $a = b = 2: 2 \div 2 = 2 \div 2$. **61. (a)** messing up your room **(b)** spending money **(c)** decreasing the volume on your CD player **62. (a)** no **(b)** no **(c)** yes **63.** identity **65.** no **68.** Both procedures yield 234 as an answer. **69.** −81 **70.** −81 **71.** 81 **72.** 81 **73.** −81 **74.** 81 **75.** −81 **76.** −81 **77. (a)** $25.1 billion **(b)** −$11.3 billion **(c)** $5.0 billion **(d)** −$2.2 billion **78. (a)** 50,395 feet **(b)** 37,486 feet **(c)** 1345 feet **(d)** 8274 feet **79. (a)** 2000: $129 billion; 2010: $206 billion; 2020: $74 billion; 2030: −$501 billion **80. (a)** $2.32 million (in the red) **(b)** $22.88 million (in the red) **(c)** $25.59 million (in the black) **81.** −$35.00 **82.** −184 meters **83.** 45° F **84.** 112° F **85.** −41° F **86.** 27 feet **87.** 14,776 feet **88.** 543 feet **89.** 365 pounds **90.** 14 feet **91.** 37 yards **92.** 34,500 feet **93.** 17 **94.** 469 B.C. **95.** $323.83 **96.** 1 minute, 28.89 seconds

6.3 Exercises (Page 279)

1. A, C, D **2.** B, C, D **3.** C **4.** A, B, C **5.** $\frac{1}{3}$ **6.** $\frac{3}{4}$ **7.** $-\frac{3}{7}$ **8.** $-\frac{1}{6}$

Answers will vary in Exercises 9–12. We give three examples in each case.

9. $\frac{6}{16}, \frac{9}{24}, \frac{12}{32}$ **10.** $\frac{18}{20}, \frac{27}{30}, \frac{36}{40}$ **11.** $-\frac{10}{14}, -\frac{15}{21}, -\frac{20}{28}$ **12.** $-\frac{14}{24}, -\frac{21}{36}, -\frac{28}{48}$ **13. (a)** $\frac{1}{3}$ **(b)** $\frac{1}{4}$ **(c)** $\frac{2}{5}$ **(d)** $\frac{1}{3}$ **14. (a)** $\frac{1}{2}$ **(b)** $\frac{1}{4}$ **(c)** $\frac{3}{4}$ **(d)** $\frac{1}{8}$ **15.** the dots in the intersection of the triangle and the rectangle as a part of the dots in the entire figure **16.** They had the same average, .400. **17. (a)** O'Brien **(b)** Bravener **(c)** Bravener **(d)** Kelly **(e)** Taylor and Britz; $\frac{1}{2}$ **19.** $\frac{1}{2}$ **20.** $\frac{8}{9}$ **21.** $\frac{43}{48}$ **22.** $\frac{41}{90}$ **23.** $-\frac{5}{24}$ **24.** $\frac{7}{30}$ **25.** $\frac{23}{56}$ **26.** $\frac{41}{60}$ **27.** $\frac{27}{20}$ **28.** $\frac{3}{28}$ **29.** $\frac{5}{12}$ **30.** $-\frac{1}{6}$ **31.** $\frac{1}{9}$ **32.** $\frac{1}{4}$ **33.** $\frac{3}{2}$ **34.** $-\frac{3}{10}$ **35.** $\frac{3}{2}$ **36.** $-\frac{3}{20}$ **37. (a)** $4\frac{1}{2}$ cups **(b)** $\frac{7}{8}$ cup **38. (a)** $\frac{1}{20}$ **(b)** $\frac{41}{50}$ **(c)** more than $1\frac{1}{25}$ million **39.** $\frac{13}{2}$ **40.** $\frac{31}{8}$ **41.** $\frac{29}{10}$ **42.** $3\frac{3}{5}$ **43.** $6\frac{3}{4}$ **44.** $6\frac{1}{3}$ **45.** $6\frac{1}{8}$ **46.** $3\frac{11}{15}$ **47.** $-17\frac{7}{8}$ **48.** $-2\frac{1}{2}$ **49.** $\frac{9}{16}$ inch **50.** $\frac{5}{16}$ inch **51.** 3 **52.** $\frac{5}{3}$ **53.** $\frac{3}{7}$ **54.** $\frac{11}{3}$ **55.** $-\frac{103}{89}$ **56.** $\frac{38}{9}$ **57.** $\frac{25}{9}$ **58.** $\frac{48}{11}$ **59.** $\frac{5}{8}$ **60.** $\frac{3}{8}$ **61.** $\frac{19}{30}$ **62.** $\frac{29}{48}$ **63.** $-\frac{3}{4}$ **64.** $-\frac{11}{4}$ **65.** $12.60 **66.** $58,355 **67.** $\frac{14}{19}$ **68.** $\frac{23}{30}$ **69.** $\frac{13}{29}$ **70.** $\frac{19}{25}$ **71.** $\frac{5}{2}$ **72.** $\frac{7}{2}$ **73.** It gives the rational number halfway between the two integers (their average). **75.** .75 **76.** .875 **77.** .1875 **78.** .28125 **79.** $.\overline{27}$ **80.** $.\overline{81}$ **81.** $.\overline{285714}$ **82.** $.\overline{73}$ **83.** $\frac{2}{5}$ **84.** $\frac{9}{10}$ **85.** $\frac{17}{20}$ **86.** $\frac{21}{200}$ **87.** $\frac{467}{500}$ **88.** $\frac{499}{625}$ **89.** repeating **90.** repeating **91.** terminating **92.** terminating **93.** terminating **94.** terminating **95. (a)** $.\overline{3}$ or .333… **(b)** $.\overline{6}$ or .666… **(c)** $.\overline{9}$ or .999… **(d)** $1 = .\overline{9}$ **96.** $3\left(\frac{1}{3}\right) = 3(.333) = .999 = 1$

6.4 Exercises (Page 290)

1. rational **2.** rational **3.** irrational **4.** irrational **5.** rational **6.** rational **7.** rational
8. rational **9.** irrational **10.** rational **11.** irrational **12.** irrational **13.** rational **14.** rational
15. (a) $.\overline{8}$ (b) irrational, rational **16.** (a) .262662666266662... (b) irrational, irrational

The number of digits shown will vary among calculator models in Exercises 17–36.
17. 6.244997998 **18.** 6.633249581 **19.** 3.885871846 **20.** 5.796550698 **21.** 29.73213749
22. 25.35744467 **23.** 1.060660172 **24.** 1.095445115 **25.** $5\sqrt{2}$; 7.071067812 **26.** $4\sqrt{2}$; 5.656854249 **27.** $5\sqrt{3}$; 8.660254038 **28.** $5\sqrt{6}$; 12.247448714 **29.** $12\sqrt{2}$; 16.97056275
30. $10\sqrt{2}$; 14.14213562 **31.** $\dfrac{5\sqrt{6}}{6}$; 2.041241452 **32.** $\dfrac{3\sqrt{2}}{2}$; 2.121320344 **33.** $\dfrac{\sqrt{7}}{2}$; 1.322875656
34. $\dfrac{2\sqrt{2}}{3}$; .9428090416 **35.** $\dfrac{\sqrt{21}}{3}$; 1.527525232 **36.** $\dfrac{\sqrt{70}}{5}$; 1.6733220053 **37.** $3\sqrt{17}$ **38.** $4\sqrt{19}$
39. $4\sqrt{7}$ **40.** $6\sqrt{3}$ **41.** $10\sqrt{2}$ **42.** $7\sqrt{3}$ **43.** $3\sqrt{3}$ **44.** $-4\sqrt{3}$
45.

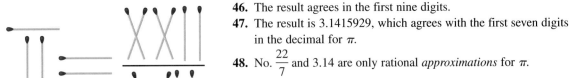

46. The result agrees in the first nine digits.
47. The result is 3.1415929, which agrees with the first seven digits in the decimal for π.
48. No. $\dfrac{22}{7}$ and 3.14 are only rational *approximations* for π.
49. 2.5 seconds **50.** 4 inches **51.** 8.2 miles
52. 1.7 amps **53.** 392,000 square miles
54. 24,900 square feet **55.** 5.4 feet **56.** .08
57. (a) 70.5 miles per hour (b) 59.8 miles per hour (c) 53.9 miles per hour
58. (a) 149.4 miles (b) 163.7 miles (c) 189.0 miles **59.** 4 **60.** 5 **61.** 7 **62.** 9 **63.** 6
64. 8 **65.** 1 **66.** 2 **67.** 4 **68.** 5 **69.** 8 **70.** 7

The number of digits shown will vary among calculator models in Exercises 71–76.
71. 3.50339806 **72.** 4.431047622 **73.** 5.828476683 **74.** 6.775106617 **75.** 10.06565066
76. 5.420827475 **77.** 5.8 feet **78.** 581 **79.** (a) 1.414213562 (b) 2.645751311 (c) 3.633180425
(d) 5 **80.** (a) $\sqrt[3]{a}$ (b) 2.5198421 (c) 2.5198421 **81.** The values seem to approach the value of e.
82. (a) 7.3890561 (b) 20.0855369 (c) 1.6487213

6.5 Exercises (Page 302)

1. true **2.** true **3.** false **4.** false **5.** true **6.** true **7.** true **8.** false **9.** false
10. false **11.** 11.315 **12.** 16.5717 **13.** -4.215 **14.** -1.587 **15.** .8224 **16.** 376.3144
17. 47.5 **18.** -2.1 **19.** 31.6 **20.** -2.45 **21.** Philadelphia; The population declined 10.6%.
22. Chicago; The population increased .6%. **23.** three (and you would have .01¢ left over) **24.** 33 cents
25. (a) $1572 (b) $2434 (c) $3296 **26.** (a) $11,042.49 (b) $8997.26 (c) $3901.35
27. (a) .031 (b) .035 **28.** 139 mph **29.** 297 **30.** 467 **31.** (a) 78.4 (b) 78.41 **32.** (a) 3689.5
(b) 3689.54 **33.** (a) .1 (b) .08 **34.** (a) .1 (b) .07 **35.** (a) 12.7 (b) 12.69 **36.** (a) 44.0
(b) 44.00 **37.** 42% **38.** 87% **39.** 36.5% **40.** 79.2% **41.** .8% **42.** .93% **43.** 210%
44. 890% **45.** 20% **46.** 40% **47.** 1% **48.** 2% **49.** $37\frac{1}{2}\%$ **50.** $83\frac{1}{3}\%$ **51.** 150%
52. 175% **54.** (a) E (b) D (c) B (d) F (e) C (f) A **55.** (a) 5 (b) 24 (c) 8 (d) .5 or $\dfrac{1}{2}$
(e) 600 **56.** (a) 15% (b) 40% (c) 20% (d) 25% **57.** No, the price is $57.60. **58.** (a) $33\frac{1}{3}\%$
(b) 25% (c) 40% (d) $33\frac{1}{3}\%$ **59.** .577; .503; .446 **60.** .626; .484; .474 **61.** 124.8 **62.** 4.56

63. 2.94 **64.** 9.234 **65.** 150% **66.** $41\frac{2}{3}\%$ or $41.\overline{6}\%$ **67.** 600 **68.** 30,000 **69.** 1.4%
70. 27% **71.** A **72.** D **73.** C **74.** A **75.** about 67% **76.** about 40% **77. (a)** .5%
(b) .6% **78.** 1.63% **79.** $1385.80 **80.** 300% **81.** about 22.7% **82.** $24,957.45 **83.** $4.50
84. $5.70 **85.** $.75 **86.** $1.20 **87.** $12.00 **88.** $8.00 **89.** $36.00 **90.** $40.00

Extension Exercises (Page 309)

1. $12i$ **2.** $14i$ **3.** $-15i$ **4.** $-20i$ **5.** $i\sqrt{3}$ **6.** $i\sqrt{19}$ **7.** $5i\sqrt{3}$ **8.** $5i\sqrt{5}$ **9.** -5
10. -3 **11.** -18 **12.** -18 **13.** -40 **14.** -99 **15.** $\sqrt{2}$ **16.** 5 **17.** $3i$ **18.** $3i$
19. 6 **20.** $6i\sqrt{2}$ **23.** 1 **24.** 1 **25.** -1 **26.** -1 **27.** $-i$ **28.** $-i$ **29.** i **30.** i

Chapter 6 Test (Page 311)

1. (a) 12 **(b)** 0, 12 **(c)** $-4, 0, 12$ **(d)** $-4, -\frac{3}{2}, -.5, 0, 4.1, 12$ **(e)** $-\sqrt{5}, \sqrt{3}$ **(f)** $-4, -\sqrt{5}, -\frac{3}{2}, -.5,$
$0, \sqrt{3}, 4.1, 12$ **2. (a)** C **(b)** B **(c)** D **(d)** A **3. (a)** false **(b)** true
(c) true **(d)** false **4.** 4 **5.** 3 **6.** 10 **7. (a)** $-11,478$ **(b)** $-17,776$ **(c)** C **(d)** C
8. 5296 feet **9.** 8929; $-10,936; -14,575; -430; 803$ **10. (a)** E **(b)** A **(c)** B **(d)** D **(e)** F **(f)** C
11. (a) Whitney **(b)** Moura and Dawkins **(c)** Whitney **(d)** Pritchard and Miller; $\frac{2}{5}$ **12.** $\frac{11}{16}$ **13.** $\frac{57}{160}$
14. $-\frac{2}{5}$ **15.** $\frac{3}{2}$ **16. (a)** .45 **(b)** $.41\overline{6}$ **17. (a)** $\frac{18}{25}$ **(b)** $\frac{58}{99}$ **18. (a)** irrational **(b)** rational
(c) rational **(d)** rational **(e)** irrational **19. (a)** 12.247448714 **(b)** $5\sqrt{6}$ **20. (a)** 4.913538149
(b) $\frac{13\sqrt{7}}{7}$ **21. (a)** -45.254834 **(b)** $-32\sqrt{2}$ **23. (a)** 13.81 **(b)** $-.315$ **(c)** 38.7 **(d)** -24.3
24. (a) 9.04 **(b)** 9.045 **25. (a)** 16.65 **(b)** 101.5 **26. (a)** $26\frac{2}{3}\%$ **(b)** $66\frac{2}{3}\%$ **27.** D
28. 345; 210; 38%; 15% **29. (a)** 200 **(b)** 90; Since 26% is twice 13%, we should expect twice the number of people to have Charlie Brown as their favorite as compared to Linus. **30.** about 13.5%

CHAPTER 7 The Basic Concepts of Algebra

7.1 Exercises (Page 322)

1. A and C **2.** B because the variable is squared, and D because there is a variable in the denominator of the second term **3.** Both sides are evaluated as 30, so 6 is a solution. **4.** It is not, because the left side is evaluated as -2 and the right side as -9. **5.** solution set **6.** Any number is a solution. For example, if the last name is Lincoln, $x = 7$. Both sides are evaluated as -48. **7.** B **9.** $\{-1\}$ **10.** $\{5\}$ **11.** $\{3\}$ **12.** $\{-3\}$
13. $\{-7\}$ **14.** $\{3\}$ **15.** $\{0\}$ **16.** $\{0\}$ **17.** $\left\{-\frac{5}{3}\right\}$ **18.** $\{-15\}$ **19.** $\left\{-\frac{1}{2}\right\}$ **20.** $\left\{\frac{4}{3}\right\}$
21. $\{2\}$ **22.** $\{-2\}$ **23.** $\{-2\}$ **24.** $\{4\}$ **25.** $\{7\}$ **26.** $\left\{-\frac{1}{2}\right\}$ **27.** $\{2\}$ **28.** $\{-6\}$ **29.** $\left\{\frac{3}{2}\right\}$
30. $\left\{-\frac{2}{5}\right\}$ **31.** $\{-5\}$ **32.** $\{0\}$ **33.** $\{3\}$ **34.** 2 (that is, 10^2, or 100) **36.** $\{4\}$ **37.** $\{-6\}$
38. $\{0\}$ **39.** $\{0\}$ **40.** $\{0\}$ **41.** $\{-5\}$ **42.** $\{2000\}$ **43.** $\{100\}$ **44.** $\{25\}$ **45.** $\{7000\}$
46. $\{40\}$ **47.** $\{60\}$ **48.** identity; contradiction **49.** A **51.** contradiction; $\emptyset$ **52.** contradiction; $\emptyset$
53. conditional; $\{-8\}$ **54.** contradiction; $\emptyset$ **55.** conditional; $\{0\}$ **56.** conditional; $\{0\}$
57. identity; {all real numbers} **58.** identity; {all real numbers} **59.** D **60.** yes; the distributive property

61. $t = \dfrac{d}{r}$ **62.** $r = \dfrac{I}{pt}$ **63.** $b = \dfrac{A}{h}$ **64.** $L = \dfrac{P - 2W}{2}$ or $L = \dfrac{P}{2} - W$ **65.** $a = P - b - c$
66. $W = \dfrac{V}{LH}$ **67.** $h = \dfrac{2A}{b}$ **68.** $r = \dfrac{C}{2\pi}$ **69.** $h = \dfrac{S - 2\pi r^2}{2\pi r}$ or $h = \dfrac{S}{2\pi r} - r$ **70.** $B = \dfrac{2A}{h} - b$ or
$B = \dfrac{2A - bh}{h}$ **71.** $F = \dfrac{9}{5}C + 32$ **72.** $C = \dfrac{5}{9}(F - 32)$ **73.** $H = \dfrac{A - 2LW}{2W + 2L}$ **74.** $h = \dfrac{3V}{B}$
75. (a) 9.65 million (b) 1992–1993 season **76.** (a) .8 million (b) 1992–1993 **77.** 1995–1996 and 1996–1997 **78.** 9.04 million; yes

7.2 Exercises (Page 334)
1. expression **2.** expression **3.** equation **4.** equation **5.** expression **6.** equation **7.** yes
9. $x - 18$ **10.** $x + 12$ **11.** $(x - 9)(x + 6)$ **12.** $\dfrac{x}{6}$ **13.** $\dfrac{12}{x}$ $(x \neq 0)$ **14.** $\dfrac{6}{7}x$ **16.** D
17. $2x + \dfrac{x}{6} = x - 8$ **18.** $x - (-4x) = x + 9$ **19.** $12 - \dfrac{2}{3}x = 10$ **20.** $6 + .75x = x + 3$
21. *Step 2:* the number of shoppers at small chain/independent bookstores *Step 3:* x; $(x - 70)$ *Step 4:* 256
Step 5: 256; 186 *Step 6:* large chain shoppers: 70; small chain/independent shoppers: 442
22. U2: $109.7 million; *NSYNC: $86.8 million **23.** Camry: 421.5 thousand; Accord: 406.5 thousand
24. wins: 61; losses: 21 **25.** Ruth: 2873 hits; Hornsby: 2930 hits **26.** Bush: 271 votes; Gore: 266 votes
27. rentals: $6.63 billion; sales: $3.18 billion **28.** 1995; rentals: $7.49 billion; sales: $7.34 billion
29. California: 59 screens; New York: 48 screens **30.** Thursday: 61 million viewers; Saturday: 41 million viewers
31. 35 milliliters **32.** 45 liters **33.** $350 **34.** $1500 **35.** $14.15 **36.** $17.50 **37.** 4 liters
38. $13\dfrac{1}{3}$ liters **39.** $18\dfrac{2}{11}$ liters **40.** 1 gallon **41.** 5 liters **42.** 20 liters **43.** $4000 at 3%;
$8000 at 4% **44.** $20,000 at 2%; $40,000 at 3% **45.** $10,000 at 4.5%; $19,000 at 3% **46.** $8000 at 3.5%;
$29,000 at 4% **47.** $58,000 **48.** $5000 **49.** 17 pennies, 17 dimes, 10 quarters **50.** 2 nickels,
2 quarters, 4 half-dollars **51.** 305 students, 105 nonstudents **52.** 450 floor tickets, 100 balcony tickets
53. 54 seats on Row 1, 51 seats on Row 2 **54.** 11 two-cent pieces, 33 three-cent pieces
55. 54 $10 coins, 26 $20 coins **56.** $.005 **57.** 3.808 hours **58.** 3.090 hours **59.** 1.715 hours
60. 2.915 hours **61.** 7.91 meters per second **62.** 7.54 meters per second **63.** 8.42 meters per second
64. 9.12 meters per second **65.** 530 miles **66.** 328 miles **68.** A **69.** $1\dfrac{3}{4}$ hours **70.** $2\dfrac{1}{2}$ hours
71. 10:00 A.M. **72.** 9:50 A.M. **73.** 18 miles **74.** 3.75 miles **75.** 8 hours **76.** $\dfrac{1}{2}$ hour

7.3 Exercises (Page 348)
1. $\dfrac{5}{8}$ **2.** $\dfrac{1}{3}$ **3.** $\dfrac{1}{4}$ **4.** $\dfrac{6}{5}$ **5.** $\dfrac{2}{1}$ **6.** $\dfrac{5}{6}$ **7.** $\dfrac{3}{1}$ **8.** $\dfrac{5}{8}$ **9.** D **10.** Answers will vary.
Examples are 12 to 9, 40 to 30, and 20 to 15. **13.** true **14.** true **15.** false **16.** false **17.** true
18. true **19.** {35} **20.** {7} **21.** {27} **22.** {16} **23.** {−1} **24.** $\left\{-\dfrac{31}{5}\right\}$ **25.** 6.875 ounces
26. 4 gallons **27.** $670.48 **28.** 74.13 francs **29.** $9.30 **30.** $9.90 **31.** 4 feet **32.** $25\dfrac{2}{3}$ inches
33. 12,500 fish **34.** 23,300 fish **35.** (a) $\dfrac{26}{100} = \dfrac{x}{350}$; $91 million (b) $112 million; $11.2 million
(c) $119 million **36.** (a) $\dfrac{22}{100} = \dfrac{x}{350}$; $77 million (b) $234.5 million **37.** 30 count **38.** 2-ounce size
39. 31-ounce size **40.** 32-ounce size **41.** 32-ounce size **42.** 16-ounce size **43.** 4 **44.** 8 **45.** 1

46. 2 **47. (a)** 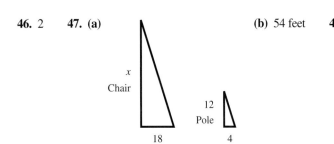 **(b)** 54 feet **48. (a)** 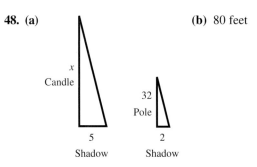 **(b)** 80 feet

49. $242 **50.** $255 **51.** $270 **52.** $281 **53.** $287 **54.** $296 **55.** 9 **56.** 15
57. 125 **58.** 147 **59.** $\frac{4}{9}$ **60.** $\frac{64}{25}$ **61.** $40.32 **62.** 15 square inches **63.** 20 miles per hour
64. 15 feet **65.** about 302 pounds **66.** about 2593 miles **67.** 100 pounds per square inch
68. $7\frac{1}{2}$ pounds per square inch **69.** 20 pounds per square foot **70.** 20 pounds **71.** 144 feet
72. $14\frac{22}{27}$ foot-candles **73.** 23 **74.** 448.1 pounds **75.** 11.8 pounds **76.** 6.2 pounds

7.4 Exercises (Page 357)

1. D **2.** C **3.** B **4.** A **5.** F **6.** E **9.** $[5, \infty)$ **10.** $[10, \infty)$
11. $(7, \infty)$ **12.** $(-\infty, 14)$ **13.** $(-4, \infty)$
14. $(-\infty, -5)$ **15.** $(-\infty, -40)$ **16.** $[3, \infty)$
17. $(-\infty, 4]$ **18.** $[.5, \infty)$ **19.** $\left(-\infty, -\frac{15}{2}\right)$
20. $\left(-\frac{28}{3}, \infty\right)$ **21.** $\left[\frac{1}{2}, \infty\right)$ **22.** $[2, \infty)$
23. $(3, \infty)$ **24.** $[6, \infty)$ **25.** $(-\infty, 4)$
26. $[-3, \infty)$ **27.** $\left(-\infty, \frac{23}{6}\right]$ **28.** $\left(-\infty, \frac{71}{150}\right)$
29. $\left(-\infty, \frac{76}{11}\right)$ **30.** $\left(-\infty, \frac{49}{2}\right]$ **31.** $(-\infty, \infty)$
32. $(-\infty, \infty)$ **33.** $\emptyset$ **34.** $\emptyset$ **36. (a)** A **(b)** D **(c)** C **(d)** B
37. $(1, 11)$ **38.** $(-2, 7)$ **39.** $[-14, 10]$
40. $[-7, 7]$ **41.** $[-5, 6]$ **42.** $(-7, -6)$
43. $\left[-\frac{14}{3}, 2\right]$ **44.** $(-6, -4)$ **45.** $\left[-\frac{1}{2}, \frac{35}{2}\right]$
46. $\left[-\frac{13}{3}, \frac{11}{3}\right]$ **47.** $\left(-\frac{1}{3}, \frac{1}{9}\right]$ **48.** $\left(-\frac{5}{2}, -\frac{1}{2}\right]$

49. April, May, June, July   **50.** April, May, June   **51.** January, February, March, August, September, October, November, December   **52.** 690   **53.** from about 2:30 P.M. to 6:00 P.M.   **54.** from about 8:00 A.M. to 10:15 P.M. and after about 9:00 P.M.   **55.** about 84°F–91°F   **56.** about 65°F–67°F   **57.** 2 miles   **58.** 2.2 miles   **59.** at least 80   **60.** at least 82   **61.** 50 miles   **62.** 628.6 miles   **63.** (a) 140 to 184 pounds (b) Answers will vary.   **64.** (a) 130 to 157 beats per minute (b) Answers will vary.   **65.** 26 tapes   **66.** 921 deliveries

7.5 Exercises (Page 369)

1. A **2.** C **3.** A **4.** C **5.** D **6.** B **7.** 625 **8.** 1000 **9.** -32 **10.** 625 **11.** -8 **12.** -9 **13.** -81 **14.** -25 **15.** $\frac{1}{49}$ **16.** $\frac{1}{4}$ **17.** $-\frac{1}{49}$ **18.** $-\frac{1}{4}$ **19.** -128 **20.** $\frac{9}{8}$ **21.** $\frac{16}{5}$ **22.** 8 **23.** 125 **24.** $\frac{9}{4}$ **25.** $\frac{25}{16}$ **26.** $\frac{5}{6}$ **27.** $\frac{9}{20}$ **28.** 1 **29.** 1 **30.** 1 **31.** 1 **32.** -1 **33.** 0 **34.** -2 **35.** reciprocal; additive inverse **37.** D **38.** B **39.** x^{16} **40.** x^8 **41.** 5 **42.** $\frac{1}{3}$ **43.** $\frac{1}{27}$ **44.** $\frac{1}{2}$ **45.** $\frac{1}{81}$ **46.** 144 **47.** $\frac{1}{t^7}$ **48.** $\frac{1}{p}$ **49.** $9x^2$ **50.** $\frac{4}{x^4}$ **51.** $\frac{1}{a^5}$ **52.** $\frac{1}{k^4}$ **53.** x^{11} **54.** $\frac{1}{p^8}$ **55.** r^6 **56.** $\frac{1}{z^8}$ **57.** $-\frac{56}{k^2}$ **58.** $\frac{30}{a^3}$ **59.** $\frac{1}{z^4}$ **60.** $\frac{1}{p^7}$ **61.** $-\frac{3}{r^7}$ **62.** $\frac{2}{y^6}$ **63.** $\frac{27}{a^{18}}$ **64.** $\frac{27}{m^{16}}$ **65.** $\frac{x^5}{y^2}$ **66.** $\frac{1}{a^6 b^{10}}$ **67.** D **68.** B **69.** 2.3×10^2 **70.** 4.65×10^4 **71.** 2×10^{-2} **72.** 5.1×10^{-3} **73.** 6500 **74.** 231,700 **75.** .0152 **76.** .000163 **77.** 6×10^5 **78.** 2.7×10^2 **79.** 2×10^5 **80.** 3×10^3 **81.** 2×10^5 **82.** 6×10^{-7} **83.** $\$1.37574 \times 10^{10}$ **84.** $\$2.07754 \times 10^8$ **85.** 1×10^9; 1×10^{12}; 2.128×10^{12}; 1.44419×10^5 **86.** 1.5×10^3; 6.8×10^5; 1.1×10^6; 2.2×10^6 **87.** 1×10^{10} **88.** 4.7×10^6 **89.** 2,000,000,000 **90.** .0000000000001 **91.** 26,000 or 2.6×10^4 **92.** \$40,045 **93.** approximately 9.474×10^{-7} parsec **94.** 63,360 inches in a mile **95.** 300 seconds **96.** approximately $3.2 \times 10^4 = 32{,}000$ hours (about 3.7 years) **97.** approximately 5.87×10^{12} miles **98.** 500 seconds, which is approximately 8.3 minutes **99.** 20,000 hours **100.** 4.6×10^7 meters or 46,000,000 meters **101.** 1.23×10^5 **102.** 3×10^3 **103.** 4.24×10^5 **104.** 1.36×10^5 **105.** 4.4×10^5 **106.** 3.74×10^5

7.6 Exercises (Page 379)

1. $x^2 - x + 3$ **2.** $m^3 - 4m^2 + 10$ **3.** $9y^2 - 4y + 4$ **4.** $5p^2 - 3p - 4$ **5.** $6m^4 - 2m^3 - 7m^2 - 4m$ **6.** $-6x^3 - 3x^2 - 4x + 4$ **7.** $-2x^2 - 13x + 11$ **8.** $10x^2y - 12xy^2$ **9.** $x^2 - 5x - 24$ **10.** $y^2 - 12y + 27$ **11.** $28r^2 + r - 2$ **12.** $15m^2 + 2m - 24$ **13.** $12x^5 + 8x^4 - 20x^3 + 4x^2$ **14.** $2b^5 - 8b^4 + 6b^3$ **15.** $4m^2 - 9$ **16.** $64s^2 - 9t^2$ **17.** $16m^2 + 16mn + 4n^2$ **18.** $a^2 - 12ab + 36b^2$ **19.** $25r^2 + 30rt^2 + 9t^4$ **20.** $4z^8 - 12z^4y + 9y^2$ **21.** $-2z^3 + 7z^2 - 11z + 4$ **22.** $12k^4 + 21k^3 - 5k^2 + 3k + 2$ **23.** $m^2 + mn - 2n^2 - 2km + 5kn - 3k^2$ **24.** $2r^2 - 7rs + 3s^2 + 3rt - 4st + t^2$ **25.** $a^2 - 2ab + b^2 + 4ac - 4bc + 4c^2$ **26.** $k^2 - 2ky + y^2 + 6km - 6ym + 9m^2$ **27.** A **28.** One example is $12x^5 - 6x^3 + 2x^2 - 1$. **31.** $2m^2(4m^2 + 3m - 6)$ **32.** $2p^3(p^2 - 5p + 8)$ **33.** $4k^2m^3(1 + 2k^2 - 3m)$ **34.** $7r^3s(4rs + 1 - 5rs^2)$ **35.** $2(a + b)(1 + 2m)$ **36.** $(y - 2)(4y - 5)$ **37.** $(m - 1)(2m^2 - 7m + 7)$ **38.** $(a + 3)(5a^2 + 31a + 46)$ **39.** $(2s + 3)(3t - 5)$ **40.** $(5a - 3)(2b + 7)$ **41.** $(t^3 + s^2)(r - p)$ **42.** $(m^4 + 3)(2 - a)$ **43.** $(8a + 5b)(2a - 3b)$ **44.** $(3 - m^2)(5 - r^2)$ **45.** $(5z - 2x)(4z - 9x)$ **46.** D **47.** $(x - 5)(x + 3)$ **48.** $(r + 6)(r + 2)$ **49.** $(y + 7)(y - 5)$ **50.** $(x - 6)(x - 1)$ **51.** $6(a - 10)(a + 2)$ **52.** $8(h - 8)(h + 5)$ **53.** $3m(m + 1)(m + 3)$ **54.** $9y^2(y - 5)(y - 1)$ **55.** $(3k - 2p)(2k + 3p)$ **56.** $(7m - 5r)(2m + 3r)$ **57.** $(5a + 3b)(a - 2b)$ **58.** $(4s + 5t)(3s - t)$ **59.** $x^2(3 - x)^2$ **60.** $(5a + m)(6a - m)$ **61.** $2a^2(4a - b)(3a + 2b)$ **62.** $3x^3(3x - 5z)(2x + 5z)$ **65.** $(3m - 2)^2$ **66.** $(4p - 5)^2$ **67.** $2(4a - 3b)^2$ **68.** $5(2p - 5q)^2$ **69.** $(2xy + 7)^2$ **70.** $(3mn - 2)^2$ **71.** $(x + 6)(x - 6)$ **72.** $(t + 8)(t - 8)$ **73.** $(y + w)(y - w)$ **74.** $(5 + w)(5 - w)$

75. $(3a + 4)(3a - 4)$ **76.** $(4q + 5)(4q - 5)$ **77.** $(5s^2 + 3t)(5s^2 - 3t)$ **78.** $9(2z + 3y^2)(2z - 3y^2)$
79. $(p^2 + 25)(p + 5)(p - 5)$ **80.** $(m^2 + 9)(m + 3)(m - 3)$ **81.** $(2 - a)(4 + 2a + a^2)$
82. $(r + 3)(r^2 - 3r + 9)$ **83.** $(5x - 3)(25x^2 + 15x + 9)$ **84.** $(2m - 3n)(4m^2 + 6mn + 9n^2)$
85. $(3y^3 + 5z^2)(9y^6 - 15y^3z^2 + 25z^4)$ **86.** $27(z + 3y)(z^2 - 3zy + 9y^2)$ **87.** $(x + y)(x - 5)$
88. $(4r + s)(2r - 3s)$ **89.** $(m - 2n)(p^4 + q)$ **90.** $(6a + 5)^2$ **91.** $(2z + 7)^2$
92. $(3p^2 - 1)(2p^2 + 3)$ **93.** $(10x + 7y)(100x^2 - 70xy + 49y^2)$ **94.** $(b + 4 + a)(b + 4 - a)$
95. $(5m^2 - 6)(25m^4 + 30m^2 + 36)$ **96.** $(q + 3 + p)(q + 3 - p)$ **97.** $(6m - 7n)(2m + 5n)$
98. $(6p + 5q)(36p^2 - 30pq + 25q^2)$ **99.** $4^2 + 2^2 = 20$ and $(4 + 2)(4 + 2) = 36; 20 \neq 36$
100. yes; $9(x^2 + 4)$

7.7 Exercises (Page 385)

1. $4, 5, -9$ **2.** 2   **3.** two   **4.** two   **5.** $\{-3, 9\}$ **6.** $\{-6, -4\}$ **7.** $\left\{\frac{7}{2}, -\frac{1}{5}\right\}$ **8.** $\left\{\frac{3}{7}, -\frac{2}{3}\right\}$
9. $\{-3, 4\}$ **10.** $\{-5, 1\}$ **11.** $\{-7, -2\}$ **12.** $\left\{-\frac{2}{3}, \frac{1}{5}\right\}$ **13.** $\left\{-\frac{1}{2}, \frac{1}{6}\right\}$ **14.** $\{-4, 1\}$ **15.** $\{-2, 4\}$
16. $\left\{-\frac{1}{3}, 2\right\}$ **17.** $\{\pm 8\}$ **18.** $\{\pm 4\}$ **19.** $\{\pm 2\sqrt{6}\}$ **20.** $\{\pm 4\sqrt{3}\}$ **21.** $\emptyset$ **22.** $\emptyset$
23. $\{4 \pm \sqrt{3}\}$ **24.** $\{-3 \pm \sqrt{11}\}$ **25.** $\left\{\frac{5 \pm \sqrt{13}}{2}\right\}$ **26.** $\left\{\frac{-1 \pm \sqrt{19}}{4}\right\}$ **27.** $\left\{\frac{2 \pm \sqrt{3}}{2}\right\}$
28. $\{-1 \pm \sqrt{6}\}$ **29.** $\left\{\frac{1 \pm \sqrt{3}}{2}\right\}$ **30.** $\left\{\frac{-1 \pm \sqrt{2}}{3}\right\}$ **31.** $\left\{\frac{1 \pm \sqrt{5}}{2}\right\}$ **32.** $\left\{\frac{2 \pm \sqrt{14}}{2}\right\}$
33. $\left\{\frac{-1 \pm \sqrt{2}}{2}\right\}$ **34.** $\left\{\frac{1 \pm 2\sqrt{5}}{2}\right\}$ **35.** $\left\{\frac{1 \pm \sqrt{29}}{2}\right\}$ **36.** $\left\{\frac{3 \pm \sqrt{73}}{2}\right\}$ **37.** $\emptyset$ **38.** $\emptyset$
43. 0; (c)   **44.** 0; (c)   **45.** 121; (a)   **46.** 25; (a)   **47.** 360; (b)   **48.** -7; (d)   **49.** 5.2 seconds
50. .7 second and 4.0 seconds   **51.** Find s when $t = 0$.   **52.** 3 minutes   **53.** (a) 1 second and 8 seconds
(b) 9 seconds after it is projected   **54.** 3.4 seconds   **55.** 2.3, 5.3, 5.8   **56.** 7.9, 8.9, 11.9   **57.** 412.3 feet
58. 470.0 feet   **59.** eastbound ship: 80 miles; southbound ship: 150 miles   **60.** 120 feet   **61.** 5 centimeters,
12 centimeters, 13 centimeters   **62.** 20 meters, 21 meters, 29 meters   **63.** length: 2 centimeters;
width: 1.5 centimeters   **64.** 5 feet   **65.** 1 foot   **66.** 2 feet   **67.** length: 26 meters; width: 16 meters
68. length: 20 inches; width: 12 inches   **69.** .035 or 3.5%   **70.** 18 inches   **71.** $.80   **72.** $11.93
73. 5.5 meters per second   **74.** 2.1 meters per second   **75.** 5 or 14   **76.** 4

Chapter 7 Test (Page 391)

1. $\{2\}$ **2.** $\{4\}$ **3.** identity; {all real numbers}   **4.** $v = \frac{S + 16t^2}{t}$ **5.** Hawaii: 4021 square miles; Maui:
728 square miles; Kauai: 551 square miles   **6.** 5 liters   **7.** 2.2 hours   **8.** 8 slices for $2.19   **9.** 2300 miles
10. 200 amps   **11.** $(-\infty, 4]$ ⟵─┤   **12.** $(-2, 6]$ ─┤   **13.** C   **14.** at least 82
15. $\frac{16}{9}$ **16.** -64 **17.** $\frac{64}{27}$ **18.** 0   **19.** $\frac{216}{p^4}$ **20.** $\frac{1}{m^{14}}$ **21.** 3×10^{-4} **22.** about 15,300 seconds
23. 1995: 1.028×10^{11}; 1997: 1.149×10^{11}; 1999: 1.236×10^{11} **24.** $4k^2 + 6k + 10$ **25.** $15x^2 - 14x - 8$
26. $16x^4 - 9$ **27.** $3x^3 + 20x^2 + 23x - 36$ **28.** One example is $t^5 + 2t^4 + 3t^3 - 4t^2 + 5t + 6$.
29. $(2p - 3q)(p - q)$ **30.** $(10x + 7y)(10x - 7y)$ **31.** $(3y - 5x)(9y^2 + 15yx + 25x^2)$ **32.** $(4 - m)(x + y)$
33. $\left\{-\frac{3}{2}, \frac{1}{3}\right\}$ **34.** $\left\{\frac{1 \pm \sqrt{29}}{2}\right\}$ **35.** .87 second

Answers to Selected Exercises   A-37

CHAPTER 8   Graphs, Functions, and Systems of Equations and Inequalities

8.1 Exercises   (Page 399)

1. (a) x represents the year; y represents the percent of women in math or computer science professions.   **(b)** 1990–2000   **(c)** 1990   **(d)** 1980   **2. (a)** x represents the year; y represents the revenue in billions of dollars.   **(b)** about 1450 billion dollars   **(c)** 1998   **3.** x **4.** y **5.** $(0,0)$ **6.** 2   **7. (a)** I   **(b)** III   **(c)** II   **(d)** IV   **(e)** none   **8. (a)** III   **(b)** I   **(c)** II   **(d)** IV   **(e)** none   **9. (a)** I or III   **(b)** II or IV   **(c)** II or IV   **(d)** I or III   **10.** One of the coordinates must be zero.   **11–20.**   **21. (a)** $\sqrt{34}$ **(b)** $\left(\dfrac{1}{2}, \dfrac{5}{2}\right)$

22. (a) $\sqrt{34}$ **(b)** $\left(\dfrac{1}{2}, -\dfrac{1}{2}\right)$ **23. (a)** $\sqrt{61}$ **(b)** $\left(\dfrac{1}{2}, 1\right)$ **24. (a)** $\sqrt{89}$ **(b)** $\left(\dfrac{7}{2}, -1\right)$ **25. (a)** $\sqrt{146}$ **(b)** $\left(-\dfrac{1}{2}, \dfrac{3}{2}\right)$ **26. (a)** $\sqrt{58}$ **(b)** $\left(-\dfrac{3}{2}, \dfrac{17}{2}\right)$ **27.** B   **28.** C   **29.** D   **30.** A   **31.** $x^2 + y^2 = 36$ **32.** $x^2 + y^2 = 25$ **33.** $(x+1)^2 + (y-3)^2 = 16$ **34.** $(x-2)^2 + (y+2)^2 = 9$ **35.** $x^2 + (y-4)^2 = 3$ **36.** $(x+2)^2 + y^2 = 5$ **37.** center: $(0,0)$; radius: r **38. (a)** There is only one point, $(4,1)$. The only way for the sum of two squares to equal zero is for both of the squares to equal zero. This is satisfied only when $x = 4$ and $y = 1$.   **(b)** There are no points because the sum of two squares cannot be negative.   **39.** center: $(-2, -3)$; radius: 2   **40.** center: $(4, 6)$; radius: 7   **41.** center: $(-5, 7)$; radius: 9   **42.** center: $(1, -2)$; radius: 3   **43.** center: $(2, 4)$; radius: 4   **44.** center: $(-5, -4)$; radius: 6

45.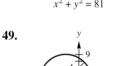

$x^2 + y^2 = 36$

46.

$x^2 + y^2 = 81$

47.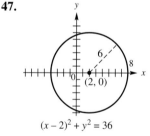

$(x-2)^2 + y^2 = 36$

48.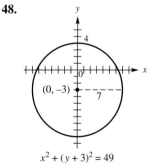

$x^2 + (y+3)^2 = 49$

49.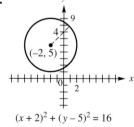

$(x+2)^2 + (y-5)^2 = 16$

50.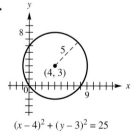

$(x-4)^2 + (y-3)^2 = 25$

51.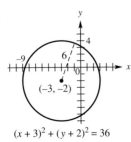
$(x+3)^2 + (y+2)^2 = 36$

52.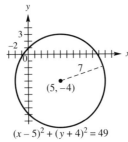
$(x-5)^2 + (y+4)^2 = 49$

53. (a) $\sqrt{40} = 2\sqrt{10}$ (b) $(-1, 4)$

54. (a) $\sqrt{205}$ (b) $\left(-\dfrac{1}{2}, 1\right)$ **55.** \$12,698.50   **56.** 1985: 4.85 million; 1995: 5.5 million

57. 1965: 3495 dollars; 1985: 10,887 dollars   **62.** $(-3, 4)$ is a solution for each of the following equations: $(x - 1)^2 + (y - 4)^2 = 16$; $(x + 6)^2 + y^2 = 25$; $(x - 5)^2 + (y + 2)^2 = 100$

63. The epicenter is $(-2, -2)$.   **67.** $(9, 18)$ **68.** B

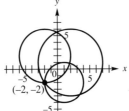

8.2 Exercises (Page 409)

1. $(0, 5)$, $\left(\dfrac{5}{2}, 0\right)$, $(1, 3)$, $(2, 1)$ **2.** $(0, -6)$, $(8, 0)$, $\left(6, -\dfrac{3}{2}\right)$, $(4, -3)$ **3.** $(0, -4)$, $(4, 0)$, $(2, -2)$, $(3, -1)$

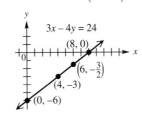 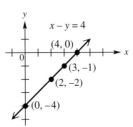

4. $(0, 4)$, $(12, 0)$, $(3, 3)$, $(-6, 6)$ **5.** $(0, 4)$, $(5, 0)$, $\left(3, \dfrac{8}{5}\right)$, $\left(\dfrac{5}{2}, 2\right)$ **6.** $\left(0, -\dfrac{12}{5}\right)$, $(6, 0)$, $(1, -2)$, $\left(-2, -\dfrac{16}{5}\right)$

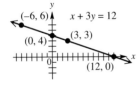 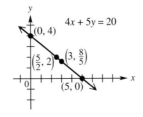 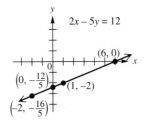

7. $(0,4)$, $\left(\frac{8}{3},0\right)$, $(2,1)$, $(4,-2)$ **8.** $(0,12)$, $\left(\frac{12}{5},0\right)$, $(3,-3)$, $(2,2)$ **11.** A **12.** 2

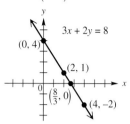
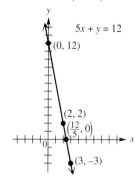

13. $(4,0)$; $(0,6)$ **14.** $(5,0)$; $(0,2)$ **15.** $(2,0)$; $\left(0,\frac{5}{3}\right)$

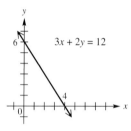
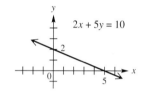
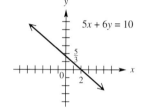

16. $(6,0)$; $(0,2)$ **17.** $\left(\frac{5}{2},0\right)$; $(0,-5)$ **18.** $\left(\frac{4}{3},0\right)$; $(0,-2)$

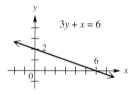
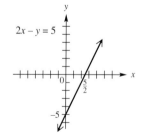
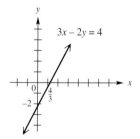

19. $(2,0)$; $\left(0,-\frac{2}{3}\right)$ **20.** $\left(-\frac{3}{4},0\right)$; $(0,3)$ **21.** $(0,0)$; $(0,0)$

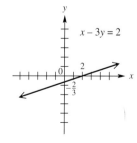
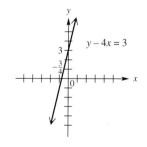
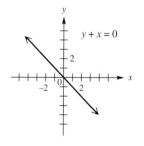

22. (0,0); (0,0)
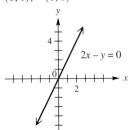

23. (0,0); (0,0)
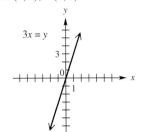

24. (0,0); (0,0)
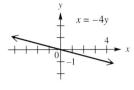

25. (2,0); none
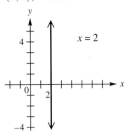

26. none; (0,−3)
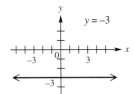

27. none; (0,4)
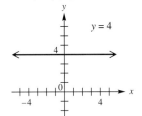

28. (−2,0); none
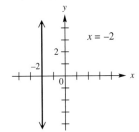

29. C **30.** C **31.** A **32.** A **33.** D **34.** D **35.** B
36. B **37.** $\dfrac{3}{10}$ **38.** $\dfrac{8}{27}$ **39.** (a) $\dfrac{3}{2}$ (b) $-\dfrac{7}{4}$
40. (a) positive (b) undefined (c) negative (d) zero
41. 8 **42.** 1 **43.** $-\dfrac{5}{6}$ **44.** $-\dfrac{9}{8}$ **45.** 0 **46.** 0
47. (a) slope = 232 (This represents enrollment in thousands.)
(b) positive; increased (c) 232,000 students (d) −1.66
(e) negative; decreased (f) 1.66 students per computer

49.
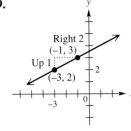

50.
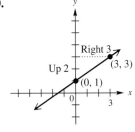

51.
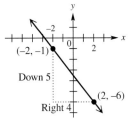

52.
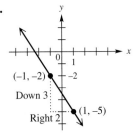

53.
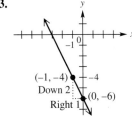

54.
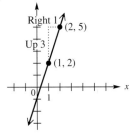

55. 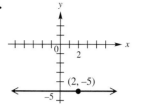 **56.** 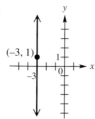 **57.** parallel **58.** parallel **59.** perpendicular

60. neither parallel nor perpendicular **61.** neither parallel nor perpendicular **62.** perpendicular **63.** $\dfrac{7}{10}$
64. 19.5 feet **65.** (a) $200 million per year **66.** (a) -1.7 million recipients per year
67. (a) 1000 million per year; 1000 million per year; 1000 million per year; The average rate of change is the same. This is true because the data points lie on a straight line. (b) All are the same. **68.** (a) 9652 thousand; 10,257 thousand; 11,269 thousand; 13,897 thousand; 16,838 thousand (b) no; no

8.3 Exercises (Page 418)

1. D **2.** C **3.** B **4.** A **5.** $y = 3x - 3$ **6.** $y = 2x - 4$ **7.** $y = -x + 3$ **8.** $y = -\dfrac{1}{2}x + 2$
9. A **10.** D **11.** C **12.** F **13.** H **14.** G **15.** B **16.** E **17.** $y = -\dfrac{3}{4}x + \dfrac{5}{2}$
18. $y = -\dfrac{5}{6}x + \dfrac{31}{6}$ **19.** $y = -2x + 18$ **20.** $y = x - 2$ **21.** $y = \dfrac{1}{2}x + \dfrac{13}{2}$ **22.** $y = \dfrac{1}{4}x - \dfrac{15}{4}$
23. $y = 4x - 12$ **24.** $y = -5x - 10$ **25.** $y = 5$ **26.** $y = -2$ **27.** $x = 9$ **28.** $x = -2$
29. $x = .5$ **30.** $x = \dfrac{5}{8}$ **31.** $y = 8$ **32.** $y = 7$ **33.** $y = 2x - 2$ **34.** $y = -2x + 8$
35. $y = -\dfrac{1}{2}x + 4$ **36.** $y = \dfrac{2}{3}x + \dfrac{19}{3}$ **37.** $y = \dfrac{2}{13}x + \dfrac{6}{13}$ **38.** $y = \dfrac{40}{7}x - \dfrac{34}{21}$ **39.** $y = 5$
40. $y = 2$ **41.** $x = 7$ **42.** $x = 13$ **43.** $y = -3$ **44.** $y = 6$ **45.** $y = 5x + 15$
46. $y = -2x + 12$ **47.** $y = -\dfrac{2}{3}x + \dfrac{4}{5}$ **48.** $y = -\dfrac{5}{8}x - \dfrac{1}{3}$ **49.** $y = \dfrac{2}{5}x + 5$ **50.** $y = -\dfrac{3}{4}x + 7$
52. D **53.** (a) $y = -x + 12$ (b) -1 (c) $(0, 12)$ **54.** (a) $y = x - 14$ (b) 1 (c) $(0, -14)$
55. (a) $y = -\dfrac{5}{2}x + 10$ (b) $-\dfrac{5}{2}$ (c) $(0, 10)$ **56.** (a) $y = -\dfrac{6}{5}x + 8$ (b) $-\dfrac{6}{5}$ (c) $(0, 8)$
57. (a) $y = \dfrac{2}{3}x - \dfrac{10}{3}$ (b) $\dfrac{2}{3}$ (c) $\left(0, -\dfrac{10}{3}\right)$ **58.** (a) $y = \dfrac{4}{3}x - \dfrac{10}{3}$ (b) $\dfrac{4}{3}$ (c) $\left(0, -\dfrac{10}{3}\right)$
59. $y = 3x - 19$ **60.** $y = -\dfrac{2}{5}x + \dfrac{13}{5}$ **61.** $y = \dfrac{1}{2}x - 1$ **62.** $y = \dfrac{1}{3}x + \dfrac{10}{3}$ **63.** $y = -\dfrac{1}{2}x + 9$
64. $y = \dfrac{2}{5}x - \dfrac{39}{5}$ **65.** $y = 7$ **66.** $y = 4$
67. (a) 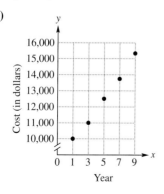 (b) $y = \dfrac{4355}{6}x + 8847.5$ (c) $18,283 **68.** (a) yes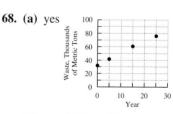

(b) $y = 1.76x + 32$
(c) $y = 49.6$, so there will be about 49,600 metric tons in 2005.

A-42 Answers to Selected Exercises

69. $y = -176.25x + 29{,}362$ **70.** $y = 15.2x + 2720$ **71. (a)** $y = 838.5x + 19{,}180.5$ **(b)** $23{,}373; It is close to the actual value.   **72. (a)** $y = 1276.5x + 17{,}692.5$ **73. (a)** $y = -.96x + 5.6$ **(b)** .8 million or 800,000 tons   **(c)** during 2001 (Since $.83333(12) = 10$, the landfill capacity will be 0 in October.)
74. (a) 32; 212   **(b)** $(0, 32)$ and $(100, 212)$ **(c)** $\dfrac{9}{5}$ **(d)** $F = \dfrac{9}{5}C + 32$ **(e)** $C = \dfrac{5}{9}(F - 32)$ **(f)** When the Celsius temperature is 50°, the Fahrenheit temperature is 122°.   **75. (a)** $Y_1 = \dfrac{3}{4}X + 1$ **(b)** $Y_1 = -4X - 2$

8.4 Exercises (Page 431)

3. It is the independent variable.   **4.** not a function; domain: $\{1, 2, 3\}$; range: $\{-9, -4, -1, 1, 4, 9\}$
5. function; domain: $\{2, 3, 4, 5\}$; range: $\{5, 7, 9, 11\}$
6. function; domain: {U.S., Japan, Canada, Britain, France, Germany, Italy}; range: $\{83.9, 90.91, 85.26, 83.79, 87.01, 83.12, 82.26\}$ **7.** not a function; domain: $(0, \infty)$; range: $(-\infty, 0) \cup (0, \infty)$
8. function; domain: $\{4, 9, 11, 17, 25\}$; range: $\{-69, 14, 32, 47\}$
9. function; domain: {Hispanic, Native American, Asian American, African-American, White}; range in millions: $\{21.3, 1.6, 8.2, 24.6, 152.0\}$ **10.** not a function; domain: $[0, \infty)$; range: $(-\infty, \infty)$
11. function; domain: $(-\infty, \infty)$; range: $(-\infty, 4]$ **12.** function; domain: $[-2, 2]$; range: $[0, 4]$
13. not a function; domain: $[-4, 4]$; range: $[-3, 3]$ **14.** not a function; domain: $[3, \infty)$; range: $(-\infty, \infty)$
15. function; domain: $(-\infty, \infty)$ **16.** function; domain: $(-\infty, \infty)$ **17.** not a function; domain: $[0, \infty)$
18. not a function; domain: $[0, \infty)$ **19.** not a function; domain: $(-\infty, \infty)$ **20.** not a function; domain: $(-\infty, \infty)$
21. function; domain: $[0, \infty)$ **22.** function; domain: $[0, \infty)$ **23.** function; domain: $(-\infty, 0) \cup (0, \infty)$
24. function; domain: $(-\infty, 0) \cup (0, \infty)$ **25.** function; domain: $\left[-\dfrac{1}{2}, \infty\right)$ **26.** function; domain: $\left(-\infty, \dfrac{9}{2}\right]$
27. function; domain: $(-\infty, 9) \cup (9, \infty)$ **28.** function; domain: $(-\infty, 16) \cup (16, \infty)$ **29. (a)** $[0, 3000]$
(b) 25 hours; 25 hours   **(c)** 2000 gallons   **(d)** $f(0) = 0$; The pool is empty at time zero.   **30. (a)** yes
(b) $[0, 24]$ **(c)** 1200 megawatts   **(d)** at 18 hours or 6 P.M.; at 4 A.M.   **31.** Here is one example: The cost of gasoline; number of gallons purchased; cost; number of gallons   **32.** B   **33.** 5   **34.** 11   **35.** 2
36. -2 **37.** -1 **38.** 7   **39.** -13 **40.** -7

41.

[Graph of $f(x) = -2x + 5$ showing line with y-intercept 5 and x-intercept $\tfrac{5}{2}$]

domain and range: $(-\infty, \infty)$

42.

[Graph of $g(x) = 4x - 1$ showing line with y-intercept -1 and passing through 3]

domain and range: $(-\infty, \infty)$

43.

[Graph of $h(x) = \tfrac{1}{2}x + 2$ showing line with y-intercept 2 and x-intercept -4]

domain and range: $(-\infty, \infty)$

44.

[Graph of $F(x) = -\tfrac{1}{4}x + 1$ showing line with y-intercept 1 and x-intercept 4]

domain and range: $(-\infty, \infty)$

45.

[Graph of $G(x) = 2x$ through origin, passing through $(1, 2)$]

domain and range: $(-\infty, \infty)$

46.

[Graph of $H(x) = -3x$ through origin, passing through $(1, -3)$]

domain and range: $(-\infty, \infty)$

47.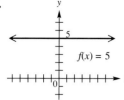

domain: $(-\infty, \infty)$; range: $\{5\}$

48.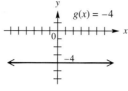

domain: $(-\infty, \infty)$; range: $\{-4\}$

49. (a) $f(x) = 3 - 2x^2$ (b) -15

50. (a) $f(x) = 2 + 3x^2$ (b) 29   **51.** (a) $f(x) = \dfrac{8 - 4x}{-3}$ 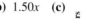 (b) $\dfrac{4}{3}$ **52.** (a) $f(x) = \dfrac{9 + 2x}{5}$ (b) 3

53. line; -2; $-2x + 4$; -2; 3; -2 **54.** A

55. (a) \$0; \$1.50; \$3.00; \$4.50 (b) $1.50x$ (c)

56. (a) 8.25 (dollars) (b) 3 is the value of the independent variable, which represents a package weight of 3 pounds; $f(3)$ is the value of the dependent variable representing the cost to mail a 3-pound package. (c) \$13.75; $f(5) = 13.75$

57. (a) 194.53 centimeters (b) 177.29 centimeters (c) 177.41 centimeters (d) 163.65 centimeters

58. (a) 1.83 cubic meters (b) 2.86 cubic meters (c) 4.11 cubic meters (d) 6.43 cubic meters   **59.** (a) 39,851 (b) 39,485 (c) 39,119 (d) In 1992, there were 39,668 post offices in the U.S.   **60.** (a) \$286,322 million (b) \$273,674 million (c) \$267,350 million (d) In 1994, the defense budget was \$279,998 million.

61. (a) $C(x) = .02x + 200$
(b) $R(x) = .04x$
(c) 10,000
(d)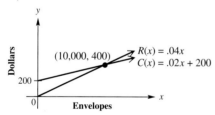

For $x < 10{,}000$, a loss
For $x > 10{,}000$, a profit

62. (a) $C(x) = .01x + 3500$
(b) $R(x) = .05x$
(c) 87,500
(d)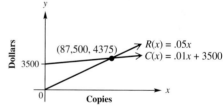

For $x < 87{,}500$, a loss
For $x > 87{,}500$, a profit

63. (a) $C(x) = 3.00x + 2300$
(b) $R(x) = 5.50x$
(c) 920
(d)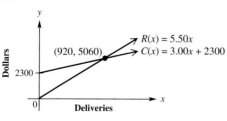

For $x < 920$, a loss
For $x > 920$, a profit

64. (a) $C(x) = 2.50x + 40$
(b) $R(x) = 6.50x$
(c) 10
(d)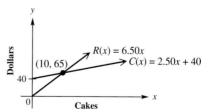

For $x < 10$, a loss
For $x > 10$, a profit

8.5 Exercises (Page 442)

1. F 2. B 3. C 4. A 5. E 6. D 9. (0, 0) 10. (0, 0) 11. (0, 4) 12. (0, −4)
13. (1, 0) 14. (−3, 0) 15. (−3, −4) 16. (5, −8) 18. downward; narrower 19. downward; narrower
20. upward; wider 21. upward; wider 22. If $|a| > 1$, the graph of $f(x) = a(x − h)^2 + k$ is narrower than the graph of $y = x^2$. If $0 < |a| < 1$, the graph is wider than the graph of $y = x^2$. If a is negative, the graph opens downward; if a is positive, it opens upward. 23. (a) I (b) IV (c) II (d) III 24. (a) 0 (b) 0 25.

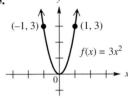

26.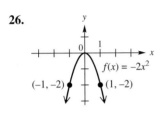

27.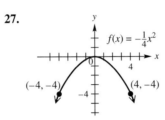

28.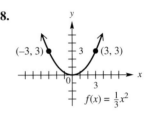

29.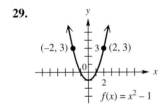

30.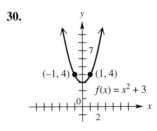

31.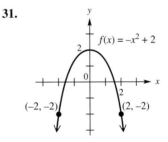

32.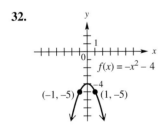

33.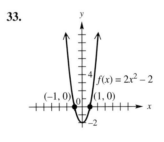

34.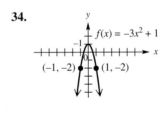

35.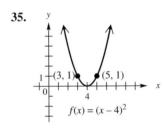

36.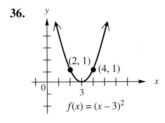

37.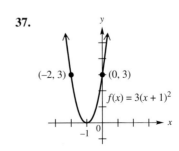

38. 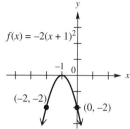 **39.** 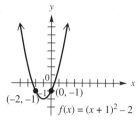 **40.**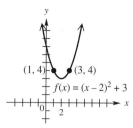

41. 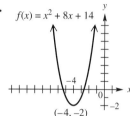 **42.** 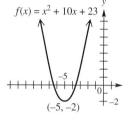 **43.**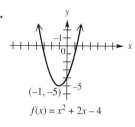

44. 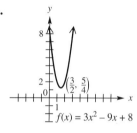 **45.** 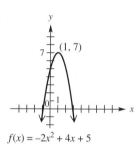 **46.**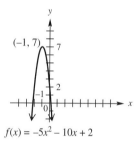

47. 25 meters   **48.** 140 feet by 70 feet; 9800 square feet (The longer side is along the highway.)   **49.** 16 feet; 2 seconds   **50.** 2500 feet; 12.5 seconds   **51.** 4.1 seconds; 81.6 meters   **52.** 573.6 feet; 664.2 seconds
53. (a) minimum   (b) 1995; 1.7%   **54.** (a) maximum   (b) 1994; 41 million   **55.** (a) The coefficient of x^2 is negative, because the parabola along the top would open downward.   (b) $(18.45, 3860)$ (c) In 2018, Social Security assets will reach their maximum of $3860 billion.   **56.** (a) No. For example, two ordered pairs correspond to the same x-value of about 15.9%, and there are other examples.   (b) 80% U.S. and 20% foreign; about 15.8%; about 14.2%   (c) the point corresponding to 100% foreign stocks; 16%
57. (a) $R(x) = (100 - x)(200 + 4x) = 20{,}000 + 200x - 4x^2$ (b)   (c) 25

(d) $22,500   **58.** (a) $R(x) = (42 - x)(48 + 2x) = -2x^2 + 36x + 2016$ (b) 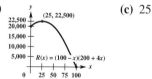   (c) 9   (d) $2178

8.6 Exercises   (Page 452)

1. rises; falls   **2.** $(0, 1)$ **3.** does not   **4.** 243   **5.** rises; falls   **6.** $(1, 0)$ **7.** does not
8. 2   **9.** 2.56425419972   **10.** 2.12551979078   **11.** 1.25056505582   **12.** .901620746784
13. 7.41309466897   **14.** 5.39357064307   **15.** .0000210965628481   **16.** .000597978996117

17. 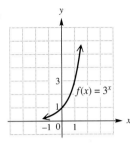 **18.** 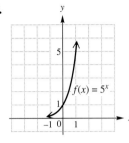 **19.**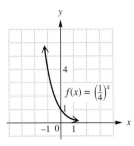

20. 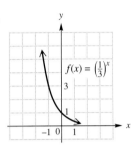 **21.** 20.0855369232 **22.** 54.5981500331 **23.** .018315638889

24. .049787068368 **25.** $2 = \log_4 16$ **26.** $3 = \log_5 125$ **27.** $-3 = \log_{2/3}\left(\dfrac{27}{8}\right)$ **28.** $-4 = \log_{1/10} 10{,}000$
29. $2^5 = 32$ **30.** $4^3 = 64$ **31.** $3^1 = 3$ **32.** $12^0 = 1$ **33.** 1.38629436112 **34.** 1.79175946923
35. -1.0498221245 **36.** .896088024557 **37.** (a) .5°C (b) .35°C **38.** (a) 1.0°C (b) .4°C
39. (a) 1.6°C (b) .5°C **40.** (a) 3.0°C (b) .7°C
41. 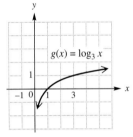 **42.** 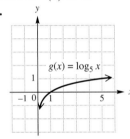 **43.**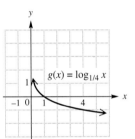

44. 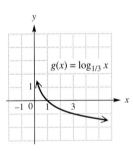 **45.** $7686.32 **46.** $13,891.16 **47.** $76,855.95 **48.** $88,585.47

49. (a) $26,281.78 (b) $32,100.64 (c) $41,218.03 **50.** 7% compounded quarterly; $800.31
51. (a) 132,359 thousand tons (b) 97,264 thousand tons (c) It is slightly less than what the model provides (93,733 thousand tons). **52.** (a) 157.28 million tons (b) 173.99 million tons (c) It is less than what the model provides (230.84 million tons). **53.** (a) 176,000 (b) 202,171 (c) It will have increased by almost 15%.
54. (a) 146,250 (b) 198,403 (c) It will have increased by almost 36%.
55. 32,044 million dollars **56.** 1,132,873 **57.** (a) 100 grams (b) 31.25 grams (c) .30 gram
58. (a) $5000 (b) $2973 (c) $1768 **59.** (a) 645 sites (b) 962 sites **60.** (a) 51.47 trillion cubic feet (b) 69.602 trillion cubic feet **61.** almost 4 times as powerful **62.** more than 300 times as powerful

8.7 Exercises (Page 470)

1. (a) 1989–1997 (b) 1997; NBC; 17% (c) 1989, share 20%; 1998, share 16% (d) NBC and ABC; (2000, 16)
2. (a) 1991; about 350 million (b) (1987, 100 million) (c) 1988–1990 (d) CD production generally increased during these years; positive **3.** yes **4.** yes **5.** no **6.** no
7. $\{(2,2)\}$ **8.** $\{(-2,-3)\}$ **9.** $\{(3,-1)\}$ **10.** $\{(1,1)\}$

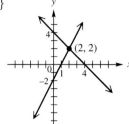
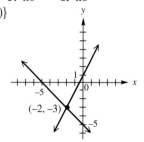

11. $\{(2,-3)\}$ **12.** $\{(4,-5)\}$ **13.** $\left\{\left(\frac{3}{2},-\frac{3}{2}\right)\right\}$ **14.** $\left\{\left(\frac{1}{4},-\frac{1}{2}\right)\right\}$ **15.** $\left\{\left(\frac{6-2y}{7},y\right)\right\}$
16. $\{(2+4y,y)\}$ **17.** $\{(2,-4)\}$ **18.** $\left\{\left(13,-\frac{7}{5}\right)\right\}$ **19.** $\emptyset$ **20.** $\emptyset$ **21.** $\{(1,2)\}$ **22.** $\{(-2,-10)\}$
23. $\left\{\left(\frac{22}{9},\frac{22}{3}\right)\right\}$ **24.** $\left\{\left(\frac{10}{7},-\frac{5}{7}\right)\right\}$ **25.** $\{(2,3)\}$ **26.** $\{(-1,4)\}$ **27.** $\{(5,4)\}$ **28.** $\{(-1,3)\}$
29. $\left\{\left(-5,-\frac{10}{3}\right)\right\}$ **30.** $\left\{\left(\frac{2}{7},\frac{8}{21}\right)\right\}$ **31.** $\{(2,6)\}$ **32.** $\{(-12,-60)\}$ **34.** Answers will vary. One example is $x+y+z=6, 2x+3y-z=7, 3x-y-z=6$.   **35.** $\{(1,4,-3)\}$ **36.** $\{(2,-1,3)\}$
37. $\{(0,2,-5)\}$ **38.** $\{(0,3,-1)\}$ **39.** $\left\{\left(-\frac{7}{3},\frac{22}{3},7\right)\right\}$ **40.** $\left\{\left(\frac{20}{59},-\frac{33}{59},\frac{35}{59}\right)\right\}$ **41.** $\{(4,5,3)\}$
42. $\{(5,1,-1)\}$ **43.** $\{(2,2,2)\}$ **44.** $\{(4,4,4)\}$ **45.** $\left\{\left(\frac{8}{3},\frac{2}{3},3\right)\right\}$ **46.** $\left\{\left(\frac{3}{4},\frac{5}{4},2\right)\right\}$
47. wins: 95; losses: 67   **48.** wins: 65; losses: 97   **49.** length: 78 feet; width: 36 feet
50. length: 94 feet; width: 50 feet   **51.** ExxonMobil: $214 billion; General Motors: $185 billion
52. Canada: $134 billion; Mexico: $77 billion   **53.** length: 12 feet; width: 5 feet   **54.** square: 12 centimeters; triangle: 8 centimeters   **55.** marker pen: $1.10; colored paper sheet: $.40   **56.** CGA monitor: $400; VGA monitor: $500   **57.** National Hockey League: $219.74; National Basketball Association: $203.38
58. Major League Baseball: $102.58; National Football League: $208.45   **59.** single: $2.09; double: $3.19
60. Tokyo: $430; New York: $385   **61.** units of yarn: 6; units of thread: 2   **62.** dark clay: $5 per kilogram; light clay: $4 per kilogram   **63.** (a) 6 ounces (b) 15 ounces (c) 24 ounces (d) 30 ounces   **64.** (a) $100 (b) $150 (c) $200 (d) $175   **65.** $.99x   **66.** $8y   **67.** 6 gallons of 25%; 14 gallons of 35%
68. $26\frac{2}{3}$ liters of 15%; $13\frac{1}{3}$ liters of 33%   **69.** 3 liters of pure acid; 24 liters of 10% acid
70. 3 liters of pure antifreeze   **71.** 50 pounds of $3.60 pecan clusters; 30 pounds of $7.20 chocolate truffles
72. 150 liters of 50% juice; 50 liters of 30% juice   **73.** 76 general admission; 108 with student identification
74. 50 pounds of $1.20 candy; 30 pounds of $2.40 candy   **75.** $1000 at 2%; $2000 at 4%   **76.** $10,000 at 4%; $5000 at 3%   **77.** freight train: 50 kilometers per hour; express train: 80 kilometers per hour
78. train: 60 kilometers per hour; plane: 160 kilometers per hour   **79.** top speed: 2100 miles per hour; wind speed: 300 miles per hour   **80.** boat: 21 miles per hour; current: 3 miles per hour   **81.** Independent: 38; Democrat: 34; Republican: 28   **82.** gold: 39; silver: 25; bronze: 33   **83.** shortest: 12 centimeters; middle: 25 centimeters; longest: 33 centimeters   **84.** shortest: 10 inches; middle: 20 inches; longest: 26 inches
85. A: 180 cases; B: 60 cases; C: 80 cases   **86.** type A: 80; type B: 160; type C: 250   **87.** $10 tickets: 350; $18 tickets: 250; $30 tickets: 50   **88.** up close: $40; in the middle: $30; far out: $20

Extension Exercises (Page 480)

1. $\{(2, 3)\}$ **2.** $\{(-3, 4)\}$ **3.** $\{(-3, 0)\}$ **4.** $\{(0, -2)\}$ **5.** $\left\{\left(\dfrac{7}{2}, -1\right)\right\}$ **6.** $\{(1, 1)\}$ **7.** $\{(1, -4)\}$
8. $\{(3, 1)\}$ **9.** $\{(-1, 23, 16)\}$ **10.** $\{(1, 0, -1)\}$ **11.** $\{(2, 1, -1)\}$ **12.** $\{(2, 2, 2)\}$ **13.** $\{(3, 2, -4)\}$
14. $\{(-1, 2, -2)\}$ **15.** $\{(0, 1, 0)\}$ **16.** $\{(0, 3, 2)\}$

8.8 Exercises (Page 485)

1. C **2.** A **3.** B **4.** D

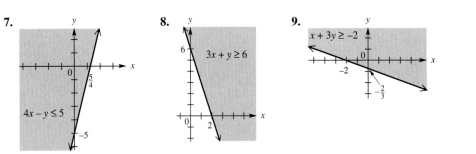

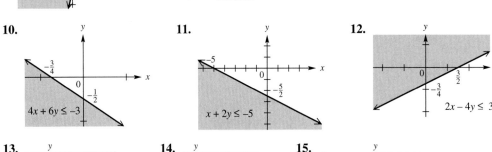

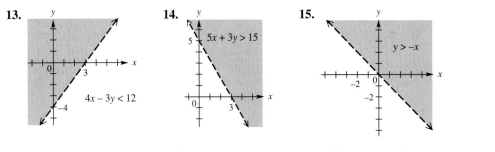

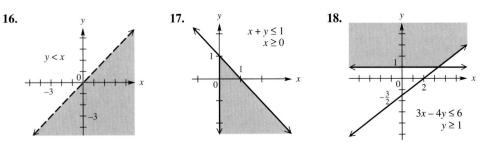

19. 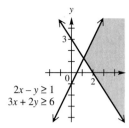   **20.** 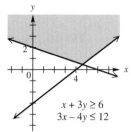   **21.**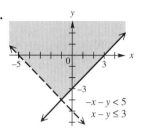

22. 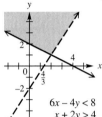   **23.** maximum of 65 at (5, 10); minimum of 8 at (1, 1)   **24.** maximum of 900 at (0, 12); minimum of 0 at (0, 0)   **25.** $\left(\frac{6}{5}, \frac{6}{5}\right); \frac{42}{5}$ **26.** $\left(\frac{10}{3}, \frac{5}{3}\right); \frac{25}{3}$ **27.** $\left(\frac{17}{3}, 5\right); \frac{49}{3}$ **28.** $\left(10, \frac{80}{3}\right); 90$

29. Ship 20 to A and 80 to B, for a minimum cost of $1040.   **30.** Use $3\frac{3}{4}$ servings of A and $1\frac{7}{8}$ servings of B, for a minimum cost of $1.69.   **31.** Take 3 red pills and 2 blue pills, for a minimum cost of 70¢ per day.   **32.** Manufacture 1600 type A and 0 type B, for a maximum revenue of $160.   **33.** Produce 6.4 million gallons of gasoline and 3.2 million gallons of fuel oil, for a maximum revenue of $16,960,000.   **34.** Make 3 batches of cakes and 6 batches of cookies, for a maximum profit of $210.   **35.** Ship 4000 medical kits and 2000 containers of water.   **36.** Ship 0 medical kits and 4000 containers of water.

Chapter 8 Test (Page 489)

1. $\sqrt{41}$ **2.** $(x + 1)^2 + (y - 2)^2 = 9$ **3.** x-intercept: $\left(\frac{8}{3}, 0\right)$; y-intercept: $(0, -4)$

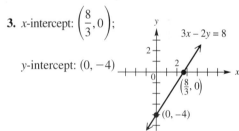

4. $\frac{2}{7}$ **5.** (a) $y = -\frac{2}{5}x + \frac{13}{5}$ (b) $y = -\frac{1}{2}x - \frac{3}{2}$ (c) $y = -\frac{1}{2}x + 2$ **6.** B   **7.** (a) $y = 1379.25x + 15,496.75$ (b) about $25,152; It is close to the actual value.   **8.** $y = .05x + .50$; (1, .55), (5, .75), (10, 1.00)   **9.** $y = \frac{2}{3}x + 1$ **10.** (a) $(-\infty, \infty)$ (b) 22   **11.** $(-\infty, 3) \cup (3, \infty)$ **12.** 500 calculators; $30,000   **13.** axis: $x = -3$; vertex: $(-3, 4)$; domain: $(-\infty, \infty)$; range: $(-\infty, 4]$

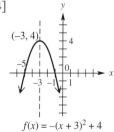

14. 80 feet by 160 feet   **15. (a)** 2116.31264888 **(b)** .157237166314 **(c)** 3.15955035878   **16.** A
17. (a) $13,521.90 **(b)** $13,529.96   **18. (a)** 1.62 grams **(b)** 1.18 grams **(c)** .69 gram **(d)** 2.00 grams
19. $\{(4, -2)\}$ **20.** $\{(2, 0, -1)\}$ **21.** $\{(1, 2z + 3, z)\}$ **22.** *Pretty Woman:* $178.4 million; *Runaway Bride:* $152.3 million   **23.** $40,000 at 10%; $100,000 at 6%; $140,000 at 5%   **24.**   **25.** Manufacture 0 VIP rings and 24 SST rings, for a maximum profit of $960.

CHAPTER 9 Geometry

9.1 Exercises (Page 499)

1. 90   **2.** 180   **3.** equal   **4.** two   **5.** true   **6.** true   **7.** false   **8.** true   **9.** true   **10.** false
11. true   **12.** false   **13.** false   **14.** false   **15. (a)** $\overleftrightarrow{AB}$ **(b)**
16. (a) $\overrightarrow{BC}$ **(b)**   **17. (a)** $\overleftrightarrow{CB}$ **(b)**
18. (a) $\overrightarrow{AD}$ **(b)**   **19. (a)** $\overleftrightarrow{BC}$ **(b)**
20. (a) $\overrightarrow{AD}$ **(b)**   **21. (a)** $\overleftrightarrow{BA}$ **(b)**
22. (a) $\overleftrightarrow{DA}$ **(b)**   **23. (a)** $\overleftrightarrow{CA}$ **(b)**
24. (a) $\overleftrightarrow{DA}$ **(b)**   **25.** F   **26.** A   **27.** D   **28.** G   **29.** B   **30.** C
31. E   **32.** H
There may be other correct forms of the answers in Exercises 33–40.
33. $\overleftrightarrow{MO}$ **34.** N   **35.** $\overleftrightarrow{MO}$ **36.** $\overleftrightarrow{MO}$ **37.** ∅   **38.** $\overleftrightarrow{OP}$ **39.** $\overleftrightarrow{OP}$ **40.** $\overleftrightarrow{NP}$ **41.** 62°
42. 58°   **43.** 1°   **44.** 45°   **45.** $(90 - x)°$ **46.** $x°$ **47.** 48°   **48.** 75°   **49.** 154°
50. 90°   **51.** $(180 - y)°$ **52.** $y°$ **53.** ∡CBD and ∡ABE; ∡CBE and ∡DBA
54. ∡TQR and ∡PQS; ∡TQS and ∡PQR   **55. (a)** 52° **(b)** 128°   **56. (a)** 126° **(b)** 54°
57. 107° and 73°   **58.** 48° and 132°   **59.** 75° and 75°   **60.** 51° and 51°   **61.** 139° and 139°
62. 105° and 105°   **63.** 65° and 115°   **64.** 26° and 64°   **65.** 35° and 55°   **66.** 49° and 49°
67. 117° and 117°   **68.** 48° and 132°   **69.** 141° and 141°   **70.** 25°   **71.** 80°   **72.** 30°   **73.** 52°
74. (a) 3 **(b)** 6 **(c)** 7 **(d)** 7; exterior   **75.** Measures are given in numerical order, starting with angle 1: 55°, 65°, 60°, 65°, 60°, 120°, 60°, 60°, 55°, 55°.   **76. (a)** 180 **(b)** 180 **(c)** 180; 180 **(d)** 0 **(e)** 0 **(f)** 3
78. $x = 10, y = 30$

9.2 Exercises (Page 506)

1. chord   **2.** radius   **3.** equilateral (or equiangular)   **4.** diameter   **5.** false   **6.** true   **7.** false
8. true   **9.** true   **10.** false   **12.** a STOP sign   **13.** both   **14.** simple   **15.** closed   **16.** both
17. closed   **18.** closed   **19.** neither   **20.** neither   **21.** convex   **22.** not convex   **23.** convex
24. convex   **25.** not convex   **26.** not convex   **27.** right, scalene   **28.** obtuse, scalene
29. acute, equilateral   **30.** acute, isosceles   **31.** right, scalene   **32.** obtuse, isosceles   **33.** right, isosceles
34. right, scalene   **35.** obtuse, scalene   **36.** acute, equilateral   **37.** acute, isosceles   **38.** right, scalene
42. For example, the triangle in Exercise 30 is an isosceles triangle, but $\sqrt{9} + \sqrt{9} \neq \sqrt{6}$, since $3 + 3 = 6 \neq \sqrt{6}$.

43. $A = 50°; B = 70°; C = 60°$ **44.** $A = 130°; B = 30°; C = 20°$ **45.** $A = B = C = 60°$
46. $A = B = C = 60°$ **47.** $A = B = 52°; C = 76°$ **48.** $B = C = 50°; A = 80°$ **49.** $165°$
50. $130°$ **51.** $170°$ **52.** $60°$ **53.** (a) O (b) $\overrightarrow{OA}, \overrightarrow{OC}, \overrightarrow{OB}, \overrightarrow{OD}$ (c) $\overleftrightarrow{AC}, \overleftrightarrow{BD}$ (d) $\overleftrightarrow{AC}, \overleftrightarrow{BD}, \overleftrightarrow{BC}, \overleftrightarrow{AB}$
(e) $\overleftrightarrow{BC}, \overleftrightarrow{AB}$ (f) $\overleftrightarrow{AE}$ **55.** (e) The sum of the measures of the angles of a triangle equals 180° (since the pencil has gone through one-half of a complete rotation).

Extension Exercises   (Page 511)

1. With radius of the compasses greater than one-half the length PQ, place the point of the compasses at P and swing arcs above and below line r. Then with the same radius and the point of the compasses at Q, swing two more arcs above and below line r. Locate the two points of intersections of the arcs above and below, and call them A and B. With a straightedge, join A and B. AB is the perpendicular bisector of PQ.   **2.** With radius of the compasses greater than one-half the length PQ, place the point of the compasses at P and swing arcs to the left and right of line r. Then with the same radius and the point of the compasses at Q, swing two more arcs to the left and right of line r. Locate the two points of intersections of the arcs to the left and right, and call them A and B. With a straightedge, join A and B. AB is the perpendicular bisector of PQ.   **3.** With the radius of the compasses greater than the distance from P to r, place the point of the compasses at P and swing an arc intersecting line r in two points. Call these points A and B. Swing arcs of equal radius to the left of line r, with the point of the compasses at A and at B, intersecting at point Q. With a straightedge, join P and Q. PQ is the perpendicular from P to line r.   **4.** With the radius of the compasses greater than the distance from P to r, place the point of the compasses at P and swing an arc intersecting line r in two points. Call these points A and B. Swing arcs of equal radius above line r, with the point of the compasses at A and at B, intersecting at point Q. With a straightedge, join P and Q. PQ is the perpendicular from P to line r.   **5.** With any radius, place the point of the compasses at P and swing arcs to the left and right, intersecting line r in two points. Call these points A and B. With an arc of sufficient length, place the point of the compasses first at A and then at B, and swing arcs either both above or both below line r, intersecting at point Q. With a straightedge, join P and Q. PQ is perpendicular to line r at P.   **6.** With any radius, place the point of the compasses at P and swing arcs above and below, intersecting line r in two points. Call these points A and B. With an arc of sufficient length, place the point of the compasses first at A and then at B, and swing arcs either both to the left or both to the right of line r, intersecting at point Q. With a straightedge, join P and Q. PQ is perpendicular to line r at P.   **7.** With any radius, place the point of the compasses at vertex A and swing an arc intersecting both sides of the angle. Call these points of intersection P and Q. With an arc of sufficient length, place the point of the compasses first at P and then at Q, and swing arcs in the interior of the angle, intersecting each other at point B. With a straightedge, join A and B. AB is the bisector of the angle.   **8.** With any radius, place the point of the compasses at vertex P and swing an arc intersecting both sides of the angle. Call these points of intersection A and B. With an arc of sufficient length, place the point of the compasses first at A and then at B, and swing arcs in the interior of the angle, intersecting each other at point Q. With a straightedge, join P and Q. PQ is the bisector of the angle.   **9.** Use Construction 3 to construct a perpendicular to a line at a point. Then use Construction 4 to bisect one of the right angles formed. This yields a 45° angle.

9.3 Exercises   (Page 518)

1. 12   **2.** 16   **3.** 6   **4.** 4   **5.** $9\sqrt{3}$ **6.** 4   **7.** 12 cm^2 **8.** 9 cm^2 **9.** 5 cm^2 **10.** 3 cm^2
11. 8 in.^{2} **12.** 10 in.^{2} **13.** 4.5 cm^2 **14.** 936 mm^2 **15.** 418 mm^2 **16.** 7.5 m^2 **17.** 8 cm^2
18. 13.5 cm^2 **19.** 3.14 cm^2 **20.** 706.5 cm^2 **21.** 1017.36 m^2 **22.** 113.04 m^2 **23.** 4 m
24. 34 in. by 54 in.   **25.** 300 ft, 400 ft, 500 ft   **26.** 6 in., 24 in., 24 in.   **27.** 46 ft   **28.** 2.5 cm
29. 23,800.10 ft^2 **30.** 9036.24 ft^2 **31.** perimeter   **32.** area   **33.** 12 in., 12π in., 36π in.^{2}
34. 18 in., 18π in., 81π in.^{2} **35.** 5 ft, 10π ft, 25π ft^2 **36.** 20 ft, 40π ft, 400π ft^2 **37.** 6 cm, 12 cm, 36π cm^2
38. 9 cm, 18 cm, 81π cm^2 **39.** 10 in., 20 in., 20π in.   **40.** 16 in., 32 in., 32π in.   **41.** 14.5   **42.** 11
43. 7   **44.** 23   **45.** 5.1   **46.** 4   **47.** 5   **48.** 8   **49.** 5   **50.** 7.5   **51.** 2.4   **52.** 1.5
53. (a) 20 cm^2 (b) 80 cm^2 (c) 180 cm^2 (d) 320 cm^2 (e) 4   (f) 3; 9   (g) 4; 16   (h) n^2 **54.** $240
55. $800   **56.** $720   **57.** n^2 **58.** n **59.** 80   **60.** 115   **61.** 76.26   **62.** 164.48   **63.** 132 ft^2
64. 868 cm^2 **65.** 5376 cm^2 **66.** 309.91 ft^2 **67.** 145.34 m^2 **68.** 100.48 cm^2 **69.** 14-in. pizza

70. 14-in. pizza **71.** 14-in. pizza **72.** 14-in. pizza **73.** 26 in. **74.** 54 in. **75.** 625 ft²
76. 256 in.² **77.** 648 in.² **78.** yes; 40 in. **79.** $\dfrac{(4-\pi)r^2}{4}$ **80.** 72 in.²

9.4 Exercises (Page 530)

1.

STATEMENTS	REASONS
1. $AC = BD$	1. Given
2. $AD = BC$	2. Given
3. $AB = AB$	3. Reflexive property
4. $\triangle ABD \cong \triangle BAC$	4. SSS congruence property

2.

STATEMENTS	REASONS
1. $AC = BC$	1. Given
2. $\angle ACD = \angle BCD$	2. Given
3. $CD = CD$	3. Reflexive property
4. $\triangle ADC \cong \triangle BDC$	4. SAS congruence property

3.

STATEMENTS	REASONS
1. $\overleftrightarrow{DB}$ is perpendicular to $\overleftrightarrow{AC}$.	1. Given
2. $AB = BC$	2. Given
3. $\angle ABD = \angle CBD$	3. Both are right angles by definition of perpendicularity.
4. $DB = DB$	4. Reflexive property
5. $\triangle ABD \cong \triangle CBD$	5. SAS congruence property

4.

STATEMENTS	REASONS
1. $BC = BA$	1. Given
2. $\angle 1 = \angle 2$	2. Given
3. $\angle DBC = \angle DBA$	3. Supplements of equal angles are equal to each other.
4. $DB = DB$	4. Reflexive property
5. $\triangle DBC \cong \triangle DBA$	5. SAS congruence property

5.

STATEMENTS	REASONS
1. $\angle BAC = \angle DAC$	1. Given
2. $\angle BCA = \angle DCA$	2. Given
3. $AC = AC$	3. Reflexive property
4. $\triangle ABC \cong \triangle ADC$	4. ASA congruence property

6.

STATEMENTS	REASONS
1. $BO = OE$	1. Given
2. $\overleftrightarrow{OB}$ is perpendicular to $\overleftrightarrow{AC}$.	2. Given
3. $\overleftrightarrow{OE}$ is perpendicular to $\overleftrightarrow{DF}$.	3. Given
4. $\angle ABO = \angle FEO$	4. Both are right angles by definition of perpendicularity.
5. $\angle AOB = \angle FOE$	5. Vertical angles are equal.
6. $\triangle AOB \cong \triangle FOE$	6. ASA congruence property

7. 67°; 67° **8.** 76° **9.** 6 in. **10.** 15 in. **13.** $\angle A$ and $\angle P$; $\angle C$ and $\angle R$; $\angle B$ and $\angle Q$; $\overleftrightarrow{AC}$ and $\overleftrightarrow{PR}$; $\overleftrightarrow{CB}$ and $\overleftrightarrow{RQ}$; $\overleftrightarrow{AB}$ and $\overleftrightarrow{PQ}$ **14.** $\angle A$ and $\angle P$; $\angle B$ and $\angle Q$; $\angle C$ and $\angle R$; $\overleftrightarrow{AC}$ and $\overleftrightarrow{PR}$; $\overleftrightarrow{CB}$ and $\overleftrightarrow{RQ}$; $\overleftrightarrow{AB}$ and $\overleftrightarrow{PQ}$ **15.** $\angle H$ and $\angle F$; $\angle K$ and $\angle E$; $\angle HGK$ and $\angle FGE$; $\overleftrightarrow{HK}$ and $\overleftrightarrow{FE}$; $\overleftrightarrow{GK}$ and $\overleftrightarrow{GE}$; $\overleftrightarrow{HG}$ and $\overleftrightarrow{FG}$ **16.** $\angle A$ and $\angle C$; $\angle E$ and $\angle D$; $\angle ABE$ and $\angle CBD$; $\overleftrightarrow{EB}$ and $\overleftrightarrow{DB}$; $\overleftrightarrow{AB}$ and $\overleftrightarrow{CB}$; $\overleftrightarrow{AE}$ and $\overleftrightarrow{CD}$ **17.** $\angle P = 78°$; $\angle M = 46°$; $\angle A = \angle N = 56°$ **18.** $\angle P = 90°$; $\angle Q = 42°$; $\angle B = \angle R = 48°$ **19.** $\angle T = 74°$; $\angle Y = 28°$; $\angle Z = \angle W = 78°$ **20.** $\angle B = 106°$; $\angle A = \angle M = 44°$ **21.** $\angle T = 20°$; $\angle V = 64°$; $\angle R = \angle U = 96°$ **22.** $\angle X = \angle M = 52°$ **23.** $a = 20$; $b = 15$ **24.** $a = 30$; $b = 60$ **25.** $a = 6$; $b = \dfrac{15}{2}$ **26.** $a = 2$ **27.** $x = 6$ **28.** $m = 18$ **29.** $x = 110$ **30.** $y = 12$ **31.** $c = 111\tfrac{1}{9}$ **32.** $m = 85\tfrac{1}{3}$ **33.** 30 m

34. 108 ft **35.** 500 m, 700 m **36.** 1.75 m **37.** 112.5 ft **38.** Phoenix to Tucson: 153.3 km; Yuma to Tucson: 325.8 km **39.** $506\frac{2}{3}$ ft **40.** 8 ft, 11 in. **41.** $c = 17$ **42.** $b = 24$ **43.** $a = 13$ **44.** $b = 7$ cm **45.** $c = 50$ m **46.** $b = 96$ km **47.** $a = 20$ in. **48.** $a = 119$ ft **49.** The sum of the squares of the two shorter sides of a right triangle is equal to the square of the longest side. **50.** For example, let $a = 3$, $b = 4$, $c = 5$. We have $3^2 + 4^2 = 5^2$ true, but $3 + 4 = 5$ false. **51.** $(3, 4, 5)$ **52.** $(5, 12, 13)$ **53.** $(7, 24, 25)$ **54.** $(8, 6, 10)$ **55.** $(12, 16, 20)$ **56.** $(15, 8, 17)$ **58.** $(20, 21, 29)$; $(119, 120, 169)$; $(696, 697, 985)$ **59.** $(3, 4, 5)$ **60.** $(5, 12, 13)$ **61.** $(7, 24, 25)$ **62.** $(9, 40, 41)$ **65.** $(4, 3, 5)$ **66.** $(6, 8, 10)$ **67.** $(8, 15, 17)$ **68.** $(10, 24, 26)$ **70.** No, since if the integer a is the length of one of the equal sides, then $\sqrt{2}\,a$ must be the length of the hypotenuse, and $\sqrt{2}\,a$ is not an integer. **71.** 24 m **72.** 8 cm, 15 cm, 17 cm **73.** 16 ft **74.** length: 4 in.; width: 3 in. **75.** 4.55 ft **76.** 12 ft **77.** 19 ft, 3 in. **78.** 24 ft, 5 in. **79.** 28 ft, 10 in. **80.** 37 ft, 9 in. **81.** (a) $\frac{1}{2}(a + b)(a + b)$ (b) PWX: $\frac{1}{2}ab$; PZY: $\frac{1}{2}ab$; PXY: $\frac{1}{2}c^2$ (c) $\frac{1}{2}(a + b)(a + b) = \frac{1}{2}ab + \frac{1}{2}ab + \frac{1}{2}c^2$. When simplified, this gives $a^2 + b^2 = c^2$. **82.** (a) b (b) k (c) cj (d) ck (e) $j + k$; $j + k$; $a^2 + b^2 = c^2$ **83.** $24 + 4\sqrt{6}$ **84.** 672 in.² **85.** 10 **86.** 9 **87.** $256 + 64\sqrt{3}$ **88.** 115° **89.** 5 in. **90.** $\overleftrightarrow{BC}$ measures 12 cm and $\overleftrightarrow{AC}$ measures 5 cm.

9.5 Exercises (Page 542)

1. true **2.** true **3.** true **4.** false **5.** false **6.** true **7.** (a) $3\frac{3}{4}$ m³ (b) $14\frac{3}{4}$ m² **8.** (a) 96 cm³ (b) 128 cm² **9.** (a) 96 in.³ (b) 130.4 in.² **10.** (a) 1436.03 m³ (b) 615.44 m² **11.** (a) 267,946.67 ft³ (b) 20,096 ft² **12.** (a) 1696.54 cm³ (b) 687.79 cm² **13.** (a) 549.5 cm³ (b) 376.8 cm² **14.** (a) 1808.64 m³ (b) 1205.76 m² **15.** (a) 65.94 m³ (b) 100.00 m² **16.** (a) 100.48 cm³ (b) 140.81 cm² **17.** 168 in.³ **18.** 160 ft³ **19.** 1969.10 cm³ **20.** 305.46 cm³ **21.** 427.29 cm³ **22.** 125.6 in.³ **23.** 508.68 cm³ **24.** 30.38 cm³ **25.** 2,415,766.67 m³ **26.** 1,461,732.8 m³ **27.** .52 m³ **28.** 200.96 in.³ **29.** 12 in., 288π in.³, 144π in.² **30.** 18 in., 972π in.³, 324π in.² **31.** 5 ft, $\frac{500}{3}\pi$ ft³, 100π ft² **32.** 20 ft, $\frac{32,000}{3}\pi$ ft³, 1600π ft² **33.** 2 cm, 4 cm, 16π cm² **34.** 4 cm, 8 cm, 64π cm² **35.** 1 m, 2 m, $\frac{4}{3}\pi$ m³ **36.** 6 m, 12 m, 288π m³ **37.** volume **38.** surface area **39.** $\sqrt[3]{2}\,x$ **40.** (a) $\frac{4}{3}\pi$ m³ (b) $\frac{32}{3}\pi$ m³ (c) 8 times (d) 36π m³ (e) 27 times (f) n^3 **41.** $8100 **42.** $19,200 **43.** $37,500 **44.** n^2 **45.** 2.5 **46.** 9 **47.** 6 **48.** 7 **49.** 210 in.³ **50.** 1 to 2 **51.** $\frac{62,500}{3}\pi$ in.³ **52.** It is multiplied by 4.5. **53.** 2 to 1 **54.** $16\sqrt{2}$ in. **55.** 288 **56.** $\frac{1 + \sqrt{5}}{2}$ (the golden ratio)

9.6 Exercises (Page 554)

1.  **2.**   **3.**   **4.**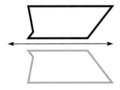

5.

6.

7. The figure is its own reflection image.

8. The figure is its own reflection image.

9.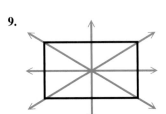

10.

11.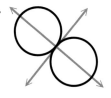

12.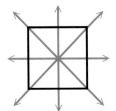

13.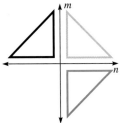

14.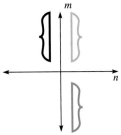

15.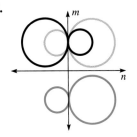

16.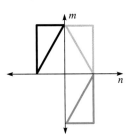

17.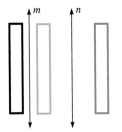

18.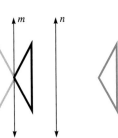

19.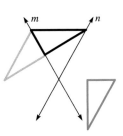

20.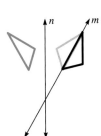

21.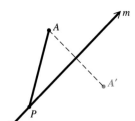

22.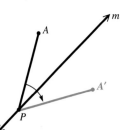

23.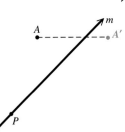

24. 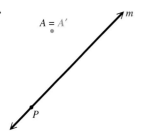   **25.** 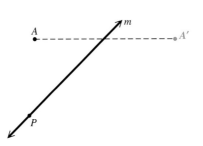   **26.**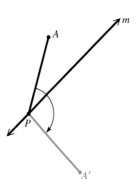

27. 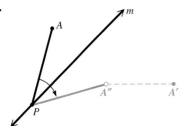   **28.**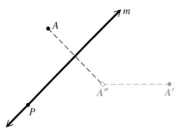

29. 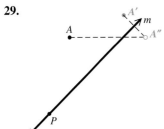   **30.** 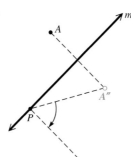   **31.**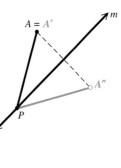

32. 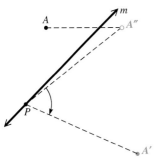   **33.** yes   **34.** no   **35.**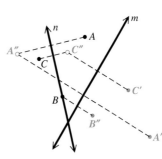

36. (a) yes   (b) yes   **37.**   **38.**

39. – **46.**

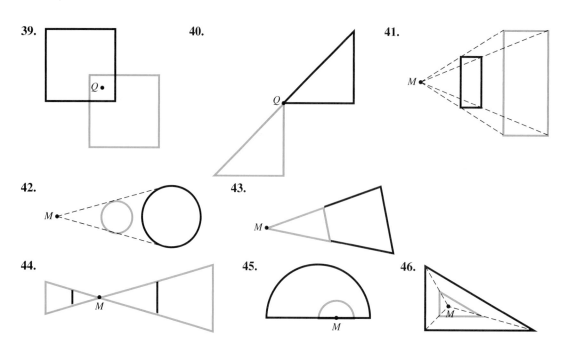

9.7 Exercises (Page 563)

1. Euclidean **2.** Riemannian **3.** Lobachevskian **4.** less than **5.** greater than **6.** Lobachevskian
7. Riemannian **8.** Riemannian **9.** Euclidean **10.**

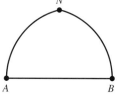

11. (g)

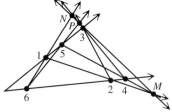

(h) Suppose that a hexagon is inscribed in an angle. Let each pair of opposite sides be extended so as to intersect. Then the three points of intersection thus obtained will lie in a straight line.

12.   **13.** C **14.** A, E **15.** A, E **16.** B, D **17.** A, E

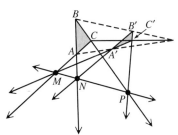

18. A, E **19.** none of them **20.** B, D **21.** no **22.** yes **23.** yes **24.** no **25.** 1 **26.** 1
27. 3 **28.** 2 **29.** 1 **30.** 0 **31.** A, C, D, and F are even; B and E are odd. **32.** A, D, E, and H are even; B, C, F, and G are odd. **33.** A, B, C, and F are odd; D, E, and G are even. **34.** A, C, and D are even; B and E are odd. **35.** A, B, C, and D are odd; E is even. **36.** All vertices $A-F$ are odd.

37. traversable **38.** not traversable **39.** not traversable **40.** traversable

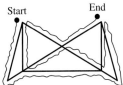
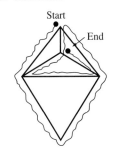

41. traversable **42.** traversable **43.** yes

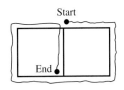

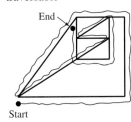
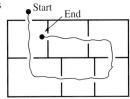

44. yes **45.** no **46.** no

9.8 Exercises (Page 570)

1. 4 **2.** 4 **3.** 2 **4.** 2/1 = 2 **5.** 4/1 = 4 **6.** 3/1 = 3; 9/1 = 9 **7.** 4/1 = 4; 16/1 = 16
8. 4, 9, 16, 25, 36, 100 **9.** Each ratio in the bottom row is the square of the scale factor in the top row. **10.** 4
11. 4, 9, 16, 25, 36, 100 **12.** Each ratio in the bottom row is again the square of the scale factor in the top row.
13. Answers will vary. Some examples are: $3^d = 9$, thus $d = 2$; $5^d = 25$, thus $d = 2$; $4^d = 16$, thus $d = 2$. **14.** 8
15. 2/1 = 2; 8/1 = 8 **16.** 8, 27, 64, 125, 216, 1000 **17.** Each ratio in the bottom row is the cube of the scale factor in the top row. **18.** Since $2^3 = 8$, the value of d in $2^d = 8$ must be 3. **19.** 3/1 = 3 **20.** 4
21. 1.262 or $\frac{\ln 4}{\ln 3}$ **22.** 2/1 = 2 **23.** 3 **24.** It is between 1 and 2. **25.** 1.585 or $\frac{\ln 3}{\ln 2}$
26. .812, .495, .812, .495, ... The two attractors are .812 and .495. **27.** .842, .452, .842, .452, ... The two attractors are .842 and .452. **28.** .540, .882, .370, .828, .506, .887, .355, .813, ... The eight attractors are these eight numbers.

Chapter 9 Test (Page 574)

1. (a) 52° **(b)** 142° **(c)** acute **2.** 40°, 140° **3.** 45°, 45° **4.** 30°, 60° **5.** 130°, 50°
6. 117°, 117° **8.** C **9.** both **10.** neither **11.** 30°, 45°, 105° **12.** 72 cm²
13. 60 in.² **14.** 68 m² **15.** 180 m² **16.** 24π in. **17.** 1978 ft **18.** 57 cm²
19.

STATEMENTS	REASONS
1. ∡CAB = ∡DBA	1. Given
2. DB = CA	2. Given
3. AB = AB	3. Reflexive property
4. △ABD ≅ △BAC	4. SAS congruence property

20. 20 ft **21.** 29 m

22. 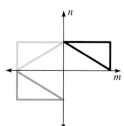   **23.** 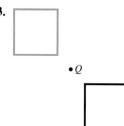   **24. (a)** 904.32 in.³   **(b)** 452.16 in.²
25. (a) 864 ft³   **(b)** 552 ft²   **26. (a)** 1582.56 m³   **(b)** 753.60 m²   **27.** Answers will vary.
28. (a) yes   **(b)** no   **29. (a)** yes   **(b)** no   **30.** The only attractor is .5238095238.

CHAPTER 10   Trigonometry

10.1 Exercises   (Page 581)

1. (a) 60° **(b)** 150°   **2. (a)** 30° **(b)** 120°   **3. (a)** 45° **(b)** 135°   **4. (a)** 35° **(b)** 125°
5. (a) 1° **(b)** 91°   **6. (a)** 88° **(b)** 178°   **7.** $90 - x$ degrees   **8.** $180 - x$ degrees   **9.** 83°59′
10. 158°47′   **11.** 119°27′   **12.** 16°12′   **13.** 38°32′   **14.** 55°9′   **15.** 17°1′49″   **16.** 53°41′13″
17. 20.900°   **18.** 38.700°   **19.** 91.598°   **20.** 34.860°   **21.** 274.316°   **22.** 165.853°
23. 31°25′47″   **24.** 59°5′7″   **25.** 89°54′1″   **26.** 102°22′38″   **27.** 178°35′58″   **28.** 122°41′7″
29. 320°   **30.** 262°   **31.** 235°   **32.** 157°   **33.** 179°   **34.** 339°   **35.** 130°   **36.** 280°
37. $30° + n \cdot 360°$ **38.** $45° + n \cdot 360°$ **39.** $60° + n \cdot 360°$ **40.** $90° + n \cdot 360°$
Angles other than those given are possible in Exercises 41–48.

41.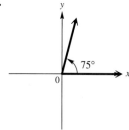
435°; −285°;
quadrant I

42.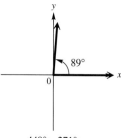
449°; −271°;
quadrant I

43.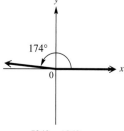
534°; −186°;
quadrant II

44.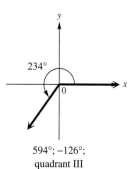
594°; −126°;
quadrant III

45.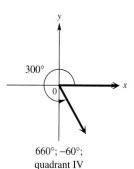
660°; −60°;
quadrant IV

46.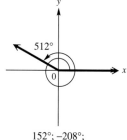
152°; −208°;
quadrant II

47.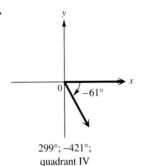
299°; −421°;
quadrant IV

48.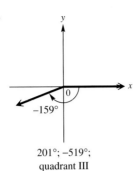
201°; −519°;
quadrant III

10.2 Exercises (Page 585)

1.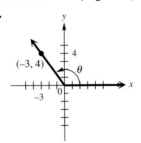

2.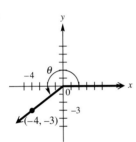

3.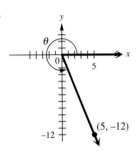

4.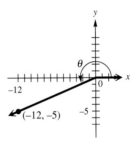

In Exercises 5–14 we give, in order, sine, cosine, tangent, cotangent, secant, and cosecant.

5. $\frac{4}{5}; -\frac{3}{5}; -\frac{4}{3}; -\frac{3}{4}; -\frac{5}{3}; \frac{5}{4}$ **6.** $-\frac{3}{5}; -\frac{4}{5}; \frac{3}{4}; \frac{4}{3}; -\frac{5}{4}; -\frac{5}{3}$ **7.** 1; 0; undefined; 0; undefined; 1

8. 0; −1; 0; undefined; −1; undefined   **9.** $\frac{\sqrt{3}}{2}; \frac{1}{2}; \sqrt{3}; \frac{\sqrt{3}}{3}; 2; \frac{2\sqrt{3}}{3}$ **10.** $-\frac{1}{2}; -\frac{\sqrt{3}}{2}; \frac{\sqrt{3}}{3}; \sqrt{3}; -\frac{2\sqrt{3}}{3}; -2$

11. $\frac{5\sqrt{34}}{34}; \frac{3\sqrt{34}}{34}; \frac{5}{3}; \frac{3}{5}; \frac{\sqrt{34}}{3}; \frac{\sqrt{34}}{5}$ **12.** $\frac{7\sqrt{53}}{53}; -\frac{2\sqrt{53}}{53}; -\frac{7}{2}; -\frac{2}{7}; -\frac{\sqrt{53}}{2}; \frac{\sqrt{53}}{7}$

13. 0; −1; 0; undefined; −1; undefined   **14.** 1; 0; undefined; 0; undefined; 1   **16.** 0   **17.** It is the distance from a point (x, y) on the terminal side of the angle to the origin.   **18.** $\tan\theta$ and $\cot\theta$ are positive; all other function values are negative.   **19.** positive   **20.** negative   **21.** negative   **22.** negative   **23.** positive   **24.** negative   **25.** negative   **26.** negative   **27.** 0   **28.** 1   **29.** undefined   **30.** 0   **31.** undefined   **32.** 1   **33.** 0   **34.** −1   **35.** 0   **36.** 0   **37.** 1   **38.** 0   **39.** 0   **40.** −1

10.3 Exercises (Page 590)

1. $-\frac{1}{3}$ **2.** $\frac{1}{5}$ **3.** $\frac{1}{3}$ **4.** $-\frac{2}{5}$ **5.** −5   **6.** $\frac{\sqrt{15}}{15}$ **7.** $2\sqrt{2}$ **8.** $-\sqrt{7}$ **9.** $-\frac{3\sqrt{5}}{5}$
10. $\frac{5\sqrt{11}}{11}$ **11.** $\frac{2}{3}$ **12.** $\frac{2}{15}$ **13.** II   **14.** I   **15.** I   **16.** IV   **17.** II   **18.** III
19. II or III   **20.** I or III

In Exercises 21–30 we give signs first of the sine and cosecant, then of the cosine and secant, and finally of the tangent and cotangent.

21. +; +; + **22.** +; −; − **23.** −; −; + **24.** −; +; − **25.** −; +; − **26.** +; +; +
27. +; +; + **28.** −; +; − **29.** −; +; − **30.** −; −; + **31.** $-2\sqrt{2}$ **32.** $-\dfrac{\sqrt{5}}{3}$ **33.** $\dfrac{\sqrt{15}}{4}$
34. $-\dfrac{\sqrt{5}}{2}$ **35.** $-2\sqrt{2}$ **36.** $-\dfrac{4}{3}$ **37.** $\dfrac{\sqrt{15}}{4}$ **38.** $-\dfrac{\sqrt{3}}{2}$

In Exercises 39–46 we give, in order, sine, cosine, tangent, cotangent, secant, and cosecant.

39. $\dfrac{15}{17}; -\dfrac{8}{17}; -\dfrac{15}{8}; -\dfrac{8}{15}; -\dfrac{17}{8}; \dfrac{17}{15}$ **40.** $-\dfrac{4}{5}; -\dfrac{3}{5}; \dfrac{4}{3}; \dfrac{3}{4}; -\dfrac{5}{3}; -\dfrac{5}{4}$ **41.** $-\dfrac{4}{5}; -\dfrac{3}{5}; \dfrac{4}{3}; \dfrac{3}{4}; -\dfrac{5}{3}; -\dfrac{5}{4}$

42. $\dfrac{7}{25}; -\dfrac{24}{25}; -\dfrac{7}{24}; -\dfrac{24}{7}; -\dfrac{25}{24}; \dfrac{25}{7}$ **43.** $-\dfrac{\sqrt{3}}{2}; -\dfrac{1}{2}; \sqrt{3}; \dfrac{\sqrt{3}}{3}; -2; -\dfrac{2\sqrt{3}}{3}$

44. $\dfrac{1}{2}; -\dfrac{\sqrt{3}}{2}; -\dfrac{\sqrt{3}}{3}; -\sqrt{3}; -\dfrac{2\sqrt{3}}{3}; 2$ **45.** $\dfrac{\sqrt{5}}{7}; \dfrac{2\sqrt{11}}{7}; \dfrac{\sqrt{55}}{22}; \dfrac{2\sqrt{55}}{5}; \dfrac{7\sqrt{11}}{22}; \dfrac{7\sqrt{5}}{5}$

46. $\dfrac{8\sqrt{67}}{67}; \dfrac{\sqrt{201}}{67}; \dfrac{8\sqrt{3}}{3}; \dfrac{\sqrt{3}}{8}; \dfrac{\sqrt{201}}{3}; \dfrac{\sqrt{67}}{8}$

10.4 Exercises (Page 598)

In Exercises 1–6, we give, in order, sine, cosine, tangent, cotangent, secant, and cosecant.

1. $\dfrac{3}{5}; \dfrac{4}{5}; \dfrac{3}{4}; \dfrac{4}{3}; \dfrac{5}{4}; \dfrac{5}{3}$ **2.** $\dfrac{8}{17}; \dfrac{15}{17}; \dfrac{8}{15}; \dfrac{15}{8}; \dfrac{17}{15}; \dfrac{17}{8}$ **3.** $\dfrac{21}{29}; \dfrac{20}{29}; \dfrac{21}{20}; \dfrac{20}{21}; \dfrac{29}{20}; \dfrac{29}{21}$ **4.** $\dfrac{45}{53}; \dfrac{28}{53}; \dfrac{45}{28}; \dfrac{28}{45}; \dfrac{53}{28}; \dfrac{53}{45}$

5. $\dfrac{n}{p}; \dfrac{m}{p}; \dfrac{n}{m}; \dfrac{m}{n}; \dfrac{p}{m}; \dfrac{p}{n}$ **6.** $\dfrac{k}{z}; \dfrac{y}{z}; \dfrac{k}{y}; \dfrac{y}{k}; \dfrac{z}{y}; \dfrac{z}{k}$

In Exercises 7–10, we give, in order, the unknown side, sine, cosine, tangent, cotangent, secant, and cosecant.

7. $c = 13; \dfrac{12}{13}; \dfrac{5}{13}; \dfrac{12}{5}; \dfrac{5}{12}; \dfrac{13}{5}; \dfrac{13}{12}$ **8.** $c = \sqrt{34}; \dfrac{5\sqrt{34}}{34}; \dfrac{3\sqrt{34}}{34}; \dfrac{5}{3}; \dfrac{3}{5}; \dfrac{\sqrt{34}}{3}; \dfrac{\sqrt{34}}{5}$

9. $b = \sqrt{13}; \dfrac{\sqrt{13}}{7}; \dfrac{6}{7}; \dfrac{\sqrt{13}}{6}; \dfrac{6\sqrt{13}}{13}; \dfrac{7}{6}; \dfrac{7\sqrt{13}}{13}$ **10.** $a = \sqrt{95}; \dfrac{7}{12}; \dfrac{\sqrt{95}}{12}; \dfrac{7\sqrt{95}}{95}; \dfrac{\sqrt{95}}{7}; \dfrac{12\sqrt{95}}{95}; \dfrac{12}{7}$

11. cot 40° **12.** tan 17° **13.** sec 43° **14.** csc 51° **15.** cot 64.6° **16.** cos 51.3° **17.** sin 76°30′
18. cot 63°50′ **19.** $\dfrac{\sqrt{3}}{3}$ **20.** $\sqrt{3}$ **21.** $\dfrac{1}{2}$ **22.** $\dfrac{\sqrt{3}}{2}$ **23.** $\sqrt{2}$ **24.** $\sqrt{2}$ **25.** $\dfrac{\sqrt{2}}{2}$
26. $\dfrac{\sqrt{2}}{2}$ **27.** $\dfrac{\sqrt{3}}{2}$ **28.** $\dfrac{1}{2}$ **29.** $\sqrt{3}$ **30.** $\dfrac{\sqrt{3}}{3}$ **31.** 82° **32.** 32° **33.** 45° **34.** 60°
35. 30° **36.** 60°

In Exercises 37–56, we give, in order, sine, cosine, tangent, cotangent, secant, and cosecant.

37. $\dfrac{\sqrt{3}}{2}; -\dfrac{1}{2}; -\sqrt{3}; -\dfrac{\sqrt{3}}{3}; -2; \dfrac{2\sqrt{3}}{3}$ **38.** $\dfrac{\sqrt{2}}{2}; -\dfrac{\sqrt{2}}{2}; -1; -1; -\sqrt{2}; \sqrt{2}$

39. $\dfrac{1}{2}; -\dfrac{\sqrt{3}}{2}; -\dfrac{\sqrt{3}}{3}; -\sqrt{3}; -\dfrac{2\sqrt{3}}{3}; 2$ **40.** $-\dfrac{\sqrt{2}}{2}; -\dfrac{\sqrt{2}}{2}; 1; 1; -\sqrt{2}; -\sqrt{2}$

41. $-\dfrac{\sqrt{3}}{2}; -\dfrac{1}{2}; \sqrt{3}; \dfrac{\sqrt{3}}{3}; -2; -\dfrac{2\sqrt{3}}{3}$ **42.** $-\dfrac{\sqrt{3}}{2}; \dfrac{1}{2}; -\sqrt{3}; -\dfrac{\sqrt{3}}{3}; 2; -\dfrac{2\sqrt{3}}{3}$

43. $-\dfrac{\sqrt{2}}{2}; \dfrac{\sqrt{2}}{2}; -1; -1; \sqrt{2}; -\sqrt{2}$ **44.** $\dfrac{\sqrt{2}}{2}; \dfrac{\sqrt{2}}{2}; 1; 1; \sqrt{2}; \sqrt{2}$

45. $\dfrac{\sqrt{3}}{2}; \dfrac{1}{2}; \sqrt{3}; \dfrac{\sqrt{3}}{3}; 2; \dfrac{2\sqrt{3}}{3}$ **46.** $\dfrac{\sqrt{3}}{2}; -\dfrac{1}{2}; -\sqrt{3}; -\dfrac{\sqrt{3}}{3}; -2; \dfrac{2\sqrt{3}}{3}$

47. $\dfrac{\sqrt{2}}{2}; -\dfrac{\sqrt{2}}{2}; -1; -1; -\sqrt{2}; \sqrt{2}$ **48.** $-\dfrac{1}{2}; \dfrac{\sqrt{3}}{2}; \dfrac{\sqrt{3}}{3}; \sqrt{3}; -\dfrac{2\sqrt{3}}{3}; -2$

49. $\dfrac{1}{2}; \dfrac{\sqrt{3}}{2}; \dfrac{\sqrt{3}}{3}; \sqrt{3}; \dfrac{2\sqrt{3}}{3}; 2$ **50.** $-\dfrac{\sqrt{2}}{2}; -\dfrac{\sqrt{2}}{2}; 1; 1; -\sqrt{2}; -\sqrt{2}$

51. $\dfrac{\sqrt{3}}{2}; \dfrac{1}{2}; \sqrt{3}; \dfrac{\sqrt{3}}{3}; 2; \dfrac{2\sqrt{3}}{3}$ **52.** $\dfrac{1}{2}; -\dfrac{\sqrt{3}}{2}; -\dfrac{\sqrt{3}}{3}; -\sqrt{3}; -\dfrac{2\sqrt{3}}{3}; 2$

53. $-\dfrac{1}{2}; \dfrac{\sqrt{3}}{2}; -\dfrac{\sqrt{3}}{3}; -\sqrt{3}; \dfrac{2\sqrt{3}}{3}; -2$ **54.** $-\dfrac{1}{2}; -\dfrac{\sqrt{3}}{2}; \dfrac{\sqrt{3}}{3}; \sqrt{3}; -\dfrac{2\sqrt{3}}{3}; -2$

55. $\dfrac{\sqrt{3}}{2}; \dfrac{1}{2}; \sqrt{3}; \dfrac{\sqrt{3}}{3}; 2; \dfrac{2\sqrt{3}}{3}$ **56.** $\dfrac{1}{2}; -\dfrac{\sqrt{3}}{2}; -\dfrac{\sqrt{3}}{3}; -\sqrt{3}; -\dfrac{2\sqrt{3}}{3}; 2$

57. $\dfrac{\sqrt{3}}{3}; \sqrt{3}$ **58.** $\dfrac{\sqrt{2}}{2}; \dfrac{\sqrt{2}}{2}; \sqrt{2}; \sqrt{2}$ **59.** $\dfrac{\sqrt{3}}{2}; \dfrac{\sqrt{3}}{3}; \dfrac{2\sqrt{3}}{3}$ **60.** $-\dfrac{1}{2}; -\dfrac{\sqrt{3}}{3}; -2$

61. $-1; -1$ **62.** $\dfrac{1}{2}; -\sqrt{3}; -\dfrac{2\sqrt{3}}{3}$ **63.** $-\dfrac{\sqrt{3}}{2}; -\dfrac{2\sqrt{3}}{3}$ **64.** $\sqrt{3}; \dfrac{\sqrt{3}}{3}$ **65.** 210°, 330°

66. 120°, 240°   **67.** 45°, 225°   **68.** 30°, 210°   **69.** 60°, 120°   **70.** 30°, 330°   **71.** 120°, 240°
72. 210°, 330°   **73.** 225°, 315°   **74.** 135°, 225°   **75.** 120°, 300°   **76.** 135°, 315°

10.5 Exercises (Page 606)
1. .5657728   **2.** .6252427   **3.** 1.1342773   **4.** .8896822   **5.** 1.0273488   **6.** 1.4267182
7. .6383201   **8.** .3095517   **9.** 1.7768146   **10.** 15.055723   **11.** .4771588   **12.** 1.4887142
13. −5.7297416   **14.** .6743024   **15.** 1.9074147   **16.** .9999905   **17.** 57.997172°   **18.** 55.845496°
19. 30.502748°   **20.** 38.491580°   **21.** 46.173581°   **22.** 68.673241°   **23.** 81.168073°
24. 16.166641°   **25.** $B = 53°40'; a = 571$ m; $b = 777$ m   **26.** $B = 58°20'; c = 68.4$ km; $b = 58.2$ km
27. $M = 38.8°; n = 154$ m; $p = 198$ m   **28.** $Y = 42.2°; x = 66.4$ cm; $y = 60.2$ cm
29. $A = 47.9108°; c = 84.816$ cm; $a = 62.942$ cm   **30.** $A = 21.4858°; b = 3330.68$ m; $a = 1311.04$ m
31. $B = 62.00°; a = 8.17$ ft; $b = 15.4$ ft   **32.** $A = 44.00°; a = 20.6$ m; $b = 21.4$ m
33. $A = 17.00°; a = 39.1$ in.; $c = 134$ in.   **34.** $B = 29°00'; a = 70.7$ cm; $c = 80.9$ cm
35. $c = 85.9$ yd; $A = 62°50'; B = 27°10'$ **36.** $c = 1080$ m; $A = 63°00'; B = 27°00'$
37. $b = 42.3$ cm; $A = 24°10'; B = 65°50'$ **38.** $a = 609$ m; $A = 70°10'; B = 19°50'$
39. $B = 36°36'; a = 310.8$ ft; $b = 230.8$ ft   **40.** $B = 76°13'; a = 306.2$ m; $b = 1248$ m
41. $A = 50°51'; a = .4832$ m; $b = .3934$ m   **42.** $A = 7°9'; b = 4.787$ cm; $a = .6006$ cm   **43.** 9.35 meters
44. 67°10'   **45.** 88.3 meters   **46.** 33.4 meters   **47.** 26.92 inches   **48.** 134.7 centimeters
49. 583 feet   **50.** 13.3 feet   **51.** 28.0 meters   **52.** 37°40'   **53.** 469 meters   **54.** 42,600 feet
55. 146 meters   **56.** 26°20'   **57.** 34.0 miles   **58. (a)** 29,008 feet   **(b)** shorter
59. $a = 12, b = 12\sqrt{3}, d = 12\sqrt{3}, c = 12\sqrt{6}$ **60.** $x = \dfrac{9\sqrt{3}}{2}, y = \dfrac{9}{2}, z = \dfrac{3\sqrt{3}}{2}, w = 3\sqrt{3}$

10.6 Exercises (Page 616)
1. $C = 95°, b = 13$ m, $a = 11$ m   **2.** $A = 99°, b = 34$ cm, $c = 21$ cm   **3.** $B = 37.3°, a = 38.5$ ft, $b = 51.0$ ft
4. $A = 37.2°, a = 178$ m, $c = 244$ m   **5.** $C = 57.36°, b = 11.13$ ft, $c = 11.55$ ft
6. $A = 36.54°, b = 44.17$ m, $a = 28.10$ m   **7.** $B = 18.5°, a = 239$ yd, $c = 230$ yd
8. $A = 49°40', b = 16.1$ cm, $c = 25.8$ cm   **9.** $A = 56°00', AB = 361$ ft, $BC = 308$ ft
10. $C = 91.9°, BC = 490$ ft, $AB = 847$ ft   **11.** $B = 110.0°, a = 27.01$ m, $c = 21.36$ m
12. $A = 65.60°, b = 1.942$ cm, $c = 2.727$ cm   **13.** $A = 34.72°, a = 3326$ ft, $c = 5704$ ft
14. $C = 109.72°, a = 955.4$ yd, $b = 2327$ yd   **15.** $C = 97°34', b = 283.2$ m, $c = 415.2$ m
16. $B = 67°45', b = 37.50$ mm, $a = 22.04$ mm   **17.** 118 meters   **18.** 448 yards   **19.** 10.4 inches
20. 111°   **21.** 12   **22.** .49 mile   **23.** 7   **24.** 5   **25.** 30°   **26.** 120°
27. $c = 2.83$ in., $A = 44.9°, B = 106.8°$ **28.** $a = 2.60$ yd, $B = 45.1°, C = 93.5°$
29. $c = 6.46$ m, $A = 53.1°, B = 81.3°$ **30.** $a = 43.7$ km, $C = 59.5°, B = 53.2°$
31. $a = 156$ cm, $B = 64°50', C = 34°30'$ **32.** $c = 348$ ft, $A = 63°50', B = 43°30'$

33. $b = 9.529$ in., $A = 64.59°$, $C = 40.61°$ **34.** $c = 4.276$ mi, $A = 48.77°$, $B = 71.53°$
35. $a = 15.7$ m, $B = 21.6°$, $C = 45.6°$ **36.** $b = 34.1$ cm, $A = 5.2°$, $C = 6.6°$ **37.** $A = 30°$, $B = 56°$, $C = 94°$
38. $A = 24°$, $B = 31°$, $C = 125°$ **39.** $A = 82°$, $B = 37°$, $C = 61°$ **40.** $A = 29°$, $B = 53°$, $C = 98°$
41. $A = 42.0°$, $B = 35.9°$, $C = 102.1°$ **42.** $A = 33.2°$, $B = 38.7°$, $C = 108.1°$ **43.** 257 meters
44. 5.2 centimeters and 8.8 centimeters **45.** 22 feet **46.** 18 feet **47.** 36° with the 45-foot cable and 26° with the 60-foot cable **48.** approximately 47.5 feet **49.** $\dfrac{\sqrt{3}}{2}$ **50.** $\sqrt{3}$ **51.** $\dfrac{\sqrt{2}}{2}$ **52.** 1
53. 46.4 m² **54.** 732 ft² **55.** 356 cm² **56.** 163 km² **57.** 722.9 in.² **58.** 289.9 m² **59.** $24\sqrt{3}$
60. $15\sqrt{3}$ **61.** 78 m² **62.** 310 in.² **63.** 12,600 cm² **64.** 228 yd² **65.** 3650 ft² **66.** 83.01 in.²
67. 100 m² **68.** 373 m² **69.** 33 cans **70.** 392,000 mi²

10.7 Exercises (Page 625)
1. G **2.** A **3.** E **4.** D **5.** B **6.** H **7.** F **8.** C
9. 2π; 2 **10.** 2π; 3 **11.** $2\pi; \dfrac{2}{3}$

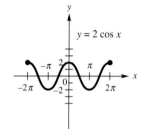
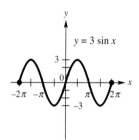
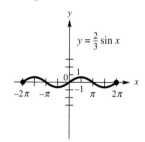

12. $2\pi; \dfrac{3}{4}$ **13.** 2π; 1 **14.** 2π; 1

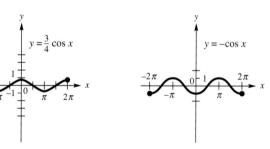
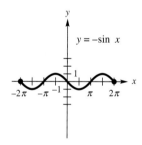

15. 2π; 2 **16.** 2π; 3 **17.** 4π; 1

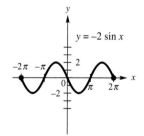
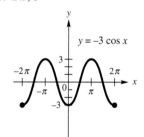
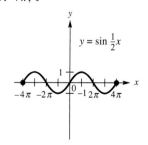

18. 3π; 1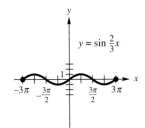

19. 6π; 1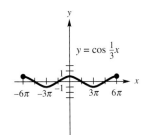

20. $\dfrac{8\pi}{3}$; 1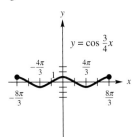

21. $\dfrac{2\pi}{3}$; 1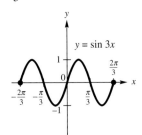

22. π; 1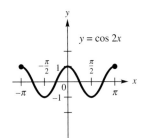

23. 8π; 2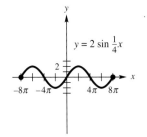

24. π; 3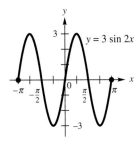

25. $\dfrac{2\pi}{3}$; 2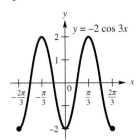

26. π; 5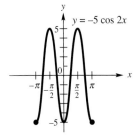

27. 2; 1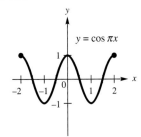

28. 2; 1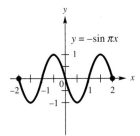

29. **(a)** 80°; 50° **(b)** 15° **(c)** about 35,000 years **(d)** downward   **30.** **(a)** 1; $\dfrac{4\pi}{3}$ **(b)** 1; $\dfrac{2\pi}{3}$

31. **(a)** 20 **(b)** 75   **32.** **(a)** about 2 hours **(b)** 1 year

Chapter 10 Test (Page 627)

1. 74.2983° 2. 203° 3. $\sin\theta = -\dfrac{5\sqrt{29}}{29}$; $\cos\theta = \dfrac{2\sqrt{29}}{29}$; $\tan\theta = -\dfrac{5}{2}$
4. III 5. $\sin\theta = -\dfrac{3}{5}$; $\tan\theta = -\dfrac{3}{4}$; $\cot\theta = -\dfrac{4}{3}$; $\sec\theta = \dfrac{5}{4}$; $\csc\theta = -\dfrac{5}{3}$
6. $\sin A = \dfrac{12}{13}$; $\cos A = \dfrac{5}{13}$; $\tan A = \dfrac{12}{5}$; $\cot A = \dfrac{5}{12}$; $\sec A = \dfrac{13}{5}$; $\csc A = \dfrac{13}{12}$ 7. (a) $\dfrac{1}{2}$ (b) 1 (c) undefined
(d) $-\dfrac{2\sqrt{3}}{3}$ (e) undefined (f) $-\sqrt{2}$ 8. (a) .97939940 (b) .20834446 (c) 1.9362132
9. 16.16664145° 10. 135°, 225° 11. $B = 31°30'$, $a = 638$, $b = 391$ 12. 15.5 ft 13. 137.5°
14. 180 km 15. 49.0° 16. 4300 km² 17. 264 square units 18. 2.7 miles 19. 5500 meters
20. distance to both first and third bases: 63.7 feet; distance to second base: 66.8 feet
21. 3; 2π 22. 1; π

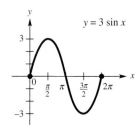

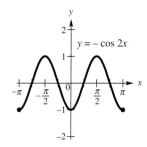

CHAPTER 11 Counting Methods

11.1 Exercises (Page 635)

1. AB, AC, AD, AE, BA, BC, BD, BE, CA, CB, CD, CE, DA, DB, DC, DE, EA, EB, EC, ED; 20 ways
2. CA, CB, CD, CE, EA, EB, EC, ED; 8 ways 3. AC, AE, BC, BE, CA, CB, CD, DC, DE, EA, EB, ED; 12 ways
4. CAE, CBE, CDE, EAC, EBC, EDC; 6 ways 5. ACE, AEC, BCE, BEC, DCE, DEC; 6 ways
6. ACB, ACD, AEB, AED, BCA, BCD, BEA, BED, DCA, DCB, DEA, DEB; 12 ways 7. ABC, ABD, ABE, ACD, ACE, ADE, BCD, BCE, BDE, CDE; 10 ways 8. ABC, ABD, ABE, ACD, ADE, BCD, BDE; 7 ways 9. 1
10. 2 11. 3 12. 4 13. 5 14. 6 15. 5 16. 4 17. 3 18. 2 19. 1 20. 18
21. 18 22. 16 23. 15 24. 6 25.

	1	2	3	4	5	6
1	11	12	13	14	15	16
2	21	22	23	24	25	26
3	31	32	33	34	35	36
4	41	42	43	44	45	46
5	51	52	53	54	55	56
6	61	62	63	64	65	66

26. 11, 13, 15, 21, 23, 25, 31, 33, 35, 41, 43, 45, 51, 53, 55, 61, 63, 65
27. 11, 22, 33, 44, 55, 66
28. 12, 24, 36, 42, 54, 66
29. 11, 13, 23, 31, 41, 43, 53, 61
30. 15, 21, 36, 45, 55, 66 31. 16, 25, 36, 64 32. 13, 21, 34, 55 33. 16, 32, 64
34. (a) hhh, hht, hth, thh (d) htt, tht, tth, ttt
    (b) hhh
    (c) hht, hth, htt, thh, tht, tth, ttt

First coin — Second coin — Third coin — Result

- h — h — h — hhh
- h — h — t — hht
- h — t — h — hth
- h — t — t — htt
- t — h — h — thh
- t — h — t — tht
- t — t — h — tth
- t — t — t — ttt

35. (a) tttt
 (b) hhhh, hhht, hhth, hhtt, hthh, htht, htth, thhh, thht, thth, tthh
 (c) httt, thtt, ttht, ttth, tttt
 (d) hhhh, hhht, hhth, hhtt, hthh, htht, htth, httt, thhh, thht, thth, thtt, tthh, ttht, ttth

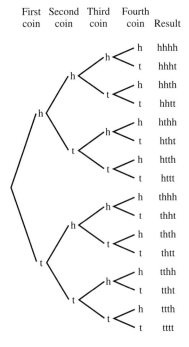

36. 10 **37.** 16 **38.** 10 **39.** 36 **40.** 30 **41.** 17 **42.** 19 **43.** 72 **44.** 8
45. 12 **46.** 14 **47.** 10 **48.** 11 **49.** 6 **50.** 3 **51.** 3 **52.** none **53.** 54
54. 49 **55.** 18 **56.** 21 **57.** 138 **58.** 13 **59.** 16 **60.** 2 **61.** 13 **62.** 8
63. 4 **64. (a)** 40 **(b)** 30 **(c)** 20 **(d)** 10 **(e)** 8 **65.** 883
66. (a) 1600 **(b)** $4k + 1$ for all positive integers k **(c)** $4k^2$ **67.** 3
Wording may vary in answers for Exercises 68–71.
68. (a) Determine the number of two-digit numbers that can be formed using digits from the set $\{1, 2, 3\}$ if repetition of digits is allowed. **(b)** Determine the number of two-digit numbers that can be formed using digits from the set $\{1, 2, 3\}$ if the selection is done with replacement.
69. (a) Determine the number of ordered pairs of digits that can be selected from the set $\{1, 2, 3, 4, 5, 6\}$ if repetition of digits is allowed. **(b)** Determine the number of ordered pairs of digits that can be selected from the set $\{1, 2, 3, 4, 5, 6\}$ if the selection is done with replacement.
70. (a) Find the number of ways to select an ordered pair of letters from the set $\{A, B, C, D, E\}$ if repetition of letters is not allowed. **(b)** Find the number of ways to select an ordered pair of letters from the set $\{A, B, C, D, E\}$ if the selection is done without replacement.
71. (a) Find the number of ways to select three letters from the set $\{A, B, C, D, E\}$ if repetition of letters is not allowed. **(b)** Find the number of ways to select three letters from the set $\{A, B, C, D, E\}$ if the selection is done without replacement.

11.2 Exercises (Page 644)
3. (a) no **4. (a)** no **5.** 24 **6.** 5040 **7.** 336 **8.** 240 **9.** 20 **10.** 360 **11.** 84
12. 210 **13.** 840 **14.** 495 **15.** 39,916,800 **16.** $3.556874281 \times 10^{14}$ **17.** 95,040
18. 3,603,600 **19.** 1716 **20.** 5,765,760 **21.** 184,756 **22.** 18,564 **23.** $4.151586701 \times 10^{12}$

24. 37,442,160 **25.** $2^3 = 8$ **26.** 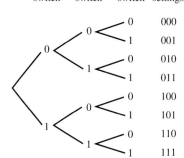 **28.**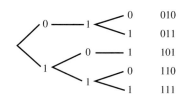

29. $6^3 = 216$ **30.** $9 \cdot 10^4 = 90,000$ **31.** $5! = 120$ **32.** $5^5 = 3125$ **33.** $3 \cdot 2 = 6$
34. $4 \cdot 3 = 12$ **35.** $3 \cdot 3 = 9$ **36.** $2 \cdot 3 \cdot 3 = 18$ **37.** $3 \cdot 2 \cdot 1 = 6$ **38.** $3 \cdot 3 \cdot 3 = 27$
39. $4 \cdot 2 \cdot 3 = 24$ **40.** $2 \cdot 3 = 6$ **41.** $2^6 = 64$ **42.** $5^{20} = 9.536743164 \times 10^{13}$
43. $3 \cdot 3 \cdot 4 \cdot 5 = 180$ **44.** $2 \cdot 3 \cdot 3 \cdot 5 = 90$ **45.** $3 \cdot 3 \cdot 4 \cdot 3 = 108$ **46.** $2 \cdot 2 \cdot 3 \cdot 5 = 60$
47. $3 \cdot 3 \cdot 1 \cdot 3 = 27$ **48.** $1 \cdot 3 \cdot 4 \cdot 1 = 12$ **49.** $2 \cdot 5 \cdot 6 = 60$ **50.** $10 \cdot 4 \cdot 6 \cdot 3 = 720$
51. $5! = 120$ **52. (a)** $3! \cdot 4! = 144$ **(b)** $5 \cdot 3! \cdot 4! = 720$ **(c)** $10 \cdot 3! \cdot 4! = 1440$ **53. (a)** 6
(b) 5 **(c)** 4 **(d)** 3 **(e)** 2 **(f)** 1; $6 \cdot 5 \cdot 4 \cdot 3 \cdot 2 \cdot 1 = 720$ **54. (a)** 5 **(b)** 2 **(c)** 4 **(d)** 3
(e) 2 **(f)** 1; $5 \cdot 2 \cdot 4 \cdot 3 \cdot 2 \cdot 1 = 240$ **55. (a)** 3 **(b)** 3 **(c)** 2 **(d)** 2 **(e)** 1
(f) 1; $3 \cdot 3 \cdot 2 \cdot 2 \cdot 1 \cdot 1 = 36$ **56. (a)** 6 **(b)** 3 **(c)** 2 **(d)** 2 **(e)** 1 **(f)** 1; $6 \cdot 3 \cdot 2 \cdot 2 \cdot 1 \cdot 1 = 72$
59. 450; no

11.3 Exercises (Page 655)
1. 360 **2.** 132 **3.** 455 **4.** 1716 **5.** 143,640 **6.** 240,240 **7.** 252 **8.** 715
9. 4.3609104×10^{10} **10.** 3,796,297,200 **15. (a)** permutation **(b)** permutation **(c)** combination
(d) combination **(e)** permutation **(f)** combination **(g)** permutation **16.** $P(6,3) = 120$
17. $P(8,5) = 6720$ **18.** $P(10,4) = 5040$ **19.** $P(12,2) = 132$ **20.** $P(10,3) = 720$
21. $P(25,5) = 6,375,600$ **22.** $P(12,12) = 479,001,600$ **23. (a)** $P(4,4) = 24$ **(b)** $P(4,4) = 24$
24. $C(24,5) = 42,504$ **25.** $C(18,5) = 8568$ **26.** $C(100,5) = 75,287,520$ **27. (a)** $C(13,5) = 1287$
(b) $C(26,5) = 65,780$ **(c)** 0 (impossible) **28.** $C(39,7) = 15,380,937$ **29. (a)** $C(6,3) = 20$
(b) $C(6,2) = 15$ **30.** $C(10,3) = 120$ **31.** $C(25,2) = 300$
32. (a) $C(47,5) \cdot C(27,1) = 1,533,939 \cdot 27 = 41,416,353$ **(b)** $C(45,3) \cdot C(27,1) = 14,190 \cdot 27 = 383,130$
33. (a) 5 **(b)** 9 **34.** $4 \cdot C(13,5) = 5148$ **35.** $P(26,3) \cdot P(10,3) \cdot P(26,3) = 175,219,200,000$
36. $2 \cdot P(25,2) = 1200$ **37.** $2 \cdot P(25,3) = 27,600$ **38.** $3 \cdot C(8,2) = 84$ **39.** $7 \cdot P(12,8) = 139,708,800$
40. (a) $8! \cdot 7! = 203,212,800$ **(b)** $8 \cdot 7 \cdot 13! = 348,713,164,800$ **(c)** $13! = 6,227,020,800$
(d) $8! \cdot 7! = 203,212,800$ **(e)** $8 \cdot 7 \cdot 6 \cdot 12! = 160,944,537,600$ **41. (a)** $7^7 = 823,543$ **(b)** $7! = 5040$
42. (a) $C(13,4) \cdot 39 = 27,885$ **(b)** $C(12,2) \cdot C(40,3) = 652,080$ **(c)** $C(26,2) \cdot C(13,2) \cdot 13 = 329,550$
43. $C(25,3) \cdot C(22,4) \cdot C(18,5) \cdot C(13,6) = 2.473653743 \times 10^{14}$ **44.** $C(7,2) = 21$ **45.** $C(20,3) = 1140$
46. $40^3 = 64,000$ **47. (a)** $P(7,2) = 42$ **(b)** $3 \cdot 6 = 18$ **(c)** $P(7,2) \cdot 5 = 210$ **48.** $6 \cdot 8 = 48$
49. $P(8,3) = 336$ **50.** $C(9,2) \cdot C(7,3) \cdot 2 \cdot 3 \cdot 4 = 30,240$ **51. (a)** $C(6,2) \cdot C(6,3) \cdot C(6,4) = 4500$
(b) $3! \cdot C(6,2) \cdot C(6,3) \cdot C(6,4) = 27,000$ **52. (a)** $6! = 720$ **(b)** 745,896 **53. (a)** $8! = 40,320$
(b) $2 \cdot 6! = 1440$ **(c)** $6! = 720$ **54.** $C(60,3) \cdot C(40,2) = 26,691,600$ **55. (a)** $2 \cdot 4! = 48$
(b) $3 \cdot 4! = 72$ **56.** $C(30,5) = 142,506$ **57.** Each is equal to 220. **58.** Each is equal to $\dfrac{n!}{r!(n-r)!}$.
59. (a) 1 **60. (a)** 1

11.4 Exercises (Page 662)

1. 4   **2.** 10   **3.** 15   **4.** 21   **5.** 28   **6.** 126   **7.** 36   **8.** 210   **9.** $C(7,1) \cdot C(3,3) = 7$
10. $C(7,2) \cdot C(3,2) = 63$ **11.** $C(7,3) \cdot C(3,1) = 105$ **12.** $C(7,4) \cdot C(3,0) = 35$ **13.** $C(8,3) = 56$
14. $C(8,4) = 70$ **15.** $C(8,5) = 56$ **16.** $C(8,6) = 28$ **17.** $C(9,4) = 126$ **18.** $C(8,4) = 70$
19. $1 \cdot C(8,3) = 56$ **20.** $56/126 = .444$ **21.** 1   **22.** 5   **23.** 10   **24.** 10   **25.** 5   **26.** 1
27. 32   **29. (a)** All are multiples of the row number.   **(b)** The same pattern holds.   **(c)** The same pattern holds.
30. …,15, 21, 28, 36, 45,…; (These are the triangular numbers.)   **31.** …,8, 13, 21, 34,…; A number in this sequence is the sum of the two preceding terms. This is the Fibonacci sequence.
32.

```
            1
          1   1
        1   0   1
      1   1   1   1
    1   0   0   0   1
  1   1   0   0   1   1
```

In rows 2 and 4, every entry, except for the beginning and ending 1s, is 0. This is because the corresponding entries in the original triangle were all even.

33. row 8   **34.** 255   **35.** The sum of the squares of the entries across the top row equals the entry at the bottom vertex.   **36.** The rows of Tartaglia's rectangle correspond to the diagonals of Pascal's triangle.

Wording may vary for Exercises 37–40.
37. sum = N; Any entry in the array equals the sum of the two entries immediately above it and immediately to its left.   **38.** sum = N; Any entry in the array equals the sum of the column of entries from its immediate left upward to the top of the array.   **39.** sum = N; Any entry in the array equals the sum of the row of entries from the cell immediately above it to the left boundary of the array.   **40.** sum = $N - 1$; Any entry in the array equals 1 more than the sum of all entries whose cells comprise the largest rectangle entirely to the left and above that entry.
41. $x^6 + 6x^5y + 15x^4y^2 + 20x^3y^3 + 15x^2y^4 + 6xy^5 + y^6$
42. $x^8 + 8x^7y + 28x^6y^2 + 56x^5y^3 + 70x^4y^4 + 56x^3y^5 + 28x^2y^6 + 8xy^7 + y^8$ **43.** $z^3 + 6z^2 + 12z + 8$
44. $w^5 + 15w^4 + 90w^3 + 270w^2 + 405w + 243$ **45.** $16a^4 + 160a^3b + 600a^2b^2 + 1000ab^3 + 625b^4$
46. $81d^4 + 540d^3f + 1350d^2f^2 + 1500df^3 + 625f^4$ **47.** $b^6 - 6b^5h + 15b^4h^2 - 20b^3h^3 + 15b^2h^4 - 6bh^5 + h^6$
48. $32n^5 - 320n^4m + 1280n^3m^2 - 2560n^2m^3 + 2560nm^4 - 1024m^5$ **49.** $n + 1$
50. $x^{n-r+1}y^{r-1}$; $\dfrac{n!}{(n-r+1)!(r-1)!}x^{n-r+1}y^{r-1}$ **51.** $n = 14, r = 5$; $\dfrac{14!}{10!\,4!}x^{10}y^4 = 1001x^{10}y^4$
52. $n = 18, r = 16$; $\dfrac{18!}{3!\,15!}a^3b^{15} = 816a^3b^{15}$

11.5 Exercises (Page 669)

3. $2^4 - 1 = 15$ **4.** $2^5 - 1 = 31$ **5.** $2^7 - 1 = 127$ **6.** $2^7 - (1 + 7) = 120$ **7.** 120
8. $2^7 - 2 = 126$ **9.** 6   **10.** $36 - 1 = 35$ **11.** $6 + 6 - 1 = 11$ **12.** $36 - 6 = 30$
13. $90 - 9 = 81$ **14.** $29 + 9 - 2 = 36$ **15. (a)** $C(8,3) = 56$ **(b)** $C(7,3) = 35$ **(c)** $56 - 35 = 21$
16. (a) $C(10,4) = 210$ **(b)** $C(8,4) = 70$ **(c)** $210 - 70 = 140$ **17.** $C(7,3) - C(5,3) = 25$
18. $C(9,2) - C(7,2) = 15$ **19.** $P(8,3) - P(6,3) = 216$ **20.** $C(8,3) - C(6,3) = 36$
21. $P(10,3) - P(7,3) = 510$ **22. (a)** $9 \cdot 10^3 - 9^4 = 2439$ **(b)** $9 \cdot 9 \cdot 8 \cdot 7 - 9 \cdot 8 \cdot 7 \cdot 6 = 1512$
23. $C(25,4) - C(23,4) = 3795$ **24.** $C(12,4) - C(8,4) = 425$ **25.** $13 + 4 - 1 = 16$
26. $12 + 26 - 6 = 32$ **27.** $30 + 15 - 10 = 35$ **28.** $50 - 35 = 15$
29. $2{,}598{,}960 - C(13,5) = 2{,}597{,}673$ **30.** $2{,}598{,}960 - 4 \cdot C(13,5) = 2{,}593{,}812$
31. $2{,}598{,}960 - C(40,5) = 1{,}940{,}952$ **32.** $2{,}598{,}960 - [C(39,5) + C(13,5)] = 2{,}021{,}916$
33. $C(12,0) + C(12,1) + C(12,2) = 79$ **34.** $C(12,10) + C(12,11) + C(12,12) = 79$ **35.** $2^{12} - 79 = 4017$
36. $2^{12} - (79 + 79) = 3938$ **37.** $26^3 \cdot 10^3 - P(26,3) \cdot P(10,3) = 6{,}344{,}000$ **38.** $1 \cdot 12 + 3 \cdot 13 = 51$
39. $C(4,3) + C(3,3) + C(5,3) = 15$ **40.** $C(12,3) - C(5,3) = 210$ **41.** $C(12,3) - 4 \cdot 3 \cdot 5 = 160$
42. $C(12,3) - (15 + 60) = 145$ **50.** $300 - (150 + 100 + 60 - 50 - 30 - 20 + 10) = 80$

Chapter 11 Test (Page 672)

1. $6 \cdot 7 \cdot 7 = 294$ **2.** $6 \cdot 7 \cdot 4 = 168$ **3.** $6 \cdot 6 \cdot 5 = 180$ **4.** $6 \cdot 5 \cdot 1 = 30$ end in 0; $5 \cdot 5 \cdot 1 = 25$ end in 5; $30 + 25 = 55$ **5.** 13   **6.**   **7.** $4! = 24$ **8.** 12   **9.** 120

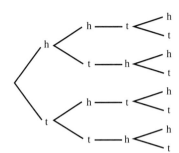

10. 336   **11.** 11,880   **12.** 35   **13.** $P(26, 5) = 7,893,600$ **14.** $32^5 = 33,554,432$
15. $C(12, 4) = 495$ **16.** $C(12, 2) \cdot C(10, 2) = 2970$ **17.** $C(12, 5) \cdot C(7, 5) = 16,632$
18. $2^{12} - [C(12, 0) + C(12, 1) + C(12, 2)] = 4017$ **19.** $2^4 = 16$ **20.** $2^2 = 4$ **21.** $2 \cdot 2^2 = 8$
22. 8   **23.** 2   **24.** $16 - (1 + 4) = 11$ **25.** $C(6, 3) = 20$ **26.** $C(5, 2) = 10$ **27.** $2 \cdot C(5, 3) = 20$
28. $C(5, 2) = 10$ **29.** $C(5, 4) + C(2, 1) \cdot C(5, 3) = 25$ **30.** $x^5 + 10x^4 + 40x^3 + 80x^2 + 80x + 32$
31. $495 + 220 = 715$ **32.** the counting numbers

CHAPTER 12 Probability

12.1 Exercises (Page 681)

1. (a) $\frac{1}{3}$ (b) $\frac{1}{3}$ (c) $\frac{1}{3}$ **2.** (a) $\frac{1}{4}$ (b) $\frac{1}{2}$ (c) $\frac{1}{4}$ **3.** (a) $\frac{1}{2}$ (b) $\frac{1}{3}$ (c) $\frac{1}{6}$ **4.** (a) $\frac{1}{2}$ (b) $\frac{1}{2}$ (c) 0
5. (a) $\{1, 2, 3\}$ (b) 2 (c) 1 (d) 3 (e) $\frac{2}{3}$ (f) 2 to 1   **6.** (a) 13 (b) 8 (c) 5 (d) $\frac{8}{13}$ (e) 8 to 5
7. (a) $\{11, 12, 13, 21, 22, 23, 31, 32, 33\}$ (b) $\frac{2}{3}$ (c) $\frac{1}{3}$ (d) $\frac{1}{3}$ (e) $\frac{4}{9}$ **8.** (a) $\{hh, ht, th, tt\}$ (b) $\frac{1}{2}$ (c) $\frac{1}{2}$
(d) $\frac{1}{4}$ (e) $\frac{1}{2}$ **9.** (a) 4 to 7 (b) 5 to 6 (c) 2 to 9   **10.** (a) 4 to 1 (b) 3 to 2 (c) 3 to 2   **11.** (a) $\frac{1}{50}$
(b) $\frac{2}{50} = \frac{1}{25}$ (c) $\frac{3}{50}$ (d) $\frac{4}{50} = \frac{2}{25}$ (e) $\frac{5}{50} = \frac{1}{10}$ **12.** $\{hhh, hht, hth, thh, htt, tht, tth, ttt\}$ (a) $\frac{1}{8}$ (b) $\frac{3}{8}$
(c) $\frac{3}{8}$ (d) $\frac{1}{8}$ **13.** (a) $\frac{1}{36}$ (b) $\frac{2}{36} = \frac{1}{18}$ (c) $\frac{3}{36} = \frac{1}{12}$ (d) $\frac{4}{36} = \frac{1}{9}$ (e) $\frac{5}{36}$ (f) $\frac{6}{36} = \frac{1}{6}$ (g) $\frac{5}{36}$
(h) $\frac{4}{36} = \frac{1}{9}$ (i) $\frac{3}{36} = \frac{1}{12}$ (j) $\frac{2}{36} = \frac{1}{18}$ (k) $\frac{1}{36}$ **14.** $\frac{170}{200} = .850$ **15.** (a) $\frac{38,550}{75,320} \approx .512$
(b) $\frac{36,770}{75,320} \approx .488$ **16.** $\frac{6}{16} = \frac{3}{8}$ **19.** $\frac{1}{4}$ **20.** $\frac{2}{4} = \frac{1}{2}$ **21.** $\frac{1}{4}$ **23.** (a) $\frac{3}{4}$ (b) $\frac{1}{4}$ **24.** $\frac{1}{2000} = .0005$
25. $\frac{1}{250,000} = .000004$ **26.** 75   **27.** $\frac{1}{4}$ **28.** $\frac{2}{4} = \frac{1}{2}$ **29.** $\frac{1}{4}$ **30.** 0   **31.** $\frac{2}{4} = \frac{1}{2}$ **32.** $\frac{2}{4} = \frac{1}{2}$
33. $\frac{1}{500} = .002$ **34.** $\frac{1}{160,000} \approx .000006$ **35.** about 160   **36.** $\frac{1}{4}$ **37.** $\frac{2}{4} = \frac{1}{2}$ **38.** $\frac{1}{4}$ **39.** (a) 0
(b) no   (c) yes   **42.** (a) 37 to 63 (b) 63 to 37   **43.** $\frac{12}{31}$ **44.** $\frac{3}{13}$ **45.** $\frac{36}{2,598,960} \approx .00001385$
46. $\frac{123,552}{2,598,960} \approx .04753902$ **47.** $\frac{1}{13} \cdot \frac{624}{2,598,960} \approx .00001847$ **48.** $\frac{1}{4} \cdot \frac{5108}{2,598,960} \approx .00049135$
49. $\frac{6^2 - 4^2 + 2^2}{8^2} = \frac{3}{8} = .375$ **50.** (a) 28 (b) 44 (c) 43 (d) 49

Answers to Selected Exercises   A-69

51. $3 \cdot 1 \cdot 2 \cdot 1 \cdot 1 \cdot 1 = 6$; $\dfrac{6}{720} = \dfrac{1}{120} \approx .0083$ **52.** $3 \cdot 3 \cdot 2 \cdot 2 \cdot 1 \cdot 1 = 36$; $\dfrac{36}{720} = \dfrac{1}{20} = .05$

53. $4 \cdot 3! \cdot 3! = 144$; $\dfrac{144}{720} = \dfrac{1}{5} = .2$ **54.** $2 \cdot 3! \cdot 3! = 72$; $\dfrac{72}{720} = \dfrac{1}{10} = .1$ **55.** $\dfrac{2}{C(7,2)} = \dfrac{2}{21} \approx .095$

56. $\dfrac{1}{P(5,3)} = \dfrac{1}{60} \approx .017$ **57.** $\dfrac{C(5,3)}{C(12,3)} = \dfrac{1}{22} \approx .045$ **58.** $\dfrac{1}{C(10,3)} = \dfrac{1}{120} \approx .008$

59. $\dfrac{1}{P(36,3)} \approx .000023$ **60.** $\dfrac{1}{C(36,3)} \approx .000140$ **61.** $\dfrac{3}{28} \approx .107$ **62. (a)** $\dfrac{2 \cdot 4!}{5!} = \dfrac{2}{5}$ **(b)** $\dfrac{2 \cdot 3!}{5!} = \dfrac{1}{10}$

63. (a) $\dfrac{8}{9^2} = \dfrac{8}{81} \approx .099$ **(b)** $\dfrac{4}{C(9,2)} = \dfrac{1}{9} \approx .111$ **64.** $\dfrac{9}{9 \cdot 10} = \dfrac{1}{10}$ **65.** $\dfrac{9 \cdot 10}{9 \cdot 10^2} = \dfrac{1}{10}$

12.2 Exercises (Page 689)

1. yes   **2.** no   **4.** $\dfrac{5}{6}$ **5.** $\dfrac{1}{2}$ **6.** $\dfrac{5}{6}$ **7.** $\dfrac{5}{6}$ **8.** 1   **9.** $\dfrac{2}{3}$ **10. (a)** $\dfrac{12}{13}$ **(b)** 12 to 1

11. (a) $\dfrac{2}{13}$ **(b)** 2 to 11   **12. (a)** $\dfrac{1}{2}$ **(b)** 1 to 1   **13. (a)** $\dfrac{11}{26}$ **(b)** 11 to 15   **14. (a)** $\dfrac{10}{13}$ **(b)** 10 to 3

15. (a) $\dfrac{9}{13}$ **(b)** 9 to 4   **16.** $\dfrac{1}{12}$ **17.** $\dfrac{2}{3}$ **18.** $\dfrac{11}{18}$ **19.** $\dfrac{7}{36}$ **20. (a)** $\dfrac{5}{12}$ **(b)** $\dfrac{5}{12}$

21. $P(A) + P(B) + P(C) + P(D) = 1$ **22.** .023094   **23.** .005365   **24.** .048522   **25.** .971285

26. .04   **27.** .76   **28.** .15   **29.** .92   **30.** .25   **31.** 6 to 19

32.

x	P(x)
red	4/11
yellow	5/11
blue	2/11

33.

x	P(x)
3	.1
4	.1
5	.2
6	.2
7	.2
8	.1
9	.1

34. Answers will vary.   **35.** $n(A') = s - a$

36. $P(A) = \dfrac{a}{s}$; $P(A') = \dfrac{s-a}{s}$ **37.** $P(A) + P(A') = 1$ **38.** complements rule of probability   **39.** 180

40. 60   **41.** $\dfrac{2}{3}$ **42.** $\dfrac{5}{6}$ **43.** 1

12.3 Exercises (Page 698)

1. independent   **2.** independent   **3.** independent   **4.** not independent   **5.** independent

6. not independent   **7.** $\dfrac{42}{100} = \dfrac{21}{50}$ **8.** $\dfrac{34}{100} = \dfrac{17}{50}$ **9.** $\dfrac{68}{100} = \dfrac{17}{25}$ **10.** $\dfrac{18}{100} = \dfrac{9}{50}$ **11.** $\dfrac{19}{34}$

12. $\dfrac{27}{42} = \dfrac{9}{14}$ **13.** $\dfrac{4}{7} \cdot \dfrac{4}{7} = \dfrac{16}{49}$ **14.** $\dfrac{1}{7} \cdot \dfrac{2}{7} = \dfrac{2}{49}$ **15.** $\dfrac{2}{7} \cdot \dfrac{1}{7} = \dfrac{2}{49}$ **16.** $\dfrac{1}{7} \cdot \dfrac{1}{7} = \dfrac{1}{49}$ **17.** $\dfrac{4}{7} \cdot \dfrac{3}{6} = \dfrac{2}{7}$

18. $\dfrac{2}{6} = \dfrac{1}{3}$ **19.** $\dfrac{1}{6}$ **20.** $\dfrac{1}{7}$ **21.** 0   **22.** 0   **23.** $\dfrac{12}{51} = \dfrac{4}{17}$ **24.** $\dfrac{13}{51}$ **25.** $\dfrac{12}{52} \cdot \dfrac{11}{51} = \dfrac{11}{221}$

26. $\dfrac{40}{52} \cdot \dfrac{39}{51} = \dfrac{10}{17}$ **27.** $\dfrac{4}{52} \cdot \dfrac{11}{51} = \dfrac{11}{663}$ **28. (a)** general multiplication rule of probability   **(b)** Divide both sides of the equation in (a) by $P(A)$.   **(c)** theoretical probability formula   **(d)** Multiply the numerator and denominator of the expression in (c) by $n(S)$.   **29.** $\dfrac{1}{3}$ **30.** 1   **31.** 1   **32.** $\dfrac{1}{2}$ **33.** $\dfrac{3}{10}$ (the same)

35. $\dfrac{1}{2} \cdot \dfrac{1}{2} \cdot \dfrac{1}{2} = \dfrac{1}{8}$ **36.** $\dfrac{1}{2} \cdot \dfrac{1}{2} \cdot \dfrac{1}{2} = \dfrac{1}{8}$ **37.** .640   **38.** .512   **39.** .008   **40.** .096   **41.** .95

42. .77   **43.** .23   **44.** .40   **45.** $\dfrac{1}{35}$ **46.** $\dfrac{1}{35}$ **47.** $\dfrac{3}{7}$ **48.** $\dfrac{3}{7}$ **49.** $\dfrac{3}{7}$ **50.** $\dfrac{8}{33}$

51. (a) $\frac{3}{4}$ (b) $\frac{1}{2}$ (c) $\frac{5}{16}$ **52.** (a) 3 (b) 9   **53.** .2704   **54.** .2496   **55.** .2496   **56.** .2304
58. $2^6 = 64$ **59.** $\frac{1}{64} \approx .0156$ **60.** .1479   **61.** 10   **62.** $\frac{2}{3}$ **63.** .400   **64.** .320   **65.** .080
66. .154   **67.** $(.90)^4 = .6561$ **68.** $C(4,1) \cdot (.10) \cdot (.80)^3 = .2048$ **69.** $C(4,2) \cdot (.10) \cdot (.20) \cdot (.70)^2 = .0588$
70. $1 - (.6561 + .2048 + .0588) = .0803$ **71.** .30   **72.** .21   **73.** .49   **74.** (a) Answers will vary.
(b) $\frac{25}{36} \approx .69$

12.4 Exercises (Page 705)

1. $\frac{1}{8}$ **2.** $\frac{3}{8}$ **3.** $\frac{3}{8}$ **4.** $\frac{1}{8}$ **5.** $\frac{3}{4}$ **6.** $\frac{7}{8}$ **7.** $\frac{1}{2}$ **8.** $\frac{7}{8}$ **9.** $\frac{3}{8}$ **11.** $x; n; n$ **12.** $\frac{1}{128}$
13. $\frac{7}{128}$ **14.** $\frac{21}{128}$ **15.** $\frac{35}{128}$ **16.** $\frac{35}{128}$ **17.** $\frac{21}{128}$ **18.** $\frac{7}{128}$ **19.** $\frac{1}{128}$ **20.** $\frac{125}{216}$ **21.** $\frac{25}{72}$
22. $\frac{5}{72}$ **23.** $\frac{1}{216}$ **25.** .041   **26.** .103   **27.** .268   **28.** .050   **30.** .302; .299   **32.** .228
33. .016   **34.** .299   **35.** .020   **36.** .058   **37.** .198   **38.** .296   **39.** .448   **40.** .246   **41.** .656
42. .002   **43.** .002   **44.** .205   **45.** .883   **46.** .209   **48.** .063   **49.** .091   **50.** .147   **51.** .018
52. $\frac{1}{1024} \approx .001$ **53.** $\frac{45}{1024} \approx .044$ **54.** $\frac{45}{1024} \approx .044$ **55.** 0   **56.** $\frac{210}{1024} = \frac{105}{512} \approx .205$
57. $\frac{210 + 120 + 45 + 10 + 1}{1024} = \frac{193}{512} \approx .377$ **58.** $\frac{1024 - 252}{1024} = \frac{193}{256} \approx .754$ **59.** $\frac{252}{1024} = \frac{63}{256} \approx .246$

12.5 Exercises (Page 713)

3. $\frac{5}{2}$ **4.** $\frac{1}{2}$ **5.** $1   **6.** $1   **7.** 50¢   **8.** unfair in favor of the player   **9.** no (expected net winnings: $-\frac{3}{4}$¢)
10. $-\frac{\$1}{37} \approx -2.7$¢   **11.** 1.72   **12.** $110   **13.** (a) $-$60 (b) $36,000 (c) $72,000
14. (a) $-$70 (b) $50,400 (c) $86,400   **15.** 50¢   **16.** $1   **17.** $2500   **18.** 1.7   **19.** 2.7
20. 6   **21.** an increase of 250   **22.** $-$20¢   **24.** Project B   **25.** Project C   **26.** Project C
27. $2200   **28.** Accept the computer system (since $2300 > $2200).   **29.** column 5: 25,000, 60,000, 16,000
30. column 6: 22,000, 51,000, 30,000, 60,000, 46,000   **31.** column 7: C, C, C, B, C, A, B   **32.** $106,500
33. about $-$15¢

12.6 Exercises (Page 719)

2. $\frac{15}{47} \approx .319$ **3.** no   **4.** Answers will vary.   **5.** $\frac{18}{50} = .36$ (This is quite close to .375, the theoretical value.)
6. $\frac{15}{50} = .30$ **7.** $\frac{11}{50} = .22$ **8.** $\frac{24}{50} = .48$ **9.** Answers will vary.   **10.** Answers will vary.
12. The walk ends $3\sqrt{2}$ blocks northwest of the starting point (that is, 3 blocks west and 3 blocks north).

Chapter 12 Test (Page 721)

3. 3 to 1   **4.** 25 to 1   **5.** 11 to 2   **6.**

	C	c
C	CC	Cc
c	cC	cc

7. $\frac{1}{2}$ **8.** 3 to 1   **9.** $\frac{7}{7} \cdot \frac{6}{7} \cdot \frac{5}{7} = \frac{30}{49}$

10. $\frac{7}{7} \cdot \frac{1}{7} \cdot \frac{1}{7} = \frac{1}{49}$ **11.** $1 - \left(\frac{30}{49} + \frac{1}{49}\right) = \frac{18}{49}$ **12.** $\frac{C(2,2)}{C(5,2)} = \frac{1}{10}$ **13.** $\frac{C(3,2)}{C(5,2)} = \frac{3}{10}$ **14.** $\frac{6}{10} = \frac{3}{5}$

Answers to Selected Exercises | **A-71**

15. $\frac{3}{10}$ **16.**

x	P(x)
0	0
1	3/10
2	6/10
3	1/10

17. $\frac{9}{10}$ **18.** $\frac{18}{10} = \frac{9}{5}$ **19.** 5 to 1 **20.** $1 - \frac{1}{36} = \frac{35}{36}$ **21.** 7 to 2

22. $\frac{4}{36} = \frac{1}{9}$ **23.** $(.78)^3 \approx .475$ **24.** $C(3,2)(.78)^2(.22) \approx .402$ **25.** $1 - (.22)^3 \approx .989$

26. $(.78)(.22)(.78) \approx .134$ **27.** $\frac{25}{102}$ **28.** $\frac{25}{51}$ **29.** $\frac{4}{51}$ **30.** $\frac{3}{26}$ **31.** 38 **32.** 1 **33.** .026

CHAPTER 13 Statistics

13.1 Exercises (Page 730)

1. (a)

x	f	f/n
0	10	10/30 ≈ 33%
1	7	7/30 ≈ 23%
2	6	6/30 = 20%
3	4	4/30 ≈ 13%
4	2	2/30 ≈ 7%
5	1	1/30 ≈ 3%

(b) 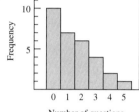 **(c)**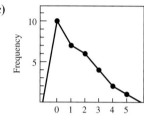

2. (a)

x	f	f/n
1	1	1/28 ≈ 4%
2	2	2/28 ≈ 7%
3	1	1/28 ≈ 4%
4	4	4/28 ≈ 14%
5	3	3/28 ≈ 11%
6	5	5/28 ≈ 18%
7	4	4/28 ≈ 14%
8	3	3/28 ≈ 11%
9	3	3/28 ≈ 11%
10	2	2/28 ≈ 7%

(b) 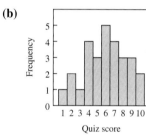 **(c)**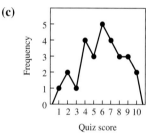

3. (a)

Class Limits	Tally	Frequency f	Relative Frequency f/n
45–49	|||	3	3/54 ≈ 5.6%
50–54	|||| |||| ||||	14	14/54 ≈ 25.9%
55–59	|||| |||| ||||		16	16/54 ≈ 29.6%
60–64	|||| |||| |||| ||	17	17/54 ≈ 31.5%
65–69	||||	4	4/54 ≈ 7.4%

Total: $n = 54$

(b)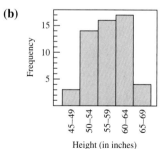

(c)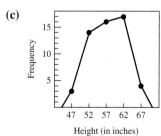

4. (a)

Class Limits	Tally	Frequency f	Relative Frequency f/n
40–59	|||	3	3/50 = 6%
60–79	~~||||~~ ~~||||~~		11	11/50 = 22%
80–99	~~||||~~ ~~||||~~	10	10/50 = 20%
100–119	~~||||~~ ~~||||~~ ||	12	12/50 = 24%
120–139	~~||||~~ |||	8	8/50 = 16%
140–159	|||	3	3/50 = 6%
160–179	|||	3	3/50 = 6%

Total: $n = 50$

(b)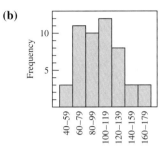

Account balance (in dollars)

(c)

Frequency plotted at 49.5, 69.5, 89.5, 109.5, 129.5, 149.5, 169.5

Account balance (in dollars)

5. (a)

Class Limits	Tally	Frequency f	Relative Frequency f/n
70–74	||	2	2/30 ≈ 6.7%
75–79			1	1/30 ≈ 3.3%
80–84	|||	3	3/30 = 10.0%
85–89	||	2	2/30 ≈ 6.7%
90–94	~~||||~~	5	5/30 ≈ 16.7%
95–99	~~||||~~	5	5/30 ≈ 16.7%
100–104	~~||||~~		6	6/30 = 20.0%
105–109	||||	4	4/30 ≈ 13.3%
110–114	||	2	2/30 ≈ 6.7%

Total: $n = 30$

(b)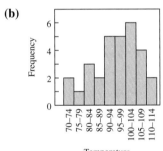

Temperature

(c)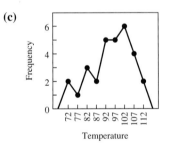

Temperature

6. (a)

Class Limits	Tally	Frequency f	Relative Frequency f/n
91–95			1	1/50 = 2%
96–100	|||	3	3/50 = 6%
101–105	~~||||~~	5	5/50 = 10%
106–110	~~||||~~ ||	7	7/50 = 14%
111–115	~~||||~~ ~~||||~~ ||	12	12/50 = 24%
116–120	~~||||~~ ||||	9	9/50 = 18%
121–125	~~||||~~ ||	7	7/50 = 14%
126–130	||||	4	4/50 = 8%
131–135	||	2	2/50 = 4%

Total: $n = 50$

(b)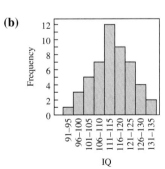

IQ

(c)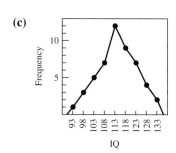

7.
```
0 | 7 9 8
1 | 1 1 2 8 9 4 3 1 0 5 0 5 5
2 | 0 9 6 6 2 5 2 3 4 4
3 | 1
```

8.
```
0 | 4 8
1 | 2 3 2 5 7 1 3 8 4 9 7 7 0
2 | 1 2 4 6 5 1 0
3 | 3 2 8
4 | 2 2
5 | 3 4
6 | 2
```

9.
```
0 | 8 5 4 9 6 9 4 8
1 | 2 0 1 8 8 2 4 0 8 8 6 3
2 | 6 6 2 5 1 3
3 | 0 4 6
4 | 4
```

10.
```
 0  | 7 3
 1  | 9 2 8 6
 2  | 8 2 4 9 9 5
 3  | 6 7 3 9 0 6 3 2
 4  | 5 3 3 0 2 1 9 6
 5  | 4 8 1 5
 6  | 7 0 2
 7  | 3 9 3
 8  | 8 6
 9  | 4
10  | 2
11  | 2
12  | 3
```

11. about 31% **12.** about 24% **13.** CBS; about 2.9%

14. from 1991–92 to 1992–93 **15.** about 1.6%; ABC; 1994–95 **16.** about $9,200,000,000,000
17. 3.4% in 2000 **18.** An increase of about .7% from 2000 to 2001 **21.** 187° **22.** .4%

23.  **24.**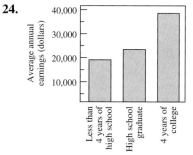

25. about 79 years **26.** about $400,000; about 76 years **27. (a)** about 6 years **28.** about four times as much

A-74 Answers to Selected Exercises

32.
(100–104)	10	0 0 4
(105–109)	10	6 9 5 9 7 6 8 5
(110–114)	11	2 0 2 4 4 2 4 0 0 3 1 1 3 0 2 4 4
(115–119)	11	8 8 7 6 8 5 8 7 8 9 7 8
(120–124)	12	0 1 2 1 0 0 0
(125–129)	12	8 5
(130–134)	13	4

34.
Letter	Probability
E	.13
T	.09
A, O	.08
N	.07
I, R	.065
S, H	.06
D	.04
L	.035
C, M, U	.03
F, P, Y	.02
W, G, B	.015
V	.01
K, X, J	.005
Q, Z	.002

35.
Letter	Probability
A	.208
E	.338
I	.169
O	.208
U	.078

36.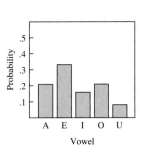

37.
Letter	Probability
A	.263
E	.288
I	.169
O	.195
U	.085

39.
Class Limits	Probability
10–19	.150
20–29	.275
30–39	.225
40–49	.175
50–59	.100
60–69	.050
70–79	.025

40. (a) .225 (b) .275 (c) .425 (d) .175

41. 7/40

42.
Sport	Probability
Sailing	.225
Hang gliding	.125
Bungee jumping	.175
Sky diving	.075
Canoeing	.300
Rafting	.100

43. (a) empirical

13.2 Exercises (Page 744)

1. (a) 12.2 (b) 12 (c) none **2.** (a) 40.2 (b) 40 (c) none **3.** (a) 201.2 (b) 210 (c) 196
4. (a) 31.4 (b) 31 (c) 31 **5.** (a) 5.2 (b) 5.35 (c) 4.5 and 6.2 **6.** (a) 16,017.5 (b) 16,210
(c) none **7.** (a) .8 (b) .795 (c) none **8.** (a) .2 (b) .22 (c) .28 **9.** (a) 47.2 (b) 45.6 (c) none
10. (a) .5 (b) .5 (c) .3 and .9 **11.** (a) 129 (b) 128 (c) 125 and 128 **12.** (a) 5.0 (b) 4.995
(c) none **13.** (a) 46,908,000 (b) 42,779,000 **14.** (a) 36,244,000 (b) 34,328,000
15. (a) 14,770,000 (b) 14,577,000 **16.** (a) 25,507,000 (b) 19,325,000 **17.** 4000 square miles
18. 70,000,000 metric tons **19.** 5.27 sec **20.** 2.39 sec **21.** 2.42 sec **22.** 2.28 sec
23. the mean **24.** the median **25.** mean = 77; median = 80; mode = 79 **26.** the median **27.** 92
28. (a) 5.22 (b) 6 (c) 6 **29.** (a) 597.42 (b) 598 (c) 603 **30.** $31,700 **31.** 2.41 **32.** 2.77

33. 773,300 square miles   **35.** 25,302,000   **37.** 716,489   **38.** 1,093,119   **39.** −162,314
40. [(1,093,119 × 7) + (−162,314 × 3)]/(7 + 3) = 716,489. This is the same as Exercise 37.   **41.** $687,333.33
42. 6.1   **43.** 5   **44.** 4 and 6   **45.** (a) 18.4 (b) 16 (c) 11 and 16   **46.** (a) $116.30 (b) $115.00
(c) none   **47.** (a) 74.8 (b) 77.5 (c) 78   **48.** 83   **50.** (a) mean = 3; median = 3; mode = 4
(b) mode   **51.** (a) mean = 13.7; median = 16; mode = 18 (b) median   **52.** (a) 7 (b) 4   **53.** (a) 4
(b) 4.25   **54.** (a) 8 (b) 6.75   **55.** (a) 6 (b) 9.33   **56.** 0   **57.** 80   **58.** three choices: 1, 6, 16
60. no   **61.** no

13.3 Exercises   (Page 753)
1. the sample standard deviation   **3.** (a) 17 (b) 5.53   **4.** (a) 18 (b) 5.74   **5.** (a) 19 (b) 6.27
6. (a) 29 (b) 9.54   **7.** (a) 23 (b) 7.75   **8.** (a) 2.8 (b) .76   **9.** (a) 1.14 (b) .37
10. (a) 7.9 (b) 2.23   **11.** (a) 8 (b) 2.41   **12.** (a) 12 (b) 3.29   **13.** 3/4   **14.** 15/16
15. 45/49   **16.** 105/121   **17.** 88.9%   **18.** 96%   **19.** 64%   **20.** 84%   **21.** 3/4
22. 8/9   **23.** 15/16   **24.** 24/25   **25.** 1/4   **26.** 4/25   **27.** 4/49   **28.** 16/81   **29.** $202.50
30. $80.38   **31.** six   **32.** all twelve   **33.** at least nine   **35.** Brand B ($\bar{x}_B$ = 48,176 > 43,560)
36. Brand A (s_B = 2235 > 2116)   **38.** 18.29; 4.35   **39.** 23.29; 4.35   **40.** 8.29; 4.35   **42.** 54.86; 13.04
45. (a) 12.5 (b) −3.0 (c) 4.9 (d) 4.2 (e) 15.5 (f) 7.55 and 23.45   **46.** (a) when the distribution is
skewed to the right (b) when the distribution is skewed to the left   **48.**
49. no   **50.** no   **53.** (a) smallest (b) largest

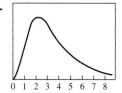

13.4 Exercises   (Page 760)
1. Lynn (since z = .48 > .19)   **2.** Michael (since z = 1.00 > .86)   **3.** Arvind (since z = −1.44 > −1.78)
4. Timothy's bass (since z = 2.50 > 2.18)   **5.** 58   **6.** 70   **7.** 62   **8.** 71   **9.** −.30   **10.** .79
11. −.50   **12.** .26   **13.** Antonio McDyess   **14.** Tim Duncan   **15.** Shawn Marion   **16.** Chris Webber
17. Tim Duncan (since z = 1.08 > .79)   **18.**

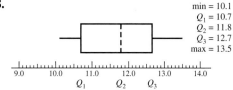

19. (a) The median is 11.8. (b) The range is 3.4. (c) The middle half of the items extend from 10.7 to 12.7.
27. the overall distribution   **28.** 5400   **29.** (a) no   **34.** mean = 51.55; standard deviation = 12.24
35. 101.4   **36.** 95.5   **37.** Q_1 = 77.7; Q_2 = 83.7; Q_3 = 94.1   **38.** Q_1 = 78.5; Q_2 = 82.9; Q_3 = 86.5
39. 94.1   **40.** 95.5   **41.**

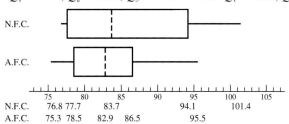

42. N.F.C.; about 1 unit   **43.** N.F.C.; about 3 units   **44.** N.F.C.; about 4 units   **45.** N.F.C.; about 8 units
46. 107.5   **47.** 114.2   **48.** high percentage of touchdown passes (about 13.9%)

13.5 Exercises (Page 769)

1. discrete **2.** discrete **3.** continuous **4.** continuous **5.** discrete **6.** continuous **7.** 50
8. 84 **9.** 68 **10.** 81 or 82 **11.** 50% **12.** 16% **13.** 95% **14.** .15% **15.** 43.3%
16. 32.1% **17.** 36.0% **18.** 48.8% **19.** 4.5% **20.** 8.6% **21.** 97.9% **22.** 82.2% **23.** 1.28
24. −1.75 **25.** −1.34 **26.** .74 **27.** 5000 **28.** 4330 **29.** 640 **30.** 9780 **31.** 9970
32. 360 **33.** 84.1% **34.** 37.1% **35.** 37.8% **36.** 79.4% **37.** 15.9% **38.** 10.6% **39.** 4.7%
40. .6% **41.** .994 or 99.4% **42.** 2150 units **43.** 189 units **44.** .433 **45.** .888 **46.** .082
47. about 2 eggs **48.** 6.7% **49.** 24.2% **50.** 38.2% **52.** 82 **53.** 79 **54.** 71 **55.** 68
56. 83.6 **57.** 90.4 **58.** 54.3 **59.** 40.3 **60. (a)** at least 36% **(b)** 78.8%

Extension Exercises (Page 776)

1. We have no way of telling. **2.** We can't tell; there is no scale. **3.** We can tell only that it rises and then falls. **4.** Again, there is no scale. **5.** 28% **6.** The actual decrease ($2.40 to $1.72) is about 28%, but the artist for the magazine reduced *each* dimension of the original figure by 28%. This causes the area to decrease by about 50%, thus giving a false impression. **7.** How long have Toyotas been sold in the United States? How do other makes compare? **8.** The tobacco is 44% fresher than what? **9.** The dentists preferred Trident Sugarless Gum to what? Which and how many dentists were surveyed? What percentage responded? **10.** So what—a Volvo is a smaller car than a Continental and should have a smaller turning radius. **11.** Just how quiet *is* a glider, really? **12.** There is no scale. We can't tell if the increase is substantial or not. **13.** The maps convey their impressions in terms of *area* distribution, whereas personal income distribution may be quite different. The top map probably implies too high a level of government spending, while the bottom map implies too low a level. **14.** It turns out that there were *three* women students, and just *one* had married a faculty member. **15.** By the time the figures were used, circumstances may have changed greatly. (The Navy was much larger.) Also, New York City was most likely not typical of the nation as a whole. **16.** When Huff's book was written, those three states had the best system for reporting these diseases. **17.** B **18.** C **19.** C **20.** A **21. (a)** $m = 1, a = 6, c = 3$
(b) There should be 1.4 managers, 5 agents, and 3.6 clerical employees. **22. (a)** $d = 2, f = 4, a = 7, s = 7$
(b) There should be 2 deans, 4 full professors, 6.5 associate professors, and 7.5 assistant professors.

13.6 Exercises (Page 783)

1. $y' = .3x + 1.5$ **2.** $r = .20$ **3.** 2.4 decimeters **4.** $y' = .556x - 17.8$ **5.** 48.9°
6. $r = 1$ (The experimental points lie perfectly along a line.) **7.** $y' = 3.35x - 78.4$
8. 123 lb **9.** 156 lb **10.** $r = .66$ **11.**   **12.** $y' = 2x - 51$
13. 79 **14.** $y' = 8.06x + 49.52$ **15.** $r = .996$ **16.** about $106,000
17.   **18.** $y' = 4.26x + 13.18$

20. .99 **21.** $y' = 1.44x - .39$ **22.** .99 **23.** The linear correlation is strong. **24.** 16; 22.7%
25. $y' = .00197x + 105.7$ **26.** .46 **27.** The linear correlation is moderate. **28.** about $135,000

Chapter 13 Test (Page 785)

1. about 1890 **2.** No. The *percentage* was 4 times as great, not the number of workers. **3.** about 3 million
4. The amount of land farmed in 1940 was about 26% to 28% greater than the amount farmed in 2000.

5.

Class Limits	Frequency f	Relative Frequency f/n
6–10	3	$3/22 \approx .14$
11–15	6	$6/22 \approx .27$
16–20	7	$7/22 \approx .32$
21–25	4	$4/22 \approx .18$
26–30	2	$2/22 \approx .09$

6. (a) 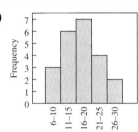 (b)

7. 8 **8.** 12.4 **9.** 12 **10.** 12 **11.** 10 **12.**

```
3 | 3 8
4 | 3 5 8 9
5 | 0 2 5
6 | 1 1 4 5 6 7 7 8
7 | 0 1 2 3 7 7 8 9 9
8 | 0 4 4
9 | 1
```

13. 35 **14.** 33 **15.** 37 **16.** 31 **17.** 49 **18.**

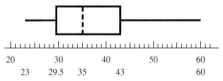

19. about 95% **20.** about .3% **21.** about 16% **22.** about 13.5% **23.** .684
24. .340 **25.** West **26.** East **27.** 80.0 **28.** *y* (scatter plot) **29.** $y' = -.98x + 10.64$
30. 7.70 **31.** $-.72$

CHAPTER 14 Consumer Mathematics

14.1 Exercises (Page 798)

1. $112 **2.** $400 **3.** $29.25 **4.** $140 **5.** $548.38 **6.** $209.32 **7.** (a) $826 (b) $835.84
8. (a) $2400 (b) $2433.31 **9.** (a) $2650 (b) $2653.02 **10.** (a) $7000 (b) $7346.64 **11.** $200

12. $30.56   **13.** $15,910   **14.** $270.40   **15.** $168.26   **16.** $587.73   **17.** $817.17   **18.** $3883.52; $1513.52   **19.** $17,724.34; $10,224.34   **20.** $4379.55; $929.55   **21.** (a) $1331   (b) $1340.10   (c) $1344.89
22. (a) $4221.30   (b) $4238.92   (c) $4247.98   **23.** (a) $17,631.94   (b) $17,762.93   (c) $17,831.37
24. (a) $27,576.89   (b) $27,862.34   (c) $28,011.11   **25.** (a) $361.56   (b) $365.47   (c) $368.13
(d) $369.44   (e) $369.48   **26.** (a) $123.17   (b) $123.62   (c) $123.92   (d) $123.96   (e) $123.97
28. $750.00   **29.** $1461.04   **31.** $747.26   **32.** $12,424.29   **33.** $4499.98   **34.** $12,351.59
35. $46,446.11   **36.** $83,761.59   **37.** $45,370.91   **38.** $82,661.67   **39.** 5.000%   **40.** 5.063%
41. 5.095%   **42.** 5.116%   **43.** 5.127%   **44.** 5.127%   **45.** 5.127%   **47.** $20,392   **48.** quarterly
49. 11 years, 233 days   **50.** 11 years, 202 days   **51.** 2.91%   **52.** (a) 3.818%   (b) Bank A; $6.52
(c) no difference; $232.38   **53.** $r = m[(Y + 1)^{1/m} - 1]$ **54.** 70 years   **55.** 35 years   **56.** 9 years
57. 8 years   **58.** 14.0%   **59.** 10.0%   **60.** 4.4%   **61.** 3.2%
62. years to triple $= \dfrac{100 \ln 3}{\text{annual inflation rate}} \approx \dfrac{110}{\text{annual inflation rate}}$
64. $190,000; $232,000; $332,000; $903,000   **65.** $5.62; $6.87; $9.85; $26.77
66. $1.90; $2.32; $3.32; $9.03   **67.** $15,500; $19,000; $27,200; $73,900   **68.** $702
69. $89   **70.** $176   **71.** $7759   **72.** $685,678.62   **73.** $81,102,828.75

Extension Exercises   (Page 805)

1. (a) $14,835.10   (b) $10,000   (c) $4835.10   **2.** (a) $62,742.85   (b) $40,000   (c) $22,742.85
3. (a) $3488.50   (b) $3000   (c) $488.50   **4.** (a) $13,125.23   (b) $9000   (c) $4125.23   **5.** (a) $3374.70
(b) $3120   (c) $254.70   **6.** (a) $9530.99   (b) $7800   (c) $1730.99   **9.** $3555.94   **10.** $13,278.98
11. $3398.08   **12.** $8623.08

14.2 Exercises   (Page 809)

1. $1650   **2.** $396   **3.** $2046   **4.** $85.25   **5.** $13,500   **6.** $3645   **7.** $17,145
8. $476.25   **9.** $1215; $158.75   **10.** $432; $130.50   **11.** $83.25; $46.29   **12.** $563.50; $100.45
13. $79.00; $165.90   **14.** $294; $171   **15.** $79.18; $38.39   **16.** $198.64; $34.37   **17.** $229.17
18. $347.46   **19.** $330.75   **20.** $180.60   **21.** $10,850   **22.** 10.3%   **23.** 6 years   **24.** 30
25. $3.26   **26.** $5.40   **27.** $2.72   **28.** $7.93

29.

Month	Unpaid Balance at Beginning of Month	Finance Charge	Purchases During Month	Returns	Payment	Unpaid Balance at End of Month
February	$319.10	$3.51	$86.14	0	$50	$358.75
March	358.75	3.95	109.83	$15.75	60	396.78
April	396.78	4.36	39.74	0	72	368.88
May	368.88	4.06	56.29	18.09	50	361.14

30.

Month	Unpaid Balance at Beginning of Month	Finance Charge	Purchases During Month	Returns	Payment	Unpaid Balance at End of Month
October	$828.63	$9.11	$128.72	$23.15	$125	$818.31
November	818.31	9.00	291.64	0	170	948.95
December	948.95	10.44	147.11	17.15	150	939.35
January	939.35	10.33	27.84	139.82	200	637.70

31.

Month	Unpaid Balance at Beginning of Month	Finance Charge	Purchases During Month	Returns	Payment	Unpaid Balance at End of Month
August	$684.17	$7.53	$155.01	$38.11	$100	$708.60
September	708.60	7.79	208.75	0	75	850.14
October	850.14	9.35	56.30	0	90	825.79
November	825.79	9.08	190.00	83.57	150	791.30

32.

Month	Unpaid Balance at Beginning of Month	Finance Charge	Purchases During Month	Returns	Payment	Unpaid Balance at End of Month
March	$1230.30	$13.53	$308.13	$74.88	$250	$1227.08
April	1227.08	13.50	488.35	0	350	1378.93
May	1378.93	15.17	134.99	18.12	175	1335.97
June	1335.97	14.70	157.72	0	190	1318.39

33. $2.50   **34.** $4.77   **35.** $12.08   **36.** $18.16   **37. (a)** $607.33   **(b)** $7.29   **(c)** $646.26
38. (a) $500.51   **(b)** $6.01   **(c)** $517.53   **39. (a)** $473.96   **(b)** $5.69   **(c)** $505.07   **40. (a)** $950.69
(b) $11.41   **(c)** $1046.85   **41. (a)** $7.20   **(b)** $6.62   **42. (a)** $11.77   **(b)** $10.12   **43. (a)** $38.18
(b) $38.69   **(c)** $39.20   **44.** $116.07   **45.** 16.0%   **46.** $20.31   **47.** $32   **55. (a)** $127.44
(b) $139.08   **56.** Bank A

14.3 Exercises   (Page 820)

1. 13.5%   **2.** 11.0%   **3.** 8.5%   **4.** 9.0%   **5.** $114.58   **6.** $125.83   **7.** $92.71   **8.** $120.00
9. 11.0%   **10.** 13.0%   **11.** 9.5%   **12.** 11.0%   **13. (a)** $8.93   **(b)** $511.60   **(c)** $6075.70
14. (a) $4.66   **(b)** $418.63   **(c)** $9766.87   **15. (a)** $2.79   **(b)** $97.03   **(c)** $4073.57   **16. (a)** $10.75
(b) $732.65   **(c)** $7129.85   **17. (a)** $206   **(b)** 9.5%   **18. (a)** $731   **(b)** 14.0%   **19. (a)** $180
(b) 11.0%   **20. (a)** $2551.44   **(b)** 10.5%   **21. (a)** $41.29   **(b)** $39.70   **22. (a)** $230.76   **(b)** $203.36
23. (a) $420.68   **(b)** $368.47   **24. (a)** $728.08   **(b)** $694.15   **25. (a)** $1.80   **(b)** $14.99   **(c)** $1045.01
26. (a) $3.53   **(b)** $156.84   **(c)** $5018.16   **27.** finance company APR: 12.0%; credit union APR: 11.5%; choose credit union   **28.** $29.07   **29.** $32.27   **30.** $1119.23   **32.** 13.5%   **34.** 8   **35.** 12   **36.** 15

14.4 Exercises   (Page 829)

1. $675.52   **2.** $568.30   **3.** $469.14   **4.** $617.78   **5.** $2483.28   **6.** $2295.88   **7.** $1034.56
8. $2224.79   **9. (a)** $513.38   **(b)** $487.50   **(c)** $25.88   **(d)** $58,474.12   **10. (a)** $755.01   **(b)** $616.25
(c) $138.76   **(d)** $86,861.24   **11. (a)** $1247.43   **(b)** $775.67   **(c)** $471.76   **(d)** $142,728.24   **(e)** $1247.43
(f) $773.11   **(g)** $474.32   **(h)** $142,253.92   **12. (a)** $1046.90   **(b)** $935.63   **(c)** $111.27   **(d)** $124,638.73
(e) $1046.90   **(f)** $934.79   **(g)** $112.11   **(h)** $124,526.62   **13. (a)** $1390.93   **(b)** $776.61   **(c)** $614.32
(d) $113,035.68   **(e)** $1390.93   **(f)** $772.41   **(g)** $618.52   **(h)** $112,417.16   **14. (a)** $1237.79   **(b)** $781.25
(c) $456.54   **(d)** $149,543.46   **(e)** $1237.79   **(f)** $778.87   **(g)** $458.92   **(h)** $149,084.54   **15.** $552.18
16. $530.29   **17.** $1127.27   **18.** $795.15   **19.** $1168.44   **20.** $1419.24   **21.** 360
22. $387,532.80   **23.** $247,532.80   **24.** Total interest is $107,532.80 more than the amount financed.
25. (a) $304.01   **(b)** $734.73   **26. (a)** 26.0%   **(b)** 1.3%   **27. (a)** $59,032.06   **(b)** $59,875.11
28. (a) $29,333.83   **(b)** $46,417.87   **29. (a)** payment 176   **(b)** payment 304

A-80 Answers to Selected Exercises

30. (a) $1.14 (b) $8.77 **31.** $25,465.20 **32.** $56,006.40 **33.** $91,030.80 **34.** $129,523.20
35. (a) $674.79 (b) $746.36 (c) an increase of $71.57 **36.** (a) $388.88 (b) $370.20 (c) a decrease of $18.68 **37.** (a) $395.83 (b) $466.07 **38.** (a) $470.73 (b) $531.09 **39.** $65.02 **41.** $140,000
42. $2800 **43.** $4275 **44.** $39,275

14.5 Exercises (Page 844)

1. $29.75 **2.** 438,600 shares **3.** 7¢ per share lower **4.** $15.60 **5.** $47.48 **6.** $17.40
7. $1.72 per share **8.** $2.23 per share **9.** 7600 shares **10.** 74¢ per share higher **11.** 10.6% higher
12. 40.1% lower **13.** 35 **14.** 18 **15.** $5140.00 **16.** $17,372.00 **17.** $1161.00 **18.** $10,815.00
19. $2273.82 **20.** $5799.20 **21.** $17,822.55 **22.** $22,259.95 **23.** $24,804.23 **24.** $17,115.00
25. $55,800.00 **26.** $56,717.15 **27.** $12,424.96 **28.** $596.07 **29.** $5709.88 **30.** $39,672.85
31. $17,441.47 **32.** $62,141.33 **33.** $56,021.45 **34.** $36,014.28 **35.** $30.56 net paid out
36. $15,941.00 net taken in **37.** (a) $800 (b) $80 (c) $960 (d) $1040 (e) 130% **38.** (a) $500
(b) $20 (c) −$60 (d) −$40 (e) −8% **39.** (a) $1250 (b) $108 (c) −$235 (d) −$127
(e) −10.16% **40.** (a) $1760 (b) $224 (c) $500 (d) $724 (e) 41.14% **41.** $275 **42.** $3200
43. $177.75 **44.** $1220 **45.** (a) $10.49 (b) 334 **46.** (a) $10.82 (b) 166 **47.** (a) $8.27
(b) 3080 **48.** (a) $12.46 (b) 6681 **49.** (a) $8.39 (b) $100.68 (c) 15.6% **50.** (a) $8.06
(b) $96.72 (c) 10.8% **51.** (a) $57.45 (b) $689.40 (c) 27.6% **52.** (a) $89.69 (b) $1076.28
(c) 21.6% **53.** (a) $1203.75 (b) $18.06 (c) 19.56% **54.** (a) $2294 (b) $52.76 (c) 31.37%
55. (a) $4179 (b) $76.48 (c) 24.31% **56.** (a) $10,159.92 (b) $225.55 (c) 30.15% **57.** (a) $10.00
(b) $35.17 (c) $29.95 **58.** (a) $1000.00 (b) $52.00 (c) $29.95 **59.** (a) $400.00 (b) $41.80
(c) $29.95 **60.** (a) $40,000.00 (b) $188.00 (c) $29.95 **61.** (a) $4000.00 (b) $91.40 (c) $120.00
62. (a) $400,000.00 (b) $595.00 (c) $120.00 **63.** large; low **66.** The annual report was available free to *Wall Street Journal* readers. **67.** Pinnacle Systems (PCLE)

	Aggressive Growth	Growth	Growth & Income	Income	Cash
69.	$ 1400	$ 8600	$ 6200	$ 2800	$ 1000
70.	$ 15,000	$ 72,500	$ 90,000	$ 60,000	$12,500
71.	$ 8000	$112,000	$144,000	$116,000	$20,000
72.	$ 16,900	$126,750	$371,800	$287,300	$42,250

73. 3.75% **74.** 4.9% **75.** 5.2% **76.** 6.0% **77.** 5.3% **78.** 5.3%

Chapter 14 Test (Page 849)

1. $130 **2.** $58.58 **3.** 3.04% **4.** 14 years **5.** $67,297.13 **6.** $10.88 **7.** $480 **8.** $145
9. 14.5% **10.** $35.63 **11.** 11.5% **13.** $1162.95 **14.** $1528.48 **15.** $3488 **17.** 12.1%
18. 112,300 shares **19.** $10 **20.** .078%; No, since the given return was not for a one-year period.

CHAPTER 15 Graph Theory

15.1 Exercises (Page 860)

1. 7 vertices, 7 edges **2.** 4 vertices, 4 edges **3.** 10 vertices, 9 edges **4.** 6 vertices, 9 edges
5. 6 vertices, 9 edges **6.** 5 vertices, 8 edges **7.** Two have degree 3. Three have degree 2. Two have degree 1. Sum of degrees is 14. This is twice the number of edges. **8.** One has degree 3. Two have degree 2. One has degree 1. Sum of degrees is 8. This is twice the number of edges. **9.** Six have degree 1. Four have degree 3. Sum of degrees is 18. This is twice the number of edges. **10.** All vertices have degree 3. Sum of degrees is 18. This is twice the number of edges. **11.** no **12.** no

13. Yes. Corresponding edges should be the same color. AB should match AB, etc.

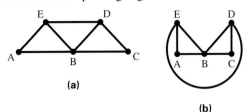

14. Yes. Corresponding edges should be the same color. AB should match AB, etc.

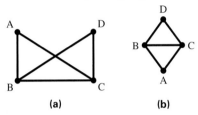

15. no   **16.** Yes. Corresponding edges should be the same color. AB should match AB, etc.

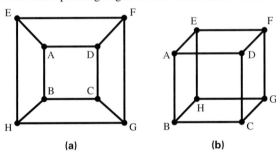

17. connected, 1 component   **18.** disconnected, 2 components   **19.** disconnected, 3 components
20. connected, 1 component   **21.** disconnected, 2 components   **22.** disconnected, 4 components   **23.** 6
24. 16   **25.** 4   **26.** 11   **27. (a)** yes **(b)** no; no edge A to D **(c)** no; no edge A to E **(d)** yes **(e)** yes
(f) yes   **28. (a)** yes **(b)** yes **(c)** no; edge B to D used more than once **(d)** no; edge GF used more than once
(e) yes **(f)** no; no edge B to E   **29. (a)** no; does not return to starting vertex **(b)** yes **(c)** no; no edge C to F
(d) no; does not return to starting vertex **(e)** no; edge F to D used more than once   **30. (a)** no; no edge J to H
(b) yes **(c)** yes **(d)** yes **(e)** yes **(f)** no; no edge J to I   **31. (a)** no; no edge B to C **(b)** no; edge I to G
used more than once **(c)** no; no edge E to I **(d)** yes **(e)** yes **(f)** no; edge A to D used more than once
32. not a walk, not a path, not a circuit   **33.** a walk and a path, not a circuit   **34.** a walk, a path, and a circuit
35. a walk, not a path, not a circuit   **36.** a walk, a path, and a circuit   **37.** a walk and a path, not a circuit
38. yes   **39.** no; no edge A to C, for example   **40.** yes   **41.** no; no edge A to F, for example
42. no; no edge A to D   **43.** yes   **44.** 16 games

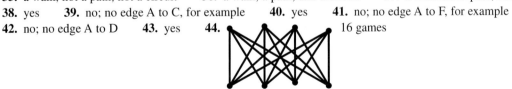

45. 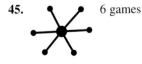 6 games   **46.**  20 edges   **47.** 36 handshakes

48. 9 handshakes   **49.** 10 telephone conversations

50. A→B→C→D→A. Corresponds to tracing around the edges of a single face. (The circuit is a square.)

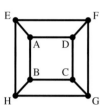

51. A→B→D→A. Corresponds to tracing around the edges of a single face. (The circuit is a triangle.)

52. (a)   **(b)** disconnected, two components **(c)**

(d)

54. (a) one answer: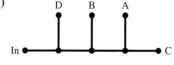

55. The degree of each vertex is 4, which shows that each face has common boundaries with 4 other faces.

56. (a)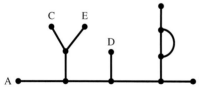

(c)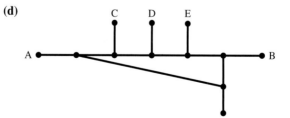

15.2 Exercises (Page 875)

1. (a) no, not a path (b) yes (c) no, not a walk (d) no, not a circuit **2.** (a) no, some edges not used (e.g., B→F) (b) no, not a circuit (c) yes (d) yes **3.** (a) no, some edges not used (e.g., B→F) (b) yes (c) no, not a path (d) yes **4.** No, some vertices have odd degree (e.g., A). **5.** Yes, all vertices have even degree. **6.** No, some vertices have odd degree (e.g., A). **7.** No, some vertices have odd degree (e.g., G). **8.** Yes, all vertices have even degree. **9.** yes **10.** no **11.** has an Euler circuit; all vertices have even degree. No circuit visits each vertex exactly once. **12.** has no Euler circuit; all vertices have odd degree. A→B→C→D→E→F→A visits each vertex exactly once. **13.** has an Euler circuit; all vertices have even degree. A→B→H→C→G→D→F→E→A visits each vertex exactly once. **14.** has no Euler circuit; some vertices have odd degree (e.g., B); has no circuit that visits each vertex exactly once **15.** has no Euler circuit; some vertices have odd degree (e.g., B); has no circuit that visits each vertex exactly once **16.** No, some vertices have odd degree. **17.** No, some vertices have odd degree. **18.** No, some vertices have odd degree. **19.** Yes, all vertices have even degree. **20.** CD, FG **21.** none **22.** all edges **23.** B→E or B→D **24.** B→D or B→F **25.** B→C or B→H **26.** A→I→H→G→F→E→K→J→H→E→D→C→B→D→I→B→A **27.** A→C→B→F→E→D→C→F→D→A **28.** A→B→C→D→E→F→D→G→F→C→G→H→J→B→H→A **29.** Graph has an Euler circuit. A→G→H→J→I→L→J→K→I→H→F→G→E→F→D→E→C→D→B→C→A→E→B→A **30.** Graph has an Euler circuit. E→F→G→H→I→G→E→H→F→I→E **31.** Graph does not have an Euler circuit; some vertices have odd degree (e.g., C). **32.** There is such a route: A→B→C→D→E→F→G→H→I→G→E→I→J→D→A→J→C→A **33.** There is such a route: A→D→B→C→A→H→D→E→B→H→G→E→F→H→J→L→C→M→A→K→M→L→K→J→A **34.** not possible **35.** not possible **36.** possible **37.** no **38.** yes; I and H **39.** yes; B and G **40.** no **41.** possible; exactly two of the rooms have an odd number of doors. **42.** possible **43.** not possible; all rooms have an even number of doors. **44.** (a) no (b) yes; Manhattan and Randall's Island **45.** A→B→D→C→A; 4 edges in any such circuit **46.** A→B→C→D→E→C→F→B→E→F→A; 10 edges in any such circuit **47.** A→B→C→D→E→F→G→H→C→A→H→I→A; 12 edges in any such circuit **48.** Only the octahedron has such a tracing; the octahedron is the only regular solid with an even number of edges meeting each vertex. **49.** those for which the number of vertices is an odd number greater than or equal to 3; In a complete graph with n vertices, the degree of each vertex is $n - 1$. "$n - 1$ is even" is equivalent to "n is odd."

15.3 Exercises (Page 889)

1. (a) no; visits vertex E twice (b) yes (c) no; does not visit C (d) no; does not visit A **2.** (a) no; for example, visits C twice (b) no; does not visit B (c) no; does not visit B or D (d) no; does not return to starting point **3.** (a) none, no edge B to C (b) all three (c) none, edge AD used twice **4.** (a) a circuit; not a Hamilton circuit; not an Euler circuit (b) a circuit; an Euler circuit; not a Hamilton circuit (c) none, no edge G to H **5.** A→B→D→E→F→C→A **6.** has none **7.** G→H→J→I→G **8.** A→B→F→C→E→D→A **9.** X→T→U→W→V→X **10.** has none **11.** Hamilton circuit: A→B→C→D→A. The graph has no Euler circuit, since at least one of the vertices has odd degree. (In fact, all have odd degree.)

12. This graph has no Hamilton circuit. A→B→C→D→E→B→F→D→A is an Euler circuit for the graph.

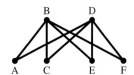

13. A→B→C→D→E→F→A is both a Hamilton and an Euler circuit.

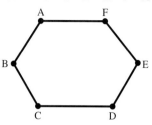

14. (a) true **(b)** true **(c)** false

A→B→C→D→A is a Hamilton circuit that does not use each edge.

(d) true **(e)** false

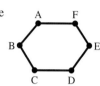

A→B→C→D→E→F→A is both a Hamilton and an Euler circuit.

(f) false

A→B→C→D→E→B→F→D→A is an Euler circuit that visits vertices B and D more than once.

15. Hamilton circuit **16.** Hamilton circuit **17.** Euler circuit **18.** Euler circuit **19.** Hamilton circuit
20. Hamilton circuit **21.** 24 **22.** 720 **23.** 362,880 **24.** $14! \approx 8.72 \times 10^{10}$ **25.** 9!
26. 14! **27.** 17! **28.** 59! **29.** P→Q→R→S→P   P→Q→S→R→P   P→R→Q→S→P
P→R→S→Q→P   P→S→Q→R→P   P→S→R→Q→P   **30.** E→F→G→H→I→E   E→F→G→I→H→E
31. E→H→I→F→G→E   E→H→I→G→F→E   **32.** E→I→H→F→G→E   E→I→H→G→F→E
33. E→F→G→H→I→E   E→F→G→I→H→E   E→F→H→G→I→E   E→F→H→I→G→E   E→F→I→G→H→E
E→F→I→H→G→E   **34.** E→I→F→G→H→E   E→I→F→H→G→E   E→I→G→F→H→E
E→I→G→H→F→E   E→I→H→F→G→E   E→I→H→G→F→E   **35.** E→G→F→H→I→E
E→G→F→I→H→E   E→G→H→F→I→E   E→G→H→I→F→E   E→G→I→F→H→E   E→G→I→H→F→E
36. E→F→G→H→I→E   E→G→F→H→I→E   E→F→G→I→H→E   E→G→F→I→H→E   E→F→H→G→I→E
E→G→H→F→I→E   E→F→H→I→G→E   E→G→H→I→F→E   E→F→I→G→H→E   E→G→I→F→H→E
E→F→I→H→G→E   E→G→I→H→F→E   E→H→F→G→I→E   E→I→F→G→H→E   E→H→F→I→G→E
E→I→F→H→G→E   E→H→G→F→I→E   E→I→G→F→H→E   E→H→G→I→F→E   E→I→G→H→F→E
E→H→I→F→G→E   E→I→H→F→G→E   E→H→I→G→F→E   E→I→H→G→F→E
37. A→B→C→D→E→A   A→C→B→D→E→A   A→B→C→E→D→A   A→C→B→E→D→A
A→B→D→C→E→A   A→C→D→B→E→A   A→B→D→E→C→A   A→C→D→E→B→A
A→B→E→C→D→A   A→C→E→B→D→A   A→B→E→D→C→A   A→C→E→D→B→A
A→D→B→C→E→A   A→E→B→C→D→A   A→D→B→E→C→A   A→E→B→D→C→A
A→D→C→B→E→A   A→E→C→B→D→A   A→D→C→E→B→A   A→E→C→D→B→A
A→D→E→B→C→A   A→E→D→B→C→A   A→D→E→C→B→A   A→E→D→C→B→A
38. Minimum Hamilton circuit is F→G→I→H→F; weight is 15.   **39.** Minimum Hamilton circuit is
P→Q→R→S→P; weight is 2200.   **40.** Minimum Hamilton circuit is A→B→D→E→C→A; weight is 20.
41. Minimum Hamilton circuit is C→D→E→F→G→C; weight is 64.   **42. (a)** F→H→I→G→F; Total weight
is 15.   **(b)** G→I→H→F→G; Total weight is 15.   **(c)** H→F→I→G→H; Total weight is 17.   **(d)** I→G→F→H→I;
Total weight is 15.   **43. (a)** A→C→E→D→B→A; Total weight is 20.   **(b)** C→A→B→D→E→C; Total weight is 20.

(c) D→C→A→B→E→D; Total weight is 23. (d) E→C→A→B→D→E; Total weight is 20.
44. (a) A→E→D→C→F→B→A; Total weight is 11.85. (b) B→F→C→D→E→A→B; Total weight is 11.85.
(c) C→D→E→A→B→F→C; Total weight is 11.85. (d) D→C→A→E→B→F→D; Total weight is 12.9.
(e) E→A→C→D→B→F→E; Total weight is 14.1. (f) F→B→C→D→E→A→F; Total weight is 12.25.
45. (a) A→C→D→E→B→A; Total weight is 87. B→C→A→E→D→B; Total weight is 95. C→A→E→D→B→C; Total weight is 95. D→C→A→E→B→D; Total weight is 131. E→C→A→D→B→E; Total weight is 137.
(b) A→C→D→E→B→A with total weight 87 (c) For example, A→B→C→D→E→A has total weight 52.
46. (a)

(b) $K_{2,3}$ and $K_{2,4}$ have no Hamilton circuit. For $K_{2,2}$: A→B→C→D→A; For $K_{3,3}$: P→Q→R→S→T→U→P; For $K_{4,4}$: E→F→G→H→I→J→K→L→E (c) For $K_{m,n}$ to have a Hamilton circuit, $m = n$. The Hamilton circuit must alternate between the two groups of vertices until the circuit is complete. This can be done without visiting any vertex more than once only if $m = n$. (d) For $K_{m,n}$ to have an Euler circuit, both m and n must be even numbers (not necessarily equal). The degree of each vertex in the graph is m or n, so by Euler's theorem we have an Euler circuit if and only if m and n are both even.   47. A→B→C→D→E→F→A A→B→C→F→E→D→A
A→B→E→D→C→F→A A→B→E→F→C→D→A A→D→E→F→C→B→A A→D→E→B→C→F→A
A→D→C→B→E→F→A A→D→C→F→E→B→A A→F→E→B→C→D→A A→F→E→D→C→B→A
A→F→C→D→E→B→A A→F→C→B→E→D→A 48. A→B→C→D→E→A A→B→C→E→D→A
A→B→D→C→E→A A→B→D→E→C→A A→C→B→D→E→A A→C→E→D→B→A
A→D→B→C→E→A A→D→E→C→B→A A→E→C→B→D→A A→E→C→D→B→A
A→E→D→B→C→A A→E→D→C→B→A 49. A→B→C→D→E→F→A A→B→C→E→D→F→A
A→B→C→E→F→D→A A→B→C→F→E→D→A A→D→E→F→C→B→A A→D→F→E→C→B→A
A→F→D→E→C→B→A A→F→E→D→C→B→A 50. A→B→C→G→F→H→D→E→A
A→E→D→H→F→G→C→B→A   51. (a) graphs (2) and (4) (b) graphs (2) and (4) (c) no; graph (1) provides a counterexample. (d) No; if $n < 3$ the graph will have no circuits at all. (e) Degree of each vertex in a complete graph with n vertices is $(n - 1)$. If $n \geq 3$, then $(n - 1) > n/2$. So we can conclude from Dirac's theorem that the graph has a Hamilton circuit.   52. A→B→C→D→E→N→M→L→K→J→I→H→G→R→S→T→U→Q→P→F→A   53. A→F→G→R→S→T→U→Q→P→N→M→L→K→J→I→H→B→C→D→E→A
54. A→B→H→G→F→P→N→M→L→K→T→U→Q→R→S→I→J→C→D→E→A

15.4 Exercises (Page 902)

1. tree   2. no; has a circuit   3. no; not connected   4. no; has a circuit   5. tree   6. no; has circuits
7. no; has a circuit   8. For example draw in edge AE.   9. not possible, since the graph has a circuit
10. For example draw in edges RS and RT.   11. tree   12. tree   13. not necessarily a tree
14. false   15. true   16. false   17. false

18.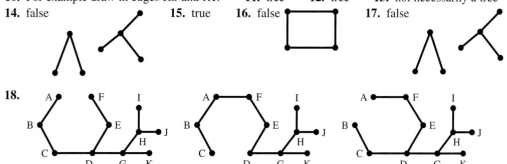

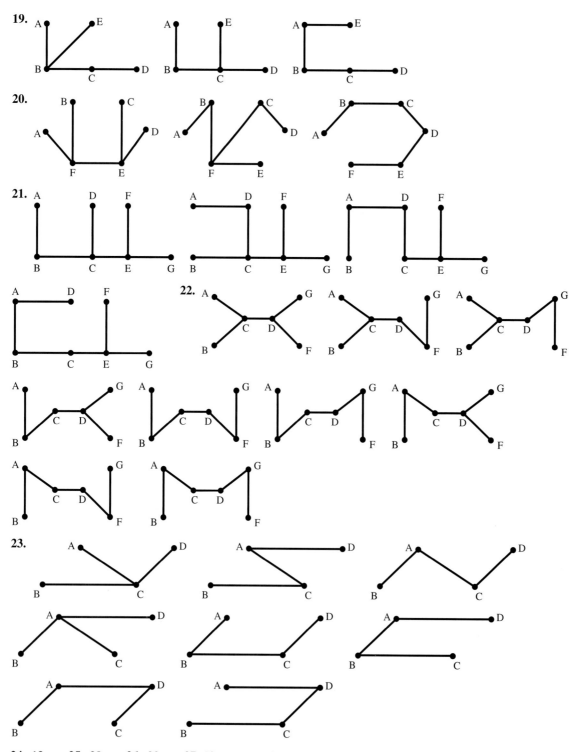

24. 12   **25.** 20   **26.** 90   **27.** If a connected graph has circuits, none of which have common edges, then the number of spanning trees for the graph is the product of the numbers of edges in all the circuits.

28. (a) (i)

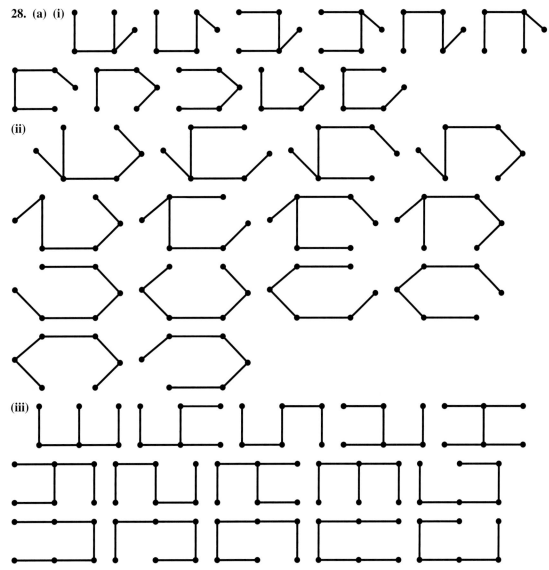

(b) (i) 17 **(ii)** 29 **(iii)** 47 **(c)** If a connected graph has two circuits having a common edge, and if the circuits have *m* and *n* edges respectively, then the number of spanning trees for the graph is $(m-1)(n-1) + (m-1) + (n-1)$ (or $mn - 1$).

29. Total weight is 51.

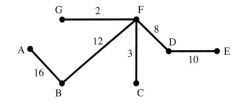

30. Total weight is 74.

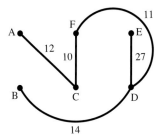

31. Total weight is 66.

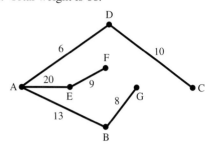

32. Total weight is 35.

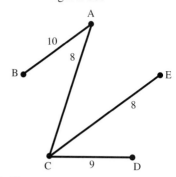

33. Total length to be covered is 140 ft.

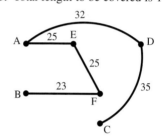

34. The total cost is 9.3 million dollars.

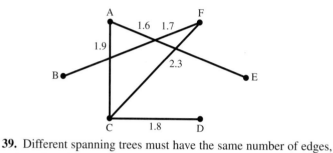

35. 33 **36.** 41 **37.** 62 **38.** 17 **39.** Different spanning trees must have the same number of edges, since the number of vertices in the tree is the number of vertices in the original graph, and the number of edges has to be one less than this. **40.** (a) 8 (b) 16 (c) 2 (d) 8 (e) $n-1$; $2(n-1)$; 2; $n-1$ **41.** (a) 9 (b) 18 (c) 0 (d) 2

42. (a) 16 (b) 32 (c) 3
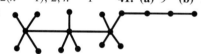

43. 22 cables **44.** 12 **45.** is possible; graph must be a tree since it has one fewer edge than vertices.
46. is not possible; if we retain connectedness of the graph, each vertex drawn has to have at least one edge associated to it. **47.** is possible; graph cannot be a tree, since it would have at least as many edges as vertices.
48. is possible; a graph such as that shown could be drawn next to the original tree.

49. 3:
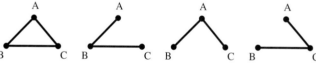

50. 16:

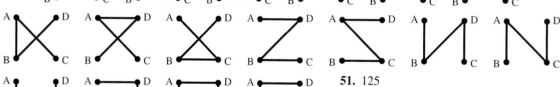
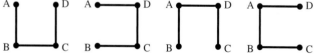
51. 125

52. 2 non-isomorphic trees: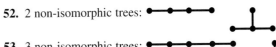

53. 3 non-isomorphic trees: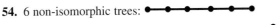

54. 6 non-isomorphic trees: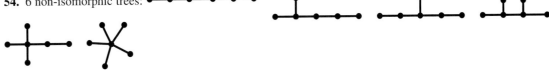

55. 11 non-isomorphic trees: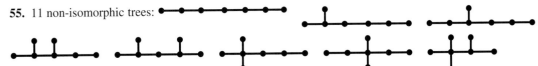

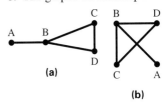

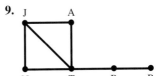

56. **(a)** 1 **(b)** The tree is connected and has more than one vertex.
(c) 2 **(d)** 3 **(e)** 4, 5 **(f)** $n + 1$ **(g)** The components are the vertices. **(h)** There are $n + 1$ vertices.

Chapter 15 Test (Page 908)

1. 7 **2.** 20 **3.** 10 **4.** (a) no; edge AB used twice (b) yes (c) no; no edge C to D
5. (a) yes (b) no; for example, no edge B to C (c) yes **6.** for example: **7.** 13 edges

8. The graphs are isomorphic. Corresponding edges should be the same color. AB should match AB, etc.

9. The graph is connected. Tina knows the largest number of other guests.

10. 28 games **11.** yes; there is an edge from each vertex to each of the remaining 6 vertices.
12. (a) no; does not use all the edges (b) no; is not a circuit (for example, no edge B to C) (c) yes
13. no; some vertices have odd degree. **14.** yes; all vertices have even degree. **15.** no; two of the rooms have an odd number of doors. **16.** F→B→E→D→B→C→D→K→B→A→H→G→F→A→G→J→F
17. (a) no; does not visit all vertices (b) no; is not a circuit (for example, no edge B to C) (c) no; visits some vertices twice before returning to starting vertex **18.** F→G→H→I→E→F F→G→H→E→I→F F→G→I→H→E→F F→G→I→E→H→F F→G→E→H→I→F F→G→E→I→H→F. There are 6 such Hamilton circuits. **19.** P→Q→S→R→P; weight is 27. **20.** A→E→D→C→F→B→A; total weight is 11.85.

21. 24! **22.** Hamilton circuit

23. Any three of these:

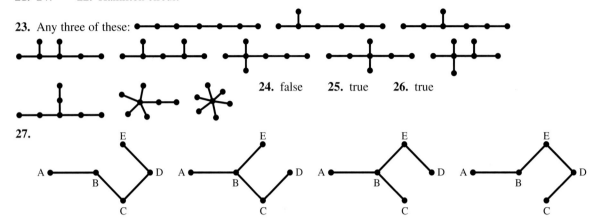

24. false **25.** true **26.** true

27.

There are 4 spanning trees.

28.

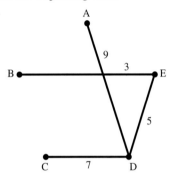

Weight is 24. **29.** 49

CHAPTER 16 Voting and Apportionment

16.1 Exercises (Page 924)

1. (a) 4! = 24
(b)

Number of Voters	Ranking
3	b > c > a > d
2	a > d > c > b
1	c > d > b > a
1	d > c > b > a
1	d > a > b > c
2	c > a > d > b
1	a > c > d > b
2	a > b > c > d

(c) Australian shepherd

2. (a) 4! = 24
(b)

Number of Voters	Ranking
3	a > b > c > d
1	b > c > a > d
2	d > b > c > a
1	c > b > a > d
2	a > d > c > b
1	b > c > d > a
1	c > b > d > a
2	a > b > d > c

(c) Australian shepherd

3. 5! = 120; 7! = 5040   **4.** 6! = 720; 8! = 40,320   **5.** For any value of $n \geq 5$ there are at least 120 possible rankings of the candidates. This means a voter must select one ranking from a huge number of possibilities. It also means the mechanics of any election method, except the plurality method, are difficult to manage.
6. C(5, 2) = 10; C(7, 2) = 21   **7.** C(6, 2) = 15; C(8, 2) = 28   **8.** The number of pairwise comparisons needed becomes large, making it quite labor intensive to determine the outcome using the pairwise comparison method.

9. With 40,320 possible rankings to be examined, finding the outcomes of 28 different pairwise comparisons seems a nearly impossible task. **10. (a)** b, with 6 first-place votes **(b)** a, with 2 (pairwise) points **(c)** b, with a total of 16 (Borda) points **(d)** a **(e)** a **11. (a)** b, with 6 first-place votes **(b)** c, with 2 (pairwise) points **(c)** b, with 16 (Borda) points **(d)** c **(e)** c **12.** Logo b is selected by all the methods. **13.** Logo a is selected by all the methods. **14. (a)** j, with 8 first-place votes **(b)** e, with 3 (pairwise) points **(c)** e, with 40 (Borda) points **(d)** h **(e)** e **15. (a)** e, with 8 first-place votes **(b)** j, with 3 (pairwise) points **(c)** j, with 40 (Borda) points **(d)** h **(e)** j **16. (a)** a, with 3 (pairwise) points **(b)** b, with 24 (Borda) points **(c)** a, Australian shepherd has 7 of 13 first-place votes—a majority—at the first round of voting. **17. (a)** c, with 3 (pairwise) points **(b)** a, with 24 (Borda) points **(c)** c **18. (a)** t, with 18 first-place votes **(b)** h, with 4 (pairwise) points **(c)** k, with 136 (Borda) points **(d)** b **(e)** h **19. (a)** t, with 18 first-place votes **(b)** h, with 4 (pairwise) points **(c)** b, with 136 (Borda) points **(d)** k **(e)** h **20.** c beats t **21.** c beats t **22.** h beats j **23.** h beats e **24.** In a runoff between b and c, activity b is selected. Activity b faces activity t, and b is selected. **25.** In a runoff between k and c, activity k is selected. Activity k faces activity t, and k is selected. **26.** Logo b **27.** Logo a **28. (a)** Joan **(b)** Joan and Lori **29. (a)** Mary **(b)** Mary and Alison **30. (a)** 1 **(b)** b, with 6 pairwise points **31. (a)** 2 **(b)** c, with 7 pairwise points **32. (a)** 8 **(b)** d, with 8 pairwise points **33. (a)** 7 **(b)** e, with 7 pairwise points **34.** 14; a **35.** 16; c **36.** 30; e **37.** 55; c **38.** e **39.** j **40.** n **41.** n **42.** a_y **43.** a_y **44.** e **45.** Answers may vary; one possible arrangement of the voters is 2, 4, 5, 7, 3. **46.** Answers may vary. One possible arrangement of the voters is 2, 4, 5, 7, 3.

16.2 Exercises (Page 939)

1. (a) Alternative a has a majority of the first-place votes (6 of 11). **(b)** b **(c)** yes **2. (a)** Alternative b has a majority of the first-place votes (6 of 11). **(b)** a **(c)** yes **3. (a)** Alternative a has a majority of first-place votes (20 of 36). **(b)** b **(c)** yes **4. (a)** Alternative d has a majority of first-place votes (19 of 36). **(b)** c **(c)** yes **5. (a)** Alternative a has a majority of first-place votes (16 of 30). **(b)** b **(c)** yes **6. (a)** Alternative b has a majority of first-place votes (16 of 30). **(b)** c **(c)** yes **7. (a)** a **(b)** b **(c)** a **(d)** b **(e)** plurality and Borda methods **8. (a)** c **(b)** b **(c)** c **(d)** b **(e)** plurality and Borda methods **9. (a)** e **(b)** j **(c)** h **(d)** e **(e)** plurality and Hare methods **10. (a)** j **(b)** e **(c)** h **(d)** j **(e)** plurality and Hare methods **11. (a)** h **(b)** t **(c)** b **(d)** k **(e)** all three methods **12. (a)** h **(b)** t **(c)** k **(d)** b **(e)** all three methods **13. (a)** m has 2 pairwise points, b and s have $1\frac{1}{2}$ pairwise points each, and c has 1 pairwise point. **(b)** b has $2\frac{1}{2}$ pairwise points, m has 2 pairwise points, s has 1 pairwise point, c has $\frac{1}{2}$ pairwise point, so b is selected. **(c)** Yes, the rearranging voters moved m, the winner of the nonbinding election, to the top of their ranking, but b wins the official selection process. **14. (a)** c has 2 pairwise points, s and b have $1\frac{1}{2}$ pairwise points each, and m has 1 pairwise point. **(b)** s has $2\frac{1}{2}$ pairwise points, c has 2 pairwise points, b has 1 pairwise point, m has $\frac{1}{2}$ pairwise point, so s is selected. **(c)** Yes, the rearranging voters moved c, the winner of the nonbinding election, to the top of their ranking, but s wins the official selection process. **15. (a)** s has 35 Borda points, b has 29 Borda points, d has 23 Borda points, c has 27 Borda points. **(b)** s has 42 Borda points, b has 43 Borda points, d has 13 Borda points, c has 16 Borda points, so b is selected. **(c)** Yes, the rearranging voters moved s, the winner of the nonbinding election, to the top of their ranking, but b wins the official selection process. **16. (a)** c has 35 Borda points, d has 29 Borda points, b has 23 Borda points, s has 27 Borda points. **(b)** c has 42 Borda points, d has 43 Borda points, b has 13 Borda points, s has 16 Borda points, so d is selected. **(c)** Yes, the rearranging voters moved c, the winner of the nonbinding election, to the top of their ranking, but d wins the official selection process. **17. (a)** d drops out after one round of votes; b drops out after Round two; in the third vote a is preferred to c by a margin of 11 to 6. **(b)** d drops out after one round of votes; c drops out after Round two; in the third vote b is preferred to a by a margin of 9 to 8. **(c)** Yes, the rearranging voters moved a, the winner of the nonbinding election, to the top of their ranking, but b wins the official selection process. **18. (a)** d drops out after one round of votes; a drops out after Round two; in the third vote b is preferred to c by a margin of 11 to 6. **(b)** d drops out after one round of votes; c drops out after Round two; in the third vote a is preferred to b by a margin of 9 to 8. **(c)** Yes, the rearranging voters moved b, the winner of the nonbinding election, to the top of their ranking, but a wins the official selection process. **19. (a)** Candidate a has 6 first-place votes. **(b)** Candidate b has 7 first-place votes. **(c)** yes

20. (a) Candidate b has 8 first-place votes. (b) Candidate c has 9 first-place votes. (c) yes **21.** (a) Candidate a has 75 first-place votes. (b) Candidate c has 80 first-place votes. (c) yes **22.** (a) Candidate a has 40 first-place votes. (b) Candidate c has 45 first-place votes. (c) yes **23.** (a) e has 3 pairwise points. (b) m has 2 pairwise points. (c) yes **24.** (a) e has 3 pairwise points. (b) m (c) yes **25.** No, however, the second pairwise comparison results in a tie. **26.** No, however, the second pairwise comparison results in a tie.
27. (a) b has 37 Borda points. (b) c wins the second Borda election. (c) yes **28.** (a) a has 25 Borda points. (b) b wins the second Borda election. (c) yes **29.** (a) Round one eliminates b; a is preferred to c in Round two, by a margin of 22 to 12. (b) b (c) yes **30.** (a) Round one eliminates b; a is preferred to c in Round two, by a margin of 32 to 22. (b) b (c) yes **31.** (a) Round one eliminates d; b is preferred to a in Round two, by a margin of 9 to 8. (b) yes **32.** (a) Round one eliminates d; a is preferred to b in Round two, by a margin of 9 to 8 (b) yes **34.** Original winner x gained a pairwise point by beating Percussionist y, so the withdrawal of y hurts x. Percussionist v was beaten originally by y, so the withdrawal of y does not alter the number of pairwise points for v.
35. Answers will vary. One possible voter profile is given.

Number of Voters	Ranking
10	a > b > c > d > e > f
9	b > f > e > c > d > a

Candidate a is a majority candidate and wins all of its pairwise comparisons by a margin of 10 to 9, earning 5 pairwise points. Candidate b wins all of its comparisons, except the one with a, earning 4 pairwise points.

36. Answers will vary. One possible way to complete the voter profile is given. b wins the Borda method election on the sample profile.

Number of Voters	Ranking
21	d > b > c > a
8	b > c > a > d
6	c > b > d > a
5	a > c > b > d

37. Answers will vary. One possible way to complete the voter profile is given. b wins the Borda method election on the sample profile.

Number of Voters	Ranking
21	a > b > c > d
5	b > c > a > d
9	c > b > d > a
5	d > c > b > a

38. Answers will vary. One possible profile is the voter profile for the animal shelter poster dog contest in Exercise 2 and Exercise 16 of Section 16.1. **39.** Answers will vary. One possible profile is the voter profile for the animal shelter poster dog contest in Exercise 1 and Exercise 17 of Section 16.1. **40.** (a) a has 2 pairwise points. (b) Answers may vary. The 3 voters in the bottom row all switch to the ranking a > z > x > y. **41.** (a) m has 25 Borda points. (b) Answers may vary. The 3 voters in the third row all switch their ranking to m > b > s > c. The 2 voters in the bottom row all switch their ranking to m > b > c > s. **42.** The 4 voters in the bottom row all switch their ranking to a > c > b. **43.** Answers will vary. Delete Candidate c. **44.** Answers will vary. Delete both b and d. **45.** Answers will vary. One possible profile is given.

Number of Voters	Ranking
21	g > j > e
12	j > e > g
8	j > g > e

16.3 Exercises (Page 959)

1. (a) 29,493; 123.4017 (b)

State park	a	b	c	d	e
Number of trees	11	70	62	54	42

(c) $md = 122$

State park	a	b	c	d	e
Number of trees	11	70	62	54	42

(d) The traditionally rounded values of Q sum to 240, which is greater than the number of trees to be apportioned.

State park	a	b	c	d	e
Traditionally rounded Q	12	70	62	54	42

(e) The value of md for the Webster apportionment should be greater than $d = 123.4017$, because bigger divisors make smaller modified quotas with a smaller total sum.

(f) $md = 124$

State park	a	b	c	d	e
Number of trees	12	70	61	54	42

(g) The Hamilton and Jefferson apportionments are the same. The Webster apportionment of the trees is different from the other two apportionments.

2. (a) $md = 20.25$

School	Applegate	Bayshore	Claypool	Delmar	Edgewater
Number of computers	16	22	14	28	29

(b) The Jefferson and Webster apportionments of the computers agree. The Hamilton apportionment is different.

(c) The Hamilton method apportions 15 computers to Claypool School (enrollment 298) and both of the other methods apportion 14 computers to Claypool. The average number of students sharing a computer drops from 21.29 to 19.87 if Claypool gets 15 instead of just 14 computers.

3. (a) 269; 24.45455 **(b)**

Course	Fiction	Poetry	Short Story	Multicultural
Number of sections	2	2	3	4

(c) $md = 20$

Course	Fiction	Poetry	Short Story	Multicultural
Number of sections	2	1	3	5

(d) The traditionally rounded values of Q sum to 10, which is less than the number of sections to be apportioned.

Course	Fiction	Poetry	Short Story	Multicultural
Traditionally rounded Q	2	1	3	4

(e) The value of md for the Webster apportionment should be less than $d = 24.45455$, because smaller divisors make larger modified quotas with a larger total sum.

(f) $md = 23$

Course	Fiction	Poetry	Short Story	Multicultural
Number of sections	2	2	3	4

(g) The Hamilton and Webster apportionments are the same. The Jefferson apportionment of the sections is different from the other two apportionments. **(h)** The Hamilton and Webster methods both apportion two sections of poetry; the Jefferson method apportions only one. If there are two sections of poetry, then the 35 enrolled students can be divided into two small sections with 17 and 18 students each instead of all 35 students being forced into one large section. **(i)** The Jefferson method apportions 5 sections of multicultural literature instead of 4. This means that the average class size will be 20 students, rather than 25 students.

4. (a) 66

Resort	Anna	Bob	Cathy	David	Ellen	Floyd
Number of boats	5	4	6	5	4	6

(b) The Hamilton and Webster apportionments are the same. The Jefferson apportionment is different.

5. (a)

State	Abo	Boa	Cio	Dao	Effo	Foti
Number of seats	15	22	6	19	14	55

(b) $md = 356$

State	Abo	Boa	Cio	Dao	Effo	Foti
Number of seats	15	22	6	19	13	56

(c) The Hamilton, Jefferson, and Webster apportionments all are different.

6.

State	Abo	Boa	Cio	Dao	Effo	Foti	Total
Number of seats	15	22	7	19	14	55	132

7.

State	Abo	Boa	Cio	Dao	Effo	Foti	Total
Number of seats	15	22	7	18	14	54	130

8. $md = 34{,}400$

State	Number of Seats	State	Number of Seats
Virginia	18	New Jersey	5
Massachusetts	14	New Hampshire	4
Pennsylvania	13	Vermont	2
North Carolina	10	Georgia	2
New York	10	Kentucky	2
Maryland	8	Rhode Island	2
Connecticut	7	Delaware	2
South Carolina	6	Total	105

9. (a) 1721; 43.025 **(b)**

Hospital	A	B	C	D	E
Number of nurses	3	5	8	11	13

(c) $md = 40$

Hospital	A	B	C	D	E
Number of nurses	3	5	8	11	13

(d) The traditionally rounded values of Q sum to 41, which is greater than the number of nurses to be apportioned.

Hospital	A	B	C	D	E
Traditionally rounded Q	3	6	8	11	13

(e) The value of md for the Webster apportionment should be greater than $d = 43.025$, because larger divisors make smaller modified quotas with a smaller total sum.

(f) $md = 43.1$

Hospital	A	B	C	D	E
Number of nurses	3	5	8	11	13

(g) All three apportionments are the same.

10. (a) 29,493; 67.18223 **(b)**

State park	a	b	c	d	e
Number of trees	21	129	113	99	77

(c) $md = 66.8$

State park	a	b	c	d	e
Number of trees	21	129	113	99	77

(d) The traditionally rounded values of Q sum to 439, which is equal to the number of trees to be apportioned.

(e) The value of md for the Webster apportionment should equal $d = 67.18223$, because the traditionally rounded values of Q already sum to exactly the number of trees to be apportioned.

State park	a	b	c	d	e
Traditionally rounded Q	21	129	113	99	77

(f) $md = 67.18223$

State park	a	b	c	d	e
Number of trees	21	129	113	99	77

(g) All three apportionments of the 439 trees agree. The governor avoided a potential political battle over which apportionment method should be used by making sure all three of the apportionment methods under consideration would result in the same apportionment of the trees.

11. Answers will vary. One possible population profile is given.

State	a	b	c	d	e	Total
Population	50	230	280	320	120	1000

12. Answers will vary. One possible enrollment profile is given.

Course	Biology	Chemistry	Physics	Mathematics	Total
Enrollment	35	40	28	95	198

13. Answers will vary. One possible ridership profile is given.

Bus route	a	b	c	d	e	Total
Number of riders	131	140	303	178	197	949

14. The modified divisor for the Adams method is always found by slowly increasing the value of the standard divisor because the modified quotas are always rounded up to the nearest integer. Rounding up clearly increases the sum. A larger divisor produces modified quotas that are small enough to sum to the number of seats to be apportioned even when rounded up.

15. (a) $md = 29$

Course	Fiction	Poetry	Short Story	Multicultural
Number of sections	2	2	3	4

(b) The Adams apportionment is the same as the Hamilton and Webster apportionments. It is different from the Jefferson apportionment.

16. (a) $md = 46$

Hospital	A	B	C	D	E
Number of nurses	3	6	8	10	13

(b) The Hamilton, Jefferson, and Webster apportionments of the nurses all agree; the Adams method apportionment is different.

17. (a) $md = 377.3$

State	Abo	Boa	Cio	Dao	Effo	Foti
Number of seats	16	22	7	18	14	54

(b) All four methods produce different apportionments of the 131 seats in Timmu's legislature.

18. $\sqrt{5 \cdot 6} = 5.477$ **19.** $\sqrt{56 \cdot 57} = 56.498$ **20.** $\sqrt{11 \cdot 12} = 11.489$ **21.** $\sqrt{32 \cdot 33} = 32.496$

22. If the sum of rounded Q values is less than the number of objects being apportioned, the modified divisor md is found by slowly decreasing the value of d, because a smaller divisor produces larger modified quotas with a larger sum.   **23.** If the sum is greater, then md is found by slowly increasing the value of d, because a larger divisor produces smaller modified quotas with a smaller sum.   **24.** If the sum is equal to the number of objects to be apportioned, then $md = d$ because the correct number of objects is already apportioned.
25. (a) $md = 24$

Course	Fiction	Poetry	Short Story	Multicultural
Number of sections	2	2	3	4

(b) The Hill-Huntington apportionment is the same as the Hamilton, Webster, and Adams apportionments. It is different from the Jefferson apportionment.

26. (a) $md = 43.3$

Hospital	A	B	C	D	E
Number of nurses	3	5	8	11	13

(b) The Hill-Huntington apportionment is the same as the Hamilton, Webster, and Jefferson apportionments. It is different from the Adams apportionment.

27. (a) $md = 367$

State	Abo	Boa	Cio	Dao	Effo	Foti
Number of seats	15	22	7	19	14	54

(b) The Hill-Huntington apportionment is the same as the Webster apportionment. The other three apportionments differ.

16.4 Exercises (Page 972)

1. $md = 595$

State	a	b	c	d
Q rounded down	28	12	81	9
Number of seats	28	12	83	9

2. $md = 626$

State	a	b	c	d
Q rounded down	105	55	23	15
Number of seats	107	55	23	15

3. $md = 356$

State	Abo	Boa	Cio	Dao	Effo	Foti
Q rounded down	15	21	6	18	13	54
Number of seats	15	22	6	19	13	56

4. $md = 50.9$

State	a	b	c	d	e
Q rounded down	88	29	38	21	21
Number of seats	90	29	39	21	21

5. $md = 48.4$

State	a	b	c	d	e
Q rounded down	52	30	164	19	22
Number of seats	53	30	166	19	22

6. $md = 295$

State	a	b	c	d	e
Q rounded down	5	11	23	80	29
Number of seats	5	11	23	82	29

7.

State	a	b	c	d
Number of seats if $n = 204$	35	74	42	53
Number of seats if $n = 205$	34	75	43	53

8.

State	a	b	c	d
Number of seats if $n = 71$	8	14	21	28
Number of seats if $n = 72$	7	14	22	29

Answers to Selected Exercises   A-97

9.

State	a	b	c	d	e
Number of seats if $n = 126$	10	29	34	28	25
Number of seats if $n = 127$	9	30	34	29	25

10.

State	a	b	c	d	e
Number of seats if $n = 45$	7	7	7	14	10
Number of seats if $n = 46$	7	6	8	14	11

11.

State	a	b	c	d	e
Number of seats if $n = 149$	27	39	28	30	25
Number of seats if $n = 150$	26	40	29	30	25

12.

State	a	b	c	d
Number of seats if $n = 160$	36	44	47	33
Number of seats if $n = 161$	36	45	48	32

13.

State	a	b	c
Initial number of seats	1	4	6
Percent growth	10.91%	18.40%	13.16%
Revised number of seats	2	4	5

14.

State	a	b	c
Initial number of seats	1	4	6
Percent growth	12.73%	20.00%	14.74%
Revised number of seats	2	4	5

15.

State	a	b	c
Initial number of seats	6	5	2
Percent growth	4.84%	1.63%	1.45%
Revised number of seats	6	4	3

16.

State	a	b	c
Initial number of seats	6	5	2
Percent growth	4.84%	1.90%	1.69%
Revised number of seats	6	4	3

17.

State	a	b	c
Initial number of seats	2	4	7
Percent growth	8.99%	16.00%	9.78%
Revised number of seats	3	4	6

18.

State	a	b	c
Initial number of seats	2	4	7
Percent growth	10.11%	16.00%	10.67%
Revised number of seats	3	4	6

19. Three additional seats are added for the second apportionment.

State	Original State a	Original State b	New State c
Initial number of seats	12	4	*****
Revised number of seats	11	5	3

20. Three additional seats are added for the second apportionment.

State	Original State a	Original State b	New State c
Initial number of seats	12	4	*****
Revised number of seats	11	5	3

21. Two additional seats are added for the second apportionment.

State	Original State a	Original State b	New State c
Initial number of seats	20	55	*****
Revised number of seats	21	54	2

22. Two additional seats are added for the second apportionment.

State	Original State a	Original State b	New State c
Initial number of seats	20	55	*****
Revised number of seats	21	54	2

23. Seven additional seats are added for the second apportionment.

State	Original State a	Original State b	New State c
Initial number of seats	36	47	*****
Revised number of seats	37	46	7

24. Seven additional seats are added for the second apportionment.

State	Original State a	Original State b	New State c
Initial number of seats	36	47	*****
Revised number of seats	37	46	7

25. Answers may vary. One possible enrollment profile is given.

Class	a	b	c	d
Enrollment	225	45	35	30

26. Answers may vary. One possible population profile is given.

County	a	b	c	d
Population	56,000	15,000	7080	9500

27. Answers may vary. One possible answer is 200 seats.
28. Answers may vary. One possible answer is 100 seats.
29. Answers may vary. One possible answer is $n = 49$ seats.
30. Answers may vary. One possible answer is $n = 17$.
31. Answers may vary. One possible answer is Course a actual enrollment = 61 and Course c actual enrollment = 213.
32. Answers may vary. One possible answer is Course a actual enrollment = 96 and Course c actual enrollment = 245.
33. The new states paradox does not occur if the new population is 531, but it does occur if the new population is 532.
34. The new states paradox occurs in favor of the first subdivision if the new population is 548, but the second subdivision itself snags a council seat from the original town if it has a population of 549.

35. $md = 208$

State	a	b	c	d	e
Q rounded down	8	16	33	118	42
Number of seats	9	17	34	117	43

36. $md = 209$

State	a	b	c	d	e
Q rounded down	8	16	33	117	42
Number of seats	9	17	34	116	43

Chapter 16 Test (Page 977)

1. Cancún **2.** Aruba **3.** the Bahamas **4.** the Dominican Republic **5.** the method that selects *your* first choice; You are the dictator. **6.** Aruba **7.** the Dominican Republic **8.** Aruba and Cancún each have two pairwise points and the Bahamas and Dominican Republic each have one pairwise point, so the pairwise comparison method vote results in a tie. **9.** Answers may vary. One possibility is a, b, d, c. **14.** Alternative a has a majority of the first-place votes, 16 of 31. The Borda method selects alternative b. **15.** c **16.** Alternative c is the Condorcet candidate. Plurality and Borda violate the Condorcet criterion and select b. Hare does not violate the criterion because it selects c. **17.** Alternative c is selected before the five voters rearrange their ranking. Alternative s is selected after they rearrange their ranking. The rearranging voters moved alternative c, the previous pairwise selection, to the top of their ranking, but c is not selected a second time. This shows that the pairwise comparison method can violate the monotonicity criterion.

18. Alternative a is selected before the five voters rearrange their ranking. Alternative b is selected after they rearrange their ranking. The rearranging voters moved alternative a, the previous Hare selection, to the top of their ranking, but a is not selected a second time. This shows that the Hare method can violate the monotonicity criterion.

19. Alternative a is selected before losing alternative c is dropped. Alternative b is selected after c is dropped. The two outcomes show that the plurality method can violate the independence of irrelevant alternatives criterion. A losing alternative was dropped from the selection process, but the original preferred alternative is not selected a second time.

20. Alternative b is selected before losing alternative d is dropped. Alternative a is selected after d is dropped. The two outcomes show that the Borda method can violate the independence of irrelevant alternatives criterion. A losing alternative was dropped from the selection process, but the original preferred alternative is not selected a second time.

21. Alternative a is selected before losing alternative b is dropped. Alternative c is selected after b is dropped. The two outcomes show that the Hare method can violate the independence of irrelevant alternatives criterion. A losing alternative was dropped from the selection process, but the original preferred alternative is not selected a second time.

23. standard divisor $d = 133.3692$ (That is, each seat represents 13,337 Smithapolis citizens.)

Ward	1st	2nd	3rd	4th	5th
Number of seats	11	65	57	50	12

24. $md = 131$

Ward	1st	2nd	3rd	4th	5th
Number of seats	10	65	58	50	12

25. $md = 133.7$

Ward	1st	2nd	3rd	4th	5th
Number of seats	11	65	57	50	12

30. $md = 347$

State	a	b	c	d
Q rounded down	6	12	15	64
Number of seats	6	12	16	66

31.

State	a	b	c	d	e
Number of seats if $n = 126$	10	29	34	28	25
Number of seats if $n = 127$	9	30	34	29	25

32.

State	a	b	c
Initial number of seats	1	4	6
Percent growth	14.55%	20.00%	15.79%
Revised number of seats	2	4	5

33. Fifteen additional seats are added for the second apportionment.

State	Original State a	Original State b	New State c
Initial number of seats	23	77	*****
Revised number of seats	24	76	15

Appendix The Metric System **(Page A-5)**

1. 8000 mm 2. 1476 cm 3. 85 m 4. .25 m 5. 689 mm 6. 32.5 mm 7. 5.98 cm
8. .3542 cm 9. 5300 m 10. 9240 m 11. 27.5 km 12. 14.592 km 13. 2.54 cm; 25.4 mm
14. 3.3 cm; 33 mm 15. 5 cm; 50 mm 16. 2.54 cm; 25.4 mm 17. 600 cl 18. 4100 ml
19. 8700 ml 20. 1250 cl 21. 9.25 L 22. .412 L 23. 8.974 L 24. 56.39 L 25. 8 kg
26. 25 kg 27. 5200 g 28. 12,420 g 29. 4200 mg 30. 389 cg 31. .598 g 32. 76.34 g
33. 30°C 34. 280°C 35. −81°C 36. −40°C 37. 50°F 38. 77°F 39. −40°F 40. 5°F
41. 200 nickels 42. 3.5 g 43. .2 g 44. 1.5 g 45. $40.99 46. $312.19 47. $387.98
48. $64.30 49. 1,082,400 cm^3; 1.0824 m^3 50. 1,076,700 cm^3; 1.0767 m^3 51. 200 bottles
52. 16 bottles 53. 897.9 m 54. 7.581 mi 55. 201.1 km 56. 1609 km 57. 3.995 lb
58. 644.7 g 59. 21,428.8 g 60. 16.90 lb 61. 30.22 qt 62. 62.77 qt 63. 106.7 L 64. 15.14 L
65. unreasonable 66. unreasonable 67. reasonable 68. reasonable 69. unreasonable
70. unreasonable 71. B 72. C 73. B 74. A 75. C 76. A 77. A 78. B 79. A
80. C 81. C 82. A 83. A 84. C 85. B 86. A 87. B 88. A 89. A 90. B
91. B 92. A 93. C 94. A 95. B 96. A

Credits

1 AP/Wide World Photos, 7 © Reuters NewMedia Inc./CORBIS, 14 Hulton Getty/Liaison Agency, Inc., 20(L) Chuck Painter/Stanford News Service, 20(C) From *How to Solve It: A New Aspect of Mathematical Method* by George Polya. Copyright 1945, © 1973 renewed by Princeton University Press. Excerpt reprinted with permission of Princeton University Press., 22 © Aaron Horowitz/CORBIS, 23 © PhotoDisc/Getty Images, 25(B) Reprinted with permission from *Mathematics Teacher*, © 2003 by the National Council of Teachers of Mathematics. All rights reserved., 25(T) Bill Greenblatt/Liaison Agency, Inc., 25–28 (Exercises 1–25) Reprinted with permission from *Mathematics Teacher*, © The National Council of Teachers of Mathematics. All rights reserved., 27 ©2000 Daniel J. Cox/Stone, 28(L) © PhotoDisc/Getty Images, 28(R) © Eyewire/Getty Images, 31 Courtesy of Texas Instruments, 32 Courtesy of Texas Instruments, 33(T) © PhotoDisc/Getty Images, 33(C) AP/Wide World Photos, 33(B) © CORBIS, 37(R) © Michael & Patricia Fogden/CORBIS, 38 © ORIGLIA FRANCO/CORBIS SYGMA, 47 © PhotoDisc/Getty Images, 47–48 (Exercises 10–15) Reprinted with permission from *Mathematics Teacher*, volume 96, Number 2, February 2003. © 2003 by the National Council of Teachers of Mathematics. All rights reserved., 49(T) © Digital Vision/Getty Images, 49(C) From Watt, *3D Computer Graphics*. © 1993 Pearson Education, Inc. Publishing as Pearson Addison Wesley., 50 Bavaria-Verlag, 54 ©Corbis-UPI/Bettmann, 77(L) © Reuters NewMedia Inc./CORBIS, 77(R) © Reuters NewMedia Inc./CORBIS, 80 Hulton Getty/Liaison Agency, Inc., 81 © Richard Hamilton Smith/CORBIS, 83 © Royalty-Free/CORBIS, 84 Silvio Fiore/Superstock, 87 ELDER NEVILLE/CORBIS SYGMA, 88 © Reuters NewMedia Inc./CORBIS, 89 © Peter Christopher Canadian Beer Brewery, Ontario, Canada, 92 © Underwood & Underwood/CORBIS, 93 © Deborah Davis/Photo Edit, 94 Courtesy of Sony Electronics, 95 ©Corbis-Bettmann Archive, 97 Erich Lessing/Art Resource, NY, 104 (Adaptation) From "Raymond Smullyan" by Ira Mothner, *Mathematical People*. Copyright © 1985 by Berkhauser Boston. Reprinted by permission., 105 Library of Congress, 107 Library of Congress, 112 © Jay Freis/Getty Images, 113 Photofest, 123 ©Bettmann/Corbis, 129 ©Corbis-Bettmann Archive, 133, 135, 136 Copyright © 1999 Dell Magazines, a division of Crosstown Publications. Reprinted by permission., 140 © Richard Hamilton Smith/CORBIS, 143(LC) © Reuters NewMedia Inc./CORBIS, 143(RC) ©Tribune Media Services, Inc. All Rights Reserved. Reprinted with permission., 147 © VCG/Getty Images, 148-149 Copyright © 1999 Dell Magazines, a division of Crosstown Publications. Reprinted by permission., 151 © PhotoDisc/Getty Images, 152(T) John Buienkant/Photo Researchers, 152(B) Courtesy, Prindle, Weber & Schmidt, Inc., 153(T) © Bridgeman Art, 153(B) The Metropolitan Museum of Art, Egyptian Expedition, Rogers Fund, 1930., 156 Transparency no. 5001(3) Courtesy Department of Library Services, American Museum of Natural History., 158 British Museum, 163 © Trinity College. Reprinted by permission of the Master and Fellows of Trinity College Cambridge., 165(TC) Courtesy, IBM Corporation, 171(T) ©Corbis-Bettmann Archive, 176 Registered trademarks of Kellogg Company, 187 Courtesy, Professor Noether, 195 © BZB/Getty Images, 199 Hugh C. Williams, University of Manitoba., 200 This program was written by Charles W. Gantner and is provided courtesy of Texas Instruments., 202 Roger Viollet/Liaison Agency, Inc., 209(T) Roger Viollet/Liaison Agency, Inc., 209(B) Princeton University Library, University Archives, Department of Rare Books and Special Collections, 213 Giraudon/Art Resource, NY, 223 Tom Wurt/Stock Boston, 234(R, L) Roy Morsch/Stock Market, 236(TL) Scala, Art Resource, NY, 236(B) ©The Walt Disney Company, 237 James Randkley/Tony Stone Images, 241(L) ©Corbis-Bettmann Archive, 241(R) Benjamin Franklin Collection/Sterling Memorial Library/Yale University, 242(L) Stadtmuseum, Aachen/ET Archive, London/SuperStock., 242(R) Metropolitan Museum of Art/Harris Brisbane Dick Fund, 1943, 247 © Eyewire/Getty Images, 255 From Lial *Intermediate Algebra*, 7e. © 2002 Pearson Education, Inc. Publishing as Pearson Addison Wesley. Reprinted with permission. (p. 185.), 267 From Lial *Introductory Algebra*, 7e. © 2002 Pearson Education, Inc. Publishing as Pearson Addison Wesley. Reprinted with permission. (p. 262.), 269 © Wayne R Bilenduke/Getty Images, 273 M. Lial, J. Hornsby, C. Miller, *Introductory Algebra* 6/e, p. 262. © 1998 Addison Wesley Longman Inc. Reprinted with permission., 274(T, B), 280(L) Photos by Doug Plasencia courtesy of Bowers and Merena Galleries., 280(R) From Lial *Introductory Algebra*, 7e. © 2002 Pearson Education, Inc. Publishing as Pearson Addison Wesley. Reprinted with permission. (p. 262.), 281(R) © PhotoDisc/Getty Images, 281(L) © PhotoDisc/Getty Images, 286 David Blatner, *The Joy of Pi*, www.joyofpi.com., 291 Courtesy www.joyofpi.com, 296 Herman is reprinted with permission from Laughing Stock Licensing, Inc., Ottawa, Canada. All rights reserved., 300 © Richard Cummins/CORBIS, 301 Courtesy of John Hornsby, 305(L) From Lial *Introductory Algebra*, 7e. © 2002 Pearson Education, Inc. Publishing as Pearson Addison Wesley. Reprinted with permission. (p. 196.), 305(R) Photography courtesy of HeritageCoin.com, 306 AP/Wide World Photos, 313(B) From Lial *Intermediate Algebra*, 7e. © 2002 Pearson Education, Inc. Publishing as Pearson Addison Wesley. Reprinted with permission. (p. 254.), 314 © Reuters NewMedia Inc./CORBIS, 321 © Reuters NewMedia Inc./CORBIS, 336(B) © Jack Hollingsworth/CORBIS, 342 © Natalie Fobes/CORBIS, 349(B) From Lial *Beginning Algebra*, 9e. © 2002 Pearson Education, Inc. Publishing as Pearson Addison Wesley. Reprinted with permission., 369 © Gary Buss/Getty Images, 372 From Lial *Intermediate Algebra*, 7e. © 2002 Pearson Education, Inc. Publishing as Pearson Addison Wesley. Reprinted with permission. (p. 321.), 393 AP/Wide World Photos, 399(T) © George Hall/CORBIS, 399(T, B) From Lial *Intermediate Algebra*, 7e. © 2002 Pearson Education, Inc. Publishing as Pearson Addison Wesley. Reprinted with permission. (p. 173.), 406 © Royalty-Free/CORBIS, 411(BR),

412(TL) From Lial *Intermediate Algebra,* 7e. © 2002 Pearson Education, Inc. Publishing as Pearson Addison Wesley. Reprinted with permission. (p. 189.), 417(R) © David Young-Wolff/Photo Edit, 420(TR, BR) From Lial *Intermediate Algebra,* 7e. © 2002 Pearson Education, Inc. Publishing as Pearson Addison Wesley. Reprinted with permission. (p. 201.), 422 © Tony Freeman/Photo Edit, 424, 432 From Lial *Intermediate Algebra,* 7e. © 2002 Pearson Education, Inc. Publishing as Pearson Addison Wesley. Reprinted with permission. (pp. 213, 225.), 438 National Optical Astronomy Observatories, 443 From Lial *Intermediate Algebra,* 7e. © 2002 Pearson Education, Inc. Publishing as Pearson Addison Wesley. Reprinted with permission. (pp. 606, 662.), 454(L) © Stephanie Maze/CORBIS, 454(R) © Barry Yee/Getty Images, 455(L) © Eyewire/Getty Images, 456 From Lial *Intermediate Algebra,* 7e. © 2002 Pearson Education, Inc. Publishing as Pearson Addison Wesley. Reprinted with permission. (p. 248.), 465 © Eyewire/Getty Images, 470(L), 472 From Lial *Intermediate Algebra,* 7e. © 2002 Pearson Education, Inc. Publishing as Pearson Addison Wesley. Reprinted with permission. (p. 261, 281.), 473 © Chuck Savage/CORBIS, 475(TR) © Owen Franken/CORBIS, 475(BR) AP/Wide World Photos, 484 Stanford University News Service, 491 Courtesy of the University of Kuopio, Department of applied physics, Inverse Problems Group, 492 From Eves' History, 506 ©Corbis-Bettmann Archive, 509 Alinari/Art Resource, NY, 510 Reprinted with permission from *The Trisection Problem,* copyright © 1971 by the National Council of Teachers of Mathematics. All rights reserved., 518 Bill Debold/Liaison Agency, Inc., 524 George Arthur Plimpton Collection, Columbia University, 526 Peter French/DRK Photo, 527 Courtesy, John Hornsby, 528 Reprinted with permission from *The Pythagorean Proposition.* Copyright © The National Council of Teachers of Mathematics., 537(R) 2072 from *On Growth and Form* by Arch Wentworth Thompson. Cambridge University Press. Abridged edition edited by John Tyler Bonner., 541 Superstock, Inc., 547(T) Courtesy of Gwen Hornsby, 547(B) © Alan Majchrowicz/Getty Images, 548 © David Trood Pictures/Getty Images, 549 © Geoff du Feu/Getty Images, 550(T) M.C. Escher's Symmetry Drawing E66 (Winged Lions) © 2003 Cordon Art B.V. - Baarn - Holland. All rights reserved., 550(B) M.C. Escher's Waterfall © 2003 Cordon Art B.V. - Baarn - Holland. All rights reserved., 551–553 The authors wish to thank The Math Forum at Drexel and Suzanne Alejandre for permission to reprint this article., 553 coolmath.com, 556 ©Corbis-Hulton/UPI Bettmann, 557(C) Courtesy, John Hornsby, 557(B) Courtesy, John Hornsby, 558 Noriyuki Yoshida/Superstock, Inc., 559 ©Corbis-Bettmann Archive, 560 Reprinted with permission from *Science News,* the weekly newsmagazine of science, copyright 1982 by Science Service Inc., 568 AP/Wide World Photos, 569(BL) NASA, 569(C) Courtesy of Philip P. Marcus, 570(B) Stephen Johnson/Tony Stone Images, 570–573 (Exercises 1–25) Pages 1–4 from *National Council of Teachers of Mathematics Student Math Notes,* November 1991. Reprinted by permission of the National Council of Teachers of Mathematics., 576 © PhotoDisc/Getty Images, 587 © Katz 1998, 605(L) Courtesy of John Hornsby, 610(B) © Dean Conger/CORBIS, 613(B) Courtesy of John Hornsby, 629 © Tony Stone Imaging/Getty Images, 631 British Museum, 634(B) (Map) From "Ride with the Wind" by Linda Garmon, *Science News,* Vol. 119, June 27, 1981. Reprinted by permission of Wind Ship Development Corporation., 640 National Museum of American Art, Washington D.C./Art Resource, NY, 649 Washington National Cathedral, 670(TR) © Randy Wells/CORBIS, 673 © PhotoDisc/Getty Images, 675 © CORBIS, 676 Lithograph by Rudolf Hoffmann, 1859, after photoprint by Maull and Polyblanck/Library of Congress Prints and Photographs Division, 684 (Exercise 64) Problem #6 from *Mathematics Teacher,* January 1989, p. 38. Reprinted by permission., 684 (Exercise 65) Problem #2 from *Mathematics Teacher,* November 1989, p. 626. Reprinted by permission., 686 Stock Montage, Inc./Historical Pictures Service, 694 ©Bettmann/CORBIS, 695(LC) © Eyewire/Getty Images, 696 Philippe Psaila/Liaison Agency, Inc., 699 Courtesy of NASA, 700 © Jonathan Nourok/Photo Edit, 702 Museum fur Volkerlunde un Schweizerishches, 706 © Reuters NewMedia Inc./CORBIS, 710 Superstock, Inc., 711 From David G. Christensens' Slot Machines, A Pictorial History 1889–1972, 712 Courtesy, John Hornsby, 715 ©LWA-Stephen Welstead/CORBIS, 723 © David Young-Wolff/Photo Edit, 724(T) Drawn by SYMVU, Harvard University Software Program, Laboratory for Computer Graphics and Spatial Analysis, Harvard Graduate School of Design, 724(L) Courtesy, Vern Heeren, 731, 732 (Exercises 11–22) From 1999 *World Book Yearbook,* © 1999 World Book, Inc., 733 (Exercises 25–28) Graph "Retirement Saving Net Worth" from *TSA Guide to Retirement Planning for California Educators,* Winter 1993. Reprinted by permission of John L. Kattman., 740(B) AP/Wide World Photos, 744 Reprinted with permission from *The World Almanac and Book of Facts 2000.* Copyright © 1999 World Almanac Education Group. All rights reserved., 752 Sovfoto, 760 AP/Wide World Photos, 764(C) North Wind Picture Archives, 768 ©Greer & Associates, Inc./SuperStock, 771–777 From *How to Lie with Statistics* by Darrell Huff, illustrated by Irving Geis. Copyright 1954 and renewed © 1982 by Darrell Huff and Irving Geis. Reprinted by permission of W. W. Norton & Company, Inc., 779 Library of Congress, 785 From the *World Almanac and Book of Facts.* Copyright © World Almanac Education Group, 2003., 788 © Peter M. Fisher/CORBIS, 790 Oriental Institute Museum, The University of Chicago, 792 Originally broadcast over the CBS Radio Network as part of the Hourly News by Charles Osgood on March 2, 1977. Copyright © 1977 CBS, Inc. All Rights Reserved. Reprinted by permission., 807 Wells Fargo Bank, 809 ©2000 David Young-Wolff/Stone, 822 Cameramann International, Ltd., 835 © Jean Miele/CORBIS, 840 ©2000 Frank Herholdt/Stone, 847 Edward Jones and Company, 2002. Used by permission., 850 © Eyewire/Getty Images, 851 ©Nicole Katano/Stone, 852 ©Scott Camazine/Photo Researchers, 854 ©Peter Cade/Stone, 855 ©Bettmann/CORBIS, 858 ©Dallas and John Heaton/CORBIS, 859 ©Michael Freeman/CORBIS, 863 © Eyewire/Getty Images, 865(T) (Exercise 58) "Poem to Autumn" by John Keats, 865(B) Courtesy of the School of Mathematics and Statistics, University of St. Andrews, Fife, Scotland, 866(L) ©Bettmann/CORBIS, 881 ©Hulton-Deutsch Collection/CORBIS, 884 ©AFP/CORBIS, 887 *What's Happening in Mathematical Sciences,* vol. 2, 1994, p. 46, American Mathematical Society. Reprinted with permission. Figure supplied by Bill Cooke of Bellcore., 897 ©Bettmann/CORBIS, 899 Maddow, Sutherland, Efstathiou & Loveday/Science Photo Library/Photo Researchers, 911 AP/Wide World Photos, 912 AP/Wide World Photos, 915 North Wind Pictures Archive, 920 THE KOBAL COLLECTION / LUCASFILM / HAMSHERE, KEITH, 921 Courtesy of NASA, 929 AP/Wide World Photos, 932 ©Bettmann/CORBIS, 936 © Tony Freeman/Photo Edit, 939 © Bettmann/CORBIS, 949 ©Bettmann/CORBIS, 953(T) ©Bettmann/CORBIS, 953(B) © Philip Gould/CORBIS, 955 AP/Wide World Photos, 956 ©CORBIS, 963 AP/Wide World Photos, 971(TL, BL) Courtesy, Michel Balinski.

Index

Pronunciation Guide

a	act, bat	j	just, fudge	Œ	as in German schön or in French feu
ā	cape, way	k	keep, token		
â	dare, Mary	KH	as in Scottish loch or in German ich	R	rolled r as in French rouge or in German rot
ä	alms, calm	N	as in French bon or un		
ch	child, beach	o	ox, wasp	sh	shoe, fish
e	set, merry	ō	over, no	th	thin, path
ē	equal, bee	o͝o	book, poor	u	up, love
ə	like a in alone or e in system	o͞o	ooze, fool	û	urge, burn
g	give, beg	ô	ought, raw	y	yes, onion
i	if, big	oi	oil, joy	z	zeal, lazy
ī	ice, bite	ou	out, cow	zh	treasure, mirage

A

Aaron, Hank, 393
Abacus [ab´ə-kəs], 164, 167
Abel, Niels Henrik [ä´bəl], 186
Abelian groups, 186, 192
Absolute value, 253–254
Abundant numbers, 207
Accumulation period of annuity, 803
Actuarial method, 815–818
Acute angles, 495, 578
Acute triangles, 503
Addition
    of decimals, 294
    identity element for, 65, 263
    in modular system, 224–225
    of polynomials, 373
    properties of, 65, 263–264
    of rational numbers, 272
    of real numbers, 257–258, 262–264
    symbol for, 259
Addition property
    of equality, 315–316
    of inequality, 352–354
Addition rule of probability
    general, 687–688
    special, 688–689

Additive inverse
    in clock arithmetic, 222
    explanation of, 252–253, 263
Additive principle of counting
    general, 667–668
    special, 666–667
Add-on interest, 806
Adjustable-rate mortgages (ARM), 826–828
Adjustment period, 826
Adleman, Leonard M., 195
Advertising, fallacies in, 143–144
Agnesi, Maria Gaetana [ag nâ´zē], 413
Alabama paradox, 965–967, 969, 971
Albert, Reka, 850
Algebra
    exponents and, 360–366
    factoring and, 375–379
    historical background of, 315, 319
    linear equations and, 315–322, 325–334
    linear inequalities and, 352–357
    polynomials and, 372–379
    proportion and, 340–343
    quadratic equations and, 381–385
    ratios and, 339–340
    scientific notation and, 366–369

    translating words into, 325–326
    variation and, 343–347
Algebraic expressions, 315
Algorithms
    alternative, 151
    approximate, 885
    brute force, 884, 885, 888
    efficient, 885
    explanation of, 155, 193
    exponential time, 888
    Fleury's, 869–873
    Kruskal's, 899–900
    nearest neighbor, 885
    origin of term for, 869
    polynomial time, 888
    speed of, 888
Alice's Adventures in Wonderland (Carroll), 40
al-Khowârizmî, Muhammad ibn Musa [äl KHwär iz´mē], 315, 319, 869
Alternate exterior angles, 498
Alternate interior angles, 498
American Association of University Women (AAUW), 691
Amicable numbers, 207–208

Amortization
    of loan, 822, 823
    negative, 828
    schedule for, 824, 825
Analyzing arguments
    with Euler diagrams, 129–131
    with truth tables, 136–143
Ancient Egyptians, 2, 153–155, 285, 317, 319
Ancient Greeks, 2, 14, 155, 199, 207, 248, 352, 492, 506, 594
Angle-angle (AA) similarity property, 526–527
Angles
    acute, 495, 503, 578
    complementary, 496, 497, 578
    corresponding, 498
    coterminal, 580–581
    degree measure for, 577–580
    of depression, 604
    of elevation, 604, 605
    explanation of, 494–498
    exterior, 498, 504
    in coordinate system, 580–581
    interior, 498, 504
    negative, 577
    obtuse, 495, 503, 578
    quadrantal, 580
    reference, 596–598
    right, 495, 503, 578, 600–605
    rotation of, 494–495
    significant digits for, 602
    in standard position, 580
    straight, 495, 578
    supplementary, 496, 497, 578
    terminology for, 577
    trigonometric function of, 582–585, 593–598
    vertical, 495–496
Angle-side-angle (ASA) congruence property, 524
Annual percentage rate (APR), 813–815
Annual rate of return, 841, 842
Annual yield, 795–796
Annuities, 802–804
Annuity due, 805
Antecedent, 113, 123
Apollonius [ap ə lō′nē əs], 441
Apportionment methods
    Adams, 961
    Alabama paradox and, 965–967, 969, 971

Balinski and Young impossibility theorem and, 971
    explanation of, 947
    Hamilton, 948–952, 954, 955, 957, 962–964, 966–971
    Hill-Huntington, 948, 961–962, 964
    historical background of, 948
    Jefferson, 948, 952–957, 962–965, 971
    new states paradox and, 969–971
    population paradox and, 967–969
    problems with, 962–963
    quota rule and, 963–965
    Webster, 948, 956–959, 962, 964, 965
Approval voting method
    examples of, 922–924
    explanation of, 922
Approximate algorithms, 885
Archimedes [är kə mē′dēz], 2, 288, 352, 597
Archimedes Palimpsest, 597
Arcs
    of circle, 506
    in networks, 561
Area
    of a circle, 517–518
    explanation of, 512
    in metric system, A-2
    of a parallelogram, 515
    of a rectangle, 514
    of a square, 514
    of a trapezoid, 515
    of a triangle, 516
Area formulas
    Heron's, 516, 616
    for triangles, 614–615
Argand, Robert [är′gand], 307
Arguments. *See* Logical arguments
Aristotle [ar′ə stot l], 97
Arithmetic
    clock, 220–223
    in Hindu-Arabic system, 160–167
    practical, 257
Arithmetic mean. *See* Mean
Arithmetic operations, 65, 69
Arithmetic sequences, 10
Arrow, Kenneth J., 939, 971
Arrow's impossibility theorem, 939
Aryabhata [är yä bä′ tə], 283
Ascher, Marcia, 156
ASCII (American Standard Code for Information Interchange), 178

Associative properties
    of addition, 65, 263
    explanation of, 179–181
    of multiplication, 65, 263
    12-hour clock system and, 221
Asymptote
    explanation of, 765
    horizontal, 445
    vertical, 449
Augmented matrix, 476
Automobile speed, 303
Average, 275, 332
Average daily balance method, 808
Average rate of change, 408
Axis, of parabola, 435

B

Babylonian mathematics, 2, 157, 158, 169, 248, 319, 524, 576, 577
Bacon, Francis, 650
Balance method
    average daily, 808
    unpaid, 807
Balinski, Michael L., 964, 969, 971
Balinski and Young impossibility theorem, 971
Banneker, Benjamin [ban′i kər], 271
Barabasi, Albert-Laszlo, 850
Barber paradox, 87
Bar graphs, 33, 34, 729–730
Bases
    for computer applications, 172–176
    conversion between, 169–176
    explanation of, 153, 160
    powers of alternative number, 169
Bear market, 834
Beckman, Petr, 287
Bernoulli, James [bər n̄oo′lē], 701, 702
Bernoulli trials, 701, 706
Bessel, Friedrich, 576
Biconditionals, 126–127
Bienaymé, Jules, 652
Bilateral cipher, 650
Bimodal distribution, 741, 743
Binary code, 896
Binary system, 171–173
Binomial coefficients, 660–661
Binomial probability
    examples of, 704–705
    explanation of, 701–703
    formula for, 704
Binomial probability distribution, 702

Binomials
    factoring, 378–379
    squares of, 375
Binomial theorem, 660–661
Birthday problem, 697
al-Biruni, 616
Blood alcohol concentration, 303
Body mass index (BMI), 351, 359
Bohr, Niels, 170
Bolyai, Janos [bô´ lyoi], 557
Bolzano, Bernard [bōlt sä´nō], 180
Bonds
    explanation of, 838–839, 842, 843
    mutual funds vs., 839–840
Bonds, Barry, 393
Boole, George, 105
Borda, Jean-Charles de, 915
Borda voting method
    examples of, 916–918, 930–931
    explanation of, 915, 917, 932, 933, 935, 936, 938, 939
Box, 538–539
Box-and-whisker plot, 759–760
Box plot, 759–760
Bragg, Sir Lawrence, 537
Brams, Steven J., 937
Break-even analysis, 430–431
Briggs, Henry, 451
Brin, Sergey, 360
Brokers, 836–837
Brown, A. Crum, 852
Brown, Robert, 676
Brute force algorithms, 884, 885, 888
A Budget of Paradoxes (De Morgan), 23
Buleo, J., 652
Bull market, 834
Burton, David M., 671
Bushoong people, 871

C

Cairo Mathematical Papyrus, 529
Calculators. *See also* Graphing calculators; Scientific calculators
    to approximate values of trigonometric functions, 600–601
    to convert bases, 170–173
    four-function, 31
    graphing, 31
    powers of *e* on, 446
    in radian mode, 621
    scientific, 31, 173
    square roots using, 284–285
    statistics using, 725

Calculus, 395, 440
Call graphs, 854
Cameron, Michael, 203
Cantor, Georg [kän´tôr], 50, 83, 639
Capital gain, 833
Capital loss, 833
Caps, interest-rate, 828
Cardano, Girolamo [kär dä´nō], 383
Cardinality
    explanation of, 51, 83
    infinite sets and their, 83–88
Cardinal number formula, 78
Cardinal numbers
    of Cartesian product, 69
    explanation of, 51, 83
    of infinite sets, 52, 88
    problem-solving techniques and, 76–79
    of sets, 50–52
    of subsets, 76
Card tricks, 227–228
Carissan, Eugene Olivier, 199
Carmen de Algorismo (de Villa Dei), 163
Carroll, Lewis, 40, 142, 143
Cartesian coordinate system [kär-tē´zhən]. *See* Rectangular coordinate system
Cartesian product of sets, 68–70
Catastrophe theory, 569
Cayley, Arthur, 895, 897
Celsius, Anders, A-4
Celsius temperature scale, A-3–A-4
Census, 911
Center, 396
Center of rotation, 548
Centigram (cg), A-3
Centiliter (cl), A-3
Centimeter (cm), A-2
Central tendency measure(s)
    box plot as, 759–760
    explanation of, 735, 743–744
    mean as, 735–738, 743
    median as, 738–740, 743
    mode as, 740–741, 743
    from stem-and-leaf displays, 741–742
Chaos, 568–570
Charts, 21–22. *See also* graphs
Chebyshev, Pafnuty Lvovich, 674, 752
Chess clock, 223
Chinese numeration, 155–158
Chronometer, 221
Chudnovsky, David [choōd näv´ skē], 286

Chudnovsky, Gregory [choōd näv´ skē], 286
Chu Shih-chieh [choō´shē-che´], 659
Circle graph. *See* Pie charts
Circles
    area of, 517–518
    circumference of, 517
    equations of, 397–399
    explanation of, 397, 441, 505–506, 538
Circuits
    Euler, 865–873
    explanation of parallel and series, 118–119
    in graphs, 857–858
    Hamilton, 880–888
    simplifying, 119–120
Circumference, 250, 517
Class mark, 727
Class width, 727
Clock arithmetic, 220–223
Closed curves, 502
Closed-end credit, 806–807
Closing costs, 828–829
Closure properties, 179, 180, 263
Codes, 195, 197, 201–202
Coefficient of variation, 761
Cofunction identities, 592–593
Coins, 274
Cole, F. N., 198
Collinearity, 546
Combinations
    explanation of, 649–650
    factorial formula for, 650
    guidelines for use of, 651–653
Combinations formula, 650
Combined variation, 347
Commissions, 836–838
Common ratio, 802
Common sense, 25
Commutative properties
    of addition, 65, 263
    explanation of, 179, 180
    of multiplication, 65, 263
Complement of a set, 57, 70
Complementary angles, 496, 497, 578
Complements principle
    of counting, 665–666
    of probability, 686–687
Complete graphs
    explanation of, 859–860
    Hamilton circuits for, 882, 883
Completing the square, 382

Complex numbers
  explanation of, 306–309
  use of, 290
Components of an ordered pair, 394
Component statements, 95
Composite numbers
  explanation of, 197–199
  prime factorization of, 199–201, 206
Compounding period, 791
Compound interest. *See also* Interest
  annual yield and, 795–796
  example of, 792
  explanation of, 791–793
  formula for, 446–447
  function of, 788
  future and present values for, 793–795
Compound statements, 113, 116. *See also* Conditional statements
  biconditional, 126–127
  circuits and, 119
  explanation of, 95
Computer analysis of images, 898
Computer coding, 197
Computers
  base systems used by, 172–176
  design elements of, 94
  DNA, 885
  palmtop, 164
  problem solving using, 1
  pseudorandom digits generated by, 718
Conclusions, of logical arguments, 3, 129–131
Conditional equations, 319, 320
Conditional probability, 691–692
Conditional statements
  alternative forms of "if p, then q" and, 125–126
  biconditionals and, 126–127
  characteristics of, 115
  converse, inverse, and contrapositive and, 123–125
  explanation of, 113–114
  negation of, 117–118
  truth tables for, 115–117, 127
Condorcet, Marquis de, 932
Condorcet candidate, 932
Condorcet criterion of voting, 932–933, 938, 939
Congruence properties, 524–525
Congruent modulo, 223–225

Congruent triangles, 523–526
Conic sections, 441
Conjectures
  explanation of, 2, 7
  Goldbach's, 208
  twin prime, 208
Conjunctions
  explanation of, 102, 118, 355
  quantifiers used with, 105
  truth table for, 127
Connectives
  explanation of, 95
  symbols for, 96–97
Consequent, 113, 123
Constant of variation, 343
Constitution, U.S., 948
Constructive solid geometry, 49
Consumer mathematics
  annuities and, 802–804
  consumer credit and, 806–809
  home purchases and, 822–829
  investments and, 832–842
  time value of money and, 789–798
  truth in lending and, 813–818
Consumer price index (CPI), 350, 396, 796
Continuous compounding
  formula for, 447, 450
  future value for, 797–798
Continuous random variables, 763
Continuous values, 768
Contradictions, 319, 320
Contrapositive of a statement, 123, 124
Converse
  of Pythagorean theorem, 529–530
  of statement, 123
Convertibility, mortgage, 828
Convex, 502
Conway, John C., 539
Coombs voting method, 928
Coordinate geometry, 395
Coordinates, 394
Coordinate system, 580–581
Correlation, 780–782
Correlation coefficients, 780, 782
Corresponding angles, 498
Cosecant, 582
Cosine(s)
  amplitude of, 622–623
  explanation of, 582
  law of, 611–614
Cost models, 430–431
Cotangent, 582

Coterminal angles, 580–581
Countable sets, 87
Counterexamples, 2, 7
Counting methods
  for counting problems involving "not" and "or," 664–668
  genetic code and, 629
  guidelines for choosing, 653
  by systematic listing, 630–634
  using binomial theorem, 660–661
  using fundamental counting principle, 637–643
  using Pascal's triangle, 658–660
  using permutations and combinations, 646–654
Counting numbers. *See also* Natural numbers
  explanation of, 2–3, 24, 51, 99
  one-to-one correspondence between, 84–85
  set of, 84, 87, 88, 152
  sum of odd, 12–13
Cox, Gertrude Mary, 724
Creative thinking, 25
Credit
  closed-end, 806–807
  explanation of, 806
  open-end, 807–809
Credit card accounts, 807, 816
Cross multiplication, 340
Cross products method
  application of, 342
  explanation of, 340, 341
Cross-product test, 270, 271
Cryptography systems, 201
Cubes
  difference of, 378
  explanation of, 536
  sum of, 378–379
Cubic centimeter (cm^3), A-2
Current yield, 835
Curve-fitting, 416–417
Curves, 502
Cut edge in graphs, 870

D

Dantzig, George B. [dant´zig], 484, 485
Data, 725
Data analysis, 724
Data sets, 742
Deadheading, 874

Deciles, 758
Decimal degrees, 579
Decimal form, 276–278
Decimal point, 301, 367
Decimals
   addition and subtraction of, 293–294
   converting between percent and, 297
   irrational numbers and, 283–285
   multiplication and division of, 294
   real numbers as, 251
   repeating, 276–278, 283
   rounding, 294–296
   terminating, 276–278, 283
Decimal system, 164
Dedekind, Richard [dā´də kint], 639
Deductive reasoning
   explanation of, 2, 3, 129
   to find sum of counting numbers, 14
   use of, 725
Deed of trust, 822. *See also* Mortgages
Deficient numbers, 207
Degree of a vertex, 852
Degrees
   decimal, 579
   explanation of, 577–580
   of polynomial, 372
de Méré, Chevalier [meR], 674
De Moivre, Abraham [de mwäv´], 671, 764
De Morgan, Augustus, 23
De Morgan's laws, 72, 110
Density property, 275–276
Dependent variables, 423–425
DeRise, George, 9
Desargues, Gerard [dā zäRg´], 558
Desargues' theorem, 564
Descartes, René [dā kärt], 68, 394, 395, 558
Descriptive statistics, 724
Deterministic phenomenon, 674
Diameter, 250, 505
Diamond, Jared, 152
Difference
   of cubes, 378
   of real numbers, 258
   of sets, 66–67, 70
   of squares, 378–379
Differential equations, 320
Digits
   explanation of, 161
   pseudorandom, 718
   significant, 601–602

Diophantus, 259
Direct statements, 123, 124
Direct variation
   explanation of, 343
   as power, 344–345
Discount brokers, 836
Discounts, 827
Discrete random variables, 763
Discrete values, 768
Discriminant of a quadratic equation, 386
Disjoint sets, 64
Disjunctions
   exclusive, 112
   explanation of, 102–103
   inclusive, 103
   quantifiers used with, 105
   truth table for, 127
Disjunctive syllogism, 140
Dispersion measures
   box plot as, 759–760
   explanation of, 748
   meaning of, 752–753
   range as, 748–749
   standard deviation as, 749–752
Disraeli, Benjamin, 771
Distance
   formula for, 332–334, 394–396
   in metric system, A-3
Distribution
   bimodal, 741, 743
   binomial probability, 702
   expected frequency, 728
   grouped frequency, 726–727
   non-symmetric, 742
   normal, 763–769
   probability, 689, 702
   skewed, 742
   symmetric, 742
Distributive properties
   of addition, 263
   explanation of, 65, 162–167, 181–183, 316
   general form of, 182
   linear equations and, 316–318
   of multiplication, 182, 263
Dividends, stock, 833
Divisibility, 196–197
Divisibility tests, 196–197
Division
   of decimals, 294
   of rational numbers, 274–275
   of real numbers, 260–261

   symbol for, 259
   of and by zero, 261
Divisor, 196
Divisor methods, 964, 971
DNA (deoxyribonucleic acid), 629, 885
Dodecahedron, 536
Dodgson, Charles [doj´s´n], 40
Domain, 425–426
Donald in Mathmagic Land, 44, 236
Double negative rule, 252–253
Dow Jones Industrial Average (DJIA), 834
Down payment, 822
du Châtelet, Emilie, Marquise, 107, 129
Dull numbers, 208
Duodecimal system, 177
Dürer, Albrecht [dyo͞or´ər], 558

E

e, 285, 445–446
E-commerce, 809
Eddin, Nassir, 613
Edward Jones and Company, 846, 847
Effective annual yield, 795–796
Egyptian mathematics, 2, 153–155, 285, 317, 319, 576
Electrical impedance tomography (EIT), 491
Electronics Frontier Foundation, 203
Elements of sets, 50, 51, 60
Elements of Euclid (Playfair), 556
Elimination method
   explanation of, 457
   solving linear systems by, 458–460, 462, 476
Eliasberg, Louis, 274
Ellipse, 438
Elway, John, 143, 144
Empirical frequencies, 728
Empirical probability, 675–676, 734
Empirical rule, 764–765
Empty set, 51
Endpoints, 494
English system, A-4–A-5
Equality
   addition and multiplication properties of, 315–316
   axioms of, 318
   explanation of, 250
   of rational numbers, 270, 271
   set, 53
   symbols for, 250, 251

Equations
  of circles, 397–399
  conditional, 319, 320
  as contradictions, 319
  differential, 320
  equivalent, 315
  first-degree, 315–318
  as identities, 320
  linear, 315–322, 325–334, 402–408, 456 (*See also* Linear equations)
  linear model, 416–417
  of lines, 412–416
  as models, 320–321
  quadratic, 381–385
  systems of, 456–470
Equilateral triangles, 503
Equivalent inequalities, 352
Equivalent sets, 84
Equivalent statements
  circuits and, 119
  explanation of, 109–110, 124
Eratosthenes [er ə tos´thə nēz], 198
Escher, M. C., 550
Escott, E. B., 203
Escrow account, 825
Estimated regression curve, 778
Estimated regression line, 778
Estimation, 33
Ethnomathematics: A Multicultural View of Mathematical Ideas (Ascher), 156
Euclid [yoo´klid], 2, 201, 207, 208, 492, 556, 557
Euclidean algorithm, 214–215
Euclidean geometry, 556–558, 563
Euclidean tools, 510
Euclid's Axioms, 556
Euclid's fifth postulate, 556–557
Euclid's postulates, 556
Euler, Leonhard [oi´lər], 129, 203, 445, 537, 561, 562, 866
Euler circuit, graphs with, 874–875
Euler circuits
  explanation of, 867–869
  Fleury's algorithm to find, 869–873
Euler diagrams, 129–131, 136
Euler path, 867
Euler's theorem
  application of, 869
  explanation of, 867–868
Events
  dependent, 693
  explanation of, 674
  independent, 693–694
  involving "and," 691–697
  involving "not" and "or," 684–689
  mutually exclusive, 688, 691
Even vertex, 561
Eves, Howard, 5
Existential quantifiers, 97, 99
Expected frequency distribution, 728
Expected value
  examples involving, 708–711
  explanation of, 707–708
  of games of chance, 712–713
Experiments, 674
Exploratory data analysis, 724
Exponential decay, 451
Exponential expressions
  evaluation of, 360–361
  explanation of, 160, 161
Exponential functions
  applications of, 446–447
  explanation of, 444–445
  graphs of, 445–446
Exponential growth, 448
Exponential growth function, 447
Exponential models, 450–451
Exponential time algorithms, 888
Exponents
  definitions and rules for, 365
  explanation of, 153, 160, 360
  negative, 362–363, 365, 366
  power rules for, 364–365
  product rule for, 361–362
  quotient rule for, 363–364
  rational number, 293
  zero, 362
Exterior angles, 498, 504
Extremes, 340

F

Face value, 157, 838
Factorial formula
  for combinations, 650
  explanation of, 642
  for permutations, 648
Factorials
  approximating, 671
  explanation of, 642
  use of, 646
Factoring
  explanation of, 375
  by grouping, 376
  trinomials, 376–377
Factoring machine, 199
Factorization, 196
Factors
  explanation of, 196, 375
  factoring out greatest common, 375–376
Factor tree, 200
Fahrenheit temperature scale, A-4
Fair game, 709
Fallacies
  in advertising, 143–144
  of the converse, 138
  example of, 130
  explanation of, 129
  forms of, 141
  of the inverse, 139
Fermat, Pierre de [feR mä´], 202, 209, 213, 448, 674
Fermat numbers, 202
Fermat's Last Theorem, 209
Fey, James T., 35
Fibonacci [fē´bo nä chē], 21, 158, 257, 289, 330
Fibonacci Association, 233
Fibonacci patterns, 233–234
Fibonacci sequence, 10, 21, 22, 232–234
Figurate numbers
  explanation of, 14–15
  heptagonal, 18, 19
  hexagonal, 18, 19
  nonagonal, 19
  octagonal, 18, 19
  pentagonal, 14, 15
  square, 14, 15, 19
  successive differences method and, 16
  triangular, 14, 15, 19
Figures
  plane, 536
  space, 536–538
Finance charges, 807, 818
Finger counting, 162
Finger reckoning, 157
Finite mathematical systems
  associative property and, 179
  closure property and, 179
  commutative property and, 179
  distributive property and, 181–183
  explanation of, 179, 220
  identity property and, 180
  inverse property and, 180
Finite sets, 51–52, 87

First component in an ordered pair, 67
First-degree equations
    explanation of, 315–318
    with two variables, 403
Fishburn, Peter C., 937
Fisher, Sir Ronald, 724
Fixed installment loans, 806
Fixed-rate mortgages, 822–826, 828
Fleury's algorithm, 869–873
Focus, of parabola, 435
FOIL method, 374, 376, 377
Formulas
    explanation of, 320
    solutions using, 320–321, 332–333
Four-color theorem, 882
Four-function calculators, 31
Fractal geometry, 569–570
Fractals, 569, 572
Fractions
    converted to percent, 297–299
    lowest terms of, 19
Franklin, Benjamin, 241
Free-falling objects, 440
Frequency distributions
    expected, 728
    explanation of, 725
    grouped, 726–727
    observed, 728
    position of median in, 739–740
Frequency polygons, 726
Friendly numbers, 207–208
Frustum of a pyramid, 2
Full-service brokers, 836
Function notation, 428–429
Functions
    explanation of, 423
    exponential, 444–447
    identifying, 426–428
    linear, 429–431
    logarithmic, 447–450
    periodic, 620
    quadratic, 434–441
    relations and, 423–425
    variations of definition of, 428
    vertical line test for, 427–428
Fundamental counting principle
    arrangement of n objects and, 643
    examples of, 639–641
    explanation of, 637–638
    factorials and, 642
    uniformity criterion for multiple-park tasks and, 638

Fundamental property of rational numbers, 270
Fundamental theorem of arithmetic, 199, 200
Future values
    of annuity, 803–804
    of annuity due, 805
    for compound interest, 793–795, 797–798
    explanation of, 790, 791

G

Galilei, Galileo [gä lē le ˉe], 84, 440
Galileo's paradox, 84
Galois, Évariste [gal wä´], 189, A-1
Galois groups, 189
Galton, Sir Francis, 724, 778
Gantner, Charles W., 200
Gaunt, John, 724
Gauss, Carl Friedrich [gous], 8, 14, 307, 557, 764, 767
Gaussian curve, 764. *See also* Normal curve
Gauss-Jordan method
    application of, 479–480
    explanation of, 477
Gematria, 85
General addition rule of probability, 687–688
General additive principle of counting, 667–668
General binomial expansion, 661
General multiplication rule of probability, 692–694
Genetics, probability in, 678–679, 682, 683
Geodesy, 495
Geometric constructions, 509–510
Geometric primitives, 49
Geometric sequences, 10, 802–803
Geometry
    angles and, 494–498
    chaos and, 567–569
    circles and, 505–506, 517–518
    constructive solid, 49
    curves and, 502
    Euclidean, 556–558, 563
    explanation of, 492
    fractal, 569–570
    networks and, 561–562
    non-Euclidean, 556–559, 563
    perimeter and area and, 512–516

    points, lines, and planes and, 492–494
    projective, 558
    space figures and, 536–538
    topology and, 559–560
    transformational, 545–551
    of triangles, 523–530
    triangles and quadrilaterals and, 502–505
    volume and surface area and, 538–541
Germain, Sophie [zheʀ men´], 209
Gigaflops, 885
Gleick, James, 570
Glide reflection, 549, 551
Goldbach, Christian, 208
Goldbach's conjecture, 208
Golden ratio, 235–236, 384
Golden rectangle, 512
Google, 360
Googol, 360
Googolplex, 360
Gossett, William S., 724
Gram (g), A-3
Graphing calculators. *See also* Calculators; Scientific calculators
    components of, 32
    explanation of, 31
    future value on, 793
    powers of e on, 446
    solving equations using, 316
    statistics using, 725
    use of, 31–33
Graphs
    bar, 33, 34, 729–730
    call, 854
    complete, 859–860, 882, 883
    components of, 853, 854
    connected, 853, 894–895
    cut edge in, 870
    directed, 896
    disconnected, 853
    with Euler circuits, 865–873
    explanation of, 33, 851–853
    of exponential functions, 445–446
    frequency distributions and, 725–730
    isomorphic, 853
    of linear inequalities, 481–482
    of linear systems, 457, 461–462
    of logarithmic functions, 449–450
    of numbers, 249
    of ordered pairs, 394, 425–426
    origin of term for, 852, 855

Graphs (*continued*)
  path in, 857–858
  of periodic functions, 620–624
  pie charts, 33, 34, 729–730
  of population density, 724
  of quadratic functions, 434–441
  of relations, 425–426
  simple, 852
  of sine and cosine functions, 623, 624
  stem-and-leaf displays, 728–729
  sum of the degrees theorem of, 856
  tree, 894–902
  types of, 33, 34
  walk in, 857
  weighted, 858
  writing equations to fit, 416–417
Graph theory, 561–562
  basic concepts of, 851–860
  Euler circuits and, 865–873
  explanation of, 850
  Hamilton circuits and, 880–888
  minimum spanning trees and, 898–902
  trees and, 894–898
Greatest common factor (GCF)
  explanation of, 212–213
  factoring out, 375–376
  methods for finding, 213–215
Great Internet Mersenne Prime Search (GIMPS), 202, 203
Greek Anthology, 327
Greek mathematics, 2, 14, 155, 199, 207, 248, 352, 492, 506, 576, 594
Greene, Bob, 143
Greenhouse effect, 450–451
Grouped frequency distribution, 726–727
Grouping
  explanation of, 153
  multiplicative, 156–157
Groups
  abelian, 186, 192
  explanation of, 186–187
  Galois, 189
  monster, 190
  permutation, 190–191
  of symmetries of a square, 187–189
Guessing and checking, 23–24, 317

H

Half-life, 451
Half-lines, 492–493

Halley, Edmund, 346
Halley's comet, 346
Hamilton, Alexander, 948, 949
Hamilton, William Rowan, 880, 881
Hamilton apportionment method, 948–952, 954, 955, 957, 962–964, 966–971
Hamilton circuits
  brute force, 884–885, 888
  for complete graphs, 882, 883
  explanation of, 881–882
  minimum, 883–884
  nearest neighbor algorithm and, 885–887
  that are the same, 882–883
Hammurabi, King, 790
Hardy, G. H., 208
Hare, Thomas, 917–918
Hare method of voting
  examples of, 918–921
  explanation of, 917–918, 932, 933, 935, 938, 939
  majority criterion and, 931
Harrison, John, 221
Heptagonal numbers, 18, 19
Heron of Alexandria [her´on], 516, 616
Heron's area formula, 516, 616
Hewitt, Paul G., A-1–A-2
Hexadecimal system, 172–175
Hexagonal numbers, 18, 19
Hexahedrons, 536
Hilbert, David, 187
Hill, Joseph, 964
Hill-Huntington apportionment method, 948, 961–962, 964
Hindu-Arabic system
  arithmetic in, 160–167, 257
  explanation of, 157–158, 169, 248
Hipparchus, 582, 587
Histograms, 726, 727, 729
A History of Pi, 287
Holyhedron, 539
Homeowners insurance, 825
Hooke's law, 344
Horizontal asymptote, 445
Horizontal lines, 406, 416
Horsepower, 303
How to Lie with Statistics (Huff), 771–775
How to Solve It (Polya), 20
Huff, Darrell, 771–775
Huntington, Edward V., 964
Huygens, Christiaan [hoi´gens], 675

Hyperbola, 440, 441
Hypotenuse, 527

I

I Ching, 170
Icosahedrons, 536
Icosian game, 880–881
Identities
  cofunction, 592–593
  example of, 320
  explanation of, 227, 319, 586
  Pythagorean, 588–589
  quotient, 589–590
  reciprocal, 586
Identity element
  of addition, 65, 265
  of multiplication, 65, 263
Identity properties
  of addition, 263
  explanation of, 180
  of multiplication, 263
  12-hour clock system and, 221
Identity translations, 548
Imaginary numbers, 308
Imaginary part, 308
Impound account, 825
Inclusive disjunctions, 103
Independence of irrelevant alternatives (IIA) criterion of voting, 935–937, 939
Independent variables, 423–425
Indirect reasoning, 139
Inductive reasoning
  examples of, 3, 5, 724–725
  explanation of, 2, 129
  false conclusions from, 4
  historical background of, 2
  number patterns and, 10–16
  pitfalls of, 5–6, 60
  problem solving using, 2–7, 10–16, 107–108
Inequalities
  addition property of, 352–353
  applications of, 356–357
  equivalent, 352
  linear, 352–357, 481–482
  multiplication property of, 354–355
  symbol for, 58
  systems of, 482–483
  three-part, 355–356
Inferential statistics, 724

Infinite sets
  cardinal numbers of, 83–88
  explanation of, 52, 85
Infinity, symbol for, 52
Inflation, 796–798
In Mathematical Circles (Eves), 5
Installment buying, 806
Integers, 51, 249
Intercepts, 404
Interest
  add-on, 806
  compound, 446–447, 788, 791–795
  explanation of, 789
  future values and present values and, 790–791
  simple, 789–790
  unearned, 815–818
Interest rates
  caps on, 828
  equations with, 328
  explanation of, 789
Interior angles
  explanation of, 498, 504
  measure of, 504–505
Internet sites, 45. *See also* World Wide Web
Interquartile range, 761
Intersecting lines, 493
Intersecting planes, 494
Intersection
  of sets, 63–64, 70
  symbol for, 63
Interval on a number line, 353
Interval notation, 353, 354
Invalid arguments. *See* Fallacies
Invariant point, 546
Inverse
  additive, 252
  of statement, 123, 124
Inverse properties
  of addition, 263
  in clock arithmetic, 222
  explanation of, 180
  of multiplication, 263
Inverse variation, 345
Investment Company Institute (ICI), 839
Investments
  bonds as, 838–839
  effect of taxes on, 846
  evaluating returns on, 841–842
  explanation of, 832
  mutual funds as, 839–841

  return on, 833
  stocks as, 833–838
  summary of, 842–843
Irrational numbers
  decimal representation and, 283–285
  explanation of, 51, 99, 249–250, 283–285
"Is greater than," 250
"Is less than," 250
Isomorphic graphs, 853
Isosceles triangles, 503, 525
Itemized billing, 807

J

Jefferson, Thomas, 271, 273, 394, 948, 952, 953
Jefferson apportionment method, 948, 952–957, 962–965, 971
Jesus Seminar, 916
Joong, Hawoong, 850
Journals, 40, 41

K

Kabbalah, 85
Kanada, Yasumasa, 285
Kaprekar number [kap´re kär], 16
Kasner, Edward, 360
Kekule, Friedrich, 895
Kepler, Johann, 346, 438
Khayyam, Omar [κHä yä´m], 659
Kilogram (kg), A-3
Kiloliter (kl), A-3
Kilometer (km), A-2
Klein, Felix, 560
Klein bottle, 560
Koch, H. von [von kook], 570
Kolmogorov, Andrei Nikolaevich, 674
Königsberg Bridge problem [kŒ´nigs bûrg], 561, 865–867
Kovalevskaya, Sofia [kov ə lef´skä yä], 320
Kruskal's algorithm, 899–900

L

Laberti, Leone Barrista, 558
Lagrange, Joseph Louis [la gRäNzh´], 209, A-1
Lambert, J. H., 286

Laplace, Pierre Simon de [la plas´], 674, 675, 685, 686, 764
Lasker, Emanuel, 222
Lasker-Noether theorem, 223
Lattice method, 164–166
Law of contraposition, 139
Law of cosines, 611–614
Law of detachment, 137
Law of hypothetical syllogism, 140
Law of large numbers, 677–678
Law of sines, 609–611
Law of tension, 293
Learning logs, 40, 41
Least common multiple (LCM)
  explanation of, 215
  formula for, 217
  methods for finding, 216–218
Least squares curves, 781
Least squares line, 778–779, 781
Leaves, of a directed tree, 896
Leblanc, M., 209
Legs of a right triangle, 527
Leibniz, Gottfried [līb´nitz], 95, 124, 170, 259, 702
Length, A-1–A-2
Leonardo da Pisa. *See* Fibonacci
Leonardo da Vinci, 558
Liber Abaci (Fibonacci), 21, 232, 330
Like radicals, 289
Limit order, 837
Linear equations
  applications of, 325–334
  in one variable, 315–318
  point-slope form of, 413–414
  slope-intercept form of, 414–416
  solutions to, 315–322
  special kinds of, 319–322
  summary of forms of, 416
  system of (*See* Linear systems)
  in two variables, 402–404
Linear functions, 429–431
Linear inequalities
  graphs of, 481–483
  in one variable, 352
  solutions to, 352–357
  three-part, 355–357
  in two variables, 481
Linear model, 416–417
Linear programming, 484–485
Linear systems
  applications of, 464–470
  elimination method to solve, 458–460, 462, 476

Linear systems (*continued*)
  graphs of, 457, 461–462
  matrix row operations to solve, 476–480
  substitution method to solve, 459
  in three variables, 461–463
  in two variables, 456–460
Line graphs, 33, 34
Line of symmetry, 547
Lines
  equations of, 412–416
  graphs of, 404
  horizontal, 406
  intersecting, 493
  parallel, 407, 493, 497
  perpendicular, 407–408
  of reflection, 545–546
  secant, 505
  skew, 494
  slope of, 405–408, 412
  vertical, 406
Liquidity, 843
Listing notation, 50
Liter (l), A-2
Loan origination fees, 829
Lobachevski, Nikolai Ivanovich [lə bu-chyef´skyē], 557–558, 563
Logarithmic functions
  explanation of, 447–449
  graphs of, 449–450
  natural, 449
Logarithms
  background of, 165
  compound interest using, 794–795
  on scientific calculators, 451–452
Logic
  circuits and, 118–120
  conditionals and, 113–118, 123–127
  equivalent statements and, 109–110
  Euler diagrams and, 129–131
  quantifiers and, 97–99
  statements and, 95–97
  symbolic, 95–97
  truth tables and, 102–109, 127, 136–143
Logical arguments
  elements of, 129, 130
  Euler diagrams and, 129–131, 136
  explanation of, 3
  truth tables and, 136–143
  valid and invalid, 129
Logical connectives. *See* Connectives

Logic puzzles
  example of, 148–149
  explanation of, 104, 133
  solutions to, 133–135
Logistica (Buteo), 652
Longitude, 221
Loomis, Elisha Scott, 528
Lower class limits, 727
Lowest terms, 270
Lucas sequence [loo kä], 238

M

Magic square sum formula, 240
Magic squares, 29, 240–241
Magnitude, 548
Majority of votes, 912
Majority criterion of voting, 930–931, 938, 939
Mandelbrot, Benoit [man´del brōt], 569, 570
Margin of a mortgage, 826
Maris, Roger, 393
Markov, A. A., 674
Mathematical expectation. *See* Expected value
Mathematical induction, 12
Mathematical statements, 104–105
Mathematical systems
  explanation of, 152
  finite, 179–183
  groups as, 186
Mathematics
  computer, 172–176
  consumer, 788–848 (*See also* Consumer mathematics)
  historical background of, 2
  Internet sites for information on, 45
  using writing to learn about, 40–42
Mathematics publications, 40–44
Mathematics Teacher, 42–43
Matrix, 476
Matrix row operations, 476–480
Mattang, 458
Maturity value, 790
Mayan Indians, 169, 248
Mazes, 858
McGwire, Mark, 393
Mean
  advantages and disadvantages of, 743, 744
  examples of, 737–738
  explanation of, 340, 735–736, 743

  method for finding, 275
  weighted, 736–737
Measures of central tendency. *See* Central tendency measures
Median
  advantages and disadvantages of, 743, 744
  explanation of, 738–739, 743
  position in frequency distribution of, 739–740
Members of a set, 50
Mendel, Gregor Johann, 678–679
Mersenne, Marin [meR sen´], 202, 558
Mersenne numbers, 202
Mersenne primes, 202, 203, 207
Meters (m), A-1
Method of least squares, 778
Metric system
  area in, A-1–A-2
  conversions to and from, A-4–A-5
  explanation of, A-1
  length in, A-1–A-2
  temperature in, A-3–A-4
  volume in, A-1–A-3
  weight in, A-3
Midpoint formula, 396–397
Midquartile, 761
Midrange, 761
Mill, John Stuart, 918
Milligram (mg), A-3
Milliliter (ml), A-3
Millimeter (mm), A-2
Minimum Hamilton circuits, 883–884
Minimum spanning trees, 898–901
Minute, as an angle measure, 579
Mixed number, 281
Möbius, August Ferdinand [mŒ´bi us], 560
Möbius strip, 560
Mode
  advantages and disadvantages of, 743, 744
  explanation of, 740–741, 743
Models
  explanation of, 320–321
  in operations research, 484
Modular systems
  cyclical changes and, 225
  explanation of, 223–227
Modus ponens, 137
Modus tollens, 139
Moiré patterns, 497
Molecular dynamics, 1

Monetary values, 331–332
Money. *See* Time value of money
Monomial, 372
Monotonicity criterion of voting, 933–935, 939
Monster groups, 190
Monte Carlo methods, 717
Mortgages
    adjustable-rate, 826–828
    explanation of, 822
    fixed-rate, 822–826
    government-backed, 827
    principal amount of, 822
Moscow Papyrus, 2
Müller, Johanne, 613
Multiple, 196
Multiplication
    of decimals, 294
    lattice method of, 164–166
    in modular system, 224–225
    of polynomials, 373–374
    properties of, 65, 263–264
    of rational numbers, 272–273
    of real numbers, 258–260, 262–264
    Russian peasant method of, 166–167
    symbol for, 259
Multiplication property
    of equality, 315–317
    of inequality, 354–355
Multiplication rule of probability
    general, 692–694
    special, 694–697
Multiplicative grouping, 156–157
Multiplicative inverse
    defining division using, 274
    explanation of, 252–253, 263
Mutual funds
    asset classes of, 846
    explanation of, 839–840, 842, 843
    net asset value of, 840–841
Mutually exclusive events, 688, 691

N

Napier, John [nāˊpē ər], 165
Napier's bones, 165
Napier's rods, 165, 166
NASDAQ (National Association of Securities Dealers Advanced Quotations), 835
Nash, John, 568
National Council of Teachers of Mathematics (NCTM), 40, 41, 691

Natural disasters, 694
Natural logarithmic function, 449
Natural numbers. *See also* Counting numbers
    abundant, 207
    amicable, 208
    deficient, 207
    divisibility of, 196–197
    divisibility tests for, 198–199
    explanation of, 2–3, 51, 99
    perfect, 206–207
    set of, 248
Nearest neighbor algorithms, 885–887
Negations
    of conditional statements, 117–118
    of statements, 96, 98
    truth table for, 103, 115, 127
Negative amortization, 828
Negative angles, 577
Negative exponents
    explanation of, 362–363, 365
    special rules for, 363, 365
Negative numbers
    applications of, 249, 264
    explanation of, 248
Net asset value of a mutual fund, 840–841
Networks, 561–562
Netz, Reviel, 597
New states paradox, 969–971
Newton, Sir Isaac, 127, 440
New York Stock Exchange (NYSE), 273, 833–835
Nines complement method, 151
Nodes, 561
Noether, Amalie [nŒ thûr], 187, 222
Nominal rate, 795
Nonagonal numbers, 19
Non-Euclidean geometry
    explanation of, 557, 563
    origins of, 557–559
Normal curve
    equivalence in, 768
    examples of, 768–769
    explanation of, 764
    properties of, 764–765
    standard, 764
Normal curve table, 765–767
Normal distribution
    examples of, 765, 767–769
    explanation of, 764–765
Null set, 51

Number patterns
    explanation of, 12–14
    figurate numbers and, 14–16
    successive differences method and, 10–11
Number sequences, 10–11
Number squares, 244–245
Number theory
    clock arithmetic and, 220–223
    deficient and abundant numbers and, 207–208
    explanation of, 196
    Fermat's Last Theorem and, 209
    Fibonacci sequence and, 232–234
    Goldbach's conjecture and, 206–209
    golden rule and, 235–236
    greatest common factor and, 212–215
    least common multiple and, 215–218
    modular systems and, 223–227
    perfect numbers and, 206–207
    prime and composite numbers and, 196–203.
Numeracy, 35
Numerals, 152
Numeration systems
    ancient Egyptian, 153–155, 285
    explanation of, 152
    Hindu-Arabic, 157–158, 248, 319
    historical background of, 152, 248
    traditional Chinese, 155–157

O

Objective function, 484
Oblique triangles, 608–609
Observed frequencies, 728
Obtuse angles, 495, 578
Obtuse triangles, 503
Octagonal numbers, 18, 19
Octahedrons, 536
Octal system, 172–175
Odd-lot differential, 837
Odd lot of shares, 837
Odds, 680–681. *See also* Probability
Odd vertex, 562
One-to-one correspondence, 84–86
On the Shoulders of Giants: New Approaches to Numeracy (Fey), 35
Open-end credit, 806–809
Operations
    in algebraic expressions, 315–316
    with decimals, 293–294

Operations (*continued*)
   explanation of, 69
   order of, 261–262
   with percent, 297–300
   with real numbers, 257–262
   on sets, 69–70
Operations research (OR), 484
Operation table, 179
Opposites, 252
Optimization model, 440–441
Order
   explanation of, 250
   of magic square, 240
   of operations, 261–262
Ordered pairs
   equal, 68
   explanation of, 67–68, 394, 395
   graphs of, 394, 425–426
   in linear equations, 403
Ordered triple, 461
Ordinary annuity, 803
Origin, 394
Orr, A. C., 288
Outcomes, 674
Outlier, 738
Overall cap, 828

P

π (pi)
   computation of, 285–287
   historical background of, 285
   value of, 288, 517
Pairwise comparison voting method
   examples of, 914–917, 919, 932
   explanation of, 913–914, 931–933, 935–939
Palindromic numbers, 27
Palmtop computers, 164
Parabolas
   explanation of, 434–435, 437, 441
   focus of, 435
   vertex of, 435, 437, 440
Paradox(es)
   barber, 87
   explanation of, 84
   Galileo's, 84
   of Zeno, 90
Parallel lines
   explanation of, 493, 497
   slopes of, 407
Parallelograms
   area of, 515
   explanation of, 503, 514–515
Parallel planes, 494
Parthenon, 236
Pascal, Blaise [pas kalˊ], 659, 674
Pascal's triangle
   activities using, 45
   explanation of, 45, 239, 658–660
Path, in a graph, 857–858
Patterns, 24, 107–108
Payment cap, 828
Payoff amount, 816–817
Pearson, Karl, 724
Peirce, Benjamin [pûrs], 285
Peitgen, H. O., 570
The Penguin Dictionary of Curious and Interesting Numbers (Wells), 210
Pentagonal numbers, 14, 15
Percent
   converting between decimals and, 297
   converting fractions to, 297–299
   equations with, 328
   explanation of, 297
   increase or decrease in, 300
Percentiles, 757–758
Percent sign (%), 297
Perfect numbers, 206–207
Perfect square trinomials, 377
Perimeter, 512, 513
Periodic cap, 828
Periodic functions
   explanation of, 620
   graphs of, 622–624
Permutation groups, 190–191
Permutations
   explanation of, 191, 647–648
   factorial formula for, 648
   guidelines for use of, 648, 649, 653
Permutations formula, 647
Perpendicular lines, 407–408
Peterson, Ivars, 199
Philosophiae naturalis principia mathematica (Newton), 127
Pi (*see* π (pi))
Pie charts, 33, 34, 729–730
Pierce, Charles Sanders, 113
Pisano, Leonardo. *See* Fibonacci
Placeholders, 158
Place value, 157
Plane figures, 536
Planes, 494
Plato [plāˊto], 97

Playfair, John, 556
Plimpton *322*, 524
Plurality of votes, 912
Plurality voting method
   examples of, 914, 919, 932
   explanation of, 913, 931, 933, 935, 938, 939
Plurality with elimination method. *See* Hare method of voting
Point reflection, 549
Points, 492–493, 829
Point-slope form, 413–414, 416
Poker, 654
Pollaiolo, Antonio de, 509
Polya, George [pōlˊyə], 20, 326
Polya's problem solving process, 20–21
Polygons, 502, 726
Polyhedrons, 536, 537
Polynomials
   addition and subtraction of, 373
   explanation of, 372–373
   factoring and, 375–379
   formulas for, 203
   in x, 372
   multiplication of, 373–375
   prime, 375
   squares of, 375
Polynomial time algorithms, 888
Poncelet, Jean-Victor [pôns leˊ], 558
Population, 724
Population paradox, 967–969
Positional numeration, 157–158
Position measures
   box plot as, 759–760
   deciles and quartiles as, 758
   explanation of, 756
   percentiles as, 757–758
   z-score as, 756–757
Positive numbers, 248
Power rules, 364–365
Powers
   explanation of, 153
   of i, 308, 309
   terms in descending, 373
Premises
   explanation of, 3
   of logical arguments, 129–131, 141
Prepayment penalty, 828
Present value(s)
   for compound interest, 793–795
   explanation of, 790–791
Price-to-earnings ratio, 835

Prime factorization
   of composite numbers, 199–201, 206
   finding greatest common factors using, 212–215
Prime numbers
   alternative definition of, 197–198
   explanation of, 197
   infinitude of, 200–201
   primorial, 209
   search for large, 201–203
   Sophie Germain, 209
   twin, 208–209
Prime polynomials, 375
Primorial primes, 209
Principal
   loan or deposit, 789
   mortgage, 822
   of stock transactions, 836
Principia Mathematica (Whitehead & Russell), 123, 124, 127
Prisms, 538
Probability
   binomial, 701–705
   classical definition of, 675
   complements rule of, 686–687
   conditional, 691–692
   empirical, 676, 734
   of events involving "and," 691–697
   of events involving "not" and "or," 684–689
   expected value and, 707–711
   explanation of, 673–675
   general addition rule of, 687–688
   general multiplication rule of, 692–694
   in genetics, 678–679, 682, 683
   law of large numbers and, 677–678
   odds and, 680–681
   properties of, 685–686
   simulation to estimate, 716–719
   special addition rule of, 688–689
   special multiplication rule of, 694–697
   theoretical, 675, 677, 684–685, 718
Probability distribution
   binomial, 702
   explanation of, 689
   theoretical, 763
Problem solving. *See also* Solutions
   high-speed computation for, 1
   importance of focus on, 22
   by inductive reasoning, 2–7, 10–16

   by reading graphs, 33–34
   strategies for, 20–25, 326
   using calculation, 31–33
   using estimation, 33
Producer Price Index (PPI), 264–265
Product
   of sum and difference of two terms, 374–375
   explanation of, 258–259
   of real numbers, 258
Product rule
   for exponents, 361–362, 365
   for square roots, 287–288
Product tables, 630–631, 634
Projective geometry, 558
Proof by contradiction, 201, 284
Proofs, 506
Proper subsets, 58–60
Property of triangle side lengths, 611
Property taxes, 825
Proportion
   examples of, 341–343
   explanation of, 340
Protractors, 495
Pseudorandom digits, 718
Pseudosphere, 558
Ptolemy, Claudius [tol′ə mē], 582, 587
Pyramid(s)
   explanation of, 538
   frustum of, 2
   volume and surface area of, 541
Pythagoras [pi thag′ər əs], 2, 14, 528
Pythagorean brotherhood, 14
Pythagorean identities, 588–589
Pythagorean theorem
   applications of, 601, 604
   converse of, 529–530
   distance formula and, 394, 395
   explanation of, 3, 209, 527–529
   proof of, 528
The Pythagorean Proposition (Loomis), 528
Pythagorean triple, 529

Q

Quadrantal angles, 580, 584, 585
Quadrants, 394
Quadratic equations
   discriminant of, 386
   explanation of, 381–384
   square root property and, 382
   zero-factor property and, 381

Quadratic formula, 382–385
Quadratic functions
   explanation of, 434–435
   graphs of, 435–439
Quadrilaterals, 502, 503
Qualitative data, 725
Quantifiers
   existential, 97, 99
   explanation of, 97–99
   universal, 97, 99
   used with conjunctions or disjunctions, 105
Quantitative data, 725
Quartiles, 758
Quipu [kē′ poo], 156
Quota rule, 963–965
Quotient identities, 589–590
Quotient of real numbers, 260
Quotient rule
   for exponents, 363–365
   for square roots, 288–290
Quotients, 260

R

Radical(s), 288
   like, 289
Radicand, 288
Radiolaria, 537
Radius, 396, 505
Ramanujan, Srinivasa [rämä noo′jən], 208
Random variables
   continuous, 763
   discrete, 763
   explanation of, 689
Random phenomenon, 674
Range
   explanation of, 425, 748–749
   interquartile, 761
   semi-interquartile, 761
Rate formula, 332, 333
Rate of return
   annual, 841, 842
   real, 847
Ratio
   explanation of, 339–340
   Golden, 235–236, 384
   price-to-earnings, 835
Rationalizing the denominator, 289
Rational number exponents, 293
Rational numbers
   addition and subtraction of, 272

Rational numbers (*continued*)
  cross-product test for equality of, 270, 271
  decimal form of, 276–278
  as decimals, 251
  density property of, 275–276
  division of, 274–275
  examples of, 249
  explanation of, 51, 99, 249, 270
  fundamental property of, 270
  multiplication of, 272–273
Raw data, 725
Rays, 493, 494
Real estate mortgage, 807
Real Estate Settlement Procedures Act (RESPA), 829
Real numbers
  absolute value of, 253
  addition of, 257–258
  additive inverse and absolute value and, 252–254
  applications of, 264
  complex numbers and, 306–309
  as decimals, 251
  division of, 260–261
  explanation of, 51, 65, 87, 99, 250, 251
  irrational numbers and, 270, 283–290
  multiplication of, 258–260
  operations with, 257–262
  operations with decimals and, 293–296
  operations with percent and, 297–300
  order and, 250
  properties of addition and multiplication of, 262–265
  rational numbers and, 270–279
  sets of, 248–251, 263
  subtraction of, 258
  use of, 247
Real part, 308
Real rate of return (RRR), 847
Reasoning
  deductive, 2, 3, 14, 129, 725
  inductive, 2–7, 10–16, 129, 724–725
Reasoning by transitivity, 140
Reciprocal, 252–253, 263
Reciprocal identities, 586
Recorde, Robert [re kôrd], 250
Rectangles
  area of, 514
  explanation of, 503
  golden, 513
  perimeter of, 512
Rectangular coordinate system, 394–396
Rectangular parallelepiped, 536
Reductio ad absurdum, 284
Refinancing, 829
Reflection(s)
  explanation of, 545–547, 551
  glide, 545–547
  product of two, 547
Reflection image, 545–546
Reflexive axiom, 318
Regiomontanus, 613
Region of feasible solutions, 484
Regression, 723, 778–779
Regression analysis, 778
Regression coefficient formulas, 779
Regression curves, 778, 782
Regular monthly payment
  explanation of, 822–823
  formula for, 824
Regular polygons, 502
Regular tessellation, 551
Relations
  explanation of, 423
  functions and, 423–425
  graphs of, 425–426
Relative frequency, 725
Relatively prime numbers, 213
Repayment schedule, 824
Repeating decimals, 276–278, 283
Reserve account, 825
Residue design, 231
Return on investment, 833
Revenue models, 430–431
Rhind Papyrus [rīnd], 153
Rhombus [räm´bus], 503
Richter, P. H., 570
Riemann, Georg Friedrich [Rē´män], 558, 559, 563
Riese, Adam, 257
Right angles, 495, 578
Right circular cones, 538, 541
Right circular cylinders
  example of, 539
  explanation of, 538
  volume and surface area of, 539–540
Right triangles
  applications of, 600–605
  explanation of, 503, 591
  function values and, 591–598
Rivest, Ronald L., 195, 197
Roman numerals, 154
Roots, of a directed tree, 896
Rotation
  explanation of, 548–549, 551
  magnitude of, 548
Roulette, 710
Round lot of shares, 836–837
RSA code, 195, 197
Rubik's Cube, 630
Rule of 70, 798
Rule of 78, 815, 817–819
Rule of False Position, 317
Russell, Bertrand, 87, 113, 123, 124, 127
Russian peasant method, 166–167
Ruth, Babe, 393

S

Saari, Donald G., 937
Saccheri, Girolamo [sa kâr´e], 556–557
Sagan, Carl, 448, 695
Sample correlation coefficient, 780
Samples, 724
Sample space, 674
Sample standard deviation, 750
Scale factor, 571
Scalene triangles, 503
Scaling, 568
Scatter diagrams, 778, 782
Scientific calculators. *See also* Calculators; Graphing calculators
  for converting bases, 173
  English-metric conversions on, A-5
  explanation of, 31
  future value on, 793
  logarithms on, 451–452
  powers of *e* on, 446
  statistics using, 725
Scientific notation
  conversion from, 368
  conversion to, 367
  examples of, 368–369
  explanation of, 366
Search for extraterrestrial intelligence (SETI), 695, 696
Secant, 582
Secant lines, 505
SEC fee, 837
Second, as an angle measure, 579
Second component in an ordered pair, 67

Securities and Exchange Commission (SEC), 837
Security deed, 822. *See also* Mortgages
Segments, line, 493, 577
Self-similarity, 568, 569
Semicircles
    angles inscribed in, 506
    explanation of, 250, 505
Semi-interquartile range, 761
Sequential pairwise comparison voting method, 921–922
Set-builder notation, 50, 249
Set equality, 53, 58
Set(s)
    cardinal numbers of, 83–88
    Cartesian product of, 68–69
    complement of, 57
    of counting numbers, 84, 87, 88, 152
    De Morgan's laws and, 72
    designation for, 50
    difference of, 66–67
    disjoint, 64
    empty, 51
    equal, 53
    equivalent, 84
    explanation of, 50–51
    finite, 51–52, 87
    infinite, 52, 83–88
    intersection of, 63–64
    null, 51
    one-to-one correspondence between, 84–86
    operations on, 69–70
    proper subset of, 58–60
    of solutions, 315
    subset of, 57–58
    union of, 64–66
    universal, 56
    Venn diagrams and, 70–72, 78
Set theory
    cardinal numbers and surveys and, 76–79
    infinite sets and their cardinalities and, 83–88
    set operations and Cartesian products and, 63–72
    symbols and terminology for, 50–53
    Venn diagrams and subsets and, 56–60
Settlement charges, 828–829
Shallit, Jeffrey, 199
Shamir, Adi, 195
Shanks, William, 286
Shannon, Claude, 118
Side-angle-side (SAS) congruence property, 524
Sides
    explanation of, 494
    initial, 577
    terminal, 577
Side-side-side (SSS) congruence property, 524
Sieve of Eratosthenes, 197–198, 208
Signed numbers, 248
Significant digits, 602
Similar triangles, 525–527
Simple curves, 502
Simple graphs, 852
Simple interest
    explanation of, 789–790
    future and present values for, 790–791
Simpson's paradox, 737
Simulation
    estimating probabilities by, 717–719
    explanation of, 716–717
Sines
    amplitude of, 622–623
    explanation of, 582
    law of, 609–611
Sine wave, 621
Single transferable voting system. *See* Hare method of voting
Sinusoid, 620
Size transformations, 549–551
Sketches, 24–25
Skewed distribution, 742
Skew lines, 494
Skewness coefficient, 755
Slope(s)
    explanation of, 405–406
    of lines, 405–408, 412
    positive and negative, 407
Slope-intercept form, 414–416, 430
Smith numbers, 206
Smullyan, Raymond, 104
Sociable numbers, 208
Solutions. *See also* Problem solving
    to applied problems, 326–334
    of equations, 315
Solution set, 315
Somerville, Mary, 686
Sonic boom shock wave, 440
Sophie Germain prime, 209
Space figures, 536–538
Spanning trees, 897–898
Special addition rule of probability, 688–689
Special additive principle of counting, 666–667
Special multiplication rule of probability, 694–697
Sperbeck, Jeff, 143
Sperry, Pauline, 613
Spheres, 540
Square
    area of, 514
    explanation of, 503
    perimeter of, 513
Square centimeter (cm^2), A-2
Square meter (m^2), A-2
Square numbers, 14, 15, 19
Square root radicals
    addition and subtraction of, 289–290
    simplified form of, 288
Square roots
    explanation of, 287
    product rule for, 287–288
    quotient rule for, 288–290
Standard deviation
    calculation of, 750–752
    explanation of, 749–750
    sample, 750
Standard form of a linear equation, 403
Standard normal curve
    areas under, 765, 766
    equivalence in, 768
    explanation of, 764
Standard position of an angle, 580
Statement(s)
    compound, 95–96, 113, 116
    conditional, 113–118, 123–127
    direct, 123, 124
    equivalent, 109–110
    explanation of, 95–96
    negation of, 96, 98
Statistics
    central tendency measures and, 735–744
    correlation and, 780–782
    descriptive, 724
    dispersion measures and, 748–753
    explanation of, 723–725
    frequency distributions and, 725–730
    inferential, 724
    lying with, 771–775
    normal distribution and, 763–769

Statistics (continued)
  position measures and, 756–760
  regression and, 778–780
Stem-and-leaf displays
  central tendency from, 741–742
  explanation of, 728–729
Stevin, Simon [stə vin´], 276
Stirling, James, 671
Stirling's formula, 671
Stock exchanges, 833
Stocks
  commissions for buying or selling, 836–838
  explanation of, 833, 842, 843
  mutual funds vs., 839–840
  prices for, 833–836
Straight angles, 495, 578
Student's *t*-test, 724
Subgraphs, 860
Subsets
  cardinal numbers of, 76
  number of, 60
  number of proper, 60
  proper, 58–60
  of sets, 57–60
  symbol for, 58
Substitution axiom, 318
Subtraction
  in clock arithmetic, 221–223
  of decimals, 294
  definition of, 258
  in modular systems, 224–225
  of polynomials, 373
  of rational numbers, 272
  symbol for, 259
Successive differences method
  explanation of, 10–11
  Fibonacci sequence and, 11
  figurate numbers and, 16
Summer Olympics, 314
Sum of cubes, 378–379
Sum of the degrees theorem, 856
Sum of real numbers, 257
Supplementary angles, 496, 497, 578
Surface area
  of a box, 538–539
  explanation of, 538
  of a pyramid, 541
  of a right circular cone, 541
  of a right circular cylinder, 539–540
  of a sphere, 540
Sylvester, James Joseph, 855, 897, 901
Symbolic logic, 95–97. *See also* Logic

Symbolic Logic (Carroll), 40
Symbol(s)
  for basic operations, 259
  for the empty set, 51
  for equality, 250, 251
  explanation of, 152
  for inequality, 58
  for infinity, 52
  for intersection, 63
  for points, lines, and planes, 492–494
  radical, 289
  for subsets, 58
  used with logic, 96–97
Symmetric axiom, 318
Symmetry
  in data sets, 742
  of normal curve, 765
  of square, 187–189
Systematic listing methods
  explanation of, 630
  miscellaneous, 634
  for one-part tasks, 630
  using product tables, 630–631, 634
  using tree diagrams, 631–634

T

Tables, 21–22, 329–331
Tally, 152
Tally sticks, 152
Tangents, 505, 582
Tangrams, 503
Tartaglia, Niccolo [täR ta´glē ə], 383
Tate, Peter, 882
Tautologies, 116
Taxes, 846
Tax Reform Act of *1986*, 823
Teaching Children Mathematics, 42–44
Temperature, A-3–A-4
Term of a mortgage, 822
Terminating decimals, 276–278, 283
Term papers, 41, 45
Tessellations, 550–553
Tetrahedrons, 536
Texas Instruments calculators
  SR-*10*, 31
  TI-*83* Plus, 200, 233, 316, 793
  TI-*1795 SV*, 31
Thales [thā´lēz], 506
Theorems, 199, 506
Theoretical frequencies, 728
Theoretical probability, 675, 677, 684–685, 718

Thom, René, 568, 569
Three-part inequalities, 355–357
Time relationship, formula for, 332, 333
Time value of money
  explanation of, 789
  inflation and, 796–798
  interest and, 789–796
Topology, 559–560
Torque approximation, 303
Torus, 538, 561
Tractrix, 558
Transformational geometry
  explanation of, 545
  reflections and, 545–547
  size transformations and, 549–551
  translations and rotations and, 547–549
Transformations
  explanation of, 546
  size, 549–551
Transitive axiom, 318
Translation
  explanation of, 548–549, 551
  of graphs, 436
  identity, 548
Transversals, 497, 498
Trapezoids, 503, 515
Traveling salesman problem (TSP), 642, 883–884, 887
Tree diagrams, 59, 631–634
Trees
  explanation of, 894–895
  minimum spanning, 898–901
  number of vertices and edges in, 901–902
  property of, 895–896
  spanning, 897–898
  unique path property of, 896, 897
Trial and error, 23
Triangle, Pascal's, 45, 239, 658–660
Triangle(s)
  angle sum of, 504, 573
  area of, 516, 615
  congruent, 523–526
  explanation of, 502
  isosceles, 503, 525
  oblique, 608–609
  perimeter of, 512
  similar, 525–527
  types of, 503
  using trigonometry to solve, 603–604

Triangular numbers, 14, 15, 19
Triangulation method, 619
Trigg, Charles, 242
Trigonometric functions
    of angles, 582–585, 587
    calculators to approximate values of, 600–601
    graphs of, 622–624
Trigonometric function values, 626–627
Trigonometric identities
    explanation of, 586–590
    Pythagorean, 588–589
    quotient, 589–590
    reciprocal, 586
Trigonometry
    angles and their measures and, 577–581
    applications of right triangles and, 600–605
    historical background of, 576, 578, 582, 587
    laws of sines and cosines and, 608–616
    right triangles and function values and, 591–598
    unit circle and graphs and, 620–624
Trinomial(s)
    explanation of, 373
    factoring, 376–377
    perfect square, 377
Truth in Lending Act, 813, 815
Truth table(s)
    analyzing arguments with, 136–143
    for biconditionals, 127
    conditional, 113–117
    for conjunctions, 102
    construction of, 105–109
    to determine truth values of mathematical statements, 104–105
    for disjunctions, 103
    explanation of, 102
    for negations, 103, 115
    summary of, 127
Truth value(s)
    of conjunctions, 102
    explanation of, 95, 104
    of mathematical statements, 104–105
Tshokwe people, 871
Tsu, Ch'ung-chih, 283

Tukey, John, 724
Tutte graph, 882
12-hour clock system, 220–223
Twin primes, 208–209

U

U.S. Metric Association, A-5
Unearned interest, 815–818
Uniformity criterion, 638
Union of sets, 64–66, 70
Unique factorization theorem, 200
Unit circle, 620
Universal Algebra (Whitehead), 123
Universal quantifiers, 97, 99
Universal sets, 56
Universe of discourse, 56
Unpaid balance method, 807
Upper class limits, 727
U.S. stock markets, 273

V

Valid arguments
    Euler diagrams and, 129–131, 136
    explanation of, 129, 131
    forms of, 141
    truth tables and, 136–143
Values, discrete vs. continuous, 768
Vanishing point, 340
Variable-rate mortgages. *See* Adjustable-rate mortgages (ARM)
Variable(s)
    dependent, 423
    independent, 423
    random, 689
    solving for specified, 320–321
Variance, 750
Variation
    coefficient of, 761
    combined, 347
    constant of, 343
    direct, 343
    explanation of, 343
    inverse, 345
    problem solving and, 344–347
Venn, John, 56
Venn diagrams
    explanation of, 56
    sets and, 70–72, 78
Vertical angles, 495–496
Vertical lines, 406, 416

Vertices
    of an angle, 494, 577
    degree of, 852
    even, 561
    explanation of, 485, 561
    formula for, 438
    odd, 561
    of a parabola, 435, 437, 440
    of a tree, 900–902
Viète, François [vē et´], 318
Vinson, Jade P., 539
Voltaire, François, 52, 107, 129
Volume
    of a box, 538–539
    explanation of, 538
    in Metric System, A-1–A-3
    of a pyramid, 541
    of a right circular cone, 541
    of a right circular cylinder, 539–540
    of a sphere, 540
von Linne, Carl, A-4
Voting methods
    approval, 922–924
    Arrow's impossibility theorem and, 937–939
    Borda, 915–918, 930–933, 935, 938, 939
    Condorcet criterion and, 932–933, 938, 939
    Coombs, 928
    explanation of, 912, 930
    Hare, 917–921, 932, 933, 935, 938, 939
    independence of irrelevant alternatives (IIA) criterion and, 935–937, 939
    majority criterion and, 930–931, 938, 939
    monotonicity criterion and, 933–935, 939
    pairwise comparison, 913–917, 919, 931–933, 935–939
    plurality, 913, 914, 919, 931–933, 935, 938, 939
    sequential pairwise comparison, 921–922

W

Walk, in a graph, 857
Wallis, John, 52
Washington, George, 952

Washington, Harold, 912
Webster, Daniel, 948, 956
Webster apportionment method, 948, 956–959, 962, 964, 965
Weight, A-3
Weighted mean, 736–737
Weighting factor, 736–737
Weights, of edges in a graph, 858
Weird numbers, 207
Wells, David, 210
The Whetstone of Witte (Recorde), 250
Whitehead, Alfred North, 123, 127
Whole numbers
    explanation of, 51, 99
    set of, 248
Wilansky, Albert, 206
Wiles, Andrew, 209
Working backward, 22–23

World Wide Web, 850. *See also* Internet sites

X

x-axis
    explanation of, 394
    as horizontal asymptote, 445
x-intercept
    explanation of, 403–404
    graphing quadratic functions and, 439

Y

y-axis, 394
Years to double, 797–798

y-intercept
    explanation of, 403–404
    graphing quadratic functions and, 439
Yin-yang, 170
Yorke, James A., 568
Young, H. Peyton, 964, 969, 971
Yu, Chinese Emperor, 240

Z

Zero
    division of and by, 261
    origins of, 248
Zero exponents, 362, 365
Zero factorial, 642
z-score, 756–757

Index of Applications

Astronomy
Aerospace industry sales, 392
Astronaut's weight, 346
Astronomy data, 371
Distance from an object to the Moon, 371
Distance to stars, 576
Elliptical orbits, 346
Filamentary structure of the universe, 899
Intergalactic Society, 921–922
Miles in a light-year, 371
NASA budget, 370
Orbit of Halley's comet, 438
Planets' distances from the sun, 371, 693, 698
Pollution from the space shuttle launch site, 699–700
Revolutions of Mercury, 38
Rocket launches, 371, 707

Automotive
Accident reconstruction, 292
Antifreeze mixture, 337, 474
Auto production, 255
Auto repair charges, 747
Automobile sales, 335
Engine control modules, 754
Getting caught when speeding, 706
License plates, 640, 670
Motor vehicle registration, 370
Number of miles driven per month, 768–769
Odometer readings, 230
Relative lifetimes of tires, 760
Skidding car, 352, 387

Biology and Life Sciences
Activity of a nocturnal animal, 626
African violets, 37
Age, 26, 30, 327, 464, 763
Birdhouse for swallows, 33
Breaching of humpback whales, 268–269
Dinosaur research, 389
Distance between atoms, 617
DNA profiling, 629
Double spiraling of a daisy, 234
Fertilizer vs. corn ear size, 783
Fish, 342, 349, 352, 771
Flea jumping, 5
Forensic studies, 433
Frog climbing up a wall, 28
Frost survival among orange trees, 706
Genetics, 678–679, 682
Growth of a lily pad, 28
Heights and weights of adult men, 783
Heights of spruce trees, 786
Herd of camels, 23–24
Identification numbers in research, 656
Languages spoken by Americans, 746
Male honeybees, 234
Mineral sample masses, 733
Neurons in the human brain, 859
Pea plant production, 716–717, 719
Pinecones, 234
Pool size for sea otters, 433
Rabbits producing offspring, 21–22
Seed germination, 682
Siblings, 30, 725–726
Size grading of eggs, 771
Weights of animals, 269, 351, 770

Business
Advertising, 143–145, 776
Book sales, 219, 312, 335, 412, 454, 651
Break-even analysis, 430–431
Broker's commission, 837–838, 845
Business failures in the U.S., 311
Carpet cleaning service, 343
CD players, 655
Cereal marketing survey, 755
Client contacts of a publisher's representative, 786
Copying service, 434
Cost and revenue models, 434
Courier delivering documents, 886–887
Delivery service, 434
Discount and markup, 304
Distribution to a wholesaler, 475
Expected value in business accounts, 716
Inspecting calculators, 219
Labeling boxes, 25
Largest U.S. companies, 472
Lengths of long-distance phone calls, 767
Managers networking computers, 905
Market share, 38, 456
Meeting in a hospitality suite, 62
Merchandise pricing, 349, 391
New company logo, 925
Producer Price Index, 256, 264–265
Production, 469–470, 473, 475, 484–485, 487, 489
Profit, 34, 714
Profit/cost analysis, 360
Real estate commission, 490
Restaurant customers, 757–758, 760
Revenue and expenditures of Yahoo, 268
Sales, 312, 434, 490, 714, 736, 783
Stuffing envelopes, 434
Tipping procedure, 306
Traveling salesman problem, 643, 884, 885, 887
Value of a copy machine, 455
Warehouse grocery shopping, 699

Chemistry
Acid mixture, 328, 329–330, 336, 466–467, 473, 474
Alcohol mixture, 336, 473
Analyzing the structure of crystals, 890

I-19

Boiling point of chlorine, 255
Chemical bonds in a molecule, 901–902
Chemical mixture, 337, 391
Decay of plutonium-241, 490
Dissolving sugar in water, A-6
Half-life, 451
Molecular dynamics, 1
Ore samples, 454–455, 770
Pressure of a gas or liquid, 351
Salt crystal, 537
Salt in seawater, A-6
Weight of helium, A-6

Construction
Air conditioner size, 34
Angle between a beam and cables, 618
Area of the Sistine Chapel, 38
Arranging new home models, 655, 656, 657
Building flagstone paths, 905
Carpeting, 514
Chain saw oil mixture, 348
Construction materials, 264–265
Contractor decisions based on expected profits, 715
Cost of paving a road, 898–899
Dimensions of a piece of sheet metal, 389
Distance between points on a crane, 618
Distance determined by a surveyor, 613
Dividing a board into pieces, 327–328
Dry-wall contractor, 227
Floor plans, 562, 878–879, 909
Grade of a ramp, 411
Hardware, 30, 475
Height of a building, 533, 607
Height of a tower, 255, 533, 534, 607
Installing an underground irrigation system, 894, 897
Königsberg bridge problem, 865–868
Leaning ladder, 388, 529, 606
Length of a wire, 387
Lumber, 219, 710
Maximum load of a cylindrical column, 347
Mazes, 858, 864
Patterns in floor tiling, 637, 876–877
Playhouse layout, 618
Required amount of paint, 619
Size of a developer's property, 388
Slope of a roof, 409

Socket wrench measurements, 281
Squaring off a floor under construction, 530, 534
Steepness of an upper deck, 411
Strength of a rectangular beam, 346–347
The Parthenon, 236
Transamerica Tower, 541
Vietnam Veterans' Memorial, 291, 516
Window side length, 519

Consumer Information
Air conditioner choices, 420
Analyzing a consumer loan, 849
Average housing costs, 282
Best buy, 342–343, 518, 522
Boat purchase and sale, 305
Buying vs. renting a house, 848
Consumer debt burden, 305
Consumer-packaged products, 747
Cost, 348, 433, 473, 487, A-6
Discount, 302
Electricity consumption, 432
Gasoline and fuel oil costs, 305, 422–423, 487
Library fines, 489
Lifetimes of lightbulbs, 770
Paying for a mint, 29
Price per acre of land, 37
Prices, 34, 306, 473
Rentals, 357, 359
Taxicab fare, 359, 433

Economics
Bond investments, 838–839, 845
Changing value of the British pound, 776
Consumer Price Index, 350, 396, 732
Decline in purchasing power of the dollar, 397
Exchange rate, 27, 348
Gross domestic product, 732
Inflation rate, 796–798, 800, 801
Influence of Spanish coinage on stock prices, 273
Interest rate, 389
Primary energy sources, 732
Property taxes, 302
Sales tax, 348
Stocks, 56, 305, 711, 833–835, 841–844, 846, 849
Supply and demand, 389
Tax returns, 74

Taxes on investment returns, 846
U.S. copper coinage, 280
U.S. petroleum imports, 425
Value of currency, 25, 28, 305, 336

Education
Absences in a math class, 714
Accumulated college units, 731
Alpha Beta Gamma fraternity, 652
Attendance, 689
Bachelor's degrees earned, 454, 666–667
Calculating a missing test score, 747
Choosing school assignments for completion, 669
Class reports, 684
College Art Association, 914–915, 916–917, 919–920, 930–931, 933–934, 935–936
College class schedule, 645, 684
College expenses, 729–730
Comparing gender and career motivation of university students, 698
Correspondence between education and earnings, 733
Course enrollment of English classes, 959, 960, 961, 962
Distances to school, 731
English composition class, 863
English essays, 643
Enrollment profile, 974
Exam scores, 730, 740, 745, 747, 757, 765, 786
Faculty categories at a small college, 777
Faculty-student marriages, 777
Fifth-grade teachers needed, 37
Financial aid for students, 81, 93
Grade point average, 359, 391, 736, 746
High school graduation year, 29
Homework, 688–689
IQ scores, 731, 770, 783
Impact of art and music education on standardized mathematics exams, 967–969
Kindergarten class voting on a class pet, 917
Likelihood of capable students attending college, 706
Lining up preschool children, 681
Lowest passing grade, 302
Multiple-choice test, 698, 706

Oral report, 694
Paths between school buildings, 904–905
Public school enrollment, 410
Questions omitted on a math exam, 730
Reading ability, 783
Relative positions on a test, 760, 761
School district apportionment of computers, 949–950, 957, 959, 966–967
School board election, 913
School bus routes, 960
Selecting answers on a test, 645
Selecting classrooms, 662
Selecting faculty committees, 669
Social interaction among preschool children, 851–852, 853, 860
Student goals, 82
Student grades, 658, 771
Student ownership of personal computers, 706
Student survey, 90–91
Students per computer, 410
Students who enjoy music and literature, 670
Study time, 726–727, 728, 734, 742, 759
Teaching assistants, 974
Test scores, 356–357
Tuition and fees, 401, 420
Two-year college enrollment, 401
Women in mathematics, 48

Engineering
Computer speed, 371
Cost to fill a spherical tank, 540, 544
Dirt in a hole, 30
Distance across a lake, river, tunnel or canyon, 607, 610–611, 617, 618, 628
Electrical circuits, 118–120, 122, 391
Electrical impedance tomography, 491
Electronics formula, 291
Engine horsepower, 303
Gates and memory cells, 94
Guy wires, 607
Measurement using triangulation, 619
Screening computer processors, 658
Settings on a switch panel, 644, 672
Time traveled for a radio signal, 391
Torque approximation, 303
Vanishing point, 340

Environment
Carbon monoxide emissions, 453
Cloud ceiling, 607
Cold fronts, 900
Depths of water and heights of mountains, 255, 268, 534, 608
Dry natural gas consumption, 455
Earthquake intensity, 455
Elevation, 255, 269
Epicenter of an earthquake, 399, 402
Forest loss and atmospheric pollution, 745
Global warming, 453
Greenhouse effect, 450–451
Hazardous waste sites, 455
Insecticide mixture, 337
Landfill capacity, 421
Municipal solid waste, 454
Nuclear waste, 420
Sun's ultraviolet rays, 247
Temperature, 265, 268, 359, 421–422, 625, 731, 734, 783, A-3, A-4, A-6
Tornado activity, 358
Weather, 39, 701
Windchill, 255

Finance
Adam and Eve's assets, 29
Add-on interest loan, 810
Adjustable rate mortgage, 826–828, 831, 832, 849
Allowance increase, 304
Analyzing a "90 days same as cash" offer, 812
Annual percentage rate, 814–815, 820, 821, 849
Average daily balance of an account, 811–812
Banking transactions, 302–303
Borrowing money, 799
Budgeting to buy a car, 310
Charge card account balances, 269, 730, 808–809, 810, 849
Charges of a bank account, 812
Checking account balance, 268
Closing costs of a mortgage, 829, 832
Comparing loan choices, 821
Comparing rates, 800, 813
Compound interest, 446–447, 450, 793–795
Cost of points in a home loan, 849
Deposits, 790–791, 794

Effective annual yield, 795, 800, 849
Finance charges, 807, 810, 811–812, 820
Financing a purchase, 809
Fixed rate mortgage, 829–830
Future value, 790, 793–794, 798, 799, 849
Half-cent coins, 338
Interest earned, 336, 351, 793, 799
Investment, 292, 328, 330–331, 337, 444, 446–447, 450, 453, 473, 474, 490, 710–711, 838–842, 844, 845, 846
Loan payoff amount, 816–818
Making change, 30
Money spent at a bazaar, 28
Monthly mortgage payments, 822–828, 831, 849
Monthly payment for a purchase, 806, 810, 820
Mutual funds, 839–841, 842, 845, 846
Net results of combined transactions, 844
Nominal rate of a savings account, 800
Ordinary annuity, 804, 805
Prepayment penalty, 821
Present value of an account, 800, 801, 849
Real rate of return, 847
Retirement plans, 255, 733, 788, 797
Savings account, 794–795, 796
Simple interest, 790, 798, 799
Total principal and interest on a mortgage, 831
Unearned interest, 816–818, 819, 820, 821
Years to double an account, 798, 800, 849

General Interest
Alphabetizing, 29
Array of nine dots, 24–25
ATM PIN numbers, 655
Beverage preference, 79
Birthdays, 23, 29, 269, 697
Books on a shelf, 30
Broken tree, 527
Buckets of water, 27
Building words from sets of letters, 672
Candles, 47, 350
Children in a circle, 28
Choosing a monogram, 656

Clocks, 27, 28
Coin or bill mixture, 331–332, 337, 338
Collections, 27, 80, 227, 230, 692–693
Combination lock, 657
Cooking habits, 81
Counting prizewinners, 655
Creature comforts, 313
Days in a month, 30
Determining a range of dates, 230
Determining day of the week, 225, 230, 669
Dimensions on Mount Rushmore, 533
Distance to the horizon, 291, 292
Dividing people into groups, 657
Donating fruit, 919
Drawers for videocassettes, 37
Error in measurement, 608
Facts about inventions, 92
Fair decisions from biased coins, 700
Favorite *Peanuts* characters, 313
Filling containers with food, 770, 771
Food and drink mixture, 474
Food stamp recipients, 412
Friday the thirteenth, 231
George Washington's chair, 350
Hazardous situations, 673
Height, 29, 388, 527, 533, 534, 575, 605, 617, 627, 628, 730
Imaginary friends, 136
Interesting property of a sentence, 28
Layout of a botanical garden, 878, 890
Length of a shadow, 607
Letter occurrence frequencies in the English language, 734
Local animal shelter, 924
Magic trick age detector, 176
Marriage, 134–135
Matching club members with tasks, 644
Matching socks, 28
Measurement of a folding chair, 617
Missing pages in a newspaper, 48
Mounting a video camera, 607
Naming a parrot, 938–939
Naming children, 28
Neckties, 136
Palindromic greeting, 31
Paradoxes of Zeno, 90
Perpetual calendar algorithm, 193–194
Phone numbers, 31, 645
Piles of ticket stubs, 230
Planning a route, 872–875, 878

Pool water level, 424–425, 432
Position of a searchlight beam, 387
Poultry on a farm, 82
Powerball lottery, 371
Programming a garage door opener, 700
Pumpkin-patch kids, 135
Puzzles, 27, 28, 29, 30, 47
Radio station call letters, 656
Raising money to purchase trees for state parks, 959, 960
Random walk, 707, 720
Recipe for grits, 280
Rhyme schemes in poetry, 865
Seating arrangements, 683
Selecting a meeting site, 941–943
Selecting items, 62, 63, 636, 645, 653, 698
Shaking hands in a group, 636, 637, 863
Size of a toy piece, 388
Slope of a hill, 409
Spreading a rumor, 863, 903
Stacking coins, 219
Swiss cheese hole sizes, 281
The barber paradox, 87
Total bill for King Solomon, 160
U.S. oyster catch, 443
Value of a biblical treasure, 160
Vertical symmetry in states' names, 26
Weddings, 658, A-6
Weekend workshop, 925–926, 941
Weighing coins, 29, A-6
Weights of coffee jars, 753
Who's telling the truth?, 29–30
ZIP codes, 645
"100 Bottles of Beer on the Wall," 89

Geometry
Altitude of a triangle, 607
Angle of elevation of the sun, 605, 607
Area of Hawaiian islands, 391
Area of lots, 520
Area of a quadrilateral, 523
Area of a shaded region, 523, 575
Area of a square, 523
Area of a trapezoid, 523
Area of a triangle, 291, 351, 619
Cardboard box dimensions, 389
Change in volume, 544
Circumference of a dome, 575
Constructive solid geometry, 49
Counting triangles, 672
Diagonal of a box, 292

Diagonal of a rectangle, 575
Diagonals of a parallelogram, 618
Diameter of a circle, 522
Dimensions of a rectangular region, 351, 388, 464–465, 472, 490, 519, 534
Dimensions of a square, 472
Dimensions of a triangle, 472, 475
Frustum of a pyramid, 2
Geometry puzzle, 30, 31
Matching triangles and squares, 26
Maximum area, 440–441, 443
Pennant side lengths, 520
Perimeter of a polygon, 523
Perimeter of a square, 545
Perimeter of a triangle, 522
Points, lines, and triangles in a plane, 657
Radius of a circular foundation, 520
Radius of an aluminum can, 291
Ratio of areas, 545
Ratio of side lengths, 545
Ratio of volumes, 544
Rectangles in a figure, 27
Side length of a cube, 543
Side lengths of a triangle, 387, 534, 607
Unfolding and folding a box, 26
Volume of a box, 544, A-6
Volume of a sphere, 544

Government
Aid to earthquake victims, 487
Amendments, 929
Appointing committees, 631, 635, 652, 653, 657, 662, 722
Approval method election, 922–924
Army housing, 83
Breaking codes, 197
California exports, 311
Campaign promises, 63
Committee selecting a chairperson, 925, 940, 941
Committees of U.S. senators, 655, 698
Council seats, 969–971, 974, 975, 978
Decoding messages, 195
Diplomatic delegation, 667
Election, 631, 635, 639, 655, 681
Federal taxes, 400, 776
House of Representatives, 911, 951–952, 955, 960, 961, 962, 972
How a new bill is processed, 928–929

INDEX OF APPLICATIONS | I-23

Legislative seats, 958–959, 960, 961, 962, 972, 973, 974, 975, 979
National health care expenditure, 369
Postage stamp pricing, 302
Ranking candidates, 944, 945
Ranking issues, 944–945
Rate of favorable media coverage of an incumbent president, 706
Social security, 268, 443
Top U.S. trading partners, 472
Town council, 905
U.S. budget, 370
U.S. defense budget, 267, 434
U.S. post offices, 420, 433
U.S. voting-age population, 431
United States presidents, 54
Voter affiliations, 475
Voter profile, 929, 932, 939, 940–947, 977–978
Votes in 2000 presidential campaign, 335
Voting for candidates, 912, 927, 928

Labor
Age vs. income, 778–781
Annual rates of employment change, 254
Committees, 652, 657
Company employment data, 257
Days off, 245, 722
Employees, 78–79, 777, 964–965
Farming, 33, 785
Flight attendant schedules, 230
Garage employing mechanics, 695–696
Hourly earnings in private industry, 282
Income for African Americans, 489
Job cost, 521
Labor union members, 275, 305, 707
Median income, 421
Night off for security guards, 219
Ordering job interviews, 700
Poverty level income cut-offs, 401
Prospects for electronics jobs in a city, 715
Salary, 737, 738–739, 745, 748
Security team visits, 655
Sources of job training, 733
Summer jobs, 680
Time on a job vs. time for a task, 782
Unemployment rate, 732
Wal-Mart employees, 371

Women in math or computer science professions, 399

Medical
AIDS, 488
Adverse effects of alcohol and tobacco, 73
Alcohol mixture in first aid spray, 336
Births in the United States, 677, 682
Blood alcohol concentration, 303
Body mass index, 351, 359
Burning calories, 56
Comparative death rates, 777
Cystic fibrosis, 682, 722
Filling bottles with medicine, A-6
Finding blood clotting times, 771
Having children, 676
Heart attack data, 372
Highly infectious disease, 903
Hospital patient symptoms, 82
Medicine doses, A-6
Nurses volunteering to treat earthquake victims, 960, 961, 962
Pharmaceutical research, 411
Pharmacy payments, 443
Recommended daily vitamin allowances, 770
Reported cases of influenza and pneumonia, 777
Retail spending on prescription drugs, 416–417
Search and rescue teams, 669
Sickle-cell anemia, 683
Side effects of prescription drugs, 706
Target heart rate, 360
Threshold weight, 292
Variation in blood pressure, 626
Vitamin pill costs, 487

Numbers
Alphametric, 27
Building numbers from sets of digits, 636, 644–645, 657, 672, 684, 690, 722
Counting, 637, 644, 645, 669
Decimal digit, 30
Determining operations, 31
Devising a correct addition problem, 47
Dice sum, 26
Difference triangle, 26
Digits puzzle, 27

Fibonacci sequence, 30, 47, 232–234, 245
Final digits of a power of seven, 28
Forming perfect square sums, 26
Froude number, 389
Integers containing the digit 2, 636
Lattice points, 636–637
Magic square, 29, 240–245
Number of inches in a mile, 371
Numbers generated by spinners, 681, 690
Octagonal number, 47
Palindromic number, 27, 684
Pascal's triangle, 46
Perfect number, 28
Perfect square, 27
Prime numbers, 684
Random numbers, 684, 715
Sums of digits, 636, 655, 658, 672
Units digit of a product, sum, or power, 28, 47, 48
Unknown numbers, 28, 30

Physics
Angle formed by radii of gears, 617
Elapsed time in a physics experiment, 745
Falling body, 345, 351
Force required to compress a spring, 351
Height of a projectile, 386, 387, 443
Hooke's law for an elastic spring, 344
Illumination of a light source, 351
Law of tensions, 293
Object thrown upward, 385
Period of a pendulum, 291
Sonic boom, 440
Speed of light, 371
Speed of sound, 332
Time an object has descended, 392

Probability
Comparing the likelihood of girl and boy births, 718, 719, 722
Drawing balls from an urn, 681, 690, 695
Drawing cards, 656, 668, 670, 683, 686, 688, 689, 694, 695, 699, 700, 713, 719, 721, 722
Drawing colored marbles from boxes, 690
Gender in sequences of babies, 682, 692, 699, 700, 704, 705, 706, 708

Lottery, 656, 709
Probability of rain, 680
Raffle, 680
Rolling dice, 644, 669, 682, 685, 689, 690, 698, 701, 704, 705, 722
Sweepstakes, 680
Tossing coins, 660, 662, 665, 669, 672, 676, 678, 681, 682, 685, 698, 704, 705, 707, 708–709, 713, 714, 717, 718, 728

Sports and Entertainment
500-meter speed skating, 321–322
Attendance at an event, 337, 338
Bandstands at a festival, 890
Baseball, 29, 218, 277, 279, 280, 335, 393, 472, 628, 656, 705, 706, 737, 754, 786–787
Basketball, 48, 83, 312, 335, 656, 672, 701, 719, 731, 760–761
Bowling, 135
Broadway shows, 324, 669
Card trick, 227–228
Chess, 223, 636, 863, 908
Children attending an amusement park, 714–715
Concert revenues, 335
Counting card hands, 657
Crawfish racing, 402
Culinary Arts Club, 913, 914, 915, 916, 918, 932, 939
Dancing at a party, 863
Dart throwing, 683
Dining out, 636, 645, 669
Diving, 749
Door prize winners, 696–697
Downhill skiing times, 269
Drive-in movie screens, 336
Dunking booth, 148–149
Expected winnings, 709, 714, 715, 716
Fan Cost Index, 473
Favorite sports among recreation students, 734–735
Fishing, 760
Five-person relay team, 890
Football, 33, 38, 218, 269, 731, 762–763
Freestyle swimming, 314, 333
Gambling, 710, 712–713, 714, 715, 716
Golf, 641, 690, 722
Hiking, 268
Horse racing, 657, 684, 746
Indianapolis *500* race, 333
Inviting friends to dinner, 908
Julia Roberts' box office hits, 490
Movies, 81, 217
Music, 77–78, 81, 470–471, 656–657, 669, 681, 684
Network news programs, 470
Olympics, 326–327, 338, 349, 359, 424, 475, 746–747
Percussionist auditions, 936–937
Poker hands, 651, 654, 656, 670, 677, 683, 690
Preschoolers at a park, 643
Racetrack betting, 22–23
Racing, 219, 297, 338, 390, 655
Reading, 657
Rock band tour, 909
Round-robin tournament, 859
Scuba diving depths, 269
Seating arrangements at a theater or concert, 633, 641, 645–646
Soccer team electing captain, 927
Sports viewing habits, 82
Star Wars movies, 920–921
Team sports equipment, 299
Television, 336, 687, 731
Ticket prices and sales, 324, 465–466, 473, 474, 475, 476
Ticktacktoe strategy, 26
Track events, 760
Trading cards, 37
Triominoes score, 269
Video rentals, 313, 336
Wine tasting, 81
Winning percentage, 301, 304
Women's *100*-meter run, 683

Statistics and Demographics
Aging U.S. population, 723
Area and population in the Commonwealth of Independent States, 746
Census, 724, 911
Comparing state populations with governors' salaries, 784
Filling out a questionnaire, 659
People per square mile, 37
Population, 255, 300, 302, 305, 454
Statistics on the western population movement, 784
U.S. immigrants, 280

Technology
Binary code, 896
Cellular telephone subscribers, 412
DNA computers, 885
Growth of e-mail, 38
Homes with multiple personal computers, 408
Most visited Internet Websites, 744–745
Personal computer prices, 39
Prices of computer monitors, 473
Search engine to search the Web, 903
Setting options on a computer printer, 633, 636
Three-dimensional computer graphics, 49
World Wide Web, 850

Travel
Crossing a river, 29
Distance between cars, 333–334
Distance between cities, 338, 348, 391, 533
Distance traveled, 230, 339, 384–385, 388, 607
Engine failures in a vintage aircraft, 701
Greyhound bus links, 856–857, 858
Highway slopes, 406
Maximum airline or bus fare revenue, 444
National monuments, 670
Number of paths from point to point, 656
Plane's altitude, 269, 311
Rate, 332
Sea Isle Vacations, 953–954, 960
Seating on an airliner, 666, 669
Selecting drivers and passengers for a trip, 657
Sleeping on the way to Grandma's house, 27
Sorority traveling for spring break, 977
Speed, 303, 338, 351, 391, 468–469, 474
Time of a round trip, 37
Travel costs, 473
Travel times, 339
Traveling between countries, 890
Vacationing in Washington D.C., 62
Voyage in a paddleboat, 351
Yearly vacation, 926, 941

Carl Gauss publishes masterpiece on theory of numbers (important to development of statistics and geometry).

Non-Euclidean geometry is developed.

Pierre Simon de Laplace works out mathematical formulas describing interacting gravitation forces in the solar system.

Augustin Louis Cauchy does important work in complex analysis.

Evariste Galois develops theory of groups.

Georg Riemann founds second non-Euclidean system.

Arthur Cayley and James Sylvester develop matrix theory.

❈

1800 A.D. to 1850 A.D.

❈

Napoleon Bonaparte attempts domination of Europe.

Georg Ohm describes principles of electric resistance.

Joseph Jacquard improves mechanical loom, allowing mass production of fabric.

Michael Faraday discovers electromagnetic induction.

Charles Babbage develops analytic engine (forerunner of computers).

Ada Augusta, Lady Lovelace, devises the concept of computer programming.

Leonhard Euler pioneers work in topology, organizing calculus and using it to describe motion of objects and forces acting on them.

❈

1700 A.D. to 1750 A.D.

❈

Age of French Enlightenment is ushered in by thinkers such as Diderot, Montesquieu, Rousseau, and Voltaire.

First suspension bridge is completed.

James Watt creates steam engine.

Benjamin Banneker, a self-taught mathematician and astronomer, compiles a yearly almanac.

Joseph Lagrange develops theory of functions; studies of moon lead to methods for finding longitude.

❈

1750 A.D. to 1800 A.D.

❈

Celsius thermometer is invented.

Benjamin Franklin does kite experiment.

American Revolution and French Revolution take place.

| Mathematical Events |
| Cultural Events |

Modern Period (Early) 1450 A.D. to 1800 A.D.
Logarithms; modern number theory; analytic geometry; calculus; the exploitation of the calculus